Essentials of
World Regional Geography
Fourth Edition

Christopher L. Salter
Joseph J. Hobbs

University of Missouri – Columbia

THOMSON

BROOKS/COLE

Australia • Canada • Mexico • Singapore • Spain • United Kingdom • United States

Geography Editor: *Keith Dodson*
Development Editors: *Marie Carigma-Sambilay, Jennifer Pine*
Assistant Editor: *Carol Benedict*
Editorial Assistant: *Heidi Blobaum*
Technology Project Manager: *Samuel Subity*
Marketing Manager: *Ann Caven*
Marketing Assistant: *Sandra Perin*
Advertising Project Manager: *Laura Hubrich*
Project Manager, Editorial Production: *Tom Novack*
Production Service: *Nancy Shammas, New Leaf Publishing Services*
Copyeditor: *Frank Hubert*
Indexer: *James Minkin*

Print/Media Buyer: *Vena Dyer*
Permissions Editor: *Mary Kay Polsemen*
Text Designer: *Patrick Devine*
Photo Researcher: *Dena Betz*
Illustrator: *Maps.com*
Cover Designer: *Patrick Devine*
Cover Images: *Michael Kelly/Getty Images (top);*
   B. Tanaka/Getty Images (bottom)
Compositor: *Progressive Information Technologies*
Cover Printer: *Lehigh Press*
Printer: *R. R. Donnelley & Sons, Willard*

For more information about our products, contact us at:
Thomson Learning Academic Resource Center
1-800-423-0563

For permission to use material from this text, contact us by:
Phone: 1-800-730-2214   **Fax:** 1-800-730-2215
Web: http://www.thomsonrights.com

Library of Congress Control Number: 2002106314

ISBN 0-03-033966-9

Brooks/Cole–Thomson
511 Forest Lodge Road
Pacific Grove, CA 93950
USA

Asia
Thomson Learning
5 Shenton Way #01-01
UIC Building
Singapore 068808

Australia
Nelson Thomson Learning
102 Dodds Street
South Melbourne, Victoria 3205
Australia

Canada
Nelson Thomson Learning
1120 Birchmount Road
Toronto, Ontario M1K 5G4
Canada

Europe/Middle East/Africa
Thomson Learning
High Holborn House
50/51 Bedford Row
London WC1R 4LR
United Kingdom

Latin America
Thomson Learning
Seneca, 53
Colonia Polanco
11560 Mexico D.F.
Mexico

Spain
Paraninfo Thomson Learning
Calle/Magallanes, 25
28015 Madrid, Spain

Brief Contents

This book arrives in a world very different from the one visited by our Third Edition. If ever there was a need for all of us — especially the youth of the United States — to comprehend how dissimilar world regions and cultures can be, that need is now. We dedicate this book to the students who will undertake the learning necessary to see the richness, the diversity, and the critical significance of global dissimilarity in world regional geographies.

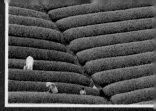

Contents

List of Maps

About the Authors

Christopher L. "Kit" Salter did his undergraduate work at Oberlin College, with a major in Geography and Geology. He spent 3 years teaching English at a Chinese university in Taiwan immediately following Oberlin. Graduate work for both the M.A. and the Ph.D. was done at the University of California, Berkeley. He taught at UCLA from 1968 to 1987, when he took on full-time employment with the National Geographic Society in Washington with his wife, Cathy. They were both involved in the Society's campaign to bring geography back into the American school system. Kit has been professor and chair of the Department of Geography at the University of Missouri Columbia since moving to the heartland in 1988.

Themes that have made Salter glad he chose geography as a life field include landscape study and interpretation in both domestic and foreign settings; landscape and literature to show students that geography occurs in all writing, not just in textbooks; and geography education to help learners at all levels see the critical nature of geographic issues. He has written more than 125 articles in various aspects of geography; has traveled to a lot of the places that he writes about in this text; and has been wise enough to have a son who is an architect in Los Angeles, a daughter who is a teacher in the San Francisco Bay Area, and a neat wife who is a writer in the heart of the heartland in Breakfast Creek, Missouri.

Joseph J. Hobbs received his B.A. at the University of California, Santa Cruz in 1978 and his Ph.D. at the University of Texas Austin in 1986. He is a professor at the University of Missouri Columbia, and a geographer of the Middle East with many years of field research on biogeography and Bedouin peoples in the deserts of Egypt. Joe's interests in the region grew from a boyhood spent in Saudi Arabia and India. His research in Egypt has been supported by Fulbright fellowships, the American Council of Learned Societies, the American Research Center in Egypt, and the National Geographic Society. He served as the team leader of the Bedouin Support Program, a component of the St. Katherine National Park project in Egypt's Sinai Peninsula. His current research interests are Bedouin participation in park development in Egypt's Eastern Desert, human uses of caves worldwide, and the global narcotics trade. He is the author of *Bedouin Life in the Egyptian Wilderness* and *Mount Sinai* (both University of Texas Press) and *The Birds of Egypt* (Oxford University Press). He teaches graduate and undergraduate courses in world regional geography, environmental geography, the geography of the Middle East, the geography of caves, and the geography of global current events. In 1994, he received the University of Missouri's highest teaching award, the Kemper Fellowship. Since 1984, he has led adventure travel trips to remote areas in Latin America, Africa, the Indian Ocean, Asia, Europe, and the High Arctic. Joe lives in Missouri with his wife, Cindy, daughters Katherine and Lily, and a wildlife menagerie of tortoises and Old World chameleons.

Preface

Approach and Scope

After the events of September 11, 2001, there was much soul-searching about how or even if the world had changed as a result of those events. As the war in Afghanistan wound down and the tempo of popular culture in the United States was revived, there was a growing sense that, despite the tragedy, perhaps things were more or less as they had been before September 11.

The authors of this text believe that must not be. We regard the world as fundamentally different today than it was in early September 2001. Most significantly for you, as the user of this text, you are more challenged than any student has been in two generations to learn about how the world works. Despite any defensive emphasis on "homeland security," no country can simply shore up its borders and restrict its worldview as a means of guaranteeing its safety and welfare. The far more urgent task is for a people to look beyond its borders to the farthest corners of the world — for in those far corners many world-changing events may occur — and seek to understand the human and environmental contexts from which all events originate.

This is not the time for shortcuts or simplifications. This is not a small book, but we call it *Essentials of World Regional Geography* because we believe that, to come to grips with global affairs, it is essential for the reader to have a thorough grounding in the environmental, political, cultural, economic, and even strategic contexts of the world's regions and nations. We have tried to offer you that complete basic training in how the world works, while being careful to avoid being exhaustive or exhausting. We hope that between this book's covers you will find the answers to many of the follow-up questions that would naturally come to you from a more basic introduction to the world. Of course, there is much more out there for you to follow up on, and we have worked with Brooks/Cole to design a rich and easy-to-use set of ancillary resources. We wish you a most satisfying and interesting journey through this world.

Features of This Edition

This edition has been updated so that all of the statistical tables reflect the most current information available in early 2002. These data have been added to all tables as well as to textual comments. In addition, map data have been revised to reflect the changing world. There are also new thematic maps, including religions and languages of each world region.

Users of the third edition will find many familiar features, including the Regional Snapshot, Problem Landscape, Regional Perspective, Landscape in Literature, Definitions & Insights, and the end-of-chapter summary, review, and discussion elements. These were retained because of favorable reviews about their usefulness in the third edition. There were many requests to have the authors do more to convey the nature and rewards of doing fieldwork, and we have responded with a new profile chapter feature called Perspectives from the Field.

Also by popular demand, we organized this edition along more thematic lines. The profile chapter for each region follows a consistent format: area and population, physical geography and human adaptations, cultural and historical context, economic geography, and geopolitical issues. There is also more attention to thematic and conceptual issues in the more detailed country chapter following each profile chapter. Our goal has been to make it easier for the text to be used in a single semester course; in some cases, the instructor may wish to assign only the profile chapter and encourage students to explore the subsequent countries chapter on their own or in supplementary assignments.

Organization of the Text

Each of the eight world regions we have built the text around has certain common features that provide for internal coherence. The regional profile opens the study of the region, outlines basic characteristics of that region, and describes the sorts of physical and cultural features that create regional unity. Not every region has the same elements of unity, but each can be identified by similar traits generally found across the region. For example, not all of the people in Middle and South America speak Spanish, but the Spanish language and Spanish cultural influences in Part 8 serve as major features of cohesion.

Chapters 1 and 2 consider how a geographer sees the world and illustrate how critical environmental assessment, modification, and management apply to this vision. In these two chapters, Chapter 2 especially, clear links between human-environmental modification and world regional problems are established.

Chapters 3, 4, and 5 explore Europe. In addition to the regional profile chapter, the next two chapters divide Europe into the Coreland that contains the very productive nations that are part of the European Union (EU). While the EU does not define the full range of geographic traits of northwest Europe, this dynamic 15-nation organization is bringing a unity to economic and even political accomplishments in Europe that is unprecedented during the past century. The realms that lie to the north, the south, and the east of the European Coreland are given their own identities so that students can leave Europe with a good sense of the economic, political, and historical diversity that characterizes this critical world region.

Chapters 6 and 7 describe and discuss the geography of Russia and the Near Abroad. This is a region in transition; indeed, there is much debate among geographers about whether or not the 15 successor states of the Soviet Union should be handled as a unit. The authors believe that the legacy of Soviet rule still lingers sufficiently strongly, particularly in economic relationships and in cultural geography (witness the populations of Russians in the Baltic states and in Kazakstan), so that at least for a few more years, there will be more benefits than drawbacks in having students introduced to these countries under the same conceptual roof. Longtime users of this text will be pleased that the historical depth that Wheeler and Kostbade gave to this region still remains in this edition. There is close attention to the dynamic and dangerous situations in the Caucasus and Central Asia. The prospects for Russia's emergence from its economic and demographic malaise are also explored.

Because it contains approximately 60% of the world's oil, the region of the Middle East and North Africa has immense relevance to people around the world. Chapters 8 and 9 take students deep into a region that is likely to make headlines throughout their lives. These chapters aspire to lead students to a well-founded understanding of the circumstances underlying strife in the Middle East, including the underpinnings for the September 2001 attacks against the United States. Understanding of the Arab-Israeli conflict begins here with comprehension of the importance of sacred places in Judaism and Islam, the exile of the Jews, the Zionist movement, and the establishment of Israel. The subsequent wars are described thoroughly because each is necessary to comprehend the difficult issues confronting would-be peacemakers in the region. The Palestinian-Israeli peace process, which is above all a set of geographic problems, is updated through its apparent collapse, and the risks of its demise are considered. These chapters also contain a wealth of material unrelated to the questions of conflict and oil, and students should come away with a strong appreciation of the cultural and environmental features of the nations that comprise the Middle East.

Monsoon Asia is the realm covered in Chapters 10 through 14. This region is home to more than half of the world's population and some of the world's most dynamic, and most troubled, economies. The Asian Tigers were the forces that led to

the 21st century being anticipated as the Pacific Century four or five years ago. These chapters chronicle the changes that have come over the region and point out why such a label is somewhat less likely now. Even while regional commonalities based on economic development are central to much of our treatment here, these chapters also craft a clear picture of unique national characteristics.

Numerous reviewers of the third edition were pleased by the thorough treatment afforded to the Pacific World. Chapters 15 and 16 have made no sacrifices in this edition. A strong foundation for the environmental setting of this vast region remains, along with several case studies of historical and contemporary problems in environmental management. There is new treatment of the perceived dilemma the majority of white Australians face in deciding whether to forge a new, multiethnic Australia oriented to the Pacific Rim and Asia or to retain ties with the Commonwealth and the British monarchy. Ancient Easter Island and modern Nauru tell much about the liabilities of island living, however romantic it may seem to continent-dwellers.

Africa has long troubled students of geography, who have had difficulty coming to terms with such a vast and complex region. The authors believe strongly that this region can be studied and comprehended as well as any other. We have used Chapter 17 to lay out the major themes that unite and differentiate the subregions and countries that make up Africa south of the Sahara. There are new case studies of compelling issues such as HIV/AIDS and "dirty diamonds." Chapter 18 is a comprehensive and thoroughly updated journey through each of the seven subregions.

Chapter 19 is the regional profile chapter for Middle and South America, the world of Latin America. Continuing the change made for the third edition, this region is organized into two chapters. Chapter 20 discusses Middle America with Mexico, Central America, and Caribbean America. The world south of Panama is comprehensively covered in Chapter 21, South America.

Part 9 assesses the United States and Canada, with the regional profile chapter (Chapter 22) looking at the broad world that lies between the northern border of Mexico and the North Pole. Chapter 23 is devoted to Canada because of the significant role that country plays in the lives of the major readership of this text, and Chapter 24 focuses exclusively on the United States. Themes as diverse as environmental perception, resource presence and development, urbanization, human mobility, and regional identities are covered in this final chapter.

The progression through the text may be taken from Chapters 1 through 24 in straight sequence, or a class may chart its own course after reading the overview in Chapters 1 and 2. The eight geographic profile chapters (Chapters 3, 6, 8, 10, 15, 17, 19, and 22) call attention to the elements that give character and identity to each of the eight regions in the text. The regional chapters that round out each section are fundamentally self-contained in their presentations. A glossary at the end of the text provides definitions of selected terms, should a class cover only a portion of the text in their particular study of world regional geography.

Also to enhance students' understanding, significant cross-references are noted in the margin. Many times they indicate an important table or figure that relates to the current text discussion or sometimes they are references to relevant text discussions elsewhere in the book. For example, F8.1, 222 indicates Figure 8.1 on page 222; T15.1, 396 indicates Table 15.1 on page 396; and just an italic number in the margin would indicate a page where the text discussion relates to the current topic.

Ancillaries

This text is accompanied by a number of ancillary publications to assist instructors and enhance student learning:

- Instructor's Manual/Test Bank
  by Christopher "Kit" Salter and Joseph Hobbs
  Written by the main text authors, this instructor's manual contains both general and curricular suggestions for the teaching of world regional geography, as well as multiple-choice questions keyed to the text. It includes chapter outlines, key terms, and demonstration ideas.
- Study Guide and Student Resources
  by Dr. Tarek A. Joseph, University of Michigan-Dearborn
  This study guide includes chapter objectives, key terms, critical thinking problems and a variety of self-test questions and answers.
- Places of the World: Place/Name Workbook and Map-Pack
  by Dr. James Lett, Indian River Community College
  This workbook provides students with organized practice for matching the names of places with their physical location on a map.
- Transparency Acetates
  100 full color transparency images from the text are available.
- Multimedia Manager: A Microsoft® PowerPoint® Link Tool
  Contains prepared lectures by Dr. Jonathan Taylor, California State University-Fullerton, or create your own lectures using the PowerPoint images and photos taken directly from the text. Search feature and DVD format are new to this edition.
- ExamView for Windows®/MacIntosh®
  This computerized test bank is the software version of the printed test bank.
- World Regions Interactive CD-ROM
  This CD-ROM is both a game and a dynamic learning tool designed to engage students in the learning process. Students will learn the fundamental concepts geographers use to understand the people and places on our planet and the complexity of their interrelationships. Approximately 200 images are included.

- CNN® Video: Geography 2002, Vol. 1
  Approximately 50 minutes of short, relevant clips from CNN are sure to engage student interest and enhance learning of the subject matter.
- World Population Data Sheets
  These data sheets provide detailed census-style quantitative information. They are included free with each new copy of the text.
- Atlas of World Geography by Rand McNally
  This comprehensive world atlas can be packaged with this text.
- World Regional Geography Companion Web Site
  The Web site includes instructor- and student-specific content and functionality. Instructors can use the Web site as a tool for classroom preparation by taking advantage of the free image downloads from the text and the Microsoft Word files from the Instructor's Manual. Students who use the World Regional Companion Web site have access to a host of content and self-assessing tools *and* can take a virtual journey with author Joseph Hobbs as he discusses his fieldwork around the world. Students can check their learning with Web quizzes containing multiple choice, true/false, and essay questions for each chapter. Interactive maps allow students to further quiz themselves on the basic geography of world regions. An online glossary and animated flashcards reinforce the key terms and concepts.

In addition, the Companion Web site contains plenty to expand students' knowledge beyond the text readings. Internet Activities launch students on Web explorations into interesting topics while InfoTrac Exercises make use of the huge database of magazine articles accessible using a passcode that comes with each new textbook. News Updates by author Joseph Hobbs tie in current world events with readings from the text. Carefully selected hypercontents allow students to do some further exploring on their own.

Acknowledgments

We would like to extend our warmest thanks and respect to the late Jesse H. Wheeler, Jr. and professor emeritus Trent Kostbade for all they have done across the decades for this book. Although they have not been a part of the last three editions, their spirit and prose continue to be a presence in the teaching of world regional geography.

We wish to acknowledge the long list of reviewers whose comments helped to bring this fourth edition to completion. We sincerely thank each and every one of these distinguished geographers and outstanding teachers and students.

Reviewers for this edition include: Richard Benfield, Central Connecticut State University; Michelle Calvarese, California State University, Fresno; David Daniels, Central Missouri State University; Matthew Ebiner, El Camino College; Jane Ehemann, Shippensburg University; John J. Hickey, Inver Hills Community College; Marcia M. Holstrom, San Jose State University; Tarek A. Joseph, University of Michigan-Dearborn; Thomas J. Karwoski, Anne Arundel Community College; David Lee, Florida Atlantic University; Richard Pillsbury, Georgia State University; William L. Preston, California Polytechnic State University; Thomas Ross, University of North Carolina at Pembroke; Daniel Selwa, Coastal Carolina University; Jennifer Speights-Binet, Louisiana State University; Philip Thiuri, William Paterson University; Jeffrey S. Torguson, St. Cloud State University; and Steve Wolfe, Chemeketa Community College.

The authors wish to extend special thanks and appreciation to the Brooks/Cole editorial staff who labored so diligently to help us all meet the tight deadlines associated with the transition from Saunders College Publishing to Thomson Learning. We are both indebted to Keith Dodson of Brooks/Cole for working with this new project and the smooth skills of Nancy Shammas of New Leaf Publishing Services. We also appreciate the rest of the Brooks/Cole team: Carol Benedict for coordinating the ancillaries, Sam Subity for getting the Web site up and running; Ann Caven for her marketing expertise, and Tom Novack for overseeing all the production issues. Thanks especially to Dena Betz for making the transition with us and continuing her work as photo researcher on this edition and to Frank Hubert for his careful copyediting. In the rear view mirror, we send thanks to Jennifer Pine, Kelley Tyner, and Anne Gibby for work they did with the third edition and the transition toward this current fourth.

Kit Salter gives special thanks to Joseph Weidinger for his steady labor with statistical updating and to Amy Hummel for her careful reading of so many varied items. Thanks also go to Bryan Crousore, Tommy Falke, Diana Lord, William Least Heat-Moon, Matt Gerike, Larry Hall, Natalie Simon, Bill Laughter, and Heather Woods for discussions of language and concept in the flow of this process. To Jim Harlan a special thanks for his crisp responses to questions of focus, to Greg Breuer a nod for his always-stunning computer knowledge, and to Fritz Gritzner, my thanks for a Dakotan who seems to know everything and share such wisdom graciously. My ultimate thanks, always, to my author lady, Cathy, for bringing such grace to our lives and labors.

Joe Hobbs also thanks Joseph Weidinger for his painstaking work on statistics, populations, and end-of-chapter materials. The teaching assistants in the Department of Geography at the University of Missouri — Amy Roust, Tim Howerton, Jason Jindrich, Denise Pass, Erin Wilson, Cheryl Morton, Victor Lozano and Nathaniel Albers — helped Joe catch some of the errata in the third edition and other important changes in this edition. Most of all, Joe thanks his ever-supportive family — especially mom, partner, and best friend Cindy, and little geographers Katie and Lily.

Kit Salter
Joseph Hobbs

Perspectives and Issues in World Regional Geography

CHAPTER 1

Engaging the Many Meanings of Geography

KIT SALTER

Definitions of geography are often varied. Although we provide an "official definition," the boundaries and definitions of geographic regions are a particularly good example of the variance of things geographic. This map was photographed on a street corner in San Francisco where a small flea market was occurring. The division of the contiguous United States into four regions and the approximate boundaries and characteristics of such regions are an evocative example of a "mental map." The image is a useful picture of geography in everyday life. The divisions are evocative: The coasts are the regions of expensive; the heartland is cheap; Texas and the noncoastal west are odd.

"Where Are You Calling From?"

Your car breaks down and your hurried phone call evokes this question, "Where are you calling from?" What does this question ask for? Its answer sets in motion a whole mindful of geographic themes — without you even thinking about the subject of geography. The nervous question wants news of exact location. What place are you in right this minute? It wants to know because of concerns for accessibility. How easily can I get there? How easily can you get here from there? It also seeks geographic information that relates to your comfort, to your safety, to your sense of frustration and anger. Are you okay while I come and get you? Can you get someone to give me clear directions on how to get to where you and your dead car are? Can you get something to eat while I get wheels organized? All sorts of issues are just below the surface of that question.

"Where are you calling from?" is the sort of geography that weaves through all of our lives but seldom makes us stop to think about the discipline of geography. It makes us think geographically as we answer the question — looking around from the phone booth trying to get visual landmarks that can help us answer the question. It does not evoke recollection of a college class or an academic discipline or considerations of a career. It just wants real information that has real utility and that tells the person you called something about your scene right this minute. The question initiates a consideration of the real heart of geographic information.

This discipline that you are now coming face-to-face with as you start to read the fourth edition of the *Essentials of World Regional Geography* is a body of content and perspectives that will touch your life in dozens of different ways. We hope to help you understand how such information will not only give you a good sense of the geographic characteristics of the regions and nations of the world but that it will also make you think about geography for life.

In thinking about how you might answer the question, "Where are you calling from?" you are beginning geographic exploration and communication. You have been on such a journey for years. It is our intention to give you new direction and new directions in how you see, utilize, and appreciate the many ways in which geography touches your life.

1.1 Geography for Life

Geography is the study of Earth's surface and the spatial distributions and patterns of its physical and human characteristics. Geographers determine what distinct forces of nature and culture have been at work in the creation of the **landscape** — the collection of physical and human geographic features on the earth's surface — and evaluate the role such a landscape will have on the economic and social development of the local area. Such evaluation involves **environmental perception** — our individual response to environmental features — and **environmental assessment** — our determination of the value and proper management of a given environment. Observation and analysis of the distribution and arrangement of given geographic phenomena lead to better understanding of the landscapes around us and are at the heart of geography.

Our hope in writing this text is that you, too, will begin to develop a sense of the geographer's mind and learn the richness and usefulness of a geographer's field of vision. The ability to come upon a scene — whether an urban intersection, a rural farmstead, a forest with deer, a photograph on the front page of a newspaper, or an image on television — and to begin to make sense of the physical and cultural geographic elements that create this image is one of the skills we hope you will achieve with this text (Fig. 1.1). Since our life is a series of continually changing scenes, it makes sense that learning landscape appreciation would enhance our understanding of the real world around us.

Because geography has so many varied perspectives, our profession has created a learning model (Table 1.1) that divides geography into six realms, or the Six Essential Elements. In confronting these varied meanings of geography, our field of vision — the way in which we each see the world — draws on each of these Essential Elements in distinct ways.

1.2 Finding and Making Order in Earth Patterns

We recognize that all phenomena of the world can be organized in spatial patterns. As random as a landscape might seem at first glance, there is a spatial order in the location of all physical and cultural phenomena (see Fig. 1.1). In our National Geography Standards, this first Essential Element is called *The World in*

DEFINITIONS + INSIGHTS

Spatial Analysis

The term **spatial** comes from the noun "space," and it relates to the ways in which space is organized and patterned. In spatial analysis, geographers examine the patterns created in the distribution of phenomena (farms, auto dealerships, port cities, Baptists, etc.). In considering spatial organization, one studies the decision-making processes and outcomes that give such distributions their particular configuration. Important to understanding this term is knowing that all phenomena possess a spatial organization of one sort or another. Even aspects of the world that appear random are generally bound by some system of ordered location and distribution.

Spatial Terms. As you open each of this text's chapters on specific world regions (beginning in Chapter 3), we will give you facts on the area, population, and characteristics of the natural environment of each region. Learning how to see the patterns and distributions of physical and cultural features in these varied landscapes is fundamental to seeing the world in spatial terms.

Maps

To think geographically is to think in terms of spatial distributions, patterns, associations, and interconnections. This inevitably means to think in terms of maps because maps portray spatial patterns and associations with a clarity beyond the reach of words. Students should refer to maps frequently as they read geographic material, use outline maps for note taking, make their own sketch maps to show important spatial relationships, and try to visualize map relationships when patterns are stated in words in text material. In short, thinking geographically is to think spatially. And as the graphics that have given geography meaning for millennia, maps are central to all of our learning.

A map is the most common artifact of geography and of the geographer. At the very beginning of someone's questioning of the meaning of geography, the map comes to mind. It is the *symbol* of geography. You have made and seen sketch maps made on the back of envelopes or business cards. But you may also have seen enormously complex, multicolored, highly detailed maps newly created by computer technology and Geographic Information Systems (GIS). In all cases, the map represents the organization of space, the patterns of recurring phenomena, and the human effort to show where we are and where we have been — that is, to show what is where on the surface of the earth.

A person's life is filled with numerous local spatial realities. The route you take when you walk, ride, or drive to school is built around spatial decisions: which road, which highway, what parking lot or structure? When you come into a classroom, particularly on the first day of a new term, you have a whole raft of spatial decisions to make: sit by the aisle, sit in the back, sit by friends, sit in new territory, sit alone? In adjusting to a new campus and town, you ask yourself: Where do I eat? What street do I explore to find a bookstore, coffee house, the Salvation Army Thrift Store, a theater? When you have the leisure or the money to seek entertainment, where do you go? Life is quilted together by the sorts of spatial decisions we make, and each of those is part of our personal geography.

One reason for studying geography is to learn where things are in the world, to acquire a framework within which coun-

Figure 1.1   Geographers are always interested in knowing how the physical environment influences the human use of the earth. This scene from China shows an elaborate terraced landscape, created to make small level lots that can be irrigated for a more productive rice harvest. Such fields represent great labor investments from the present farmer and maybe from others in the past.

KIT SALTER

TABLE 1.1   The Six Essential Elements and the Eighteen Geography Standards

I. The World in Spatial Terms

Geography studies the relationships between people, places, and environments by mapping information about them into a spatial context.

The geographically informed person knows and understands

  1. how to use maps and other geographic representations, tools, and technologies to acquire, process, and report information from a spatial perspective

  2. how to use mental maps to organize information about people, places, and environments in a spatial context

  3. how to analyze the spatial organization of people, places, and environments on Earth's surface

II. Places and Regions

The identities and lives of individuals and peoples are rooted in particular places and in those human constructs called regions.

The geographically informed person knows and understands

  4. the physical and human characteristics of places

  5. that people create regions to interpret Earth's complexity

  6. how culture and experience influence people's perceptions of places and regions

III. Physical Systems

Physical processes shape Earth's surface and interact with plant and animal life to create, sustain, and modify ecosystems.

The geographically informed person knows and understands

  7. the physical processes that shape the patterns of Earth's surface

  8. the characteristics and spatial distribution of ecosystems

IV. Human Systems

People are central to geography; human activities, settlements, and structures help shape Earth's surface, and humans compete for control of Earth's surface.

The geographically informed person knows and understands

  9. the characteristics, distribution, and migration of human populations

10. the characteristics, distribution, and complexity of Earth's cultural mosaics

11. the patterns and networks of economic interdependence

12. the processes, patterns, and functions of human settlement

13. how the forces of cooperation and conflict among people influence the division and control of Earth's surface

V. Environment and Society

The physical environment is influenced by the ways in which human societies value and use Earth's natural resources, while at the same time human activities
    are influenced by Earth's physical features and processes.

The geographically informed person knows and understands

14. how humans modify the physical environment

15. how physical systems affect human systems

16. the changes that occur in the meaning, use, distribution, and importance of resources

VI. Uses of Geography

Knowledge of geography enables people to develop an understanding of the relationships between people, places, and environments over time — that is,
    of Earth as it was, is, and might be.

The geographically informed person knows and understands

17. how to apply geography to interpret the past

18. how to apply geography to interpret the present and plan for the future

Source: *Geography for Life: National Geography Standards 1994.* National Geographic Research and Exploration, Washington, D.C., pp. 34–35. Reprinted by permission.

tries, important cities, rivers, mountain ranges, climatic zones, agricultural and industrial areas, and other features can be related to each other. However, it is not enough simply to know the facts of location; one must also develop an understanding and appreciation of the *significance* of location.

Two Distinct Island Systems: An Example of Location Dynamics

Perhaps an illustration will clarify these remarks. Let us compare two distinct countries and island systems. Both are island coun-

tries. Westerly winds, which blow off the surrounding seas, bring abundant rain and moderate temperatures throughout the year. Their climates are remarkably similar, even though these areas are in opposite hemispheres and are about as far from each other as it is possible for two places on Earth to be (Fig. 1.2).

Great Britain is located in the Northern Hemisphere, which contains the bulk of the world's land and most of the principal centers of population and industry; New Zealand is on the other side of the equator, in the Southern Hemisphere. Great Britain is located near the center of the world's land

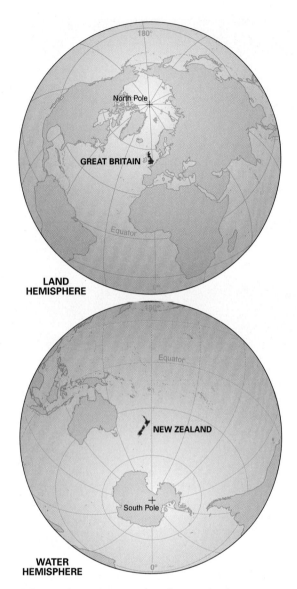

Figure 1.2 In the top map, note how the major land masses are grouped around the margins of the Atlantic and Arctic Oceans. The British Isles and the northwestern coast of Europe lie in the center of the "land hemisphere," which constitutes 80% of the world's total land area and has approximately 91% of the world's population. New Zealand lies near the center of the opposite hemisphere, or "water hemisphere," which has only 20% of the land and 9% of the population.

masses and is separated by only a narrow channel from the densely populated industrial areas of Western continental Europe; New Zealand is surrounded by vast expanses of ocean. Great Britain is located in the western seaboard area of Europe, where many major ocean routes of the world converge; New Zealand is far away from the centers of world commerce. For more than four centuries, Great Britain has shared in the development of northwestern Europe as a great organizing center for the world's economic and political life; New Zealand, meanwhile, has existed in comparative isolation.

Great Britain, in other words, has a central location within the existing frame of human activity on Earth, whereas New Zealand has a peripheral location. Centrality of location is a highly important factor to consider in assessing

the economic as well as political geography of any country, region, or other place. Location, in other words, is of major significance (Fig. 1.3).

The Language of Maps

Because maps are so basic to all geographic understanding, it is important to learn the language of maps. Cartographers are the geographers and skilled technicians who create maps through the careful design and skillful presentation of geographic information. Increasingly, **cartography** — the art and science of making maps — utilizes computer manipulation of spatial data, a field increasingly dependent on computers and known as computer cartography. Cartography is a graphic portrayal of location (Fig. 1.4).

The cartographer shows where things and places on Earth are located in relation to each other. Thus, maps are indispensable tools for discovering, examining, and portraying spatial relationships and associations. Some maps show the distribution and interrelation of things that are fixed in place, such as structures and terrain features, whereas other maps portray the flow of people, goods, and ideas from place to place. A map communicates locational facts and concepts to us by means of a specialized "language" that has to be learned so that the map can be properly understood and used. Major elements of this language of maps (beyond title and date) are **scale, coordinate systems, projections,** and **symbolization.**

Scale. A map is a reducer; it enables us to comprehend an extent of Earth-space by reducing it to the size of a single sheet. The amount of reduction appears on the map's scale, which shows the actual distance on Earth as represented by a given linear unit on the map. A common way of denoting scale is to use a representative fraction, such as 1:10,000 or 1:10,000,000. In other words, one linear unit on the map (e.g., an inch or centimeter) represents 10,000 or 10,000,000 such real-world units on the ground. A large-scale map is one with a relatively large representative fraction (e.g., 1:10,000 or even 1:100) that portrays a relatively small area in more detail. A "global" or small-scale map has a relatively small representative fraction (e.g., 1:1,000,000 or 1:10,000,000) that, in contrast, portrays a relatively large area in more generalized terms (see Fig. 1.4).

Coordinate Systems. Another concept relating to maps is **location:** You have to know where something "is" and how it relates spatially to other things. There are two general types of locational information: relative location and absolute location. **Relative location** defines a place in relationship to other places. It is a dynamic of accessibility. It may be used as a simple locational feature (Oregon is north of California; Sumatra is west of Java). It may also deal with accessibility in other ways, such as, "It is easier to get into the central city if you go in at 7 P.M. rather than 7 A.M." **Absolute location,** also known as mathematical location, uses different constructions to label places on the earth so that every place has its own unique location or "address." Coordinate systems are a way of determining absolute location that are common on maps. Different types of coordinate systems create grids consisting of horizontal and vertical lines covering the entire globe. The intersec-

This excerpt from Beryl Markham's *West with the Night* is a powerful note to anyone who has ever used maps in a serious way. She was the flyer made famous in a richly produced Sydney Pollack film with Robert Redford and Meryl Streep entitled *Out of Africa* (1985). Her small plane was the center of her life. This excerpt illustrates her belief in the power and significance of a map.

A map in the hands of a pilot is a testimony of a man's faith in other men; it is a symbol of confidence and trust. It is not like a printed page that bears mere words, ambiguous and artful, and whose author, perhaps — must allow in his mind a recess for doubt.

A map says to you, 'Read me carefully, follow me closely, doubt me not.' It says, 'I am the earth in the palm of your hand. Without me, you are alone and lost.'

And indeed you are. Were all the maps in this world destroyed and vanished under the direction of some malevolent hand, each man would be blind again, each city be a stranger to the next, each landmark become a meaningless signpost pointing to nothing.

Yet, looking at it, feeling it, running a finger along its lines, it is a cold thing, a map, humourless and dull, born of calipers and a draughtsman's board. That coastline there, that ragged scrawl of scarlet ink, shows neither sand nor sea nor rock; it speaks of no mariner, blundering full sail in wakeless seas, to bequeath, on sheepskin or a slab or wood, a priceless scribble to pos-

terity. This brown blot that marks a mountain has, for the casual eye, no other significance, though twenty men, or ten, or only one may have squandered life to climb it. Here is a valley, there a swamp, and there a desert; and there is a river that some curious and courageous soul, like a pencil in the hand of God, first traced with bleeding feet.

Here is your map. It is only paper. It is only paper and ink, but if you think a little, if you pause a moment, you will see that these two things have seldom joined to make a document so modest and yet so full with histories of hope or sagas of conquest.[a]

[a]Beryl Markham, *West with the Night.* San Francisco: North Point Press, 1942, 1983, 245–246.

---

tion of such lines creates the address in the global coordinate system, giving a specific, mathematical, and unique location.

The most common coordinate system on maps uses **parallels of latitude** and **meridians of longitude.** The term latitude denotes position with respect to the equator and the poles. Latitude is measured in degrees (°), minutes ('), and seconds ("). The equator, which circles the globe east and west midway between the poles, has a latitude of 0°. All other latitudinal lines are parallel to the equator and to each other and therefore are called parallels. Every point on a given parallel has the same latitude. Places north of the equator are in north latitude; places south of the equator are in south latitude. The highest latitude a place can have is 90°N or 90°S latitude. Thus, the latitude of the North Pole is 90°N, and that of the South Pole is 90°S. Places near the equator are said to be in low latitudes; places near the poles are in high latitudes. The Tropic of Cancer and the Tropic of Capricorn, at 23.5°N and 23.5°S, respectively, and the Arctic and Antarctic Circles, at 66.5°N and 66.5°S, respectively, form convenient and generally realistic boundaries for the low and high latitudes. Places occupying an intermediate position with respect to the poles and the equator are said to be in middle latitudes.

Meridians of longitude are straight lines connecting the poles. Every meridian is drawn due north and south. They converge at the poles and are farthest apart at the equator, where the distance between lines of longitude is approximately 69.15 statute miles (111.29 km). Longitude, like latitude, is measured in degrees, minutes, and seconds. The meridian most often used as a base (starting point) is the one at the Royal Astronomical Observatory in Greenwich, England. It is known as the meridian of Greenwich, or the prime meridian, and has a longitude of 0°. Places east of the prime meridian are in east longitude; places west of it are in west longitude.

The meridian of 180°, exactly halfway around the world from the prime meridian, is the other dividing line between places east and west of Greenwich called the International Date Line. However, because the meridians converge toward the poles, a degree of longitude at the Arctic Circle is equivalent to only 27.65 miles (44.50 km). The combination of latitude and longitude gives us mathematical or **absolute location.**

Projections. Transferring images from a globe to a flat map creates distortions because it is not possible to represent the three-dimensional curved surface of Earth with complete accuracy on a two-dimensional flat sheet of paper. If the area represented is very small (e.g., a large-scale map), the distortion may be slight enough to be disregarded, but maps that represent larger spatial areas (small-scale maps) may introduce very serious distortions. Distortions affect four different properties of a globe: direction, distance, shape, and area. A line drawn from point A to point B on a map may not have the correct distance. The purpose of the various map projections is to minimize distortion in one or more of these four properties.

Conformal map projections preserve shape over small areas. The Mercator projection (Fig. 1.4a) accomplishes shape preservation for small areas and retains direction for straight lines. To achieve this, and to serve the original purpose of aiding navigation, the parallels of latitude and meridians of longitude are straight and meet at right angles. In keeping the lines of longitude straight, instead of curving toward the poles as on a globe, the Mercator projection greatly exaggerates the east–west dimension of areas near the poles. It also exaggerates the north–south dimension because the parallels are not spaced evenly between the equator and poles, as on a globe, but gradually draw farther and farther apart with increasing distance from the equator. The angular array of parallels and meridians makes

Figure 1.3 Map of major world regions that form the basic framework of this text. Some regions overlap others. Russia and the Near Abroad refers to the total expanse occupied by the former Soviet Union (Union of Soviet Socialist Republics) prior to the period of dissolution that began with independence for the Baltic Republics, 1990–1991. Names and boundaries of the 15 new independent states in the Russian Realm are shown in Chapter 6.

MAJOR REGIONS — A Global View

the absolute latitudinal and longitudinal location of every place accurate, but it distorts the size and shape of polar regions so that Greenland may appear larger than all of South America, although in reality it is only slightly larger than Mexico.

Equal-area projections, such as the Lambert Equal-area projection (Fig. 1.4b) and Goode's Interrupted Homolosine projection (Fig. 1.4c), minimize distortions relating to how big an area looks. The area of Kansas on a Lambert Equal-area projected map represents the real-world area of that state. Equal-area projections must distort shape to retain equal-area characteristics. Goode's Interrupted Homolosine projection tries to minimize some of the shape distortion by segmenting the map into lobes (the cartographic equivalent of orange sections if Earth were an orange), thereby minimizing shape distortion in midlatitudes while preserving area over all. Compromise projections,

such as the Robinson projection (Fig. 1.4d), do not eliminate distortion in any of the four properties. Rather, the Robinson projection allows different degrees of shape, areal, scale, and distance distortion to minimize the severity of the other distorted properties, leading to an aesthetically pleasing map.

Symbolization. Maps enable us to extract certain information from the totality of things, to see patterns of distribution, and to compare these patterns with each other. No map is a complete record of an area; it represents instead a selection of certain details, shown by symbols, that a cartographer utilizes to accomplish a particular purpose. Unprocessed data must be classified to provide categories that the symbols will represent. The details may be categories of physical or cultural forms (rivers, roads, settlements, etc.), aggregates such as 100,000

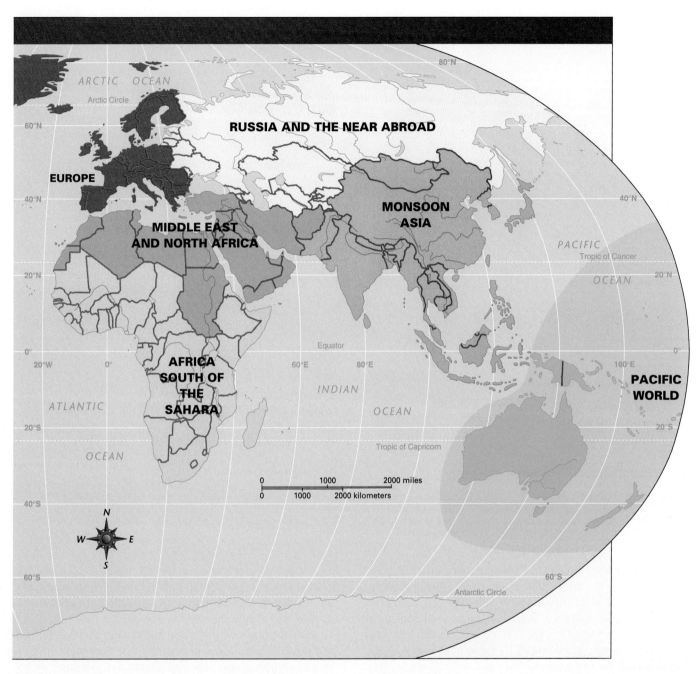

people, or 1 million barrels of crude oil, or averages such as population density (the number of persons per square mile within a defined area). Aggregates or averages are frequently ordered into ranked categories in a graded series (e.g., population densities of 0–49, 50–99, or 100 or more people per square mile) for portrayal on the map. The categories are not self-evident; they are selected by the cartographer, sometimes by elaborate statistical procedures. To represent varied phenomena, the cartographer has available a wide range of symbols: lines, dots, circles, squares, shadings, and others. Color, increasingly, is used in the production of maps.

Dots usually portray quantities, and the dot map is one of the most common types of maps used to show distributions of people or things on Earth. On a dot map of population, for example, each dot represents a stated number of people and is placed as near as possible to the center of the area that these people occupy. In interpreting such a map, we are not greatly concerned with the individual dots but with the way the dots are arranged. We look for the pattern and distribution of a given phenomenon.

One complexity of cartography is that different scales and ways of showing distributions can give different impressions. For example, methods of symbolization other than dots could be used to show the distribution of world population and might convey rather different ideas from those presented in the dot map. Some alternative methods might include the use of a single symbol for each country, with the size of the symbol proportional to the country's population, or drawing a map on which the area of each country is proportional to the item being shown, such as the number of doctors per 100,000 population.

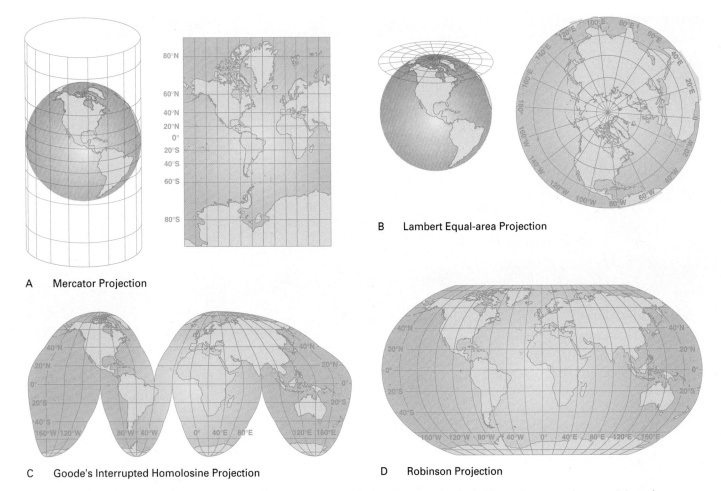

A   Mercator Projection

B   Lambert Equal-area Projection

C   Goode's Interrupted Homolosine Projection

D   Robinson Projection

Figure 1.4a–d   These assorted map projections each attempt to represent the three-dimensional surface of the earth on a two-dimensional sheet of paper. Note the name of each different projection included in this set of map figures and try to determine what particular map function that projection deals with. Look at the maps at the front of any good college atlas (these are from Goode's 20th edition published by Rand McNally) and you will be able to find examples and a detailed explanation of a variety of projections.

This sort of a proportional map is called a cartogram (Fig. 1.5a). As you compare the distribution of any given entity by differing gradients (colors or patterns) in choropleth maps (Fig. 1.5b), you can see a graphic distinction from the cartogram. In looking at the two maps in Figure 1.5, you can see how the different goals of the cartographer, as evidenced by the style of map selected, will influence the maps selected for a text or for a presentation. Even if symbolization by dots is retained, one map might use one dot per 100,000 people, whereas another uses one dot per 1 million people, giving the two maps a very different look.

Once a pattern is perceived, the question arises: Why is this located or distributed the way it is? The search for answers generally involves an attempt to explain relationships between the mapped subject and other phenomena that apparently influence its distribution or people's perceptions of it. In seeing the world in spatial terms, we can begin to see how geographers use maps as tools in understanding patterns of spatial distribution.

Mental Maps

Our understanding of location, however, is not completely objective, for each of us has our own personal sense of geography (i.e., a set of ever-changing mental maps), which include relative location, geographic and cultural characteristics, and accessibility. A **mental map** is the collection of personal geographic information that we use to order the images and facts we have about places, both local and distant. The mental map also serves to accommodate changes that one might imagine for a landscape (chapter opening photo). In this course, we will confront and reshape your mental map repeatedly as we introduce new images and facts about places that you have possibly never seen but about which you may have some impressions.

Geographers are keenly involved in attempting to understand the nature of place. Given a list of places, one's mind tends to produce an image for each of them, whether or not you have been there. For example, think about your reactions to these place names: New Orleans; Denver; Paris; Paris, Missouri; Surf City; your hometown; your neighborhood. Each of these place words conjures up a mental image in your mind — that is, draws information from your mental map of the world. Those images include both physical and cultural components. They provide a sense of place identity, which is an essential aspect of geographic learning.

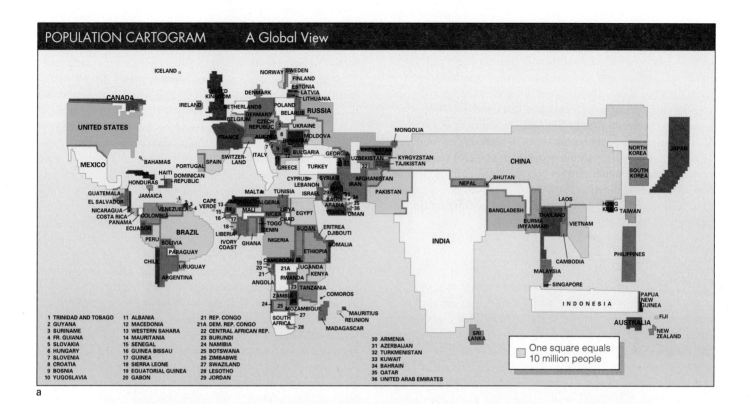

POPULATION CARTOGRAM · A Global View

1	TRINIDAD AND TOBAGO	11	ALBANIA	21	REP. CONGO	30	ARMENIA
2	GUYANA	12	MACEDONIA	21A	DEM. REP. CONGO	31	AZERBAIJAN
3	SURINAME	13	WESTERN SAHARA	22	CENTRAL AFRICAN REP.	32	TURKMENISTAN
4	FR. GUIANA	14	MAURITANIA	23	BURUNDI	33	KUWAIT
5	SLOVAKIA	15	SENEGAL	24	NAMIBIA	34	BAHRAIN
6	HUNGARY	16	GUINEA BISSAU	25	BOTSWANA	35	QATAR
7	SLOVENIA	17	GUINEA	26	ZIMBABWE	36	UNITED ARAB EMIRATES
8	CROATIA	18	SIERRA LEONE	27	SWAZILAND		
9	BOSNIA	19	EQUATORIAL GUINEA	28	LESOTHO		
10	YUGOSLAVIA	20	GABON	29	JORDAN		

One square equals 10 million people

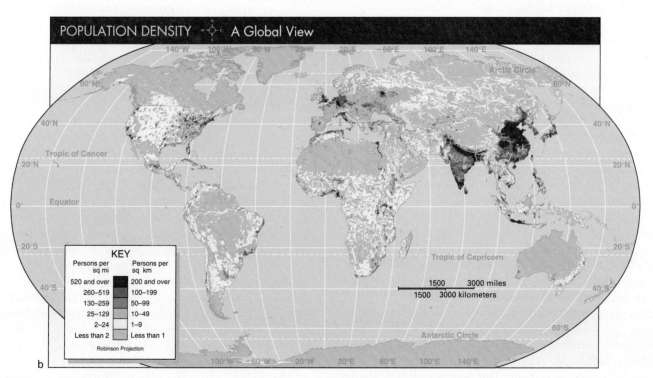

POPULATION DENSITY · A Global View

Figure 1.5 (a) This cartogram illustrates one of the ways cartographers represent different patterns on Earth. In a cartogram such as this one of world population by country, the shape and size of each country named differ in direct relationship to the size of its population. The cartogram provides the viewer with a graphic and geometric sense of a world ranking of national population or of any other item that the cartogram is designed to represent. (b) This is a choropleth map designed to show the approximate distribution of global population. The colors define six size units. By comparing 1.5a and 1.5b, you can see the different visual messages sent by cartograms and choropleth maps. The choropleth map is a common cartographic convention used to show the generalized distribution of everything from hospital beds to religious preferences through distinctive patterns of color or shading.

You will begin to add places to your own mental map of the world as you become increasingly confident in your study of the world's spatial relationships — the recurring patterns of roads and settlements, intersections with businesses, parking spots with less busy streets. Places most often take part of their identity from the human influences that have been important in shaping such settings. In this text, we will look continually at the interaction between human influences and the environmental setting. In the geographer's field of vision, phenomena such as settlement, economic activity, lines of transport and communication, and human movement all relate closely to the physical nature of place and to differing human perceptions of even the same places.

1.3   Worlds Within Worlds

One of the features that lends color and interest to geography is the opportunity to observe and interpret the extraordinary variety of landscapes that the world affords. This study carries intellectual as well as aesthetic rewards. A great deal of both the human record and nature's works is inscribed on the earth's surface, and both must be subjected to careful mapping and rigorous analysis if one is truly to understand and appreciate what is "written" on the landscape. From the earliest times, the landscape has been a data bank of human-environmental interaction and transformation (Fig. 1.6a).

The landscapes that humankind has created serve as a primary document — the record of where we have lived and traveled and of what we have done to try to make Earth more responsive to human needs. The process of reading theses landscapes to understand the intent of earlier peoples in their creation of cultural landscapes is called landscape analysis. In this exploration, we begin to better see and understand the worlds within worlds as we understand the nature of places nested within larger places, or regions. This process is focused on Places and Regions as the second of the Essential Elements in the National Geography Standards.

Some landscapes are intricate, irregular, and hard to characterize, whereas others are regular, sharply defined, and conducive to precise geographic description and characterization (see Fig. 1.1). In the study of landscapes, geographic analysis is most often drawn to the features that are typical — landscape units that repeat themselves and that are subject to mapping and orderly analysis. American small towns, for example, generally share the characteristics of a Main Street, small cafés, corner markets and bars, local churches, and adjacent farmland. Yet, even with such similarities, most small towns take pride in claiming that they possess individuality and uniqueness. Geographers are engaged by the aspects that are representative of a general, repetitive pattern, but they also find the idiosyncratic landscape characteristics fascinating. Often, present-day landscapes are described as suffering from visual blight and, thus, are uninteresting (or even menacing) to tourists. But these same landscapes convey significant geographic messages to someone trying to understand the geographic genesis and implications of a given scene (Fig. 1.6b).

Regions as Human Constructs

We have divided the world into regions (see Fig. 1.3) for this text because it is impossible to introduce order and logic to something as massive, complex, and diverse as the earth's surface without a system and organized framework built around smaller spatial units. The need for analysis of smaller units is the very genesis of the regional concept (Table 1.2). Regions are human constructs, and as such, they are collections of relatively homogeneous characteristics defined and connected through the process of human observation and analysis. In the Essential Elements of the National Geography Standards, Places and Regions are listed as the second of the Elements because of the central importance of these two spatial units to geographic understanding.

Think of your own view of the world. Many of you have probably planned vacation trips in part because of your perception of a location's distinct regional environmental characteristics. You have, in all likelihood, a reasonably well-developed sense of the regional characteristics of your home territory or the region where you have lived the longest. In your own personal mental maps, you have a feeling for climate, population characteristics, economic activity, cultural traits, and perhaps even for vegetation and landforms in your home region. But what of the same considerations for, say, upland Philippines? Libya? Iceland? The Texas-Mexico borderlands? We continually use the regional concept because of the desire we have to bring order to our mind and its collection of geographic images and information, but this concept is not always expressed in formal and objective ways (see chapter opening photo).

We have used an arbitrary but fairly common system of regionalization in this text. We have divided the world into eight regions. Each of these regions has varying features that provide internal cohesion and coherence. The opening chapters of the discussions of these eight regions — the Geographic Profile chapters (Chapters 3, 6, 8, 10, 15, 17, 19, and 22) — outline basic geographic characteristics and describe the sorts of physical, cultural, economic, and political features that create regional unity and identity. Not every region has the same elements of homogeneity, but each region can be identified by characteristics such as language, religion, location, or some other blend of features that connect the parts. Not all of Latin America, for example, speaks Spanish, but the Spanish language and Spanish cultural influence serve as a major feature of cohesion in the Latin American world region.

The route by which you explore the world in a textbook is, like regional borders, also selective. Once you read Chapters 1 and 2, however, you can really chart any route you wish through the text, preferably beginning with the profile chapter each time you go to a new world region and then reading all the subsequent regional chapters before turning to yet another region. As you grow to comprehend the physical and cultural features that develop regional identity, you are beginning to think like a geographer.

a

DAVID DOUBILET

b

KIT SALTER

Figure 1.6   (a) As geographers look at a physical setting from above, the question, "What sort of human activity is evident?" is often asked. In this aerial scene from a thinly settled region of southern Australia, the only evidence of human impact is the thin line of dirt road that winds along above the rugged coast. Yet that road adds a clear marker of human intent. Being part of a network of road communications means that life is touched by peoples and places very distant from this isolated setting. (b) This family is on the slopes of eastern Taiwan, opening up government mountain flanks for short-term farming. Their technology is simple, but their intensity of effort and ambition reflects on the human goals of making the landscape more productive and more responsive to a family's, or a community's, needs. The bicycle and walking are their link to lowland markets and the supplies that they will need before a crop comes in.

DEFINITIONS + INSIGHTS

The Regional Concept

Regions are human constructs. There are no "natural regions" on Earth. There are areas that are deserts and rain forests, but we are the creators of regions because it is the human decision-making process that puts boundaries on geographic space. In the real world, you could have one foot in a desert and one foot in a semi-arid zone. The regional boundary is a function of decisions made by cartographers and geographers and politicians as they segment the earth into worlds within worlds. In general, a region is a distinct segment of Earth space that is defined by the presence of one or more distinctive characteristics.

A **formal region** (also called a **uniform** or **homogeneous region**) is one in which all the population shares a defining trait. The most easily defined example is an administrative political unit such as a county or a state, where the regional boundaries are made explicit on a map. A **functional region** (also called a **nodal region**) is a spatial unit characterized by a central focus on some (often economic) activity. At the center of a functional region, the activity is most intense (e.g., the marketing area for a metropolitan

newspaper), and as you move toward the edges of the region, the defining activity diminishes in importance. A **vernacular region** (or **perceptual region**) is a region that exists in the mind of a large number of people (southern California), and it may play an important role in cultural identity (Dixie) but often does not possess explicit or objective borders (see chapter opening photo). The Rust Belt in the American Northeast is such a region. It has economic and cultural connotations, but 10 people might have 10 different definitions as to the northern or southern or western boundaries of such a vernacular region. The power of vernacular or perceptual regions is that the media make active use of such places and they become a shorthand for regional perception and place identity, even though the popular images of such regions may be highly inaccurate as defining characteristics of the locale. In all cases, however, regions are human constructs because whether we create formal regions with uniform characteristics or begin to allude to vernacular regions, human use and language become the defining force in such regional systems.

1.4 Physical Systems and the Forces of Nature

In the Beryl Markham excerpt from *West with the Night* earlier in this chapter, she wrote of some of the physical features that mapmakers have long used to orient people to the findings of explorers and earlier travelers. We have always used landmarks to give some visual cues to the landscape and its distinctiveness. Chapter 2 of our text introduces, explains, and illustrates the major elements of the physical systems that are fundamental to understanding geography. In virtually all scenes that you will be exposed to in the text, and certainly the great majority of the ones you will begin to wonder about as you add geographic perspectives to your life, the role of the physical setting is significant. Your learning from the next chapter will equip you with the skills to make better sense of the physical systems and the forces of nature that underlie our world. Physical Systems is the third of the six Essential Elements in the National Geography Standards.

TABLE 1.2   The Major World Regions: Basic Data

Political Unit	Area (thousand/sq mi)	Area (thousand/sq km)	Estimated Population (millions)	Annual Rate of Increase (%)	Estimated Population Density (sq mi)	Estimated Population Density (sq km)	Human Development Index[a]	Urban Population (%)	Arable Land (%)	Per Capita GNP (SUS)
Europe	2728	7067	511.7	0.06	270.9	104.6	0.894	74.4	26	19,036
Russia and the Near Abroad	8601	22,276	289.3	−0.21	33.6	13.0	0.754	64.4	10	5822
Middle East and North Africa	5920	15,334	445.4	2	75.2	29.0	0.667	54.7	6.7	5008
Monsoon Asia	8014	20,757	3377.7	1.4	421.5	162.7	0.657	35.4	19	4120
Pacific World	3306	8563	30.7	1.1	9.3	3.6	0.857	68.6	5.8	17,729
Africa South of the Sahara	8412	21,787	640.9	2.6	76.2	29.4	0.444	30.1	6.3	1591
Latin America	7941	20,568	525.1	1.7	66.1	25.5	0.757	74.6	6.1	6930
USA and Canada	7567	19,600	315.5	0.57	41.7	16.1	0.934	75.3	12	35,080
World Regional Totals	**52,491**	**135,951**	**6137**	**1.3**	**118.0**	**45.6**	**0.716**	**46**	**10**	**7200**
USA (for comparison)	3718	9629	284.5	0.6	77.0	29.7	0.934	75	19	36,200

[a]This is a numeric index developed by the United Nations that reflects gross domestic product per capita, literacy rate, level of education on average, and life expectancy. The maximum HDI possible is 1.0.

Sources: *World Population Data Sheet*, Population Reference Bureau, 2001; *U.N. Human Development Report*, United Nations, 2001; *World Factbook*, CIA, 2001.

1.5 Human Systems and Landscape Transformation

Human systems are the powerful mechanisms that people have created to make the earth more satisfying, more productive, and more clearly reflective of human needs and desires. The driving force in such modification of the environment is the power of culture and its associated technology. Cultural forces shape not only the political interactions of groups but very often also affect the physical appearance of landscapes from a local to global scale. This shaping act is called landscape transformation. Such a process is increasingly central to ever more people as cultural diffusion and globalization have accelerated the pace of change for almost all cultures. A group's culture includes the values, beliefs, aspirations, modes of behavior, social institutions, knowledge, and skills that are learned and transmitted within the group. It also includes the **material culture** — the group's tangible possessions and products that play a role in cultural expression and daily life.

Each culture group alters the environment in its own way, thereby creating a distinctive cultural landscape — the landscape as transformed by human action (see Figs. 1.6a and 1.6b). Cultures are dynamic and constantly changing. They change by learning and adapting traits from the cultures of other groups with whom they come in contact, a process known as cultural diffusion (Fig. 1.7). The theory of **environmental determinism,** in which the role of culture was seen as subordinate to the power of the physical environment and its influence on human actions, had its origin in the early geographies of the Greeks and Romans. It came to its strongest expression in the 19th century through the work of German geographer Friedrich Ratzel (1844–1904) and his American student, Ellen Churchill Semple (1863–1932). **Environmental possibilism** — the belief that humans always had a range of opportunities for cultural expression and economic activity in any given environment — grew out of the work of French geographer Paul Vidal de la Blache (1845–1918). In the early years of the 20th century, particularly through the work of Carl O. Sauer (1889–1975), culture began to be seen as a more powerful engine in the shaping of human landscape through differing social preferences about what development might occur in a specific environment. In this text, you will see that culture and environment are in continual tension simply because there are so many alternative ways to create a setting for human activity.

Because of the complexity of human systems and their central role in shaping cultural landscapes and world regional geography, this element of the National Geography Standards is divided into a series of themes that appear throughout this text. All influences of Human Systems (the fourth of the six Essential Elements) appear in each region but in differing force and impact. The listing of the Standards numbers in this next series relates to Table 1.1.

Human Populations (Standard 9). Central to all geographic analysis in human geography is the nature of population and population distribution (see Fig. 1.5b). The dimensions of

KIT SALTER

Figure 1.7 Cafés, restaurants, and even small eating stalls often are quick to evoke a regional or ethnic connection whether or not it is genuine. This sandwich and coffee stand in downtown Los Angeles promises "Kosher-style Burritos," and the people behind the counter are as likely to be Korean or black or Latino or white. The people who show up to test and taste this special burrito or any of the other fast foods will see that only the evocation is kosher, but the regional or ethnic image is effective in identifying the place to others.

growth, mortality, health, education, economic vitality, demographic patterns, migration, and mobility are some of the most frequently used indices of population activity. The next chapter introduces population themes; in your own efforts to make sense of the world, you will find that human population plays a dominant role in the geographic activities of settlement, landscape transformation, and resource utilization.

Cultural Mosaics (Standard 10). Recognizing the broadly differing cultural and architectural mosaics that occur in farming villages and urban settlements, the geographer recognizes landscape variety, ethnic differentiation, and cultural dissimilarity as part of daily reality. The political, cultural, and economic dynamics that are set in motion by these mosaics are part of the geographic patterns that are central to an understanding of the personality and identity of all places and regions you encounter, not only in this text, but at any scale — your dorm, neighborhood, or distant countries and regions of the world (Fig. 1.8).

Patterns of Economic Interdependence (Standard 11). Were ours a world of abundant resources, small populations, and rigid physical geographic mountain, water, and desert barriers set between peoples, there would be the potential for imagining worlds of really distinctive cultural and economic isolation. Although even today some populations do exist in relative isolation in tropical rain forests, on islands, or in mountain highlands, such isolation is unusual. The regions that this text explores are characterized by profound and increasing global interconnectedness. Economic geography is at the heart of this linkage: The extraction of raw materials from local resources, the movement of partially or fully processed goods to centers of manufacturing, and the critical movement of finished goods to markets are integral components both of economic and world regional geography. The geographer sees not only the characteristic elements that can be mapped in the study of this economic interdependence but also the human actions that created these specific patterns. Virtually every change associated with the development of human society arises from the interaction of human and physical systems; geography focuses on just such interactions.

Here in the beginnings of the 21st century, we tend to think of economic interdependence as reflecting our own country's intricate involvement in global economic trade networks. However, economic systems are interdependent constructions that can be as local as a grocery store selling apple cider from a nearby farmer or as global as the same store selling Swiss chocolates. The nature of this global interdependence depends on the level of economic development of the area. This concept of interdependence weaves through every chapter of our text.

Globalization is a term that will become increasingly common as we grow into the new century. When you see basketball shoes being designed in Europe, fabricated in East Asia, marketed in the United States, and traded as stock all across the world's markets, you begin to comprehend how absolutely

global economic connections are. Add to that reality the nature of political entanglements, human migration streams, and telecommunication linkages and you see how much of our future will be deeply influenced by economic, political, and demographic activities that take place not only halfway across the world, but all across the globe. The concept of globalization is yet another aspect of geography that is beginning to touch all of our lives.

As countries and regions industrialize and networks of geographic interdependence expand in extent and complexity, their occupational structure changes markedly. The number of farmers decreases, and as global population grows, their proportion in the population also steadily decreases. With improving technology and increasing farm consolidation (larger farmers buying up the farmland of smaller farmers who are retiring or leaving the land), the still remaining farmers often produce more. Employment in manufacturing, trade, and services becomes more dominant. Per-person industrial output and average incomes rise, and industrial and commercial cities grow in size and complexity.

Countries or regions begin to specialize in producing and exporting certain types of goods and services. Countries even become known for their distinctive landscapes and earn significant amounts of tourist income from such settings (Fig. 1.9). Such items are products of the basic industries that are the main elements of the area's economy. Understanding the geographic and other conditions relative to the growth of such industries (or their failure to develop) casts much light on both the economic and social character of countries and regions.

Since the end of World War II, trends have shown that some industrial societies are becoming postindustrial; that is, the industrial labor force begins to decrease in importance, and employment in trade and services becomes more dominant. In some countries, industrial output may substantially decrease, but in most industrial societies, productivity increases through technological innovations despite a smaller labor force. This has been true both in manufacturing and in agriculture. The overall welfare of such a society comes to depend largely on productivity and incomes in the trade and service sectors. Those human activities lead to the exchange of goods and the provision of services such as transportation, banking, and information management. The same forces, however, also lead to rapid change and industrial abandonment (Fig. 1.10). The geographer is always trying to identify and comprehend economic patterns on the landscape and must be continually conscious of the change in the scale from local to global in the study of any of these phenomena. It may help the student of geography to think of countries and regions as being either preindustrial, industrializing, industrial, or to some extent, postindustrial and to relate other geographic circumstances to current stages of economic development.

Human Settlement (Standard 12). One of the most important maps you can study is that of human settlement. It might be a sketch map given to you by a friend who owns a mountain cabin that you are trying to find in a snowstorm, or it could be a map showing the locations of the Midwest farm villages and

Figure 1.8   Regions are human constructs, and regional borders are decided on by human decision making. In any regional system, such as the three distinct anticipations of these three people, the criteria that define the region, or the land use, depend on the person doing the regional definition or the economic development. The potential future scenes, for example, envisioned by three people looking at undeveloped land may be quite distinct. Geography is concerned with understanding where these images come from, what might be necessary to bring such images to reality, and what the environmental consequences of such development might be. Knowing that, geography also will enable a person to define the regional characteristics likely to identify the locale after development.

SOURCE: THE KEY TO THE NATIONAL GEOGRAPHY STANDARDS. ILLUSTRATION BY SUZANNE DUNAWAY.

Figure 1.9   Rio de Janeiro, Brazil, occupies a spectacular setting of bays, peninsulas, islands, low mountains, and world-famous beaches. This landscape not only serves as a hallmark for world tourist trade, but it also demonstrates the smooth blend of physical and cultural landscape features often found in an urban settlement nested in a dramatic physical setting.

towns that were flooded in the Great Flood of '93. Where people settle, where they prosper, where they fail and leave, and where they seem just to persist are all elements of spatial and cultural significance. Settlement points on a map are beginning points in trying to understand a place, a region, and a human pattern. As Beryl Markham says in her earlier praise to the concept of the map, "each city [would] be a stranger to the next . . . " if maps were taken away and we were denied our cartographic icon for the networks of human settlements that exist.

Millions — in fact, billions — of people have been faced with location decisions about human settlement, and for most of history, rural settlements in different patterns have been selected (see Fig. 1.1). Studies of such decisions and the resulting patterns are central to the discipline of geography and a useful understanding of the world's globally interconnected regions and nations.

Forces of Human Cooperation and Conflict (Standard 13). Humankind seems not to be a peaceful species. From the beginning, there have been geographic tensions over space. Boundaries, resources, settlements, population movement, and population growth are some of the most contentious aspects of the human use of the earth. However, while the geography of military conflict seems to gain the attention of historians and journalists, there in fact has been a great deal of cooperation in human use of the earth. To better understand how regions and nations have created and maintained the places and patterns they possess, the significance of cooperation and conflict must be kept in mind continually. Many of the efforts of ongoing landscape transformation require cooperation in the reassignment of boundary patterns from the scale of neighbors dividing up a new driveway to countries recharting the channel of a river that flows as the border between them.

We live in a world that has been divided into a multitude of geographic units — nations (groups of people who possess common traits and occupy a specific area), sovereign states (autonomous political units), subdivisions of these (provinces,

Figure 1.10   All across the industrialized world, the melding of new technologies with traditional industrial ambition has begun to shift the location of major steelworking activities. This steel mill is the Katowice steelworks in southern Poland, the largest steel plant in the country. A century ago, scenes such as these would have characterized Pittsburgh, Pennsylvania, which has now closed down all of its steel plants except for some small specialty firms.

states, counties, etc.), and dependent political entities that re-main as legacies from a past colonial age. Separated by boundaries drawn on maps and in many cases shown by mark-ers on the ground, these political divisions form a compli-cated patchwork enclosing most of the land surface of Earth. The political status of any area is an important feature of its geography. In a broad sense, political status includes not only the political organization of the area but also the area's impact on regional and global politics and economics. Although these elements are at times difficult to define precisely and of-ten change in their significance, their influence cannot be ig-nored in a geographic study of regions and peoples.

Just as globalization is a term of importance in the context of economic interdependence, it is also central to understand-ing world patterns of cooperation and conflict. For example, North Korea's firing of ballistic missiles across the Sea of Japan in 1998 is an event that was not limited in impact to East Asia. Political centers and associated military bases all across the globe took quick note of such action. We are, as will be noted often in our text, all part of a global world. Such reality gener-ates both good and bad outcomes. The challenge to the geog-rapher is to determine what forces are at work at the different levels of human interaction because virtually all the world's landscapes reflect and exhibit cooperation as well as conflict.

1.6   Human-Environment Interaction

The relationship between environment and society represents one of the most critical interfaces examined in world regional geography. The ever-increasing influence that humankind has on the environment, not only in technology but also in population growth and patterns of consumption, has meant a rapid and serious reappraisal of human-environment interac-tion. At the same time, nature has shown that people have grown too confident of their authority over nature (Fig. 1.11). Almost a century and a half ago, George Perkins Marsh (1801–1882) authored *Man and Nature* (1864), setting forth his almost unprecedented view that the human species had been a quite considerable force for change in nature.[1] The in-fluence of technology and population size are the major causes of Marsh's nontraditional perspective. It has been in the last several thousand years, and especially in the most re-cent century, that humankind has had a truly powerful impact on the earth's surface. In the decades since author Rachel Carson (1907–1964) wrote *Silent Spring* (1962), the world has become ever more concerned about the mounting impact of the search for and depletion of resources, the transfer of wa-ter, the relentless human modification of the environment, and our increasing dependence on petrochemical fixes.[2]

Figure 1.11   Erosion caused by storm waves and rainfall can undermine whole settlements that have been placed close to the sea because of the human fascination with view and proximity to water. Even though human response to the damage caused in such settings is one of regret, the odds are that people will rebuild in the same environmental setting after this particular storm and erosional event because of continued attraction to the geographic features of such a setting.

1.7   Putting Geography to Use in Understanding Our World

Geography is not intended to be simply a class idea for one se-mester, or even for a full year. It is our hope that the perspec-tives you develop in this text and class become Geography for Life: what you want in the way of a place to live; where you go to vacation; how you react to potential environmental changes in your neighborhood or city or state; and how you learn to conserve resources and appreciate the environment and land-scapes around you. This is all geography. This perspective is central to our text. Knowledge of geography is not just "state capitals" or "capes and bays" (the two most commonly used ca-sual references to the content of geography). It is understanding the ways in which humankind perceives and interacts with Earth's surface, its resources, and its peoples in the present and over time, both locally and globally. This broad perspective leads to a wide variety of careers and professions within the field of geography. For the National Geography Standards, it is the Uses of Geography that makes up the final of the six Essen-tial Elements, and those final two Standards help illustrate the ways in which a geographic perspective can add real dimen-sions of meaning to events of the past and can serve as tools for more intelligent planning of the future.

1.8   Taking Geography to the Street: Careers in Geography

This textbook makes use of specialized work by geographers in many regional, systematic (the topical study of various physical or cultural geographic elements), and technical fields (Fig. 1.12). Such specializations have gradually devel-oped as geographic knowledge about world complexities has increased. Today, the Association of American Geographers

---

[1] George Perkins Marsh, *Man and Nature*, David Lowenthal (ed.). Cam-bridge, Mass.: Harvard University Press, 1965.
[2] Rachel Carson, *Silent Spring*. Boston, Mass.: Houghton Mifflin, 1962.

Box
292
(AAG) lists more than 50 specialty groups ("proficiencies") among its approximately 6,500 members. Regional specialties of the members show a preference for Western rather than non-Western areas. The relative dearth of U.S. geographers with professional expertise in major developing areas is a disadvantage, but it does open opportunities for younger geographers willing to undertake the challenge of language study and gaining genuine understanding of non-Western areas.

Most U.S. geographers emphasize systematic specialties over regional specialties. Specialists in physical geography study spatial patterns and associations of natural features. Prominent subfields include geomorphology (the study of landforms and a field in which geography intersects with geology), climatology (climatic processes and patterns), biogeography (the study of biotic resources), and soils geography. Closely related to physical geography is the large interdisciplinary field of environmental studies, which is concerned with reciprocal relationships between society and the environment. There is a whole subfield here related to natural hazards because of their impact on the interface of human activity and environmental processes. Another allied specialty is medical geography, which focuses on spatial associations between the environment and human health and on locational aspects of disease and health-care delivery.

Specialists in economic geography study spatial aspects of human livelihood. Major occupations and products are the emphasis, and there is an active theoretical component that intersects with the field of economics. Agricultural geography, manufacturing geography, and transportation geography are important subfields that deal with these major aspects of our networks of inter-

action and the transformation of land and resources. Marketing geography is a relatively small but active applied offshoot of economic geography and is of particular importance in the continuing expansion of residential areas on the edges of cities of all sizes. The study of economic development is an extremely important interdisciplinary field in which many geographers specialize. This text pays much attention to contrasts (expressed or implied) between "more developed countries (MDCs)" and "less developed countries (LDCs)" with respect to the economy. Urban geography studies the locational associations, internal spatial organization, and functions of cities.

Geographers specializing in cultural geography concern themselves with places of origin ("culture hearths"), diffusion, interactions, landscape evidence and analysis, and regionalization of culture. Closely allied to cultural geography is the field of cultural ecology, which focuses on the relationship between culture and environment. Social geography deals with spatial aspects of human social relationships, generally in urban settings; population geography assesses population composition, distribution, migration, and demographic shifts. Political geography studies topics such as spatial organization of geopolitical units, international power relationships, nationalism, boundary issues, military conflicts, and regional separatism within states. Historical geography studies the geography of past periods and the evolution through time of geographic phenomena such as cities, industries, agricultural systems, and rural settlement patterns.

Technical specialties in geography include cartography, computer cartography, remote sensing, quantitative methods (mathematical model building and analysis), air photo interpretation, and Geographic Information Systems (GIS). The last of these — GIS, which involves the computer manipulation and analysis of spatial data — is a very successful interface between the analysis of spatial data and the computer (Fig. 1.13). There is increasing use of this technique in everything from spatial analysis to model building to sophisticated cartography. Because so much of the data utilized in the management of cities, businesses of all scales, government, and even agriculture are **digital data,** the computer has become central to all of these fields. With the capacity to array such data in a geographic and cartographic form, an understanding of GIS has become a highly marketable skill.

Increasingly, departments of geography are developing curricula and laboratories to train students in the study of spatial statistics, GIS classes, computer cartography, remote sensing, and digital image processing so that geographers can be active players in this field of professional growth and expanding significance. All of these technique-oriented specialties center on the cartographic (mapped) expression of spatial data, while remote sensing and air photo interpretation utilize various kinds of satellite imagery or photo coverage to assess land use or other geographic patterns. The expanding use of the computer offers geographers greater speed, accuracy, and especially, facility in incorporating new and changing data into the creation of maps, graphs, and charts. This has become an area of major professional activity for geographers.

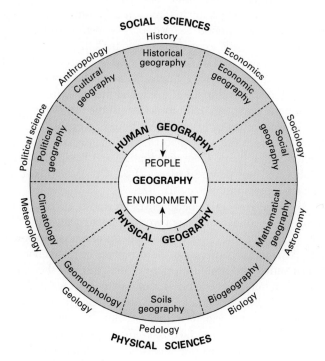

Figure 1.12   Selected subfields of geography. This diagram identifies the main subject-matter areas within human geography and physical geography and links them with the most closely related disciplines in the social sciences and the physical sciences.

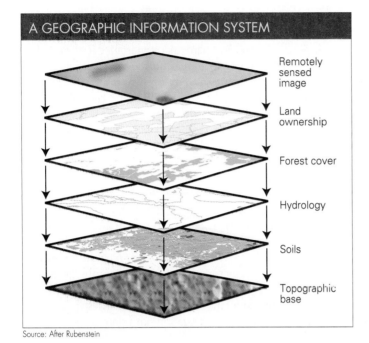

A GEOGRAPHIC INFORMATION SYSTEM

Remotely sensed image

Land ownership

Forest cover

Hydrology

Soils

Topographic base

Source: After Rubenstein

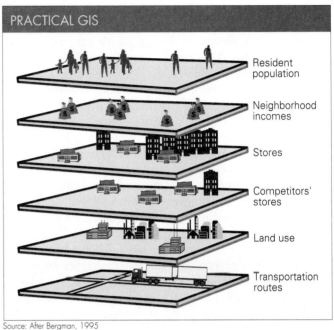

PRACTICAL GIS

Resident population

Neighborhood incomes

Stores

Competitors' stores

Land use

Transportation routes

Source: After Bergman, 1995

Figure 1.13   The use of the computer in geographic analysis and the mapping sciences has been an area of rapid development in the past three decades. There is increasing career potential in Geographic Information Systems (GIS), and many departments of geography have developed computer labs for research and instruction in map design and production, for data display and analysis, and for the easier and more immediate updating of various layers of geographic information.

Career opportunities in geography are found in government agencies at the local, state, regional, and federal levels (see Fig. 1.12). Firms working with land-use decisions at all scales use geography and geographers, particularly in the areas of GIS, remote sensing, and computer cartography. Retail firms find the problem-solving skills inherent in geographic analysis useful in management, survey design and implementation, estimation of public response to innovations or expansion to new locations, and policy shifts. Issues of environmental perception, management, shifts in utilization, and description all represent career options for geographers. The most essential lesson to learn as one pursues a career or profession that utilizes skills gained in studying geography is that, although there may be few classified ads that proclaim "Geographer wanted!" there are worlds of opportunity in those same pages that speak to the very skills and learning represented in the themes we bring to you in this world regional geography text.

Human history has been shaped by the interplay of geographic influences and human ambitions, whether at the scale of a peasant farmer deciding to irrigate his fields to better feed his family or of an army commander trying to expand his realms of military authority through blasting open a new mountain pass for troop movement. Knowing about the spatial distribution of resources, mountain passes, ethnic populations, animal herds, and settlements, for example, means that one can achieve a clearer understanding of the past. One of the prime uses of geography, then, is to be able to determine and understand the map of the locale being studied at the time of the event you are attempting to comprehend.

Understanding one place helps prepare you for seeing the meaning of yet other places. In this respect, no place is truly an island, although some locales are more insular than others; nor are places the same in the magnitude, intensity, and reach of the geographic effects they engender. For example, a hurricane in Florida will have an impact on citrus prices in Alaska. The return of Hong Kong to China in 1997 influenced financial markets all around the world. The regional electricity shortages California experienced early in 2001 had a definite impact on the potential for further growth and development of the state's Silicon Valley. Such effects may be physical, economic, political, cultural, or social, and they arise from trade, investment, migration, military action (or the threat of it), and other mechanisms of spatial interaction. Especially large and complex effects are generated by such major nations as the United States, Japan, Germany, France, and Brazil, but many smaller and less powerful places (Iraq, Cuba) may have an impact disproportionate to their size.

All of the geographic perspectives previously discussed play a role in our understanding of world regional geography. The role of the map, the distributional patterns of human resources and physical properties, and the dynamics of interaction between varied and significant players and places in the drama have keen importance to our making sense of the world. Many of the problems, for example, that your generation will face are problems that are evident in today's cultural landscape — that is, the landscape transformed by human action. We want you to see a benefit in exploring the world's varied regions and nations through a geographer's eyes and in developing a geographer's field of vision. Studying the many engaging meanings of geography will help achieve such understanding and will lead to an appreciation of geography for life.

CHAPTER SUMMARY

- Geography is the study of the earth's surface and the analysis of its spatial distributions and patterns of human and physical characteristics, brought to light not only through objective analysis but also through the study of landscapes in literature and regional popular culture.
- The 1994 National Geography Standards are a recent tool created for spatial analysis. These are built around six Essential Elements (including the world in spatial terms, places and regions, physical systems, human systems, environment and society, and the uses of geography) and 18 geography standards, all of which are designed to illustrate what American students at grades 4, 8, and 12 should know and be able to do in geography.
- Central to such learning are maps, map projections, map scale, map symbols, mental maps, latitude and longitude, cartograms, and other aspects of cartography. Such tools are seen as the language of maps, and such language is vital to both geographic analysis and the geographer's field of vision.
- Regions are human constructs, and as such, they have an arbitrary and fluid quality. Critical to their existence is a clear definition of the criteria that are utilized to define such regions and the resultant regional maps that are created. For this text, we have created eight world regions.
- The human systems incorporated in the National Geography Standards include themes of human population, cultural mo-

saics, patterns of economic interdependence, human settlement, and forces of cooperation and conflict. The interplay of these human realms is constant and operates at all scales in world regional geography.
- Environment and society are themes of human-environmental interaction that focus on the ways in which humans modify the physical environment, how the physical environment influences human activity, and how the human definition and utilization of resources change over time.
- Human history has been shaped continually by the nature of our geographic setting. Geography serves as a useful tool not only in understanding past history but also in the geographic analysis of contemporary human activities and in planning for a better-thought-out future development.
- Increasing sophistication in the human capacity to understand, shape, and analyze spatial patterns through digital data has expanded career options for people working in geography. Careers in Geographic Information Systems (GIS) and many other realms of landscape analysis, environmental management, and government work in land management utilizing computers and digital imaging are continually expanding in number and significance. Growing attention to the location, careful stewardship, and conservation of resources at both the local and global level has also provided an expanding base for careers in geography.

REVIEW QUESTIONS

1. Why is author Beryl Markham such a believer in the importance of a map?
2. List the six Essential Elements in the National Geography Standards.
3. What geographic features make Great Britain and New Zealand different?
4. List the eight world regions that are utilized in this text.
5. Define map projection and list four kinds shown in the text.
6. Define and explain the use of latitude and longitude.
7. Define cartogram and explain why it has such a different look to it than a world map.
8. Define mental map and give an example of a place in your own mental map.
9. List the five themes noted under Human Systems in the National Geography Standards.
10. List examples of scenes that might have appeared in Figure 1.11 that are from your own region in the United States. Why do people live in such threatening environmental settings?
11. Name five professions that do not have the word "geography" in them but that clearly use some of the skills that are part of the geographer's field of vision.

DISCUSSION QUESTIONS

1. What kinds of geographic questions might occur in the conversation that opens Chapter 1?
2. What sort of discussion might have taken place in the creation of the map that is in the chapter opening photo?
3. Talk about the ways in which your daily life is involved with spatial decisions and spatial analysis.
4. Discuss the ways in which the relative isolation of New Zealand compared to Great Britain has been significant for both nations.
5. List and discuss possible world regional systems other than the one used in creating the eight world regions in this text. What might make such alternative regional systems more or less useful than the one used in this text?
6. Paraphrase the three distinctive perception systems illustrated in

Figure 1.8. Can you explain what is driving each one? Why are they competitive? Can you think of other ways in which that landscape might be viewed in that cartoon?
7. Discuss maps as the language of geography. What sorts of things can you say with that language?
8. Select any two of the six Essential Elements associated with the National Geography Standards and show how landscape reflects their influence.
9. Discuss at least one way you can define and utilize each of the 18 National Geography Standards.
10. Looking at the components of Figure 1.12, discuss the ways in which geography relates to the various subfields.

CHAPTER 2

Physical and Human Processes That Shape World Regions

Early morning on the Amazon River near Leticia, Colombia. The fuel powering the boat motors may have been imported from the Middle East. Global processes of change are affecting even Earth's most remote and wild locales.

Earth has an astonishing variety of natural landscapes and a wealth of natural resources. These exist within dynamic systems. We know, for example, that natural processes have brought about the extinction of more than 90% of the plant and animal species that have ever lived on Earth. Other forms have taken their place including, within the recent geological past, the primates known as hominids. One of these species, *Homo sapiens*, emerged from the pack only within the last 40,000 years to become the most ecologically dominant animal species of all time. Just about 10,000 years ago, this species called human began changing the genetic makeup and behaviors of plants and other animals to meet its needs. People thus began a process of unprecedented landscape change. That process has accelerated exponentially in the past three centuries, and the most rapid, far-reaching changes in landscapes and natural systems that have ever taken place are happening now, in your lifetime.

Above all, geographers are concerned with the interaction of people and nature, with a spatial and often historical approach. This chapter is an introduction to geography's basic vocabulary about how the natural world works and how people have interacted with it to change the face of the earth. We begin by identifying patterns of climate and vegetation that set the global environmental stage for human activities. Then we consider how that stage may be changing as a result of human impacts on the atmosphere. We look at modern trends in people-land associations as the products of revolutionary changes in the past: the arrival of agriculture and industrialization. We see where rich and poor countries are located on Earth's surface and explain some of the causes of these patterns of prosperity and poverty. We consider where and why populations are increasing and what the implications of that growth are. Finally, we look at ideas about how to solve some of the most important global problems of our time.

2.1   Patterns of Climate and Vegetation

As you experience a warm, dry, cloudless summer day or a cold, wet, overcast winter day, you are encountering the **weather** — the atmospheric conditions occurring at a given time and place. Climate is the average weather of a place over a long time period. Along with surface conditions such as elevation and soil type, climatic patterns have a strong correlation with patterns of natural vegetation and in turn with human opportunities and activities on the landscape. Precipitation and temperature are the key variables in weather and climate.

Precipitation

Water is essential for life on Earth, and it is vital that we consider the processes of weather and climate that distribute this precious resource (see Fig. 2.1). Warm air holds more moisture than cool air, and precipitation — rain, snow, sleet, and hail — is best understood as the result of processes that cool the air to release moisture. Precipitation results when water vapor in the atmosphere cools to the point of condensation, changing from a gaseous to a liquid or solid form. The amount of cooling necessary depends on the original temperature and the amount of water vapor in the air.

For this cooling and precipitation to occur, generally air must rise in one of several ways. In equatorial latitudes or in the high-sun season (summer, when the sun's rays strike Earth's surface more vertically) elsewhere, air heated by intense surface radiation can rise rapidly, cool, and produce a heavy downpour of rain, an event that occurs often in summer over much of the United States. Precipitation that originates in this way and is released from tall cumuloform clouds is called **convectional precipitation** (Fig. 2.2). **Orographic (mountain-associated) precipitation** results when moving air strikes a topographic barrier and is forced upward. Most of the precipitation falls on the windward side of the barrier, and the lee (sheltered) side is likely to be excessively dry. Such dry areas — for example, Nevada, which is on the lee side of the Sierra Nevada in the western United States — are in a **rain shadow.** Rain shadows are the primary cause of arid and semiarid lands in some regions.

Cyclonic or **frontal precipitation** is generated in traveling low-pressure cells, called **cyclones,** that bring air masses with different characteristics of temperature and moisture into contact (Figs. 2.2 and 2.3). A cyclone may overlie hundreds or thousands of square miles of Earth's surface. In the atmosphere, air moves from areas of high pressure to areas of low pressure. Air from masses of different temperature and moisture characteristics is drawn into a cyclone, which has low pressure. One air mass is normally cooler, drier, and more stable than the other. Such masses do not mix readily and

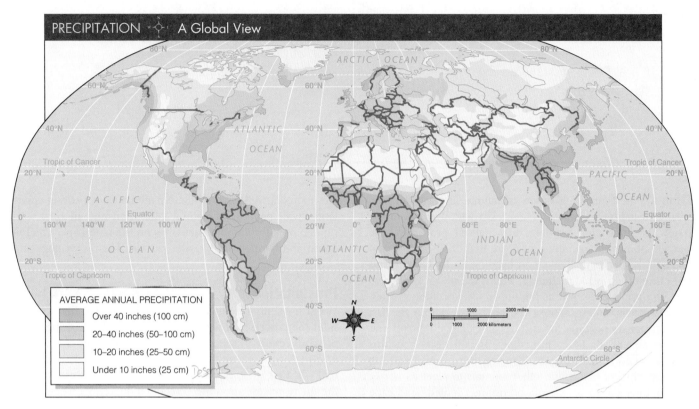

Figure 2.1 World precipitation map, modified from a map by the U.S. Department of Agriculture.

tend to retain distinctive characteristics. They come in contact within a boundary zone 3 to 50 miles wide (c. 5 to 80 km) called a **front**. A front is named according to which air mass is advancing to overtake the other. In a cold front, the colder air wedges under the warmer air, forcing it upward and back. In a warm front, the warmer air rides up over the colder air, gradually pushing it back. But whether a warm or cold front, precipitation is likely to result because the warmer air mass rises and condensation takes place.

Large areas of the earth receive very little precipitation because of subsiding air masses having high atmospheric pressure, called **high-pressure cells** or **anticyclones** (see Fig. 2.4). In a high-pressure cell, the air is descending and so becomes warmer as it comes under the increased pressure (weight) of the air above it. As it warms, its capacity to hold water vapor increases, its relative humidity decreases, and the result is minimal condensation and precipitation. Streams of dry, stable air moving outward from the anticyclones often bring prolonged drought to the areas below their path. Most famous of these are the **trade winds** — streams of air that originate in semipermanent anticyclones on the margins of the tropics and are attracted equatorward (becoming more moisture-laden as they do so) by a semipermanent low-pressure cell, the equatorial low. As Figure 2.3 illustrates, high-pressure cells rotate clockwise in the Northern Hemisphere and counterclockwise in the Southern Hemisphere. Conversely, low-pressure systems rotate counterclockwise in the Northern Hemisphere and clockwise in the Southern Hemisphere.

Cold ocean waters are responsible for the existence of coastal deserts in some parts of the world, such as the Atacama Desert in Chile and the Namib Desert in southwestern Africa. Here, air moving from sea to land is warmed. Instead of yielding precipitation, the air's capacity to hold water vapor is increased and its relative humidity is decreased; the result is little precipitation. But many areas of extremely low precipitation result from a combination of influences. The Sahara of northern Africa, for example, seems to be primarily the result of high atmospheric pressure. The rain-shadow effect of the Atlas Mountains and the presence of cold Atlantic Ocean waters along its western coast also contribute to the Sahara's dryness.

Temperature

In the middle and high latitudes, the most significant factor in determining temperatures is seasonality, which is related to the inclination of Earth's polar axis as the planet orbits the sun over a period of 365 days (Fig. 2.5). On or about June 22, the first day of summer in the Northern Hemisphere, the northern tip of Earth's axis is inclined toward the sun at an angle of $23\frac{1}{2}$ degrees from a line perpendicular to the plane of the ecliptic (the great circle formed by the intersection of Earth with the plane of Earth's orbit around the sun). This is the **summer solstice** in the Northern Hemisphere and the **winter solstice** in the Southern Hemisphere. A larger portion of the Northern Hemisphere than of the Southern Hemisphere remains in daylight, and warmer temperatures prevail. On or about September 23, and again on or about March 20, Earth

TYPES OF PRECIPITATION

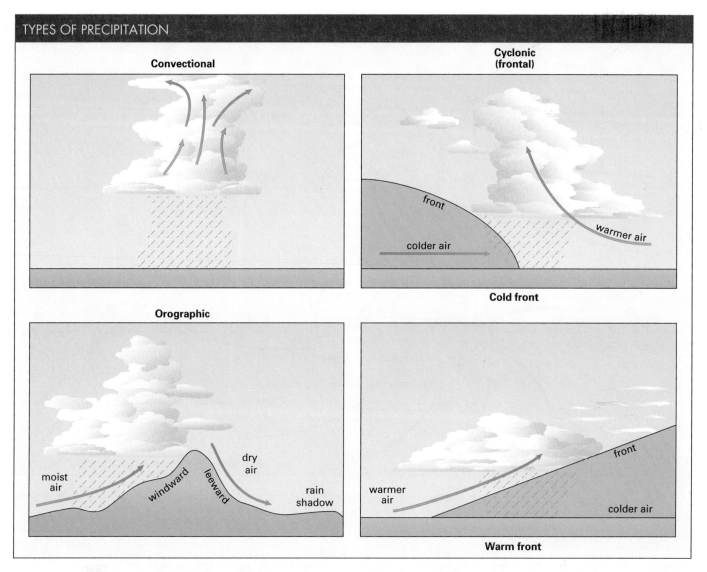

Figure 2.2 Diagrams showing origins of convectional, orographic, and cyclonic precipitation. Note that cyclonic precipitation results from both cold fronts and warm fronts.

reaches the **equinox** position. Its axis does not point toward or away from the sun, so days and nights are of equal length at all latitudes on Earth. On or about December 22, the first day of winter in the Northern Hemisphere, the southern tip of

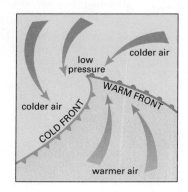

Figure 2.3 Diagram of a cyclone in the Northern Hemisphere.

Earth's axis is inclined toward the sun at an angle of 23 1/2 degrees from a line perpendicular to the plane of the ecliptic. This is the winter solstice in the Northern Hemisphere and the summer solstice in the Southern Hemisphere. A larger portion of the Southern Hemisphere than of the Northern Hemisphere remains in daylight, and warmer temperatures prevail.

Great regional differences exist in the world's annual and seasonal temperatures. In lowlands near the equator, temperatures remain high throughout the year, while in areas near the poles, temperatures remain low for most of the year. The intermediate (middle) latitudes have well-marked seasonal changes of temperature, with warmer temperatures generally in the summer season of high sun and cooler temperatures in the winter season of low sun (when sunlight's angle of impact is more oblique and daylight hours are shorter). Intermittent incursions of polar and tropical air masses increase the variability of temperature in these latitudes, bringing unseasonably cold or warm weather.

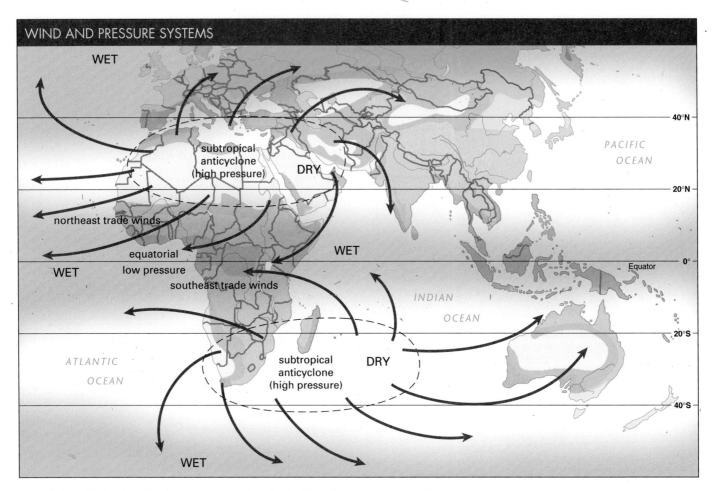

Figure 2.4 Idealized wind and pressure systems. Irregular shading indicates wetter areas.

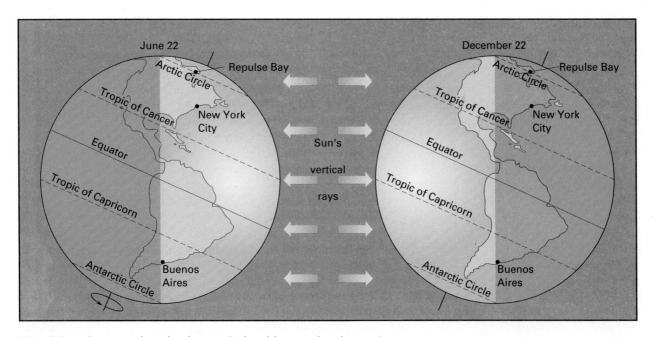

Figure 2.5 Geometric relationships between Earth and the sun at the solstices.

Climate and Vegetation Types

The varied combinations of precipitation, temperature, latitude, and elevation are characteristic of a great variety of local climates. Geographers group these local climates into a limited number of major climate types (Fig. 2.6), each of which occurs in more than one part of the world and is associated closely with other types of natural features, particularly vegetation. Geographers recognize 10 to 20 major types of terrestrial ecosystems, called **biomes,** which are categorized by dominant type of natural vegetation (Figs. 2.7 and 2.8). Climate plays the main role in determining the distribution of biomes, but differing soils and landforms may promote different types of vegetation where the climatic regime is essentially the same. Vegetation and climate types are sufficiently related that many climate types take their names from vegetation types — for example, the tropical rain forest climate and the tundra climate. You may easily see the geographic links between climate and vegetation by comparing the maps in Figures 2.6 and 2.7. The spatial distributions of climate and vegetation types do not overlap perfectly, but there is a high degree of correlation.

In the **icecap, tundra,** and **subarctic climates,** the dominant feature is a long, severely cold winter, making agriculture difficult or impossible. The summer is very short and cool. Perpetual ice in the form of glaciers may be found at very high elevations (even in equatorial regions) and in the polar realms; the icecap biome (Fig. 2.8a, p.32) is devoid of vegetation, except in those very few spots where enough ice or snow melts in the summer to allow tundra vegetation to grow. Tundra vegetation (Fig. 2.8b) is composed of mosses, lichens, shrubs, dwarfed trees, and some grasses. Needleleaf evergreen coniferous trees can stand long periods when the ground is frozen, depriving them of moisture. Thus, **coniferous forests,** often called boreal forests or *taiga,* their Russian name (Fig. 2.8c), occupy large areas where there is subarctic climate.

In **desert** and **steppe** climates, the dominant feature is aridity or semiaridity. Deserts and steppes occur in both low and middle latitudes. Agriculture in these areas usually requires irrigation. Earth's largest dry region extends in a broad band across northern Africa and southwestern and central Asia. The deserts of the middle and low latitudes are generally too dry for either trees or grasslands. They have **desert shrub vegetation** (Fig. 2.8d), and some areas have practically no vegetation at all. The bushy desert shrubs are **xerophytic** (literally, "dry plant"), having small leaves, thick bark, large root systems, and other adaptations to absorb and retain moisture. Grasslands dominate in the moister steppe climate, a transitional zone between very arid deserts and humid areas. The biome composed mainly of short grasses is also called the **steppe** or **temperate grassland** (Fig. 2.8e). The temperate grassland region of the United States and Canada originally supported both tall grass and short grass vegetation types, known in those countries as prairies.

Rainy low-latitude climates include the **tropical rain forest climate** and the **tropical savanna climate.** The critical difference between them is that the tropical savanna type has a pronounced dry season, which is short or absent in the tropical rain forest climate. Heat and moisture are almost always present in the tropical rain forest biome (Fig. 2.8f), where broadleaf evergreen trees dominate the vegetation. In tropical areas with a dry season but still having enough moisture for tree growth, **tropical deciduous forest** (Fig. 2.8g) replaces the rain forest. Here, the broadleaf trees are not green throughout the year; they lose their leaves and are dormant during the dry season and then add foliage and resume their growth during the wet season. The tropical deciduous forest approaches the luxuriance of tropical rain forest in wetter areas but thins out to low, sparse **scrub and thorn forest** (Fig. 2.8h) in drier areas. **Savanna** vegetation (Fig. 2.8i), which has taller grasses than the steppe, occurs in areas of greater overall rainfall and more pronounced wet and dry seasons.

The humid middle-latitude regions have mild to hot summers and winters ranging from mild to cold, with several types of climate. In the **marine west coast climate,** occupying the western sides of continents in the higher middle latitudes (e.g., in the Pacific Northwest region of the United States), warm ocean currents moderate the winter temperatures, and summers tend to be cool. Coniferous forest dominates some cool, wet areas of marine west-coast climate; a good example is the redwood forests of northern California.

The **Mediterranean climate** (named after its most prevalent area of distribution, the lands around the Mediterranean Sea) typically has an intermediate location between a marine west-coast climate and the lower latitude steppe or desert climate. In the summer high-sun period, it lies under high atmospheric pressure and is rainless. In the winter low-sun period, it lies in a westerly wind belt and receives cyclonic or orographic precipitation. **Mediterranean scrub forest** (Fig. 2.8j), known locally by such names as *maquis* and *chaparral,* characterizes Mediterranean climate areas. Because of hot, dry summers, the natural vegetation consists primarily of xerophytic shrubs.

The **humid subtropical climate** occupies the eastern portion of continents between approximately 20° and 40° of latitude and is characterized by hot summers, mild to cool winters, and ample precipitation for agriculture. The **humid continental climate** lies poleward of the humid subtropical type; it has cold winters, warm to hot summers, and enough rainfall for agriculture, with the greater part of the precipitation in the summer. In middle-latitude areas with these two climate types, a **broadleaf deciduous forest** or mixed broadleaf coniferous forest (as in northeastern North America; Fig. 2.8k) is found. As cold winter temperatures freeze the water within reach of plant roots, broadleaf trees shed their then-colorful leaves and cease to grow, thus reducing water loss. They then produce new foliage and grow vigorously during the hot, wet summer. Coniferous forests can thrive in some hot and moist locations where porous sandy

Figure 2.6   World distribution of the types of climate discussed in this text.

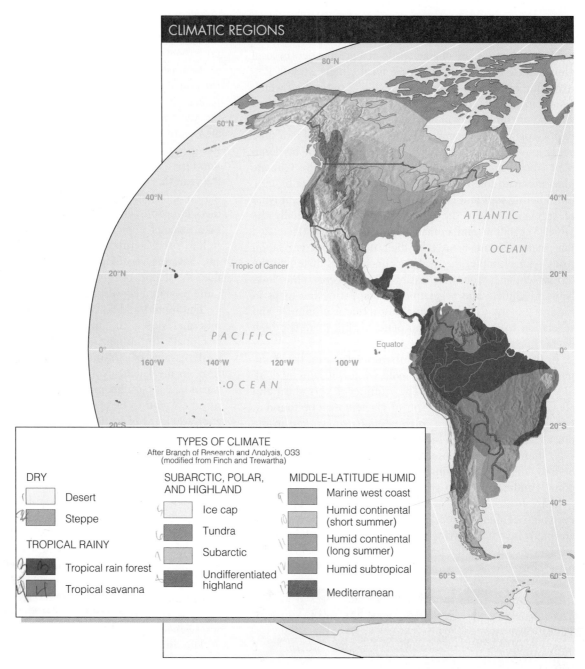

CLIMATIC REGIONS

TYPES OF CLIMATE
After Branch of Research and Analysis, O33
(modified from Finch and Trewartha)

DRY
- Desert
- Steppe

TROPICAL RAINY
- Tropical rain forest
- Tropical savanna

SUBARCTIC, POLAR, AND HIGHLAND
- Ice cap
- Tundra
- Subarctic
- Undifferentiated highland

MIDDLE-LATITUDE HUMID
- Marine west coast
- Humid continental (short summer)
- Humid continental (long summer)
- Humid subtropical
- Mediterranean

---

soil allows water to escape downward, giving conifers (which can withstand drier soil conditions) an advantage over broadleaf trees. Pine forests on the coastal plains of the southern United States are an example.

Undifferentiated highland climates exhibit a range of conditions according to elevation and exposure to wind and sun. Undifferentiated highland vegetation (Fig. 2.8l) types differ greatly depending on elevation, degree and direction of slope, and other factors. They are "undifferentiated" in the context of world regional geography because a small mountainous area may contain numerous biomes, and it would be impossible to map them on a small scale. The world's mountain regions

have a complex array of natural conditions and opportunities for human use. Increasing elevation lowers temperatures by a predictable "lapse rate," on average about 3.6°F (2.0°C) for each increase of 1,000 feet (305 m) in elevation. In climbing from sea level to the summit of a high mountain peak near the equator (e.g., in western Ecuador), a person would experience many of the major climate and biome types to be found in a sea level walk from the equator to the North Pole!

Landforms — and such factors as their elevation, latitude, soil type, and the processes that change them — have enormous relevance to the distributions of plants and animals and to human uses of resources. This chapter does not introduce

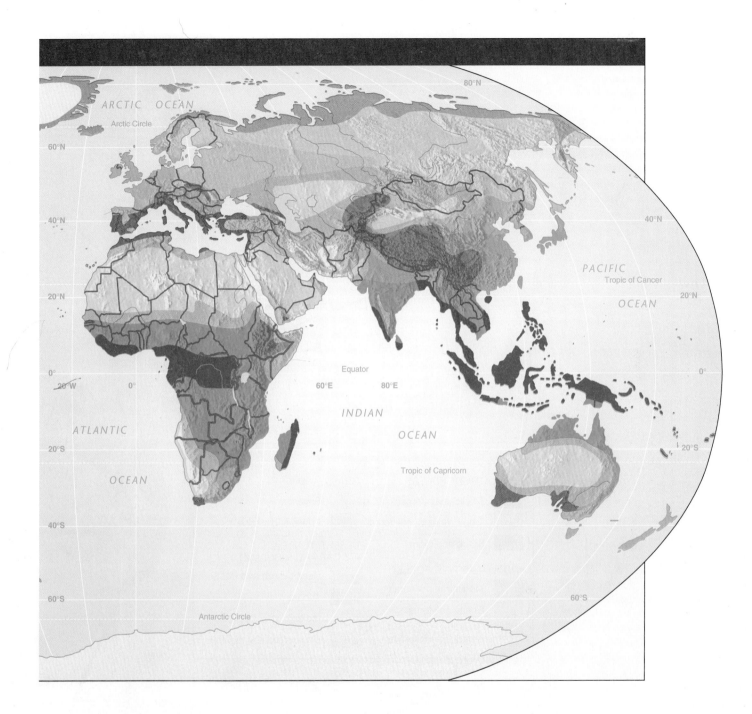

the world's landforms; these are dealt with in the following regional chapters, as are processes such as plate tectonics. For a global view, you may wish to look at the map of Earth's landforms inside the book's front cover.

2.2 Biodiversity

Anyone who has admired the nocturnal wonders of the "barren" desert or appreciated the variety of the "monotonous" Arctic can vouch for an astonishing diversity of life even in the biomes least hospitable to life. But geographers and ecologists recognize the exceptional importance of some biomes because of their **biological diversity** (or **biodiversity**) — the number of plant and animal species present and the variety of genetic materials these organisms contain.

The most diverse biome is the tropical rain forest. From a single tree in the Peruvian Amazon region, entomologist Terry Erwin recovered about 10,000 insect species. From another tree several yards away, he counted another 10,000, many of which differed from those of the first tree. Alwyn Gentry of the Missouri Botanical Garden recorded 300 tree species in a single-hectare (2.47-acre) plot of the Peruvian rain forest. Just 40 years ago, scientists calculated that there were 4 to 5 million species of plants and animals on Earth.

Figure 2.7 World biomes (natural vegetation) map.

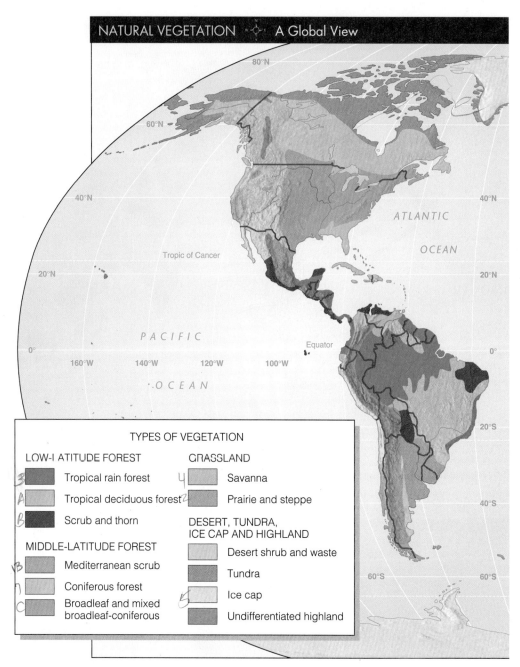

NATURAL VEGETATION ⟶ A Global View

TYPES OF VEGETATION

LOW-LATITUDE FOREST

Tropical rain forest

Tropical deciduous forest

Scrub and thorn

MIDDLE-LATITUDE FOREST

Mediterranean scrub

Coniferous forest

Broadleaf and mixed broadleaf-coniferous

GRASSLAND

Savanna

Prairie and steppe

DESERT, TUNDRA, ICE CAP AND HIGHLAND

Desert shrub and waste

Tundra

Ice cap

Undifferentiated highland

Now, however, their estimates are much higher, in the range of 40 to 80 million. This startling revision is based on research, still in its infancy, on species inhabiting the rain forest.

Such diversity is important in its own right, but it also has vital implications for nature's ongoing evolution and for people's lives on Earth. Humankind now relies on a handful of crops as staple foods. In our agricultural systems, the trend in recent decades has been to develop high-yield varieties of grains and to plant them as vast **monocultures** (single-crop plantings). This trend, which characterizes the **Green Revolution,** may render agriculture more vulnerable to pests and diseases.

In evolutionary terms, we have reduced the natural diversity of crop varieties that allows nature and farmer to turn to alternatives when adversity strikes. As we remove tropical rain forests and other natural ecosystems to provide ourselves with timber, agriculture, or living space, we may be eliminating the foods, medicines, and raw materials of tomorrow, even before we have collected them and assigned them scientific names. "We are causing the death of birth," laments biologist Norman Myers.[1]

[1] In Edward O. Wilson, "Threats to Biodiversity." *Scientific American*, 261(3): 1989, 60–66.

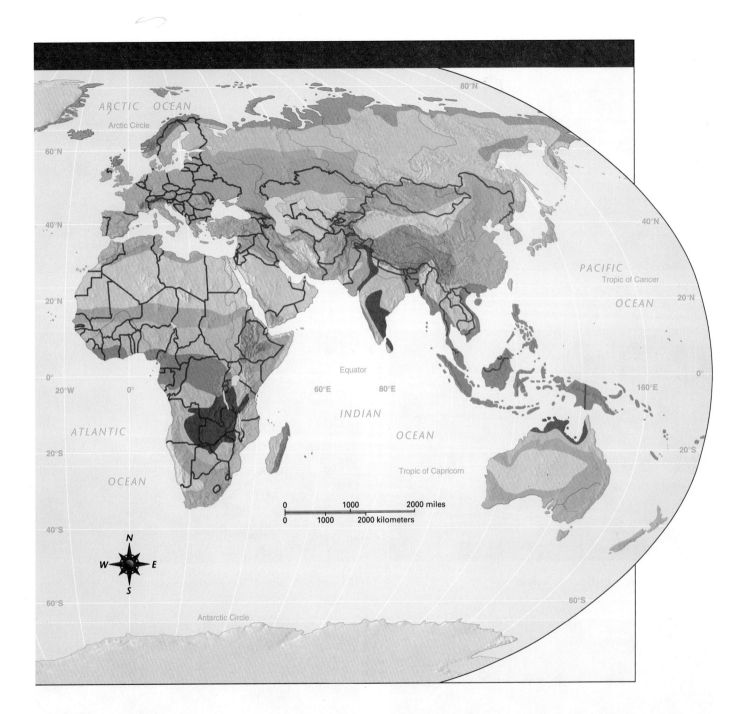

Regions where human activities are rapidly depleting a rich variety of plant and animal life are known as **biodiversity hot spots,** which scientists believe deserve immediate attention for study and conservation (Fig. 2.9). These 25 priority regions, as identified by Conservation International, are in South, Central, and North America: the tropical Andes, the Choco-Darien Western Ecuador region, central Chile, the Atlantic coastal forest of Brazil, the Brazilian cerrado, the Caribbean, Mesoamerica, and the California floristic province; in Asia: Sri Lanka and the Western Ghats of India, the mountains of south-central China, Indo-Burma, the Philippines, Sundaland, and Wallacea; in the Pacific: southwest Australia, New Zealand, New Caledonia, and Polynesia/Micronesia; in

Africa and adjacent regions: the Guinean forests of West Africa, the Cape floristic province of South Africa, the succulent Karoo of South Africa and Namibia, the Eastern Arc Mountains and coastal forests of Tanzania and Kenya, and Madagascar and the Indian Ocean islands; and in Europe, North Africa, and Southwest Asia: the Mediterranean Basin and the Caucasus.

By referring to the map of Earth's biomes in Figure 2.7, you will see that most of these hot spots are within tropical rain forest areas. Many, too, are islands that tend to have high biodiversity because species on them have evolved in isolation to fulfill special roles in these ecosystems and because human pressures on island ecosystems are particularly intense. Efforts

(a) Icecap, glacier in British Columbia, Canada

(b) Tundra, northern Norway

(c) Coniferous forest, British Columbia, Canada

(d) Desert shrub, southern Sinai Peninsula, Egypt

(e) Steppe, eastern Turkey

(f) Tropical rain forest, Dominica, West Indies

Figure 2.8 An album of Earth's biomes.

(g) Tropical deciduous forest, Gir Forest, western India

(h) Scrub and thorn forest, northern Zimbabwe

(i) Savanna, southern Kenya

(j) Mediterranean scrub forest, southern California, U.S.

(k) Broadleaf deciduous forest, southern Missouri, U.S.

(l) Undifferentiated highland vegetation, San Juan Mountains, Colorado, U.S.

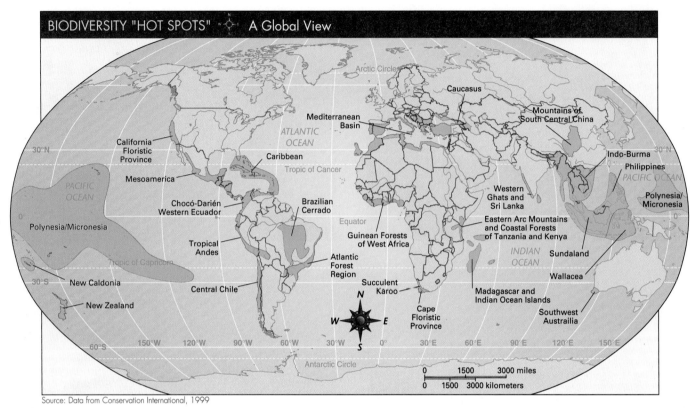

Source: Data from Conservation International, 1999

Figure 2.9   World biodiversity hot spots.

are already underway in most of these hot spots to establish national parks and other protected areas.

2.3   Global Environmental Change

Efforts to preserve biodiversity, especially with the development of protected areas, are often hard-fought. Ironically, once established, these islands of wilderness in a sea of humanized landscapes might end up losing the species they were established to defend, not as a direct result of human activities within them, but because of what is happening overhead. The global climate is, or may be, changing, and if it is, the spatial distributions of plants and animals will be changing, too.

There is a polarized and dynamic debate between those who insist that human activities are responsible for a documented warming of Earth's surface and those who either insist the warming is not occurring or, if it is, that a natural climatic cycle is responsible. However, an emerging consensus, supported by improvements in computer modeling, is that Earth's atmosphere is warming because of human production of "greenhouse gases" such as carbon dioxide (Fig. 2.10). In 2000, an international team of atmosphere scientists known as the Intergovernmental Panel on Climate Change (IPCC) abandoned their previously neutral stance on the question of whether or not people were to blame for global warming. "There is now stronger evidence for a human influence on the climate," the IPCC concluded. "There is increasing evidence from many sources that the signal of human influence on the climate has emerged from natural variability, sometime around 1980 . . . Man-made greenhouse gases have contributed substantially to the observed warming over the last 50 years."[2]

In 1827, French mathematician Baron Jean Baptiste Fourier established the concept of the **greenhouse effect,** noting that Earth's atmosphere acts like the transparent glass cover of a greenhouse (Fig. 2.11) (for modern purposes, think of a car's windows). Visible sunlight passes through the glass to strike Earth's surface. Ocean and land (the floor of the greenhouse, or the car upholstery) reradiate the incoming solar energy as invisible infrared radiation (heat). Acting like the greenhouse glass or car window, Earth's atmosphere traps some of that heat.

In the atmosphere, naturally occurring greenhouse gases such as carbon dioxide (CO_2) and water vapor make Earth habitable by trapping heat from sunlight. Concern over global warming focuses on human-made artificial sources of greenhouse gases, which trap abnormal amounts of heat. Carbon dioxide released into the atmosphere from the burning of coal, oil, and natural gas is the greatest source of concern, but methane (from rice paddies and the guts of

[2] Quoted in "Panel: Earth to Become Hotter." *Columbia Daily Tribune,* October 26, 2000, 4A, and in Andrew C. Revkin, "A Shift in Stance on Global Warming Theory." *New York Times,* October 26, 2000, A18.

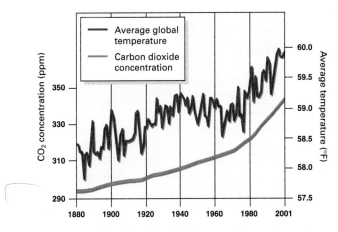

Figure 2.10 Industrialization and the burning of tropical forests have produced a steady increase in carbon dioxide emissions. Many scientists believe these increased emissions explain the corresponding steady increase in the global mean temperature.

ruminating animals like cattle), nitrous oxide (from the breakdown of nitrogen fertilizers), and chlorofluorocarbons, or CFCs (from cooling and blowing agents), are also human-made greenhouse gases (CFCs also destroy stratospheric ozone, a gas that has the important effect of preventing much of the sun's harmful ultraviolet radiation from reaching Earth's surface). In the car-as-Earth metaphor, continued production of these greenhouse gases has the effect of rolling up the car windows on a sunny day with the result of increased temperatures and physical problems for the occupants.

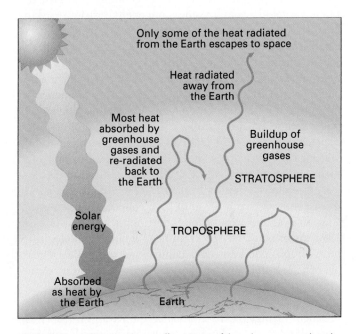

Figure 2.11 The greenhouse effect. Some of the solar energy radiated as heat (infrared radiation) from Earth's surface escapes into space, while greenhouse gases trap the rest. Naturally occurring greenhouse gases make Earth habitable, but carbon dioxide and greenhouse gases emitted by human activities accentuate the greenhouse effect, making Earth unnaturally warmer.

Although formal records of meteorological observations began just over a century ago, past climates left evidence in the form of marine fossils, corals, glacial ice, fossilized pollen, and annual growth rings in trees. These indirect sources, along with formal records, indicate that the 20th century was the warmest century in the last 600 years, and the warmest years in that 600-year period were 1990, 1995, 1997, 1998, and 1999. Direct sources indicate that the 1990s was the hottest decade, and 1998 the hottest year, since instrument recording began in 1861.

In 1896, Swedish scientist Svante Arrhenius feared that Europe's growing industrial pollution would eventually double the amount of carbon dioxide in Earth's atmosphere and as a consequence raise the global mean temperature as much as 9° Fahrenheit (5°C). Computer-based climate change modelers today also use the scenario (which assumes that nothing will be done to reduce greenhouse gases) of a doubling of atmospheric carbon dioxide, and they conclude that the mean global temperatures might warm from 3° to 11° Fahrenheit (1.7° to 6.1°C) by 2100; this would be in addition to the increase of just over 0.5° Celsius (1°F) since 1975. The estimated range and consequences of the increases vary widely because of different assumptions about the little-understood roles of oceans and clouds. Most models concur that because of the slowness with which the world's oceans respond to changes in temperature, the effects of this rise will be delayed by decades.

Geographers are most concerned with where the anticipated climatic changes will occur and what the impacts of these changes will be. Computer models conclude that there will not be a uniform temperature increase across the entire globe but rather that the increases will vary spatially and seasonally. Differing models produce contradictory results about the timing and impacts of warming. Two teams of scientists publishing their results in the journal *Nature*, for example, first assumed a doubling of CO$_2$. Both predicted rises in temperature, but because they differed on what season of year the increases would be greatest, they reached different conclusions about the effects on agriculture in the United States. One predicted warmer summers and forecast that crop productivity would decline by 20%. The other predicted warmer winters and forecast a 10% increase in crop yields.

Geographically, the impacts of global warming appear to be greatest at the higher latitudes. The volume of the north polar ice declined by 40% between 1958 and 2000, and the ice shelves surrounding Antarctica have also retreated substantially. Sailing on a Russian nuclear-powered icebreaker, the author of this chapter saw open water at the North Pole in August 1996 (Fig. 2.12), and a similar observation from the same vessel in August 2000 produced worldwide headlines about polar ice melting. Some quarters are cheering the trend. A further retreat of polar ice would bode well for maritime shipping through the long-icebound "Northwest Passage" across Arctic Canada and through the Northern Sea

JOE HOBBS

Figure 2.12 The passengers and crew of the Russian icebreaker *Yamal* found open water at the North Pole on August 16, 1996. Much of the voyage from northeast Siberia was ice-free. Similar observations in 2000 led to widespread concern about global warming. The ship's GPS instrument indicated that the vessel's bow was on 90°N latitude when this photo was taken.

F7.1
190
route across the top of Russia, which has been navigable only in summer and only with the aid of icebreakers.

There is general agreement that with global warming the distribution of climatic conditions typical of biomes will shift poleward geographically and upward in mountainous regions. Many animal species will be able to migrate to keep pace with changing temperatures, but plants, being stationary, will not. Conservationists who have struggled to maintain islands of habitat as protected areas are particularly concerned about rapidly changing climatic conditions. Ecologists have documented that two-thirds of Europe's butterfly species have already shifted their habitat ranges northward by 22 to 150 miles (35 to 240 km), coinciding with the continent's warming trend. The World Wildlife Fund issued a report in 2000 warning that due to global warming as much as 70% of the natural habitat could be lost, and 20% of the species could become extinct, in the arctic and in subarctic regions of Canada, Scandinavia, and Russia. The report predicted that more than a third of existing habitats in the American states of Maine, New Hampshire, Oregon, Colorado, Wyoming, Idaho, Utah, Arizona, Kansas, Oklahoma, and Texas would be irrevocably altered by global warming.

There is also general agreement that if global temperatures
410
rise, so will sea levels as glacial ice melts and as seawater warms and thus occupies more volume. Sea levels have already risen 8 inches (20 cm) in the past century, possibly due to global warming, and are continuing to rise at a rate of one-tenth of an inch (2 mm) each year. Decision makers in coastal cities throughout the world, in island countries, and in nations with important lowland areas adjacent to the sea are worried about the implications of rising sea levels. The coastal

lowland countries of the Netherlands and Bangladesh have for many years been outspoken advocates for reductions of greenhouse gases. The president of the Maldives, an Indian Ocean country comprised entirely of islands barely above sea level, pleaded, "We are an endangered nation!"

There is a growing international political will to take the strong measures that may be necessary to prevent or reverse global warming. Present efforts to reduce the production of greenhouse gases focus on the world's wealthiest nations, where about 85% of the emissions originate. The United States alone now accounts for 25% of all emissions, China for 14%, and more than 70 less developed countries together for only about 5%. However, this balance is likely to change as the wealthier countries adopt stricter emissions standards and as coal-wealthy China matures into a major industrial power, possibly with less stringent standards.

In 1997, 84 countries (of 160 countries represented at the conference) signed the **Kyoto Protocol,** a landmark international treaty on climate change. The agreement requires 38 more developed countries (known as the Annex I countries) to reduce carbon dioxide emissions to at least 5% below their 1990 levels by the year 2012. The United States pledged to cut its emissions to 7% below 1990 levels by that date. This would be a huge cut; in 2000, the United States produced 22% more carbon dioxide than it did in 1990. The European Union made a promise of 8% below 1990 levels, and Japan promised 6%.

Meeting these pledges would require substantial legislative, economic, and behavioral changes in these more affluent countries. Transportation and other technologies that use fossil fuels would have to become more energy efficient, making these technologies at least temporarily more expensive. Gasoline prices would rise, so consumers would feel the pinch. Advocates of the protocol argue that the initial sacrifices would soon be rewarded by a more efficient and competitive economy powered by cleaner and cheaper sources of energy derived from the sun. Opponents, however, feel that higher fuel prices will be too costly for the United States and other industrialized economies to bear. Their position will make it difficult for some signatories of the Kyoto Protocol to ratify and then implement the treaty. For the protocols to take effect, at least 55 countries must ratify them, and the industrialized Annex I countries that ratify must have collectively produced at least 55% of the world's total greenhouse gas emissions in 1990. As of mid-2002, 68 countries—including all 15 European Union members—had ratified, and the critical threshold of 55% of the 1990 emissions by Annex I countries was being approached.

Much to the dismay of the European Union countries, President George W. Bush rejected the Kyoto Protocol soon after taking office, making the treaty's ratification by the Senate a very unlikely prospect. The Bush administration had two objections: the potential economic costs of implementing the treaty and the fact that China, along with all the world's less developed countries, is not required by the Kyoto Protocol to take any steps to reduce greenhouse gas emissions (it is anticipated that China will pass the United States as the world's greatest producer of carbon dioxide emissions by the year 2018). The European Union countries and Japan dismissed these concerns and expressed outrage that the United States broke ranks with them on global warming.

There are, however, reasons for optimism that countries can unite in effective action against global environmental change. Thanks to the **Montreal Protocol** and its amendments signed by 37 countries in the late 1980s, the production of CFCs worldwide has all but ceased; the wealthier countries no longer produce them, and the less developed countries are scheduled to cease production by 2010. The anticipated result is that there will be a marked reduction in the size of stratospheric ozone "holes" that have been observed seasonally over the southern and northern polar regions since about 1985. However, CFC molecules live extremely long (up to 110 years). The formation of the largest ozone hole to date over Antarctica in the spring of 2000 suggests that many years will pass before the lasting impact of CFC elimination will be felt. The current forecast is that the effects of the Montreal Protocol will be noticeable by 2010, with the ozone layer recovering to pre-1980 levels by 2050.

Compared with the richer nations, the less developed countries, which tend to use fuelwood to complement scarce fossil fuel supplies, release relatively little CO_2 into the atmosphere, but they do contribute to the greenhouse effect. When burned, trees not only release the CO_2 they stored in the process of photosynthesis, but trees are also eliminated as organisms that in the process of photosynthesis remove CO_2 from the atmosphere. However, because these countries produce relatively little CO_2, policymakers are considering innovative ways to reward them and encourage them to develop clean industrial technologies. One system being considered in the wake of the Kyoto Protocol is that of **tradable permits,** in which each country would be assigned the right to emit a certain quantity of CO_2, according to its population size, setting the total at an acceptable global standard. The United States, already an overproducer by this scale, could purchase emission rights from a populous country like India, which currently "underproduces" CO_2. India would be obliged to use the income to invest in energy-efficient and nonpolluting technologies.

Most notable about this system of emissions trading is that it views global warming as a global problem. However, international talks in the Hague, Netherlands, in 2000 on how to implement the Kyoto Protocol broke down over this concept. Although the United States argued that it should be able to buy emissions rights from poorer countries, the European Union countries responded that in doing so the United States would simply be avoiding the important work of reducing carbon dioxide emissions in its own country. The Europeans were also angry over U.S. insistence that it should not have to reduce carbon dioxide emissions as much as demanded in the Kyoto Protocol because the United States has huge areas of forest and farmland that absorb this gas and act as carbon

dioxide "sinks." Here again, the United States was viewed as trying to find the easy way out.

2.4 Two Revolutions That Have Shaped Regions

Geography has long been concerned with people as agents of change in the global environment. The geographer's approach to understanding a current landscape (in almost all cases, a cultural landscape that has been fashioned by human activity) is sometimes deeply historical, involving study of its evolution from the primordial natural landscape. With a perspective of great historic depth, the current spatial patterns of our relationship with Earth may be seen as products of two "revolutions": the Agricultural, or Neolithic (New Stone Age), Revolution that began in the Middle East about 10,000 years ago and the Industrial Revolution that began in 18th-century Europe. Each of these revolutions transformed humanity's relationship with the natural environment. Each increased substantially our capability to consume resources, modify landscapes, grow in number, and spread in distribution.

Hunting and Gathering

Until about 10,000 years ago, our ancestors practiced a hunting and gathering livelihood (also known as foraging). Joined in small bands consisting of extended family members, they were nomads practicing extensive land use in which they covered large areas to locate foods such as seeds, tubers, foliage, fish, and game animals (Fig. 2.13). Moving from place to place in small numbers, they had a relatively limited impact on natural environments compared with human impact today. Many scholars praise these preagricultural people for the apparent harmony they maintained with the natural world in

Figure 2.13   Until the relatively recent past—just a few thousand years ago—people were exclusively hunters and gatherers. This rock art in Egypt's Sinai Peninsula was created over a long time span, as it depicts Neolithic period hunting of ibex, later uses of camels and horses, and writing from the Nabatean period (first century A.D.).

both their economies and their spiritual systems. Hunters and gatherers have even been described as the "original affluent society" because after short periods of work to collect the foods they needed, they enjoyed long stretches of leisure time. Studies of those few hunter-gatherer cultures that lingered into modern times, such as the San ("Bushmen") of southern Africa and several Amerindian groups of South America, suggest that although their life expectancy was low, they suffered little from the mental illnesses and broken family structures that characterize industrial societies.

Hunters and gatherers were not always at peace with one another or with the natural world, however. With upright posture, stereoscopic vision, opposable thumbs, an especially large brain, and no mating season, *Homo sapiens* was from its earliest times an **ecologically dominant species** — one that competes more successfully than other organisms for nutrition and other essentials of life or that exerts a greater influence than other species on the environment. Using fire to flush out game or create new pastures for the herbivores they hunted, preagricultural people, unlike other animals, shaped the face of the land on a vast scale relative to their small numbers. Many of the world's prairies, savannas, and steppes where grasses now prevail originated from hunters and gatherers setting fires repeatedly. These people also overhunted and in some cases eliminated animal species. The controversial **Pleistocene overkill hypothesis** states that rather than being at harmony with nature, hunters and gatherers of the Pleistocene Era (which ended 10,000 years ago) hunted many species to extinction, including the elephantlike mastodon of North America.

The Agricultural Revolution

Despite these excesses, the environmental changes that hunters and gatherers could cause were limited. Humankind's power to modify landscapes took a giant step with **domestication,** the controlled breeding and cultivation of plants and animals. This was the **Agricultural Revolution,** also known as the Neolithic Revolution and the Food-Producing Revolution. Why people began to produce rather than continue to hunt and gather plant and animal foods — first in the Middle East and later in Asia, Europe, Africa, and the Americas — is uncertain. Two theories prevail. Climatic change in the form of increasing drought and reduced plant cover may have forced people and wild plants and animals into smaller areas, where people began to tame wild herbivores and sow wild seeds to produce a more dependable food supply. Another theory is that their own growing populations in areas originally rich in wild foods compelled people to find new food sources, so they began sowing cereal grains and breeding animals. The latter process apparently began about 8000 B.C. in the Zagros Mountains of what is now Iran. The culture of domestication spread outward from there but also developed independently in several world regions.

In choosing to breed plants and animals, people slowly abandoned the nomadic, **extensive land use** of hunting and gathering for the sedentary, **intensive land use** of agriculture

and animal husbandry. The new system enabled them to create large and reliable surpluses of food. Through **dry farming,** or planting and harvesting according to the seasonal rainfall cycle, population densities could be 10 to 20 times higher than they were in the hunting and gathering mode. By about 4000 B.C., people along the Tigris, Euphrates, and Nile Rivers began crop **irrigation** — bringing water to the land artificially — an innovation that allowed them to grow crops year-round, independently of seasonal rainfall or river flooding (Fig. 2.14). Irrigation technology allowed even more people to make a living off the land; irrigated farming yields about five to six times more food per unit area than dry farming. In ecological terms, the expanding food surpluses of the Agricultural Revolution raised Earth's **carrying capacity** — that is, the size of a species' population (in this case, humans) that Earth's ecosystems can support.

This steep increase in food surpluses freed more people from the actual work of producing food, and they undertook a wide range of activities unrelated to subsistence needs. Irrigation and the dependable food supplies it provided thus set the stage for the development of **civilization,** the complex culture of urban life characterized by the appearance of writing, economic specialization, social stratification, and high population concentrations. By 3500 B.C., for example, 50,000 people lived in the southern Mesopotamian city of Uruk, in what is now Iraq. Other **culture hearths** — regions where civilization followed the domestication of plants and animals — emerged between 8000 and 2500 B.C. in China, Southeast Asia, the Indus River Valley, Egypt, West Africa, Mesoamerica, and the Andes.

The agriculture-based urban way of life that spread from these culture hearths had larger and more lasting impacts on the natural environment than either hunting and gathering or

Figure 2.14 The Tigris River in southeastern Turkey. The brown areas are rainless in the long, hot summers and are capable of producing only one crop a year through dry (unirrigated) farming. The green areas along the river are irrigated and can produce two or three crops a year. Irrigation was thus a revolutionary technology that greatly increased the number of people the land could support.

early agriculture. Acting as agents of humankind, domesticated plants and animals proliferated at the expense of the wild varieties that people came to regard as pests and competitors. Agriculture's permanent and site-specific nature magnified the human imprint on the land, while the pace and distribution of that impact increased with growing numbers of people.

The Industrial Revolution

The human capacity to transform natural landscapes took another giant leap with the **Industrial Revolution,** which began in about A.D. 1700. This new pattern of human-land relations was based on breakthroughs in technology that several factors made possible (Fig. 2.15). First, western Europe had the economic capital necessary for experimentation, innovation, and risk. Much of this money was derived from the lucrative trade in gold and slaves undertaken initially in the Spanish and Portuguese empires after 1400. Second, in Europe prior to 1500, there had been significant improvements in agricultural productivity, particularly with new tools such as the heavy plow, and with more intensive and sustainable use of farmland. Crop yields increased, and human populations grew correspondingly. A third factor was population growth itself. More people freed from work in the fields represented a greater pool of talent and labor in which experimentation and innovation could flourish. As agricultural innovations and industrial productivity improved, an increasingly large proportion of the growing European population was freed from farming, and for the first time in history, a region had more city-dwellers than rural folk. The process of industrialization continues to promote urbanization today.

Most geographers see population growth today as a drain on resources, but the Industrial Revolution illustrates that, given the right conditions, more people do create more resources. Innovations such as the steam engine tapped the vast energy of fossil fuels — initially, coal and, later, oil and natural gas. This energy, the photosynthetic product of ancient ecosystems, allowed Earth's carrying capacity for humankind to be raised again — this time into the billions.

As they began to deplete their local supplies of resources needed for industrial production, Europeans began to look for these materials abroad. As early as their Age of Exploration, which began in the 15th century, Europeans probed ecosystems across the globe to feed a growing appetite for innovation, economic growth, and political power. The process of European colonization was thus linked directly to the Industrial Revolution. Mines and plantations from such faraway places as central Africa and India supplied the copper and cotton that fueled economic growth in Belgium and England.

No longer dependent on the foods and raw materials they could procure within their own political and ecosystem boundaries, European vanguards of the Industrial Revolution had an impact on the natural environment that was far more extensive and permanent than that of any other people in history. There are many measures of the unprecedented changes that the Industrial Revolution and its wake have wrought on Earth's landscapes. Between 1700 and 2000, the total forested area on Earth declined by more than 20%. During the same period, total cropland grew nearly 500%, with more expansion in the period from 1950 to 2000 than in the 150 years from 1700 to 1850. Human use of energy increased more than a hundredfold from 1700 to 2000. Today, a whopping 40% of Earth's land-based photosynthetic output is dedicated to human uses, especially in agriculture and forestry. Of particular interest to geographers is how the costs and benefits of such expansion are distributed in unequal patterns across the earth.

2.5 The Geography of Development

One of the most striking characteristics of human life on Earth is the large disparity between wealthy and poor people, both within and among countries. At a high level of generalization, the world's countries can be divided into "haves" and "have-nots" (Fig. 2.16 and Table 2.1). Writers refer to these distinctions variously as "developed" and "underdeveloped," "developed" and "developing," "more developed" and "less developed," "industrialized" and "nonindustrialized," and "north" and "south" based on the concentration of wealthier countries in the middle latitudes of the Northern Hemisphere and the abundance of poorer nations in the Southern Hemisphere. This text uses the terms **more developed countries (MDCs)** and **less developed countries (LDCs).** It must be emphasized that this framework is an introductory tool and cannot account for the tremendous variations and ongoing changes in economic and social welfare that characterize the world today. Some countries, such as the "Asian Tigers," are best described as **newly industrializing countries (NICs)** because they do not fit either the MDC or LDC idealized type. We will describe these cases in the relevant regional chapters.

Figure 2.15 A tweed mill in Stornoway on Scotland's Lewis Island. The process of industrialization that began in 18th-century Europe has rapidly transformed the face of the earth.

Figure 2.16 Wealth and poverty by country. Note the concentration of wealth in the middle latitudes of the Northern Hemisphere.

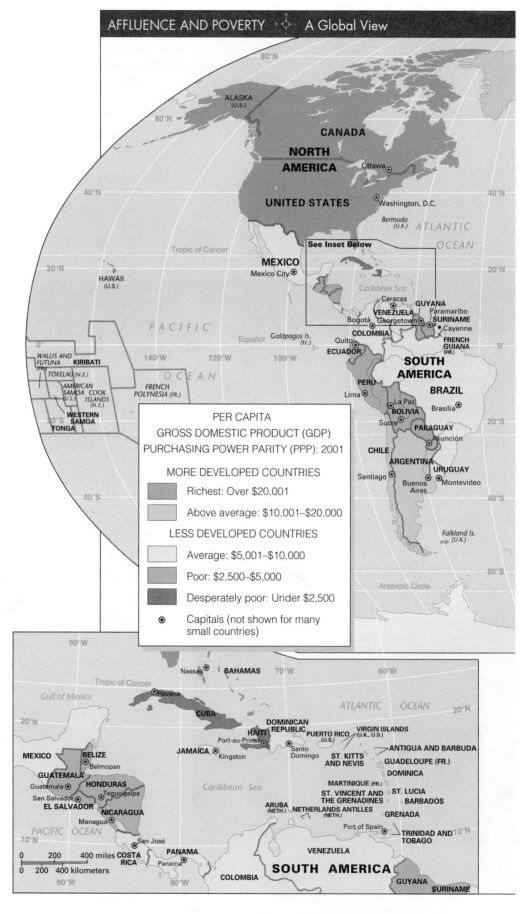

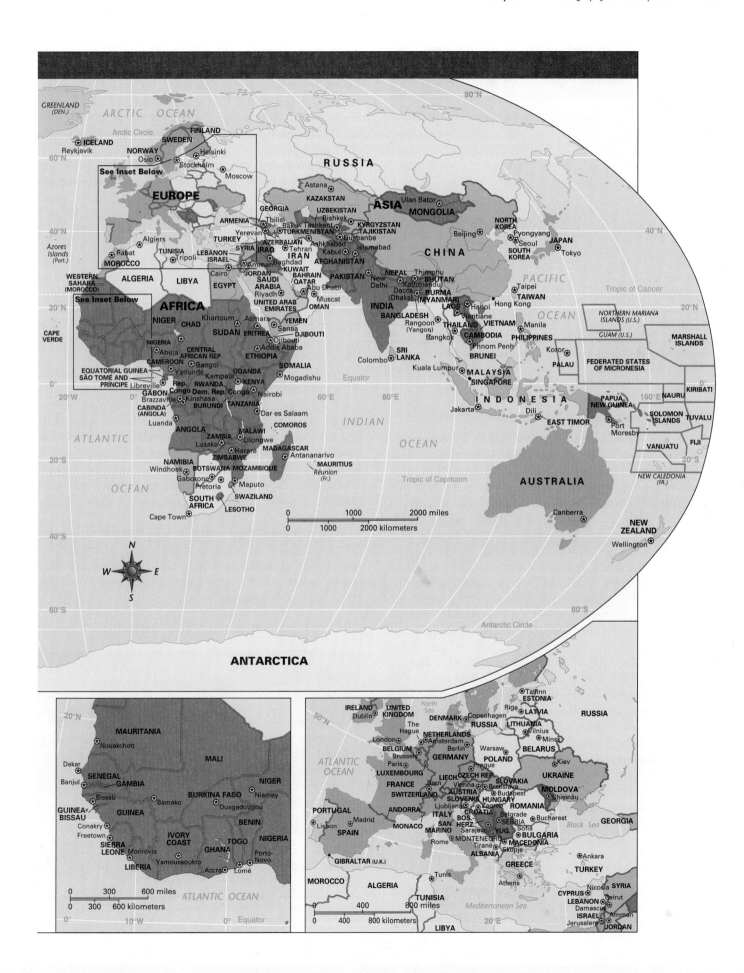

TABLE 2.1 Characteristics of More Developed Countries (MDCs) and Less Developed Countries (LDCs)

Characteristic	MDC	LDC
Per capita GNP and income	High	Low
Percent in middle class	High	Low
Percent in manufacturing	High	Low
Energy use	High	Low
Percent urban	High	Low
Percent rural	Low	High
Birth rate	Low	High
Death rate	Low	Low[a]
Population growth rate	Low	High
Percent under age 15	Low	High
Percent literate	High	Low
Leisure time available	High	Low
Life expectancy	High	Low

[a] Although death rates are high in the LDCs relative to the MDCs, they are quite low compared to what they were previously in the LDCs.

Measures of Development

MDCs may be distinguished from LDCs by comparing countries' annual **per capita gross domestic product (GDP)** — the total output of goods and services a country produces for home use in a year divided by the country's population. Even more useful for comparative purposes is that figure adjusted for **purchasing power parity (PPP)**, which involves the use of standardized international dollar price weights that are applied to the quantities of final goods and services produced in a given economy. The resulting measure, **per capita GDP PPP**, which is the one used in this text, provides the best available starting point for comparisons of economic strength and well-being between countries. The gulf between the world's richest and poorest countries is startling (Table 2.2). The average per capita GDP PPP in the MDCs is nearly six times greater than in the LDCs. In 2001, with a per capita GDP PPP of $36,400, Luxembourg was the world's richest country, whereas Sierra Leone was the poorest with only $510. By another measure, in 2001, 1.5 billion people, or 25% of the world's population, were "abjectly poor," living on less than $1 per day. These raw numbers suggest that economic productivity and income alone characterize **development**, which, according to a common definition, is a process of improvement in the material conditions of people through diffusion of knowledge and technology.

Such economic definitions reveal little about measures of well-being such as income distribution, gender equality, literacy, and life expectancy. Recognizing the shortcomings of strictly economic definitions, the United Nations Development Programme created the **Human Development Index (HDI)**, a scale that considers these attributes of quality of life.

According to this index, Norway is "the world's best place to live," although it is 17th in per capita GDP PPP ranking. Following Norway, in descending order, are Australia, Canada, Sweden, Belgium, the United States, Iceland, the Netherlands, Japan, and Finland. In HDI terms, as in per capita GDP PPP terms, Sierra Leone in West Africa is the world's worst place to live.

On the basis of per capita GDP PPP, 1.2 billion, or 20%, of the world's 6.1 billion people inhabit the MDCs. Most citizens of these countries, such as the United States, Canada, Japan, Australia, New Zealand, and the nations of Western Europe, enjoy an affluent lifestyle with freedom from hunger. Employed in industries or services, most of the people live in cities rather than in rural areas. Disposable income, or money that people can spend on goods beyond their subsistence needs, is generally high. There is a large middle class. Population growth is low as a result of low birth rates and low death rates (these demographic terms are explained in the section on population that follows). Life expectancy is long, and the literacy rate is high.

Life for the planet's other 80%, or about 4.9 billion people, is very different. In the LDCs, including most countries in Latin America, Africa, and Asia, poverty and often hunger prevail. The leading occupation is subsistence agriculture, and the industrial base is small. The middle class tends to be small, with an enormous gulf between the vast majority of poor and a small wealthy elite, which owns most of the private landholdings. With high birth rates and falling death rates, population growth is high. Life expectancy is short, and the literacy rate is low.

With four-fifths of the world's people living in the poorer, less developed countries, it is important to understand the root causes of underdevelopment and to appreciate how wealth and poverty affect the global environment in very different but equally profound ways.

TABLE 2.2 Top 10 and Bottom 10 Countries in Per Capita GDP PPP (U.S. Dollars, 2001)

Top 10 Country	GDP PPP, per capita	Bottom 10 Country	GDP PPP, per capita
1. Luxembourg	36,400	1. Sierra Leone	510
2. United States	36,200	2. Dem. Republic Congo	600
3. Switzerland	28,600	3. Somalia	600
4. Norway	27,700	4. Ethiopia	600
5. Singapore	26,500	5. Tanzania	710
6. Denmark	25,500	6. Eritrea	710
7. Belgium	25,300	7. Comoros	720
8. Austria	25,000	8. Burundi	720
9. Japan	24,900	9. Afghanistan	800
10. Canada	24,800	10. Madagascar	800

Note: The table excludes the city-states of San Marino and Monaco, whose per capita GDP PPP would be in the top 10.
Source: *World Factbook,* CIA 2001.

Causes of Disparities

Many theories attempt to explain the disparities between MDCs and LDCs. The most widely debated, and also the most embraced in the LDCs, is **dependency theory,** which argues that the worldwide economic pattern established by the Industrial Revolution and the attendant process of colonialism persists today. In his book *Ecological Imperialism*, historian and geographer Alfred Crosby explains how dependency led to the rich-poor divide by depicting the two very different ways in which European powers used foreign lands during the Industrial Revolution. In the pattern of "settler colonization," Europeans sought to create new Europes, or "neo-Europes," in lands much like their own: temperate midlatitude zones with moderate rainfall and rich soils where they could raise wheat and cattle. Thus, between 1630 and 1930, more than 50 million Europeans emigrated from their homelands to create European-style settlements in what are now Canada, the United States, Argentina, Uruguay, Brazil, South Africa, Australia, and New Zealand. These lands were destined to become some of the world's wealthier regions and countries.

In contrast to their preference to settle familiar middle latitude environments, Europeans viewed the world's tropical lands mainly as sources of raw materials and markets for their manufactured goods. The environment was too different from home to make settlement attractive. In establishing a pattern of "mercantile colonialism," Crosby explains, Europeans were less inhabitants than conquering occupiers of the colonies, overseeing indigenous peoples and resettled slaves in the production of primary or unfinished products: sugar in the Caribbean; rubber in Latin America, West Africa, and Southeast Asia; and gold and copper in southern Africa, for example. Colonialism required huge migrations of people to extract the earth's resources, including 30 million slaves and contracted workers from Africa, India, and China to work mines and plantations around the globe.

In the mercantile system, the colony provided raw materials to the ruling country in return for finished goods; thus, people in India would purchase clothing made in England from the raw cotton they themselves had harvested. The relationship was most advantageous to the colonizer. England, for example, would not allow its colony India to purchase finished goods from any country but England. It prohibited India from producing any raw materials the empire already had in abundance, such as salt (India's Mohandas Gandhi defiantly violated this prohibition in his famous "March to the Sea"). Finished products (or "value-added" products) are worth much more than the raw materials they are made from, so manufacturing in the ruling country concentrated wealth there while discouraging industrial and economic development in the colony.

The colony was obliged to contribute to, but was prohibited from competing with, the economy of the ruling country — a relationship that dependency theorists insist continues today. Dependency theory states that to participate in the world economy, the former colonies but now-independent countries continue to depend on exports of raw materials to, and purchases of finished goods from, their former colonizers, and this disadvantageous position keeps them poor. Dependency theorists call this relationship **neocolonialism**. With independence, the former colonies needed revenue. To earn that money, they continued to produce the goods for which markets already existed — generally the same unprocessed primary products they supplied in colonial times. Dependency theorists argue that when former colonies try to break their dependency by becoming manufacturers, the former colonizers impose trade barriers and quotas to preclude that step. To support their argument, they point out that countries like Thailand and South Korea, which the Europeans never colonized and which never developed these dependencies, are among the most prosperous LDCs or are in that select group of NICs.

Many geographers, however, view dependency theory as too simplistic and politicized. Rather, they consider a wider and more complex set of factors, including culture, location, and natural environment, to explain why some countries are wealthy and others are poor. For example, because it is situated close to a great mainland with which to trade, the island of Great Britain enjoys a central location favorable for economic development. Japan has a similar location relative to the Asian landmass. In contrast, landlocked nations such as Bolivia in South America and numerous nations in Africa have locations unfavorable for trade and economic development, and they have not overcome this disadvantage. But it is important to recognize that geographic location is never the sole decisive factor in development. Like Japan and Great Britain, Madagascar and Sri Lanka are island nations situated close to large mainlands, but neither has experienced prosperity.

A superabundance of one particular resource (e.g., oil in the Persian/Arabian Gulf states) or a diversity of natural resources has helped some countries to become more developed than others. The former Soviet Union and the United States developed superpower status in the 20th century in large part based on the enormous natural resources of both countries. In other cases, human industriousness has helped to compensate for resource limitations and promote development. For example, Japan has a rather small territory with few natural resources (including almost no petroleum). Yet, in the second half of the 20th century, it became an industrial powerhouse largely because the Japanese people united in common purpose to rebuild from wartime devastation, placing priorities on education, technical training, and seaborne trade from their advantageous island location. Conversely, cultural or political problems like corruption and ethnic factionalism can hinder development in a resource-rich nation, as in the mineral-wealthy Democratic Republic of Congo.

Whether because of neocolonialism (as many dependency theorists argue) or a more complex array of variables, many

GLOBAL PERSPECTIVE : : : Globalization: The Process and the Backlash

One of the most remarkable international trends of recent decades has been **globalization** — the spread of free trade, free markets, investments, and ideas across borders and the political and cultural adjustments that accompany this diffusion. There has been much debate and sometimes violent conflict over the pros and cons of globalization.

Advocates of globalization argue that the newly emerging global economy will bring increased prosperity to the entire world. Innovations in one country will be transferred instantly to another country, productivity will increase, and standards of living will improve. They propose that one obvious solution to the problem of the LDCs' inability to compete in the world economy, and therefore escape their dependency, is to reduce the trade barriers that MDCs have erected against them. (One such effort to mitigate the trade imbalance is the Lomé V Convention, an agreement signed between the European Union and more than 70 LDCs in 2000. The Lomé deal provides European Union aid and market access to former colonies in Africa, the Pacific, and the Caribbean.) With free trade, free enterprise will prosper, pumping additional capital into national economies and raising incomes

for all. Much of the support for globalization comes from **multinational companies** (also called transnational companies, meaning companies having operations outside their home countries), as they increase their investments abroad. Most of the companies are based in the MDCs, but multinational corporations also have grown in LDCs such as Mexico.

Opponents of globalization argue that the process will actually increase the gap between rich and poor countries; a selected few developing nations will prosper from increased foreign investment and resulting industrialization, but the hoped-for "global" wealth will bypass other countries altogether. And the increasing interdependence of the world economies will make all the players more vulnerable to economic and political instability. The multinational companies will recognize huge profits at the expense of poor wage laborers. In addition, environments will be harmed if environmental regulations are reduced to a lowest common denominator (e.g., the high standards of air quality demanded by the U.S. Clean Air Act would be deemed noncompliant with World Trade Organization rules because they make it harder for countries with

"dirtier" technologies to compete in the marketplace). Such concerns have already led to massive protests against "corporate-led globalization," notably at the meeting of the World Trade Organization in Seattle, Washington, in 1999, and at the meetings of the International Monetary Fund and the World Bank in Prague, Czech Republic, in 2000. Many more confrontations like these can be anticipated. Typical protestors' demands are that working conditions be improved in the foreign "sweatshops," where textiles and other goods are produced at low cost for U.S. corporations, and that Starbucks Coffee should sell only "fair trade" coffee beans bought at a price giving peasant coffee growers a living wage rather than at the "exploitive" price typically paid.

The problems of **hot money** and the **digital divide** are also cited as drawbacks to globalization. "Hot money" refers to short-term (and often volatile) flows of investment that can cause serious damage to the "emerging market" economies of less developed countries. Individual and corporate investors in the MDCs can invest such money heavily in stock market securities of the LDCs, reasoning that these developing nations have much greater economic growth

developing countries continue to rely heavily on income from the export of a handful of raw materials. This makes them vulnerable to the whims of nature and the world economy. The economy of a country heavily dependent on rubber exports, for example, may suffer if an insect pest wipes out the crop or if a foreign laboratory develops a synthetic substitute. When demand for rubber rises, that country may actually harm itself trying to increase its market share by producing more rubber because in the process it drives down the price (or consider oil: OPEC countries drove oil prices to all-time lows in the mid-1980s when they overproduced oil in a bid to get more revenue). If the country withholds production to shore up rubber's price, it provides consuming countries with an incentive to look for substitutes and alternative sources (again look at oil: After OPEC embargoed shipments of oil to the United States in the 1970s, the United States

began developing domestic oil supplies and becoming more energy efficient, thus reducing oil prices and OPEC's revenues). The developing country is in a dependent and disadvantaged position.

Environmental Impacts of Underdevelopment

Of particular interest to geographers are the environmental impacts of relations between MDCs and LDCs, especially the impacts on the poorer countries. LDCs generally lack the financial resources needed to build roads, dams, energy grids, and the other infrastructure assets they perceive as critical to development. They turn to the World Bank, International Monetary Fund (IMF), and other institutions of the MDCs to borrow funds for these projects. Many borrowers are unable to pay even the interest on these loans, which is sometimes huge (e.g., in 2000, Tanzania spent 40% of its an-

potential than the mature economies of the MDCs. This investment can bring rapid wealth to at least some sectors of the economies of LDCs. The problem, however, is that the capital can be withdrawn as quickly as it is pumped in, resulting in huge economic consequences.

This is what happened in August 1998, when Russia's ruble was devalued. Institutions such as foreign banks that had bought the country's debt in the form of bonds saw the value of their holdings fall to around 10% of their worth. Investors fretted about what would happen in Latin America and other emerging markets if, as in Russia, there were to be a collapse in the prices of emerging market bonds. Such a collapse would make it much more expensive for the LDCs to borrow to meet their financial needs. With borrowing becoming more expensive, these countries might not be able to pay off their debts and cover their high trade deficits. Worried investors from the MDCs therefore rushed their money out of LDCs. In a single day, stock markets in Venezuela, Brazil, and Argentina each lost about 10% of their value, and Russia's declined by 17%.

It is often assumed that the rapid growth and spread of computer and wireless technologies are benefiting all of humankind, but the notion of the digital divide challenges this view. The divide is between the handful of countries that are the technology innovators and users and the majority of nations that have little ability to create, purchase, or use new technologies. The United States, most European countries, and Japan are leaders in **information technology,** and the LDCs are well behind. For example, more than 25% of U.S. citizens use the World Wide Web, compared with 3% of Russians, 4 one-hundredths of 1% of the population of South Asia, and two-tenths of 1% of the Arab countries (Figure 2.A). Statistically, an American must save a month's salary to buy a computer, but a Bangladeshi must save 8 years' wages to buy one. It was estimated that in 2000, U.S. companies earned 85% of the revenues from the global Internet business and represented 95% of the stock market value of Internet corporations. The fear is that the growth of e-commerce will concentrate the wealth generated by that commerce in the technology leadership countries, enabling them to make further advances in technology and realize even bigger economic gains, while the technology laggards fail to catch up and simply get poorer. Will globalization prove to be good or bad? It may depend on where and whom you are.

Figure 2.A   An Internet café in Mérida, Yucatán, Mexico. Information technology—especially computers, the Internet, and cellular telephones—is spreading rapidly around the world. Some people believe there is now an unfolding "Information Revolution," similar to the Agricultural and Industrial Revolutions in its capacity to change humanity's relationship with Earth.

nual revenue on interest payments to the IMF, World Bank, and other lenders — more than it spent on education and health combined). When lender institutions threaten to sever assistance, borrowing countries often try to raise cash quickly to avoid this prospect. One method is to dedicate more quality land to the production of cash (commercial) crops, luxuries such as coffee, tea, sugar, coconuts, and bananas exported to the MDCs. Governments or foreign corporations often displace or "marginalize" subsistence farmers in the search for new lands on which to grow these commercial crops. In the process of **marginalization,** poor subsistence farmers are pushed onto fragile, inferior, or marginal lands that cannot support crops for long and that are degraded by cultivation. In Brazil's Amazon Basin, for example, peasant migrants arrive from Atlantic coastal regions where government and wealthy private landowners cultivate the best soils for sugarcane and other cash crops. The newcomers to Amazonia slash and burn the rain forest to grow rice and other crops that exhaust the soil's limited fertility in a few years. They move on to cultivate new lands, and in their wake come cattle ranchers whose land use further degrades the soil.

National decision makers often face a difficult choice between using the environment to produce more immediate or more long-term economic rewards. In most cases, they feel compelled to take short-term profits and, by cash cropping and other strategies, initiate a sequence having sometimes tragic environmental consequences. Most LDCs have resource-based economies that rely not on industrial productivity but on stocks of productive soil, forests, and fisheries. The long-term economic health of these countries could be assured by the perpetuation of these natural assets, but to

DEFINITIONS + INSIGHTS

The Fuelwood Crisis

The removal of tree cover in excess of sustainable yield (e.g., as in highland Nepal) illustrates the many detrimental effects that a single human activity can have in the LDCs (Fig. 2.B). These impacts comprise the complex phenomenon, widespread in the world's LDCs, known as the **fuelwood crisis**. As people remove trees to use as fuel, for construction, or to make room for crops, their existing crop fields lose protection against the erosive force of wind. Less water is available for crops because, in the absence of tree roots to funnel water downward into the soil, it runs off quickly. Increased salinity (salt content) generally accompanies increased runoff so that the quality of irrigation and drinking water downstream declines. Eroded topsoil resulting from reduced plant cover can choke irrigation channels, reduce water delivery to crops, raise floodplain levels, and increase the chance of floods destroying fields and settlements. As reservoirs fill with silt, hydroelectric generation, and therefore industrial production, is diminished. Upstream, where the problem began, fewer trees are available to use as fuel.

Owing to depletions of wood, people living in rural areas must now change their behavior (the fuelwood crisis often refers only to this part of the problem). Women, most often the fuelwood collectors, must walk farther to gather fuel. Turning to animal dung and crop residues as sources of fuel deprives the soil of the fertilizers these poor people most often use when farming. Reduced food output is the result. A family may eventually give up one cooked meal a day or tolerate colder temperatures in their homes because fuel is lacking—measures that negatively impact the family's health.

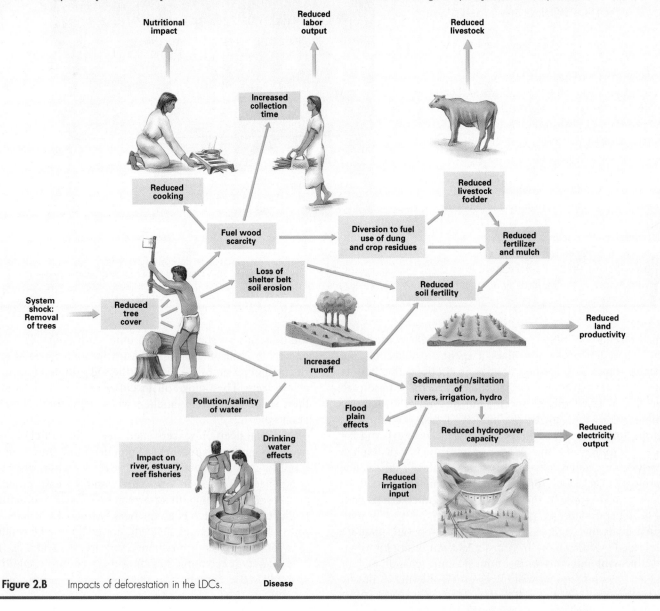

Figure 2.B   Impacts of deforestation in the LDCs.

pay off international debts and meet other needs, the LDCs generally draw on their ecological capital faster than nature can replace it. In ecosystem terms, they exceed **sustainable yield** or the **natural replacement rate,** the highest rate at which a renewable resource can be used without decreasing its potential for renewal. In tropical biomes, for example, people cut down 10 trees for every tree they plant. In Africa, that ratio is 29 to 1. In 1950, about 30% of Ethiopia's land surface was forested. Today, less than 1% is in forest. Such countries are ecologically bankrupt; they have exhausted their environmental capital. Many political and social crises result from this bankruptcy: Revolutions, wars, and refugee migrations in developing nations sometimes have underlying environmental causes. Such problems are becoming more central to the national interests of the United States and other powerful MDCs. The U.S. Central Intelligence Agency, for example, is increasingly using Geographic Information Systems (GIS) and other geographic techniques to analyze the natural phenomena and human misuses of the environment that are root causes of war and threats to global security.

People in the LDCs feel the impacts of environmental degradation more directly than do people in the MDCs. The world's poor tend to drink directly from untreated water supplies and to cook their meals with fuelwood rather than fossil fuels. They are more dependent on nature's abilities to replenish and cleanse itself, and thus, they suffer more when those abilities are diminished (see Fig. 2.B).

Two Types of Overpopulation

High rates of human population growth intensify the environmental problems characteristic of the LDCs. More people cut more trees, a phenomenon that suggests there is a problem of **people overpopulation** in the poorer countries. Many persons, each using a small quantity of natural resources daily to sustain life, may add up to too many people for the environment to support.

There is another type of overpopulation, **consumption overpopulation,** which is characteristic of the MDCs. In the wealthier countries, a few persons, each using a large quantity of natural resources from ecosystems across the world, may also add up to too many people for the environment to support. With less than 5% of the world's population, the United States accounts for about a third of the world's annual energy consumption. Americans use more fossil fuels, iron, and copper than do the inhabitants of Latin America, Africa, and Asia combined. One analyst calculated that the average U.S. citizen, in his or her lifetime, will consume more than 250 times as many goods as the average person in Bangladesh. Such disparities suggest that, just as Bangladesh is "underdeveloped," the United States is "overdeveloped." Much of this overconsumption is unnecessary for sustenance or well-being and has large impacts on the global environment.

If the vast majority of the world's population were to consume resources at the rate that U.S. citizens do, the environ-

mental results might be ruinous. Even if consumption levels in the LDCs do not rise substantially, the sheer increase in numbers of people in those countries suggests that degradation of the environment will accelerate in the coming decades. The basic attributes of the world's population geography provide an insight into this dilemma.

2.6 Population Geography

Earth's human population 10,000 years ago, before the Agricultural Revolution, was probably about 5.3 million. By A.D. 1, it was probably between 250 and 300 million, or about the population of the United States today. The first billion was reached about 1800. Then, a staggering population explosion occurred in the wake of the Industrial Revolution. The second billion came in 1930, the fourth in 1975, and the sixth in 1999. As one measure of ecological dominance, humankind is now by far the most populous large mammal on Earth and has succeeded where no other animal has in extending its range to the world's farthest corners. Some world regions and parts of some countries have especially dense populations. Notable are China and India, where a long history of productive agriculture is the main reason for high density, and Western Europe and the northeastern United States, where industrial productivity is the main factor.

Figure 2.17 illustrates the population explosion as seen in the classic J-shaped curve of exponential growth in the human population. Most striking is the upward curve of growth after the dawn of the Industrial Revolution, with the greatest increase in the population growth rate taking place since 1950. At the 2001 rate of 1.3% annual growth, our population would double in 54 years.

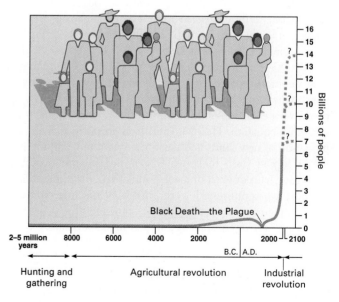

Figure 2.17 The J-shaped curve tracing exponential growth of the human population.

It is useful to understand how energy flows through ecosystems because this process is central to the question of how many people Earth and its regions can support and at what levels of resource consumption. Food chains (e.g., in which, an antelope eats grass and a leopard eats an antelope) are short, seldom consisting of more than four feeding (trophic) levels. The collective weight (known as biomass) and absolute number of organisms decline substantially at each successive trophic level (there are fewer leopards than antelopes). The reason is a fundamental rule of nature known as the **second law of thermodynamics,** which states that the amount of high-quality usable energy is lost as energy passes through an ecosystem. In living and dying, organisms use and lose the high-quality, concentrated energy that green plants produce and that is passed up the food chain. As an organism is consumed in any given link in the food chain, about 90% of that organism's energy is lost, in the form of heat and feces, to the environment. The amount of animal biomass that can be supported at each successive level thus declines geometrically, and little energy remains to support the top carnivores. So, for example, there are many more deer than there are mountain lions in a natural system.

The second law of thermodynamics has important consequences and implications for human use of resources. The higher we feed on the food chain (the more meat we eat), the more energy we use. The question of how many people Earth's resources can support thus depends very much on what we eat. The affluent consumer of meat demands a huge expenditure of food energy because that energy flows from grain (producer) to livestock animal (primary consumer, with a 90% energy loss) and then from livestock animal to human consumer (secondary consumer, with another 90% energy loss). Each year, the average U.S. citizen eats 100 pounds of beef, 50 pounds of pork, and 45 pounds of poultry. By the time it is slaughtered, a cow has eaten about 10 pounds of grain per pound of its body weight; a pig, 5 pounds; and a chicken, 3 pounds. Thus, in a year, a single American consumes the energy captured by two-thirds of a ton of grain (1,330 lb) in meat products alone. By skipping the meat and eating just the corn that fattens these animals, many people could live on the energy required to sustain just one meat-eater. The point is not to feel guilty about eating meat, but to realize there is no answer to the question: How many people can the world support? The question needs to be rephrased: How many people can the world support based on people consuming _____? with the blank filled in, whether as a specific number of calories per day or as a general lifestyle.

Development means a more affluent lifestyle, including eating more meat (feeding higher on the food chain). Some analysts fear the burden that the changing dietary habits accompanying development will pose to the planet's food energy supply. Lester Brown of the WorldWatch Institute wrote a provocative essay entitled "Who Will Feed China?" Keeping in mind the second law of thermodynamics, he argued that if China continues on its present course of growing affluence, with a move away from reliance on rice as a dietary staple to a diet rich in meats and beer, the entire planet will shudder. How could more than a billion people be sustained so high on the food chain?

Dense settlement and growing numbers pose a fundamental question: Will we exceed Earth's carrying capacity for our species, and what will happen if we do? Early in the Industrial Revolution, an English clergyman named Thomas Malthus (1766–1834) postulated that human populations, which can grow geometrically or exponentially, would exceed food supplies, which usually grow only arithmetically or linearly. He predicted a catastrophic human die-off as a result of this irreconcilable equation. He could not have foreseen that the exploitation of new lands and resources, including tapping into the energy of fossil fuels, would permit food production to keep pace with or even outpace population growth for at least the next two centuries. However, this **Malthusian scenario** of the lost race between food supplies and mouths to feed remains a source of constant and important debate today.

Population Concepts and Characteristics

Two principal variables determine population change in a given village, city, country, or over the entire Earth: birth rate and death rate. The **birth rate** is the annual number of live births per 1,000 people in the population. The **death rate** is the annual number of deaths per 1,000 people in the population. The **population change rate** is the birth rate minus the death rate in that population. On a worldwide average (supposing there were a perfect sample of 1,000 people representing the world's population) in 2001, the birth rate was 22 per 1,000 and the death rate was 9 per 1,000. By year's end, among the 1,000 people, 22 babies had been born but 9 people of varying ages had died, resulting in a net growth of 13. That figure, 13 per 1,000, or 1.3%, represents the 2001 population change rate (population growth rate) for the world. As no one leaves or enters Earth for or from other planets, there is no need to consider a third variable, **migration,** at this scale. However, at the scale of a village, city, or country, migration does affect the population. Compulsory and voluntary migrations are strong forces in the world's regions and countries, and this text cites frequent examples of both.

Many factors affect birth rates. Better educated and wealthier people have fewer children. Conversely, less educated and poorer people generally want and have more children. Poor parents view additional children as an economic asset rather than a burden because they represent additional labor to work in fields or factories and will care for them in their old age. People in cities tend to have fewer children than those in rural areas. Those who marry earlier generally have more children. Couples with access to and understanding of contraception may have fewer children. Cultural norms are important because even where contraception is available, for religious or social reasons a couple may decide not to interfere with what they perceive as God's will or the social status associated with a larger family (Fig. 2.18).

Death rates are affected mainly by health factors. Improvements in food production and distribution help reduce death rates. Better sanitation, hygiene, and drinking water eliminate infant diarrhea, a common cause of infant mortality in LDCs. The availability of antibiotics, immunizations, insecticides, and other improvements in medical and public health information and technologies have a marked correlation with declining death rates.

It is vital to understand that the explosive growth in world population since the beginning of the Industrial Revolution is the result not of a rise in birth rates but of a dramatic decline in death rates, particularly in the LDCs. The death rate has fallen as improvements in agricultural and medical technologies have diffused, or spread, from the MDCs to the LDCs. Until recently, however, there were no strong incentives for people in LDCs to have fewer children. With birth rates remaining high and death rates falling quickly, the population has grown sharply.

With about 9 of 10 babies worldwide born in the LDCs, the current and projected rates of population growth are distributed quite unevenly between the MDCs and LDCs. This phenomenon is apparent in the age-structure diagrams typical of these countries. An **age-structure diagram** (often called a "population pyramid") classifies a population by gender and by 5-year age increments (Figs. 2.19 and 2.20). One important index these profiles show is the percentage of a population under age 15. The very high proportion typical of LDCs, about 33%, is remarkable, for it means that populations will continue to grow in these poor countries as these children enter their reproductive years. The bottom-heavy, pyramid-shaped age-structure diagram of LDCs contrasts markedly with the more rectangular structure of the MDCs. The wealthier countries have a much more even distribution of population through age groups, with a modest share of 18% under age 15. Such profiles suggest that, not considering migration, their population growth will be low in the near future.

The Demographic Transition: Will the LDCs Complete It?

The history of population change in the MDCs provides a model whose usefulness in predicting the population future for LDCs is still uncertain. Known as the **demographic transition,** this model depicts the change from high birth rates and high death rates to low birth rates and low death rates that accompanied economic growth in the MDCs. Four stages comprise the model (Fig. 2.21). In the first or preindustrial stage, birth rates and death rates were high, and population growth was negligible. In the second or transitional stage, birth rates remained high, but death rates dropped sharply after about 1800 due to medical and other innovations of the Industrial Revolution. In the third or industrial stage, beginning around 1875, birth rates began to fall as affluence spread. Finally, after about 1975, some of the industrialized countries entered the fourth or postindustrial stage, with both low birth rates and low death rates and, therefore (once again, as in stage one), low population growth. Some MDCs, including in 2001, Greece, Spain, and Austria, officially registered zero population growth (ZPG), a rate of equal birth rates and death rates, and many in Europe (Sweden, Italy, Croatia, Slovenia, Estonia, Latvia, Lithuania, Germany, Moldova, Belarus, Bulgaria, the Czech Republic, Hungary, Romania, Russia, and Ukraine) were experiencing a negative growth rate with more deaths than births.

Viewed as a group, the LDCs are in the latter part of stage two of the demographic transition. Birth rates in the LDCs are now falling and, consequently, so is the world's rate of population growth (the actual numbers are dropping, too, from an all time annual growth of 88 million per year in 1994 to 85 million in 2001). In view of declining birth rates worldwide, the United Nations in 1997 revised its projection for future population growth downward. The agency predicted 9.4 billion by 2050 (half a billion fewer than that projected only 2 years earlier) and a stabilization at 10.7 billion in 2120 (the earlier projection was 12.6 billion). This revision prompted a rash of popular and academic articles proclaiming "the population explosion is over" and even some essays arguing there would soon be too few people on Earth, particularly in the MDCs. Other sources cautioned that it was too soon to declare the population bomb defused. "World population growth turned a little slower," a Population Institute report argued. "The difference, however, is comparable to a tidal wave surging toward one of our coastal

Figure 2.18   These graffiti on the back of a road sign in Agra, India, sum up the attitude many of the world's people, particularly the rural poor, have about children. For many, family planning thus means trying to go against divine will.

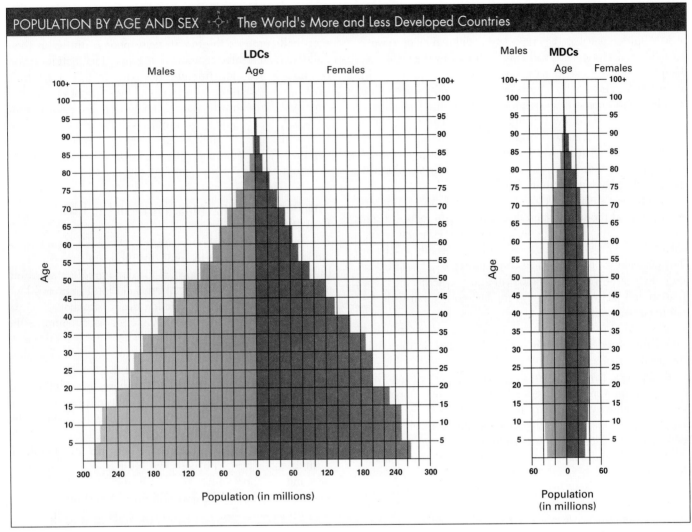

POPULATION BY AGE AND SEX ⟡ The World's More and Less Developed Countries

Source: United Nations Population Division, 2000 estimates; Population Reference Bureau.

Figure 2.19 Age-structure profiles of the LDCs and MDCs in 2000.

cities. Whether the tidal wave is 80 feet or 100 feet high, the impact will be similar."[3] Not content with its own projected figure for stabilization, the United Nations has pledged to do its best to help stabilize the global population at no more than 9.8 billion after the year 2050. The organization's plan is to focus on enhancing the education and employment of women in the LDCs as a means of bringing down birth rates.

There are two very distinct points of view on where these LDCs will proceed from stage two of the demographic transition, with its low death rates, high but falling birth rates, and rapid population growth. Some observers believe that the poorer countries will follow the wealthier ones through the transition, achieving prosperity and a stabilized population. Most recognizable among the optimists are the so-called **technocentrists** or **cornucopians**, who argue that human history

[3] Quoted in Steven A. Holmes. "Global Crisis in Population Still Serious, Group Warns." *New York Times*, December 31, 1997, A7.

provides insight into the future: Through their technological ingenuity, people always have been able to conquer food shortages and other problems, and therefore always will. The late Julian Simon, a University of Maryland economist, argued that far from being a drain on resources, additional people create additional resources. Technocentrists thus insist that people can raise Earth's carrying capacity indefinitely and that the die-off that Malthus predicted will always be averted. Our more numerous descendants will instead enjoy more prosperity than we do.

In contrast, the **neo-Malthusians** (heirs of the reasoning of Thomas Malthus) argue that although successful so far, we cannot indefinitely increase Earth's carrying capacity. There is an upper limit beyond which growth cannot occur, with calculations ranging from 8 to 40 billion. Neo-Malthusians insist that LDCs cannot remain indefinitely in the transitional stage. Either they must intentionally bring birth rates down further and make it successfully through the

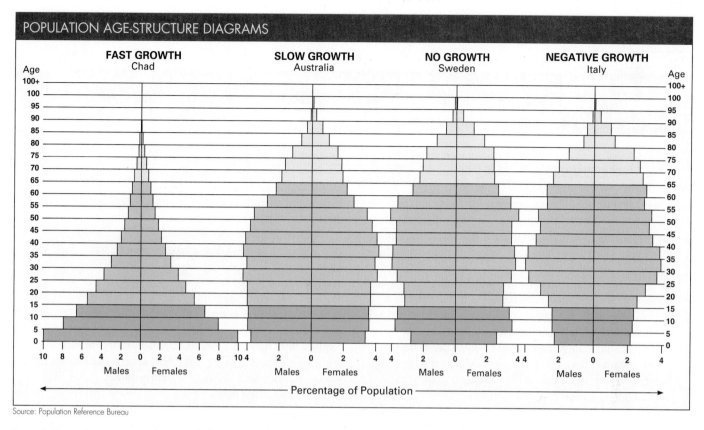

Figure 2.20 Age-structure profiles typical of countries with rapid, slow, zero, and negative rates of population growth.

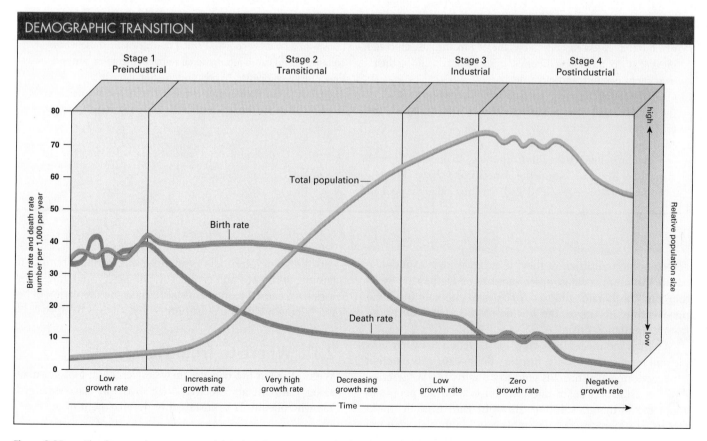

Figure 2.21 The demographic transition model. In the MDCs, economic development has resulted in a transition from high birth rates and high death rates to low birth rates and low death rates. Population growth rates at stages 1 and 4 are low because birth rates and death rates are nearly equal.

Neo-Malthusian ecologist Garrett Hardin introduces **lifeboat ethics** — the question of whether or not the wealthy should rescue the "drowning" poor — in an essay so distressing but challenging that we are obliged to consider and respond to it. "People turn to me," writes Hardin, "and say 'my children are starving. It's up to you to keep them alive.' And I say, 'The hell it is. I didn't have those children.'"[a] He describes our world not as a single "spaceship earth" or "global village" with a single carrying capacity, as many environmentalists do, but as a number of distinct "lifeboats," each occupied by the citizens of single countries and each having its own carrying capacity. Each rich nation is a lifeboat comfortably seating a few people. The world's poor are in lifeboats so overcrowded that many fall overboard. They swim to the rich lifeboats and beg to be brought aboard. What should the passengers of the rich lifeboat do? The choices pose a dilemma.

Hardin proposes the following: There are 50 rich passengers in a boat with a capacity of 60. Around them are 100 poor swimmers who want to come aboard. The rich boaters have three choices. First, they could take in all the swimmers, capsizing the boat with "complete justice, complete catastrophe." Second, as they enjoy an unused excess capacity of 10, they could admit just 10 from the water. But which 10? And what about the margin of comfort that excess capacity allows them? Finally, the rich could prevent any of the doomed from coming aboard, ensuring their own safety, comfort, and survival.

Translating this metaphor into reality, as an occupant of the rich lifeboat United States, for example, what would you do for the drowning refugees from lifeboat Haiti, Sierra Leone, or Afghanistan?

Hardin's choice is the third: drowning. To preserve their own standard of living and ensure the planet's safety, the wealthy countries must cease to extend food and other aid to the poor and must close their doors to immigrants from poor countries. "Every life saved this year in a poor country diminishes the quality of life for subsequent generations," Hardin concludes. "For the foreseeable future, survival demands that we govern our actions by the ethics of a lifeboat."[b]

In the United States and other Western countries, there is now a growing sense of **donor fatigue,** a trend that Garrett Hardin would applaud as beneficial for the global environment. Wearied by constant images of people in need and feeling their contributions might be only marginally useful, people who can afford to give are simply tired of thinking about giving. Governmental support has flagged, too; development aid fell by a third between 1990 and 2000. That decline is widely attributed to the end of the Cold War, when it began to be regarded as less critical to national security to win and maintain influence in foreign lands. Critics of foreign aid point out that corrupt governments often steal or squander the assistance money, and that too much of the aid is "tied," so that the recipient country is obliged to buy goods or services from the donor country.

So why shouldn't we be fed up with giving? Recently, U.S. foreign policy analysts have begun to identify certain strategically "pivotal countries" for enhanced aid, while assistance to other nations would be drawn down. **Pivotal countries** are those whose collapse would cause international refugee migration, war, pollution, disease epidemics, or other international security problems. According to one model, nine countries are clearly pivotal: Mexico, Brazil, Algeria, Egypt, South Africa, Turkey, India, Pakistan, and Indonesia.

[a] Garrett Hardin, "The Tragedy of the Commons." Science, 162: 1968, 1243–1248. His original lifeboat essay was entitled "Living on a Lifeboat" and was published in *Psychology Today* in September 1974.
[b] Ibid.

demographic transition, or they must unwillingly experience nature's solution, a catastrophic increase in death rates. By either the "birth rate solution" or "death rate solution," the neo-Malthusians argue, the less developed world must confront a population crisis.

Whereas technocentrists view the equation between people and resources passively, insisting that no corrective action is needed, neo-Malthusians tend to be activists who describe dire scenarios of a death rate solution to motivate people to adopt the birth rate solution. Biologist Paul Ehrlich thinks that we are increasingly vulnerable to a Malthusian catastrophe, particularly as HIV and other viruses diffuse around the globe with unprecedented speed. "The only big question that remains," wrote Ehrlich, "is whether civilization will end with the bang of an all-out nuclear war, or the whimper of famine, pestilence and ecological collapse."[4] Such dire warnings have earned many neo-Malthusians the reputation of being "gloom and doom pessimists."

2.7   Where to from Here?

Within the last two decades, new concepts and tools for managing Earth in an effective, long-term way have emerged. Known collectively as **sustainable development** or ecodevel-

[4] Paul R. Ehrlich. "Populations of People and Other Living Things." In *Earth '88: Changing Geographic Perspectives*, Harm De Blij, ed. Washington, D.C.: National Geographic Society, 1988, 302–315.

opment, these ideas and techniques consider what both MDCs and LDCs can do to avert the possible Malthusian dilemma and improve life on the planet. Sustainable development offers an activists' agenda without the peril of doom depicted by the neo-Malthusians.

The World Conservation Union defines sustainable development as "improving the quality of human life while living within the carrying capacity of supporting ecosystems."[5] Sustainable development refutes what its proponents perceive as the current pattern of unsustainable development, whereby gross natural product (GNP) and other measures of economic growth are based in large part on exceeding carrying capacity and overusing resources. According to these measures, a country that depletes its resource base for short-term profits gained through deforestation increases its GNP and appears to be more "developed" than a country that protects its forests for a long-term harvest of sustainable yield. Deforestation appears to be beneficial to a country because it raises GNP through the production of pulp, paper, furniture, and charcoal. However, GNP does not measure the negative impacts of deforestation, such as erosion, flooding, siltation, and malnutrition. These consequences are known as **external costs** or **externalities,** and are not priced into goods and services. Advocates of sustainable development argue that these externalities should be added or "internalized" before a good or service is marketed because their true high costs would be recognized. The lower true costs of less destructive practices would then be evident, providing stronger incentives for individuals, companies, and nations to invest in sustainable practices and technologies.

Sustainable development is a complex assortment of theories and activities, but its proponents call for eight essential changes in the way people perceive and use their environments:

1. People must change their worldviews and value systems, recognizing the finiteness of resources and reducing their expectations to a level more in keeping with Earth's environmental capabilities. Proponents of sustainable development argue that this change in perspective is needed especially in the MDCs, where, instead of trying to "keep up with the Joneses," people should try to enjoy life through more social rather than material pursuits.

2. People should recognize that development and environmental protection are compatible. Rather than viewing environmental conservation as a drain on economies, we should see it as the best guarantor of future economic well-being. This is especially important in the LDCs, with their resource-based economies.

3. People all over the world should consider the needs of future generations more than we do now. Much of the wealth we generate is in effect borrowed or stolen from our descendants. Our economic system values current environmental benefits and costs far more than future benefits and costs, and thus we try to improve our standard of living today without regard to tomorrow.

4. Communities and countries should strive for self-reliance, particularly through the use of appropriate technologies. For example, remote villages could rely increasingly on solar power for electricity rather than be linked into national grids of coal-burning plants.

5. LDCs need to limit population growth as a means of avoiding the destructive impacts of "people overpopulation." Advances in the status of women, improvements in education and social services, and effective family planning technologies can help limit population growth.

6. Governments need to institute land reform, particularly in the LDCs. Poverty is often not the result of too many people on too little total land area, but of a small, wealthy minority holding a disproportionably high share of quality land. To avoid the environmental and economic consequences of marginalization, a more equitable distribution of land is needed.

7. Economic growth in the MDCs should be slowed to reduce the effects of "consumption overpopulation." If economic growth, understood as the result of consumption of natural resources, continues at its present rate in excess of sustainable yield, Earth's environmental "capital" will continue to diminish rapidly.

8. Wealth should be redistributed between the MDCs and LDCs. Because poverty is such a fundamental cause of environmental degradation, the spread of a reasonable level of prosperity and security to the LDCs is essential. Proponents argue that this does not mean rich countries should give cash outright to poor countries. Instead, the lending institutions of MDCs could forgive some existing debts owed by LDCs or use such innovations as **debt-for-nature swaps,** in which a certain portion of debt is forgiven in return for the borrower's pledge to invest that amount in national parks or other conservation programs. Reducing or eliminating trade barriers that MDCs impose against industries in LDCs would also help redistribute global wealth.

Some geographers and other scientists believe that sustainable development (rather than information technology) will bring about the "Third Revolution," a shift in human ways of interacting with Earth so dramatic that it will be compared with the origins of agriculture and industry. The formidable changes called for in sustainable development are attracting increasing attention, perhaps because at present there are no comprehensive alternative strategies for dealing with some of the most critical issues of our time.

---

[5] World Wildlife Fund. *Sustainable Use of Natural Resources: Concepts, Issues and Criteria.* Gland, Switzerland: World Wildlife Fund, 1995, 5.

CHAPTER SUMMARY

- Weather is the atmospheric conditions prevailing at one time and place. Climate is a typical pattern recognizable in the weather of a region over a long period of time. Climatic patterns have a strong correlation with patterns of vegetation and in turn with human opportunities and activities in relation to the environment.

- Warm air holds more moisture than cool air; therefore, precipitation is best understood as the result of processes that cool the air to release moisture. Convectional precipitation is the result of air being heated by intense surface radiation and rising rapidly to cool and produce rain. Orographic precipitation results when moving air confronts a topographic barrier and is forced upward to cool and produce rain. Cyclonic (frontal) precipitation is generated in cyclones or traveling low-pressure cells that bring air masses with different characteristics of temperature and moisture into contact.

- Most of the sun's visible short-wave energy that reaches Earth is absorbed, but some of it returns to the atmosphere in the form of infrared long-wave radiation, which generates heat and helps to warm the atmosphere. Although the sun is the primary source of heat, the air is mainly heated by reradiation from the land and water.

- Geographers group local climates into major climate types, each of which occurs in more than one part of the world and is associated with other natural features, particularly vegetation. Geographers recognize 10 to 20 major types of ecosystems or biomes, which are categorized by the type of natural vegetation. Vegetation and climate types are so sufficiently related that many climate types are named from the vegetation types — for example, the tropical rain forest climate and the tundra climate.

- Geographers and ecologists recognize the importance of some biomes because of their biological diversity — the number of plant and animal species and the variety of genetic materials these organisms contain. Regions where human activities are rapidly depleting a rich variety of plant and animal life are known as biodiversity hot spots, a ranked list of places scientists believe deserve immediate attention for study and conservation.

- Computer-based climate change models use the scenario of a doubling of atmospheric carbon dioxide and indicate that the mean global temperature might warm up to 11° Fahrenheit by the year 2100 due to the greenhouse effect. The general agreement is that with global warming the distribution of climatic conditions typical of biomes will shift poleward and upward in elevation.

- A useful perspective on the current spatial patterns of our relationship with Earth is to view these patterns as products of two "revolutions" in the relatively recent past: the Agricultural or Neolithic (New Stone Age) Revolution and the Industrial Revolution. Each of these transformed humanity's relationship with the natural environment.

- High rates of human population growth intensify the environmental problems characteristic of the LDCs. More people use more resources, a phenomenon that suggests there is a problem of people overpopulation in the poorer countries. Consumption overpopulation is more characteristic of the MDCs, where fewer people use large quantities of natural resources from across the world.

- Considering population projections, the geographer must ask how and where global resources can support 10 billion or more people. Our demands for more food and other photosynthetic products are growing, and it is questionable whether supplies can keep pace. The Malthusian scenario, which has not yet been realized, insists that a catastrophic die-off of people will occur when their numbers exceed food supplies.

- The explosive growth in world population since the beginning of the Industrial Revolution is not the result of a rise in birth rates but of a dramatic decline in death rates, particularly in the LDCs.

- Within the last two decades, new concepts and tools for managing Earth in an effective, long-term way have emerged. Known collectively as sustainable development or ecodevelopment, these ideas and techniques consider what humanity can do to avert the Malthusian dilemma and improve life on the planet.

REVIEW QUESTIONS

1. What is the difference between weather and climate?
2. List the major climate types and their associated natural vegetation. Identify the general locations of these climate and vegetation types on maps (Figs. 2.6 and 2.7).
3. List Earth's biodiversity hot spots and identify their locations in association with Earth's biomes.
4. What apparent long-term effects on Earth's atmospheric processes can be attributed to modern technology?
5. Why, perhaps, did the Agricultural Revolution come about? What factors made the Industrial Revolution possible?
6. List and define the measures of development used to distinguish MDCs from LDCs.
7. According to the dependency theory, what are the causes of disparities between MDCs and LDCs?
8. What are the primary environmental impacts of underdevelopment?
9. What are the two types of overpopulation and how do they differ?
10. What has been the main cause of the world's explosive population growth since the beginning of the Industrial Revolution?
11. What variables distinguish the four stages of the demographic transition and what explains them?
12. Explain Garrett Hardin's lifeboat metaphor.

DISCUSSION QUESTIONS

1. Explain the high degree of spatial correlation between the distributions of Earth's climate and vegetation types.
2. What are the indicators that support the notion that Earth is getting warmer? What might these changes mean for agriculture, plant and animal life, and human adaptations?
3. Discuss the Kyoto Protocol. What would be the advantages and drawbacks of its implementation? Do you agree or disagree with the position taken against the treaty by the administration of U.S. President George W. Bush?

4. Using the album of Earth's biomes in Fig. 2.8, discuss the variables of precipitation, latitude, and elevation that likely give rise to each characteristic type.

5. What are some of the possibly negative ramifications of the Green Revolution on agriculture and Earth's biodiversity? Why does Norman Myers lament "we are causing the death of birth"?

6. Explain how *Homo sapiens* has always been an ecologically dominant species.

7. What characteristics other than productivity and income are considered in the Human Development Index (HDI) and why?

8. Why are some countries rich and some poor? Discuss and debate the arguments presented in this chapter, and use your own perspectives.

9. Discuss the graph of the fuelwood crisis on page 46. Can you explain the causes and effects of the initial system shock?

10. What is overdevelopment? Is it a useful concept for describing the way of life in the country you live in? What would you regard as indicators of overdevelopment?

11. What would happen to the global environment if China became a prosperous country with personal habits to match?

12. Compare and contrast the birth rate solution and the death rate solution. Is one or the other really inevitable? Which is Earth's population heading for?

13. Debate Garrett Hardin's lifeboat ethics. Do you agree or disagree with his view and why?

14. Discuss the nine pivotal countries. What defines a country as pivotal and therefore deserving of more foreign aid? What nine countries have been identified as pivotal and why? Prior to study of the following chapters in this text, what did you know about these nations deemed vital to the interests of the United States and other countries?

15. Discuss sustainable development. Are these notions pie in the sky or are they achievable? Should achieving them be attempted? Why or why not?

Europe

Europe continues to be a major global engine of innovation, manufacturing, and consuming. In its patterns of marketing and international visibility, this region — especially the Coreland of Western Europe — is still a powerful worldwide culture broker. Even while the European Union is attempting to further regional unity with the introduction of the euro, there are political forces that are tending to break up the fragile unity the EU represents. At the same time, countries lying east of the EU are trying to gain membership in this fifteen-country alliance. From the very origins of Western civilization to the modernity of today's transport technologies, Europe has been a world region worthy of study and understanding.

With the changing nature of warfare in the past two decades, Europe has spawned new industries. This plywood tank is made to be sold to countries that may be visited by smart bombs or air raids focusing on military equipment rather than human settlements. There is enormous cost in blasting this wooden model to pieces with a million dollar missile.

Although Europe is seldom thought of as a major energy source, the oil being pumped from beneath the North Sea has changed both the political and energy dynamics of this region in the past three decades. Even as oil has become a more important regional resource, coal has lost importance, showing that resource equations are in continual flux in Europe as well as the rest of the world.

Although it was in Europe that the Industrial Revolution came alive and, with that powerful force, changed the face of the world through colonization and global trade, older industrial plants are now being closed and replaced by modern factory complexes in Asia, Latin America, and other parts of the world.

All through Europe rivers have been major travel and trade arteries. They have been bridged, dammed, barged upon, and bordered deeply with dense human settlements from east to west, north to south.

CHAPTER 3

A Geographic Profile of Europe

This view of Europe is a mosaic composed of a number of weather satellite images taken on cloud-free summer days. The colors approximate natural tones and are a good indication of different ground cover, with light and midgreen colors indicating pasture and cultivated land and the bluer greens showing greater aridity and less lush vegetation. Most evident is the geographic fact that Europe is more a peninsula of the massive Eurasian landmass than a freestanding continent.

Many regions of the globe have contributed to shaping today's world but none more powerfully than Europe. The influence of Europe on modern nation building, democracy, the development of science and technology, the rise of advanced capitalist economies, and colonial expansion overseas is central to world history, even though there are both positive and negative perspectives to these areas of influence. In some views, it is the very essence of this Western economic and cultural influence that played a role in the September 11, 2001, attack on New York City's World Trade Towers. Such a centrality in the development of global attitudes — and resource use — has become an enemy target for international terrorist activities. Europe plays a vital role in the world's economic, political, cultural, and intellectual life — all the while trying to maintain its independence from American influences — as it attempts to craft an ever-closer regional compact through the slowly expanding map of the **European Union (EU).**

3.1   Area and Population

Europe is a culture region made up of the countries of Eurasia lying west of Russia and the Near Abroad and Turkey. Thus defined, Europe is a great peninsula, fringed itself by lesser peninsulas and islands and bounded by the Arctic and Atlantic Oceans, the Mediterranean Sea, Turkey, the Black Sea, and Russia and the Near Abroad (Fig. 3.1). Europe achieves the label of continent largely because fundamental geographic standards for such language were determined and defined by Europeans when they had enormous global authority.

We segment Europe into these several regions: The Coreland is Northwest and North Central Europe. It includes the United Kingdom, France, the Benelux Countries (Belgium, the Netherlands, and Luxembourg), Switzerland, Austria, and Germany. These are countries that have the largest populations and play major economic and political roles in contemporary Europe. The North, a second region, is made up of Denmark, Greenland, Iceland, Norway, Sweden, and Finland. The South, the third region, is made up of Portugal, Spain, Italy, Greece, and Malta. The East completes the regional unit and includes Poland, the Czech Republic, Slovakia, Hungary, Romania, Bulgaria, Albania, Yugoslavia, Bosnia-Herzegovina, Croatia, Macedonia, and Slovenia. Table 3.1 provides detailed statistical information on these countries and the European subunits.

There are also the **microstates** of Andorra, Monaco, and Liechtenstein. In addition, there is the curious role of Greenland, the world's largest island. It has an area of 840,000 square miles and a total population of just over 60,000, giving it a population density of less than 0.1 person per square mile. It lies more than 2,000 miles west of the European nation with which it is associated, Denmark (a country of 16,600 sq mi). The linkage between Denmark and Greenland is shown dramatically by the fact that annually the European country subsidizes the massive island with an average support of more than $8,000 a person. Even though there is increasing talk of independence among the Greenland population, the reality of this financial dependence on Denmark makes such discussion mostly academic. F22.1 584

Understanding the geographic reality of Europe requires a grasp of some basic magnitudes that are very different from those of the United States. For example, all of Europe has an area only about two-thirds as large as that of the 48 contiguous United States. This means that the areas of Europe's many countries tend to be rough equivalents of those of U.S. states. One of the striking things about Europe is the large impact that societies nourished on some of these small pieces of land have had on the world.

A different set of comparative magnitudes emerges when comparing the populations of the European countries with those of the individual United States. California, with its population of nearly 34 million, would be a rather populous country if it were European, but even California is considerably exceeded in population by six European countries (Table 3.1). Four major countries — Germany, the United Kingdom, France, and Italy — far outsize all others in Europe in population. Their respective populations range from about 82 million for Germany to approximately 60 million each for France, Italy, and the United Kingdom. Together, the four countries represent about one-half of Europe's population, or nearly as many people as there are in the entire United States.

Another useful comparison between European countries and the United States is by density of population. Even excluding great disparities among small units such as Malta and the microstates of Andorra and Monaco, Europe

Western civilization and related terms, such as the **West** and **Westernization,** are indispensable to a study of world geography because they connote innumerable traits that give geographic distinctiveness to places. In general, Western civilization refers to the sum of values, practices, and achievements that had roots in ancient Mesopotamia — and Palestine, Greece, and Rome — that subsequently flowered in Europe. These traits are still being developed and modified there and in other areas to which they have been diffused. The term incorporates a set of languages, religious practices (associated with "Western" Christian churches, including the Roman Catholic Church and the Protestant churches but not the Orthodox Eastern churches), systems of law, and systems of social, economic, and political organization. Among the core concepts of modern Western civilization are strong commitments (not

always effective) to education, experimental science, technological progress, economic development, democratic representative government, and explicit protection of individual rights and liberties. A strong reliance on private capitalism developed prior to and during the Industrial Revolution, although this has been modified by experiments in communism, socialism, and fascism during the 20th century. Westernization today refers to the process whereby non-Western societies acquire Western traits, adopted with varying degrees of completeness. As used today, "the West" refers primarily to Europe, the United States and Canada, Australia, and New Zealand, although the concept embraces other countries (e.g., the Latin American countries, Israel, and South Africa) to a smaller degree.

exhibits an extreme range of population density — from over 1,222 persons per square mile (over 472 per sq km) in the intensively developed Netherlands to about 8 per square mile (3 per sq km) in Iceland (Table 3.1 and Fig. 3.2). Comparisons with the United States reveal about the same range, although in general the European countries are more densely populated. Table 3.1 provides a set of data that allows a comparison within Europe and with other world regions and nations as well.

This is a region worthy of some detailed study because many of the cultural characteristics and patterns that give image, personality, and tension to the world today derive from the influences of Western civilization in its many manifestations.

Europe's Population Dynamics

A major reason for Europe's importance in the global economy lies in the region's large and technically skilled population. Europeans represent one of the great masses of world population (see Table 3.1). Europe's approximately 512 million people in 2001 were nearly twice that of the United States. One of every 11 people in the world live in a space half the size of the United States. Birth rates are low in this region of relative affluence and high urbanization, and the whole region is increasing in numbers much more slowly (or in some cases, is losing population) than most of the world.

Major Population Belts and Their Significance

Except for some northern, rugged, or infertile areas, European population density is everywhere greater than the world average (107/sq mi; 41/sq km). However, the greatest densities appear

along two belts of industrialization and urbanization near coal or hydroelectric power sources. One belt extends north–south from Great Britain to Italy (see Fig. 3.2). In addition to large parts of Great Britain, it includes extreme northeastern France, most of Belgium and the Netherlands, Germany's Rhineland, northern Switzerland, and across the Alps, northern Italy. The second belt of dense population extends west–east from Great Britain to southern Poland and continues into the Ukraine. It corresponds to the first belt as far east as the Rhineland, where it forms a relatively narrow strip eastward across Germany, the western Czech Republic, and Poland. Although the two belts represent a minor part of Europe's total extent, they contain more large cities and a greater value of industrial output than the rest of Europe combined. Only in eastern North America and Japan are there urban-industrial belts of the same order of complexity and importance.

The European belts coincide with major route ways that were in use very early for migration, trade, and military movement. Many cities along these routes are very old. With the rise of modern industry, coal fields along the west–east belts became important, and in the southern part of the north–south belt, the age of electricity saw the development of many industries based on the hydroelectric resources of the Alps. Both belts benefit from the relatively good soil they encompass.

An issue of growing concern in Europe is the tension caused by high rates of domestic unemployment in many EU nations and the steady increase in rates of legal and illegal immigration. From 1990 to 1995, Germany alone absorbed 2.4 million foreigners, most of them from Central and Eastern Europe, during the bloodshed of the Bosnian chaos in the breakup of the former Yugoslavia. By 1996, 8.8% of Ger-

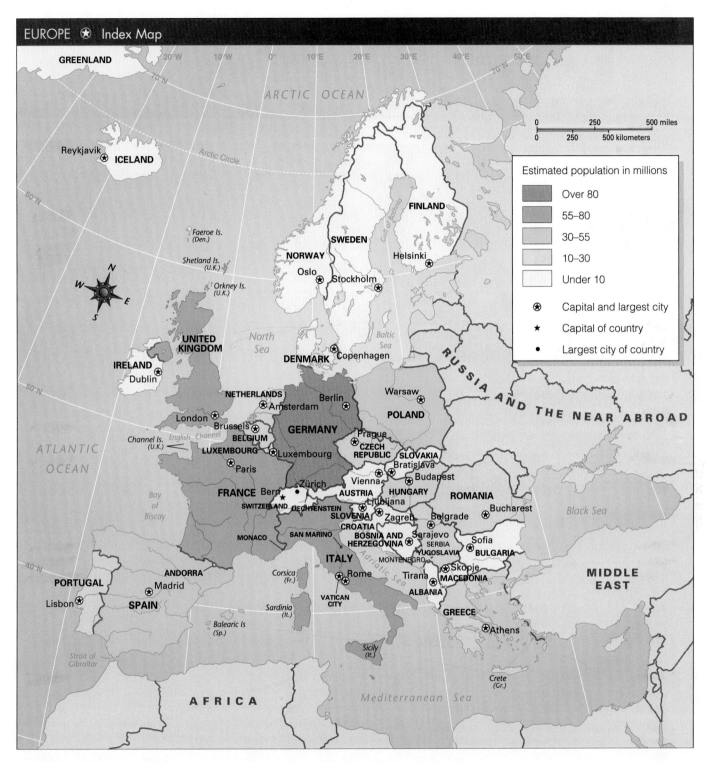

Figure 3.1 Europe showing political units as of 2001. Note that in nearly all European countries, the political capital is also the largest and most important city (metropolitan area).

many's population were foreigners; in France it was 6.3%; in Italy, 6%; and in the United Kingdom, 3.4%. Switzerland was 19.5% foreign. In that same period, the United States was 7.9% foreign born. The EU has eliminated border controls as one aspect of the new openness of flow among its member nations. This means, in essence, that Italy — with its 5,000 miles (9,000 km) of coastline — plays the gatekeeper role for much of the annual migrant flow into Western Europe. The EU's open borders, however, are intended for legal and official citizens of the EU nations. As it is, a significant migrant

TABLE 3.1   Europe: Basic Data

Political Unit	Area (thousand/sq mi)	Area (thousand/sq km)	Estimated Population (millions)	Annual Rate of Increase (%)	Estimated Population Density (sq mi)	Estimated Population Density (sq km)	Human Development Index	Urban Population (%)	Arable Land (%)	GDP PPP Per Capita ($US)
British Isles										
United Kingdom	94.5	244.9	60	0.1	635	245	0.923	90	25	22,800
Ireland	27.1	70.3	3.8	0.6	142	55	0.916	58	13	21,600
Total	**121.6**	**315.2**	**63.8**	**0.13**	**525**	**202**	**0.923**	**88**	**22**	**22,729**
West Central Europe										
France	212.9	551.5	59.2	0.4	278	107	0.924	74	33	24,400
Germany	137.8	357.0	82.2	−0.1	597	231	0.921	86	33	23,400
Belgium	11.8	30.5	10.3	0.1	827	337	0.935	97	24	25,300
Netherlands	15.8	40.8	16	0.4	1018	393	0.931	62	25	24,400
Liechtenstein	0.06	0.16	0.03	0.6	534	206	NA	23	24	23,000
Monaco	0.001	0.003	0.03	0.3	45,333	17,503	NA	100	0	27,000
Luxembourg	1.0	2.6	0.4	0.4	446	172	0.924	88	24	36,400
Switzerland	15.9	41.3	7.2	0.2	453	175	0.924	68	10	28,600
Austria	32.4	83.9	8.1	0	251	97	0.921	65	17	25,000
Total	**427.7**	**1107.8**	**183.46**	**0.13**	**429**	**166**	**0.924**	**79**	**30**	**24,220**
Northern Europe										
Denmark	16.6	43.1	5.4	0.2	322	124	0.921	72	60	25,500
Norway	125.1	323.9	4.5	0.3	36	14	0.939	74	3	27,700
Sweden	173.7	450.0	8.9	0	51	20	0.936	84	7	22,200
Finland	130.6	338.2	5.2	0.2	40	15	0.925	60	8	22,900
Iceland	39.8	103.0	0.3	0.8	7	3	0.932	93	0	24,800
Greenland (Denmark)	840	2175.6	0.056	0.1	0.07	0.03	NA	NA	0	20,000
Total	**1325.8**	**3433.8**	**24.4**	**0.15**	**18**	**7**	**0.931**	**74**	**8**	**24,134**
Southern Europe										
Andorra	0.17	0.45	0.1	0.9	380	147	NA	93	4	18,000
Italy	116.3	301.3	57.8	0	497	192	0.909	90	31	22,100
Spain	195.4	506.0	39.8	0	204	79	0.908	64	30	18,000
Portugal	35.5	92.0	10	0.1	282	109	0.874	48	26	15,800
Greece	51.0	132.0	10.9	0	214	83	0.881	59	19	17,200
Malta	0.12	0.32	0.4	0.3	3157	1219	0.866	91	32	14,300
San Marino	0.023	0.06	0.03	0.4	1166	450	NA	89	17	32,000
Total	**398.5**	**1032.1**	**119**	**0.01**	**299**	**115**	**0.903**	**75**	**29**	**19,724**
East Central Europe										
Poland	124.8	323.3	38.6	0	310	120	0.828	62	47	8500
Czech Republic	30.5	78.9	10.3	−0.2	337	130	0.844	77	41	12,900
Slovakia	18.9	49.0	5.4	0	286	110	0.831	57	31	10,200
Hungary	35.9	93.0	10	−0.4	278	107	0.829	64	51	11,200
Romania	92.0	238.4	22.4	−0.1	243	94	0.772	55	41	5900
Bulgaria	42.8	110.9	8.1	−0.5	190	73	0.772	68	43	6200
Albania	11.1	28.7	3.4	1.2	310	120	0.725	46	21	3000
Yugoslavia	39.4	102.2	10.7	0.1	270	104	NA	52	40	2300
Bosnia-Herzegovina	19.7	51.1	3.4	0.4	173	67	NA	40	14	1700
Croatia	21.8	56.5	4.7	−0.2	197	76	0.803	54	21	5800
Macedonia	9.9	25.7	2	0.5	205	79	0.766	60	24	4400
Slovenia	7.8	20.3	2	−0.1	256	99	0.874	50	12	12,000
Total	**454.6**	**1178.0**	**121**	**−0.05**	**266**	**103**	**0.809**	**60**	**40**	**7530**
Summary Total (Land area and population includes Greenland)	**2728.4**	**7066.6**	**511.7**	**0.06**	**270.9**	**104.6**	**0.894**	**74.4**	**26**	**19,036**

Sources: *World Population Data Sheet,* Population Reference Bureau, 2001; *U.N. Human Development Report,* United Nations, 2001; *World Factbook,* CIA, 2001.

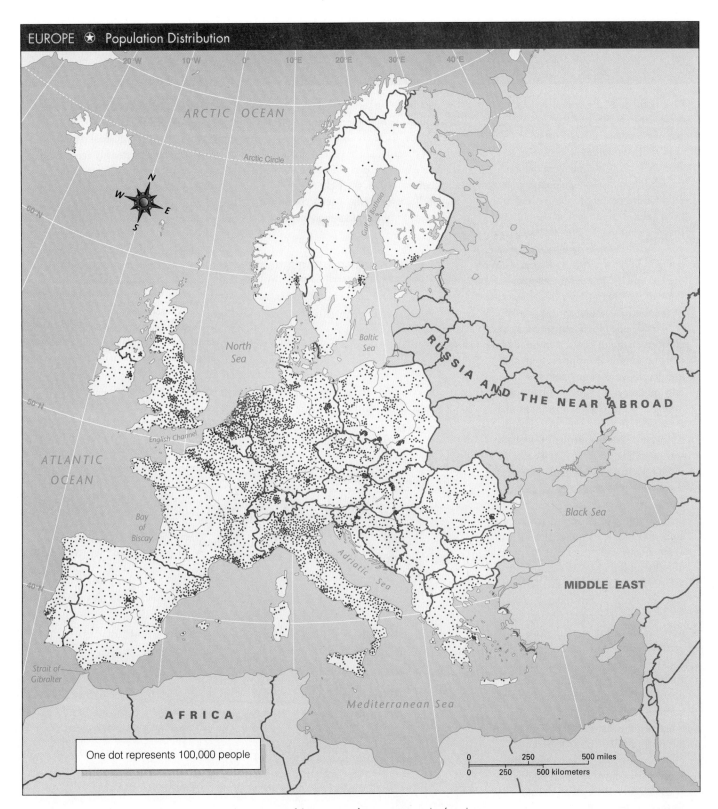

EUROPE ✪ Population Distribution

One dot represents 100,000 people

Figure 3.2   The distribution of the human population is one of the most significant patterns to be found in geography. In this map of European population distribution, the strong role of rivers, coasts, and lowlands is evidenced. In addition, the long-time concentration of urban industrial population on the island of Great Britain, along the coastlines of northwest Europe, on the North European Plain, and along both margins of the Italian peninsula is apparent. The dot map clearly distinguishes high population areas from areas that are relatively less densely populated, and it also highlights the urban clustering of major city settings.

One of the most contentious geography topics in an assessment of the current population dynamics of Europe has to deal with population flows. As can be seen in Table 3.1, annual population growth rates of European populations tend toward zero or, in some cases, drop even below zero. In fact, at the outset of the new century, the Population Reference Bureau now designates Europe as being in negative population growth. As the EU and the rest of the European nations attempt to capture a larger share of global markets, manufacturing, and commerce, the need for labor continues to grow. At the same time, countries that lie east and south of Europe are generally less developed economically and provide only modest employment opportunities for their youth. This comparative imbalance of labor opportunity in the north and west (Europe, especially the Coreland) and labor surplus in the east and south (North Africa particularly) serves as a catalyst for very dynamic migration streams. In the year 2000, some 500,000 illegal immigrants are estimated to have come to Europe. Compare this with an approximate 40,000 in 1993! This migration stream was given a keen impetus in the 1989 tearing down of the Berlin Wall and in the subsequent collapse of the Soviet Union. Migration streams that were profoundly dangerous prior to 1990 are still chancy, but less dangerous. This pattern of human drift toward the EU and toward perceived better economic opportunities — and oftentimes, toward family members who made these migrations earlier — has stimulated a trade in false passports and other papers and, more problematic, people smuggling. Particularly active "jumping off points" are in Morocco (only 9 miles from Spain), Tunis in Tunisia, and most ac-

tive, Istanbul, Turkey. Even in the face of increasingly tight border controls in Eastern Europe, there continues to grow a migrant stream that has set its sights on residence in some part of Europe. And as a further confusion, the EU estimates that by 2025, there may need to be an additional 35 million (another Working

Paper suggests 75 million by 2050) immigrants in Europe, especially in the Coreland region, to meet industrial requirements in the face of a steadily aging local population and nearly universal slow growth populations.

Source: The *New York Times*, Dec. 24, 2000, pp. 1, 9 and Dec. 25, 2000, pp. 1, 6.

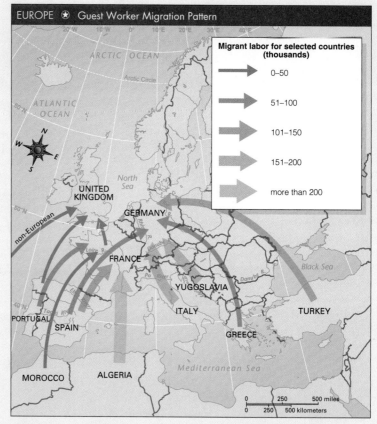

Figure 3.A   In the 1970s an active pattern of labor migration into the European Coreland developed. As the European Union (EU) experienced steady industrial growth and economic development, there was a continuing labor shortage. *Guest workers*—the term used for young male workers from the margins of Eastern Europe, North Africa, and earlier colonial holdings—began to come to Germany, France, and the U.K. in great numbers. Guest workers became an increasingly evident part of the economic and cultural landscape. These demographic changes have had a significant impact on the urban scene in these EU nations. The social unrest associated with these large numbers of unattached males when industrial slowdowns have thrown them out of work has grown into a social issue of considerable concern.

flow into Europe comes either in the form of migrants seeking asylum or as illegal immigrants hoping to keep a low profile long enough to take advantage of periodic amnesties granted by most EU nations. Such an amnesty generally leads to legal citizenship.

During periods of more rapid economic growth a decade ago, the immigrant pool (which often provided needed labor resources) was more quickly welcomed. In fact, major flows of **guest workers**—workers encouraged to come to industrial and agricultural centers that were not able to meet labor needs with domestic populations—were a part of steady immigration to Germany and France and other countries as well. But current high levels of unemployment (Italy, 10%; France 10%; Germany, 10%; and the EU average, 9%) have changed the social climate toward immigrants, both legal and illegal, considerably.

Another factor that has influenced this demographic equation has been the governmental strain resulting from the increasingly large welfare benefits provided by European governments. As the European population gets older, lives longer, and expects a continuation of the generous welfare programs that have evolved since the end of World War II, governments are faced with a significant budget requirement. Such a mind-set makes the vision of an immigrant population standing in need of benefits a particularly unsettling one, even for this region that has such a long history of active human migration.

3.2 Physical Geography and Human Adaptations

There is a broad range of environmental settings in Europe. The application of human creativity to this varied environment is an important feature of Europe's long record of struggle and accomplishment. The roots of this development within Europe far antedate the **Age of Discovery** (see Chapter 1). Later, profits from exploitation of a worldwide colonial realm stimulated and funded technological advances and major economic growth. It should not be forgotten, however, that Europeans also exploited each other: The unbelievably primitive working conditions of early English coal mines is only one of many possible illustrations. The marshaling of human energy to extract resources by Europeans to transform the environment over centuries has created landscapes of considerable significance and served as the foundation for patterns of landscape transformation and resource use that have diffused globally.

Physical Attributes of Europe
Irregular Outline

A noticeable characteristic of Europe is its extremely irregular coastal outline. The main peninsula of Europe is fringed by numerous smaller ones, most notably the Scandinavian, Jut-

land, Iberian, Italian, and Balkan Peninsulas (see Fig. 3.4). Offshore are numerous islands, including Great Britain, Ireland, Iceland, Sicily, Sardinia, Corsica, and Crete. Around the indented shores of Europe, arms of the sea penetrate the land in the form of significant **estuaries** (the tidal mouth of a river), and countless harbors offer protection for shipping. This complex mingling of land and water provides many opportunities for maritime activity, and much of Europe's history has focused on maritime trade, sea fisheries, and sea power.

Northerly Location

Another striking environmental characteristic of Europe is its northerly location: Much of Europe lies north of the 48 conterminous United States (Fig. 3.3). Despite their moderate climate, the British Isles are at the same latitude as Hudson Bay in Canada. Athens, Greece, is only slightly farther south than St. Louis, Missouri. One effect of a northerly latitude that visitors to such cities as Berlin and Stockholm notice is the long duration of daylight during summer (giving a boost to the flow of tourist dollars into the economy) and the brevity of daylight in the winter.

Temperate Climate

The overall climate of Europe is much more temperate than its northerly location would suggest. Winter temperatures, in particular, are mild for the latitude. For example, London has approximately the same average temperature in January as Richmond, Virginia, which is 950 miles (c. 1,500 km) farther south. These anomalies of temperature are caused by relatively warm currents of water (the **Gulf Stream** and the **North Atlantic Drift**) that originate in tropical western parts of the Atlantic Ocean, drift to the north and east, and cause the waters around Europe to be much warmer in winter than the latitude would warrant (Fig. 3.4). Comparing Europe with other world regions of approximately the same latitude will show that Europe is considerably milder

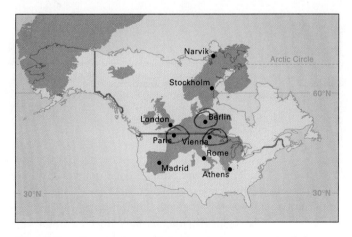

Figure 3.3 Europe in terms of latitude and area compared with the United States and Canada. Most islands have been omitted.

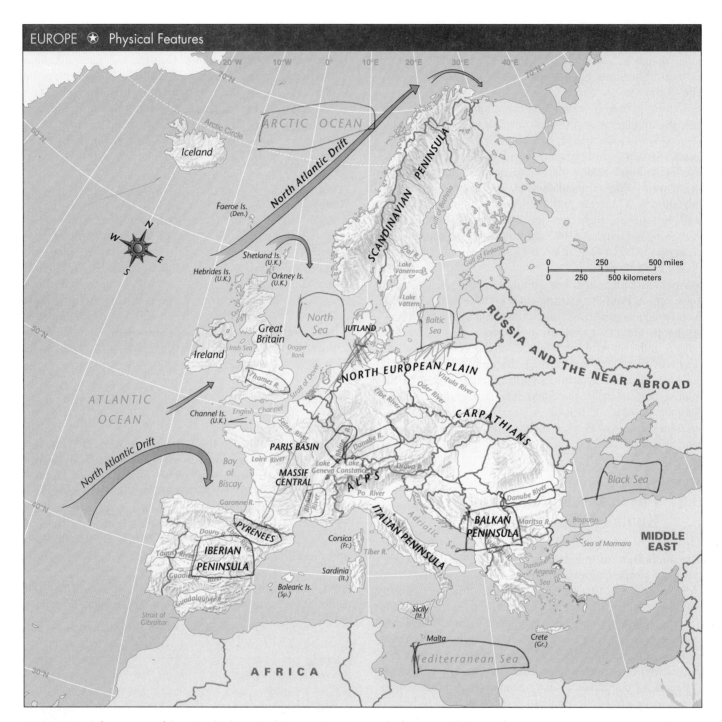

Figure 3.4   Reference map of the main islands, peninsulas, seas, straits, topographic features, and rivers of Europe. A few lakes are also shown.

in its winter temperature and climate conditions than other regions lying on the same parallels of latitude. This highly important factor in the physical environment of Europe derives from this North Atlantic Drift. With winds blowing predominantly from the west (from sea to land) along the At-

lantic coast of Europe, the moving air in winter absorbs heat from the ocean and transports it to the land, making winter temperatures abnormally mild for this latitude, especially along the coast. In the summer, the climatic roles of water and land are reversed: Instead of being warmer than the

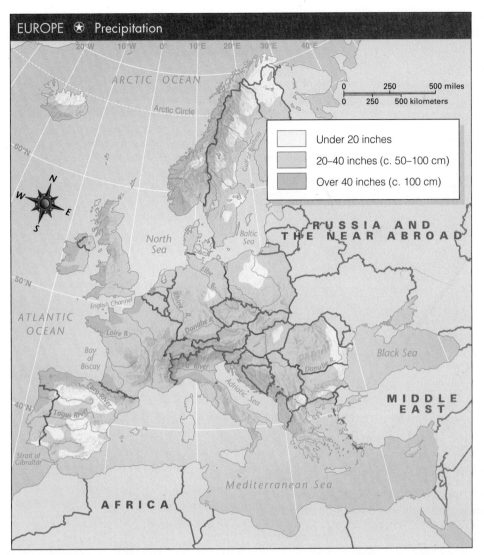

EUROPE ⊛ Precipitation

Under 20 inches

20–40 inches (c. 50–100 cm)

Over 40 inches (c. 100 cm)

Figure 3.5 Average annual precipitation in Europe. Most parts of Europe receive sufficient total precipitation for crop production, although in areas that have a Mediterranean climate, the concentration of rain occurs in the colder half of the year, thereby reducing its usefulness. Highlands along windward coasts often receive heavy orographic precipitation. International boundaries are not shown for areas outside Europe as defined in this text.

land, the ocean is now cooler; hence, the air brought to the land by **westerlies** in the summer has a cooling effect. These influences are carried as far north as the Scandinavian peninsula and into the Barents Sea, giving the Russians a warm-water port at Murmansk, even though it lies within the Arctic Circle.

The same winds that bring warmth in winter and coolness in summer also bring abundant moisture. Most of this falls as rain, although the higher mountains and more northerly areas have considerable snow. Ample, well-distributed, and relatively dependable moisture has always been one of Europe's major assets. Actually, the average precipitation in most places in the European lowlands is only 20 to 40 inches (c. 50 to 100 cm) per year (Fig. 3.5). In some parts of the world nearer the equator, this amount would be distinctly marginal for agriculture. But in most of Europe, the moisture is sufficient for a wide range of crops because mild tempera-

tures and high atmospheric humidity reduce the rate of evapotranspiration.

Advantages and Limitations of a Varied Topography

Europe's topographic features are highly diversified. Each class of features — plains, plateaus, hill lands, mountains, and water bodies — is well represented (Fig. 3.6). The whole physical assemblage, overspread with varied plant life and enriched by the human associations and constructions of an eventful history, presents a series of distinctive and often highly scenic landscapes (Fig. 3.7).

One of the most prominent surface features of Europe is a plain that extends without a break from the Pyrenees Mountains at the French-Spanish border, across western and northern France, central and northern Belgium, the Netherlands, Denmark, northern Germany, and Poland, and far into Russia. Known as the **North European Plain** (see Fig. 3.6), it

EUROPE ✹ Natural Regions

Legend:
- Humid mid-latitude plains
- Humid mid-latitude hill lands (including small areas of low mountains and of plains)
- Mediterranean (dry-summer) subtropical hills, tablelands, and small plains
- Arctic tundra
- Glacially scoured subarctic plains and hills forested in conifers
- Mountains
- Approximate boundaries between marine west coast climate (west of line) and humid continental climate (east of line) between the Alps and the Baltic Sea

Selected cities are shown as reference points.

Figure 3.6 Natural regions. The "natural regions" of the map are attempts to show natural features as composite habitats. Each color symbol represents a distinctive association of landforms and climate, with natural vegetation also stated or implied. The result is a broad-scale view of the *natural settings* for human activity in Europe.

Figure 3.7 The long-time European talent and inclination to blend water, canals, outdoor cafés, classic architecture, and wonderful people-watching plazas have been hallmarks of many European cities for ages. This scene of Venice, Italy, presents such a landscape. This pattern has promoted tourism for decades, even centuries. The massive cruise ship moored in the background represents an expression of the major importance of tourism to Europe in almost all of its countries. However, there are also many lesser inns and pensions that will accommodate travelers at a lower cost and differing level of comfort.

has outliers in Great Britain, the southern part of the Scandinavian peninsula, and southern Finland. For the most part, the plain is undulating or rolling. Flat stretches are generally on alluvial land. The North European Plain contains the greater part of Europe's cultivated land, and it is underlain in some places by deposits of coal, iron ore, potash (used in the production of soaps, glass, and the manufacture of other potassium compounds), and other minerals that have been important in the region's industrial development. Many of the largest European cities, including London, Paris, and Berlin, developed on the plain. From northeast France eastward, a band of especially dense population extends along the southern edge of the plain; this concentration apparently has been present since prehistoric times. It coincides with (1) extraordinarily fertile soils formed from deposits of **loess** (a windblown soil of generally high fertility), (2) an important natural transportation route skirting the highlands to the south, and (3) a large share of the mineral deposits of the region.

South of this northern plain, Europe is predominantly mountainous or hilly, although the hills and mountains enclose many plains, valleys, and plateaus. The hill lands and some low mountains are geologically old. Exposed to erosion for a long time, these uplands are often rather smooth and round in outline. But many mountains in southern Europe are geologically young, and they are often high and ruggedly spectacular, with jagged peaks and snowcapped summits. They reach their peak of height and grandeur in the Alps.

Glaciation has played a major role in the shaping of the landscape of Europe as well. During the periods of ice advance in the Great Ice Age of the Pleistocene (c. 2,000,000 to 10,000 years B.P.), continental ice sheets formed over

Europe, beginning on the Scandinavian Peninsula and Scotland. Glacial evidence suggests that there were four major periods of widespread glacial coverage. Today, landscapes of **glacial scouring** — the erosive action of ice masses in motion — characterize most of Norway and Finland, much of Sweden, parts of the British Isles, Iceland, and Greenland. These changes have often created favorable sites for hydroelectric installations. **Glacial deposition** — the process of offloading rock and soil in glacial retreat or lateral movement — had a major effect on the present landscape. Glacial deposits of varying thickness were left behind on most of the North European Plain. Most of the deposits still reflect their glacial origins, although careful farming and the use of fertilizers have made them productive. Loess soils, created in part by glacial action and carried south by wind, play an important role in the productivity of contemporary European farming.

The North European Plain is bordered by glaciated lowlands and hill lands in Finland and eastern Sweden and by rugged, ice-scoured mountains and fjords in western Sweden and most of Norway. The British Isles also have considerable areas of glacially scoured hill country and low mountains, along with lowlands where glacial deposition occurred.

Climate Patterns

Climate patterns tend to be fairly uniform over wide areas of Earth. This allows classification of Earth's surface into climatic regions, each characterized by a particular type of climate. Certain ranges and combinations of temperature and precipitation conditions define the climate types. Associated with and strongly influenced by each type are certain vegetation and soil conditions. One common classification is shown in this text on the map of world climate types (pp. 28–29). An examination of this map reveals that Europe, despite its modest spatial dimensions, has remarkable climatic diversity.

Marine West Coast Climate

This is the type of climate in which Atlantic influences dominate. It extends from the coast of Norway to northern Spain and inland to West Central Europe, or the Coreland. The main characteristics are mild winters, cool summers, and ample rainfall, with many drizzly, cloudy, and foggy days. Throughout the year, changes of weather follow each other in rapid succession as different air masses temporarily dominate or collide with each other along weather fronts. Most precipitation is frontal in origin or results from a combination of frontal and orographic (highland) influences (see Chapter 2). In lowlands, winter snowfall is light, and the ground is seldom covered for more than a few days at a time. Summer days are longer, brighter, and more pleasant than the short, cloudy days of winter, but even in summer, there are many chilly and overcast days. The frost-free season of

175 to 250 days is long enough for most crops grown in the middle latitudes to mature, although most areas have summers that are too cool for heat-loving crops such as corn (maize) to ripen.

Humid Continental Climates

Inland from the coast, in Western and Central Europe, the marine climate gradually changes. Winters become colder and summers are hotter; cloudiness and annual precipitation decrease. Influences of maritime air masses from the Atlantic diminish and are modified by continental air masses from inner Asia. At a considerable distance inland, conditions become sufficiently different that two new climate types are apparent: the humid continental short-summer climate in the north (principally in Poland, Slovakia, and the Czech Republic) and the humid continental long-summer climate in the warmer south (principally in Hungary, Romania, Serbia, and northern Bulgaria). The natural vegetation is mostly forest, and soils vary greatly in quality. Among the best soils are those formed from alluvium and loess along the 1,776-mile (2,842-km) valley of the Danube River.

Mediterranean Climate

F5.6
141
Southernmost Europe has a distinctive climate: the Mediterranean (dry summer-wet winter) subtropical climate. This pattern of precipitation results from a seasonal shifting of atmospheric belts. In winter, the belt of westerly winds shifts southward, bringing precipitation at a time of relatively low evaporation. In summer, the belt of subsiding high atmospheric pressure over the Sahara shifts northward, bringing desert conditions. Mediterranean summers are warm to hot, and little precipitation occurs during the summer months when temperatures are most advantageous for crop growth. Winters are mild; frosts are few. Drought-resistant trees originally covered Mediterranean lands, but little of that forest remains. It has been replaced by the wild scrub that the French call *maquis* (called *chaparral* in the United States and Spain) or by cultivated fields, orchards, or vineyards. Much of the land consists of rugged, rocky, and badly eroded slopes where thousands of years of deforestation, overgrazing, and excessive cultivation have taken their toll. The subtropical temperatures make possible a great variety of crops, but irrigation is necessary to counteract the dry summers. However, with irrigation, an enormous variety of crops can be raised productively. In addition, this climate is a boon for summer travelers, and the money made from its exploitation for tourism is of immense economic importance in Mediterranean Europe. The rains that generally fall in the winter season are particularly useful because of lower evaporation rates, allowing more of the precipitation to find its way into the water table for use during the late spring and summer.

Subarctic and Tundra Climates

Some northerly sections of Europe experience the harsh conditions associated with subarctic and tundra climates. The subarctic climate, characterized by long, severe winters and short, rather cool summers, covers most of Finland, the greater part of Sweden, and some of Norway. A short frost-free season, coupled with thin, highly leached, acidic soils, handicaps agriculture. Human settlement is scanty, and a forest of needle-leaf conifers, such as spruce and fir, covers most of the land. In the tundra climate of northernmost Norway and much of Iceland, cold winters combine with brief, cool summers and strong winds to create conditions hostile to tree growth. An open, windswept landscape results, covered with lichens, mosses, grass, low bushes, dwarf trees, and wildflowers. Wildlife is present, but human inhabitants are few. Agriculture in a normal sense is not feasible. In the taiga region, south of the tundra, a slight increase in moisture and temperature allows some forest growth.

Highland and Icecap Climates

The higher mountains of Europe, like high mountains in other parts of the world, have an undifferentiated highland climate varying with elevation and differential exposure to sun, wind, and precipitation. Given enough height, the variety can be startling. The Italian slope of the Alps, for instance, ascends from subtropical conditions at the base of the mountains to tundra and icecap climates at the highest elevations. The icecap climate experiences temperatures that average below freezing every month of the year, generally enabling ice fields and glaciers to be preserved. Greenland is a good example of this sort of environment.

The Importance of Rivers and Waterways

As one might expect in such a humid area, Europe has numerous river systems that are very important economically for transport and water supply, the generation of electricity, and the creation of regional images. Geographic discussion of these rivers requires an initial understanding of certain terms. A river system is a river together with its tributaries, and a river basin is the whole area drained by a river system. The source of a river is its place of origin; the mouth is where it empties into another body of water, often forming an estuary. As they near the sea, many rivers become sluggish, depositing great quantities of sediment to form **deltas** and often dividing into a number of separate channels, known as **distributaries.** The management of a river requires a high level of cooperation throughout the river system because major change upstream, such as taking water out for irrigation or putting in industrial pollutants, will have a significant impact on downstream populations and their water use. Control of waterways, therefore, is always dependent on a combination of not only engineering skills but also political authority and control. This is particularly true when the waterway flows through a number of countries like the Rhine River and its several countries, but especially for the Danube which intersects or borders 11 different countries.

Rivers were an important part of Europe's transport system at least as far back as Roman times and probably earlier; today, they are still central to the transporting of large quantities of bulk cargo at low cost, mostly in motorized (self-propelled) barges. The more important rivers for transportation are largely those of the highly industrialized areas in the Coreland area of West North Central Europe. An extensive system of canals connects and supplements the rivers. During the time of the Roman Empire, canals were built throughout northern Europe and Britain. These were used primarily for military transport, although some systems in southeastern Britain were also used to help field drainage.

The Dutch developed the pound (pond) lock for canals in the late 14th century, and this led to continued expansion of regional connections of waterways through canals, as well as the use of canals to link farmlands with the coasts. In the late 1700s, Britain became especially active in such construction, as it was moving raw materials toward factory locations in the early years of the Industrial Revolution. This period of active growth continued until the middle of the 19th century, when the expanding influence of railroads and the steam engine provided real competition for canals. The longest canal constructed in this period of British investment in canals was the 140-mile (230-km) Leeds-Liverpool Canal that was begun in 1770 and completed in 1816. Although canals continue to play a significant transport role, the combination of railroad and highway development since the 1950s has taken over the dominant goods transport role all through Europe.

Important seaports have developed along the lower courses of many rivers, and some have become major cities. London on the Thames, Antwerp on the Scheldt, Rotterdam in the delta of the Rhine, and Hamburg on the Elbe are outstanding examples. Often, the river mouths are wide and deep, allowing ocean ships to travel a considerable distance upstream and inland. This is true even of short rivers such as the Thames and Scheldt.

The Rhine and the Danube are European rivers of particular importance, both touching or crossing the territory of many countries (see Fig. 3.4). In the case of the Rhine, these countries include Switzerland — where the river rises in the Alps — Liechtenstein, Austria, France, Germany, and the Netherlands. Highly scenic for much of its course, the Rhine River is Europe's most important inland waterway. Along or near it, a striking axis of intense urban-industrial development and extreme population density has developed. At its North Sea end, the Rhine axis connects to world commerce by the world's most active seaport, Rotterdam, in the Netherlands.

The Danube (German: Donau) touches or crosses more countries than does any other river in the world. From its source in the Black Forest of southwestern Germany, not far from Lake Constance in Switzerland, it flows eastward through Austria, with Vienna on its banks, and serves as the border between Slovakia on the north and south through Hungary, with Budapest on its banks. Farther south, it serves as the border between Croatia on the west and Yugoslavia on the east. It then turns east, passing by Belgrade, and then becomes the border between (what is left of) Yugoslavia and Romania on the north and, farther east, the border between

Bulgaria on the south and Romania on the north. Continuing farther east, the Danube enters Romania, turns north, and makes one more sharp turn to the east before pouring into the Black Sea. Within Europe, the Danube is also unusual in its southeasterly direction of flow (see Fig. 3.4).

3.3 Cultural and Historical Context

Patterns of language and religion in Europe helped establish the foundation for Europe's dominance in a large part of the world's economic, cultural, and political development during the past five centuries.

Language

Language is one of the most highly identifiable elements in a culture group's personality. In learning a language, one also learns a great deal about tradition, history, cultural mores, belief systems, and the spirit of a people. Language also plays a major role in the tensions frequently at work in the geography of borders and ethnicity in Europe (Fig. 3.8).

Europe emerged from prehistory as the homeland of many different peoples. In ancient and medieval times, certain of these peoples experienced periods of vigorous expansion, and their languages and cultures became widely diffused. First came the expansion of Greek and Celtic peoples, and later that of the Romans, Germanic (Teutonic) peoples, and the Slavic peoples. As each expansion occurred, traditional languages persisted in some areas but were displaced in others. Each of the important languages eventually developed many local dialects. With the rise of centralized nation-states in early modern times, particular dialects became the bases for standard national languages.

The first millennium B.C. witnessed a great expansion of the Greek and Celtic peoples. In peninsulas and islands bordering the Aegean and Ionian Seas, the early Greeks evolved a civilization that reached unsurpassed heights of philosophical inquiry and literary and artistic expression. Greek adventurers, traders, and colonists used the Mediterranean Sea as their highway to spread classical Greek civilization and its language along much of the Mediterranean shoreline, although there was already a strong Phoenician influence along the shores of the eastern Mediterranean before the Greeks arrived. Evidence of the early geographic range and subsequent influence of the Greek language and culture is apparent in the many Greek elements in modern European languages. But over time, the use of Greek in most areas disappeared as new peoples and languages were introduced, expanded, and gained authority.

Europe's Celtic languages expanded at roughly the same time as Greek and, like Greek, are represented today only by remnants. The expansion of preliterate Celtic-speaking tribes radiated from **hearth areas** — regions of original development — in what is now southern Germany and Austria,

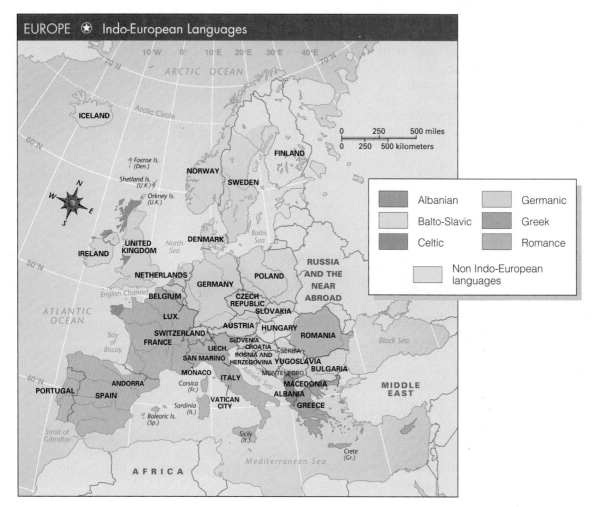

EUROPE ✶ Indo-European Languages

Albanian	Germanic
Balto-Slavic	Greek
Celtic	Romance
Non Indo-European languages	

Figure 3.8   Branches of the Indo-European language family. Most Europeans speak languages from the Indo-European language family; the three most important branches are Germanic (north and west), Romance (south and west), and Balto-Slavic (east). The fourth major branch, Indo-Iranian (not shown in the European map here) is clustered in southern Asia and the Middle East and North Africa.

eventually occupying much of continental Europe and even the British Isles. Conquest and cultural influence by later arrivals, mainly Romans and Germans, eventually eliminated the Celtic languages except for a few traces that survive today in isolated pockets in the British Isles and in French Brittany.

In present-day Europe, the overwhelming majority of the people speak Romance, Germanic, or Slavic languages. The Romance languages evolved from Latin, originally the language of ancient Rome and a small district around it. In the few centuries before and just after the birth of Christ, the Romans subdued territories extending through western Europe as far west as Great Britain, and the use of Latin spread to this large empire. Latin had by far the greatest impact in the less developed and less populous western parts of the empire in southwest Asia. Over a long period, extending well beyond the collapse of this western part of the empire in the fifth

century, regional dialects of Latin survived and evolved into Italian, French, Spanish, Catalan (still spoken in the northeast corner of Spain), Portuguese, Romanian, and other Romance languages of today. It's worth noting here that language frontiers do not always coincide exactly with political frontiers. For example, the French language extends to western Switzerland and also to southern Belgium, where it is known as Walloon.

In the middle centuries of the first millennium A.D., the power of Rome declined and a prolonged expansion by Germanic and Slavic peoples began. Germanic peoples first appear in history as a group of tribes inhabiting the coasts of Germany and much of Scandinavia. They subsequently expanded southward into Celtic lands east of the Rhine. Roman attempts at conquest were repelled, and the Latin language had little impact in Germany. In the fifth and sixth centuries A.D., Germanic incursions overran the western Roman Empire, but in

F21.4
556

many areas, the conquerors eventually were absorbed into the culture and language of their Latinized subjects.

However, the German language expanded into, and remains, the language of present-day Germany, Austria, Luxembourg, Liechtenstein, the greater part of Switzerland, the previously Latinized part of Germany west of the Rhine, and parts of easternmost France (Alsace and part of Lorraine). The Germanic languages of Europe include many languages other than German itself. In the Netherlands, Dutch developed as a language closely related to dialects of northern Germany, and Flemish—almost identical to Dutch—became the language of northern Belgium. The present languages of Denmark, Norway, Sweden, and Iceland descended from the same ancient Germanic tongue, although Finnish did not.

English is basically a Germanic language, but it has many words and expressions derived from French, Latin, Greek, and other languages. Originally, English was the language of the Germanic tribes known as Angles and Saxons who invaded England in the fifth and sixth centuries A.D. The Norman Conquest of England in the 11th century established French for a time as the language of the English court and the upper classes. Modern English retains the Anglo-Saxon grammatical structure but borrows great numbers of words from French and other languages. English is now the principal language in most parts of the British Isles, having been imposed by conquest or spread by cultural diffusion to areas outside England.

Slavic languages are dominant in most of Eastern Europe. The main ones today include the Russian and Ukrainian of the adjoining Russia and the Near Abroad, Polish,

Czech and Slovak, and Serbian, Croatian, and Bulgarian in the Balkan Peninsula. They apparently originated in eastern Europe and Russia and were spread and differentiated from each other during the Middle Ages as a consequence of migrations and cultural and trade contacts involving various peoples.

A few languages in present-day Europe are not related to any of the groups just discussed. Some are ancient languages that have persisted from prehistoric times in isolated (usually mountainous) locales. Two outstanding examples are Albanian (Indo-European in its origin) in the Balkan Peninsula and Basque, spoken in or near the western Pyrenees Mountains of Spain and France. Some languages unrelated to others in Europe have relatives in Russia. They reached their present locales through migrations of peoples westward. The prime examples are Finnish and Hungarian, also called Magyar.

Although it is easy to understand the importance of language as a medium of communication, it is just as essential to realize that a language represents a whole universe of culture, tradition, and history. The kaleidoscope of languages in Europe has contributed to the patterns of cooperation, affiliation, tension, and distrust that characterize this region today. Language plays this same role in all other parts of the world, especially where a great number of cultures and languages are resident in a relatively small area. The current tensions that are so significant in the former Yugoslavia and the Balkan Peninsula area derive in part from language differences, and their associated religious and other cultural distinctions, within the region.

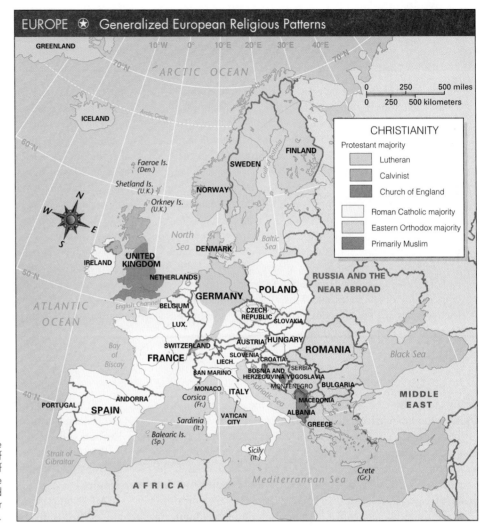

Figure 3.9   The religious patterns of Europe are largely expressions of distinct sects of Christianity. There are also significant areas of non-Christian populations, particularly in the Balkan Peninsula (especially in Kosovo and Albania) and in other parts of the former Yugoslavia.

Religious Belief Systems

Europe's diverse patterns of religious belief systems are powerful indicators of culture, badges of nationality, and repositories of achievement and myth. Along with language, they stand very high among the factors that differentiate Europe geographically (Fig. 3.9). During the long course of European history, both of these cultural systems have played an important role in political conflict, including warfare, repression, discrimination, and terrorism. They have also served as foundations of and for cooperation.

In the Roman Empire, Christianity arose, spread, survived persecution, and became a dominant institution in the late stages of the empire. It survived the empire's fall and continued to spread, first within Europe and then to other parts of the world. In total number of adherents, the Roman Catholic Church, headed by the pope in the Vatican, is Europe's largest religious division, as it has been since the Christian Church was first established. Today, it remains the principal

faith of a highly secularized Europe. The main areas that are predominantly Roman Catholic include Italy, Spain, Portugal, France, Belgium, the Republic of Ireland, large parts of the Netherlands, Germany, Switzerland, Austria, Poland, Hungary, the Czech Republic, Slovakia, Croatia, and Slovenia.

During the Middle Ages, a center of Christianity evolved in Constantinople (now Istanbul, Turkey) as a rival to Rome. From that seat of power, the Orthodox Eastern Church spread to become, and remains, dominant in Greece, Bulgaria, Romania, Serbia, and Macedonia, as well as in the Ukraine and Russia to the east.

In the 16th century, the Protestant Reformation took root in various parts of Europe, and a subsequent series of bloody religious wars, persecutions, and counterpersecutions left Protestantism dominant in Great Britain, northern Germany, the Netherlands, Denmark, Norway, Sweden, Iceland, and Finland. Except for the Netherlands, where a higher Catholic

HOWARD GARNETT/DEMBINSKY PHOTO ASSOCIATES

Figure 3.10 Although media attention in recent years has focused on Bosnia and Kosovo in the former Yugoslavia as the setting for expressions of Muslim influences in southern and eastern Europe, the Moorish influence is more geographically widespread in Europe. This mosque is in the countryside near Oberammergau, Germany, and demonstrates the breadth of such cultural diffusion.

birth rate has since reversed the balance, these areas are still mainly Protestant.

The Islamic (Muslim) faith, which is the principal religion of Albania and is an important faith in other parts of the Balkan Peninsula, plays a major influence in the contemporary struggle of European states in south central Europe to achieve both political and religious stability. Islam was once widespread on the Balkan Peninsula, where it was established by the Ottoman Turks during their period of rule from the 14th to the 19th centuries. It was also the religion of the Moors, a powerful cultural presence on the Iberian Peninsula from the 8th century until the end of the 15th century. Its presence has served as a dominant factor in the ethnic unrest in the former Yugoslavia since 1991, especially in the Kosovo conflict. Religious differences continue to surface as a source of discord in Bosnia and Herzegovina (Fig. 3.10).

Under the Roman Empire, Jewish minorities spread from Palestine into Europe, where they have since persisted as a significant minority population in many countries despite recurrent persecution. New streams of migration, both legal and undocumented, are now washing across Europe. Each of these migrants carries with him or her a whole collection of cultural baggage, including — often depending on specific age and origin — religion, language, political conventions and convictions, and societal patterns. So, even though our text has maps that illustrate the spatial extent of language and religion, it is important to realize that things are in continual flux all across the European landscape, especially on the margins of North Africa, the western edge of Russia and the Near Abroad, and the western edge of the Balkan Peninsula.

The Impact of Europe on the World in the Colonial Age

There is a global imprint of European influence expressed through language, custom, economic systems, and technology diffusion. Far more than any other region, Europe has shaped the human geography of the modern world. Before the late 15th century, Europe played a minor role in world trade patterns. Western Europe saw goods moving from the southern coast of France and northern Italy northwest to Britain. A more active European trade connected Italy with the margins of the Mediterranean Sea and beyond to Southwest Asia. However, major trade routes had centers much farther east and south (Fig. 3.11). The longest trade link was the centuries-old Silk Road that moved goods overland (and on the Mediterranean Sea) from Ch'ang-an (Xi'am) in China to Venice. The route was more than 5,000 miles (9,000 km) long.

From the beginning of the Age of Discovery in the 15th century, European seamen, missionaries, traders, soldiers, and colonists burst upon the world scene. By the end of the 19th century, Europeans had created a world in which they and their descendants were culturally dominant. This influence is seen in everything from landscape to philosophical mind-set.

The process of exploration and discovery by which Europeans filled in the world map began with 15th-century Portuguese expeditions down the west coast of Africa. In 1488, a Portuguese expedition headed by Bartholomeu Dias rounded the Cape of Good Hope at the southern tip of Africa and opened the way for subsequent European voyages eastward into the Indian Ocean. Then, in 1492, North America was brought into contact with Europe when a

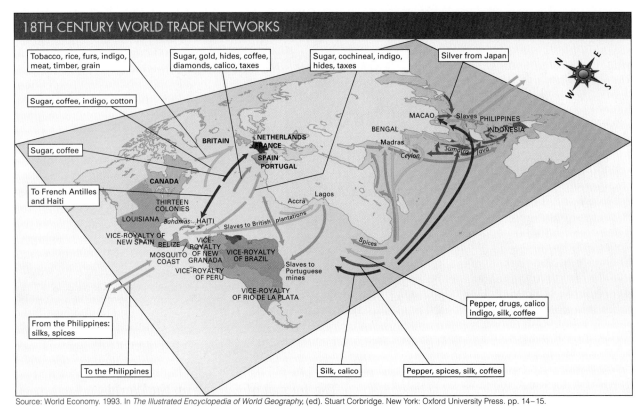

18TH CENTURY WORLD TRADE NETWORKS

Tobacco, rice, furs, indigo, meat, timber, grain

Sugar, gold, hides, coffee, diamonds, calico, taxes

Sugar, cochineal, indigo, hides, taxes

Silver from Japan

Sugar, coffee, indigo, cotton

Sugar, coffee

To French Antilles and Haiti

From the Philippines: silks, spices

To the Philippines

Silk, calico

Pepper, spices, silk, coffee

Pepper, drugs, calico indigo, silk, coffee

BRITAIN

NETHERLANDS
FRANCE
SPAIN
PORTUGAL

CANADA

THIRTEEN COLONIES

LOUISIANA   *Bahamas*   HAITI

VICE-ROYALTY OF NEW SPAIN   BELIZE   VICE-ROYALTY OF NEW GRANADA

MOSQUITO COAST   VICE-ROYALTY OF BRAZIL

VICE-ROYALTY OF PERU

VICE-ROYALTY OF RIO DE LA PLATA

Slaves to British plantations

Slaves to Portuguese mines

Spices

Accra   Lagos

MACAO   Slaves   PHILIPPINES

BENGAL   INDONESIA

Madras

Ceylon   *Sumatra*   *Java*

Source: World Economy. 1993. In *The Illustrated Encyclopedia of World Geography*, (ed). Stuart Corbridge. New York: Oxford University Press. pp. 14–15.

Figure 3.11   In the 18th century, the global flow of goods, slaves, and information was carried all around the world by European merchant fleets. From the small countries in Europe came the decisions, the marine power, the mercantile systems, and the military force to promote the global movement of sugar, spices, and human cargo. The details provided in the items along the sides of the map show how many goods flowed to and from how many places.

Box 518

Spanish expedition commanded by a Genoese (Italian), Christopher Columbus (Cristóbal Colón), crossed the Atlantic to the Caribbean Sea. Less than half a century later, these early feats of exploration culminated in the first circumnavigation of the globe by a Spanish expedition commanded initially by a Portuguese, Ferdinand Magellan, who named the Pacific Ocean. Worldwide exploration, both coastal and inland, continued apace for centuries, with leadership being wrested from the early Spanish and Portuguese by the ascending Dutch, French, and English.

The discovery of new peoples and places by the Europeans generally meant that missionaries, soldiers, and traders were seldom far behind the explorers, often traveling with the first ships. Trading posts, Christian missions, and military garrisons were established along many coasts or, in some cases, inland. Inland posts were especially notable in Latin America, where they were often a response to the highland location of native people who could be Christianized, robbed of precious metals, and put to work in mines or on the newly developed and highly profitable agricultural plantations or more generalized farming operations. Many European outposts came to domi-

nate the areas where they were established and eventually became bases from which complete colonial control was often extended.

This process of discovery inevitably set in motion a whole ensemble of changes for both the people discovered and those doing the discovering. As researchers look more at the lives of common people rather than only at social patterns and **material culture** — the artifacts of a group — of the wealthy and politically powerful, it is increasingly clear that being "discovered" extracts a high cultural cost. Every exploring, discovering, exploiting, and conquering people has had its own explanations about why such geographic acts were meant to bring benefits to those being found. However, the recipients of such discovery have long complained about the ruinous impact such discovery had on their lives and cultures.

Subsequent settlement by European colonists, on differing scales and at different times, took place in a great many non-European areas. In well-populated or environmentally difficult areas, European populations remained small minorities, but in other circumstances, local versions of European societies took such firm root that they became numerically dominant. Today,

TABLE 3.2 Diffusion of Foodstuffs Between New World and Old World: The Columbian Exchange	
This table shows some of the most significant foodstuffs diffused from the New World to the Old World after 1492, as well as the major New World foodstuffs introduced into world trade after 1492 by the ships and crews that came to the newly found lands of North, Middle, and South America.	
Diffused from Old World to New World	**Diffused from New World to Old World**
Wheat	Maize
Grapes	Potatoes
Olives	Cassava (manioc)
Onions	Tomatoes
Melons	Beans (lima, string, navy, kidney, etc.)
Lettuce	Pumpkins
Rice	Squash
Soybeans	Sweet potatoes
Coffee	Peanuts
Bananas	Cacao
Yams	

the descendants of transplanted Europeans make up most of the population in the United States, Canada, and parts of Latin America, Australia, and New Zealand, and they continue to be the financially dominant group in South Africa.

In carrying on their various overseas enterprises, Europeans not only migrated themselves but often transferred non-Europeans from place to place. The most monumental instance of this was the massive forced migration of Africans to the New World in the slave trade that began in the 16th century. Such transfers greatly influenced the ultimate racial and cultural makeup of sizable areas.

Also of great importance in shaping the world's geography was the transfer of plants and animals from one place to another. A few major examples include the introduction of hogs and cattle and the reintroduction of horses to the Americas from Europe; of tobacco and corn (maize) from the Americas to Europe and other parts of the world; of rubber from South America to Asia; of the potato from the Americas to Europe, where it became a major food; and of coffee from Africa to Latin America, where it became the principal basis of a number of national economies (Table 3.2).

A worldwide system of trade, with Europe at its core, was a major outcome of European expansion. The system principally involved the movement of raw materials and food products from the rest of the world to Europe and the return sale of European manufactured goods to greatly varied global trade destinations. In general, this exchange favored Europe, and wealth from the exchange mainly accumulated there, particularly in Seville, Spain, during the Age of Discovery (late 15th to early 19th centuries). The consequent infusions of money enabled certain European cities, particularly London, to grow and become centers of world finance. Today, other centers of commerce and finance, notably in the United States and in Japan and Hong Kong in East Asia, have risen to challenge Europe's dominance.

Europe's Rise to Global Centrality, and Potential Decline

What factors account for the rise of this Eurasian peninsula as the center of world trade and industrialization? Many interpretations are possible, with various scholars emphasizing factors such as the following:

1. *Capitalism.* Parts of Europe had already developed capitalist institutions by the end of the Middle Ages, and the Colonial Age saw further rapid development of capitalism in the region. Profit became an acceptable motive, and the relative freedom of action afforded by the capitalistic system provided opportunities for gaining wealth by taking risks. Energetic entrepreneurs found it possible to mobilize capital into large companies and to exploit both European and overseas labor (Fig. 3.12).

2. *Science.* Europe's scientific prowess must not be overlooked as an explanation for the region's rise to world dominance. In part, technology is applied science, and the foundations of modern science were constructed almost entirely in Europe during the centuries when European influence was becoming paramount. Until recently, the great names of science were overwhelmingly the names of Europeans. As late as World War II, the development of atomic fission and associated space science in the United States was carried out by a team composed largely of prominent refugee European scientists. Nobel laureates continue to reflect the scientific capacities of the peoples of, and from, this region.

3. *Technology.* By the end of the Middle Ages, Europeans had reached a level of technology generally superior to that of non-Europeans with whom they came in contact during the early Age of Discovery. In particular, achievements in shipbuilding, navigation, and the manufacture and handling of weapons gave them decided advantages. Nations that had earlier made significant progress in the development of technology (e.g., China) were disinclined to compete at a global level during the Age of Discovery. This gave Europe a virtual free rein in energetic expansion and control of technological innovation for several centuries.

In the 20th century, however, the preeminence in world trade and industry of leading European countries diminished. What caused the relative decline in their fortunes? Some important reasons and circumstances include:

1. *War dislocation.* Europe suffered enormous casualties and damage in World Wars I and II, which were initiated and fought mainly in Europe. Recovery was eventually achieved with U.S. aid, but the region's altered position

Figure 3.12   Paris, France, has been one of the most innovative metropolitan centers in terms of urban showcase architecture. These public housing towers surround one side of the impressive new Grand Arch La Defence, a new structure that puts Paris in a competition with Berlin for having some of the most striking urban structures of the late 20th century.

could not be reversed. The wars destroyed Europe's ability to maintain its predominance in the face of such trends as anticolonialism and thereby accelerated the capacity of the United States and Russia and the Near Abroad to rise to world power.

2. *New nationalism.* Rising **nationalism** — the quest of a nationality to possess its own homeland — in the colonial world during the 20th century resulted in the formal end of the European colonial empires. Taking advantage of a weakened Europe and a mounting disapproval of colonialism in the world at large, one European colony after another gained independence quickly in the decades following the end of World War II. Opposition to continued European control was often spearheaded by colonial leaders who had been educated in Europe and had absorbed nationalistic ideas there. Nationalism is a sentiment that has had its most pervasive expression in Europe. This same drive received an additional boost with the collapse of the Soviet Union in 1991. Much of the bloodshed and warfare in the former Yugoslavia and in the Balkan Peninsula is fueled by a rising sense of Serb, non-Serb, and anti-Serb nationalism.

3. *Ascendancy of the United States and Russia.* Europe's predominance was seriously eroded by the rising economic and political stature of the United States and, to some degree, Russia and its neighboring states. These enormous countries, each far larger than any European country, outstripped Europe in military power, economic resources, and world influence, particularly in the rebuilding years of 1950–1970. With the collapse of communism and the USSR in the early 1990s, Russian influence diminished considerably.

4. *Shift in world manufacturing patterns.* Europe once enjoyed almost a near monopoly in exports of manufactured goods, but in the past three decades, manufacturing has developed in many countries outside Europe. Japan, China, South Korea, Hong Kong (now part of China), and Taiwan (a contested province of China) are prime examples just in East Asia alone. In the decades following World War II — although Japan began its transformation in the 1870s — these Asian countries reinvented themselves and changed from more traditional rural societies to become industrial forces capable of competing with Europe in markets all over the world. The United States began its industrialization earlier than Japan, but it, too, reached industrial maturity in the 20th century and also became a vigorous competitor of Europe.

5. *Energy factors.* Europe's ability to assert itself in world affairs has been weakened by the region's new dependence on outside sources of energy. The region's traditionally significant coal fields (Fig. 3.13) have become increasingly costly to exploit, and despite the recent development of North Sea oil and gas resources (which benefit primarily Great Britain and Norway), Europe is now very dependent on Middle East oil and other imported sources of energy.

The speed of technology transfer and the complex global interdependence of capital flow make it difficult to pinpoint current, much less future, centers of manufacturing. However, current patterns of industrial activity in Europe show that the coalescence of resources, skilled labor, cultures that promote entrepreneurship, and spatial networks of transport and communication are all still in place in this region. If all of the productive volume of Europe were added together,

These include grains, butter, cheese, olive oil, and table wines.

As a region, Europe is by no means agriculturally self-sufficient. Imports of supplemental agricultural commodities are necessary. Increasingly, policies dealing with agricultural goods are taken under consideration by the EU. With the early 1990s collapse of Soviet control of Eastern Europe, lines of trade with the rest of Europe — the world to the west — began to increase.

Throughout history, fishing has been an important part of the European food economy, and the coasts are still thickly dotted with fishing ports. At times, control of fishing grounds has been a major commercial and political objective of nations, even resulting in warfare. Fisheries are particularly important in shallow seas that are rich in the small organisms known as plankton. These organisms are the principal food for schools of herring, cod, and other fish of commercial value. The Dogger Bank (to the east of Great Britain) in the North Sea and the waters off Norway and Iceland are major fishing areas, and Norway and Denmark are Europe's leading nations in total catch. Iceland, which has few other resources and few people, ranks fourth in total catch. Its economy depends mainly on exports of fish and fish products, which represented 75% of all Iceland's exports by value in 1998.

Figure 3.13   The mining of coal has been an important European occupation for centuries. Seen here are miners coming off shift at a British colliery. The children of these miners, however, will not find their fathers' jobs as easy to secure as in the past. Industrial changes and the ever-diminishing role played by coal in Europe, especially Western Europe, make this a vanishing occupation. The landscape of Western Europe has grand evidence of earlier industrial efforts built of brick and stone, left behind now as manufacturing has so changed its setting and its location.

JAKE SUTTON/GAMMA LIAISON

Europe would be shown to be the most powerful economic engine in the world today. It is certain that Europe will continue to play a major role in manufacturing and global trade for a long time to come, even though the very urban centers that began these patterns are losing relative importance as new manufacturing areas emerge both within and beyond Europe.

3.4 Economic Geography
Agriculture and Fisheries

Agriculture was the original foundation of Europe's economy, and it is still a very important component. Food provided by the region's agriculture allowed Europe to become a relatively well-populated area at an early time. After about 1500, a period of steady agricultural improvement began. Introduction of important new crops such as the potato played a part (see Table 3.2 on crop diffusion), but so did such practical improvements as new systems of crop rotation and scientific advances that produced a better knowledge of the chemistry of fertilizers. The continuing expansion of industrial cities provided growing markets for European farmers, who received some protection through tariffs or direct subsidies to encourage production and support rural incomes. Both of these governmental benefits are now enjoyed by farmers throughout the European Union (EU). Large surpluses of many products have emerged as a result of the EU's Common Agricultural Policy (CAP).

Geographic Patterns of Economic Activity

Diversity is a keynote of the European economy, just as it is of the European environment. The economic vitality that has characterized the region for the past two centuries plays a major role in the current global centrality of Europe. Numerous forms of economic activity are highly developed, and much variety exists from one country and region to another. Most European countries rank high in productivity and affluence compared with other countries of the world. Western Europe was the first world region to evolve from an agricultural society into an industrial economy. The series of events, commonly called the **Industrial Revolution** (see Chapter 4), were followed by vast advances in productive (and in terms of weaponry, sometimes destructive) knowledge, technology, and environmental transformation. Even military devastation did not stop the economic dynamism of Europe, which capitalized on U.S. aid after World War II — primarily, the Marshall Plan that began in 1947 — to rapidly reorganize, rebuild, and attain new heights of production, affluence, and global influence.

One of the most significant facts of global geography is the high proportion of the world's manufacturing capacity and output that is European. Today, this proportion is decreasing, but approximately one-quarter to one-third of the world output of most major industrial products originates in Europe. Industrialization goes far toward explaining the highly urban character of most European countries as well as their high standing in overall productivity and average income.

In traveling abroad as an American college student, one of the rules of the road is to avoid tourist landscapes. There is a persistent paradox in such a rule because the very places that have made it to the tourist maps often are landscapes that a geographer might find really interesting. The paradox comes in the fact that such locales generally have a commercial component developed on site, and the goods featured in such a shop are often not what the college student is seeking to bring home. So, the trick is to learn how to use the landscape benefits of a tourist trap (the landmark itself, the bathrooms, the maps, the film availability, and the other travel information) while not getting trapped in the so-called Friendship Stores that can pull money out of your wallet like a siphon.

A mechanism that can work well in trying to deal with this situation is using the field resource that virtually all travelers possess: your legs and the capacity to leave a tour and say, "I'll meet you back at the hotel." This brings us to the first axiom of uncharted urban exploration: *Axiom One: Always have a hotel card with you as you explore foreign cities on your own.* Armed with this card and a willingness to explore a foreign city, a tourist travel tour can serve really well as a morning trip to some church, museum, or factory. Such a formal trip can then be followed by a 4–8-hour walk through the city. The best size for such an exploration is a group of two, although traveling alone is also pretty fine if you are willing to let your eyes do a lot of walking as you wander back to your home base.

Consider, for example, the payoff of such an exploration in Venice, Italy. A tour bus might take you anywhere in that spectacular city, and all you have to do at the farthest point from your hotel is ask to be let off or explain to the tour guide (if there is one) that you are going to walk back to the hotel. You find yourself going into ever smaller lanes as you end up canal surfing as you search out either your hotel (unless it is on Lido Island) and then you just search for the grand campanile at San Marcos Square and the Doges Museum. This brings up the second axiom: *Axiom Two: Do your uncharted urban exploration with a soft clock; do not promise an exact return time.* By being somewhat indifferent to the time, your exploration of the small canals and their associated streets and lanes lets you see a part of Venice that is seldom studied by the tourist population (which is often busy in leather and glass factories and shops). Your sense of the city is enhanced by the fact that your landscapes are unusual ones.

On any given day of such unplanned walking, determine a landscape focus as you do your hours of undirected walking. *Axiom Three: Determine a landscape focus for each day of exploration.* Make yourself consider one or more of these questions: What is the ethnic makeup of the people you see? What sorts of window flower boxes are there? What sorts of cottage industry and repair shops do you see in the small shops in the minor streets of the city? Are there free-ranging cats and dogs? How are the churches used during the day? You could select any small set of descriptive characteristics and arm yourself with a few specifics to key your attention as you walk. By highlighting your perceptual goals on each day of exploration, you'll see the landscape, both physical and cultural, a little more clearly than if you just wander.

Finally, punctuate your exploration with recordkeeping. *Axiom Four: Sit down for a coffee or something every couple of hours with a journal (not just postcards) to record at least some of what you are seeing.* As you are writing, think of what your focus is and think about what has caught your attention and caused you to ponder a scene. You may think that such journal keeping is too obvious to be of any use to you, but you'll find exactly such notes powerfully evocative in the future as you try to put your travels in order and recall what you have learned. Using words rich in imagery will give you a resource in the future that will surprise you with the evocative power of noted building colors or textures, shop shutters, road textures, shopkeeper attitudes and looks, and your own continually changing reaction to the city.

These four axioms can be your friends as you engage in uncharted urban explorations in not only the European landscape but anywhere in the world. Try such exploration in London or Paris or Berlin, but know that the same process works just as well in Shanghai or St. Tropez or Sydney. Walking is your friend. Just have the hotel card with you always!

Kit Salter

Europe was where a large-scale manufacturing industry, using machines driven by inanimate power, first arose. Notable for its surge of inventiveness, the Industrial Revolution began to be felt in the first half of the 18th century. Industrial innovations, many of which British inventors developed, made water power — and then coal-fueled steam power — increasingly available to turn machines and drive gears in the new factories. The invention of a practical steam engine by James Watt in 1769 in Great Britain suddenly made coal a major resource and greatly increased the

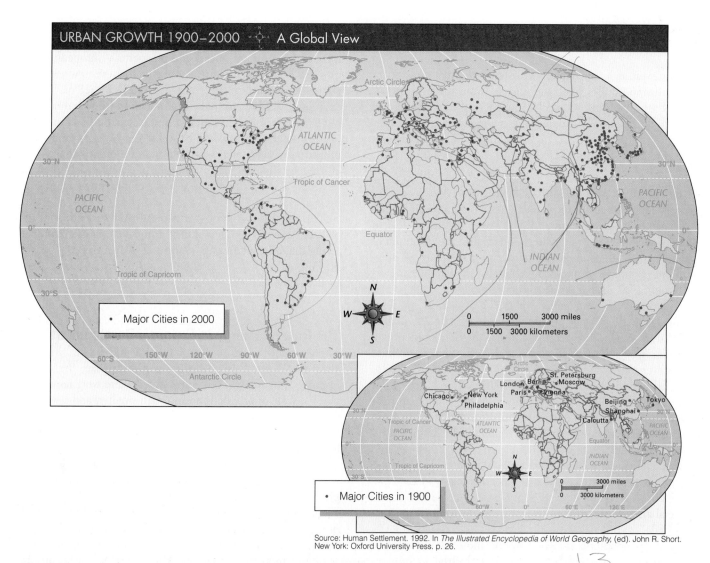

URBAN GROWTH 1900–2000   A Global View

- Major Cities in 2000

- Major Cities in 1900

Source: Human Settlement. 1992. In *The Illustrated Encyclopedia of World Geography*, (ed). John R. Short. New York: Oxford University Press. p. 26.

Figure 3.14   In the past century, the growth of cities all across the globe has been a true urban explosion. A century ago, there were 13 cities that were larger than 1 million inhabitants. A study of this map shows the increase in the number of large cities. Europe has gained much population growth overall, and in city growth, the increase in "million cities" has touched all parts of the region. The steady inflow of guest workers during the decades since the conclusion of World War II has been a significant factor in such population shift and growth.

amount of power available. New processes and equipment, including the development of coke (coal with most of its volatile constituents burned off), made iron smelting relatively cheaper and, hence, more abundant. The invention of new industrial machinery made of iron and driven by steam engines multiplied the output of manufactured products, principally textiles at first. Then, in the 19th century, British inventors developed processes that allowed steel, a metal superior to iron in strength and versatility, to be made on a large scale and cheaply for the first time. Such developments led the world into the modern age of massive, mechanized industrial production. At the same time, this shift to new industrial activity also stimulated sig-

nificant human migration from the countryside into the cities. Europe was the primary hearth for all of these changes.

Europe has one of the highest regional averages of urban population in the world (see Fig. 3.2 and Table 3.1). In the United Kingdom and Belgium, more than 90% of the population is urban, with Belgium reaching 97% urban in 1998. More than 85% of the population of Germany, Denmark, and Luxembourg resides in cities, while Sweden is 83% urban. These high levels of urbanization complement the industrial nature of these nations' economies, but they also have an influence on levels of population growth (Fig. 3.14). These nations generally have the lowest levels of crude birth rate (the total number of live births per 1,000 population per year) and often the highest per capita **gross national product (GNP)** levels.

However, as later chapters will show, the high rates of industrialization and urbanization are far from uniform in Europe. From the Western and Northwestern subregions toward Eastern and Central Europe, there is a gradient toward lower levels of industrial activity, higher ratios of nonurban population, and lower per capita GNP levels.

Moving Toward a Postindustrial Society

Scholars have traced a slowly rising trend of European technical inventiveness since the Middle Ages, involving devices such as clocks, gears, waterwheels, windmills, the invention of printing in 15th-century Germany, improvements in shipbuilding, drainage of mines by steam-driven pumping to permit mining at deeper levels, improvements in artillery and other armaments, and many others. The practical improvements leading to greater production were furthered by mathematical and scientific discoveries and by the development of a scientific viewpoint toward nature and the further possibilities for its ever more productive use. At the same time, still prior to the 18th century, capitalist values, practices, and institutions were evolving and becoming prominent — for example, private ownership of property; favorable attitudes toward profit making, including interest from loans; commercial banking, insurance, and credit institutions; double-entry bookkeeping; and corporate forms of business organization. Increased law and order imposed by newly centralized states, especially in England, the Netherlands, France, and Spain, protected the rising commercial and industrial interests that profited steadily from overseas trade.

The developments spawned by the Industrial Revolution continue to the present time. Today, industrial technology has advanced from the crude steam engines and textile machines of 18th-century Britain to modern electronics, computer-controlled factory robots, and the human exploration of space. As it has advanced, it has diffused geographically. The early coal-based and steam-powered industry first spread from Britain to Belgium and then more widely in Europe and the eastern United States. It began to expand in influence to some non-European areas like Japan in the 1870s. Today's more modern technology is continuing its expansion to Monsoon Asia, Latin America, and parts of Africa with a rising tempo.

Much evidence exists that the older and more advanced industrial societies are in the process of becoming "postindustrial"; that is, they will eventually be societies in which industrial workers are nearly as uncommon as farmers are now in most "industrial" societies. The introduction and improvement of **computer-controlled robots** (machines programmed to do specific jobs without further direction) on industrial production lines are moving this possibility along. The recent occurrence of serious **technological unemploy**ment (the supplanting of workers by labor-saving technology) in Western Europe is, in part, related to this trend. Hence, the more advanced and prosperous European countries are beset by concerns such as (1) where new jobs will come from (most likely from the "service" sectors of the economy), (2) how to make the transition from manufacturing to service roles without too much social dislocation, and (3) how to allocate income between a few highly productive industrial workers who tend robots and a predominant mass of service employees whose output per worker will often be less than a robot's and whose basic employment is increasingly at risk.

This process is moving across northwestern Europe at the present, and it is a major shift in employment and manufacturing that characterizes the entire developed world. It is significant not only as an economic transformation but also as a societal change.

The Role of Energy in Europe's Development

Energy resources are central to any program of industrialization. When Europe began to industrialize, the energy resource situation was quite favorable. There were fewer people, the scale of production was much smaller, and raw materials were generally sufficient for the simpler requirements of the time. A relatively primitive technology was served by extensive forests, numerous small-scale sources of water power, and a large variety of minerals in small deposits found in widespread locations. When coal became the main source of power and heat in the 18th century, Europe — especially England — was favored by its many coal fields. These fields proved to be the main localizing factor for much of Europe's initial industrial and urban development: During the early period of industrialization, electricity was still in the future, and coal was expensive and cumbersome to transport; consequently, the new factories were generally built on or near the coal fields. The working population in turn clustered near the factories and coal mines (see Fig. 3.13) and developed a group of industrial districts that incorporated mines, factories, and urban areas and provided both labor and markets in a new spatial pattern.

These older districts are still very prominent on the map of Europe today. Notable examples include most major cities in Great Britain, although not London; the east–west line of cities strung across Belgium south of Brussels; the huge urban agglomeration in western Germany called the Ruhr; and the Polish industrial cities near the Czech border. Economic depression and urban blight are increasingly common in these places, which had the advantage of an early start but now suffer from steadily changing industrial circumstances, including increasing competition from non-European industries, competition from newer products made elsewhere (e.g., new synthetic fibers competing with the cottons and woolens of old textile centers), and decreas-

Box 136

ing requirements for labor due to the increasing automation of factories.

Industrial Resources and Landscapes Today: The Impact of New Patterns

A few countries, notably Sweden, Finland, Norway, and Austria, currently have surpluses of wood. Hydroelectricity has been developed in the Alps, the Scandinavian Peninsula, and along the Rhone and Danube Rivers. Most deposits of metallic ores are too depleted or too small to be important today. But the most far-reaching change is the altered situation of coal. Worked long and intensively, most European coal fields are now expensive to mine and yield less coal. Many were phased out as Europe shifted to cheaper imported oil for power after World War II and as it became apparent that coal had a major negative environmental impact on the quality of city air. Indeed, new environmental demands have further increased the cost of coal as a fuel source because of its implication in acid precipitation as well as in the generation of smog. North Sea oil production started only in the 1970s; prior to that, much of the oil used in Europe after World War II was imported from the Middle East or Russia. In a broader sense, Europe, which was once relatively self-sufficient in energy resources, must now buy and import energy and raw materials from a global base, and most European countries now enter global trade actively to acquire energy resources.

This shift away from relative energy self-sufficiency has had implications for economic well-being, factory employment levels, and the vitality of the economic landscape in Europe. The transition is an ongoing example of the globalization of manufacturing, trade, and political interdependence. What Europe has achieved in developing a much wider network of raw material imports and the export of high value-added manufactured products is a pattern that is occurring in all world regions. Because of its colonial legacy and earlier worldwide trade dominance, Europe has more experience in such global trade patterns than most of the world's manufacturing centers.

A number of European countries do have mineral resources of considerable importance. As already mentioned, in the 1970s, the United Kingdom became a large producer and net exporter of oil and also a large producer of natural gas when its offshore fields in the North Sea were brought into production (Fig. 3.15). Norway's oil and gas fields in the North Sea likewise yield a large production. For several decades, the Netherlands has been a major producer and exporter of natural gas. Large iron ore fields are worked in eastern France and northern Sweden. Coal from southern Poland is a mainstay of that country's economy, and Germany is still a sizable coal producer.

3.5 Geopolitical Issues

Geopolitics is a term that connects basic spatial aspects of geography like borders in association with the politics of administration and expansion. It also has, from geographic theory of the early 20th century, a similarity in spelling to the German term **geopolitiks**. Geopolitiks relates to empire building and arguments for spatial expansion that favor one but never both of the countries or regions involved in geopolitical discussions. Our use of the term geopolitics in this text identifies issues that are a blend of geographic factors that are heavily influenced by political decision making. Geopolitical Issues is the final segment of the eight Geographic Profile chapters in this text. In these sections, we will highlight geographic issues whose solutions have real political significance for the region and its neighbors.

Figure 3.15 The discovery of major oil resources in the North Sea in the last three decades has both changed the appearance of the seascape and given Norway, for one, major economic independence. This photo shows how these oil production platforms have become a much more commonplace feature of the North Sea.

MARK A. LEMAN / STONE/GETTY

The consideration of environmental issues serves as an excellent example of geopolitical issues. As you consider all of the various forces that have an influence on environmental change, it can soon be seen that political attitudes and policies are as much responsible for environmental degradation as factors of air movement, groundwater flow, or precipitation patterns. Human modification of the setting in which we live is always created by a combination of human action, human indifference, and the extraordinary human belief that we have the right to change any and all elements into whatever form provides us with a more productive, more convenient, and more profitable product. Such attitudes and actions have been particularly manifest during the past century. In this section of this European Profile, the discussion of the Danube River and the European Union is one example of the interplay of politics and environmental control. The Field Essay on Venice is another window on the close relationship between a location's environmental conditions and the multiplicity of forces working to sustain or modify an environmental situation. As you put these materials into the context of Europe, realize that the societal dynamics of such environmental issues are, almost always, geopolitical as well as environmental.

Problems of Europe's Environmental Pollution

F5.17
154
A much-publicized problem in some European areas (particularly the former Eastern Europe) is **environmental pollution.** For example, pollutants pour into the Mediterranean Sea from urban-industrial districts in Europe, Africa, and Southwest Asia. Wastes discharged into this most historic of seas damage both the coastal tourist industry and fisheries. The Mediterranean, almost totally enclosed by land, cannot cleanse itself by flushing wastes into the ocean. By contrast, the North Sea is far more able to interchange water with the ocean, thanks to strong currents that move in and out through the broad opening to the Atlantic between Scotland and Norway. But even the North Sea has become alarmingly contaminated with pollutants discharged by great industrial metropolises, the undersea oil industry, and a heavy volume of shipping (Fig. 3.16).

Still another famous water feature, the Rhine River, is under stress from untreated wastes. Stringent legislation requiring municipalities and industries to maintain waste-treatment plants has reduced the pollution, but the Rhine, from Basel, Switzerland, through a major German-French and a Low Countries industrial district, and to the North Sea, is still so dangerous to human health that swimming is generally forbidden. Aquatic life in the Rhine continues to be damaged, not only by toxic wastes but also by excessive warmth when water withdrawn for industrial use is returned to the river in a heated state. Nuclear stations, for example, use large volumes of Rhine water to cool radioactive cores and produce steam to generate electricity. The return of this water heats the river. Similarly, steel mills and coke works withdraw water to cool red-hot metal or to quench flaming coal in the coking process, and this water is still warm when it comes back to the river. (Look at the Definitions & Insights box on the Danube (p. 86) to see another example of the politics of environmental management.)

Atmospheric pollution is another widespread environmental affliction in Europe, as it is in North America and various other parts of the world where there are massive metropolises and industrial concentrations. One form of this pollution is known as **acid rain** (sometimes called acid precipitation). This is precipitation that has interacted with airborne industrial pollutants, causing the moisture to become highly acidic. Especially publicized have been the ravages of pollutants that have killed great numbers of trees in some of Europe's finest forests. Emissions containing sulfur, when combined with precipitation, produce weak sulfuric acid, and those containing nitrogen yield weak nitric acid. Both can damage trees, soils, and aquatic life in lakes that receive acid precipitation and its runoff. There has been much alarm about lakes that are "dying." Germany and Scandinavia are among the European areas particularly affected.

Another form of atmospheric pollution caused worldwide alarm in 1986 when significant radioactive fallout from a nuclear disaster — a partial meltdown and resulting fire — at the **Chernobyl** nuclear power plant in the Ukraine (western part of Russia and the Near Abroad) was deposited on parts of Europe as well as on a large area of Russia and the Near Abroad itself. Sections of Scandinavia and Finland were particularly affected; in fact, some northerly areas used by Lapp reindeer herders became so radioactive that the wild forage could no longer be used and the animals that had grazed on it had to be destroyed.

Many other examples of environmental damage can be drawn from Europe. For instance, old mining and industrial regions contain numerous waste heaps, sometimes centuries old, that are full of poisonous substances. Seepage of these poisons into groundwater may cause cancers and a variety of health problems. Today, there is a very active environmental consciousness in Europe, as in many other parts of the world, but many problems deriving from environmental neglect are so deep-seated that solutions will be slow and very costly. This is particularly true of former Communist bloc areas in Eastern Europe, where high rates of unemployment make it particularly difficult to close factories for environmental reasons.

The Drive for New Patterns of Regional Cooperation

Since World War II, the countries of Western Europe have moved toward greater economic, military, and political cooperation. Searching for development and security, the countries have banded together in a series of organizations designed to foster unity. In Eastern Europe, there were efforts to create economic and security pacts under the former Soviet leadership, but the collapse of communism in the early 1990s aborted such cooperative arrangements.

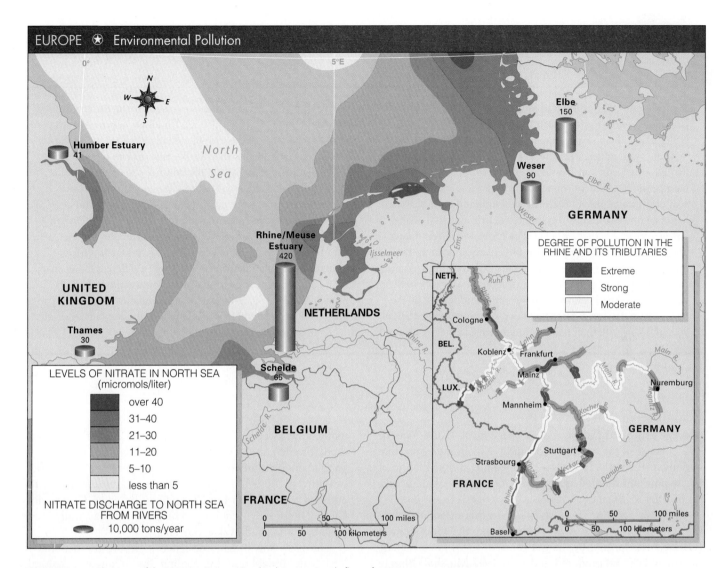

EUROPE ⊛ Environmental Pollution

North Sea

Humber Estuary
41

Elbe
150

Weser
90

GERMANY

Rhine/Meuse
Estuary
420

Ijsselmeer

DEGREE OF POLLUTION IN THE
RHINE AND ITS TRIBUTARIES

Extreme
Strong
Moderate

**UNITED
KINGDOM**

NETHERLANDS

NETH.

Ruhr R.

Cologne

Thames
30

BEL.

Koblenz   Frankfurt

Mainz

Nuremburg

LEVELS OF NITRATE IN NORTH SEA
(micromols/liter)

Schelde
65

LUX.

Mannheim

over 40
31–40
21–30
11–20
5–10
less than 5

BELGIUM

Stuttgart

GERMANY

Strasbourg

NITRATE DISCHARGE TO NORTH SEA
FROM RIVERS
10,000 tons/year

FRANCE

FRANCE

0   50   100 miles
0   50   100 kilometers

Basel

0   50   100 miles
0   50   100 kilometers

Figure 3.16   The rivers of the Western Europe Coreland carry a steady flow of nitrates into various parts of the North Sea. Although the Rhine River is by far the biggest contributor to this pollution of the North Sea, there is a continual residue of nitrates brought seaward from heavily farmed lands and varied industrial centers on the banks of the Rhine and virtually all other rivers in this region. The European Union and other international organizations make continuing efforts to diminish such runoff and consequent pollution but levels of human activity along these rivers make it impossible to bring the waterways back to earlier levels of more modest pollution.

What is responsible for the new age of cooperation in a region too well-known for national quarrels and strife? Important among the many motivations are a longstanding ideal of European unity dating from the Roman Empire and the medieval Catholic Church; the experience of two World Wars (1914–1918; 1939–1945); perceived military and political threats from the then-USSR after World War II; and a search for economic betterment through the concerted action of nations working in an elusive unity. After World War II, economic recovery was slow, and there were fears in Western Europe of military aggression by Russia and the Near Abroad and/or a political takeover by powerful Communist parties in

Italy and France. In the meantime, Russia and the Near Abroad and the Communist governments of the satellite countries east of the Iron Curtain (the term Winston Churchill devised for the western boundary of the Soviet bloc in a 1956 speech in Fulton, Missouri) saw U.S. military power in Europe as a major threat to be countered by economic and military cooperation under Soviet leadership. Thus, the desire for economic recovery and military security in both the western and eastern nations set the stage for a remarkable period of supranational organization.

The alliances developed in this era of regional cooperation have played a major role in both the economic and political

DEFINITIONS + INSIGHTS

The Politics of River Management

Approximately 80% of the **wetlands** on the margins of the Danube River have been lost in the steadily increasing levels of settlement and landscape change along this river that passes through 11 countries. The forces of landscape transformation that have replaced marshlands and floodplains along the Danube have been the traditional agricultural expansion onto the good soils of river margins. This act of **environmental modification** has been one of the classic examples of environmental change to a river system. Another aspect of such change has been the use of the river, for the past half-century particularly, as a dump bin for industrial and municipal wastes. On the Danube, there has been steady difficulty in getting upstream nations to worry about their environmental policies, which have a profound impact on the downstream nations. A force that has taken on new potential for changing this pattern of political indifference to the peoples and

settlements on other parts of the river system of the Danube is the **European Union.** The **World Wide Fund for Nature (WWF)** has taken advantage of the wish of virtually every nation on the Danube to be admitted to the EU. Because of such interest in becoming members of the EU, in the non-EU countries from Slovakia to Bulgaria, there is a growing willingness to attempt to manage the river, better manage its wetland resources, and control the urban and industrial effluents discharged into this international waterway. A river system presents complexities for environmental management because of the fact that upstream river users have traditionally been relatively indifferent to the environmental impacts of their actions on downstream river users. The WWF and the Danube experience are showing new river management as yet another impact of the expanding influence of the EU all across the European region.

F10.12
290

development of the region since the end of World War II. The collapse of the Soviet Union and the Communist satellite nations in the early 1990s has changed the political fabric of the region and opened new possibilities, and new demands, for broader regional cooperation.

In 1948, a major step was taken in this direction when the **Organization for European Economic Cooperation (OEEC)** was formed by many European states to coordinate the use of U.S. aid proffered under the 1947 **Marshall Plan** to bolster Europe against communism. The Marshall Plan was terminated in 1952, but the OEEC continued to function. In 1960, its name was changed to the **Organization for Economic Cooperation and Development (OECD)**. Membership now includes the United States and several other countries outside Europe.

But the most significant economic organization among Europe's western nations was the **European Economic Community (EEC)**, also called the **Common Market,** formed in 1957 and now officially called the **European Union** or **EU** (since November 1, 1993) (Fig. 3.17). It was initially established by six countries: France, the former West Germany, Italy, Belgium, Luxembourg, and the Netherlands. In 1967, this group was reorganized into the **European Community.** These countries had already eliminated barriers to trade in coal and steel and now looked toward removal of all economic barriers. Later, England, Denmark, Ireland, Greece, Portugal, and Spain became members. Cyprus, Malta, and Turkey are currently associate members. As of 2000, the 15 members of the European Union included the United Kingdom, Ireland, Portugal, Spain, France, Luxembourg, Germany, Italy, Greece, Austria, Belgium, the Netherlands, Denmark, Sweden, and Finland. These 15 countries have a total population of more than 378 million people and produce two-fifths of the world's global exports. Countries that are requesting admission to the EU include Bulgaria, Cyprus, the Czech Republic, Estonia, Hungary, Latvia, Lithuania, Poland, Romania, Slovakia, Slovenia, and Turkey.

The EEC was initially designed to secure the benefits of large-scale production by pooling the resources — natural, human, and financial — and markets of its members. Hence, tariffs were eliminated on goods moving from one member state to another, and restrictions on the movement of labor and capital between member states were greatly reduced. Monopolistic trusts and cartels that formerly restricted competition were discouraged. Meanwhile, a common set of external tariffs was established for the EEC's entire area to regulate imports from the outside world, and a common system of price supports for agriculture replaced the individual systems of member states. The founders of the EEC anticipated that free trade within such a populous and highly developed bloc of countries would (1) stimulate investment in mass-production enterprises, which could, wherever they were located, sell freely into all EEC countries and (2) encourage a productive geographic specialization, with each part of the Community expanding lines of production for which it was best suited. Thus, each country might achieve greater production, larger exports, lower costs to consumers, higher wages, and a higher level of living than it could achieve on its own.

It is possible that this success would have been attained without the formal organization of an international eco-

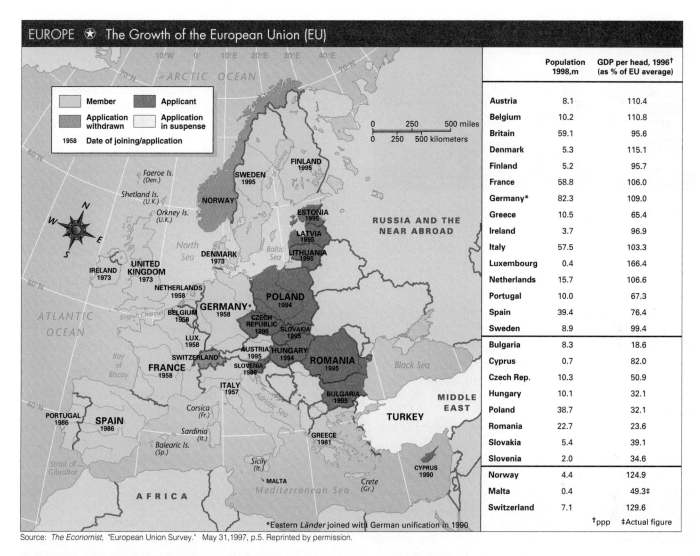

EUROPE ✪ The Growth of the European Union (EU)

	Population 1998,m	GDP per head, 1996† (as % of EU average)
Austria	8.1	110.4
Belgium	10.2	110.8
Britain	59.1	95.6
Denmark	5.3	115.1
Finland	5.2	95.7
France	58.8	106.0
Germany*	82.3	109.0
Greece	10.5	65.4
Ireland	3.7	96.9
Italy	57.5	103.3
Luxembourg	0.4	166.4
Netherlands	15.7	106.6
Portugal	10.0	67.3
Spain	39.4	76.4
Sweden	8.9	99.4
Bulgaria	8.3	18.6
Cyprus	0.7	82.0
Czech Rep.	10.3	50.9
Hungary	10.1	32.1
Poland	38.7	32.1
Romania	22.7	23.6
Slovakia	5.4	39.1
Slovenia	2.0	34.6
Norway	4.4	124.9
Malta	0.4	49.3‡
Switzerland	7.1	129.6

†ppp   ‡Actual figure

*Eastern *Länder* joined with German unification in 1990

Legend: Member | Applicant | Application withdrawn | Application in suspense | 1958 Date of joining/application

Source: *The Economist*, "European Union Survey." May 31,1997, p.5. Reprinted by permission.

Figure 3.17 The European Union has grown steadily from 6 nations in 1952 to 9 in 1973, 10 in 1981, 12 in 1986, and 15 in 1997, with another dozen interested in becoming part of this extraordinary collection of diverse but economically ambitious European countries. The table to the right of the map provides population data and comparative economic efficiency.

nomic community. For example, both Switzerland, which is a close neighbor of the European Union but not a member, and Norway, which voted in 1995 not to join the EU, have among the highest GNPs and per capita incomes in the world. Sweden, which joined the EU in 1995, also ranks near the top of the world scale in affluence. However, Switzerland, Norway, and Sweden have all depended heavily on trade and other dealings with the Common Market, and thus, their prosperity has been closely tied to that of the Market. Six nations that were not initially enrolled in the EEC—Iceland, Norway, Sweden, Switzerland, Finland, and Austria—belonged for many years to a separate trading organization called the **European Free Trade Association (EFTA).** It has worked closely with the Common Market, and both are increasingly integrated within a single market

area, particularly for nonagricultural products. The Common Market made preferential trade and economic aid treaties with many other countries, mainly ex-colonies of Common Market members, in areas outside Europe. By late 1996, however, three of the EFTA nations (Sweden, Finland, and Austria) had joined the EU, and Norway had rejected membership.

The European Union, under provisions of the **Maastricht Treaty of European Union** that went into force on November 1, 1993, is now trying to achieve major new steps in unification. These steps include the removal of nontariff trade barriers, such as "quality standards," that still exist between member countries. Major energy focused on the implementation of a single EU currency, the **euro,** in 1999. Progress has been slow because of questions concerning

Figure 3.18 Tourism and the global diffusion of American popular culture have brought countless landscape scenes of American influences in the European landscape. This scene gives graphic evidence of McDonalds in Italy and is also evocative of other fast-food companies that have brought aspects of the American throwaway culture to societies seemingly all too willing to adopt such culture habits, particularly in cities.

national sovereignty and occasional protest by populations through the EU who do not want to lose the autonomy of having their own unique currency. On January 1, 1999, the euro became the currency of record for 12 (U.K., Sweden, and Denmark elected not to do the conversion, and Greece was required to wait because it had not yet met the EU budgetary criteria) of the 15 nations in the EU. On January 1, 2002, euro notes and coins became "coin of the realm." Supposedly, by March 2002, traditional national coins and currency will cease to serve as legal tender within the EU nations that have converted their currencies to the euro. This coinage shift will involve the replacement of more than $15 billion in bills and $56 billion in coins.[1] This change will have direct impact on the more than 300 million people resident in those nations. Part of the significance of this monetary innovation is the fiscal criteria that were decided on in the December 1991 meetings in Maastricht, Netherlands. The euro and this move toward an ever more closely linked European Union community of nations represent a major shift in the body politic of the dominant economic and cultural forces of Europe, although in tourist settings and activities, the scenes seem largely unchanged (Fig. 3.18).

In the sphere of military defense, the key international body in non-Communist Europe became the **North Atlantic Treaty Organization (NATO).** This military alliance was formed in 1949 and now includes the United States, Canada, a majority of the European nations west of the former Iron Curtain, and Turkey. In 1997, NATO membership was extended to Poland, Hungary, and the Czech Republic. The accession process to full membership was intended to take 2 years. The NATO membership offer to three of the coun-

tries that had been part of the Union of Soviet Socialist Republics for nearly half a century led to considerable anger from Russia. It saw this eastward expansion of NATO as particularly threatening. After considerable negotiation, Russia finally agreed to accept this act because of its own deeper involvement in the NATO-sponsored Partnership for Peace program. Russia also played a role in the NATO-led Stabilization Force (SFOR) that enforced the terms of the 1995 Dayton Accords between Bosnia-Herzegovina and Serbia.

NATO member countries pledged to settle disputes among themselves peacefully, to keep their individual and joint defense capacities in good order, to consider an attack on any of them in Europe or Anglo America (e.g., the Falkland Islands) as an attack on all, and to come to the aid of the country or countries being attacked. Questions arise today as to whether NATO is necessary any longer and about what its role should be in view of the dissolution of the Soviet Union and the retreat of Soviet power from Eastern Europe. The Partnership for Peace program of military and political cooperation (including joint military exercises and peacekeeping operations) with Russia continues to be a response to such questions. Eighteen ex-Soviet Iron Curtain allies joined the program on June 22, 1994.

The response of the former Communist East to the formation of the Common Market and NATO was the creation in 1955 of an international economic organization called the **Council for Mutual Economic Assistance (COMECON)** and a military alliance called the **Warsaw Pact.** The former Soviet region dominated the Communist "satellite" states in these organizations. The new Eastern Europe still needs freely entered agencies of economic cooperation and collective security, but since COMECON and the Warsaw Pact were not freely entered, they have thus been terminated. Currently, NATO and the EU are working with a variety of potentially useful and productive collaboratives with the Eastern European nations. For example, in the former Yugoslavia — once an East European nation — warring factions have thrown NATO into a new role as a peacekeeping force. The 1995 Dayton Accords tried to bring stability to a European region already known for centuries of discord. In the winter of 1999, NATO played a military role in attempting to hold back Serbian military advances against ethnic Albanians in the Kosovo region of the former Yugoslavia. There continues to be a NATO peacekeeping presence in Bosnia, Kosovo, and a continual potential for expanded engagement in other parts of the Balkan Peninsula and in other reaches of the former Yugoslavia.

Finally, one additional facet of regional interaction is the European purchase of more than $260 billion worth of U.S. firms in the year 2000 alone. The companies bought range from telecommunications to foods to financial management firms. The German purchase of Chrysler in 1998 and the creation of the Daimler-Chrysler firm reflected a major upswing in such international acquisition. To have a European

---

[1] "Permanent Revolution for Europe's Union?" *Economist*, February 3, 2001, 46.

INTERNATIONAL ORGANIZATIONS OF EUROPE, MARCH 2002

Sources: EUROSTAT (Statistical Office of the European Communities) (1996). CIA, *The World Fact Book* (1995).

Figure 3.19 Some of the key organizations in Europe's intricate framework of political, strategic, and economic relationships. The sequence of the creation and elaboration of the various steps leading to the contemporary European Union (EU) is chronicled in the text. The critical fact to remember is that this region has a long history of conflicts that have been political, religious, and economic. The EU marks a very momentous innovation in European regional cooperation.

firm buy outright one of the major American auto firms in the Daimler-Chrysler deal illustrates the way in which the balance of fiscal and manufacturing is in continual change. However, these new corporations, new political alliances, and new perspectives on globalization are all evidence of a changing geographic network of economic, political, and consequently, cultural interchange. Hence, it can be clearly seen that patterns of traditional geographic isolation, independence, and relative autonomy are being modified by these new regional and increasingly global linkages. Because of the enormous economic strength of the giant nations of Europe, and the market associated with those countries with the largest populations, the new profile created by the EU sets the stage for still additional innovations in regional efforts to unify the historically individualistic nations of Europe (Fig. 3.19).

CHAPTER SUMMARY

- Europe in topographic terms is, in reality, a Eurasian peninsula. However, it is generally labeled a continent today because of the political, religious, and economic authority held in Europe for the past two millennia. It has a series of other smaller peninsulas, rivers, mountain systems, and plains, which are generally much more densely settled than the United States. The European population is about twice the size of the U.S. population. Western civilization has its origins in this region, and its influences have reached all areas of the world.

- Northerly location, temperate climate, the North Atlantic Drift, and varied topography are all factors in the productivity and long history of productive settlement of this region. The North European Plain is a major belt of settlement and agricultural productivity. The marine west coast climate and the Mediterranean climate are the two most characteristic climates of Europe. Major rivers flow out from mountain highlands in West Central Europe and spill into the North Atlantic, the Mediterranean, the Baltic, the Black Sea, and other proximate seas. The Rhine and the Danube are the two most important rivers.

- European languages derive primarily from Indo-European roots and are expressed primarily in Romance, Germanic, and Slavic languages. English is a Germanic language and, like most European languages, it is enriched by many other tongues. The languages of Europe reflect a long history of human migrations from east to west and a pattern of continual mixing. The dominant religious influence in Europe has been Christianity, with the major three religions being Protestant, Roman Catholic, and Eastern Orthodox. There are significant pockets of Islam and Judaism as well as many regional variants on all of these belief systems.

- Before the Age of Discovery, Europe played only a minor role in world trade networks. From the beginning of the 16th century until late in the 19th century, Europe was at the center of global patterns of colonization and foreign settlement, long-distance trade, slave trade, and agricultural innovation. One of the major outcomes of the maritime trade and travel such patterns engendered was the Columbian Exchange, or the diffusion of crops and animals from the Old World of Europe and Asia to the New World of the Western Hemisphere. Major crops, including maize, coffee, cacao, rice, grapes, and potatoes, were taken to new locales and became major food crops.

- The Industrial Revolution originated in Europe, with energy coming from coal and factory technology focused on textiles and iron manufacture. Rivers, canals, and railways all played a role in making Europe strong, and a shipbuilding industry and skill in maritime endeavors from exploration to navigation gave Europe a dominance in the 18th and 19th centuries that it has only slowly given up.

- Europe rose to dominance in world trade and politics because of factors such as capitalism, technology, and science. However, recent decades have seen a global change in the patterns of power, and factors such as war dislocation, rising nationalism, the ascendancy of the United States, a shift in world manufacturing patterns, and new energy sources have combined to diminish Europe's global centrality. Europe continues to be a dominant force in world economic, political, and social realms, but there is a steady increase in competition from many other world regions.

- Europe has a history of human migrations coming from the east and going to all parts of not only Europe but the world beyond. Contemporary migration streams from eastern Europe to western and northwestern Europe, and from northern Africa to southern and southwestern Europe, are aspects of migration that shape the current scene. Guest workers, very important in the rapid economic growth of Europe in the 1970s and 1980s, have recently become a demographic problem, as has technological unemployment. Population growth rates vary a great deal from country to country and region to region in Europe, but overall, Europe has the lowest fertility rates of any world region.

- The European Union (EU) is a collection of 15 nations that has grown from 6 member states that came together to gain economic strength in 1957. Now the EU is leading the way in the creation of a Europe characterized by open borders, limited or no tariffs, and a common currency called the euro. The North Atlantic Treaty Organization (NATO) is another example of a supranational organization that attempts to create a military and political force that gives Europe a more powerful voice than it would achieve through its individual states.

REVIEW QUESTIONS

1. What defines the continent of Europe?
2. What are its major physical and environmental characteristics?
3. Define the major types of climates and differentiate between the two dominant ones.
4. List four major rivers in Europe and explain their historic and contemporary importance.
5. What are the dominant languages and religions in Europe and what have been the sources of these cultural features?
6. Define the factors that led to Europe's central role in the Age of Discovery and the subsequent patterns of colonization and world trade patterns of the 18th and 19th centuries.
7. Explain the origins and significance of the Industrial Revolution, including the role of resources in that revolution.
8. List some of the environmental problems that Europe is faced with; describe how they came about and explain why they are difficult to fix.
9. Describe patterns of human migration in the history and in the contemporary world of Europe. Make a sketch map that shows these spatial patterns.
10. Recall the six Essential Elements in Chapter 1 and describe a scene from Europe that would be suitable as an example of each element.
11. Describe the efforts Europe has been undertaking since World War II to build a stronger regional Europe.

DISCUSSION QUESTIONS

1. What is the impact of the North Atlantic Drift (Gulf Stream) on European patterns of agriculture, settlement, and economy?
2. Try to explain the reasons for the relatively low population growth rates for European countries, taking care to point out the significant differences from country to country.
3. What factors made Europe the center of the world for three centuries after the beginnings of European settlement of North, South, and Middle America?
4. Considering the importance of spatial analysis, how would you describe and explain the patterns of European cultural influences around the world today?
5. What factors may lead to a steady decline in the relative importance of Europe in the global economy in the next two decades? Are there other realms in which Europe is likely to lose relative importance?
6. Consider the factors that make up an agricultural scene in Europe. Use a photograph or recall a trip you or someone you know has taken, and explore the features of that scene that are natural versus the features that have been modified by human effort. From that observation, make some comments about agriculture in Europe today.
7. What factors are working against the success of the EU in Europe today? Are they likely to be overcome?
8. If you were a major Russian leader, describe how you would have responded when NATO invited Poland, Hungary, and the Czech Republic to join NATO. What could have changed your mind and changed your reaction?
9. Discuss some of the ways and the cities in which you would like to use the axioms presented in Perspectives from the Field on p. 80.

Europe's Dynamic Coreland

The resurgence of Germany from the ruins of World War II is dramatically illustrated by this photo of Frankfurt-am-Main contrasted with the photo of the shattered city in 1949.

H.P. MERTEN/CORBIS STOCK MARKET.

Chapter Outline

AP/WIDE WORLD.

West Central Europe, which is the Coreland of this world region, is now undergoing a transformation toward more service industries and less traditional manufacturing. Fewer jobs are opening in smokestack-laden industrial zones, and more people are bringing computer skills to their employment. Carry these images with you as you learn in the pages that follow of the increasing abandonment of 19th- and early-20th-century industrial landscapes in the Coreland of West Central Europe. Such industrial space is partially replaced with modern information-centered office complexes. This European Coreland is rightfully studied in connection with the process and ramifications of the Industrial Revolution, European colonization, and major political, economic, and cultural shifts as a consequence of World War II. All of these phenomena help give image to regional landscapes in this economic hub of Europe. This monumental three-century process of regional economic change has also led to major demographic shifts, social change, and ultimately, European concerns about the environmental deterioration that seems to travel with industrial activity. In these pages, look for the geographic factors that have served as foundation blocks for the dynamic economic and social transformations that characterize the images and realities of West Central, or Coreland, Europe.

4.1   The Geographic Foundation Blocks of European Global Influence

Three of Europe's major countries and a series of smaller nations and microstates occupy the British Isles and the west central portions of the European mainland (Fig. 4.1). The United Kingdom and the Republic of Ireland share what is called the British Isles. Germany; France; their smaller neighbors of the Netherlands, Belgium, Luxembourg, Switzerland, and Austria; and the microstates of Liechtenstein, Monaco, and Andorra may all be conveniently grouped as the countries of West Central Europe. Because of the historic and economic role this region has played for the past centuries, it is called the Coreland. Considered as a whole, these units are highly developed, and they form Europe's economic and political core. Development is far more intensive in some areas than in others, but the overwhelming majority of people in the Coreland live in areas that are among the world's most geopolitically significant and intensively developed cultural landscapes (see Fig. 4.1 and Table 3.1).

T3.1 62

Even given the complexity of Europe, the geography of the Coreland is unusually intricate. Spatial variations are great, different layers of history appear at every turn, and functional linkages are extraordinarily complicated. The individual countries differ sharply in their human geographies and national perspectives, and the whole area is differentiated in a mosaic of regions and localities exhibiting great physical and cultural variety (Fig. 4.2).

This cultural variety is matched by the range of environmental niches and the resources base of the area as well. In the 19th and early 20th centuries, the largest coal and iron ore deposits in Europe provided bases for large-scale industrial development (see Figs. 4.1 and 4.2). Consequently, a series of major industrial concentrations now stretches from the United Kingdom to Poland along the main belt of mineral deposits. Many individual industrial cities lie outside the main coal and steel axis, and some of these (e.g., Paris and London) are sufficiently important to be considered major industrial concentrations in their own right. In the typical industrial agglomeration, coal mining and iron and steel production have been the traditional economic foundations, with chemicals, textiles, and heavy engineering also common. Forces of change, however, have diminished the manufacturing importance of many of these centers, as the European Coreland has steadily been transformed into postindustrial scenes that feature much less industrial activity and much more service and information-centered commerce.

4.2   Tensions of Past and Present in the United Kingdom

Located off the northwest coast of Europe, two countries occupy what are called the British Isles: the Republic of Ireland, with its capital at Dublin, and the larger United Kingdom of Great Britain and Northern Ireland, with its capital at London. The United Kingdom (U.K.) incorporates the island of Great Britain, consisting of England, Scotland, and Wales plus the northeastern corner of Ireland and most of the smaller out-islands. Northern Ireland separated from Ireland in 1922 and maintained its link with the U.K. while Ireland moved toward full independence. Altogether, the U.K. encompasses about four-fifths of the area and well over nine-tenths of the population of the British Isles, which are defined as the islands of Great Britain and Ireland (see Table 3.1).

T3.1 62

Political Subdivisions and World Importance of the United Kingdom

England, which lies only 21 miles (34 km) from France at the closest point (see Fig. 4.1), is the largest of the four main subdivisions of the United Kingdom (the others being Scotland, Wales, and Northern Ireland). These all were originally independent political units. England conquered Wales in the Middle Ages but preserved some cultural distinctiveness associated with the Welsh language.

The international influence once attained by the United Kingdom and the imprint it left on the world are a legacy of the 19th century. When British power was at its peak in the century, between the defeat of Napoleonic France in 1815 and the outbreak of World War I in 1914, the United Kingdom was generally considered the world's greatest power. Its overseas empire covered a quarter of the earth at the time of its maximum spatial extent, and the influence of English law, education, and culture spread still farther. Until the late 19th century, the United Kingdom was the world's foremost manufacturing and trading nation. The English Royal Navy dominated the seas, and the British merchant marine carried half or more of the world's ocean trade. London grew to be the center of a free trade and financial system that invested its profits from industry and commerce around the world.

From the late 19th century onward, a relative decline in the U.K.'s economic importance set in. The free trade and financial system based in London were damaged by World War I, and decline accelerated after World War II, although London remains a major world financial center. Relative economic decline has been accompanied by a drop in political stature. The large colonial countries once included in the British Empire have all gained independence, and Britain's remaining overseas possessions are scattered and small. The United Kingdom, however, is still a country of consequence in many fields. It plays a major role in the European Union and is associated with many of its former colonies in the worldwide Commonwealth of Nations.

F3.19 89

Regional Variations on the Island of Great Britain

The physical base of the United Kingdom is the island of Great Britain, about 600 miles (somewhat under 1,000 km) long by 50 to 300 miles (80 to nearly 500 km) wide. For its small size, this island provides a homeland with a notable variety of

Figure 4.1 Industrial concentrations and cities, seaports, internal waterways, and highlands of the Core-land of West Central Europe. Older industries such as coal mining, heavy metallurgy, heavy chemicals, and textiles cluster around highland margins in the congested districts shown as "major industrial concentrations." Local coal deposits provided fuel for the Industrial Revolution in most of these districts, which have shifted increasingly to newer forms of industry as coal has lost its economic significance and older industries have declined. But the new and more diversified industries have also proliferated in metropolitan London, Paris, and smaller industrial cities away from the traditional industrial concentrations. Relative sizes of dots and circles indicate approximate groupings of industrial cities and seaports according to economic importance. Rivers and canals shown on the map are navigable by barges and, in some instances, by ships. Give careful eye to the canal symbol, for these waterways continue to have a major economic significance, particularly in France, the Benelux region, and Germany. The Channel Tunnel (Eurotunnel) will change forever the sense of proximity, isolation, and access perceived by the political units on the map.

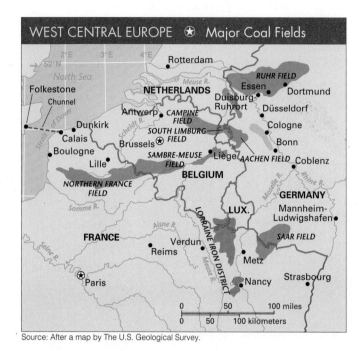

WEST CENTRAL EUROPE ✷ Major Coal Fields

Source: After a map by The U.S. Geological Survey.

Figure 4.2 Major coal fields of the European Coreland. Note the line of fields stretching across western Germany, the Low Countries, and northeastern France. The Northern France and South Limburg coal fields have been closed. The early importance of these fields as centers of 19th-century industrialization (and in some places, earlier) is being eclipsed now by the increasing importance of imported natural gas and petroleum, but the footprint of significant European Coreland industrial origins is shown in this map.

landscapes and modes of life. Some of this variety can be attributed to a broad division of the island into two contrasting physical areas: Highland Britain and Lowland Britain (see Fig. 4.3).

Highland Britain

The principal highlands include the Scottish Highlands, the Southern Uplands of Scotland, the Pennine Chain and the Lake District of northern England, the mountains that occupy most of Wales, and the uplands of Cornwall and Devon in southwestern England (Fig. 4.3). The mountains are low—Ben Nevis in the Scottish Highlands is the highest at 4,406 feet (1,343 m)—but many slopes are steep, and the mountains often look high because they rise precipitously from a base at or near sea level. Highland Britain's most important low-lying area is the Scottish Lowlands, a densely populated industrialized valley separating the rugged Scottish Highlands from the gentler and more fertile Southern Uplands of Scotland. The Scottish Lowlands compose only about one-fifth of Scotland's area, but they incorporate more than four-fifths of its population and its two main cities: Glasgow (with a population of 1.1 million in the metropolitan area, including 610,000 in Glasgow city proper plus suburban and satellite districts nearby) on the west, and Edinburgh, Scotland's political capital (with a population of 665,000 in the metropolitan area and 382,600 in the city proper), on the east. In the summer of 1997, the Scots voted overwhelmingly to create their own parliament, changing the nature of their 290-year union with England. This new body politic will give Scotland more independence in its domestic matters. These

highland landscapes have long been popular for tourist populations because of the open hillsides, the quaint villages, and the evidences of pastoralism, although the crisis over the potential for mad cow disease in the late 1990s reduced interest in this area markedly.

Lowland Britain

Practically all of Lowland Britain lies in England, and it is often called the English Lowland. It has better soils than the highlands and a gentler topography (Fig. 4.4). Most of it exhibits a mosaic of well-kept pastures, meadows, and crop fields fenced by hedgerows and punctuated by closely spaced villages, market towns, and industrial cities. The largest industrial and urban districts, aside from London, lie in the midlands and the north, around the margins of the Pennine Chain. Here, clusters of manufacturing cities such as Manchester, Leeds, and Sheffield rise in the midst of coal fields.

Geographic Elements and Impact of Britain's Industrial Revolution

Development in the Industrial Revolution

Four industries were of major significance in Britain's industrial rise: coal mining, iron and steel manufacture, cotton textiles, and shipbuilding. The advantage of an early British start in these enterprises gave the United Kingdom industrial preeminence through most of the 19th century. Exported surpluses paid for massive food and raw material imports and made Britain the hub of world trade and very wealthy for several centuries. Current difficulties in these same industries — due to growing international competition, changing patterns of demand, and shifting urban populations — are fundamental to many of the problems that Britain faces in the 21st century. These industrial innovations basically began with 17th-century coal mining and associated steam technology.

The British Coal Industry

Coal became a major industrial resource between 1700 and 1800, when it fueled the blast furnace and the steam engine. The link between the two technologies related back to coal. Coal mining had gone into deeper shafts by the early decades of the 1700s. Steam power became the best force to enable those shafts to reach new depths. Steam was also used for extracting water from those deeper shafts. James Watt's improvements in 1765 on Thomas Newcowmen's earlier design of a steam engine led to a steady expansion of the British role in the Industrial Revolution. Coal was coked for use in blast furnaces, and coke replaced the charcoal previously used for smelting iron ore. With Britain's development of an economically practical steam engine, coal supplanted human muscle and running water as the principal source of industrial energy.

Coal production rose in England and Wales until World War I. Then, between 1914 and the 1990s, output fell by about two-thirds, the number of jobs in coal mining was cut by more than four-fifths, and practically all of the coal export trade disappeared. Major factors contributing to this decline included (1) increasing competition from foreign coal in world markets, (2) depletion of Britain's own coal, which was

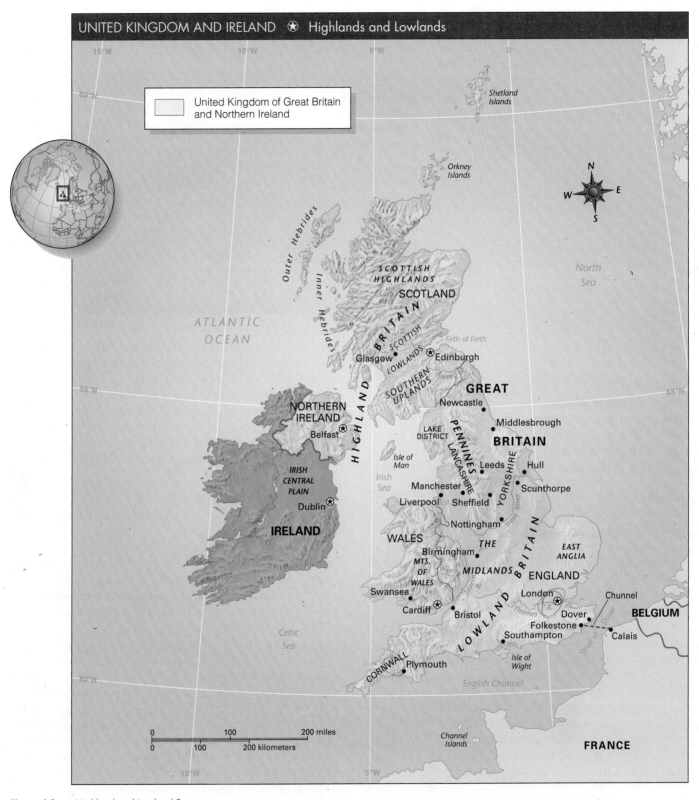

UNITED KINGDOM AND IRELAND ✪ Highlands and Lowlands

United Kingdom of Great Britain
and Northern Ireland

Figure 4.3   Highland and Lowland Britain.

the easiest and cheapest to mine (although large reserves of coal still remain), (3) the environmental unattractiveness of coal, and (4) increasing substitution of other energy sources for coal, both in overseas countries and in Britain itself. In the late 1960s and the 1970s, large-scale development of newly discovered oil and gas fields underneath the North Sea revolutionized Britain's energy situation. By the 1980s, once-dominant coal's principal remaining role in Britain had been reduced to supplying fuel for generating domestic electricity, and even that is under steady competition from natural gas

Figure 4.4　Fields in the English Lowland normally are fenced by hedgerows, seen here in southwestern England near the mouth of the River Severn. Hedgerow trees blend in the distance to give a misleading impression of forest. This landscape was created centuries ago by the Enclosure Movement, in which large landowners appropriated local "commons," or unfenced land tilled by villagers. Hedges were planted so that sheep could be raised under a system of controlled breeding. This closure of lands that had historically provided a safety valve for marginal peasant farmers helped promote the migration of peasant youth to the newly industrializing English cities.

JESSE H. WHEELER, JR.

and oil. This shift was also promoted by ever-increasing alarm over coal and industrial-environmental pollution in the U.K.

Iron and Steel

Steel was a scarce and expensive metal before the Industrial Revolution. Iron, not too cheap or plentiful itself, was more commonly used. It was made in small blast furnaces by heating iron ore over a charcoal fire, with temperatures raised by an air blast from a primitive bellows. Supplies of iron ore in Great Britain were ample for the needs of the time, but the island was largely deforested by the 18th century, and charcoal was in short supply. By 1740, Britain was importing nearly two-thirds of its iron from countries with better charcoal supplies, notably the American colonies, Sweden, and Russia.

This heavy import dependence then changed when 18th-century British ironmasters, led in the early 18th century by Abraham Darby when he was able to replace charcoal with coal and coke, made revolutionary changes in iron production. There was improvement of the blast mechanism; the invention and use of a refining process called puddling, which made iron more malleable; and the adoption of the rolling mill in place of the hammer for final processing. By 1855, Britain was producing about one-half the world's iron, much of which was exported. Further advances in production resulted when British inventors developed the Bessemer converter and the open-hearth furnace in the 1850s. Steel, a metal superior in strength and versatility to iron, could now be made on a large scale and inexpensively for the first time.

The early iron industry in the United Kingdom used domestic ores (mostly located in the Midlands of Great Britain), but these eventually became inadequate, and the U.K. now relies almost entirely on high-grade imported ores. And although British inventors and entrepreneurs led the world into the age of cheap, mass-produced, and extensively used iron and steel, the country lost its leadership in global totals of

steel production by the late 19th century. The United States and Germany surpassed the British output before 1900. By 2000, the EU produced a total of more than 176 million tons of steel, with Germany producing 51.1 million tons — the fourth highest global output. China led world production with 127.2 million tons, the United States had 101.5 million tons, and Japan produced 106.4 million tons in 1998. The United Kingdom had fallen far down the rankings, producing 15.2 million tons of steel in 2000.

Cotton Milling in Lancashire

During the period of Britain's industrial and commercial ascendancy, cotton textiles were its leading export by a wide margin, amounting to almost one-quarter of all exports by value in 1913. The enormous cotton industry eventually became concentrated in Lancashire, in northwestern England, with Manchester and Liverpool now seen as one metropolitan area with a total population of 3.6 million in 2001, with 395,000 in Manchester city proper (Fig. 4.5). Imported raw cotton was the basis for this industry, and manufactured textiles were exported through the Mersey River port of Liverpool (population: city proper, 468,300), although Manchester itself became a supplementary port after completion of a ship canal to the Mersey Estuary in 1894.

A hand-labor textile industry in Lancashire dated from the Middle Ages and found itself involved in the Industrial Revolution in several lines of development. First, a series of inventors, in good part Lancashire men, mechanized and sped up the spinning and weaving processes. Then, power was applied to drive enlarged and improved machines, first using falling water from Pennine streams and then steam fueled by coal from underlying Lancashire coal fields. In 1793, the invention of the cotton gin by American Eli Whitney provided a machine that separated seeds from raw cotton fibers economically, making cotton a relatively cheap material from which

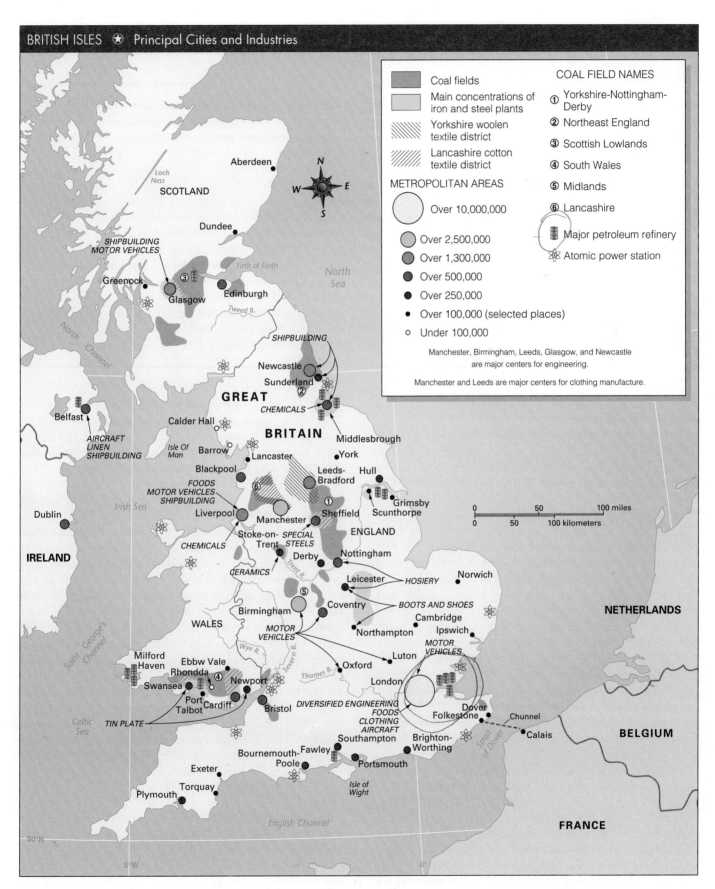

BRITISH ISLES ★ Principal Cities and Industries

Coal fields

Main concentrations of iron and steel plants

Yorkshire woolen textile district

Lancashire cotton textile district

METROPOLITAN AREAS

Over 10,000,000

Over 2,500,000

Over 1,300,000

Over 500,000

Over 250,000

Over 100,000 (selected places)

Under 100,000

COAL FIELD NAMES

① Yorkshire-Nottingham-Derby

② Northeast England

③ Scottish Lowlands

④ South Wales

⑤ Midlands

⑥ Lancashire

Major petroleum refinery

Atomic power station

Manchester, Birmingham, Leeds, Glasgow, and Newcastle are major centers for engineering.

Manchester and Leeds are major centers for clothing manufacture.

Loch Ness

SCOTLAND

Aberdeen

Dundee

SHIPBUILDING MOTOR VEHICLES

Greenock

③ Glasgow

Edinburgh

Firth of Forth

Tweed R.

North Sea

North Channel

SHIPBUILDING

Newcastle

Sunderland ②

GREAT

CHEMICALS

Belfast

AIRCRAFT LINEN SHIPBUILDING

Calder Hall

BRITAIN

Middlesbrough

Isle Of Man

Barrow

Lancaster

York

Blackpool

⑥

Leeds-Bradford

Hull

Dublin

FOODS MOTOR VEHICLES SHIPBUILDING

Irish Sea

Liverpool

Manchester

① Sheffield

Grimsby

Scunthorpe

ENGLAND

IRELAND

CHEMICALS

Stoke-on-Trent

SPECIAL STEELS

Derby

Nottingham

CERAMICS

Trent R.

Leicester

HOSIERY

Norwich

⑤

Coventry

BOOTS AND SHOES

Cambridge

Birmingham

Northampton

Ipswich

WALES

MOTOR VEHICLES

Luton

MOTOR VEHICLES

Saint George's Channel

Milford Haven

Ebbw Vale

Rhondda

④

Oxford

Thames R.

London

Wye R.

Swansea

Newport

Severn R.

NETHERLANDS

Port Talbot

Cardiff

Bristol

DIVERSIFIED ENGINEERING FOODS CLOTHING AIRCRAFT

Dover

Folkestone

Chunnel

Celtic Sea

TIN PLATE

Bournemouth-Poole

Fawley

Southampton

Brighton-Worthing

Portsmouth

Calais

BELGIUM

Exeter

Isle of Wight

Strait of Dover

Torquay

Plymouth

English Channel

FRANCE

50°N

5°W

Figure 4.5   Principal cities, coal fields, and industries of the British Isles.

One of the rural changes that played a role in the stimulation of rural youth migration to the city was a shift in rural land use called enclosure (also spelled inclosure). This was the establishment of fences, hedges, ditches, or other barriers around the perimeters of farm fields (see Fig. 4.4). The process began in the 12th century in England, but it became widespread in the mid-18th and early 19th centuries. Landowners realized that, with enclosure, they could make more money on their animal herds, especially sheep, by controlling breeding and, hence, producing better wools for the rapidly expanding international textile trade.

Enclosure in the countryside also provided abundant, cheap labor for new city industries because this act eliminated or severely restricted peasant access to lands traditionally held in commons for those farmers who had little or no land for themselves. The commons had served traditionally as unfenced lands that all

farmers could run animals on or even plant. For the farmers who had counted on access to the commons, enclosure meant that an already marginal agricultural operation became even more untenable.

This marginal peasant population became very important to the growing urban textile centers. These rural youth, particularly the youngest children of land-poor families, came to the factory centers and found themselves pressed into factory work in difficult situations. Although the full set of influences on this critical demographic shift is quite complex, the act of enclosing the rural commons did stimulate and expand a major new migration stream — the demographic shift from the countryside to the city — in Britain as farm youth moved to the cities. It also led to higher levels of agricultural productivity because the enclosed fields were farmed more thoroughly and intensively. Enclosure, in various forms, spread to other nations of the European Coreland.

inexpensive cloth could be manufactured. Industrial growth acted like a magnet to bring farm youth to the cities in quest of factory labor. These new factory complexes that shone in the soot of England's expanding cities were difficult environments. They did, however, provide alternatives to unemployment and underemployment in the countryside, and they were seen by many youth as the only means of breaking patterns with the past and getting into the city.

For more than a century, until World War I, Lancashire dominated world trade in cotton and cotton textiles. But 1913 saw the peak of British production; after that, the decline of the industry was nearly as spectacular as its earlier rise. The root of the trouble was increased foreign competition. Throughout the 20th century, country after country surpassed the United Kingdom in cotton textile production and exports, with a resulting slide in employment and great economic distress in Lancashire. Part of the shift comes from the relatively lower labor costs in less developed parts of the globe. Another factor in the shift is the ease with which new textile machinery can be imported and set into productivity. This has been particularly true in China, South Korea, and other parts of East and Southeast Asia. Lancashire's future as an industrial area lies mainly with other industries such as engineering (a general name for a wide range of industries that produce machinery, vehicles, and tools), chemicals, and clothing.

Shipbuilding on the Tyne and Clyde

When wood was the principal material used in building ships, Great Britain, as a major seafaring and shipbuilding force of global significance, eventually created serious difficulties for itself due to deforestation. Large quantities of timber were imported from the Baltic area and from North America, and the importation of completed ships from the American colonies,

principally New England, was so great that an estimated one-third of the British merchant marine was American-built by the time of the American Revolution. During the first half of the 19th century, American wooden ships posed a serious threat to Britain's commercial dominance on the seas.

In the later 19th century, iron and steel ships propelled by steam transformed ocean transportation. By the end of the century, Britain led the way and built four-fifths of the world's seagoing tonnage. Two districts developed as the world's greatest centers of shipbuilding: the Northeast England district, with a great concentration of shipyards along the lower Tyne River between Newcastle (population: metropolitan 1.2 million; city proper, 187,500) and the North Sea; and the western Scottish Lowlands, with shipyards lining the River Clyde for miles downstream from Glasgow. Both districts had a seafaring tradition (Newcastle in fishing and the early coal trade and Glasgow in trade with America), both were immediately adjacent to centers of the iron and steel industry, and both had suitable waterways for location of the shipyards. Their ships were built mainly for the United Kingdom's own merchant marine, as they still are, but even the minority built for foreign shipowners were an important item in Britain's exports. Like the other three industries of early importance and current decline in the U.K. shipbuilding has also lost major economic significance.

The Industrial Revolution and the Growth of Cities

The Industrial Revolution gave a powerful impetus to the growth of cities at the same time that Britain's total population was increasing rapidly. In 1851, the United Kingdom became the world's first predominantly urban nation, and it remains so today. Most of the large cities, with the notable exception of

London — and to some degree, Birmingham — are on or near the coal fields that supplied the power for early industrial growth. The contemporary transportability of electricity, natural gas, and oil has made it unnecessary for industries to locate near coal now, and recent development tends to favor the relatively more pleasant environments of southern England.

Food and Agriculture in Britain

A major characteristic of the United Kingdom's agriculture is its high concentration on grass farming for milk and meat production. Meat and, to an even greater degree, fresh milk are traditionally more difficult and expensive to transport than grains and many other vegetable products; consequently, they afford local producers a degree of advantage over competitors located greater distances from the market.

Nearly 2% of the working population of Great Britain continues to be employed in agriculture, forestry, and fishing. A little more than a quarter of the total land area is given over to crops, with nearly half used for pasture and rough grazing. Even with such a broad arable base, more than 60% of Britain's food supply is imported.

Total food production has increased sharply since World War II, mainly an outcome of various government policies aimed at reducing imports. A large increase in wheat and barley production has been fostered by subsidies to farmers and has been expensive. However, despite the high cost of subsidies, Britain's membership in the Common Market, and now the European Union, probably will continue to foster the tendency to expand British agricultural production. Up to the late 1990s, the agricultural policy of the EU emphasized subsidies to agriculture, including generally high import tariffs on farm products from non-EU countries.

As long as this policy holds, British farmers will continue to receive incentives in the form of relatively high and protected prices, and the British people, like the other peoples in the European Union, will eat more expensive food, produced to a greater degree within the home country or at least within the EU. This agricultural aspect of EU policies will continue to foment political tension as other countries continue to try to compete with EU-produced farm products.

Great Britain's fishing industry has been in steady decline since the 1960s. EU policies have led to fishing restrictions and a decline in catch, but even with this, Britain possesses the largest fishing fleet in the EU. Major ports still central to fisheries include Hull, Grimsby, Fleetwood, and Plymouth in England and Peterhead in Scotland.

The English Channel Tunnel. The 31-mile (50-km) three-tube underwater tunnel that links southern Britain with Calais, France, is also called the Chunnel or Eurotunnel. It opened for commercial travel in 1994 (a year and a half later than its scheduled opening), but discussion about the possibility of a tunnel link between the British Isles and the European continent began 140 years earlier during the reign of Britain's Queen Victoria. The estimated cost for this structure, which houses two commercial rail lines and a service shaft, was set at $7.7 billion in the initial 1987 plans. The final cost came to more than $15

billion in 1994, making it the most expensive single building project in human history. It is now possible to board a train in London and arrive in central Paris in 3 hours, traveling at speeds that can reach 200 miles per hour. In 1997, design started on the 16-mile (26-km) twin-tube tunneling that will fully complete the high-speed connection between Paris and London.

The Channel Tunnel has opened a whole new era of speculation about the impact of reducing the insularity of Britain, a geographic characteristic that the British have long seen as central to their traditional ability to chart a history somewhat independent of Western Europe. However, this opening to the vital landscapes of France, Germany, and the Low Countries has the potential of significant economic expansion not only for London but for Birmingham and other British manufacturing (and tourist) centers as well (Fig. 4.6).

4.3 Ireland: Technology as Change Agent in National Prosperity

Ireland is a land of hills and lakes, marshes and peat bogs, cool dampness and verdant grassland. England completed an effective conquest of Ireland in the 17th century and subsequently made the island a formal part of the United Kingdom. Ireland's status, in reality, was virtually colonial. Most of the land was expropriated and divided among large estates held by English landlords with Irish peasants functionally tenants on these estates. Whatever income was gained in this rule generally found its way to Britain, and Ireland became a land of deep poverty, sullen hostility, and periodic violence.

The island consists of a central plain surrounded on the north, south, and west by hills and low rounded mountains. The Republic of Ireland, which occupies a little over four-fifths of the island and is included in the definition of the term "British Isles," is in the midst of a transition from an agricultural to an industrial economy. Manufacturing is now a larger employer than agriculture and supplies more exports by value. But the transition is far from complete. However, agriculture is still more important as an employment sector in Ireland than in most countries in Western Europe, and the Republic is a considerable exporter of agricultural commodities and an importer of manufactures.

In per capita income, Ireland ranked near the bottom of Western Europe's countries for the past decades, but recent economic growth has begun to change this pattern. In the most recent data, Ireland has gone beyond the U.K. in per capita GDP/PPC (see Table 3.1). By the end of 2000, Ireland's unemployment rate had been dropped from double digits to approximately 8%. At the same time, the per capita GNP had been raised to $17,900, compared to just under $5,000 in 1986 (see Table 3.1). The expanding role of newly developed industrial facilities, tourism, and high-technology innovation and production all worked together to give at least temporary stimulus to Ireland's economic growth. In 1998, for example, the export of software from Ireland had a value of more than $3 billion dollars, with "high tech industry

T3.1 62

In geography, the factor of location is almost always a significant element of a city's, or of a business's, success or failure. Here the city of London is studied in terms of both location and changing production technologies. The factors that are most relevant to the London details have global perspectives. As you read the script for London, learn this set of patterns in a broader way because the same conditions operate all across the European scene, and even farther.

Perspective I: The Role of Specific Location. Cities are almost always where they are because of the presence of some particular geographic characteristic(s). These might include a ford across a river, a strategic pass, the location of a resource, the intersection of significant migration paths, or even a religious or historic site. London, as one of the most dominant of European cities, is characterized by a complex set of urban location dynamics that are worth some detailed consideration. Although what follows is the story of London, it is important to realize that many aspects of this development would apply to Paris or Brussels or Berlin. Use this chronicle as a set of specifics that has broad implications for urban positioning across the globe.

The London metropolitan area, with more than 11 million people and 6,574,000 in the city proper, sprawls over a considerable part of southeastern England (Fig. 4.A). Its historical background is unique in that it was already a major city before the Industrial Revolution, and its economy is not based so much on industry. Although the metropolitan area contains a great deal of manufacturing, this sector does not dominate. Instead, it shares leadership with several other major elements, some of which developed before the Industrial Revolution and have greatly expanded since then. These include commerce, government, finance and insurance, and corporate administration. Basic elements more recently devel-

oped include communications media (publishing, television, films) and tourism.

London's early function as a seaport and commercial center laid the original foundation for its development. Trade generated population growth, profits, and financial operations, including insurance (which began with marine insurance covering risks to ships and cargoes). Population, trade, and wealth all responded to and further attracted government. Finance, trade, and government attracted corporate offices. The large, growing, and increasingly prosperous population provided both a labor force and a large market for manufacturing. Dominant geographic factors that have shaped this urban port development include:

1. Location relatively near major trading partners. London is located in the corner of Britain nearest the European continent, which was the most important area with which England traded before the Age of Discovery and which is still of dominant importance to British trade today.
2. Location in a productive agricultural area. London is located in the fertile English Lowland, which was the most productive and populous part of Great Britain in the preindustrial period, supplying most of the wool that was Britain's main export and providing most of the British market for imports.
3. Location facilitating transport connections with the surrounding region. From London, a major route to the interior, with branches, was provided by the Thames River and its tributaries. Before the age of the 19th-century railroad and the 20th-century truck, water transport had greater advantages over land transport than it has today, and these natural inland waterways were used extensively for transporting goods to and from Lon-

don. During the Industrial Revolution, their utility was enhanced by canals (little used today) that connected the Thames system with other British rivers. Furthermore, London became a major junction for roads as early as Roman times, and again the city's location provided advantages. Originally, tidal marshes fringed the lower Thames, forming a major barrier to transportation. Highway networks could avoid the marshes by converging on London and crossing the river there.
4. Location on a river (and estuary) permitting relatively deep land penetration by ocean ships. A seaport set inland in such a manner can serve more trading territory in the surrounding countryside, within a given radius, than can a port on a comparatively straight stretch of coast or on a promontory. Dense urban development surrounds the port of London on all margins of the Thames (Fig. 4.B). Not only does the river provide good channel access for oceangoing ships, but London's network of highway connections to the rest of Britain facilitates port activities that link the island with global trade. The inland location of London's port tends to maximize the distance goods can be carried toward many destinations by relatively cheap water transportation (ocean vessels) and by minimizing the more expensive land transport.

Although London, like many other ports, originally developed well inland, later growth of the port has been downstream toward the ocean. Although ships have been growing larger for centuries, in the industrial age they have been growing at an accelerated pace, especially with the advent of oil supertankers and other very large ships in the last few decades. In addition, more efficient loading and unloading methods, such as containerization, require

(Continued)

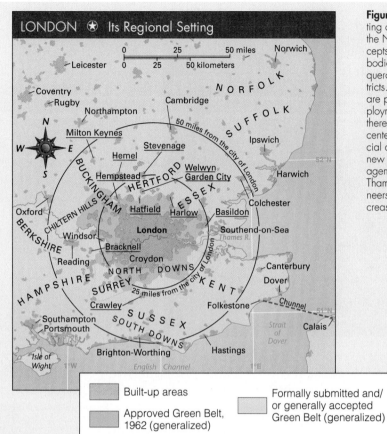

LONDON ✪ Its Regional Setting

Figure 4.A These maps of London illustrate the structure and setting of a major world metropolis. Look at them in conjunction with the New York maps panel in Chapter 24 to examine general concepts of urban geography and the relationships of urban location to bodies of water. The City of London, built originally by Roman conquerors of Britain, contains one of the world's greatest financial districts. The Green Belt limits further urban sprawl, and the new towns are planned communities built to provide residential areas and employment for migrants from London proper to reduce congestion there. The West End is the governmental, hotel, and entertainment center and houses the Millennium Dome. The docks, built as artificial anchorages for shipping, have been closed but are finding new life as tourist and artist venues. There is an ingenious river management construction called the Thames Barrier that lies across the Thames just upstream from Tilbury Docks. These gates enable engineers to close the river to rising floodwaters on command, a task increasingly significant because of subsidence in the Thames Estuary.

Built-up areas

Approved Green Belt, 1962 (generalized)

Formally submitted and/ or generally accepted Green Belt (generalized)

New towns are underlined.

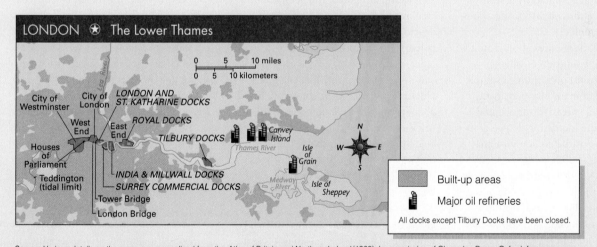

LONDON ✪ The Lower Thames

Built-up areas

Major oil refineries

All docks except Tilbury Docks have been closed.

Source: Various details on the maps are generalized from the *Atlas of Britain and Northern Ireland* (1963), by permission of Clarendon Press, Oxford; from a government map of the green belt on an ordnance survey base, Crown Copyright; and from maps in J. T. Coppock and High C. Prince (eds.), *Greater London*, Faber and Faber, 1964.

large areas of relatively open ground along the waterfront for tracking and moving containers, and such areas cannot generally be found in the congested older ports built up for earlier ships and methods. Instead, new wharfing areas must be sought in the areas downstream from the original port. Thus, a large share of the port and industrial facilities in the older part of the London port have closed, with much loss of employment, whereas new port areas, using modern methods that require fewer people, are being developed even closer to the mouth of the Thames estuary.

Perspective II: London and the Example of Industrial Shift. As in the case of urban location dynamics, the city of London serves well as an example for

(Continued)

Figure 4.B   London on the Thames River in southeastern England is a major node in the worldwide system of places that geographers study. Tower Bridge is at the lower left, with new London Bridge next upstream. The financial district (City of London) is at the right center, adjacent to London Bridge.

another common economic and geographic dynamic characterizing nearly the past half century: industrial and commercial relocation.

Industrial difficulties in older industrial areas and countries, of which Britain is the most outstanding example, are a product of many circumstances. The most vital, at least for the United Kingdom, is the increasing degree of international industrial competition that developed in the 20th century, especially since World War II. Intensified competition is partly the logical outcome of the continued spread of industrialization around the world, accentuated since the 1960s by the rapid improvement of communications and transportation. This brought a precipitous drop in the costs both of maintaining contact with distant production facilities and markets and of shipping materials, subassemblies, and finished products for long distances. Distance from competitors no longer offers much protection, and distance from markets is no longer much of a handicap. Competition has also been accentuated by the systematic and scientific development of new products with the potential to compete with and replace older ones. The most significant 20th-century example is the replacement of older means of transport, particularly rail connections, by the automobile and truck and the attendant replacement of coal by oil. Another is the development of synthetic fibers, made from wood or petroleum, to compete with cotton and wool.

In the strenuous competition for global markets, older industrial areas tend to lose to newer ones. Major reasons for this decline include:

1. Loss of a resource advantage. A common example is the diminishing significance of local coal. Location on a coal field was the main basis for the early development of many older districts, but that advantage has been lost because of (a) depletion of the local coal, (b) the rise of lower cost coal fields elsewhere, or (c) replacement of the coal by oil or natural gas.

2. Cheaper labor in newer areas. In older districts, there is time for unionization to occur and to increase wages, whereas a new area is likely to have an unorganized work force to whom even low wages seem a great improvement compared to urban unemployment or rural underemployment. With lower wages, prices of the manufactured product can be lower.

3. Technological advantages of newer areas. Such areas can be developed with the newest and most efficient technology. In older areas, there is a large investment in older plants and equipment, which is difficult to write off, scrap, and replace with the latest technology.

4. Unemployment resulting from the efforts of older industries to survive. Even if an industry in an older area

(Continued)

survives the competitive struggle, unemployment and hardship in the community often result because the industry is likely to survive only through increased mechanization, automation, and the use of robots. In some cases, strong unions may oppose the new technology because it provides fewer jobs.

5. Conservatism of management. After years or decades of success, there is often a tendency for the management as well as the workers in an older industrial area to be understandably resistant to compelling change in manufacturing. There is a human tendency to be complacent in the belief that procedures that worked in the past will continue to succeed in the future. Many charges of this nature have been leveled against established management teams not only in Britain but anywhere such dislocation occurs.

As the process of dislocation takes place in one sector of the economy, there is often a parallel expanding employment sector. This is currently taking place in the service sector — a term embracing all workers not directly involved in producing material goods. Apparently, this trend must continue if an increase in employment is to be enjoyed: Agriculture has fallen to low levels of employment, and industrial jobs are being lost to competition and to automation. Increasingly, Britain — along with other industrial nations — has become a country of white-collar workers.

The expanding service sector that employs them is highly varied in its makeup and includes government, finance, transportation, insurance, health services, management, sales, education, research, personal services, law, tourism, and communications media.

There are many unanswered questions about this service sector-oriented, "postindustrial" society toward which the industrialized Western nations are moving. Such questions are focused on:

1. The role of exports. In a global trading world, will this transition be fully competitive?

2. Capacity of the service sector. Can the service industries provide an adequate supply of exports needed in international trade? The United Kingdom long supported itself in part by "invisible exports" such as capital, financial services, insurance services, ocean shipping, airline services, and tourism, and other Western nations are now involved in such activities. But a productive industrial sector remains important to thriving economies. Just as important, can modern service jobs, which are sometimes described as knowledge-worker jobs, be expanded rapidly enough to supply high employment in the face of declining industrial employment?

3. Income potential in the service sector. Can service jobs be made productive enough to provide high incomes to those holding them? There is reason to doubt the ability of many parts of the service sector to increase productivity at rates comparable to those of large-scale and highly mechanized agricultural or manufacturing industries. If such doubts prove justified, the countries with a high dependence on services are likely to experience relatively slow economic growth unless they are able to create a service sector that is based more on information management and less on fast food and tourism.

4. Location of the service sector. Will new service employment locate in the afflicted old industrial areas where it is especially needed, or will it grow mainly elsewhere? In Britain, the primary growth of such employment, as with employment in newer forms of manufacturing, has taken place in London and surrounding parts of southeast England rather than in the depressed coalfield industrial areas. This has accentuated the sharp contrast between the relatively prosperous London region and the rest of the country.

The answers to the foregoing questions will go far toward determining the future of Western countries both within and outside Europe. The United Kingdom will be a key country to watch because the process of change is relatively far along there, as it has been progressing for more than two and a half centuries, and is proceeding rapidly.

---

account[ing] for between a quarter and a third of Ireland's gross domestic product." This total exceeded software exports from the United States in the same year.[1]

This major surge of industrial development has been accomplished under a governmental program designed to attract foreign-owned plants by a combination of cheap labor, tax concessions, and help in financing plant construction. Hundreds of plants have been attracted, although most of them are rather small. They are owned mainly by U.S., British, and German companies, and their major products are processed foods, textiles, office machinery, organic chemicals, and clothing. Most of the new plants have been built in and around the two main cities—Dublin (population: metropolitan area, 986,600) and Cork (population: 186,400)—and in the vicinity of Shannon International Airport in western Ireland.

[1] Peter Ford, "Irish Prosperity Has Its Downside." *Christian Science Monitor,* February 6, 2001, 1, 7.

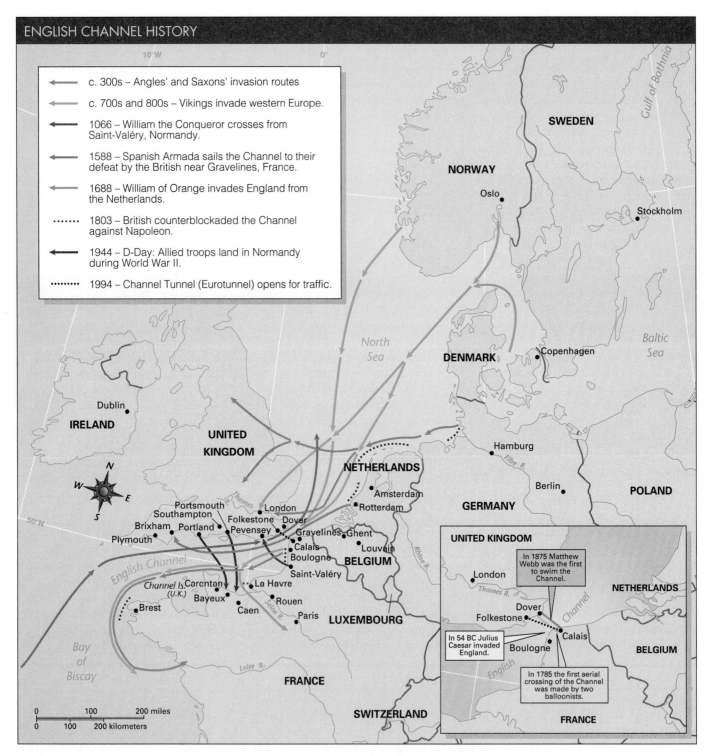

ENGLISH CHANNEL HISTORY

◄── c. 300s – Angles' and Saxons' invasion routes

◄── c. 700s and 800s – Vikings invade western Europe.

◄── 1066 – William the Conqueror crosses from Saint-Valéry, Normandy.

◄── 1588 – Spanish Armada sails the Channel to their defeat by the British near Gravelines, France.

◄── 1688 – William of Orange invades England from the Netherlands.

••••••• 1803 – British counterblockaded the Channel against Napoleon.

◄── 1944 – D-Day: Allied troops land in Normandy during World War II.

•••••••• 1994 – Channel Tunnel (Eurotunnel) opens for traffic.

UNITED KINGDOM

In 1875 Matthew Webb was the first to swim the Channel.

In 54 BC Julius Caesar invaded England.

In 1785 the first aerial crossing of the Channel was made by two balloonists.

Figure 4.6   This map shows the great variety of ways in which the English Channel has been challenged and, generally, overcome across the centuries. The current rail traffic through the Channel Tunnel (also called Eurotunnel and Chunnel) adds a new dimension to this map of interaction between the Continent and the island of Great Britain.

The development of new high-tech industries has been a critical aspect of Ireland's rapid economic growth in the 1990s. Partly because of a highly developed work force, some 25% of the country's exports in the late 1990s came from electronic products. In terms of overall growth, Ireland increased its manufacturing output during the 1990s on a par with the strongest global growth patterns. In terms of West Central Europe, it became the showcase of growth as its GDP increased above the levels of more fully developed nations. The remarkable increase of per capita GDP reflects the fact that Ireland has had one of the most impressive rates of economic growth in the European Coreland in the past half dozen years.

With now only 9% of its labor force on farms, the Republic of Ireland is loosening its traditional dependence on agriculture to a marked degree. Although the large estates of the period of English rule are gone and the Irish farmer now tends to be a landowner, there continues to be classic rural poverty. Farms tend to be both small and undermechanized, but even they have taken on some additional value as Ireland has become a major tourist destination. Most important, though, have been the changes worked by the growth of high-tech industries and services in Ireland during the past decade. Such an economic shift, although slowed by the global technology slowdown in the first years of this century, has accelerated the continuing migration of Irish youth away from the countryside and into the city.

Northern Ireland

The dominantly Protestant North of Ireland, now known as Northern Ireland, is a part of the United Kingdom. Shortly after the end of World War I, the island of Ireland was divided into a largely Protestant Northern Ireland and the Republic of Ireland, which was almost entirely Catholic. The Protestants were the descendants of Scottish Presbyterians and other British Protestants who were brought to the north of Ireland in the 17th century under the auspices of the British crown. They were intended to form a nucleus of loyal population in a hostile, conquered, largely Roman Catholic country. Protestant entrepreneurs were able to develop shipbuilding and linen textile industries in and near the North's main city, Belfast (population: metropolitan area, 493,500; city proper, 257,400). It was in Belfast that the English White Star Line had its luxurious ocean liner, *Titanic*, constructed and outfitted. These industries have declined recently, but Northern Ireland has been effective in attracting new industries seeking cheap labor despite recent political violence.

Ever since the 1960s, the religious tensions in Northern Ireland, also known as Ulster, have been manifest in car bombings, the shooting of British police and soldiers, and street bloodshed through clandestine bombings. In the spring of 1998, former Senator George Mitchell of the United States was central to an intense effort to move the Northern Ireland peace talks to a new level of productive interaction between the Catholics and Protestants on this corner of the island of Ireland. The three nations of Ireland, the United Kingdom, and the United States led an effort to finally resolve the bloody tensions that exist between Catholic Ireland and Protestant Northern Ireland. A document was signed by the antagonistic factions on Good Friday 1998. This document was given a majority vote of support by both the Irish Republican Army (IRA) and Ulster populations, leading people to think that perhaps the disquiet that had been so characteristic of Northern Ireland since its 1922 inception might be coming to an end.

However, such hopes were premature. A bloody bombing in a crowded marketplace in Omagh, 60 miles (96 km) due west of Belfast, in August 1998 was, in fact, the event that most completely set the stage for cautious talks between the two sides split by long-time religious enmity. Both sides declared a cease-fire after the Omagh bombing, and political progress toward open discussions and resolution of the problem of clandestine arms and violence began in earnest.

On December 2, 1999, there was further political movement when the U.K. yielded direct rule of Northern Ireland. On this day, the Northern Ireland Assembly, which had been established with the signing of the 1998 Good Friday Agreement, took over wider powers of self-government. While there is no certainty yet that the lingering issues of religion, tradition, and political autonomy will be clearly resolved in Northern Ireland, by mid-2001 there was greater optimism about such an outcome than had been felt for decades, although the profound change in world attitudes toward terrorism and bloody ethnic/religious activities after September 11, 2001, put every such act on a different platform than before.

4.4   The Many Worlds of France

France has long been one of the world's most celebrated countries. It is the largest country by area in Europe as defined in our text (see Table 3.1) and is Europe's leading nation in value of agricultural output. The country is one of the world's major industrial and trading nations and a nuclear power. Although France has lost its former significance as a major colonial power, it still maintains special relations with a large group of former French colonies. More than a dozen are now Departments of France and send representatives to the Assembly in Paris. Still another important foundation for France's international importance is the high prestige of French culture in many parts of the world. It is a country of high definition and high self-esteem.

Topography and Frontiers

France lies mostly within an irregular hexagon framed on five sides by seas and mountains (Fig. 4.7). In the south, the Mediterranean Sea and Pyrenees Mountains form two sides of the hexagon; in the west and southwest, the Bay of Biscay and the English Channel form two sides; in the southeast and east, a fifth side is formed by the Alps and the Jura Mountains, and farther north, by the Vosges Mountains, with the Rhine River a short distance to the east. Part of the sixth or northeastern side of the hexagon is formed by the low Ardennes Upland, which lies mostly in Belgium and Luxembourg, but which has broad lowland passageways leading into France from Germany both north and south of the Ardennes.

Aside from the large Massif Central in south central France, all the country's major highlands are peripheral. Even the Massif Central, a hilly upland with low mountains in the east, is skirted by lowland corridors that connect France's extensive plains of the north and west with the Mediterranean littoral. These corridors include the Rhône-Saône Valley between the Massif to the west and the Alps and Juras to the east and the Carcassonne Gap between the Massif and the Pyrenees. Thus, France is a country with no serious internal barriers to movement, however, it does have barriers at its borders.

The state that emerged as modern France was to a considerable degree a natural fortress, as two of its three land frontiers lay in rugged mountain areas. The Pyrenees are

FRANCE ✵ Index Map

GREAT BRITAIN

NETHERLANDS

GERMANY

Thames R.   Strait of Dover   Chunnel

Folkestone   Dunkirk

FLANDERS

BELGIUM

Calais

Boulogne   Bruay   Lille

Lens-Hénin-Liétard   Valenciennes

Douai

Sambre R.

LUXEMBOURG

ARDENNES

Meuse R.

Dieppe   PICARDY   *Somme R.*

Cherbourg   Rouen

Bristol Channel

English Channel

COTENTIN

Channel Islands (U.K.)

Caen   Le Havre   *Seine R.*   *Oise R.*   *Aisne R.*   Reims   Verdun   Thionville   SAAR   Metz

48° N

Brest   NORMANDY   PARIS   BASIN   Paris   *Marne R.*   Nancy   LORRAINE   SAVERNE GAP   VOSGES   Strasbourg   UPPER RHINE PLAIN

BRITTANY   Rennes   Chartres   *Yonne R.*   CHAMPAGNE   *Moselle R.*   BLACK FOREST

Concarneau   *Rance R.*   Le Mans   *Aube R.*

Lorient   NORTHERN LIMITS OF VITICULTURE   Orléans   BURGUNDY   Mulhouse   *Rhine R.*

Vilaine R.   Belfort   BELFORT GAP   Basel

Saint-Nazaire   *Loire R.*   Tours   Dijon   ALSACE

Nantes   *Vienne R.*   *Loire R.*   SWITZERLAND

POITOU HILLS   Poitiers   Le Creusot   JURA MTS.

Bay of Biscay   *Charente R.*   Vichy   *Allier R.*   RHONE-SAONE VALLEY   Mt. Blanc 15,771 ft. (4807 m)   ALPS

Limoges   Clermont-Ferrand   Lyon   SAVOY

Cognac   Saint-Étienne   Grenoble

Gironde R.   *Dordogne R.*   MASSIF CENTRAL   ITALY

Bordeaux   *Lot R.*

SOUTHWESTERN

100 miles
100 kilometers

44° N   Roquefort   LOWLAND   *Garonne*   Nîmes   Avignon   Nice   MONACO   44° N

4° W   *Adour R.*   Arles   Cannes   RIVIERA

Biarritz   Toulouse   PROVENCE   Marseilles

BASQUE COUNTRY   Carcassonne   LANGUEDOC   Golf of Fos   Toulon

PYRENEES   CARCASSONNE GAP

SPAIN   Pic d'Aneto 11,168 ft. (3404 m)   Mediterranean Sea

ANDORRA   4° E   8° E

URBAN AREAS

✵ Over 10,000,000 (national capital)

◉ Over 1,000,000

● Over 600,000

○ Over 275,000

∘ Over 100,000 (selected places)

• Selected smaller places

City-size symbols are based on metropolitan area population estimates.

LANDFORMS

Lowlands

Hills and uplands

Low mountains

Swiss Plateau

High mountains

Figure 4.7   General location map of France.

LANDSCAPE IN LITERATURE

The Walking Drum

This excerpt from Louis L'Amour's *The Walking Drum* is presented to provide a sense of how big a role trade and markets played in European life almost 1,000 years ago. Although set in France, the images and the lore captured in this scene spill over into the rest of Mediterranean Europe and other regions as well. As you read these paragraphs, think about the routes that had to be traveled to bring all of these goods and people together in this market setting. Think about the role of change agent played by the vendors and the soldiers who traveled with the merchant caravans from the distant source points for many of these goods.

The fair at Provins was one of the largest in France during the twelfth century. There was a fair in May, but the most important was that in September. Now the unseasonably cold, wet weather had disappeared, and the days were warm and sunny.

Long sheds without walls covered the display of goods. Silks, woolens, armor, weapons, leather goods, hides, pottery, furs, and every conceivable object or style of goods could be found there.

Around the outer edge of the market where the great merchants had their displays were the peasants, each with some small thing for sale. Grain, hides, vegetables, fruit, goats, pigs, and chickens, as well as handicrafts of various kinds.

Always there was entertainment, for the fairs attracted magicians, troupes of acrobats, fire-eaters, sword-swallowers, jugglers, and mountebanks of every kind and description.

The merchants usually bought and sold by the gross; hence, they were called grossers, a word that eventually came to be spelled grocer. Dealing in smaller amounts allowed too little chance for profit, and too great a quantity risked being left with odds and ends of merchandise. The White Company had come from Spain with silk and added woolens from Flanders. Our preferred trade was for lace, easy to transport and valued wherever we might go.

Merchants were looked upon with disdain by the nobles, but they were jealous of the increasing wealth and power of such men [as the merchants in this chapter], or a dozen of others among us.

The wealth of nobles came from loot or ransoms gained in war or the sale of produce from land worked by serfs, and there were times when this amounted to very little. The merchants, however, nearly always found a market for their goods.

At the Provins fair there were all manner of men and costumes: Franks, Goths, Saxons, Englanders, Normans, Lombards, Moors, Armenians, Jews, and Greeks. Although this trade was less than a century old, changes were coming into being. Some merchants were finding it profitable to settle down in a desirable location and import their goods from the nearest seaport or buy from the caravans.

Artisans had for some time been moving away from the castles and settling in towns to sell their goods to whoever passed. Cobblers, weavers, coppers, and armorers had begun to set up shops rather than doing piecework on order. The merchant-adventurers were merely distributors of such goods. . . .

The Church looked upon the merchants with disfavor, for trade was considered a form of usury, and every form of speculation considered a sin. Moreover, they were suspicious of the fartraveling merchants as purveyors of freethinking.

Change was in the air, but to the merchant to whom change was usual, any kind of permanence seemed unlikely. The doubts and superstitions of the peasants and nobles seemed childish to these men who had wandered far and seen much, exposed to many ideas and ways of living. Yet often the merchant who found a good market kept the information for his own use, bewailing his experience and telling of the dangers en route, anything to keep others from finding his market or his sources of cheap raw material.[a]

[a]Louis L'Amour. *The Walking Drum*. Toronto: Bantam Books, 1984, 256–258.

F21.4
556

a formidable barrier, lowest in the west, where the fiercely independent Basques, who are neither French nor Spanish in language, occupy an area extending into both France and Spain. In southeastern France, the Alps and Jura Mountains follow France's boundaries with Italy and Switzerland. The sparsely populated Alpine frontier between France and Italy is even higher than that of the Pyrenees: Mont Blanc, the highest summit, reaches 15,771 feet (4,807 m). The number of important routes through the mountains is limited, and the Alps have been an important defensive rampart throughout France's history. The Jura Mountains, on the French-Swiss frontier, are lower but also difficult to cross, as they are arranged in long ridges separated by deep valleys.

France's Flourishing Agriculture

Agriculture is a more important part of the economy in France than in all but a very few of the most highly developed countries. Some 60% of the French landscape is agricultural, and another 30% is in woodland. In terms of global agricultural importance, France, in 1999, was the second largest exporter of wine, third of milk and butter, third of meat, sixth of eggs, second of wheat, first of barley, and fourth of sugar (from the sugar beet). Since World War II, rapid economic change has been accompanied by a massive migration from farms to cities. The number of farmers has fallen to the point that less than 5% of the country's labor force was in agriculture as of 1999. But this is still higher than in many developed countries, and although their numbers have fallen, France's farmers have become more efficient. Hence, agricultural output has increased greatly, which is a common experience in industrializing countries.

For a country with only four-fifths the area of Texas to attain such a large volume of farm exports, some human and physical advantages in agriculture must be present. One advantage often cited is the country's relatively small population compared to its land area. But one must remember that a density of 281 people per square mile is low only by Western European standards; it is almost four times the density of the United States (80 per sq mi). Also, France's membership in the

PROBLEM LANDSCAPE

The European Union and the Horror of Mad Cow Disease

Maybe this is how people behaved during the great plagues of the Middle Ages. Maybe one stricken community would hurl its infected bedclothes over the walls of the next town and then exult in pointing out that its neighbors were unclean. . . . Both the British and French press have taken pleasure in the news that there is one confirmed case of bovine spongiform encephalopathy (BSE) in a German-born cow. That is one case France has now had 190 cases of mad cow disease, more than half of them this year. Britain has had 180,000 cases, including 1,000 this year.[a]

This harsh alarm at the most recent news of the spread of "mad cow disease" is just one of many hundreds of articles in the European press on this frightening sickness. First discovered in cattle in Britain in 1986, it was thought that the disease had come from the presence of sheep offal in some of the protein and bone meal fed to British cows. Sheep have long been known to get a disabling disease called *scrapie*. It had been rare in cows. This sickness, although it had apparently made the transition from sheep to cows, was not known to be transmissible to humans. In 1986, it became illegal in Britain to use or export any feed made up of the remnants of BSE-infected cattle.

The most frightening aspect of this mad cow disease — and it is this aspect that is evoked in the opening lines of the newspaper quoted earlier — is that it is now believed that aspects of BSE can be transmitted to humans and the disease becomes manifest as a variant of a degenerative brain disorder in humans called Creutzfeldt-Jakob disease (CJD). The European Commission, the administrative arm of the European Union, banned all exports of cattle, beef, or beef products from the U.K. in 1996. Also banned were milk and milk products from BSE-infected cows. As might be expected, levels of beef consumption plummeted in the U.K., and even though public health officials continued to say that there was no danger to humans, beef became a flat item in markets, butcheries, and agricultural trade all through the EU.

With the increasing likelihood that BSE can cause a variant form of CJD in humans who have consumed beef, or even animals that have been fed infected beef remains, there arose a general alarm all across the European landscape. By the time the 16th human victim linked to BSE died in Britain, medical teams began a rush to figure out what the vectors were in this transmission. What role did travel, diet, eating patterns, or residence have in putting a person at risk for this offshoot of mad cow disease? The complexity of such a search leads to observations like this: "Then, says the doctor, you have to plot a map showing the supply lines for meat and retrace the paths that the supplies followed. This rarely leads us back to the local farmers. For example, the cafeteria did not get its supplies from the surrounding farms but from institutional suppliers that operate throughout England. The central slaughterhouses processed the meat, which was then distributed to the supermarkets, and so on. This is why we are fighting for a far-reaching investigation on a national scale. This is a very urgent matter. . . . . This will be helpful for other countries which, like France, will soon uncover the disease.'"[b]

By the beginning of 2001, mad cow disease had hit France, Germany, Switzerland, Portugal, Spain, Poland, Greece, and other nations. Every country was creating some sort of ban, but as one agricultural minister said, "Mad cow disease knows no borders but is moving from one member state to another." Another said, "Is it the cows, or have we gone mad?"

The panic that rippled through the EU as new testing showed a higher incidence of BSE came in part from the fear of the disease itself, partly from the hidden ways in which the spores of such illness could be delivered into one's life by everything from milk to meat to sausage, even to cosmetics and candy. A third facet of the fear was the realization that this ease of movement and diffusion through the European Union — in fact, through all of Europe — was a new and very real dimension of the global linkages that are ever more common in the contemporary world. The problem landscape for this fear was not simply the millions of cattle that have been purposefully killed to eradicate the disease, but the knowledge that there really was no fully local world anywhere anymore.

[a]John Lichfield, *The Independent*, London, November 27, 2000 (in *World Press Review*, February 2001, 22). Reprinted by permission.
[b]Annick Cojean, *Le Monde*, Paris, November 15, 2000, in *World Press Review*, February 2001, 24. Reprinted by permission.

European Union is a major advantage. This gives French farmers free access to a huge and affluent consuming market protected by tariffs from non-EU producers. In addition, the European Union budget has generally provided generous price supports at rates above world prices for many agricultural products, with French farmers being the leading beneficiaries.

France's physical advantages include superior topographic, climatic, and soil conditions. The main topographic advantage is the country's high proportion of plains. Large areas, especially in the north and west, are level enough for cultivation. Climatically, most of the country has the moderate temperatures and year-round moisture of the marine west coast climate. The wettest lowland areas are in the northern part of the country, and it is here that crop and dairy production is especially intense. Southern France often experiences notably warmer temperatures, allowing a sizable production of corn in the southwestern plains. Productive vineyards are also concentrated in the south and southwest, although important vineyard areas are found as far north as Champagne, east of Paris. Finally, northeastern France from the frontier to beyond Paris has fine soils developed from loess. As in much of Europe, these exceptionally fertile soils are used principally to produce wheat and sugar beets. This wheat belt in northeastern France is the country's chief breadbasket; the beet production from the same area provides not only sugar but also livestock feed (beet residue from processing) for meat and milk production. Wheat and sugar beets tend to occupy the best soils because both crops are unusually sensitive to soil fertility.

Since 1995, more than 4 million farm jobs have been lost in France, and the country is now approximately 80% urban. The average French farm size has grown to nearly 30 hectares (75 acres) through the remaining farmers expanding to absorb the unworked farmland. Though this total is about one-eighth the size of a U.S. farm, it does mean that it takes increasing capital to be able to stay afloat in such an operation. For the French, to see all farming becoming capital intensive, export-market-oriented, and controlled by the EU is quite unacceptable. The French continue to view themselves as an agricultural nation, holding the customs of the family farm central to their identity and their moral well-being. The economic growth resulting from EU membership has extracted its cost by increasing the migration of French youth away from family farms toward Paris and other cities. Meanwhile, the aging farm population goes farther into debt by purchasing new farm machinery that might enable them to increase productivity and stay competitive in the domestic, EU, and global marketplaces.

The family farm in France is much more than just an agricultural unit. It has long been a cultural institution and carries its tradition with an energy that is reflected both in the productivity of the fields and in the politics that surround French agriculture. One of the most celebrated efforts to diminish the forces of globalization, an economic force that is perceived as particularly ruinous to the family farm in France, was the late 1999 destruction of a new McDonald's being constructed outside Paris. Jose Bove, a French sheep farmer, long active in efforts to slow down the French accommodation and support of global expansion of American influences in agricultural issues, led a militant farming union called the Peasants Confederation in this civil disobedience. The French have made Bove a national folk hero, and he has become international in his battle against expanding global influences on farming and the family farm. Although virtually all nations in Europe are seeing a steady decline in the number of family farms and farmers, this demographic shift has been fought with the most energy and effectiveness in France.

French Cities and Industries
The Primacy of Paris

Paris is the greatest urban and industrial center of France, completely overshadowing all other cities in both population and manufacturing development. With an estimated metropolitan population of more than 11 million (and 2,115,400 in the city proper), it is by far the largest city on the mainland of Europe (excluding Moscow in Russia). Paris is located at a strategic point relative to natural lines of transportation, but it is especially the product of the growth and centralization of the French government and of the transportation system created by that government (Figure 4.8). Like London, it has no major natural resources for the industry in its immediate vicinity, yet it is the greatest industrial center of the country.

The Primate City: The Example of Paris

By formal geographic definition, a primate city is one that is many times larger in population than all other cities in the country. In a more common usage, an urban center is the primate city when its demographic, economic, political, and cultural importance far outshines all other cities. Such a city is considered primate by the hierarchal rule, meaning that its cultural and economic influence overshadows all other urban centers. In the rank order of urban-center size in France (using city-proper population figures), Paris leads with 2,115,400 followed by Marseilles with 809,400, Lyon with 444,300, Toulouse with 400,700, and Nice with 336,300.

The concept of the primate city is, however, a more culturally defined term, and in that sense, there is a singularity in the authority and dominance of Paris, making it an excellent example of a contemporary and popular primate city. The unusual aspect of the Paris example is that a primate city is more often a geographic characteristic in developing nations, where the dominant city is the center of everything — industry, politics, culture, and government. In France, the less important cities have considerable industrial and cultural strength, but they all continue to exist in the shadow of Paris culturally, economically, and politically.

In the Middle Ages, Paris became the capital of the kings who gradually extended their effective control over all of France. As the rule of the French monarchs became progressively more absolute and centralized, their capital, housing the administrative bureaucracy and the court, grew in size and came to dominate the cultural as well as the political life of France. When national road and rail systems were built in the last century and a half, their trunk lines were laid out to connect Paris with the various outlying sections of the country. The result was a radial pattern, with Paris at the hub. As the city grew in population and wealth, it became an increasingly large and rich market for goods.

The local market, plus transportation advantages and proximity to the government, provided the foundations on which a huge industrial complex developed. Speaking broadly, this development involves two major classes of industries. On the one hand, Paris is the principal producer of the high-quality luxury items (fashions, perfumes, cosmetics, jewelry, etc.) for which that city and France have long been famous. The trades that produce these items are very old. Their growth before the Revolution of 1789 was based in considerable part on the market provided by the royal court; since the Revolution, the production of specialty goods has been favored by the continued concentration of wealth in the city. On the other hand, Paris is now the country's leading center of engineering industries, secondary metal manufacturing, and diversified light industries. They are concentrated in a ring of industrial suburbs that sprang up in the 19th and 20th centuries. Automobile manufacturing is the most important single industry, but a great variety of other goods is produced. In addition, the city's economic base includes an endless variety of services for a local, regional, national, and worldwide clientele. In terms of national industrial productivity, France has now climbed to the fourth most productive nation in the world after the United States, Japan, and Germany. In addition to automobiles, the Airbus and the Concorde as aircraft, armaments, rubber, and chemicals make

Figure 4.8   Notre Dame rests on a small island in the Seine River in Paris, standing majestically on its site as a major landscape signature of this most famous European city. The Seine functions as a major transportation artery for the city and the region, but it is better known as a landscape hallmark of the French capital.

FRANCOIS DUMONT/STONE/GETTY

up major products. Textiles, food processing, and world-famous prestige products are also nationally significant.

With the more recent growth of Marseilles, Lyon, Arras, and Lille, the dominance of Paris has been steadily modified and diminished. Two ports near the mouth of the Seine handle much of the overseas trade of Paris: Rouen (population: metro, 522,000; city, 105,700), the medieval capital of Normandy, is located at the head of navigation (the farthest inland a ship can go on a river) for smaller ocean vessels using the Seine. However, with the increasing size of ships in recent centuries, more and more of Rouen's port functions have been taken over by Le Havre (population: metro, 253,675; city, 195,932), located at the entrance to the wide estuary of the Seine.

French Urban and Industrial Districts Adjoining Belgium

France's second-ranking urban and industrial area, located in the north near the Belgian border, stretches along the western edge of the Ruhr District coal and industrial region. Lille is the largest city there and gained its initial importance by being on France's once-leading coal field (now closed). The cities form a metropolitan agglomeration of more than 2 million people. This region, called the Nord, is both an important contributor to, and a major problem in, the economy of France. In the 19th and early 20th centuries, it developed a rather typical concentration of the coal-based industries of that period, with close juxtaposition of coal mining, steel production, textile plants, coal-based chemical production, and heavy engineering. Such operations were not among France's growth industries in the second half of the 20th century, and the region suffers now with extraordinary employment problems, including classic decline in heavy industries and only modest labor replacement by other industries and service employment (see Regional Perspectives on Shifting Factors of Economic Success earlier in this chapter).

Urban and Industrial Development in Southern France

France's third largest urban-industrial cluster centers on the metropolis of Lyon (population: metropolitan area, about 1.7 mil-

lion) in southeastern France. The valleys of the Rhône and Saône Rivers join here, providing routes through the mountains to the Mediterranean, northern France, and Switzerland. Since prehistoric times, the valley has been a major connection between the North European Plain and the Mediterranean. Today, it forms the leading transportation artery in France, providing links between Paris and the coast of the Mediterranean via superhighways, rails that carry some of the fastest trains in the world, and a barge waterway formed by the Rhône, Saône, and connecting rivers and canals to the north.

France has stressed hydroelectric and nuclear energy to compensate for the inadequacy of its fossil fuel supplies, and the Rhône Valley plays a major role in both. A series of massive dams on the Rhône and its tributaries produce about one-tenth of France's electricity, and further power comes from several large nuclear plants in the Rhône Valley using uranium from the nearby Massif Central and foreign sources.

France's principal Mediterranean cities and metropolitan areas — Marseilles, Toulon, and Nice — are strung along the indented coast east of the Rhône delta. Marseilles (population: metropolitan area, 1,527,200) ranks in a class with Lyon and Lille in metropolitan population and is France's leading seaport. It is located on a natural harbor far enough from the mouth of the Rhône to be free of the silt deposits that have clogged and closed the harbors of ports closer to the river mouth. Recently, rapid growth of port traffic and manufacturing in the Marseilles area has been particularly associated with increasingly large oil imports and related development of the petrochemical industry. Toulon (population: 568,900), southeast of Marseilles, is France's main Mediterranean naval base, and Nice (population: metropolitan area, 939,800), near the Italian border, is the principal city of the French Riviera, the most famous resort district in Europe. Along this easternmost section of the French coast, the Alps come down to the sea to provide a spectacular shoreline dotted with beaches. Near Nice, a string of resort towns and cities along the coast includes the tiny principality of Monaco, which is nominally independent but closely related to France economically, culturally, and administratively.

Western France and the Massif Central

As a microcosm of the range of French landscapes, this south central wedge of France is characterized by a rich mix of viticulture, manufacturing, varied agriculture, and tourism. Western France and the Massif Central are considerably less populous, urban, and industrial than the parts of France to the northeast and east. The largest urban center, Toulouse, has 971,700 people in its metropolitan area, and the next two larger cities, Bordeaux and Nantes, have only about 932,000 and 716,200, respectively. Bordeaux is a seaport on the Garonne River, which discharges into the Gironde Estuary on the Bay of Biscay. The city passed its name to the wines from the surrounding region, exported through Bordeaux for centuries. Nantes is the corresponding seaport where the other main river of western France, the Loire, reaches the Atlantic. Toulouse, a very old and historic city with modern machinery, chemical, and aircraft plants, developed on the Garonne well upstream from Bordeaux, at the point where the river comes closest to the Carcassonne Gap. Lying between the Pyrenees and the Massif Central, the Gap provides a lowland route from France's Southwestern Lowland to the Mediterranean.

Recent Dynamism in French Urban and Industrial Development

France shows striking contrasts in the nature of its development before and since World War II. Since the dawn of the Industrial Revolution until recent decades, the country tended to fall behind its European neighbors, except for Spain and Italy, in industrialization and modernization. In the decades following World War II, France's old patterns were rapidly modified with new development.

Population growth was encouraged after World War II by government subsidies for children. Traditionally inadequate coal resources became less of a handicap as other energy sources increased in availability and importance. Protection of inefficient producers from competition was weakened or abolished by France's membership in the European Union. The French government promoted the combination of companies into larger units capable of higher efficiency and international competitiveness. France was well placed, in both site and situation, to move into newly developing industries, technologies, and markets. It had a large pool of labor available for transfer from an overmanned agriculture sector, a relatively modest share of its labor force and capital tied up in older industries, and an excellent educational system.

Tourism.   Tourist flows of the late 1990s have now made France the world's number one tourist destination. In the late 1990s, annual income from tourist activity in France amounted to more than $30 billion. Major global tourist destinations included Paris, which is the world's leading urban tourist destination, and its traditional and highly imaged landmarks. Tourist dollars flowed as well to the redone Disneyland Paris (Fig. 4.9), the Côte d'Azur in the south of France, and to the Cannes Film Festival. The real impact of tourism in France, however, can be seen in the small-scale ways in which it has reached the

Figure 4.9   Euro Disneyland outside of Paris has had a very slow start. Opening as an imported theme park over a decade ago, it has only been in the past few years that it has become more European and, consequently, more successful.

Mediterranean coast, the mountain flank towns and villages, and the urban sidewalks and ceremonial spaces all through the country. The widely popular image of romance and conversation at outdoor cafés, and of leisurely meals with fine wine over which memorable travels are recounted, is evidence of the major impact of tourism on the well-being of the French economy. These landscape elements that may seem so Parisian or French in their origin have now been widely disseminated across the globe. Tourism is a truly global activity that generates significant economic impact wherever it flourishes.

F3.18
88

The ever-expanding images of France and other major tourist destinations come, in part, from these café venues enjoyed — and perpetuated — by tourists and journalists alike. It remains to be seen what ultimate impact the September 11 bombing of the World Trade Towers and the subsequent Western efforts to defuse fundamental Islamic terrorism will have on general tourist activity. There are political movements in France, as well, that are keenly anti-American or antiglobal. They, too, show another face of the way in which a region of high tourist activity can become distressed with the diffusion of popular culture that is so often associated with significant tourist flows.

French National and Regional Planning

The new dynamism of France has been achieved with deliberate planning. Since 1947, the country has engaged in national and regional planning with the stated objectives of improving national output and living standards, reducing the considerable differences in prosperity of different sections of the country, and making France a pleasant place to live in a technological age. In addition to fostering overall national growth and improvement, French planning is directed toward reducing imbalances of long standing among the regions of France. In the mid-1960s, the government called particular attention to the contrasting levels of development among four major regions: Paris, the Northeast, the Southeast, and the West. The most significant contrasts were (1) between the Paris region and the rest of the country and (2) between the less developed and poorer West and the rest of the country. Because modern economic growth is concentrated strongly in cities, the French planners are attempting to divert development from Paris to the main cities of other regions. They use restrictions and penalties on various types of new development in the Paris region, a varied array of financial rewards for firms locating facilities in regions of greater need, and direct location or relocation of government-owned facilities in the less concentrated regional centers.

Although there has been a real decentralization of growth in France during the planning period, some analysts believe that decentralization might have occurred simply as a normal response to changing spatial patterns with or without any planning policy. They maintain that after manufacturing industries reach a certain stage of development, their work becomes so standardized and routine that little skill is required from the work force, and such businesses then tend to relocate from industrial cities to previously nonindustrial places where lower wages can be paid to unskilled and nonunionized labor. This point of view sees French economic decentralization as a normal result of the evolution of industries (see Regional Perspective, p. 102). Whether such transformation comes from official planning or more natural spatial changes associated with new technologies, some decentralization of jobs and prosperity in France seems to have been essential for political stability and national cohesion.

4.5   The Dominant Importance of Germany

Germany reappeared on the map of Europe as a unified country in 1990. Between 1949 and 1990, there had been two Germanys: West Germany (German Federal Republic), formed from the occupation zones of Britain, France, and the United States after Germany's defeat in World War II in 1945; and Communist East Germany (German Democratic Republic), in territory occupied by the former Soviet Union. The urban landscape and political climate of West Berlin were a part of West Germany for those decades, although the city of Berlin overall was entirely surrounded by East Germany. The withdrawal of Soviet support for East Germany in 1989, and the destruction of the Berlin Wall, led to the rapid collapse of its Communist government in 1990, the reunification of Germany in 1991, and the de facto inclusion of East Germany in the EU.

The new Germany is, by a wide margin, the most dominant country of Europe. Its population of 81 million is much greater than that of any other nation in the region. Economic considerations are even more important. The former West Germany alone, with 61 million people, has been Europe's leading industrial and trading state for most of the past seven decades and was the principal focus of the economy of the Coreland of Europe. The former East Germany was an advanced country by Soviet bloc standards but an economic disaster by West German standards. Many billions of dollars and much time have been required to fully rehabilitate Germany's East, and the task is far from done. This area, however, can be expected to contribute increasingly to overall German dominance in the European and global economy, even though there is a continuing irritation felt by the (once) West Germans because of the tax costs they are having to bear for the national efforts to bring (once) East Germany into fuller employment and more productive economic levels of activity.

The Significance of the German Environmental Setting

Germany arrived at its favorable economic position by skillfully exploiting the advantages of its centrality within Europe, together with certain key features of a diverse environment (Fig. 4.10). Broadly conceived, the terrain of Germany can be divided into a low-lying, undulating plain in the north and higher country to the south. The lowland of northern Germany is a part of the much larger North European Plain. The central and southern countryside is much less uniform. To the extreme south are the moderately high German (Bavarian) Alps, fringed by a flat to rolling piedmont or foreland that slopes gradually down to the east-flowing upper Danube River. Between the Danube and the northern plain is a complex series of uplands, highlands, and depressions. The higher lands are predominantly composed of rounded, forested hills or low mountains. Interspersed with these are agricultural lowlands draining to the Rhine, Danube, Weser, or Elbe Rivers, which drain into three different seas.

The climate of Germany is maritime in the northwest and increasingly continental toward the east and south. The annual precipitation is adequate but not excessive, with most places in the lowlands receiving an average of 20 to 30 inches (c. 50 to 75 cm) per year. Farmers, however, must contend with soils that generally are rather infertile. Two areas are major exceptions:

1. The southern margin of the North German Plain forms a belt with soils of extraordinary fertility formed from deposits of loess. This belt continues beyond Germany both east and west.
2. The long, nearly level alluvial Upper Rhine Plain, extending from Mainz to the Swiss border and mostly enclosed by highlands on both sides, is also an area with fertile agricultural soils.

Locational Relationships Among German Cities

Within the environmental framework just described, Germany's major cities tend to be located in the most economically advantageous areas and those most strategic with respect to transportation. Canals have long played both a transport and a recreational role in Germany) (Fig. 4.11). A string of major metropolitan areas is located in the western part of Germany, along and near the Rhine River. This waterway has been a major route since prehistoric times. Four of the eight largest German metropolises—Düsseldorf, Cologne, Wiesbaden-Mainz, and Mannheim-Ludwigshafen-Heidelberg—are on the Rhine itself. The other four—Essen, Wuppertal, Frankfurt, and Stuttgart—are on important Rhine tributaries. This north–south belt of cities also includes several other large metropolitan centers, such as Dortmund, Duisburg, and Aachen, plus numerous smaller places. The concentration of seven of the forenamed cities at or just north of the boundary between the

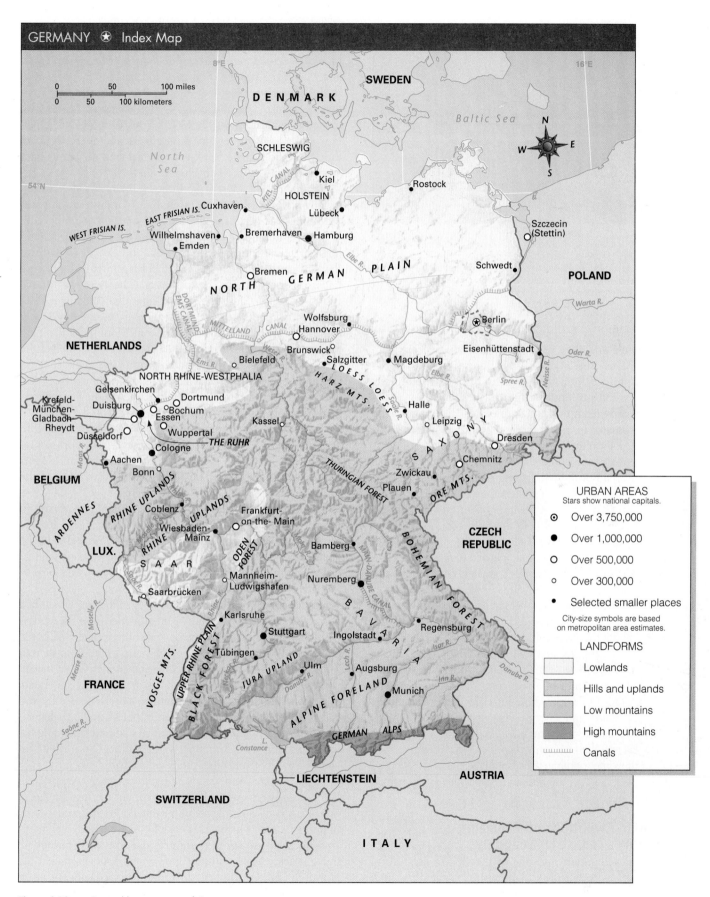

Figure 4.10 General location map of Germany.

JANA R. JIRAK, VISUALS UNLIMITED

Figure 4.11   In the Rhine River valley of Germany, there is an order to the landscape that begins at river's edge, courses through the narrow band of residential and commercial settlement, and then marches up the farmed mountain flanks, all with an eye toward making the scene both productive and lovely. Waterways and canals in Europe have been sources both of vital transportation and of scenic beauty that have long had this region function as a major tourist destination.

Rhine Uplands and the North German Plain reflects additional conditions (see Fig. 4.10). Historically, some of the seven cities profited from their location on the agriculturally excellent loess soil just north of the Uplands. This concentration of large and medium-sized urban centers differentiates the German landscape from the French world, in which Paris plays such a profoundly singular and dominant role.

Other German cities developed as early industrial centers before the Industrial Revolution, drawing on the Uplands for ores, waterpower, wood for charcoal, and surplus labor. Then, in the 19th and early 20th centuries, all seven cities grew in connection with the development of both the Ruhr coal field, located at the south edge of the Plain and east of the Rhine, and the smaller Aachen field. Except for Aachen and Cologne, the cities became part of "the Ruhr," Europe's greatest concentration of coal mines and heavy industries within a network of active river transport. Still other important German cities lie apart from the Rhine and Ruhr. These include (1) Hanover, Leipzig, and Dresden, spread along the loess belt across Germany; (2) the North Sea ports of Hamburg, located well inland on the Elbe estuary, and Bremen, up the Weser estuary; (3) two cities in the southern uplands — Munich, north of a pass route across the Alps, and Nuremberg; and (4) Berlin, whose large size belies its unlikely location in a sandy and infertile area of the northern plain. Berlin is mainly an artificial product of the central governments, first of Prussia and then of Germany, which developed it as their capital. Berlin's metropolitan area population is 3,951,400 and Hamburg's is 3,258,500; the remaining cities previously named vary between nearly 1 and 3 million.

Geographic Evolution and Problems of Germany's Industrial Economy

Germany's development into one of the world's greatest industrial areas has roots in its distant past. Certain areas, such as the Harz Mountains, were exceptionally important centers of European mining in the Middle Ages, and many urban handicraft industries were also well established in medieval times. The most powerful connection between urban merchants and the cottage industries and handicraft artisans was in the Hanseatic League developed in the late 1100s. Goods that were traded by the League, and all across Europe, included silks, leather goods, woolens, armaments, hides, pottery, and all sorts of jewelry. The growth and effectiveness of the Hanseatic League were a strong factor in the subsequent growth of cities in Germany and beyond.

By about 1800, there were three German regions with unusually high concentrations of small-scale industries. One was the Rhine Uplands, in which industry spread along both sides of the river north of Wiesbaden and also away from the river in both directions, but especially toward the east. A second was Saxony, north of and including the Erzgebirge (Ore Mountains) on the German-Czech border. The third was Silesia (now in Poland), consisting of the plains along the upper Oder River, which contained the major mineral resources of Upper (southeastern) Silesia, and the Sudeten Mountains adjoining the plains. All three regions mined metallic ores, smelted the ores with charcoal, used the waterpower of mountain streams, and produced textiles as well as metals and metal goods.

In the 19th and early 20th centuries, each of the foregoing regions developed into a major center of coal-based industry. As the Industrial Revolution spread into Germany in the early 19th century, each area discovered coal resources that were conveniently located and, in the case of the Rhine area and Upper Silesia, extremely large. The Ruhr coal field, located at the north edge of the Rhine Uplands, is the greatest coal field in Europe, and the Upper Silesia field is also of major importance. In this region, the firms with the names of Krupp, Thyssen, and Stinnes had their origins. These same firms continued to be the global players in the mergers and the industrial expansion so characteristic of the globalization of the European economy in the late 20th century. In both areas, the abundance of coal favored the growth of large iron and steel industries, which had become increasingly dependent on imported ores. Secondary development (e.g., metal goods, machinery, and coal-based chemicals) evolved (Fig. 4.12) much more in the Ruhr than in Silesia, perhaps because of the Ruhr area's much better market position. Today, "the Ruhr" denotes an area of somewhat indefinite extent beyond the coal field proper. The "Inner Ruhr," also called Rhein Ruhr (total population: about 11 million), denotes the more concentrated urban-industrial zone incorporating Essen, Duisburg, and Dortmund, whereas the "Outer Ruhr" includes Düsseldorf and other more widely spaced cities.

In Saxony, coal resources were much poorer. Brown coal and lignite fields were eventually developed, but the heat produced by these resources is only about one-quarter of that produced by bituminous coal, which yields coke for the iron and steel industry. Hence, an iron and steel complex did not arise in Saxony, although factory-type industries grew rapidly and promoted the growth of Leipzig, Dresden, and other cities. Machinery, precision instruments, textiles, and chemicals were (and are) the

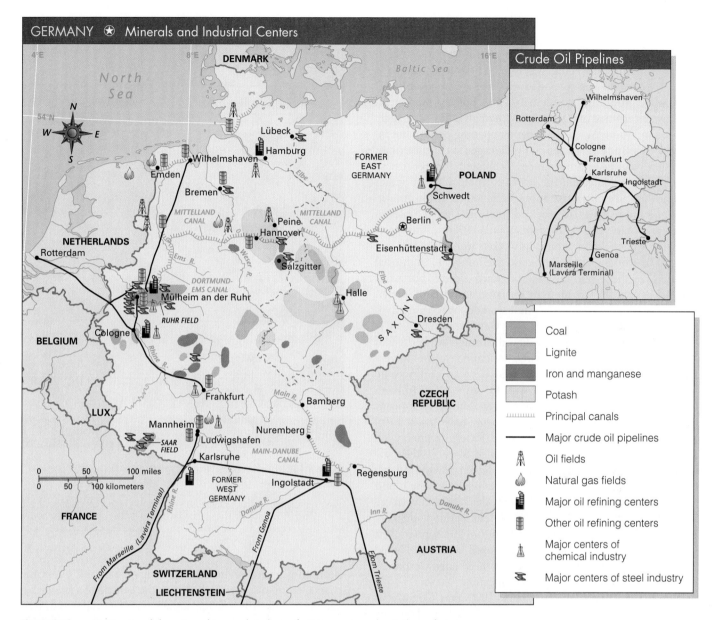

Figure 4.12 Major mineral deposits and iron and steel manufacturing centers, chemical-manufacturing centers, and oil-refining centers in Germany. Dashes show the boundary between former East Germany and former West Germany.

main emphases. Saxony's industrial area is roughly triangular and is often called the Saxon Triangle (see Fig. 4.1).

Much of Germany's present industry, however, is not within the Ruhr or Saxony. A wide scatter of industry and urban centers is characteristic of the region. Many cities are large, but none is as gigantic and nationally dominant as London or Paris. This pattern comes from the influence of two circumstances:

1. Germany was divided for a long time into petty states, a number of which eventually became internal states within German federal structures. These units often promoted the growth of their own capitals, and many of the large cities of today were once the capitals of independent German states.

2. German cities were able to secure power from a distance during the age of industrialization. In the sequence of German development, railways generally reached major cities before coal became practically an industrial necessity. Thus, when coal was needed, the cities away from the fields could generally get it and continue to industrialize and grow. Waterways also helped many cities get coal; power grids then began to provide electricity from a distance.

The former West Germany recovered rapidly from World War II. The U.S. Marshall Plan (officially called the European Recovery Program, but better known for its namesake, U.S. General George Marshall) was authorized in 1948.

From 1948 until 1952, the United States funneled more than $13 billion in capital goods, as well as management expertise and political stability, into West Germany in an effort to keep communism from spreading west into that war-torn region.

A new currency was provided in 1948, and a government empowered to act was formed in 1949. Former production levels were regained and surpassed. German cities were badly bombed in World War II (see chapter opening photo), but industrial capacity was not as badly damaged as is sometimes suggested. Various circumstances facilitated the process of economic rebuilding, not the least of which were the rather freewheeling free-enterprise approach followed by the government and the massive aid received from the United States in those early years. The skills of the population were still there, and labor was cheap for a time, as workers were abundant and eager to be working again. Markets were hungry for industrial products. By the 1960s, the pace of growth slowed, but it was still fast enough to provide jobs for large numbers of foreign workers (sometimes known as **guest workers**) who came in to take jobs vacated by upwardly mobile German workers.

The pace of development slowed during the 1970s and 1980s, although West Germany continued to be a very wealthy and productive country. West German coal became increasingly unable to compete with the imported petroleum that replaced it as Germany's main energy source. New plants to generate electricity with nuclear energy have sparked violent antinuclear protests, and the need for cheap energy led West Germany to quarrel with the United States in the early 1980s in order to complete a deal for Soviet newly available natural gas from a new Russia-to-Europe pipeline. The export of German industrial products to world markets became more difficult as Japanese and other foreign competition stiffened. However, despite the various difficulties of the former West Germany's "economic miracle," it ranked with the United States and Japan as one of the world's three largest exporters of goods at the time of the 1990 reunification.

The former East Germany (1949–1990) was Communist ruled and Soviet dominated, but it was still German. Even Soviet exploitation and the inefficiencies of bureaucrats attempting to plan and operate a "command economy" were unable to destroy completely a historical and traditional German capacity to be productive. East Germany was the most productive of the Soviet-launched and closely managed countries that lay between Russia and Western Europe. When East Germany collapsed and extensive outside investigation became possible, it was revealed to be a poor and inefficient country by Western standards. Images of life in West Germany, and the West in general, led to an upwelling of desire in its population to leave East Germany, stimulating major migration streams from East to West Germany after the 1989 opening of the Berlin Wall (Fig. 4.13). Many billions of dollars of investment and considerable economic pain over a number of years will be needed to bring the economy of what is now the eastern part of reunified Germany, closer to the levels of productivity and living standards in the western part. Huge costs are also involved in the

92–93

Figure 4.13   Berlin citizens from both sides of the Berlin Wall watch its demolition in June 1990. The wall had been opened in November 1989, and there had been a daily flow of some 2,000 East Germans into the former West Germany. This wall had served as one of the most powerful icons of the Cold War since its construction in 1961. Its destruction literally paralleled the collapse of the Soviet Union and the near conclusion of the Cold War contest between communism and capitalism.

cleanup of severe environmental damage incurred under the East's Communist rule. Even within the context of a keenly felt traditional link between the former East Germany and West Germany, there continues today to be a strong sense of frustration by the people of the former West Germany at the financial costs being paid to try to bring the former Communist state up to economic and environmental par with the other Germany.

Germany as a Migrant Destination. In June 1997, the EU countries signed the Amsterdam Treaty, which committed the European Union to creating an "area of freedom, justice, and security" within 5 years. This was intended to mean that frontiers among the 15 members of the EU would be open and allow free movement. However, increases in migration from Turkey, which was turned down for EU membership in 1998, and from Albania, Bosnia, and other Balkan areas caught in ethnic warfare have made Germany particularly uneasy about the EU open border policy (Table 4.1). The current pattern in the EU is to accept migrants who seek political asylum; if they are turned down, they are given 2 weeks to leave the EU country. Since Italy and Austria are the gateway EU countries

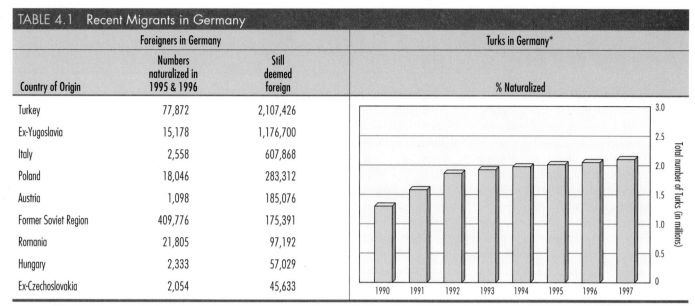

TABLE 4.1 Recent Migrants in Germany

Country of Origin	Foreigners in Germany		Turks in Germany*
	Numbers naturalized in 1995 & 1996	Still deemed foreign	% Naturalized
Turkey	77,872	2,107,426	
Ex-Yugoslavia	15,178	1,176,700	
Italy	2,558	607,868	
Poland	18,046	283,312	
Austria	1,098	185,076	
Former Soviet Region	409,776	175,391	
Romania	21,805	97,192	
Hungary	2,333	57,029	
Ex-Czechoslovakia	2,054	45,633	

Sources: German Interior Ministry. *The Economist.* Vol. 348. July 4, 1998. Reprinted by permission.   *Estimate

for major migrant streams, when immigrants are turned away there, they often are officially "lost" and then tend to move north and west, into the heart of the EU nations.

This regional role as a major migrant destination has been positive in one sense because of the relatively low (or even negative) population growth in most of the EU nations, particularly in Germany and France. Immigration was very important to the dramatic economic growth of the 1960s and 1970s. With the slowdown and the high rates of domestic unemployment in the 1990s, however, Germany has found the migration scenario to be more troublesome. Table 4.1 shows the sources of current immigration in Germany and the changing magnitude of the Turkish migration stream.

The Political and Spatial Shift from Bonn to Berlin

When Bonn was made the capital of West Germany in 1949, it was thought that this was only a temporary shift necessitated by the enclosure of the traditional capital city of Berlin by hostile Soviet control. In 1991, after the political reunification of Germany, it was decided to return the national capital to Berlin. This process was slated to be completed by the end of 1999 so that the year 2000 would see Germany governed, again, from Berlin. This has been only partly achieved. At the final end of this political shift, Bonn will be left with the Bundesrat (federal council) as well as eight federal ministries.

Part of this political shift is being set in motion by Europe's largest building project: the Potzdamer Platz (Fig. 4.14). In late 1998, it was said that there were six Berlins visible in the 17-acre construction site in the heart of old Berlin: (1) the baroque city that was fashioned in the last years of the 18th century by Frederick the Great; (2) the imperial Berlin of Bismarck and Kaiser Wilhelm II (spanning the creation of modern Germany in the 1870s through World War I); (3) the

architectural remnants of the hyperinflated Weimar Republic of the 1920s; (4) Nazi Berlin of 1933–1945; (5) the postwar Berlin that was divided into its western enclave and the eastern city run by Moscow until 1989; and (6) the unified Berlin that has been preparing itself to stand again as the capital of a unified Germany. Because of the combination of enormous construction and reconstruction activity, as well as political centrality, Potzdamer Platz has emerged as a major tourist destination and serves as a bellwether of Germany's continuing industrial and financial development.

F4.14
121

The Berlin that is promoting the massive rebuilding project with its profound images of a new European unity ("We are the laboratory of unity," said Berlin's mayor, Eberhard Diepgen, in 1988) has gained a visibility that only a few anticipated. The city has created a vivid red observation platform overlooking the Potzdamer Platz site, which has received more than 2 million visitors, who, the mayor claims, are drawn by "construction site tourism." The unusual red platform that was the observation platform for this 19-building complex was finally closed in early 2001 as the last of the major new buildings was completed.

Berlin has nearly 4 million people in its metropolitan area and 3,317,000 in the city proper. Bonn has a population of 299,000 in the city proper. The shift of the capital from west to east, from the present to the past (and future), is a geographic shift of major political and cultural magnitude.

4.6 Benelux: Trade Networks as the Key to Prosperity

Belgium, the Netherlands, and Luxembourg have been closely associated throughout their long histories. Today, they are often referred to collectively as "the Benelux Countries." An older name often applied to the three is "the Low Countries." This

REGIONAL PERSPECTIVE   ::: The Geography of Discontent

With the steady expansion of the European Union over the past decade and the number of nations that have already asked to be considered for membership in the EU, it seems strange that there is a simultaneous devolution of European political units from larger to smaller. Yet, looking at the tensions in the fall of 1998, one can see that political and ethnic forces are at work seeking new political autonomy for places of all sizes. In Germany, for example, there has been such a focus on not letting too much political power become centered in one locale again that there has been major power given to the 16 German Lander (states). This has enabled these separate political units to play significant roles in modifying or even blocking reform or federal projects. This devolution — the dispersal of political power and autonomy to smaller spatial units — of the power structure of the federal government has generally been coupled with democratic reforms.

In the fall of 1997, Scotland voted 3 to 1 in favor of establishing its own parliament, making a move toward greater independence. Scotland is similar in size and population to Denmark, Finland, or Ireland, so the success of an independent Scotland will have increasing attraction to those who are seeking more freedom from the governments that have controlled their economies for decades, even centuries.

Ever since the 1975 death of Generalissimo Franco, Spain has given more latitude to 17 new regions, including those of the Basques and the Catalans, who have been famous for seeking such

F5.13
149

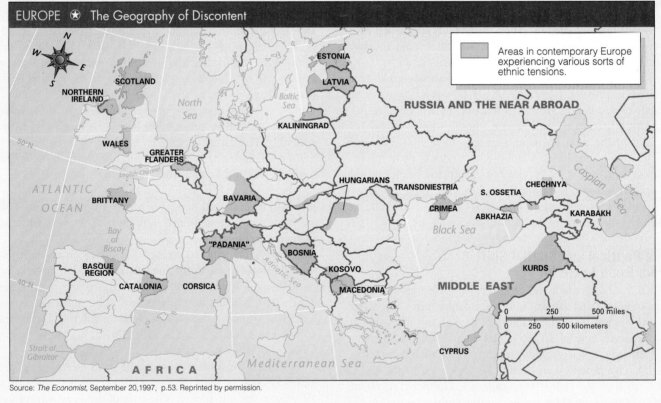

EUROPE ✪ The Geography of Discontent

Areas in contemporary Europe experiencing various sorts of ethnic tensions.

Source: *The Economist*, September 20, 1997, p.53. Reprinted by permission.

Figure 4.C   Areas in contemporary Europe experiencing various sorts of ethnic tensions.

term is properly applied primarily to Belgium and the Netherlands because approximately the northern two-thirds of the former and practically all of the latter consist of a very low plain facing the North Sea (Fig. 4.15). This plain is the narrowest section of the North European Plain. However, Luxembourg is also included in the Low Countries designation because of the long historic connection between all three countries. Large sections near the coast, especially in the Netherlands, are indeed below sea level and have been made into major farming and urban landscapes through the engineering of polders—land that lies below sea level but that has been protected from the sea by dikes, canal systems, and pumps (Fig. 4.16).

The economic life of the Benelux countries is characterized by an intensive and interrelated development of industry, agriculture, and trade. An especially distinctive feature of their economies is a remarkably high development of, and

authority at the cost of bloodshed and continuing belligerence. France has been following the same theme with its own expansion of independence to 22 new mainland regions outside Paris. These central governments — not just in Spain and France but elsewhere in Europe as well — hope that this devolution, accomplished in an ordered and peaceful manner, will quell the strong separatist movements that have sparked fights and bad blood through much of Europe.

Another aspect of this process took place in the former Yugoslavia for most of the 1990s. In the winter of 1999, Kosovo, a subsection of Serbia in the former Yu-

goslavia, made world news after the North Atlantic Treaty Association (NATO) undertook to bomb Slav forces out of the countryside of Kosovo, where the population is 90% ethnic Albanian. This devolution through efforts to develop new political units bonding similar peoples to other similar peoples, called ethnic cleansing in some contexts, has the potential in Europe of continuing as a subtheme of political geography for the indefinite future.

This potential for atomization of places in Europe into ever smaller, more homogeneous units is in stark, but not unrealistic, contrast to the continuing fiscal union being sought by the member na-

tions of the EU, particularly as the euro currency became an electronic currency in early 1999 and coin and bill in 2002. In some situations it is felt that the two forces — as contradictory as they seem in terms of simple geography — may even support each other. For example, some feel that the EU's economic growth may help nudge Ireland and Northern Ireland closer together, even though religion and politics have kept them distant and unwelcoming to each other for most of the 20th century.

The map in Figure 4.C shows the hot spots for the geography of discontent.

Figure 4.14   This 17-acre construction site in Berlin, Germany, is one of the most dramatic new urban plazas in Europe's Coreland. With the German decision to remove the national capital from its Cold War placement in Bonn in the west back to Berlin in the east, an enormous investment in construction and image enhancement was undertaken. The nearly decade-long rebuilding of this ceremonial, political, and financial locale became so central to the ever-changing Berlin that there was a very popular observation site established (the red structure on the right), and it was not closed until 2001 when the rebuilding was fundamentally completed.

Figure 4.15   General location map of the Benelux countries: Belgium, the Netherlands, and Luxembourg. Note the extensive area of polder land reclaimed from the sea.

Ports of the Netherlands and Belgium

dependence on, international trade. These Low Countries have had their economies linked to global trading patterns for more than four centuries.

Exploitation of commercial opportunities by the Low Countries is reflected in the presence of three of the world's major port cities within a distance of about 80 miles (c. 130 km): Rotterdam and Amsterdam in the Netherlands and Antwerp in Belgium. Amsterdam (population: metropolitan area, 1.2 million) is the largest metropolis and the constitutional capital of the Netherlands, although the government is actually located at The Hague (population: 773,000). Amsterdam was the main port of the Netherlands during the 17th and 18th centuries and for most of the 19th century, while the Dutch colonial empire was expanding. The principal element in this empire was the Netherlands East Indies (now Indonesia), where the Royal Dutch Shell Company originated. Until the loss of Indonesia after World War II, Amsterdam built a growing trade in imported tropical specialties. Many of these imports were reexported, often after processing, to other European countries. Profits accumulated during these centuries

supplied funds for the large overseas investments of the Netherlands and made Amsterdam an important financial center, which it continues to be.

Amsterdam also remains an important port, but it has lost much of its relative position in this respect during the past 50 years. The original sea approach via the Zuider Zee, a landlocked inlet of the North Sea, proved too shallow for the larger ships of the 19th century. The opening in 1876 of the North Sea Canal solved this problem for a time, but even successive expansions of the canal did not give Amsterdam access to the sea comparable to that of Rotterdam. And the Dutch loss of Indonesia in 1949 greatly decreased the port's importance as an entrepôt, although continuing ties with its former colony have given Amsterdam a large Indonesian community that imparts a distinctive ethnic and cultural flavor to the city.

Rotterdam (population: metropolitan area 1,125,500; city proper, 589,500), the world's largest port in tonnage handled, is better situated for Rhine shipping than either Amsterdam or Antwerp, being located directly on one of the navigable distributaries of the river rather than to one side. Accordingly, Rotterdam controls and profits from the major portion of the river's transit trade, receiving goods by sea and dispatching

Figure 4.16   Windmills are a common feature in the polder landscape of the Low Countries. This scene from Alkmaar, Hoorn, in the Netherlands shows the fertile farmland resulting from the poldering process. This resultant cultural landscape is a very productive blend of engineering skills, continuing agricultural ambition, and continual efforts to enhance the land base of the Low Countries.

them upstream by barge, and receiving goods downstream by barge and dispatching them to global destinations. Two developments are mainly responsible for the port's tremendous expansion. First was the opening in 1872 of the New Waterway, an artificial channel to the sea far superior to the shallow and treacherous natural mouths of the Rhine and the sea connections of either Amsterdam or Antwerp. Second, and more fundamental, is the increasing industrialization of areas near the Rhine (particularly the Ruhr district), for which Rotterdam has long been the main sea outlet. A large addition to Rotterdam's facilities known as Europoort lies along the New Waterway at the North Sea entrance to the Rotterdam port area. Europoort is specially designed to handle supertankers and other oversized container carriers.

Antwerp (population: metropolitan area, 945,800; proper, 445,500), located about 50 miles (80 km) up the Scheldt River, is primarily a port for Belgium itself but also accounts for an important share of the Rhine transit trade. Belgium's coast is straight and its rivers are shallow so that the deep estuary of the Scheldt gives Antwerp the best harbor in the country, even though it must be reached through the Netherlands.

Industrial Patterns in Benelux

All the Benelux countries are highly industrialized. Before major discoveries of natural gas were made in the northeastern Netherlands in the 1950s, Belgium and Luxembourg were better provided with mineral resources. This helped them become somewhat more industrialized than the Netherlands. Now, however, the Netherlands, with its advantageous trade position and natural gas, is the leading country within Benelux in total value of industrial output.

Heavy Industry in the Sambre-Meuse District and Luxembourg

One of Europe's major coal fields crosses Belgium in a narrow east–west belt about 100 miles (161 km) long (see

Fig. 4.2). It follows roughly the valleys of the Sambre and Meuse Rivers and extends into France on the west and Germany on the east. Liège (population: metropolitian area, 621,000; proper, 185,400) and smaller industrial cities along this Sambre-Meuse field account for most of Belgium's metallurgical, chemical, and other heavy industrial production. A sizable iron and steel industry developed here even before the Industrial Revolution, using local iron ore and charcoal. Subsequently, local coal and the early adoption of British techniques made Liège the first city in continental Europe to develop modern large-scale iron and steel manufacturing. The Sambre-Meuse district also specializes in smelting imported nonferrous ores, with much of the metal being exported, though it now suffers from the customary problems of old, coal-based heavy industrial districts. Coal production has declined drastically in the face of competition from imported oil and even coal imported from better and less depleted fields in North America and elsewhere. International competition has also adversely affected the iron and steel industry.

Newer industrial plants have generally located in northern Belgium, where unions are weaker and wages lower. Belgium's capital and largest city, Brussels (population: metropolitan area, 2,390,000), is the leading center of a breed of new, less energy-dependent cities, where industry is able to utilize less skilled, nonunion employees. Note that this pattern reflects the same general process that we outline in the Regional Perspective on Shifting Factors of Economic Success (p. 102). Wherever industries can trace their roots back to the beginnings of the Industrial Revolution and vital coal fields that fueled those innovations and growth, today's picture is generally bleak as local coal loses utility.

Luxembourg's most important exports come from the iron and steel industry, located in several small centers near the southern border where the Lorraine iron ore deposits of France overlap into Luxembourg. Production is on a large scale, and the very small home market both necessitates and permits export of most of the metal. Luxembourg continues to follow the European pattern in diversifying its manufacturing industries and exports. The mark of its success is shown in part by the fact that its annual per capita income of more than $45,000 makes it the richest country in the world.

Industries in the Netherlands

Having traditionally lacked local coal supplies of any significance, the Netherlands has long been dependent on imported fuels. The country's dense population provided cheap labor, and its location allowed for the cheap import of fuel and cheap export of products. Thus, the Netherlands developed no heavy industrial region but had a rather scattered development of light industries such as textiles and electrical equipment. This situation changed after World War II, as imported oil increasingly supplanted coal in the industries of Western Europe. More oil imports now enter Europe via Rotterdam-Europoort than any other port, with some arriving as crude oil and some refined further in the Rotterdam

One of the most definitive elements of the Low Countries landscape is the presence of land reclaimed from the North Sea. This pattern exists in other parts of the world, but the leading example of this expansion of the cultural landscape is the polder land in the Netherlands. These are lands that have been surrounded by dikes and artificially drained. The process of turning former swamps, lakes, and shallow seas into agricultural land has been going on for more than seven centuries. An individual polder, of which there are a great many of various sizes, is an area enclosed within dikes and kept dry by constant pumping into the drainage canals that surround it. About 50% of the Netherlands now consists of an intricate patchwork of polders and canals, and Belgium has a narrow strip of such lands behind its coastal sand dunes. Even though the polders are the best agricultural lands in the Netherlands and production from them is the heart of Dutch farming, this land has become increasingly important in accommodating the continuing growth of the urban landscape.

These highly manipulated landscapes are perfect examples of a combination of tradition and modern engineering to expand a limited farmland resource. They represent the conflict created by the modification of a natural landscape, the shallow seascape, into a more productive cultural landscape. In a major 1953 flood, more than 1,800 people drowned, providing further evidence of nature's capacity to remind engineers of the potential danger in massive landscape transformation. Polders not only underlie much of the extremely productive agriculture of the Low Countries, but they are also central to tourist images of Holland: windmills, dikes, and gardenlike farmscapes. The entire scene, however, is always subject to threats from the sea at times of storm. Like all modified and artificial landscapes, polders require strict human maintenance to continue the unusually high levels of productivity that have been wrested from nature.

area. This activity has generated one of the world's largest concentrations of refineries. Petrochemical plants near the refineries represent the main branch of the sizable Dutch chemical industry. Port locations and dependence on port functions also characterize many other important Dutch industries and activities, particularly the operation of a good-sized merchant fleet that is actively engaged in the expansion of global trade networks. Port activities reflect not only the country's present emphasis on trade but also a long Dutch maritime tradition.

The Dutch played a prominent part in European discoveries and colonization overseas and, for a brief time in the 17th century, were probably the world's greatest maritime power. In the early part of the 20th century, they were still able to capitalize on their earlier dominance, and by about 1910, the Netherlands had achieved their maximum extent of control over what they called the Dutch East Indies — later Indonesia. Major petroleum reserves were discovered and exploited during the following four decades, and when Indonesia gained independence in 1949, the Dutch had significant oil holdings in the East Indies.

Certain other industries of consequence are in the major ports but also in smaller Dutch cities. Among them are the food processing, electrical machinery, textile, and apparel industries. Some food industries process imported foods for European distribution, but more of them handle products of the remarkably productive Dutch agriculture, which produces everything from dairy products to sugar beets to salad vegetables. Certain industries tend toward interior locations having relatively cheap, productive labor and good transport connections. Among the best known are the electrical and electronics industries.

The Distinctive Agricultural Patterns of the Low Countries

Both the Netherlands and Belgium are very productive agriculturally and are characterized particularly by extremely high yields per unit of land. In fact, yields are so high that the total output of farm products is surprisingly great for such small countries. Such productivity results mainly from the following circumstances:

1. *Fertile soils.* Parts of each country have very fertile soils, the largest such areas being the polders (diked and drained lands) in the Netherlands and the loess lands of Belgium.
2. *Farm labor.* Despite rapid decreases of farm population in recent times, there is still an intensive use of labor on the small farms that characterize these countries.
3. *Highly modified landscape.* There is an intensive application of capital in such forms as water control, agricultural machines, knowledge, technical expertise, and especially fertilizer. Both countries are export producers of nitrogen fertilizers made from oil, natural gas, or coal, and they have developed specialty crops and very productive greenhouse agriculture.
4. *Role of the EU.* Agriculture in the Low Countries has profited from Common Market tariff protection and subsidies for agricultural producers. These have stimulated spectacular increases in output to the extent that overproduction of certain products such as milk and butter has become a more worrisome problem than food supply, even in such crowded countries as these.
5. *Trade networks.* Belgium and the Netherlands profit agriculturally from trading relationships that enable the two

Figure 4.17   The commerce that the Dutch have brought to the color of flowers is evident all around the country. This photo of Kaukenhot gardens near Lisse, Holland, shows the intensity of the colors that now serve as the basis for international flower and bulb raising. These same swaths of color are also woven through parks, urban gardens, and tourist attractions for both local and foreign visitors.

countries to specialize in labor-intensive farm commodities such as commercial flowers and dairy products, particularly suited to their conditions (Fig. 4.17). The dense and affluent populations in and near the Low Countries provide a large market for such specialties. Requirements for other commodities such as grains are met increasingly by imports.

Population and Planning Problems

Countries as intensively and effectively developed as the Netherlands and Belgium must consider how development can be guided to sustain traditional environmental amenities for their crowded populations. In the Netherlands, such planning problems are especially acute because this country has a population density higher than virtually any country of Europe but with relatively small urban concentrations. The annual population growth rate is 0.3% compared to the European average of 0%, or by some estimations, negative growth. In this situation, the Dutch are trying to limit urban sprawl, provide more outdoor and waterfront recreational areas, and in general, maintain an environment that is pleasant for its people and attractive to potential tourists. Planners are especially concerned with maintaining the open space that is semiencircled by Amsterdam, Utrecht, Rotterdam, The Hague, and other cities. These urban areas are tending to coalesce into what is referred to as Randstad Holland, the "Ring City" of Holland.

4.7   Switzerland and Austria

With respect to physical environment, Switzerland and Austria have much in common. More than half of each country is occupied by the high and rugged Alps (Fig. 4.18). North of the Alps, both countries include strips of the rolling morainal foreland of the mountains. The Swiss section of the Alpine foreland, often called the Swiss Plateau, extends between Lake Geneva on the French border and Lake Constance (German: Bodensee) on the German border. The Austrian section of the foreland, slightly lower in elevation, lies between the Alps and the Danube River from Salzburg on the west to Vienna on the east. The Swiss and Austrian sections are separated by a third portion of the foreland in southern Germany. On their northern edges, both Switzerland and Austria include mountains that are much lower and smaller in extent than the Alps. These are the Jura Mountains, on the border between Switzerland and France, and the Bohemian Hills of Austria, on the border with the Czech Republic.

The Disparity Between Resources and Economic Success

Historical assessment of economic success has often suggested a close link between such success and a generous resource base. In a comparison of Austria and Switzerland, however, it can be seen that natural resources and raw materials play only a partial role in the distinct levels of economic achievements in the two nations. Based on natural resources alone, one might expect Austria to be the more successful country economically. It has more arable land than Switzerland and more per person, primarily as a result of Austria's greater proportion of nonmountainous terrain. It also has more forested land than Switzerland and more per person. This results mainly from the fact that the Austrian Alps have somewhat lower elevations than the Swiss Alps, and hence, a smaller proportion of land is above the tree line. The same contrast holds true for minerals. Austria has no mineral resources of truly major size, but it does have a varied output of minerals that possess value collectively. Switzerland, on the other hand, is almost devoid of significant mineral resources. Both countries depend heavily on abundant hydroelectric potential, but Austria's resource base is greater.

In economic success, however, the advantage lies with Switzerland, which in 2000 had a per capita GDP PPP of $28,600; in contrast, Austria had a per capita GNP of GDP PPP of $25,000. But this must be seen in perspective, as Austria is quite well off today, largely as a result of economic growth since World War II.

Historical developments that favored Switzerland far outweighed Austria's natural resource advantage. The primary reason seems to be the long period of peace enjoyed by Switzerland. Except for some minor internal disturbances in the 19th century, Switzerland has been at peace inside stable boundaries since 1815. The basic factors underlying this long period of peace seem to be (1) Switzerland's position as a buffer between larger powers, (2) the comparative defensibility of much of the country's terrain, (3) the relatively small value of Swiss economic production to an aggressive state, (4) the country's historic and political value as an intermediary between belligerents in wartime and as a refuge for people and money, and (5) Switzerland's policy of strict and heavily armed neutrality.

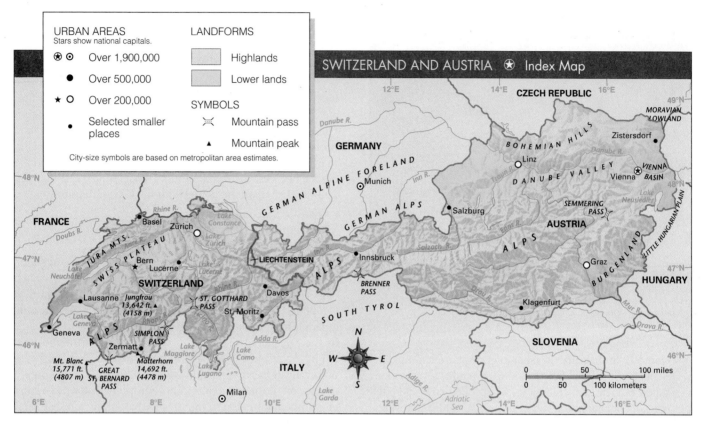

Figure 4.18   General location map of Switzerland and Austria.

Switzerland

During well over a century and a half of peace, the Swiss have had the opportunity to develop an economy finely adjusted to the country's potentials and opportunities. In particular, they have skillfully exploited the fields of banking, tourism, manufacturing, and agriculture. Favored by Swiss law, the country's banks enjoy a worldwide reputation for security, discretion, and service. The result has been a massive inflow of capital, including some of dubious origins, to Swiss banks. Switzerland's largest city, Zürich (population: metropolitan area, 958,100; city proper, 334,300), is one of the major centers of international finance, and other Swiss cities are also heavily involved.

Switzerland has scenic resources in abundance, and the country has long been a major tourist destination. Development of winter sports makes snow an important resource, and Alpine ski resorts such as Zermatt, Davos, and St. Moritz are world famous (Fig. 4.19). The same mountains that make skiing such a recreational resource also make surface links with the rest of Europe, forcing the construction of numerous tunnels and costly highway engineering marvels. Less publicized than the landscape of relative isolation is Switzerland's development of the manufacturing of delicate timepieces and high-technology instrumentation, based primarily on hydroelectricity from mountain streams and the skills of Swiss workers and management. Because most of its raw materials and fuels must be imported, industrial specialties in Switzerland minimize the importance of bulky materials and derive value from skilled design and workmanship.

Reliance on hydroelectricity has facilitated the development of many small industrial centers, mostly in the Swiss Plateau. Five of the country's six largest cities are located on

Figure 4.19   The topography of Switzerland provides both a burden and an asset. The complexity of building and maintaining roads and shelter is a continual cost and concern. But the mountain landscapes and the winter snows that drape across peaks, flanks, and valleys are seen by millions of tourists annually as worlds worth trying to conquer or at least enjoy. This is the village of Laax, Switzerland.

the Plateau. Except for Zürich, these cities are under 650,000 in metropolitan area population, although Geneva approaches that figure. The remaining city, Basel (population: 552,600), lies beyond the Juras at the point where the Rhine River turns north between Germany and France. Basel is the head of navigation for Rhine barges that connect Switzerland with ocean ports and other destinations via river and canal. Mention of the importance of Swiss cheese and Swiss chocolate must be made because the images of such products are part of the global identity of this landlocked country.

The National Unity of Switzerland

The internal political organization of Switzerland expresses and makes allowances for linguistic and ethnic diversity, with German being the primary language of 65% of the population, French for 19%, and Italian for 8%. The country was originally a loose alliance of small sovereign units known as cantons. When a stronger central authority became desirable in the 19th century, not only were the customary civil rights of a democracy guaranteed, but governmental autonomy was retained by the cantons except for limited functions specifically assigned to the central government. Although the responsibilities allotted the central government have tended to increase with the passing of time, each of the local units (now 26 in number) has preserved a large measure of authority. Local autonomy is supplemented by the extremely democratic nature of the central government. In no country are the initiative and the referendum mechanism more widely used to submit important legislation directly to the people. Thus, guarantees of fundamental rights, local autonomy, and close governmental responsiveness to the will of the people are used to foster national unity despite the potential handicaps of ethnic diversity and local particularities.

The central government in turn has pushed economic development vigorously. Two outstanding accomplishments have been the construction and operation of Switzerland's railroads and the development of the hydroelectric power generating system. Despite the difficult terrain, the Swiss government has built a railway network of great density, which carries enough international transit traffic to earn important revenues. The hydroelectric system, utilizing the many torrential streams of the mountains, yields about 59% of the country's electric output and places Switzerland very high among the world's nations in hydroelectricity produced per capita.

Austria

Austria emerged in its present form as a remnant of the Austro-Hungarian Empire when that empire disintegrated in 1918 as a consequence of defeat in World War I. Austria's population is essentially German in language and culture — and the countries have been closely linked historically — but unification with Germany was forbidden by the victors. The economy was seriously disoriented by the loss of its empire, and the country limped through the interwar period until it was absorbed by Nazi Germany in 1938. In 1945, after defeat in World War II, Austria was reconstituted as a separate state but was divided into occupation zones administered, respectively, by the United States, the United Kingdom, France, and the Soviet Union. During 10 years of occupation, the Austrian Soviet Zone in the east (which included Vienna), already badly damaged in the war, was subjected to extensive eastward removal of industrial equipment to the former Soviet Union for reparations. It was only in 1955 that this occupation was ended by agreement among the four powers, and Austria became an independent, neutral state.

As the core area of the earlier Austro-Hungarian (Hapsburg) empire in southeastern Europe, Austria developed a diversified industrial economy prior to 1914. Its industries depended heavily on resources and markets provided by the broader empire that covered much of the North European Plain. Austrian iron ore was smelted mainly with coal drawn from Bohemia and Moravia, now in the Czech Republic. Austria's textile industry specialized to a considerable extent on spinning, leaving much of the weaving to be done in Bohemia. Austrian industry was not outstandingly efficient, but it had the benefit of a large protected market in which one currency was in use. In turn, the producers of foodstuffs and raw materials also had a market within the empire protected by external tariffs.

When the empire disintegrated at the end of World War I, the areas that had formed Austria's protected markets were incorporated into independent states. These states, motivated by a desire to develop industries of their own, began to erect tariff barriers, and other industrialized nations began to compete with Austria in their markets. The resulting decrease in Austria's ability to export made it more difficult for the country to secure the imports of food and raw materials that its unbalanced economy required. Such difficulties were further increased after World War II by the absorption of East Central Europe into the Soviet Communist sphere. The necessary 1990s reorientation of Austrian industries toward new markets, principally in Western Europe, was made more difficult because Austria did not join the EU until 1995. As a consequence, Austrian exports of goods were consistently inadequate to pay for imports. However, the country has been able to compensate for much of this deficit by revenues from a growing tourist industry.

Austria made notable progress after World War I, and has made especially rapid progress since World War II, in building a successful economy suited to its status as a small independent state. Major elements in the economy include forest industries, tourism, some iron and steel production based on Austrian iron ore and imported ore and fuel, increased hydroelectric development, traditional manufacturing of clothing and crafts, and larger scale engineering and textile industries. The latter industries contribute quite importantly to exports, but development is not sufficient to satisfy the internal market, and Austria is now a net importer of machinery and textiles.

Since the end of the empire, Austrian agriculture has moved toward greater specialization in dairy and livestock farming. Such production is a logical use of the large upland areas of Austria that are best adapted to pasture. An enlarged acreage devoted to feed crops such as barley and corn accompanies the increased emphasis on animal products. But Austria

F5.15
151

is still far behind Western Europe's leading countries in agricultural efficiency. Many small and poor farms still exist, and 6% of the labor force was still in agriculture as of 1998 — a high proportion by Western European standards. Meanwhile, substantial imports of food and feed remain necessary.

A significant aspect of Austrian readjustment since World War I has been the reduced importance of the famous Austrian capital, Vienna (population: 1,609,600). Although Vienna is still by far the largest city in Austria, there has been a pronounced tendency for population and industry to shift away from the capital toward the smaller cities and mountain districts. From a position as the capital of a great empire, Vienna has declined to only the capital of a small country. The city's present metropolitan area population is less than the population of the city proper in 1918.

Vienna's importance, however, has persisted since Roman times and is based on more than purely Austrian circumstances. The city is located at the crossing of two of the European continent's major natural routes: the Danube Valley route through the highlands separating Germany from the Hungarian plains and southeastern Europe and the route from Silesia and the North European Plain to the head of the Adriatic Sea. The latter follows the lowland of Moravia to the north and makes use of the passes of the eastern Alps, especially the Semmering Pass, to the south. Thus, Vienna has long been a major focus of transportation and trade and also a major strategic objective in time of war. In the geography of the early 21st century, Vienna finds itself particularly well-positioned for the potential expansion and development of the Eastern European nations and the western tiers of the states emerging from the former Soviet region.

CHAPTER SUMMARY

- This chapter gives attention to the Industrial Revolution, particularly the forces that led to its genesis and development in Great Britain. Coal, steam power, and the railroad all took new roles in an assemblage of industrial changes, demographic shifts, and growing urbanism in England and Scotland in the middle 17th century as part of this revolution.

- The nations that are considered part of West Central Europe in this chapter include the United Kingdom, France, Germany, the Low Countries (the Netherlands, Belgium, and Luxembourg), Austria, and Switzerland, as well as a number of smaller nations and microstates. The giant nations in this region, however, are considered to be Germany, the United Kingdom, and France. They are the primary engines for the economic momentum in this part of the world.

- The United Kingdom is made up of England, Scotland, Wales, and Northern Ireland. It forms the greater part of two major islands: Great Britain and Ireland, which are called the British Isles. The island of Great Britain has a highlands and a lowlands with distinctive landscapes and mineral resources. Most of the 18th- and 19th-century industrial and urban development of Britain related to the location of coal and iron, which were the two most dominant resources of the Industrial Revolution. Later, the location of London on the Thames Estuary enabled that city to rise to vital importance with neither coal nor iron as a local resource.

- The four major factors that led to Britain's importance in the genesis of the Industrial Revolution were coal, iron and steel, cotton textiles, and shipbuilding. The process of enclosure also helped promote a demographic shift that provided urban factory workers for the new textile mills. London is a major example of a dominant urban center that reflects the forces of innovation and change that characterize this era of industrial growth and development. Central to London's development were location on the Thames, proximity to regional trading partners, the high level of productivity of the London region (even without coal or iron ore), and the significant role of English shipping for almost all of the 18th and 19th centuries. London also illustrates late 20th-century industrial relocation. Central to this locational shift are a loss of resource advantage, cheaper labor in new areas, technological and employment changes, and managerial conservatism.

- The island of Ireland has long experienced a tense relationship between Roman Catholics, who make up the major population on the island, and Protestants, who came from Scotland and Britain in the 17th century and settled in Ulster, the locale now called Northern Ireland. Economically, Ireland has encouraged high-tech industries, as well as tourism, to play an important role in its recent development.

- France is the largest of the West Central European countries. The nation has an internal coherence and reasonably clear national borders made up of rivers, mountains, and seacoasts. Agriculture is more important in France than in any other West Central European country. It plays a role not only in economic but also in social terms. Markets for French agricultural products are not only domestic and regional but global, especially for wines, olives, and wheat. Paris, an example of a primate city, is the most dominant city in all of France and more singular than are other national capitals in this region. The growth of Paris from a small settlement on an island in the Seine River into a world capital is an example of many geographic features coming to play in a single location. These factors include location, transportation, human migration, markets, cultural uniqueness, and colonialism. France is made up of four regions, each less economically vibrant than the prior one. The regions, in order, are the Paris region, the Northeast, the Southeast, and the West. Tourism continues to be a strong factor in the economic health of France, with Paris the most popular tourist city destination in the world.

- Germany, re-formed when West Germany reunified in 1990 with East Germany at the collapse of the former Soviet Union and its collection of satellite nations lying between Russia and West Central Europe, has the largest economy and population of all Europe. German urban centers have a much broader distribution than do the industrial cities of France and even Britain. These cities, generally, have been located in direct response to availability of coal, river systems, iron ore, and the soils of the North European Plain. The German adoption of the Industrial Revolution occurred much later than the British or Belgian experience, but by the beginning of the 19th century, Germany had developed major industrial zones in the Rhine uplands, Saxony, and Silesia (now in Poland). All three of these centers had coal, iron ore, hydroelectric power, charcoal, labor, and the technology for textile manufacturing and production of metal goods. The geography of German manufacturing is more widespread than is common in West Central Europe. There has been considerable immigration of guest workers or foreign laborers. Germany has also been a target destination for migrating peoples from the Balkan Peninsula, Turkey, Russia, and various countries neighboring Russia. At the end of the 1990s, Germany moved its federal capital from Bonn, where it had been since 1949, to Berlin, as a result of reunification.

- The Netherlands, Belgium, and Luxembourg are called the Low Countries, or Benelux. They have developed global trading networks, busy river systems, and active expansion of a land base through polders, and continue to derive significant income from agriculture. Three major port cities in Benelux are Rotterdam and Amsterdam in the Netherlands and Antwerp in Belgium.

- Rotterdam, in the Netherlands, serves as an example of the ways that the river systems of West Central Europe have played and continue to play a central role in the economic growth and importance of port cities on the North Sea. The reasons for the design,

engineering, and construction of polders in the Low Countries derive from some of the same influences that make Rotterdam and other port cities so vital to current economic health.

- Switzerland and Austria represent two nations with somewhat similar geographic features that have had very distinct historical geographies. They are both in the Alps, have played the buffer state role in the heart of Europe, and both are significant tourist destinations. Neither has particularly abundant natural resources, but scenery, hydroelectricity, and a strong cultural tradition have all played a role in their development. Switzerland has gained more stability with its traditional neutrality and financial management, whereas Austria has had its fate tied more closely to Germany and battles on the North European Plain.

REVIEW QUESTIONS

1. List the major nations of West Central Europe (the Coreland) and locate the most important beds of coal and iron ore, both in the 18th century and now.

2. What are the four industries that were central to the Industrial Revolution? Discuss each of them in relation to the development of industrial growth in the United Kingdom.

3. Discuss the role of agriculture in France, explaining why it is so much more central a role in that country than in the United Kingdom.

4. Using maps and the text, and reflecting on Thinking of the World in Spatial Terms from Table 1.1, Chapter 1, why is Paris the city that it is? What makes it not only central to France and the French population but enables it to be the most popular tourist city in the world?

5. Explain why West Germany went to the great expense in 1990 of bringing East Germany back into the single state of Germany.

6. Compare and contrast the maps of industrial centers in the United Kingdom, France, and Germany. What role do resources play in those distinct maps?

7. Looking at the population map of Europe in Chapter 3, suggest reasons for the active migration from Eastern Europe and Russia into West Central Europe, especially Germany.

8. What are the reasons for, and possibly against, the move of the capital of Germany from Bonn in the west to Berlin in the eastern part of Germany?

9. What has made Rotterdam the most active and prosperous port in West Central Europe? What are the geographic features that have led to this and what are the cultural features?

10. Explain the concept of the polder. Where is it important and how do you think its construction cost and maintenance costs have been justified?

11. What role has geography had in the distinct histories of Switzerland and Austria?

DISCUSSION QUESTIONS

1. What is the impact of the Eurotunnel (the Chunnel or the English Channel Tunnel) on Britain? On France? On West Central Europe?

2. Discuss the locales that have been caught up in the Geography of Discontent (see Regional Perspective, pp. 120–121) that is cited in this chapter. Where are they? How would you characterize the features that put them in this category?

3. Consider the growth of the German economy and the return of the federal capital from Bonn to Berlin. What are the various conversations that such a move might raise? Give a French, a Dutch, an English, a German, and a Russian perspective in such conversations.

4. Discuss the importance of European rivers and of European coal resources shown on the maps in this chapter. What can you say about the role of geographic resources and landscapes?

5. Discuss how the National Geography Standards' six Essential Elements that deal with Places and Regions would be helpful in making an analysis of the geographic features that are discussed in questions 3 and 4.

6. What is the pattern of population growth in Europe? What is the significance of the strong regional differences in growth rates?

7. Think of your mental maps of European cities. Which cities most quickly come to mind as places you would like to visit? How many of them are in the Coreland? What are the qualities that make them so appealing to you? Talk about how your mental maps of these cities have been shaped. How might they be changed?

8. Discuss your migration decision-making process if you are living in the Balkan Peninsula. What forces might lead you to leave? What news and images affect the decisions you make about where you want to go?

9. If you were asked to create a Wait Disney World complex or theme park in Berlin, would you have it be the same as the one outside Paris? Why or why not?

10. Discuss the likelihood of the United Kingdom, Germany, and France being able to bring old industrial cities into the 21st century as economically viable urban centers. Give examples of where this has worked or not worked.

Europe: Northern, Southern, Eastern

In Spain's Costa del Sol, on the coast of the Mediterranean Sea, you see the elements that are so characteristic of the landscape in Europe. The raw presence of the natural landscape is shown in clear light. In the background, well-tended agricultural fields butt against tourist accommodations of hotels and cafés that capitalize on the southern European landscape. Images like this are increasingly important to Eastern Europe as it tries to capture some of the traditional flow of tourists. This locale is east of Malaga.

Chapter Outline

Present-day Europe's northwestern Coreland, which was the focus of Chapter 4, is ringed by three outer groups of countries: Northern Europe (sometimes called Norden or Fennoscandia), Southern Europe (best known as Mediterranean Europe), and Eastern Europe. Although these countries all possess distinctive traits and unique histories, today they tend to play a more minor role than do the Coreland nations in influencing the other regions and nations of the world. In this chapter, we explore the geographic personalities and economic and political dynamics of these countries outside the European center. Area and population data can be found in Table 3.1.

5.1　Northern Europe: Prosperous, Isolated, and Distinct

Denmark, Norway, Sweden, Finland, and Iceland are the countries of Northern Europe (Fig. 5.1). The peoples of these lands recognize their close relationships with each other and group their countries geographically under the regional term Norden, or "the North." A more common term to describe these countries is *Scandinavia*, or the Scandinavian countries. But this regional name is ambiguous: It occasionally refers only to the two countries that occupy the Scandinavian Peninsula: Norway and Sweden. More often, it includes these countries plus Denmark, and sometimes it includes these three plus Iceland. When Finland is included in the group, the term *Fennoscandia*, or Fennoscandian countries, is used, thus acknowledging the cultural and historical differences between the Finns and the other northern nations.

Dominant Traits of Northern Europe as a Region

No other highly developed countries in the world have their principal populated areas so near the North Pole (see Fig. 3.2). Located in the general latitude of Alaska, the countries of Northern Europe represent the northernmost concentration of advanced industrial nations in the world.

Westerly winds from the Atlantic, warmed in winter by the North Atlantic Drift, moderate the climatic effects of this northern location. Temperatures average above freezing in winter over most of Denmark and along the coast of Norway. But away from the direct influence of the west winds, winter temperatures average below freezing and are particularly severe at elevated, interior, and northern locations. In summer, the ocean tends to be a cooling rather than a warming influence, and most of Northern Europe's Fahrenheit temperatures in July average no higher than the 50s or low 60s (10° to 18°C). The populations of the various countries tend to cluster in the southern sections, and all countries except Denmark have considerable areas of sparsely populated terrain where the problems of development are largely those of overcoming the rigors of a northern environment.

Historical interconnections and cultural similarities are important factors in the regional unity of Northern Europe. Historically, each country has been more closely related to others of the group than to any outside power. In the past, warlike relations often prevailed, and for considerable periods, some countries ruled others of the group. But since the early 19th century, relations among the Northern European peoples have been peaceful and have come to be expressed in close international cooperation among the five countries.

The languages of Denmark, Norway, and Sweden descend from the same ancient tongue and are mutually intelligible. Icelandic, although a branch of the same root, is more difficult for the other peoples because it has changed less from the original Germanic language and has borrowed less from other languages. It derives from Old Norse, the language of the Vikings, and Icelanders have been proud of their linguistic conservatism, which has diminished the adoption of new terms from other tongues. Only Finnish, which belongs to an entirely distinct language family, is fully different from the others. Even in Finland, however, about 6% of the population speaks Swedish as a native tongue, and Swedish is recognized as a second official language.

Among all these countries, cultural unity is also embodied in a common religion. The Evangelical Lutheran Church is the church of approximately 90% of the population, at least nominally. It is a state church, supported by taxes, and is probably the most all-embracing organization outside of the state. The countries of Northern Europe also exhibit basic similarities with respect to law and political institutions. They all have a long tradition of individual rights, broad political participation, limited governmental powers, and democratic control. Today, they are recognized as strongholds of democratic institutions, not just within Europe but globally.

Small size and resource limitations have made it necessary for each of the countries of Northern Europe to build a highly specialized economy in pursuit of a high standard of living.

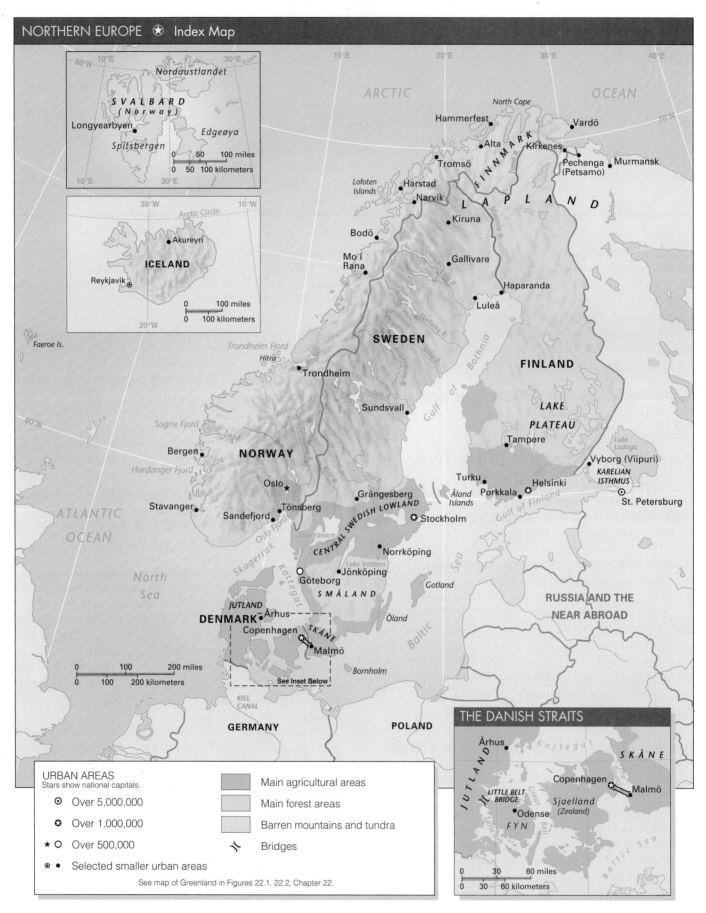

Figure 5.1   Index map of Northern Europe. Small tracts of agricultural land are scattered through the areas of forest, mountain, and tundra, and smaller forest tracts occur within the agricultural areas. Nearly one-third of Northern Europe lies within the Arctic Circle. The major island of Greenland (840,000 sq mi) and the smaller Faeroe and Svalbard Islands are not shown.

Figure 5.2   Farmland in the North often comes down to the edges of the fjords that provide both transport for farm surpluses and a connection with the more densely settled world. This photo of Aurlandfjord, Norway, shows how little of the mountain flank landscape has been opened for farming.

Landscape images of the North often revolve around small, picturesque villages nestled on the flanks of mountains rising out of placid fjords (Fig. 5.2). Success of settlements in such environments has been so marked that these countries are probably known as much for high living standards as for any other characteristic. High standards of health, education, security for the individual, and creative achievement are characteristics common to all five of the Northern nations.

Denmark

Denmark has the distinction of possessing the largest city in Northern Europe, but at the same time, it is the most dependent on agriculture of any country in the region. The Danish capital of Copenhagen (population: metropolitan area, 2,313,500; proper, 1,086,000) has one-third of Denmark's population in its metropolitan area, allowing it — like Paris — to play the role of a primate city. Denmark has a far greater density of population than the other countries of Northern Europe, a fact accounted for by the size of Copenhagen, the greater productivity of the land, and the lack of any sparsely populated zone of frontier settlement.

Copenhagen and the Danish Straits

Copenhagen lies on the island of Sjaelland (Zealand) at the extreme eastern edge of Denmark (see Fig. 5.1). Sweden lies only 12 miles (c. 20 km) away across the Sound, the main passage between the Baltic and the North Seas. The city grew beside a natural harbor well placed to control traffic through the Sound. Before the 17th century, Denmark controlled adjacent southern Sweden as well as the less favored alternative channels to the Sound: the Great Belt on the east of the island of Fyn and the Little Belt to the west. Tolls were levied on all shipping passing to and from the Baltic. Although the days of levying tolls are now long past, Copenhagen still does a large transit and entrepôt business in North Sea–Baltic Sea trade.

In mid-2000, the Oresoundbridge (also called the Oresund fixed link) was opened. This 10-mile (16-km) bridge and tunnel is the first link that has connected Denmark and Sweden. The eastern anchor will rest in Malmo, Sweden. The western terminus will lead to the Copenhagen (Denmark) airport. This link is somewhat like the Channel Tunnel in that it breaks down a very significant insularity that Sweden and Denmark have experienced for more than 7,000 years. The project was built at a cost exceeding $2 billion. F5.3 136

Trade led to industry, and Copenhagen is the principal industrial center of Denmark as well as its chief port and capital. Outstanding industries include the processing of both exported and imported foods, the manufacture of chemicals, brewing, and engineering.

Factors in Denmark's Agricultural Success

About three-fifths of Denmark is arable, and this is the largest proportion of any European country. It is fortunate that so much of the nation has this agricultural potential because the country is greatly lacking in other natural resources. Danish agriculture is so efficient that less than 4% of the working population is employed on the land. Agriculture, however, is basic to the country's economy. Many additional workers engage in processing and marketing the country's agricultural products and in supplying the needs of the farms. Approximately one-fifth of Denmark's exports relate to agriculture (brewing, meat canning, fish processing), and this is second only to machinery (shipbuilding, engineering, metal working). LEGO® building blocks originated in Denmark and continue to have both economic and cultural significance.

Few countries or areas that depend so heavily on agriculture are as materially successful as Denmark. Danish agriculture is based on the highly specialized and carefully fostered development of animal husbandry, which was emphasized when the competition of cheap grain from overseas brought ruin to the previous Danish system of grain farming in the latter part of the 19th century. Traditionally, Danish animal husbandry focused on dairy farming, but since the rapid expansion of the impact of global marketing since the 1970s, the production and sale of animals for meat have become more important than dairying.

A number of factors have been responsible for the general success the Danes have had in agriculture. Membership in the EU and a location adjacent to the largest country in Western Europe (Germany) have provided a ready market for Danish agricultural products. Two other factors have had a major impact as well: The Danes have developed agricultural cooperatives to a high level of efficiency and have given keen attention to adult education in addition to traditional schooling, leading to a skilled labor force in agriculture as well as the industrial sector. Denmark has a high level of overall literacy and an effective skilled labor force.

A final note on Denmark. Greenland has been under Danish control for centuries. In 1979, the Danes granted the large island home rule. In 1985, Greenland withdrew from the European Communities (EC) because of its frustration over EC fishing policies and the lack of any EC development aid.

Norway

Norway stretches for well over 1,000 miles (c. 1,600 km) along its west side and around the north end of the Scandinavian Peninsula. The peninsula, as well as Finland, occupies part of the Fennoscandian Shield, a block of ancient and very stable metamorphic rocks. The western margins of the block in Norway are extremely rugged. Most areas have little or no soil to cover the rock surface, which was scraped bare by the continental glaciation that centered in Scandinavia. About 70% of Norway is classified as wasteland, less than 3% as arable or pasture land, and the remainder as forest (see Fig. 5.2).

The nature of the terrain hinders not only agriculture but also transportation. Glaciers deepened the valleys of streams flowing from Norway's mountains into the sea, and when the ice melted, the sea invaded the valleys. Thus were created the famous **fjords** — long, narrow, and deep extensions of the sea into the land, usually edged by steep cliff faces and valley walls. It is difficult to build highways and railroads parallel to such a coast. The fjords provide some of the finest harbors in the world, but most of them have practically no **hinterland** or productive rural landscape dependent on the central places established on the fjords. Much of Norway's population is scattered along these waterways in many relatively isolated clusters, most of which are small and only moderately interconnected. The linkage provided by the sea is often more functional than surface roads.

More than half of the population of Norway lives in the southeastern core region, which centers on the capital, Oslo (population: 758,949). Located at the head of Oslo Fjord, where several valleys converge, Oslo is the principal seaport and industrial, commercial, and cultural center of Norway as well as its largest city. In this region, valleys are wider and the land is less rugged. The most extensive agricultural lands and the largest forests in the country are located here. Hydroelectricity powers sawmilling, pulp and paper production, metallurgy, electrochemical industries, and industries that produce various consumer goods for Norwegian consumption and some export trade.

Apart from its scanty agricultural base, Norway has rich natural resources relative to the size of its population, which is small and increasing at a very slow rate (0.3% annually). Key resources include spectacular scenery, a vast potential for the generation of hydroelectricity, forests, and a variety of minerals. The country's resource position was greatly enhanced by the discoveries and development, in the 1960s and 1970s, of large oil and natural-gas deposits far offshore in the Norwegian sector of the North Sea floor. In just a few years, Norway's oil and gas exports grew from nothing to the country's largest by value, and Norway advanced from being reasonably well-off to being one of Europe's richest countries as measured by per capita GNP (see Table 3.1).

T3.1
62

Meanwhile, older economic emphases remain important. Manufacturing (petroleum products, chemicals, aluminum, and machinery) is a large employer and provides exports of goods related to Norway's resources. The major export of wood products reflects such a pattern. And Norway has had a major fishing industry for centuries. An ancient seafaring tradition going back to the Vikings is expressed in the current possession of one of the world's largest merchant fleets, and Norway enjoys great success in cruise shipping as well.

The real giant in the Norway economy is not only the income derived from the energy resources of the North Sea, but the political independence associated with that. For example, in 1995, Norway was second only to Saudi Arabia in the export of petroleum worldwide. In addition, Norway gains more than 99% of its electricity from hydroelectricity so that its oil wealth is almost entirely an exportable resource. Currently, Norway gets 99% and Iceland 93% of their power from hydroelectricity. This power-generating resource is perfect in the upland areas of all of Northern Europe because there is so little stream silt and because snowfall and precipitation patterns tend to be relatively stable.

F3.15
83

In economic growth, Norway experienced a rate of 2.7% in 2000 but had an average growth rate of 3.9% for the 1990s, one of the highest in Europe. Norway's 3% unemployment rate is one of the lowest in Europe. The combination of an effective local energy source free of major environmental shortcomings and the new Norway authority in the sale of petroleum products has led to an important political development that has parallels to the successes in Norwegian North Sea oil exploration.

In November 1994, Norwegians voted, by a margin of 53 to 47%, not to join the European Union (EU). This negative vote came about largely because of the scale of the recent economic stability that these energy resources provided. This allowed Norway the capacity to decline the interference of the EU as well as the benefits of membership. Sweden, during the same time period, voted to join the EU by the reverse percentages. In the European Economic Area (EEA), this means that only Iceland, Liechtenstein, and Norway are not members of the increasingly powerful EU. As the regional influence of the EU continues to expand, Norway's decision becomes even more significant.

The forces that led the battle against membership in the EU, at a time when even most of the Eastern European nations are petitioning for EU membership, were coalitions of farmers, environmentalists, fishermen, and nationalists. Their call was to preserve the relative independence of Norway and not slip into the situation where uniquely Norwegian situations would be tampered with by bureaucrats in Brussels, the administrative "capital" of the EU. On the other hand, the people nervous about the "no" vote warned that Norway's petroleum income will surely decline in the future and that the aging of the population will change the budget picture profoundly as ambitious promises for retirement and medical benefits become more costly to honor. However, the farmers and fishermen wanted to retain their Norwegian state subsidies, which are even more generous than those offered by the EU. In the end, it was Norwegian suspicions of transferring political power to Brussels that resulted in the unusual "no" vote. As are so many aspects of the European Union, this vote and associated membership will be revisited in the years to come.

In the late 1990s, Norway began an ambitious investment in transportation infrastructure. Bridges, tunnels, and highways have received much attention, especially in the more remote parts of the country. The government is trying to find a way to stem the increasing migration stream from isolated and remote settlements to small towns and cities in the south. Planners feel that if, for example, a tunnel to Hitra Island (see Fig. 5.1), some 200 miles (320 km) northwest of Oslo, can make the 4,100 inhabitants feel more closely connected to the mainland, it will reduce the number of youth who leave the island permanently in quest of jobs and a more urban lifestyle. Hundreds of millions of dollars were spent in 1998 in just such capital construction projects, even though relatively small numbers of people were directly affected by such work. These efforts represent another example of the demographic flow away from rural places to towns and cities all across the Northern European landscape.

Sweden

Sweden is the largest in area and population (8.9 million) and the most diversified of the countries of Northern Europe. In the northwest, it shares the mountains of the Scandinavian Peninsula with Norway; in the south, it has rolling, fertile farmlands like those of Denmark; in the central area of the great lakes lies another block of good farmland. To the north, between the mountains and the shores of the Baltic Sea and Gulf of Bothnia, Sweden consists mainly of ice-scoured forested uplands. A smaller area of this type occurs south of Lake Vättern, just west of Norrkoping, south of Stockholm.

Economy and Resources

The high development of engineering and metallurgical industries is the feature that most distinguishes the Swedish economy from that of the other countries of Northern Europe. These industries have evolved out of a centuries-old tradition of mining and smelting. The iron ores of the Bergslagen region, just north of the Central Swedish Lowland, have been made into iron and steel for more than a millennium. Originally, the fuel used to smelt the ore was charcoal made from the surrounding forest. When coal and coke began to replace charcoal as fuel in Europe's iron industries in the late 17th century, Sweden was handicapped by its scarcity of coal. But in the later age of electricity, the Swedes responded to this problem by a heavy reliance on electric furnaces using current from well-sited hydropower plants. In 1996, Sweden made the decision to decommission all of its nuclear power plants, standing as an enormously significant environmental decision. The modern Swedish steel industry continues to focus on the high esteem held for "Swedish steel" — quality achieved in part because of the exactness of temperature control associated with Swedish steel production. Traditional small steel centers still exist in many small towns, but output in recent years has been supplemented by production from some large conventional plants built on the Baltic and Bothnian coasts which use imported coal and coke. And although Sweden now produces large quantities of ordinary steel, the ultrafine steels and the products made from them still impart a special character to Swedish metallurgy.

Among the specialties of the Swedish steel industry that yield a large volume and value of exports are automobiles (Volvo and Saab), industrial machinery (including robots), office machinery, telephone equipment, medical instruments and machinery, agricultural machinery, and electric transmission equipment. The country's reputation for skill in design and quality of product also extends to items such as ball bearings, cutlery, tools, and the glassware and furniture crafted in Småland, the infertile and rather sparsely populated plateau south of the Central Lowland (which lies west of the coast from Stockholm to Norrkoping).

Overall, Swedish manufacturing occurs principally in numerous centers in the Central Lowland. These places, ranging in population from about 125,000 to mere villages, have grown up in an area favorably situated with respect to minerals, forests, water power, labor, food supplies, and trading possibilities. The Central Lowland is the historic core of Sweden

In the generation of electricity, there is no means more environmentally benign and geographically sensitive than water power. Water that is ponded — or captured behind dams — to develop **water head** or the volume of water necessary to turn turbines possesses the potential for **multiple uses**. Once the water moves through a complex of turbines and dynamos, generating power with virtually no waste product of any sort, it returns to a stream system that exists downstream from the dam. In small systems, power is generated by diverting flowing streams into small generating units and then returning water directly to the stream system. Thus, water is a **flow resource** or a **renewable resource**.

In hydroelectric complexes that require the construction of dams, there is the inevitable loss of some settlement land as rivers are stopped in their flow, ponded to depths of sometimes scores or hundreds of feet, and then released through major turbine and penstock networks. This system of power generation works best where there is little or only modest silt carried in the water being utilized; a heavier silt load in the reservoir behind the dam tends to settle, reduces the water head obtainable at that site, and thus shortens the life of the hydroelectric project. In the larger projects built around massive river systems, major flood control components are commonly designed into the project. This has great potential benefit to downstream farming and urban populations. Currently, Norway gets 99% and Iceland 93% of their power from hydroelectricity. It is a perfect resource in the upland areas of the north and an important resource in this region particularly.

and has long maintained important agricultural development and a relatively dense population.

The introduction of new labor-saving technologies in the forestry industries has reduced job opportunities in this most traditional occupation in the country's remote lands. Recent national promotions, however, have touted remote villages in the Swedish north as excellent settings for computerized telephone call-centers. Because of a tradition of well-educated, keyboard-literate youth, even in the remote reaches of the country, the Swedish government is attempting to lure telecommunication firms to these places. The hope is also that the far north and upland will seem less distant and remote if a young population will come there for good jobs and global computer linkages. Thus, Sweden — like Norway with its transportation investments — is hoping to slow down the steady migration of young populations from villages to the cities in the south.

Unlike Denmark, Sweden has not developed a specialized export-oriented agriculture; and unlike Norway, Sweden commands enough good land to supply all but a minor share of its small population's food requirements. More than 92% of Sweden is nonagricultural, but there is some good land, largely in Skåne at the southern tip of the peninsula (Fig. 5.3) and in the Central Lowland. Skåne, with relatively fertile soils like those of adjacent Denmark and with Sweden's mildest climate, is the prime agricultural area. This fertile agricultural and relatively well-settled area is the source for the term "Scandinavia."

Stockholm and Göteborg

Stockholm (population: metropolitan area, 1,620,900; proper, 1,226,200) is the third largest city in Northern Europe. The location of the capital reflects the role of the Central Lowland as the center of the Swedish state and the early orientation of that state toward the Baltic and trans-Baltic lands. Stockholm is the principal administrative, financial, and cultural center of the country and shares in many of the manufacturing activities typical of the Central Lowland. But in the past century, as Sweden has traded more via the North Sea, Stockholm has been displaced by Göteborg (population: metropolitan area, 744,300; proper, 500,800), located at the western edge of the Central Lowland, as the leading port of the country and, in fact, of all Northern Europe. In addition to its North Sea location, Göteborg has a harbor that is ice-free year round, whereas icebreakers are needed to keep Stockholm's harbor open in winter.

The Swedish Policy of Neutrality and Preparedness

In the Middle Ages and early modern times, Sweden was a powerful and imperialistic country. In the 17th century, the Baltic Sea functioned almost like a Swedish lake. During the

Figure 5.3   In the final stages of the completion of the 10-mile (16-km) Oresund Fixed Link (bridge) that now links the Danish capital of Copenhagen with Sweden's third largest city (Malmo, located in Skåne), 35,000 cyclists participated in a 3-day celebration of this new link between Sweden and Denmark that stretches across The Sound.

18th century, however, the rising power of Russia and Prussia put an end to Swedish imperialism, aside from a brief campaign in 1814 through which Sweden won control of Norway from Denmark. Since 1814, Sweden has not engaged in a war, and it has become known as one of Europe's most successful neutrals. The long period of peace has undoubtedly been partially responsible for the country's success in attaining a high level of economic welfare and a reputation for social advancement. However, Sweden stays heavily armed while maintaining its traditional policy of neutrality.

Finland

Conquered and Christianized by the Swedes in the 12th and 13th centuries, Finland was ceded by Sweden to Russia in 1809 and was controlled by Russia until 1917. Under both the Swedes and the Russians, the country's people developed their own culture and feelings of nationality. The opportunity for independence provided by the collapse of Tsarist Russia in 1917 was eagerly seized. This independence has been also promoted by the fact that the Finnish language is entirely distinct from the other Scandinavian languages, showing a closer historical link with Hungarian Magyar and lands to the east, as opposed to the countries of the Scandinavian Peninsula.

Most of Finland is a sparsely populated, glacially scoured, subarctic wilderness of coniferous forest, ancient igneous and metamorphic rocks, thousands upon thousands of lakes, and numerous swamps. One-third of Finland lies within the Arctic Circle. Despite this environment, Finland was primarily an agricultural country until very recently, and the majority of the population resides in the relatively fertile and warmer lowland districts scattered through the southern half of the country. Hay, oats, and barley are the main crops of an agriculture in which livestock production, especially dairying, predominates.

To pay for a wide variety of imports, the nation depends heavily on exports of forest products. About two-thirds of Finland is forested, mainly in pine or spruce, and about two-fifths of the country's exports generally consist of wood products such as paper, pulp, and timber. Forest production is especially concentrated in the south central part of the country, often referred to as the Lake Plateau. A poor and rocky soil discourages agriculture here, but a multitude of lakes, connected by streams, aid in transporting the logs.

Helsinki (population: metropolitan area, 1,131,200; proper, 566,800), located on the coast of the Gulf of Finland, is Finland's capital, largest city, main seaport, and principal commercial and cultural center. It is also the country's most important and diversified industrial center. Hydroelectricity originally powered Finland's industries, but this source has now been surpassed by both thermal and nuclear power. The thermal stations use imported fossil fuels.

The Finns were able to make a rapid recovery after suffering great devastation in World War II. The economy drastically changed in some respects. One of the most striking changes was the rise of metalworking industries. This was made necessary when the Soviets required that a large part of the reparations required of Finland to be paid to the USSR were to be in metal goods. Several small steel plants and other metalworking establishments were built to meet this demand, and they continue to operate today. Finland currently gains nearly half of its export income from machinery. Nokia Corporation, one of the world's most successful mobile-phone firms, is an example of high-tech manufacturing in Finland. The second largest export sector is paper and forestry products.

Iceland

Iceland is a fairly large (39,699 sq mi/102,819 sq km), mountainous island in the Atlantic Ocean just south of the Arctic Circle. Its rugged surface shows the effects of intense glaciation and vulcanism. Some upland glaciers and many active volcanoes and hot springs remain. The vegetation consists mostly of tundra, with considerable grass in some coastal areas and valleys. Trees are few, their growth discouraged by summer temperatures averaging about 52°F (11°C) or below, as well as by the prevalence of strong winds. The cool summers also handicap agriculture. Mineral resources are almost nonexistent.

Despite the deficiencies of its environment, however, Iceland has been continuously inhabited at least since the ninth century and is now the home of a progressive and democratic republic with a population of about 300,000. Practically the entire population lives in coastal settlements, with the largest concentration in and near the capital, Reykjavik, which has a city population of just over 112,600. Because of the proximity of the relatively warm North Atlantic Drift, the coasts of Iceland have winter temperatures that are mild for the latitude. Reykjavik (64°N latitude) has an average January temperature that is practically the same as New York City's (41°N latitude). The island's topography and abundant water support significant generation of hydroelectric power (Fig. 5.4).

ROBERT GILL, PAPILIO/CORBIS

Figure 5.4 The capacity to generate electricity from controlled water is not limited only to major hydroelectric complexes. This small-scale operation in southern Iceland is an example of how a small river can be harnessed by a low dam and serve to produce power for local uses. All this is done with minimal landscape modification.

LANDSCAPE IN LITERATURE

The Cloud Sketcher

In this novel about the short life of a strong and creative Finnish architect, we are given a scene in which he determines his first major design. Having seen him grow up in the lean and uncertain years of Finland at the end of the second decade of the last century, we here see him determining the site and the overall design for his creation to be later known as The Church of the Shadow Cross. It is from this design that he gained the confidence necessary to undertake other minor Finnish efforts and then to cross the Atlantic and attack the growing and flexing New York City architectural frontier. As you try to envision the structure that Esko is contemplating, think of the Western traditions that worked on his mind and look at the ways his design of space and function were unique. The novel is loosely related to the life of Eliel Saarinen, a major Finnish architect and the father of Eero Saarinen, the Finnish designer of the Gateway Arch in St. Louis, commemorating westward migration across the United States.

Esko stopped, turned, looked back toward the barn at the distant end of the meadow; it was a fine barn, built in the old Finnish style, with chinks between the horizontal logs that let in wind and light; it was in a fine location, well protected and with good drainage; it was an admirable barn in every way, red and sturdy

and with years of use in it yet; it was going to have to go, for it lay on the precise spot where he knew he must build the church.

Under the shade of his wide straw hat Esko's face was humorous and determined. He doffed the hat, sipped cool water from a bottle that he'd brought with him, and smoothed his hair with the attitude of a man satisfied to have come to terms with a problem. . . .

Esko turned his face to the sun, its warmth making him shudder with pleasure; he put his hat back on, took another gulp of water, and sat down, opening his sketch pad. He began to draw, his hand scratching and swirling across the paper. Now that he'd finally decided on the site, everything else came easily, in a breath, in a burst, as if the church already existed in his mind and the unearthing of this one last detail opened a door to reveal it.

The space *within* the building would be its primary reality, not the exterior. The altar, he saw, must be a simple white oblong struck by light. Behind it would hang a cross, and behind the cross, recessed into the wall to a depth of three or four inches, would be the shape of a cross, a cross of exactly the same dimensions, the shadow cross. In front of the altar, and level with it, would stand pews of white pine, leading back toward tall doors of oak. The walls would be plain, whitewashed; the white floors would be of concrete. The ceiling would be barrel-vaulted, with openings cut out of it at unpredictable intervals through its heights and curve. Thus, during each day, as the sun moved above the open Ostrobothnian plain, there would be a constant rotation of light and shadow. The light within the space would never be

still; it would shift and dance, from brilliance to shadow, and back again. When the sun went down, and throughout the winter, interior light would come from deep-hanging lamps of glass and copper; the barrel ceiling would reside in mysterious dark, with perhaps the stars and the northern lights flashing through, while the space below would be warm, welcoming.

Esko was trying for something bold. Hitherto churches had developed since Roman times in many different ways but according to one architectural law that had been universally applied: a church always had two naves crossing each other at right angles, forming a cross. Esko was ignoring that, using the cross motif in an entirely different and original way. This in itself would make his church striking, modern; some might even call it ungodly, just on principle. He knew he had to counter that possible charge by really making sure that this was a holy and peaceful place. After all, whatever its architectural pretensions, it would still be a church, and people would be bringing the most religious and wounded parts of themselves here; accordingly it should evoke both what was unknowable, eternal, those lofty ideas to which man aspires in his dreams, and what was real, intimate, the drama of changing fortune and destiny with which every human being struggles every day.[a]

[a]From *The Cloud Sketcher*, by R. Rayner, pp. 162–163. Copyright © 2000 HarperCollins. Reprinted by permission.

Questions: What role does environment play in design decisions? Does architecture reflect the environment in which it occurs?

F24.14
644

The economy of Iceland depends basically on fishing, farming, and the processing of their products. Agriculture focuses on cattle and sheep raising and a limited production of potatoes and hardy vegetables. The real backbone of the economy is fishing, which supplied almost three-quarters of all exports by value in 1997.

In 1944, Iceland declared itself a republic. Iceland has strategic importance because of its position along major sea and air routes across the North Atlantic; the country is also a member of the North Atlantic Treaty Organization (NATO).

Greenland, the Faeroes, and Svalbard

Denmark and Norway possess outlying islands of some significance. Greenland was earlier a colony of Denmark, became a Danish province in 1953, and in 1979 was granted home rule. It is the world's largest island (840,000 sq mi/2,175,600 sq km) and lies off the coast of North America. Greenland and the Faeroe Islands, between Norway and Iceland, are considered integral, although self-governing, parts of Denmark, and their peoples have the rights of Danish citizens. Although Greenland has nearly one-fourth the area of the United States, about 85% is covered by an icecap, and

90% of the total island population is distributed along the southern coast. Total population amounts to only about 56,000. Nearly all of this population is of mixed Eskimo, Inuit, and Scandinavian descent. The principal means of livelihood are fishing, hunting, trapping, sheep grazing, and the mining of zinc and lead. Under the auspices of NATO, joint Danish-American air bases are maintained there, and the United States has a giant radar installation at Thule in the remote northwest. (See maps in Chapter 22.)

The Faeroes are a group of treeless islands where some 46,000 people of Norwegian descent make a living by fishing and grazing sheep. Norway controls the island group of Svalbard, located in the Arctic Ocean and commonly known as Spitsbergen, which is also the name of its main island. Although largely covered by ice, the main island has the only substantial deposits of high-grade coal in Northern Europe. Mining is carried on by Norwegian and Russian companies.

5.2 Southern Europe: The Mediterranean World

On the south, Europe is separated from Africa by the Mediterranean Sea. Three large peninsulas extend southward from the continent of Europe into the Mediterranean (Fig. 5.5). To the west, south of the Pyrenees Mountains, is the Iberian Peninsula, unequally divided between Spain and Portugal. In the center, south of the Alps, is Italy and its southern offshoot, the island of Sicily. To the east, between the Adriatic and Black Seas, is the Balkan Peninsula, from which the Greek subpeninsula extends still farther south between the Ionian and Aegean Seas. The four main countries of Southern Europe, as considered in this text, are Portugal, Spain, Italy, and Greece. But Southern Europe also includes the small island country of Malta, the microstates of Vatican City (enclosed within Rome), San Marino, and the British colony of Gibraltar. Three of the peninsular countries include islands in the

Figure 5.5 Southern Europe. General location map of Mediterranean Europe. Certain Southern European traits mark southern France and other countries named on the map but are discussed elsewhere.

Mediterranean, the largest of which are Italy's Sicily and Sardinia, the Greek island of Crete, and the Spanish Balearic Islands. Some Mediterranean islands lie outside these countries, the most notable being Corsica, which is a part of France, and Cyprus, a contested landscape claimed by both Turkey and Greece, but we include it as part of Russia and the Near Abroad.

Topography and Population Distribution

In terrain and population distribution, as well as in climate and agriculture, the countries of Mediterranean Europe present various points of similarity. Rugged terrain predominates in all four major countries. Plains tend to be small and face the sea, separated from each other by mountainous areas. Population distribution corresponds in a general way to topography, with the lowland plains being densely populated and the mountainous areas much less so, although a number of rough areas attain surprisingly high densities. The Alps and the Pyrenees have long served as a permeable barrier between the Coreland and Southern Europe.

The Ecology of Mediterranean Agriculture

A distinctive natural characteristic of Southern Europe is its **Mediterranean climate,** which combines mild, rainy winters with hot, dry summers. Characteristics associated with this climate become increasingly pronounced toward the south. The northern extremities of both Italy and Spain have atypical climatic characteristics. Except for a strip along the Mediterranean Sea, northern Spain has a marine climate like that of northwestern Europe. It is cooler and wetter in summer than the typically Mediterranean climate areas, and the basin of the Po River in northern Italy is distinguished by a relatively wet summer and cold-month temperatures in the lowlands averaging just above freezing. Much of the high interior plateau of Spain, called the Meseta, is cut off from rain-bearing winds and has somewhat less precipitation and colder winters than is typical of the Mediterranean climate, although the seasonal regimen of precipitation with dry summers is characteristically a Mediterranean climate trait. Thus, the Meseta is enormously productive agriculturally, and the agriculture is based principally on crops that are naturally adapted to winter rainfall and summer drought.

Although all agricultural patterns create a mosaic of crops and land-use patterns in agricultural geography, winter wheat deserves special note here. Although it is a common grain crop in much more than the Mediterranean world, it has great importance in this region, and has for thousands of years. This grain crop is generally planted at the end of the harvest of a summer crop. In September or as late as October, farmland is cleared of crops that were planted in spring, cultivated through the summer, and harvested before the first frost. Wheat is planted as fall temperatures drop and daylight shortens. Winter wheat is an unusually hardy grain, for it has to endure any snow cover that comes, although this is not common

in the South. Winter wheat tends to give a higher yield than does spring wheat, which is planted in late spring and harvested in early fall. Tillers (grain heads) form on winter wheat, and when cold weather and snow come, the plant goes dormant. After mild, rainy winters and the spring snows melt, wheat begins growing again, taking advantage of the moisture, the return of the longer hours of daylight, and the new warmth. In May or June, before any other crops have grown long enough — or have even been planted — to bring the farmer a harvest, winter wheat comes to fruition. Such timing produces a very popular and usable crop at the outset of the busy summer farming season.

Barley, a more adaptable grain, tends to supplant wheat in some areas that are particularly dry or infertile. Other typical crops in the South are olives, grapes, citrus, and vegetables. The olive tree and the grapevine have extensive root systems and certain other adaptations that allow them to survive the summer droughts. Olive oil is a major source of fat in the typical Mediterranean diet, and virtually all of the world supply is produced in countries that touch or lie near the Mediterranean Sea or are in a Mediterranean climate zone. The principal use of grapes is for wine, a standard household beverage in Southern Europe and a major export product. Where irrigation is lacking, vegetables are grown that will mature during the wetter winter or in the spring. The most important are several kinds of beans and peas. These are a source of protein in an area where meat animals make only a limited contribution to the food supply; feedstuffs are not available to fatten large numbers of animals, and parched summer pastures further inhibit the meat supply. Extensive areas too rough for cultivation are used for grazing, but the carrying capacity of such lands is relatively low. Sheep and goats, which can survive on a sparser pasturage than cattle, are favored. They are kept only partially for meat, and the total amount they supply is relatively small. In many places, grazing depends on **transhumance** — utilizing lowland pastures during the wetter winter and mountain pastures during the summer. In some areas, nonfood crops supplement the basic Mediterranean products. An example is tobacco, which is an export of Greece.

Mediterranean agriculture comes to its peak of intensity and productivity in areas where the land is irrigated (Fig. 5.6). Irrigation on a small scale is found in many parts of Southern Europe, but a few irrigated areas stand out from the rest in size and importance. Among these are northern Portugal, the Mediterranean coast of Spain, the northern coast of Sicily, and the Italian coastal areas near Naples. The largest of all, the plain of the Po River in northeast Italy, uses irrigation to supplement year-round rainfall. In northern Portugal, irrigated corn (maize) replaces wheat as the major grain, and some cattle are raised on irrigated meadows. Grapes are the other agricultural mainstay of Southern Europe, with wines of the region providing significant income for all four of the major nations, as well as providing a very major source of profitable tourist images.

F10.11
290

DEFINITIONS + INSIGHTS

Irrigation

The act of irrigation — the watering of land by controlled methods — is one of the most significant aspects of the agricultural landscape. The Chinese talk about "teaching water" when they devise systems of irrigation. In Europe, and especially in the Mediterranean climate zone, hydraulic control is the factor that has taken farming from a marginal economic activity to a more prosperous one. Elements central to the irrigation process include canals, dams, ponds, pumps, gravity flow, wells, and evaporation. The process is thousands of years old and is at the center of being able to increase a region's arable land. In the Mediterranean world, irrigation is particularly significant because of the summer drought climate pattern. There is good sunlight and often good soils in this climate zone, but there is never enough rainfall in the main summer growing season. For this reason, labor, technology, and tradition are all brought into play to bring summer water to orchards, vineyards, and even grains. Wherever farmers have invested labor, capital, and ingenuity in designing effective irrigation systems, they realized increased agricultural yields. Irrigation is an innovation that has brought even more significance to the agricultural sector in Southern Europe.

In the Mediterranean coastal regions of Spain, irrigation makes possible the development of extensive orchards. Oranges are the most important product, with growth concentrated around the city of Valencia, which has given its name to a type of orange. Lemons are particularly important on the island of Sicily. The district around Naples, known as Campania, is more intensively farmed, productive, and densely populated than any other agricultural area of comparable size in Southern Europe. Irrigation water applied to exceptionally fertile soils formed of volcanic debris from Mt. Vesuvius supports a remarkable variety of production that includes almost every crop grown in Southern Europe. Despite the intensive

Figure 5.6 The stone town of San Gimignano on the hilltop in this scene overlooks small farm homes and tightly worked farmland. The close-packed row of the vineyards (the most intensely green) and the neatly spaced orchard trees represent the productive organization of Mediterranean agriculture, especially where irrigation has been introduced and maintained. This scene is from Tuscany, Italy.

cultivation and productivity of many locales in this area, the overcrowded population on its tiny farms is relatively poor, though resident in very picturesque settings. Fruits and vegetables, a large proportion of which are often destined for export, tend to supplement, and sometimes largely displace, other types of agricultural products.

Spain

On the Iberian Peninsula, the greater part of the land consists of a plateau, the Meseta, with a surface lying between 2,000 and 3,000 feet (c. 600–900 m) in elevation. The plateau surface is interrupted by deep river valleys and ranges of mountains. Population density is generally light, between 25 and 100 per square mile (c. 10–40 per sq km), due a lack of rainfall, compared to the European average of approximately 280 per square mile (c. 108 per sq km). The plateau edges are mostly steep and rugged. The Pyrenees and the Cantabrian Mountains border the Meseta on the north, and the Betic Mountains, culminating in the Sierra Nevada, border it on the south. Most of the population of Iberia is distributed peripherally on discontinuous coastal lowlands.

In Spain, the three largest cities are the capital, Madrid (population: metropolitan area, 5,028,700; proper, 2,894,100), and the two Mediterranean ports of Barcelona (population: metropolitan area, 3,855,300; proper 1,497,000) and Valencia (population: city proper, 740,500). Madrid was chosen as the capital of Spain in the 16th century because of its location near the geographic center of the Iberian Peninsula, approximately equidistant from the various peripheral areas of dense population and of sometimes separatist political tendencies. Located in a poor region, Madrid has had little economic excuse for existence during most of its history. But its position as the capital has made it the center of the Spanish road and rail networks and thus has given it nodal business advantages. In Spain's recent economic boom, the city has begun to attract industry as well as tourists on a considerable scale.

Barcelona is Spain's major port and the center of a diversified industrial district. The main elements of Barcelona's industrialization have been (1) the early formation of a commercial class that supplied financing, (2) the availability of hydroelectric power, mainly from the Pyrenees, (3) cheap and sufficiently skilled labor, and (4) imported raw materials.

Valencia, Spain's third city in population, is the business center and port for an unusually productive section of coastal Spain. An extensive irrigation system in the area was developed by the Moors in the Middle Ages. Today, the density of population on irrigated land resembles that of Italy's Campania, but here, much of the agricultural effort goes into producing oranges, which are Spain's largest single agricultural export.

The main period of Spanish power and influence reached a new zenith in 1492, when a centuries-long struggle, the Reconquista, ended with Spain's defeat of the Kingdom of Granada, the last part of Iberia held by the Muslim Moors. In that same year, Columbus, a Genoese Italian navigator in the service of Spain, landed in the Americas in search of a sea route to India

KIT SALTER

Figure 5.7   A doorway to La Mezquita, or the Great Mosque of Cordoba, Spain. This Moorish temple was built in the eighth century and turned into a Christian cathedral in 1236. It exists today in southern Spain as one of the most visually stunning examples of the blending of Moorish and Christian sacred space. The tension and contest that characterized the Moorish presence from 711 to 1492 in the European world have left elements of language, custom, religion, cuisine, and architecture all across the landscape.

and East Asia. For a century thereafter, Spain was the greatest power in Europe and probably in the world (Fig. 5.7). It rapidly built an empire in areas as widespread as Italy, the Low Countries, Latin America, Africa, and the Philippines. These possessions were subsequently lost at various times during a long, slow decline of Spanish power over the next three centuries. The Canary Islands, however, off the Atlantic coast of Morocco, are still Spanish-held and governed as an integral part of Spain.

Portugal

Portugal also played a part in expelling the Moors from Iberia and, in the 15th century, took the lead in seeking a sea route around Africa to the Orient. Henry the Navigator of Portugal set in motion the exploration of the coast of northwest Africa between 1418–1460. The accomplishment of rounding the Cape of Good Hope and reaching the Muslim-controlled major trading port of Calicut near the southwest tip of the peninsula of India became even more important than the 1492 Columbian voyage in terms of a shift in the patterns of spice trade. The first Portuguese expedition to succeed in the quest for the spice lands of south and southeast Asia was the voyage, headed by Vasco da Gama, that returned from India in 1499. For the better part of a century thereafter, Portugal dominated European trade with the East and built an empire there and across the Atlantic in Brazil. Then, there was a rapid decline in Portuguese fortunes, although they controlled Brazil until 1822, when it gained independence. Portugal also held large African territories until the 1970s.

The two large cities of Portugal are both seaports on the lower courses of rivers that cross the Meseta and reach the Atlantic through Portugal. Lisbon (population: metropolitan area, 2,606,300; proper, 556,800), on a magnificent natural harbor at the mouth of the Tagus River, is both the leading seaport of the country and its capital. The smaller city of Oporto (population: metropolitan area, 1,207,000), at the mouth of the Douro River, is the regional capital of northern Portugal and the commercial center for trade in Portugal's famous port wine, which comes from terraced vineyards along the hills overlooking the Douro Valley.

Italy

Although Italy is a crucial segment of Southern Europe as outlined in this chapter, it is also important to see this country as a major part of the European industrial engine. As can be seen from Table 3.1, Italy's income level exceeds that of the U.K. and stands clearly above any other country in Southern Europe. The multiplicity of roles played by this country reminds us again of how many worlds exist in Europe and of how many distinct roles most countries play.

In Italy, the Alps, and the Apennines are the principal mountain ranges. Northern Italy includes the greater part of the southern slopes of the Alps. The Apennines, lower but often rugged, form the north–south trending backbone of the peninsula, extending from their junction with the southwestern end of the Alps to the toe of the Italian boot, appearing again across the Strait of Messina in Sicily, where Mt. Etna, a volcanic cone, reaches 10,902 feet (3,323 m). Near Naples, Mt. Vesuvius, at 4,190 feet (1,277 m), is one of the world's most famous volcanoes and is a continual threat to the city of Naples with more than 3 million people. West of the Apennines, most of the land between Florence on the north and Naples on the south is occupied by a tangled mass of lower hills and mountains, often of volcanic origin. All through the Mediterranean region, the monumental architecture that is so much a part of the ancient and current landscape comes from Italian marble (Fig. 5.8).

Italy overall has a population density of nearly 508 per square mile (c. 190 per sq km), more than twice the density of any other large Mediterranean nation. Parts of the Italian high-

(margin: T3.1 62)

(vertical credit: DAVID PATERSON, WILD COUNTRY/CORBIS)

Figure 5.8   Much of the monumental architecture in Western Europe is created from marble, alabaster, or smooth grained limestone that is abundant in most of the countries that surround the northern Mediterranean. These blocks of gleaming Carrara marble come from the region that gives this rock its name, Carrara in the Tuscany region of Italy.

lands have population densities of more than 200 per square mile (c. 75 per sq km), but even these areas are much less densely populated than most Italian lowlands. The largest lowland, the Po Plain, has about two-fifths of the entire Italian population, with nonmetropolitan densities frequently reaching 500–700 per square mile (c. 190–310 per sq km). Some other lowland areas with extremely high densities include the narrow Ligurian Coast centering on Genoa, the plain of the Arno River inland to Pisa and Florence, the lower valley of the Tiber River including Rome, and Campania around Naples.

The Po Basin

The Po Basin is the economic heart of modern Italy and the most highly developed and productive part of Southern Europe. Comprising just under a fifth of Italy's area, this slightly moister Mediterranean climate zone contains about two-fifths of Italy's population (who are more prosperous by far than the rest) and accounts for approximately one-half of its agricultural production and two-thirds of its industrial output.

A very rapid expansion and diversification of Po Basin industry, previously dominated by textiles, occurred after World War II. Prominent components of this expansion were (1) the growth of engineering and chemical industries, (2) the development of a sizable iron-and-steel industry, mainly in northern Italy but partly in peninsular Italy, and (3) a shift to domestic natural gas and imported oil as the predominant sources of power, with hydroelectric energy continuing to play an important role.

Many cities in the basin share in the region's industry, but the leading ones are Milan (population: metro area, 4,047,500; proper 1,301,600), Turin (population: metro area, 1,619,400; proper, 901,100), and Genoa (population: metro area, 682,100; proper, 632,500). Milan is Italy's largest industrial city and a dominant center of finance, business administration, and railway transportation. It is advantageously located between the port of Genoa and important passes, now shortened by railway tunnels through the Alps to

Switzerland and the North Sea countries. Turin is the leading center of Italy's automobile industry. It is particularly associated with FIAT SpA, which is by far the largest Italian auto manufacturer. The port of Genoa functions as a part of the Po Basin industrial complex, although it is not actually in the basin. It is reached from the Milan-Turin area by passes across the narrow but rugged northwestern end of the Apennines. The city dominates the overseas trade of the Po area and is Italy's leading seaport. Genoa is now much larger and economically more important than its old medieval commercial and political rival, Venice (population: 425,000), whose location on islands at the eastern edge of the Po Plain is relatively unfavorable for serving present-day industrial centers.

In Italy, the largest cities outside of the northern industrial area are the capital, Rome (population: metro area, 3,546,500; proper, 2,646,700), and Naples (population: metro area, 3,620,300; proper, 997,700). Rome (Fig. 5.9) has a location that is roughly central within Italy and is predominantly a governmental, religious, and tourist center. Naples, located farther south on the west coast of the peninsula, is the port for the populous and productive, but very poor, Campanian agricultural region described earlier. It is also the main urban center of one of Italy's major tourist regions, with attractions such as Mt. Vesuvius, the ruins of Pompeii, and the island of Capri in the vicinity.

Italy's main period of preeminence occupied approximately five centuries, during which the Roman Empire ruled over the whole Mediterranean basin and some lands beyond. Within the Empire, Christianity began and spread, and by the time the Empire collapsed in the fifth century A.D., Rome had become the seat of the popes and the Catholic church. During the later Middle Ages, a number of Italian cities, notably Venice, Genoa, Milan, Bologna, and Florence, became powerful independent states. Venice and Genoa controlled maritime empires within the Mediterranean region, and all the cities profited as traders between the eastern Mediterranean and Europe north of the Alps. After unification in the 19th century, Italy attempted to emulate past glories, but the empire that it acquired in northern Africa — on the eastern Mediterranean and Ethiopia — was lost when Italy was defeated in World War II.

Domestic Imbalances in Italy's Economic Growth, and Implications for the EU

There has long been an imbalance in the pace and scale of growth between northern Italy (Milan, Turin, the Po Valley) and southern Italy (Rome, Naples, and islands in the Mediterranean). In the 1990s, northern Italy had labor shortages while the south had approximately 20% unemployment. The productivity of labor and industry in the south is approximately 80% that of the north. Per capita income in the south is less than three-fifths that of the north. Even with considerable aid from Rome, this regional imbalance continues apace.

Shortly after World War II, the Italian government began efforts to develop the southern part of its peninsula. Agricultural development, land reform, and the introduction of subsidized industries were among the methods used to stimulate development. But the differential between north and south remains and is widely perceived in Italy as a major national problem. This was accentuated by the 1996 efforts of the Po Valley region to draw attention to the possibility of secession and the creation of a new country, to be called Padania, made up of the northern third of the Italian Peninsula. This region is home to most of Italy's heavy industry and considerable agricultural and resource wealth. Given the history of politi-

Figure 5.9   Panorama of Rome. The low skyline and uniform architecture are characteristic of older European cities. The gleaming white structure in the left foreground is the Victor Emmanuel Monument, celebrating the 19th-century unification of Italy. The ancient Roman Coliseum is at the upper right. The blend of massive ceremonial structures that span millennia with parks, gardens, and densely built-up residential and commercial space is representative of much of Europe. This combination is important to the tourist flow in such places.

cal flux in Europe, even seemingly facetious suggestions like this have to be given some consideration.

Italy's domestic imbalance is of particular geographic concern because one of the prime goals of the European Union, with its new euro currency, is to bring all of the member nations of the EU into a greater common prosperity. It is hoped that with the full achievement of **open borders,** which is a treaty-guaranteed goal of the EU, and with a common currency built around the euro, the wealth of some of the member nations will transcend historic borders and regional differences and expand prosperity regionwide. However, if a single EU country cannot achieve this prosperity, what will be the impact on the future scenarios of the EU? In the case of Italy, it is felt that this regional differentiation is both historic and not particularly unusual. This rather common pattern of relatively more wealth in highly urbanized regions with the associated concentration of industrial activity, good harbors, and networks of transportation leads to a regional dominance. Add to that the early geographic centrality of these wealthy regions in the major industrial growth of Europe and you find asymmetry in the wealth of countries. In the case of Italy, it is the wealthier north and less prosperous south. Such imbalances occur all through Europe and in most other parts of the world as well. The lesson to be learned from the Italian example is that the broad expansion of wealth across the full realm of the European Union may not happen nearly as quickly as anticipated — if it happens at all.

Greece

Box 60 In Greece, most of the peninsula north of the Gulf of Corinth is occupied by the Pindus Mountains and the ranges that branch from them. Extensions of these ranges form islands in the Ionian and Aegean Seas. Greece south of the Gulf of Corinth, commonly known as the Peloponnesus, is composed mainly of the Arcadian mountain knot. Along the coasts of Greece, many small lowlands face the sea between mountain spurs and contain the majority of the people. A particularly famous lowland, although far from the largest, is the Attica Plain, still dominated as in ancient times by Athens and its seaport, Piraeus. Centuries before the Christian era, Greek artists, architects, authors, philosophers, and scholars produced many of the ideas and works that laid the foundation for Western civilization. Between about 600 and 300 B.C., the sailors, traders, warriors, and colonists of the Greek city-states spread their culture throughout the Mediterranean area. A second period of Greek power and influence occurred in the Middle Ages, when Constantinople (modern Istanbul, Turkey) was the capital of the large Byzantine Empire. This Greek empire developed and diffused the Eastern Orthodox branch of Christianity.

Athens, the largest city in Greece by an overwhelming margin with a metropolitan population of 3,192,600 (proper, 759,100) (including the port of Piraeus), contains nearly a third of the population of Greece, is growing rapidly, and has

Figure 5.10   In so much of Europe, and particularly in Western Europe in the areas of high tourist interest, there is a continual interplay of the past and the present. In this photo of the theater of Herod Atticus in the Acropolis, a stage has been created and set for a concert among the stone remains of a theater built in approximately A.D. 200. In the background spreads Athens, Greece, blanketed in part by the smog that has traveled into the scene with the growth of the city and the use of the automobile.

no national rival. It is the capital, main port, and main industrial center. It is a classic primate city. Its development is based on imported fuels, and manufactured products are sold almost entirely in a domestic market whose main center is Athens itself.

Tourism is the mainstay of the Greek economy, with nearly $4 billion a year coming from this trade. Greece has not only the beauty of the Mediterranean Sea to bring people to its shores but also the ruins of early Greek culture and ongoing festivals in art and drama, all combining to make this an economic and cultural environment of major significance (Fig. 5.10).

Geographic Factors Influencing Southern Europe's Uneven Development

Today, Southern Europe is, as it has long been, one of the relatively poor parts of Europe, along with Eastern Europe (see Table 3.1). The main factor contributing to Southern European poverty seems to have been a lagging development of industry related to unfavorable social and resource conditions during the 19th and early 20th centuries, when many European countries were industrializing rapidly. Lack of industrial employment has left too many people in crowded small-farm areas where output per person (although not always per acre) is low. Among the inimical social conditions in Mediterranean Europe were:

1. *Trade deficiency.* The absence of large-scale trade and therefore of capital accumulation to invest in manufacturing created a steady lack of development capital.
2. *Low educational levels.* The preponderance of poor and uneducated populations offered little in the way of skilled

T3.1 62

labor or dynamic markets and, in combination with the trade deficiency, worked to slow economic growth.

3. *Land tenure patterns.* The control of land and often government by wealthy landowners whose interests lay in maintaining the agrarian societies that they dominated meant that there was little drive for a significant change in the way that society and the economy were organized.

4. *Ineffectual government assistance.* The prevalence of unstable, undemocratic, and ineffective governments little oriented toward economic development served as a continuing impediment to real economic growth.

5. *Resource deficiencies.* Resource deficiencies included a lack of good coal and iron ore deposits, except in northern Spain; deficient wood supplies due to dryness and centuries of deforestation; and lack of water power in the areas of Mediterranean climate, where streams often dry up in summer.

6. *Transportation difficulties.* The rugged topography of much of the region made construction of adequate rail and road systems difficult and expensive.

A variety of these factors worked in different combinations as major impediments to economic development in the region. Only three sizable industrial areas developed in Southern Europe before World War II. The most important was in the Po Plain of northern Italy. Here, imported coal and raw materials, hydroelectric power from the surrounding mountains, and cheap labor formed the basis for an area specializing in textiles but including some heavy industry, automobile production, an aircraft industry, and engineering (Fig. 5.11). A second area very similar in its foundations and specialties, but developed on a much smaller scale, came into existence in the northeastern corner of Spain in and around Barcelona. A third area, focused on heavy industry, developed in the coal and iron mining section of northern coastal Spain. All three areas lie near the foot of mountains with humid conditions that provide the basis for hydroelectric power development.

Economic Progress in the Last Half Century

In the later 20th century, the economies of Southern Europe began to show a new dynamism. Although the South is still relatively poor compared to the North and West Central Europe, recent growth has been shaped by both industrial modernization and an expanding role of the EU. Industrial and service activities led the way, and there was a sharp decline in the proportion of the population still employed in agriculture. A number of factors played a part in this revitalization:

1. *Transportation changes.* Improved transportation, particularly in ocean transport, made the importation of fuels and materials for manufacturing more economical.

2. *Energy shift.* Coal ceased to be the dominant source of power in Southern Europe, replaced by oil or, in Italy, by

Figure 5.11 The western Po Valley around Milan, Italy, is a hub of varied manufacturing in this most industrial part of Italy. This Turin-based firm specializes in the fabrication of full-scale plastic decoy tanks and fighter planes. Such products as these were used in the Gulf War by Iraq in an effort to have the United Nations forces expend missiles on these plastic shells at considerable financial cost to the United Nations war effort. A similar use may have been made of such dummy tanks in the 1999 Kosovo bombing and the fall 2001 bombing of Afghanistan.

natural gas and oil. The countries in Southern Europe became major importers of oil, as did their industrial competitors. Thus, Southern Europe moved to a more even footing in securing power than when it had to depend on large imports of coal. In Italy, the power situation also improved with the discovery of large natural gas deposits, principally under the Po Plain, as well as development of an active geothermal program.

3. *Market expansion.* Market conditions for Southern European exports improved greatly in the decades after World War II, largely as a result of the rising prosperity of northwestern Europe and the expansion of European markets. These exports included goods such as subtropical fruits and vegetables, Italian-built automobiles, and less obviously, labor. Streams of temporary migrants from Southern Europe found jobs in booming, labor-short industries to the north. Absence of the migrants relieved unemployment, and their remittances added purchasing power in their home areas.

4. *Capital infusion.* There were other infusions of foreign money into the Southern European countries. Each country received American economic and military aid, and favorable economic circumstances and government policies attracted a considerable amount of foreign investment in industrial and other facilities (Fig. 5.11). Southern Europe also saw a tremendous boom in tourism. To increasingly prosperous Europeans and Americans, Southern Europe offers a combination of spectacular

scenery (see Fig. 5.10), historical depth exceeding that of even the other parts of Europe, a sunny subtropical climate, the sea, and the beaches. The ancient cities, now modern and bustling with activity around their monuments of the past, supplemented attractions with new or greatly expanded resort areas such as Spain's Costa Brava and Costa del Sol, the Balearic Islands, the Italian Riviera, and various other places.

Economic and Demographic Trends: The Shift from Agriculture to Industry

Since World War II, Italy, Spain, Portugal, and Greece have all evolved from agricultural to industrial and service-oriented countries. Agricultural exports are still important, but manufactured exports have become more valuable. There has also been a steady flow of youth from the farm and village world into cities in quest of factory positions. Large tourist industries "marketing" sun, scenery, historic sites, and beaches supplement exports with an increasing economic significance.

5.3 Eastern Europe: Struggles for Economic and Political Transformation

As defined in this chapter and text, Eastern European countries include Poland, the Czech Republic, Slovakia, Hungary, Romania, Bulgaria, Slovenia, Croatia, Bosnia-Herzegovina, Yugoslavia, Albania, and Macedonia, occupying an arc that drops from the Baltic Sea in the north to the Balkan Peninsula in the south and the Black Sea on the east (Fig. 5.12). There is a major effort in these countries to "reinvent" themselves after more than four decades of control by the Soviet Union. Prior to German reunification in 1990, the former East Germany was also part of Eastern Europe, and it will be brought into the present discussion when appropriate. Many of these countries (including East Germany) became parts of the former Soviet empire during World War II, when Soviet troops entered them in pursuit of retreating German armies in 1944 and 1945. Exceptions were Albania and the former Yugoslavia, where national Communist resistance forces took power on their own as German and Italian power collapsed. In the others, Soviet occupation forces supported local Communists, and all had Communist "satellite" governments subservient to the Soviets by 1948.

Following their takeover by communism, the nations of Eastern Europe (and East Germany) were reshaped along Communist lines, with one-party dictatorial governments; national economies planned and directed by organs of the state; abolition of private ownership (with some exceptions) in the fields of manufacturing, mining, transportation, commerce, and services; abolition of independent trade unions; and vary-

ing degrees of socialization of agriculture. Attempts by East Germany in 1953, Hungary in 1956, and Czechoslovakia in 1968 to break away from Soviet control were crushed energetically by Soviet military force.

In 1989 and 1990, Soviet backing for the region's Communist order was abruptly withdrawn because of the inability of the Russian government to continue financial support and because of the energy of the rapidly expanding calls for greater autonomy by the former Soviet satellite nations. Several Communist regimes (including East Germany's) quickly collapsed and were replaced by democratically elected non-Communist governments. However, "reformed" and renamed Communist parties and ex-Communist officials still play a significant role in the region. In late 1995, a Communist was elected president of Poland, and in that same year, the Communists gained more than 20% of the common vote — the largest of any single party — in a parliamentary election in Russia. Governments are now struggling to build democracies with capitalist economies out of the economic wreckage of frequently nonproductive, nonprofitable, noncompetitive, state-owned enterprises left as a Communist legacy. The struggle to make Western free-market economies work in these nations of Eastern Europe has been a contest of a varied and often serious nature. There is no place that proclaims full success as yet, but that is not surprising given the scale of change required in such a transformation.

Political Geography of the "Shatter Belt"

The extension of Soviet power into Eastern Europe at the end of World War II was in keeping with the history of the region (Fig. 5.13). In the Middle Ages, several peoples in the region — the Poles, Czechs, Magyars, Bulgarians, and Serbs — enjoyed political independence for long periods and at times controlled extensive territories outside their homelands. Their situation then deteriorated as stronger powers — Germans, Austrians, Ottoman Turks, and Russians — pushed into East Central Europe and carved out empires. These empires frequently collided, and the local peoples were caught in numerous wars that devastated great areas, often resulted in a change of authority, and sometimes brought about large transfers of population from one area to another. There was also the change effected by the expansion of the role of Islam. People began to become Muslim as the Moorish Islamic influence that had been very successful in the Iberian Peninsula since early in the eighth century began to diffuse through the northern Mediterranean region. This influence continued to gain in importance even after the fall of the Ottoman Empire at the end of World War I.

A longer view of regional political and historical geography reminds us that the civil war of the 1990s fought in the former Yugoslavia has long been characteristic of this shatter belt of Eastern Europe. There have been large population transfers in the region since the beginning of World War II.

EASTERN EUROPE ⭐ Index Map

Baltic Sea

POMERANIA

Gdynia
Gdansk (Danzig)

Szczecin (Stettin)

Berlin

GERMANY

SAXONY INDUSTRIAL CONCENTRATION

MASURIA

Bydgoszcz

Bialystok

Poznan (Posen)

Plock

Warsaw

Brest (Brest-Litovsk)

POLAND

Lodz

Lubin

Wroclaw (Breslau)

SILESIA

Czestochowa

Katowice

Nowa Huta

Krakow

ORE MTS. SUDETEN MTS.

Walbrzych (Waldenberg)

BOHEMIA INDUSTRIAL CONCENTRATION

Prague

Pilsen

BOHEMIA

CZECH REPUBLIC

MORAVIAN GATE

Ostrava

UPPER SILESIA INDUSTRIAL CONCENTRATION

GALICIA

Lvov

BOHEMIAN FOREST

Brno

MORAVIA

TATRA MTS.

SLOVAKIA

Kosice

RUTHENIA

C A R P A T H I A N   M T S.

BUKOVINA

Vienna

Bratislava

LITTLE HUNGARIAN PLAIN

Ozd

Miskoic

Eger

Debrecen

Iasi (Jassy)

AUSTRIA

Gyor

Budapest

GREAT

Cluj

ALPS

BAKONY FOREST

HUNGARIAN PLAIN

Szeged

TRANSYLVANIA

BIHOR MTS.

ROMANIA

Ljubljana

HUNGARY

Pecs

Subotica

Timisoara

Hunedoara

Brasov

Braila

Galati

ITALY

SLOVENIA

Zagreb

Trieste

CROATIA

Rijeka-Susak

ISTRIAN PEN.

DRAVA R.

VOJVODINA

Vukovar

BANAT

Resita

Novi Sad

Belgrade

IRON GATE

TRANSYLVANIAN ALPS

Ploesti

Bucharest

WALACHIA

DOBRUJA

Constanta

Mouths of the Danube

Adriatic Sea

DINARIC ALPS

BOSNIA-HERZEGOVINA

Sarajevo

SERBIA

Danube R.

Split

DALMATIAN COAST

YUGOSLAVIA

Nis

BULGARIA

Varna

Black Sea

MONTENEGRO

Dubrovnik

Titograd

Cetinje

Bar

Shkoder

KOSOVO

Sofia

Mt. Musala 9592 ft. (2924 m) ▲

BALKAN RANGE

Plovdiv

Burgas

Skopje

RHODOPE MTS.

Istanbul (Constantinople)

Bosporus

30° E

Tirana

MACEDONIA

Sea of Marmara

ALBANIA

Thessaloniki (Salonika)

Dardanelles

TURKEY

GREECE

Aegean Sea

40° N

Ionian Sea

RUSSIA AND THE NEAR ABROAD

Dniester R.

Southern Bug R.

Prut R.

Inset map (upper right):

LITHUANIA, LATVIA, ESTONIA (off map) absorbed by former USSR; now independent

LATVIA

LITHUANIA

Gdansk (Danzig)

EAST PRUSSIA

To former USSR from Germany

To former USSR from Poland

Free City of DANZIG to POLAND

To POLAND from Germany

GERMANY

POLAND

To former USSR from Czecho-slovakia

CZECH REPUBLIC

SLOVAKIA

AUSTRIA

HUNGARY

ROMANIA

To former USSR from Romania

Trieste

To YUGOSLAVIA from Italy

"FORMER YUGOSLAVIA" (see main map for successor nations)

To BULGARIA from Romania

ITALY

ALBANIA

MACEDONIA

BULGARIA

GREECE

TURKEY

Legend:

URBAN AREAS
Stars show national capitals

⊛ ⊙ Over 3,500,000

✪ ● 1,000,000–3,500,000

★ ○ 500,000–1,000,000

☆ ○ 300,000–500,000

⊛ ● Selected smaller places

The independent successor nations of former Yugoslavia are underlined.

City-size symbols are based on metropolitan area estimates.

Mountainous areas

Major industrial concentrations

Main territorial changes since 1938

Main agricultural areas

0   50   100 miles
0   50   100 kilometers

Figure 5.12 Eastern Europe. The "mountainous areas" are broadly generalized. City-size symbols are based on metropolitan area estimates.

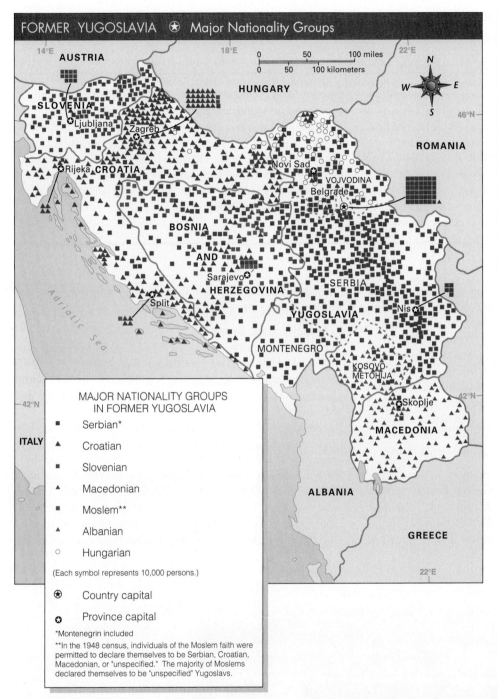

FORMER YUGOSLAVIA ✪ Major Nationality Groups

MAJOR NATIONALITY GROUPS
IN FORMER YUGOSLAVIA

- ■ Serbian*
- ▲ Croatian
- ■ Slovenian
- ▲ Macedonian
- ■ Moslem**
- ▲ Albanian
- ○ Hungarian

(Each symbol represents 10,000 persons.)

✪ Country capital

✪ Province capital

*Montenegrin included

**In the 1948 census, individuals of the Moslem faith were
permitted to declare themselves to be Serbian, Croatian,
Macedonian, or "unspecified." The majority of Moslems
declared themselves to be "unspecified" Yugoslavs.

Figure 5.13   This map has been redrawn from a map in George W. Hoffman and Fred Warner Neal, *Yugoslavia and the New Communism* (New York: Twentieth Century Fund, 1962), p. 30, by permission of the authors and the Twentieth Century Fund. Names in the largest type denote the six republics that constituted the former Socialist Federal Republic of Yugoslavia. The Vojvodina and Kosovo-Metohija areas are former "autonomous provinces" brought under tight control by Serbia in 1988. They continue, like most of the political units within the former Yugoslavia, to have a most uncertain future, even in early 2002. A map such as this one provides a solid base of demographic reality to be considered as one attempts to understand contemporary patterns of Balkan tensions and their causes. The tensions that are primary in the locales of former Yugoslavia are not only powerfully expressed in Eastern Europe, but they are replicated in much of Southwest Asia and other parts of the world as well.

During that war and the immediate postwar years, transfers involving millions of Germans, Poles, Hungarians, Italians, and others "simplified" the ethnic distribution patterns but at enormous human cost. People often were uprooted without notice, losing all their possessions, and were dumped as refugees in a so-called "homeland" that many had never seen. The prewar populations of Germans in Poland and Czechoslovakia were expelled at the end of World War II and forced into East and West Germany. Many died in this process of forced migration. Millions of Jews were systematically killed throughout East Central Europe during the German occupation in World War II. Ethnic minorities now constitute only 2% of the population in East Central Europe's largest country, Poland, which transferred most of its German population to Germany and whose territory containing Lithuanians, Russians, and Ukrainians was taken away.

A number of countries still have sizable minorities — for example, Hungarian minorities in the Slovak Republic, the

DEFINITIONS + INSIGHTS

Shatter Belt

In regions characterized by frequent or even continual shifting of boundaries due to warfare, rapidly changing roles of authority, or frequent friction between two major nations, there tends to be a fluidity in the alignment and independence of the adjacent smaller nations. The result is a **shatter belt.** Historically, the term has referred to the Eastern European nations from Poland in the north to Albania in the south. In broader use, however, this term connotes any region that experiences continual demographic shifts through ongoing migrations, efforts at **ethnic cleansing** —

the often bloody removal of ethnic populations from a state or region controlled by different and more powerful ethnic groups — and continual internecine warfare and border disputes. This pattern of geopolitical splintering is at the base of enormous instability, tension, ongoing enmity, and very often, local warfare. In the case of the former Yugoslavia, the shatter-belt characteristic has a history of leading to much broader wars, and it carries that same potential still today.

peoples of Vojvodina in Serbia, and the mountainous area called Transylvania in Romania. However, while Communist control put a quieting blanket over most of this region, its collapse beginning in 1989 was accompanied by a rapid reemergence of some old ethnic quarrels. For example, fighting among diverse ethnic elements has fractured the former federal state of Yugoslavia and served as a catalyst in military actions in Bosnia-Herzegovina, Kosovo, and has produced a heightened tension in Montenegro as well (Fig. 5.14).

The Slavic Realm of Europe

The expulsion of Germans from East Central Europe and the extermination or flight of more than 6 million Jews in total during the period of Nazi control in the 1930s and 1940s intensified the Slavic character of the region. Practically all of

Europe's Slavic people live there, and they form a large and ambitious majority within the region.

The Major Slavic Groups

The Slavic peoples are often grouped into three large divisions: (1) East Slavs of the former Soviet region, (2) West Slavs, including Poles, Czechs, and Slovaks, and (3) South Slavs, including Serbs, Croats, Slovenes, Bulgarians, and Macedonians.

Although the various Slavic peoples speak related languages, such languages may not be mutually intelligible. For example, the Polish language is not easily understood by a Czech, although Poles and Czechs are customarily grouped as West Slavs. Other significant differences also exist. Serbian and Croatian, for example, are essentially one language in spoken form, but the Serbs, like the Bulgarians and most of the East Slavs, use the Cyrillic alphabet (based on the Greek

DARIKO DOZET/AP, THE CHRISTIAN SCIENCE MONITOR, MAY 11, 2000

Figure 5.14   One of the major outcomes of the 78 days of NATO bombing in Yugoslavia in the spring of 1999 was the destruction of regional infrastructure. This photo shows an opening ceremony of a pontoon bridge newly constructed across the Danube River near Novi Sad, northwest of Belgrade, Yugoslavia, in the fall of 1999. The Danube serves as border, transport avenue for barge traffic, for tourism, as an industrial corridor, and as a water source in each of the 11 countries it touches from its source in the Alps and its confluence with the Black Sea.

language), whereas the Croats use the Latin alphabet, as do the Slovenes and the West Slavs. The Romanian language is classed with the Romance languages derived from Latin; however, the contemporary Romanian language contains many Slavic words and expressions. The various Slavic tongues are generally described as being more mutually intelligible than are the Romance languages. Romanians are conventionally classified as non-Slavs and are an additional example of the ethnic and linguistic variety of Eastern Europe.

A religious division exists between the West Slavs, Croats, and Slovenes, who are Roman Catholics for the most part, and the East Slavs, Romanians, Bulgarians, Serbs, and Macedonians, who adhere principally to various branches of the Eastern Orthodox Christian Church.

Non-Slavic Peoples

The principal non-Slavic elements in East Central Europe are the Hungarians, the Albanians, and the Romanians. The Hungarians, also known as Magyars, are the descendants of nomads from the east who settled in Hungary in the ninth century. They are distantly related to the Finns and speak a language entirely distinct from the Slavic languages. Roman Catholicism is the dominant religious faith in Hungary, although substantial Calvinist and Lutheran groups have long existed there. The Albanians speak an ancient language that is not related to the Slavic languages except in a very distant sense. Excluding Turkey, Albania is the only European state in which Muslims form a majority of the population, although Slavic Muslims (Moslems) are the largest ethnic group in Bosnia and Herzegovina (in the former Yugoslavia). Serbia, Macedonia, Montenegro, and Bulgaria have many Albanian or Turkish Muslims. These groups are legacies from former Turkish rule, and the continuing efforts to achieve even a modest level of regional cooperation among all of these groups consume enormous energy and generate only an uncertain stability (Fig. 5.15).

Physical Geography: A Checkerboard of Habitats

A glance at Figure 5.12 shows that East Central Europe forms a kind of crude checkerboard of mountains and plains. The mountains, located in the center and south, come in a wide range of elevations and degrees of ruggedness. The plains, concentrated in the north and in the center along the Danube, exhibit varying degrees of flatness, fertility, warmth, and humidity, and agriculturally, they range from some of Europe's better farming areas to some of the poorest. A handful of Europe's largest rivers carries the drainage of the region to the Black Sea (via the Danube), the Baltic Sea (via the Vistula), or the North Sea (via the Elbe).

The Northern Plain

Most of Poland lies in a large northern plain between the Carpathian and Sudeten Mountains on the south and the Baltic Sea on the north. The plain is a segment of the North

Figure 5.15   The task of trying to create a regional unit that shares space, resources, and ambitions but does not have common ethnic roots is most difficult. However, it has been thus for decades, perhaps centuries. Today, it is a continuing balancing act between political maneuvering and military presence.

European Plain, which extends westward from Poland into Germany and eastward into Ukraine in the former Soviet region. At the south, the plain rises to low uplands. Here, where Poland touches Europe's loess belt, are the country's most fertile soils. Central and northern Poland exhibit land that is rather sandy and infertile, with many swamps, marshes, and lakes.

Poland's largest rivers, the Vistula (Polish: Wisla) and Oder (Polish: Odra), wind across the northern plain from Upper Silesia to the Baltic. The Oder is the main internal waterway, carrying some barge traffic between the important Upper Silesian industrial area (to which it is connected by a canal) and the Baltic. Poland's main seaports — Gdansk (formerly, Danzig), Gdynia, and Szczecin (formerly, Stettin) — are along or near the lower courses of these rivers. Gdansk (population: city, 457,400) and Gdynia form a metropolitan area of just under 900,000 people, and Szczecin has a metropolitan area of about 500,000.

The Central Mountain Zone

A central mountain zone is formed by the Carpathian Mountains and lower ranges farther west along the Czech frontiers. The Carpathians, considerably lower and less rugged than the Alps, extend in a giant arc for about a 1,000 miles (c. 1,600 km) from Slovakia and southern Poland to south central Romania. West of the Carpathians, lower mountains enclose the hilly and highly industrialized Bohemian basin in the Czech Republic. On the north, the Sudeten Mountains and Ore Mountains separate this republic from Poland and the former East Germany. Between these ranges, at the Saxon Gate, the valley of the Elbe River provides a lowland connection and a navigable waterway from Bohemia to the industrial region of Saxony in Germany and, farther north, to the German seaport of Hamburg and the North Sea. To the southwest, the Bohemian forest occupies the border zone between Bohemia and the former West Germany.

Between mountain-rimmed Bohemia and the Carpathians of Slovakia, a busy passageway is provided by the lowland corridor of Moravia in the Czech Republic. Through this corridor run major routes connecting Vienna and the Danube valley with the plains of Poland. Near the Polish frontier, the corridor narrows at the Moravian Gate between the Sudeten Mountains and the Carpathians. Just beyond this gateway, East Central Europe's most important concentration of coal mines and iron and steel plants developed in the Upper Silesian–Moravian coal field of Poland and the Czech Republic.

The Danubian Plains

Two major lowlands, bordered by mountains and drained by the Danube River and its tributaries, compose the Danubian plains. One of these, the Great Hungarian Plain, occupies two-thirds of Hungary and smaller adjoining portions of Romania, Serbia, and Croatia. It is very level in most places and contains much poorly drained land near its rivers. The second major lowland is composed of the plains of Walachia and Moldavia in Romania, together with the northern fringe of Bulgaria. The Danubian plains contain the most fertile large agricultural regions in East Central Europe.

The Danube River, which supplies a vital navigable water connection between these lowlands and the outside world, rises in the Black Forest of southwestern Germany and follows a winding course of some 1,750 miles (2,800 km) to the Black Sea, making it the longest river in Europe. Upstream from Vienna, the river is swift and hard to navigate, although large barges go upstream as far as Regensburg, Germany. Below Vienna, the Danube flows leisurely across the Hungarian and Serbian plains past Budapest and Belgrade. In the border zone between Serbia and Romania, the river follows a series of gorges through a belt of mountains about 80 miles (c. 130 km) wide. The easternmost gorge is the Iron Gate gorge (also called Portile de Fier, a 2-mile-long gorge with rapids and serving as the border between Romania and Yugoslavia). Beyond it, the Danube forms the boundary between the plains of southern Romania and northern Bulgaria. The river then turns northward into Romania and enters the Black Sea through a marshy delta. Traffic on the Danube is modest compared with the tonnage carried by barges on the Rhine or by road and rail in Eastern Europe. The fact that the Danube flows into the Black Sea while the Rhine pours into the North Sea has made a powerful difference in the economic centrality of the two waterways. The industrial world served by the Rhine, and the very active trading world into which it flows, is very distinct from the more agricultural world served by the Danube. The Danube flows mostly through agrarian areas, and heavy industries — the most important generators of barge traffic — are found in only a few places along its banks.

The Southern Mountain Zone

East Central Europe's southern mountain zone occupies the greater part of the Balkan Peninsula. Bulgaria, the Yugoslav successor states, and Albania, which share this zone, are very mountainous, although important lowlands exist in some places. In Bulgaria, the principal mountains are the Balkans, extending east–west across the center of the country, and the Rhodope Mountains in the southwest. These are rugged mountains that attain heights of over 9,000 feet (c. 2,800 m) in a few places. Between the Rhodope Mountains and the Balkan Mountains is the productive valley of the Maritsa River, constituting, together with the adjoining Sofia Basin, the economic core of Bulgaria. North of the Balkans, upland plains, covered with loess and cut by deep river valleys, slope to the Danube.

In southern Serbia and neighboring areas to the west and south, a tangled mass of hills and mountains constitutes a major barrier to travel. Through this difficult region, a historic lowland passage connecting the Danube valley with the Aegean Sea follows the trough of the Morava and Vardar Rivers. At the Aegean end of the passage is the Greek seaport of Salonika (Thessaloniki). The Morava-Vardar corridor is linked with the Maritsa Valley by an east–west route that leads through the high basin in which Sofia, the capital of Bulgaria, is located. Along the rugged, island-fringed Dalmatian Coast, the Dinaric Alps rise steeply from the Adriatic Sea. The principal ranges run parallel to the coast and are crossed by only a few significant passes. These numerous physical impediments to movement around the margins of and coursing through the heart of the Balkan Peninsula add yet another major geographic feature to the character of Eastern Europe.

The Varied Range of Climates

In most of East Central Europe, winters are colder and summers are warmer than in Western Europe. In the Danubian plains, the humid continental long-summer climate is comparable to that of the corn and soybean region (Corn Belt) in the United States Midwest, but the plains of Poland have a less favorable humid continental short-summer climate, comparable to that of the American Great Lakes region. The most exceptional climatic area is the Dalmatian Coast. Here, temperatures are subtropical and precipitation, concentrated in the winter, reaches 180 inches (about 260 cm) per year on some slopes facing the Adriatic. This is Europe's greatest precipitation. The climate of the Dalmatian Coast is classed as Mediterranean (dry-summer subtropical). Together with spectacular scenery, this environmental attribute has fostered a sizable tourist industry. All through this region, tourist dollars are not only becoming more important than ever before — and are seen as a sign of welcomed economic change — but that same upswing places Western cultural influences right in the center of these countries. The capital cities are the most frequently sought tourist destinations, and it is from these same centers that political and cultural influences radiate.

Patterns of Erratic Agricultural Development Under Communism

Following World War II, the new Communist governments liquidated the remaining large private holdings in East Central Europe, and programs of collectivized agriculture on the Soviet model were introduced. Some farmland was placed in large state-owned farms on which the workers were paid wages, but most land was organized into collective farms owned and worked jointly by peasant families who shared the proceeds after operating expenses of the collective had been met. **Collectivization** — the bringing together of individual land holdings into a government-organized and government-controlled agricultural unit — met with strong resistance, and it was discontinued in the 1950s in former Yugoslavia and Poland. Today, all but a minor share of the cultivated land in Poland is privately owned. The remaining countries are pursuing programs to reprivatize their farmland, most of which was still in collective ownership at the outset of the new century. Conversion to private farming after four decades of col-

lectivization presents painful obstacles, and progress is slow, even though images of this transformation continue to fuel hopes of the region's small farmers.

Despite recent attempts to diversify and intensify agriculture, farming in Eastern Europe remains primarily a crop-growing enterprise based on corn and wheat, the leading crops in the region from Hungary and Romania south. In the Czech and Slovak republics and Poland, with their cooler climates, corn is unimportant, but wheat is a major crop as far north as southern Poland's belt of loess soils. Most of Poland lies north of the loess belt and has relatively poor sandy soils; here, rye, beets, and potatoes become the main crops. Raising livestock is a prominent secondary part of agriculture throughout East Central Europe. The region's meat supply is much less abundant than in Western Europe and comes primarily from pigs. The more southerly areas contain poor uplands that pasture millions of sheep, the main source of meat in Bulgaria, Albania, and parts of former Yugoslavia. This great landscape variety has led to steadily growing tourist activity all through Eastern Europe (Fig. 5.16).

Urban and Industrial Development in Eastern Europe

Eastern Europe has few cities of major size, although small regional centers, often old and historic, are widespread. Only two areas — the Upper Silesian–Moravian coal field of Poland and the Czech Republic, and the industrialized Bohemian basin — exhibit a closely knit web of urban places (Fig. 5.17). Even the agglomeration of mining and metallurgical centers in Upper Silesia-Moravia, a Ruhr in miniature, contains only about 3.4 million people. The Polish part of Upper Silesia-Moravia is somewhat larger than Poland's capital, Warsaw (population: metro area, 2,201,900; proper,

Figure 5.16   Brasov is Romania's second largest city and an important economic, cultural, and tourist center. The center of the city dates back to medieval times. The Coca Cola and the Camel Cigarettes signs reflect the rapid diffusion of commercial Western influence into Eastern Europe, particularly in urban centers.

BILL BACHMAN/PHOTO RESEARCHERS, INC.

Figure 5.17   Industrial development in Eastern Europe has lagged behind the post-World War II buildup that characterized so much of Western Europe. In this scene of the Nowa Huta steelworks near Krakow, Poland, can be seen an industrial plant that was intended to replicate some of the industrial power that Pittsburgh, Pennsylvania, developed nearly a century earlier. Plants such as this become difficult to maintain because of the competition of steel from new mills, but to close these older plants creates a social dislocation that is always significant. It was mill scenes such as this that led to the creation of the very successful late 1990s movie and stage play, *The Full Monty*, built around the idea of out-of-work steel workers turning to entertainment to try to regain some income after the closing of the local steel mills that had been their lives for decades.

1,611,800). In all the other countries, the largest cities are national capitals. None of the capitals, including Warsaw, ranks in Europe's top 10 cities in metropolitan population. The largest, Budapest, has a metro population of about 2.6 million (proper, 1,864,000); Bucharest has a metro area of 2.3 million; Prague, Sofia, and Belgrade each have between 1.1 and 1.7 million; Zagreb and Skopje each have more than 500,000; and Ljubljana, Sarajevo, and Tirane have 250,000 to 500,000. In every instance, the capital is its country's leading center of diversified industries and also serves as a primate city.

Industrial Trends in the Communist Era

The ascension of communism in East Central Europe following World War II inaugurated a new industrial and urban era in this region. With minor exceptions, the existing industries were taken out of private hands, and national economic plans were developed. Communist planners did not aim at a balanced development of all types of industry. Instead, they stressed the types that were deemed most essential to regional industrial development as a whole. Investment was channeled heavily into a few fields: mining, iron and steel, machinery, chemicals, construction materials, and electric power. These were favored at the expense of consumer-type industries and agriculture.

The planned expansion of mining and industry in East Central Europe under communism produced notable increases in the total output of minerals, manufactured goods, and power. But it also produced inefficient, overmanned industries ineffective in their later competition for the world markets of the post-Communist period. The central planning agencies of the Communist governments maintained rigid control over indi-

vidual industries. Plant managers were directed to produce certain goods in quantities determined by state governmental planners, and the success of a plant was judged by its ability to meet production targets rather than its ability to sell its products competitively and at a profit. Political reliability and conformity to the central plan were qualities much desired in plant managers. This system frequently resulted in shoddy goods that were often in short supply or in mountainous oversupply because production was not being driven or keyed by buyer demand and satisfaction.

A key element in East Central Europe's industrialization in the Communist era was its heavy dependence on the Soviets for vital minerals. These included iron ore, coal, oil, natural gas, and other minerals, which were imported in great quantities and over long distances by rail or pipeline from Soviet mines and wells. In return, the Soviet Union was a large importer of East Central Europe's industrial products.

During the 1970s and 1980s, however, the region's trade relationship with the Soviet Union weakened. Trade and financial relations with Western countries, companies, and banks expanded rapidly. With relatively little to export, the East Central European countries borrowed massively from Western governments and banks to pay for imports from the West. This course was followed particularly by Poland and Romania. New industrial plants and equipment acquired in this way were supposed to be paid for by goods that would be produced for export to the West by industries that had learned efficient production and marketing techniques from the West. But then the Western economies began to slump, lowering the demand for imports. This led to a situation by the 1980s in which large debts to Western governments and banks needed to be repaid if countries were to maintain any credit

Figure 5.18 This 1989 anti-Communist rally of ship-yard workers in Gdansk, Poland, reflects the surge of protest that rapidly overthrew the Communist order in most of Eastern Europe and in the former East Germany as well. This meeting was addressed by former U.S. President George Bush and Solidarity leader Lech Walesa. This process of re-creating political structures and patterns after the 1989 collapse of the former Soviet Union served to unleash a seemingly endless series of political efforts at many scales to reinvent political and economic futures.

at all. However, with the help of mismanagement by Communist bureaucrats, exports with which to pay were not being produced or, if produced, could not be sold. The result was a severe impediment to economic expansion, falling standards of living as the governments squeezed out the needed money from their people, and in Poland, social action through the formation of an independent trade union called Solidarity, which paralyzed the country and threatened to upset Communist dominance (Fig. 5.18). These conditions, along with the USSR's own growing economic and political distress, set the stage for the withdrawal of Soviet control and de-Communization in 1989 and 1990.

The Political Revolutions of 1989 and the Recession of Communism

As late as the early summer of 1989, the former East Germany and the East Central European countries seemed firmly in the control of totalitarian Communist governments. Then, in the late summer and autumn, the Communist order began to crumble. Public demands for freedom, democracy, and a better life gathered momentum in country after country. Communist dictators who had ruled for many years were forced out, and reformist governments took charge. This liberalizing process continued into the 1990s. By mid-1991, democratic multiparty elections had been held in all countries.

In a process that has had bold implications for the political as well as the economic landscape of Eastern Europe, the whole geography of economic activity underwent profound change in the 1990s. Ten highlights of the economic restructuring of the region included:

1. the privatization of state-owned enterprises;
2. increasing the efficiency of state-run enterprises by allowing noncompetitive enterprises to fail and removing diseconomic support of poor performers;
3. encouraging new private enterprises to develop, in many cases with foreign capital and management playing a role in the growth of these new units;
4. ending price controls and allowing prices to reflect competition in the market — a shift that has been particularly difficult for the firms that paid little attention to markets, the need for their product, or quality;
5. developing new institutions required by a market-oriented economy (banks, insurance companies, stock exchanges, accounting firms, etc.);
6. fostering joint enterprises between state-owned firms and foreign firms;
7. eliminating bureaucratic restrictions on the private sector;
8. making currencies internationally convertible to increase trade and thus increase competition;
9. converting farmland from state and collective ownership to private ownership; and
10. expanding access to international communications media of all sorts to help local entrepreneurs learn from foreign examples, such as encouraging travel of potential foreign investors to plant sites to assess the prospects for success in joint ventures (see Fig. 5.18).

This economic transformation has been highly varied in its success in Eastern Europe. The period from 1989 to the mid-1990s saw only spasmodic economic progress in the region.

Economic hardship, uncertainty, and uneasiness were widespread. Loans by the International Monetary Fund (IMF) and other outside sources were helpful but modest in scale. In all of the countries, unemployment was high. Freeing of prices as a move toward a market economy often resulted in more and better goods in stores but at prices few local customers could afford. The increased prices created great hardship for pensioners and others living on low fixed incomes. Meanwhile, a high proportion of all industries remained state-owned despite various efforts to dispose of them to private owners. Such privatization was slowed not only by the decrepitude of many properties but also by resistance to change on the part of workers and managers whose jobs, security, and power were at stake. Similarly, the privatization of state-owned farmland moved very slowly. Most prospective farmers were reluctant to exchange the relative security of employment on collectivized farms for the potential hazards of private farming. The uncertainty generated by all these changes was further promoted by the success of Communist party candidates in the region in elections in the second half of the 1990s.

A Minisketch of Regional Development in Eastern Europe

Poland. Poland is nearly 50% arable, sited as it is on the North European Plain (although most of Poland lies north of the fertile loess belt), with another 13% in meadow and some 28% in forest. Currently, Poland has a favored position among the countries of the post-Soviet breakup because of the catalyst role it played in the early 1980s with Solidarity's first rising in the Gdansk shipyards (see Fig. 5.18). The Gdansk Accords gave Polish workers the right to strike and to form free trade unions. This, in essence, began the end of Soviet control of Eastern Europe. Warsaw, Poland's capital and primate city, has given much attention to the reconstruction of the structures that survived heavy bombing toward the end of World War II.

With Germany as its major trading partner, Poland gains more than two-thirds of its export dollars from the sale of machinery, chemicals, and manufactured goods. In 1997, Poland was offered membership in NATO and accepted. Poland also suffers from continued environmental problems spawned by the Soviet promotion of economic development and industrial growth between the 1950s and the end of the 1980s. By the end of the 1980s, 75% of Poland's rivers had been declared biologically dead.

The Czech Republic. The economically diversified Czech Republic, where a long tradition of economic accomplishment was undercut by communism, is also enjoying more than usual success in breaking out of the Communist mold, although, like Hungary, it is still a poor area by broader European standards. It is 43% arable and one-third forested. The 1993 split-up of Czechoslovakia into the Czech Republic and Slovakia was called the Velvet Revolution, which began in 1989, led to the 1993 break, and finally, resulted in the 1996 delineation of new borders between the two independent countries. Prague, like Warsaw in Poland, is both the capital and the primate city of the Czech Republic. Proximity

to Germany and to European tourists plays a major role in its economic growth, with more than $3.7 billion in tourist dollars earned in 1998. By the mid-1990s, more than 90% of the Czech economy had been privatized.

The Slovak Republic. With the 1993 creation of the Slovak Republic, Bratislava (on the Danube River) became its national capital. The country is one-third arable and two-fifths forested and is somewhat more ethnically diverse than either the Czech Republic or Poland. As a result, Slovakia has a potential problem of further political separatism within its minority community of about 600,000 ethnic Hungarians. Slovakia still suffers from the fact that many professionals, and much potential foreign investment, stayed with the Czech Republic after the 1993 split. In the late 1990s, Slovakia had an unemployment rate of 15%, but there is hope that the upswing in European economic growth might enlarge the tourist flow to landscapes along the Danube, to the Tatra Mountains in the east, and to the country's many caves and thermal spas.

Hungary. Hungary has been a relatively bright spot economically in Eastern Europe. Its move toward free-market capitalism began early and has been aided by Western and Japanese investments larger than those received by any other nation in this subregion. A relatively small but growing class of "newly rich" private entrepreneurs has emerged, often as participants in joint business ventures with foreign companies. This new "business bourgeoisie" contrasts, however, with a far larger class of "newly poor" persons living near or below the poverty line. Hungary has a manufacturing edge because of local bauxite and aluminum production capabilities. More than 50% of its land is classified as arable, giving a strong staying power to agriculture and, unfortunately, a poor peasantry.

Romania. In 1993, Romania applied for membership in the EU and was turned down. It also was not included in the first set of Eastern European countries asked to join NATO. Part of the reason that the country has fared so poorly in its quest for greater recognition from Western Europe and the world beyond is because of its poor economic performance in the collection of Eastern European nations. Currently, the country has to import oil and gas to provide its own energy needs. The export of textile, minerals (coal), chemicals, and machinery make up approximately 50% of Romania's export income. Germany and Italy are the primary export destinations.

With a long eastern coastline on the Black Sea and hundreds of miles of Danube River landscape, Romania has hoped for an influx of foreign exchange from the tourist trade. But the tourist traffic has largely been from the east, mostly Russia, and recently has grown so small that no foreign exchange is gained from the trade. Even in the mid-1990s, Romania did not gain a net positive income from tourist activity. Industry accounts for one-third and agriculture for one-fifth of Romania's annual gross domestic product (GDP).

Bulgaria. Bulgaria has the Danube River and its flood plain as its northern border, the Balkan Mountains coursing from east to west through the middle of the country, and mountains separating it from Macedonia and Greece in the south and southwest. To the southeast lie the lowlands that separate Bulgaria from Turkey. Some 34% of the country is arable, another 30% is forested, and 16% is meadows and pastures. Traditionally, the population was largely agricultural, but in recent decades, there has been an expanded effort in industrialization and much demographic shift. Manufacturing and mining are responsible for more than 40% of Bulgaria's current GDP, with machinery and chemicals responsible for 25% of the industrial sector. Agricultural products now account for only 13% of the total annual GDP. Currently, Bulgaria is 68% urban.

One of the most important economic resources the country possesses is the rail route that links Istanbul, Turkey, with points in Central and Western Europe. The Maritsa River that flows from west to east and drains into the Aegean Sea also provides an important travel corridor.

The Breakup of Yugoslavia

Yugoslavia. Economic progress in Yugoslavia was disrupted by ethnic warfare following the dissolution of the Yugoslav federal state in 1991. Previously, under the Communist regime instituted by Marshal Josip Broz Tito during World War II, Yugoslavia had been organized into six Socialist People's Republics: Serbia (composition: Serbs 63% in 1997); Croatia (Croats 78%, Serbs 12%); Slovenia (Slovenes 91%); Bosnia and Herzegovina (Slavic Muslims 38%, Serbs 40%, Croats 22%); Montenegro (Montenegrins 69%, Slavic Muslims 13%, Albanians 6%); and Macedonia (Macedonians 65%, Albanians 22%, Turks 5%). In addition, two "autonomous provinces" were created: Kosovo (Albanians 77%, Serbs 13%) and Vojvodina (Serbs 54%, Hungarians 19%) (Fig. 5.19).

In 1988, Serbia took direct control of Kosovo and Vojvodina as part of Serbia's push (under an elected Communist president, Slobodan Milosevic) for greater influence within federal Yugoslavia. All of these quasi-independent states demanded a looser federation with more autonomy for each republic, and in 1991, they all declared independence. Slovenia, Croatia, and Bosnia and Herzegovina were admitted to the United Nations in 1992.

The secessions and subsequent political independence left federal Yugoslavia a rump state composed only of Serbia and Montenegro. The subsequent political moves to segment the already patchwork nation of Yugoslavia into smaller, more homogeneous political units was simply a renewed attempt toward **"balkanization"** (this very term for political breakup

RESHAPING THE WESTERN BALKAN PENINSULA

0 50 100 miles
0 50 100 kilometers

—— International boundary
- - - - Republic boundary
········· Provincial boundary
—— Inter-entity boundary

PEACEKEEPING OPERATIONS
UNTAES Transitional Administration for Eastern Slavonia*
SFOR Stabilization Force
UMOP Mission of Observers in Prevlaka
UNPREDEP Preventive Deployment Force

*Until January 15th, 1998.
† Withdrawn August 1997

Source: *The Economist*, January 24, 1998.

Figure 5.19 Ever since the 1980 death of Josip Tito, the fiercely independent leader who had held Yugoslavia together for more than four decades in the face of both German and Soviet aggression, Yugoslavia has been the victim of powerful forces of fragmentation. This map shows evidence of both boundary and demographic shifts in the past two decades. It also shows evidence of continuing international efforts to keep open warfare to a minimum. The North Atlantic Treaty Organization (NATO) has had to allocate major resources and considerable military personnel and material in the past years in an effort to retard or eliminate the process of "ethnic cleansing" that has caused the removal of hundreds of thousands of ethnic Albanians from Kosovo and, in 2001, from Macedonia at the hands of Slavic forces. One of the most astute cartoons on the Balkan tensions showed a very senior person saying, "We don't need NATO . . . we need Tito!" The fragmentation of this region of the Balkan Peninsula is clearly with us in this new century as well.

PROBLEM LANDSCAPE

"Bosnia, Sort Of"

Zvornik, Bosnia

What is the best way to sum up the situation in Bosnia today, five years after NATO came to its rescue? Very simple: Bosnia is sort of a country—it's sort of united, sort of divided, sort of peaceful, sort of corrupt, sort of coming back, sort of not coming back—a place where NATO armies feel sort of good about what they've done to stabilize the country, but where NATO diplomats feel sort of bad that Bosnia's Serbs, Muslims and Croats still have not come together enough to function on its own.

It's sort of like that.[a]

It was in the early 1990s that warfare became a regular part of the Serb efforts to achieve what became known as "ethnic cleansing," — that is, the relocation of Croatian and Bosnian Muslims. Patterns that had allowed these varied and diverse populations in the former Yugoslavia to live in at least moderate peace had been ruptured by the death of Josip Tito (the long-time and very powerful leader of Yugoslavia) in 1980 and then steady economic decline in the waning years of the Soviet Union and its control over Eastern Europe. NATO led a bombing attack against the Serb efforts in 1995, and this was followed by peace talks outside of Dayton, Ohio, which produced what have been called the Dayton Accords.

Now, 7 years after the signing of the 1995 Dayton Accords, there has been achieved a marginal peace. Most overt Serb efforts to eliminate the Muslim populations in the northwest sector of what was once called The Land of the South Slavs (Yugoslavia) have been suppressed. It has taken a NATO force of some 60,000 troops to keep the peace.

The 4,900 U.S. troops within that larger number view this exercise, which has lasted some 4 years longer than was initially promised, as a prime exercise in achieving battle readiness. Author Friedman goes on in his article to say, "You hear this from all U.S. troops: We're keeping the peace and getting experience in everything that an army does — patrols, intelligence, transport, operations — except killing people. It beats marching around a flagpole in Germany. It is no accident that the highest rate of re-enlistment in the U.S. Army today comes from troops in Bosnia and Kosovo."

The problem in this landscape is that in both Bosnia and Kosovo (which was a parallel operation that was affected in NATO bombing in 1998 and subsequent allocation of NATO troops

to keep the peace) massive numbers of foreign troops have now taken the role of keeping a lid on ethnic, religious, land ownership, and governmental tensions that have been part of the Balkan Peninsula for centuries. There is a continual effort at the village level to reclaim houses and farm fields that both Serb and non-Serb peoples feel were wrongly taken from them. While such tensions are absolutely not unique to this region, the potential for monumental bloodshed to derive from peasant and political efforts to deal with these issues makes this Balkan scene more critical than most.

And at the heart of the problem lies the basic reality that village and urban populations are increasingly seeking more autonomy, more access to private lands and forces of production, and greater political independence. So, as in so many aspects of world regional geography, what is discussed as an unusual and regionally unique reality in one place turns out to be in truth simply a variant on global wants and patterns of achieving such desires.

[a]Thomas Friedman, "Bosnia, Sort Of." *New York Times,* January 26, 2001, A23.

comes from this region initially because of the continuing history of tension and political shift in its political organization) or the generally "bloody" breakup of larger political units into smaller ethnic units. As soon as Croatia made its declaration in 1991, the Croatian Serbs took over one-third of the new country and began an "ethnic cleansing," the forced emigration—or killing—of peoples different from the ethnic group in power. As a result of this act, the tensions between the Serb peoples and the Muslim peoples of the former Yugoslavia have become a major focus of the world press. It is critical to realize that the Balkan Muslim population is made up mostly of local peoples who have converted to the Islamic religion and are not just a relic population left over from the Ottoman Turk rule of southeastern Europe in the last five centuries. In 1992, the United Nations withdrew recognition of Yugoslavia

because of the government's failure to stop Serbian atrocities occurring in Croatia and Bosnia-Herzegovina.

U.N. troops have been deployed in Bosnia since late 1995 as part of the Dayton Accords signed in December 1995 by the presidents of Bosnia, Croatia, and Serbia on behalf of Serbia and the Bosnian Serb Republic. The Dayton Accords and the positioning in this region of 60,000 NATO troops brought a slowdown to the Serb military's attempts to continue ethnic cleansing, and in a sense, political efforts have replaced military efforts. Three years later, the 78 days of NATO bombing of Serbia in 1998 was another strike to force Serbia to discontinue its bloody efforts to separate the Slavic peoples from all others who had distinct religious or ethnic characteristics in the Kosovo region. The goal was to bring Muslims back to their homes in cities that were controlled by Serbs and to stimulate government cooperation at all levels in this effort at civil reunification. As of mid-2001, only modest success has been achieved.

This Balkan pattern of ever smaller ethnic, religious, and political groups—all generally demanding autonomy, political independence, and widespread political recognition—fighting for such independence is centuries old. The warfare in Kosovo in 1997–1999 is a variant of this same theme of Serb desire for more homogeneous nation-states and for more autonomy in the realignment of the tortured landscapes that were held together by Tito for decades before his 1980 death. The efforts of the Kosovo Liberation Army to gain full autonomy and independence are more of the continuing Balkan, perhaps global, effort to turn an ethnic realm into a nation-state. The ouster of Serbian maximum leader, Slobodan Milosevic, and his 2001 capture and transfer to The Hague for a war crimes trial early in 2002 are yet another aspect of the political uncertainty associated with this southeastern peninsula of Europe. It is highly likely that the region will need to overcome the enormous tradition of infighting before it will be able to play a more stable and productive role in the nations of Europe.

Albania. This small country has gained most of its visibility since the end of World War II by linking itself economically with the People's Republic of China. For China, Albania became a unique locus to show off its political capacity to relate to a distant non-Asian country. The Albanians had been on the bottom rungs of economic and cultural development for so long that even such a distant ally seemed very beneficial. However, even that linkage did not enable the Communist government of Albania to design and achieve a pattern of productive economic growth in the years following Italy's wartime occupation.

In 1944, Communist Enver Hoxha took charge of Albania, and as the dominant leader for four decades, he oversaw first the Soviet linkage, then the China connection, and then harsh isolation. With Hoxha's ouster in 1984 and the opening up of Eastern Europe to new political, economic, and social winds, there was a possibility that Albania would grow into a position of greater strength. This has largely been unrealized;

Albania is one of the few countries in Eastern Europe in which a majority of its population continues to live in the rural sector (63%). There is the hope that tourism might be able to capitalize on the more than 100 miles (160 km) of Albanian coastline on the Adriatic Sea, but there is no infrastructure yet established to begin to highlight that locale or deal with the demands of a foreign tourist trade. The coast has been more important as a point of departure for Albanians and others trying to gain access to the eastern shores of Italy as part of the growing refugee flow from the Balkan Peninsula. There has also been continued ethnic strife among the Muslim and non-Muslim populations that are particularly present in northern and eastern Albania.

Macedonia. Macedonia is another of the new political units carved, perhaps ruptured, from the Yugoslavia that had been held together so forcefully by Josip Tito for nearly four decades. With the breakup of Yugoslavia and the warfare in Bosnia and, later, Kosovo, Macedonians also began to search for more independence. Like the whole southwestern corner of the Balkan Peninsula, Macedonia is home to a number of ethnic and religious groups, and the history of cooperative living in this difficult environment has not been peaceful. The demographic element leading to the most difficulty is the tension between ethnic Albanians (with a large proportion of that group Muslim) and the Macedonians.

Macedonia is 24% arable and another quarter of its land is meadow and pasture. It is landlocked, and the political need to maintain positive relations with Greece, from which it is partially cleaved, to have predictable access to the Greek port of Thessaloniki further complicates the autonomy of Macedonia. In 1995, it brokered a treaty with Greece over the use of the name Macedonia, which Greece has claimed historically. By the middle of 2001, it had become involved in its own warfare, and there were efforts to quell the political ambitions of the ethnic Albanians who make up 23% of its population and who, like their analogs in Kosovo, would like much more political autonomy.

CHAPTER SUMMARY

- Present-day Europe is ringed with three sets of nations called the Northern, the Southern, and the Eastern. These represent nations that are more peripheral to the European Union nations that lie in the core of Europe. The Northern countries include the Scandinavian countries of Norway, Sweden, generally Denmark, and sometimes Iceland. If Finland is added, this cluster is called Fennoscandia, or the Fennoscandian countries.
- The Northern countries are the world's most northerly highly settled, economically developed region. This is because of the North Atlantic Drift, a northward extension of the Gulf Stream that comes from the Caribbean Sea and the Gulf of Mexico, bringing unusual warmth to the North Sea and even the Baltic Sea, and allowing Murmansk, Russia, to be an ice-free port, even though it is within the Arctic Circle. The winter climate of the

whole of Western Europe is much more moderate than places in North America at the same latitude, directly because of the geographic influence of the North Atlantic Drift.
- Countries in Fennoscandia have close connections through language (except for Finland) and religion. The Evangelical Lutheran Church is the church of 90% of the people. Finnish has its origin, unlike the other languages of this area, in languages of Central Asia. Other qualities that link these countries are settlements in the southern reaches of the countries, very sparse populations in the north and in the mountains of all the countries (except Denmark, which has no mountains to speak of), high levels of literacy and education, and the major economic importance in forestry, fishing, skilled manufacturing, and in Denmark and part of Sweden, very productive agriculture.

Norway also has very significant petroleum reserves in the North Sea. Hydroelectricity is also a major resource in all of the Northern countries except Denmark.

- The Southern region is the term used for the countries on the northern margin of the Mediterranean Sea (excluding France) including Portugal, Spain, Italy, and Greece, as well as a number of major islands and some microstates including San Marino, Vatican City, the British colony of Gibraltar, and the small state of Malta. The countries on the eastern edge of the Aegean Sea are also part of the Eastern region. The Iberian Peninsula and the Italian Peninsula are of major importance to the Southern region and the Balkan Peninsula is seen as part of the Eastern region.

- The countries in the South are much influenced by the Mediterranean climate, which is characterized by summer drought and winter precipitation. Irrigation has been an important factor in this region's agricultural history. Important crops are winter wheat, olives, citrus, grapes and associated winemaking, and vegetables.

- The Moors and the Islamic/Moorish influence have been an important part of Spanish history. Also central was Christopher Columbus because of his voyage to America in 1492. Portugal had great fame through the voyages of Vasco da Gama and with the control of Brazil. Italy had some colonial influence, and is currently the most industrial and economically significant nation of the South. The Po Valley is central to industrial northern Italy and houses two-fifths of the nation's population. There is a classic and costly economic imbalance between northern Italy and southern Italy. Padania is the name suggested for a new country that some residents hope might be created out of the prosperous north of Italy.

- Diffusion, migration, the Roman Catholic church, merchants, and trade are themes that provide images of 12th-century Mediterranean Europe (and beyond) in *The Walking Drum*. This literature excerpt (see p. 109) deals with periodic markets and major agricultural fairs. The variety of languages, customs, and trade and travel patterns help provide images of medieval Europe.

- The Southern region has lagged behind West Central Europe in development because of trade deficiency, low educational levels, land tenure patterns, and resource deficiencies. These traditional characteristics have changed somewhat in recent decades due to transportation changes, energy shifts, market expansion, and capital infusion.

- The Eastern region is made up of Poland, the Czech Republic, Slovakia, Hungary, Romania, Bulgaria, Slovenia, Croatia, Bosnia-Herzegovina, Yugoslavia, Albania, and Macedonia. These countries have been shaped powerfully by shifting human migrations and power struggles between various empires over the past millennium. Most critical has been the Communist control of the Soviet Union after the end of World War II until the collapse of the USSR in 1989. The term "shatter belt" is useful in explaining the shifting political affiliations of these nations.

- Slavs have been a dominant population group in the East, especially in the Balkan Peninsula. The three divisions that still operate in the political tensions of this region are East Slavs of the former Soviet Union; South Slavs, including Serbs, Croats, Slovenes, Bulgarians, and Macedonians; and West Slavs, including Poles, Czechs, and Slovaks. The current geopolitics of the region still are responding to this Slavic population and its relationship to Russia and to other non-Slavic local populations.

- The environment of the Eastern countries includes the northern plain, with Poland dominant; the central mountain zone, with the Czechs dominant; the Danubian plains, including major landscapes in Hungary, Romania, Serbia, and Croatia; and the southern mountain zone, with Bulgaria, part of Yugoslavia, and Albania dominant. Varied climates exist.

- Collectivization under Communist rule during the Soviet era from 1947 until nearly 1990 provided the primary agricultural motif. Corn and wheat are the primary crops, and farmland has now been largely privatized but is still less productive than croplands to the west in Europe. Raising livestock and sheep flocks in the uplands of these countries are part of the farming pattern. There is a steady rural-to-urban migration all through the region.

- Mining, iron and steel, machinery production, construction materials, and electrical power were the highlights of the industrial effort during the Soviet years, with the majority of raw materials coming from Russia. The shift to the market economy has left many firms uncompetitive and many have been abandoned. There has been a major effort to make the industrial base of the East more productive, including increasing the efficiency of state-run enterprises (not everything has been privatized), encouraging new private enterprises, fostering joint enterprises, eliminating price controls and bureaucratic restrictions, expanding international connections, and increasing the quality of the region's manufactured products.

- Politically, the disintegration of the former Yugoslavia has been the most disabling phenomenon of the post-Soviet era, leading to the Bosnian conflict, the Dayton Peace Accords of 1995, and in 1998, the emergence of the Kosovo Liberation Army and its active military action against the Serbian forces controlling Yugoslavia. These various campaigns of ethnic cleansing have led to NATO involvement, including bombing and the stationing of forces in Bosnia as well as Kosovo. In 1998, NATO extended an invitation to Poland, Hungary, and the Czech Republic to join NATO, much to the expressed disapproval of Russia. In March 1999, Hungary, Poland, and the Czech Republic joined the alliance.

REVIEW QUESTIONS

1. List the countries of the European North and outline their similarities and their distinctive differences.

2. Which of the countries in the European North has a language fully unlike the other four? Explain the historical geography that explains this distinction. What are other examples of a highly distinctive national image characteristic of this region?

3. Describe the climate patterns of the European North, comparing this region with North America at the same latitudes. Explain what geographic feature is most responsible for the regional warmth of the Northern region. What economic impact does this lead to?

4. Reconsidering the National Geography Standards of Chapter 1, explore your mental map of the European North and try to determine the source of your images. Do the same for the Southern and the Eastern regions. Consider the urban images and their relative strength compared to environmental or agricultural images.

5. List the countries of the European South, and the islands related to these countries.

6. What is the ecology of Mediterranean agriculture? Outline the climate and soil patterns that relate to it, explaining why such agriculture is so historically significant. Give some detail to the role of winter wheat in your discussion.

7. Inventory the major urban centers that appear from west to east in the European South, noting the comparative roles of industry, governmental function, tourism, and agriculture in defining their major images and geographic characteristics.

8. What factors have influenced the European South's relatively uneven economic development, and what changes have been realized in these patterns since World War II?

9. List the nations that make up the European Eastern region as defined in this text and relate them to broader patterns of European historical geography.

10. Define "shatter belt" and cite examples of such a phenomenon both in the European East and in the North. In what regions and in what country or countries does the concept play the most significant role?

11. Outline the breakup of the former Yugoslavia and use the aspects of change in the last decade there to discuss aspects of the National Geography Standards, especially relating to cultural mosaics, forces of cooperation and conflict, and human settlement.

12. What has been the Slavic presence in the European East and what role does that play in the stability of the region today? Use Kosovo and other nations carved from the former Yugoslavia to define your observations.

13. Using river systems, topography, and soils, define the physical geography of the European Eastern region, linking such characteristics to history where possible.

14. Utilizing Human Systems from Table 1.1, show the ways that essential elements are reflected in the landscape of the European East in this chapter. T1.1 5

DISCUSSION QUESTIONS

1. What physical and cultural characteristics have blended to create the lifestyle of the European North?

2. Looking at the larger map of Europe and western Russia, discuss the ways in which basic physical geography has had a strong role in historical development of the region.

3. Try to imagine the European North without the North Atlantic Drift. How would the reality of that region change, and what significance would such changes have?

4. Think about growing up in rural or isolated lands of the European North. What factors might pull you toward other regions? What can governments do to reduce the steady rural-to-urban migration in that region?

5. Consider the tourist images of the European South. What are the geographic elements in such images? What is the economic importance of such images, and how are they promoted?

6. Why do you think countries in the European South played such a significant role in cultural development two millennia ago, but play such a relatively modest role now? What influence have geographic factors had in this shift?

7. Discuss efforts in the north of Italy to secede from the rest of the country. Why do such thoughts emerge? What political, economic, and cultural significance might they have?

8. Discuss the influence of 50 years of Soviet control on the nations of the European Eastern region. What might have been patterns of post-World War II development had the Cold War not been so strongly developed?

9. If you had money to invest in a place that was going through a political change from collectivization to privatization, where in the European East would you invest your money? What sort of industry would you promote? What sort of enterprise would you avoid?

10. Discuss the patterns of ethnic cleansing that have emerged in the past decade in the Balkan Peninsula. Are there any other parts of Europe where there is a potential for similar conflict? What dynamics have made this process so characteristically Balkan? What might lead to a change in such patterns?

North Pole

CHUKCHI

KORYAK

NENETS

TAIMYR

YAKUT

KARELIAN KOMI-
PERMYAK
MARI EL   KOMI
MORDOVIAN   YAMAL
                         NENETS   EVENKI
CHUVASH   KHANTY-
            TATAR   MANSI
                    UDMURT   UST-ORDYN-
    KALMYK   KAKHASS   BURYAT   BURYAT   JEWISH
        BASHKIR   GORNO-ALTAI
    DAGESTAN   TUVAN   AGIN-BURYAT
        KARAKALPAK

GORNO-BADAKSHAN

Russia and
the Near Abroad

Seen from Russian eyes, the fourteen other successor states
of the Soviet Union comprise the *Near Abroad*, a strategically vital perimeter retaining important economic and ethnic ties with
Russia. Large petroleum reserves, the world's greatest forest, nuclear weapons, and many other issues make this a critical region
in world affairs.

The Soviet Union experienced unimaginable suffering and loss of life in the two world wars of the 20th Century. The stamina of Russians and other peoples in the region has been tested repeatedly in troubled times.

The world's largest country, Russia is endowed with abundant natural resources but, paradoxically, is handicapped by huge transportation distances, high elevations, and natural hazards in its most productive areas. Jesse Wheeler described Siberia as a "bleak storehouse of resources."

The Soviet Union was, and Russia is, a "land empire" built when a single group imposed its political will and established economic sovereignty over disparate peoples in a vast hinterland. Ethnic diversity made the long-term survival of the Soviet Union an unlikely prospect and now threatens the unity of the Russian Federation.

The prospects for Russia, so recently the core of a superpower state, are uncertain. Russia appears to be a "misdeveloped" country torn between conflicting economic, political, and cultural choices. There is a serious health crisis, and with falling birth rates and rising death rates, the population is undergoing a profound decline. A recent uptick in Russia's economic growth brings hope that these trends may be reversed.

CHAPTER 6

A Geographic Profile of Russia and the Near Abroad

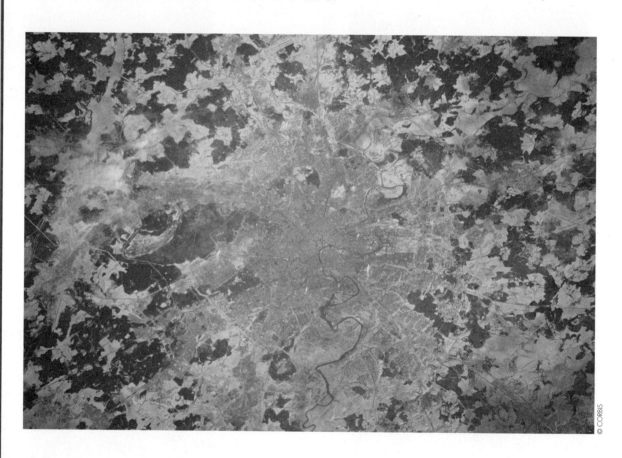

Moscow in winter. From this modest core
Russia, then the Soviet Union, then just Russia
again, maintained a mighty land empire. This is a
satellite view of the Russian capital.

Chapter Outline

From 1917 to 1991, the huge region labeled in this text as Russia and the Near Abroad was a single Communist-controlled country called (after 1922) the Union of Soviet Socialist Republics (USSR). After World War II, the government based in Moscow exercised control over Communist "satellite" countries in Eastern Europe and maintained a strong influence in other Marxist states, particularly in less developed countries (LDCs) not aligned with the capitalist West. For four decades, the **Cold War** between the Soviet bloc of nations and the Western bloc led by the United States dominated world politics. The perspectives and actions of the Soviet Union had major impacts on world events.

Suddenly, in 1991, the country split into 15 independent nations (see Table 6.1 and Regional Perspective, p. 169). Russia is by far the largest in area, population, and political and economic influence. From the Russian perspective, the remaining 14 countries comprise the "Near Abroad," in which Russia's special interests and influence must be recognized. Because of Russia's prominence, and its real and potential influence on all the successor states of the Soviet Union, the region is known in this text as "Russia and the Near Abroad." Twelve of the units are associated loosely in the Commonwealth of Independent States (CIS), but further political fragmentation and decentralization, rather than integration, are the dominant political themes in this region. Without the USSR's Red Army to impose order, long-simmering ethnic conflicts have boiled over. The future of this huge territory is uncertain. There are even indications that many of these nations are moving backward economically and now resemble less developed countries rather than component parts of a former superpower.

This chapter offers geographic and historical perspectives on a major world region in which momentous changes are underway. Placing the 15 former republics of the Soviet Union, now 15 independent countries, in a single region is a somewhat awkward but still useful designation early in the 21st century. These countries have lasting and often uncomfortable strategic and economic associations. For example, the Soviet-era oil refinery and pipeline system still links Russia with Kazakstan, Azerbaijan, Ukraine, Belarus, and other countries. Since the breakup of the Soviet Union, Russia has periodically shut off the pipelines supplying natural gas to Ukraine, Belarus, and Moldova to obtain political concessions

from these nations. The needs of these countries to buy and sell oil provide an incentive to hold them together in some kind of economic or political union.

The Soviet government failed in its vision of creating a single, unified "Soviet people" from the multitude of ethnic groups living within the boundaries of the USSR. However, it did succeed in transforming the human geography of the vast Soviet empire. For many years to come, the 15 countries of that crumbled empire will be attempting to sever or to reestablish ties with the former imperial hub of Moscow. For its part, the Russian government in Moscow will try vigorously to maintain influence in the former republics. Some analysts even fear that Russia may attempt to reexert control over some of the recently independent nations, especially those vital to the economic and political security of the Russian Federation.

6.1 Area and Population

With an area (including inland waters) of 8.6 million square miles (22.3 million sq km), the region of Russia and the Near Abroad is the second largest (after Africa South of the Sahara) of the world regions recognized in this text. A good indication of its staggering size is the fact that this region (even Russia alone) spans eleven time zones; by comparison, the United States from Maine to Hawaii spans seven.

The eventful geopolitical history of Russia and the Near Abroad has given this region land frontiers with 12 countries in Eurasia (Fig. 6.1). Between the Black Sea and the Pacific, the region borders Turkey, Iran, Afghanistan, China, Mongolia, and North Korea. Pakistan and India also lie very close by. In the Pacific Ocean, narrow water passages separate the Russian-held islands of Sakhalin and the Kurils from Japan. If India, Pakistan, and Japan are considered, the Asian part of Russia and the Near Abroad is a neighbor or near neighbor of about half the world's people.

In the west, the region has frontiers with Romania, Hungary, Slovakia, Poland, Finland, and Norway. These differ from the sparsely populated, arid, and mountainous frontiers in Asia. On the frontier between the Black and Baltic Seas, international boundaries pass through populous lowlands that have long been disputed territory between Russia and other countries. During World War II (1939–1945), the Soviet Union expanded its national territory westward in several stages.

Figure 6.1 General reference map of the 15 independent successor states that emerged from the dissolution of the Soviet Union. The inset map shows subordinate ethnic political units that did not have Union Republic status in the Soviet era (one unit, Karelia, was downgraded from that status). Names of the titular nationalities are shown. Some of these designations have carried over into the post-Soviet Russian Federation.

KEY TO SMALL UNITS IN CAUCASUS
1 ABKHAZ
2 ADJAR
3 SOUTH OSSETIAN
4 NORTH OSSETIAN
5 KABARDIN-BALKAR
6 CHECHNYA
7 INGUSHETIA
8 NAKHICHEVAN
9 NAGORNO-KARABAKH

SUBORDINATE NATIONALITY UNITS OF THE FORMER SOVIET UNION

LEGEND

Autonomous republic (Former ASSR: see text)

Autonomous region (Oblast)

Autonomous area (Okrug)

Units are labeled with nationality names.

———— Boundaries of independent states (for names, see main map)

- - - - Boundaries of subordinate units (in color)

Table 6.1   Russia and the Near Abroad: Basic Data

Political Unit	Area (thousand/sq mi)	Area (thousand/sq km)	Estimated Population (millions)	Estimated Annual Rate of Increase (%)	Estimated Population Density (sq mi)	Estimated Population Density (sq km)	Human Development Index	Urban Population (%)	Arable Land (% of total area)	Per Capita GDP PPP ($US)
Slavic States and Moldova										
Russia	6592.8	17,075.4	144.4	−0.7	22	8	0.775	73	8	7700
Ukraine	233.1	603.7	49.1	−0.7	211	81	0.742	68	58	3850
Belarus	80.2	207.6	10.0	−0.4	125	48	0.782	70	29	7500
Moldova	13.0	33.7	4.3	−0.1	328	127	0.699	46	53	2500
Total	6919.0	17,920.4	207.8	−0.67	30	12	0.766	71	10	6673
Trancaucasia										
Georgia	26.9	69.7	5.5	0	203	78	0.742	56	9	4600
Armenia	11.5	29.8	3.8	0.3	330	127	0.745	67	17	3000
Azerbaijan	33.4	86.6	8.1	0.9	243	94	0.738	51	18	3000
Total	71.8	186.1	17.4	0.48	242	93	0.741	56	14	3506
Central Asia										
Kazakstan	1049.2	2717.3	14.8	0.5	14	5	0.742	56	12	5000
Uzbekistan	172.7	447.4	25.1	1.7	145	56	0.698	38	9	2400
Turkmenistan	188.5	488.1	5.5	1.3	29	11	0.73	44	3	4300
Kyrgystan	76.6	198.5	5.0	1.3	65	25	0.707	35	7	2700
Tajikistan	55.3	143.1	6.2	1.4	112	43	0.66	27	6	1140
Total	1542.3	3994.4	56.6	1.3	37	14	0.709	42	10	3153
Baltic States										
Estonia	17.4	45.1	1.4	−0.4	78	30	0.812	69	25	10,000
Latvia	24.9	64.6	2.4	−0.6	95	37	0.791	69	27	7200
Lithuania	25.2	65.2	3.7	−0.1	147	57	0.803	68	39	7300
Total	67.5	174.9	7.5	−0.32	111	43	0.801	69	31	7772
Summary Total	8600.7	22,275.8	289.3	−0.21	33.6	13.0	0.754	64	10	5822

Sources: *World Population Data Sheet*, Population Reference Bureau, 2001; *U.N. Human Development Report*, United Nations, 2001; *World Factbook*, CIA, 2001.

So much of this vast region is sparsely populated that its estimated population of 289 million in 2001 ranked it only seventh among the eight world regions. The average population density of 34 per square mile (13 per sq km) is only half that of the United States. Great stretches of economically unproductive terrain separate many outlying populated areas from one another.

Russia and the Near Abroad comprise a complex population and linguistic mosaic. There are about 30 major ethnic groups and more than 100 languages (Figs. 6.2 and 6.3). In Russia, Belarus, and Ukraine, the majority are Slavs who originated in east-central Europe as speakers of an ancestral Slavonic language (a member of the Indo-European language family). In Moldova, the majority ethnic Moldovans speak a Romance language. Native Latvians and Lithuanians speak a Baltic branch of the Indo-European family. Their northern neighbor Estonia is populated by a distinct group, the Estonians, who speak a language in the Finnic branch of Uralic languages (it is closer to the language spoken in Finland than to the languages spoken in Estonia's

neighboring countries). The Armenians have a unique language within the Indo-European group. Their neighbors in the Caucasus — the Georgians and Azerbaijanis — speak, respectively, a Caucasian and a Turkic language.

Smaller ethnic groups in the Caucasus speak Caucasian languages related to Georgian. The Turkic languages, within the Altaic language family, are spoken by most of the ethnic groups in the Central Asian countries of Turkmenistan, Kazakstan, Uzbekistan, and Kyrgyzstan. In Tajikistan, however, the dominant group of Tajiks speaks an Iranian language within the Indo-European family. In extreme northeastern Russia, there are speakers of Paleo-Siberian languages, including Chukchi and Koryak, related to the languages carried across an ancient land bridge by the ancestral Native Americans. This is just a brief glimpse of the region's tongues; there are many smaller languages. As will be seen in this and the next chapter, this rich multiethnicity has at once bestowed great cultural wealth on the region and threatened to tear apart the nations within it.

Here are some useful geographic terms to know when studying this region. The region of Russia and the Near Abroad is outlined by the borders of what used to be the Union of Soviet Socialist Republics, also known as the Soviet Union or USSR. This state came into existence in 1922 following the overthrow of the last Romanov tsar in the Russian Revolution of 1917 and the subsequent civil war. Prerevolutionary Russia is known as Old Russia, Tsarist Russia, Imperial Russia, or the Russian Empire.

The name Russia now refers to the independent country of Russia. It is known politically as the Russian Federation, which was the largest of the 15 Soviet Socialist Republics (Union Republics or SSRs) that made up the Soviet Union. The full name of the Russian Federation during the Communist period was the Russian Soviet Federated Socialist Republic or RSFSR; the name appears on many older maps.

The loosely aligned Commonwealth of Independent States (CIS) was formed by 12 of the 15 former Union Republics late in 1991. Estonia, Latvia, and Lithuania are not members of this organization.

The area west of the Ural Mountains and north of the Caucasus Mountains has been known historically as European Russia. The Caucasus and the area east of the Urals have been called Asiatic Russia, Soviet Asia, or the eastern regions. Here, Transcaucasia is the region of the mountains plus the area south of the Caucasus Mountains; Siberia is a general name for the area between the Urals and the Pacific; and Central Asia is the arid area occupied by the five countries with large Muslim populations immediately east and north of the Caspian Sea.

F7.1
190

Russian nationals are by far the largest group, with a population of nearly 150 million. The next largest country is Ukraine, with about 50 million people. Kazakstan (this is the country's official spelling, though Kazakhstan is also widely used) is the most populous Central Asian nation, with about 25 million people. All the other countries in the region trail far behind, each with populations generally under 15 million. Population growth rates are highest, around 1.5% annually, among the predominantly Muslim populations of the Central Asian countries. At the other end of the spectrum, several countries — Russia, Ukraine, Estonia, and Latvia — are losing populations at a rate of about 0.5% per year. In Russia and Ukraine, this trend is due to the plummeting quality of life since the breakup in the Soviet Union in 1991 (see Definitions & Insights, p. 172). In an atmosphere of economic and political uncertainty, couples have chosen to have fewer children, and there are more broken marriages. More disturbingly, death rates have soared due to declining health care, increasing alcoholism, violence, suicide, and more frequent botched abortions.

6.2 Physical Geography and Human Adaptations

Stretching nearly halfway around the globe in northern Eurasia, the region of Russia and the Near Abroad has formidable problems associated with climate, terrain, and distance. Most of the region is handicapped economically by excessive cold, infertile soils, marshy terrain, aridity, and ruggedness. Natural conditions are more similar to those of Canada than to those of the United States. Interaction with a complex and demanding environment was a major theme in Russian and Soviet expansion and development. Nature continues to provide large assets but also poses great problems for the 15 countries that have succeeded the Soviet Union.

The Role of the Climatic and Biotic Environment

The region of Russia and the Near Abroad has a harsh climatic setting. Severe winter cold, short growing seasons, drought, and hot summer winds that shrivel crops in the steppes are major handicaps. But there are also advantages in the form of good soils, the world's largest forests, natural pastures for livestock, and diverse wild fauna. However, many of these resources are hard to access because of the logistical difficulties posed by an unfavorable climate and the vast distances involved.

Most parts of the region have continental climatic influences, with long cold winters, short warm summers, and low to moderate precipitation. Severe winters, for which Siberia is particularly infamous, prevail because the region is generally at a high latitude and is cut off from the moderating influences of oceans by continental periphery and mountain barriers. Four-fifths of the total area is farther north than any point in the conterminous United States (Fig. 6.4). The most extreme continental climate of the world is here. The lowest official temperature ever recorded in the Northern Hemisphere (−90°F) occurred in the Siberian settlement of Verkhoyansk (67°N, 135°E). Westerly winds from the Atlantic moderate the winter temperatures somewhat in the west, but their effects become steadily weaker toward the east, where it is much colder.

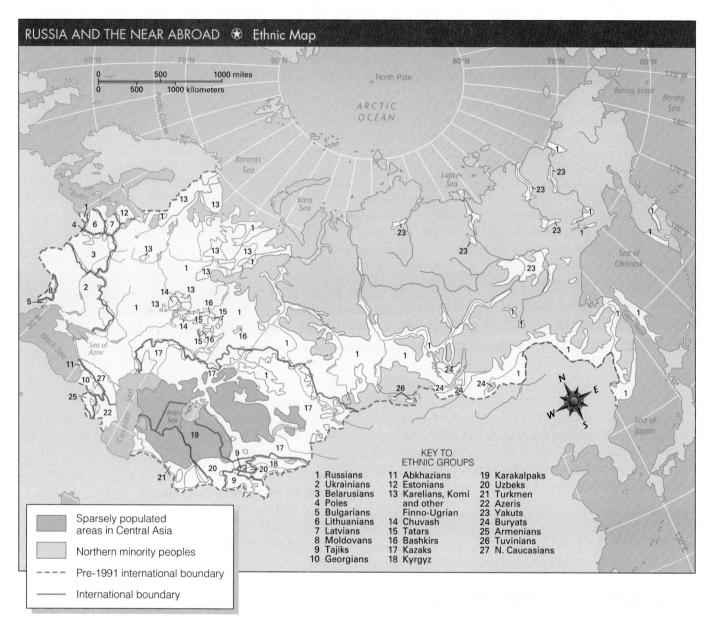

KEY TO
ETHNIC GROUPS

1 Russians	11 Abkhazians	19 Karakalpaks
2 Ukrainians	12 Estonians	20 Uzbeks
3 Belarusians	13 Karelians, Komi	21 Turkmen
4 Poles	and other	22 Azeris
5 Bulgarians	Finno-Ugrian	23 Yakuts
6 Lithuanians	14 Chuvash	24 Buryats
7 Latvians	15 Tatars	25 Armenians
8 Moldovans	16 Bashkirs	26 Tuvinians
9 Tajiks	17 Kazaks	27 N. Caucasians
10 Georgians	18 Kyrgyz	

Sparsely populated
areas in Central Asia

Northern minority peoples

- - - - Pre-1991 international boundary

——— International boundary

Figure 6.2   The Soviet Union had difficulties trying to hold together such a vast collection of ethnic groups. The Russian Federation faces many of the same challenges.

The average frost-free season of 150 days or less in most areas (except the extreme south and west) is too short for many crops to mature. Most places are relatively warm during the brief summer, and the southern steppes and deserts are hot. A factor partially offsetting the brevity of the summer is the long summer daylight period that encourages plant growth. This is also a time of accelerated human activities outdoors. Even more than Americans, Russians are passionate campers and sun-worshippers, and at every opportunity, they get away from it all to enjoy the fleeting pleasures of summer. Urban residents never lose the traditional Russian passion for nature, and those who can afford them purchase country cabins called *dachas*, to which they retreat on weekends and longer holidays.

For agriculture, a general shortage of moisture and the periodic occurrences of drought are even greater handicaps than low temperatures. The annual average precipitation is less than 20 inches (50 cm) nearly everywhere except in the extreme west, along the eastern coast of the Black Sea and the Pacific coast north of Vladivostok, and in some of the higher mountains.

There are five main climatic belts in the region of Russia and the Near Abroad: tundra, subarctic, humid continental, steppe, and desert. In this order, these belts, each with its associated vegetation and soils, succeed each other from north to south. There are also smaller scattered areas of subtropical, Mediterranean, and undifferentiated highland climates.

F2.6
28–29
F2.7
30–31

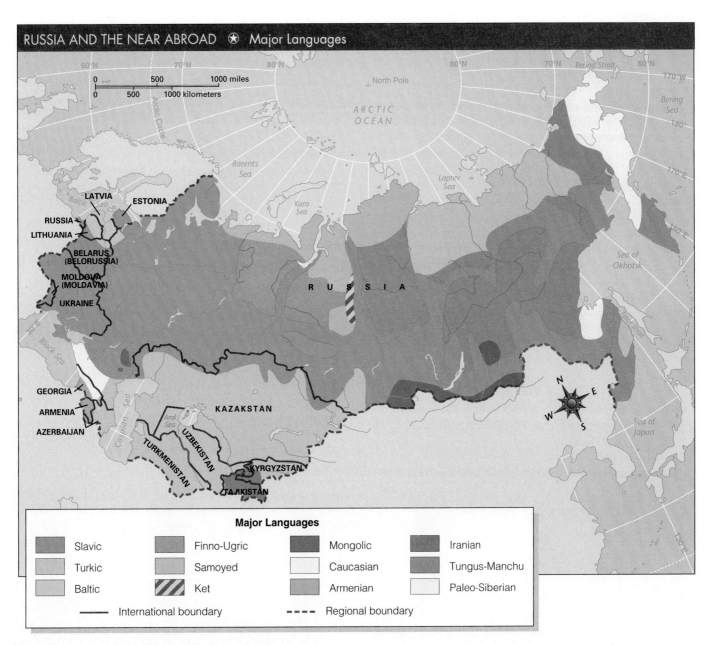

RUSSIA AND THE NEAR ABROAD ✪ Major Languages

Major Languages

Slavic	Finno-Ugric	Mongolic	Iranian
Turkic	Samoyed	Caucasian	Tungus-Manchu
Baltic	Ket	Armenian	Paleo-Siberian

——— International boundary   - - - - Regional boundary

Figure 6.3 Languages of Russia and the Near Abroad.

Huge areas in the tundra and subarctic climatic zones are permanently frozen at a depth of a few feet. The frozen ground, called **permafrost,** makes construction difficult. Heat generated by buildings melts the upper layers of permafrost and causes foundations and walls to sink and tilt. Most buildings are elevated on pilings to reduce this risk (Fig. 6.5). Pipelines carrying crude oil, which is hot when it comes from the ground, would likewise melt the permafrost and sink, and so they are heavily insulated or built on elevated supports.

The most productive human activities take place in the subarctic, humid continental, and steppe climatic zones. The subarctic climate is essentially coextensive with the Russian **taiga,** or northern coniferous forest, the largest continuous area of forest on Earth. The main trees are spruce, fir, larch,

and pine, which are useful for pulpwood and firewood but often do not make good lumber. Large reserves of timber suitable for lumber do exist in parts of the taiga, and this is one of the major areas of lumbering in the world (Fig. 6.6). Many observers fear that Russia, in its drive to advance the economy, will deplete the taiga. There is some marginal farming in the taiga, especially toward the south. The dominant soils are the acidic **spodosols** (from the Greek word for "wood ash" and known in Russian as *podzols*), which have a grayish, bleached appearance when plowed, lack well-decomposed organic matter, and are low in natural fertility.

A humid continental climate occupies a triangular area south of the subarctic climate, narrowing eastward from the western border of the region of Russia and the Near Abroad to

DEFINITIONS + INSIGHTS

The Russian Cross

Russia's economic health has plummeted since the breakup of the Soviet Union (as discussed in Section 6.4). The accompanying unemployment and decline in health and other services, increasing crime rates, and a growing sense of despair at the individual level have given rise to some very unhealthful habits among Russians. The rate of alcoholism has soared; the national average for annual vodka consumption is an eye-blearing 4.4 gallons (17 liters). Smoking is a national pastime. Organized and petty crime, and simple drunken brawls, result in an alarmingly high incidence of physical violence.

These and other factors have led to a drop in Russians' life expectancy, from 68 in 1990 (average for men and women) to 66 in 2001, the same as in the developing nations of Guatemala, Egypt, and Vietnam. Between 1990 and 2001, the death rate rose almost one-third, to one of the world's highest levels (15 per 1,000), surpassed only in such desperate LDCs as Afghanistan, Laos, and many nations in Africa south of the Sahara. The leading causes of death, in descending order, are heart disease, accidents, violence, and cancer. The universal system of health care under the Soviets was not efficient, but a health-care system close to collapse has replaced it, and hospitals are ill-equipped to help stem the rising tide of deaths.

As death rates have soared, birth rates have plummeted. The birth rate in 2001 was 9 per 1,000, down 25% from 12 per 1,000 in 1990. The disturbing trend lines of growing death rates and falling birth rates in Russia intersected in the mid-1990s, forming the grim graphic that demographers dubbed "the Russian cross" (Fig. 6.A). It is an ominous intersection. Russia's rate of annual population loss is the highest in the world and has seldom been seen on Earth except in times of warfare, famine, and epidemic. One United Nations estimate puts Russia's population at 121 million in 2050, down 17% from its 2001 figure of 144 million.

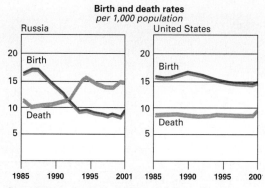

Source: Data from Russian Health Ministry; U.S. Census Bureau International Database.

Figure 6.A   The Russian Cross.

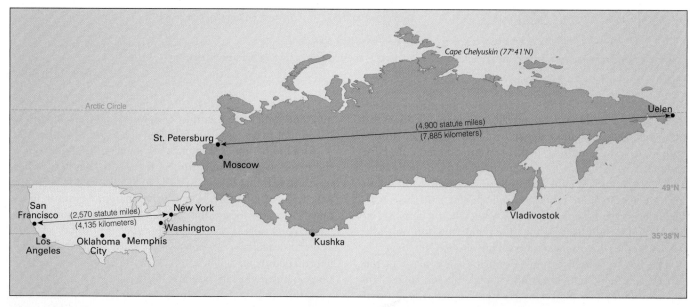

Figure 6.4   Russia and the Near Abroad compared in latitude and area with the conterminous United States.

Figure 6.5 In the High Arctic, buildings must be erected on pilings so that they do not melt the permafrost below. This is the Russian coal-mining settlement of Barentsburg in Norway's Svalbard (Spitsbergen) Archipelago.

the vicinity of Novosibirsk. This area has the short-summer subtype of humid continental climate, comparable to that of the Great Lakes region and the northern Great Plains of the United States and adjacent parts of Canada. Broadleaf deciduous forest supplants the evergreen taiga forest; oak, ash, maple, elm, and other broadleaf trees alternate with conifers (this is "mixed forest"). Both climate and soil are more favorable for agriculture in the humid continental climate than in the subarctic. Soils developed under broadleaf deciduous or mixed forest are normally more fertile than spodosols, although less so than grassland soils.

The steppe climate characterizes the grassy plains south of the forest in Russia and is the main climate of Ukraine, Moldova, and Kazakstan. The average annual precipitation is 10 to 20 inches (c. 25 to 50 cm), barely enough for unirrigated crops. This is the setting of the **black-earth belt,** the most important area of crop and livestock production in the region of Russia and the Near Abroad. The main soils of this belt are the **chernozem,** meaning in Russian "black earth." Among the best soils to be found anywhere, these are thick, productive, and durable soils from the soil order known as **mollisols.** A similar belt of mollisols occupies the eastern portion of the Great Plains of North America. Their great fertility is due to an abundance of humus in the topsoil. The major natural threat to productive agriculture here is the occurrence of severe droughts and hot, desiccating winds (*sukhovey*) that damage or destroy crops. The steppe also includes extensive areas of **chestnut** or **alfisol** soils in the zones of lighter rainfall. These are lighter in color than chernozems and lack their superb fertility but are among the world's better soils.

F7.1 190

The most characteristic natural vegetation of the steppe is short grass. Pastoralists, including the Scythians who in the fourth century B.C. produced astonishing artwork of gold in the area that is now Ukraine, grazed their herds from an early time on the treeless steppe grasslands, which stretched over a vast area between the forest and the southern mountains and deserts. Today, much of this area is cultivated, with wheat the main crop. The steppe is also a major producer of sugar beets, sunflowers grown for vegetable oil, various other crops, and all the major types of livestock (Fig. 6.7).

Growing crops is difficult in the desert climate areas east and immediately north of the Caspian Sea in the Central Asian states. Cotton is the most important crop where irrigation water is available from the Syr Darya, Amu Darya, and other rivers flowing from high mountains to the south and east.

212

The Role of Rivers

The Moscow region lies on a low upland from which a number of large rivers radiate like the spokes of a wheel (Fig. 6.8). The longest ones lead southward: the Volga to the landlocked

Figure 6.6 The taiga in Russia is Earth's largest continuous forest biome, and Russia's economic growth is based in part on its exploitation. These milled conifers are awaiting shipment from St. Petersburg's docks.

Figure 6.7 A rancher grazing his cattle on the steppe near the Don River in southern Russia.

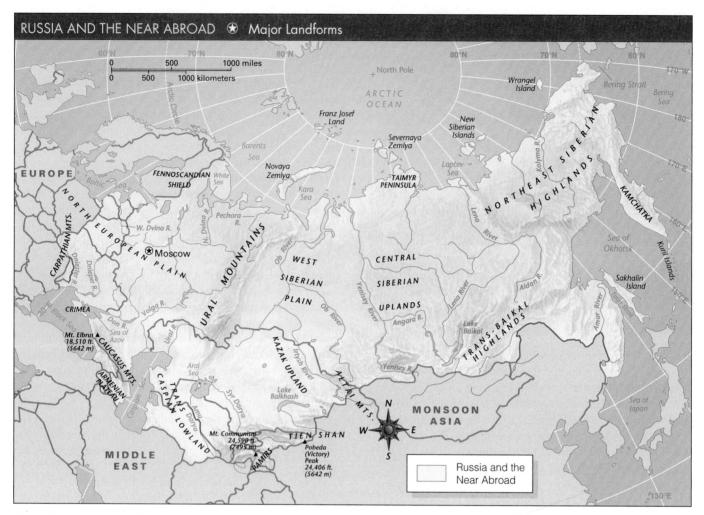

Figure 6.8 Map of landforms and water features in the region of Russia and the Near Abroad.

Caspian Sea, the Dnieper (Dnepr) to the Black Sea, and the Don to the Sea of Azov, which connects with the Black Sea through a narrow strait. Shorter rivers lead north and northwest to the Arctic Ocean and Baltic Sea. These river systems are accessible to each other by portages. In the early history of Russia, the rivers formed natural passageways for trade, conquest, and colonization. They were especially crucial in the settlement of Siberia, which is drained by some of the greatest rivers on Earth: the Ob, Yenisey, Lena, and Kolyma, all flowing northward to the Arctic Ocean, and the Amur, flowing northward to the Pacific. By following these rivers and their lateral tributaries, the Russians advanced from the Urals to the Pacific in less than a century.

The Volga is the most important river; Russians call it *Matushka* — "Mother." It rises about 200 miles (320 km) southwest of St. Petersburg and flows 2,300 miles (3,680 km) to the Caspian Sea. Before railroads supplemented river traffic, the Volga was Russia's premier commercial artery. Traditionally, wheat, coal, and pig iron from Ukraine, fish from the Caspian Sea, salt from the lower Volga, and oil from Baku, on the western shore of the Caspian, traveled upriver toward Moscow

and the Urals. Timber and finished products moved downriver to the lower Volga and the Ukraine (Fig. 6.9). Boatmen towed barges upstream on a 70-day journey from Astrakhan, near the Volga mouth, to Kazan, on the middle Volga. The steamboat arrived in the late 1800s, bringing an end to the way of life recalled in the famous Russian song "The Volga Boatmen." Despite changing times, the ethnic mosaic of the middle Volga region remains colorful. In the autonomous regions of Tatarstan, Chuvashia, and Mari-El, non-Russian groups of Turkish and Finnish origin make up the majority of the population.

The Volga was difficult to navigate until the latter half of the 20th century. There were shoals and shallows, especially in very dry summers, and ice still closes the waterway each winter for 120 to 160 days. During the Soviet era, the "Great Volga Scheme" transformed the river. The goal was to control completely the flow of the river with a stairway of huge reservoirs, each of which reaches upstream to the dam forming the next reservoir, thus assuring complete navigability during the 6 months when the river is ice-free and supplying hydroelectric power and water for irrigation. Behind the dams that are

Figure 6.9 Timber moving on the Volga, a major transportation artery in Russia. This scene is near Saratov on the lower Volga.

the backbone of the Great Volga Scheme, the reservoirs are so vast that they are generally called "seas." With these improvements in navigation, the Volga secured its place as Russia's most important internal waterway. Large numbers of barges and log rafts, as well as a fleet of passenger vessels that carry tourists, use it. Alteration of rivers for power, navigation, and irrigation became an important component of economic development under the Soviets. Large dams were also raised across the main courses or major tributaries of the Dnieper, Don, Kama, Irtysh, Ob, Yenisey, and Angara Rivers.

A major link in the inland waterway system is the Volga-Don Canal, opened in 1952 to tie together the two rivers where they approach each other in the vicinity of Volgograd (Figs. 6.1 and 6.10). The completion of the canal meant that the White Sea and Baltic Sea in the north were linked to the Black Sea and Caspian Sea in the south in a single water transport system. The accomplishment was no small task. The Don River, which empties into the Black Sea via the Sea of Azov, is about 150 feet (45 m) higher than the Volga. The ground that separates these two rivers rises to almost 300 feet (90 m). Engi-

neers solved this discrepancy with 13 locks, each with a lift of about 30 feet (9 m), which carry water 145 feet (44 m) above the Don and drop it 290 feet (87 m) to the Volga.

The Role of Topography

Most of the important rivers of Russia and the Near Abroad wind slowly for hundreds or thousands of miles across large plains. Such plains, including low hills, compose nearly all the terrain from the Yenisey River to the western border of the region. The only mountains in this lowland are the Urals, a low and narrow range (average elevation: less than 2,000 feet/610 m) that separates Europe from Asia and European Russia from Siberia. The Urals trend almost due north–south but do not occupy the full width of the lowland. A wide lowland gap between the southern end of the mountains and the Caspian Sea permits uninterrupted east–west movement by land. Cut by river valleys offering easy passageways, the Urals are not a serious barrier to transportation.

Between the Urals and the Yenisey River, the great lowland of western Siberia (West Siberian Plain) is one of the flattest

Figure 6.10 A lock in the Volga-Don Canal, a vital link for trade between the Volga watershed and the Black Sea.

areas on Earth. Immense wetlands, through which the Ob River and its tributaries slowly wind their way, cover much of the plain. This waterlogged country, underlain by permafrost that blocks downward seepage of water, is a major barrier to land transport and is uninviting to settlement. Tremendous floods occur in the spring when the breakup of ice in the upper basin of the Ob releases great quantities of water while the river channels farther north are still frozen, acting as natural dams. In spring 2001, Russian aircraft bombed numerous ice jams on the rivers in an innovative attempt to relieve severe flooding.

The area between the Yenisey and Lena Rivers is occupied by the hilly Central Siberian Uplands, which have a general elevation of 1,000 to 1,500 feet (c. 300 to 450 m). Mountains dominate the landscape east of the Lena River and Lake Baikal. Extreme northeastern Siberia is an especially bleak and difficult country for human settlement.

High mountains rim the region of Russia and the Near Abroad on the south from the Black Sea to Lake Baikal, and lower mountains rim the region from Lake Baikal to the Pacific. Peaks rise to over 15,000 feet (c. 4,500 m) in the Caucasus Mountains between the Black and Caspian Seas and in the Pamir, Tien Shan, and Altai Mountains east of the Caspian Sea. From the feet of the Pamir, Tien Shan, and Altai ranges, and the lower ranges between the Pamirs and the Caspian Sea, arid and semiarid plains and low uplands extend northward and gradually merge with the West Siberian Plain and the broad plains and low hills west of the Urals.

6.3 Cultural and Historical Context

The ethnic and cultural compositions of Russia and the Near Abroad are complex. The cultural histories of these peoples have always been influenced and often dominated by the Russians, with whom we begin; the non-Russian peoples are described in more detail in Chapter 7. Russian political and cultural origins reach more than 1,000 years into the past. The central figures were Slavic peoples who colonized Russia from the west, interacted with many other peoples, stood off or outlasted invaders, and acquired the giant territory that would become Russia and the Soviet Union.

Early Scandinavian and Byzantine Influences

Slavic peoples have inhabited Russia since the early centuries of the Christian era. During the Middle Ages, Slavic tribes living in the forested regions of western Russia came under the influence of Viking adventurers from Scandinavia known as Rus or Varangians. The newcomers carved out trade routes, planted settlements, and organized principalities along rivers and portages connecting the Baltic and Black Seas. In the ninth century, the principality of Kiev, ruled by a mixed Scandinavian and Slavic nobility, achieved mastery over the others and became a powerful state. The culture it developed was the foundation on which the Russian, Ukrainian, and Belarusian cultures later arose.

Contacts with Constantinople (now Istanbul, Turkey) much affected Kievian Russia. Located on the straits connecting the Black Sea with the Mediterranean, Constantinople was the capital of the Eastern Roman or Byzantine Empire, which endured for nearly 1,000 years after the collapse of the Western Roman Empire in the fifth century A.D. Constantinople became an important magnet for Russian trade, and the Russians borrowed heavily from its culture. In 988, the ruler of Kiev, Grand Duke Vladimir I, formally accepted the Christian faith from the Byzantines and had his subjects baptized. Following its cleavage from the Roman Catholic faith in A.D. 1054 Orthodox Christianity became a permanent feature of Russian life and culture (see Fig. 6.11 for the region's map of religions). Moscow eventually came to be known as the "Third Rome" (after Rome itself and Constantinople) for its importance in Christian affairs. The Bolshevik Revolution in 1917 began a 75-year period of official repression and neglect of the Church. Since the breakup of the Soviet Union, however, there has been a renaissance in religious observation across the vast region of Russia and the Near Abroad (Fig. 6.12).

The Tatar Invasion

Yet another cultural influence reached the Russians from the heart of Asia. The steppe grasslands of southern Russia had long been the habitat of nomadic horsemen of Asian origin. During the later days of the Roman Empire and in the Middle Ages, these grassy plains, stretching far into Asia, provided the Huns, Bulgars, and other nomads a passageway into Europe. In the 13th century, the Tatars took this route. Many steppe and desert peoples were in their ranks, but Mongols led them. In 1237, Batu Khan ("Batu the Splendid"), grandson of Genghis Khan, launched a devastating invasion that brought all the Russian principalities except the northern one of Novgorod under Tatar rule.

The Tatars collected taxes and tributes from the Russian principalities but generally allowed the rulers of these units latitude in local government. Even the princes of Novgorod, who were not technically under Tatar control, paid tribute to avoid trouble. The Tatars established the Khanates (governmental units) of Kazan, Astrakhan, and Crimea. Russians called the Kazan Tatars "the Golden Horde," after the brightly colored tents in which they lived. When Tatar power declined in the 15th century, the rulers of the Moscow principality were able to begin a process of territorial expansion that resulted in the formation of present-day Russia. Today, oil-rich Tatarstan, with its capital at Kazan, is one of Russia's most important "autonomous" political units.

The Empire of the Tsars

The Russian monarchy reached outward from Muscovy, its original domain in the Moscow region. From the 15th century until the 20th, the tsars built an immense Russian empire by accretion around this small nuclear core. This imperialism by land (see Definitions & Insights, p. 180), in the era when the maritime powers of Western Europe were expanding by sea, brought

Perspectives from the Field

Into the Russian Arctic

In the summers since 1992, I have served as "expedition leader" aboard vessels carrying Western tourists to the high Arctic regions of Russia and Norway. For its extraordinary beauty, the surprising abundance of wildlife, and the bizarre rhythms of a land where the sun never sinks below the horizon in summer nor rises in winter, I have come to love the Arctic. But the greatest pleasures come from serving on these vessels with Russian officers and crew, being embraced by their unending warmth and hospitality, and gaining insight into their lives and personalities, at once brimming with soulful joy—what they call dokhovnost, *the spiritual dimension of life—and suffering from reflection on the difficult circumstances their loved ones face at home, south of the Arctic Circle. Following are some excerpts from my journal written at the end of my first Arctic voyage aboard a small Russian passenger vessel, the* Alla Tarasova. *She (ships are always referred to in the feminine) was not capable of penetrating fast ice, so to reach Franz Josef Land, we enlisted the aid of a Russian nuclear-powered icebreaker, the* Vaygach. *These passages offer flavors of Arctic land, sea, and wildlife and glimpses of the Russian people working in this remote region. They also show how tourism has invaded Earth's most remote corners.*

Tromso, Norway, July 6. On board the ship I immediately met with Captain Vashislav ("Slava") Vasyuk, who expressed his eagerness for this to be a most successful voyage. I found it very easy to work with the captain. He is very reasonable and flexible, and his judgment is sound. Another key individual in the success of this cruise was Bogdan Gavrilchuk, the ship's navigator. His facility with English, his good humor, and his willingness to share his culture with the passengers kept him in constant demand. He always said "yes" to any favor we asked of him. Like the captain, his at-

titude was, "No problem — we want you to be happy and we'll do whatever we can." The captain informed me there were no guns on the ship. It was too late in the day for us to purchase these in Tromso. [Norwegian law requires that visitors to Spitsbergen, where we sailed first, carry weapons to defend themselves against polar bears, and in Franz Josef Land, we would also need guns for this reason.]

At Sea en route to Franz Josef Land, July 11. At about 0400, the ship began striking ice. Most passengers awoke. The ship again had to slow as we worked slowly through the ice. About 0930, a polar bear was sighted swimming. All passengers had a chance to view her as she emerged from the water and stood briefly on floating ice. They were thrilled! At 1545, we reached a long line of floating ice, about 2 miles wide, and filled with a vast number of seals — the animals were everywhere. Surely, at that moment, it was the largest concentration of pinnipeds anywhere on Earth. They must have congregated here to feed, but I don't know why so many fish would be in this spot. At 1745, we had our rendezvous with the icebreaker Vaygach, another

highlight for the passengers, who had been concerned about our slow pace through the ice. Slava talked me out of trying to get on board the icebreaker until we were closer to Franz Josef Land and so in a better position to make decisions. At this time, we learned there were no guns aboard the icebreaker — a problem that lingered as a major concern in polar bear country.

Hooker Island, Franz Josef Land, July 12. Slava woke me at 0500 saying we needed to get to the icebreaker to meet with her captain and do a helicopter reconnaissance. I woke Chantal [Chantal Langley, my assistánt] to come along, and we had a most exciting morning in the helicopter. On the icebreaker, Chantal presented Captain Valerie Ivanovich Chestapolov with a bottle of whiskey and a tie, and we sat down to business. Captain Chestapolov, like our own captain, was very eager to see that our passengers enjoyed themselves and insisted the only way to know what we could do was to get up and have a look — and so we did, in the helicopter. We landed briefly at Cape Flora and then flew northward over pack ice and intermittent open water (where several pods of beluga whales sheltered),

Figure 6.B   In Franz Josef Land on the *Alla Tarasova*, August 1992, from left to right: Chantal Langley, Bogdan Gavrilchuk, Vashislav Vasyuk, and Dmitri Vasyuk.

(Continued)

viewing the characteristic landscape of glacial cover broken occasionally by basaltic cliffs. We flew on to Rubini Rock and Calm Bay by Hooker Island. We circled Rubini Rock and landed at the abandoned station the Russians call *Tikhiya Papanyin* on Calm Bay. Chantal and I were enchanted by this place and agreed the passengers must come here. Returning to the icebreaker, we passed over walruses resting on an ice floe . . . Each group of the passengers had 90 minutes ashore. Katerina Salén arranged to have Markus Lehofer receive each arriving helicopter with champagne and sandwiches. He wore black jacket and tie — he was very funny and the passengers were delighted.

There was one very large problem with the excursion today. I took what I feel was a terrible risk in putting passengers ashore without any adequate protection against polar bears. Everything I have read on the subject of people-polar bear interaction suggests that polar bear attacks are not unlikely and that there is no sure protection against an attacking bear but to kill the animal. In a thorough briefing on the excursion ashore, I promised passengers that Bogdan and the ship's doctor would be positioned with flare guns as sentinels at either end of our excursion area. A flare meant that a bear had been sighted and you should take cover immediately. I showed them the body position to assume if a bear should attack. Once ashore though, our two sentinels abandoned their responsibility and playfully fired their flares, which about gave me a heart attack. It was all the more distressing to find polar bear tracks in the snow on the beach. We did not have a problem with bears today — but we could have, and been totally unprepared for the result.

Hays Island, July 13. During the night, the two ships sailed through the pack ice of the British Channel and Markham Strait and reached Hays (Kheysa) Island at 0500. My wakeup call to the passengers was an announcement at 0500 that they could get up and see a polar bear. By 0545, the ships were moored together. There were some delays as I sent our interpreter, Tatya, to negotiate at the station to have some of the researchers show us around. The answer at first was a firm "no." Tatya was informed that this was a secure, secret facility, where no tourists had ever visited, and none ever would. Apparently, she was treated rather roughly there until Katerina arrived and fixed the situation by arranging to have most of the research staff flown to our ship to enjoy our bars; we had both bars opened by 0700 and the scientists began having a merry time. We also shuttled in some brandy for the research staff hosting our visitors, and they became happy hosts to the groups we flew ashore. We saw the highlight of the station, the place where rockets for weekly atmospheric soundings are assembled.

Our few hours here made it clear how difficult and remote a place this is. They call this "queen of the Arctic research stations," but as it sleeted and snowed on us here in the height of summer and we peeked into the scientists' threadbare accommodations, we all reflected on how lonely and isolated these people must feel. There are no spouses or children here, and there is only work. Bogdan, who said he makes only $10 per day, said the researchers and staff of Hays Island are paid only $1 per day.

All passengers were back on ship at 0930, and by 1030, we had disembarked the cheery researchers and were underway. The weather took an extraordinary turn to clear blue skies. The icebreaker created a path of broken ice that glistened like diamonds. We saw several more polar bears, including two sets of mothers with cubs. The passengers were by now nearly ecstatic about being here. At 1500, after passing through Abedaire Strait, we left Franz Josef Land behind.

Joe Hobbs

Figure 6.C   Some of the scientific research staff of the station on Hays Island, Franz Josef Land, August 1992.

a host of alien peoples under tsarist control. The motivations were diverse. Quelling raids by troublesome neighbors, particularly nomadic Muslim peoples in the southern steppes and deserts, was one objective. Many Russian cities, such as Volgograd (originally, Tsaritsyn and then Stalingrad) on the Volga River, were founded as fortified outposts on the steppe frontier. In the wilderness of Siberia, the search for valuable furs and minerals, especially gold, stimulated early expansion, and the missionary impulse of Orthodox priests also played a role. Land hunger and a desire to escape serfdom and taxation led to the flight of many peasants into the fertile black-earth belt of southwestern Siberia, and many landlords also moved there with their serfs.

The initial outward thrust from Muscovy under Ivan the Great (reigned 1462–1505) was mainly northward and east-

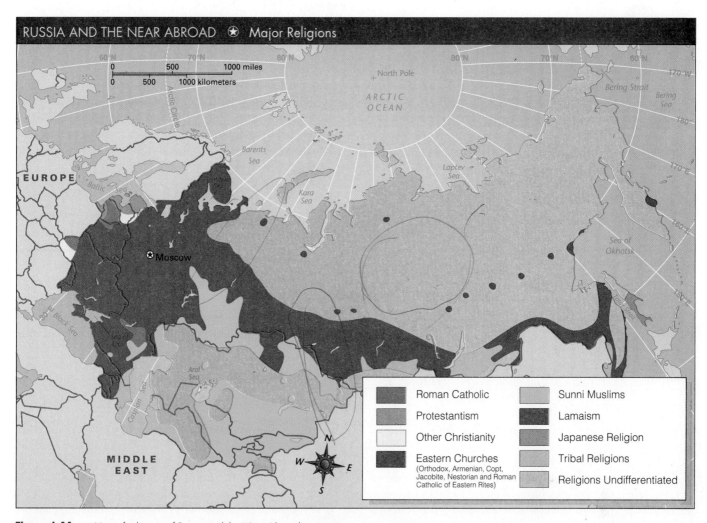

RUSSIA AND THE NEAR ABROAD ✷ Major Religions

■ Roman Catholic	▨ Sunni Muslims
■ Protestantism	■ Lamaism
□ Other Christianity	■ Japanese Religion
■ Eastern Churches (Orthodox, Armenian, Copt, Jacobite, Nestorian and Roman Catholic of Eastern Rites)	▨ Tribal Religions
	▨ Religions Undifferentiated

Figure 6.11   Map of religions of Russia and the Near Abroad.

Figure 6.12   Some of the faithful in a Russian Orthodox church in Vyborg, Russia. Since the breakup of the USSR, there has been a resurgence of organized worship throughout the region.

ward. The rival principality of Novgorod was annexed and a domain secured that extended northward to the Arctic Ocean and eastward to the Ural Mountains. Ivan the Terrible (reigned 1533–1584) added large new territories to Russia by conquering the Tatar khanates of Kazan and Astrakhan, thus giving the Russian state control over the entire Volga River. Later tsars pushed the frontiers of Russia westward toward Poland and southward toward the Black Sea. Peter the Great (reigned 1682–1725) defeated the Swedes under Charles XII to gain a foothold on the Baltic Sea, where he established St. Petersburg as Russia's capital and its "window on the West." Catherine the Great (reigned 1762–1796) secured a frontage on the Black Sea at the expense of Turkey.

Meanwhile, the conquest of Siberia proceeded rapidly. At the end of Ivan the Terrible's reign, traders and Cossack military pioneers were already penetrating this wild and lonely territory, where there were only scattered indigenous inhabitants. The Cossacks were peasant-soldiers in the steppes, formed originally of runaway serfs and others fleeing from tsardom. They eventually gained special privileges as military communities serving the tsars.

Some geographers have characterized the Soviet Union and now Russia alone as a **land empire.** Rather than establishing its colonies overseas, as did imperial powers such as Spain and Great Britain, Russia founded its colonies in its own vast continental hinterland. Many of the colonized peoples had little in common with the Russian ethnic majority, which ruled from faraway Moscow. Like former colonies of overseas empires such as Great Britain, non-Russian regions of the Soviet Union's periphery were drawn into a relationship of economic dependence on the imperial Russian core. The Central Asian republics, for example, followed Moscow's demands to grow cotton, which was shipped to Moscow to be manufactured into expensive clothing marketed in Central Asia and elsewhere in the empire. This pattern contributed to the growth of the USSR's economy but often inhibited local development. Other entities characterized historically as land empires include the United States, China, Brazil, and the Ottoman, Mogul, Aztec, and Inca empires.

In 1639, a Cossack expedition reached the Pacific. Russian expansion toward the east did not stop at the Bering Strait but continued down the west coast of North America as far as northern California, where the Russian trading post of Fort Ross was active between 1812 and 1841. In 1867, however, Russia sold Alaska to the United States and withdrew from North America. Russia's selling price was two cents per acre. In the United States, the transaction was widely derided as "Seward's Folly," after William Henry Seward, the American secretary of state who negotiated it. The value of the oil pumped from Alaska each day now exceeds what the United States paid for its vast new territory.

The relentless quest of Russian fur traders for sable, sea otter, and other valuable furs brought great cultural changes to aboriginal peoples in Siberia and Pacific North America. These effects became increasingly pronounced, especially in the last decades of the Russian empire when millions of Russians moved into Siberia.

In the Amur River region near the Pacific, the Russians were not able to consolidate their hold for nearly two centuries as a result of opposition by strong Manchu emperors who claimed this territory for China. They were thus barred from the east Siberian area best suited to grow food for the fur-trading enterprise and best endowed with good harbors for maritime expansion. This situation changed in 1858–1860 following the victory of European sea powers over China in the Opium Wars. Russia was able to add the Amur region to its earlier gains in the Ob, Yenisey, and Lena basins. In a series of military actions during the 19th century, Russian tsars annexed most of the Caucasus region and Turkestan, the name given to the arid Central Asia region east of the Caspian Sea.

Russia and the Soviet Union, Tempered by Revolution and War

Russia and the Soviet Union repeatedly triumphed over powerful invaders, notably the Swedish forces led by King Charles XII in 1709, the French and associated European forces under Napoleon I in 1812, and the German and other European forces sent by Hitler into the Soviet Union in World War II. In each case, the Russians lost early battles and much territory but eventually inflicted a crushing and decisive defeat on the invaders. The success of Old Russia and the Soviet Union in withstanding invasions by such formidable armies was due in part to the environmental rigors (particularly, the brutal Russian winter) invaders faced, the overwhelming distances of a huge country with poor roads, and the defenders' love of their homeland. It was also due to talented Russian military leadership and to the willingness of good and poor generals alike to lose great numbers of men in combat. Finally, the successful defense used the **scorched earth** strategy to protect the motherland; rather than leave Russian railways, crops, and other resources to fall into the invaders' hands, the defenders destroyed them.

The Russian Revolution of 1917, which set the stage for the formation of the Soviet Union, was really two revolutions that occurred against the backdrop of World War I. In 1914, when a Serb in Sarajevo assassinated the heir to the throne of Austria-Hungary, a complicated series of alliances required Tsar Nicholas II to commit Russian troops to fight with Serbia, France, and Great Britain against Austria-Hungary and Germany in World War I. The first revolution in Russia began early in 1917 as a general protest against the terrible sacrifices of Russian forces on the Eastern Front during this war. That revolt overthrew Nicholas II, the last of the Romanov tsars. The Bolshevik Revolution came later that year. Led by Vladimir Ilyich Lenin (1870–1924), the Bolshevik faction of the Communist party seized control of the government. The new regime made a separate peace with Germany and its allies and survived a difficult period of civil war and foreign intervention between 1917 and 1921. Lenin presided over the establishment of Russia's successor state, the Soviet Union, in 1922.

World War II found Russians again allied with France and Britain in a far more ferocious war against Germany. The Soviet Union's success in withstanding the German onslaught that began as Operation Barbarossa in June 1941 surprised many outside observers who had predicted that the Soviets would prove too weak and disunited to resist for more than a few weeks or months. The crucially important cities of Leningrad

JOE HOBBS

Figure 6.13   The Mamayev Hill monument to the memory of the defenders of Stalingrad. Within the hill lie the bodies of tens of thousands of the city's defenders. This sword-wielding figure called "the Russian Motherland" is the world's largest statue, 279 feet (85 m) high and weighing 8,000 tons.

(now St. Petersburg) and Moscow held out through brutal sieges. Late in 1942, Soviet forces halted the German push eastward at the Volga River in the huge Battle of Stalingrad (Fig. 6.13). This engagement, which cost 2.7 million Soviet and 800,000 German lives, was the turning point in the war; thereafter, Soviet and Allied forces reversed the Nazi advance.

The failure of powerful Germany to conquer the Soviet Union was a clear indication that the strength of the Soviet Union had been underrated and that the country's power would henceforth be a major feature of world affairs. But the war's impact on the USSR would linger. The German invasion took an estimated 20 million or more Soviet lives and caused the relocation of millions of people. It did enormous damage to settlements, factories, and livestock. As a strategic precaution during the war, Lenin's successor, Josef Stalin, directed major Soviet industries to relocate eastward away from the front, a move that has had a lasting imprint on the region's economic geography.

Russians and Other Nationalities

The Soviet Union was a mosaic of over 100 languages and cultures, many of which were officially recognized as non-Russian nationalities. The Soviets established 16 Autonomous Soviet Socialist Republics (ASSRs) as homelands for large ethnic minorities, in theory endowing them with limited self-governing

powers. Smaller nationalities were organized into politically subordinate "autonomous" units, including five "autonomous regions" (autonomous *oblasts*) and ten "autonomous areas" (autonomous *okrugs*). Soviet labeling of these units as "autonomous" came back to haunt the nation of Russia in the 1990s. As a result of the devolution of power since 1991, within the Russian Federation there are now 89 areal "subjects" divided into 6 categories: 49 *oblasts* (regions), 6 *krays* (territories), 21 republics, 10 autonomous *okrugs* (ethnic subdivisions of *oblasts* or *krays*), 2 federal cities, and 1 autonomous oblast.

The Soviet Union permitted the different nationalities to retain their own languages and other elements of their traditional cultures. Alphabets were created for nationalities that previously had no written language. In spite of such cultural concessions to non-Russians, however, the Russian-dominated regime implemented a deliberate policy of **Russification,** an effort to implant Russian culture in non-Russian regions and make non-Russians more like Russians. Large numbers of Russians migrated to work in the factories and state farms in non-Russian republics. Russians were prominent in positions of responsibility even within the non-Russian republics; throughout the USSR, they held the majority of top posts in the Communist party, the government, and the military. But Russian ways had little impact on non-Slavic cultures such as the Georgians and Armenians. Even Slavic kinspeople of the Russians, such as the Ukrainians, insisted on retaining their cultural autonomy. In general, the policy of Russification was a failure because of strong nationalist sentiments throughout the Soviet Union.

6.4 **Economic Geography**
The Communist Economic System

The Soviet Union's Communist economic system was an attempt to put into practice the economic and social ideas of the 19th-century German philosopher Karl Marx. According to Marx, the central theme of modern history is a struggle between the capitalist class (bourgeoisie) and the industrial working class (proletariat). He forecast that exploitation of workers by greedy capitalists would lead the workers to revolt, overthrow the capitalists, and turn over ownership and management of the means of production to new workers' states. In the classless societies of these states, there would be social harmony and justice, with little need for formal government.

Marx's utopian vision did not materialize anywhere, but it did provide guidelines for antigovernment revolts and Communist political systems in many countries. The ideas of Marx and Lenin (who died in 1924), as Lenin's successor Josef Stalin interpreted and implemented them, provided the philosophical basis for the Soviet Union's centrally planned **command economy.** Beginning in 1928, a series of 5-year economic plans demanded — thus, the term "command economy" — quotas of production for the nation such as types and quantities of minerals, manufactured goods, and agricultural commodities to be produced; factories, transportation links, and dams to be built or improved; and locations of new residential areas to house industrial workers. The goals were to abolish the old aristocratic and capitalist institutions of Tsarist Russia and to develop a strong socialist state of equal stature with the major industrial nations of the West.

In the command economy, an agency in Moscow called Gosplan (Committee for State Planning) formulated the national plans, which were then transmitted downward through the bureaucracy until they reached individual factories, farms, and other enterprises. This was an unwieldy and inefficient process. First, the planners in Moscow were essentially required to act as CEOs of a giant corporation that would manage the economy of an area larger than North America — a gargantuan, impossible task. Further, the Soviet planning bureaucracy had no free market to guide it, so the system produced goods that people would not buy or failed to produce goods that people would have liked to buy. Third, Gosplan stated production targets in quantitative rather than qualitative terms, churning out abundant but substandard products. Finally, fearing reprisals from people higher up the ladder, no one wanted to suggest ways to increase quality and efficiency.

Although cumbersome, this system in fact propelled the Soviet Union to superpower status, improved the overall standard of living, prompted rapid urbanization and industrialization, and altered the landscape profoundly. The principal goal of Soviet national planning after 1928 was a large increase in industrial output, with emphasis on heavy machinery and other capital goods, minerals, electric power, better transportation, and military hardware. Masses of peasants were converted into factory workers, new industrial centers were created, and old ones were enlarged. There were huge investments in defense, and the country grew strong enough to survive Germany's onslaught in World War II. After the war, the Soviets maintained large armed forces and accumulated a massive arsenal of conventional and nuclear weapons in the **arms race** against the world's only other superpower, the United States (Fig. 6.14). By the height of the Cold War in

the early 1980s, 15 to 20% of the country's GNP was dedicated to the military (in contrast to less than 10% in the United States), representing a considerable diversion of investment away from the country's overall economic development.

Farmers and farming had troubled careers under the Soviet system of collectivized agriculture. Between 1929 and 1933, about two-thirds of all peasant households in the Soviet Union were collectivized. Their landholdings were confiscated and reorganized into two types of large farm units: the collective farm (*kolkhoz*) and the factory-type state farm (*sovkhoz*). The consolidation of individual farmsteads and villages into fewer but larger communities on the collective farms was supposed to permit the government to more efficiently and cheaply administer, monitor, and indoctrinate the rural population and to provide them with services, including education, health care, and electricity. But the rural people resisted collectivization fiercely. In their own version of scorched earth, peasants and nomads slaughtered millions of livestock and burned crops to avoid turning them over to the "socialized sector." Government reprisals followed, including wholesale imprisonments and executions, together with confiscation of food at gunpoint (often including the peasants' own food reserves and seed). The more prosperous private farmers, known as *kulaks*, were killed, exiled, sent to labor camps, or left to starve. Famines took millions of lives. Soviet leaders disregarded these costs, and collectivization was virtually complete by 1940.

The drive to increase the national supply of farm products also demanded an enlargement of cultivated area. In the 1950s, the Soviet Union began a program to increase the amount of grain (mainly spring wheat and spring barley) produced by bringing tens of millions of acres of virgin and idle lands or "new lands" into production in the steppes of northern Kazakstan and adjoining sections of western Siberia and the Volga region.

Soviet enterprises harnessed the energies and resources of the whole country to achieve specific objectives. The government called on the people to sacrifice and to make the country strong, especially by working on so-called "Hero Projects" such as construction of tractor plants, dams, railways, and land reclamation. The government in turn operated as a collectivized welfare state, providing guaranteed employment, low-cost housing, free education and medical care, and old-age pensions. Social services were often minimal but in some sectors were quite successful; the literacy rate, for example, rose from 40% in 1926 to 98.5% in 1959. However, military-industrial superpower status was generally achieved at the expense of Soviet consumers, whose needs were slighted in favor of heavy metallurgy and the manufacture of machinery, power-generating and transportation equipment, and industrial chemicals. Lines of consumers waited in stores to purchase scarce items of clothing and everyday conveniences. The exasperation of shoppers confronted by long lines and empty shelves was an important factor generating dissatisfaction with the economic system and leading to demands that the system be redesigned.

Figure 6.14   A nuclear-powered aircraft carrier in Russia's port of Murmansk. The arms buildup during the Cold War was unbearably costly for the Soviet Union.

Economic Roots of the Second Russian Revolution

Internal freedoms and prosperity did not accompany geopolitical power in the Soviet Union. Near paralysis gripped the flow of goods and services in the late 1980s, when large demonstrations and strikes underscored public anger at a political and economic system that was sliding rapidly downhill. The Communist system came under open challenge on the grounds that it stifled democracy, failed to provide a good living for most people, and thwarted the ambitions of the country's many ethnic groups for a greater voice in running their affairs. The outpouring of dissent was unprecedented in Soviet history. Worsening economic conditions led to official calls for fundamental reform in the mid-1980s.

The revamping of the economic system became an urgent priority during the regime of Mikhail Gorbachev that began in 1985. Gorbachev initiated new policies of *glasnost* ("openness") and *perestroika* ("restructuring") to facilitate a more democratic political system, more freedom of expression, and a more productive economy with a market orientation (Fig. 6.15). At the same time, there were increasing problems of disunity as the different republics and the ethnic groups organized by the Soviet system into nominally "autonomous" units took advantage of new freedoms to resurrect old quarrels and demand that their units be given more autonomy. Fighting among ethnic groups erupted in several republics. Meanwhile, in all of the republics, declarations of sovereignty and in some cases outright independence challenged the authority of the central government. In 1991, the USSR and most other nations recognized the independence of the three Baltic republics of Estonia, Latvia, and Lithuania. The process of the empire's disintegration had begun.

At the center of the crumbling empire, having failed to reverse the downward slide of the economy, Gorbachev faced growing sentiment to scrap the command system and move as rapidly as possible to a market-oriented economy. The reformers' cause was championed by Boris Yeltsin, a leading advocate of rapid movement toward a market-oriented economy and greater control by individual republics over their own resources, taxation, and affairs. Yeltsin's political stature grew when, in June 1991, the citizens of Russia chose him president of the giant Russian Federation in the Soviet Union's first open democratic election. Gorbachev, himself elected president of the USSR (but by a Congress of People's Deputies rather than by the people as a whole), opposed Yeltsin's proposals for radical reform. He advocated more gradual movement toward free-market orientation within the state-controlled planned economy and insisted on the need for unity and continued political centralization within the Soviet state.

Matters came to a head in August 1991, when an attempted coup by hardliners failed. Yeltsin gained enhanced stature from his defiant opposition to the coup. Gorbachev resigned his position as head of the Communist party and began to work with Yeltsin to reconstruct the political and economic order. But his efforts to preserve the Soviet Union and a modified form of the Communist economic system could not stand against the growing tide of change. During the autumn of 1991, the Communist party was disbanded and the individual republics seized its property. On December 25, 1991, Gorbachev resigned the presidency, and the national parliament formally voted the Soviet Union out of existence on the following day. A powerful empire had quickly and quietly faded away, to be replaced by 15 independent countries.

Russia, the Misdeveloped Country

The unprecedented experiment to transform the colossal Communist state into a bastion of free-market democracy has produced strange and often unfortunate results, leading some observers to classify Russia not as a "less developed" or "more developed" country but as a "misdeveloped" country. The

Figure 6.15 Freedom of expression erupted all over the USSR during the Gorbachev era. This was an artist's kiosk in Leningrad (now St. Petersburg) in 1991.

disturbing demographic trend of a plummeting population — the outcome of the "Russian cross" — is perhaps the best indicator of Russia's backward momentum, and the roots of this trend are mainly economic.

Despite heightened public expectations for a better life in Russia, criticism of the triumphant President Yeltsin and his policy of economic reform toward freer markets increased throughout the course of his two terms in office (1991–1999). Early in his first term, Yeltsin introduced a program of rapid economic reform, known as **economic shock therapy,** designed to replace the Communist system with a free-market economy. It removed price controls and encouraged privatization of businesses. Products and services from the private sector had already been indispensable during the Communist era, but now they were to be the centerpiece of the economy (Fig. 6.16).

The results of shock therapy were generally poor. Former employees of the Communist party resented the loss of their jobs and privileges. Printing of money to meet state obligations caused inflation. Poor people living on fixed incomes struggled to survive high prices for basic necessities. A new consumer-oriented society developed along class lines. There was growing unemployment and homelessness and a widening gap between rich and poor (Fig. 6.17). Access to choice goods and services was no longer forbidden to ordinary citizens as it was during the Communist era, but prices were so high that most people could not afford them. Most Russians grew economically worse off; an estimated 70% lived below the poverty line by 2000. After a 7-year absence, American journalist Michael Dobbs returned to Russia in 1999 and found a country transformed. "I am amazed by the gleaming new face of Moscow," he wrote: "the American-style supermarkets and shopping malls, the repainted facades, the gentrification of crumbling tenement buildings, the bustling highways and beltways that are the pride and joy of Mayor Yuri Luzhkov. At the same time, I am saddened by the vulgarity of the new rich, the sight of beggars being splattered with

Figure 6.17   Having fallen through the cracks in Russia's transition from a command economy to free enterprise, these women are begging for help on a St. Petersburg sidewalk. The message of their placards may be summarized as: "Help us dear brothers and sisters. Please give us money for food, for God's sake. We are invalids with cancer. We have given of ourselves but are now forgotten by the state. We are poor and hungry, with no protection, just left to rot."

mud by Toyota Land Cruisers, the seemingly all-pervasive corruption, the stunning gap between the haves and have-nots."[1]

In its persistent conflict between capitalist reformers and old-school Communists, Russia's political leadership has been powerless to prevent the deterioration of the country's economy. Suffering under an enormous debt burden, the Russian government is trying to keep spending down in part by not paying salaries to federal employees. Many privately owned industries are too unproductive to pay their employees. Unpaid, workers cannot pay taxes to the government. The government has therefore turned to other sources for the tax revenues it desperately needs — for example, by cracking down on the street vendors who have been among the vanguards of free-market enterprise in Russia. The widespread failure of economic reforms in Russia, in both the agricultural and industrial sectors, points toward a possible reversal of the reform process toward more state control of the economy.

One of the major components that has emerged from Russia's new economic geography is the underground economy, also known as the "countereconomy" or "second economy" (or economy *na levo,* meaning "on the left"). Widespread barter has resulted from the declining value of the ruble (see Definitions & Insights, p. 185). Many people resort to selling personal possessions to buy high-priced food and other necessities. Rampant black-marketeering has developed since prices were decontrolled. Much of this traffic is still officially illegal, but there is a general tendency to overlook such transactions because they are so essential to the economy.

Russia's new private entrepreneurs include a large criminal "mafia" that preys on government, business, and individuals.

Figure 6.16   Free enterprise thrives throughout what was once the Soviet Union. Customers are lined up to buy vegetables from a street vendor in Vyborg, Russia. Note the seller's abacus "calculator."

[1] Michael Dobbs, "Were the Bad Old Days Better?" *Washington Post National Weekly Edition,* August 21, 2000, 21.

DEFINITIONS + INSIGHTS

Barter

Barter is the exchange of goods or services in the absence of cash. Bartering was a common practice in the USSR and has increased in importance since Russia became independent; it accounts for about half the transactions in Russian industries and 40% of national tax payments. Barter is even a medium of foreign exchange. When companies do not have cash to buy equipment, pay utility bills, pay workers, or pay taxes, they resort to barter. For example, a Novgorod-based company that mainly manufactures valves also produces railroad-car brake pads, which it uses to pay rail cargo

fees, which have been kept artificially high as a means of subsidizing passenger travel.

Companies that owe unpaid taxes also like to use barter as a means of avoiding state seizure of their cash assets. Although critics argue that barter corrupts free enterprise by distorting prices and making a mockery of tax collection, others emphasize that it allows individuals and businesses some insulation against severe economic turmoil.

192

Organized crime and corruption are all pervasive. If you give money to a beggar, you are paying the mafia; that beggar must give some income to the local mafia to keep his or her place on the street corner. If you want to build a factory, you pay off officials at every stage on constructions; without your payouts, they would terminate your project. Russia has become a "kleptocracy," an economic and political system based on crime. The biggest beneficiaries are the so-called **oligarchs** (from the Greek *oligarchy*, meaning "rule of the few"), Russia's leading businessmen who in most cases are essentially crime bosses. These tycoons wield enormous political power and helped usher Vladimir Putin, a former KGB (state security) officer of the Soviet Union into Russia's highest office, the presidency, in 2000. Once in office, Putin vowed to crack down on the oligarchs.

Russia's criminal way of life has raised public fears and prompted some people to call for a return to the relative stability and security of the Soviet state. The journalist Dobbs was astounded by many Russians' longings for what they had despised for decades: the grim way of life under communism. "Everywhere I went, from the Caucasus to Central Asia, I encountered a deep-seated nostalgia for the old Soviet Union. The nostalgia assumes different forms in different people, but the common thread is a yearning for a way of life that may have been drab and regimented, but was nevertheless secure and predictable."[2] Many have emigrated, including 380,000 from the former Soviet Union to the United States in the decade 1990–2000. But most have stayed, struggling in creative and often brilliant ways with limited resources to make ends meet and, ideally, help build the middle class and the normal way of life that communism's collapse was supposed to bring.

6.5 Geopolitical Issues

With the collapse of the Soviet Union, the Russian Federation took over the property of the former government within the Federation's borders, including Moscow's Kremlin (see Fig. 6.18). Russia also took custody of the international functions of the USSR, including its seat at the United Nations. In early

[2]Ibid.

December 1991, the republics of Russia, Ukraine, and Belarus (formerly Belorussia) formed a loose political and economic organization called the Commonwealth of Independent States (CIS) headquartered in Minsk, Belarus. Except for Estonia, Latvia, and Lithuania, the other republics eventually joined the CIS, in which Russia took a strong leadership role. Disputes erupted, notably between Russia and Ukraine, over such questions as ownership of property inherited from the defunct USSR, particularly the Black Sea naval fleet based in Ukraine's Crimea region, and disposition and control of nuclear weapons. These were eventually resolved. To date, the Commonwealth of Independent States has been a mostly moribund organization. However, Russia's President Putin has proposed that its members unite in security matters (e.g., by forming an antiterrorism alliance).

Since the Soviet Union dissolved in 1991, many important links between Russia and the other 14 successor states have persisted. Economic, ethnic, and strategic ties are especially crucial. For example, one major economic link is the vital flow of Russian oil and gas to other states such as Ukraine,

Figure 6.18 Moscow's Red Square, seat of power for the USSR and now the Russian Federation. The Lenin Mausoleum, the red building below the spire with the star, backs up to the Kremlin wall at the extreme left of the photo.

Belarus, and the Baltic states, which are highly dependent on this supply of energy. About 25 million ethnic Russians lived in the 14 smaller states at the time of independence. Although a considerable number have migrated to Russia since 1991, many have had difficulties finding housing and employment, and the great majority still live in the other countries. In certain areas, notably eastern Ukraine and Crimea, many ethnic Russians are causing political instability because they want to secede and form their own state or join Russia. Russians in the 14 non-Russian nations complain that governments and peoples discriminate against them.

In response, the Russian government has said that it has a right and a duty to protect Russian minorities in the other countries. Russia's perspective on the other 14 former members of the USSR — the Near Abroad — is that they constitute a special foreign policy region. Some observers contend that Russia's diplomatic, economic, and limited military involvement in these countries is an effort to establish a buffer zone between Russia and the "far abroad."

Russia sometimes forcefully asserts claims to a special sphere of influence in the Near Abroad. Along the outer frontiers of the cordon of successor states, Russia maintains a chain of military bases where, in 2002, an estimated 100,000 Russian soldiers were stationed. Within the Near Abroad, notably in Moldova, Tajikistan, Georgia, Armenia, and Azerbaijan, Russian troops have become engaged frequently as "peacekeepers" in local ethnic conflicts. In Armenia and Tajikistan, at least, the governments fear what would happen without Russian military support. There are some in the West and among Russia's neighbors, however, who fear that "peacekeeping" is merely a euphemism for a plan to restore Russian imperial rule. Some observers contend that Russia wants access to important resources and economic assets in its former republics, such as uranium in Tajikistan, aviation plants in Georgia, military plants in Moldova, and the Black Sea coast and naval fleet in Ukraine's Crimea. Noting Moscow's reluctance to recall Russian troops from the Near Abroad, some analysts fear that Russia's reasoning may be that Moscow's empire will again be expanded, and it is not worth withdrawing troops and dismantling bases only to have to redeploy and rebuild them later.

Within Russia, there is in fact a body of public opinion favoring reassertion of Russian control over its former empire, but the strength of this feeling is unknown. President Putin has spoken of Russia's urgent priority to draw the countries that were republics of the former Soviet Union closer to Russia. Such reintegration could, of course, take place through peaceful political and economic means. In one case it has; in 2000, Russia and Belarus officially established a political union and elected a joint parliament. It is uncertain what this union will mean. Some observers believe it represents the beginning of a push of Russian influence even farther to the west. What is certain is that several of the former republics continue to be economically intertwined with Russia. Turkmenistan will be exporting natural gas to Russia, for example, and Kyrgyzstan maintains a huge energy debt to Russia.

Russia and the "lesser" members of the former Soviet Union have generally enjoyed improving relations with the West since the USSR dissolved. The Warsaw Pact, a military alliance formed around the Soviet Union to take on the West's North Atlantic Treaty Organization, has dissolved. Both sides have reduced their nuclear arsenals, with Russia taking the lead in calling for even deeper cuts. Russia's economic problems have led to a dramatic decline in the size and quality of its armed forces and arsenal, tragically symbolized by the explosion and subsequent loss of all hands aboard the nuclear submarine *Kursk* in the Barents Sea in 2000.

Western governments and their financial institutions have extended large amounts of economic aid to many of these countries, particularly to Russia during a severe financial crisis in 1998. It has been difficult to gauge how useful this assistance has been, and there is resentment on both sides about it. Many critics complain that such aid has failed to reach its intended targets and instead lined the pockets of the oligarchs. One commented that while under communism the arrangement between the worker and the state was "they pretend to pay us, we pretend to work," the new arrangement between the West and Russia is "You Russians pretend to be creating a law-based, market-friendly liberal democracy. We westerners pretend to believe you — and what's more, we pay you for it."[3] For their part, many Russians perceive their leaders as having "sold out" to the West and resent their growing dependence on its institutions, exports, and media. The Russian poet Yevgeny Yevtushenko lamented what he called "the McDonaldization of Russia."

CHAPTER SUMMARY

- It is still useful to place the 15 former republics of the Soviet Union into a single region. These 15 countries will be attempting to sever or reestablish ties to Moscow for many years to come. Some analysts believe that Russia may even attempt to reexert control over some of the recently independent nations.
- The area west of the Ural Mountains and north of the Caucasus Mountains has been known as European Russia. The Caucasus and the area east of the Urals have been called Asiatic Russia.

Siberia is the general name for the area between the Urals and the Pacific. Central Asia is the general name for the arid area occupied by five states immediately east and north of the Caspian Sea.
- Russia and the Near Abroad span 11 time zones. Stretching nearly halfway around the globe, the region has formidable problems associated with climate, terrain, and distance.
- The most productive human activities take place in the subarctic, humid continental, and steppe climatic zones.

[3] "Fuelling Russia's Economy." *Economist*, August 28, 1999, 13.

- The most productive agricultural area is located in the steppe region characterized by the grassy plains south of the forest in Russia and in Ukraine, Moldova, and Kazakstan. Here the low and variable rainfall is partially offset by the fertility of the mollisols, which are among the best soils to be found anywhere.
- In early Russia, rivers formed natural passageways for trade, conquest, and colonization. In Siberia, the Russians followed tributaries of some of Earth's greatest rivers to advance from the Urals to the Pacific in less than a century. More recently, alteration of rivers for power, navigation, and irrigation became an important aspect of economic development under the Soviet regime.
- Large plains dominate the terrain from the Yenisey River to the western border of the country. The area between the Yenisey and Lena Rivers is occupied by the hilly Central Siberian Uplands. In the southern and eastern parts of the region, mountains, including the Caucasus, Pamir, Tien Shan, and Altai ranges, dominate the landscape.
- Slavic and Scandinavian peoples were greatly influential in early cultural development, which was the foundation for the Russian, Ukrainian, and Belarusian cultures.
- Early contacts with the Eastern Roman or Byzantine Empire led to the establishment of Orthodox Christianity as a feature of Russian life and culture. Although it was repressed and neglected after the Bolshevik Revolution, this and the region's other religions have been experiencing a rebirth since the breakup of the Soviet Union.
- From the 15th century until the 20th, the tsars built an immense Russian empire around the small nuclear core of Muscovy. This imperialism brought great areas of land and hosts of alien peoples under tsarist control. Cossack expeditions extended to the Pacific, across the Bering Strait, and down the west coast of North America as far as northern California.
- Russia has often triumphed over powerful foreign invaders. These achievements have been due in part to the environmental obstacles the invaders faced, the overwhelming distances involved, and the defenders' willingness to accept large numbers of casualties and implement a scorched earth strategy to protect the motherland.

- Policies of the Soviet Union permitted the various nationalities to retain their own languages and other elements of traditional cultures. However, the Soviet regime implemented a deliberate policy of Russification in an effort to implant Russian culture in non-Russian regions. Millions of Russians settled in non-Russian areas, but local cultures were not converted to Russian ways.
- To meet the needs of its people, the USSR became a collective welfare state, with the government providing guaranteed employment, low-cost housing, free education and medical care, and old-age pensions. However, military-industrial superpower status was generally achieved at the expense of the ordinary consumer.
- Current efforts to privatize industry and agriculture are slowed by the difficulties inherent in a complete overhaul of the old inefficient state-run systems such as the collectivized agriculture of the Communists.
- The sudden transition from a command economy to capitalism in Russia and other countries in the region has led to a widening gap between rich and poor people. Birth rates have fallen and death rates have risen dramatically, leading to steep declines in population.
- Many important links between Russia and the other 14 successor states — the Near Abroad — have persisted. Economic, ethnic, and strategic links are especially crucial, and the Russian government has said that it has the right and duty to protect Russian minorities in the other countries. Some observers contend that Russia is attempting to reestablish the "near abroad" as buffers between Russia and the "far abroad."
- Because of nuclear weapons and other critical issues, relations between Western countries and the countries of the former Soviet Union, particularly Russia, are vital to global security. The West gives much economic aid to Russia and some of the other countries but worries that it is misspent. For their part, some of the countries of Russia and the Near Abroad resent growing dependence on the West.

REVIEW QUESTIONS

1. List the five climatic belts of Russia and the Near Abroad and briefly describe their attributes.
2. Using maps and the text, locate and list the major river systems.
3. Why is Russia's population declining?
4. What significant conflicts have occurred in this region? Why has Russia so often won them?
5. What were the perceived advantages of "collectivized agriculture"?
6. Name and describe some Soviet Communist projects that changed the landscape of this region.

7. List some of the successful elements of the drive to industrialize the Soviet Union.
8. What were some of the results of Russia's "shock therapy" economic reform and the subsequent conditions of the country and its people?
9. What are the areal names that have been used for Russia and the Near Abroad?
10. What are the major ethnic, cultural, and political external forces that have helped shape Russia and the Near Abroad?

DISCUSSION QUESTIONS

1. Examine and discuss the significance of the area and population of Russia and the Near Abroad.
2. Use the location 67°N, 135°E to find the Siberian settlement of Verkhoyansk and discuss why it has recorded the lowest official temperature known in the Northern Hemisphere.
3. Locate, examine, and discuss the important physiographic features of Russia and the Near Abroad.
4. What roles have the natural environment played in this region's history and development?
5. Is tourism of the kind described in the Russian Arctic good or bad for local environments and people?

6. What cultures were important in Russia's early years and what were the major contributions of these cultures?
7. Examine and discuss how warfare has been such a major influence on this region.
8. What were the foundations for the Communist economic system and what is this system's legacy for today's economy?
9. Why did the "Second Russian Revolution" occur? What are some of the difficulties that the present government is facing?
10. Are Russia and some of the other countries in this region important to you, your community, or your country? Discuss the reasons why or why not.

CHAPTER 7

Fragmentation and Redevelopment in Russia and the Near Abroad

Boys of Kazan on the upper Volga. They are growing up in the new and uncertain world that follows decades of Soviet rule.

JOE HOBBS

Few geographers foresaw the collapse of the Soviet Union. Like most other observers, we assumed that, when threatened with disintegration, the core in Moscow would assert its military strength to retain the union at any cost. From the geographer's perspective, the survival of the Soviet Union and its status as a global superpower always depended on Russia's ability to control non-Russian resources such as the oil of the Caucasus, the soil of Ukraine, and the cotton of Central Asia. Asked prior to 1991 whether Russia would let go of these, most of us answered "no," not without a violent struggle.

But the union of so many disparate peoples was so unlikely that, in retrospect, it seems absurd. National unity was, in hindsight, an oxymoron in the Soviet Union. It was not supposed to be; Marxism-Leninism had assumed that the cultural and linguistic differences among more than 200 separate groups would disappear as the ideal, universal proletarian populace emerged. In effect, Moscow's rulers hoped, they would all become Russian. Russification was supposed to level out the differences, but it could not and did not.

Now Russia has lost all but the core: Russia herself and her somewhat endangered Federation, itself an amalgam of many non-Russian entities cemented artificially to the core. The other former republics have gone their own ways as independent countries. How are all of these entities faring today and what is their future? What has happened to the Russian and non-Russian resources, agricultures, and industries that once supported a global power? This chapter will answer these questions and more.

The most outstanding feature of the economic geography of this huge region is that resources are distributed unevenly, and they favor certain subregions and countries. Agricultural and industrial resources are in fact clustered in a core region encompassing western Russia, northern Kazakhstan, Ukraine, Belarus, Moldova, and the Baltic countries of Estonia, Latvia, and Lithuania. Nearly three-fourths of the region's people — the great majority of them Slavic — and an even larger share of the cities, industries, and cultivated land of this immense region are packed into this roughly triangular Fertile Triangle composing about one-fifth of the region's total area (Fig. 7.1). Also known as the Agricultural Triangle and the Slavic Coreland, this is the distinctive functional core of the region.

That there is a core also suggests there is a periphery, and there is — the world's largest. The remaining four-fifths of the region of Russia and the Near Abroad, lying mostly in Asia, consists of land supporting only spotty human settlements in conditions of sometimes huge environmental diversity and adversity. Most of the region's non-Slavic peoples live in these lands, but many immigrant Slavs also reside in them, generally in cities. Although these areas are marginal relative to the well-endowed Fertile Triangle, they do include particular resource and production centers that represent real or potential wealth. Kazakstan and several countries in the Caspian Sea region are rich in oil and other minerals. There are pockets of diverse and productive agriculture in the Caucasus. Russia's Siberia is a bleak storehouse of minerals, timber, and waterpower. The geographic pattern that has emerged is a series of discrete production nodes widely separated by taiga, tundra, mountains, deserts, wetlands, and frozen seas.

The second prominent feature of the region's economic geography is that, with few exceptions, the countries that succeeded the Soviet Union have been moving backward. Here, ironically, Russia has continued its traditional role as pacesetter, seemingly inspiring its neighbors to follow examples of rampant corruption, growing poverty, and sluggish agricultural and industrial activity. The common challenge facing the countries of Russia and the Near Abroad is to reverse these dismal trends.

7.1 Peoples and Nations of the Fertile Triangle

Slavic peoples are the dominant ethnic groups in most of the region of Russia and the Near Abroad in both numbers and political and economic power. The major groups are the Russians (whose cultural geography is described in Chapter 6), Ukrainians, and Belarusians (Byelorussians). Most of these Slavs live in the coreland extending from the Black and Baltic Seas to the neighborhood of Novosibirsk in Siberia. This area contains about one-half the area and more than four-fifths the population of the United States. Moscow (population: metropolitan area, 12,230,000; city proper, 8,383,000) and St. Petersburg (formerly Leningrad; population: metropolitan area, 4,948,000) are the largest cities, followed by the Ukrainian capital Kiev (population: metropolitan area, 3,241,200). Lowlands predominate — the only mountains are the Urals, a small segment of the Carpathians, and a minor range in the Crimean Peninsula. The original vegetation was mixed coniferous and deciduous forest in the northern part and steppe in the south.

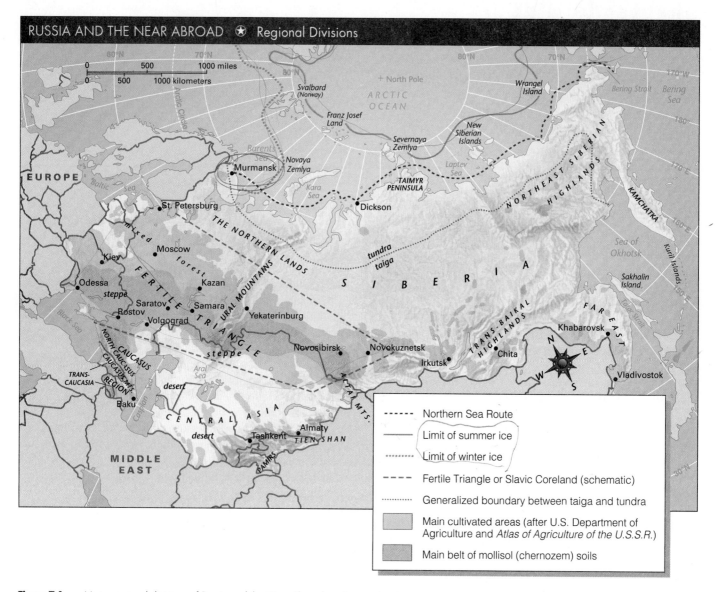

RUSSIA AND THE NEAR ABROAD ✪ Regional Divisions

Legend:
- – – – – Northern Sea Route
- ——— Limit of summer ice
- ·········· Limit of winter ice
- – – – Fertile Triangle or Slavic Coreland (schematic)
- ············ Generalized boundary between taiga and tundra
- Main cultivated areas (after U.S. Department of Agriculture and *Atlas of Agriculture of the U.S.S.R.*)
- Main belt of mollisol (chernozem) soils

Figure 7.1   Major regional divisions of Russia and the Near Abroad as discussed in this text.

Ukraine

Although closely related to the Russians in language and culture, Ukrainians — the second largest ethnic group in the coreland — are a distinct national group. The name Ukraine translates as "at the border" or "borderland," and this region has long served as a buffer between Russia and neighboring lands. Here, armies of the Russian tsars fought for centuries against nomadic steppe peoples, Poles, Lithuanians, and Turks, until the 18th century when the Russian empire finally absorbed Ukraine. For three centuries, Ukrainians were subordinate to Moscow, first as part of the Russian empire and then as a Soviet republic. Ukraine's industrial and agricultural assets were always vital to the Soviet Union; the Bolshevik leader Lenin once declared, "If we lose the Ukraine, we lose our head."

Ukraine (population: 49.1 million) lies partly in the forest zone and partly in the steppe. On the border between these biomes, and on the banks of the Dnieper River, is the historic city of Kiev, Ukraine's political capital and a major industrial and transportation center (Fig. 7.2). Although its agricultural output has plummeted since the USSR's demise, Ukraine was traditionally a great "breadbasket" of wheat, barley, livestock, sunflower oil, beet sugar, and many other products, grown mainly in the black-earth soils of the steppe region south of Kiev. North of Kiev is Chernobyl, where an explosion at a nuclear power station in 1986 rendered parts of northern Ukraine and adjoining Belarus incapable of safe agricultural production for years to come. Neurological, endrocrinological, gastrointestinal, and urinary illnesses are 15 times greater in the parts of Ukraine contaminated by this accident than in other regions of the country. Having shut down the Chernobyl plant in 2000, Ukraine has promised to close all of its Chernobyl-type nuclear plants within a few years. This will be

Figure 7.2   Kiev, the historic capital of Ukraine.

a costly proposition, considering the country's shortage of affordable alternative fuels.

Ukraine has paid dearly for independence. Previously, the Soviet economy subsidized the provision of Russian oil, gasoline, natural gas, and uranium to Ukraine. But with independence, Ukraine was compelled to pay much higher world market prices for these commodities. As of 2002, Ukraine held a $1.5 billion gas debt to Russia. Ukraine is, however, well-endowed with mineral resources, including large deposits of coal, iron, manganese, salt, and even its own natural gas (see Fig. 7.4). These were vital in the Soviet Union's heavy industries but have lain underutilized during the economic slowdown following independence. Coal is concentrated in the Donets Basin (Donbas) coal field near the industrial center of Donetsk, where most of Ukraine's iron and steel plants are also located. Iron ore comes by rail from central Ukraine in the vicinity of Krivoy Rog, the most important iron-mining district in the region of Russia and the Near Abroad. Most of the ore is used in Ukrainian iron and steel plants. This area is also a center of chemical manufacturing. Surrounding the inner core of mining and heavy-metallurgical districts in Ukraine is an outer ring of large industrial cities, including Kiev, the machine-building and rail center of Kharkov (east of Kiev), and the seaport and diversified industrial center of Odessa on the Black Sea.

Like most Russians, most Ukrainians are poorer today than they were under communism. There are problems of corruption, and as in Russia, much business and accompanying wealth are in the hands of oligarchs and organized crime. As in Russia, workers frequently go for months without wages. Power outages are common, and consumers stockpile as many foodstuffs and dry goods as possible.

Despite hard times, Ukraine's demographic, military, and potential economic strength demands attention on the world stage. Since independence, Ukraine and the West have courted carefully. Ukraine became a nuclear weapon-free nation by disposing of the 1,800 nuclear warheads that had briefly made it the world's third largest nuclear military power. The United States rewarded this move with a much-increased foreign aid package to Ukraine, which in 2001 was the fourth largest recipient of U.S. economic assistance after Israel, Egypt, and Russia. Ukraine's growing orientation to the West may also be seen in its membership in the North Atlantic Treaty Organization's Partnerships for Peace program and in its application to become a member of the European Union. Many European countries are dependent on natural gas shipped by pipeline across Ukrainian land from Russia.

Ukraine has improved relations with Russia that had soured after the collapse of the USSR. President Leonid Kuchma, who favored economic union and closer political ties with Russia, resolved some of the problems that followed Ukraine's efforts to distance itself from post-Soviet Russia. One issue was the fate of Russia's Black Sea fleet, which traditionally operated from the Ukrainian port of Sevastopol. Kuchma and Russia's then-President Yeltsin signed a treaty allowing Russian ships to continue using the port.

Another problem was the Crimean Peninsula, a picturesque and verdant resort region. Russia's Catherine the Great annexed the Crimea in 1783, but Soviet leader Nikita Khrushchev returned it to Ukraine in 1954. After the Soviet Union collapsed, Russia agitated to get back the Crimea, with its 70% Russian ethnic population. Ukraine yielded a bit by granting the Crimea special status as an autonomous republic. Crimeans then elected their own president, a pro-Russian politician who advocated the Crimea's reunification with Russia. But for now, Russia seems resigned to Crimea's belonging to Ukraine.

Many Ukrainians would like to cultivate their Slavic distinctiveness and pursue closer ties with their Slavic Russian kin rather than seek a closer orbit around the West. Along with Belarus, Ukraine is the former Soviet republic most like Russia — "Russia writ small," as one author described it.[1] Since achieving independence from Moscow in 1991, Ukraine has restored the Ukrainian language to its educational system. Still, a majority of Ukrainians wants Russian to be a second official language.

Belarus

The much smaller country of Belarus (formerly called Belorussia and "White Russia;" population: 10 million) adjoins Ukraine on the north. Minsk (population: city proper, 1,751,300) is the political capital and main industrial city. Belarus developed slowly in the past, mainly because of a lack of fuels other than peat. Recently, however, oil and gas pipelines from Siberia and other areas have provided raw material for oil-refining and petrochemical industries. Some oil production has begun in Belarus itself. Under Soviet rule, Belarus became an industrial powerhouse, shipping televisions, farming equipment, automobiles, textiles, and machinery to both Moscow and its Eastern European satellites. Belarus' foreign

[1] "The Ukrainian Question." *Economist*, November 29, 1999, 20.

trade is now almost exclusively with Russia, much of it in the form of barter. In the late 1990s, for example, Belarus paid off a debt to Russia's natural gas monopoly (Gazprom) by supplying it with $300 million worth of refrigerators, tractors, and other hard goods. Unlike Russian industries, most manufacturing enterprises in Belarus are still state-run. Industrial output is low, however, and Belarus has become increasingly impoverished since its independence.

Belarus is too cool, damp, and infertile to be prime agricultural country, but traditionally, it had a substantial output of potatoes (particularly for the production of vodka), small grains, hay, flax, and livestock products. Southern Belarus contains the greater part of the Pripyat (Pripet) Marshes, large sections of which were drained for agriculture. Agricultural productivity declined after 1986, when much ground was contaminated by the nuclear accident in nearby Chernobyl. To compensate for these losses, Belarus stepped up production of agricultural machinery for export.

Belarus' strategic geographic location between the Russian coreland and Western Europe brought it enormous suffering in World War II. En route to Moscow and Leningrad, the Nazi army laid waste to Belarus. Three-quarters of its urban centers were utterly destroyed. One and a half million Belarusians fled eastward to escape the fighting. The Jews of Belarus were decimated by Hitler's "final solution."

Of all the former Soviet republics, the Slavic nation of Belarus has the closest ties with Russia. The two countries unified their monetary systems in the mid-1990s. In this agreement, Belarus gave up sovereignty over its currency and banking system in exchange for preferential access to Russian energy and other resources. Soon after this, an overwhelming majority of Belarusians voted for even closer economic integration with Russia and for the adoption of Russian as their official language. This led to the official designation of Belarus and Russia as a "union state" with plans to integrate their economies, political systems, and cultures. During its subsequent financial crisis, Russia evaluated the costs of absorbing some of the financial liabilities of Belarus, and based on the evaluation, it decided not to implement. Militarily, however, the two nations are already intertwined; Belarus has reintegrated its air force, ground forces, intelligence, and arms production with those of Russia. And apparently to prepare for absorption into Russia, the government has removed Belarusian history and language from school curricula.

Moldova

Tiny Moldova (known formerly as Moldavia and, in the 19th and early 20th centuries, as Bessarabia; population: 4.3 million), made up largely of territory that the Soviet Union took from Romania in 1940, adjoins Ukraine at the southwest. The indigenous Moldovan majority (about two-thirds of the country's population) is ethnically and linguistically Romanian, not Slavic. There are problems of ethnic antagonism and political separatism within Moldova's borders. In the eastern part of the country, in a sliver of territory along the Dniestr River known as Transdniestria, ethnic Ukrainians and Russians have been agitating, sometimes violently, to form a separate state. Russia maintains troops there, ostensibly to protect the ethnic Russians, but has promised to withdraw them in 2003. Elsewhere in the country, irredentist Moldovans, who claim that Moldovans are in fact Romanians, want Moldova to join Romania.

Industrial production in Moldova jumped after World War II, especially around the capital of Chisinau (also known as Kishinev; population: metropolitan area, 723,000). Mainly a fertile, black-earth steppe upland, however, Moldova is primarily agricultural. The Moldovan specialty is cultivation of vegetables and fruits, especially grapes for winemaking.

The Baltic States and Kaliningrad

Estonia, Latvia, and Lithuania, known as the Baltic States or the Baltics, lie between Belarus and the Baltic Sea. These countries share many environmental, economic, and historical traits with the other nations of the Fertile Triangle. However, the indigenous inhabitants of the Baltics are not Slavs, and they have long struggled to safeguard their national identities against Russian encroachment. They were part of the Russian empire before World War I, but became independent following the Russian Revolution. The Soviet Union reabsorbed them in 1940.

Today, they are again independent states. They are striving to become more European, to be recognized as part of the up and coming new Central Europe, and to join European economic and military organizations. But their geographic past has been hard to shake, and they are still linked closely to the remainder of the former Soviet Union in various ways. Russia uses Baltic ports to transport goods abroad and exports fossil fuels and a variety of other goods and services to the Baltic nations. But Estonia, Latvia, and Lithuania want to minimize their ties with their giant neighbor to the east. They have refused to become members of the Commonwealth of Independent States (CIS). They would like to become members of the European Union (EU), but only Estonia, the most prosperous of the three, has been accepted for "fast track" EU incorporation. The Baltics would also like to become members of the North Atlantic Treaty Organization (NATO), but Russia considers that option a serious threat to its security, and resists it. So, although the Baltic governments prefer to portray themselves as European, as many outside observers do (the French Geological Institute recently calculated that a spot in eastern Lithuania is geographically the center of Europe), vital and difficult links with Russia suggest that these nations, particularly Latvia, will in many ways belong in a sphere of "former Soviet" countries for some years to come.

Dairy farms alternate with forests in the hilly Baltic landscape. Mineral resources are scarce; the most notable are deposits of oil shale in Estonia, mined for use as fuel in electric power plants. Like Belarus, the Baltics benefit from natural gas and petroleum brought from Russian fields located hundreds and even thousands of miles away. Expansion of industry after World War II helped make the area one of the more prosperous and technically advanced parts of the former USSR. The largest city and manufacturing center is the port of Riga (population: metropolitan area, 893,700), the capital of Latvia (population: 2.4 million).

About 85% of Latvia's non-Latvian ethnic minorities are Russian-speaking; they are mostly retired Soviet military officers and their offspring. While independent Lithuania and Estonia established procedures early on to grant citizenship to ethnic Russians, Latvia delayed taking such measures until the mid-1990s and gave in only under pressure from Russia and the West. The Latvian government feared that enfranchised Russians would destabilize the country in future elections or even try to reannex Latvia to Russia. But Moscow's economic leverage prevailed; it threatened to cut off the shipment of its oil through Latvia (the transit fees generate important revenue for Latvia) and to stop buying Latvian agricultural exports. Latvia relented, in part by agreeing to allow all children born on Latvian soil since 1991 to become Latvian citizens and by allowing more non-Latvians to apply for citizenship.

The presence of Russian troops in the Baltic countries was a serious foreign relations problem after the countries gained independence. With independence, only Lithuania (population: 3.7 million; capital, Vilnius, population: city proper, 577,800) successfully negotiated the immediate withdrawal of Russian troops. A large number of these Russian forces resettled in neighboring Kaliningrad, an anomalous Russian enclave two countries away from Russia. As the northern half of the former German East Prussia, the territory — then known as Konigsberg — was transferred to the USSR at the end of World War II to become the Kaliningrad Oblast. Soviet authorities expelled nearly all of Kaliningrad's Germans to Germany, and Russians now make up more than three-fourths of the population of 424,800. Kaliningrad remains a massive military and naval establishment, with defense workers and their dependents and retired military families comprising most of the population. Moscow would like Kaliningrad to continue to be principally a military state, but many in Kaliningrad have other ideas about how to capitalize on their strategic geographic location. Seeing themselves mainly as "Euro-Russians," they want Kaliningrad to become a more vibrant Central European economy. Some dream of making Kaliningrad one of the world's great free-trade zones, the "Hong Kong on the Baltic." Kaliningrad does have some geographic and other factors in its favor: Its port seldom freezes, labor is inexpensive, and it is a logical way station for east–west trade. Already much legitimate trade and smuggling are carried on through Kaliningrad, helping to raise standards of living for its people. The fear, however, is that when neighboring Lithuania and Poland join the European Union, Kaliningrad will be left out to drift in a state of underdevelopment.

Figure 7.3   Always defiant of Soviet rule, the Baltic states set the pace for the dissolution of the USSR and continue to seek distance from Russia. Here, the Estonian flag flies defiantly over Tallinn in July 1991, almost 6 months before the Soviet Union collapsed.

Since independence, Estonia (population: 1.4 million; capital, Tallinn, population: city proper, 401,100; Fig. 7.3) has developed close economic ties with Finland, which quickly replaced Russia as the country's dominant trading partner. It should be remembered that although Estonia is situated geographically in a largely Slavic region, its indigenous people are closely related to the ethnic Finns across the Gulf of Finland. It is only a 2.5-hour ferry ride between Tallinn and Helsinki, and this is a busy route.

7.2 Agriculture in the Russian Coreland

Throughout the second half of the 20th century, the Fertile Triangle was a global-scale producer of farm commodities such as wheat, barley, oats, rye, potatoes, sugar beets, flax, sunflower seeds, cotton, milk, butter, and mutton. The overall high output, however, did not reflect high agricultural productivity per unit of land and labor. There were many inefficiencies in the state-operated and collective farms that dominated the agricultural sector. On the state-operated farms, workers received cash wages in the same manner as industrial workers. Bonuses for extra performance were paid. Workers on collective farms received shares of the income after obligations of the collective had been met. As on the state farms, there were bonuses for superior output.

These methods, however, failed to provide enough incentives for highly productive agriculture. Farm machinery stood idle because of improper maintenance; since no individual owned the machine, the incentive to repair it was diminished. There were also shortages of spare parts. Poor storage, transportation, and distribution facilities, plus wholesale pilfering, caused alarming losses after the harvest. Younger people were deserting the farms for urban work, often living with the family in the countryside but commuting to their city jobs. With the shortage of younger male workers, women, children, and the elderly did a large share of the farm work.

In an effort to reverse such disturbing trends, all of the countries of the Fertile Triangle are promoting land reform by privatizing collective and state farms and developing more independent or "peasant" farming. Privatization of the collective and state farms occurred rapidly in the Baltic States and Armenia. Agricultural reform began in Russia but then ground to a halt. After 1991, the Russian government held about 30% of the country's farms in reserve as state farms (*sovkhozes*) and technically allowed the remaining 70% to be privatized. However, arguing that only rich land barons would be able to purchase land available on auction, Communist opposition in Russia's legislature (Duma) has largely prevented the privatization of the collective farms (*kolkhozes*) that make up half of Russia's approximately 50,000 farms. Only about 6% of Russia's farmland is privately owned and worked by family farmers, and that share is barely growing.

There is simply no remembered tradition of family farming in Russia, and with so many economic uncertainties, there is little incentive for the would-be farmer to innovate. At least two-thirds of the so-called privately owned land is still under collective share holding. The farms are quite large; more than 80% are larger than 100 hectares (about 250 acres). In theory, this scale should promote more efficient farming techniques and higher output. But in Russia's new economic environment, the collective agricultural system has faltered and flagged. As the incomes of Russian consumers have fallen, so has their demand for agricultural produce. This has meant falling prices that could be fetched by farmers, who have had to pay increasingly higher costs for such inputs as fertilizers and pesticides. Unfortunately, farmers rarely try to cultivate the crops most suited to local conditions or use the most effective planting and harvesting methods. Instead, they choose the crops and growing methods supported by their regional governments, which are often ill advised. There is no concept of specialization based on competitive advantage.

The failure thus far to reform Russia's inefficient, still largely collectivized agricultural system is exacting a high cost. In 2002, the share of agriculture in Russia's economy was less than half what it was in 1990. Exports have fallen and imports have surged. Overall, Russia is a net importer of agro-food products from its trading partners.

7.3 The Industrial Potential of the Russian Coreland

The Soviet Union's emphasis on heavy industry and armaments required the use of huge quantities of minerals, most of which existed in the Soviet Union. From the beginning of its 5-Year Plans, the Soviet Union stressed the need for heavy industry, which it developed in three principal areas of mineral abundance: Ukraine, the Urals, and the Kuznetsk Basin (Fig. 7.4). One of the most problematic legacies of the breakup of the Soviet Union is that production of industrial commodities is not evenly spread across the region. The main production of each commodity tends to come from only a few areas, such as textiles from the Moscow region; steel from Ukraine, the Urals, and areas immediately south and north of Moscow; automobiles from the Volga and Moscow areas; and raw cotton from Central Asia. The manufacture of machinery is widespread, but a given city specializes only in particular kinds of machinery, such as grain harvesters at Rostov-on-Don and textile machinery near Moscow.

This regional specialization, which the 15 successor countries inherited from Communist planning, should make continued cooperation essential among the now-independent states. For example, a single factory in Belarus produces all the electric motors required by certain types of appliance factories throughout the region of Russia and the Near Abroad. For the short term at least, it is in the interest of the producers and the consumers in the former republics to maintain trading relations.

In the case of Russia, the industrial resources are there; the country has a large and highly skilled work force; and the markets both within and outside its borders are vast. Sadly, however, Russia has squandered its industrial promise in recent years. The process of privatizing formerly state-controlled industries has been corrupted by the Russian mafia and by "cronyism," the deal making and favoritism among Russia's new tycoons that thwart the emergence of a genuine free market. The largest industries are extremely inefficient and continue the Soviet practice of churning out products that no one needs. A single steel factory typically employs more than 10,000 workers — far more than are needed — in great contrast to the lean and efficient plants of the West. Such inefficient enterprises are propped up by the barter system and by a round-robin cycle in which everyone owes money to someone but no one ever pays up. "The effect," wrote one analyst, "is to keep alive concerns that chew up $1.50 worth of resources in order to turn out a product that is worth only $1 to consumers. Economists call this 'negative value added.' Ordinary folks call it economic suicide."[2] Russia's gross domestic product (GDP) shrunk by almost half in the 1990s, the largest fall in production any industrialized country has ever experienced in peacetime.

Russia is one giant "rust belt," where aging industries are suffering mechanical breakdowns that are seldom fixed. Russia is simply finding it too expensive to buy new equipment or maintain Soviet-era equipment. The systems that should function to move raw materials, finished goods,

185

---

[2] Michael M. Weinstein, "Russia Is Not Poland, and That's Too Bad." *New York Times*, August 30, 1998, 5.

REGIONAL PERSPECTIVE ::: The Sex Trade

One of the outcomes of the collapse of communism and the dissolution of the Soviet Union has been the unleashing of long-restrained expressions of sexuality. Entrepreneurs in Russia and other former Soviet countries quickly adopted an axiom of Western advertising: Sex sells. Russian television has surpassed even the spiciest fare in standard American television. On one Moscow network, there is a prime-time news show called *The Naked Truth* in which an attractive female anchor begins the newscast fully clothed but ends it topless. The public delights in watching how politicians interviewed on this show interact with the interviewer at each stage of undress. The station's ratings have soared.

Many sexual awakenings in the region are applauded or accepted as good clean fun, but others have a decidedly dark side, and some send ripples far abroad. Women of the former Soviet Union have become commodities. Russian, Ukrainian, and Kazak partners or brides can, in effect, be purchased by anyone in the world with access to the Internet. Dot com agencies act as brokers for these New Age "mail order brides." Some of the subsequent unions end happily, of course, but others collapse with physical abuse and heartache. Frequently, the woman ends up abandoned and destitute in a strange land.

The sex trade takes the greatest toll. Women are driven into it often unwittingly and almost always out of poverty. Women have paid the greatest price economically since communism's demise; two-thirds of Russia's unemployed, for example, are women. Criminal gangs use false promises of conventional employment to lure young women into the sex trade. The typical victim is an 18- to 30-year-old Russian, Ukrainian, Belarusian, or Moldovan — often with children to support — who is approached by a trafficker promising her a job as waitress, barmaid, babysitter, or maid in Central or Western Europe. She pays him a fee of $800 to $1,000 for travel and visas costs. Arriving in Czechoslovakia or Germany, she is met by another trafficker who confiscates her passport and identity papers and then sells her for several thousand dollars to a brothel owner. She is escorted to her guarded apartment. There she is compelled to reside with several other women who have taken a similar journey, and she is informed of her real job: prostitute.

Berlin papers advertise the merchandise: "The Best from Moscow!" "New! Ukrainian Pearls!" The workingwomen meet their clients in clubs or brothels or are driven to their homes. The going rate for their services is $75 per half hour, but less than 10% of that goes to the women. Their pimps and other gang members take the lion's share, and the women use their meager earnings to buy food and pay rent. Most end up in debt. Unable to pay off her debts, lacking identification papers, fearful of going to the police, and terrified the gang might harm her family members back home if she tries to escape, the woman is trapped. There are rife accounts of depression, isolation, venereal disease, beatings, and rape.

This is a big business with striking demographics. The sex industry in Europe generates revenues of about $9 billion per year. Ukrainian sources estimate that since the country's independence in 1991, 400,000 Ukrainian emigrant women have ended up in the European sex trade. An estimated 300,000 to 600,000 women of the former Soviet Union immigrate to Central and Western Europe each year, the great majority of them becoming prostitutes. There is an overflow into other regions. These so-called "Natashas" work in Turkey, Israel, Egypt, and other countries in the Middle East in similar deplorable circumstances.

JOE HOBBS

Figure 7.A Russian women and those of other former Soviet republics are some of the main participants in and victims of the global sex trade.

energy supplies, and people all seem to be on the verge of collapse. Accounts are rife of pilots bribed to overload planes with cargo, causing the planes to crash. Apartment buildings explode when lack of maintenance causes their natural gas lines to rupture. People in rural areas hack holes into oil pipelines to siphon off fuel, causing fires and explosions. Every year, hundreds of people die trying to steal hot electrical cable, and residential and industrial customers suffer power failures. Others plunder trains and planes for spare parts to be sold as scrap metal.

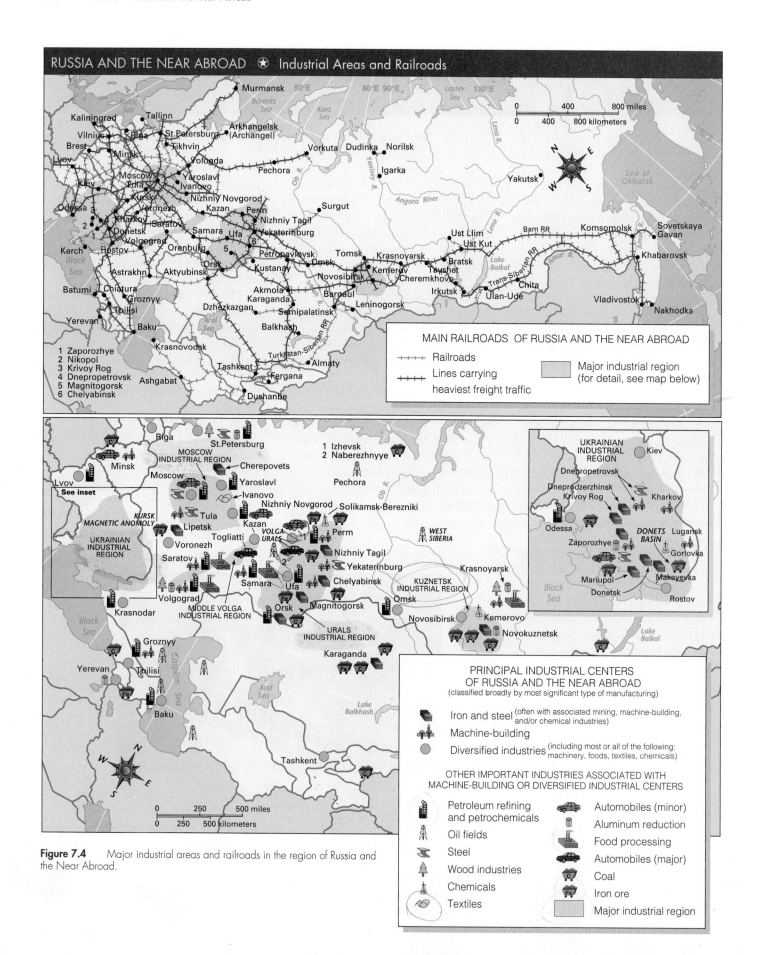

Figure 7.4   Major industrial areas and railroads in the region of Russia and the Near Abroad.

There is some good news, and the first decade of the 21st century may see the resurrection of Russian industry. Russia's economy grew 2% in 1999 and then 6% in 2000 and in 2001. Ironically, the improvement resulted from the severe economic crisis Russia experienced in 1998. When Russia's currency was devalued that year, sales of imported goods plummeted, and Russian entrepreneurs boosted production of textiles, clothing, and automobiles. A tripling in the price of oil on world markets between April 1999 and October 2000 also stimulated the economy. Russia has enormous energy reserves; its 50 billion barrels of proven oil reserves represent 5% of the world total, and its 50 trillion cubic meters of natural gas represent a third of proven global reserves. After years of belt tightening due to low world market prices, Russia has finally begun to realize energy revenues. Energy exports in 2000 and 2001 accounted for about half of Russia's export earnings. To help support the United States in its military effort in Afghanistan, Russia boosted production even more in 2001 and 2002, thereby helping to offset any potential spike in oil prices. There was also considerable growth in exports of aluminum and other metals and minerals. The Russian companies selling these resources have used their growing profits to purchase coownership in automobile and other industries, thereby perhaps sparking a broader revival of Russian manufacturing. One area in which Russia hopes to reestablish dominance is in weapons exporting. Even with 90% of its 1,700 Soviet-era military factories shut down, Russia was the world's fourth largest arms exporter in 2001.

While Russian industry is in flux, it is worthwhile to study its considerable potential, boding perhaps a brighter future for the country. Industries and industrial resources are concentrated in four areas: the regions of Moscow and St. Petersburg, the Urals, the Volga, and the Kuznetsk region of southwestern Siberia. The industrialized area surrounding Moscow is known as the Central Industrial Region, Old Industrial Region, or Moscow-Tula-Nizhniy Novgorod Region. The region has a central location physically within the populous western plains and is functionally the most important area of Russia and the Near Abroad. It lies at the center of rail and air networks reaching Transcaucasia, Central Asia, and the Pacific and is connected by river and canal to the Baltic, White, Azov, Black, and Caspian Seas (Moscow is known as the "Port of Five Seas"). Moscow, capital of Russia before 1713 and of the Soviet Union from 1918 to 1991, remains the most important manufacturing city, transportation hub, and cultural, educational, and scientific center in the region. It is the political capital of the Russian Federation.

This region ranks with Ukraine as one of the two most important industrial areas in the region of Russia and the Near Abroad. South of Moscow lies the metallurgical and machine building center of Tula and to the east is the diversified industrial center of Nizhniy Novgorod (formerly Gorkiy) on the Volga. Textile milling, mainly using imported U.S. cotton and Russian flax, was the earliest form of large-scale manufacturing to be developed in the Moscow region. The area still has the main concentration of textile plants (cottons, woolens,

linens, and synthetics) in the region of Russia and the Near Abroad, with Moscow and Ivanovo the leading centers. The breakup of the Soviet Union has done much damage to this textile industry, however. The price of Central Asian cotton quadrupled with independence, forcing mills to close and consumers to turn to cheap clothing imports.

The Moscow region's favorable geographic connections help compensate for its lack of mineral resources. The area's well-developed railway connections, partly a product of political centralization, provide good facilities for an inflow of minerals, materials, and foods and for a return outflow of finished products to all parts of Russia and the Near Abroad.

The St. Petersburg area (Fig. 7.5) is not as great an industrial center as the Moscow region, Ukraine, or the Urals, but it is still a significant presence in industrial development and is a critical port. As in the case of Moscow, local minerals have not formed the basis of this industrialization. Instead, metals from outside the area provide material for the metal-fabricating industries that are the leaders in St. Petersburg's diversified industrial structure. Supported by university and technological-institute research workers, the city's highly skilled labor force played an extremely significant role in early Soviet industrialization. They pioneered the development of many complex industrial products, such as power-generating equipment and synthetic rubber, and supplied groups of experienced workers and technicians to establish new industries in other areas.

Other industrial cities are strung along the Volga River from Kazan southward, in the region Russians know as the Povolzhye. The largest are Samara (formerly Kuybyshev); Volgograd (formerly Stalingrad); Saratov, heart of the Volga German Autonomous Republic until World War II; and Kazan. These four cities in the middle Volga industrial region have diversified machinery, chemical, and food-processing plants. Under the Soviets, Samara was one of the military-industrial "closed cities" (closed to foreigners because of security concerns) whose

Figure 7.5   St. Petersburg, established on the Baltic Sea at the mouth of the Neva River to serve as Russia's "window on the West." For its canals and distinctive architecture, it is also known as the "Venice of the North."

industries Stalin commanded to be hastily built as a defensive measure against German attack in World War II. Like Nizhniy Novgorod upriver on the Volga, the city is undergoing a process known officially as "conversion" — that is, the retooling of military enterprises for civilian-oriented manufacturing. Kazan has a major university whose alumni include Vladimir Lenin and Leo Tolstoy. Maxim Gorki wrote about Kazan in *My Universities:* Down along the Volga wharves, he found "a whirling world where men's instincts were coarse and their greed was naked and unashamed."

Industries in these cities were powered by the construction of dams and hydroelectric stations along the Volga River and its large tributary, the Kama. The Volga region took leadership in the automobile industry. Russia's "Detroit" is the Volga city of Togliatti (Fig. 7.6), by far the leading center for the production of passenger cars in the region of Russia and the Near Abroad. The cars come from the Volga Automobile Plant, which was built and equipped for the Soviet government by Italy's Fiat Company; the city itself takes its incongruous name from a former leader of Italy's Communist party.

Large-scale exploitation of petroleum in the nearby Volga-Urals fields has contributed to the industrial rise of the Volga cities. Prior to the opening of fields in western Siberia during the 1970s, the Soviet Union's most important area of oil production was the Volga-Urals fields, stretching from the Volga River to the western foothills of the Ural Mountains. Ufa and Samara are the region's principal oil-refining and petrochemical centers.

The Ural Mountains contain a diverse collection of useful minerals, including iron, copper, nickel, chromium, manganese, bauxite, asbestos, magnesium, potash, and industrial salt. Low-grade bituminous coal, lignite, and anthracite are mined. The important Volga-Urals oil fields lie partly in the western foothills of the Urals, and a major gas field lies at the southern end of the mountains.

The former Communist regime fostered the development of the Urals as an industrial region well removed from the exposed western frontier of the Soviet Union. The major industrial activities are heavy metallurgy, emphasizing iron and steel and the smelting of nonferrous ores; the manufacture of heavy chemicals based on some of the world's largest deposits of potassium and magnesium salt; and the manufacture of machinery and other metal-fabricating activities.

Yekaterinburg (formerly Sverdlovsk), located at the eastern edge of the Ural Mountains, is the largest city of the Urals and the region's preeminent economic, cultural, and transportation center. The second most important center is Chelyabinsk, located 120 miles (193 km) to the south. This grimy steel town was infamous in Old Russia as the point of departure for exiles sent to Siberia. Today, areas in the vicinity are highly polluted with radioactive waste from Soviet nuclear operations. In western Siberia, between the Urals and the large industrial and trading center of Omsk, rail lines from Yekaterinburg and Chelyabinsk join to form the Trans-Siberian Railroad, the main artery linking the Far East with the coreland (see Fig. 7.7 and the railroad map, Fig. 7.4). Omsk is a major metropolitan base for Siberian oil and gas development. It is also Siberia's most important center of oil refining and petrochemical manufacturing.

From Omsk, the Trans-Siberian Railroad leads eastward to the Kuznetsk industrial region, the most important concentration of manufacturing east of the Urals. The principal localizing factor for industry here is an enormous reserve of coal. The manufacture of iron and steel is also a major

Figure 7.6 Renowned for its drab functionality, Russian mass housing, such as these apartment blocks for auto workers in the factories at Togliatti, also offers some amenities—like fountains on a hot summer day.

LANDSCAPE IN LITERATURE

Spring Comes to the Urals

With its continental climate, Russia has some of the fiercest weather in the world: stifling, humid (but very short) summers and infamously long, cold, snowbound winters. Russians embrace springtime's thaw. In Boris Pasternak's *Doctor Zhivago*, the doctor (Yurii Andreievich) and his family take a journey by train from Moscow to the Urals. It is also a journey from winter into spring:

Suddenly everything changed — the weather and the landscape. The plains ended, and the track wound up hills through mountain country. The north wind that had been blowing all the time dropped, and a warm breath came from the south, as from an oven.

Here the woods grew on escarpments projecting from the mountain slopes, and when the track crossed them, the train had to climb sharply uphill until it reached the middle of the wood, and then go steeply down again.

The train creaked and puffed on its way into the wood, hardly able to drag itself along, as if it were an aged forest guard walking in front and leading the passengers, who turned their heads from side to side and observed whatever was to be seen.

But there was nothing yet to see. The woods were still deep in their winter sleep and peace. Only here and there a branch would rustle and shake itself free of the remaining snow, as though throwing off a choker.

Yurii Andreievich was overcome with drowsiness. All these days he lay in his bunk and slept and woke and thought and listened. But there was nothing yet to hear.

While Yurii Andreievich slept his fill, the spring was heating and melting the masses of snow that had fallen all over Russia, first in Moscow on the day they had left and since then all along the way — all that snow they had spent three days clearing off the line at Ust Nemdinsk, all that thick, deep layer of snow that had settled over the immense distances.

At first the snow thawed quietly and secretly from within. But by the time half the gigantic labor was done it could not be hidden any longer and the miracle became visible. Waters came rushing out from below with a roar. The forest stirred in its impenetrable depth, and everything within it awoke.

There was plenty of room for the water to play. It flung itself down the rocks, filled every pool to overflowing, and spread. It roared and smoked and steamed in the forest. It streaked through the woods, bogging down in the snow that tried to hinder its movement, it ran hissing on level ground or hurtled down and scattered into a fine spray. The earth was saturated. Ancient pine trees perched on dizzy heights drank the moisture almost from the clouds, and it foamed and dried a rusty white at their roots like beer foam on a mustache.

The sky, drunk with spring and giddy with its fumes, thickened with clouds. Low clouds, drooping at the edges like felt, sailed over the woods and rain leapt from them, warm, smelling of soil and sweat, and washing the last of the black armor-plating of ice from the earth.

Yurii Andreievich woke up, stretched, raised himself on one elbow, and looked and began to listen.[a]

[a]From *Doctor Zhivago*, by B. Pasternak, pp. 232–233. Copyright © 1958 Pantheon Books, an imprint of Knopf Publishing Group. Reprinted by permission.

Figure 7.7   View of Omsk and the Trans-Siberian Railway from space. The railway draws a line through the April snow, and Osmk lies where it crosses the Irtysh River.

industrial activity of the Kuznetsk region, whose main center is Novokuznetsk. The industry draws its iron ore from various sources among Russia's Central Asian neighbors.

The largest urban center of the Kuznetsk region is Novosibirsk, a diversified industrial, trading, and transportation center located on the Ob River at the junction of the Trans-Siberian and Turkestan-Siberian (Turk-Sib) Railroads. Sometimes called the Chicago of Siberia, Novosibirsk has developed from a town of a few thousand at the turn of the 20th century to more than a million today. In Soviet times, hundreds of factories produced mining, power-generating, and agricultural machinery, tractors, machine tools, and a wide range of other products.

The city of Akademgorodok (Academy Town or Science Town), Siberia's main center of scientific research, is an outlying satellite community of Novosibirsk. The city was established in 1957 as a parklike haven where some of the Soviet Union's greatest scientific minds could research and develop both civilian and military-industrial innovations. Reflecting a general crisis in the post-Soviet military-industrial establishment, Akademgorodok's funding has shrunk dramatically, forcing many prominent scientists to seek menial jobs. Internationally, there are fears that underpaid Russian nuclear scientists will sell their know-how or materials to governments or terrorist organizations such as al-Qa'ida seeking to develop atomic weapons. There have already been hundreds of thefts of radioactive substances at nuclear and industrial institutions in the former USSR. The destination points of these materials remain largely unknown. Former Soviet nuclear experts have applied for work or are already working in Iran, Iraq, Algeria, India, Libya, and Brazil.

7.4   **The Russian Far East**

The Far East is Russia's mountainous Pacific edge. Most of it is a thinly populated wilderness in which the only settlements are fishing ports, lumber and mining camps, and the villages and camps of aboriginal peoples. Port functions, fisheries, and forest industries provide the main support for most Far Eastern communities, and the output of coal, oil, and a few other minerals is small. Most of the Russians and Ukrainians who make up the majority of its people live in a narrow strip of lowland behind the coastal mountains in the southern part of the region. This lowland, drained by the Amur River and its tributary, the Ussuri, is the region's main axis of industry, agriculture, transportation, and urban development.

Several small- to medium-sized cities form a north–south line along two important arteries of transportation: the Trans-Siberian Railroad and the lower Amur River. At the south on the Sea of Japan is the port of Vladivostok, which is kept open throughout the winter by icebreakers. About 50 miles (80 km) east of the city, the main commercial seaport area of the Far East has developed at Nakhodka and nearby Vostochnyy (East Port). Both ports are nearly ice-free and well positioned for trade in goods manufactured in Korea, China, and Japan. Seaborne shipments of cargo containers from northeast Asia to Scandinavia require 40 days, but the cargo can make the 6,000-mile (9,600-km) journey from Vostochnyy to Finland by rail in only 12 days. Business in Vostochnyy is picking up.

North of Vladivostok lies a small district that is the most important center of the Far East's meager agriculture, producing cereals, soybeans, sugar beets, and milk for Far Eastern consumption. The Far East is far from self-sufficient in food and consumer goods; large shipments from the Russian coreland and from overseas supplement local production.

The diversified industrial and transportation center of Khabarovsk is located at the confluence of the Amur and Ussuri Rivers, where the main line of the Trans-Siberian Railroad turns south to Vladivostok and Nakhodka and the Amur River turns north toward the Sea of Okhotsk.

Before World War II, the Soviet Union and Japan held the northern and southern halves, respectively, of the large island of Sakhalin, which today has important forest and fishing industries and some coal, petroleum, and natural gas reserves. At the end of the war, the USSR annexed southern Sakhalin

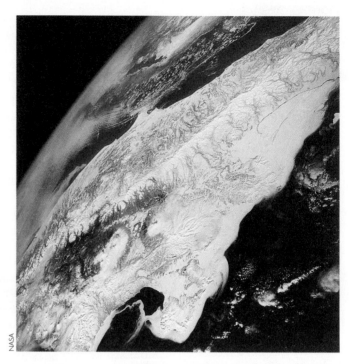

Figure 7.8   Russia's wild Kamchatka Peninsula seen from space. Deep snows accentuate the relief of its many volcanoes. At left is the Sea of Okhotsk and in the right is the Bering Sea.

and the Kuril Islands and repatriated the Japanese population. Russian control of these former Japanese territories continues to be a major problem in relations between the two countries.

The Kurils are small volcanic islands that screen the Sea of Okhotsk from the Pacific. Fishing is the main economic activity. At the north, the Kurils approach the mountainous peninsula of Kamchatka (total population: 360,000, with 230,000 living in two cities). Like the islands, Kamchatka is located on the Pacific Ring of Fire — the geologically active perimeter of the Pacific Ocean–and has 23 active volcanoes (Fig. 7.8). Soviet authorities protected Kamchatka for decades as a military area, and no significant development activities occurred there. A land of rushing rivers, extensive coniferous forests, and gurgling hot springs, Kamchatka is now one of the last great wilderness areas on Earth, resembling the U.S. Pacific Northwest landscape of a century ago. A struggle is now underway between those who want to develop Kamchatka's gold and oil resources and those who wish to set the land aside in national parks and reserves where ecotourism would be the only significant source of revenue.

7.5   **The Northern Lands of Russia**

North and east of the Fertile Triangle and west of (and partially including) the Pacific littoral lie enormous stretches of coniferous forest (taiga) and tundra extending from the

Finnish and Norwegian borders to the Pacific. These outlying wilderness areas in Russia may for convenience be designated the Northern Lands, although parts of the Siberian taiga extend to Russia's southern border.

These difficult lands form one of the world's most sparsely populated large regions. The climate largely prevents ordinary types of agriculture, but hardy vegetables, potatoes, hay, and barley are grown in scattered places, and there is some dairy farming. A few primary activities, including logging, mining, reindeer herding, fishing, hunting, and trapping, support most people. Significant towns and cities are limited to a small number of sawmilling, mining, transportation, and industrial centers, generally along the Arctic coast to the west of the Urals, along the major rivers, and along the Trans-Siberian Railroad between the Fertile Triangle and Khabarovsk.

The oldest city on the Arctic coast is Arkhangelsk (Archangel), located on the Northern Dvina River inland from the White Sea. Tsar Ivan the Terrible established the city in 1584 for the purpose of opening seaborne trade with England. Today, the city is the most important sawmilling and lumber-shipping center in the region of Russia and the Near Abroad. Despite its restricted navigation season from late spring to late autumn, it is one of the more important seaports.

Another Arctic port, Murmansk (population: 361,000; Fig. 7.9; the world's largest city within the Arctic Circle), is located on a fjord along the north shore of the Kola Peninsula west of Arkhangelsk and is the headquarters for important fishing trawler fleets that operate in the Barents Sea and North Atlantic. Murmansk also has a major naval base and cargo port and is home port to the icebreakers that escort cargo vessels and carry Western tourists to the North Pole and other High Arctic destinations (Fig. 7.10). It is connected to

Figure 7.9 The ice-free port of Murmansk is situated on a fjord at the northern end of the Kola Peninsula.

the coreland by rail. The harbor is open to shipping all year thanks to the warming influence of the North Atlantic Drift, an extension of the Gulf Stream. Murmansk and Arkhangelsk played a vital role in World War II, continuing to receive supplies by sea from the Soviet Union's Western allies after Nazi forces captured and closed the other ports of the western Soviet Union.

Murmansk and Arkhangelsk are western termini of the Northern Sea Route (NSR), a waterway the Soviets developed to provide a connection with the Pacific via the Arctic Ocean (see Fig. 7.1). Navigation along the whole length of the route is possible for only up to 4 months per year, despite the use of powerful icebreakers (including nine nuclear-powered vessels; see Problem Landscape, p. 202) to lead convoys of ships. Areas along the route provide cargoes such as the timber of Igarka and the metals and ores of Norilsk. Some supplies are

Figure 7.10 The Russian nuclear-powered icebreaker *Yamal* at the North Pole. The former Soviet icebreaker fleet still escorts commercial ships on the Northern Sea Route but during the summer carries Western tourists such as these to the High Arctic.

On the map, the Russian islands of Novaya Zemlya ("New Land"), separating the Barents and Kara Seas in the high Arctic, appear remote, wild, and untouched. These ice- and tundra-covered islands are indeed an isolated wilderness, which is precisely why Soviet authorities perceived them as prized grounds for nuclear weapons testing and nuclear waste dumping.

Soon after the breakup of the Soviet Union in 1991, a former Soviet radiation engineer stepped forward with disturbing information: Soviet authorities had ordered the dumping of highly radioactive materials off the coast of Novaya Zemlya for the past 30 years (Fig. 7.B). From the 1950s through 1991, the Soviet navy and icebreaker fleet secretly and illegally used the shallow waters off the eastern shore of Novaya Zemlya as a nuclear dumping ground. At least twelve nuclear reactors removed from submarines and warships, six of which still contained their highly radioactive nuclear fuel, were sent to the shallow seabed. Some of these individual reactors contain roughly seven times the nuclear material contained in the Chernobyl reactor that exploded in 1986. Between 1964 and 1990, 11,000 to 17,000 containers of solid radioactive waste were also dumped here at shallow depths between 200 and 1,000 feet (61 and 305 m). Soviet seamen reportedly cut holes in those "sealed" containers, which otherwise would not have sunk. This dumping occurred even though the Soviet Union joined other nations in a 1972 treaty that allowed only low-level radioactive waste to be dumped at sea, and then only at depths greater than 12,000 feet (3,657 m).

The danger posed by the nuclear material is not limited to the islands' shores but extends through the entire ecosystem of Russia's Arctic seas and possibly beyond. Contamination can begin at the base of the food chain (the phytoplankton that "bloom" in the area each spring) and pass upward to higher levels, where fish, seals, walrus, polar bears, and people eat. Alaskan authorities are worried, and Norwegians are particularly concerned. Norway's prime minister described the dumping as a "security risk to people and to the natural biology of northern waters." Western European and North American markets for Norwegian and Russian fish, which now represent significant exports for these countries, may refuse purchases if radiation levels rise.

Due to the long half-life of the nuclear materials, the dangers posed by 30 years of haphazard dumping near Novaya Zemlya may last thousands of years. The threat can be reduced greatly by retrieving the waste and disposing of it in terrestrial sites deep underground, which are considered safer. (The Russian military supports a proposal to convert some of the shafts used in 1950s atomic weapons tests to permanent radioactive waste repositories.) The major obstacle to this step is the extreme costliness of such an operation—estimated to be in the hundreds of billions of dollars—which is well beyond Russia's economic means. An arms-agreement obligation for Russia to decommission about 80 nuclear submarines based in Murmansk is also costly and, because of the radioactive materials involved, potentially dangerous. Confessing and confronting the nuclear legacy of the Soviet Union, Russia has asked the United States and other nations to help pay for an environment with fewer nuclear hazards. The United States in turn is asking Russia not to export its nuclear power technologies to Iran and other countries where the West fears they may be used to develop weapons.

The predicament of Novaya Zemlya and the future of the vast wilderness in Russia's far East reflect important questions about environment and development all across Russia and the Near Abroad. Like the United States and other industrialized countries, the Soviet Union experienced problems of environmental pollution caused by rapid economic growth. Russia and the other countries must now deal with the Soviet legacy of decades of environmental neglect. Environmental cleanup is hindered by the massive cost of repairing past damage and a reluctance to bear the expenses of stringent enforcement of pollution-abatement measures. Some Russian sources estimate that 80% of all industrial enterprises in Russia would go bankrupt if forced to comply with environmental laws. A desperate search for hard currency and widespread corruption have led to other environmental problems, including the wide-scale selling of natural resources like timber and the poaching of animals. With the breakup of the USSR, many environmentalists had predicted more protection for nature. They have been disappointed.

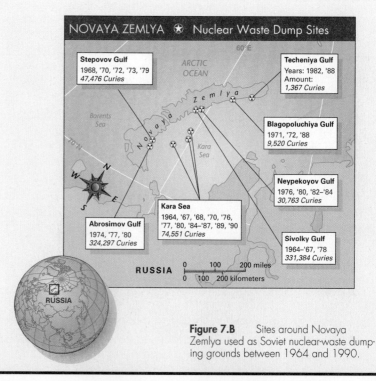

Figure 7.B   Sites around Novaya Zemlya used as Soviet nuclear-waste dumping grounds between 1964 and 1990.

shipped north on Siberia's rivers from cities along the Trans-Siberian Railroad and then loaded onto ships plying the Northern Sea Route for delivery to settlements along the Arctic coast. The Northern Sea Route is now open to international transit shipping, and if not for the capricious sea ice, it would be an attractive route. In recent winters, the Arctic ice cover has retreated substantially — some authorities fear because of global warming — and the Russian shipping industry is hoping this trend will continue. The distance between Hong Kong and all European ports north of London is shorter via the Northern Sea Route than through the Suez Canal. Russia is anxious to develop the Northern Sea Route more for its own exports because Russian exports by land to Western Europe must now pass over the territories of Ukraine, Belarus, and the Baltic countries.

The railroad that connects Murmansk with the coreland serves important mining districts in the interior of the Kola Peninsula and logging areas in Karelia, which adjoins Finland. The Kola Peninsula is a diversified mining area, holding nickel, copper, iron, aluminum, and other metals, as well as phosphate for fertilizer. The peninsula and Karelia are physically an extension of Fennoscandia, located on the same ancient, glacially scoured, granitic shield that underlies most of the Scandinavian Peninsula and Finland. Karelia, with its thousands of lakes, short and swift streams, extensive softwood forests, timber industries, and hydroelectric power stations, bears a close resemblance to nearby parts of Finland (which annexed some of this area during World War II).

Important energy resources, including good coking coal, petroleum, and natural gas, are present in the basin of the Pechora River, east of Arkhangelsk. The main coal-mining center is Vorkuta, infamous for the labor camp where great numbers of political prisoners died during the Stalin regime. Russian, U.S., and Norwegian oil companies are working together to explore for and develop oil for export from the Pechora region. The proven reserves of this region are enormous. The Timan Pechora field contains an estimated 126 billion barrels of oil, one of the world's largest reserves and enough oil to supply the world at current levels of oil use for 4 years. Natural gas reserves in this area and nearby in the Barents Sea are also huge. One of the big challenges is to find a way to transport these hydrocarbons to market. Plans are underway to construct a pipeline to send oil and gas south to the Gulf of Finland near St. Petersburg.

East of the Urals, the swampy plain of western Siberia north of the Trans-Siberian Railroad saw a surge of petroleum and natural gas production after the 1960s. The West Siberian fields are the largest producers in the region of Russia and the Near Abroad. However, an estimated half of the reserves of these fields have already been pumped, and so Russia looks to the Arctic for its energy future. Extraction and shipment of oil and gas take place in West Siberia under frightful difficulties caused by severe winters, permafrost, and swampy terrain. Huge amounts of steel pipe (Fig. 7.11) and pumping equip-

Figure 7.11   Laying an oil pipeline in Russia's Northern Lands. Immense quantities of steel have been required for thousands of miles of oil and gas pipes that have been laid in Russia and other parts of Russia and the Near Abroad since World War II.

ment have been required to connect the remote wells with markets in the coreland and Europe.

Still farther east, the town of Igarka, located about 425 miles (c. 680 km) inland on the deep Yenisey River, is an important sawmilling center accessible to ocean shipping during the summer and autumn navigation season. Northeast of Igarka lies Norilsk, the most northerly mining center of its size on the globe. Rich ores yielding nickel, copper, platinum, and cobalt have justified the large investment in this Arctic city, including a railway connecting the mines with the port of Dudinka on the Yenisey downstream from Igarka.

In central and eastern Siberia, the taiga extends southward beyond the Trans-Siberian Railroad. A vast but sparsely settled hinterland is served by a few cities spaced at wide intervals along the railroad, notably Krasnoyarsk on the Yenisey River and Irkutsk on the Angara tributary of the Yenisey near Lake Baikal. From east of Krasnoyarsk, a branch line of the Trans-Siberian leads eastward to the Lena River. Years of construction in the 1970s and 1980s under very difficult conditions resulted in continuation of this line to the Pacific under the name Baikal-Amur Mainline, or BAM. The railroad was built to open important mineralized areas east of Lake Baikal and to reduce the strategic vulnerability caused by the close proximity of the Trans-Siberian Railroad to the Chinese frontier. The Pacific terminus is located on the Tatar Strait, which connects the Sea of Okhotsk with the Sea of Japan (see Figs. 6.1 and 7.4).

The BAM railway passes through Bratsk, site of a huge dam and hydroelectric power station on the Angara River. Lake Baikal, which the Angara drains, lies in a mountain-rimmed rift valley and is the deepest body of fresh water in the world at more than a mile deep in places (Fig. 7.12). Lake Baikal is at the center of an environmental controversy over

Figure 7.12   Fishermen at work in Lake Baikal, Earth's deepest lake.

the estimated 8.8 million cubic feet of wastewater that pour into it daily from two large paper and pulp mills on its shores. The waste jeopardizes Baikal's unique natural history, as the lake contains an estimated 1,800 endemic plant and animal species. "Nature as the sole creator of everything still has its favorites, those for which it expends special effort in construction, to which it adds finishing touches with a special zeal, and which it endows with special power," wrote Valentin Rasputin, a native of Siberia. "Baikal, without a doubt, is one of these. Not for nothing is it called the pearl of Siberia."[3]

Siberia's development is based in large part on hydroelectric plants located along the major rivers (Fig. 7.13). The largest hydropower stations are on the Yenisey (at Krasnoyarsk and in the Sayan Mountains) and its tributary the Angara (at Bratsk and Ust Ilim). These power stations and dams are among the largest in the world. They supply inexpensive electricity, consumed mainly in mechanized industries that use power voraciously. These include cellulose plants that process Siberian timber and aluminum plants making use of aluminum-bearing material (alumina) derived from ores in Siberia, the Urals, and other areas and imported from abroad via the Black Sea.

Scattered throughout central and eastern Siberia are important gold-mining centers, mainly in the basins of the Kolyma and Aldan Rivers and in other areas in northeastern Siberia. (In Stalin's time, the Kolyma mines, located in one of the world's harshest climates, developed an evil reputation as death camps for great numbers of political prisoners forced to work there.) Large coal deposits are mined along the Trans-Siberian Railroad at various points between Irkutsk and the Kuznetsk Basin.

Aside from irregularly distributed centers of logging, mining, transportation, and industry peopled mostly by Slavs, the

[3] Valentin Rasputin, "Baikal." In *Siberia on Fire: Stories and Essays* by Valentin Rasputin. DeKalb: Northern Illinois University Press, 1989, pp. 188–189.

Figure 7.13   Siberia's hydroelectric power potential is vast. This river in extreme northeastern Siberia is one of the abundant northward-flowing drainages in the region.

Figure 7.14 Yakut herders and reindeer, Sakha.

Northern Lands are home mainly to non-Slavic peoples who make a living by reindeer herding, trapping, fishing, and in the more favored areas of the taiga, by cattle raising and precarious forms of cultivation (Fig. 7.14). The domesticated reindeer is especially valuable to the tundra peoples, providing meat, milk, hides for clothing and tents, and draft power. The Yakuts, a Turkic people living in the basin of the Lena River, are among the most prominent of the non-Slavic ethnic groups in the Northern Lands. Their political unit, Sakha (formerly Yakutia), is larger in area than any former Union Republic except the Russian Federation, within which it lies. Sakha has only about 1 million people inhabiting an area of 1.2 million square miles (3.1 million sq km). The capital city of Yakutsk on the middle Lena River has road connections with the BAM and Trans-Siberian Railroads and with the Sea of Okhotsk. Riverboats on the broad and deep Lena provide a connection with the Northern Sea Route and the southern rail system during the warm season. Sakha will figure prominently in future mineral and timber development, with its large reserves of diamonds, natural gas, coking coal, other minerals, and timber.

7.6 The Future of the Russian Federation

Russia's internal republic of Sakha is the largest in area of many autonomies (nationality-based republics and lesser units) in the Russian Federation. They are successors to the autonomous ASSRs, autonomous *oblasts*, and autonomous *okrugs* of the Soviet era (some of these are in other former republics of the Soviet Union; see inset map, Fig. 6.1). These autonomies occupy over two-fifths of Russia's total area and contain about a fifth of the country's population.

Nearly half of the people in Russia's autonomies are ethnic Russians, who are a majority in many units. Titular nationalities of the autonomies are ethnically diverse (see Fig. 6.2): Some, such as the Karelians, Mordvins, and Komi, are Finnic (related to the Finns and Magyars); others are Tur-

kic (e.g., Tatars, Bashkirs, and Yakuts), Mongol (e.g., Buryats near Lake Baikal), and members of many other ethnic groups. Russian expansion brought these peoples into the Russian Empire at different times over a period of several centuries, and the Communist rulers of the USSR organized them into "autonomous" units ranked on a nationalities ladder with the former SSRs (Union Republics) at the top. Some of the lesser units eventually climbed the ladder to become Union Republics.

The largest areal units form a nearly solid band stretching across the northern part of the Northern Lands from Karelia, bordering Finland, to the Bering Strait. The Karelian, Komi, and Sakha (Yakut) Republics are in this group. This band also includes several large but thinly inhabited units of aboriginal peoples that the Soviets designated as "autonomous *okrugs.*" There is more pastoralism than agriculture in the autonomies. There are few large cities, with only Kazan in Tatarstan, Ufa in Bashkiria, and Perm in Komi-Permyak having more than 1 million people. There is little manufacturing. However, there are major mineral resources in some of the autonomies: oil and gas in the Volga-Urals fields in Tatarstan and Bashkiria, high-grade coal deposits at Vorkuta in the Komi Republic, and a quarter of all diamonds produced in the world in Sakha.

Such resources have great potential significance in the geopolitical realm. After the breakup of the USSR, many of the former Soviet ASSRs, recalling Moscow's longstanding promises of self-rule for them, issued declarations of sovereignty asserting their right to greater self-direction of their internal affairs and greater control over their own resources. Moscow has relented to some of them. The Sakha (Yakut) Republic, for example, won the right to keep 45% of hard currency earnings from foreign sales of Sakha diamonds, compared with only a small fraction during the Soviet era. In turn, Sakha must now pay for the government subsidies that Moscow previously paid to local industries. Without credits from Moscow, Sakha no longer pays taxes to Moscow. Salaries and other indexes of living standards in the Sakha Republic have risen since this agreement was implemented.

Such regional demands for greater self-rule threaten the unity of the Russian Federation. Eighteen of Russia's twenty republics (the sixteen Soviet-era ASSRs plus four autonomous regions upgraded to republic status) signed a 1992 Federation Treaty that granted them considerable autonomy. The treaty called for the devolution of power centralized in Moscow and for more cooperation between regional and federal governments. Each republic of the Russian Federation was legally entitled to have its own constitution, president, budget, tax laws and other legislation, and foreign and domestic economic partnerships. But many republics, and the *oblasts* and *krays* that have had their own elected governors since 1997, are complaining that Moscow has not honored the treaty, and they are looking for regional solutions to their economic and other problems. Moscow in turn has created seven of its own administrative regions (North-West, Central, North Caucasus,

Volga, Urals, Siberia, and far East) and appointed a presidential envoy (known as a governor general) to manage national defense, security, and justice matters in each. The message is that Moscow rules. It remains to be seen whether yet another layer of bureaucracy will do anything to keep the country's vast periphery tied to its core.

The Caucasus autonomy of Chechnya (then joined with Ingushetia) and the upper Volga autonomy of Tatarstan did not sign the 1992 Russian Federation Treaty and insisted on independence from Moscow. Oil-rich Tatarstan now has its own constitution, parliament, flag, and official language and has since signed the treaty. Other autonomous regions are keeping eyes on Tatarstan and Chechnya. The oil-rich Turkic-speaking Bashkirs and the Chuvash might push for more independence. This could begin a process that would virtually cut Russia in half. Although some analysts argue that this process of devolution in Russia may actually bring more stability and prosperity to the country, the government in Moscow worries that what happened to the USSR might happen to Russia itself. The situation in Chechnya (population: 544,400) is of particular concern.

The Chechens, who are Sunni Muslims, have periodically resisted Russian rule, and Moscow has punished them. Russian troops attempted but failed to seize control over Chechnya soon after its 1991 declaration of independence. Beginning a second offensive in 1994, and at a cost of an estimated 30,000 to 80,000 lives on both sides, Russian troops succeeded in exerting physical control over most of Chechnya. But late in 1996, Chechen forces recaptured the capital city of Groznyy (renaming it Jokhar-Gala). At that time, the Chechen war was deeply unpopular among Russians, and the prospect of further Russian losses led to a peace agreement with the Chechens. The pact required an immediate withdrawal of Russian forces from Chechnya and deferred the question of Chechnya's permanent political status until 2001. Russia allowed Chechens to elect their own president (they did, choosing Aslan Maskhadov in 1997) but insisted that Chechnya was, for the time being, still part of the Russian Federation.

The cessation of hostilities was short-lived. Chechen rebels kidnapped and tortured hundreds of Russian citizens in southern Russia and, in September 1999 carried their campaign to the heart of Russia. They detonated bombs in Moscow and other vulnerable civilian centers, killing more than 300 Russians. The Chechen resistance by this time had become internationalized, with Islamic militant groups from abroad, including from Osama bin Laden's notorious al-Qa'ida organization based in Afghanistan, supplying men and war material to their Islamic guerilla brethren in Chechnya. No longer controlled by Chechen President Maskhadov, in August 1999 these Chechen Islamists invaded the neighboring province of Dagestan with the hope of creating an independent Islamic state.

That invasion, and the terrorist bombings in Moscow, led to a surge in anti-Chechen sentiment among Russians and popular calls for resolute action against Chechnya. Russia's newly appointed prime minister, Vladimir Putin, decided to pursue an all-out second Chechen War. In September 1999, Russian forces launched a furious attack that this time put them firmly in control of most of Chechnya. Putin's military success helped secure his victory in the Russian presidential election that followed, but the reaction abroad was mostly negative. Russian troops were documented to have carried out atrocities against Chechen civilians, and Western governments came under pressure to apply sanctions against Russia. None did, however. With Chechyna's president in hiding and Chechen guerilla resistance continuing, Russia is maintaining a vice on Chechnya, and prospects for its independence from Russia look remote.

7.7 The Caucasus

Chechnya is in the northernmost Caucasus, the isthmus between the Black and Caspian Seas. In the north of the isthmus is the Greater Caucasus Range, forming an almost solid wall from the Black Sea to the Caspian (Fig. 7.15). It is similar in age and character to the Alps but is much higher, up to 18,510 feet (5,642 m) on Mt. Elbrus, on the Russia/Georgia border. The southern Caucasus is known as Transcaucasia. In southern Transcaucasia is the mountainous, volcanic Armenian Plateau. Between the Greater Caucasus Range and the Armenian Plateau are subtropical valleys and coastal plains where the majority of people in Transcaucasia live. A humid subtropical climate prevails in coastal lowlands and valleys of Georgia, south of the high Caucasus Mountains. Mild winters and warm to hot summers combine with the heaviest precipitation to be found anywhere in the region of Russia and the Near Abroad — 50 to 100 inches (c. 125 to 250 cm) a year. The lowland area bordering the Black Sea in Georgia is the most densely populated part of Transcaucasia. Specialty crops such as tea, tung oil, tobacco, silk, and wine grapes, together with some citrus fruits, grow there.

JOHN CORBETT/ECOSCENE/© CORBIS

Figure 7.15   Mountains and glaciers seen from Mt. Elbrus in the Caucasus.

Figure 7.16 Ethnic Georgians selling produce in a market at Tbilisi, the main city and capital of Georgia.

The Caucasian isthmus has been an important north–south passageway for thousands of years, and the population includes dozens of ethnic groups who have migrated into this region. Most of these are small ethnic groups confined to mountain areas that became their refuges in past times. The republic of Dagestan, within Russia on its southeastern border with Georgia and Azerbaijan, is home to 34 different ethnic groups. Some mountain villages, particularly in southeastern Azerbaijan, have achieved international fame as the abodes of the world's longest lived people — up to the much-challenged claim of 168 years! In addition to Russians and Ukrainians, most of who live north of the Greater Caucasus Range, the most important nationalities are the Georgians (Fig. 7.16), Armenians, and Azerbaijanis, each represented by an independent country. Throughout history, all have defiantly maintained their cultures in the face of pressure by stronger intruders.

These nationalities have ethnic characteristics and cultural traditions that are primarily Asian and Mediterranean in origin. Their religions differ. The Azerbaijanis (also known as Azeri Turks) are Muslims. The Georgians belong to one of the Eastern Orthodox churches. The Armenian Apostolic Church is a very ancient, independent Christian body that is an offshoot of Eastern Orthodoxy; in A.D. 301, Armenia became the world's first Christian country when its government pronounced the young religion to be its national faith (Fig. 7.17). There has been a history of animosity between the Armenians and Azeri Turks (including a war between them in 1905), growing out of the Turks' persecution of Armenians in the Ottoman Empire prior to and during World War I. Turkey and Azerbaijan still have not accepted responsibility for their roles in the Armenian genocide, in which an estimated 1.5 million Armenians died between 1915 and 1918. The descendants of Armenians who fled this holocaust have established themselves in many places around the world (e.g., there is a

Figure 7.17 The Armenian Church of the Holy Cross on the island of Akhtamar. Armenia was the world's first Christian country.

large population in Hollywood's film industry) and today outnumber Armenians living within the country by almost two to one. One indicator of Armenia's declining economic fortunes following independence in 1991 is that 40% of the country's population has moved abroad since then.

The complex and often bitter world of the Caucasus is summarized in unrestrained terms in Table 7.1, adopted from the British publication *The Economist*. Here is a more detailed presentation of the problems and events underlying that table.

Early in the 1990s, historical animosity flared into massive violence between Armenians and Azeris over the question of Nagorno-Karabakh, a predominantly Armenian 1,700 square mile (4420 sq km) enclave within Azerbaijan, then governed by Azerbaijan but claimed by Armenia. From 1992 until 1994, when Armenian forces finally secured the region, fighting over Nagorno-Karabakh took thousands of lives and created nearly a million refugees. Large numbers of Armenians fled from Azerbaijan to Armenia as refugees. There was also a refugee flight of Azeris from Armenia to Azerbaijan; today, fully one-eighth

TABLE 7.1   Summary of Ethnic and Political Problems in the Caucasus

Where	What	Neighbors' view	What they say back	Main ally	Mafia involvement	Importance
Abkhazia	Unrecognized separatist republic	Murderous, idiotic Russian stooges	Georgian imperialists tried to destroy us and our culture	Russia	Lightly supervised port offers great potential	Blocks railway to Russia, 200,000 angry refugees in Georgia
Ajaria	Autonomous republic within Georgia	Got rich by stealing customs revenues	Rest of Georgia is wild, badly run	Russia, probably	Russian mobsters have been spotted there	Ignores Georgian government, Russian base
Armenia	Independent country	Arrogant, untrustworthy Russian stooges	Genocidal maniacs (Turkey, Azerbaijan)	Russia (guns), America (money), Iran (trade)	War profiteers, corrupt officials	Large Russian military presence
Azerbaijan	Independent country	Primitive, murderous, American lackeys	Russian-backed aggressors (Armenia)	America, Turkey	Ruling elite very corrupt	Oil, gas. Helps Chechnya, says Russia
Georgia	Independent country	Wild, corrupt, horrid to minorities	Russian-backed separatist stooges	America	Rampant cronyism at all levels	Could disintegrate or implode violently
Javakheti	Armenian part of Georgia	Disloyal, potential separatists	Georgia neglects us	Armenia, Russia	Very poor	Potential flashpoint, large Russian base
Nagorno-Karabakh	Self-proclaimed republic	Illegal Armenian aggressor puppet state	Refuge from homicidal neighbors	Armenia, Russia indirectly	War profiteers, veterans' groups	Cause of the largest conflict in the region
Nakhichevan	Exclave of Azerbaijan	Autocratic, primitive	Our guys run the whole country	Azerbaijan, Turkey	Officialdom does conspicuously well	Birthplace of Azerbaijan's president; border with Turkey
Pankisi Gorge	Chechen-populated part of Georgia	Wild	Wimpy	Chechnya	Don't ask	Russia complains Chechen fighters shelter there
South Ossetia	Unrecognized separatist republic	Murderous, ungrateful, recent immigrants, Russian stooges	Georgian imperialists terrorized us	Russia, North Ossetia	Tunnel to Russia offers huge potential	Blocks road route to Russia

Source: After "The Caucasus: Where Worlds Collide." *Economist*, August 19, 2000, 18.

of Azerbaijan's people are war refugees, many living in holes in the ground covered by thatch. As this book went to press, there were hopes of a peace deal between Armenia and Azerbaijan. The Armenians would return to Azerbaijan six of the seven regions they captured. Nagorno-Karabakh and the adjacent Lachin region would be granted self-governing status. Azerbaijan's compensation would be the construction of an internationally protected road linking Azerbaijan with its enclave of Nakhichevan, now cut off from Azerbaijan by Armenian territory.

During the conflict of the early 1990s, tens of thousands of Azeris fled into Iran. Iranian Azeris, who make up a third of Iran's population, feel that Iran has leaned too far toward Armenia as a de facto ally against Turkish influence in the Caucasus and Central Asia and should move closer to Azerbaijan. But Azerbaijan fears that Iran might try to export a more militant Islam into the country.

Russia negotiated a cease-fire between Armenia and Azerbaijan in 1994, but the countries are still technically at war.

Armenians are angry that Moscow did not restore to Armenia the enclave of Nagorno-Karabakh, which the Bolsheviks had put under Azeri control. But the Armenians also embrace Russia as a strategic ally against their historic enemy, the Turks, who virtually surround them.

In Georgia, there also have been recent episodes of ethnic and political violence, particularly between the Georgians and the primarily Muslim South Ossetians, who would like to free themselves from Georgian control and establish an Ossetian nation. The North Ossetians, whom they wish to join, live adjacent to them in Russia. South Ossetia now exists as a Russian puppet state within Georgia's borders. Its major source of income is from transit fees charged for a little-used tunnel running into Russia.

Also seeking independence from Georgia is another Muslim people, the Abkhazians, who make up about 2% of Georgia's population and are concentrated in the province known as Abkhazia. In 1993, Abkhazian separatists captured the Georgian Black Sea port of Sukhumi in Abkhazia. Russian forces

F6.1 167

initially aided them, both to regain access to Black Sea resorts and to take revenge on Georgian President Eduard Shevardnadze, the former Soviet foreign minister whom many Russians hold partly responsible for the breakup of the USSR. Russia was also putting pressure on Georgia to rejoin the CIS. Shevardnadze responded by having Georgia rejoin the CIS and agreeing to allow four Russian military bases to remain for a while on Georgian soil and to allow Russian troops to be stationed on Georgia's border with Turkey. Russian forces then put a stop to the Abkhazian offensive but did not drive the rebels from the territory they had captured. In 2001, when Georgia insisted that it was time for Russia to scale back its military presence in the region, Russia responded by cutting off supplies of natural gas to Georgia and insisting that Georgians working in Russia obtain proper visas. Those expatriate workers are a vital source of revenue for Georgia, where 60% of the people live below the poverty line, so Russia has strong leverage in hinting they might be sent home.

Georgians see Russia as using the Abkhazian problem, energy supplies, and the visa issue as means of retaining leverage in the strategically vital Caucasus region and, above all, securing access to Abkhazia's 110 miles (180 kms) of Black Sea coastline. For its part, Georgia wants to loosen ties with Russia in favor of better relations with the West and its allies (e.g., it wants to join the European Union and the North Atlantic Treaty Organization). Most road and rail links between the countries have been severed, and Turkey has already replaced Russia as Georgia's main trading partner. Georgia also wants to reestablish its historically important position in overland trade with the Central Asian countries, particularly by serving as an outlet for the export of oil from the Caspian Basin.

The Caspian Sea, which is actually the world's largest freshwater lake in area, is an important fishery, especially for its sturgeon, which supply over 70% of the world's caviar (sturgeon eggs). When only the USSR and Iran bordered the Caspian, this resource was managed sustainably. Now, however, the smaller, recently independent countries bordering the shallow Caspian are exerting tremendous pressure on this resource in an attempt to boost their market shares in competition with Russia and Iran. The result has been a rapid drop in the lake's Beluga sturgeon population (from 1.5 million in 1986 to 300,000 in 2000) and soaring prices of caviar on the world market (a pound goes for $2,000 in Paris and New York). Despite police patrols, sturgeon poaching is big business, netting five to ten times the licensed harvest of these fish. There are growing fears about what an oil spill would do to the caviar industry.

Oil and gas are the most valuable minerals of the Caucasus region. The main oil fields are along and underneath the Caspian Sea around Azerbaijan's capital, Baku (population: metropolitan area, 2,073,900) and north of the mountains at Groznyy in Chechnya. Dry and windswept Baku, the largest city of the Caucasus region, was the leading center of petroleum production in Old Russia and, at the dawn of the 20th century, was the world's leading center of oil production. Un-

der the Soviets, its leadership was decisively surpassed in the 1950s by the Volga-Urals fields and again in the 1970s by the West Siberian fields. However, petroleum development is now reestablishing the Caucasus as a vital world region (see Regional Perspective, p. 210).

Under the Soviets, diverse manufacturing industries developed in Transcaucasia. Most of these factories, which now produce only a fraction of their Soviet-era output, are located in and near the political capitals: Azerbaijan's Baku, Georgia's Tbilisi (population: metropolitan area, 1,449,600), and Armenia's Yerevan (population: metropolitan area, 1,432,000).

7.8 The Central Asian Countries

Across the Caspian Sea from the Caucasus region lie large deserts and dry grasslands, bounded on the south and east by high mountains. Like Transcaucasia, this Central Asian region is inhabited by ethnic groups who outnumber Slavs. The largest populations are four Muslim peoples speaking closely related Turkic languages — the Uzbeks, Kazaks, Kyrgyz, and Turkmen — and a Muslim people of Iranian origins, the Tajiks (Fig. 7.18). Each has its own independent nation (see Table 6.1).

Division and Conquest

The peoples of this region are heirs of ancient oasis civilizations and of the diverse cultures of nomadic peoples. Small-scale social and political units such as clans and tribes were traditionally associated with particular oases. Conquerors have repeatedly sought to possess this area because of its strategic location on the famed Silk Road, an ancient route across Asia from China to the Mediterranean. Among them were Alexander the Great, the Arabs who brought the Muslim faith in the eighth century, the Mongols, and the Turks. By the time of the Russian conquests in the 19th century, Turkestan (as this

Figure 7.18 The elders of this community in Tajikistan are about to partake in the traditional meal celebrating Novruz, the Muslim New Year.

REGONAL PERSPECTIVE ::: Oil in the Caspian Basin

Estimated oil reserves beneath and adjacent to the Caspian Sea total as much as 200 billion barrels, or about one-tenth of the world's total (and there are an estimated 600 billion cubic meters of natural gas in the region) Reserves in Azerbaijan, Kazakstan, and Turkmenistan represent the world's third most important oil region, behind the Persian/Arabian Gulf and Siberia. The potential wealth of this resource will be realized only through a difficult and delicate resolution of several geographic and political problems (Fig. 7.C).

One of the first to be dealt with was the question of which countries owned the fossil fuels under Caspian waters. With relatively small reserves of oil lying adjacent to their Caspian shorelines, Iran and for a while Russia argued that

the Caspian should be regarded as a lake, meaning its lake-bed resources would be treated as common property, with shares divided equally among the five surrounding states. Eventually, Russia conceded that the Caspian should instead be regarded as a sea consisting of five separate sovereign territorial waters, with each country having exclusive access to the reserves in its domain. Iran unhappily accepted this outcome, which gives Azerbaijan the largest share of Caspian oil. But in 2001, Iran and Azerbaijan skirmished over rights to an oil-field below the Caspian. Joined by Turkmenistan, Iran, and Russia again began insisting that the Caspian should be treated as a lake rather than a sea.

Azerbaijan's oil situation epitomizes the economic prospects and

obstacles facing the southern countries of the region of Russia and the Near Abroad. The Azeri resource is large; there are huge reserves of oil and gas under the bed of the Caspian Sea. Planners have studied Azerbaijan's oil with a difficult question in mind: What is the best way to get this oil to market? They are considering a variety of export pipeline routes, each of which has a different set of geopolitical, technical, and ecological problems. Since the Caspian is a lake, Azerbaijan is landlocked, and Azeri oil must cross the territories of other nations to reach market. These territories, however, are embroiled in wars and political unrest left in the wake of the collapse of the USSR. Russia wants to retain strong influence over Azerbaijan in part by insisting that

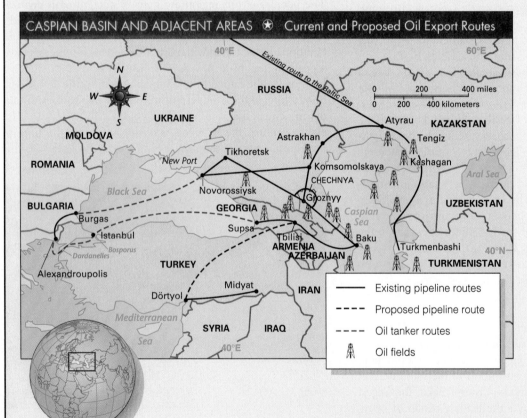

Figure 7.C Numerous physical and political obstacles stand in the way of oil exports from the Caspian region.

Azeri oil reach export terminals on the Black Sea by passing through Russian territory, as it now does, through a pipeline to the Russian port of Novorossiysk.

Turkey opposes this strongly on the grounds that oil-laden ships navigating the narrow Bosporus and Dardanelles Straits imperil Turkey's environment (and Turkey would not receive transit fees for those shipments). Environmental concerns have not, however, kept energy-hungry Turkey from planning to import new supplies of natural gas from Russia via the Blue Stream pipeline (the world's deepest) being built across the Black Sea floor.

Iran has offered a swap with Azerbaijan: Iran would import Azeri oil for Iran's domestic needs and would export equivalent amounts of its own oil from established ports on the Persian Gulf. Iran also argues that its territory offers the shortest and safest pipeline route for fossil fuel exports from Turkmenistan, Kazakstan, and Azerbaijan. Iran is also in an excellent geographic position to facilitate transportation of other goods to and from the Central Asian nations, using its Gulf of Oman and Persian/Arabian Gulf ports. To stimulate this trade, Turkmenistan and Iran have begun linking their rail systems.

Azerbaijan's leaders, however, want more ties with the West (which is generally hostile to Iran) and independence from Moscow. The United States and its Western allies anticipate that by about 2005 Azerbaijan and Kazakstan will become major counterweights to the volatile oil-producing states of the Middle East, and they want to be able to obtain Caspian Sea oil without relying on Russia's or Iran's goodwill. Therefore, in 1995, Azerbaijan signed an agreement with a U.S.-led Western oil consortium to export oil through a combination of new and old pipelines to the Georgian port of Supsa on the Black Sea. To avoid a direct confrontation with Russia over this sensitive issue, some of the initial production passed through the existing Russian pipeline to Novorossiysk.

This compromise applied only to the first stage of production, through 1997. A permanent route had yet to be established, with both Turkey and Russia vying for the pipeline to pass through their territories. Russia insisted that the environmental hazard of navigating the Turkish straits could be averted by offloading oil at the Bulgarian Black Sea port of Burgas, shipping it overland through a new pipeline to the Greek Aegean Sea port of Alexandroupolis, and reloading it there. The alternative, a 1,080-mile (1,725-km) Turkish route, would be far from the straits, running across Azerbaijan from Baku, through Georgia (skirting hostile Armenia), and then across Turkey to the Mediterranean port of Dörtyol, near Ceyhan. The Western powers prefer this route, mainly as a political means to block either Russian or Iranian reassertion of power in the region. Proponents of the Turkish route do, however, have security concerns: The pipeline would have to take a detour to skirt the volatile Kurdish-dominated region of southeastern Turkey. And the project is costly: If chosen, the Turkish line would take about 3 years to build at an estimated cost of $3 billion.

Other Caspian Basin countries face similar questions about how best to export their resources. Turkmenistan, on the lake's southeast shore, has large oil deposits and the fourth largest natural gas reserves in the world. At present, these are exported mainly in pipelines passing around the Caspian Sea through Russia, and Turkmenistan is seeking an alternative through Iran to Turkey (it already exports some natural gas to Iran via pipeline), across Afghanistan to Pakistan (a hazardous route, given conditions in Afghanistan), or under the Caspian to link up with Azerbaijani export routes. The country's leadership is in favor of that third alternative, the export route promoted by the United States, with oil and gas traveling westward through pipelines on the bed of the Caspian Sea to Baku and then through Georgian and Turkish pipelines to the Mediterranean Sea.

Of the five Central Asian countries, Kazakstan has the largest fossil fuel endowment. With recent discoveries of huge reserves in the Kashagan oil field of the Caspian Sea, some have predicted that Kazakstan will eclipse Saudi Arabia as the world's leading oil producer by 2015. In the late 1990s, Kazakstan began exporting its oil from fields around Tengiz to join the Baku-to-Novorossiysk pipeline at Komsomolskaya. Unwilling for the time being to find an alternative to a Russian route, the Kazaks worked with Omani, Russian, and American firms to build a pipeline bypass from Komsomolskaya to Novorossiysk. Kazakstan is also studying the possibility of building a 2,000-mile (c. 3,000-km) pipeline to export oil eastward into China, but the construction costs are daunting.

region was then known) was contested between feuding, tradition-bound Muslim khanates with political ties to China, Persia, Ottoman Turkey, and British India. Cultural and political fragmentation continued up to the Russian Revolution and into the period of civil war that followed.

Nationalist identities existed historically in Central Asia, but nation-states did not. During World War I, Central Asian peoples revolted against Russia's tsarist government, which had begun to draft the region's Muslims for menial labor at the front. The Bolshevik government gained control in the early 1920s and organized the area into nationality-based units, effectively setting the boundaries for what are now independent countries.

During the 20th century, Russians and Ukrainians poured into Central Asia, mainly into the cities. Meant to Russify this non-Russian area, these were political dissidents banished by the Communists, administrative and managerial personnel, engineers, technicians, factory workers, and in the north of Kazakstan, farmers in the "new lands" wheat region. Most of the Slavic newcomers lived apart from the local Muslims, as most of those who remain today still do.

Kazakstan provides a good example of the dilemma facing ethnic Russians in Central Asia. At the time it became independent in 1991, Kazakstan's population of 17 million was roughly 40% Kazak and 39% Russian, with the rest a mix of nationalities. Ethnic Kazaks, who generally believe they were treated as second-class citizens until independence, have been working to diminish the Russian cultural footprint. They are increasingly predominant in government and business. Kazak is now the official language, and Russian is the language of "interethnic communication." Even ethnic Kazaks have a lot of catching up to do; half of them cannot speak Kazak. With rising ethnic tensions throughout the 1990s, about 1 million Russians and a half million ethnic Germans emigrated from Kazakstan. Meanwhile, the government is calling for an estimated 4.5 million ethnic Kazaks who live abroad to come home.

Environment and Agriculture

The five Central Asian states are mainly flat, with plains and low uplands, except for Tajikistan and Kyrgyzstan, which are spectacularly mountainous and contain the highest summits in the region of Russia and the Near Abroad in the Pamir and Tien Shan Ranges (see Fig. 6.8). Central Asia is almost entirely a region of interior drainage. Only the waters of the Irtysh, a tributary of the Ob, reach the ocean; all the other streams either drain to enclosed lakes and seas or gradually lose water and disappear in the Central Asian deserts.

Most people live in irrigated valleys at the base of the southern mountains (Fig. 7.19). There, the wind-blown loess and other soils are fertile, the growing season is long, and rivers flowing from mountains provide irrigation water. The principal rivers in the heart of the region are the Amu Darya (Oxus) and Syr Darya, both of which empty into the enclosed Aral Sea.

Only northern Kazakstan, which extends into the black-earth belt, receives enough moisture for dry farming (unirrigated agriculture). Irrigation is needed everywhere else, providing water to mulberry trees (grown to feed silkworms), rice, sugar beets, vegetables, vineyards, and fruit orchards. Commercial orchard farming is focussed on the area around Almaty ("City of Apples"; population: metropolitan area, 1,222,400), the former capital of Kazakstan. But most of the irrigated land is in Uzbekistan. The most important farming area is the fertile Fergana Valley on the upper Syr Darya, nearly enclosed by high mountains, and the oases around Uzbekistan's cities of Tashkent (the nation's capital and Central Asia's largest city; population: metropolitan area, 3,331,700), Samarkand, and Bukhara. These old cities became powerful political and economic centers on the Silk Road. Samarkand was the capital of Tamerlane's empire, which encompassed much of Asia in the 14th century. Samarkand and nearby Bukhara have superb examples of Islamic architecture that the government of Uzbekistan has recently restored at great cost (Fig. 7.20).

Cotton began to be cultivated here because during the Civil War in the United States, U.S. exports abroad virtually ceased. It remains the region's major crop. This area grows over nine-tenths of the cotton produced in the region of Russia and the Near Abroad. Production supplies textile mills in the Moscow region, Tashkent, and other cities in far-flung locations, including many outside the region. However, a significant environmental price was paid to develop agriculture here. Driven by the need for foreign exchange from export sales, Communist planners expanded irrigated cotton so much that the rivers and the Aral Sea became highly polluted with agricultural chemicals draining from the fields. Diversion of water from the rivers caused the Aral Sea to shrink rapidly in volume and area, virtually destroying a vast natural ecosystem and an important regional fishery (Fig. 7.19). Since the 1960s, the Aral Sea has lost 50% of its surface area, 75% of its volume, and has all but divided into two small lakes. Kazakstan is trying to muster the financial resources to build a dam to contain the smaller lake, known as the Little Aral Sea, completely within its borders, in an effort to reduce its salinity and restore its fishery.

Another environmental disaster threatens on the island of Vozrozhdeniye, on Uzbekistan's side of the Aral Sea. In 1988, as the Soviet Union hastened to build new bridges with the West, Soviet biological warfare specialists buried hundreds of tons of living anthrax bacteria, encased in steel canisters, on the island. With the Aral Sea shrinking, the island may soon become a peninsula, and there are fears these biohazards will be exposed and blown into populated areas or carried away by would-be terrorists. In 2001, following a number of incidents of anthrax-tainted letters reaching U.S. Senate offices and other facilities, the American government pledged the funds needed to clean up "Anthrax Island."

Many of this region's people previously were pastoral nomads who grazed their herds on the natural forage of the steppes and in mountain pastures in the Altai and Tien Shan Ranges. Over the centuries, there was a slow drift away from nomadism. The Soviet government accelerated this process by forcibly collectivizing the remaining nomads and settling them in permanent villages. Raising livestock became a mainly sedentary form of ranching in the process. Sheep and cattle are the principal livestock animals, followed by goats, horses, donkeys, and camels. Horses remain central to the national personality, if not the livelihoods, of the indigenous people of Kyrgyzstan. In the post-Soviet period, they have resurrected the oral epics that tell the story of Manas, a great warrior horseman who, 1,000 years ago, fought off the enemies of the Kyrgyz and unified their disparate tribes for the first time. The Kyrgyz are passionate horsemen, and even townspeople keep horses in the countryside for weekend races and other games.

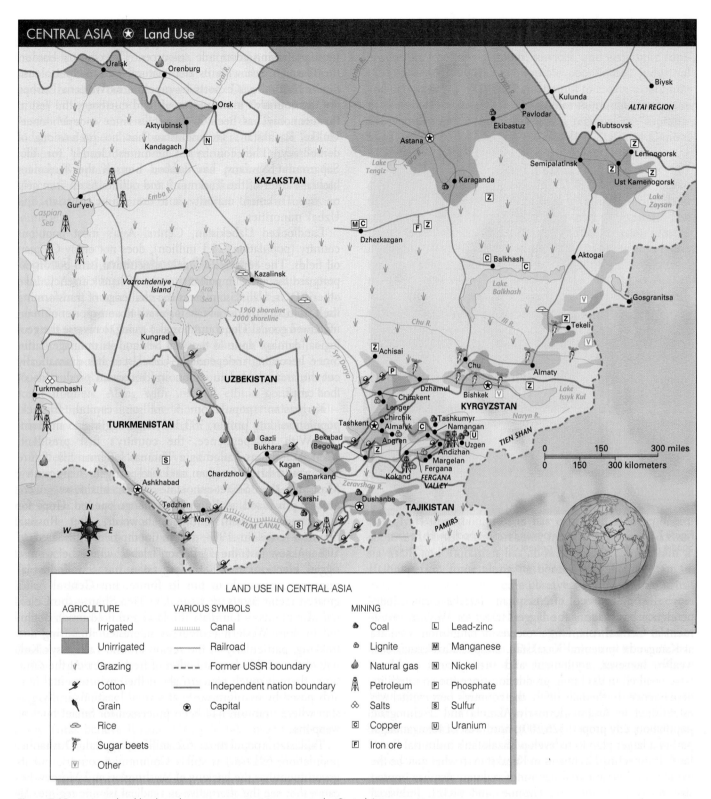

CENTRAL ASIA ✯ Land Use

LAND USE IN CENTRAL ASIA

AGRICULTURE
- Irrigated
- Unirrigated
- Grazing
- Cotton
- Grain
- Rice
- Sugar beets
- Other

VARIOUS SYMBOLS
- Canal
- Railroad
- Former USSR boundary
- Independent nation boundary
- Capital

MINING
- Coal
- Lignite
- Natural gas
- Petroleum
- Salts
- Copper
- Iron ore
- Lead-zinc
- Manganese
- Nickel
- Phosphate
- Sulfur
- Uranium

Figure 7.19   Agricultural lands and some important mining areas in the Central Asian countries.

Economic and Political Prospects

In a familiar pattern, the economies of the five Central Asian countries have worsened after an initial period of exuberance and growth following independence. The long-term prospects for some of the countries are bright, however.

Central Asia has a generous endowment of minerals, particularly natural gas, coal, oil, iron ore, nonferrous metals, and ferroalloys. Kazakstan (population: 14.8 million), the region's richest country, has the greatest natural resource wealth. Political stability under the nominally democratic (but quite

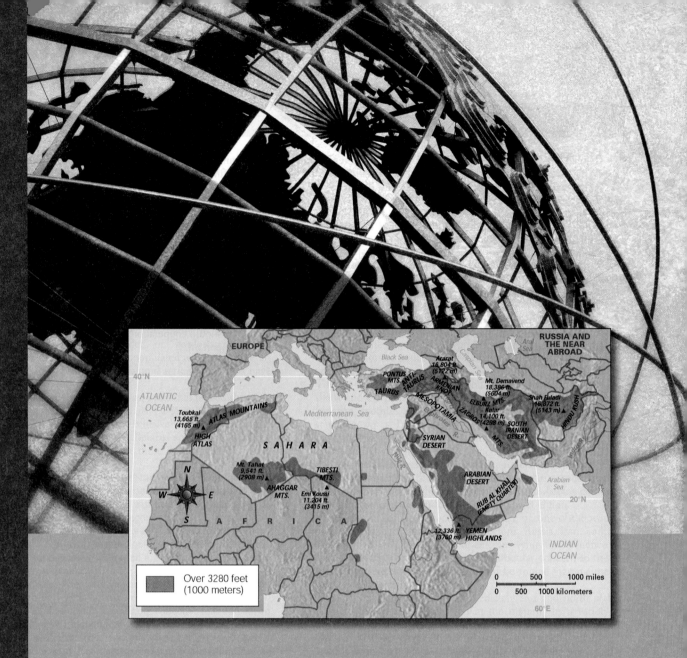

The Middle East and North Africa

The Middle East and North Africa are pivotal global crossroads linking Asia, Europe, Africa, the Mediterranean Sea, and the Indian Ocean. More than two-thirds of the world's proven oil reserves are in the region. Global and regional powers have frequently contested over and controlled strategically and economically important areas of the Middle East and North Africa. The region's inhabitants include Jews, Arabs, Turks, Persians, Berbers, and other ethnic groups who practice a wide variety of ancient and modern livelihoods.

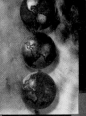

Many of the world's earliest and most accomplished civilizations emerged in the Middle East, contributing much to humankind's scientific and artistic legacies.

Villagers, pastoral nomads, and city-dwellers make up a distinctive "ecological trilogy" within which they have generally symbiotic interactions.

The three great monotheistic faiths of Judaism, Christianity, and Islam were born in the Middle East. Muslims far outnumber Jews and Christians in the region. In Jerusalem, the holy places for these religions are close together, and some are even the same. Animosity between Christians and Jews has a long history, but between Arabs and Jews is fairly recent.

Generally the region is arid, but great river systems, freshwater aquifers, and copious rains in some mountainous regions sustain large human populations.

CHAPTER 8

A Geographic Profile of the Middle East and North Africa

This extraordinary view from the U.S. Space Shuttle *Atlantis* shows some of the world's most storied places: in the lower left, the green ribbon of the Nile Valley; in the center, the Sinai Peninsula, pointing southward down the Red Sea; the Great Rift Valley, traced by the Red Sea and the Gulf of Aqaba leading northward to the world's lowest place, the Dead Sea shore; and between the Dead Sea and the Mediterranean, the verdant land of Israel.

Chapter Outline

"The Middle East" is a fitting designation for the location and the cultures of this vital world region because they are literally in the middle. This is a physical crossroads, where the continents of Africa, Asia, and Europe meet and the waters of the Mediterranean Sea and the Indian Ocean mingle. Its peoples—Arab, Jew, Persian, Turk, Kurd, Berber, and others—express in their cultures and ethnicities the coming together of these diverse influences. Occupying as they do this strategic location, the nations of the Middle East and North Africa have through five millennia been unwilling hosts to occupiers and empires originating far beyond their borders. They have also bestowed upon humankind a rich legacy that includes the ancient civilizations of Egypt and Mesopotamia and the world's three great monotheistic faiths of Judaism, Christianity, and Islam.

People outside the region tend to forget about such contributions because they associate the Middle East and North Africa with war and terrorism. These negative connotations are often accompanied by muddled understanding; this is arguably the most inaccurately perceived region in the world. Does sand cover most of the area? Are there camels everywhere? Does everyone speak Arabic? Are Turks and Persians Arabs? Are all Arabs Muslims? The answer to each of these questions is "no," yet chances are that if television and other sources of popular culture were your only guide, you would answer some of them "yes." This and the following chapter attempt to make the misunderstood, complex, sometimes bloody, but often hopeful events and circumstances in the Middle East and North Africa more intelligible by presenting the geographic context within which they occur.

8.1   Area and Population

What and where are the Middle East? The term itself is Eurocentric, created by Europeans and their American allies who placed themselves in the figurative center of the world. They began to use the term prior to the outbreak of World War I, when the "Near East" referred to the territories of the Ottoman Empire in the Eastern Mediterranean region, the "East" to India, and the "Far East" to China, Japan, and the western Pacific Rim. With "Middle East," they designated as a separate region the countries around the Persian Gulf (known to Arabs as the Arabian Gulf, and in this text as the Persian/Arabian

Gulf and simply The Gulf). Gradually, the perceived boundaries of the region grew.

Today, depending on what source you consult, you might find that the Middle East includes only the countries clustered around the Arabian Peninsula or that it spans a vast 6,000 miles (9,700 km) west to east from Morocco in northwest Africa to Afghanistan in central Asia and 3,000 miles (4,800 km) north to south from Turkey, on Europe's southeastern corner, to Sudan, which adjoins East Africa (Figs. 8.1 and 8.2). This is the region covered in these chapters, where it is referred to as "The Middle East and North Africa" because the North African peoples of Morocco, Algeria, and Tunisia generally do not perceive themselves as Middle Easterners; they are, rather, from what they call "the Maghrib." Primarily because of the issues that involve Turkey, Cyprus is also included in the region in this text. Thus defined, the region incorporates 22 countries, the Palestinian territories of the West Bank and Gaza Strip, and the disputed Western Sahara (Table 8.1), occupying 5.79 million square miles (14.99 million sq km) and inhabited by about 445 million people as of mid-2001.

The Middle East and North Africa are sometimes mistakenly referred to as the "Arab World"; in fact, the region has huge populations of non-Arabs. It is true that most of the region's inhabitants are Arabs. An Arab is anyone whose first language is Arabic, and that language is spoken by about 238 million or 53% of the region's people (see the language map, Fig. 8.3). Originally, the Arabs were inhabitants of the Arabian Peninsula, but conquests after their majority conversion to Islam took them, their language, and their Islamic culture as far west as Morocco and Spain. The region is also the homeland of the Jews. It is important to recognize that although political circumstances made them enemies in the 20th century, Arabs and Jews lived in peace for centuries, and they share many cultural traits. Both recognize Abraham as their patriarch. Arabic is a Semitic language, in the same Afro-Asiatic language family as Hebrew, which is spoken by most of the 5.1 million Jewish inhabitants of Israel.

There are other very large populations of non-Semitic ethnic groups and languages in the region. The greatest are the 58 million Turks of Turkey, who speak Turkish, an Altaic language, and the 30 million Persians of Iran, who speak Farsi or Persian in the Indo-European language family. Both Persian and Arabic are written in Arabic script and so appear related,

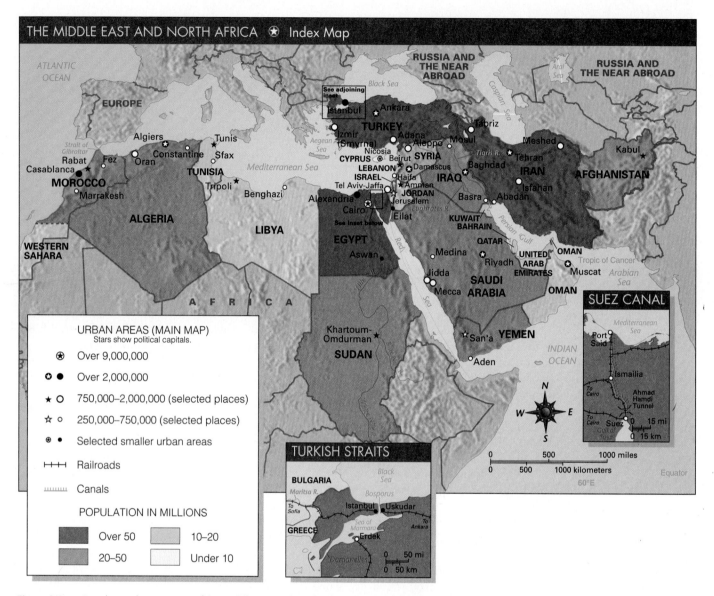

THE MIDDLE EAST AND NORTH AFRICA ★ Index Map

URBAN AREAS (MAIN MAP)
Stars show political capitals.

⊛   Over 9,000,000

✪ ●   Over 2,000,000

★ ○   750,000–2,000,000 (selected places)

☆ ○   250,000–750,000 (selected places)

⊛ •   Selected smaller urban areas

┝┿┥   Railroads

╓╙╙   Canals

POPULATION IN MILLIONS

Over 50	10–20
20–50	Under 10

Figure 8.1   Introductory location map of the Middle East and North Africa showing political units.

F2.1
24

but they are not. Turkish also was written in Arabic script until early in the 20th century, but since then, it has been written in a Latin script. About 20 million Kurds—a people living in Turkey, Iraq, Iran, and Syria—speak Kurdish (in the Indo-European language family). Many people in North Africa speak Berber (in the Afro-Asiatic language family).

These diverse peoples are not distributed evenly across the region but are concentrated in major clusters. Three countries contain the lion's share of the region's population: Turkey, Iran, and Egypt, each with more than 65 million people. One look at a map of precipitation explains why people are clustered this way. Where water is abundant in this generally arid region, so are people. Egypt has the Nile River, and parts of Iran and Turkey have bountiful rain and snow. Conversely, where rain seldom falls, as in the Sahara of North Africa and in the Arabian Peninsula, people are few.

The Middle East and North Africa as a whole have a high rate of population growth, a general indication that this is a developing rather than industrialized region. The average annual rate of population change for the 22 countries, the Palestinian territories, and Western Sahara was 2% in 2001. Excluding Cyprus, the lowest rate of population growth (1.2%) is in Iran. The highest is a staggering 3.7% in the Palestinian territories, followed closely by 3.5% in Oman. These are the highest population growth rates in the world. In the Palestinian territories, such rapid growth may be ascribed in large part to the Palestinians' poverty and perhaps to the wishes of many Palestinians to have more children to counterbalance the demographic weight of their perceived Israeli foe.

Between these extremes are countries with modest rates of population growth of 1.5 to 2% per year; these include Israel, Morocco, Algeria, Bahrain, Kuwait, Lebanon, Qatar, and

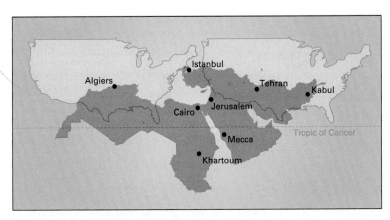

Figure 8.2   The Middle East and North Africa compared in latitude and area with the conterminous United States.

TABLE 8.1 Middle East: Basic Data

Political Unit	Area (thousand/sq mi)	Area (thousand/sq km)	Estimated Population (millions)	Estimated Annual Rate of Increase (%)	Estimated Population Density (sq mi)	Estimated Population Density (sq km)	Human Development Index	Urban Population (%)	Arable Land (% of Total Area)	Per Capita GDP PPP (SU.S.)
Arab States										
Egypt	386.7	1001.4	69.8	2.1	181	70	0.635	43	2	3600
Sudan	967.5	2505.8	31.8	2.4	33	13	0.439	27	5	1000
Tunisia	63.2	163.6	9.7	1.3	154	59	0.714	62	19	6500
Algeria	919.6	2381.7	31	1.9	34	13	0.693	49	3	5500
Morocco	172.4	446.5	29.2	2	169	65	0.596	55	21	3500
Lebanon	4.0	10.4	4.3	1.7	1061	410	0.758	88	18	5000
Syria	71.5	185.2	17.1	2.6	231	89	0.7	50	28	3100
Jordan	34.4	89.2	5.2	2.2	150	58	0.714	79	4	3500
Iraq	169.2	438.3	23.6	2.7	139	54	NA	68	12	2500
Saudi Arabia	830.0	2149.7	21.1	2.9	25	10	0.754	83	2	10500
Kuwait	6.9	17.8	2.3	1.8	297	115	0.818	100	0	15000
Bahrain	0.3	0.7	0.7	1.9	2688	1038	0.824	88	1	15900
Qatar	4.2	11	0.6	2.7	139	54	0.801	91	1	20300
United Arab Emirates	32.3	83.6	3.3	1.4	103	40	0.809	84	0	22800
Yemen	203.8	528	18	3.3	88	34	0.468	26	3	820
Oman	82.0	212.5	2.4	3.5	29	11	0.747	72	0	7700
Totals	3948.0	10225.4	270.1	2.3	68	26	0.628	51	5.0	4289
Other Units										
Libya	679.4	1759.5	5.2	2.4	8	3	0.77	86	1	8900
Turkey	299.2	774.8	66.3	1.5	221	85	0.735	66	32	6800
Cyprus	3.6	9.2	0.9	0.6	247	95	0.877	66	12	11700
Iran	630.6	1633.2	66.1	1.2	108	42	0.714	64	10	6300
Afghanistan	251.8	652.1	26.8	2.4	106	41	NA	22	12	800
Israel	8.1	21.1	6.4	1.6	791	305	0.893	91	17	18900
Western Sahara	97.3	252.1	0.3	2.9	3	1	NA	95	0	NA
Palestinian Territory	2.4	6.3	3.3	3.7	1365	527	NA	NA	27	1300
Totals	1972.4	5108.3	175.3	1.6	89	34	0.735	60	10.004	6108
Summary Totals	5920.4	15333.8	445.4	2.0	75.2	29.1	0.667	54.7	6.7	5008

Sources: *World Population Data Sheet, Population Data Sheet, 2001; U.N. Human Development Report, United Nations, 2001; World Factbook, CIA, 2001.*

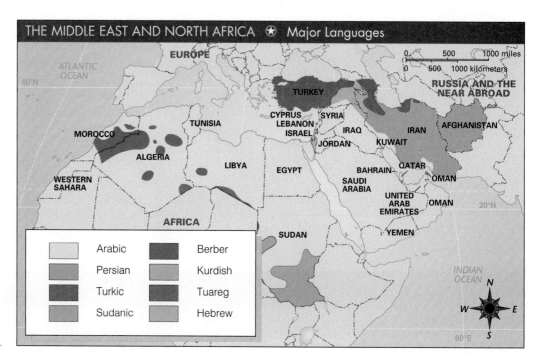

Figure 8.3 Language map of the Middle East and North Africa.

Turkey. Their governments have regarded most of these countries as too populous for their resource and industrial base and encouraged family planning; they have been successful in lowering birth rates. On the other hand, oil-rich Oman is a good example of how rapid population growth is not always a sign of poverty, particularly in this region. The United Arab Emirates, too, has a very high growth rate of 3.3% per year. In these two cases, oil-rich nations have encouraged their citizens to create more citizens so that in the future they do not need to import foreign laborers and technicians and will therefore become more self-sufficient in their development.

Developing countries generally have economies largely dependent on subsistence agriculture and have low percentages of urban inhabitants. Perhaps surprisingly, however, the Middle East and North Africa have more urbanites than country folk. The average urban population among the 24 countries and territories is 55%. The most prosperous countries are also the most urban. Essentially a city-state, Kuwait is 100% urban. The other oil-wealthy Gulf states also have urban populations over 70%. Consistent with its profile as a Western-style industrialized country without oil resources, Israel is 90% urban. At the other end of the spectrum, desperately poor Afghanistan, Yemen, and Sudan have urban populations of less than 28%.

8.2 Physical Geography and Human Adaptations

The margins of the Middle East and North Africa are mainly oceans, seas, high mountains, and deserts. To the west lies the Atlantic Ocean; to the south, the Sahara and the highlands of East Africa; to the north, the Mediterranean, Black, and Caspian Seas, together with mountains and deserts lining the

southern land frontiers of Russia and the Near Abroad; and to the east, the Hindu Kush mountains on the Afghanistan-Pakistan frontier and the Baluchistan Desert straddling Iran and Pakistan. The land is composed mainly of arid plains and plateaus, together with considerable areas of rugged mountains and isolated "seas" of sand. Despite the environmental challenges, this region has given rise to some of the world's oldest and most influential ways of living.

Hot, Dry, and Barren but in Places Damp and Forested

Aridity dominates the Middle East and North Africa (see the deserts in Fig. 8.4). At least three-fourths of the region has an average yearly precipitation of less than 10 inches (25 cm), an amount too small for most types of unirrigated agriculture (dry farming). Sometimes, however, localized cloudbursts release moisture that allows plants, animals, and small populations of people — the Bedouins, Tuaregs, and other pastoral nomads — to live in the desert. Even the vast Sahara, the world's largest desert, supports a surprising diversity and abundance of life. Plants, animals, and even people have developed strategies of **drought avoidance** and **drought endurance** to live in this harsh biome. Plants either avoid drought by completing their life cycle quickly wherever rain has fallen or endure drought by using their extensive root systems, small leaves, and other adaptations to take advantage of subsurface or atmospheric moisture. Animals endure drought by calling on extraordinary physical abilities (e.g., a camel can sweat away a third of its body weight and still live) or by avoiding the worst conditions by being active only at night or by migrating from one moist place to another. (Migration to avoid drought is also the strategy pastoral nomads use.) Popu-

F2.6
28–29
F2.7
30–31

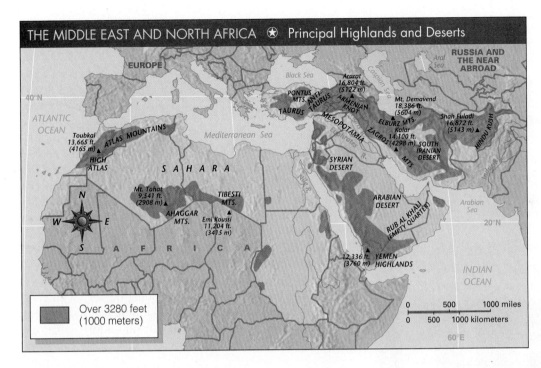

THE MIDDLE EAST AND NORTH AFRICA ✹ Principal Highlands and Deserts

Figure 8.4 Many of the region's deserts are mountainous rather than sandy deserts, but in some places—especially Turkey—high mountains help to create abundant rains, and deserts are absent.

lations of people, plants, and animals are all but nonexistent in the region's vast **sand seas,** including the Great Sand Sea of western Egypt (Fig. 8.5) and the Empty Quarter of the Arabian Peninsula.

The region's climates exhibit the comparatively large diurnal (daily) and seasonal ranges of temperature that are characteristic of dry lands. Summers in the lowlands are very hot almost everywhere. The hottest shade temperature ever recorded on Earth, 136°F (58°C), occurred in Libya in September 1922. Many places regularly experience daily maximum temperatures over 100°F (38°C) for weeks at a time. Human settlements located near the sand seas often experience the unpleasant combination of high temperatures and hot, sand-laden winds, creating the sandstorms known locally by such names as *simuum* ("poison") and *sirocco.* Only in mountainous sections and in some places near the sea do higher elevations or sea breezes temper the intense midsummer heat. The population of Alexandria explodes in summer as Egyptians flee from Cairo and other hot inland locations. In Saudi Arabia, the government relocates from Riyadh to the highland summer capital of Taif to escape the lowland furnace.

Lower winter temperatures bring relief from the summer heat, and the more favored places receive enough precipitation for dry farming of winter wheat, barley, and other cool-season crops. In general, winters are cool to mild. But very cold winters and snowfalls occur in the high interior basins and plateaus of Iran, Afghanistan, and Turkey. These locales generally have a steppe climate. Only in the southernmost reaches of the region, notably Sudan, do temperatures remain consistently high throughout the year. A savanna climate and biome prevail there.

Most areas bordering the Mediterranean Sea have 15 to 40 inches (38 to 102 cm) of precipitation a year, falling al-

most exclusively in winter, while the summer is dry and warm—a typical Mediterranean climate pattern. Throughout history, people without access to perennial streams have stored this moisture to make it available later for growing those crops that require the higher temperatures of the summer months. The Nabateans, for example, who were contemporaries of the Romans in what is now Jordan, had a sophisticated network of limestone cisterns and irrigation channels (Fig. 8.6). Rainfall sufficient for dry farming during the summer is concentrated in areas along the southern and northern margins of the region. The Black Sea slope of Turkey's Pontic Mountains is lush and moist in the summer, and tea grows well there (Fig. 8.7). In the southwestern Arabian Peninsula, a monsoonal climate brings summer rainfall

Figure 8.5 Sand "seas" cover large areas in Saudi Arabia, Iran, and parts of the Sahara. This is the edge of the Great Sand Sea near Siwa Oasis in Egypt's Western Desert.

The hinge of this crustal movement is the Bekaa Valley of Lebanon, and the widening fault line that follows the Jordan River valley southward to the Dead Sea. This valley is the deepest depression on Earth's land surface, lying about 600 feet (183 m) below sea level at Lake Kinneret (the Sea of Galilee or Lake Tiberius) and nearly 1,300 feet (400 m) below sea level at the shore of the Dead Sea, the lowest point on Earth.

The rift then continues southward to the very deep Gulf of Aqaba and Red Sea before turning inland into Africa at Djibouti and Ethiopia. In tectonic processes, one consequence of rifting in one place is the subsequent collision of Earth's crustal plates elsewhere. There are several such collision zones in the Middle East and North Africa, particularly in Turkey and Iran. These are tectonically active zones — meaning earthquakes occur in them — and rarely does a year go by without a devastating quake rocking Turkey or Iran.

A larger area of mountains, including the region's highest peaks, stretches across Turkey, Iran, and Afghanistan. On the eastern border with Pakistan, the Hindu Kush Range has

Figure 8.6 The Treasury, a temple carved in red sandstone, probably in the first century A.D., by the Nabateans at their capital of Petra in southern Jordan. Note man at lower right for scale.

and autumn harvests to Yemen and Oman, probably accounting for the Roman name for the area: *Arabia Felix*, or "Happy Arabia" (Fig. 8.8).

Mountainous areas in the region, like the river valleys and the margins of the Mediterranean, play a vital role in supporting human populations and national economies. Due to orographic or elevation-induced precipitation, the mountains tend to receive much more rainfall than surrounding lowland areas.

There are three principal mountainous regions of the Middle East and North Africa (see Fig. 8.4). In northwestern Africa between the Mediterranean Sea and the Sahara, the Atlas Mountains of Morocco, Algeria, and Tunisia reach over 13,000 feet (3,965 m) in elevation. Mountains also rise on both sides of the Red Sea, with peaks up to 12,336 feet (3,760 m) in Yemen. These are the result of tectonic processes that are pulling the African and Arabian plates apart, creating the northern part of the Great Rift Valley.

Figure 8.7 Rainfall is heavy on the Black Sea side of Turkey's Pontic Mountains. Note that roofs are pitched to shed precipitation; in the region's drier areas, most roofs are flat. The crop shown here is corn.

374
443

Figure 8.8   The mountainous landscape near Sana'a, Yemen's capital. The only crop under cultivation in this scene is the stimulant known as *qat* (*Cathya edulis*).

peaks over 25,000 feet (7,600 m). The loftiest mountain ranges in Turkey are the Taurus and Anti-Taurus, and in Iran, the Elburz and Zagros Mountains. These chains radiate outward from the rugged Armenian Knot in the tangled border country where Turkey, Iran, and the countries of the Caucasus meet. Mt. Ararat is an extinct, glacier-covered volcano of 16,804 feet (5,122 m) towering over the border region between Turkey and Armenia. Many biblical scholars and explorers think the ark of Noah lies high on the mountain (and some go in search of it), for in the book of Genesis, this boat was said to have come to rest "in the mountains of Ararat."

Extensive forests existed in early historical times in the Middle East and North Africa, particularly in these mountainous areas, but overcutting and overgrazing have almost eliminated them. Since the dawn of civilization in this area, around 3000 B.C., people have cut timber for construction and fuel faster than nature could replace it. Egyptian King Tutankhamen's funerary shrines and Solomon's Temple in Jerusalem were built of cedar of Lebanon. So prized has this wood been through the millennia that only a few isolated groves of cedar remain in Lebanon. Described in ancient times as "an oasis of green with running creeks" and "a vast forest whose branches hide the sky," Lebanon is now largely barren (Fig. 8.9). Lumber is still harvested commercially in a few mountain areas such as the Atlas region of Morocco and Algeria, the Taurus Mountains of Turkey, and the Elburz Mountains of Iran, but supply falls far short of demand.

The Middle Eastern Ecological Trilogy: Villager, Pastoral Nomad, and Urbanite

In the 1960s, American geographer Paul English developed a useful model for understanding relationships between the three ancient ways of life that still prevail in the Middle East and North Africa today: villager, pastoral nomad, and urbanite.

Each of these modes of living is rooted in a particular physical environment. Villagers are the subsistence farmers of rural areas where dry farming, or irrigation, is possible; pastoral nomads are the desert peoples who migrate through arid lands with their livestock, following patterns of rainfall and vegetation; and urbanites are the inhabitants of the large towns and cities, generally located near bountiful water sources, but sometimes placed for particular trade, religious, or other reasons. Describing these ways of life as components of the Middle Eastern ecological trilogy, English explained how each of them has a characteristic, usually mutually beneficial, pattern of interaction with the other two (see Fig. 8.10).

The peasant farmers of Middle Eastern and North African villages (the villagers) represent the cornerstone of the trilogy. They grow the staple food crops such as wheat and barley that feed both the city-dweller and the pastoral nomad of the desert. Neither urbanite nor nomad could live without them. The village also, often unwillingly, provides the city with tax revenue, soldiers, and workers. And before the mid-20th century, villages provided pastoral nomads with plunder as the desert-dwellers raided their settlements and caravan supply lines. Generally, however, the exchange is beneficial. The nomads provide villagers with livestock products, including live animals, meat, milk, cheese, hides, and wool, and with desert herbs and medicines. Educated and progressive urbanites provide technological innovations, manufactured goods, religious and secular education and training, and cultural amenities (today, e.g., films and music).

There is little direct interaction between urbanites and pastoral nomads, although some manufactured goods such as clothing travel from city to desert, and some desert-grown folk medicines pass from desert to city. Historically, the exchange has been violent, as urban-based governments have sought to control the movements and military capabilities of the elusive and sometimes hostile nomads. Pastoral nomads once

Figure 8.9   One of the twelve groves of cedars remaining in Lebanon. People of the Mediterranean Basin harvested this valuable resource for about 5,000 years, nearly depleting it. This symbol of Lebanon now has complete protection.

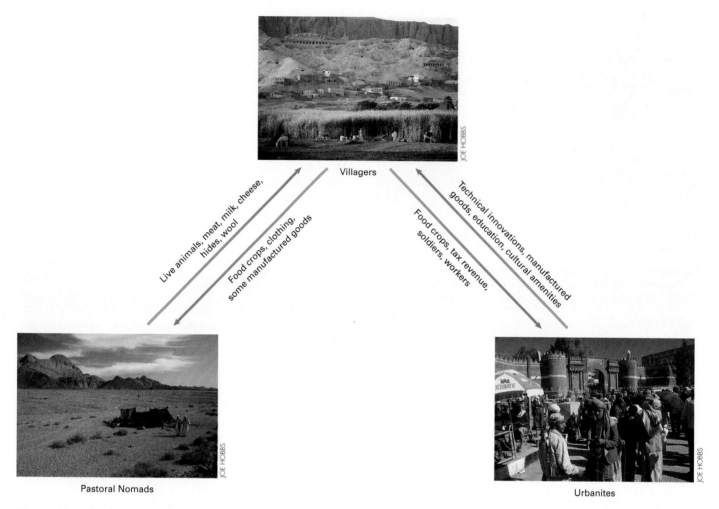

Villagers

Live animals, meat, milk, cheese, hides, wool

Food crops, clothing, some manufactured goods

Technical innovations, manufactured goods, education, cultural amenities

Food crops, tax revenue, soldiers, workers

JOE HOBBS

Pastoral Nomads

JOE HOBBS

Urbanites

JOE HOBBS

Figure 8.10   People, environments, and interactions of the ecological trilogy (here, villagers harvesting sugarcane in Upper Egypt, Bedouin at camp in Egypt's Eastern Desert, and shoppers at the main gate to the historic city of San'a, Yemen). This relationship is generally symbiotic, although historically both urbanites and pastoral nomads "preyed" on the villagers who are the trilogy's cornerstone.

plundered rich caravans plying the major overland trade routes of the Middle East and North Africa. Governments did not tolerate such activities and often cracked down hard on the nomads they were able to catch.

In the 1970s, Paul English wrote an article marking the "passing of the ecological trilogy." He noted that cities were encroaching on villages, villagers were migrating into cities and giving some neighborhoods a rural aspect, and pastoral nomads were settling down — thus, the trilogy no longer existed. In reality, although the makeup and interactions of its parts have changed somewhat, the trilogy model is still valid and useful as an introduction to the major lifeways of the Middle East and North Africa. It is especially significant that a given man or woman in the region strongly identifies himself or herself as either a villager, a pastoral nomad, or an urbanite. This perception of self has an important bearing on how these people of very different backgrounds interact, even when they live in close proximity. Urban officials may work in rural village areas, but they remain at heart and in their perspectives city people and usually live apart from farmers. Extended families of pastoral nomads may settle down and become farmers

but continue to identify themselves by affiliation with the nomadic tribe. Many continue to harvest desert resources on a seasonal basis and retain marriage and other ties with desert-dwelling relatives.

The Village Way of Life

Agricultural villagers historically represented by far the majority populations in the Middle East and North Africa; only within recent decades have urbanites begun to outnumber them. In this generally dry environment, the villages are located near a reliable water source with cultivable land nearby. They are usually made up of closely related family groups, with the land of the village often owned by an absentee landlord. Most often, the villagers live in closely spaced flat-roofed houses made of mud brick or concrete blocks. Production and consumption focus on a staple grain such as wheat, barley, or rice. As land for growing fodder is often in short supply, villagers keep only a small number of sheep and goats and rely in part on nomads for pastoral produce. Residents of a given village usually share common ties of kinship, religion, ritual, and custom, and the changing demands of agricultural seasons regulate their patterns of activity.

Village life has been increasingly exposed to outside influences since the mid-18th century. Contacts with European colonialism brought significant economic changes, including the introduction of cash crops and modern facilities to ship them. Improved and expanded irrigation, financed initially with capital from the West, brought more land under cultivation. Recent agents of change have been the countries' own government doctors, government teachers, and land reform officers. Modern technologies such as sewing machines, motor vehicles, gasoline powered pumps, radio, television, and even the Internet and cell phones have modified old patterns of living. The young and more ambitious have been drawn to urban areas. Improved roads and communications in turn have carried urban influences to villages, prompting villagers to become more integrated into national societies.

The Pastoral Nomadic Way of Life

Pastoral nomadism emerged as an offshoot of the village agricultural way of life not long after plants and animals were first domesticated in the Middle East (about 7000 B.C.). Rainfall and the wild fodder it produces, although scattered, are sufficient resources to support small, mobile groups of people who migrate with their sheep, goats, and camels (and in some locales, cattle) to take advantage of this changing resource base. In mountainous areas, they follow a pattern of **vertical migration** (sometimes referred to as **transhumance,** moving with their flocks from lowland winter to highland summer pastures. In the flatter expanses that comprise most of the region, the nomads have a pattern of **horizontal migration** over much larger areas where rainfall is typically far less reliable than in the mountains. In addition to selling or trading livestock to obtain foods, tea, sugar, clothing, and other essentials from settled communities, pastoral nomads also hunt, gather, work for wages, and where possible, grow crops. Their multifaceted livelihood has been described as a strategy of "risk minimization" based on the exploitation of multiple resources so that some will support them if others fail.

Although renowned in Middle Eastern legends and in popular Western films like *Lawrence of Arabia,* pastoral nomads have been described as "more glamorous than numerous." It is still impossible to obtain adequate census figures on the number living in the deserts of the Middle East and North Africa, though estimates range from 5 to 13 million. Recent decades have witnessed the rapid and progressive settling down, or **sedentarization,** of the nomads — a process attributed to a variety of causes. In some cases, prolonged drought virtually eliminated the resource base on which the nomads depended. Traditionally, they were able to migrate far enough to find new pastures, but modern national boundaries now prohibit such movements. Some have returned with the rains to their desert homelands, but others have chosen to remain as farmers or

wage laborers in villages and towns. On the Arabian Peninsula, the prosperity and technological changes prompted by oil revenues made rapid inroads into the material culture — and then the livelihood preferences — of the desert people; many preferred the comforts of settled life. Some governments, notably those of Israel and pre-Revolutionary Iran were unable to count, tax, conscript, and control a sizable migrant population and compelled nomads to settle.

Pastoral nomads of the Middle East and North Africa identify themselves primarily by their tribe, not by their nationality. The major ethnic groups from which these tribes draw are the Arabic-speaking Bedouins of the Arabian Peninsula and adjacent lands, the Berber and Tuareg of North Africa, the Kababish and Bisharin of Sudan, the Yoruk and Kurds of Turkey, the Qashai and Bakhtiari of Iran, and the Pashtun of Afghanistan. Members of a tribe claim common descent from a single male ancestor who lived countless generations ago; their kinship organization is thus a patrilineal descent system. It is also a segmentary kinship system, so called because there are smaller subsections of the tribe, known as clans and lineages, that are functionally important in daily life. Members of the most closely related families comprising the lineage, for example, share livestock, wells, trees, and other resources. Both the larger clans, made up of numerous lineages, and the tribes possess territories. Members of a clan or tribe typically allow members of another clan or tribe to use the resources within its territory on the basis of "usufruct," or nondestructive mutual use.

Although some detractors have depicted pastoral nomads as the "fathers" rather than "sons" of the desert, blaming them for wanton destruction of game animals and vegetation, there are numerous examples of pastoral nomadic groups who have developed indigenous and very effective systems of resource conservation. Most of these practices depend on the kinship groups of family, lineage, clan, and tribe to assume responsibility for protecting plants and animals.

The Urban Way of Life

The city was the final component to emerge in the ecological trilogy, beginning in about 4000 B.C. in Mesopotamia (modern Iraq) and 3000 B.C. in Egypt. Unlike the villages they resembled in many ways, the early cities were distinguished by their larger populations (more than 5,000 people), the use of written languages, and the presence of monumental temples and other ceremonial centers. The early Mesopotamian city and, after the seventh century, the classic Islamic city, called the **medina,** had several structural elements in common (Fig. 8.11). The medina was characterized by a high surrounding wall built for defensive purposes. The congregational mosque and often an attached administrative and educational complex dominated the city center. Although Islam is often characterized as a faith of the desert, religious life has always been focused in, and diffused from, Middle Eastern and North African cities. The importance of the city's congregational mosque in religious and everyday life

Perspectives from the Field

Wayfinding in the Desert

The research for my doctoral dissertation in geography at the University of Texas—Austin was an 18-month journey with the Khushmaan Ma'aza, a clan of Bedouin nomads living in the northern Eastern Desert of Egypt. I studied their perceptions, knowledge, and uses of their desert resources and tried to understand how their worldviews and kinship patterns apparently helped them to devise ways of protecting these scarce resources. I have worked with these people from the early 1980s to the present and have always been astonished by how different they are from me—how much better they are—in their ability to "read" the ground and recall landscape details. In the following passage from my book Bedouin Life in the Egyptian Wilderness, *I attempt to explain why.*

The process by which the Khushmaan nomads have developed roots in their landscape, fashioning subjective "place" from anonymous "space," and the means by which they orient themselves and use places on a daily basis deserve special attention: these are essential parts of the Bedouins' identity and profoundly influence how they use resources and affect the desert ecosystem.

The Bedouins have little need or knowledge of maps. The Mercator projection is meaningless. One evening as three of the men watched television news in Hurghada, they argued over whether the world map behind the anchorman was a map or a depiction of clouds. On the other hand, when shown a map or aerial image of countryside they know,

Figure 8.A Saalih Ali, a Bedouin of the Khushmaan Ma'aza.

the nomads accurately orient and interpret it, naming the mountains and drainages depicted. Saalih Ali (Fig. 8.A) especially enjoyed my star chart. He would tell me what time the Pleiades would rise, for example, and ask me what time the chart indicated. The two were almost always in agreement.

Many desert travelers have been astonished by the nomads' navigational and tracking ability, calling it a "sixth sense." The Khushmaan are exceptional wayfinders and topographical interpreters able, for instance, to tell from tracks whether a camel was carrying baggage or a man; whether gazelle tracks were made by a male or female; which way a car was traveling and what make it was; which man left a set of footprints, even if he wore sandals; and how old the tracks are. Bedouins are proud of their geographical

skills which, they believe, distinguish them from settled people. Saalih told me, "Your people don't need to know the country but we do, to know exactly where things are in order to live." Pointing to his head, he said, "My map is here."

The difference in the way-finding abilities of nomads and settled persons may be due to the greater survival value of these skills for nomads and to the more complex meanings they attribute to locations. Bedouin places are rich in ideological and practical significance. The nomads interpret and interact with these meanings on a regular basis and create new places in their lifetimes. Theirs is an experience of belonging and becoming with the landscape, whereas settled people are more likely to inherit and accommodate themselves to a given set of places.

The nomads' homeland is so vast, and the margin of survivability in it so narrow, that topographic knowledge must be encyclopedic. Places either are the resources that allow human life in the desert, or are the signposts that lead to resources. A Khushmaan man pinpointed the role of places in desert survival: "Places have names so that people do not get lost. They can learn where water and other things are by using place names." Tragedy may result if the nomad has insufficient or incorrect information about places: many deaths by thirst are attributed to faulty directions for finding water.[a]

[a]Joseph Hobbs, *Bedouin Life in the Egyptian Wilderness.* Austin: University of Texas Press, 1989, pp. 81–82.

is often emphasized by its large size and outstanding artistic execution.

A large commercial zone, known as a *bazaar* in Persian and *suq* in Arabic, and recognizable as the ancestor of the modern shopping mall, typically adjoined the ceremonial and administrative heart of the city. Merchants and craftspeople of different commodities occupied separate spatial areas within this complex, and visitors to an old medina today can still expect to find streets where spices, carpets, gold, silver, traditional medicines, and other goods are sold exclusively. Smaller clusters of shops and workshops were located at the city gates.

Residential areas were differentiated as quarters not by income group but by ethnicity; the medina of Jerusalem, for example, still has distinct Jewish, Arab, Armenian, and non-Armenian Christian quarters. Homes tended to face inward

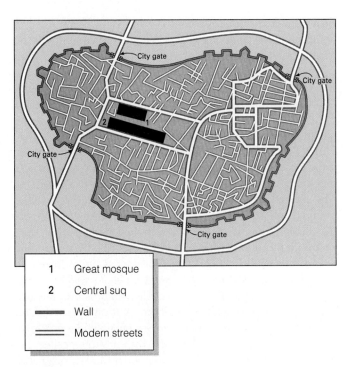

1	Great mosque
2	Central suq
▬▬	Wall
══	Modern streets

Figure 8.11 An idealized model of the classic *medina*, or Muslim Middle Eastern city.

toward a quiet central courtyard, buffering the occupants from the noise and bustle of the street. The narrow, winding streets of the medina were intended for foot traffic and small animal-drawn carts, not for large motor vehicles, a fact that accounts for the traffic jams in some old Middle Eastern and North African cities today and for the wholesale destruction of the medina in others.

The medinas that survive today are gently decaying vestiges of a forgotten urban pattern. Periods of European colonialism and subsequent nationalism changed the face and orientation of the city. During the colonial age, resident Europeans preferred to live in more spacious settings at the outer edges of the city, and later, the national elite followed this pattern. In recent times, independent governments have adopted Western building styles, with broad traffic arteries cutting through the old quarters and large central squares near government buildings. This opening up of the cityscape has scattered commercial activity along the wide avenues, thus diluting the prime importance of the central bazaar as the focus of trade.

Rural-urban migration and the city's own internal growth contribute to a rapid rate of urbanization that puts enormous pressure on services in the region's poorer countries. Governments often build high-rise public housing to accommodate the growing population, contributing to a cycle in which the urban poor move into the new dwellings only to leave their old quarters as a vacuum to draw in still more rural migrants. In Cairo, millions of former villagers now live in the "City of the Dead," an extraordinary urban landscape composed of multistory dwellings erected above graves — a last resort for the poor who have no other place to go. The overwhelmingly largest

city, or **primate city,** so characteristic of Middle Eastern and North African capitals, thus grows at the expense of the smaller city. Much of the rural-urban migration and subsequent urban gridlock and squalor could probably be avoided if governments invested more in the development of villages and smaller cities. The oil-rich countries with relatively small populations generally enjoy an urban standard of living equaling that of affluent Western countries. Modern industrial cities such as Saudi Arabia's Jubail and others founded on oil wealth were built virtually overnight, providing fascinating contrast to the region's colorful, complex ancient cities.

8.3 Cultural and Historical Context

Cultures of the Middle East and North Africa have made many fundamental contributions to humanity. Many of the plants and animals upon which the world's agriculture is based were first domesticated in the Middle East between 5,000 and 10,000 years ago in the course of the Agricultural Revolution. The list includes wheat, barley, sheep, goats, cattle, and pigs, whose wild ancestors were processed, manipulated, and bred until their physical makeup and behavior changed to suit human needs. The interaction between people and the wild plants and animals they eventually domesticated took place mainly in the well-watered Fertile Crescent stretching from Israel to western Iran.

By about 6,000 years ago, people sought higher yields by irrigating crops in the rich but often dry soils of the Tigris, Euphrates, and Nile River valleys. Their efforts produced the enormous crop surpluses that allowed civilization — a cultural complex based on an urban way of life — to emerge in Mesopotamia (literally, "the land between the rivers" Tigris and Euphrates) and Egypt. Accomplishments in science, technology, art, architecture, language, mathematics, and other areas diffused outward from these centers of civilization. Egypt and Mesopotamia are thus among the world's great culture hearths.

The Middle East also gave the world the closely related monotheistic faiths of Judaism, Christianity, and Islam. It is impossible to consider the human and political geographies of this region without attention to these religions, the ways of life associated with them, and their holy places, so here is an introduction.

The Promised Land of the Jews

The world's first monotheistic faith, Judaism is today practiced by about 17 million people worldwide, mostly in Israel, Europe, and North America. The year 2002 is 5762 in the Jewish calendar, but unlike its kindred faiths, Christianity and Islam, Judaism does not have an acknowledged starting point in time. Also unlike Christianity, Judaism does not have a fixed creed or doctrine. Jews are encouraged to behave in this life according to God's laws, which He gave to Moses on Mt. Sinai as a covenant with His people. Those laws are part of the Jewish Bible, which Christians know as the Old Testament. Jews do not accept the Christian New Testament because they do not recognize Jesus Christ as the Messiah ("Anointed One" in Hebrew) or savior prophesied in the Jewish scriptures.

Another distinction from both Christianity and Islam is that Judaism is not a proselytizing religion. It does not seek converts. In fact, practicing the Jewish faith is not the basic definition of what a Jew is; there are many nonpracticing Jews, but they are Jews nonetheless. A non-Jew can technically become a practicioner of Jewish beliefs, and thus be known as a proselyte or "immigrant," but cannot become a Jew. A Jew cannot cease to be Jewish. To be Jewish is to be a member of the Semitic ethnic group known as Jews who originated in the Middle East and shared a common historical experience over thousands of years. That historical experience included deep-seated geographic associations with particular sacred places in the Middle East — particularly to places in Jerusalem, capital of ancient Judah, the province from which Jews take their name. Tragically, the Jewish history also has included unparalleled persecution.

Depending on one's perspective, the Jewish connection with the geographic region known as Palestine, essentially the area now composed of Israel, the West Bank, and the Gaza Strip, is most significant on a time scale of either about 4,000 years or about 100 years. According to the Bible, around 2000 B.C., God commanded Abraham and his kinspeople, known as Hebrews (later as Jews), to leave their home in what is now southern Iraq and settle in Canaan. God told Abraham that this land of Canaan — geographic Palestine — would belong to the Hebrews after a long period of persecution. The Bible says that the Hebrews did settle in Canaan, until famine struck that land. At the command of Abraham's grandson Jacob, the Hebrews — known then as Israelites — relocated to Egypt, where grain was plentiful. That began the long sojourn of the Israelites in Egypt, which, according to the Bible, ended in about 1200 B.C. when Moses led them out (the Exodus).

According to Jewish history, the prophecy of Abraham was first fulfilled when the Israelites settled once again in their "promised land" of Canaan. The Jewish King Saul unified the 12 tribes who were Jacob's descendants into the first united Kingdom of Israel in about 1020 B.C. In about 950 B.C. in Jerusalem — the capital of a kingdom enlarged by Saul's successor, David — King Solomon built Judaism's First Temple. He located it atop a great rock known to the Jews as Even HaShetiyah, the "Foundation Stone," plucked from beneath the throne of God to become the center of the world and the core from which the entire world was created.

The united Kingdom of Israel lasted only about 200 years before splitting into the states of Israel and Judah. Empires based in Mesopotamia destroyed these states: The Assyrians attacked Israel in 721 B.C., and the Babylonians sacked Judah in 586 B.C. The Babylonians destroyed the First Temple and exiled the Jewish people to Mesopotamia, where they remained until conquering Persians allowed them to return to their homeland. In about 520 B.C., the Jews who returned to Judah, the land from which they take their name, rebuilt the temple (the Second Temple) on its original site. A succession of foreign empires came to rule the Jews and Arabs of Palestine: Persian, Macedonian, Ptolemaic, Seleucid, and around the time of Christ, Roman. Herod, the Jewish king who ruled under Roman authority and was a contemporary of Christ, greatly enlarged the temple complex.

The Jews of Palestine revolted against Roman rule three times between A.D. 64 and 135. The Romans quashed these rebellions in a series of famous sieges, including those of Masada and Jerusalem. The Romans destroyed the Second Temple, and a third has never been built. All that remains of the Second Temple complex is a portion of the surrounding wall built by Herod. Today, this Western Wall, known to non-Jews as the Wailing Wall, is the most sacred site in the world accessible to Jews (Fig. 8.12). Some religious traditions prohibit Jews from ascending the Temple Mount above, the area where the temple actually stood, because it is too sacred. After the temple's destruction, that site was occupied by a Roman temple and then in 691 by the Muslim shrine called the Dome of the Rock, which still stands today (also in Fig. 8.12). The mostly Muslim Arabs know the Temple Mount as *al-Haraam ash-Shariif*, meaning "The Noble Sanctuary." Supercharged with meaning, this place has in modern times often been the spark of conflagration between Jews and Arabs.

The victorious Romans scattered the defeated Jews to the far corners of the Roman world. Thus began the Jewish exile, or **diaspora.** In their exile, the Jews never forgot their attachment to the Promised Land. The Passover prayer ends with the words "Next year in Jerusalem!" In Europe, where their numbers were greatest, Jews were subjected to systematic discrimination and persecution and were forbidden to own land or engage in a number of professions. Known as **anti-Semitism,** the hatred of Jews developed deep roots in Europe. This

Figure 8.12   Jerusalem's Temple Mount/Noble Sanctuary. In this view are some of Judaism's and Islam's holiest places. At left, below the golden dome, is the Western Wall, all that remains of the structure that surrounded the Jews' Second Temple. The dome is the Muslims' Dome of the Rock. At far right, with the black dome, is the al-Aqsa Mosque, another very holy place in Islam.

sentiment in part grew out of the perception that Jews were responsible for the murder of Christ and the fact that Christian Europeans were prohibited from practicing usury, or money lending. They assigned this role to Jews but then resented paying debts to the despised moneylenders.

In the 1930s, anti-Semitism became state policy in Germany under the Nazis, led by Adolph Hitler. Many German Jews, including Albert Einstein, fled to the United States, and others emigrated to Palestine in support of the **Zionist movement,** which aimed at establishing a Jewish homeland in Palestine with Zion (a synonym for Jerusalem) as its capital. Most Jews were not as fortunate as the emigrants. Within the boundaries of the Nazi empire that dominated most of continental Europe during World War II, Hitler's regime executed its "final solution" to the Jewish "problem." Nazi Germans and their allies killed an estimated 6 million Jews, along with other "inferior" minorities, including Gypsies and homosexuals. It was this **Holocaust** that prompted the victorious allies of World War II, from their powerful position in the newly formed United Nations, to create a permanent homeland for the Jewish people in Palestine.

Christianity: Death and Resurrection in Jerusalem

Nearly 1,000 years after Solomon established the Jewish Temple, a new but closely related monotheistic faith emerged in Palestine. This was Christianity, named after Jesus Christ. Jesus, a Jew, was born near Jerusalem in Bethlehem, probably around 4 B.C. Tradition relates that when he was about 30, Jesus began spreading the word that he was the Messiah long prophesied in Jewish doctrine (thus, after his death, his surname became Christ, meaning "Anointed One" in Greek). A small group of disciples accepted that he was, and followed him for several years as he preached his message. He explained that he was the Son of God, a living manifestation of God Himself, and that the only path to eternal life was by accepting his divinity. He taught that love, sacrifice, and faith were the keys to salvation. He said he had come to redeem humanity's sins through his own death.

Christ's teachings denied the validity of many Jewish doctrines, and in about A.D. 29, a growing chorus of Jewish protesters called for his death. Palestine was then under Roman rule, and Roman administrators in Jerusalem placated the mob by ordering that Christ be put on trial. He was found guilty of being a claimant to Jewish kingship, and Roman soldiers put him to death by the particularly degrading and painful method of crucifixion. The cornerstone of Christian faith is that Christ was resurrected from the dead after 3 days and ascended into heaven. Christians believe that he continues to intercede with his Father on their behalf and that he will come again on Judgment Day, at the end of time.

After a period of relative tolerance, the Romans began actively persecuting Christians. Nevertheless, Christian ranks and influence grew. The turning point in Christianity's career came

Figure 8.13 Jerusalem's Church of the Holy Sepulcher, containing the spots where many Christians believe Christ was crucified and buried.

after A.D. 324 when the Roman Emperor Constantine embraced Christianity and quickly established it as the official religion of the empire. The Christian Byzantine civilization that developed in the "New Rome" Constantine established–Constantinople, now Istanbul, Turkey — created fine monuments at places associated with the life and death of Jesus Christ. These include the place long acknowledged as the center of the Christian world, Jerusalem's Church of the Holy Sepulcher (Fig. 8.13). This extraordinary, sprawling building, now administered by numerous separate Christian sects, contains the spots where tradition says Jesus was crucified and buried.

Christianity has seldom been the majority religion in the land where it was born; only until Islam arrived in Palestine in A.D. 638 was the region primarily Christian. From then until the 20th century, most of Palestine's inhabitants were Muslim, and since the mid-20th century, Muslims and Jews have been the major groups. Between the 11th and 14th centuries, European Christians dispatched military expeditions to recapture Jerusalem and the rest of the Holy Land from the Muslims. These bloody campaigns, known as the **Crusades,** resulted in a series of short-lived Christian administrations in the region.

There are significant minority populations of Christians throughout the Middle East, including members of distinct sects such as the Copts of Egypt and Maronites of Lebanon. Their population percentages have generally been declining, both because of emigration and lower birth rates than those of the majority Muslims. Nevertheless, the Middle East remains the cradle of their faith for Christians the world over, and Jerusalem and nearby Bethlehem are the world's premier Christian pilgrimage sites.

The Message of Islam

Islam is by far the dominant religion in the Middle East and North Africa; only Israel within its pre-1967 borders and Cyprus have non-Muslim majorities (Jews and Christians, respectively). Because of Islam's powerful influence not merely as a set of religious practices but as a total way of life, an understanding of the religious tenets, culture, and diffusion of Islam is vital for appreciating the region's cultural geography. Islam is a monotheistic faith built upon the foundations of the region's earliest monotheistic faith, Judaism, and its offspring, Christianity. Indeed, Muslims (people who practice Islam) call Jews and Christians "People of the Book," and their faith obliges them to be tolerant of these special peoples. Muslims believe that their prophet Muhammad was the very last in a series of prophets who brought the Word to humankind. Thus, they perceive the Bible as incomplete but not entirely wrong—Jews and Christians merely missed receiving the entire message. Muslims disagree with the Christian concept of the divine trinity and regard Jesus as a prophet rather than as God.

Muhammad was born in A.D. 570 to a poor family in the western Arabian (now Saudi Arabian) city of Mecca. Located on an important north-south caravan route linking the frankincense-producing area of southern Arabia (now Yemen and Oman) with markets in Palestine (now Israel) and Syria, Mecca was a prosperous city at the time. It was also a pilgrimage destination because more than 300 deities were venerated in a shrine there called the Ka'aba (the Cube; Fig. 8.14). Muhammad married into a wealthy family and worked in the caravan trade. Muslim tradition holds that when he was about 40 years old, Muhammad was meditating in a cave outside Mecca when the Angel Gabriel appeared to him and ordered him to repeat the words of God that the angel would recite to him. Over the next 22 years, the prophet related these words of God (Allah) to scribes who wrote them down as the *Qur'an* (or Koran), the holy book of Islam.

During this time, Muhammad began preaching the new message, "There is no god but God," which the polytheistic people of Mecca viewed as heresy. As much of their income depended on pilgrimage traffic to the Ka'aba, they also viewed Muhammad and his small band of followers as an economic threat. They forced the Muslims to flee from Mecca and take refuge in Yathrib (modern Medina), where a largely Jewish population had invited them to settle. There were subsequent skirmishes between the Meccans and Muslims, but in 630, the Muslims prevailed and peacefully occupied Mecca. The Muslims destroyed the idols enshrined in the Ka'aba, which became a pilgrimage center for their one God.

The Ka'aba is Islam's holiest place, and Mecca and Medina are its holiest cities. Jerusalem is also sacred to Muslims. Muslim tradition relates that on his "Night Journey," the Prophet Muhammad ascended briefly into heaven from the great rock now beneath the Dome of the Rock. Nearby on the Temple Mount/al-Haraam ash-Shariif is the al-Aqsa Mosque, a sacred congregational site. The proximity, even the duplication, of holy places between Islam and Judaism came to be the most difficult issue in peace negotiations between Palestinians and Israelis, as explained in Chapter 9.

After Muhammad's death in 632, Arabian armies carried the new faith far and quickly. The two decaying empires that then prevailed in the Middle East and North Africa—the Byzantine or Eastern Roman Empire, based in Constantinople (now Istanbul), and the Sassanian Empire, based in Persia (now Iran) and adjacent Mesopotamia (now Iraq)—put up only limited military resistance to the Muslim armies before capitulating. Local inhabitants generally welcomed the new faith, in part because administrators of the previous empires had not treated them well, while the Muslims promised tolerance. Soon the Syrian city of Damascus became the center of a Muslim empire. Baghdad assumed this role in A.D. 750.

Figure 8.14   The black-shrouded cubical shrine known as the Ka'aba (just right of center) in Mecca's Great Mosque is the object toward which all Muslims face when they pray and is the centerpiece of the pilgrimage to Mecca required of all able Muslims.

A schism occurred very early in the development of Islam and persists today. The split developed because the Prophet Muhammad had named no successor to take his place as the leader (caliph) of all Muslims. Some of his followers argued that the person with the strongest leadership skills and greatest piety was best qualified to assume this role. These followers became known as **Sunni,** or orthodox, Muslims. Others argued that only direct descendants of Muhammad, specifically through descent from his cousin and son-in-law Ali, could qualify as leaders. They became known as **Shi'a,** or Shi'ite, Muslims. The military forces of the two camps engaged in battle south of Baghdad at Karbala in

A.D. 680, and in the encounter, Sunni troops caught and brutally murdered Hussein, a son of Ali. Thereafter, the rift was deep and permanent. The martyrdom of Hussein became an important symbol for Shi'ites, who still today regard themselves as oppressed peoples struggling against cruel tyrants, including some Sunni Muslims.

Today, only two Muslim countries, Iran and Iraq, have Shi'ite majority populations. Significant minority populations of Shi'ites are in Syria, Lebanon, Yemen, and the Arab states of the Persian/Arabian Gulf.

Arab science and civilization flourished in the Baghdad immortalized in the legends of *The Thousand and One Nights.* There were important accomplishments and discoveries in mathematics, astronomy, and geography. Scholars translated the Greek and Roman classics, and if not for their efforts, many of these works would never have survived to become part of the modern European legacy. It was an age of exploration, when Arab merchants and voyagers visited China and the remote lands of southern Africa. Many important discoveries by the Arab geographers were recorded in Arabic, a language unfamiliar to contemporary Europeans, and had to be rediscovered centuries later by the Portuguese and Spaniards. Arab merchants carried their faith on the spice routes to the East. One result, surprising to many today, is that the world's most populous Muslim country is not in the Middle East and North Africa, and its people are not Arabs; it is Indonesia, 5,000 thousand miles (8,000 km) east of Arabia.

Whether in Arabia or Indonesia, whether they be Sunni Muslims or Shi'ite Muslims (see Definitions & Insights above, all believers are united in support of the five fundamental precepts, or **pillars of Islam.** The first of these is the profession of faith: "There is no god but God, and Muhammad is His Messenger." This expression is often on the lips of the devout Muslim, both in prayer and as a prelude to everyday activities. The second pillar is prayer, required five times daily at prescribed intervals. Two of these prayers mark dawn and sunset. Business comes to a halt as the faithful prostrate themselves before God. Muslims may pray anywhere, but wherever they are, they must turn toward Mecca. There also is a congregational prayer at noon on Friday, the Muslim sabbath.

The third pillar is almsgiving. In earlier times, Muslims were required to give a fixed proportion of their income as charity, similar to the concept of the tithe in the Christian church. Today, the donations are voluntary. Even Muslims of very modest means give what they can to the needier.

The fourth pillar is fasting during Ramadan, the ninth month of the Muslim lunar calendar. Muslims are required to abstain from food, liquids, smoking, and sexual activity from dawn to sunset throughout Ramadan. The lunar month of Ramadan occurs earlier each year in the solar calendar and thus periodically falls in summer. In the torrid Middle East and North Africa, that timing imposes special hardships on the faithful, who, even if they are performing manual labor, must resist the urge to drink water during the long, hot days.

The final pillar is the pilgrimage (*hajj*) to Mecca, Islam's holiest city. Every Muslim who is physically and financially capable is required to make the journey once in his or her lifetime. A lesser pilgrimage may be performed at any time, but the prescribed season is the 12th month of the Muslim calendar. Those days witness one of Earth's greatest annual migrations, as about 2 million Muslims from all over the world converge on Mecca. Hosting these throngs is an obligation the government of Saudi Arabia fulfills proudly and at considerable expense, but with some trepidation in recent years because of the security threat foreign visitors may pose to the host country and because accidents such as tent city fires have cost many lives. Many pilgrims also visit the nearby city of Medina, where Muhammad is buried. Most Muslims regard the hajj as one of the most significant events of their lifetimes. All are required to wear simple seamless garments, and for a few days, the barriers separating groups by income, ethnicity, and nationality are broken. Pilgrims return home with the new stature and title of "hajj" but also with humility and renewed devotion.

All Muslims share the five pillars and other tenets, but they vary widely in other cultural practices related to their faith, depending on what country they live in, whether they are from the desert, village, or city, and how much education and income they have. The governments and associated clerical authorities in Saudi Arabia and Iran insist on strict application of Islamic law (*shari'a*) to civil life; in effect, there is no separation between church and state. The Qur'an does not state that women are required to wear veils, but it does urge them to be modest, and it portrays their roles as different from those of men. Clerics in Saudi Arabia insist that women wear floor-length, long-sleeved black robes and black veils in public, that they not travel unaccompanied by a male member of their families, and that they not drive cars. In Egypt, by contrast, Muslim women are free to appear in public unveiled if they choose. However, in most Muslim countries, conservative ideas about the role of women are still very strong: They should be modest, retiring, good mothers, and keepers of the home. The Qur'an portrays women as equal to men in the sight of God, and in principle, Islamic teachings guarantee the right of women to hold and inherit property.

Most Muslim women argue that what others often see as "backward" cultural practices are in fact very progressive. For example, their modest dress compels men to evaluate them on the basis of their character and performance, not their attractiveness. Segregation of the sexes in the classroom makes it easier for both women and men to develop their confidence and skills. Sexual assault is rare. A married woman retains her maiden name. These apparent advantages can be weighed against the drawbacks that women are generally subordinate to men in public affairs and have fewer opportunities for education and for work outside the home.

8.4 Economic Geography

Overall, this is a poor region; per capita GDP PPP for the 24 countries and territories averages only $5,008. This may seem surprising in view of the "rich Arab" stereotype. Only the oil-endowed states of the Persian/Arabian Gulf and Libya and Algeria in North Africa deserve reputations for wealth, and only non-Arab Israel is truly a more developed country (MDC) by measures other than per capita wealth. Israel's prosperity comes from its innovation in computer and other high-technology industries, the processing and sale of diamonds, large amounts of foreign (mostly U.S.) aid, and investment and assistance by Jews and Jewish organizations around the world.

Vital to the industrialized countries as a source of fuels, lubricants, and chemical raw materials, petroleum is one of the world's most important natural resources. A crucial feature of world geography is the concentration of approximately two-thirds of the world's proven petroleum reserves in a few countries that ring the Persian/Arabian Gulf. By coincidence, the countries rich in oil tend to have relatively small populations, whereas the most populous nations have few oil reserves; Iran is an exception. All but about 1% of the Persian/Arabian Gulf oil region's proven reserves of crude oil are located in Saudi Arabia, Iraq, Kuwait, Iran, and the United Arab Emirates (UAE), with smaller reserves in Oman and Qatar. Saudi Arabia, by far the world leader in reserves, has about 26% of the proven crude oil reserves on the globe. Iraq

has an additional 11%, UAE 10%, Kuwait 9%, and Iran 9%. By comparison, Venezuela has 7% of world reserves, Russia another 5%, and the world's largest oil consumer, the United States, only 2% (these are the top eight in the world's known oil reserves as of 2002).

Production, export, and profits of Middle Eastern and North African oil were once firmly in the hands of foreign companies. That changed after 1960 when most of the Gulf countries and other exporting nations formed the Organization of Petroleum Exporting Countries (OPEC), with the aim of joint action to demand higher profits from oil. It changed again after 1972 when the oil producing countries began to nationalize the foreign oil companies. OPEC was relatively obscure until the Arab-Israel war of 1973, after which the organization began a series of dramatic price increases. In 1981, the organization's price reached $34 (U.S.) per barrel, compared with $2 a barrel in early 1973.

These events had enormous repercussions for the world economy. Immense wealth was transferred from the MDCs to the OPEC countries to pay for indispensable oil supplies. The skyrocketing cost of gasoline and other oil products helped cause serious inflation in the United States and many other countries and contributed to the 1973 energy crisis in the United States. Desperately poor, less developed countries (LDCs) found that high oil prices not only hindered the development of their industries and transportation but also reduced food production because of high prices for fertilizer made from oil and natural gas. In the Gulf countries, the oil bonanza produced a wave of spending for military hardware, showy buildings, luxuries for the elite, and ambitious development projects of many kinds. Per capita benefits to the general populace were greatest in the oil states of the Arabian Peninsula, where small populations and immense inflows of oil money made possible the abolition of taxes, the development of comprehensive social programs, and heavily subsidized amenities such as low-cost housing and utilities, including water distilled from the Gulf by desalination plants.

Then, at the beginning of the 1980s, the era of continually expanding oil production, sales, and profits by the OPEC states came to an end. After 1973, the high price of oil stimulated oil development in countries outside OPEC. Oil-conservation measures such as a shift to more fuel-efficient vehicles and furnaces were instituted. Substitution of cheaper fuels for oil increased. Coal replaced oil in many electricity-generating stations. Oil refineries were converted to make gasoline from cheaper "heavy" oils rather than the more expensive "light" oils previously used. Meanwhile, the world entered a period of economic recession, due in part to high oil prices. Decreased business activity reduced the demand for oil. Profits of the world oil industry (and taxes paid to governments) were severely cut, large numbers of refineries had to close, and much of the world tanker fleet was idled.

Oil prices rose temporarily in 1990–1991 when the flow of Iraqi and Kuwaiti oil was cut off following Iraq's military takeover of Kuwait in August 1990, but the prices soon fell

REGIONAL PERSPECTIVE ::: Islamic Fundamentalism

A wave of Islamic "fundamentalism" has recently swept the Islamic world. Arguing "Islam is the solution," Islamists (as they are more accurately known) reject what they view as the materialism and moral corruption of Western countries and the political and military support these countries lend to Israel. Both Sunni and Shi'ite Muslims have advanced a wide range of Islamic movements, notably in Iran, Lebanon, Egypt, Afghanistan, Sudan, and Algeria.

Although nominally religious, the more radical of these movements have political and cultural aims, particularly the destabilization or removal of U.S. and Israeli interests in the region and abroad. In 1993, followers of the radical Egyptian cleric Sheikh Umar Abdel-Rahman bombed New York City's World Trade Center as a protest against American support of Israel and Egypt's pro-Western government. In an attempt to destabilize and replace Egypt's government, which they viewed as an illegitimate regime too supportive of the United States, another Egyptian Islamist group attacked and killed foreign tourists in Egypt in the 1990s. In the 1980s, members of the pro-Iranian Hizbullah, or Party of God, in Lebanon kidnapped foreigners as bargaining chips for the release of comrades jailed in other Middle Eastern countries.

Within Israel and the autonomous Palestinian territories of the West Bank and Gaza Strip, Palestinian members of HAMAS (an Arabic acronym for the Islamic Resistance Movement) have carried out terrorist attacks on Israeli civilians and soldiers in an effort (apparently successful) to derail implementation of the peace agreements reached between the Israeli government and the Palestine Liberation Organization (PLO). In Algeria, years of bloodshed have followed the government's annulment of 1991 election results that would have given the Islamic Salvation Front (FIS) majority control in the parliament. Muslim sympathizers carried the battle to France, bombing civilian targets in protest against the French government's support for the Algerian regime.

In 1998, the world began to hear about Osama bin Laden, a former Saudi businessman living in exile in Afghanistan, whose al-Qa'ida organization bombed U.S. embassies in Kenya and Tanzania as part of an avowed worldwide struggle against American imperialism and immorality. Bin Laden's organization was also suspected in the 2000 bombing of the American naval destroyer the U.S.S. *Cole* in Yemen's harbor of Aden. However shocking those assaults were, they pale in comparison to al-Qa'ida's attacks against targets in the United States on September 11, 2001. In the most ferocious terrorist actions ever undertaken to that date, members of al-Qa'ida cells in the United States hijacked four civilian jetliners and succeeded in piloting three of them into New York City's World Trade Center towers and Washington, D.C.'s Pentagon. More than 3,000 people, mainly civilians, perished. Bin Laden and his followers cheered the carnage as justifiable combat against an infidel nation whose military troops occupied the holy land of Arabia, where Mecca and Medina are located.

In all of these situations, a tiny minority of Muslims carried out terrorist actions that the great majority condemned. Islamic scholars and clerics pointed out in each case that the murder of civilians is prohibited in Islamic law and that the attacks had no legitimate grounds in the religion. Mainstream Islamic movements are not military or terrorist organizations but have distinguished themselves through public service to the needy and through encouragement of stronger moral and family values. For most Muslims, the growing Islamist trend means a reembrace of traditional values like piety, generosity, care for others, and Islamic legal systems, which have proven effective for centuries, values which now pose a reasonable alternative to Western cultural influences and often repressive political and administrative systems. For many people outside the region, however, "Muslim" and "terrorist" have become synonymous — an erroneous association that can be overcome in part by careful study of the complex Middle East and North Africa.

---

again. In the 1990s, Saudi Arabia and other oil-rich Gulf states implemented internal economic austerity measures for the first time. However, the immense oil and gas reserves still in the ground guaranteed that the Gulf region would continue to have a major long-term impact on the world and would remain prosperous as long as these finite resources are in demand in the MDCs. The lasting economic clout of the region was apparent early in the 21st century as the price of oil once again climbed to near record levels (thanks mainly to OPEC decisions to reduce production), contributing to economic slowdowns in many nations.

8.5 Geopolitical Issues

This has long been a vital region in world affairs and a target of outside interests. Its strategic crossroads location often has made it a cauldron of conflict. From very early times, overland caravan routes, including the famous Silk Road, crossed the Middle East and North Africa with highly prized commodities traded between Europe and Asia. The security of these routes was vital, and countries on either end could not tolerate threats to them. Then, in 1869, one of the world's most important waterways opened in the Middle East. Slicing

Figure 8.15 The strategically vital Suez Canal zone saw bitter fighting in the Middle East wars of 1956, 1967, and 1973.

107 miles (172 km) through the narrow Isthmus of Suez, the British- and French-owned Suez Canal linked the Mediterranean Sea with the Indian Ocean, saving cargo, military, and passenger ships a journey of many thousands of miles around the southern tip of Africa (see Figs. 8.1 and 8.15). Egyptian President Gamal Abdel Nasser's nationalization of the canal in 1956 led immediately to a British, French, and Israeli invasion of Egypt and a conflict known as the Suez Crisis and the 1956 Arab-Israeli War. Egypt effectively won that war when international pressure caused the invading forces to withdraw, leaving the Canal in Egypt's hands.

The Canal, the cotton of the Nile Delta, and the strategic location of the region were important during and since colonial times, but oil has been, and will remain (as long as fossil fuels drive the world's economies), what keeps the rest of the world interested in the Middle East and North Africa. The region's oil is sold to many countries, but most of it is marketed in Western Europe and Japan. The United States also imports large amounts of Gulf oil but has a much smaller relative dependence on this source than do Japan and Europe. However, the Gulf region is very important to the United States because of the heavy dependence of close American allies on Gulf oil and the importance of the oil as a future reserve. There is also great involvement of American companies in oil operations and oil-financed development in Gulf countries and the large economic impact of Gulf "petrodollars" spent, banked, and invested in the United States. Maintaining a secure supply of Gulf oil has therefore been one of the longstanding pillars of U.S. policy in the Middle East.

The United States has always maintained a precarious relationship with the key players in the Middle Eastern arena. On the one hand, the United States has pledged unwavering support for Israel, but on the other, it has courted Israel's traditional enemies such as oil-rich Saudi Arabia. That Arab kingdom and its neighbors around the Persian/Arabian Gulf possess more than 60% of the world's proven oil reserves and are thus vital to the long-term economic security of the Western industrial powers and Japan. The United States and its Western allies made it clear they would not tolerate any disruption of access to this supply when Iraqi troops directed by Iraqi President Saddam Hussein occupied Kuwait on August 2, 1990. U.S. President George Bush drew a "line in the sand," proclaiming "we cannot permit a resource so vital to be dominated by one so ruthless — and we won't." The United States and a coalition of Western and Arab allies fielded a massive array of military might that ousted the Iraqi invaders within months and secured the vital oil supplies for Western markets.

The Gulf War was not the first time the United States expressed its willingness to use force if necessary to maintain access to Middle Eastern oil. In the wake of the revolution in Iran in 1979, the Soviet Union invaded neighboring Afghanistan. U.S. military analysts feared the Soviets might use Afghanistan as a launch pad to invade oil-rich Iran. The United States deemed this prospect unacceptable, and then-President Jimmy Carter issued the policy statement that came to be known as the **Carter Doctrine:** The United States would use any means necessary to defend its vital interests in the region. Vital interests meant oil, and any means necessary meant the United States was willing to go to war with the Soviet Union, presumably nuclear war, to defend those interests.

Middle Eastern wars had already become proxy wars for the superpowers: In the 1967 and 1973 Arab-Israeli wars, for example, Soviet-backed Syrian forces fought U.S.-backed Israeli troops. American support of Israel in this war prompted Arab members of OPEC to impose an embargo of sales of their oil to the United States, precipitating the nation's first energy crisis. During the 1973 war, the United States put its forces on an advanced state of readiness to take on the Soviets in a nuclear exchange if necessary. All of these events illustrate that the Middle East and North Africa comprise, in political geography terms, a **shatter belt** — a large, strategically located region composed of conflicting states caught between the conflicting interests of great powers. The following chapter offers more insight into the peoples and nations of this vital region.

CHAPTER SUMMARY

- Events in the Middle East and North Africa profoundly affect the daily lives of people around the world, yet this region is often misunderstood. Misleading stereotypes about its environment and people are common, and people outside the region often associate it soley with military conflict and terrorism.

- The region has bestowed upon humanity a rich legacy of ancient civilizations, including those of Egypt and Mesopotamia, and the three great monotheistic faiths of Judaism, Christianity, and Islam.

- Middle Easterners include Jews, Arabs, Turks, Persians, Berbers, and other ethnic groups who practice a wide variety of ancient and modern livelihoods.

- Arabs are the largest ethnic group in the Middle East and North Africa, and there are also large populations of ethnic Turks, Persians (Iranians), and Kurds. Islam is by far the largest religion. Jews live almost exclusively in Israel, and there are minority Christian populations in several countries.

- Population growth rates in the region are moderate to high. Oil wealth is concentrated in a handful of countries, and as a whole, this is a developing region.

- The Middle East has served as a pivotal global crossroads, linking Asia, Europe, Africa, and the Mediterranean Sea with the Indian Ocean. These countries have historically been unwilling hosts to occupiers and empires originating far beyond their borders.

- The margins of this region are occupied by oceans, high mountains, and deserts. The land is composed mainly of arid and semi-arid plains and plateaus, together with considerable area of rugged mountains and isolated "seas" of sand.

- Aridity dominates the environment, with at least three-fourths of the region receiving less than 10 inches of yearly precipitation. Plants, animals, and people have developed strategies of drought avoidance and drought endurance to live here. In addition, great river systems and freshwater aquifers have sustained large human populations.

- Many of the plants and animals upon which the world's agriculture depends were first domesticated in the Middle East in the course of the Agricultural Revolution. The interaction between people and the wild plants and animals that were domesticated took place mainly in the Fertile Crescent.

- The Middle Eastern "ecological trilogy" consists of peasant villagers, pastoral nomads, and city-dwellers. The relationships among them have been mainly symbiotic and peaceful, but city-dwellers have often dominated the relationship, and both pastoral nomads and urbanites have sometimes preyed upon the villagers, who are the trilogy's cornerstone.

- About two-thirds of the world's oil is here, making this one of the world's most vital economic and strategic regions.

- Since World War II, several international crises and wars have been precipitated by events in the Middle East. Strong outside powers depend heavily on this region for their current and future industrial needs.

REVIEW QUESTIONS

1. Using maps and the text, list and locate the countries of the Middle East.
2. What are the major climatic patterns of the Middle East?
3. Explain how residents have culturally adapted to heat and aridity in the Middle East.
4. Where were the extensive forests of the Middle East found during early historical times? What nation supplied the cedar for King Tutankhamen's funerary shrines?
5. List some of the non-Arab nations of the Middle East and define their cultural composition.
6. Using maps and the text, locate the two great culture hearths of Mesopotamia and Egypt.
7. Using maps and the text, explain why the Middle East has held such a strategic importance in international conflict.
8. Define vertical migration (transhumance) and horizontal migration in relation to the Middle East.
9. What are the main elements that make up the classic medina, or Muslim Middle Eastern city? List them and then draw a model city.
10. What sites in the Middle East are sacred to Jews, Christians, and Muslims?
11. What countries in the Middle East have how much of the world's oil?

DISCUSSION QUESTIONS

1. Explain the origin of the term "Middle East."
2. List the major ethnic groups and the countries in which they are found. Do the same with the major religions.
3. What is an Arab? A Jew? A Turk? A Kurd?
4. Examine and discuss the important land and water features of the Middle East.
5. Explain how heat and aridity have helped shape environments and human cultures in the Middle East.
6. Why is the Middle East considered such an important culture hearth for the rest of the world?
7. Islam is often considered more a way of life than just a religion. Why?
8. What is the difference between Shi'ite and Sunni Islam?
9. What are the five pillars of Islam and what do they require of a Muslim?
10. What are the mainly non-Arab nations of the Middle East?
11. Three ancient ways of life—making up the ecological trilogy—prevail in the Middle East. What are these and what are some of the important characteristics and interrelationships of each? This may be answered in an exercise dividing students into three groups, with each group representing one livelihood.
12. Is the old city of Jerusalem a medina? Research this question, being sure to locate maps of the city.
13. Has there historically been animosity among Jews, Christians, and Muslims? When, where, and why?
14. How important is the Middle East in the global economy?

CHAPTER 9

Countries of the Middle East and North Africa: Modern Struggles in an Ancient Land

Bedouin of the southern Sinai Peninsula, Egypt, and Egypt's second highest mountain, Umm Shumar.

Conflict in the Middle East has been one of the most persistent and dangerous problems in global affairs since the end of World War II. This chapter begins with an introduction to the key players and issues in conflict. The reader should come away with a well-informed, detailed understanding of why the Palestinian-Israeli dispute is so difficult to solve and how beneficial its resolution would be. The chapter continues with a survey of all of the major subregions and nations of the diverse, fascinating Middle East.

9.1   The Arab-Israeli Conflict and Its Setting

The Arab-Israeli conflict persists as one of the world's most intractable disputes. It has not been resolved because the central issues are closely tied to such life-giving resources as land and water and to deeply held religious beliefs. The Arab-Israeli conflict is above all a conflict over who owns the land — sometimes very small pieces of land — and is therefore of extreme interest to geographers and geography. A geographic understanding of the Middle East today requires familiarity with the events leading up to the creation of the state of Israel and with the many skirmishes and wars that have followed, particularly as they have rearranged the boundaries of nations and territories.

The modern state of Israel was carved from lands whose fate had been undetermined since the end of World War I. The Ottoman Empire, based in what is now Turkey, had ruled Palestine (roughly the area now made up of Israel and the Occupied Territories) and surrounding lands in the eastern Mediterranean since the 16th century. After the British and French defeated the Turks in World War I and destroyed their empire, they divided the region between them. The British received the mandate for Palestine, Transjordan (modern Jordan), and Iraq, while the French received the mandate for Syria (now Syria and Lebanon). During World War I, British administrators of Palestine had made conflicting promises to Jews and Arabs. On the one hand, they implied they would create an independent Arab state in Palestine and, on the other, vowed to promote Jewish immigration to Palestine with an eye to the eventual establishment of a Jewish state there. The Palestinians — Arabs who historically formed the vast majority of the region's inhabitants — did not welcome the ensuing Jewish immigration.

Placing themselves in a no-win position with these conflicting promises, and under increasing attack from both Jews and Arabs, in 1947 the British decided to withdraw from Palestine and leave the young United Nations with the task of determining the region's future. The United Nations responded in 1947 with a "two-state solution" to the problem of Palestine. It established an Arab state (which would have been called Palestine) and a Jewish state (Israel). The plan was deeply flawed. The states' territories were long, narrow, and almost fragmented, giving each side a sense of vulnerability and insecurity (Fig. 9.1). War broke out in May 1948 between newborn Israel and the armies of the neighboring Arab countries of Transjordan, Egypt, Iraq, Syria, and Lebanon. The smaller but better organized and more highly motivated Israeli army defeated the Arab armies, and Israel acquired its "pre-1967" borders (see Fig. 9.1).

Prior to and during the fighting of 1948, approximately 800,000 Palestinian Arabs were both forced and chose to flee from the new state of Israel to neighboring Arab countries (Fig. 9.2). The United Nations established refugee camps for these displaced persons in Jordan, Egypt, Lebanon, and Syria. Little was done to resettle them in permanent homes, and both the Arab governments and the refugees themselves continued to insist on the return of the refugees to Israel and the restoration of their properties there. This Palestinian **right of return** is one of the central problems of the current peace process, which will be discussed further.

Other countries assumed control of those parts of the proposed Palestine that Israel did not absorb. Egypt occupied the Gaza Strip, a piece of land on the Mediterranean shore adjacent to Egypt's Sinai Peninsula and inhabited mostly by Palestinian Arabs. Transjordan occupied the West Bank, the predominantly Arab hilly region of central Palestine on the west side of the Jordan River, and the entire old city of Jerusalem, including the Western Wall and Temple Mount, Judaism's holiest sites. The Palestinian Arab state envisioned in the U.N. Partition Plan was thus stillborn.

The 1967 war fundamentally rearranged the region's political landscape in Israel's favor, setting the stage for subsequent struggles and the peace process. This conflict was precipitated in part when, by positioning arms at the Strait of Tiran chokepoint (see Definitions & Insights, p. 245), Egypt closed the Gulf of Aqaba to Israeli shipping. Egypt's President Nasser and his Arab allies took several other belligerent but nonviolent

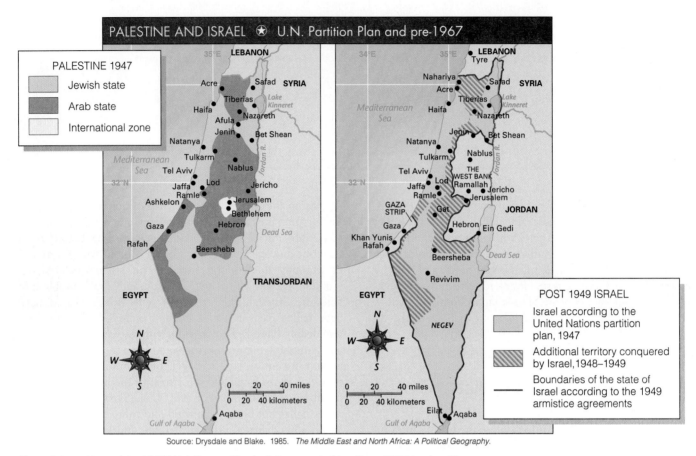

Source: Drysdale and Blake. 1985. *The Middle East and North Africa: A Political Geography.*

Figure 9.1 Maps of the 1947 U.N. Partition Plan for Palestine and of Israel's pre-1967 borders. The war, which began as soon as Britain withdrew from Palestine and Israel proclaimed its existence, aborted the U.N. Partition Plan and created a tense new political landscape in the region.

steps toward a war they were ill prepared to fight. Israel elected to make a preemptive strike on its Arab neighbors, virtually destroying the Egyptian and Syrian air forces on the ground. Israel gave Jordan's King Hussein an opportunity to stay out of the conflict. However, Jordan went to war and quickly lost the entire West Bank and the historic Old City of Jerusalem. The entire nation of Israel was transfixed by the news that Jewish soldiers were praying at the Western Wall. The Israeli army (Israeli Defense Forces, or IDF) also seized the Gaza Strip, Egypt's Sinai Peninsula (thus closing the Suez Canal), and the strategic Golan Heights section of Syria overlooking Israel's Galilee region. Israel had tripled its territory in 6 days of fighting; this conflict is known as The Six Day War (see Fig. 9.3).

The Palestinian refugee situation became more complex as a result of the war. Israel took over the areas where most of the camps were located. Many persons in the camps again took flight, and Palestinians fleeing villages and towns in the newly Occupied Territories joined them as refugees (see Fig. 9.2). Some later returned to their homes, but an estimated 116,000 either did not attempt to return or were denied permission by Israel authorities to do so.

The 1973 war was a multifront Arab attempt to reverse the humiliating losses of 1967. On October 6, 1973, the Jewish holy day of Yom Kippur, Egypt and Syria launched a surprise attack on Israel with the hope of rearranging the stalemated political map of the Middle East. Egypt's army initially showed surprising strength, penetrating deep into the Sinai Peninsula and overturning Israel's image of invulnerability. Israeli troops soon reversed the tide and surrounded Egypt's army, but in the ensuing disengagement talks, Egypt won back the eastern side of the Suez Canal and by 1975 was able to reopen it to commercial traffic. In 1979, Israel returned Sinai to Egypt under the U.S.-sponsored Camp David Accords, and Egypt recognized Israel's rights as a sovereign state. The Gaza Strip, West Bank, and Golan Heights continued to be held by Israel and came to be known as the **Occupied Territories.** Israel has variously held them as lands that rightfully belong to Jews and as cards to be played in the regional game of peacemaking.

Arabs and Jews: The Demographic Dimension

As well as issues of land, water, politics, and ideology, the Palestinian-Israeli conflict is about sheer numbers of people. Each side has wanted to maximize its numbers to the disadvantage of the other. To realize the Zionist dream of establish-

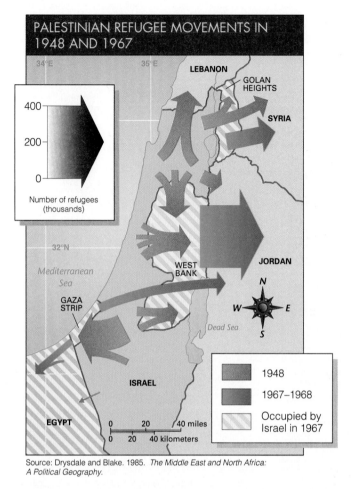

PALESTINIAN REFUGEE MOVEMENTS IN 1948 AND 1967

Number of refugees (thousands)

400
200
0

LEBANON
GOLAN HEIGHTS
SYRIA
34°E
35°E
32°N
Mediterranean Sea
WEST BANK
JORDAN
GAZA STRIP
Dead Sea
ISRAEL
EGYPT

N
W   E
S

0   20   40 miles
0   20   40 kilometers

1948
1967–1968
Occupied by Israel in 1967

Source: Drysdale and Blake. 1985. *The Middle East and North Africa: A Political Geography.*

Figure 9.2   Palestinian refugee movements in 1948 and 1967. Many who fled in the first conflict relocated again in the second.

Another 3.3 million people live in the West Bank, Gaza Strip, and Golan Heights — areas occupied by Israel in the wars of 1967 and 1973. About 205,000 of those are Jewish settlers, and most of the rest are Palestinian Arabs (in the West Bank and Gaza Strip) and Druze and other Arabs (in the Golan Heights). Another 110,000 Jewish settlers live in East Jerusalem, which is also part of the territories Israel conquered in 1967. There are 4.2 million more Palestinians living outside Israel and the Occupied Territories. Jordan has the largest number (2.3 million), followed by Syria (466,000) and Lebanon (430,000). The remaining million are in other Arab countries and in numerous nations around the world, comprising a sizable diaspora of their own. In this respect, the Palestinians are much like the Jews prior to Israel's creation.

After 1967, and especially after the election of its conservative Likud party to power in 1977, Israel moved to strengthen its grip on the Occupied Territories through security measures and the government-sponsored establishment of new Jewish settlements. Most of these settlements were, and are still, spread through the West Bank, where 205,000 Jewish settlers live (see Fig. 9.A, p. 247). These are generally not frontier farming settlements but communities inhabited by relatively prosperous middle-class Jews who commute to jobs in Israel. They are not built as temporary encampments but as permanent fixtures meant to create what have been described as "facts on the ground" — an Israeli presence so entrenched that its withdrawal would be almost inconceivable (Fig. 9.5, p. 248). Some of the motivation for this settlement is ideological, as the West Bank is composed of historical Judea and Samaria, which many devout Jews regard as part of Israel's historic and inalienable homeland.

The proliferation of Jewish settlements worsened relations between Israel and its Arab neighbors and even drove a wedge into the Jewish population of Israel itself. Many Israelis strongly opposed further settlement in the Occupied Territories and annexation of these territories to the Jewish state because if the territories were annexed, Israel would become a country whose population was only about 60% Jewish. With an Arab minority that has a higher birth rate than that of the Jews, Jews could eventually become the minority population in their own country. There is also international pressure against the settlements because they violate Geneva Conventions on activities allowed in territories captured in war. Several U.N. Resolutions, notably 242 and 338, have called on Israel to withdraw completely from the Occupied Territories.

The settlements and other aspects of Israeli occupation, including the periodic "closing off" of the territories which prevents many Palestinians from reaching their jobs inside Israel proper, have fostered widespread and long-lasting Palestinian resistance against Israel. In 1987, Palestinians instigated a popular uprising in the Occupied Territories known as the **Intifada** (Arabic for "the shaking"). Initially, they relied on rocks and bottles to engage Israeli troops, who answered with rubber and metal bullets. International media coverage of a Palestinian "David" fighting an Israeli

ing Israel, Jews began immigrating to Palestine early in the 20th century. Jews were 11% of Palestine's population in 1922, 16% in 1931, and 31% in 1946, on the eve of Israel's creation. In keeping with national legislation known as the **Law of Return,** the Jewish state of Israel has always granted citizenship to any Jew who wishes to live there. Following the 1948–1949 war, waves of new immigrants from Europe (Ashkenazi Jews) joined the Middle Eastern (Sephardic) Jews who had inhabited Palestine and other parts of the Middle East (and, until 1492, Spain) since early times. In the 1990s, more waves of Jewish immigration followed the dissolution of the Soviet Union and a change of government in Ethiopia, where an ancient Jewish group called Falashas had lived in isolation for thousands of years. In 2001, Jews made up 79% of the population of 6.4 million people living within Israel's pre-1967 borders. About 950,000 Palestinians, or 15% of the country's population, are now Israeli citizens residing within the pre-1967 boundaries of Israel; they are known as "Israeli Arabs." Palestinian Arabs without Israeli citizenship, and a variety of other ethnic groups, comprise the balance of the population within Israel's pre-1967 borders.

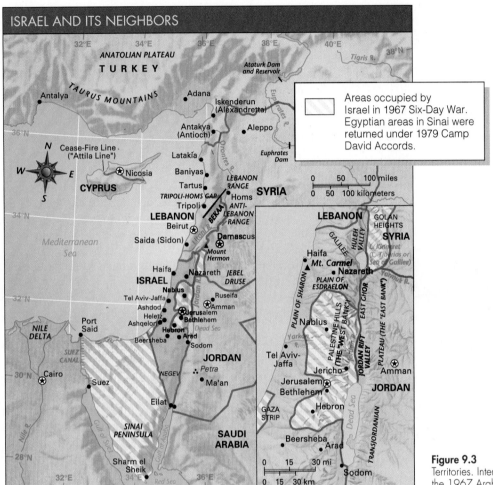

ISRAEL AND ITS NEIGHBORS

Areas occupied by Israel in 1967 Six-Day War. Egyptian areas in Sinai were returned under 1979 Camp David Accords.

Figure 9.3   Israel and its neighbors, with the Occupied Territories. International boundaries show the status prior to the 1967 Arab-Israeli War.

"Goliath" did much damage to Israel's image abroad, and many Israelis came to question the justice and importance to security of Israel's continued occupation. Popular opinion and, by 1992, a new Labor government began to consider the previously unthinkable: giving the Palestinians at least some control over land and internal affairs in the Occupied Territories.

A Geographic Sketch of Israel

Israel is much more than a country in a perpetual state of conflict. It is a physically and culturally diverse and vibrant nation with a promising future, especially if relations with its Arab neighbors improve.

One of Israel's greatest achievements since independence has been the expansion and intensification of agriculture (Fig. 9.6, p. 248). Production concentrates on citrus fruits (including the famous Jaffa oranges), which provide export revenue, and on dairy, beef, and poultry products, as well as flowers, vegetables, and animal feedstuffs. Israel's intensive, mechanized agriculture resembles the agriculture of densely populated areas in Western

Europe. Collectivized settlements called *kibbutzim* (singular, *kibbutz*) are a distinctively Israeli feature on the agricultural landscape. Many of these lie near the frontiers and have defensive as well as agricultural and industrial functions. Far more numerous and important in Israel's agricultural economy are other types of villages, including the small farmers' cooperatives called *moshavim* (singular, *moshav*) and villages of private farmers.

Agricultural development is concentrated in the northern half of the country, with its heavier rainfall (concentrated in the winter in this Mediterranean climate) and more ample supplies of surface and underground water for irrigation. Annual precipitation averages 20 inches (50 cm) or more in most sections north of the city of Tel Aviv and in the interior hills where Jerusalem lies (see Fig. 9.3). Due to the rain shadow effect, very little rain falls in the deep rift valley of the Jordan River. From Tel Aviv southward to Beersheba, annual rainfall decreases to about 8 inches (20 cm), and in the central and southern Negev, it drops to 4 inches (10 cm) or less. To expand agriculture along the coastal plain and in the northern Negev, Israel transfers large quantities of water from the north by the aqueduct and pipeline network of the Na-

DEFINITIONS + INSIGHTS

Chokepoints

The Strait of Tiran at the mouth of the Gulf of Aqaba is one of the Middle East's many **chokepoints** — strategic narrow passageways on land or sea that may be easily closed off by use, or even the threat, of force. Other notable chokepoints in the area are the Strait of Gibraltar at the mouth of the Mediterranean Sea, the Dardanelles and Bosporus Straits (together known as the Turkish Straits; see Fig. 9.4) linking the Black Sea and Aegean Sea, the Bab el-Mandeb at the mouth of the Red Sea, the Strait of Hormuz at the mouth of the Persian/Arabian Gulf, and the Suez Canal. Notable events in military history and the formation of foreign policy in the Middle East focus on these strategic places. For example, Iran's plans to station Chinese-made silkworm missiles on the Strait of Hormuz and thus threaten international oil shipments led to a new level of U.S. involvement late in the Iran-Iraq War of 1980–1988. And Egyptian closure of the Strait of Tiran chokepoint to Israeli shipping in 1967 helped precipitate the Six Day War.

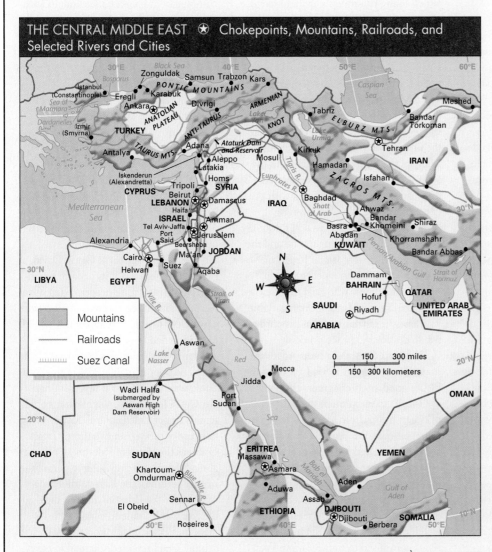

Figure 9.4   Some of the world's most critical waterways, and their vulnerable chokepoints, are located in the Middle East.

tional Water Carrier. The largest source for this network is Lake Kinneret (Lake Tiberias, or Sea of Galilee), which is fed by the upper Jordan River.

Jerusalem (population: 671,700, excluding East Jerusalem in occupied territory) is the country's capital. Tel Aviv-Jaffa (population: metropolitan area, 834,000; city proper, 348,000) on the Mediterranean coast is Israel's greatest industrial center. The leading seaport and center of heavy industry is Haifa in northern Israel. The ancient town of Beersheba is the administrative and industrial center of the Negev Desert region.

President Anwar Sadat of Egypt set the precedent in 1979: Arabs could make peace with Israel. Egypt's Arab neighbors scorned this influential Arab country for a decade for its peacemaking, and Sadat was assassinated because of it. However, in the process, Israel restored the Sinai to Egyptian control and the United States rewarded Egypt handsomely with about $2 billion in aid annually. Along with about $3 billion awarded Israel annually, for about two decades these countries have been the two largest recipients of U.S. foreign aid.

Israel and its remaining Arab adversaries stand to gain much more through the implementation of a comprehensive framework for peace. The United States, Russia, and Norway initiated negotiations in the 1990s that made such a prospect appear possible. The process began in earnest in 1993, when Israel recognized the legitimacy of and began to negotiate with the **Palestine Liberation Organization** (PLO), long recognized by the Palestinians as their legitimate governing body. In return, the PLO, under the leadership of Yasir Arafat, recognized Israel's right to exist and renounced its long-standing use of military and terrorist force against Israel. U.S. President Clinton orchestrated a historic handshake between Arafat and the Israeli Prime Minister Yitzhak Rabin, and the leaders signed an agreement known as the Gaza-Jericho Accord (in its implementation, it came to be known as the Oslo I Accord, after the Norwegian capital where it was negotiated). It was designed to pave a pathway to peace by Israel's granting of limited **autonomy,** or self-rule, in the Gaza Strip and in the West Bank town of Jericho. Autonomy meant the Palestinians were responsible for their own affairs in matters of education, culture, health, taxation, and tourism. Israel also pulled its troops out of these areas. The accord established a 5-year timetable for the resolution of much more difficult matters, the so-called **Final Status issues.** These included the political status of Jewish and Muslim holy places in Jerusalem and of the city itself, the possible return of Palestinian refugees, and implicitly,

Palestinian statehood.

The calendar proved impossible to abide by as each side struggled with lingering and new problems. Arafat's PLO had to respond to a series of violent attacks on Jewish civilians and soldiers by the radical Palestinian organization HAMAS. While these terrorist attacks caused many Israelis to reject the peace process, subsequent PLO detention and prosecution of Palestinian extremists also hurt Palestinian support for the PLO, which many Palestinians came to see as an agent for Israeli interests. Many Palestinians turned against the peace process after February 1994, when a right-wing Israeli extremist opposed to Israel's negotiations with the Palestinians shot and killed Muslim Palestinian worshipers in the Tomb of the Patriarchs, a holy place in the West Bank town of Hebron where the prophet Abraham is said to be buried. Many of the Jewish settlers living in the West Bank are vehemently opposed to the peace process. Regarding Jewish presence in the area as a God-given right, they fear the peace process will result in an independent Palestinian state and the forced withdrawal of Israeli Jews. On November 4, 1995, after a peace rally in Tel Aviv, an Israeli Jew who held this view assassinated Prime Minister Yitzhak Rabin. The assassin proclaimed that his purpose was to destroy the peace process.

Nevertheless, the peace process remained on track. In 1995, Israel and the PLO signed another agreement (known as Oslo II) to implement the withdrawal of Israeli troops from several large Palestinian cities in the West Bank. Palestinians then acquired self-rule over about 30% of the West Bank, with the rest, including all Jewish settlements, remaining in Israeli hands. In a subsequent deal known as the Wye Agreement, two Israeli prime ministers (Netanyahu and Barak) promised to transfer more West Bank lands from Israeli to Palestinian control, and Arafat promised increased Palestinian efforts to crack down on Palestinian terrorists and so guarantee the security of Israelis in the Palestinian territories and in Israel. Most critically, Israel pledged to transfer an additional

13% of the territory it controlled in the West Bank from the area designated in the peace process as Zone C (complete Israeli control) to Zone B (Palestinian civilian control, Israeli security control) and pledged to transfer 14% of the West Bank from Zone B to Zone A (complete Palestinian control). If finally implemented (it had not been as of 2002), this arrangement would have brought the total of West Bank lands under complete Palestinian control to 17%, leaving 57% completely in Israeli hands and 26% under joint control (see Fig. 9.A). Israel also released some Palestinian prisoners from Israeli jails, allowed the opening of an international Palestinian airport in Gaza, and accepted the establishment of a corridor of "safe passage" for Palestinians between the West Bank and the Gaza Strip.

During his final year in office in 2000, U.S. President Clinton sought to solidify his legacy as peacemaker by brokering a historic final settlement between Israelis and Palestinians. PLO Chairman Arafat and Israeli Prime Minister Barak huddled with Clinton and his advisers in the presidential retreat at Camp David, a site chosen because of its historic significance in Middle East peacemaking. Over weeks of tough negotiations, the two sides came close to a deal. It apparently would have included the creation of an independent Palestinian country on the Gaza Strip and approximately 95% of the West Bank, with the other 5% representing clusters of Jewish settlements that would remain under Israeli control. However, the talks broke down over two thorny issues. One was the status of Palestinian refugees abroad; Israel was unwilling to permit the right of return of the large numbers demanded by the Palestinians. But the deal breaker was Jerusalem, which both Israelis and Palestinians proclaim as their capital. Israel reportedly agreed to have the Old City of Jerusalem divided, with Israel assuming sovereignty over the Jewish and Armenian quarters and Palestinians assuming sovereignty over the Arab and Christian quarters. These issues were not as problematic as were issues surrounding the places held holy by each side. Israel reportedly agreed to allow the Palestinians "soft sovereignty" over the Temple Mount/al-Haraam ash-Shariif, meaning Palestinians could administer the area and fly a flag over it, but Israel would technically "own" it. Israel saw this point as essential because, after all, the First and Second Temples had stood on this spot. The Palestinians, however, insisted on full sovereignty over the Temple Mount/al-Haraam ash-Shariif, and both sides walked away from the deadlock.

Tragically, the near-peace evaporated and was replaced by a state of war within weeks. On September 28, 2000, the right-wing Israeli opposition leader (subsequently, the prime minister) Ariel Sharon walked up to Temple Mount/al-Haraam ash-Shariif under heavily armed escort. He meant to make the point that this sacred precinct belonged to Israel. Many observers claim he also meant to disrupt the peace process. Whether or not it was his intention, that disruption resulted. Beginning that day, and throughout the following months, Palestinians took to the streets in the "al-Aqsa Intifada" to challenge Israeli forces with

rocks, bottles, and guns. Hundreds of Palestinians and scores of Israelis died. In a ceaseless cycle of revenge killings, HAMAS stepped up its shootings and suicide bombings against Israeli civilians and soldiers, while Israelis assassinated suspected Palestinian militants. New militant Palestinian groups emerged including the al-Aqsa Martyrs' Brigade, which was affiliated with Yasir Arafat's Fatah wing of the PLO. Along with HAMAS, the al-Aqsa group carried out a series of ferocious suicide bombings of civilian Israeli targets, including a seaside hotel where Jews were beginning their Passover dinner in March 2002. Prime Minister Sharon reacted to the "Passover Massacre" with "Operation Defensive Shield." This full-scale Israeli army invasion of several autonomous Palestinian towns in the West Bank, including Bethlehem, Ramallah, Nablus, and Jenin, destroyed much of the infrasructure of Arafat's Palestinian Authority and leveled entire sections of some towns. The Israeli assault on Jenin was particularly devastating. Palestinian claims that a massacre of civilians had taken place there could not be readily investigated be-

cause of Israel's success in keeping media and diplomatic observers at bay. The escalating violence prompted the United States to call for an international peace conference, and Saudi Arabia fielded a plan by which all the Arab countries would recognize Israel's soveignty if Israel withdrew from all of the Arab territories it occupied in 1967.

The peace process has both progressed and faltered on other fronts. In October 1994, following the PLO's lead, Jordan signed a treaty with Israel. It called for an end to hostility between the two nations; establishment of full diplomatic relations; the opening of exchanges in trade, tourism, science, and culture; and an end to economic boycotts. Meanwhile, Syria and Israel have failed to reach a peace agreement because of a deadlock on the issues of recognition of Israel and return of occupied land: Israel refuses to return the Golan Heights to Syria until Syria recognizes the sovereignty and legitimacy of Israel, and Syria refuses to recognize the sovereignty and legitimacy of Israel until Israel returns the Golan Heights. Syrian President Hafez al-Assad (who died in 2000) and his son, Bashar, who succeeded him, apparently wanted to see how successful Israel's existing peace treaties would be before concluding Syria's own with Israel. And because Syria is the dominant power in Lebanon, with tens of thousands of troops stationed in that tiny, politically weak country, peace between Israel and Lebanon is improbable unless Israel and Syria first come to terms.

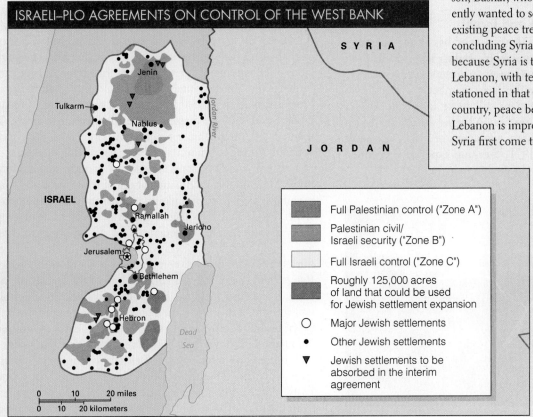

ISRAELI–PLO AGREEMENTS ON CONTROL OF THE WEST BANK

Full Palestinian control ("Zone A")

Palestinian civil/ Israeli security ("Zone B")

Full Israeli control ("Zone C")

Roughly 125,000 acres of land that could be used for Jewish settlement expansion

○ Major Jewish settlements

• Other Jewish settlements

▼ Jewish settlements to be absorbed in the interim agreement

Source: *Yediot Ahronot* newspaper, *AP* research. *Columbia Missourian*, 11/20/98, p. 3A. AP/Wm. J. Castello

Figure 9.A   Israeli and Palestinian areas of control in the West Bank (as of 2002). This patchwork quilt may be rearranged by the violence and political uncertainty that have followed years of peacemaking.

Figure 9.5   An aerial view shows a Jewish settlement built in 1998 on a hilltop north of the Jewish settlement Eli in the Samarian hills, southeast of Nablus, on the West Bank.

To the south lies Israel's small Red Sea port and resort town of Eilat. Thanks to the successful peace accords with Egypt and Jordan, open doors now link Eilat with the neighboring resort towns of Taba in the Egyptian Sinai and Aqaba in Jordan. Whenever conflict subsides tourism booms, and developers are building and touting a three-nation "Red Sea Riviera" to satisfy mostly European sun-worshippers.

Jordan: Between Iraq and Israel

Jordan's economic and political situation has always been precarious. Throughout his long reign, ending with his death in 1999, the country's King Hussein withstood various crises. Many of these related to the large numbers of Palestinian refugees in Jordan. Palestinians still make up about 60% of Jordan's population of 5.2 million. After the 1967 war, Jordan was the chief base for military operations of the PLO against Israel, but in 1970 and 1971, the king expelled the armed Palestinian forces, which relocated to Lebanon.

Figure 9.6   Avocado orchard in the Upper Hula Valley, Galilee region, northern Israel.

Jordan is a land of desert and semidesert. The main agricultural areas and settlement centers, including the largest city and capital, Amman (population: metropolitan area, 2,443,500; city proper, 1,213,800), are in the northwest. Prior to the 1967 war with Israel, northwestern Jordan included two upland areas — the West Bank on the west (captured during the war) and the East Bank, or Transjordanian Plateau, on the east — separated by the deep rift valley occupied by the Jordan River, Lake Kinneret, and the Dead Sea (see Fig. 9.3). Most of the valley bottom lies in deep rain shadow and receives less than 4 inches (10 cm) a year. The bordering uplands, rising to more than 2,000 feet (610 m) above sea level and lying in the path of moisture-bearing winds from the Mediterranean, receive winter precipitation averaging 20 inches (50 cm) or more annually. But overall, rainfall is so scarce that only small areas can be cultivated without irrigation, and the water available for irrigation is limited. Only about 4% of Jordan is cultivated, although larger areas are used to graze sheep and goats. Irrigated vegetables, together with olives, citrus fruits, and bananas, provide substantial exports, but some food has to be imported. Natural resources are limited, and there is little manufacturing. Rock phosphate and potash extracted from the Dead Sea are shipped out of Jordan's small port of Aqaba on the Gulf of Aqaba. Tourism is a growing industry, with the Nabatean rose-red city of Petra the chief attraction (see Fig. 8.6).

Lebanon and Syria

Israel's immediate neighbors on the north are Lebanon and Syria (see Fig. 9.3). Their physical and climatic features are similar to those of Israel and Jordan. As in Israel, narrow coastal plains backed by highlands front the sea. The highlands of Lebanon and Syria are loftier than those of Israel and Jordan. Mt. Hermon and the Lebanon and Anti-Lebanon Ranges rise above 10,000 feet (c. 3,050 m). Cyclonic precipitation brought by air masses off the Mediterranean is supplemented by orographic precipitation, the rain and snow generated along the mountain chains paralleling the coast.

Lebanon and Syria have agricultural patterns that conform to the Mediterranean climatic pattern of winter rain and summer drought. The coastal plains and seaward-facing mountain slopes are settled by villagers growing Mediterranean crops such as winter grains, vegetables, and drought-resistant trees and vines. Lebanon's Bekaa Valley, drained in the south by the Litani River, lies between the Lebanon and Anti-Lebanon Ranges and is a northward continuation of the Jordan rift valley and the African Great Rift Valley. Baalbek, its chief city and the site of an important ancient Roman temple, became a Syrian military stronghold in 1976 and a center of pro-Iranian Hizbullah activities against Israel after 1982. The political and military volatility of the Bekaa was conducive to the emergence of a flourishing drug industry, producing both hashish (from marijuana plants) and opium (from poppies), from the mid-1970s until the 1990s. In the early 2000s, in appears that at least marijuana cultivation is making a comeback as alternative crops have brought farmers far less income.

Syria is more industrialized than Lebanon, but neither country compares with Israel industrially. Textile milling, agricultural processing, and other light industries predominate. Both countries are deficient in minerals, although Syria has an oil field in the east. Its reserves are expected to be exhausted by 2010.

There are three large, historically important cities in Syria and Lebanon. Damascus (population: metropolitan area, 2,232,800; city proper, 1,747,200), Syria's capital, is in a large irrigated district at the foot of the Anti-Lebanon Range. Its Umayyad Mosque, originally a Byzantine cathedral, is one of the great monuments of the Middle East. The main city of northern Syria, Aleppo, is an ancient caravan center located in the Syrian Saddle between the Euphrates Valley and the Mediterranean. The old Phoenician city of Beirut (population: metropolitan area, 1,858,600; city proper, 1,135,800), on the Mediterranean, is Lebanon's capital, largest city, and main seaport. Equipped with modern harbor facilities and an important international airport, Beirut before 1975 was one of the busiest centers of transportation, commerce, finance, and tourism in the Middle East.

For its prosperity and its natural beauty, pre-1975 Lebanon had earned the name "Switzerland of the Middle East." Then a precarious political balance broke down, and the country was plunged into civil war in 1975 and 1976. The war's origins were rooted in Lebanon's mixture of strongly localized religious communities. Seventeen separate ethnic and religious groups, known as "confessions," are recognized officially (Fig. 9.7). Major ones include the Maronite Christians, who predominate in the Lebanon Range north of Beirut; Sunni Muslims close to the northern border with Syria; Shi'ite Muslims in the Bekaa Valley and along Lebanon's southern border with Israel; and Druze Muslims in the Shouf Mountains, part of the Lebanon range south of Beirut. Eastern Orthodox, Greek Catholic, and nearly a dozen other Christian sects are also present in Lebanon. Beirut itself is divided between predominantly Muslim West and Christian East. Conflict between the communities broke out from time to time, and foreign powers that gained influence with one or another group heightened the antagonisms.

What the French carved out of greater Syria and put together as Lebanon in 1920 was a collection of small geographic areas, each dominated by an ethnic/religious group. A 1932 census revealed a majority of Christians in Lebanon, and the country's 1946 constitution distributed power according to those census figures. Maronite Christians held a majority of seats in parliament, followed by Sunni Muslims and Shi'ite Muslims. Top leadership positions were allocated by a formula still followed today, with the president a Maronite Christian, the prime minister a Sunni Muslim, and the speaker of parliament a Shi'ite Muslim.

There was never an effort to unify the disparate geographic and ethnic enclaves of this mountainous land. Muslim populations grew and eventually came to outnumber the Christians, who were unwilling to relinquish their political and economic privileges. Palestinians became part of the volatile mix. Pushed from Jordan, PLO guerrillas established a virtual "state within a state" in the southern slums of Beirut and in Shi'ite-controlled areas of southern Lebanon, from which they launched attacks on Israel. The war started in 1975 as the "have-not" Sunnis, soon joined by Shi'ites and Palestinians, took on the more prosperous Maronite Christians. Syria and Israel supported opposing sides. By the end of the war in 1976, as many as 100,000 people, most of them civilians, had been killed (Fig. 9.8). The war accomplished little. Lebanon effectively split into a number of geographically distinct microstates or fiefdoms—a politically unstable situation that set the stage for the next war in 1982.

The 1982 conflict was one facet of a larger war in Lebanon involving both civil strife and international intervention. Lebanon had become the main stronghold of the Palestine Liberation Organization, which used the country as a base for guerrilla operations against Israel. In June 1982, Israel invaded Lebanon with the avowed intention of smashing the PLO and establishing security for the northern Israeli frontier. Syria

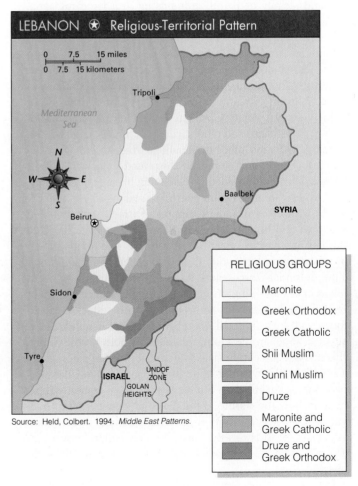

LEBANON ✪ **Religious-Territorial Pattern**

0   7.5   15 miles
0   7.5   15 kilometers

Mediterranean Sea

Tripoli

Baalbek

Beirut

SYRIA

Sidon

Tyre

ISRAEL   UNDOF ZONE
GOLAN HEIGHTS

RELIGIOUS GROUPS

Maronite

Greek Orthodox

Greek Catholic

Shii Muslim

Sunni Muslim

Druze

Maronite and Greek Catholic

Druze and Greek Orthodox

Source: Held, Colbert. 1994. *Middle East Patterns.*

Figure 9.7   Generalized map of the distribution of religious groups in Lebanon.

Figure 9.8   The Lebanese Civil War of 1975–1976 devastated the capital city of Beirut. This is the lobby of the Holiday Inn, where a fierce tank battle had taken place a few days earlier.

resisted the Israeli advance and gave support to the PLO. Israeli troops routed and inflicted heavy casualties on both the Syrians and the PLO. The Israeli push continued northward to the suburbs and vicinity of Beirut, culminating in a ferocious aerial and artillery bombardment of the city — the Battle of Beirut — that ended only when U.S. President Ronald Reagan exerted pressure on Israel to cease fire. Under supervision by U.S. and other international peacekeeping troops, the PLO withdrew from Beirut to Tunisia. After the multinational troops withdrew, alleged Muslim assailants killed Lebanon's Christian President Bashir Gemayal. Angry Christian forces then massacred hundreds of Palestinians and other civilians in refugee camps south of Beirut. U.S. and other multinational forces were again deployed to Beirut, and American ships used firepower to assist the new Christian president in widening his area of authority.

Lebanese Shi'ite Muslim guerrillas, seeking to avenge U.S. attacks on Lebanon and its support of Israel, soon bombed American and French targets in Beirut, killing over 200 U.S. Marines in a barracks near Beirut's airport. The multinational forces withdrew, but Syrian troops remained in Lebanon and allowed Iranian forces and Iran's Shi'ite Muslim allies in Lebanon to establish a stronghold in the Bekaa Valley town of Baalbek. Subjected to persistent guerrilla attacks, Israeli troops withdrew from all of Lebanon except a self-declared "security zone" in the south, inhabited by both hostile Shi'ite Muslims and Israel's de facto Christian Lebanese allies. Israel exerted control over this zone and engaged in regular skirmishes with Hizbullah, a Shi'ite Muslim organization sympathetic to Iran and hostile to the peace process between Israel and its Arab neighbors. With growing numbers of casualties among Israeli forces in the zone, the occupation became increasing unpopular in Israel, and troops were finally withdrawn in 2000.

Lebanon is now on a steady course of reconciliation. Beirut is rising from the ashes in an ambitious, expensive re-

construction program. Political power has been redistributed to reflect the new demographic composition in which Muslims outnumber Christians. Once dominated by Christians, the Lebanese army is now comprised of soldiers from all of Lebanon's ethnic confessions. There are numerous social and education programs underway that mix children from different ethnic groups, with the hope they will learn tolerance. Lebanon is still not in control of its own destiny, however. The government in Damascus has never accepted the 1920 division of greater Syria, which gave a portion of Syria to Lebanon, and as of 2002, 30,000 Syrian soldiers remained in Lebanon. Surprisingly, perhaps, many Lebanese from all the major ethnic groups want the Syrians to remain and keep the peace while the country's former enemies continue the long, slow healing process.

Syria, like Lebanon, is a country beset by religious factionalism. In Syria, however, Muslims outnumber Christians nine to one. In 1970, a coup overthrew the ruling Sunni Muslim establishment and brought to power Hafez al-Assad, a member of the Alawite sect of Shi'ite Islam. Under his authoritarian rule as president, dominance of the government quickly passed to the Alawites, who make up only 12% of the country's population. The police and army crushed opposition within other communities, notably Sunni Islamists who rejected Alawites as heretics. For three decades, until his death in 2000, President Assad was a skilled politician who played Middle Eastern affairs cautiously. His regime survived the economic and political setbacks accompanying the downfall of its benefactor, the Soviet Union, and slowly adopted a more accommodating relationship with the United States and the West. His son and successor Bashar al-Assad has continued the cautious and generally unaccommodating approach to Israel established by Hafez al-Assad.

9.2 Egypt: The Gift of the Nile

At the southwest, Israel borders the Arab Republic of Egypt, the most populous Arab country (69.8 million). This ancient land, strategically situated at Africa's northeastern corner between the Mediterranean and Red Seas, is utterly dependent on a single river. The Nile has created the floodplain and delta on which more than 95% of all Egyptians live (see map, Fig. 9.9). The Greek historian and geographer Herodotus called Egypt "the gift of the Nile." The river's bountiful waters and fertile silt helped Egypt become a culture hearth that produced remarkable achievements in technology and cultural expression over a period of almost 3,000 years until about 500 B.C., when a succession of foreign empires came to rule the land.

The Nile Valley may appropriately be described as a "river oasis," for stark, almost waterless desert borders this lush ribbon (Fig. 9.10). Only 2% of Egypt is cultivated, and nearly all this land lies along and is watered by the great river. The conversion of the original papyrus marshes and other wetlands along the Nile to the thickly settled, irrigated landscape of to-day is a process that has been unfolding for more than 50 cen-turies. It is difficult to imagine that, in the time of the pharaohs, crocodiles and hippopotami swam in the Nile and game animals typical of East Africa roamed the nearby plateaus.

Ancient Egyptian civilization was based on a system of **basin irrigation,** in which the people captured and stored in built-up embankments the floodwaters that spread over the Nile floodplain each September. Egyptian farmers allowed the water to stand for several weeks in the basins, where it su-persaturated the soil and left a beneficial deposit of fertile vol-

Figure 9.10 The Nile Valley in southern Egypt. The Nile River snakes through the left side of the cultivated area. Few other places on Earth exhibit such a stark contrast between productive, populated landscapes and those virtually devoid of water, people, and vegetation.

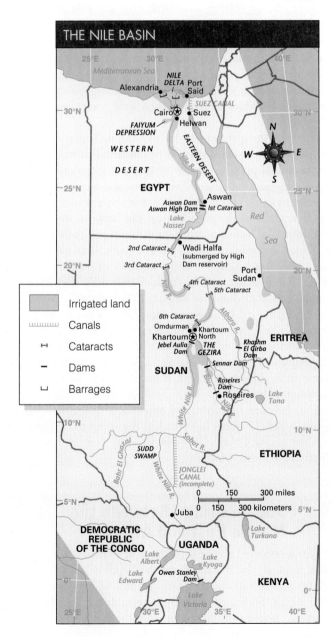

Figure 9.9 The Nile Basin. The widths of the irrigated strips along the Nile are exaggerated.

canic silt washed down by the Blue Nile from Ethiopia. They drained the excess water back into the Nile and planted their winter crops in the muddy fields. They were able to harvest one crop a year with this method. They could produce a sec-ond or third crop on limited areas of land using water-lifting devices such as a weighted lever, or *shaduf* (a bucket on a counterweighted pole, manually operated to raise water from canal to field), and an Archimedes' screw (a cylinder contain-ing a screw, also manually turned to raise the water).

However, achieving **perennial irrigation** on a vast scale re-quired the construction of barrages and dams, which were in-novations of the 19th and 20th centuries in this part of the world (see Fig. 9.9). Late in the 19th century, French and British colonial occupiers of Egypt, anxious to boost Egypt's exports of cotton (a summer crop demanding perennial irriga-tion), began Egypt's conversion from basin to perennial irriga-tion by constructing a number of barrages, or low barriers designed to raise the level of the river high enough that the water flows by gravity into irrigation canals. Barrages are not designed to store large amounts of water, a function now per-formed by two dams in Upper Egypt (the old Aswan Dam, completed in 1902, and the huge Aswan High Dam, com-pleted in 1970) and several others along the Nile and its tribu-taries in Sudan and Uganda.

The enormous Aswan High Dam (Fig. 9.11) is located on a stretch of the Nile known as the First Cataract (the Nile's many cataracts are areas where the valley narrows and rapids form in the river). Its reservoir, Lake Nasser, stretches for more than 300 miles (480 km) and reaches into northern Sudan.

Like all giant hydrological schemes, the Aswan High Dam has generated benefits and liabilities, and its construction was controversial. On the plus side, by storing water in years of high flood for use in years of low volume, the High Dam pro-vides a perennial water supply to the Nile Valley's 6 million

JOE HOBBS

Figure 9.11 Egypt's controversial Aswan High Dam (at top center, just beyond the white monument). Lake Nasser stretches toward the right. In the foreground is the settlement that was inhabited by the Soviet technicians who worked on the project.

irrigated acres (2.4 million hectares), thus extending the cultivation period. The dam also made possible the "reclamation" of previously uncultivated land on the desert fringe adjoining the western Nile delta. Navigation of the Nile downstream from (north of) Aswan is much easier, and the country's industries rely heavily on the hydroelectricity produced at the dam. Egypt weathered a 2-year drought in the late 1980s thanks to the excess waters stored in Lake Nasser. The dam also eliminates the prospect of catastrophic flooding, which sometimes occurred before it was built.

On the other hand, the dam has caused the water table to rise, making it harder for irrigated soil to drain properly. When farmers use too much water, standing water evaporates and leaves a deposit of mineral salts so that the once-fertile soil loses its productivity — a problem known as **salinization.** In addition, perennially available canal waters are ideal breeding grounds for the snail that hosts the parasite that causes schistosomiasis or bilharzia, a debilitating disease affecting a large proportion of Egypt's rural population. Mediterranean sardine populations, now deprived of the rich silt that nurtured their feeding ground, have plummeted off the Nile delta, and the sardine industry has faltered. Without the free fertilizer the silt offered, Egypt's bill for artificial fertilizers has increased. Generally, however, Egyptians are very proud of the High Dam, and it has increased food output dramatically for Egypt's mushrooming population.

Egypt's food crops include corn, wheat, barley, rice, millet, fruit, vegetables, and sugarcane. The country was once a great food exporter, but its population has grown so explosively that food now comprises over 30% of Egypt's import expenditures. Long-staple Egyptian cotton has long been the main commercial export crop for the hard currency Egypt needs desperately. Rural population densities already are among the highest in the world, and the crowding is growing worse despite a steady migration to Egypt's primate city, Cairo (population: metropolitan area, 15,206,000; city proper, 7,594,800) — the largest metropolis in Africa and the Middle East — and other cities. So far, the government's efforts to lure urban-dwellers and factories out of the Nile Valley to desert "satellite cities" have borne

little fruit. Egypt is much more successful in exporting its brains and brawn abroad: Egyptians are prominent among the teachers, doctors, lawyers, engineers, and laborers working in the wealthy Gulf countries, and the remittances they send home pump much-needed cash into Egypt's economy.

Egypt's industries are concentrated in Cairo and in the Mediterranean port of Alexandria. They include cotton textiles, food processing, clothing manufacture, chemical fertilizers, and cement. Automobiles and other consumer durables are assembled primarily from imported components. The country is self-sufficient in some vital minerals, including oil, and exports some oil. But Egypt lacks the mineral wealth to support extensive industrialization, and for its development, the country depends heavily on foreign loans and grants. Income from tourism, particularly to the outstanding temples and tombs of ancient Egypt and to the world-famous dive sites on the Red Sea and Gulf of Aqaba, helps redress the unfavorable trade balance, as do transit fees from ships using the Suez Canal. However, tourism is a very vulnerable resource, as the number of visitors plummets each time violence rocks the Middle East.

Egypt is the giant of the Arab World, its largest country in terms of population, and its most influential in regional and international politics. The Western powers recognize its key role in mediation between Palestinians and Israelis. The West is anxious to see that Egypt does not become poorer than it already is and that the already huge gap between the rich and poor does not widen, perhaps throwing its population into the embrace of militant Islam. And Egypt's strategic location astride the Suez Canal is also a major concern for the West. The United States is Egypt's chief non-Arab ally, lavishing huge sums of civilian and military aid in support of the autocratic regime led by President Hosni Mubarak. To quell real and perceived threats by Islamists, Mubarak has maintained a state of emergency for two decades and effectively stifled most opposition. A grass-roots Islamic movement has grown at the same time. Mostly peaceful, it has involved the widespread adoption of traditionally Islamic values in dress, education, relations between the sexes, and other social and cultural practices. A radical fringe has attacked the interests of Egypt's 6 million strong Coptic Christian minority and foreign targets such as the tourism industry.

9.3 Sudan: Bridge Between the Middle East and Africa

Egypt is bordered on the south by Africa's largest country, the vast, tropical, and sparsely populated republic of Sudan (population: 31.8 million). Formerly controlled by Great Britain and Egypt, Sudan received its independence in 1956. This southern neighbor is vitally important to Egypt, for it is from Sudan that the Nile River brings the water that sustains the Egyptian people. The Nile receives its major tributaries in Sudan (see Fig. 9.9), and storage reservoirs exist there on the

Blue Nile, the White Nile, and the Atbara River, all of which benefit Egypt as well as Sudan itself.

The Nile's main branches — the White Nile and Blue Nile — originate respectively at the outlets of Lake Victoria in Uganda and Lake Tana in Ethiopia and flow eventually into Sudan. In southern Sudan, the river passes through a vast wetlands called the Sudd, where much water is lost by evaporation. An artificial channel to straighten the river's course and increase the volume and velocity of its flow through the Sudd was under construction for a long time. Called the Jonglei Canal, its future is uncertain. Civil war in Sudan has halted its progress since 1985. Predicting the death of the Sudd wetlands if the canal is completed, many environmentalists would be pleased if work was not resumed.

Sudan is overwhelmingly rural. The largest urban district, formed by the capital, Khartoum (population: metropolitan area, 5,524,500; city proper, 1,191,800), and the adjoining cities of Omdurman and Khartoum North, is located toward the center of the country at the junction of the Blue and White Niles. Transportation within Sudan as a whole is poorly developed. Several cataracts prohibit navigation on the Nile north of Khartoum (see Fig. 9.9), but the White Nile is continuously navigable at all seasons from Khartoum to Juba in the deep South. Most of the country has no railways or good highways, although there is widespread airplane service and much automobile traffic despite the lack of surfaced roads. Khartoum itself has rail connections with several widely separated sections (see Fig. 9.4). The main route for rail freight connects Khartoum and the Gezira with Port Sudan, a modern, well-equipped port that handles most of Sudan's seaborne trade.

Approximately 6% of Sudan is cultivated, but only about 10 to 20% of the cultivated land is irrigated. The largest block of irrigated land is found in the Gezira (Arabic for "island") between the Blue and White Niles. The most important cash crop and export of Sudan is irrigated Egyptian-type cotton. Many Sudanese support themselves by raising cattle, camels, sheep, and goats. The country's industries are relatively meager, and there are few mineral reserves, except for some oil in southern Sudan.

Most of Sudan is an immense plain, broken frequently by hills and in a few places by mountains. Climatically and culturally, the area is transitional between the Middle East and Africa South of the Sahara. A contrast exists between Sudan's arid Saharan North, peopled mainly by Arabic-speaking Muslims, and the seasonally rainy equatorial South, with its grassy savannas, papyrus wetlands, and non-Arabic-speaking peoples of many tribal and ethnic affiliations who practice both Christian and animistic faiths. About three-fourths of Sudan's total population is Muslim.

There has been much political friction between the two sections. An all-out civil war has raged between northern and southern Sudan since 1983, with southern factions of the Sudanese People's Liberation Army (SPLA) engaging government army troops from the north. Some casualty estimates exceed 2 million people, mostly civilians, and 4.5 million more have been driven from their homes. Among the reasons for this conflict have been historical antagonisms and economic disparities between the more developed North and the less developed and poorer South, along with efforts by the Arab-dominated government to impose Islamic law (*shari'a*) and the Arabic language on the South and to exploit the South economically. The conflict has helped keep Sudan mired in poverty. The country could become a modest oil exporter, but SPLA rebels have repeatedly attacked the country's single export pipeline leading from the oil producing area around Bentui to the coast at Port Sudan. Bentui is perilously close to the southern SPLA stronghold, and the county's largest oil reserves are actually within SPLA territory.

Warfare and prolonged drought have devastated southern Sudan and adjacent Ethiopia during recent years, and Sudan in particular has hovered near catastrophic famine. Huge tides of refugees have flowed back and forth between Sudan and Ethiopia. International relief agencies have been allowed to provide food and medical aid in only a small portion of the war zone. Sudan's military regime has meanwhile been accused of gross human rights violations, including **ethnic cleansing** of the Nuba, Nuer, Dinka, and other tribespeople in the South. Human rights groups have accused the government of promoting the abduction and enslavement of non-Muslim, mostly Dinka, southerners, who usually end up serving Arab masters in the North.

Sudan's foreign relations have changed since 1989, when an Islamic military regime overthrew an elected government. The new rulers allied themselves with militant Iran. In 1993, after obtaining information linking Sudanese officials with the first bombing of New York City's World Trade Center, the United States branded Sudan a "terrorist" nation and halted economic aid. Following the analysis of intelligence information about bombings of U.S. embassies in Kenya and Tanzania, in August 1998 American cruise missiles obliterated a pharmaceutical factory, alleged to be a chemical weapons factory, in Sudan's capital.

9.4 Libya: Deserts, Oil, and a Defiant Survivor

Egypt's neighbor to the west, Libya — formerly an Italian colony — became an independent kingdom in 1951. In 1969, Libya became a republic after Colonel Muammar al-Qaddafi and other army officers led a coup against the monarchy. Some 97% of the country's people are Muslim Arabs and Berbers.

Libya is made up of three major areal divisions: Tripolitania in the northwest, Cyrenaica in the east, and the Fezzan in the southwest (Fig. 9.12). The principal cities are the capital, Tripoli (population: metropolitan area, 2,188,800; city proper, 1,948,400) in Tripolitania, and Benghazi in Cyrenaica.

Most of Libya lies in the Sahara and is too dry to support cultivation except at scattered oases watered by wells and springs. The country is sparsely inhabited; its 5.2 million people are heavily concentrated in coastal lowlands and low

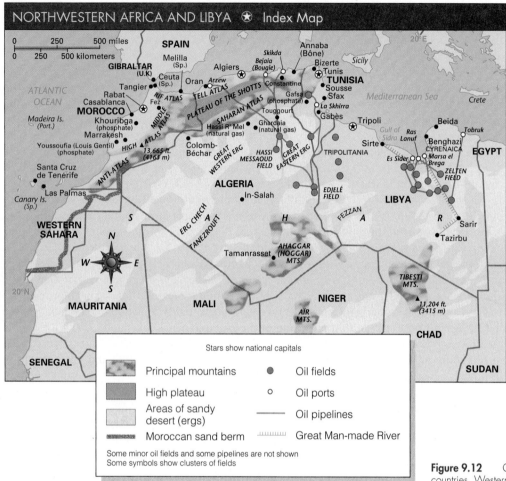

NORTHWESTERN AFRICA AND LIBYA ✪ Index Map

Figure 9.12   General reference map of the Maghreb countries, Western Sahara, and Libya.

highlands along the Mediterranean, where limited winter rains and irrigation with groundwater allow small-scale cultivation of typical Mediterranean crops. Center-pivot irrigation of wheat is carried on in southern Libya. Deep wells there tap **fossil waters** that accumulated in deep limestone aquifers in ancient times when rains fell abundantly in what is now the Sahara (Fig. 9.13). In an ongoing project called the Great Man-made River, these same fossil waters are diverted from 1,155-feet (347-m) deep wells around Sarir and Tazirbu in east central Libya by a pipeline 13 feet (4 m) in diameter stretching 560 miles (900 km) to Sirte and Benghazi on the Mediterranean (see Fig. 9.12). The goal of this effort, which Colonel Qaddafi calls "the world's largest civil engineering project," is agricultural self-sufficiency through the irrigation of fruits, vegetables, and wheat in northern Libya.

Crude oil, petroleum products, and natural gas represent nearly all of Libya's exports (primarily to Western Europe) and have brought prosperity to the country since the 1960s. The government spends much of the oil revenue on public works, education, housing, aid to agriculture, and other projects for social and economic development. The country has also spent large sums on weapons.

Libya's revolutionary government has long been involved in an exchange of hostile rhetoric and sometimes hostile action with the United States, its Western allies, and Israel. In 1986, American warplanes bombed Tripoli and Benghazi in reprisal for alleged terrorist acts against American citizens

Figure 9.13   By tapping into "fossil waters" in aquifers beneath the Sahara Desert, Libyan farmers are able to use center-pivot irrigation to produce a surprisingly abundant yield of wheat and other crops.

instigated by Libya, which is on the U.S. list of "terrorist states." After Western intelligence authorities linked two Libyans to the 1988 bombing of Pan American Airlines' Flight 103 over Lockerbie, Scotland, Libya was subjected to an international air embargo. Following Libya's release of the suspects for trial (in which they were convicted) in the Netherlands, the embargo was lifted in 1999, and Libya's relations with the West have softened considerably. Even U.S. oil companies are gradually getting back to work in this oil-blessed country. International tourists are beginning to be drawn to Libya's outstanding Roman sites, pristine beaches, and desert wilderness.

Figure 9.14   Looking north from Morocco across the Strait of Gibralter to Spain. The Mediterranean is on the right and the Atlantic Ocean is to the left.

9.5 Northwestern Africa: The Maghreb

Arabs know the northwestern fringes of Africa as the Maghreb (meaning "Place of the West") of the Arab world. Most of this area's people are Muslim Arabs. The ethnically distinct Berbers, who are most numerous in Morocco and Algeria, converted to Islam after the Arabs brought the new faith into the region in the seventh century A.D. Many Berbers now speak Arabic, and most have adopted Arab customs.

The Maghreb includes four main political units: Morocco, Algeria, Tunisia, and the disputed Western Sahara (see Fig. 9.12). Morocco is geographically and culturally the closest to Europe; only the 11-mile-wide (18-km) Strait of Gibraltar separates it from Spain (Fig. 9.14). Islamic influences crossed the strait beginning early in the eighth century and prevailed in southern Spain until 1492, when Spain's monarchs ordered both Jews and the Muslim Moors to be expelled from Spain. France became the dominant power in the Maghreb during the 19th and early 20th centuries. France invested large sums to develop mines, industries, irrigation works, power stations, railroads, highways, and port facilities, but French rule was very unpopular. In 1956, after long agitation by local nationalists, Tunisia and Morocco secured independence. In Algeria, independence in 1962 came only after a bitter civil war lasting 8 years.

Most people in the Maghreb live in the coastal belt of Mediterranean climate (called "The Tell" in Algeria), with its cool-season rains and hot, rainless summers. Here, in a landscape of hills, low mountains, small plains, and scrubby Mediterranean vegetation, people grow the familiar products associated with the Mediterranean climate (Fig. 9.15). The region's major cities are on the coasts: on the Mediterranean, Tunisia's capital, Tunis (population: metropolitan area, 1,625,400; city proper, 693,700), and Algeria's capital, Algiers (population: metropolitan area, 3,717,100; city proper, 1,650,300), and on the Atlantic, Morocco's famed Casablanca (population: city proper, 3,292,100).

Inland, the Atlas Mountains extend in almost continuous chains from southern Morocco to northwestern Tunisia, with the highest peaks in the High Atlas Mountains of Morocco.

They block the path of moisture-bearing winds from the Atlantic and the Mediterranean, and some mountain areas receive 40 to 50 inches (c. 100 to 130 cm) or more of precipitation annually. The precipitation nourishes forests of cork oak or cedar in some places. In Algeria, the mountains form two east-west chains, the Tell Atlas nearer the coast and the Saharan Atlas (see Fig. 9.12) farther south.

All the countries reach the Sahara to the south. Some pastoral nomadic tribepeople, including the legendary Tuareg, still migrate through the Sahara with their camels, sheep, and goats. Clusters of oases exist in a few places, such as in the dramatic sandstone Ahaggar Mountains of southern Algeria, which rise high enough to catch moisture from the passing winds. A line of oases fed by springs, wells, mountain streams, and *foggaras* (tunnels from mountain water sources, known as

Figure 9.15   The western Maghreb has breathtaking physical landscapes, from the Mediterranean regions in the north to the Atlas Mountains and Sahara Desert in the south. Traditional village and urban architecture lends a unique character to the human environment. This is Vallée du Dades in Morocco, with the Atlas Mountains in the background.

qanats in Iran) lies along or near the southern base of the Atlas Mountains. Several large areas of sandy desert (*ergs*) and a barren gravel plain, the Tanezrouft, occupy portions of the Algerian Sahara.

The Maghreb economies benefit from a valuable endowment of minerals. Algeria's oil, located in the Sahara, is the region's greatest asset (see Fig. 9.12). The oil flows through pipelines to shipping points on the Mediterranean in Algeria and Tunisia. Petroleum products, crude oil, and natural gas make up almost all of Algeria's exports by value. Morocco's exports also include some minerals, notably phosphate, but agricultural exports and clothing comprise a much larger total. Consumer industries, including plants that assemble foreign-made components, are the second leading industries in the Maghreb countries.

Algeria's oil boom went bust in the 1980s, but population growth continued, fueling an economic crisis and shortages of food and consumer goods that contributed to widespread unrest in the 1990s. Algeria also has built up a staggering foreign debt that has contributed to instability. Some economic relief might be provided by a pipeline that will carry Algerian natural gas to Europe. Meanwhile, political instability in Algeria has reverberated through the entire Maghreb and across the Mediterranean to the region's former colonial power, France. For almost 8 years after 1992, when Algeria's military government canceled the results of parliamentary elections that would have given majority seats to the Islamic Salvation Front (FIS), civil strife claimed an estimated 80,000 lives within Algeria. Violent sympathizers of the FIS also carried their insurgency to France in the 1990s, bombing government and civilian targets to protest France's support of the military government in Algiers.

Morocco and Tunisia view events in neighboring Algeria with anxiety. Citing the violence in Algeria, Tunisia banned Islamist political parties in 1994. Tunisia's government has promoted family planning and women's rights and is often cited as an "oasis" of openness and progressive change in the region. Tunisia also has the Maghreb's strongest economy. In Morocco, King Muhammad VI rules over a constitutional monarchy that promises political accommodation of Islamic groups, in contrast with Algeria's experience, but human rights monitors are increasingly alarmed by the young king's muzzling of opposition voices. Morocco is the poorest of the Maghreb countries and, as of 2002, had suffered several years of drought that drove large numbers of former farmers into the cities. But there has been some progress in recent years. The country has shown exceptional success in bringing down birth rates; while the average woman had seven children in 1980, she now has three.

Morocco and Algeria have long been at odds over the future of Spain's former dependency known as Western Sahara. This sparsely populated coastal desert (estimated population: 300,000, a figure that includes 170,000 refugees living in Algeria), which contains rich phosphate deposits, adjoins Morocco on the south and is claimed by that country. When Spain relinquished control in 1975, the territory was partitioned between Morocco and Mauritania, but in 1979, Mauritania withdrew its presence. Algeria objected to Moroccan control and, together with other outside parties, gave assistance to an armed resistance movement in Western Sahara against Morocco called the Polisario Front. In 1989, Morocco and the Polisario Front agreed to a referendum on independence, and a peace treaty was signed in 1991. To influence the outcome of the referendum in its favor, Morocco relocated 40,000 Moroccans into the region in 1991; they remain there today under Moroccan guard. The Moroccan government has also used tax breaks and other subsidies to attract new Moroccan settlers there. Such questionable actions led the United Nations to repeatedly suspend the referendum and finally call it off in 2000. The U.N. is studying the prospect of helping to establish limited Western Saharan autonomy under Moroccan rule. Whereas the status of Western Sahara remains unresolved, more than 170,000 Saharawis (as the indigenous people of Western Sahara are known) live in refugee camps in the Algerian desert, as they have for decades.

Morocco has aggressively asserted its claim to Western Sahara. It has sold Western Sahara's offshore fishing rights to European countries. To stave off Algerian interests, Moroccan technicians have also erected a 700-mile (1,120-km) sand barrier parallel to Western Sahara's border with Algeria, reinforcing it with land mines and 150,000 troops (see Fig. 9.12). West of this berm, in Morocco-held territory, is the world's largest reserve of phosphate.

9.6 The Gulf Oil Region

Foreign companies, principally British and American, originally developed the oil industry in the Persian/Arabian Gulf. These firms made huge profits from their oil concessions, but with a few exceptions, after 1972 the companies saw their properties nationalized by the Gulf governments. However, many outside companies still work closely with national oil companies in the Gulf countries, providing them with technical and managerial expertise, skilled workers, equipment, marketing channels, and varied services.

Gulf oil is shipped out by a combination of pipelines and ships (Fig. 9.16). Most oil produced in the Gulf region is transported to market by oceangoing tankers, generally the huge vessels called supertankers. These ships load oil at terminals spaced along the Gulf. Some crude oil moves by pipelines from oil fields in the Gulf countries to tanker terminals on the Mediterranean and Red Sea coasts. A growing proportion of extracted crude oil is refined in the Gulf countries for local use and export. Petrochemical industries exist in most of the producing countries, with the largest development in Saudi Arabia. The industries use local natural

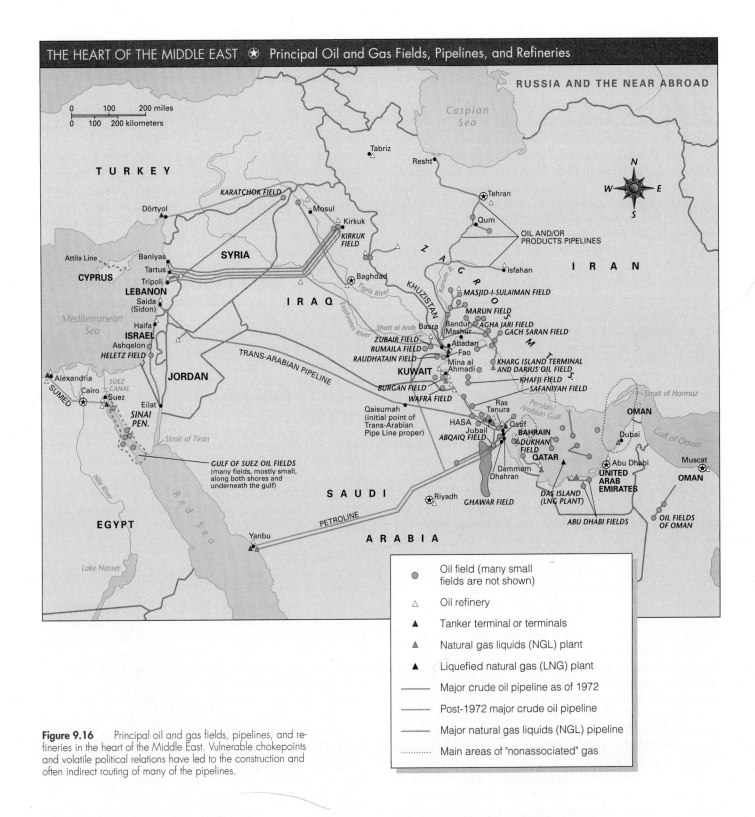

THE HEART OF THE MIDDLE EAST ✶ Principal Oil and Gas Fields, Pipelines, and Refineries

Legend:

- ● Oil field (many small fields are not shown)
- △ Oil refinery
- ▲ Tanker terminal or terminals
- ▲ Natural gas liquids (NGL) plant
- ▲ Liquefied natural gas (LNG) plant
- —— Major crude oil pipeline as of 1972
- —— Post-1972 major crude oil pipeline
- —— Major natural gas liquids (NGL) pipeline
- ·········· Main areas of "nonassociated" gas

Figure 9.16 Principal oil and gas fields, pipelines, and refineries in the heart of the Middle East. Vulnerable chokepoints and volatile political relations have led to the construction and often indirect routing of many of the pipelines.

gas as their primary source of energy. Huge gas reserves are associated with oil deposits, and other major reserves exist apart from the oil fields, in "nonassociated" deposits. Growing amounts of gas are exported in liquefied form; Saudi Arabia is the world's largest exporter.

Saudi Arabia: The Oil Colossus

With an area of roughly 830,000 square miles (2.1 million sq km), Saudi Arabia occupies the greater part of the Arabian Peninsula, the homeland of Arab civilization and the faith of Islam. The interior of the country is an arid plateau, fronted on the west by mountains that rise steeply from a narrow coastal plain bordering the Red Sea. Most of the plateau lies at an elevation of 2,000 to 4,000 feet (c. 600 to 1200 m), with the lowest elevations near the Gulf and in the eastern part of the gigantic sea of sand called the Empty Quarter (al-Rub al-Khali). Mountains (Arabic: *jebels*) rise locally from the plateau, and the plateau surface is cut by many valleys (Arabic: *wadis*) that carry water whenever it happens to rain; Saudi Arabia has no permanent rivers. Most interior sections average less than 4 inches (10 cm) of precipitation annually, but years may pass at a given locale with no precipitation at all. Rainfall is greatest along the mountainous western margin of the peninsula, especially in the southwestern Asir region, near the border with Yemen, which benefits from summer monsoon rains. About 95% of Saudi Arabia's present crop production depends on irrigation. Cultivated land makes up only 2% of the country's area, yet surprisingly, Saudi Arabia is self-sufficient in many agricultural staples and is an exporter of wheat, irrigated mostly in the central part of the country by gigantic center-pivot sprinkler systems.

The estimated population of Saudi Arabia is 21.1 million (as of 2001). The population is young and growing; half of all Saudis are younger than 19, and the annual growth rate is 2.9%. About one-quarter of that population consists of foreign workers. Saudi Arabia's best-known city is Mecca, Islam's holiest city. The interior city of Riyadh (population: city proper, 3,533,500) is Saudi Arabia's capital. The Red Sea port of Jidda, some 40 miles (64 km) from Mecca, is the main point of entry for foreigners who make the pilgrimage to Mecca.

Most of Saudi Arabia's oil fields, including the world's largest, the Ghawar field, are located in the eastern margins of the country, including its adjacent seabed. The great thickness of Saudi Arabia's oil-bearing strata and high reservoir pressures have made it possible to secure an immense amount of oil from a small number of wells. Such advantages are widespread in the Gulf oil states. The productivity of each well makes each barrel inexpensive to extract and makes it simple to increase or reduce production quickly in response to world market conditions. Gulf oil is thus "cheap" to produce unless expenditures to maintain huge military forces to defend it are factored in. In 2002, the United States maintained about 20,000 military personnel in the region at a cost of more than $1.5 billion per year. A U.S. Navy secretary once remarked that the real price of oil, with military expenditures factored in, is about $100 per barrel, not the typical market price of around $25.

Some Saudi Arabian oil is processed in refineries within the country and in nearby Bahrain, but most of the production leaves Saudi Arabia in crude form by tankers, using loading terminals that jut into both the Gulf and the Red Sea. To meet anticipated future demand, huge refining and petrochemical complexes have been developed at Jubail on the Gulf and Yanbu on the Red Sea. These cities sprang up almost literally overnight in the 1980s.

Most Saudi factories produce and assemble consumer goods for domestic consumption, and the country is not a significant industrial power. The country's transition to a standard of living comparable to the more developed countries (MDCs) is not complete. For example, infant mortality remains high. And there are striking discrepancies in income; although the sharing of oil wealth has brought once undreamed-of prosperity to most of the population, the Shi'ite Muslim minority in the Eastern Province is poor. Saudis in general no longer take their prosperity for granted; the population is growing much faster than the economy, and years of declining oil revenues have compelled many Saudis to take menial jobs once reserved for foreigners.

Despite the luxuries that oil wealth has brought to Saudi Arabia, the conservative Saudi family that rules this absolutist monarchy has struggled to maintain local Muslim traditions. There are restrictions on the freedoms of women, who are prohibited from driving automobiles, for example. The legal system is based on an interpretation of Islamic law that does not flinch at severe corporal and capital punishment. The Saudi royal family keeps a wary eye on many internal and external threats. Political acts of violence have included bombings of U.S. military facilities, probably carried out by Islamist organizations intent on purging the region of secular and "imperialist" Western interests and on replacing the monarchy with an Islamic government. Countercurrents are now buffeting the desert kingdom. Just as strong as efforts to prevent outside influences from changing Saudi ways are efforts to invite the world in by allowing tourists to visit, encouraging foreign investment, and signing international trade and human rights treaties.

Bahrain, Qatar, and the UAE

The sheikdom of Bahrain (ruled by a monarch called a *Sheikh* or *Emir*; population: 700,000) occupies a small archipelago in the Gulf between the Qatar Peninsula and Saudi Arabia. A modern four-lane causeway connects the main island to nearby Saudi Arabia. In addition to being an oil producer, Bahrain is a center of oil refining, seaborne trade, ship repair, air transportation, business administration, international banking, and aluminum manufacturing using imported alumina and local natural gas. Like Saudi Arabia, Bahrain is an oil-financed welfare state. However, in contrast to its giant neighbor, Bahrain's petroleum reserves are expected to be exhausted within two decades.

East of Bahrain along the Gulf lie the small emirates of Qatar (population: 600,000) and the United Arab Emirates (UAE) (population: 3.3 million). The UAE is a loose confederation of seven units, three of which produce oil; only Abu Dhabi and Dubai pump it in major quantities. Qatar and the UAE are extremely arid and, prior to the oil era, supported only a meager subsistence economy based on fishing, seafaring, herding, and a little agriculture, augmented by smuggling and piracy. Today, the UAE is capitalizing on its central loca-

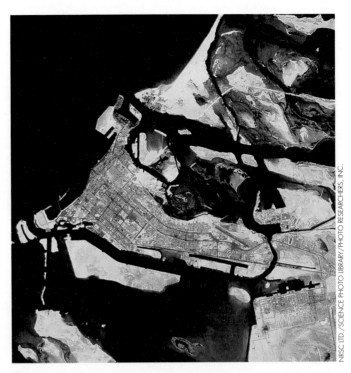

Figure 9.17   These extraordinary before-and-after satellite images reveal how oil wealth has transformed some Middle Eastern landscapes. The first shows Abu Dhabi, capital of the UAE, in 1972. The city occupies much of the natural island in the center. The second image of the same area, recorded in 1984, reveals the outcome of huge investments, including human-made islands and deepwater port facilities.

tion between Asia, Africa, and Europe with the construction of sophisticated air and seaport facilities. The recently built seaports of Abu Dhabi and Dubai are colossal facilities that have dramatically changed the coastal landscapes and are easily visible from space (Fig. 9.17). The UAE is successfully diversifying its economies beyond the oil industry. Dubai is a well-established international trade and banking center (Fig. 9.18). Abu Dhabi, the largest and richest unit of the UAE, is hoping to establish regional supremacy as a financial center through the construction of a massive commercial development on Saadiyat Island. Unlike its more conservative neighbors, the UAE has promoted itself as a destination for sun-worshipping international tourists and shoppers. Dubai is the most liberal unit within the UAE and, in marked contrast with its teetotaling neighbors, tolerates bars and brothels.

The Southern Margins of Arabia

Known to the ancient Romans as *Arabia Felix*, or "Happy Arabia," because of their verdant landscapes, Yemen (population: 18 million) and Oman (population: 2.4 million) occupy, respectively, the mountainous southwestern and southeastern corners of the Arabian Peninsula.

The present Republic of Yemen was created in 1990 by the union of the former North Yemen and South Yemen. They re-

main united despite a brief civil war in 1994. Its political capital and largest city is San'a (population: city proper, 1,565,100), located in the former North Yemen. Until the 1960s, North Yemen was an autocratic monarchy, extremely isolated and little known. In the mid-1960s, a republican

Figure 9.18   Like most cities in the oil-producing countries of the Gulf, Dubai is a thoroughly modern metropolis with only hints of the past—here, the storied wooden sailing vessels known as *dhows*.

government supported by Egypt and opposed by Saudi Arabia took control of large sections at a cost of much bloodshed.

The former South Yemen, known until 1967 as Aden, a British colony and protectorate, lies along the Indian Ocean between the former North Yemen and Oman. The main city, Aden, is located on a fine harbor near the straits. Aden is perfectly located for refueling vessels that transit the Suez Canal and is close to the strategic chokepoint of Bab el-Mandeb connecting the Red Sea with the Indian Ocean. Possession or use of this port has therefore long been prized by outside powers. After British colonial times, South Yemen became a Marxist state supported by the Soviet Union and other Eastern-bloc nations. Since the two Yemens were unified, Aden has been an important refueling station for the United States and its Western allies. Security at the port has been stepped up since the bombing there of the U.S.S. *Cole* in 2000, and Western governments have pressured Yemen's government to crack down on Islamist groups that advocate struggle against Western interests in the region.

Yemen receives the heaviest rainfall in the Arabian Peninsula, enjoying a summer rainfall maximum derived from the same monsoonal wind system that brings summer rain to the Horn of Africa and the Indian subcontinent. Its main subsistence crops are millet and sorghums. Most of the country's best agricultural land consists of terraces and is dedicated to the cultivation of the valuable cash crop *qat* (*Cathya edulis*), a stimulant shrub chewed by virtually all adults in the country on a daily basis (see Fig. 8.8). Most of the country's people live in the rainier highlands, and the coastal plain is extremely arid and sparsely populated.

Yemen's meager agricultural output and lack of industrial production have been compensated for during recent decades by earnings sent home by Yemeni workers in the oil fields and services of the wealthy Gulf countries. The support of Yemen's government for Saddam Hussein's regime in Iraq during the 1991 Gulf War prompted Saudi Arabia to expel millions of these expatriate Yemeni workers, leading to a precipitous decline in Yemen's economy. Oil production began in Yemen itself only in the 1980s. The fields are located in a contested border area between Yemen and Saudi Arabia and subjected periodically to raids and kidnappings by groups opposed to Yemen's government, so production has been slow to increase.

Most of the Sultanate of Oman lies along the Indian Ocean and the southern shore of the Arabian Sea, with a small detached part bordering the Strait of Hormuz and the Gulf. As in Yemen, the coastal region is arid, and most people live in interior highlands where mountains induce a fair amount of rain (Fig. 9.19). For about 2,000 years, beginning around 1000 B.C., Oman was the source of most of the world's frankincense, a prized commodity that made the region a hub of important oceanic and continental trade routes. Crude oil now furnishes almost all of Oman's exports.

The country's strategic position on the Strait of Hormuz,

through which much of the oil produced by the Gulf countries is transported, has made it an object of concern in industrial nations, particularly Great Britain, with which it has historic ties, and the United States. These Western powers have rights of access to base facilities in Oman in the event of any military emergency that threatens the continued flow of oil from the Gulf region. Allied forces used the bases after Oman joined the United Nations coalition against Iraq in the Gulf War of 1991.

Kuwait and the Gulf War of 1991

With 9% of the world's proven oil reserves within its borders, tiny Kuwait (6,900 sq mi/17,900 sq km — about the size of the U.S. state of New Jersey) has the world's fourth largest store of proven petroleum resources (after Saudi Arabia, Iraq, and the UAE). Associated with the oil is much natural gas, which Kuwait both uses and exports. Oil revenues have changed Kuwait (population: 2.3 million) from an impoverished country of boatbuilders, sailors, small traders, and pearl fishermen to an extraordinarily prosperous welfare state. Generous benefits have flowed to Kuwaiti citizens but not to the Asian and other non-Kuwaiti expatriates drawn by employment opportunities to this emirate. (By 1990, foreign workers actually outnumbered Kuwaiti citizens in the country.) To reduce possible security threats and become more self-reliant, the government is trying to reduce dependence on foreign workers. A huge challenge will be to provide jobs for its young population (60% of Kuwaitis are younger than 18) in non-oil-related industries and services. Kuwait produces only this one commodity, which makes it vulnerable to developments such as alternative energy.

Kuwait took center stage in world affairs in August 1990, when Iraqi military forces invaded Kuwait. Claiming that

Figure 9.19 The southern margins of Arabia are surprisingly verdant. Thanks to summer monsoonal rains, this Jebel Qara region, above the city of Salalah in southwestern Oman, sustains cattle.

Figure 9.20 Iraqi troops set more than 600 oil wells ablaze before retreating from Kuwait in 1991. This image shows some of the fires on April 28, 1991. Kuwait City occupies much of the peninsula in the top center.

A formal cease-fire came in April 1991, with harsh conditions imposed on Iraq by the U.N. Security Council. Particularly important was the abolition of Iraqi chemical, biological, and nuclear "weapons of mass destruction" capabilities. Iraq agreed to pay reparations for the damage its forces had caused, and it renounced its claims to Kuwait. Saddam Hussein's surviving forces then mercilessly put down rebellions by Shi'ite Arabs in southern Iraq and by Kurds (members of a non-Arab ethnic group; see p. 262) in northern Iraq. To deter Iraq's military and provide a safe haven for the Kurds in the north and the Shi'ites in the south, the United Nations established "no-fly zones" north of the 36th parallel and south of the 32nd (later, 33rd) parallel, where Iraqi aircraft were prohibited from operating. But there was widespread condemnation of the Western coalition for apparently encouraging the Kurds and Shi'ites to rebel and then failing to assist them. Iraq's Shi'ites fared badly.

One of the great environmental and cultural tragedies of the late 20th century was Iraq's systematic draining after 1991 of the marshes adjoining the southern floodplains of the Tigris and Euphrates Rivers. To quell any opposition by the ancient Shi'ite culture of the Maadan, or "Marsh Arabs," the Iraqi military built massive embankments and canals that drained and destroyed the marshes, forcing the Maadan to abandon their historic homeland.

Throughout the 1990s, the regime of Saddam Hussein played a "cat and mouse" game with U.N. inspectors whose duty was to eliminate Iraq's ability to produce weapons of mass destruction. United States, British, and other forces were repeatedly deployed to strike against Iraq unless the country allowed unimpeded access by the U.N. inspectors and, in 1998, attacked scores of military targets when access was denied. Since then, the Baghdad regime has not allowed U.N. inspectors back into the country, has flaunted the no-fly zones, and has restated its wartime position that Kuwait is an illegitimate country. U.N.-sponsored economic sanctions against Iraq continue, with shortages of food and medical supplies bringing great hardship to the country's population, but international support for the sanctions has weakened greatly, particularly in the Arab world. President Saddam Hussein endured, much to the chagrin of his foes.

Kuwait was historically part of Iraq and calling it Iraq's "19th province," the Baghdad regime attempted to annex the country. Saudi Arabia and other Gulf oil states seemed in danger of invasion. Worldwide opposition to Iraq's aggression resulted in a sweeping embargo on Iraqi foreign trade imposed by the Security Council of the United Nations. Led by the United States, a coalition of 28 U.N. members (including several Arab states) staged a rapid buildup of armed forces in the Gulf region.

Beginning in January 1991, the coalition's crushing air war drove the Iraqi air force from the skies, massively damaged Iraq's infrastructure, and hammered the Iraqi forces in Kuwait and adjacent southern Iraq with intensive bombardment. In the ensuing ground war, the Iraqis were driven from Kuwait within 100 hours. Overall, an estimated 110,000 Iraqi soldiers and tens of thousands of Iraqi civilians were killed, while the technologically superior coalition forces suffered 340 combat deaths, of which 148 were American. During the conflict, which came to be known as the **Gulf War,** the Iraqis also made war on the environment, setting fire to hundreds of oil wells in Kuwait (Fig. 9.20) and allowing damaged wells to discharge huge amounts of crude oil into the Gulf.

Iraq: Troubled Oil State of Mesopotamia

Iraq occupies a broad, irrigated plain drained by the Tigris and Euphrates Rivers (see Fig. 9.4), together with fringing highlands in the north and deserts in the west. This riverine land of Mesopotamia has a complex geopolitical history, having been a seat of empires, a target of conquerors, and in the 20th century, a focus of oil development and political and military contention involving many other nations.

Iraq was a province of the Ottoman Empire for almost 400 years. British forces defeated the Turks in Iraq in World War I, and Great Britain had a League of Nations mandate over the

REGIONAL PERSPECTIVE   :::   The Kurds

A Sunni Muslim people of Indo-European origins, the Kurds are by far the largest of several non-Arab minorities in Iraq. They have revolted against Iraqi central authority on numerous occasions, most recently during the Gulf War of 1991. The Kurdish question has international implications because the area occupied by an estimated 22 million Kurds also extends into Iran, Turkey, Syria, and the southern Caucasus region (Fig. 9.B); an estimated 8 million more live outside the Middle East. European powers promised an independent state of Kurdistan at the end of World War I, but the

governments concerned never took steps to create it. Kurds today remain the world's largest ethnic group without a country.

A Kurdish rebellion against Turkish rule has smoldered and sometimes flamed since 1990. The main Kurdish resistance comes from the Kurdistan Workers Party, known by its Kurdish acronym PKK. The organization suffered a major setback in 1999 when Turkey caught, jailed, and sentenced to death its leader, Abdullah Ocalan. Following the Gulf War, Iraq's Kurds enjoyed a decade of relative autonomy in

northern Iraq. However, Iraq is presently trying to "Arabize" its Kurds by forcing them to renounce their ethnicity and sign forms saying they are Arabs. Turkish officials have also downplayed the identity of Kurds (who make up 15% of Turkey's population), often referring to them as "Mountain Turks," and Kurdish language media in Turkey have long been banned or restricted. A book about the Kurds is appropriately entitled *No Friends but the Mountains*.[a]

[a]John Bullock, 1993. *No Friends but the Mountains*. London: Oxford University Press.

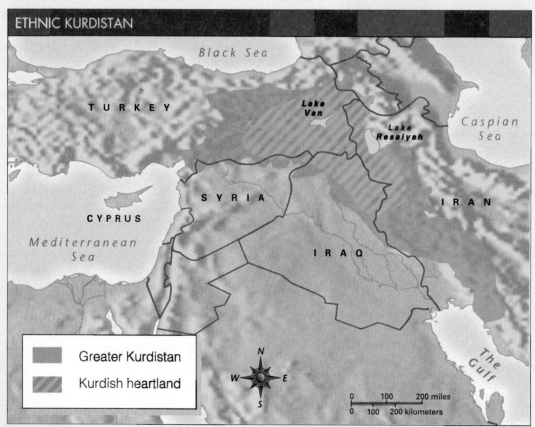

ETHNIC KURDISTAN

Legend:
- Greater Kurdistan
- Kurdish heartland

Source: Alasdair Drysdale and Gerald Blake, 1995. *The Middle East and North Africa: A Political Geography,* London: Oxford University Press, p. 61.

Figure 9.B   The Kurds have a distinct national territory but have never had a country.

country until it gained its independence in 1932. Iraq's government was a pro-Western monarchy from 1923 to 1958, when an army coup made the country a republic. By the mid-1970s, Iraq had become an authoritarian one-party state ruled by President Saddam Hussein in the name of the Arab Baath Socialist Party. After 1973, increasing oil revenues from the country's huge reserves (about 11% of the world's total) underwrote new social programs, large construction projects, and the creation of new businesses. New high-rise buildings, busy expressways, and an active commercial and social life gave metropolitan luster to ancient Baghdad. Subsequent wars resulted in destruction in the city and in Iraq's economy. This country should be rich from its oil (in 2001 it produced 4% of the world's total) but, instead, is forced to use its oil revenues to try to make up ground lost through hapless military adventures.

In 1980, perceiving internal weakness in the neighboring fledgling Islamic Republic of Iran and a means of securing land and mineral resources on the Iranian side of the Shatt al-Arab waterway (the river formed by the confluence of the Tigris and Euphrates Rivers), President Hussein launched an invasion of his more populous neighbor. Thus began an 8-year war that proved enormously costly to both sides, with an estimated 500,000 people killed or wounded. Iran mounted a far better defense than Iraq had anticipated, expending its greater military personnel in bloody "human wave" assaults to counter Iraq's greater firepower from tanks, artillery, machine guns, and warplanes. A complex web of geopolitical relationships made the Iran-Iraq War a storm center in the political world, particularly as the superpowers and other countries sold sophisticated arms to both sides. The war ended in stalemate in 1988, with neither side gaining significant territory or other assets.

An estimated 70% of Iraq's 23.6 million people (about 60% of whom are Shi'ite Muslims) are urban-dwellers. Most of the rest are agricultural villagers. Iraq's government has expanded irrigation through the construction of dams on the Tigris and Euphrates and their tributaries for multiple uses of flood control, hydroelectric development, and irrigation storage. Agricultural output in Iraq is poor, however, due mainly to the problem of overwatering, which causes the salinization of once-fertile soils. Small bands of nomadic Bedouins still raise camels, sheep, and goats in the western deserts. Some of the Kurdish tribespeople in the mountain foothills of the north are seminomadic grazers, although most Iraqi Kurds are sedentary farmers.

Baghdad (population: metropolitan area, 6,241,500; city proper, 4,869,500), Iraq's political capital, largest city, and main industrial center, is located strategically on the Tigris near its closest approach to the Euphrates River. The country invested heavily in rapid reconstruction of its capital city after the 1980–1988 war and again after the 1991 conflict. Basra is Iraq's main seaport, although it is located upriver on the Shatt al-Arab. Iraq suffers from a fundamental geographic handicap of not having natural deep-water ports in its tiny 35-mile (55-km)

window on the Gulf. One of the main reasons Iraq attacked Iran in 1980 and Kuwait in 1990 was to widen this window.

Oil and Upheaval in Iran

Iran, formerly Persia, was the earliest Middle Eastern country to produce oil in large quantities. The main oil fields form a line along the foothills of the Zagros Mountains in southwestern Iran. British-built pipelines linked the large refinery at Abadan (on the Shatt-al Arab waterway) with the inland oil fields. Now nearly all exports of crude oil from Iran pass through a newer terminal at Kharg Island (see Fig. 9.16), which allows supertankers to load in deep water.

Oil is the leading source of revenue for Iran, a country where a dry and rugged habitat makes it hard to provide a good living for its population of 66.1 million. Arid plateaus and basins, bordered by high, rugged mountains, are Iran's characteristic landforms (Fig. 9.21). Encircling mountain ranges prevent moisture-bearing air from reaching the interior, creating a classic rain-shadow desert. Only an estimated 10% of Iran is cultivated. Enough rain falls in the northwest during the winter half-year to permit unirrigated cropping (dry farming), but rainfall sufficient for intensive agriculture is confined mainly to a densely populated lowland strip between the high, volcanic Elburz Mountains and the Caspian Sea. The western half of this lowland receives more than 40 inches (c. 100 cm) annually and produces a variety of subtropical crops such as wet rice, cotton, oranges, tobacco, silk, and tea.

Agriculture elsewhere in Iran depends mainly on irrigation. Many irrigated districts lie on gently sloping alluvial

Figure 9.21 View southwest from the space shuttle *Atlantis* of the Strait of Hormuz and the Zagros Mountains of Iran. Most of Oman and the United Arab Emirates can be seen in the upper right.

fans (triangular deposits of stream-laid sediments) at the foot of the Elburz and Zagros Mountains; the capital, Tehran (population: metropolitan area, 10,737,300; city proper, 7,555,000), is in one of these areas south of the Elburz. Qanats (underground sloping tunnels) supply water to a large share of the country's irrigated acreage. However, they are expensive to build and maintain, and an increasing share of water comes from stream diversions and from deep wells. Wheat is the most important irrigated crop. The leading revenue earner, and Iran's third most important export after oil and its famous carpets, is the pistachio nut. Cattle are raised in areas where there is enough moisture to grow forage for them, but drier areas support sheep and goats. Many livestock belong to seminomadic mountain peoples—the Qashqai, Baktiari, Lurs, Kurds, and others—whose independent ways have long been a source of friction between these tribespeople and the central government. Before the Islamic revolution in Iran in 1979, it was government policy to sedentarize Iran's nomads so that they did not threaten the state.

Prior to World War I, Iran was a poor and undeveloped country, in marked contrast to its imperial grandeur when Persian kings ruled a great empire from Persepolis. In the 1920s, a military officer of peasant origins, Reza Khan, seized control of the government and began a program to modernize Iran and free it from foreign domination. He had himself crowned as Reza Shah Pahlavi, the founder of the new Pahlavi dynasty. Influenced by the modernizing efforts of Mustafa Kemal Ataturk in neighboring Turkey, the new shah (king) introduced social and economic reforms. In 1941, his young son took the throne as Mohammed Reza Shah Pahlavi and set out to expand the program of modernization and Westernization begun by his father. In the 1970s, Iran played a leading role in raising world oil prices. Mounting oil revenues underwrote explosive industrial, urban, and social development.

In the countryside, the shah's government attempted to upgrade agricultural productivity through land reform and better technologies. Iran made major efforts to increase the supply of irrigation water and electricity by building storage dams and hydropower stations. But agricultural reforms generally failed, and poverty-stricken families from the countryside poured into Tehran and other cities. From this devoutly Muslim group of new urbanites came much of the support for the revolution that ousted the shah in 1978–1979.

Oil money funded Iran's development under the Pahlavi dynasty, but the vast sums the regime spent on the military undercut the benefits to Iran's people. Arms came primarily from the United States. In 1978 and 1979, a revolution overthrew the shah and abruptly took Iran out of the American orbit. The country became an Islamic republic governed by Shi'ite clerics, who included a handful of revered and powerful ayatollahs ("signs of God") at the head of the religious establishment, and an estimated 180,000 priests called mullahs. These religious leaders continue to supervise all aspects

of Iranian life and perform many functions allotted to civil servants in most countries. They base their authority on Shi'ite interpretations of the Islamic faith. (About 93% of Iran's population is Shi'ite.)

The central figure among the revolutionaries who overthrew the shah was the Ayatollah Ruhollah Khomeini, an elderly critic of the regime who had been forced into exile by the shah in 1964. From Paris, Khomeini sent repeated messages to Iran's Shi'ites that helped spark the revolution. A furious tide of Shi'ite religious sentiment rose against corruption, police heavy-handiness, modernization, Westernization, Western imperialism, the exploitation of underprivileged people, and the institution of monarchy, which Shi'ite beliefs regard as illegitimate. Growing numbers of people staged strikes and demonstrations in 1978, forcing the shah to flee the country and abdicate his throne. (After hospitalization in the United States and life in exile in Panama and Egypt, the shah died in Cairo in 1981 and is buried there; Fig. 9.22.) The Ayatollah Khomeini returned in triumph to Tehran in 1979, where he soon was able to take control of the government. His followers held 52 U.S. diplomatic personnel as hostages in Tehran for 444 days. His inability to resolve this crisis led to U.S. President Jimmy Carter's defeat in his 1980 reelection bid.

What has emerged since the revolution is a theocratic state, guided by edicts of the Ayatollah Khomeini (who died in 1989) and his successors, who have undertaken to create a society governed by Islamic law. In contrast to their status during the shah's reign, when they were encouraged to have stronger public roles, women in postrevolutionary Iran have lost many freedoms. Under the new regime, "polluting" influences of Western thought and media were purged. However,

Figure 9.22 The late 1970s and early 1980s were turbulent times in the Middle East. The exiled shah of Iran was welcomed only in the United States and in a single sympathetic country in the region: Egypt. Here a line of dignitaries, besieged by crowds, is following the casket of the shah to its resting place in a Cairo mosque in July 1981. At the far right center, wearing a military uniform, is Egypt's President Anwar Sadat, who was assassinated by Islamist militants 3 months later.

recent years have seen a general softening of restrictions on personal freedom of expression, and the Internet and other media have introduced broader worldviews to Iran's youth. Generally, there is a tug of war between the moderate, democratically elected, and reform-minded President Muhammad Khatami, who favors more openness, and the less tolerant supreme religious leader, Ayatollah Ali Khamanei, who has lifetime control over the military and the judicial branch of government. Western analysts alternatively perceive Iran to be either developing a more accommodating attitude toward the international community or exporting ever more dangerous notions of revolutionary Islam to currently pro-Western countries in the Middle East. Since 1998, relations have improved between Iran and the United States and between Iran and Great Britain.

9.7 Turkey: Where East Meets West

The Turks who organized the Ottoman Empire beginning in the 14th century were pastoral nomads from Central Asia, where Turkic cultures still prevail. From the 16th to the 19th century, their empire, based in what is now Turkey, was an important power. Modern Turkey was created from the wreckage of the old empire after World War I. Its founder, Mustafa Kemal Ataturk, was determined to Westernize the country, raise its standard of living, and make it a strong and respected national state. He inaugurated social and political reforms designed to break the hold of traditional Islam and open the way for modernization and Turkish nationalism.

Islam had been the state religion under the Ottoman Empire, but Ataturk separated church and state, and to this day, Turkey is the only Muslim Middle Eastern country to officially cleave them. Wearing the red cap called the *fez*, an important symbolic act under the Ottoman caliphs, was prohibited, and state-supported secular schools replaced the religious schools that had monopolized education. To facilitate public education and remove further traces of Muslim culture, Latin characters replaced the Arabic script of the Qur'an. Slavery and polygamy were outlawed, and women were given full citizenship. Legal codes based on those of Western nations replaced Islamic law, and forms of democratic representative government were instituted, although Ataturk ruled in dictatorial fashion.

Turkey has continued to have trouble establishing a fully democratic system, and there have been frequent periods of military rule (Fig. 9.23). Turkey's constitution invests the military with the responsibility of protecting the country's secular government, and the army has used its powers. After decades of secularization, there was, for 2 years (1996–1997), a democratically elected government coalition comprised of avowedly secular politicians and Islamist leaders, including a prime minister who wanted to see Turkey reestablished on Islamic foundations. In 1998, however, Turkey banned govern-

Figure 9.23   The military has a strong presence in secular Turkey. The soldiers are guarding an archeological site in eastern Turkey, a region of Kurdish rebellion against government rule.

ment participation by the pro-Islamic Welfare (Refah) party that had put the prime minister in office. Backed by the army, the new government reaffirmed Ataturk's vow that Turkey remain secular. According to many observers, the regime's perception that the Islamist tendencies threaten the state represents a setback for democracy and human rights in Turkey — there are even restrictions against women wearing Islamic-style head scarves — and raises the specter of violent dissent.

From a physical standpoint, Turkey is composed of two units: the Anatolian Plateau and associated mountains occupying the interior of the country and the coastal regions of hills, mountains, and small plains bordering the Black, Aegean, and Mediterranean Seas. Turkey in general has a physical presence quite unlike the popular image of the Middle East as a desert region; it is, in fact, the only country in the region that has no desert, and cultivated land occupies about a third of Turkey's land area. The Anatolian Plateau, an area of wheat and barley fields and grazing lands, is bordered on the north by the lofty Pontic Mountains and on the south by the Taurus Mountains.

The annual precipitation, concentrated in the winter half-year, is barely sufficient for grain. The plateau ranges from 2,000 to 6,000 feet (c. 600 to 1800 m) and is highest in the east where it adjoins the mountains of the Armenian Knot. Its surface is rolling and windswept, hot and dry in summer and cold and snowy in winter, with a natural vegetation of short steppe grasses and shrubs. Production of cereals (especially barley, wheat, and corn) and livestock (especially sheep, goats, and cattle), employing both traditional and mechanized means, prevails in Anatolia (Fig. 9.24). Coastal plains and valleys along and near the Aegean Sea, Sea of Marmara, and Black Sea are generally Turkey's most densely populated and

REGIONAL PERSPECTIVE ::: Water in the Middle East

In the arid Middle East, where most water is available either from rivers or from underground aquifers that cross national boundaries, control over water is an especially difficult and potentially explosive issue. An estimated 90% of the usable fresh water in the Middle East crosses one or more international borders.

Water is one of the most problematic issues in the Palestinian-Israeli conflict. Fresh-water aquifers underneath the West Bank supply about 40% of Israel's water. Palestinians point to Israeli control over West Bank water as one of the worst elements of its occupation. The average Jewish settler on the West Bank uses 74 gallons (278 litres) per day, whereas the average West Bank Palestinian uses 19 (72 litres). The World Health Organization calculates that 27 gallons (102 litres) per person per day is needed for minimal health and sanitation standards, but Israeli policies prohibit Palestinians from increasing their water usage. Many Israeli policymakers insist that water resources on the West Bank must remain under

strict Israeli control and, on these grounds, oppose the creation of a Palestinian state on the West Bank. Critically, the West Bank aquifers do not contain enough water to support the region's population at current levels of consumption for more than a few more years (even taking into account anticipated replenishment from rainfall).

More promisingly, Jordan and Israel are working on agreements to share waters from the Jordan River (which forms a portion of their common border) and its tributary, the Yarmuk River. They are discussing a joint venture to build the Dead-Red Canal, which would connect the Gulf of Aqaba with the Dead Sea. The gravity flow of seawater to the Dead Sea would spin turbines and run generators to produce electricity to be shared between the two nations.

Water is a critical issue blocking a peace treaty between Israel and Syria. If Syria were to recover all of the Golan Heights area, it would have shorefront on the Sea of Galilee and therefore, presum-

ably, rights to use its water. That prospect is unacceptable to Israel, as this is its principal supply of fresh water. In occupying the Golan Heights, Israel also controls some of the northern bank of the Yarmuk River on the border with Jordan. For many years, Israel indicated it would never allow Syria and Jordan to construct the Unity Dam that would store waters of the Yarmuk River to be shared between those countries. Israel thus implied it would bomb the dam rather than allow it to deprive Israel of Jordan River water.

As a "downstream" riverine, or riparian, nation, completely dependent on water originating outside the country, Egypt considers relations with upstream countries vital to its long-term national security. Egypt and Sudan are signatories of a Nile Waters Agreement that apportions water between them, but Egypt does not have similar agreements with other countries. Ethiopia's diversion of water from vital Nile headstreams is a potential future source of friction with Egypt and Sudan, but at present, Ethiopia has little money

---

productive areas. Here are grown cash crops such as hazelnuts, tobacco, grapes for sultana raisins, and figs. The slopes of the Pontic Mountain facing the Black Sea have copious rains in both summer and winter and support prolific tea harvests (see Fig. 8.7).

Figure 9.24   The wheat harvest on the Anatolian Plateau in east central Turkey.

Relative poverty compared with the nearby European countries is a striking characteristic of present-day Turkey. There remains a strong rural and agricultural component in Turkish life, with about one-third of the population of 66.3 million still classified as rural. However, the country has embarked on a course of change that promises to modernize its agriculture, expand industry, and raise its general standard of living. The value of output from manufacturing (mainly textiles, agricultural processing, cement manufacturing, simple metal industries, and assembly of vehicles from imported components) is already greater than that from agriculture. Turkey does produce some oil.

Early in the 1990s, Turkey began an impressive $32 billion agricultural scheme called the Southeast Anatolia Project, known by its Turkish acronym as GAP. Its aim is to convert the semiarid southeast quarter of the country into the "Breadbasket of the Middle East." Surpluses of cotton, wheat, and vegetables are to be exported, along with valued-added manufactures such as cotton garments and textiles. The centerpiece of this massive irrigation project is the Ataturk Dam on the Euphrates River (Fig. 9.25). About 20 other dams on the Turkish Euphrates and

to finance the dams, canals, and pumps that large-scale diversions would require. Meanwhile, Egypt's demands on Nile waters are increasing. Egypt is now excavating the multibillion-dollar Tushka Canal that will transport water from Lake Nasser over a distance of 300 miles (500 km) to the Kharga Oasis of the Western Desert. Proponents of the scheme insist it will result in the cultivation of nearly 1.5 million acres (600,000 hectares) of "new" land and provide a living for hundreds of thousands of people. Critics argue that it is a waste of money and that salinization and evaporation will take a huge toll on the cultivated land and the country's water supply.

Turkey is an "upstream" state and is the source of four-fifths of Syria's water and two-thirds of Iraq's (Fig. 9.C). Turkey's position has long been that water in Turkey belongs to Turkey, just as Saudi Arabian oil belongs to Saudi Arabia. Not surprisingly, downstream Syria and Iraq reject this position and are distraught by the diminished flow and quality of water resulting from Turkey's comprehensive Southeast Anatolia Project. When completed, the project is expected to reduce Syria's share of the Euphrates waters by

40% and Iraq's by 60%. Also increasing the likelihood of serious future tension is a history of strained relations among Turkey, Syria, and Iraq, accompanied by the fact that no commonly accepted body-of-water law governs the allocation of water in such international situations. Turkish leaders have said they will never use water as a political weapon, but Turkey has wielded water to its advantage.

In 1987, for example, Turkey increased the Euphrates flow into Syria in exchange for a Syrian pledge to stop support of Kurdish rebels inside Turkey. Turkey now says it wants to use water to promote peace in the Middle East by shipping it in converted supertankers for sale to such thirsty (and therefore potentially combative) countries as Israel, Jordan, Saudi Arabia, Tunisia, and Algeria.

Figure 9.C   The Tigris and Euphrates Rivers rise in Turkey, giving this non-Arab country control over a resource vital to the lives of millions of Arabs in downstream Syria and Iraq. This waterfall is on a tributary of the Tigris in far eastern Turkey.

Tigris are also part of the ongoing project to provide hydroelectricity as well as water. The GAP is controversial. It is inundating historic settlements and priceless archeological treasures and is compelling people to move. It is being developed in the part of the country where a restive Kurdish population is seek-

Figure 9.25   Where there had been a bridge across the Euphrates River, there is now a ferry across the Ataturk Reservoir, upstream from the huge Ataturk Dam.

ing recognition, autonomy, and among some factions, independence from Turkey. In addition, Turkey's Tigris and Euphrates waters flow downstream into neighboring states, raising serious questions about downstream water allocation and quality (see Regional Perspective above).

Turkey is an "in-between" country. Economically, it is well below the level of MDCs but above most of the world's LDCs. Culturally, it is between traditional Islamic and secular European ways of living. Part of Turkey — the section called Thrace — is actually in Europe, and Turkey as a whole aspires to become more European. It is a member of the North Atlantic Treaty Organization (NATO). It is an associate member of the European Union and is a candidate for full membership. The EU, however, would first like to see Turkey become more European politically, for example, by diminishing the army's influence in government and by improving its human rights record.

Istanbul (population: metropolitan area, 9,433,500; city proper, 9,018,500), formerly Constantinople, is Turkey's main metropolis, industrial center, and port. One of the world's most historic and cosmopolitan cities, Istanbul was for many centuries the capital of the Eastern Roman (Byzantine)

Figure 9.26 When completed, this bridge across the Bosporus in Istanbul would link Asia (left) with Europe (right).

Empire. It became the capital of the Ottoman Empire when it fell to the Turks in 1453. However, the capital of the Turkish Republic was established in 1923 at the more centrally located and more purely Turkish city of Ankara (population: metropolitan area, 3,362,700; city proper, 3,258,000) on the Anatolian Plateau. Turkey's third largest city is the seaport of Izmir on the Aegean Sea.

Istanbul is located at the southern entrance to the Bosporus Strait, the northernmost of the three water passages (the Dardanelles, Sea of Marmara, and Bosporus) that connect the Mediterranean and Black Seas and are known as the Turkish Straits (Fig. 9.26). The straits have long been a focus of contention between Turkey and Russia, who fought several wars in the past; their relations are much better now. Turkey's military posture has also been shaped by fragile relations with its NATO neighbor, Greece. The main questions recently at issue have been the status of Cyprus and conflicting claims to hundreds of islands in the Aegean Sea and that sea's oil and gas reserves.

9.8 Ethnic Separation in Cyprus

The large Mediterranean island of Cyprus, located near southeastern Turkey (see Fig. 9.3), came under British control in 1878 after centuries of Ottoman Turkish rule. In 1960, it gained independence as the Republic of Cyprus. The critical problem on the island is the division between the Greek Cypriots, who are Greek Orthodox Christians comprising about three-fourths of the estimated population of 900,000, and the Turkish Cypriots, who are Muslims comprising about one-fourth. Agitation by the Greek majority for union (*enosis*) with Greece was prominent after World War II and led in the 1950s to widespread terrorism and guerrilla warfare by Greek Cypriots against the occupying British. Violence also erupted between Greek advocates of enosis and the Turkish Cypriots,

who greatly feared a transfer from British to Greek sovereignty.

In 1974, a major national crisis erupted when a short-lived coup by Greek Cypriots, led mainly by Greek military officers, temporarily overthrew Cyprus' President Makarios, who had followed a conciliatory policy toward the Turkish minority. Turkey then launched a military invasion that overran the northern part of the island. Cyprus was soon partitioned between the Turkish North and the Greek South. A buffer zone (the Attila Line or Green Line) sealed off the two sectors from each other, and even the main city of Nicosia was divided. A separate government was established in the North, and in 1983, an independent Turkish Republic of Northern Cyprus was proclaimed. Only Turkey has recognized this state. Meanwhile, the internationally recognized Republic of Cyprus functions in the Greek-Cypriot sector, which comprises about three-fifths of the island. Both republics have their capitals in Nicosia (population: city proper, 93,700).

Prior to the partitioning of Cyprus, the North had dominated the economy, but since then, the North has had severe economic difficulties while Greek Cyprus has prospered. Most outside nations refused to trade directly with Turkish Cyprus after the invasion, and the trade and economic assistance offered by an economically weak Turkey were insufficient to provide much momentum. The depressed North remains tied to the struggling economy of the Turkish mainland.

The Greek sector was able to make effective use of economic aid from Greece, Britain, the United States, and the United Nations. A construction program provided new housing, commercial buildings, roads, and port facilities. Tourism based on both beach and mountain resorts flourished. New businesses were attracted. Greek Cyprus is expected to become a member of the European Union, pending the outcome of negotiations concerning some of its social and military policies. Already a European Union member, Greece is pressuring Turkey to revise its policy toward Cyprus and to settle the Aegean Sea dispute as preconditions for Turkey joining the organization.

9.9 Rugged, Strategic, Devastated Afghanistan

High and rugged mountains dominate Afghanistan (population: 26.8 million), the only landlocked nation in the Middle East (Fig. 9.27). Historically, it has occupied an important strategic location between India and the Middle East. Major caravan routes crossed it, and a string of empire-builders sought control of its passes. Today, it has limited resources, poor internal transportation, little foreign trade, and much conflict; Afghanistan is one of the poorest of the world's LDCs.

Most Afghans live in irrigated valleys around the fringes of

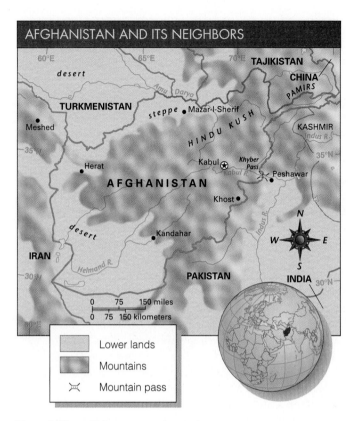

AFGHANISTAN AND ITS NEIGHBORS

Figure 9.27 Afghanistan in its regional setting.

the high mountains that occupy a large part of the country. The country's second most populous area is the northern side of the central mountains, where people live in oases in foothills and steppes. Northern Afghanistan borders three of the five Central Asian countries, and millions of people on the Afghan side are related to peoples of those countries.

The most heavily populated section is the northeast, particularly the fertile valley of the Kabul River, where the capital and largest city, Kabul (population: 2,080,000; elevation: 6,200 ft/1,890 m) is located. Most of the inhabitants of the southeast are Pushtuns (also known as Pashtuns or Pathans). Their language, Pushtu, is related to Farsi (Persian), which is the main language of administration and commerce in Afghanistan. The Pushtuns are the largest and most influential of the numerous ethnic groups that make up the Afghan state. The independent-minded tribal Pushtun people have always been loath to recognize the authority of central governments. In the days of Great Britain's Indian Empire, the area saw warfare among tribes, tribal raids on British-controlled areas, and British punitive expeditions against the tribes. In those days, Peshawar — on the Indian (now Pakistan) side of the Khyber Pass into Afghanistan — became the hotbed for British, Russian, and other agents playing what came to be known as the "Great Game" of vying for strategic influence in this part of the world.

Afghanistan is overwhelmingly a rural agricultural and pastoral country. It is so mountainous and arid that only about

12% is cultivated. Enough rain falls in the main populated areas during the winter for dry farming of grains. A variety of cultivated crops, fruits, and nuts are important locally. Raising of livestock on a seminomadic and nomadic basis is widespread. In general, Afghanistan's agriculture bears many of the customary Middle Eastern earmarks: traditional methods, simple tools, limited fertilizer, and low yields. The most successful crop has been the opium poppy. A succession of rebel armies and then Afghanistan's ruling Taliban used revenues from opium and heroin to obtain their arms. Afghanistan recently was the world's largest producer of opium (its 2000 harvest was three-fourths of the global total). Prior to their eviction from power, the Taliban imposed a ban on cultivation, and there was virtually no opium harvest in 2001. While there are fears that the Taliban's demise might allow the widespread resumption of poppy farming, U.S. and other foreign military and aid agencies have pledged to try to prevent its reemergence as Afghanistan rebuilds.

Through most of the 20th century, this highland country was remote from the main currents of world affairs. After the Islamic revolution in Iran in 1979, however, Afghanistan's location next to that oil-rich country made it once again the target of foreign interests. The Soviet military intervention of 1979 and the ensuing devastation catapulted the country into world prominence. The USSR's motives for its invasion of Afghanistan may have included a desire to prevent by force the spread of Iranian-style Islamic fundamentalism into Afghanistan (which is 84% Sunni Muslim and 15% Shi'ite Muslim) or into the Central Asian Soviet republics deemed vital to the superpower's security. For its part, the United States warned the Soviet Union that it would not tolerate further Soviet expansionism.

In the ensuing civil war, there was widespread killing and maiming of civilians, sowing of land mines over vast areas, destruction of villages, burning of crops, killing of livestock, and destruction of irrigation systems. Soviet ground and air forces caused several million Afghan refugees to flee into neighboring Pakistan and Iran (Fig. 9.28). Meanwhile, arms from various foreign sources filtered into the hands of the *mujahadiin*, the anti-Soviet rebel bands who kept up resistance in the face of heavy odds. The United States was one of the powers supporting the rebels and, in that sense, waged a proxy war against the Soviet Union in Afghanistan. In its waning years, the USSR recognized it could not win its "Vietnam War," and its troops withdrew from Afghanistan by 1989. It is estimated that, of the 15.5 million people who lived in Afghanistan when it was invaded in 1979, 1 million died, 2 million were displaced from their homes to other places within the country, and 6 million fled as refugees into Pakistan and Iran. As of 2002, fewer than half of these refugees had returned.

Support lent to the *mujahadiin* and sympathetic Arab fighters by the United States and "moderate" Arab states such as Egypt and Saudi Arabia soon came back to haunt those countries. Intelligence analysts credit some of the

The scars of war will last a long time in Afghanistan. When peace does come, the country must confront the problem of clearing the millions of land mines sown across the landscape during wartime.

Land mines are antitank and antipersonnel explosives, usually hidden in the ground, designed to kill or severely wound the enemy. There are many kinds: Soviet troops planted at least nine varieties in Afghanistan, including seismic mines, which are triggered by vibrations created by passing horses or people; mines with devices that pop up and explode when a person approaches; "butterfly" mines meant to maim rather than kill; and explosives disguised as toys, cigarette packs, and pens.

There are an estimated 10 million land mines in Afghanistan, or 40 per square mile (15 per sq km). Even after the fighting ends, reconstruction of this agricultural country cannot begin in earnest until the mines are cleared. The obstacles are enormous in rugged Afghanistan. Mines typically remain active for decades after they are planted and tend to last longer in arid places. The mountainous terrain prevents systematic clearance of the explosives. Many mines were scattered indiscriminately from the air and will be difficult to locate. Heavy rains move the mines downhill and away from known locations. And it is expensive to clear mines; Kuwait spent $800 million to rid its landscape of them after the Gulf War. So far, Western countries have donated $12 million toward the expense of clearing Afghanistan's land mines; Russia has contributed nothing.

Land mines are a global problem. Up to 70 million mines contaminate the land in 64 countries, and they kill or maim more than 25,000 civilians each year. The numbers and concentrations in some countries are staggering. The leader in numbers is Egypt, with 23 million mines; in density of mines, Bosnia-Herzegovina leads with 152 per square mile (58 per sq km). The United Nations estimates that with current technology, it would take $33 billion and 1,100 years to clear the world's land mines.

Several recent events may portend an eventual end to the land mine scourge. The world's most visible critic of land mines, Princess Diana of Wales, was killed in an automobile accident in Paris in 1997. In her wake, there were renewed calls for a moratorium on the manufacture and use of antipersonnel land mines. Two months after Diana's death, anti-land mine activist Jody Williams won the Nobel Peace Prize. Soon after, representatives from 125 nations met in Ottawa, Canada, to sign a treaty strictly banning the use, production, storage, and transfer of antipersonnel land mines. Many of the largest producers, users, and exporters of these explosives, including the United States, Russia, China, India, Pakistan, Iran, Iraq, and Israel, refused to sign. The United States argued that land mines were necessary to protect South Korea against an invasion from North Korea and failed to win an exemption on this count that would have allowed it to sign the treaty.

Too many countries regard these as legitimate weapons to make the prospect of a mine-free world likely. Costing as little as $2 each, they are very affordable weapons for cash-strapped warring LDCs (and so have been dubbed the "Saturday-night specials of warfare"). For now, these hidden horrors continue to be planted and to be cleared, as one demining specialist put it, "one arm and one leg at a time."

groups armed and trained by these countries in the 1980s with some spectacular acts of terrorism in subsequent years, including an assassination attempt on Egyptian President Hosni Mubarak, the bombing of U.S. military barracks in Saudi Arabia, the bombing of U.S. embassies in Kenya and Tanzania, and the horrific 2001 airliner attacks on New York City and Washington, D.C. The East African bombings in 1998 led within days to a cruise missile strike by the United States on six guerrilla camps near Khost, Afghanistan. These camps were allegedly funded by Osama bin Laden, who was also suspected of masterminding the 2000 attack on the U.S.S. *Cole* and the 2001 airliner attacks on the United States.

The *mujahadiin* succeeded in overthrowing the Communist government of Afghanistan in 1992, but after that, rival factions among the formerly united rebels engaged in civil warfare. By 1996, one of the factions, the Taliban (backed by Saudi Arabia and Pakistan), gained control of most of the country, including the capital of Kabul. Proclaiming itself the sole legitimate government of Afghanistan, the Taliban imposed a strict code of Islamic law in the regions under its control and gained international notoriety for its austere administration. The Taliban removed almost all women from the country's work force, forbade public education of girls, and outlawed "un-Islamic" practices such as dancing, kite-flying, television, bird-keeping, and beard-trimming. The Taliban continued to make advances against its opponents inside Afghanistan (particularly the Northern Alliance, whose leader, Ahmed Shah Massoud, was assassinated just days before and in apparent preparation for the terrorist

JOB HOBBS

Figure 9.28 When this photo was taken in 1987, the *mujahadiin* fighters pictured here in an Afghan refugee camp outside Peshawar, Pakistan, were regularly involved in firefights with Soviet troops across the border in Afghanistan. More recently, many of the various guerrilla factions that made up the anti-Soviet resistance have been involved in conflicts with one another.

strikes against New York and Washington) and, by 2001 controlled 95% of the country's territory. The neighboring Central Asian countries, Russia, and even Iran grew increasingly fearful of the spread of the Taliban's extreme interpretation of Islam into their nations. Russia ended up supporting some of the rebels it once fought because they were fighting the Taliban. Russia's once unlikely ally in supporting those rebels was the United States, which successfully lobbied the United Nations to levy economic sanctions against Afghanistan in 1999. The United States hoped that economic losses would pressure the Taliban into turning over bin Laden for prosecution and also announced a $5 million

reward for information that would lead to bin Laden's capture or death.

The bin Laden bounty rose into the tens of millions of dollars after September 11, 2001. Named as the mastermind of the attacks against New York and Washington, bin Laden became the "world's most wanted" as U.S. President George W. Bush evoked a wild west vow to have him captured dead or alive. In crafting its pronounced war against terrorism (which it promised to carry anywhere in the world necessary), the U.S. administration identified the Taliban as al-Qa'ida's mentor and targeted both organizations for elimination. Within a month of the attacks on the United States, American warplanes and special forces were operating against Taliban and al-Qa'ida facilities and personnel around the country. The British military assisted in the air campaign, and most of the ground fighting was carried on by the Northern Alliance, the Taliban's rival and nemesis. The United States successfully persuaded Pakistan to drop its support of the Taliban and join the effort, and much of the military campaign was based on U.S.-Pakistani cooperation. One after another of the Taliban's urban strongholds fell, including Kabul and the Taliban spiritual capital, Kandahar. Northern Alliance forces filled the void, and an international commission appointed an interim leader, Hamid Karzai, to rule Afghanistan during the country's political reconstruction.

Withering assaults from U.S. warplanes apparently succeeded in crushing the caves and other refuges used by al-Qa'ida guerillas, and many al-Qa'ida prisoners were taken to the U.S. naval base in Guantanamo Bay, Cuba. The elusive bin Laden, however, along with the Taliban leader Mullah Omar, apparently slipped out of Afghanistan, perhaps seeking safe harbor in Pakistan, Yemen, Somalia, or even Indonesia. As this book went to press, as so often before, the world held its breath over events happening in and emerging from the Middle East.

CHAPTER SUMMARY

- The Arab World stretches from Morocco to the Indian Ocean. Although an ethnically diverse region, most inhabitants speak dialects of Arabic and are known as Arabs. The wide distribution of these peoples is due primarily to the rise and spread of Islam and its empire.
- The United Nations Partition Plan of 1947 attempted a "two state solution" to the Arab/Jewish conflict. However, it produced almost fragmented states, making both sides vulnerable to the other and complicating the issue further. The Arab/Israeli conflict has continued from that time through numerous wars, crises, and terrorist actions.
- The famous Bekaa Valley, shared by Lebanon and Syria, has been a center of political activity since before Roman times and

continues to be a center for pro-Iranian Hizbullah activities today.
- Lebanon and Syria are the home of the historic and modern urban centers of Damascus, Aleppo, and Beirut. Before Lebanon's Civil War, the country was known as the "Switzerland of the Middle East."
- Sudan is a nation of marked environmental diversity and is considered a transition zone. This diversity has led to long-term military and political conflict between the northern and southern peoples.
- Most of Libya lies in the Sahara, and its primary natural resource is oil. The revenues generated from the exported oil are used to support social and economic projects along with large

purchases of weapons and weapons development. Muammar al-Qaddafi's brand of socialism and his support for international terrorism have earned the nation a dubious reputation.

- The Maghreb (Place of the West) is composed of Morocco, Algeria, Tunisia, and the disputed Western Sahara. This area is closer, geographically and culturally, to Europe than is any other of the Middle East.

- Two-thirds of the world's proven petroleum reserves are concentrated in a few countries that ring the Persian/Arabian Gulf. All but 1% of the Gulf's reserves are located in Saudi Arabia, Iraq, Kuwait, Iran, and the United Arab Emirates (UAE). Saudi Arabia controls approximately 26% of the world's oil production.

- Iraq, the region of Mesopotamia ("Land Between the Rivers"), lies on the eastern end of the Fertile Crescent. It is the home of the first known civilization and culture hearth. The Baath (Arab Socialist) party under Saddam Hussein has driven for Arab solidarity, freedom from the West, and economic development, but it has also caused the Iraqis to suffer heavy casualties in the long Iran/Iraq War and a political, economic, and military thrashing in the Gulf War for its attempt at expansion (for oil and seaports) into Kuwait and other areas of the Persian Gulf.

- Iran (formerly known as Persia) was the earliest Middle Eastern country to produce oil in large quantities. The oil revenues were used to modernize and Westernize the nation during the Pahlavi dynasty. The rapid changes in the culture precipitated a revolutionary movement, which forced the shah to abdicate and flee. The new theocratic state, since the death of the Ayatollah Khomeini, is alternately perceived to be either developing a more accommodating attitude toward the international community or exporting ever more dangerous notions of revolutionary Islam to the region.

- Turkey was formerly the seat of power for the Islamic Ottoman Empire. It is now an independent, secular nation with fewer ties to traditional Islam. Turkey has become a rather unique Middle Eastern nation based on such attributes as official separation of church and state, democratic government, membership in NATO, and full citizenship for women.

- Cyprus has a history of foreign rule by the Greeks, Romans, Ottoman Turks, and the British. Although independent today, the ethnic composition of Cyprus is three-fourths Greek and one-fourth Turkish. Despite hopes for enosis (unification), several internal conflicts have occurred in efforts to gain control of this large island near southeast Turkey.

- Afghanistan is one of the world's poorest nations, being landlocked as well as having limited resources, poor internal transportation, and little foreign trade. Its historic location along the ancient caravan routes made it the target of empires. Its role in the September 11, 2001, attacks against the United States brought another round of warfare to this beleaguered land.

REVIEW QUESTIONS

1. What precipitated the respective conflicts between Israel and her Arab neighbors and how did each of these conflicts rearrange the political map?
2. What did the peace process promise to Israelis and Arabs? What powers do Palestinians now have that they lacked before the peace process began? What are the prospects for peace now?
3. Using maps and the text, identify the major physical and cultural features of Israel, Jordan, and Syria.
4. What are some of the positive and negative outcomes of the construction of Egypt's Aswan High Dam?
5. In what ways is Sudan's North different from its South?
6. What are some of the major factors driving U.S. interests in the Gulf Oil Region?
7. Using maps and the text, locate the countries of the Gulf Oil Region. Which of these are more important in petroleum production?
8. List some of the unique aspects of Turkey in relation to most of the other countries of the Middle East.
9. What have been the primary goals of the Baath regime in Iraq? List some of the Baath regime's results.
10. What were some of the grievances against the Pahlavi dynasty that led to revolution in Iran?
11. What were Afghanistan's shifting fortunes between 1979 and 2001?

DISCUSSION QUESTIONS

1. Where are the chokepoints of the Middle East?
2. Examine and discuss the birth of Israel and its lasting legacy for the rest of the world.
3. Define, discuss, and debate the following terms and issues in the Arab-Israeli conflict: diaspora, right of return, autonomy, Intifada, Final Status issues.
4. Discuss the sectarian problems that ripped apart Lebanon's delicate fabric between 1975 and the late 1980s.
5. What is the origin of the phrase "the gift of the Nile"? How does this relate to modern Egypt?
6. What is the difference between basin and perennial irrigation in Egypt and what technologies are associated with each?
7. What differences exist between Sudan's North and South and how have these contributed to civil conflict?
8. Examine and discuss the political problems associated with Sudan and Libya and the rest of the world.

9. Explain how the Maghreb is closely associated with Europe.
10. What regions or nations of the world are primarily dependent on the Gulf Oil Region for petroleum? What is the U.S. interest?
11. Why did the United States and other nations feel driven to engage Iraq in the Gulf War?
12. Examine and discuss Iran's complicated struggle between traditional Islamic culture and modern society.

13. Why has Turkey, once the home of the Ottoman Empire, evolved as a comparatively Westernized nation in the Middle East?
14. Examine and discuss the conflict in Afghanistan after September 11, 2001.

Monsoon Asia

With more than three-fifths of the world's population living in its cities, villages, and remote places, Monsoon Asia is quickly known for its abundance of population. This world region, however, is also economically dynamic, culturally significant, and worthy of study for many additional reasons. From the technological innovations in East Asia to the religious traditions in South Asia, this region represents a way of life that will be increasingly significant to all world regions as this new century unfolds.

This region has a history rich in human effort to remake the landscape into an ever more productive and distinctive setting for its large populations. The Great Wall in China, the Taj Mahal in India, and the thousands of village temples in Monsoon Asia all reflect the long history of political authority and civil engineering at many scales.

The growth of new city space in Monsoon Asia has had architectural and structural distinctiveness. With the world's tallest buildings in this region, you also have flashy design innovations in East and Southeast Asian urban centers. The city in Monsoon Asia has become not only a migration destination, but also a setting for global innovation.

Rivers are a major factor in the geographic character of Monsoon Asia. They are a resource that absorbs enormous amounts of human labor, accommodates massive settlements, and has served as imagery for poetry and painting for thousands of years.

One of the constants in Monsoon Asia is the centrality of the family to all culture and all tradition. Even where population growth is an anxiety to economic and urban planners, there continues to be keen pleasure in the continuity of family growth, although new ideas about the optimum size of families are increasingly evident.

A Geographic Profile of Monsoon Asia

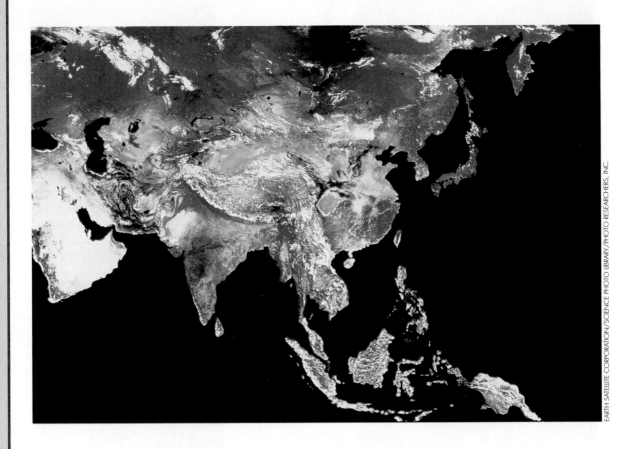

This composite satellite image from the National Oceanic and Atmospheric Administration (NOAA) shows the full landscape of Asia, with its great mountain highlands and plateaus, desert expanses, ranges of borderland mountains, verdant river valleys, and the plains and river systems with dense accumulations of farming and urban populations.

Chapter Outline

Monsoon Asia extends, in a broad, sweeping crescent, from Pakistan and the Indus River in the west through an arc of mountains, rivers, plains, islands, and seas to the northern island of Japan in the east (Fig. 10.1). In this landmass of less than 8 million square miles (20,761,000 sq km) lives more than half of the world's population on less than one-quarter of Earth's land mass. The cultural makeup of these 3.4 billion people is as diverse and expressive as are the physical environments they inhabit. The world's highest mountain peaks, some of the longest rivers, and Earth's most highly trans-

formed and densely settled river plains are the setting for some of our oldest civilizations and our most modern developing economies. Just as this region has played a major role in societal development over the past millennia, so, too, will it be a major force in defining the world of the 21st century.

"The Orient" is the term traditionally used to refer to the countries occupying the southeastern quarter of Eurasia. The term *orient* comes from the Latin meaning "to rise" or "to face the east." *Occident* means "to set" (as does the sun, in the west). Early on, Western Europe was the point from which

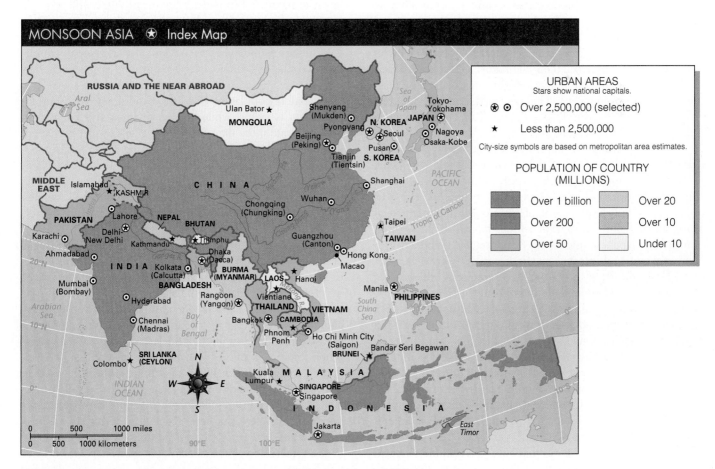

Figure 10.1   Introductory location map of Asia, showing political units as of early 2002. Macao was returned to China from Portugal in December 1999.

this daily orbit of the sun was described. Hence, the Orient and the Occident became European terms for spatial regional identification. In this text, we use the term Monsoon Asia for this region rather than the Orient or the Far East, terms which tend to disinclude South Asia (see Fig. 10.1).

10.1 Area and Population

Monsoon Asia encompasses the following regions: East Asia, which includes Japan, North and South Korea, China, Mongolia, Taiwan, and countless nearshore islands; South Asia, which includes Pakistan, India, Sri Lanka, Bangladesh, and the mountain nations of Bhutan, Nepal, and offshore islands; Southeast Asia, which is the term used for the dominant peninsula jutting out from the southeast corner of the Asian continent and includes the countries of Burma (Myanmar), Thailand, Laos, Cambodia, Vietnam, Malaysia, and Singapore; and the island world that rings this peninsula, which includes the countries of Indonesia, the Philippines, Brunei, and East Timor.

This crescent of Asia has been home to some of the most important cultural developments of humankind in landscape transformation, settlement patterns, religion, art, and political innovations. It is not the magnitude of population alone that makes understanding Monsoon Asia central to our study of world regional geography; it is also the pivotal role of past and present Asian cultural innovations and a significant global economic role that make this region a major building block in the process of better understanding the nature of the world today.

Table 10.1 portrays the demographic and geographic magnitude of Monsoon Asia. The region is home to approximately 55% of the world's population. To gain a feeling for the diversity of this region, it is important to see it today as a collection of distinctive approaches to achieving economic growth and, in the face of the severe economic decline of 1997–1998, sustaining such growth. This challenge is set against a backdrop of widely varying distinctive histories, particularly in the development of technology, farming, literature, and religion. At the same time, images in the West of prosperous Japanese and Hong Kong tourists festooned with cameras and sophisticated electronic gadgetry must be tempered with the reality of a large Asian peasantry still working the land with traditional tools and only modest evidence of recent change in agricultural technology. Monsoon Asia, to a greater extent than other regions chronicled in this text, is a restless amalgam of tradition and innovation, rural industriousness and urban experimentation, and societal continuity and tense assimilation of Western popular culture.

This is also a region of monumental physical hazards: the earthquakes of India, China, and Japan; the massive floods of Bangladesh and India; and the perennial threat of drought through much of South Asia and Inner Asia. There are mountain heights in the Himalayas that have been central to global images of isolation for centuries. There are river system floodplains that play both agricultural and religious roles as they course through upland and lowland South Asia (Fig. 10.2).

Deserts in Inner Asia have been sources of mystery, trade, and adventure for millennia (Fig. 10.3). And the fabled urban ghettos of Calcutta, Guangzhou (Canton), Jakarta, and Manila are settings and landscapes that hold image banks for Asians and non-Asians alike. In turning to Monsoon Asia in our attempt to better understand the global nature of geography and its linkages across the past and present, the richness of this region's historical geography and economic importance will come into focus.

Population, Settlement, and Economy

In Monsoon Asia, populations range from small tribal groups with locally distinct cultures in the hills and uplands of Southeast Asia and southwest China to major and expansive culture groups like the Chinese and the Indians. Mongoloid peoples form a majority in China, Japan, Korea, Burma, Thailand, Cambodia, Vietnam, and Laos. But the majority of the people in India, although darker skinned than Europeans, are considered in many classification systems to belong to the Caucasian race; similar peoples form a majority of the native inhabitants of the Malay Peninsula, the East Indies, and the Philippines.

The birth rate ranges from a 0.2 annual rate of increase in Japan to nearly 3.0 in parts of Southeast and Southwest Asia (see Table 10.1). In Hong Kong, Macao (both now returned to China), and Singapore, some of the world's highest urban densities are found, whereas in Mongolia and Nepal, population densities are extremely low. Singapore is one of the few nations to claim a 100% urban population, whereas 89% of the Nepalese population is still rural and nonurban. Only 2% of Bhutan is arable, while farmers in India and Bangladesh both cultivate more than 50% of their land, with Bangladesh defining two-thirds of its area to be arable. The densest Monsoon Asian populations are found on river and coastal plains (Fig. 10.4), although surprisingly high densities occur in some hilly or mountainous areas. Higher mountains, steppes, deserts, and some areas of tropical rain forest are very sparsely inhabited.

Most of the countries in Monsoon Asia have experienced large increases in population during recent centuries, especially since the beginning of the 19th century, and most of the additional people have accumulated in areas that were already the most crowded. Food production levels remain precarious, although factors such as increased grain supplies from new high-yielding varieties through the success of the Green Revolution, the ability of Thailand and Burma to continue to export rice, and the availability of surplus grain from overseas countries are generally making it possible to maintain adequate levels of nutrition. Through irrigation, farmland expansion, and the increasing use of chemical fertilizers and new seed stock (much of it derived from the new strains developed in the Green Revolution), Asian agriculture has grown to be much more productive than it has been historically. It is not known how long this can continue in the face of continuing population increases—in India particularly—and just as

TABLE 10.1 Monsoon Asia: Basic Data

Political Unit	Area (thousand/sq mi)	Area (thousand/sq km)	Estimated Population (millions)	Annual Rate of Increase (%)	Estimated Population Density (sq mi)	Estimated Population Density (sq km)	Human Development Index	Urban Population (%)	Arable Land (%)	Per Capita GDP PPP ($US)
Indian Subcontinent										
India	1148.0	2973.2	1033.0	1.7	899.9	347.4	0.571	28	56	2200
Pakistan	300.7	778.7	145.0	2.8	482.3	186.2	0.498	33	27	2000
Bangladesh	51.7	133.9	133.5	2.0	2582.3	997.0	0.47	21	73	1570
Nepal	52.8	136.8	23.5	2.4	444.9	171.8	0.48	11	17	1360
Bhutan	18.1	47.0	0.9	3.1	49.6	19.1	0.477	15	2	1100
Sri Lanka	25.0	64.7	19.5	1.2	780.6	301.4	0.735	22	14	3250
Maldives	0.1	0.3	0.3	3.2	2590.0	1000.0	0.739	25	10	2000
Total	**1596.4**	**4134.6**	**1355.7**	**1.9**	**849.2**	**327.9**	**0.554**	**27**	**49**	**2116**
Southeast Asia										
Myanmar (Burma)	253.9	657.7	47.8	1.6	188.2	72.7	0.551	27	15	1500
Thailand	197.6	511.8	62.4	0.8	315.8	121.9	0.757	30	34	6700
Vietnam	125.6	325.4	78.7	1.4	626.4	241.9	0.682	24	17	1950
Cambodia	68.1	176.5	13.1	1.7	192.2	74.2	0.541	16	13	1300
Laos	89.1	230.8	5.4	2.5	60.6	23.4	0.476	17	3	1700
Malaysia	126.9	328.6	22.7	2.0	178.9	69.1	0.774	57	3	10,300
Singapore	0.2	0.638	4.1	0.9	16,644.2	6426.3	0.876	100	2	26,500
Brunei	2.0	5.27	0.3	2.0	147.4	56.9	0.857	67	1	17,600
Indonesia	705.2	1826.4	206.1	1.7	292.3	112.8	0.677	39	10	2900
Philippines	115.1	298.2	77.2	2.2	670.5	258.9	0.749	47	19	3800
East Timor	NA	NA	0.8	1.6	NA	NA	NA	8	NA	NA
Total	**1683.9**	**4361.3**	**518.6**	**1.6**	**308.0**	**118.9**	**0.687**	**36**	**14**	**3685**
Chinese Realm										
China (PRC)	3600.9	9326.4	1273.3	0.9	353.6	136.5	0.718	36	10	3600
China, Hong Kong	0.4	1.0	6.9	0.3	17,150.7	6621.9	0.88	100	6	25,400
China, Macao	0.008	0.02	0.4	0.6	49,333.4	19,047.6	NA	99	0	17,500
Taiwan	12.5	32.3	22.5	0.8	1804.2	696.6	NA	77	24	17,400
Mongolia	604.2	1565.0	2.4	1.4	4.0	1.5	0.569	57	5.7	1780
Total	**4218.1**	**10,924.8**	**1305.5**	**0.9**	**309.5**	**119.5**	**0.719**	**37**	**9**	**3954**
Japan and Korea										
Japan	144.7	374.7	127.1	0.2	878.5	339.2	0.928	78	11	24,900
Korea, North	46.5	120.4	22	1.5	473.3	182.7	NA	59	14	1000
Korea, South	37.9	98.2	48.8	0.9	1287.1	496.9	0.875	79	19	16,100
Total	**229.1**	**593.3**	**197.9**	**0.5**	**863.9**	**333.6**	**0.913**	**76**	**13**	**20,073**
Summary Total	**7727.4**	**20,014.0**	**3377.7**	**1.4**	**437.1**	**168.8**	**0.657**	**35.4**	**19**	**4120**

Sources: *World Population Data Sheet, Population Data Sheet, 2001. U.N. Human Development Report, United Nations, 2001 World Factbook, CIA, 2001.*

important, in the face of changing diet patterns. In China, the newly prosperous population is moving away from rice and toward increased meat and wheat consumption, and this is going to have an impact on not only regional but global patterns of grain movement. Although Japan has lowered its rate of population increase to below that of the United States and Canada, and China has considerably reduced its growth rate in the past two decades by stringent birth control measures built around a limit of one child per couple, many Asian countries continue to have high fertility rates characteristic of less developed countries (LDCs) discussed earlier in the text (see Table 10.1).

Rural-to-Urban Demographic Shift
Even though the village is central to Monsoon Asian demography, this era of unprecedentedly rapid cultural diffusion and change since World War II has rapidly accelerated migration. Rural peoples have been leaving the farming life and going to the cities at a pace never before experienced.

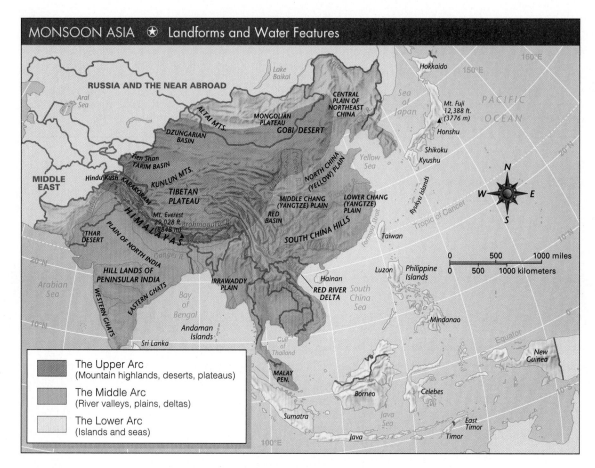

MONSOON ASIA ✶ Landforms and Water Features

Legend:

- **The Upper Arc** (Mountain highlands, deserts, plateaus)
- **The Middle Arc** (River valleys, plains, deltas)
- **The Lower Arc** (Islands and seas)

Figure 10.2   Major landforms and water features of Asia. The upper, middle, and lower arcs define major regions of landscape distinction and the middle and lower arcs are very heavily settled.

Figure 10.3   Dunhuang, in China's western Gansu Province, is an ancient stop on the Silk Road that linked northern China with trading ports as far west as Baghdad and various Mediterranean ports. In the early 1970s, a whole series of grottoes was discovered carved into the cliffsides of the river valley seen in the distance. These caves were used for ceremonial and religious displays by travelers either relieved to have successfully crossed the deserts of western China (especially the Takla Makan) or nervous about now beginning such a traditionally difficult crossing. In the foreground can be seen small ceremonial structures built by the farmers of this wheat-growing and carpet-weaving market stop on this ancient route between China and the West.

Figure 10.4 In India, the great rivers play a role not only in the flow of agricultural life but also in religious life. Here, the holy act of washing in the waters of the Ganges River is both a sacred and a social act. Rivers in Monsoon Asia play many roles as they carry barge and boat traffic, serve as sources for irrigation and urban settlement waters, and flow through the spiritual lives of many of the millions who live along their banks or come to them for ceremonial cleansing.

In China, for example, in the late 1940s, approximately 85% of the population lived in the countryside. Although there were massive cities such as Shanghai, Beijing, and Guangzhou (Canton), the great bulk of the Chinese population was still deeply involved in rural activity, and a vast majority of these hundreds of millions were living in relatively small villages. However, in the 1990s, China experienced a demographic shift that doubled the proportion of urban population it now has, giving the country a ratio of approximately 64% rural and 36% urban. India is 72:28, and Japan is very nearly the opposite (23:77). In the village nations of China and India, especially, this demographic shift has meant enormous problems for city management as these tens of millions of rural migrants (all through Monsoon Asia generally) seek space, jobs, services, food, and goods. At the same time, there has been new pressure on the agricultural sectors to further increase food production and surpluses for these new urban populations.

The magnitude of these demographic shifts in China and India is so great that such moves can mean the transfer of hundreds of millions of people within a decade in those two Asian countries alone. Consequently, those left behind in the countryside have been prompted to introduce more agricultural mechanization to fill the resulting labor gaps. At the same time, cities have been energetically absorbing this labor force into everything from small sweatshop activities to major industrial enterprises.

The lure of the city, so powerfully expressed via today's widely available television and movies, has never played its siren song so successfully before. The whole demography of Monsoon Asia is shifting from a traditional rural past to an evolving urban present. In this region, as all across the globe, this population shift is seldom planned (except in some very specific efforts such as those of the Indonesians to shift Javan populations to Sumatra over the past six decades in their campaigns of transmigration). The shift comes from a pattern of young males deciding that there are too few options in the countryside, particularly if they are a second son or farther down the line. Filled with optimism and perhaps some lore or feedback from cousins or villagers who tried the rural-to-urban migration stream earlier and may have been successful, the break from the countryside is undertaken. The ramifications of this demographic shift are felt in both the city and the countryside, and in East as well as South Asia. Such changes are keenly felt among the elderly.

A final aspect to the population issue in Monsoon Asia is a general note on population control and family planning. The generally above average fertility rates in the first half of the 20th century have seen a substantial reduction. The Chinese population had doubled in less than 30 years in the 1930s, and in 60 years in 1985; it now has a doubling time of 82 years. The Indian population doubled in 31 years in 1985 and in 42 years in 2001. Indonesia had a birth rate of 31 children/1,000 people in 1985. By 2001, this rate had been reduced to 23 children/1,000 population. Although the pattern for each country has been different, there has generally been a major impact in Asian efforts at family planning and a

Figure 10.11   The terraced gardens and small plots of the Monsoon Asian farmer serve to remind observers and travelers of how energetically the Asian peasant farmer has remade the landscape to gain better and more predictable productivity. Step terraces are designed to allow water to flow by gravity through all the fields, generally reentering a stream at a lower level. By creating small terraces like these, the farmer is replacing sloping farmland with level minifields. The labor costs for such transformation are high and influence the continuing relatively high rural birth rates in Monsoon Asia. The labor demands of the farming life in a setting like this also help to explain the ever-increasing rate of rural-to-urban migration. This scene is from Sichuan Province in western China.

abundant, useful landscape modifications allowing irrigation can be designed and created by manual labor. The benefit of such hydraulic control has always led to higher yields, more certainty of crop success, and labor opportunities for marginally underemployed farm populations. The addition of irrigation facilities also always adds to the value of farmland. This penchant for the control of water has grown from household gardens and plots all the way to massive river projects. The greatest of these is China's Three Gorges Project (Fig. 10.12).

Rural Migration to Asian Cities

Like garden farming and irrigation control, this pattern of rural-to-urban migration is certainly not unique to Asia.

However, the speed with which long traditions of having two-thirds, three-quarters, or even four-fifths of a country's population living in the countryside is changing profoundly. Sons (especially) who see no likelihood of getting land in the expansion of agricultural businesses and ever more intense farming patterns are picking up and slipping into a demographic stream that is heading for middle sized and massive urban centers. Such migration is generally unapproved by official governmental policy, but the reality of such a demographic shift is occurring all across Monsoon Asia.

Figure 10.12   China's massive Three Gorges Dam Hydroelectric Project has been underway more than 5 years already. The project will create a reservoir more than 400 miles long, causing the displacement of approximately 3 million people. The dam face will be nearly 200 meters high and the turbines are expected to generate more than 80 billion kilowatt-hours a year. This scene is from work in the building of the coffer dam that allowed the major channel of the Yangtze (Chiang Jiang) River to be closed for this dam. It is a project that has been both praised and damned for years because of its scale, cost, and potential environmental consequences.

Figure 10.13   In China, it is still very difficult to leave the countryside and move smoothly into urban employment. The pervasive myth, however, of opportunities that exist in the city continually draws farmers who have no chance of getting land and little chance of earning enough to have a family. Men coming to the city—in this case Shanghai—with no guaranteed job, no residential permit, and possibly no real connections of any predictable nature become "floaters" in the urban population. This crowd of people, not all of whom are unemployed, are hanging around on the margin of the Hwang Po River that feeds from Shanghai to the Yangtze just above its confluence with the East China Sea. The men wait in hope of picking up some type of job that might not only provide cash for living but also lead to other connections that could provide residence and maybe even a courtship.

Often, these migrants have neither jobs nor accommodations in the cities they head for (Fig. 10.13). This is a classic migration stream in the sense that there is always a lore about some cousin or villager who found work, made money, and has undergone a considerable change in lifestyle through this sequence. This pattern occurs all over the world, but in the Asian context, it is vital to realize that the migration stream encompasses literally scores of millions of ill-educated, undercapitalized, and probably illegal (in China at least) migrants. They come to cities with good ambition, but they also come with needs for shelter, food, jobs, and training as they molt from rural traditions into an urban environment. This is never an easy migration transition, but when it is done with little or no support, it is particularly disruptive for both the city scene and the ever growing body of unemployed recent village migrants hanging around day-labor lines or wharf and rail-yard areas.

10.4   Economic Geography

The Asian population is characteristically attentive to family productivity, education, agricultural tradition and success, and increasingly, industrial productivity. India has launched a silicon revolution over the past two decades and now provides both services and highly trained migrants for high-tech industries both within Asia and beyond. China and Taiwan have become producing sources for literally thousands of consumer goods ranging from cheap prizes in food boxes to highly complex electronics products. Clothes of both a simple and a highly fashionable style come from sweatshops in Southeast Asia, South Asia, and East Asia. In the early 1990s, patterns of economic development were exceptional until there occurred some significant slowdown in 1997, especially in Thailand and South Korea. Overall, however, Monsoon Asia came through that difficult time reasonably well.

In setting your sights on the economic geography of Asia, refer again to Table 10.1 and look at the factors dealing with population, arable land, percentage of urban population, and GNP per capita. Add these dimensions to these several characteristics that follow, and you will be prepared to see why Asia is so important in world regional geography.

Settlement Patterns

Although an estimated 1.2 billion people in Asia live in urban settlements, the main unit of Asian settlement is the village. About two-thirds of the region's people are residents of an estimated 1.9 million villages. Highly urbanized Japan, Singapore, Macao, and Brunei do not fit this pattern, nor in smaller measure do Taiwan, South Korea, or even Mongolia. But in most other Asian countries, the typical inhabitant is a villager. The Asian village is essentially a grouping of farm homes, although some villages house other occupational groups, such as miners or fishers. Clusters of houses bunched tightly together are typical, and cheap and simple structures — often made of local clays and other building materials — are characteristic. Piped water and indoor plumbing continue to be somewhat exceptional in village homes, although the availability of electricity has been steadily expanding all through Asia. Details of village life vary according to culture and place; for instance, Hindu villages have segregated quarters for different castes, even though the caste system has been officially outlawed. In China, the village has, during the past two decades, witnessed a steady expansion of private farmland, local, private cottage industries, and low-tech plants as well.

As in all corners of the agricultural world, the original siting of villages was closely adapted to natural conditions. For example, villages are slightly elevated in floodplains located on natural levees (raised river banks built up by deposition of sediments during floods), dikes, or raised mounds. Early villages in Indonesia were often built in defensible mountain sites, although the Dutch colonial administration gradually required the building of villages along main roads and trails in the lowlands to make it easier to exercise control, collect taxes, and draft soldiers or laborers for roadwork or other projects.

One of the landscape signatures associated with village settlement in the crescent of Asia is the continuing tradition of shifting cultivation (slash-and-burn agriculture), particularly in the island world of Southeast Asia. In 1997 and 1998, Indonesia suffered through two seasons of the worst forest fires in two decades, and perhaps of the century. At least a portion of the widespread fires was caused by runaway fires associated with the traditional slash-and-burn farming. The other major cause was a very severe drought, influenced by El Niño. An additional factor was the steadily expanding forest clearance for commercial forestry and the creation of new plantations for the oil palm. Such clearing often results in attendant piles of forest debris that can serve as crisp fuel for forest fires. Estimates of the amount of land burned in the Indonesian fires of 1997–1998 run as high as 2 million hectares.

Means of Livelihood

Japan was the first Asian country to develop modern cities and modern types of manufacturing on a large scale, even while maintaining a broad-based village structure. For more than a century, it has been a major industrial power. But China and India also have important and dynamically expanding industrial bases. These two nations are the largest in the world in population, and they are much better supplied with mineral resources than Japan. They also have cheaper labor. But Japan's labor force is more skilled, and Japan has shifted increasingly to types of industry requiring such labor and advanced technology. Not only does it lead Asia in high value-added manufacturing — the process of refining and fabricating more valuable goods from raw or semiprocessed materials — but Japan continues to be the leader in the provision of financial services and other activities essential to steady economic growth.

The remaining Asian countries present highly varied patterns of urbanism and associated development of manufacturing and services (see Table 10.1). The urban centers of Hong Kong and

In the People's Republic of China from the mid-1960s until late in the 1970s, China experimented with a national model for agricultural village building. Because Chairman Mao Zedong and Premier Chou Enlai (the two dominant political forces in China in the mid-1960s) had become fascinated with a small village in north central China, it rose to national prominence. The village of Dazhai had about 500 people, was nested in the flanks of a mountain system in Shanxi Province, immediately west of the densely settled North China Plain south of Beijing. Like all of the villages in those mountains, it had been a long-time exporter of farm youth fearful of never being able to support themselves or a family in that relatively unproductive landscape. However, in the revolutionary years of the Chinese Revolution, this particular village changed patterns dramatically.

Chairman Mao had told the Chinese people that "there are no unproductive regions, there are only unproductive people" and warned the Chinese peasants that the government was not going to be able to help all of the poor villagers improve their settings and their lives. They would have to make a bold stab at undertaking such change on their own.

In Dazhai, a coal miner turned farmer named Chen Yonggui took Mao to heart. For nearly a decade before 1964, Chen had cajoled the villagers into numerous labor-intensive landscape modification projects. The biggest ones were terrace making and irrigation networks. The prime result of all of this work was that in 1964 Zhou Enlai was convinced to come and visit the newly sculpted Dazhai village. He returned to Beijing and told Mao that he had witnessed something truly remarkable. Always fond of national campaigns, Mao proclaimed, "Nongye, shui Dazhai: In agriculture, learn from Dazhai!" This became the clarion call all across China from the end of 1964 until my first visit to the village in 1977. The little place had become China's single most popular and significant model village and a major peasant-tourist (and occasional professional geographer) touring destination.

In August 1977, I was traveling with the first American delegation of geographers asked to visit China. Ten of us from the States spent some 30 days traveling as a professional delegation with Chinese hosts from the Chinese Academy of Sciences, and we saw a wide range of urban, rural, and tourist-scenic settings all across China. A few of us had lobbied very hard to see Dazhai, and in mid-August, we were taken by train, truck, and car to see this little "Appalachian" village southwest of Taiyuan in Shanxi Province.

We got to the village midmorning. Since we had been traveling for many hours, everyone was ready for a breather and a chance to put their gear in their rooms for the overnight visit. I opted to

---

Singapore are outstandingly productive in proportion to their size. South Korea and Taiwan also have a relatively high level of development of skilled labor. This pattern of developing major urban sectors skilled in finance, services, and global networking has become a very significant hallmark of East Asia and an indicator of aggressive economic development.

In Asia as a whole, agriculture remains the major source of livelihood. In most countries — Japan is a conspicuous exception — the majority of the people continue to be farmers. Two major types of agriculture — plantation agriculture and shifting cultivation — are discussed in Chapter 12 on Southeast Asia, the subregion in which these forms of farming are the most prominent. In the steppes and deserts, nomadic or seminomadic herding and oasis farming are practiced. Over large sections of Asia, most farmers make a living by cultivation of small rain-fed or irrigated plots worked by family labor. In Communist-held areas, particularly in China, the past four decades have seen a variety of experiments in agricultural organization. The current pattern has largely reverted to family farming operations with markets controlled, at least in part, by local and provincial governments or, increasingly, by an expanding free-market system.

Although there has been some farm mechanization in Asia, the general practice (outside Japan) continues to be characterized by the steady input of large amounts of very arduous hand labor. Production is often of a semisubsistence character. This type of agriculture, which is often referred to as intensive subsistence agriculture, is built around the growing of cereals. Where natural conditions are not suitable for irrigated rice, grains such as wheat, barley, soybeans, millet, sorghums, or corn (maize) are raised. However, irrigated rice yields the largest amount of food per unit of area where conditions are favorable for its growth, and this crop is the agricultural mainstay in the areas inhabited by a large majority of Asian farmers. Because of the role of rice in the traditional Asian diet, it is the crop of choice in areas with adequate rainfall, other things being equal.

In response, in part, to these distinctive landscapes and growing cities, tourism has also grown enormously in Asia. In Hong Kong, Thailand, South Korea, and Japan, for example, tourist income increased five- to sevenfold in the 1990s. The attraction to touring this region of Asia has long been fueled by the images of classic and traditional landscape treasures of art, architecture, shrines, gardens, and markets in Asia. The upswing in these recent years has been stimulated by the construction of world-class hotels in major cities all across Asia and by the continuing attention the region has given to extending services and

break from the group and spend a couple of hours walking up and down the terrace paths of this handmade landscape that had driven millions of Chinese peasants into frenzies of terrace building. Imagine the scene. Because I had written about Dazhai before the trip, I had come to the model village with myriad images of terrace construction, rock digging, earth hauling, stream channeling, and the continual creation of ponds and spillways. These labor-intensive acts all combined to help me anticipate this remarkable landscape, the reported product of these self-reliant bootstrap efforts to turn this traditionally underproductive landscape into a series of level, well-watered, minifarm fields. It was claimed that these new terrace fields were gaining crop yields that were on a par with the major agricultural region to the east in the level North China Plain. I had conjured up panoramas of the work teams, of the peasant meetings, even of the touring villagers from all parts of China making the politically correct pilgrimage to Dazhai.

Yet, in all of the scores of acres of fields I walked through, and in the additional areas I could see from my upslope walking, I saw not a soul. I did not see a single person. I acknowledge that it was midday, but the reading about the fervor of building Dazhai never allowed for my consideration of normal meal breaks. Someone, some crew, some socialist model peasant, was *always* working. How could I possibly explain the total absence of dirt movers, field hands, or rock breakers on this stunning August day? It was, as far as I knew, no special holiday, and besides, Dazhai did not observe such days. The only good day was a working day in the village lexicon.

I walked slowly down the slope and finally found the building where our delegation had been housed. This was trickier than I thought because I saw a number of flatbed trucks and buses in steady motion now, offloading visitors with their banners, notebooks, and some cameras. When I got a bowl of noodles and some hot tea, I went and sat with our Chinese delegation who were in counsel with Dazhai leadership. I was eager to ask my questions. But from the expressions on their faces, it appeared that something serious was afoot, so I held off on my questions about the phantom crews and their magic Dazhai landscape.

"Hsiao Shensheng (my Chinese name), I am afraid that we have serious news to tell you. Someone very important has died in the United States." Having lived in China 14 years earlier when John F. Kennedy had been killed, my heart tightened as I anticipated grave news. "Elvis Presley, The King, has died."

And that news, that event — told to me in a mountain village in north China — basically captured the attention of our group for the afternoon in Dazhai. I did see some workers in the fields later in the day but never, ever, was there the mass of landscape-changing peasants that I had so keenly anticipated after having read of Dazhai's monumental success in creating a single model landscape for a nation bigger than all but three nations in the world. They were surely not on holiday because of The King's death, but they were nowhere to be seen. It is always difficult to get all your field questions answered!

Kit Salter

---

welcome to foreign visitors. This inflow of tourist dollars has not only been an important source of investment capital for the countries, but tourist traffic has led to the upkeep and significant refurbishing of traditional landscape features (e.g., temples, sacred burial sites, and monumental structures) that might otherwise have been given less attention in the current rush to modernize and, in many cases, Westernize.

10.5 Geopolitical Issues

Monsoon Asia has been in the center of the world geopolitical stage ever since September 11, 2001. However, before the destruction of the World Trade Towers and a segment of the Pentagon by hijacked commercial airliners by Islamic terrorists, India and Pakistan had captured significant press attention with their parallel nuclear test explosions in the spring of 1998. The armed response from the United States and other allies against Afghanistan after the World Trade Towers horror became a whole new focus on the western edge of Monsoon Asia. The bombing of Afghanistan and the deployment of U.S. Special Forces in Afghanistan, Pakistan, the Philippines, and potentially other Asian countries have changed the geopolitical reality of this part of the world profoundly. The detailed impact of this shift in world attention and military action will be dealt with in more detail in the appropriate chapters, but know at the outset that at least a portion of Monsoon Asia will be more completely discussed and wondered about in this first decade of the 21st century than it has been for decades in the past.

In the category of geopolitical issues for Asia, we are reminded again about how acts such as innovation, adoption, and technological diffusion are never simply acts of geographic diffusion. When a culture, a village, or a country decides to undertake a major cultural change through the switch to, for example, irrigated agriculture, the process of change touches all facets of life. Political control must be gained of upstream as well as downstream villages and culture groups. Such control means a loss of some independence, even though there will probably be a yield increase because of the added water to the crop calendar. This means that such transformation is, absolutely, a geopolitical issue.

In this final segment of this Geographic Profile of Monsoon Asia chapter, we look particularly at two geographic events. One is an innovation that has been developed across decades. The other is more fully encapsulated in 4 months of very difficult times. We first turn to one of the most significant agricultural innovations in Asia during that past century. It is the development

DEFINITIONS + INSIGHTS

Shifting Cultivation

An early, traditional agricultural system important in Asia is one called **shifting cultivation.** It is also known as *slash-and-burn* and *swidden*. Its dynamics are built around low technology, farming unowned land generally, multicrop planting and harvesting, low farm density, and short-term use of the prepared plots. The pattern runs something like this. At the end of the rainy season, farm families will take to the forest that they had access to and select perhaps as much as 1 hectare (2.47 acres) as their plot. By axe or saw or girdling the trees, the major tree growth of the plot was brought down and left where it fell. There was no effort to have rectangular lot shapes. No fences were ever put up. The wood, once "slashed" (as in slash-and-burn) was left to dry during the nonrain season. Then, just prior to the coming of the monsoons, fire would be set to the dried vegetative matter, and the farmers would burn as much of the trunks, branches, leafy matter as they could. The ash cover that was left after the burn then served as the soil base into which the farmers would plant not one or two crops but as many as a dozen or even more. Ceremonies would surround such planting, and the hope was that the rains would come once the seeds were covered with the still warm ash.

The only technology in the process up to this step was the knowledge of fire and a fire-hardened digging stick (something like a short javelin) that would be used to drive a hole into the still warm soil for a handful of seeds. After the rains came, crop growth would be set in motion, and over the next 2 to 4 months, everything from beans to maize to spice plants to vegetables and more would be harvested. Little attention was given to cultivation during the growth period, and harvest was done in ceremony and at the end of the basic capture of all the food. The field would be

abandoned (the "shifting" of shifting cultivation) except for some late maturing crops, and the swidden family would begin looking for a new plot for the following year. The land just farmed would become home to volunteer grasses and, ultimately, secondary forest return.

From Taiwan through the Philippines and all through insular Southeast Asia and the peninsula as well, shifting cultivation was the farming force that first began forest clearance in the river valley floodplains. It then began to get pushed upslope as permanent field farming took over the prime lowlands. The environmental impact of shifting cultivation in early history of human use of Asian land was minimal because the system built in adequate time for regrowth. However, as populations began to increase and tensions developed between differing groups, regrowth (sometimes called fallow) time was diminished and, instead of forest replacing forest, unproductive grasses came in and took over the soil base.

Today, shifting cultivation continues to be important on the large islands of Sumatra and Borneo and other outer islands of Indonesia as well as other islands going north. Dense island populations see the relatively unsettled nature of such areas, and this serves as a catalyst for both formal and informal efforts at resettlement. In March 2001, the bloody battles between the Dayaks of Borneo and new settlers from Madura, a small island east of Java, came largely from such migration and forest clearance efforts. This means that although the actual agricultural impact of swidden is smaller today than ever before, the geopolitical implications of the landscapes that have and might accommodate such farming continue to be very real and very significant.

---

of new seed types and planting, cultivating, harvesting, and marketing patterns known as the **Green Revolution.**

The Green Revolution

As a step toward eliminating hunger by using science to increase yields of rice, the single most important crop in Asia, the International Rice Research Institute (IRRI) was founded in the Philippines in 1962. The institute is one facet of a worldwide research effort involving the use of new varieties and strains of popular grains in association with a modification of traditional farming practices. Governments have also invested increased capital in building better roads and bridges to enable farmers to get their surpluses to markets and to bring new farming tools and materials to the countryside more efficiently.

Notable success has been achieved in breeding the new high-yielding varieties of seed stock, and there has been a large upsurge of production in certain areas where the new

strains have been widely introduced (Fig. 10.14). Peasant farmers have changed their crop calendars — the dates for planting, cultivating, and harvesting crops — to accommodate an increasing dependence on chemical fertilizers as well. The entire effort has come to be known as the Green Revolution.

For Asian farmers to capitalize fully on the Green Revolution, they must overcome many obstacles. For example, success requires levels of capital that are often beyond the present means or inclination of peasant farmers, landlords, and governments to provide. Such expenditures are needed for water-supply facilities (e.g., the tubewells that have burgeoned by hundreds of thousands in the Indo-Gangetic Plain of the northern Indian subcontinent), chemical fertilizers, and chemicals to control weeds, pests, and diseases. And as agriculture becomes more mechanized, considerable increases in the costs of machinery, fertilizers, and fuel have to be borne by farmers who have, in many cases, had little experience with the cash economy needed to achieve a positive return on their

SYLVAN H. WITTWER/VISUALS UNLIMITED

Figure 10.14   Rice was the primary focus crop for the so-called Green Revolution. The International Rice Research Institute (IRRI) was established in 1962 in the Philippines. Through comprehensive and thorough investigation of a great variety of native strains of rice, and the creation of hybrids as well, levels of yield, pest resistance, and grain quality all improved. However, as the rice plants became potentially more productive, the actual act of farming changed. Farmers had to follow a more exact crop calendar. Chemical as well as organic fertilizers had to be applied more exactly, and farm villages had to have good access to all-season roads to get their new surpluses to markets. The overall impact of the IRRI researches and innovations has been very positive although some farmers continue to lament the loss of some independence as they find that their new cropping patterns pull them more completely into the marketing world and associated urban culture. These rice beds are in Sri Lanka.

investment. Governments must improve transportation so that the large quantities of fertilizer required by the new seed varieties can be delivered in a timely fashion. Not only must there be these associated infrastructural changes to support this "revolution," but the crop calendar of the farmer becomes much less forgiving because many of the newest seed grains demand more precise water, fertilizer, and cultivation requirements than traditional grains. Grain-storage facilities, now subject to plundering by rats, must be improved.

Overcoming the financial obstacles is rendered more difficult by the widespread system of share tenancy. Farmers who are share tenants generally have no security of tenure on the land, and thus cannot be sure that money they invest in the Green Revolution will actually benefit them in the future. They may not wish to assume any additional risk, even if credit on reasonable terms is available. Even if they remain on their holdings, landlords may take up to half of the increased crop while bearing little or none of the additional expense. Landlords in turn may be content to collect their customary rents without expending the additional capital necessary in this new mode of farming, or they may endeavor to turn tenants off the land to create larger spatial units that they themselves can farm more profitably with machinery and hired labor. In fact, landowners with large holdings often become the chief beneficiaries of the new technology, with many smaller farmers becoming a class of landless workers hired for low wages on a seasonal basis or migrants in the rural-to-urban migration stream.

Other problems associated with the Green Revolution include damage to ecosystems by large infusions of agricultural

chemicals and the economic dislocations that result when rice-importing countries become more self-sufficient, thus causing hardships for rice exporters. The Asian experiment with the Green Revolution provides yet additional evidence of the widespread ramifications that are set in motion when any major technological shift, either agricultural or industrial, is introduced into societies. Indian or Filipino farmers, for example, who had developed patterns of reasonable self-sufficiency through traditional agricultural practices, can find their lives turned upside down as they try to adopt some of the standards of the Green Revolution. Although there has generally been no overt pressure to get all farmers to become part of the agricultural innovations associated with the Green Revolution, in reality, the changes caused by this program tend to make it very difficult for isolated peasants not to adopt the new changes. Everything from local crop calendars to marketing systems changes as a region's farmers begin to shape their lives and landscapes in accordance with such changes.

The scale of culture change that must accompany the pattern shifts in a farmer's involvement in the Green Revolution may seem acceptable to an urbanite, whether from Asia or Latin America or the world beyond, but to the traditional Asian farmer, such changes are not only financially costly but serve to absolutely rupture long-held traditions. As you explore the various nations and landscapes of the crescent of Monsoon Asia, keep in mind how much impact the global flow of information, fads, and images has on human ambitions for an improved livelihood and lifestyle. The ways in which peoples of Asia have responded to these pulsating influences of change — from sophisticated mechanization to maintenance of traditional customs and costumes and growth — are significant geopolitical issues. We conclude with one other set of problem issues. They all relate to environmental hazards and the uncertainty of shelter, livelihood, and even life in Asia.

Impact of Natural Events

Even with Asia actively involved in experiments, technology transfer, and diffusion of new agricultural practices as part of the Green Revolution, there are parts of the region that are virtually untouched by all this change. The most significant example is North Korea, where its population of some 23 million experienced 4 years of major food deficiency in the 1990s. The country experienced a 100-year flood — a flood of such a magnitude that it is expected to occur only once in a century — in 1995–1996, and it suffered through a 1997 drought that was partially caused by El Niño. In 1995–1996, the food crop yield in North Korea was less than 2.8 million tons, or 2 million tons less than what was deemed necessary to feed its population at even a minimal level. Yields and total output have been down dramatically ever since mid-1990, leading to the unprecedented North Korean acceptance of grain from South Korea.

The major flooding of 1995–1996 also had other impacts on North Korea. The country's major mining area was flooded, and mining itself was disrupted significantly. This led

F19.B
523

not only to major industrial unemployment in mining and associated industries but also to a major drop in the generation of electricity, which then pulled the plug on other manufacturing. This combination led to a reduction in winter heating, overall electricity availability, and general industrial activity. At the same time, agriculture had seen its output drop to about half of the normal productivity. So, even while parts of Asia are experiencing steady increases in agricultural yields through the Green Revolution and other changes in farming patterns, other locales are either slow or unable to adopt the changes and are often harmed by unpredictable weather conditions, which are part of the nature of residence in the Asian world.

The January 2001 Gujarat Earthquake

When considering natural hazards in Asia, the so-called "ring of fire" that courses along the island margins of East Asia is most often thought of as the home of earthquakes and volcanoes. However, in January 2001, a major quake hit southwest India in the state of Gujarat. The epicenter was near the city of Bhuj, a town of 150,000. It is estimated that between 20,000 and 25,000 were killed in the quake, and some 750,000 rendered homeless (Fig. 10.15). The rubble of concrete buildings, still bound together by steel bars and electrical wiring, is expressed in great cluttered piles on the landscape even months after the quake because of the enormous financial and social costs represented by such physical hazards.

Floods, earthquakes, volcanic explosions, and droughts represent some of the geographic conditions that are always political as well as physical. From the scale of an individual family's response up to central governmental action, all decisions that are made in an effort to offset the problems that come with these crises must pass through social and religious filters even as fiscal and technological solutions are being considered. As will be seen in the various chapters on Monsoon Asia, the ways in which these "filters" are utilized change from place to place and from time to time. They are virtually always operative, however.

Figure 10.15   In January 2001, massive earthquakes hit Gujarat Province in southwest India. Rolling across a number of days, it is estimated that more than 20,000 people died and hundreds of thousands were left homeless or living in highly damaged shelters. There is a tyranny in natural disasters such as earthquakes, typhoons, and floods for they have a way of undoing enormous human effort at trying to improve standards of living and levels of economic development. This scene is from the city of Ahmedabad in Gujarat Province.

CHAPTER SUMMARY

- Monsoon Asia includes the countries of Japan, North and South Korea, China, Taiwan, Macao, Pakistan, India, Sri Lanka, Bangladesh, Bhutan, Nepal, Burma (Myanmar), Laos, Thailand, Cambodia, Vietnam, Malaysia, Singapore, Indonesia, the Philippines, Brunei, East Timor, and numerous islands scattered along the edges of this major continental block. The adjective "monsoon" is used because of the central role this precipitation pattern plays all through this region.

- Three concentric arcs make up the broad physiography of the region. The arcs include the highest mountain ranges of the Himalaya, Karakoram, and Hindu Kush. The next arc, going eastward, is made up of major floodplains, deltas, and relatively low mountains. These include the Indus, Ganges, and Brahmaputra Rivers in South Asia; the Irrawaddy, Chao Praya (Menam),

Mekong, and Red Rivers in Southeast Asia; and the Chiang Jiang (Yangtze) and Huang He (Yellow) in East Asia.

- In areas of dense settlement, climates are warm and monsoonal. Varied types include tropical rain forest, savanna, humid subtropical, humid continental, steppe, desert, and undifferentiated highland. Although there is significant desert landscape, there has been a long cultural history of caravans trekking across it, connecting East Asia with Europe for more than 2000 years.

- Landscape signatures of many sorts exist in Monsoon Asia, including indigenous ones (e.g., the Great Wall in China, the Taj Mahal in India, and the intensely worked garden plots supporting peasant households), but there are also signatures of the colonial period. Great Britain, the Netherlands, France, and Portugal were the most important colonial powers in this region. Most of these domains were relinquished by the middle of the 20th century, with the British return of Hong Kong and the Portuguese return of Macao (both to China) fundamentally closing the colonial period.

- Major religions include Hinduism, Islam, Buddhism, Confucianism, Taoism, and Christianity. There continue to be tensions between populations of the different belief systems, especially in the case of Pakistan (Islam) and India (Hinduism). Wherever Buddhism and Confucianism are predominant, there tends to be less tension because of the capacity of those religions to accommodate the existence of other belief systems.

- Garden agriculture, rural-to-urban migration, and "teaching water" (irrigation) are three very significant cultural features of Monsoon Asia. The combination of concern for intensive farming and water management has created distinctive landscapes both in mountains and on the plains. The importance of education has enabled Asians to adapt productively to technology transfer.

- Nearly three-fifths of the world's population lives in Monsoon Asia. Because of the importance of rural settlement and associated farm labor and partly because of cultural support of large families, China and India have enormous populations. These two countries and Indonesia are three of the four most populated nations in the world. Birth control programs and family planning have been successful in reducing fertility rates, and increasing migration from villages to cities has also been important in reducing population growth rates.

China has been the most efficient in wide-scale reduction of population growth, and India will possibly become the most populous nation in the world by the middle of the 21st century or even sooner.

- Villages are the most common settlement form in Monsoon Asia, although the percentage of people resident in urban places has been steadily increasing all through the region. Japan was the first Monsoon Asian nation to adopt Western industrialization. In the past three decades, there has been widespread economic development, especially in East Asia and Southeast Asia (Taiwan, South Korea, Hong Kong before it was returned to China, Singapore, Indonesia, Thailand, and Malaysia), but economic decline since 1997 has been rapid and significant, particularly in Japan. China and Taiwan have been particularly diligent in avoiding the decline.

- The Green Revolution is the name of a broad effort to increase agricultural productivity in dominant crops through expanded use of chemical fertilizers, new seed stock, a more exacting crop calendar, and better access to good market roads, markets, and credit. Although there has been some success with the innovations associated with the Green Revolution, it has been difficult to overcome the entrenchment of strong traditional agricultural practices. North Korea represents one nation that has had virtually no productive encounter with these innovations. In addition, it has been plagued by both floods and drought, bringing the nation to the end of the 20th century with a great deal of failure to mark its half century of experimentation with communism.

REVIEW QUESTIONS

1. What nations make up Monsoon Asia?
2. Name the largest plains and river valleys in Monsoon Asia and comment on their role in human settlement.
3. What are the characteristics of the monsoon climate? How does agriculture relate to such patterns?
4. Where does the term "the Orient" come from? The "Far East"? What is the problem in using such names now?
5. What percentage of the world's population lives in Monsoon Asia? Describe the range of population growth rates in the region. What areas have had the greatest success in reducing fertility rates?
6. Discuss the reasons that rural-to-urban migration has an impact on population growth rates. Give examples of where this trend is significant in Monsoon Asia.
7. List the seven different climate classifications found in Monsoon Asia. Take three and discuss the types of landscapes and settlements they support.

8. Using some of the maps in Chapter 10, discuss the colonial period and explain where and why colonial influences are the most pronounced.
9. What are the primary religions in Monsoon Asia? What landscape features would they be likely to create or influence?
10. Explain the significance of garden agriculture, "teaching water," and rural-to-urban migration in Monsoon Asia. In what way are these three things part of geographic perspective? What three National Geography Standards would reflect and support your answer?
11. Look at the world population dot map in this chapter and Chapter 2 and discuss the population settlement patterns that are evident in Monsoon Asia.
12. Discuss tourism, industrialization, and rural-to-urban migration in Monsoon Asia. How are these distinct processes and activities related? Where in this region would you see the clearest presence of any or all three?

DISCUSSION QUESTIONS

1. Discuss the role that the natural environment has played in the patterns of settlement in Monsoon Asia.
2. Discuss the ways in which patterns of human settlement relate to climate patterns. Use various maps to support your perception and answers.
3. What are the hallmarks of colonialism in Monsoon Asia? Name the good and bad influences that this process has had on the region.
4. What is the consequence of the explosion of nuclear devices by India and Pakistan in May 1998? What sort of impact will this act possibly have beyond the region?

5. Name the countries that have not been colonized in Monsoon Asia. In what ways have their histories been distinct from those of the nations that underwent colonization?
6. Discuss the sorts of changes that take place in migrating from the countryside to the city in Monsoon Asia. What economic and demographic changes does such a migration set in motion?
7. Discuss the role of natural disasters in Monsoon Asia.

CHAPTER 11

Complex and Populous South Asia

JOE HOBBS

Tamil mother and child: faces in the rich cultural fabric of South Asia. They are Hindu in religion and Dravidian in ethnicity—quite distinct from other major faiths and ethnic groups of the Subcontinent.

Chapter Outline

A triangular peninsula thrusts southward 1,000 miles (c. 1,600 km) from the Asian landmass, splitting the northern Indian Ocean into the Bay of Bengal and the Arabian Sea. The peninsula is bordered on the north by the alluvial plain of the Indus and Ganges (Ganga) Rivers, and north of these rise the highest mountains on Earth. The entire unit — peninsula, plain, and fringing mountains — is known as the Indian subcontinent. It contains the five countries of India, Bangladesh, Pakistan, Nepal, and Bhutan in an area about half the size of the 48 conterminous United States (Figs. 11.1 and 11.2). Off the southern tip of India, across the narrow Palk Strait, lies the island nation of Sri Lanka, which shares many physical and cultural traits with the subcontinent. These six countries make up the subregion of South Asia, home to about 20% of the world's people.

A world-class giant, India outranks the other South Asian nations in both population (1.033 billion) and area (1.1 million sq mi/3 million sq km, about the combined size of the U.S. states of Alaska, Texas, California, and Colorado). Until 1947, it consisted of the entire area of the present India, Pakistan, and Bangladesh. For more than a century, India was the most important possession of the British Empire — the "jewel in the crown." In 1947, it gained freedom but quickly divided along religious lines into two countries: the secular but predominantly Hindu nation of India and the Muslim nation of Pakistan. Pakistan had two parts, West Pakistan and East Pakistan, separated by Indian territory. In 1971, an Indian-supported revolt in East Pakistan led to the birth there of the new independent country of Bangladesh.

Bangladesh (population: 133.5 million) and Pakistan (population: 145 million) are much smaller than India in area and population, but both are among the world's 10 most populous countries. Nepal and Bhutan, on the southern flank of the Himalaya Mountains, are small, rugged, and remote, with fewer people (23.5 and 0.9 million, respectively). They serve as buffer states between India and China, which have engaged in sometimes violent border disputes since the early 1950s. Strife has not spared Sri Lanka, whose 19.5 million people have experienced a civil war for almost two decades. Altogether, about 1.36 billion people, more than one of every five on Earth, lived in the complex and conflicted subcontinent and Sri Lanka in 2001.

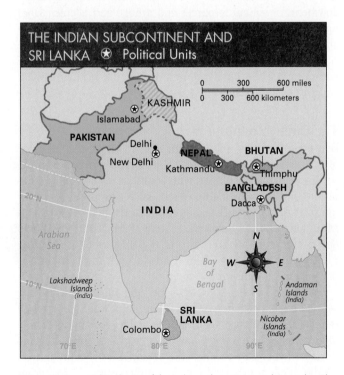

Figure 11.1   Political units of the Indian subcontinent and Sri Lanka. The future status of Kashmir, disputed between India and Pakistan, with some parts occupied by China, remains unresolved.

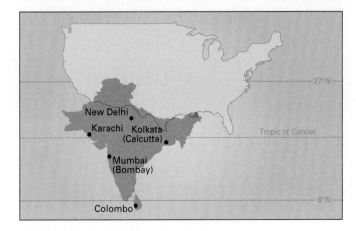

Figure 11.2   The Indian subcontinent and Sri Lanka compared in latitude and area with the conterminous United States.

11.1 The Cultural Foundation

The Indian subcontinent is one of the world's culture hearths. Among the animals first domesticated in the region were zebu cattle, which today are important livestock in many world regions. Before 3000 B.C., some of the world's earliest cities (notably Mohenjo Daro) developed on the banks of the Indus River in what is now Pakistan. Their inhabitants were probably the ancestors of the Dravidians, a dark-skinned people whose stronghold is now southern India.

Between 2000 and 1000 B.C. came the Aryans (or Indo-Aryans), tribal cultivators and pastoral nomads from the inner Asian steppes who were the source of many ethnic and cultural attributes that are dominant in the subcontinent today. These include the Caucasoid racial majority, the numerically dominant Aryan languages such as Hindi (most of which were derived from the Sanskrit used by the conquerors), and the faith of Hinduism, which developed from the beliefs and customs of the Aryans. In the eighth century, Arabs brought the Islamic religion into India from the west to what is now Pakistan. Waves of Muslim conquerors who came through mountain passes from Afghanistan and central Asia spread Islam farther afield in later centuries.

The last great Islamic invasion, by a Mongol-Turkish dynasty called the Moguls (Mughals), began in the early 16th century, soon after the Europeans first came to India by sea. Before it succumbed to rebellious local Hindus and Sikhs and to the British Empire, the Mogul empire ruled nearly the entire subcontinent from its capital at Delhi. Some of the region's most spectacular artistic and scientific achievements date to the peak of Mogul power in the 16th and 17th centuries, including the Red Fort in Delhi and the tomb of Mogul Emperor Shah Jahan and Empress Mumtaz Mahal — this is the astonishing Taj Mahal in Agra (Fig. 11.3). Muslims now make up about 12% of India's population, seemingly a small number except that it equals more than 120 million people, ranking India third among the world's countries with the largest Muslim populations, behind Indonesia and Pakistan.

British colonialism left an indelible mark on the subcontinent's cultural landscape. To serve the needs of their empire, the British developed the great port cities of Karachi, Bombay (now officially known as Mumbai), Madras (now Chennai), and Calcutta (now Kolkata), and India's current inland capital of New Delhi. The existence today of English as the de facto national language of both India and Pakistan and the presence in Great Britain of large numbers of immigrants from the subcontinent testify to the close relations between these countries.

This simple sketch of Dravidian, Aryan, Islamic, and British influences can only begin to shed light on the extreme social diversity of the Indian subcontinent. There is an enormous range of ethnic groups, social hierarchies, languages, and religions among regions of the subcontinent and even within single settlements. The subcontinent is the most culturally complex area of its size on Earth. Some of the problems related to this complexity are discussed later in the chapter.

11.2 Regions, Resources, and Settlements

Physical conditions are very important in the lives of the rural majorities who live in the Indian subcontinent and Sri Lanka. With little foreign aid reaching the region, people's survival depends largely on local agricultural resources, and future industrialization will have to draw mainly from available natural resources. Fortunately, the subcontinent has some generous endowments of natural resources and a huge and talented labor force to work with them. It also has colossal natural hazards, including cyclones, floods, droughts, landslides, and earthquakes.

The subcontinent may be roughly subdivided into three natural areas: the outer mountain wall, the northern plain, and peninsular India.

The Outer Mountain Wall

Between 40 and 50 million years ago, a triangle-shaped piece of the earth's crust completed its tectonic journey across the Indian Ocean and collided with southern Asia. The collision resulted in the thrusting up of a massive, almost solid, wall of mountains. That wall, made up mainly of the Himalaya Mountains and the Karakoram Range, forms an inverted U to define the northern boundary of the subcontinent (see Fig. 11.4). From each end of this massive wall, lower mountain ranges trend southward to the sea.

Pakistan's borders with Iran and Afghanistan traverse the rugged western section of the mountain wall. Desert and steppe climates reach upward to high elevations. These mountains provide only a few crossing places, including the storied Khyber Pass through the Hindu Kush Mountains on the border between Pakistan and Afghanistan. These passes have long been gateways into India for conquerors from the outer world, including the Aryans and the Arabs.

Figure 11.3 For Indians and foreigners alike, the Taj Mahal in Agra symbolizes the cultural achievements of the Indian subcontinent.

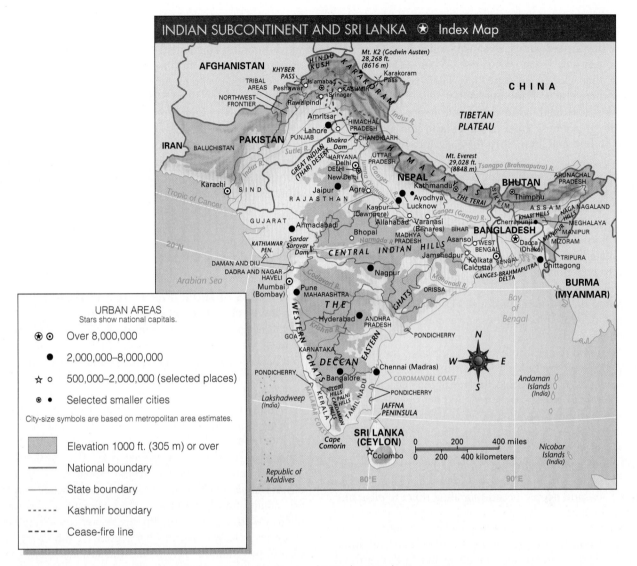

INDIAN SUBCONTINENT AND SRI LANKA ✪ Index Map

URBAN AREAS
Stars show national capitals.

✪ ⊙ Over 8,000,000

● 2,000,000–8,000,000

☆ ○ 500,000–2,000,000 (selected places)

⊛ • Selected smaller cities

City-size symbols are based on metropolitan area estimates.

▨ Elevation 1000 ft. (305 m) or over

—— National boundary

—— State boundary

------ Kashmir boundary

– – – – Cease-fire line

Figure 11.4 General reference map of the Indian subcontinent and Sri Lanka. Note how mountainous areas frame the northwestern, northern, and northeastern portions of this region, while ocean waters set the boundaries in the south.

The towering northern segment of the mountain wall has been much less passable. Across the north of the subcontinent, the Himalaya extend about 1,500 miles (c. 2,400 km) from Kashmir to the northeastern corner of India (Fig. 11.5). Paralleling the Himalaya on the northwest, and separated from them by the deep gorge of the upper Indus River, lies another lofty range called the Karakoram. The two ranges contain 92 of the approximately 100 world peaks greater than 24,000 feet (7,315 m), including the two highest: Mt. Everest (29,035 ft/8,859 m) in the Himalayas on the border between Nepal and China and Mt. K2, or Godwin Austen (28,268 ft/8,616 m), in the Karakoram within the Pakistani-controlled part of Kashmir. Geographic surveys recently quieted a

Figure 11.5 The roof of the world: looking across Nepal's boulder-strewn Gokyo Glacier toward Mt. Everest (hidden by clouds in the upper left).

brewing debate over whether K2 might actually be higher than Everest: The Himalayan peak is the highest. Both great ranges, however, are young, dynamic, and still rising, so K2 could conceivably outgrow Everest. While the peaks are climbing, their glaciers are retreating. There is more ice here than anywhere else on Earth outside Greenland and the polar regions, but it is melting quickly. Some fingers point to human-made greenhouse gases as the culprit, and by one estimate, at current rates of melting these glaciers would be gone by 2035.

Although most passes through these ranges are higher than any peak in the Alps of Europe, a trickle of trade and cultural exchange has crossed these mountains for millennia. However, neither commerce nor military force has ever crossed this barrier on a large scale.

The kingdoms of Nepal (capital, Kathmandu; population: metropolitan area, 1,169,800; city proper, 616,900; Fig. 11.6) and Bhutan (capital, Thimphu; population: 55,700) are perched on this mountain wall. Their landlocked locations and difficult topographies have contributed to their slow economic development. Both countries have productive agricultural regions in the "middle ranges," the bands of foothills between the vast lowland of the subcontinent and the towering Himalayan peaks. In Nepal, the capacity of this productive area has been saturated, and poor peasants "marginalized" by rapid population growth have sought new ground on steeper slopes in nearby Bhutan and in Nepal's lowland region of the Terai. Once a malaria-plagued region of dense forest and savanna, the Terai helps serve as a "safety valve" for Nepal's overpopulation. Control of malaria began in the 1950s, and people began migrating there to clear and cultivate the land.

Nepal's Himalayan region is a popular destination for international tourists. An estimated 460,000 trekkers visit annually, bound especially for the Everest region in the east and the Annapurna area of the west. They bring in much-needed

Figure 11.6 Nepal's capital, Kathmandu, sprawls across a plain in the country's "middle" mountain range. In the foreground is the Royal Palace.

foreign currency to this poor country but also create problems by inducing social change in traditional cultures. The conservative Buddhist kingdom of Bhutan has been more careful in opening its doors to tourism, placing restrictions on the numbers permitted in and on the routes they may follow.

The mountains along and near the subcontinent's border with Burma (Myanmar) are much lower than the Himalayan ranges but, being cloaked with dense vegetation and soaked with rain, are also nearly impenetrable. Cherrapunji in the Khasi Hills of India has an average annual rainfall of 432 inches (1,097 cm), more than nine-tenths of it falling in the summer. It ranks with a station on Kauai in the Hawaiian Islands as one of the two spots with the greatest annual rainfall ever recorded on Earth. Cherrapunji reportedly has the highest alcoholism rate in South Asia because so many people stay inside and drink on rainy days.

The Plains of India, Pakistan, and Bangladesh

Just inside the outer mountain wall lies the subcontinent's northern plain. This large alluvial expanse contains the core areas of the three major countries. In the west, the plain is split between Pakistan and India. The Indus River and its tributaries and **distributaries** cross the portion lying in Pakistan and water this country's two most populous provinces, Punjab and Sind. Pakistan's major cities are here: the seaport of Karachi (population: city proper, 10,013,400) at the western edge of the Indus River delta in Sind and the cultural center of Lahore (population: city proper, 5,470,000) in the Punjab. North of Lahore, at the foot of the Himalaya, is the capital city of Islamabad ("City of Islam"), a much smaller, very modern administrative metropolis. The climate of the western plains is desert and steppe; the dry region straddling the two countries is known as the Thar Desert, or the Great Indian Desert. Irrigation water from the rivers sustains highly productive agriculture and millions of people here.

East of the Punjab lies the part of the northern plain traversed by the Ganges (Ganga) River and its tributaries, especially the Jumna (Yamuna). The western portion of this region is known as the Doab, "the land between the two rivers." The northern plain is India's core region, containing more than two-fifths of the country's population as well as its capital, many other major cities, and sacred places in the religions of Hinduism, Buddhism, and Jainism.

This region of India is relatively well watered, especially toward the east, where agriculture based on irrigated rice supports high population densities. At the narrow western end of the Ganges-Jumna section of the plain, the old fortified capital of Delhi arose as a bastion against invaders from beyond the western mountains (Fig. 11.7). The present adjoining capital of New Delhi was founded after 1912 as the capital of British-controlled India. Today, the population of Delhi–New Delhi is an estimated 16.5 million, making it India's second largest metropolis after Mumbai.

The southeastern part of the northern plain lies in the region of Bengal, which is essentially the huge delta formed by

Figure 11.8 Bangladesh is both a low-lying coastal land and a downstream country of several large watersheds. As a result of storms blowing in from the Bay of Bengal and of accelerated runoff from rainfall and snowmelt on deforested mountains upstream, it is subjected to devastating floods. This is an aerial view following the cyclone (hurricane) that struck Bangladesh on April 29, 1991, leaving more than 120,000 people dead and inundating most of the country's precious farmlands.

Figure 11.7 The Mogul Emperor Shah Jahan established Delhi as India's capital early in the 17th century. His monumental Red Fort still dominates this old city. Adjacent New Delhi is India's capital today.

the combined flow of the Ganges and Brahmaputra Rivers. Bengal is divided between India and Bangladesh and has an extremely dense population. Its chief city is the English-founded port of Kolkata (Calcutta) in the Indian state of West Bengal. A city of 13.8 million (metropolitan area), Kolkata has a worldwide reputation for teeming slums and dire poverty — Rudyard Kipling called it "The City of Dreadful Night" — and it was this Indian city to which the extraordinary nun Mother Teresa dedicated her life's work of aiding the poor. Within India, Kolkata is recognized as a great center for culture and the arts.

Across the border, Bangladesh's industries are concentrated in Dacca (Dhaka; population: metropolitan area, 10,168,600; city proper, 8,539,500), the capital and largest city, and Chittagong, the main seaport. The Bangladesh portion of the delta ranks, along with the Indonesian island of Java, as one of the two most crowded agricultural areas on Earth. This area is subject to catastrophic flooding (Fig. 11.8). Massive hurricanes,

known locally as cyclones, ravage Bangladesh regularly in September and October. Deforestation upstream on the steep slopes of the Himalayas has increased water runoff and sediment load in the Ganges, also contributing to Bangladesh's flood problems. Bangladesh must also keep a wary eye on the potential for the sea level to rise if the world's temperatures increase according to many climate change models. At the turn *36* of the century, the country's environment minister insisted that 20% of Bangladesh would be under water by 2015 if nothing was done to control global warming.

Peninsular India

The southern peninsular portion of the subcontinent, which is entirely within India, consists mainly of a large volcanic plateau called the Deccan. It is generally low, less than 2,000 feet (c. 600 m) above sea level. Its rivers run in valleys cut well below the plateau surface. This topographic problem, together with the rivers' seasonal flow pattern, makes it difficult to use their waters for irrigation of the upland surface without large inputs of capital and technology. Wells and rain-catchment ponds called tanks are used, but water is in short supply and population density is low compared with the better watered parts of the northern plain (Fig. 11.9).

Hills and mountains outline the edges of the Deccan. In the north, belts of hills (the Central Indian Hills in Fig. 11.4) separate the plateau from the northern plain. On both east and west, the plateau is edged by ranges of low mountains called Ghats ("steps") overlooking alluvial plains along the coasts of the Bay of Bengal and the Arabian Sea. The low and discontinuous eastern mountains are known as the Eastern Ghats. The higher, almost continuous western mountains are the Western Ghats (Fig. 11.10). The two ranges merge near the southern tip of the peninsula in clusters known locally as

Figure 11.9   Water is at a premium in the dry period preceding the summer monsoon in South Asia. It is usually the task of women to draw water from communal supplies such as this one in a small settlement near Mumbai.

the Nilgiri, Palni, and Cardamon "hills." They are really rather high mountains, with some peaks rising above 8,000 feet (c. 2,400 m). The colonizing British founded several "hill stations" in them for recreation and administration during summer months, when the climate on the subcontinent's low-lands is stifling. They are now popular tourist destinations for Indians during that season.

India's largest city, Mumbai (metropolitan area, 13,790,000; city proper, 5,066,600), is on the peninsula's northwest coast. The coastal plains between the Ghats and the sea support very dense populations, especially in the far south along the Malabar Coast of western India and on the corresponding Coromandel Coast in the east. The largest metropolis of the far south, Chennai (formerly Madras; population: metropolitan area, 7,046,700; city proper, 4,476,400) is a seaport on the Coromandel Coast. The main metropolises of the interior Deccan are Hyderabad and Bangalore.

Sri Lanka

Sri Lanka (population: 19.5 million), formerly called Ceylon, is a tropical island country with affinities to the subcontinent.

Figure 11.10   Southwest India's mountains, known as the Western Ghats, have been widely cleared of native forest by logging and agricultural expansion. This is a tea plantation in eastern Kerala state.

Its name conveys its physical beauty: Sri Lanka means "Resplendent Isle" in the native Sinhalese language, and medieval Arabs knew it as "Serendip," the island of serendipity (Fig. 11.11). The island consists of a coastal plain surrounding a knot of mountains and hill lands (see Fig. 11.4). Most people live in the wetter southwestern portion of the plain, in the south central hilly areas, and in the drier Jaffna Peninsula of the north. Coconuts and rice are the major crops of the low southwestern coast and Jaffna Peninsula, while tea and rubber plantations dominate the economy of the uplands. Colombo (population: metropolitan area, 2,381,900; city proper, 770,000), in the southwest, is the capital, chief port, and only large city.

Sri Lanka's economy is highly commercialized. Three cash crops — rubber, coconuts, and the world-famous Ceylon tea — supply about 20% of export earnings. Sri Lanka is the world's largest exporter of tea. Clothing manufactured in the Colombo area is exported. Sri Lanka is nearly self-sufficient in

Figure 11.11   Verdant landscapes are typical of all but northernmost Sri Lanka. This view is near Kandy in the south central part of the country.

DEFINITIONS + INSIGHTS

The Sacred Cow

Many attitudes, beliefs, and practices associated with cattle make up the world-famous but often poorly understood "sacred cow" concept of India. There are nearly 200 million cattle in India, representing about 15% of the world total and the largest concentration of domesticated animals anywhere on Earth. India's dominant religion of Hinduism forbids the slaughter of cows but allows male cattle and both male and female water buffalo to be killed. Reverence for the cow is well founded. Indians favor cow's milk, ghee (clarified butter), and yogurt over dairy products from water buffalo. They value cows as producers of male offspring, which serve as India's principal draft animal. Both cows and bullocks provide dung, an almost universal fuel and fertilizer in rural India.

Reverence for the cow in particular and cattle in general pervades the Hindu religion and mythology. The bull Nandi is associated with the Hindu god Shiva. People allow cattle to freely roam the streets of Indian cities (Fig. 11.A). As a symbol of fertility, cows are associated with (but not worshiped as) several deities. The mother of all cows, Surabhi, was one of the treasures churned from the cosmic ocean. Hindus honor cows at several special festivals. They use cow's milk in temple rituals. They believe the "five products of the cow" — milk, curds, ghee, urine, and dung — have unique magical and medicinal properties, particularly when combined. All over India, there are goshalas, or "old folks' homes," for aged and infirm cattle.

Remarkably, however, the subcontinent countries of Pakistan, Bangladesh, and India are major exporters of leather and

leather goods. In India, the leather industry is mainly in the hands of Muslims, who do not share Hindu's restrictions on killing or eating cattle. Muslims are forbidden to eat pork; therefore, pigs, a major food resource in many developing countries, are of little importance here. They are eaten mainly by Christians, very low-caste Hindus (some of whom are pig breeders), and tribal peoples.

Figure 11.A   People and cows share street space across India.

rice production, and with declining rates of population growth, its development prospects are good, particularly if its civil conflict can be ended.

External influences have diversified the island's population and sown seeds of unrest. Centuries of recurrent invasion from India were followed by Portuguese domination in the early 16th century, Dutch in the 17th, and British from 1795 until the country was granted independence in 1948 as a member of the British Commonwealth. This eventful history, plus the longstanding commercial importance of the sea route around southern Asia, has given Sri Lanka a polyglot population that includes Burghers (descendants of Portuguese and Dutch settlers) and Arabs.

The two major ethnic groups, distinguished from each other by language and religion, are the predominantly Buddhist Sinhalese, making up about 74% of the population, and the Hindu Tamils, about 18% (Fig. 11.12). The light-skinned Sinhalese are an Indo-Aryan people who settled in Sri Lanka about 2,000 years ago. The dark-skinned Tamils, whose main area of settlement is the Jaffna Peninsula and adjoining areas in the north, are descendants of early invaders and more recent imported tea plantation workers from southern India.

The migrations of Tamil tea workers from India to Sri Lanka, and their subsequent travails there, are part of the legacy of British colonial rule.

Figure 11.12   Buddhism is the religion of most Sri Lankans. These are the Colossi of Buddha and Ananda at Polonnaruwa.

11.3 Climate and Water Supply

Climatic conditions in the subcontinent and Sri Lanka range from Himalayan ice to the tropical heat of peninsular India and from some of the world's driest climates to some of the wettest. Climatic and biotic types in the subcontinent include undifferentiated highland climates in the northern mountains; desert and steppe in Pakistan and adjacent India; humid subtropical climate in the northern plains; tropical savanna in the peninsula; and rain forest along the seaward slopes of the Western Ghats and in parts of the Ganges-Brahmaputra Delta and the eastern mountains near Burma (see Figs. 2.6 and 2.7).

Heat is nearly constant except in the mountains. The tropical peninsula is hot all year. The subtropical north has stifling heat before the break of the wet monsoon, and warm conditions even in the winter. Some of the warmest temperatures on the globe occur in the plains of Pakistan and northwestern India. Delhi averages 94°F (34°C) in both May and June. The highest temperatures over most of the subcontinent occur in May and June, and high humidities produced by the impending monsoon rains combine with the heat to create almost insufferable conditions. Then the monsoon rains break, bring cooling relief — but sometimes catastrophic flooding — to the subcontinent's people. Of a particularly violent onset of the monsoon in India, Kamala Markandaya wrote: "That year the monsoon broke early with an evil intensity such as none could remember before. It rained so hard, so long and so incessantly that the thought of a period of no rain provoked a mild wonder. It was as if nothing had ever been but rain, and the water pitilessly found every hole in the thatched roof to come in, dripping onto the already damp floor."[1]

The **monsoons** are winds and include both the wet southwest monsoon that brings the short, wet summer between June and September and the dry northeast monsoon that creates mostly arid conditions the rest of the year (Fig. 11.13). There are two main arms of the wet monsoon. One, approaching from the west off the Arabian Sea, strikes the Western Ghats and produces heavy rainfall on these mountains and the coastal plain. The amount of rain diminishes in the interior Deccan rain-shadow region to the east of the mountains. Here, there is barely enough rainfall for dry farming, and drought sometimes brings crop failures. The second arm approaches from the Bay of Bengal, bringing moderate rainfall to the eastern coastal areas of the peninsula and heavy rains to the Ganges-Brahmaputra Delta region and northeastern India. This is the monsoon that feeds cyclones. Wet monsoon winds pass up the Ganges Valley to drop moisture that diminishes in quantity from east to west. Both arms of the monsoon bring some rain to Pakistan, but it is so little that semiarid or desert conditions prevail in most areas (Fig. 11.14). Blowing mainly from land to sea, the dry monsoon brings dry and cooler weather to most parts of the subcontinent. An exception is in the far south of the peninsula, where the heaviest rainfall of the year falls along the eastern coast and in adjacent uplands from October through January.

[1] Kamala Markandaya, *Nectar in a Sieve*. New York: John Day, 1954, p. 57.

Figure 11.13 The torrential monsoon rains are a regular and welcome feature of land and life in South Asia. The skies cleared after copious rains flooded this road and many villages in south central Sri Lanka.

With precipitation so concentrated within a short part of the year, it is essential to extend the crop season by irrigation. People in the subcontinent therefore expend an enormous amount of labor to get additional water onto the land. Modern technology does the job in places where huge dams impound rivers and where networks of canals distribute water from these reservoirs. Most dams are in the steppe and desert areas of the Punjab and the lower Indus Valley (Sind Province) in Pakistan. A massive irrigation and hydroelectric power project is now underway in the north central part of peninsular India, where the Sardar Sarovar Dam is being constructed on the Narmada River (see Fig. 11.4). Largely because it is expected to displace 320,000 tribal and rural people and inundate 28,000 acres of cropland and 32,000 acres of forest, environmentalists and human rights advocates both in India and abroad oppose the dam. The World Bank withdrew its financial support of the project in 1993 in response to these protests, but asserting that its benefits will greatly outweigh its

Figure 11.14 Aridity and deforestation are among the urgent environmental problems faced by Pakistan. Foraging by goats and cutting trees for fuel in an already dry environment have created a lunar landscape in the country's Karakoram Mountain region.

REGONAL PERSPECTIVE ∴ "Upstream" and "Downstream" Countries

The region of the subcontinent called the Punjab, meaning "Five Waters" (referring to the five Indus River tributaries), is split between two countries, with its rivers flowing through Indian territory before entering Pakistan (see Fig. 11.4). In geographic and geopolitical terms, India is an "upstream country," and Pakistan is a "downstream country" where Punjab waters are concerned. There are no internationally accepted laws regarding water sharing between upstream and downstream states, and the upstream country often exercises its geographic advantage to siphon off what the downstream country views as too much water. **Riparian** (river-owning) countries often come close to, or actually go to, war over the problem. In many cases, formal treaties have averted

what could have become major international conflicts.

After 1947, relations between India and newly independent Pakistan deteriorated in part because of the disputed waters of the Indus and the five Punjab rivers. For many years, the parties could not agree on how much of the water India could divert and how much should remain for Pakistan. The countries finally reached a satisfactory resolution in the 1960 Indus Waters Treaty. The agreement allocates the water of the three eastern rivers of the Punjab to India, which in return allows unrestricted and undiminished flow of the two others, and the Indus River itself, into Pakistan. The two countries have abided by the agreement ever since, despite their periodic wars and constant state of tension.

Another landmark water-sharing agreement was signed between India and Bangladesh in 1996. For two decades, these nations had a bitter disagreement about the allocation of water from the Ganges River. In the 1996 accord, which is meant to be in effect for 30 years, upstream India guarantees a larger share of water to downstream Bangladesh. This achievement may open the way to the integrated development of what is known as the Ganges-Brahmaputra-Barak (GBB) river basin, a watershed that is home to 500 million people in Bangladesh, Bhutan, China, India, and Nepal. Cooperation to settle the vital issue of water allocation among these countries could promote peace and reduce poverty in the region.

costs, India has pressed ahead with the dam's construction. It is the centerpiece of the massive Narmada Valley Development Project, where 30 dams will ultimately displace about 1 million people by the completion date of 2025.

11.4 Subsistence and Software

The subcontinent of India is a paradox. The true subcontinent is to be found in the villages, where 7 of 10 persons live. But the true subcontinent is also to be found in its megacities: Mumbai, Kolkata, Karachi, and Dacca are central to the identities and economies of the region's people. The subcontinent of India is an ancient, modern, rural, urban land.

Food for a Billion People

Both staple food crops and population densities correlate spatially with the amount of water available in the subcontinent. Rice is the basis of life in the wetter areas and, with its high caloric yields per acre, is associated with the highest population densities. Rice dominates in the delta area of Bengal (in both India and Bangladesh), in the adjacent lower Ganges valley, and in the coastal plains of the peninsula. India's East and West Punjab and Pakistan's Sind and Punjab Provinces produce surplus rice, including the Basmati variety, which is an important export (Fig. 11.15). Irrigated wheat is the staple crop and food in the drier upper Ganges valley of India and in the dry Punjab of India and Pakistan. Here, the caloric yield and

population densities are intermediate. Unirrigated sorghums and millets prevail over most of the Deccan plateau and in other areas where low rainfall cannot be supplemented much by irrigation. Caloric yields from these crops are low, and so are population densities. Maize is an important grain in the lower regions of the Himalaya. However, rice is the preferred food almost everywhere, and where enough water is available, there are patches of irrigated rice. A host of minor crops supplements the staple grains.

Industrial and Export Crops

Cotton, jute, tea, and rice are the main crops grown in the subcontinent for industrial use and export. Irrigated cotton in the Punjab and lower Indus Valley makes Pakistan the world's fourth largest cotton producer (after China, the United States, and India). Textiles, yarn, ready-made garments, and raw cotton contribute 77% of Pakistan's exports. In India, cotton is grown mainly in the interior Deccan. India's large textile industry absorbs most of the production.

The Ganges-Brahmaputra Delta is the world's greatest producing area of jute, the raw material for burlap. Both India and Bangladesh grow and mill jute in this region of Bengal. Of the four major commercial crops in the subcontinent, only tea is exported on a sizable scale. Unlike cotton, jute, and most other crops of the subcontinent, tea is a plantation crop. The plantations were developed, and are still largely owned, by British interests. Production is greatest in the northeastern

48

Poverty and marginal human health are already problems in the Indian subcontinent and Sri Lanka, and the prospect of continued high population growth raises the question of whether growth in food supplies can avert the proverbial Malthusian crisis.

Population growth in India has accelerated rapidly since independence as birth rates have remained relatively high and death rates have continued to fall. Given increases in urbanization and industrialization, and the dividends from India's aggressive family planning programs, future population growth in the region is expected to slow. The population base is already so vast, however, that even modest growth will add huge numbers. India is predicted to overtake China as the world's most populous country by 2015.

Already for many, the name "India" invokes an image of grinding poverty, and with reason: An estimated 20% of the population is abjectly poor, defined as living on less than $1 per day. A few people in India are very wealthy, and there is an emerging middle class of as many as 300 million people (Fig. 11.B). Perhaps in de-

fiance of Malthus, an increased population has so far succeeded in producing more economic resources. Recent statistics show that India's economy is growing impressively (about 8% per year through the 1990s). The problem is that the benefits of the growing economy are unevenly distributed; during that decade, the number of Indians living below the poverty line dropped only 1% to 34%. Apparently, most of the economic growth benefits those who are already better off, widening the gap between rich and poor. Population growth in the long run may threaten a more equitable distribution of wealth.

The issue of population growth is also critical for Pakistan, which has never mounted an effective family planning program, mainly because of opposition from Muslim religious authorities who regard birth control as an intervention against God's will (Pakistan's birth rate in 2001 was 39 per thousand, compared with 26 per thousand in India).

Agricultural output in South Asia has increased since independence. Most notably, despite its huge and growing population, India has managed to remain self-sufficient in food production. The successes of agriculture in the subcontinent have been due mainly to the increased use of artificial fertilizers, the introduction of new high-yield varieties of wheat and rice associated with the Green Revolution, more labor provided by the growing rural population, better irrigation, the spread of education (Fig. 11.C), and the development of government extension institutions to aid farmers.

Social conditions and services in rural areas are improving in some parts of

South Asia, notably in India, suggesting that development is progressing and that agricultural production has so far been sufficient to keep up with population growth. These are signs that a Malthusian catastrophe long predicted for the region is not imminent.

One of the challenges confronting India is raising the status of women. Despite a 1961 ban on dowries (the money and gifts given by a bride's parents to the groom), the practice is continuing. So is the rate of killing brides who do not provide enough dowry. The burden placed on the bride's family has prompted many parents to abort female fetuses, which can be detected with ultrasound technology.

Technology is thus a mixed blessing in this traditional society. Generally, it plays a constructive role because, where face-to-face efforts are not possible, radio, television and now the Internet can encourage positive change in the remotest villages. South Asia is developing. The challenge will be for technology and human resources to keep agricultural growth apace with or ahead of population growth.

Figure 11.B   This Indian family visiting the Red Fort in Agra is among the growing ranks of the subcontinent's middle class.

Figure 11.C   Education and agricultural production are advancing in South Asia despite great odds. These schoolboys are studying in a mosque in Peshawar, Pakistan.

state of Assam, with another center in the mountainous far south of peninsular India (see Fig. 11.10). Coffee, rubber, and coconuts are minor plantation crops, and India exports products made from coir (coconut fiber).

Opium is a major illegal export crop of Pakistan. Cultivation is restricted to the rugged mountain region of the North-

west Frontier Province, inhabited mainly by Pushtun (Pathan) tribespeople, along the frontier with Afghanistan. Processing of the raw opium into heroin is also a significant industry in the region. The government of Pakistan is under pressure from the United States and other heroin-consuming nations to crack down on the illegal trade. Domestic heroin addiction

Figure 11.15   These meticulously crafted and maintained terraces produce a high yield of rice in Pakistan's northern Punjab.

is a growing problem, with an estimated 5 million Pakistanis dependent on the drug in 2002.

Space Age Industries, and Others, in Lands of Tradition

In the partition following independence from Britain, India received almost all of the subcontinent's modern industries and most of its mineral wealth and energy supplies. Pakistan, including the present Bangladesh, was left with a small industrial infrastructure and fewer natural resources. In global terms, the subcontinent today has a small but growing industrial base. Handicraft workers still exceed factory workers by millions, and handicrafts supply many everyday needs for the population. Visitors to the subcontinent are astounded to see human hands performing, on a vast scale, the tasks routinely done by machines in more developed countries (MDCs).

Pakistan and Bangladesh are still very far behind India in both high-tech and traditional industrial output. Both have developed successful cotton-textile industries. Chances are that the American reader of this text owns clothing made in Bangladesh. India's leading factory industries are cotton textiles, jute products, food processing, and iron and steel. Modern cotton mills are concentrated mainly in the metropolitan area of Mumbai and neighboring areas. India is both a major user and exporter of cotton textiles and clothing. Favoring cotton manufacturing are a huge home market, cheap labor, available hydroelectric power, and the large domestic production of cotton.

The Bengal is an important industrial region. Both India and Bangladesh make jute products here. India's iron and steel industry is concentrated in the Bengal. The most valuable assemblage of mineral resources in India, including iron ore, coal, manganese, chromium, and tungsten, is found in this region. These resources have led to the rise of small- to medium-sized industrial and mining cities (notably Jamshedpur) in an area that otherwise is one of the least developed parts of the country.

Until the 1990s, India was intent on keeping out most foreign manufacturing in favor of industrial self-sufficiency. The economy was largely socialized, with many state-owned industries. With these policies, India's industries expanded and diversified rapidly. But lately, India has opened up to more imports, privatized many state-run industries, and boosted exports. The result has been remarkably strong economic growth.

India's diverse manufactures include old-economy products like cars, trucks, motorcycles, bicycles, and farm machinery, along with high-tech data-processing equipment, silicon chips for electronics, and computer software. New technology venture firms are racing to establish satellite telephone service across the subcontinent. Improved phone service is closely linked to government and private efforts to wire India to the World Wide Web. Internet cafés are turning up even in remote villages, where their operators use diesel generators to keep computer terminals from failing during frequent power outages. Increasing access by the poor to the Net could bring significant social and economic change in this poor country. For example, farmers could get information on crop prices in different towns, improving their bargaining positions during harvests. Villagers would have unbridled access to land records rather than having to bribe officials to get them. Those seeking work abroad could download passport applications instantly rather than spending time and money working through intermediaries.

There is great optimism in India that the country's future lies with information technology. Bangalore is known as India's Silicon Valley, and its space-age character presents a startling contrast to the largely rural and unchanging image that many outsiders have of India. India's software exports (two-thirds of which go to the United States) have been growing in recent years at rates of up to 50% per year and earned $5.7 billion in revenue in 2000. There are other Indian counterparts with profitable Western industries. Mumbai is India's Wall Street and also its Hollywood (it calls itself "Bollywood") — the center of a prolific and accomplished film industry (and the world's largest) that produces movies of international stature as well as the regular fare that a huge domestic market consumes passionately (Fig. 11.16).

The downside of India's intellectual productivity is that so much of it is lost to other countries. India suffers from a **brain drain,** the emigration of highly trained and highly educated individuals abroad. Chances are that the American reader of this text has been or will be treated in a hospital or clinic by a physician who immigrated from India. The United States' gain, in the form of highly qualified medical personnel, is

Figure 11.16 Mumbai produces the romance and action films that are the staples of Indian cinema.

Figure 11.17 The central rail station in New Delhi, India, is one of the hubs in a vast network dating to the British colonial period. Note New Delhi's modern skyline in the distance.

inevitably India's loss because better development requires better human resources. Indian computer scientists, engineers, and other professionals have in recent years followed the trend in "exports" of doctors to the United States and other countries. There is a kind of global bidding war for Indian brains.

India produces a wide range of chemical products, from vitamins to fertilizers. There is a flourishing pharmaceutical industry. Much to the chagrin of Western drug companies, their Indian counterparts routinely replicate foreign drug formulas and produce copycat medications for sale at a fraction of their cost in the West. Arguing that India is a poor country that needs low-cost medicines, the Indian government does not crack down on its drug "pirates." India's success in providing lifesaving medications to its population cannot be disputed; life expectancy has risen by 20 years since the country became independent.

Inadequate energy resources are a problem for further industrialization in the subcontinent. The major domestic sources of commercial energy are India's coal deposits and the hydropower generated in all the countries but lowland Bangladesh. All the countries are heavily dependent on imported energy, especially oil, and would probably be more so if their levels of industrial development and consumption were higher. India has a modest but rapidly growing oil output from fields in Assam, Gujarat, and offshore of Bombay. Natural gas from domestic wells is especially important to Pakistan, and smaller quantities are in India and Bangladesh. India derives about 2% of its electricity from nuclear power plants.

On the plus side, the existence of a superior rail transportation network favors industrial development in India, Pakistan, and Bangladesh (Fig. 11.17). Railways are universally regarded as one of the positive legacies of British colonialism in the subcontinent. The vast network created by the British has been expanded and improved on by the independent nations. In addition to transporting goods to internal markets and to ports for export, this system carries millions of passengers daily.

11.5 Social and Political Complexities

On a physical map, the Indian subcontinent looks like a discrete unit, marked off from the rest of the world by its mountain borders and seacoasts. But the social complexity of this area is so great that it has never been unified politically, except when Britain ruled. As soon as the British withdrew, divisive forces split the subcontinent into Muslim and Hindu countries. Within a quarter century, other forces split Pakistan into two nations. All countries of the subcontinent and Sri Lanka have serious problems in internal or international relations, and sometimes both. Seldom is there talk of reconciliation.

Religion and Sectarianism

The major social divisions within the subcontinent relate to religion and language. Religious differences are particularly troubling. The most serious division has been between the two major religious groups: Hindus and Muslims. Indigenous Hinduism was the dominant religion at the time Islam made its appearance in this region. It continues to have the largest number of adherents: an estimated 81% of the population of India, 11% in Bangladesh, and 2% in Pakistan.

The faith of Islam is described in Chapter 8. It is much more difficult to characterize the complex and regionally varied faith of Hinduism. It lacks a definite creed or theology. It is very absorptive, encompassing an unlimited pantheon of deities (all of which, most Hindus say, are simply different aspects of the one God) and an infinite range of types of permissible worship (Fig. 11.18). Briefly, it has the following notable elements distinguishing it from the region's other religions: Most Hindus recognize the social hierarchy of the caste system (see Definitions & Insights, p. 311), deferring authority to the highest (Brahmin) caste. They practice rituals to honor one or another of five principal deities, attending temples to worship them in

DEFINITIONS + INSIGHTS

The Caste System

One ancient Hindu belief is that every individual is born into a particular **caste**, a socioeconomic subgroup that determines the individual's rank and role in the society. Castes form a hierarchy, with the Brahmin caste at the top (comprising about 5% of all Hindus) and three others (Kshatriya, Vaisya, and Sudra) below. Caste membership is inherited and cannot be changed. Particular castes are associated with certain religious obligations, and their members are expected to follow traditional caste occupations. Marriage outside the caste is generally forbidden, and meals are usually taken only with fellow caste members.

At the bottom of the social ladder are the Dalits, or "untouchables" (about 20% of all Hindus), also known as the "scheduled castes." They are not part of the caste system but are "outcaste" and according to Hindu belief, they are not twice-born. The untouchables are so called because they traditionally performed the worst jobs, such as handling of corpses and garbage, and therefore, their touch would defile caste Hindus.

This ancient, rigid social system is changing. Brahmin privileges are being increasingly challenged and restricted by law. Indian law now forbids recognition of untouchables as a separate social group. Confronting the fact that upper castes account for less than one-fifth of India's population but command more than half of the best government jobs, the Indian government has instituted an affirmative action program to allocate more jobs to members of lower castes. Indian leaders of high caste have championed the Dalits' cause for both moral and practical reasons. Politicians use caste divisions to their advantage to bring out voters, promising if elected to act in the interests of their respective large social blocs. In 1997, an untouchable was elected president of India for the first time in the country's history. Since then, untouchables have won more regional and local offices.

The Dalits are voting in greater numbers than ever before with the conviction they can use India's democratic system to overcome ancient discrimination against them. They are making progress. Disintegration of the caste system has been especially rapid in the cities, where people of different castes must mingle in factories, public eating places, and public transportation. In India's vast rural areas, the caste system is more entrenched.

---

the form of sanctified icons in which the gods' presence resides. They believe in reincarnation and the transmigration of souls. They are supposed to be tolerant of other religions and ideas. They participate in folk festivals to commemorate legendary heroes and gods. To earn religious merit and to struggle toward liberation from the bondage of repeated death and rebirth, they make pilgrimages to sacred mountains and rivers. The pilgrimage to celebrate the festival of Purna Kumbh Mela at Allahabad, on the Ganges River, drew 30 million pilgrims in January 2001. It was the largest pilgrimage in human history; 30 million is the conservative estimate, and some sources put the number as high as 70 million!

The Ganges is a particularly sacred river to Hindus, who believe it springs from the matted hair of the god Shiva. The most enduring Hindu pilgrimage destination is the Ganges city Varanasi (Benares) in the state of Uttar Pradesh (Fig. 11.19). Many elderly people go to die in this city and to be cremated where their ashes may be thrown into the holy waters. The Indian government is now attempting to clean up the Ganges, polluted in part by incompletely cremated corpses; many of the faithful poor cannot afford to buy the fuel needed for thorough immolation. Officials in Varanasi have released scavenging water turtles to dispose of the cadavers.

Seldom in history have two large groups with such differing beliefs lived in such close association with each other as have Hindus and Muslims in the subcontinent. Islam holds to an uncompromising monotheism and prescribes uniformity in religious beliefs and practices. Hinduism is monotheistic for some believers but polytheistic for others and asserts that a variety of religious observances is consistent with the differing natures and social roles of humans. The exuberant and noisy celebrations of the Hindu faith are a striking contrast to the austere ceremonies of Islam. Islam has a mission to convert others to the true religion, whereas most Hindus regard proselytizing as essentially useless and wrong — one is born or reborn as a Hindu, so conversion is not an issue. Islam's concept of the essential equality of all believers is a total contrast to the inequalities of the caste system endorsed in Hinduism. Islam's use of the cow for food is anathema to Hinduism (see Definitions & Insights, p. 305).

Figure 11.18 Hinduism has a complex and very colorful pantheon. This is the preserver god, Vishnu, and his consort, Lakshmi, goddess of wealth and prosperity.

Figure 11.19 The Hindu faithful pray and bathe on the banks of the sacred Ganges River in Varanasi.

Practical differences have sometimes added to doctrinal and cultural differences and produced outright conflict between the groups. During the British occupation, the formerly subordinate Hindus came to dominate the civil service and most businesses. Many Muslims feared the results of being incorporated into a state with a Hindu majority, and their demands for political separation led to the creation of the two independent states — India and Pakistan — from the British colony.

Immediately preceding and following partition in 1947, violence broke out between the two peoples, and hundreds of thousands of lives were lost in wholesale massacres. More than 15 million people migrated between the two countries. Pakistan today has a large population known as *Mohajirs*, or "migrants," the Muslim immigrants and their descendants who poured into the region from India when partition occurred. Particularly in Karachi and elsewhere in the southern province of Sind, violent conflict frequently occurs between rival factions among the Mohajirs and between the Mohajirs and other ethnic groups, including Pathans (Pushtuns) and Biharis.

Since 1947, India and Pakistan have been in conflict over the status of Kashmir (called Jammu and Kashmir in India), a disputed province straddling their northern border. Before independence, Kashmir was a princely state administered by a Hindu maharajah. About three-fourths of Kashmir's estimated population of 13.5 million is Muslim, which is the basis of Pakistan's claim to the territory. But under the partition arrangements, the ruler of each princely state was to have the right to

join either India or Pakistan, as he chose. Kashmir's Hindu ruler chose India, which is the legal basis of India's claim.

After partition, fighting between India and Pakistan led to a cease-fire line leaving eastern Kashmir, with most of the state's population, in India and the more rugged western Kashmir in Pakistan. Beginning in the mid-1950s, China pressed its own claims to remote northern mountain areas of Kashmir and occupied some of the border territories. Pakistan ceded part of the occupied territory to China and established friendly relations with its giant northern neighbor, while India rejected China's claims (these countries still have chilly relations). Until now, India's recurrent small-scale military actions have failed to dislodge Chinese forces. India now holds about 55% of the old state of Kashmir, Pakistan 30%, and China 15%.

Three wars between India and Pakistan have effected little change in Kashmir. In 1965, conflict began in Kashmir, spread to the Punjab, and escalated to a brief but indecisive fullscale war involving tanks, airborne forces, and widespread air raids. Renewed hostilities in Kashmir in 1971, as part of the war in which India supported the revolt of Bangladesh against Pakistan, again did not alter the political landscape of Kashmir. During that conflict, the Siachen Glacier in the Karakoram, at about 20,000 feet (2,800 m), earned the title of "world's highest battlefield."

In 1989, Kashmir's Muslim majority escalated the campaign for secession from India, and the strife has continued ever since, claiming an estimated 35,000 lives by 2002. December 2001 and January 2002 saw India and Pakistan perched perilously on the brink of war over the Kashmir issue. In mid-December, armed insurgents attacked India's parliament in New Delhi in a failed attempt to assassinate Indian politicians. India's government identified the assailants as members of two Muslim rebel movements (Lashkar-e-Taiba and Jaish-e-Muhammad) based in Kashmir. India further accused Pakistan's government of supporting these rebel groups and therefore being ultimately responsible for the attack on India's parliament. Indian military forces began amassing on the line of control in Kashmir, and Pakistani troops responded in kind. Tremendous international diplomatic pressure was applied to both sides in an effort to prevent all-out war. The pressure was particularly intense for Pakistan. Pakistani President Musharraf responded by outlawing and beginning to crack down on five militant groups, including the two blamed for the December attack. It was a risky move, especially because many people within Pakistan believed their leader was forsaking a legitimate struggle in Kashmir and "selling out" to the United States and other Western powers in the wake of September 11. While Musharraf's moves did at least temporarily defuse the explosive situation in Kashmir, many observers remained concerned that Pakistan would prove unable to rein in Muslim attacks against Indian targets in Kashmir. There are also many questions about how much control Musharraf has over his own intelligence service, the ISI (Inter-Service Intelligence Agency), which may contain "rogue" elements intent on continued support of the militants fighting India.

Figure 11.20   Kashmir, "Paradise on Earth" but for the blood shed there. The mountain is Rakaposhi (25,550 ft/7,788 m), on the Pakistani side of the line of control.

The Kashmir conflict has been very costly to both sides. India's international reputation has suffered because of its unyielding position on the region. Pakistan spends about a third of its budget on defense, mostly focused on a potential engagement with India over Kashmir. There would be economic benefits for both India and Pakistan if peace returned. "Kashmir is a Paradise on Earth," wrote Samsar Kour.[2] The stunning snow-crowned peaks and flower-laden valleys of this western Himalayan region once lured millions of Indian tourists and ranks of Western trekkers each year, and the tourism development potential is enormous (Fig. 11.20).

Pakistan has internal problems centering on religious divisions, mainly between the majority Sunni Muslims (75% of the country's population) and Shi'ite Muslims (20%). Shi'ite Muslims opposed the government's efforts in the 1980s to establish an avowedly Sunni Islamic state, and there was violent conflict between members of the two sects. Pakistani factions who preferred a Western-style secular democracy also rioted in protest against the Islamization of the government. Democratic elections in the 1980s and 1990s twice installed Benazir Bhutto, a liberal, Western-educated woman, as the country's prime minister. Her regimes were so riddled with corruption that she was twice voted out of office to be replaced by Nawaz Sharif, who promised to cleanse the country of corruption and embrace the traditional Islamic system of law known as *shari'a*. Sharif in turn was ousted in a military coup in 1999.

His successor, General Pervez Musharraf, took the helm of a country torn between the secular wishes of a Western-oriented middle class and popular calls for greater Islamic influences and fewer ties with the West. The growing Islamist sentiment is focused both on ousting India from disputed Kashmir and on resisting the U.S. and other Western intervention in the region that stepped up in the wake of the terrorist attacks on the

United States in September 2001. Much of the dissent is based in the North West Frontier Province, especially in and around its principal city, Peshawar. At the time of the attack on the United States, there were already 3 million mainly ethnic Pashtun refugees from Afghanistan living in camps outside the city. Hundreds of thousands more poured in from Afghanistan in the weeks following the attack. Within these impoverished, desolate encampments simmer much anti-U.S. and generally anti-Western sentiment.

An even larger breeding ground for anti-Western (and anti-India) thought and action in Pakistan is the institution known as the *madrasa*, which may be translated as "Islamic school" or "seminary." There are as many as 8,000 *madrasas* in Pakistan, where as many as 700,000 children (mainly boys) are educated. Their curriculum is based largely on study of the Qur'an and other Islamic texts, and some instructors promote an interpretation of the Islamic concept of *jihad* or "struggle" in which Muslims are urged to fight India, the United States, and other infidel entities. Many of the *jihadis* or "holy warriors" who join groups like al-Qa'ida, Lashkar-e-Taiba, and Jaish-e-Muhammad have been educated in these schools. But *madrasas* also offer much legitimate instruction and help to compensate for Pakistan's poor public educational system. Pakistani President Musharraf is thus once again in a difficult position, at once called upon by the outside world to cleanse the *madrasas* of militant content and by his own people to uphold the long-cherished institution of the religious school.

In India, sectarian violence (known in the region as "communal" violence) between Hindus and Muslims is a constant threat to the country's social and political fabric. Contention over places sacred to both faiths sometimes ignites widespread violence. A 16th-century mosque in the city of Ayodhya in Uttar Pradesh was a place of prayer for Muslims and also revered by the Hindus as the birthplace of the god-king Ram, an incarnation of the god Vishnu. Backed by Hindu nationalists in the provincial government, a mob of about 250,000 Hindu fundamentalists demolished the mosque in December 1992 with the intention of building a Hindu temple on the site. The ensuing communal violence between Hindus and Muslims throughout India, notably involving the cosmopolitan and usually tolerant urbanites of Bombay, left thousands dead. Ever since this incident, as a means of garnering votes from Hindu Indians, Hindu politicians have made veiled wishes to proceed with construction of the temple. His apparent support for the temple helped sweep India's Prime Minister Vajpayee into power in 1998. The conflict over this multireligious sacred site in India has parallels with the Temple Mount *247* dispute between Israel and the Palestinians.

In addition to the Muslims who make up 12% of the population in India, there are other significant religious minorities in the country. About 19 million Sikhs are concentrated mainly in the unusually prosperous state of Punjab. Theirs is India's fourth largest religion, after Hinduism, Islam, and Christianity. Their faith emerged in the 15th century as a religious reform movement intent on narrowing the differences

---

[2] Samsar Chand Kour, *Beautiful Valleys of Kashmir and Ladakh.* Mysore, India: Wesley Press, 1942, p. 2.

PROBLEM LANDSCAPE

Nationalism and Nukes in the Subcontinent

Since independence more than 50 years ago, India has been an avowedly secular democratic state. But an explicitly sectarian political party captured the highest number of the country's parliamentary seats for the first time in the national election of 1998. This was the Bharatiya Janata party, or BJP, a Hindu nationalist party that promised India's Hindu majority the strongest say in the country's affairs. The BJP platform included a pledge to build a Hindu temple atop the ruins of the mosque at Ayodhya and the adoption of a code of civil law that would strip Muslims of separate laws in matters of marriage, divorce, and property rights. However, to attract the support of non-Hindu parties and thereby gain enough parliamentary seats to ensure his place as prime minister, the BJP leader Atal Vajpayee dropped many of the Hindu-oriented elements of his agenda. He also suggested he would make peace with Pakistan. At the same time, though, he declared that India would reevaluate its nuclear arms policy. Among BJP supporters, there was hope for a "Hindu bomb," a counterweight to a long-feared "Islamic bomb" in neighboring Pakistan.

On May 11, 1998, much to the surprise of the U.S. Central Intelligence Agency and other Western intelligence agencies, India conducted three underground nuclear tests in the Thar Desert. With the blasts, India's government seemed to be trying to stake India's claim as a great world power, exhibit its military muscle to Pakistan and China,

and garner enough political support for the BJP to form an outright parliamentary majority in the future. Initial reactions among India's vast populace were highly favorable; 91% of those polled within 3 days of the event supported the tests. Outside India, there was alarm. India had defied an informal worldwide moratorium on nuclear testing that went into effect in 1996, when 149 nations (not including India and Pakistan) signed the Comprehensive Test Ban Treaty (known as the Nuclear Non-Proliferation Treaty, or NPT), which prohibits all nuclear tests.

The world's eyes quickly turned to India's neighbor to the west. Governments pleaded with Pakistan to refrain from answering India with nuclear tests of its own, arguing that Pakistan would have a public relations triumph it if exercised restraint: It would appear to be a mature and responsible power while India revealed itself as a dangerous rogue state. But Pakistan's leaders felt obliged by their population's demands for a tit-for-tat response to India's blasts and soon followed with six nuclear tests. The United States and other nations reacted with economic sanctions against both countries, and these were particularly hard on Pakistan's smaller economy. Within 2 years, however, the United States began courting India diplomatically, as a counterweight against China (India's longtime foe in the region) and as a means of expressing displeasure with Pakistan (China's ally) for helping Afghanistan's Taliban to harbor

accused terrorist Osama bin Laden. The United States and Pakistan did an abrupt diplomatic about-face after the 2001 terrorist attacks against New York and Washington. Pakistan agreed to allow the United States to use its territory in staging attacks against purported bin Laden bases in Afghanistan, and in return, the United States lifted its sanctions against Pakistan and forgave much of its debt to the United States. To keep from isolating India, the United States also lifted its post-nuclear test sanctions against that country.

India and Pakistan have joined only five other nations — the United States, Russia, China, the United Kingdom, and France — in acknowledging that they possess nuclear weapons. There is disagreement about what this new regional and global shift in the balance of power means. Some analysts fear an escalating nuclear arms race that, perhaps ignited by a border skirmish in Kashmir, could lead to the mutually assured destruction of Pakistan and India. There are concurrent fears of nuclear war between China and India. Others, particularly within South Asia, argue that the weapons represent the best deterrent against conflict, as they did for decades between the countries of NATO and the Warsaw Pact. However, few people anywhere argue with the contention that the nuclear rivalry between India and Pakistan has taken an enormous economic toll in two nations that need to wage war on poverty.

---

between Punjab's Hindus and Muslims. They regard men and women as equals and disavow the caste system recognized by Hindus. Sikh men are recognizable by their turbans, which contain their never-shorn hair, by a bracelet called a "bangle" worn on the right arm, and by their ubiquitous surname Singh, meaning "lion."

In India's Punjab, where Sikhs make up 60% of the population and Hindus 36%, the 1980s and early 1990s were violent years during which about 20,000 people died in armed clashes. Sikh factions intent on establishing their own home-

land of Khalistan, a movement prompted in part by New Delhi's plans to divert water from the Punjab, challenged Indian authority and turned Amritsar's Golden Temple, the holiest site in the Sikh faith, into their military stronghold (Fig. 11.21). In a controversial move to quell the revolt, Prime Minister Indira Gandhi ordered troops to storm the Golden Temple in June 1984. The resulting deaths and desecration led directly to Mrs. Gandhi's assassination by her own Sikh bodyguards later that year. Indian authorities accused Pakistan of arming the Sikh militants.

PAOLO KOCH/PHOTO RESEARCHERS, INC.

Figure 11.21 The Golden Temple (established in 1577, rebuilt in 1764) in Amritsar is the holiest shrine of the Sikh religion.

There are about 22 million Christians in India, most of whom live in the south of the peninsula; Buddhists, Jains, Parsis, and members of a variety of tribal religions make up the numerous smaller remaining religious minorities. Most of the estimated 8 million Buddhists are recent converts from among India's lowest castes. Numbering perhaps 200,000 and concentrated mainly in Mumbai, the famously entrepreneurial Parsis have attained wealth and economic power far out of proportion to their number. Their religion is the pre-Islamic Persian faith of Zoroastrianism, known mainly for its reverence of fire. Zoroastrians follow the teachings of the prophet Zarathustra, who was born around 600 B.C. in Persia. He preached that people should have good thoughts, speak good words, do good deeds, and believe in a single god, Ahura Mazda ("Wise Lord"). Persia's Zarathustras, as the Zoroastrians are also known, fled from Persia in two waves, first during the conquest by Alexander the Great around 300 B.C. and then again when Islam swept into Persia in the seventh century. They settled in China, Russia, and Europe, where they became assimilated with other cultures, and in India, where they maintained a separate existence. Today, there are an estimated 18,000 in the United States.

India's 4 million Jains also have influence beyond what their numbers suggest, as they control a significant share of India's business. Their faith, founded on the teachings of Vardhamana Mahavira (a contemporary of Buddha, c. 540–468 B.C.), is renowned for its respect for geographical features and animal life. Jains believe that souls are in people, plants, animals, and nonliving natural entities such as rocks and rivers. Jainism has taken the principle of nonviolence (*ahimsa*, also present in Hinduism) to mean they should not even harm microbes. Therefore, Jain worshippers often wear masks to prevent inhalation of microscopic organisms, and Jains cannot be farmers because they would have to destroy plant life and living organisms in the soil (Fig. 11.22).

Religious and political troubles exist in what otherwise looks like Shangri-la, the mountain country of Bhutan. This is a monarchy ruled from the capital of Thimphu by a Buddhist

JOE HOBBS

Figure 11.22 A priest in a Jain temple in Mumbai wears a mask so as not to breathe in and thereby harm microbes.

king of the Drukpa tribe, to which most of the government ministers also belong. The government forbids political parties and is being challenged by a number of outlawed organizations, including the Bhutan People's party (BPP). The BPP represents ethnic Nepalis, who are Hindus and who for decades have been migrating out of their own overcrowded Himalayan kingdom to work the productive soils of Bhutan. Ostensibly as a means of promoting national unity, Bhutan's king has outlawed the use of the Nepali language in Bhutan's schools and insisted that all inhabitants of Bhutan wear the national dress and hair style of the Drukpa tribe. Violent clashes ensued between ethnic Nepalis and Drukpas during the 1990s.

Nepal itself has suffered violent consequences of one-party rule and of palace intrigue. The country was an absolute monarchy until 1991, when prodemocracy riots and demonstrations forced King Birendra to accept a new role as constitutional monarch and permit elections for a national parliament. Although greater democracy has been established, the government in Kathmandu faces opposition from a Maoist insurgency based mainly in Nepal's far west. When King Birendra and many other royal family members were shot to death at a dinner gathering in 2001, many Nepalis believed the Maoists were responsible. Numerous reports, however, corroborated the official explanation that Nepal's crown prince did the killing in a drunken rage over his mother's refusal to allow him to take the bride of his choice.

In Sri Lanka, internal violence reflects antagonisms between ethnic groups and discontent with economic and political conditions, especially among the minority Tamils (18% of the population). Since the mid-1980s, more than 60,000 people have been killed and many more made homeless as the

Tamils of the country's north have fought for independence from Sri Lanka's majority Sinhalese government. Tamils complain that the Sinhalese treat them as second-class citizens and deprive them of many basic rights. Tamil fighters want to establish their own homeland, Tamil Eelam, in the north and west of the country. In 1987, guerrilla attacks by the organization of Tamil separatists called the Tamil Tigers, or the Liberation Tigers of Tamil Eelam (LTTE), led to military action by Indian troops invited to help restore order. India withdrew in 1990, but the troubles have continued, and India is worried that Tamil war refugees from Sri Lanka could flood southern India or that Sri Lankan infiltrators could inspire Tamil separatism within India itself.

The 1991 assassination in southern India of Indian Prime Minister Rajiv Gandhi, son of the slain Indira Gandhi, was linked to Indian sympathizers of the Tamil separatist movement in Sri Lanka. In 1995, Sri Lankan government troops succeeded in retaking the Tamils' geographic stronghold on the Jaffna Peninsula. The Tamil guerrillas retreated to the wild forests just south of the Jaffna and continued their insurgency with the sinking of Sri Lankan naval ships, a devastating car bombing of downtown Colombo, an attack on the Sinhalese Buddhists' holiest site, the Temple of the Tooth in Kandy, an assassination attempt on the Sri Lankan prime minister, and a spectacular assault on Colombo's airport that destroyed most of the national airline's planes. Prime Minister Kumaratunga is being variously advised to crush the revolt militarily or to pursue a course of devolution that would turn over federal powers to outlying states — a possible track for granting autonomy in the north and east to the Tamils.

Language Problems

Language is yet another divisive factor in India and Pakistan. India alone has 17 major languages and more than 22,000 dialects. The languages of the subcontinent fall into two principal families: in the north, the Aryan languages of the Indo-European language family and, in the southern third of peninsular India, the languages of the Dravidian language family. Languages of major importance, each with at least tens of millions of speakers, include four languages of the Dravidian group and eight Aryan languages.

Many people speak two or even three languages, but overall literacy rates are low and communication is often a problem.

Since British colonial times, English has been the lingua franca and the de facto official language of the subcontinent. However, only a small percentage of the population, mainly the educated, is literate in English. This fact, along with nationalist sentiments, has prompted calls for indigenous languages to be adopted as official languages. After independence and partition, disputes arose in each country over which language should be chosen. India decided on Hindi, an Aryan language spoken by the largest linguistic group, and made it the country's national language in 1965. This government decision resulted in strong protests, especially in the Dravidian south. The result has been a proliferation of official languages to satisfy huge populations; India now has 15, all of which appear on the country's paper currency. Rioting between speakers of different languages has occurred repeatedly, and demands by language groups have reshaped the boundaries of several Indian states.

In Pakistan, Urdu, a language similar to Hindi but written in Perso-Arabic script, is the official language. As in India, English is widely used in Pakistan's business and government. In Bangladesh, Bengali is the predominant and official language.

Despite linguistic, social, and religious differences, the subcontinent's accomplishments in recent decades have been remarkable. India has lowered its birth rate, and its economy has grown. Despite all its internal divisions and frictions, India has retained a representative democracy. But sectarian violence between Hindus and Muslims, a growing Hindu nationalist movement that threatens India's remarkable diversity, Sikh desire for autonomy, and the problem of Kashmir all pose long-term difficulties for this South Asian giant. Pakistan must walk the tightrope between support for its own majority Muslim interests and support for the secular West in its campaign against terrorism in South and Central Asia while also avoiding war with India. Both Pakistan and Nepal must confront the problems of rapid population growth, deforestation, and land degradation where the balance between people and resources is already precarious. Bangladesh faces the huge challenge of dealing with the environmental consequences of upstream deforestation and repeated flooding assaults from the Bay of Bengal. Sri Lanka must accommodate its Tamil minority. And all of these nations struggle with the chronic problem of poverty, which ultimately is the source of many of the region's most severe environmental and political dilemmas.

CHAPTER SUMMARY

- Most of South Asia is a triangular peninsula that thrusts from the main mass of Asia and splits the northern Indian Ocean into the Bay of Bengal and the Arabian Sea. The entire unit is often called the Indian subcontinent.
- South Asia contains the five countries of India, Bangladesh, Pakistan, Nepal, and Bhutan in an area slightly larger than one-half of the 48 conterminous United States. The island nation of Sri Lanka lies off the southern tip of India and shares many physical and cultural traits with the subcontinent.

- Mountains enclose the subcontinent on its northern borders. There lie the Himalaya Mountains and the Karakoram Range. The countries of Nepal and Bhutan, on the southern flank of the Himalaya, are small, rugged, and remote.
- Until 1947, the entire area, with the exception of the small Himalayan states and Sri Lanka, was referred to as "India." For over a century, it was the most important unit in the British colonial empire — the "jewel in the crown."

- The Indian subcontinent is one of the world's culture hearths. Some of the world's earliest cities developed on the banks of the Indus River before 3000 B.C. The Aryans arrived between 2000 and 3000 B.C. and were the source of many ethnic and cultural attributes of the subcontinent today.
- The subcontinent's northern plain contains the core areas of the three major countries of India, Bangladesh, and Pakistan. Three great rivers water this plain: the Indus River, the Ganges River, and the Brahmaputra River. Although mostly desert and steppe in Pakistan, the plain is relatively well-watered in India where it supports unusually high population densities.
- The southern portion of the subcontinent is mainly a large volcanic plateau called the Deccan. Generally low in elevation, it is rimmed by highlands: the Western Ghats, the northern Central Indian Hills, and the Eastern Ghats.
- Climates vary, with undifferentiated highland in the northern mountains to desert and steppe in Pakistan and western India; humid subtropical occurs in the northern plain; tropical savanna is in the peninsula with the exception of the rain shadow in the lee of the Western Ghats; and there are rain forests in the Western Ghats' seaward slopes, the coastal plain to the south, and part of Bangladesh.
- The southwest or wet monsoon is at its height from June to September, blowing in from the Arabian Sea and Bay of Bengal. The dry or northeast monsoon occurs in the winter, with dry, cooler winds from the northeast coming from high-pressure systems over the continent.
- Although it is perceived as an overwhelmingly poor and rural country, India has many modern industries that have contributed to its substantial rate of economic growth in recent years. These include the software and computer industries of Bangalore and the film industry of Mumbai.
- The subcontinent is home to many different faiths, including Hinduism, Sikhism, Christianity, Islam, Buddhism, Zoroastrianism, and Jainism. Religious, ethnic, and other differences underlie serious and often violent conflicts. The principal ones have been the Hindu versus Muslim in India, Sikhs versus the government of India, and Tamils versus Sinhalese in Sri Lanka.
- The hundreds of languages and dialects spoken in the region fall into the two principal families of Aryan and Dravidian. English is spoken often for business, but only a small percentage of the people, generally the well educated, are literate in English.
- There has been animosity between India and Pakistan since the two countries became independent in 1947. Several wars have been sparked or fought in the still-volatile frontier province of Kashmir, which, despite its Muslim majority, was joined within India in 1947. Both India and Pakistan are nuclear weapons powers.

REVIEW QUESTIONS

1. What four groups of peoples have been most significant in forming the cultural foundation of South Asia?
2. Using maps and the text, locate the outer mountain wall, the northern plain, peninsular India, and Sri Lanka. What are the major climates, resources, and other natural attributes of these areas?
3. Using maps and the text, locate the three major rivers of the northern plain.
4. Using maps and the text, locate the main agricultural production areas of rice, irrigated wheat, sorghum, millet, jute, and tea.
5. What are the main industrial and export crops of South Asia? Where are they concentrated?
6. What are the major nonagricultural industries of South Asia? Where is India's Silicon Valley? Its Hollywood? Its Wall Street? Where are the other major industries located?
7. Identify the areas of high, intermediate, and low population densities and relate these to climate.
8. What is the cause of civil unrest in Sri Lanka?
9. What major internal and international problems does Pakistan face?
10. What are the roots and expressions of India's many sectarian conflicts?

DISCUSSION QUESTIONS

1. Explain why India could be considered one of the most important nations to have gained independence since World War II.
2. Why is the Indian subcontinent considered the most culturally complex area of its size on Earth?
3. Discuss the impact that the outer mountain wall has had on the subcontinent's development.
4. Why are the wet and dry monsoons so significant to South Asia?
5. What legacies were left to South Asia by the British Empire? Which are "good" and which "bad" as seen through local eyes?
6. What disputes center on Kashmir and the Punjab?
7. Discuss how language and religion appear to be divisive in South Asia. In the case of Hinduism and Islam, are religious differences the reason for conflict? In Sri Lanka, is ethnicity the reason Tamils and Sinhalese have been fighting?
8. What are the major contributors to Bangladesh's flood problems?
9. What have been the results of India's attempt at industrial self-sufficiency?
10. What are the major points of contention in the nuclear rivalry between India and Pakistan?
11. What was the political, military, and economic fallout in South Asia following the terrorist attacks on the United States on September 11, 2001?

CHAPTER 12

Southeast Asia: From Slash-and-Burn to Semiconductors

JOE HOBBS

The peoples of Southeast Asia are modernizing rapidly but not sacrificing cherished traditions. These Sarawakian women are attending a jubilee festival in the city of Kuching on the island of Borneo.

Chapter Outline

Southeast Asia, from Burma (Myanmar) in the west to the Philippines in the east, is a fragmented region of peninsulas, islands, and seas. East of India and south of China, between the Bay of Bengal and the South China Sea, the large Indochinese Peninsula projects southward from the continental mass of Asia (Fig. 12.1). From it, the long, narrow Malay Peninsula extends another 900 miles (c. 1,450 km) toward the equator. To the south and east are thousands of islands, the largest of which are Sumatra, Java, Borneo, Celebes (Sulawesi), Mindanao, and Luzon. Another large island, New Guinea, is culturally a part of the Melanesian archipelagoes of the Pacific World, but its western half, held by Indonesia, is politically part of Southeast Asia.

Southeast Asia is one of the world's culture hearths, having contributed domesticated plants (including rice) and animals (including chickens) and the achievements of civilizations to

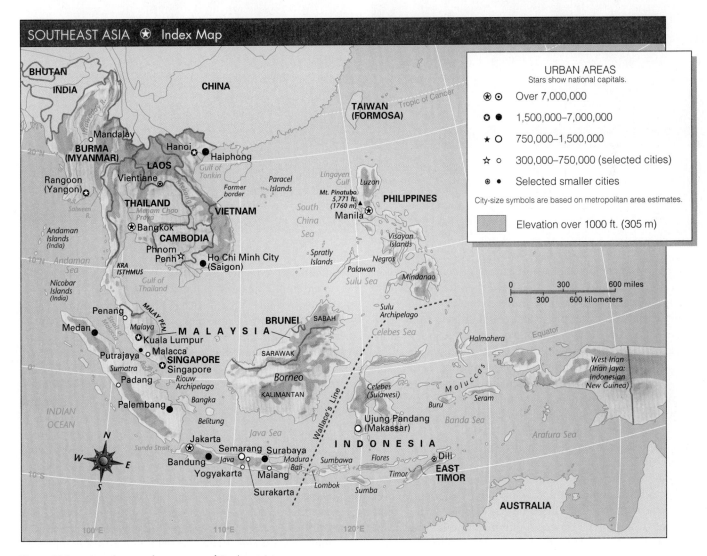

Figure 12.1   Introductory reference map of Southeast Asia.

a

JERRY ALEXANDER/STONE/GETTY IMAGES

JOE HOBBS

b

Figure 12.2   (a) Angkor Wat, the 12th-century capital of the Khmer empire, is a symbol of the rich cultural history of Southeast Asia. (b) Modernity and prosperity characterize some nations in Southeast Asia today. The twin Petronas Towers in Kuala Lumpur are the world's tallest buildings.

a wider world. Innovations and accomplishments diffused from several centers, which themselves were influenced by foreign cultures. Indian and Chinese traits have been especially strong in shaping the region's cultural geography. Hinduism came from India, and Hinayana (Theravada) Buddhism came from India through Ceylon (Sri Lanka). The Chinese cultural influences of Confucianism, Taoism, and Mahayana Buddhism were particularly strong in Vietnam. China periodically demanded and received tribute from states in Burma, Sumatra, and Java. The Islamic way of life came with Arab traders from the west from the 13th through the 15th centuries and took permanent root; Southeast Asia is the eastern margin of the world of Islam, and Indonesia has more Muslims than any other country in the world today.

The historic and contemporary monumental architectures of Southeast Asia reflect these diverse origins. In the ninth century, the advanced Sailendra culture of the Indonesian island of Java built Borobudur, the world's largest Buddhist *stupa* (shrine) and still the largest monument in the Southern Hemisphere. In the 12th century, a Hindu king built the

extraordinary complex of Angkor Wat as the capital of the Khmer empire based in Cambodia (Fig. 12.2a). The first independent Thai kingdom emerged in the 13th century, and in the 14th century, Thai kings ruling from the great city of Ayuthaya (near modern Bangkok) extinguished the Khmer empire and extended their own control into the Malay Peninsula and Burma. The world's largest mosque is in modern Brunei, a tiny oil-rich country near the northern tip of the island of Borneo, and in Malaysia stands the world's tallest buildings (1,462 ft/446 m) — the Petronas Towers in Kuala Lumpur (Fig. 12.2b).

Southeast Asia today is composed politically of 11 countries: Burma (called Myanmar by its rulers, but this name is not widely used), Thailand, Laos, Cambodia, Vietnam, Malaysia, Singapore, Indonesia, East Timor, Brunei, and the Philippines. With the exception of Thailand, which was never a colony, all of these states became independent from colonial powers after 1946. These 11 nations together occupy a land area less than one-half that of China and are fragmented both politically and topographically. Many are also culturally fragmented and have experienced strife between differing ethnic groups. Outside intervention has sometimes complicated and worsened local discord, producing enormous suffering in the region and lasting trauma among the foreign soldiers who fought the determined inhabitants of Southeast Asia.

12.1 Area, Population, and Environment

It is about 4,000 miles (c. 6,400 km) from western Burma to central New Guinea and 2,500 miles (c. 4,000 km) from northern Burma to southern Indonesia (Figs. 12.1 and 12.3). Despite these great distances, the total land area of Southeast Asia is only about 1.7 million square miles (4.4 million sq km), or about half the size of the United States (including Alaska). The estimated population of the region was 518.6 million in mid-2001, giving it an overall density of 308 per square mile (119 per sq km). To put this in perspective, Southeast Asia has nearly four times the overall population density of the United States but well under half that of the Indian subcontinent and China Proper (the humid eastern part of China south of the Great Wall).

Southeast Asia has been less populous than India and China Proper throughout history. Only about 10 million people lived in the region in the year 1800, and there were no widespread dense populations. Environmental obstacles probably kept population growth low in Southeast Asia until colonialism and new technologies boosted food production and lowered death rates. By land, the region is rather isolated, cut off to some extent by the rugged mountains of the northern peninsula of Indochina. Over many centuries, the ancestors of most of the present inhabitants entered the area from the north as recurrent thin trickles of population, crossing this mountain barrier as refugees.

There are formidable environmental challenges in this region. Southeast Asia is tropical, with continuous heat in the lowlands, torrential rains, prolific vegetation difficult to clear and keep cleared, soils that are generally leached and poor, and a high incidence of disease. A 4- to 6-month dry season on the Indochinese Peninsula and on some of the islands has discouraging agricultural effects. The lush forest nutrients are mainly in the vegetation itself, not the soil, so when people clear forest and plant crops, soil quality deteriorates rapidly.

More spectacular environmental difficulties result from the location of part of the region on the Pacific Ring of Fire, subjecting especially Indonesia and the Philippines to earthquakes, the tidal waves (tsunamis) that earthquakes create, and volcanic eruptions. Violent wind-and-rain storms known locally as typhoons (known elsewhere as hurricanes and cyclones) often strike, particularly Vietnam and the Philippines.

Typical of less developed countries (LDCs), the introduction of modern medical and other technologies into the region has softened the blows of environmental adversity and lowered death rates, producing a tremendous population increase in the last century and a half. This increase, which accelerated after World War II, is now slowing but is still rapid. Between 1973 and mid-2001, population grew from an estimated 316 million to 519 million, a 28-year increase of 203 million people, or 64%. During recent times, the race between population growth and food supply has seen food supply winning in some countries but losing to population growth in others. From the 1970s through the 1990s, increases in food production outpaced population growth in Malaysia and Indonesia. At the other end of the scale were Cambodia and Laos, in which war, repression, and inefficient communalization of agriculture slowed progress.

12.2 The Economic Pattern

Southeast Asia as a whole is among the world's poorer regions, yet some of its countries are rather prosperous. As of 2001, the mean GDP PPP per capita of $3,685 of its countries was about average for the world's less developed countries. In contrast, Singapore, Malaysia, the Philippines, Brunei, and Thailand were well above average for LDCs. Because of their high rates of industrial productivity and economic growth, Singapore, Malaysia, and Thailand are described either as "Asian Tigers" or "Tiger Cubs" (along with Hong Kong, South Korea, and Taiwan). The commercial-industrial city-state of Singapore, for example, has about the same GDP PPP per capita as the United Kingdom. In 1997, a series of events precipitated an economic crisis that roared through the region and at least temporarily declawed these emerging economies (see pp. 324–325).

This region's economic activities span the entire continuum from Stone Age hunting and gathering (on Irian Jaya) to slash-and-burn farming—one of humankind's earliest agricultural techniques—to the commercial agriculture typical of colonial times and from the export of raw materials like

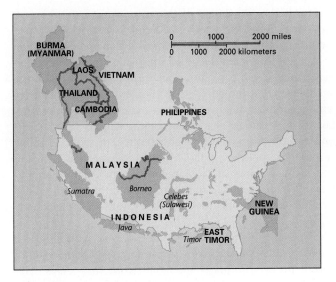

Figure 12.3   Southeast Asia compared in area (but not latitude) with the conterminous United States.

DEFINITIONS + INSIGHTS

Rubber

The earliest major cash crop introduced into Southeast Asia was rubber, first cultivated there in the 1870s. The Englishman Sir Joseph Priestley gave the name "rubber" to the latex of the *Hevea brasiliensis* tree in 1770, when he discovered he could use it to rub errors off the written page. Extensive exploitation of this tree in its native Brazil gave rise to the "rubber boom" of the late 19th century, when Brazilian rubber traders were so wealthy that they used bank notes to light cigars and sent their shirts to be laundered in Europe. In 1876, an Englishman named Henry Wickham smuggled 70,000 rubber seeds out of Brazil. Delighted botanists at London's Kew Gardens cultivated them and shipped them on to Ceylon (Sri Lanka) and Southeast Asia for experimental commercial planting. Rubber trees proved exceptionally well adapted to the climate and soils of South, and later Southeast, Asia (Fig. 12.A). Back in Brazil, however, a fungus known as leaf blight wiped out the rubber plantations, and by 1910, bats and lizards inhabited the mansions of Brazil's rubber barons. Fortunately, the Kew Gardens stock was free of the fungus; by 1940, 90% of the world's natural rubber came from Asian plantations.

Figure 12.A   Rubber latex bleeding down a fresh incision and into the collection cup.

---

rubber to value-added goods like clothing and shoes. As Southeast Asians see it, their economic future will depend not so much on rubber, coffee, or even oil, which some countries have abundant supplies of, but on high-tech exports like semiconductors and software. But despite islands of technological innovation, rapid urbanization, and industrialization, the majority of the people in Southeast Asia are still farmers, with 64% of the people classified as rural.

Shifting and Sedentary Cultivation, Plantation Agriculture, and Commercial Rice Farming

Permanent agriculture has been difficult to establish in many parts of this mountainous tropical realm. Much land is used only for a shifting subsistence form of cultivation, in which a migratory farmer clears and uses fields for a few years and then allows them to revert to secondary growth vegetation while he moves on to clear a new patch. Since an abandoned field needs about 15 years to restore itself to forest that can again be cleared for farming, only a small portion of the land can be cultivated at any one time. This extensive form of land use can support only sparse populations. Unirrigated rice is the principal crop, with corn, beans, and root crops such as yams and cassava also grown. In some forested areas, hunting and gathering supplement agriculture. These migratory subsistence farmers often differ ethnically from adjacent settled populations, having historically sought refuge in the back country from stronger invaders.

Most of Southeast Asia's people live in widely scattered, dense clusters of permanent agricultural settlements. These core regions stand in striking contrast to the relatively empty spaces of adjoining districts. Superior soil fertility is the main locational factor in these areas. However, there are a few districts of dense populations without superior soils to support them. Most of these rely on plantation agriculture of rubber and other less demanding cash crops (see Definitions & Insights above).

The typical inhabitant of the areas of permanent sedentary agriculture is a subsistence farmer whose main crop is wet rice, grown with the aid of natural flooding or irrigation (Fig. 12.4). In some drier areas, unirrigated millet or corn replaces rice. Secondary crops that can be grown on land unsuitable for rice and other grains supplement the major crop. These secondary food crops include coconuts, yams, cassava, beans, and garden vegetables. Some farmers grow a secondary cash crop such as tobacco, coffee, or rubber. Fishing is important both along the coasts and along inland streams and lakes. Because of their importance in the food supply, many fish are harvested from artificial fishponds and flooded rice fields, in which they are raised as a supplementary "crop."

Subsistence agriculture in Southeast Asia exists alongside tropical plantation cash crops grown for export. The region is one of the world's major supply areas for such crops. This highly commercial type of production is a legacy of Western colonialism, which began in the 16th century. Until the 19th

JOE HOBBS

Figure 12.4 These terraced padis in Borneo are ready to be planted in rice; a crop is already growing on the terrace between the two huts.

century, Westerners were interested mainly in the region's location on the route to China. They were content to control patches of land along the coasts and to trade for some goods produced by local people. During the 19th century, however, the Industrial Revolution in the West greatly enlarged the demand for tropical products. Europeans therefore extended their control over almost all of Southeast Asia. European colonists used their capital and knowledge, along with indigenous and imported labor, to increase export production rapidly. They focused on exporting some local produce such as copra (coconut meat) and spices and introduced some new commercial crops.

The usual method of introducing commercial production was to establish large estates or plantations managed by Europeans but worked by indigenous labor or labor imported from other parts of Monsoon Asia. There were often problems finding available local hands, so Europeans imported contract labor from India and China. Large migrations from these countries complicated an already complex ethnic mixture, setting the stage for today's unusual distribution of ethnic groups and their sometimes discordant relations.

Many export cash crops, some produced by small landowners, are produced in Southeast Asia. The major ones are rubber, palm products, and tea. Nearly three-fourths of the world's natural rubber is produced in the region, mainly in Malaysia, Indonesia, and Thailand. However, synthetic rub-

ber, most of which is made from petroleum, now meets the greater part of the world's demand. Palm products include palm oil, coconut oil, and copra (dried coconut meat from which oil is pressed). Malaysia is the world leader in producing palm oil (a popular and inexpensive vegetable oil used in cooking throughout Asia), and Indonesia dominates the world output of coconut palm products. Indonesia is also the region's leading producer of tea and is among the world's top five producers. Other export crops in Southeast Asia include coffee from Indonesia, cane sugar from the Philippines and Indonesia, pineapples (of which the Philippines and Thailand have become the world's leading producers and exporters), and many others. However, the countries of Southeast Asia are generally coming to depend less on agricultural exports and more on mining and manufacturing.

One impact of Western colonialism on the economy of Southeast Asia was the stimulation of commercial rice farming in formerly unproductive areas. Three areas became prominent in commercial rice growing: the deltas of the Irrawaddy, Chao Praya, and Mekong Rivers, located in Burma, Thailand, and Vietnam and Cambodia, respectively. Almost impenetrable wetlands and uncontrolled floods had kept these deltas thinly settled, but incentives and methods for settlement have turned them into densely populated areas within the past century. In the 1990s, Thailand, Burma, and the resurgent Vietnam became the region's leading rice exporters. Other Southeast Asian nations import much of this produce.

Agricultural Growth and Deforestation

Near areas of dense settlement in Southeast Asia, there are still some large, undeveloped forested areas, but they are shrinking as population and development progress. Many environmentalists view the destruction of the region's tropical rain forest as an international ecological problem. In 2000, only 10% of Thailand's original forest cover remained. Neighboring, poorer Burma had a 50% forest cover. Vietnam lost one-third of its forest between 1985 and 2000, by which time only 17% of the original forest cover remained. Deforestation is advancing most rapidly in Malaysia, where 45% of the land area is still forested, and in Indonesia, where 50% is forested. Malaysia is now the world's largest exporter of tropical hardwoods, and at current rates of deforestation, it will have logged all of its principal forest reserves in Sarawak on the island of Borneo by the year 2005 (Fig. 12.5).

Although Southeast Asia's tropical forests are smaller in total area than those of central Africa and the Amazon Basin, they are being destroyed at a much faster rate, most significantly by commercial logging for Japanese markets rather than by subsistence farmers. Environmentalists fear the irretrievable loss of plant and animal species and the potential contribution to global warming that deforestation in this region may cause. Malaysia and Indonesia have moved to slow the destruction, with Indonesia taking the aggressive step of banning the use of fire to clear forests. However, enforcing such legislation has thus far proven impossible. Hundreds of

JOE HOBBS

Figure 12.5   Commercial logging has removed all but small portions of the tropical rain forest in East Malaysia's states of Sarawak (seen here) and Sabah. The same process is underway in neighboring Kalimantan on Indonesia's side of the island of Borneo.

Indonesian and Malaysian companies, most of them large agricultural concerns (e.g., palm oil plantations and pulp and paper companies) with close ties to the government, continue to use fire as a cheap and illegal method of clearing forests. In 1997, a widespread drought attributed to the El Niño warming of Pacific Ocean waters turned the annual July–October burning season into a human-made holocaust. Deliberately set fires raced out of control over large areas of Sumatra and Borneo, resulting in a choking haze that shut down airports, closed schools, deterred tourists, and caused respiratory distress to millions of people in Indonesia, Malaysia, Singapore, the Philippines, Brunei, and Thailand. Less widespread but still severe smog crises struck Malaysia and Indonesia, particularly the island of Sumatra, in subsequent years, and Indonesian authorities fought back by revoking the licenses of plantation companies.

Many of the species that inhabit these forests are endemic (found nowhere else on Earth). Indonesia, which contains 10% of the world's tropical rain forests, is known, after Brazil, as the world's second most important "megadiversity" country, with about 11% of all the world's plant species, 12% of all mammal species, and 17% of all bird species within its borders. The Southeast Asian region is also particularly significant in biogeographic terms because of the so-called

Wallace's Line — named for its discoverer, English naturalist Alfred Russel Wallace — that divides it (see Fig. 12.1). Nowhere else on Earth is there such a striking local change in the composition of plant and animal species in such a small area on either side of this divide. East of the line (e.g., separating the Indonesian islands of Bali and Lombok), marsupials are the predominant mammals, and placental mammals prevail west of the line. Similarly, bird populations are remarkably different on either side of the line.

Mineral Production and Reserves

Petroleum is the most important mineral resource in Southeast Asia. Seven of the region's countries — Indonesia, Malaysia, Brunei, Burma, Thailand, Vietnam, and the Philippines — produce oil. Collectively, their proven oil reserves amount to about 1.1% of the world total. Their combined output is about 4% of world oil production, with Indonesia the leading producer. Indonesia's main oil fields are in Sumatra and Indonesian Borneo (known as Kalimantan).

Although it does not loom large on the world scene, Southeast Asian oil is very important to the countries that own and produce it. Oil and associated natural gas provide nearly all the exports of Brunei and make that tiny Muslim country one of the world's wealthiest nations as measured by GDP PPP per capita (Fig. 12.6). Oil, gas, and refined products supply almost one-fourth of Indonesia's exports and about one-tenth of Malaysia's. Singapore profits as a processor of oil. The main customer for Southeast Asian oil and natural gas is Japan, the leading trade partner of several Southeast Asian countries.

There is other mineral wealth in Southeast Asia. Tin is the most important ore mined, with about one-quarter of world output from Indonesia. The Philippines produces a variety of metals — copper, chromite, nickel, silver, and gold — in significant amounts. The most serious mineral deficiency in the region is the near absence of high-grade coal. However, more energy could come from the region's additional potential for hydroelectric power.

The Meltdown and Its Aftermath

Following years of spectacular growth, in which economies across the region grew at rates up to 8% annually, 1997 saw a stunning reversal of fortunes in Southeast Asia. Up to 1997, the region's economic Tiger Cubs were praised for high rates of savings, balanced budgets, and low inflation. Export-driven growth rates were climbing, particularly with the increased manufacturing of computers and other electronics. Gradually, however, the appetite of the main consumer for the products, the United States, began to wane, and trade deficits developed in the producing countries. At the same time, the emerging economies of Southeast Asia pursued expensive public works projects, such as Malaysia's new capital of Putrajaya.

In some countries, corruption added to the potential for economic crisis as insolvent companies were propped up with infusions of cash from sympathetic politicians (in a phenomenon known as "crony capitalism"). Banks extended risky loans

The countries of Southeast Asia have sought to strengthen their economic and security interests by forming a regional organization. Known as the Association of Southeast Asian Nations (ASEAN), the coalition was formed in 1967 to promote regional trade and minimize alignment with the two world superpowers, the United States and the Soviet Union. Singapore, Malaysia, Thailand, Indonesia, Brunei, and the Philippines were founding members. Vietnam joined in 1995 with the promise of strengthening regional ties and adopting the more open economic system that the Vietnamese call "doi moi," or renovation. Burma and Laos joined in 1997, and finally, Cambodia in 1999.

After U.S. troops withdrew from bases in the Philippines in 1991, ASEAN members grew concerned that they lacked a security "umbrella" and that China or possibly even Japan might become more aggressive in the region. Several ASEAN members, notably Indonesia, Malaysia, and Singapore, therefore strengthened ties with the U.S. military to deter other powers from attempts to dominate the region. The organization thus aspired to become both a large trading bloc and a substantial security counterweight to China. By the end of the century, however, ASEAN members moved closer to China and created an informal economic cooperation alliance with China, Japan, and South Korea known as ASEAN+3. The bloc's main goal is to adopt policies that would avert the type of economic meltdown that shook all the group's members in 1997 and 1998. The countries are also trying to reduce dependence on the International Monetary Fund and other Western monetary agencies and countries in the event of future economic crises.

for both private and public businesses and construction projects. The loans were generally made in U.S. dollars. When local currencies rapidly lost their values, the borrowers were unable to repay these loans.

Economies melted down. Stock markets, which had seen explosive growth, imploded; Malaysia's market, for example, lost 80% of its value between September 1996 and September 1998. Among the region's stronger economies (excluding Vietnam, Cambodia, Laos, and Burma, which had far less to lose), Malaysia, Thailand, and Indonesia fell into economic recession, characterized by a relatively short fall in economic output and a moderate rise in unemployment, and as the crisis dragged on, it threatened to tip into outright depression. Only the Philippines and Singapore escaped the bloodletting.

Economies gradually began to improve, but the economic meltdown had lasting impacts. Largely unchallenged in prosperous times, Southeast Asian ways of running governments and businesses have been under strong scrutiny since the hard times. Indications of this shift run the gamut from large-scale revolts against and/or impeachments of national leaders to parents fighting favoritism in school admissions, farmers protesting dam building projects, peasants protesting industrial pollution, and patients suing doctors for malpractice. Examples of these events are described in the discussions of the individual countries.

Figure 12.6   Brunei is a tiny, prosperous Muslim nation. This mosque of Omar Ali Saifuddin was designed by an Italian architect.

PAUL HARRIS/STONE/GETTY IMAGES

12.3 The Countries

Although the countries of Southeast Asia have many similarities, each has its own distinctive qualities, with unique environmental features, indigenous and immigrant peoples, culture traits, and economic activities. Many of the characteristics and problems of these countries are the outcome of European imperialism, but in no two countries have the results of colonialism been the same. Here are portraits of the countries.

Burma (Myanmar)

Burma (2001 population: 47.8 million) is centered in the basin of the Irrawaddy River and includes surrounding uplands and mountains. Within the basin are two areas of dense population: the Dry Zone, around and south of Mandalay (population: city proper, 1,057,600), and the Irrawaddy Delta, which includes the capital and major seaport of Rangoon (Yangon; population: city proper, 3,938,900; Fig. 12.7). The

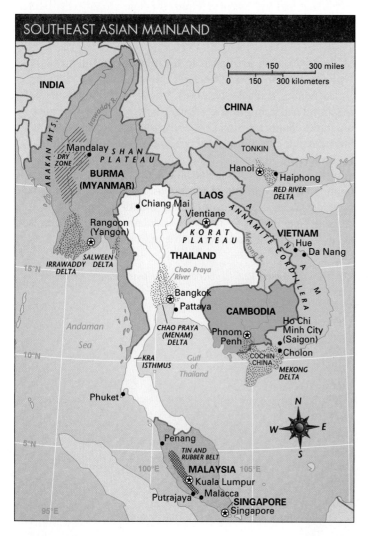

Figure 12.7   Index map of the Southeast Asian mainland. Stars show political capitals and area symbols show river deltas, the Dry Zone of Burma (Myanmar), and the tin and rubber belt of Malaysia.

Dry Zone is the historical nucleus of the country. The annual rainfall of this area—34 inches (87 cm) at Mandalay—is very low for Southeast Asia. The people are supported by mixed subsistence and commercial farming, with millet, rice, and cotton the major crops. During the past century, the Dry Zone was surpassed in population by the Delta, where a commercial rice-farming economy now provides one of the country's two leading exports (forest products is the other). The Irrawaddy forms a major artery of transportation uniting the two core areas.

The indigenous Burman people, most of whom live in the Irrawaddy Basin, number an estimated 68% of the country's population. About 90% of Burma's citizens are Buddhists. Distinct ethnic groups live in the Arakan Mountains of the west and the northern highlands. The Shan Plateau of eastern Burma is inhabited by the ethnic Shans to the north and ethnic Karens to the south and in the delta of the Salween River. They account, respectively, for about 9 and 7% of the country's population. The Shan Plateau in northeastern Burma is part of the Golden Triangle, which was long the world's leading source area for opium, until Afghanistan pushed it to number two in the 1990s (see Regional Perspective on next page).

Great Britain conquered Burma in three wars between 1824 and 1885. It became an independent republic outside the British Commonwealth in 1948. There has been almost constant civil war since independence. Communism, targeted persecution, and ethnic separatism have motivated rebellions by ethnic Karens, the Shans, a number of less numerous hill peoples, and Burmese Muslims who fear the creation of a pure Buddhist state. Beginning around 1999, however, Burma's ruling military government reached cease-fire agreements with most of these ethnic groups (except the Karen) and allowed what it called "contingent sovereignty" to these ethnic groups, offering them more civil rights and economic opportunities than they had known previously. The government also encouraged profits made by some of these groups in the drug business to be "invested" in national development.

The economy has been badly damaged by fighting and for years was mismanaged under a form of rigid state control promoted as "The Burmese Way to Socialism." Burma became the poorest non-communist country in Southeast Asia, a sad contrast with its reputation in early independence years for having the best health care, highest literacy rate, and most efficient civil service in Southeast Asia. In 1962, Burma exported 1 million tons of rice. In 2000, it exported 70,000 tons of rice and ranked second from the bottom (just above Sierra Leone) in global quality of health care. In the 1990s, Burma began to abandon socialism in favor of a free-market economy and enjoyed considerable economic growth, with China its main trading partner and military ally. Most of the fruits of this economic growth, however, lined the pockets of the ruling military.

The military government that seized power in Burma in 1988 has yielded little to popular pressure for democratization. Burma remains one of the world's most repressive places to live.

Southeast Asia has long been one of the world's leading source areas for opium and its highly addictive semisynthetic derivative, heroin. For decades, global concern focused on combating the drug trade originating in this region and on dealing with the problem of heroin addiction in the consuming nations. Now, because of the HIV virus, the other strong Southeast Asian industries of prostitution and tourism have woven drugs into a leviathan problem of global concern.

The region's drug production is centered in the Golden Triangle, the remote region where Laos, Thailand, and Burma meet. None of these countries exercises successful control over its frontier in this area. As in Afghanistan and Pakistan, the absence of a strong central government presence and ideal growing conditions have made the Golden Triangle one of the world's greatest centers of opium poppy cultivation (Fig. 12.B). China White, or Asian-grown heroin, originates here and commands approximately 18% of the global heroin market. About 100 tons of heroin leave the Golden Triangle each year, bound for markets in the United States, Australia, and Western Eu-

rope. International authorities succeed in stopping only an estimated 10% of that flow. In recent years, with new heroin-producing areas cropping up in Mexico and elsewhere, the price of heroin for the consumer has dropped and its quality has risen. In the United States, these trends are noticeable in the growing appeal of heroin among middle and upper income groups, by smoking and other previously rare forms of heroin consumption, and by the increasing numbers of deaths due to heroin overdose.

The U.S. Drug Enforcement Agency and other international antidrugs authorities focus their regional concerns on Burma. Nearly all of the region's refining of heroin from opium is done in remote Burmese laboratories. In northeastern Burma, armed ethnic minority groups specialize in the production of heroin and a potent methamphetamine known as *yabba*. They smuggle raw opium into both China and India, and amphetamines into Thailand, providing large amounts of investment capital to contribute to the country's "development." Laundered drug money has paid for the country's recent construction boom, and the government encourages such investment from the drug sector to make up for shortfalls in legitimate export revenue.

Not all of the associated problems are exported. A flood of cheap and potent heroin hit the Burmese market after the current regime came to power in 1988. Heroin addiction has soared ever since. The primary delivery is by needles shared in tea stalls. The numbers of users continue to grow; in 2000, an estimated 4% of adult men and 2% of adult women were addicts. Accompanying the heroin use is an AIDS epidemic paralleled only in neighboring Thailand and Cambodia, in Africa south of the Sahara, and in the Caribbean, with an estimated 2% of the Burmese adult population infected in 2000. The 2000 infection rate among heroin users in Burma stood at 57%. The epidemic that began among heroin users spread quickly to the sex industry. Year 2000 estimates put the HIV infection rate among prostitutes at 47%, three times the

rate in neighboring Thailand. Many young men with AIDS are shunned by their families and go to monasteries to die. Tragically, their virus spreads among the populations of Buddhist monks (there are now 400,000 in Burma) who share razors to shave their heads. AIDS prevention is virtually unheard of. The government had outlawed condoms until 1993. They are now legal, but along with HIV tests, they are too expensive for most people. There are almost no anti-HIV drugs in Burma.

Burma's prosperous, cosmopolitan neighbor Thailand also has the scourge, with a 2.2% HIV infection rate among adults in 2000. Its liberal social climate has given rise to the unique phenomenon of sex tourism, with package tours catering to an international clientele, particularly Japanese men. Thailand's overall annual income from prostitution is estimated to be between $22 and $27 billion, and prostitutes send as much as $300 million each year to relatives in the villages from which they migrated. However, the spread of HIV infection among intravenous drug users and among prostitutes in the red-light districts of Bangkok and the popular beach resorts of Pattaya and Phuket now threatens both the country's public health and its lucrative tourist trade.

There are striking differences in how this region's countries are dealing with HIV/AIDS. Thailand's government has responded with an aggressive and increasingly successful anti-AIDS public awareness campaign, resulting in a decrease in the number of new HIV infections every year since. Other Southeast Asian countries lag far behind. The government of the Philippines has been unsuccessful in mounting an anti-AIDS campaign because of opposition from the country's powerful and anticontraception Catholic clergy. Church officials have denounced the government's anti-AIDS program as "intrinsically evil" and have set boxes of condoms on fire at antigovernment demonstrations. Similarly, in the conservative and largely Islamic nations of Indonesia and Malaysia, Muslim clerics denounce their governments' anti-AIDS campaigns as efforts to encourage promiscuity.

JOE HOBBS

Figure 12.B Raw opium oozes from slices in a poppy pod. Heroin, which is derived from opium, has a hearth in Southeast Asia's Golden Triangle region.

Access to the Internet is prohibited. Foreign journalists are banned, and Burmese may not allow foreigners into their homes. It is illegal to gather outside in groups of more than five. Giant green billboards across the country read: "Crush all internal and external destructive elements as the common enemy."

In 1995, the government released the opposition leader and Nobel Peace Prize laureate Aung San Suu Kyi, who had been under house arrest since leading her antimilitary political party to overwhelming victory in the 1990 elections (the results of which the military government nullified). After her release, "The Lady," as Ms. Suu Kyi is popularly known, called for dialogue with the military junta and appealed for the parliament that was elected in 1990 to be convened. The government responded by again placing her under house arrest until 2002. She has been successful in soliciting support abroad for boycotts of American and other Western companies doing business with Burma. American corporations including Pepsi and Phillips Petroleum have withdrawn from Burma, and the U.S. government has banned any new investments. There is pressure on the American firm Unocal to suspend its contract to join a French firm in developing Burma's considerable offshore natural gas reserves in the Yadana Field of the Andaman Sea. The two companies have already built a pipeline from the field to a port west of Bangkok, Thailand, where the gas is used to generate electricity. Human rights groups complain that slave labor was used to build the pipeline and that profits from gas sales help prop up Burma's repressive government. The pipeline has to be well-defended in view of the Burmese government's many domestic opponents who might try to destroy it as a means of weakening the regime.

Thailand

Governed today as a democracy in which military influence is strong and in which a traditional monarchy has mainly symbolic functions, Thailand (formerly Siam; population: 62.4 million) was the only Southeast Asian country to preserve its independence throughout the colonial period. This unique status was due to Siam's position as a buffer between British and French colonial spheres.

Thailand is located in the delta of the Chao Praya River, known to Thais as the Menam ("The River"). Annual floods of the river irrigate the rice that is the most important crop in this Southeast Asian land of relative abundance and progress. Thailand is the world's largest exporter of rice.

On the lower Chao Praya is Bangkok (population: metropolitan area, 8,578,400; city proper, 6,416,200), the capital, main port, and only large city. It is a notoriously crowded and smoggy city; air pollution is responsible for as many as 400 deaths yearly. But it is also a city of stunning architectural marvels, particularly Buddhist temples. Bangkok's networks of canals (klongs), thronged with both commercial and residential activities, give it a distinctively amphibious character; it has been dubbed the Venice of the East. Its "floating market" is especially renowned (Fig. 12.8). Bangkok has filled in many of its canals with paved roads and buildings, thereby making the city vulnerable to flooding. Due to the pumping out of groundwater

beneath the city, Bangkok's soils are settling downward, and the entire capital is sinking at a rate of about 2 inches yearly. Thai officials forecast that Bangkok will be below median sea level by 2050. If nearby sea levels rise according to many climate change models, the city would be imperiled.

Areas outside the river's delta are more sparsely populated and include the mountainous territories in the west and north, inhabited mainly by Karens and other ethnic groups, and the dry Korat Plateau to the east, populated by the Thai, related Laotian, and Cambodian peoples. To the south in the Kra Isthmus, a part of the Malay Peninsula, live more than 2 million Malays. Most Malays are Muslims, but an estimated 95% of the people of Thailand are Buddhists whose faith includes elements borrowed from Hinduism and local spiritualism (Figs. 12.9 and 12.10). Spectacular Buddhist temples are a prime attraction for one of Southeast Asia's largest tourist industries, as are ceremonies celebrating Thailand's monarchy and other historical traditions.

The main ethnic minority of Thailand is Chinese (Fig. 12.11). An estimated 8.7 million ethnic Chinese, or 14% of the population, reside in the country. They control much of the country's business, a situation that many Thais resent and that makes the loyalty of the Chinese an important issue.

Figure 12.8 The floating market of Damnern Saduak near Bangkok, Thailand, typifies the amphibious nature of life in Southeast Asia. Water transportation is vitally important in the river deltas and islands where most Southeast Asians live.

GEORGE CHAN/PHOTO RESEARCHERS, INC.

Thailand has enjoyed more internal tranquillity than most of Southeast Asia but has had its troubles. Since the late 1960s, guerrilla insurrections motivated by communism and ethnic separatism have persisted in the northeast and south. In the late 1970s, the country dealt with a mass migration of refugees from adjoining Cambodia. Malay guerrillas seeking independence for Thailand's Muslim south have sometimes burned schools and ambushed government troops in that area.

In spite of conflict and refugee problems, Thailand has had one of the strongest economies in Southeast Asia. From 1993 to 1996, the economy grew a robust 8% annually. The country shifted away from its traditional exports of garments, textiles, cut gems, and seafood in favor of "medium-tech" industries such as computer assembly. Tourism was the leading source of foreign exchange. Then came 1997, when Thailand was the first "domino" to fall in the Asian economic crisis (see p. 324). Thailand's currency, the baht, was "pegged" to the U.S. dollar, meaning that its value fluctuated in a narrow range relative to the dollar. In 1996 and 1997, the currencies of Thailand's main competitors, Japan and China, fell, putting Thailand at an increasingly competitive disadvantage as the U.S. dollar rose in value. As the costs of Thailand's exports grew, its economy weakened and its banks, which had funneled billions of dollars into questionable investments, be-

Figure 12.10   Piya Chat, a Buddhist nun, lives in a cave in the Malay Peninsula. Devout Buddhists having spiritual or family problems seek her counsel; they make appointments with her by cell phone and travel to her refuge for the consultations.

gan to look vulnerable. Currency speculators, who profit by taking advantage of declining currencies, bet that the baht was due for a steep decline. The government exhausted its reserves of foreign currency and raised interest rates in a failed effort to prop up the currency. The baht fell precipitously when its peg to the dollar was removed, and the increase in interest rates put a virtual halt to the country's economic development activities.

Figure 12.9   Most ethnic Malays, like this woman, are Muslim.

Figure 12.11   Ethnic Chinese are a significant and generally prosperous minority in most countries of Southeast Asia. Pam and Zee Yoong, a lawyer and engineer, respectively, in the Malaysia city of Ipoh, are about to have a favorite Chinese fruit, durian.

While the currency crisis and subsequent economic recession spread from Thailand throughout the region, Thailand's emerging consumer society suddenly found itself bereft. Economic growth had attracted tens of thousands of the rural poor to the factories of Bangkok, where they made garments for such major U.S. retailers as Gap, Nike, and London Fog. Now they were laid off and joined the swelling ranks of Thailand's poor. Even before the slump, Thailand had a huge gap between the wealthy and the poor, with more than 50% of the country's wealth in the hands of the richest 10% of its population. By the beginning of the new century, Thailand's economy was mostly back on its feet, but it has yet to resume its precrisis record of strong growth.

Vietnam, Cambodia, and Laos

Vietnam, Cambodia, and Laos are the three states that emerged from the former French colony of Indochina. Vietnam is the largest and most geographically complex, with three main regions. Tonkin, the northern region, consists of the densely populated Red River delta and surrounding, sparsely inhabited mountains; Annam, the central region, includes the wild Annamite Cordillera and small, densely populated pockets of lowland along its seaward edge; and Cochin China, the southern region, lies mainly in the delta of the Mekong River and has a high population density.

Almost nine-tenths of Vietnam's people are ethnic Vietnamese, closely related to the Chinese in many cultural respects, including language and the shared religious elements of Confucianism, Buddhism, and ancestor veneration. A Catholic minority reflects the French occupation, and there are small indigenous religions. Other minorities complicate the region's ethnic makeup. Lao people form a minority in the mountains of northern Vietnam, and Cambodians are a minority in southern Vietnam. A large Chinese minority lives in the major cities of the three countries.

Cambodia and Laos occupy interior areas except for Cambodia's short coast on the Gulf of Thailand. They are very different countries physically, and a particular culture dominates each. Cambodia is mostly plains along and to the west of the lower Mekong River, with mountain fringes to the northeast and southwest. Except for half a million ethnic Vietnamese, almost all the 13.1 million people of Cambodia are ethnic Cambodians, also known as Khmer people. Their language is Khmer, and their predominant religion is Buddhism.

Laos is mountainous, sparsely populated, and landlocked. It has 68 ethnic groups, the largest of which is the ethnic Lao, who are linguistically and culturally related to the Thais and whose dominant religion is Buddhism. Their largest numbers are in the lowlands, especially in and near the capital, Vientiane. Most of the ethnic minorities of Laos live in the mountain regions, including the country's second largest ethnic group, the Hmong.

The Vietnam War

France conquered Indochina between 1858 and 1907. The French first extinguished the Vietnamese empire, which covered approximately the territory of today's Vietnam, and de-

feated the Chinese, whom the Vietnamese called upon for help. Later, the French took Laos (in 1893) and western Cambodia (in 1907) from Thailand. France's administration of Indochina, centered in Hanoi and Saigon, left a big cultural imprint. Its major economic success lay in opening the lower Mekong River area to commercial rice production, converting both Vietnam and Cambodia into important exporters of rice.

Japanese forces overran French Indochina in 1941, initiating five decades of warfare in the area. In World War II, a Communist-led resistance movement was formed to carry on guerrilla warfare against the Japanese. When French forces attempted to reoccupy the area after the end of World War II, these forces fought to expel them. After 8 years of warfare, the Vietnamese, with Chinese materiel support, destroyed a French army in 1954 at Dien Bien Phu in the mountains of Tonkin. France withdrew, ending a century of colonial occupation.

Four states came into existence with France's departure. Laos and Cambodia became independent non-Communist states. North Vietnam, where resistance to France had centered, emerged as an independent Communist state. South Vietnam gained independence as a non-Communist state that received many anti-Communist refugees from the north. Vietnamese Catholics were prominent among those who fled southward. This partition of Vietnam into two states (see Fig. 12.1) was largely the work of United States' power and diplomacy.

South Vietnam was from the outset a client state of the United States, whereas North Vietnam became allied first with Communist China and then with the Soviet Union. Warfare in Vietnam gained momentum. Between 1954 and 1965, an insurrectionist Communist force supported by North Vietnam and known as the Viet Cong achieved increasing successes in South Vietnam in their attempt to reunify the country. A similar situation existed in Laos. Increasing intervention by the United States in South Vietnam escalated from a small scale to a large scale in 1965 and eventually involved the full-scale commitment of half a million U.S. military personnel (Fig. 12.12a). North Vietnam responded by committing regular army forces against American and South Vietnamese troops. The United States never invaded North Vietnam, although American air strikes there were devastating. American bombs, napalm, and defoliants such as Agent Orange caused enormous damage to the natural and agricultural systems of Vietnam, destroying an estimated 5.4 million acres (2.1 million hectares) of forest and farmland in an effort described in military parlance as "denying the countryside to the enemy."

The United States avoided leading an invasion of the North in part because of the risks posed by Chinese and Soviet support of the North. In limiting the theater of ground warfare, however, the United States found itself unable to expel from the South a North Vietnamese army that was determined, skillfully commanded, increasingly better equipped by its allies, accomplished in guerrilla tactics, and willing to bear heavy losses. More than 3 million Vietnamese soldiers and civilians died in the conflict, and about 58,000 U.S. soldiers and support staff perished. In 1973, almost all American forces were withdrawn from the costly war, which was extremely divisive at home. There are still 2,238 Americans listed as missing in action in Indochina.

North Vietnam completed its conquest of South Vietnam in 1975. Saigon was renamed Ho Chi Minh City (though many of its people still call it Saigon), after the original leader of Vietnam's Communist party. Vietnam entered a period of relative internal peace, but there were ongoing conflicts, including Vietnam's 1979 conquest of Cambodia, continued warfare against Cambodian and Laotian resistance groups, and a brief border war with China to resist Chinese aggression in 1979. Within its borders, a reunited Vietnam unleashed a repression strong enough to cause a massive outpouring of refugees, including "the boat people" (see Definitions & Insights). This situation intensified, especially for the country's ethnic Chinese, after China invaded Vietnam briefly in 1979. China and Vietnam gradually came to terms, however, finally normalizing relations in 1991 and solving longstanding disputes such as the demarcation of their territorial waters and corresponding exclusive economic zones in the Gulf of Tonkin.

Vietnam Today

Vietnam (2001 population: 78.7 million; capital, Hanoi, population: metropolitan area, 2,462,600; city proper, 1,349,400) is now restoring its war-torn landscape through large-scale reforestation, agricultural reclamation, and nature conservation programs, aided by international organizations such as the World Wildlife Fund. Animal species new to science have been discovered in the Vu Quang Nature Reserve in the rugged Annamite Cordillera region on the border with Laos. These include two hoofed mammals, the Vu Quang ox and the giant muntjac, a new species of barking deer. Such natural treasures, along with Vietnam's tigers, rhinoceroses, and elephants, are helping to lure international ecotourists who provide much-needed revenue. The protected areas that house these species are clustered mainly in the highlands near the Laotian border.

The picture of Vietnam's economy is generally positive (Fig 12.12b). The poverty rate fell by half in the 1990s. The Communist government halted collectivized farming in 1988, turn-

ing control of land over to small farmers under 20-year lease agreements. The results have been impressive; Vietnam since 1988 has become a major exporter of rice, second in the world only to Thailand. This is a vulnerable commodity, however, especially because production is concentrated along the Mekong River, which is prone to flooding in the May–October rainy season. Upstream China has argued that it should dam the Mekong to the benefit of both nations, but Vietnam is wary of its giant neighbor and has refused to support the plan.

In addition to joining ASEAN, Vietnam in 1995 restored full diplomatic relations with the United States, and U.S. businesses began scrambling for consumers in this large, promising market. After the two countries signed a trade agreement in 2000, Vietnam began exporting shoes, finished clothing, and toys to the United States. The dual personalities of North and South are being perpetuated in the new economic climate, with most investment and infrastructural improvement focused in the South. The North is a more austere economic landscape, and northerners dominate the civil service. Along with a growing gap between the rich and the poor, the differences between North and South could slow the country's overall development.

In an effort to better integrate the nation, Vietnam has begun a costly and ambitious effort to build a new road that will run the entire 1,000-mile (1,600-km) length of the country, along the historic Ho Chi Minh Trail. Environmentalists fear the highway's impacts on wildlife. Anthropologists are worried about its effects on the traditional peoples inhabiting remote areas along the route. And many in Vietnam's growing private business sector lament the road as a costly white elephant; the country already has a north-south highway and railway, they complain, and resources would be better spent on "new economy" infrastructure such as the Internet.

Like that of many LDCs, Vietnam's progress is threatened by the specter of uncontrolled population growth. Since the war with the United States ended in 1975, Vietnam's population has increased by more than 60%. It will probably reach 168 million by 2025, a staggering figure in view of the country's limited size and resource base. The government in Hanoi

a

b

Figure 12.12 (a) U.S. troops in combat in a Vietnamese village in 1972. The war inflicted enormous suffering on all sides. (b) Vietnam opened its doors to capitalism and Western goods in the mid-1990s.

The desperation of many people in Vietnam and its neighboring war-torn countries created the crisis of the **boat people** — refugees who fled onto the open ocean on large rafts, with chances of survival variously estimated at 40 to 70%. After 1974, almost 2 million Vietnamese fled Vietnam.

The United States provided permanent refuge for more than half of those who resettled. France, Canada, and Australia also received sizable numbers. Large numbers of Vietnamese and Laotian refugees occupied camps in China near its borders with those countries. Boat people also filled refugee camps in Hong Kong, Malaysia, Indonesia, Thailand, and the Philippines.

The United States and other Western governments long ago reduced the size of the welcome mat for Southeast Asian refugees, but Vietnamese continue to immigrate to the United States at a rate of about 26,000 yearly (Vietnam is in the top six of immigrant populations into the United States, behind Mexico, China, India, the Philippines, and the Dominican Republic). Most of these immigrants are typical of an "expanding pyramid," in which family members in the United States bring extended family members over from Vietnam. About 2,000 are classified as political refugees

who survived Vietnam's postwar "education camps," many of whom are Amerasians, the offspring of American soldiers who served in the Vietnam War. Vietnam is also among the leading foreign sources of children adopted by American couples. Altogether, the United States is now home to about 2 million ethnic Vietnamese. Approximately 100,000 visit Vietnam each year, bringing much welcome hard currency with them into Vietnam.

In 1996, Vietnamese refugee camps around Asia closed and about 100,000 Vietnamese were compelled to return to their homeland. Just as those countries were closing their doors to Vietnam, Vietnam was undertaking political and economic reforms to provide incentives for the expatriates to return home and for resident Vietnamese to stay home. Since 1990, almost a million overseas Vietnamese, mostly from the United States and France, have returned to Vietnam to stay. Their repatriation has also pumped economic capital into Vietnam's fledgling free-market economy. Many Vietnamese who once aspired to leave their home country have heard accounts of the difficulties refugees have faced in adjusting to their adopted countries and have chosen to remain in the hopeful nation of Vietnam.

has mounted a family planning campaign in an effort to reduce birth rates, but so far, most Vietnamese have ignored it and continue to prefer large families.

Cambodia's Killing Fields

Cambodia for a time escaped the kinds of ills known to Vietnam but then suffered even worse catastrophes. After 5 years of fighting in the country, a Communist insurgency overcame the American-backed military government in 1975. Known as the Khmer Rouge and led by a man with the nom de guerre of Pol Pot, this Communist organization ruled the country for 4 years with exceptional savagery. In 1973, the population had been estimated at 7.5 million, but by 1979, it was estimated at only about 5 million; a third of the population had been murdered. Some intended victims managed to escape as refugees from Cambodia's "killing fields" to adjoining Thailand. The Cambodian Communists' stated policy was to build a "new kind of socialism" by eradicating the educated and the rich, emptying the cities, and breaking up the family. The population of Phnom Penh, the capital and largest city, was expelled on a death march to the countryside. The city's population dropped from more than 1 million in 1975 to an estimated 40,000–100,000 in 1976. By 2001, it had recovered to an estimated 1,098,900 (city proper).

In 1979, the Communist/Nationalist imperialism of Vietnam came to dominate Cambodia's political landscape. There had been escalating border conflicts between the two countries, and in 1979, Vietnamese troops quickly conquered

most of Cambodia and installed a puppet Communist government. Then the Khmer Rouge, supported by China (Vietnam had by then become a close ally of China's rival, the Soviet Union), began guerrilla resistance against the Vietnamese. Two non-Communist Cambodian guerrilla forces also took the field. These three resistance forces often quarreled among themselves, but all maintained the conflict with the Cambodian/Vietnamese army of the Phnom Penh government, which periodically pursued them into Thailand and engaged Thai border military units. At first, the Vietnamese rejected international food aid for Cambodia and followed a policy of systematically starving the country, apparently to weaken resistance. Later, they attempted to promote recovery under their puppet government.

Peace negotiations among Cambodia's contending factions led, in 1991, to an agreement resulting in the restoration of democratic government under United Nations supervision. A new constitution with new leaders took effect in 1993, and Cambodia's former King Norodom Sihanouk returned to the throne. In 1997, Hun Sen, a former Khmer Rouge leader, ousted the popularly elected royalist leader Prince Norodom Ranariddh in a violent coup. Under much international pressure, Hun Sen permitted elections to be held in 1998. The vote, which most international observers judged to be fair, left Hun Sen's political party with a majority of the parliament's seats but not the two-thirds needed to form a government. Even while Hun Sen's political opponents took to the streets to declare the elections unfair, Hun Sen began an effort to form a coalition government that would include members of the opposition. Finally succeeding in forming a government legitimately and peacefully, he secured Cambodia's membership in ASEAN and regained Cambodia's seat at the United Nations.

Even in this relatively peaceful time, the people of Cambodia must deal with a persistent scourge of war: land mines. Sown by hostile forces during years of warfare, the antipersonnel explosives remain active indefinitely. Mines have maimed more than 35,000 Cambodians, and 200 to 300 more casualties occur each month. People sow new mines, imported mainly from China and Singapore, to protect their property in Cambodia.

The Lao Way

Laotians also suffer from ordnance dating to the Vietnam War. Between 1964 and 1973, U.S. warplanes dropped more than 2 million tons of bombs — more than the United States dropped on Germany in World War II — on the Laotian frontier with Vietnam in an effort to disrupt Communist supply lines on the nearby Ho Chi Minh Trail and to prevent Communist troops from entering Laotian cities. This Plain of Jars region retains the distinction of being the most heavily bombed place on Earth. An estimated 30% of the bombs failed to detonate, and today, millions of unexploded American cluster bombs remain. Smaller than a tennis ball, the cluster bomb contains more than 100 steel ball bearings. Many curious Laotian children who play with them are killed or dismembered. About 200 people are killed or maimed yearly by these explosives. Of Laos' 17 provinces, 15 contain unexploded ordnance (UXO), according to United Nations studies, creating a serious deterrent to farming in a country with inadequate food supplies. Efforts are underway to clear the explosives, with only minimal economic assistance from the United States. The agency in charge of the cleanup hopes by the year 2005 to have cleared enough land to produce 10,000 tons of rice yearly, enough to feed 50,000 people in a year.

The administrative capital of Laos is Vientiane (with only 632,200 people), located on the Mekong River where it borders Thailand. Vientiane replaced the royal capital of Luang Prabang ("Royal Holy Image"), located upriver on the Mekong when, in 1975, Laos came under Vietnamese occupation. That year, the Pathet Lao organization overthrew the Laotian monarchy and established a new People's Democratic Republic. The Communist dictatorship now rules Laos in a benign fashion, encouraging a market economy and foreign investment. The government calls its conservative approach to economic reform and social change "the Lao Way." To slow the potential flood of foreign worldviews and materials into the country, Laos has built only one bridge to Thailand across the Mekong River and has resisted appeals for the construction of a second bridge. However, there is an already important and growing gambling casino industry in Laos, patronized mainly by visitors from neighboring Thailand.

Despite gambling revenues, landlocked Laos is now one of the world's least developed countries, ranking 131 of the 162 countries on the United Nations Human Development Index.

About 80% of its people are peasant farmers. Exports are limited to timber, hydroelectric power (to Thailand), and garments and motorcycles assembled by cheap Laotian labor using imported materials. Without significant industries, trade, or a stock exchange, Laos was already too poor to have felt the severe impact of the economic crisis that shook Asia beginning in 1997. However, the crisis in neighboring Thailand did prevent Laos from embarking on projects to build more than 20 new dams that would have produced electricity for sale to Thailand.

Like those of Vietnam and Cambodia, the Laotian economy has also suffered from the brain drain prompted by years of warfare, political turmoil, and underdevelopment — conditions that caused the best educated and most talented of the country's inhabitants to flee. After the 1975 takeover, 343,000 Laotians, mostly members of the Hmong tribe who had U.S. support to fight Laotian government troops during the war, fled to neighboring Thailand. Many Hmong refugees subsequently immigrated to the United States, where 250,000 ethnic Hmongs now live. A low-level Hmong insurgency continues against the Vientiane government.

Given the conditions of warfare, political turmoil, oppression, and inefficient economic systems, it is not surprising that the three Indochinese countries are, with Burma, the poorest in Southeast Asia. However, their outlook seems to be improving as tensions with the outer world relax, privatization of Communist economies increases, the supply of goods and services improves, and the number of returning refugees exceeds emigrants.

Malaysia and Singapore

The small island of Singapore (Fig. 12.13) is just off the southern tip of the mountainous Malay Peninsula. It lies at the eastern end of the Strait of Malacca, the major passageway for sea traffic between the Indian Ocean and the South China Sea. For centuries, European sea powers competed to control this passageway. Britain gained control in 1824 and used the newly established settlement of Singapore to monitor the strait. With its central position among the islands and peninsulas of Southeast Asia, British Singapore developed into a large naval base and the region's major entrepôt. Singapore is the world's largest port in tonnage of goods shipped.

From Singapore, the British extended their political hold over the adjacent southern end of the Malay Peninsula. This expansion gave Britain control over the part of the peninsula that is now included, along with northwestern Borneo (except Brunei), in the independent country of Malaysia. This southern end of the peninsula is known as Malaya, Peninsular Malaysia, or West Malaysia. The Borneo section includes the states of Sarawak and Sabah, together known as East Malaysia.

During the British period, Malaya developed a commercialized and prosperous economy. Tin and rubber were the major products, with oil palms and coconuts of secondary importance. Chinese and British companies pioneered the tin-mining industry in the late 19th century. The building of railroads gave access to a line of tin mines and rubber plantations (see Fig. 12.7) in the foothills between Malacca and the hinterland of Penang (George Town). Along this line, the inland city of Kuala Lumpur (population: metropolitan area, 3,889,600; city proper, 1,419,400) became Malaya's leading commercial center and would become Malaysia's capital (see Fig. 12.2b).

Although Kuala Lumpur will remain Malaysia's official capital, all government administration is to be transferred to Putrajaya, a city under construction about 25 miles (40 km) to the south (Fig. 12.14). It is a high-tech capital stitched together by fiberoptic cable. Prime Minister Mahathir Mohammed, who relocated his office there in 1999, boasts that the Internet and other forms of digital communication will replace paper in the country's new administrative center. Nearby is Cyberjaya, a new city that, as its name suggests, hopes to lure technology companies and become Asia's Silicon Valley. These cities are symbols of how Malaysia sees itself, a country leaving behind its colonial legacy as a "rubber republic" to become an engine of technical innovation and export. Putrajaya's $5 billion construction cost has drawn much criticism in the wake of Malaysia's financial crisis, however.

The development of a colonial economy gave Malaya an ethnically mixed population. The native Muslim Malays played only a minor role in this development. Chinese and Indian immigrants and their descendants became the principal farmers, wage laborers, and businesspeople of the tin and rubber belt. So heavy was the immigration that 27% of Malaysia's population of 23 million is Chinese and 8% is Indian. The growth of Singapore brought even heavier immigration so that now over three-fourths of Singapore's present population of 4.1 million is Chinese.

Ethnic antagonisms between Malays and Chinese shaped the political geography of Malaysia and Singapore. Malaya accepted independence in 1957 only on the condition that Singapore not be included. This was to ensure that ethnic Malays would be a majority in the new state. In 1963, however, the Federation of Malaysia was formed, composed of Malaya, Singapore, and the former British possessions of Sarawak and Sabah in sparsely populated northern Borneo. The non-Chinese majorities in Borneo were counted on to counterbalance the admission of Singapore's Chinese. This experiment in union lasted only until 1965, when Singapore was expelled from the federation and left to go its own way as an independent state.

Within Malaysia, the government passed legislation to ensure that at least 60% of the university population was ethnic Malay, with similar measures to ensure business opportunities went primarily to Malays. Although ethnic Chinese and Indians still complain that such policies continue to shut doors of opportunity to them, the economic gap between these minorities and the majority Malays has narrowed considerably since the 1970s. But tensions between Muslims and non-Muslims in this country bear monitoring. Among Malaysia's majority Muslims, there are growing complaints that the government is too secular, though it is nominally Islamic. The leading champion of the adoption of Islamic law (*shari'a*) as the law of the land is the Pan-Malaysian Islamic party (PAN), which administers the state governments of Kelantan and Terengganu in Peninsular Malaysia.

Malaysia and Singapore are among the more economically successful of the world's formerly colonial states that received independence after World War II. Singapore is an outstanding center of manufacturing, finance, and trade along the Pacific Rim. Its industries include oil refining, machine building, and many others. The tourists it attracts each year far outnumber its resident population. Singapore suffered less than any other Southeast Asian nation during the region's economic crisis of the late 1990s. This was due in part to a geographic advantage: its ability to conduct trade with a vast regional hinterland belonging to other countries without having to attend to the struggles of a large and poor rural population of its own. Singapore's prosperity is accompanied by a strict sense of propriety; for example, chewing gum is banned and graffiti artists are flogged.

Figure 12.13   Singapore, located at a crossroads where the Pacific and Indian Oceans meet, exhibits new skyscrapers of a rising business center, older buildings once occupied by the former British colonial administration (foreground), and a spacious harbor (background).

Figure 12.14   The parliament building and national mosque in Putrajaya, Malaysia's brand new, controversial administrative capital.

PROBLEM LANDSCAPE

The Spratly Islands: Hot Zone of the South China Sea

About 60 islands make up the Spratly Island chain, which lies in the South China Sea between Vietnam and the Philippines (see Fig. 12.1). They are an idyllic tourist destination, where divers can hire luxury boats to explore the coral reefs and palm-lined beaches of remote atolls. However, the islands are much more significant for their strategic location between the Pacific and Indian Oceans. During World War II, Japan used the islands as a base for attacking the Philippines and Southeast Asia. Still more significantly, as much as $1 trillion in oil and gas may lie beneath the seabed around the Spratlys.

Not surprisingly, many nations covet control of the Spratlys. Six nations claim some or all of the islands: China, Vietnam, Taiwan, Malaysia, Brunei, and the Philippines. During the Cold War, the competing claimants felt it was too hazardous to push their claims on the islands. As the Cold War drew to a close, however, the situation became more volatile. All the contenders except Brunei placed soldiers, airstrips, and ships on the islands. In 1988, the Chinese Navy invaded seven of the islands occupied by Vietnam, killing about 70 Vietnamese soldiers. In 1992, China again landed troops in the islands and began exploring for oil in a section of the seabed claimed by Vietnam. In 1995, China moved to expand its territorial claims on islands already claimed by the Philippines and, since then, has built what it calls "shelters," but which look like fortifications to Filipinos. Late in the 1990s, Malaysia occupied two disputed Spratly islands and began building on one of them. Indonesia, which has no claims on the Spratlys, is sponsoring unofficial workshops on joint efforts in oil exploration among the six claimants in an effort to defuse the emerging crisis. But so far, China, which claims not only the Spratlys but also all the South China Sea as its own, has refused to discuss the issue (China and Vietnam have disputing claims to the Paracel Islands, which China took from Vietnam by force in the early 1970s). There are fears that as the countries' petroleum needs grow, they will seek more aggressively to gain control of the Spratlys. An incident in these remote islands could trigger a much wider and more serious conflict in Asia.

Malaysia still exports rubber and tin but has a diversified export portfolio including crude oil (from northern Borneo), timber, palm oil, automobiles, and electronic components. Malaysia is the world's largest exporter of semiconductors and refrigerators. Its annual GDP PPP per capita is only about one-third that of Singapore's but, within Southeast Asia, falls only behind Singapore and Brunei and is about triple that of the average figure for the world's LDCs (see Table 10.1).

Malaysia has a national mantra of "Wawasan 2020" (Vision 2020) referring to the year by which national economic planners intend Malaysia to be a fully developed country. The slogan has been less forceful since 1997, however, when, following a decade in which the economy grew at a robust 8% annually, Malaysia began falling into recession. The country quickly froze its ambitious and expensive development projects and adopted economic austerity measures, including even an appeal to individuals to cut down on the number of sugar lumps added to their tea. Like the other "emerging markets," Malaysia had been benefiting from huge and often speculative foreign capital investments. And like the others, Malaysia suffered in 1997 and 1998 when worried foreign investors rapidly withdrew their capital.

In 1998, Malaysia effectively pulled out of the orbit of globalization and took drastic actions to protect itself from the whims of foreign investment. It required foreign investors to wait a year before repatriating their funds. Rejecting the notion that free markets would continue to benefit his country, Malaysia's prime minister imposed strict currency controls, pegging the value of the Malaysian ringgit to that of the dollar. This made Malaysian money worthless overseas but ensured that Malaysian money abroad would return home. The fixed rate of the ringgit made it easier for businesses to plan for the future because they knew the value of the goods they exported and imported would not be subjected to wide fluctuations. Malaysian politicians blamed the International Monetary Fund and other institutions in the West for Asia's financial troubles, and in turn these institutions condemned Malaysia for retreating from the principles of the global free market. Despite the ominous downturn, by the end of the decade Malaysia was once again registering economic growth — particularly as exports of computer chips and other electronics resumed — and was pulled out of the abyss in which neighboring Indonesia was still mired.

Indonesia and East Timor

Indonesia is by far the largest and most populous country of the region (2001 population: 206.1 million) and is the fourth most populous country on Earth (after China, India, and the United States). Stretching across 3,000 miles (4,800 km), its 13,600 islands comprise an area about three times the size of Texas, representing over two-fifths of Southeast Asia's land area and about two-fifths of its population. The large population of Indonesia results from an enormous concentration of people on the island of Java. About 122 million people lived there in 2001, representing nearly 60% of Indonesia's population and about 24% of that of all Southeast Asia. The island's population density is about 2,388 per square mile (922 per sq km). In contrast, the remainder of Indonesia, which is about 13 times larger than Java in land area, has an average density of only about 292 per square mile (c. 113 per sq km).

This extraordinary concentration of population on one island owes partly to the superior fertility of Java's volcanic soils (Fig. 12.15). Other Indonesian islands (with the notable exception of Borneo) have areas of fertile volcanic soil but lack the extremely high population densities found on Java. Their unfriendly coastlines, with their coral reefs, cliffs, and extensive wetlands, have handicapped the economic development of some islands.

Java's extreme population density also has cultural and historical roots in the concentration of Dutch colonial activities there. The Dutch East India Company, after lengthy hostilities with Portuguese and English rivals and with indigenous states, secured control of most of Java in the 18th century. Large sections of the other islands, however, were not brought under colonial control until the 19th and early 20th centuries. The Netherlands exploited the natural wealth of Java with the introduction of the **Culture System** in 1830. Under this system, Dutch colonizers forced Javanese farmers to contribute land and labor for the production of export crops. From the Dutch point of view, this harsh system was successful. The Culture System was abolished in 1870, but commercial agriculture continued to expand, and its success supported increasing numbers of people. The introduction of the Culture System thus coincided with an enormous increase in Java's population, which is more than 20 times larger than it was a century and a half ago.

Java's historical leadership and the island of Sumatra's current supremacy in export production are seen in the distribution of Indonesia's larger cities. Of 17 cities having estimated metropolitan populations of more than 500,000, 9 are on Java. These include four "million cities," headed by Jakarta, the country's capital, main seaport, and the largest metropolis in Southeast Asia (population: metropolitan area, 15,149,800; city proper, 10,226,200). Java's second largest city is Bandung. Of the remaining Indonesian cities with more than half a million people, four are in Sumatra.

The Indonesian state has had a turbulent career (see p. 338). Increasing nationalism during the Japanese occupation of World War II led to a bitter struggle for independence from the Netherlands following the war. This struggle finally succeeded in 1949. Another confrontation with the Netherlands in 1962 brought western New Guinea (West Irian) out of Dutch control and into the Indonesian state. A United Nations plebiscite in 1965 ratified Indonesian control, and relatively friendly relations have been reestablished between Indonesia and the Netherlands.

In 1965 (known as The Year of Living Dangerously), an attempted Communist coup against the longstanding regime of Indonesian President Sukarno was unsuccessful and resulted in the massacre of about 300,000 Communists and their supporters by the Indonesian army and by Islamic and nationalist groups. For more than three decades, this government's president was an army general named Suharto, whose regime was regularly criticized for human rights abuses.

Figure 12.15   Molten lava illuminates the morning sky as it flows down the side of Mt. Merapi in central Java on February 19, 2001. Some have linked the mountain's growing rumblings to the rising political tension in the country. Mbah Maridjan, a sprightly 73-year-old Indonesian who was made the volcano's "gatekeeper" nearly 20 years ago by the Sultan of Yogyakarta—head of the royal family in Indonesia's nearby ancient capital—said mystics had been climbing the mountain to pray that it did not explode.

Suharto ruled Indonesia undemocratically and often repressively but argued that he was a benevolent figure acting on behalf of his people. Indonesians generally tolerated his authority because the country's economy was advancing rapidly. The economic boom of the 1980s and early 1990s centered in the oil industry (in the hands of the state firm Pertamina) and in food production. Oil fields located mainly in Sumatra and Kalimantan produce 6% of the world's oil, and their production has been expanding. Crude oil and liquefied natural gas account for about 25% of Indonesia's exports and have tied its economy very closely to that of Japan, which is the main market for these products. Indonesia is a member of OPEC, the Organization of Petroleum Exporting Countries. In addition to oil and gas, clothing, wood products, various tropical agricultural products, and increasingly, electronics are important contributions to the export trade. Tourism has also brought welcome foreign exchange earnings to Indonesia. The country's principal attraction by far is the legendary island of Bali, home of Indonesia's only remnant of the India-born Hindu culture that permeated the islands from the 4th to 16th centuries (Fig. 12.16). Bali is world-famous for its unique culture and forms of art and dance, which so far have endured and become even richer despite a massive influx of international tourism to this legendary island.

Indonesia's economy collapsed under the force of the spreading Asian economic crisis of the late 1990s. The rupiah, Indonesia's currency, lost 80% of its value in less than a year. The Suharto government removed expensive subsidies on gasoline and kerosene. Resulting protests quickly grew into vast riots. The economic troubles were compounded by a prolonged drought attributed to El Niño's warming of Pacific waters, which caused significant declines in the country's rice harvest. Skyrocketing prices for rice and widespread rice shortages helped fuel widening popular discontent, which focused on two targets. First were the ethnic Chinese who represent 3% of Indonesia's population but control half the economy, and second was President Suharto. Rioters looted and destroyed hundreds of Chinese-owned shops in Indonesia's major cities, forcing thousands of ethnic Chinese to flee the country. Since the Chinese had dominated the food production business in Indonesia, food shortages intensified in the wake of the disturbances. Protestors and rioters also called increasingly for the resignation of President Suharto, on whom they also blamed their economic troubles. Bowing to the relentless pressure, and averting a potential bloodbath, Suharto resigned from office in May 1998, to be succeeded by his vice president, B. J. Habibie, who in turn stepped down in 1999 following large protests.

Indonesia's first truly democratic elections in 40 years subsequently brought to power President Abdurrahman Wahid, a widely respected, nearly blind Islamic scholar. His "honeymoon" with the public was short-lived, as by 2000 an official investigation revealed that he had embezzled millions of dollars of government funds. Wahid denied the allegations and insisted he would finish his term of office in 2004. He did not. Against a backdrop of massive anti-Wahid protests across Indonesia, the president was impeached by Indonesia's Supreme Court and fled the country in 2001. He was replaced in office by Megawati Sukarnoputri, daughter of former President Sukarno.

The scale and pace of Indonesia's recent economic decline have been dizzying. Millions of people slipped below the poverty line, and unemployment surged. The turmoil created a new generation of boat people, economic refugees who fled Indonesia for what they perceived as better lands, notably Malaysia. When Malaysia boomed, it had welcomed Indonesians to do much of the country's work, but with its own economic downturn, Malaysia tried to turn the unwelcome masses back.

Rapid population growth in Indonesia compounded the effects of the economic crisis. Indonesia's population grew by 77 million, or more than 60%, in the 26 years from 1975 to 2001. The government lacks an aggressive family planning program and instead is exporting its "surplus" population by promoting emigration from Java and other densely populated islands to some of the sparsely inhabited outer islands of the vast Indonesian archipelago.

The Philippines

The Philippine archipelago includes over 7,000 generally mountainous islands. The two largest islands, Luzon and Mindanao, make up two-thirds of the total area.

The Philippines were a Spanish colonial possession governed from Mexico from the late 16th century until 1898, when the United States achieved victory in the Spanish-American War and gained control of the Philippines. With the exception of the Japanese occupation of 1942 to 1944, the Philippines were controlled by the United States from 1898 until 1946, when independence was granted.

Spanish missionary activity in the Philippines created the only Christian nation in Asia. The country's Spanish legacy is still important. Ever since its founding in 1571, the Spanish-oriented capital of Manila (population: metropolitan area, 13,220,300; city proper, 10,032,900) has been the major metropolis and only large city of the islands. The society created by Spain was composed of a small upper class of Hispanicized Filipino landowners and a great mass of landless peasants. Problems created by this uneven distribution of agricultural land have persisted in the Philippines. Discontented peasants supported a Communist-led revolt after World War II. It was eventually suppressed, in large part through granting land (generally on sparsely populated Mindanao) to surrendered rebels. Revolts flared up again in the 1960s, however, and have continued on a reduced scale ever since.

American economic, military, and other ties with the Philippines continued after independence. Preferential treatment in the U.S. market stimulated the growth of major Philippine export industries: coconut products, sugarcane, and abaca (a fiber used in making rope and fabrics). Education in English grew so that even now English is spoken by about 45% of the people and serves as a bridge between many of the diverse linguistic groups of the population. After independence, Tagalog, one of the most widely used indigenous languages, became the principal base for the official national language of Pilipino, spoken by 29% of the populace. English also remains an official language. Both Tagalog and English are second languages for most Spanish speakers.

The Philippine economy has expanded considerably since independence. Until the 1970s, the most striking growth was in agricultural exports such as coconut products, sugar, bananas, and pineapples and in copper and other metal exports. More recently, there has been growth in manufacturing and food production. Labor-intensive manufacturing, such as that of electronic devices and clothing, now accounts for about 80% of all exports. Japanese and American capital, together with American agricultural science and technology, have

Figure 12.16 The cultural arts of Bali's Hindu people are world-renowned, making this Indonesian island a popular tourist destination.

GEORGE HUNTER/STONE/GETTY IMAGES

been very important in Philippine economic development. The United States and Japan buy more than half of all Philippine exports.

Despite encouraging progress in recent decades, the Philippine republic remains poor and potentially explosive politically and has not joined the ranks of Asian Tigers or Tiger Cubs. The country has taken a conservative approach to its development and has borrowed money carefully to invest in modest projects, avoiding the aircraft industry, the building of a new national capital, and other such showy and expensive enterprises undertaken in other Southeast Asian nations in the 1990s. Since the mid-1970s, the Philippines has also adhered closely to the terms of loans and financial reforms imposed by the International Monetary Fund. The country did not boom, but neither did it bust in the economic crisis that rolled across Asia in 1997 and 1998. During the regional economic recovery that followed, however, the Philippines' growth continued to lag. An estimated 40% of the population lived below the poverty line, and unemployment stood at 10% in 2001.

Economic expansion in the Philippines has been hard-pressed to stay ahead of the more than 70% growth in population between 1975 and 2001. The predominant Roman Catholic culture (Catholics make up 83% of the population, with 10% adhering to other Christian denominations) encourages large families, and the Philippines has a high annual population growth rate of 2.2% (birth rates have been soaring, but food production has been increasing at only about 1% annually since 1990). Church officials have labeled the government's current promotion of family planning as "demographic imperialism" masterminded by the United States.

Compounding the problem, land and wealth continue to be very unevenly distributed, as Philippine society continues to be dominated by only a few hundred wealthy families. The gap between haves and have-nots has inspired rebellion. On the island of Mindanao, rebels of the Moro Islamic Liberation Front (MILF) and a splinter organization, the Abu Sayyaf, have long fought the government in an effort to establish a separate Islamic state. In 2001, the MILF and the government reached a cease-fire, but Abu Sayyaf accelerated its campaign, employing terrorist tactics such as kidnapping and beheading civilians. In its declared worldwide war on terrorism, the United States in 2002 began providing military advisors and weaponry to Philippine troops fighting Abu Sayyaf.

This unrest has economic, political, and religious roots. Despite a wealth of oil, natural gas, fisheries, and forests, Mindanao is the Philippines' poorest region, and the mainly Christian government invests little development aid in this mainly Muslim area. Hundreds of thousands of Mindanao's Muslims have fled as illegal immigrants to more prosperous Sabah, in the Malaysian portion of the island of Borneo.

The Philippines has seen some momentous political changes in recent years. In 1986, the country's dictator, Ferdinand Marcos, was ousted in a popular uprising and replaced by a democratically elected president, Corazón Aquino. Marcos and his wife, Imelda, had looted the country of billions of dollars over a period of 20 years and headed a corrupt and abusive regime virtually ignorant of the needs of the ordinary Filipino. The Marcos path seemed to be followed by Philippine President Joseph Estrada, who allegedly enriched himself in the 1990s through bribes, kickbacks, and embezzlement of public funds. Impeachment proceedings against Estrada crawled along, but he resisted their inevitable conclusion. Only massive street protests succeeded in forcing his resignation in January 2001, when Gloria Macapagal-Arroyo replaced him as president (Fig. 12.17).

The United States has maintained a concerned watch on the country's struggles. During the 1980s, many Filipino legislators resisted the prospect of renewing the leases under which the United States held the two military bases of Subic Bay and Clark Field on Luzon. Then, in 1991, the eruption of volcanic Mt. Pinatubo put an end to the debate over the fate of these two American outposts; volcanic ash forced their closure. The Philippines has 22 active volcanoes. With volcanic eruptions, plus an average of 20 typhoon strikes yearly and regular monsoonal flooding, the country earned the distinction of having more natural disasters in the 20th century than any other. Mt. Pinatubo also ejected enough ash high into Earth's atmosphere to cool temperatures globally over the following several years, at least temporarily confounding efforts to track the trend in global warming. Meanwhile, the U.S. base at Subic Bay was converted into a thriving free-trade port that the Filipinos hope will become Asia's "New Hong Kong."

The Philippines is Southeast Asia in a nutshell: There are humanized landscapes never free of nature's whims, struggles in transforming a mercantile colonial economy to one based on high-tech, global trade, seemingly intractable odds to overcome in unifying diverse peoples under one flag, and wars on poverty that have yet to be fought.

Figure 12.17 Filipinos cheer after Gloria Arroyo was sworn in as the Philippines new president on January 20, 2001, in Manila. Arroyo took over as the new president after disgraced incumbent Joseph Estrada was ousted amid mass street protests.

CHAPTER SUMMARY

- Southeast Asia is composed politically of 11 countries: Burma, Thailand, Laos, Cambodia, Vietnam, Malaysia, Singapore, Indonesia, East Timor, Brunei, and the Philippines. With the exception of Thailand, all of these states were formerly colonies of foreign powers.
- Southeast Asia is one of the world's culture hearths, having contributed several domesticated plants (including rice) and animals and the achievements of civilization to a wider world.
- This region poses numerous environmental challenges for its inhabitants. These include tropical climates, poor soils, disease, long dry seasons, deforestation, heavy erosion, tectonic processes, and monsoon climate patterns.
- The majority of the populace of this region are still farmers, with manufacturing less developed. The land is often used for shifting subsistence cultivation. The densest populations exist where there are better than average soils.
- Plantation cash crops are grown for export and are the legacy of Western colonialism. The climate, cheap transportation, and abundance of land aided in the development of plantation agriculture.
- Southeast Asia's tropical forests are being destroyed at a faster rate

than others of the world, mostly for commercial logging for Japanese markets.
- Southeast Asia as a whole is among the world's poorer regions. The poorest countries — Burma, Cambodia, Laos, and Vietnam — have suffered from warfare and, in the case of Burma, from unwise political management. Peace has finally come to Cambodia, Laos, and Vietnam, and Vietnam is advancing economically. Singapore, Malaysia, and Thailand, because of their high rates of industrial productivity and economic growth, have often been described as Asian Tigers or Tiger Cubs.
- Considerable mineral wealth, including oil, gas, and a wide variety of metal-bearing ores, are extracted in many locations, and many unexploited reserves remain. The most serious deficiency is the near absence of high-grade coal.
- There has recently been much unrest in the region because of economic problems and tensions between ethnic groups. Following the regional economic crisis of 1997–1998, frustrations in Indonesia were directed against the prosperous Chinese minority. Indonesia is in danger of fragmentation. Muslim insurgents in the southern Philippines have stepped up their campaign against the government.

REVIEW QUESTIONS

1. What are the countries of Southeast Asia and what are their physiographic features?
2. How have Indian and Chinese ethnic groups and cultures shaped Southeast Asia?
3. Who are the boat people, why did they leave their homelands, and where are they now?
4. Why has Southeast Asia been historically rather isolated?
5. What are the leading producers and consumers of Southeast Asian tropical hardwoods?
6. What are the members of ASEAN and why was this coalition formed?
7. What are the areas of densest population in Southeast Asia and how do these relate to agriculture?
8. What are Southeast Asia's typical plantation cash crops and where are these crops produced? What are the prominent commercial rice production areas?
9. What are the major mineral resources of Southeast Asia and where are they produced?
10. What is the unique product of the Golden Triangle? What impacts does it have on health and society in Southeast Asia and abroad?
11. What lasting impacts has colonialism had on Southeast Asia? Which country was never colonized?
12. Why have the remote Spratly Islands become the object of contention among several countries?

DISCUSSION QUESTIONS

1. What happened to Brazil's economy when rubber seeds were smuggled out to Southeast Asia? Can you think of other examples of dramatic economic effects from the introduction of exotic commercial plant species?
2. What are the two main regions of Burma? What is the political situation in Burma?
3. What is the ethnic population makeup of Thailand? Why is tourism such a strong economic draw in Thailand? What are Bangkok's main environmental problems?
4. What are the three major regions of Vietnam? Describe its present political atmosphere. How is the economy doing? What is the ethnic population makeup of Vietnam?
5. How are the Southeast Asian countries dealing with problems of drugs and HIV/AIDS?
6. How do the governments of the Southeast Asian countries regard the issue of population growth?
7. How have Vietnamese immigrants been accommodated in the United States?
8. What is the ethnic population makeup of Cambodia? What was the political situation there in the 1970s and what are conditions like there now?
9. What is the ethnic population makeup of Laos? What is the Lao Way? What problems, mostly resulting from warfare, are still being suffered in Laos?
10. Why are Malaysia and Singapore considered more successful than most formerly colonial states? What ethnic considerations shaped the political geographies of these countries?
11. Why did the economic meltdown of the late 1990s occur and what impacts did it have on the respective Southeast Asian countries?
12. What foreign countries have been influential in the Philippines' development?
13. Where and how did the Asian economic crisis begin?
14. Which countries have been most and least affected by Southeast Asia's economic turmoil?
15. Why is the United States concerned about political instability in Indonesia?

chapter 13

China: Forever Ambitious, Forever Significant

KIT SALTER

China has elevated shopping to an entertain-ment in all of the major coastal and some interior cities. This shopping center in Hong Kong has created three floors of shopping variety built around an elaborate centerpiece that is intended to make this commercial experience as much social as it is economic.

So often, images of China are built almost entirely around population issues. While there is real dimension to having the world's largest population, as China does, the real drama of China relates to the ways in which its people have made their landscape accommodate such populations. From the earliest dynasties dating back to the second millennium B.C.— and the monumental construction of China's Great Wall in the third century B.C.— to the present, the story of China has been one of steady manipulation of a not particularly fertile landscape. China is a country that has maintained a distinctly Chinese perspective for nearly 4,000 years, despite major foreign infusion through Mongol, Manchu, and Western control of all or part of the country.

Outsiders have long had a fascination with China (Fig. 13.1). The West has sent missionaries, businesspeople, young scholars hopeful of understanding this giant of the East, and now, countless tourists. There is a grandeur to China that we have seldom failed to appreciate. It has long been on the horizon of interest and opportunity for people of the West. Since the 1970s, it has had a special importance for merchants of the United States.

China currently engages in active industrial and economic development in an effort to pull away from its past as a monumental agricultural nation. At the same time, the need to feed its population and raise fiber for its enormous textile base and market requires it to stay attentive to growth in the rural sector as well. With the reclaiming of Hong Kong in 1997 after a century and a half of British control, and with relations with the United States ever uncertain, China faces a broad array of demands on its social and economic fabric. Not only is China continually contested by the forces of nature, such as the annual rise and fall of the Chang Jiang (Yangtze River) and the typhoons hugging the face of the East Asian coast, but it must deal with the arid China ethnic insurgencies in the west that have played an ongoing role in fomenting instability in this enormous country, especially in recent decades.

The unfolding Chinese drama represents an effort to resolve the tension between enormous ambition and enormous struggle. And particularly in light of the late-1990s Asian economic slowdown and potential for serious recession, China has continued to make monumental efforts to maintain its continuing economic growth and market development on both domestic and foreign fronts. This is a country both ambitious and highly significant.

13.1   Area and Population

China has the world's largest total national population, estimated at 1.27 billion (excluding Taiwan) in 2001 — more than one-fifth of the world's people — and increasing by 11 million per year. It is also the world's fourth largest country. In the year 2001, China marshaled 6 million census takers to try to track down and record all of the hundreds of millions of people living in the country. It is now certain that the pace of growth has declined and that China actually stands the possibility of being surpassed by India as the world's largest country by 2050.

Population is a very serious matter for an industrializing but underdeveloped country whose area of about 3.7 million square miles (9.4 million sq km) is only slightly larger than that of the United States (Fig. 13.1), but whose inhabitants outnumber the United States' population nearly 4.6 to 1. This population lives in a state officially called the People's Republic of China (PRC), which plays a major role in world affairs (Fig. 13.2). It is, nonetheless, still relatively poor and largely agricultural despite major industrial development and economic growth under the Communist regime that came to power in 1949 (see Table 10.1).

Over 815 million Chinese (about 64% of China's population) live in rural settlements and are largely engaged in and supported by agriculture. This accumulation of peasant humanity must wrest food for itself and for the ever-expanding urban population from a landscape where nine-tenths of the land is not in cultivation because of steep slopes, dryness, or short growing seasons. China's arable (cultivable land) area, concentrated in the plains and river valleys of humid China, is about half the arable area of the United States, but through multiple-cropping practices (growing successive crops on a given field throughout the year), the sown acreage may actually be larger than the area farmed in the United States. China's arable area provides, on the average, about one-third of an acre of arable land per person to the agricultural population and only one-fourth of an acre per person to the total population.

In the 1990s, China has grown at an average of 9% annually in terms of economic development. In fact, it has now created the world's third largest economy after the United States and Japan. This growth has a major urban component, but it is also built around 25 million rural industrial enterprises with more

Figure 13.1 *China compared in latitude and area with the conterminous United States.*

than 125 million employees. This base amounts to about 40% of China's industrial labor force. These rural-based, small-scale industries generate some 50% of China's annual industrial output, 25% of its total economic value, and more than one-third of China's exports. The magnitude of this growth is shown by the fact that in 1979, trade between the United States and China amounted to $2.4 billion while in 1998 it had grown to $49.7 billion. With numbers like this, even with the regional downswing of economic health in the late 1990s, China has a possibility of becoming the world's largest economy sometime after 2010. China is a nation of major economic, demographic, and increasingly, political magnitude.

Population Policy

China's success in reducing population growth in the 1990s has been one of the global success stories in population management (Fig. 13.3), though not without controversy and serious social impact. The drive for smaller families, which began seriously in the late 1970s, was motivated by the fact that even with China's impressive recovery from war in the 1950s, there was only a modest increase in China's per capita food output for two decades. Increases in agricultural output were largely matched by rapid population growth during the time when the Mao Zedong regime (Fig. 13.4) made little or no effort to curb the birth rate.

In the early years of the revolution, Mao encouraged family growth because he felt that every new pair of hands born to China could be a productive addition to the country's economic infrastructure. Then, toward the end of Mao's life in 1976, the regime instituted one of the most stringent programs of birth control in the world. This has culminated in the "one child campaign," which aims to limit the number of children per married couple to one. Accordingly, the government takes note of individual family birthing patterns — especially in the cities — and maintains surveillance through local authorities. It dispenses free birth control devices, free

sterilization operations, free hospital care in delivery, free medical care for the child, free education for the child, an extra month's salary each year for the parents, and other favors and preferences to induce compliance with the one child per family norm. Those who violate the norm are subjected to constant social and political pressure, denial of the privileges accorded one-child families, pay cuts, and fines. Women who become pregnant without permission are pressured to have a free state-supplied abortion, with a paid vacation provided.

Under such programs, China's birth rate declined by 1995 to 1%, the lowest in any major less developed country (LDC), and its rate of natural population increase dropped to about 60% of the average for other LDCs. However, China's base population is so large that even the relatively low birth rate achieved by 1995 still meant a net increase of approximately 1 million new mouths a month, or a gain of approximately the urban population of New York City annually! This is in a country where the death rate has also been drastically lowered by better food availability, better transportation, and improved and more readily available public health facilities.

The regime hopes to continue to decrease childbearing to the extent that population in the 21st century will stabilize at 1.3 billion people, but some population experts are projecting a total of 1.4 billion by the year 2010 or 1.5 billion by 2025. The higher number may be more likely because, all through the 1990s, government officials were reporting widespread disregard of the "one couple, one child" policy. This was particularly true in the countryside as peasants became wealthier and hence more ready to pay the fines of reportedly 5,000–10,000 yuan ($600–$1,200) per extra child in a continuing traditional interest in larger families, or to produce a son if their one child was a female, or simply because governmental policies seem never to be enforced as closely in the countryside as in the city.

Urban populations have been more responsive to the continual governmental exhortation for couples to have only one child. At the same time, however, the rapid growth of the urban population in China and the increasing number of women in the urban work force have led to decreasing birth rates in the cities because of the increasing importance of material goods and new — for China — lifestyle patterns that do not necessarily revolve around parenting.

There is also a new class of child called "the little emperor," who is a single child, most often male, much loved, and often spoiled. Historically, there were usually many siblings to diffuse parental care across the lives of a number of children. Now, with few (or no) siblings and an improvement in the standard of living, many in China are fearful of the power and consequence of having so many children growing up without the social benefit of an extended family and sibling interaction.

There is also the growing social concern about who will care for the aging parents who may not have a family of children to support them in their old age. As family size diminishes, and as the one couple, one child policy is ever more

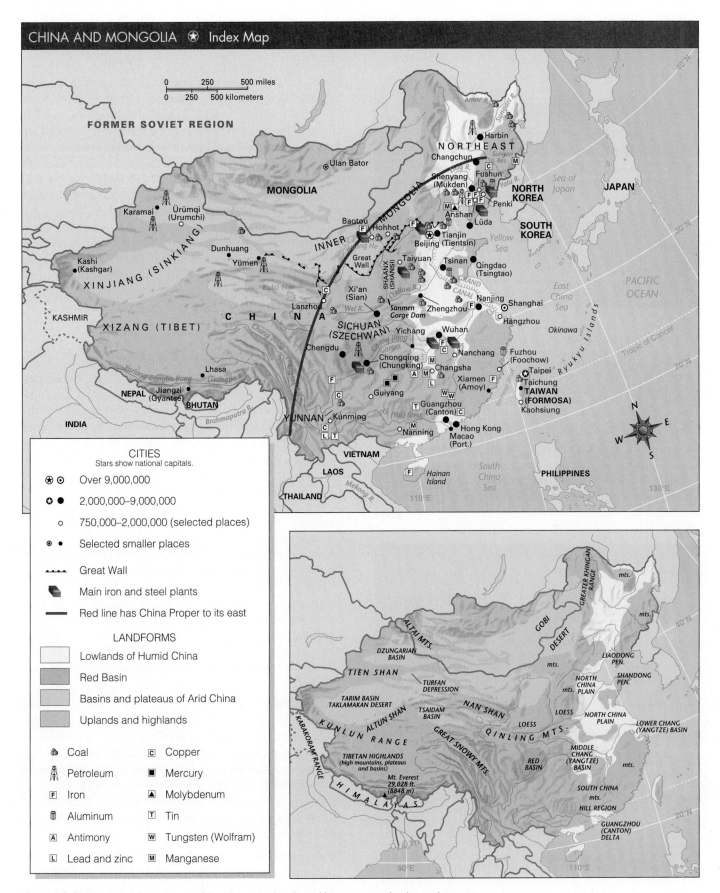

0   250   500 miles

0   250   500 kilometers

FORMER SOVIET REGION

Amur R.

Sungari R.

NORTHEAST

Harbin

Changchun

Ulan Bator

MONGOLIA

Shenyang
(Mukden)

Fushun

NORTH
KOREA

JAPAN

Sea of
Japan

Karamai

Ürümqi
(Urumchi)

Baotou

Hohhot

Penki

Anshan

Lüda

SOUTH
KOREA

Dunhuang

Yümen

INNER MONGOLIA

Great
Wall

Tianjin

Beijing (Tientsin)

Yellow
Sea

PACIFIC
OCEAN

Kashi
(Kashgar)

XINJIANG (SINKIANG)

Huang He

SHAANXI (SHANSI)

Taiyuan

Tsinan

Qingdao
(Tsingtao)

KASHMIR

XIZANG (TIBET)

Koko Nor

Lanzhou

C   H   I   N   A

Xi'an
(Sian)

Wei R.

(Yellow R.)

GRAND CANAL

Zhengzhou

Nanjing

Shanghai

East
China
Sea

Okinawa

Tropic of Cancer

Sanmen
Gorge Dam

Hangzhou

Chang Jiang

SICHUAN
(SZECHWAN)

Yichang

Wuhan

Ryukyu Islands

Chengdu

Chang Jiang
Gorges

Nanchang

Fuzhou
(Foochow)

Chongqing
(Chungking)

Changsha

Xiamen
(Amoy)

Taipei

Taichung

TAIWAN
(FORMOSA)

Guiyang

Lhasa

Yarlung Zangbo Jiang (Tsangpo R.)

NEPAL

Jiangzi
(Gyantse)

BHUTAN

Brahmaputra R.

INDIA

YUNNAN

Kunming

Guangzhou
(Canton)

Kaohsiung

Formosa Strait

N
W   E
S

Nanning

Hong Kong

Macao
(Port.)

VIETNAM

(Xi) River

LAOS

THAILAND

Mekong R.

Hainan
Island

South
China
Sea

PHILIPPINES

110°E   130°E

CITIES
Stars show national capitals.

⭐ ⊙   Over 9,000,000

✪ ●   2,000,000–9,000,000

○   750,000–2,000,000 (selected places)

⊛ •   Selected smaller places

·····   Great Wall

◤   Main iron and steel plants

——   Red line has China Proper to its east

LANDFORMS

Lowlands of Humid China

Red Basin

Basins and plateaus of Arid China

Uplands and highlands

🜨 Coal   C Copper

⛭ Petroleum   ■ Mercury

F Iron   ▲ Molybdenum

🝁 Aluminum   T Tin

A Antimony   W Tungsten (Wolfram)

L Lead and zinc   M Manganese

GREATER KHINGAN RANGE

mts.

mts.

ALTAI MTS.

GOBI DESERT

DZUNGARIAN
BASIN

TIEN SHAN

TURFAN
DEPRESSION

LIAODONG
PEN.

TARIM BASIN
TAKLAMAKAN DESERT

ALTUN SHAN

TSAIDAM
BASIN

NAN SHAN

mts.

NORTH
CHINA
PLAIN

SHANDONG
PEN.

KUNLUN RANGE

LOESS

LOESS

QINLING MTS.

NORTH CHINA
PLAIN

KARAKORAM RANGE

TIBETAN HIGHLANDS
(high mountains, plateaus
and basins)

GREAT SNOWY MTS.

RED
BASIN

MIDDLE
CHANG
(YANGTZE)
BASIN

LOWER CHANG
(YANGTZE) BASIN

mts.

HIMALAYAS

Mt. Everest
29,028 ft.
(8848 m)

SOUTH CHINA

mts.

HILL REGION

GUANGZHOU
(CANTON)
DELTA

90°E   110°E   20°N

Figure 13.2   General location map of the Chinese realm. The red line represents the division between so-called China Proper (to the east) and outer or Frontier China. China Proper has the great majority of the Chinese population, and Frontier China is home to many of the minority peoples of China, as well as to still-undetermined resources and some petroleum and natural gas.

345

FORREST ANDERSOON/GAMMA LIAISON

Figure 13.3 The Chinese government has been promoting "one couple, one child" for over two decades in an effort to slow down the annual population increase. This poster exhorts young couples to observe this constraint with the phrase: "For the sake of a prosperous today and a beautiful tomorrow, limit your family to one child." The campaign has been helped by China's rapid pace of economic growth since the early 1990s. A steady increase in the availability of consumer goods has tended to reduce fertility rates, at least among urban populations.

fully followed in urban China, the traditional safety net of extended families and children to care for seniors vanishes. For example, if a young urban couple has the allowed one child and that child is female, the likelihood is quite good that she will marry and become — by Chinese tradition — a part of the

KIT SALTER

Figure 13.4 Tiananmen Square in the center of Beijing is a 100-acre plaza that was opened after the 1949 Chinese Communist Party's military defeat of the Chinese Nationalists, two million of whom fled to Taiwan. The old housing and one- and two-story structures that were torn down were seen as expendable in the new government's quest for significant ceremonial space in the center of their capital city. Today the openness is modified only by the presence of the Mausoleum, erected in 1977, in honor of Chairman Mao Zedong who lies in state there. Tiananmen Square itself is of enormous importance as a tourist destination for foreigners as well as for millions of Chinese domestic tourists annually. This scene shows the continuing presence of Chairman Mao and also some of the wall of the Forbidden City, the fortress-like imperial dwelling that is reputed to have 5,000 rooms and is now a must walkthrough for people who come to the square. Cameras, photographs, posing, ogling, are all part of the Tiananmen Square scene in today's China.

husband's family to such an extent that she (or they) may not be available to care for her parents as they get older. Since China has traditionally used the family as the accommodation of choice for senior citizens, either a major shift in social services and associated taxation will be necessary or a full generation of seniors will find no easy setting for their final years. By 1998, China already had more than 115 million people over the age of 60, and that population segment was growing annually as the health of China's people improved steadily.

These unexpected social outcomes of a successful family planning campaign serve to remind people that virtually every change in one social domain has the capacity to send ripples of influence across all or at least many other areas of social concern. Population planning is particularly likely to have such varied impacts.

13.2 The Physical Setting

China is vast in size but, like all extremely large countries, contains a great deal of unproductive and thinly settled land. In China, such land lies mainly in the western half of the country. Here are huge outlying areas principally composed of high mountains and plateaus, together with arid or semiarid plains, where rainfall is generally insufficient for agriculture. This dry, sparsely settled country probably contains about 5% of China's population (see Fig. 13.5).

The arid interior area is in marked contrast to the better watered, more densely settled eastern core of the country. Thus, China is divided approximately into a western half, or Arid China, and an eastern core region, or Humid China, in which the country's population, developed resources, and productive capacity are heavily concentrated. A rough boundary between the two major divisions is an arc drawn from the northeastern corner of Yunnan to the northern tip of China's Northeast. This line corresponds in a general way to the stretch of land whose average annual rainfall is about 20 inches (c. 50 cm), with Arid China to the west and Humid China to the east.

Arid China

The principal regions of Arid China are the Tibetan Highlands, Xinjiang (Sinkiang), and Nei Mongol (Inner Mongolia). All three are identified historically with non-Han populations, although now — after nearly five decades of politically promoted domestic migration away from the east coast urban centers toward the borders with Russia — Han Chinese comprise the great majority in Nei Mongol and may become a majority in Xinjiang.

Tibetan Highlands

The very thinly inhabited Tibetan Highlands (see Fig. 13.5) occupy about one-fourth of China's area. Included are Tibet itself and fringes of adjoining provinces. Most of this vast area is a very high, barren, and mountainous plateau averaging nearly 3 miles (c. 5,000 m) in elevation. To the northwest, high basins of internal drainage are common, some containing large salt

DEFINITIONS + INSIGHTS

Chinese Minorities

China has some 94 million non-Han peoples divided into 56 minority groups. The Han Chinese are the ethnically "pure" people who see themselves as descending from the Han Dynasty (206 B.C. to A.D. 200), the originators of Chinese culture. Non-Han are the many varied peoples who generally live in mountain settings, especially in the southwest, or the harsh arid environments west of the margin of China Proper. Although this total of nearly 100 million Chinese amounts to under 9% of the total population, they occupy 50 to 60% of the Chinese land. They are particularly important in Inner Asia, especially Xinjiang Province. They are major percentages of the population in the very thinly settled Inner Asian Chinese interior, and they live in lands that continue to have significant resource potential. The Chinese government has launched numerous programs over the past five decades in an effort both to bring them into potential assimilation or at least to have them caught up in Chinese urban society enough to be watched by the Han Chinese.

The Muslim non-Han (the Hui) are a particularly unsettling population in western China because of the ever-increasing militancy of the Islamic minority groups. As in so many cultures, minority populations develop distinctive cultural settings with distinctive patterns of language, custom, lifestyle, and religion. This uniqueness serves both as a cohesive bond for the group itself, but also as a sometimes disturbing barb in the side of the larger culture. In China, these 56 minority groups represent a continuing source of uneasiness for the mother culture, especially in the resource-rich western region.

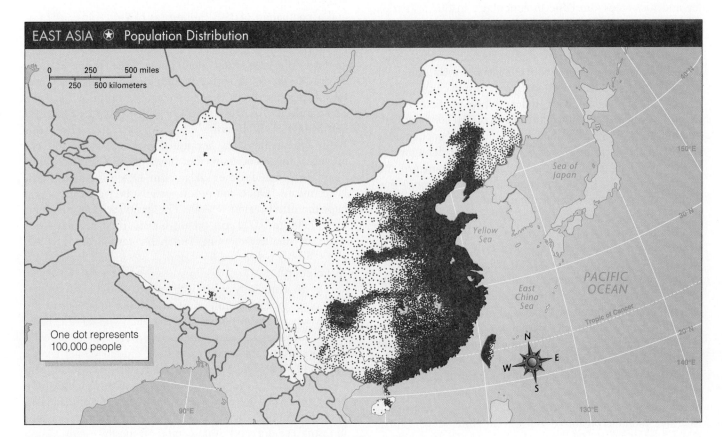

EAST ASIA ✪ Population Distribution

One dot represents 100,000 people

Figure 13.5   This dot map of population distribution in China shows the influence of waterways, coastlines, and mountain systems in the shaping of human activities. Coastal areas with open access to the adjacent seas get washed with monsoonal precipitation, as do mountain flanks on the windward side. These areas of relatively high precipitation have long favored settlement and agricultural activity. River systems have historically provided transportation, fishing resources, and water resources for settlement and irrigation. Upland and interior areas are often too arid for dense settlement but such locales may play pivotal roles as market places, transportation hubs, and sometimes religious sites. Population distribution in China reflects these very basic geographic realities all across the map.

lakes. To the southeast, the plateau is cut into ridge and canyon country by the upper courses of great rivers such as the Tsangpo (the Brahmaputra of India), the Mekong, and China's own Chang Jiang (Yangtze Kiang, or Yangtze River) and Huang He (Hwang Ho, or Yellow River). It is along the Chang Jiang's flow through the eastern edge of this plateau, in movement toward the well-settled middle Chang Jiang basins of central China, that the Chinese have begun to focus their plans to capture and control the power of this greatest river of China, which travels more than 3,400 miles (5,440 km) in its descent from the Tibetan Highlands through the Three Gorges Project landscape to its delta in the Shanghai area in the east.

Lower elevations, warmer temperatures, and greater precipitation in parts of the southeastern plateau result in some extensive grasslands and stands of conifers. Around the edges of the Tibetan Highlands are huge mountain ranges: the Himalayas on the south, the Karakoram and Kunlun Shan to the north and northwest, and the Qinling Shan and others to the east.

The people of the Highlands, numbering approximately 5.2 million (2.62 million in the political area called the Tibetan Autonomous Region) are predominantly sedentary farmers and animal herders. A few hardy crops, especially barley and root crops, are basic to agriculture in this restrictive environment. In grasslands at higher elevations, a nomadic minority graze their flocks of yaks, sheep, and goats. Yaks are particularly important not only as durable beasts of burden in these high elevations but also as providers of meat, milk and butter, leather, and hair traditionally woven into cloth.

The Tibetan Highlands are under Chinese political control, and Lhasa, the capital, now has more than 310,000 Han Chinese residents. Beyond the capital, this province's population is comprised predominantly of Tibetans, who are distinguished by their own language and by their adherence to Lamaism, the Tibetan variant of Buddhism. Prior to the Communist era, China had sometimes exerted loose control over Tibet, but Chinese authority vanished with the overthrow of the Manchus and the Qing Dynasty in 1911. From 1912 until the Chinese Communists' conquest of 1951, Tibet existed as an independent state, with its capital and main religious center at Lhasa (Fig. 13.6). After the Chinese completed roads to Tibet in late 1954, an increase in restrictive measures by the Communist government contributed to the rise of guerrilla warfare. This culminated in a large-scale Tibetan revolt in 1959 and the flight of the Dalai Lama (the spiritual and political head of Lamaism) and many other refugees to India. Subsequently, the Chinese drove most of the monks from their monasteries, expropriated the large monastic landholdings, and made institutional changes similar to those designed to implement socialism and prohibit organized religion in the rest of China. Tibetan resistance to Chinese rule and aggressive repressions by the Chinese government continue.

Xinjiang (Sinkiang)

Xinjiang adjoins the Tibetan Highlands on the north. It has an area of roughly 635,000 square miles (c. 1.6 million sq km)

Figure 13.6 The Potala Palace in Lhasa, Tibet, is the ceremonial home of the 14th Dalai Lama. It continues to stand as a monumental expression of the importance of Tibetan Buddhism in Tibet. The prayer flags that stream from the left side of the palace all the way down to the camera are representations of people who wish to have their prayers answered, and who perceive the sacred power of this place to be of particular significance. The current Dalai Lama is in exile, residing in India.

and a population of 19.3 million (as of 2001). It consists of two great basins: the Tarim Basin to the south and the Dzungarian Basin to the north (see Fig. 13.2). These basins are separated by the lofty Tien Shan Range. The Tarim Basin is rimmed to the south by the mountains bordering Tibet, and the Dzungarian Basin is enclosed on the north by the Altai Shan and other ranges along the southern border of Russia and Mongolia. Both basins are arid or semiarid. The Tarim Basin is particularly dry because it is almost completely enclosed by high mountains that block off rain-bearing winds. This basin includes the Taklamakan Desert — which, translated, means "Once you get in, you'll never get out" — perhaps the driest region in Asia. The basin varies in elevation from 2,000 to 6,000 feet (about 600 to 1800 m) above sea level; the smaller adjoining Turfan Depression drops to 928 feet (283 m) below sea level and is the second lowest land spot on Earth.

The great majority of Xinjiang's population is concentrated near oases located mainly at points around the edges of the basins where streams from the mountains enter the basin floors. For many centuries, these oases were stations on caravan routes crossing central Asia from Humid China toward the Middle East and Europe on the historic Silk Road (Fig. 13.A). Under the People's Republic, the ancient routes have been superseded by modern transportation. A railroad connects Lanzhou (Lanchow; population: 1.9 million), a major industrial center and supply base in northwestern China Proper, with Urumqi (Urumchi; population: 1.4 million), the capital of Xinjiang. Several major roads link the oasis cities, and several major airfields have been built in the region.

The transport links are key elements in a drive to expand the economic significance of this remote part of China and

PROBLEM LANDSCAPE

The Unsettled Resettlement of Xinjiang, China's Far West

The human-rights group Amnesty International in June [2000] identified Xinjiang as one of five global trouble spots where violence is escalating. Between January 1997 and April 1999, Amnesty documented 210 death sentences and 190 executions in the region. It said most of those executed were Uighurs convicted of terrorist or subversive activities. . . . Reports of violence that have filtered out of Xinjiang are incomplete, but give a sense of the scale of the problem. Take Yining [in the far northwest of the province]. According to a 1999 state-published guidebook to Yining county, police there in recent years smashed 26 terrorist groups; confiscated guns, ammunition, explosives and detonators; arrested terrorist leaders; and shot dead 15 terrorists.[a]

Of China's 56 minority groups, the Chinese Muslims — called the Hui — represent the most problematic. Although most of the non-Han populations are scattered about mountain highlands in the southwest of China, the Hui expand across the wide open spaces of Xinjiang Autonomous Region in China's far west. All minority peoples, in fact, amount to less than 6% of China's total population, but they are distributed across nearly 60% of China's land area. The Hui are the most spatially expansive.

Susan Lawrence titles her article, from which the box's opening quote comes, "Where Beijing Fears Kosovo" because of the particularly delicate nature of this problem landscape. Although the official Chinese police reports talk about the killing and capturing of terrorists, the enemies of the state in this case are Hui

leaders who are trying to pull at least a part of Xinjiang out of China's orbit to create an independent Chinese Muslim state. Whether we are talking about Turkey and the Kurds, the Serbs and Kosovo, the Basques in Spain, or the Hui in western China, this issue of breaking up larger nations into smaller, often religiously and ethnically distinct, new states is a fearful problem for state leaders at many levels.

In 1996, China defined such a threat in these terms: "It said that illegal religious activity was inciting disturbances, attacks on party and governmental offices, bombings, and terrorist and other destructive actions. It warned of a danger of future riots and turmoil affecting the stability of Xinjiang and the entire country."

From the Hui's perspective, there is time pressure in their case. At the beginning of the Chinese Communist takeover of China in 1949, the Han population in Xinjiang amounted to 6.3% of the province's population. Today, the percentage is closer to 40%, and some 250,000 more Han Chinese immigrate to the cities (especially Urumqi and Yining) annually, enticed in part by special opportunities for Han Chinese and salary differentials compared to the eastern provinces.

It is not just the demographic swing that brings tension to this scene. The Chinese governmental officials (almost all Han) bar students from participating in religious activities, attending

scripture schools, fasting, or wearing religious garb, such as Islamic head coverings.

The Chinese government has developed an ambitious Develop the West Campaign, which is partly responsible for the steady inflow of Han Chinese, and the plan is that economic development will improve people's livelihoods enough to diminish the drive for ethnic and political separatism. China has plans of spending more than $12 billion by 2004. The focus of such investment will be 70 major infrastructural improvements, including airports, factory refurbishments, irrigation and agriculture investment, and the improvement of transportation networks. In the Chinese plan, demographic dilution in combination with diminished isolation in the oases and small towns and the few cities of the province will create a Xinjiang that is more closely aligned with mother China to the east and much less with the roiling Muslim nations to the west.

As a problem landscape, this combination of strong ethnic differentiation and a global climate of escalating efforts to gain religious and ethnic independence is keenly felt in western China, just as it is in every other continent of the world.

[a] Susan V. Lawrence, "Where Beijing Fears Kosovo." *Far Eastern Economic Review*, September 7, 2000, pp. 22–24.

bring it under firmer political control. Early in the Communist era, demobilized soldiers and urban youth from eastern China were organized into quasi-military production and construction divisions to achieve expansion of irrigated land on state-operated farms in this arid landscape, allowing cotton to become a major crop. The region also has deposits of coal, petroleum, iron ore, and other minerals, and a considerable expansion of mining and manufacturing has occurred under the auspices of the People's Republic. In addition, it has been

a major focus in Chinese efforts to relocate significant numbers of Han Chinese away from the coasts in the east to the thinly settled far west in Xinjiang Autonomous Region.

This region has been valuable for both its isolation and its resources. It was in Xinjiang that the Chinese exploded their first nuclear bomb in 1964. It continues to be the most mineral-rich province in China and has the potential for considerable political as well as economic activity in the decades to come.

Nei Mongol

North and northwest of the Great Wall, rolling uplands, barren mountains, and parched basins stretch into the arid interior of Asia. Here lie slightly more than 1 million square miles (c. 2.6 million sq km) of dry, thinly grassed terrain that is divided about equally between Nei Mongol (Inner Mongolia), the area nearest the Great Wall, and the now-independent country called Mongolia. An overwhelming majority of this Autonomous Region's present population (23.6 million in 2000) is Han Chinese. This population is concentrated in irrigated areas along and near the great bend of the Huang He. By far, the greater amount of territory of Nei Mongol is still the habitat of a very sparse population of Mongol herders. From these severe, arid landscapes, the Mongols emerged to control China and an expanse of land that ranged from the Korean Peninsula in the east to the margins of Poland in the west in the 13th century. The desert landscape displays the shadows of an extraordinary history of Asia through its remnants of the Chinese Great Wall, caravan oases and routes, and temples hidden in desert caves at the end (or the beginning) of the long treks to and from distant markets (Figs. 13.A and 13.7).

The traditional picture of nomadic Mongol tribesmen herding their flocks of sheep and goats and using camels and horses for riding and pack purposes is fast disappearing. The economic core of Nei Mongol today is composed of the agricultural areas near the Huang He (Yellow River). The irrigated areas, which have been expanded by the People's Republic, produce mainly oats and spring wheat, while unirrigated fields are used mainly for drought-resistant millet and kaoliang, a sorghum grain used for animal feed and liquor. The main city of the area is Baotou (Paotow; population: 1.42 million), an expanding industrial

Figure 13.7   A scene from Dunhuang in the arid interior of China in western Gansu Province. This oasis setting was a major stop on the more than 4,000-mile (6,000-km) Silk Road that has connected northern China with the Mediterranean region for more than 2,000 years. Oasis structures like the one shown are under continual threat from encroaching sands as the desert landscape shifts with constantly blowing winds. The town of Dunhuang has now become an important tourist destination because of the relatively recent discovery of scores of caves with delicate religious paintings on their walls.

center on the Huang. One of China's major iron and steel works was opened at Baotou with Soviet help in the 1950s using nearby resources of coal and iron ore. The areas with industrial bases have grown to be major population centers of northwest China as well.

Humid China

Humid China, sometimes referred to as Eastern China, Monsoon China, or China Proper — the core region of the country — includes the densely settled parts of China south of the Great Wall and in the Northeast. The ancient provinces south of the Wall are often referred to by outsiders as "China Proper," although the Chinese seem not to employ the term. China Proper includes two major divisions, North China and South China, which differ from each other in various physical, economic, and cultural respects. As in the United States, there is a strong sense of regional identity that radiates from these words and worlds.

North China

Affected mightily by the monsoon wind patterns that characterize so much of this Asian region, North China has hot, humid, and rain-filled summers and correspondingly cold and dry winters. When the winter monsoon blows out from interior Asia, long, dark, cold winter days dominate North China. Because of the historic disinclination to have interior heating in homes in this region, winters are particularly bitter as people attempt to break the cold with just the heat from the cooking stove or, traditionally, the heated bed called the kang.

Although the region receives between 15 and 21 inches (c. 35–55 cm) of precipitation annually, it is plagued with relatively high variability from year to year (Fig. 13.8). Such a rainfall pattern has led to a history of droughts and floods, reminding one again of the burden of being a densely settled area of low precipitation and high variability. The droughts that have swept across North China have been instrumental in the initiation of the massive Chang Jiang Water Transfer Project, which is designed to bring surplus water from the Three Gorges region of the Chang Jiang through aqueducts to the heavily farmed but water-deficient North China Plain.

Floods in North China almost all relate to the continual problem of managing the Huang He (Yellow River). It is around this 2,500-mile (4,000-km) river and the Wei He tributary (east of Xi'an) that Chinese civilization was founded. The Huang drops down from its source on the Tibetan Plateau and courses for more than 1,000 miles (1,600 km) through the loess soils of North China. These airborne soils blown from the Gobi Desert to the west and northwest are very poorly structured, and they collapse easily into the river system. The river carries this yellow loess soil as silt to the Yellow Sea in the east. This physical process has built up the broad, fertile, and relatively level North China Plain. Chinese recorded history chronicles 26 major channel changes and different discharge points to the Yellow Sea in the past 1,500 years.

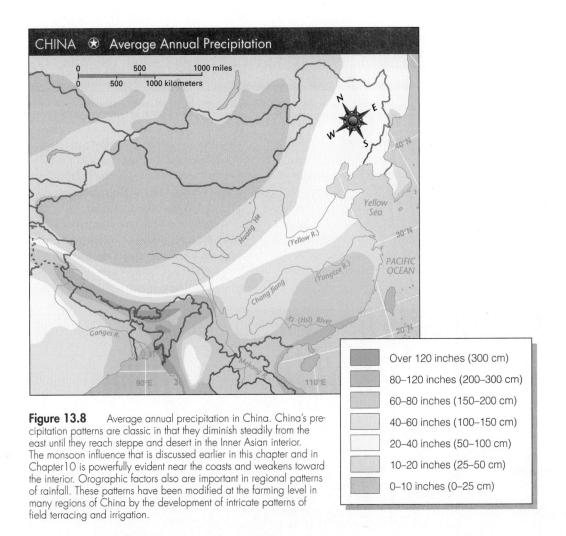

Figure 13.8 Average annual precipitation in China. China's precipitation patterns are classic in that they diminish steadily from the east until they reach steppe and desert in the Inner Asian interior. The monsoon influence that is discussed earlier in this chapter and in Chapter 10 is powerfully evident near the coasts and weakens toward the interior. Orographic factors also are important in regional patterns of rainfall. These patterns have been modified at the farming level in many regions of China by the development of intricate patterns of field terracing and irrigation.

Over 120 inches (300 cm)
80–120 inches (200–300 cm)
60–80 inches (150–200 cm)
40–60 inches (100–150 cm)
20–40 inches (50–100 cm)
10–20 inches (25–50 cm)
0–10 inches (0–25 cm)

Floods are particularly significant on the Huang He. The river carries an enormous quantity of silt that is deposited along the final 500–600 miles of the river's flow through the Tai Hang Mountains on the western edge of the North China Plain to its delta. This sedimentation has elevated the actual stream channel of the river above the surrounding landscape. The Huang flows within a critical diking system all the way from the Tai Hang Mountains to the current delta north of the Shandong Peninsula. When the dikes break during times of flood, the river pours out of the channel and down to the lower, densely settled farmlands and cities of the North China Plain. Even after the ruptured dike is repaired and the river is contained once again, the great flood of water that issued forth from the elevated stream bed has nowhere to go. Such widespread ponding and floods have been associated with the Huang He and the North China Plain for centuries. For this reason, the Huang He is also called "China's Sorrow."

The major crops in North China are winter wheat, millet, and kaoliang, with some land given over to summer rice crops and corn and a wide variety of kitchen vegetables. North China has exported wheat to South China since the seventh century A.D. Historically, much trade has been carried along the Grand Canal, a monumental civil engineering project of the Sui Dynasty (A.D. 581–618). This 1,050 mile (1,700 km) inland waterway linked Hangzhou in the Chang Jiang delta region to just southwest of Beijing. Wheat and coal were barged southward, and rice was carried north. Although the Grand Canal has never attained a status equal to the image of the Great Wall, it has come much closer to achieving its initial design goals than has the much more widely known Great Wall.

South China

The Qinling Mountains, at approximately 10,000 feet (3,048 m), have historically served as the accepted dividing line between North and South China. These mountains run east to west and divide stream drainage between the Huang He to the north and the Chang Jiang to the south. The agricultural dominance of South China comes from the 3,400-mile (c. 5,440-km) Chang Jiang that flows from its origin in the Tibetan Plateau to the delta just north of Shanghai on the east. This river plays a role in China quite similar to that of the Missouri River in the United States. Historically, the Chang Jiang has served as a major east–west transit corridor for China. It can accommodate freighters that draw 10 feet all the way from the East China Sea to Wuhan in the central basin.

In the early Han Dynasty (from 206 B.C. to about A.D.8), the Chinese began a steady exchange of goods for items from the Mediterranean trading ports. From China went silk (not raised in the West until the sixth century) and to China came wool, gold, silver, and glass. The overland route for this exchange — called the Silk Road — was approximately 4,000 miles (6,000 km) long, going from the early Chinese capital of Chang'an (now Xi'an) to ports of the eastern Mediter-ranean. The Road went along the Great Wall and the Taklamakan Desert, over the Pamir Mountains, and through Baghdad to the Mediterranean Sea. From there, goods were shipped to various European ports.

This route played a role in the diffusion of Buddhism into China in the first century A.D. Until the Venetian explorer and trader Marco Polo in the 1270s, no one ever followed the whole route because travel was dangerous and every car-avan load had to pass numerous checkpoints and pay considerable tolls. Polo is sometimes identified as the first person to travel the complete route, and it was in part the power of his description of this grand overland route of trade, diffusion, and cultural exchange that made Christopher Columbus so determined to find a sea route to replace it. The route is currently being considered as a possible base for a United Nations trans-Asian international highway (see Fig. 13.A).

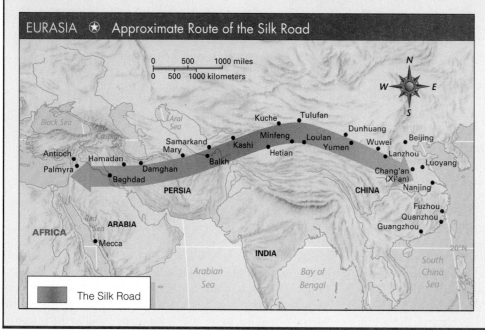

EURASIA ★ Approximate Route of the Silk Road

The Silk Road

Figure 13.A   The Silk Road (sometimes called the Silk Route) is the approximately 4,000-mile link that connected China with European markets and civilization since before the time of Christ. This series of partial trails and routes was almost never traveled by a single person or caravan, but rather was a trade route for many caravan groups each crossing its own specific segment of the Road. Silks, paper, art goods, and spices went west from China, while gold and silver bullion, glass goods, and metalware were major goods going east. The Silk Road has been a major diffusion route of religions, languages, customs, and technology as well.

The Three Gorges Project is designed to extend the Yangtze's freighter capacity from Shanghai all the way west to Chongqing in the Suchuan Basin.

The Chiang Jiang is central to China's sense of identity and has been for well over 2,000 years. The "Great River" — the Chang Jiang — has long been pivotal to a Chinese sense of place. This river drains more than 700,000 square miles and has seasonal shifts in river depth of well over 100 feet in the gorges that lie between Sichuan Province in the west and the broad agricultural basins of central China to the east. The current Chang Jiang Water Transfer Project — called the Three Gorges Dam, for short — is the latest and largest effort made by the Chinese to control the enormous power and potential of that river.

In 1919, Sun Yat-sen (sometimes called the George Washington of China) suggested that a major dam be built to help control the annual rise and fall of the Chang Jiang. One goal of the contemporary project is to build a series of dams so strong and high that the flow of the river water through the heavily populated provinces east of the gorges can be evened out, reducing or eliminating flood threats to a region that has suffered major inundations for thousands of years. There is also the goal to create a hydroelectric project that would produce more power than many of China's dams combined. The Three Gorges Dam as designed will be 610 feet high, 1.3 miles (2.1 km) long, and the reservoir will extend upstream for 385 miles (620 km). The boldness and scale of this landscape modification project make it akin to the building of the Great Wall and the Grand Canal in the eyes of the Chinese.

But there has been much disapproval, both domestic and foreign, of the plan because of its potential for environmental disturbance to a whole network of water and plant ecosystems that rely on the Chang Jiang. The World Bank, after expending $8.7 million for a feasibility study, decided *not* to fund any part of the project. In May 1996, the U.S. Export-Import Bank also refused to provide any funding for the megadam project. In view of China's major flooding in the summer of 1998, there is also the claim that the major project has already taken so much capital and administrative energy that ordinary projects that would have helped diminish the impact of these recent floods had not been undertaken.

The project is also disfavored because of the number of villages and archeological sites that will be drowned by the reservoir that will develop behind this tallest of all China dams. It is estimated that completion of the project will displace some 1.2 to 2.8 million people and submerge 13 cities, including the lower half of Wanxian (a city of more than 300,000), 350 to 700 villages, and more than 115,000 acres of arable farmland (see Fig. 13.9). The Three Gorges Dam project must be seen as a political statement — as well as an engineering hallmark — further manifesting China's wish to

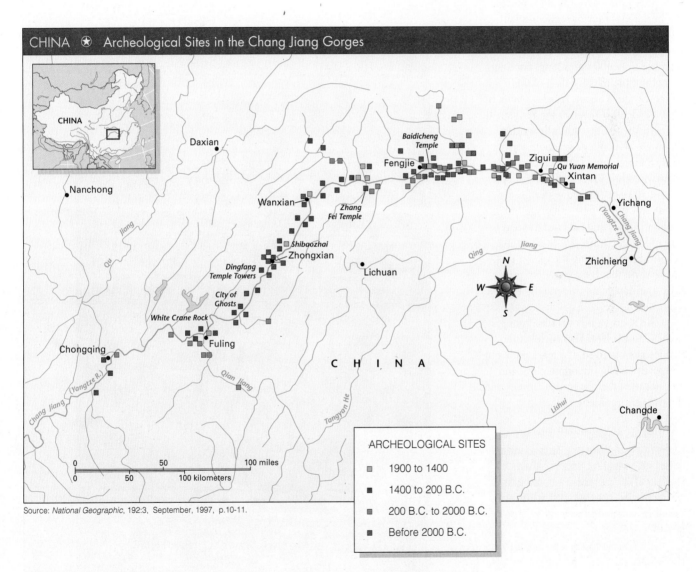

CHINA ✷ Archeological Sites in the Chang Jiang Gorges

ARCHEOLOGICAL SITES

- 1900 to 1400
- 1400 to 200 B.C.
- 200 B.C. to 2000 B.C.
- Before 2000 B.C.

Source: *National Geographic*, 192:3, September, 1997, p.10-11.

Figure 13.9   Some call the Chinese damming of the Chang Jiang (Yangtze River) "China's New Great Wall Project." The scale and landscape implications of the multiple dam project along hundreds of miles of the river as it winds through the mountains west of the Middle Basin of the Yangtze make it far overshadow any water project ever undertaken in China or anywhere else. This map shows the extent of the project and the variety of archeological sites to be lost from the 400-mile ponding of the river's water behind the world's tallest dam. The planning, initial construction, budget, and environmental ramifications of the Three Gorges Dam Project have been hotly contested for more than a decade, and some would say for nearly a century. However, China's government continues to stick to its plan to bring the Yangtze under more complete control, to increase China's capacity to generate electricity, and to possibly develop a water transfer scheme that might move some of the Yangtze flow to the water-hungry North China Plain. The project is not scheduled to be completed until 2009.

LANDSCAPE IN LITERATURE

River Town: Two Years on the Yangtze

Peter Hessler served in China as a Peace Corps volunteer for 2 years from 1996 to 1998. He and one other American young man were posted in the city of Fuling, at the confluence of the Wu River and the Yangtze River in Sichuan Province. Fuling is one of the hundreds of towns and cities that will have its geography changed profoundly by the creation of a body of water more than 400 miles long, pooling up behind the massive dam of the Three Gorges Project. Hessler's book is *River Town: Two Years on the Yangtze*. It is a colorful and thoughtfully written chronicle of a college graduate suddenly being dropped off in a river town where no other American has lived for at least 50 years. Armed with minimal initial skill in Chinese language but keen to learn it and find out the real nature of life in China, this book does a beautiful job of letting the reader look into a world that is at once both powerfully traditional and undergoing massive change.

This short segment comes from part of the Chinese New Year's celebration that Hessler observed. It offers a midnight perspective on an important event and reminds us of the power of ceremony and the river environment.

Fireworks at midnight called in the new year. I had left the Huang home early, because I was a little tired, and I was getting ready for bed when the sound started, low and steady like thunder rolling over the hills. The noise grew louder, echoing across the river valley, and I went out on my back balcony to watch.

The Wu River [a tributary to the Yangtze Chiang, China's most massive river] looked sullen in the night. The city was also dark, but as midnight approached the fireworks increased; I could see them flaring and flashing among the streets and stairways. The intensity of the sound doubled, tripled; explosions joined in from Raise the Flag Mountain, and in the distance across the Yangtze, there were flashes on White Flat Mountain. At the stroke of midnight the entire city gathered itself and roared, its voice reverberating back and forth across the Wu, the windows of the buildings flickering in reflections of sparks and bursts of fire. The old year died; evil spirits fled; deep in the valley's heart the Wu trembled, its water colored by the bright shadow of the blazing city. And finally midnight passed, and the fireworks faded, and we were left with a new year as empty and mysterious as the river that flowed silently through the valley.[a]

[a]Peter Hessler, *River Town: Two Years on the Yangtze*. New York: HarperCollins Publishers, 2001, pp. 303–304.

NEWSMAKERS/GETTY IMAGES

Figure 13.B   Upstream from where author Peter Hessler spent his years as a Peace Corps volunteer and writing *River Town* is the city of Wanxian. It is the largest city on the Yangtze that will be covered with water as the full reservoir behind the Three Gorges Dam comes to full depth. In this photo, cement factories are working full-tilt producing cement and other construction materials for the ongoing work on the Three Gorges Dam complex.

but it is not evenly distributed. There continue to be tens, perhaps many tens, of millions of urban poor who are underemployed and unemployed.

- Regional imbalances in economic activity and opportunity continue to plague China. The most prosperous areas are in the southeast, east coast, and farmlands near large cities, while regions in the west, southwest, uplands, and Arid China have dramatically less evidence of economic development.

- Special Economic Zones (SEZs) and Taiwan are two special landscapes that have China's attention. The SEZs have been developed to showcase China's ability to absorb and make productive use of Western-style urban growth and development. Taiwan, in its ever-increasing prosperity, is being closely watched as it becomes more democratic, more prosperous, and more inclined to seek full political and economic independence from China.

REVIEW QUESTIONS

1. List the area and population of China and explain the importance of both these statistics.
2. What are the characteristics of China's population growth and what governmental policy is at the heart of its contemporary birth control programs?
3. What is the meaning of "the little emperor," and what implications does that phrase and the associated concept have for future generations in China?
4. Define and find on the map Arid China. Discuss some of the associated landscapes.
5. Define and find on the map Humid China. Discuss some of the associated landscapes.
6. What is the economic and political significance of the western highlands in China? Of the Northeast?
7. Explain the importance of the Great Wall, the Grand Canal, and the Three Gorges Project. Differentiate their economic importance from their historical and political significance.
8. Who was Mao Zedong and what role did he play in China in the 20th century? What campaigns is his government associated with? What role do his ideas now play in China?
9. Outline the different eras of development in China between 1949 and 1999. Which of these eras seem to still have importance?
10. Who was Deng Xiaoping? What contemporary programs is he associated with? What is their contemporary importance?
11. What is the relationship between China and Taiwan? Explain potential future scenarios between those two places.

DISCUSSION QUESTIONS

1. What are the geographic features that give China such a monumental image in East Asia? In all of Asia? In world affairs?
2. Discuss the magnitude and significance of the Chinese population and of the government's efforts to control its growth.
3. What are the features of Arid China that support or diminish settlement and economic development?
4. Compare North China and South China in terms of physical features, demographic patterns, and cultural significance. Looking at essential element 2 in Chapter 1, what are the regional forces that make the two regions so distinct?
5. How much of China's future is linked with China Proper and how much will come from the world of Western China?
6. What is the cause and the solution to the imbalanced economic development in China? How old is this pattern?
7. In China's efforts to industrialize, what role has been played by Mao Zedong and what role by Deng Xiaoping?
8. What was the attraction of the Dazhai village model in the 1960s and 1970s? Could such a model occur again in China? Under what conditions?
9. Discuss the origins, the expected outcomes, and the environmental ramifications of the Three Gorges Project.
10. What are the geographic patterns of urban distribution in China? What do such patterns reflect in terms of China's physical landscape and demographic history?

CHAPTER 14

Japan and the Koreas: Adversity and Prosperity in the Western Pacific

Japan's most precious assets are its children.
Birth rates have been plummeting, and
these youngsters will bear the burden of looking
after an elderly nation.

Chapter Outline

On a cluster of islands off the eastern coast of Asia, the 127 million people of Japan operate an economy second only to that of the United States. This powerful economic engine sits on mountainous, resource-poor, and crowded lands about the size of California, yet holding almost half as many people as the entire United States. Japan's dramatic rise from the ashes of World War II has drastically changed the world's economic geography. This chapter explains that phenomenal achievement and the country's recent malaise.

Although the Korean Peninsula has a seemingly marginal position on the Pacific Rim and the Eurasian landmass, it, too, has a central place in geopolitical affairs, mostly because of its strategic location between Japan, China, and the former Soviet Union, the prosperity of South Korea, and the sometimes dangerous backwardness of North Korea. This chapter examines the problems of a divided Korea and its prospects for reunification.

14.1   The Japanese Homeland

There are four main islands in the Japanese archipelago (Fig. 14.1). On the west, the Sea of Japan and the Korea Strait separate the main islands from Russia and Korea. Japan faces the open Pacific on the east. Hokkaido, the northernmost island, is close to Russian-controlled islands to its north. The scenic and busy Inland Sea separates Honshu, the largest island, from Shikoku and Kyushu. South of Kyushu, the smaller Ryukyu Islands run southwestward almost to Taiwan.

All of the main islands are mountainous, with the higher peaks between 5,000 and 9,000 feet (c. 1,500 to 2,750 m) above sea level. The islands are volcanic (Fig. 14.2) and are subject to frequent and sometimes severe earthquakes (see Problem Landscape, p. 374). Located west of Tokyo, the volcano Mt. Fuji, whose name means "Fire Goddess," is Japan's highest mountain (12,388 ft/3,761 m; Fig. 14.3). An active volcano (one that has had an eruption in recorded history), Fuji began experiencing "swarms" of earthquakes in 2001 that many geologists interpreted as signs of an approaching eruption; it would be the first since 1707. Fuji was sacred to the aboriginal Ainu people of Japan as long as 2,000 years ago. It is still a holy mountain in the national Shinto faith, and about 300,000 pilgrims and tourists visit its lofty summit each July and August, when it is free of snow. The journey to the peak of this most prominent symbol of Japan is a difficult one, as suggested in the Japanese proverb, "It is as foolish not to climb Fujisan as to climb it twice in a lifetime."

Climate and Resources

Japan's location east of the Eurasian landmass and at latitudes comparable to those from South Carolina to Maine (Fig. 14. 4, p. 376) strongly influences its climates. A humid subtropical climate extends from southernmost Kyushu to north of Tokyo, a humid continental long-summer climate characterizes northern Honshu, and Hokkaido has a humid continental short-summer climate. In lowland locations, where most of the people live, the average temperatures for the coldest month range from about 20°F (−7°C) in northern Hokkaido to about 40°F (4°C) in the vicinity of Tokyo and about 44°F (7°C) in southern Kyushu. Averages for the hottest month are only about 64°F (18°C) in northern Hokkaido but rise southward to about 80°F (27°C) around Tokyo and above 80°F farther south. Conditions are cooler in the mountains, with snowfalls common in winter in Honshu and Hokkaido. Hokkaido has twice been the venue for the Winter Olympics.

Japan is a humid country. Annual precipitation of more than 60 inches (c. 150 cm) is common from central Honshu south, and 40 inches (c. 100 cm) or more falls in northern Honshu and Hokkaido. Their location off the east coast of Eurasia puts the islands in the path of the inward-blowing summer monsoon from the Pacific, and summer is the main rainy season in eastern Japan. But the islands are also subject to the seaward-blowing winter monsoon after it has crossed the Sea of Japan, so much precipitation also occurs in the winter in western Japan. Japan is also subjected in summer and fall to powerful typhoons, the great low-pressure storm systems known as hurricanes in the western Atlantic region. These add to the woes of this hazard-prone country.

Japan's natural resource base is not large enough to meet the country's growing needs. Up to the middle of the 19th century, the country was nearly self-sufficient economically, although the standard of living was low. Japan was already crowded with about 30 million people, but disease, starvation, abortion, and infanticide slowed the rate of natural

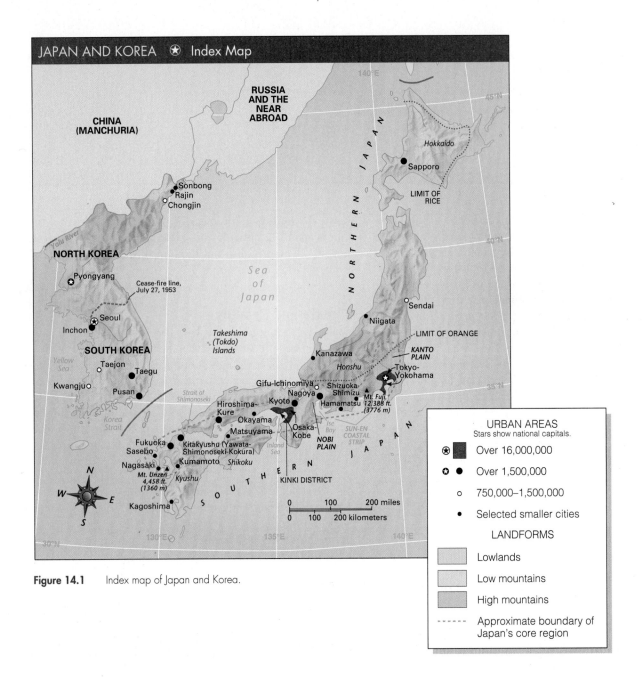

JAPAN AND KOREA ✪ Index Map

URBAN AREAS
Stars show national capitals.

⊛ ⬛ Over 16,000,000

✪ ● Over 1,500,000

○ 750,000–1,500,000

• Selected smaller cities

LANDFORMS

◻ Lowlands

◻ Low mountains

◻ High mountains

- - - - - Approximate boundary of Japan's core region

Figure 14.1   Index map of Japan and Korea.

increase. Industrialization stimulated rapid population growth in the second half of the 19th century and later. Population growth has recently almost ceased, however, as delayed marriages and the rising costs of child rearing have lowered birth rates. Meanwhile, affluence has increased dramatically. The material demands of so many people enjoying a high standard of living in a country with limited natural resources require Japan to import an astonishing 97% of its raw materials and energy supplies and a large share of its food.

Forests, which now cover about 60% of the land surface, illustrate the limitations of Japan's natural resources. In a preindustrial and early industrial society, with a much smaller population, the nation's heavily forested mountainsides supplied the needed wood, but if today's demand for forest products were met with domestic reserves, the country would be rapidly deforested. There is a wide range of valuable trees, mainly broadleaf deciduous toward the south and needleleaf coniferous toward the north. Many mountainsides are quilted with rectangular plots of trees planted in rows. Most first-growth timber is long since gone, but some virgin stands are still protected in preserves. Forests are important in Japanese culture, and this affluent country does not wish to lose them. With its enormous demand for wood products, Japan is therefore the world's largest importer of wood, mostly hardwoods from the tropical rain forests of Southeast Asia.

Figure 14.2   Japan has numerous volcanoes—some active, others intermittently active, and still others extinct. In 1991, Mt. Unzen, near Nagasaki in Kyushu, erupted for the first time since 1792. At least 35 people were killed. The photo shows firefighters fleeing for their lives as smoking red-hot lava hurtles down the side of the volcano.

THE YOMIUMI SHIMBO

JOE HOBBS

Figure 14.3   Volcanic Mt. Fuji loses its snow cover by late summer.

Kerosene stoves have largely replaced the charcoal used formerly for household cooking and heating, but the Japanese still use wood for many purposes (Fig. 14.5). Japan's large publishing industry and other industries and handicrafts require a great deal of paper. Single-family dwellings continue to be built largely of wood, which provides more safety from earthquake shock waves than more rigid construction. However, wood makes the dwellings more vulnerable to the fires that typically sweep through Japanese cities after earthquakes rupture gas lines.

Small deposits of minerals, including coal, iron ore, sulfur, silver, zinc, copper, tungsten, and manganese, supported early industrialization in Japan. These minerals are in short supply today, and like most other raw materials, they must be imported. Japan's mountain streams provided much of the power for early industrialization, but they are so swift, shallow, and rocky that they are of little use for navigation. Most of the water used for irrigation in Japan comes from these streams.

Only about one-eighth of this small country's area is arable; Japan is not one of the world's major crop-growing nations. Most cultivable land is in mountain basins and in small plains along the coast. Terraced fields on mountainsides supplement this level land. Irrigated rice, the country's basic food and most important crop, was once nearly everywhere on arable land (Fig. 14.6). Production of this grain was heavily subsidized by the Japanese government and promoted by a powerful farm lobby, which aimed at rice self-sufficiency. The country's farmers usually produce a net export of rice by intensive use of scarce cropland. However, in the 1990s, a scarcity of rice throughout East Asia and pressure on Japan to reduce its massive trade surplus with the United States prompted Japan to open its rice market to foreign imports for the first time. Now, with silos overflowing with rice, and with the Japanese diet more dependent on wheat (in the form of bread and noodles) than on rice, the government is paying Japanese farmers not to grow rice.

Mild winters and ample moisture permit **double cropping**—the growing of two crops a year on the same field—on most irrigated land from central Honshu south.

PROBLEM LANDSCAPE

Living on the Ring of Fire

The people of Japan must cope with one of the most hazard-prone regions on Earth. By far the greatest threats are related to Japan's location on the Pacific Ring of Fire, the name given to the boundaries of the Pacific plate of Earth's crust where earthquakes and volcanic eruptions are commonplace and often catastrophic (Fig. 14.A). These impressive geologic events are the results of slow but steady worldwide movements of the massive plates of the earth's mantle and crust that collectively are known as **plate tectonics.** The processes of plate tectonics have been dominant forces in the creation of continents, mountain systems, and ocean troughs and account for the past, present, and future locations of the continents. In the tectonic scheme, Japan is in an area of "subduction" where one plate (the Pacific) is being pushed, as if by a conveyor belt, under another plate (the Eurasian), creating the country's characteristic volcanoes, earthquakes, and related features.

Japan is most threatened by earthquakes on and near its islands but must also be on guard for the effects of distant earthquakes, even as far away as the shore of North America. A strong quake there can produce a *tsunami* (meaning "harbor wave" and commonly called a tidal wave) that may travel thousands of miles to strike and flood coastal Japan.

Japan has a long history of seismic assaults. One of the most significant recent events took place on January 17, 1995, when a devastating earthquake released the tension that had built up on a transverse fault near historic Kobe, a port

city of 1.5 million people. The Japanese soon labeled the 6.9-magnitude episode the Great Hanshin Earthquake, establishing its place in the annals of Japanese disasters alongside such events as the Great Kanto Earthquake (magnitude 7.9 on the Richter scale), which killed 143,000 people in and around Tokyo in 1923. Killing 6,400 people and injuring thousands more, damaging 190,000 buildings, displacing 300,000 people, and causing at least $100 billion worth of damage, the Kobe quake was the worst disaster to hit Japan since World War II. Kobe's remarkable lack of disaster preparedness and relief focused attention on a common Japanese complaint: that the government invests too much in commerce-oriented development and not enough in the welfare of its people.

Japanese authorities did not predict the Great Hanshin Earthquake. After the fact, there were reports that large schools of fish swam uncharacteristically close to the surface of the sea off Kobe in the days preceding the quake. Crows and pigeons reportedly were noisier than usual and flew erratically in the hours before the ground moved. In the weeks preceding the quake, levels of radon gas rose in a water aquifer below the city and then declined sharply after January 9. Animal behavior, water level and composition, and many other variables are often debated but still unproven barometers of an impending earthquake.

Since 1965, Japan has spent more than $1.3 billion in an effort to develop means of predicting earthquakes. During the 1990s, more than $100 million yearly

went to this effort. But by 1998, no earthquake had been successfully predicted, and Japan suspended this expensive research program. Its critics argue that earthquakes are chaotic events and will always be impossible to predict. They also point out the hazards of issuing an earthquake warning that is not followed by a real quake. A false alarm in the Tokyo region would cost an estimated $7 billion a day in lost business and other economic disruptions. The bigger question is how such a great city could be evacuated safely, efficiently, and with minimal panic.

With the prediction program scrubbed, Japan has redoubled its efforts to prepare for future big ones. New structures must meet a tough safety code that aims at "earthquake-proof" construction. Much of the focus is on Tokyo, which is underlain by a complex system of active fault lines, and which typically suffers a big quake every 70 years (as of the time this book went to press, it had been 79 years since the last one). Current estimates are that a big one nearing 8.0 on the Richter scale would kill 6,717 people and destroy 300,000 structures (mostly by fire) in Tokyo. These numbers do not count the impacts of the estimated 27-foot-high (8-m) tsunami that would strike low-lying Tokyo within minutes of an offshore earthquake. With such threats in mind, the government has resolved to build, by 2010, a new national capital inland, where tsunamis are not an issue and where mountain bedrock would reduce the devastating impacts of seismic shock waves. Much of

---

Rice is grown in summer, and wheat, barley, or some other winter crop is planted after the rice harvest. Overall, about one-third of Japan's irrigated rice fields are sown to a second crop, and more than half of the unirrigated fields are double cropped. **Intertillage** — the growing of two or more crops simultaneously in alternate rows — is also common. Dry farming also produces a variety of crops: sugar beets in Hokkaido;

all sorts of temperate-zone fruits and vegetables; potatoes, peas, and beans, including soybeans; and in the south, tea, citrus fruits, sugarcane, tobacco, peanuts, and mulberry trees to feed silkworms.

Changes in Japan's agricultural sector have accompanied a surge in urban-industrial employment and general affluence in recent decades. Farmers have been moving to towns, and

the devastation in Kobe resulted from the city's location on soft coastal soils and reclaimed lands, which also underlie great parts of Tokyo. Japan's struggle against the seismically inevitable is a national priority and a symbol of the nation's determination to maintain economic success despite environmental adversity.

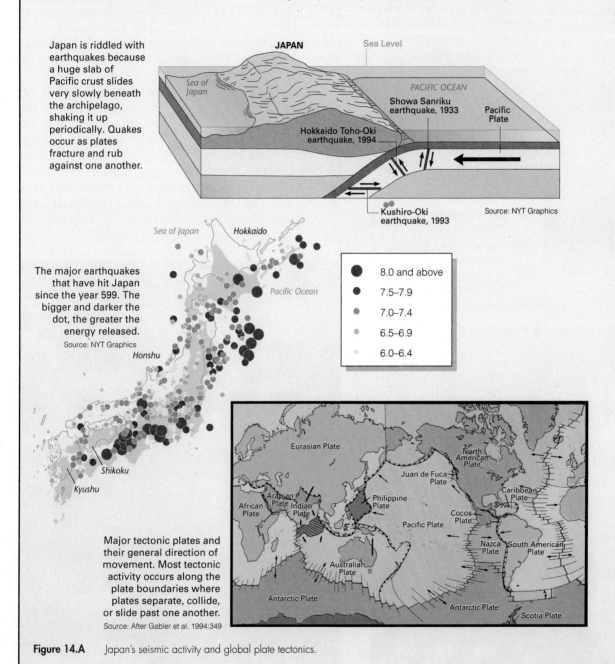

Japan is riddled with earthquakes because a huge slab of Pacific crust slides very slowly beneath the archipelago, shaking it up periodically. Quakes occur as plates fracture and rub against one another.

JAPAN Sea Level

Sea of Japan

PACIFIC OCEAN

Showa Sanriku earthquake, 1933

Pacific Plate

Hokkaido Toho-Oki earthquake, 1994

Kushiro-Oki earthquake, 1993

Source: NYT Graphics

The major earthquakes that have hit Japan since the year 599. The bigger and darker the dot, the greater the energy released.

Source: NYT Graphics

Sea of Japan　Hokkaido

Pacific Ocean

Honshu

Shikoku

Kyushu

8.0 and above
7.5–7.9
7.0–7.4
6.5–6.9
6.0–6.4

Major tectonic plates and their general direction of movement. Most tectonic activity occurs along the plate boundaries where plates separate, collide, or slide past one another.

Source: After Gabler et al. 1994:349

Eurasian Plate
North American Plate
Juan de Fuca Plate
Caribbean Plate
Arabian Plate
Indian Plate
African Plate
Philippine Plate
Cocos Plate
Pacific Plate
Nazca Plate
South American Plate
Australian Plate
Antarctic Plate
Antarctic Plate
Scotia Plate

Figure 14.A　Japan's seismic activity and global plate tectonics.

the number of agricultural workers has dropped dramatically (from about 25% of the labor force in 1962 to 5% in 2002). A huge increase in farm mechanization has made it possible to till Japan's tiny farms with far less human labor. With this technology and heavy applications of fertilizers, plus improved crop varieties, Japanese agriculture has achieved the world's highest yields per unit of land.

The island people of Japan view the ocean as an important resource and promising frontier. The country has an ambitious "inner space" program of deep-sea exploration to mine the sea's living and mineral riches. Many food fish species are in the waters around Japan, and Japanese fishermen range widely through the world's major ocean fishing grounds. Japan is the leading nation in ocean fisheries, and the

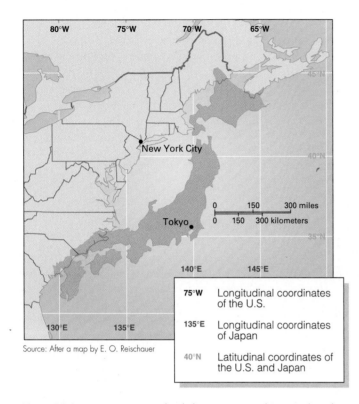

75°W	Longitudinal coordinates of the U.S.
135°E	Longitudinal coordinates of Japan
40°N	Latitudinal coordinates of the U.S. and Japan

Source: After a map by E. O. Reischauer

Figure 14.4   Japan compared with the eastern United States in latitude and area.

Japanese have the world's highest per capita consumption of fish, including *sashimi* (raw fish) and other marine foods (Fig. 14.7). Historically, the Japanese have also enjoyed whale meat, but pressure from international environmentalists compelled Japan in 1985 to support a worldwide moratorium on commercial whaling. However, Japan has resumed small-scale whaling of minke whales for what are called "scientific research" purposes. This loophole allows whale meat to be widely sold.

In recent years, the Japanese have begun to eat less fish and rice in favor of a more Western diet. Although sea fish continue to be the main source of animal protein in the Japanese diet, increased consumption of meat, milk, and eggs is reflected in the growing number of cattle, pigs, and poultry on Japanese farms. The animals require heavy imports of feed, and the increasing variety in people's tastes means many more food imports. The United States is Japan's largest supplier of

Figure 14.6   Rice fields near Tokyo. Note how people maximized the amount of land dedicated to rice by confining their settlements to the hills, and how forests on those hills have been saved.

Figure 14.5   Wood is a vital resource in Japan. This wooden home is in Shirakawa, north of Tokyo on the island of Honshu.

imported feedstuffs and food, and the United States exports more food to Japan than to any other country.

Japan's Core Area

Most of Japan's economic activities and people are packed into a corridor about 700 miles (c. 1,100 km) long. This megalopolis extends from Tokyo, Yokohama, and the surrounding Kanto Plain on the island of Honshu in the east, through northern Shikoku, to northern Kyushu in the west (see Fig. 14.1). This highly urbanized core area contains more people than live in the Boston to Washington, D.C., megalopolis on the United States' east coast. All of Japan's greatest cities and many lesser ones have developed here, on and near harbors along the Pacific and the Inland Sea. They are sited typically on small, agriculturally productive alluvial plains between the mountains and the sea.

The largest urban complex in the world—Tokyo-Yokohama and surrounding suburbs (population: 31 million by one estimate; other figures differ according to how the metropolitan area is defined)—has developed on the Kanto Plain, Japan's most extensive lowland. Tokyo grew as Japan's political capital and Yokohama as the area's main seaport. This urban complex functions today as Japan's national capital, one of its two main seaport areas (the other is Osaka-Kobe), its leading industrial center, and the commercial center for northern Japan (Fig. 14.8).

Beyond Tokyo, in the entire northern half of Japan, the only one major city is Sapporo (population: metropolitan area, 2,279,700; city proper, 1,770,000) on Hokkaido. About 200 miles (c. 320 km) west of Tokyo lies Japan's second largest urban cluster, in the Kinki District at the head of the Inland Sea. Here, three cities—Osaka, Kobe, and Kyoto—form a

Figure 14.7   Ocean fish are a major element in the Japanese diet. This is the Tsukiji fish market in Tokyo.

Figure 14.8   Tokyo, the capital of an industrious and affluent nation, is the world's largest and one of its most vibrant cities. This is rush hour at the Shinjuko Station.

metropolis, together with their suburbs, of nearly 18 million people.

The inland city of Kyoto was the country's capital from the eighth century A.D. until 1869 (when it was moved to Tokyo) and is now preserved as a shrine city where modern industrial disfigurement is prohibited (Fig. 14.9). Some U.S. military planners wanted to target Kyoto with a nuclear weapon in August 1945, but the decision that it should be avoided because of its historical importance spared the city. Kyoto continues to be the destination of millions of pilgrims and tourists annually.

Osaka developed as an industrial center, and Kobe as the district's deepwater port. Both cities now combine port and industrial functions. Another large metropolis, Nagoya, is situated directly between Tokyo and Osaka on the Nobi Plain at the head of a bay. It is both a major industrial center and a seaport.

West of the Kinki District, smaller metropolitan cities spot the coreland. The largest are metropolises of over 1.5 million people: the resurrected Hiroshima on the Honshu side of the Inland Sea, Kitakyushu (a collective name for several cities) on the Strait of Shimonoseki between Honshu and Kyushu, and Fukuoka on Kyushu. Hiroshima was destroyed on August 6, 1945, when the United States dropped the first atomic bomb ever used in warfare, detonating it directly over the cen-

ter of the city. With no topographic barriers to deter the effects of the explosion, the city was obliterated, and an estimated 140,000 died by the end of 1945, when radiation sickness had taken its greatest toll.

Among the several smaller cities of northern Kyushu is Nagasaki, which, on August 9, 1945, was hit by a second U.S. nuclear bomb, the last used in warfare. Nagasaki was chosen during the flight mission after clouds obscured the primary target, Kokura. The bomb detonated over the suburbs rather than the city center, and hills helped to diminish the explosion's impact, but even so, an estimated 70,000 died by year's end. The bombing of Nagasaki brought Japan's surrender and the end of World War II.

14.2 Historical Background

Japan has seen successive periods of isolationism and expansionism and of economic and military accomplishment and defeat. Today, it is near the top of most indicators of prosperity in the more developed countries (MDCs). The remarkable success story of Japan can best be appreciated by considering its turbulent history and the difficult home environment in which the Japanese have lived.

The Japanese are descended mainly from Mongoloid peoples who reached Japan from other parts of eastern Asia in the distant past. An earlier non-Mongoloid people, the Ainu, were driven into outlying areas where some still live, mainly in Hokkaido. The oldest surviving Japanese written records date from the eighth century A.D. These depict a society strongly influenced by Chinese cultural and ethnic traits, often reaching Japan by way of Korea. A distinct Japanese culture gradually evolved, and today, it is linguistically and in other ways different from its Chinese and Korean antecedents. Traditions about Japanese origins extend back to the reign of Jimmu, the first emperor, whose accession is ascribed to the year 660 B.C., but which probably took place centuries later. These traditions are important in the Japanese worldview, which says that the islands and their emperor have a divine origin. Jimmu was said to have descended from the Sun Goddess, and the Japanese have known their homeland as the "Land of the Gods." The Japanese concept of Japan as unique and invincible developed early, shaped by legends such as that of the kamikaze, the "divine wind" that repelled a Mongol attack on Japan in the 13th century.

The early emperors gradually extended control over their island realm. A society organized into warring clans emerged. By the 12th century, powerful military leaders called *shoguns*, who actually controlled the country, diminished the emperor's role. Meanwhile, nobles, or *daimyo*, whose power rested on the military prowess of their retainers, the *samurai*, ruled the provinces. This structure resembled the European feudal system of medieval times.

Adventurous Europeans reached Japan early in the 16th century, beginning with the Portuguese in the 1540s. Most of

Figure 14.9 Kyoto has a unique status in Japanese history and religious life. This is the 14th-century Kinkakuji Temple, also known as the Golden Pavilion.

BRUCE BURKHARDT/WESTLIGHT

these early arrivals were merchants and Roman Catholic missionaries. Japanese administrators allowed the merchants to trade and the missionaries to preach freely. There were an estimated 300,000 Japanese Christians by the year 1600.

Contacts with the wider world were curtailed as Japan entered a period when military leaders imposed central authority on the disorderly feudal structure. After winning the battle of Seikigahara in 1603 with arms purchased from the Portuguese, warlord Ieyasu Tokugawa proclaimed himself "Nihon Koku Taikun" (Tycoon of All Japan) and became the first shogun to rule the entire country. During the era of the Tokugawa Shogunate, from 1600 to 1868, the Tokugawa family acquired absolute power and shaped Japan according to its will. To maintain power and stability, the early Tokugawa shoguns wanted to eliminate all disturbing social influences, including foreign traders and missionaries. They feared that the missionaries were the forerunners of attempted conquest by Europeans, especially the Spanish, who held the Philippines. The shoguns drove the traders out and nearly eliminated Christianity in persecutions during the early 17th century. After 1641, a few Dutch traders were the only Westerners allowed in Japan. Their Japanese hosts segregated them on a small island in the harbor of Nagasaki, even supplying them with a brothel so that they would not be tempted to venture out and pollute Japan's ethnic integrity. Under Tokugawa rule, Japan settled into two centuries of isolation, peace, and stagnation.

Foreigners succeeded in reopening the country to trade two centuries later, when Japan was ill prepared to resist Western firepower. Visits in 1853 and 1854 by American naval squadrons of Black Ships under Commodore Matthew Perry (who wanted to establish refueling stations in Japan for American whaling ships) resulted in treaties opening Japan to trade with the United States; this was "gunboat diplomacy." The major European powers were soon able to obtain similar privileges. Some feudal authorities in southwestern Japan opposed accommodation with the West. United States, British, French, and Dutch ships responded in 1863 and 1864 by bombarding coastal areas under the control of these authorities. These events so weakened the Tokugawa Shogunate that a rebellion overthrew the ruling shogun in 1868. The revolutionary leaders restored the sovereignty of the emperor, who took the name Meiji, or "Enlightened Rule." The revolution of 1868 is known as the **Meiji Restoration.**

The men who came to power in 1868 aimed at a complete transformation of Japan's society and economy. They felt that if Japan were to avoid falling under the control of Western nations, its military impotence would have to be remedied. This would require a reconstruction of the Japanese economy and social order. They approached these tasks energetically and abolished feudalism, but only after a bloody revolt in 1877. The Meiji leaders forged the model of a strong central government that would modernize the country while resisting foreign encroachment. Under this system, which lasted until 1945, democracy was strictly limited, with small groups of powerful men manipulating the government's machinery and the emperor's prestige. Military leaders were prominent in the power structure.

Following their fact-finding tours of Europe and the United States, Meiji leaders pressed their people to learn and apply the knowledge and techniques that Western countries had accumulated during centuries of Japan's isolation. Foreign scholars were brought to Japan, and Japanese students were sent abroad in large numbers. Japan adopted a constitution modeled after that of imperial Germany. The legal system was reformed to be more in line with Western systems. The government used its financial power, which it obtained from oppressive land taxes, to foster industry. New developments included railroads, telegraph lines, a merchant marine, light and heavy industries, and banks. Wherever private interests lacked capital for economic development, the government provided subsidies to companies or built and operated plants until private concerns could acquire them. Rapid urbanization accompanied a process of rapid industrialization.

So spectacular were the results of the Meiji Restoration that within 40 years Japan had become the first Asian nation in modern times to become a world power. Outsiders often spoke of the Japanese in derogatory terms as mere imitators. The Japanese, however, knew which elements of Western technology they wanted, and they adapted them successfully to Japanese needs.

The shift from Tokugawa isolationism to Meiji openness put Japan on a pathway to empire and eventually to war. Japan was bent on expansion between the early 1870s and World War II and by 1941 controlled one of the world's most imposing empires (Fig. 14.10). Between 1875 and 1879, Japan absorbed the strategic outlying archipelagoes of the Kuril, Bonin, and Ryukyu Islands. Japan won its first acquisitions on the Asian mainland by defeating China in the Sino-Japanese War of 1894–1895, which broke out over disputes in Korea. In the humiliating 1895 Treaty of Shimonoseki, China ceded the island of Taiwan (Formosa) and Manchuria's Liaotung Peninsula to Japan and recognized the independence of Korea. Feeling threatened by Japan's emerging strength so close to its vital port of Vladivostok, Russia demanded and won a Japanese withdrawal from the Liaotung Peninsula and leased the strategic peninsula from China. Japan went to war with Russia in 1904 by attacking the peninsula's main settlement of Port Arthur. Russia lost the ensuing Battle of Mukden, which involved more than a half-million soldiers, making it history's largest battle to that date. Russia's loss of the Russo-Japanese War of 1904 represented the first defeat of a European power by a non-European one. Victorious Japan regained the Liaotung Peninsula, secured southern Sakhalin Island, and established Korea as a Japanese protectorate; Korea was formally annexed to Japan in 1910.

At the end of World War I, the Caroline, Mariana, and Marshall Islands were transferred from defeated Germany to Japan as mandated territory. Japanese encroachments on

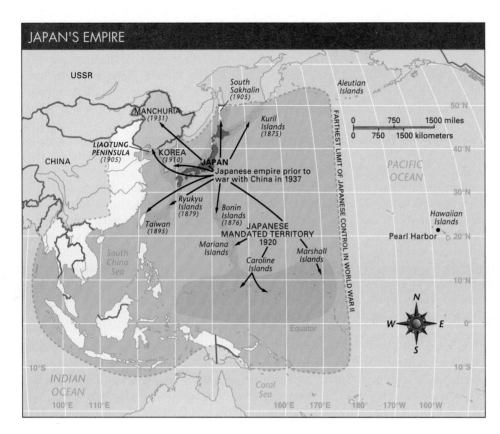

JAPAN'S EMPIRE

Figure 14.10 Map showing overseas areas held by Japan prior to 1937 and the line of maximum Japanese advance in World War II.

China's region of Manchuria followed in the 1930s. In 1937, Japan began an all-out attack that overran the most populous parts of China. Japanese troops committed many atrocities in the assault on China, even employing biological warfare by airdropping plague-infested fleas on the city of Ningbo. The most notorious was the "Rape of Nanking" (now Nanjing) in 1937, when the invaders took control of China's temporary capital, killing an estimated 300,000 Chinese civilians and soldiers, and raping about 20,000 Chinese women. All told, an estimated 2.7 million Chinese civilians were killed in Japan's effort to "pacify" China.

The encroachments continued. Japan seized French Indochina in 1940. After Japan's carrier-based air attack crippled the American Pacific Fleet at Pearl Harbor, Hawaii, on December 7, 1941 ("a day which will live in infamy," declared U.S. President Franklin Roosevelt), Japanese forces rapidly overran Southeast Asia as far west as Burma, as well as much of New Guinea and many smaller Pacific islands.

The motives for Japanese expansionism were mixed. They included a perception of national superiority and "manifest destiny," a desire for security, recognition by the great powers, the wishes of military leaders to inflate their importance and gain control of the Japanese government, and ambitions of industrialists to gain sources of raw materials (e.g., Manchurian and Korean coal) and markets for Japanese industries. The desire for materials and markets was based

solidly on need. Expansion of industry and population on an inadequate base of domestic natural resources had made Japan dependent on sales of industrial products outside the homeland. Such sales provided, as they do now, the principal funds for buying necessary imported foods, fuels, and materials. As worldwide depression hurt the trade-dependent prewar Japanese economy, Japan chose to create a Japanese-controlled Asian realm that would insulate the Japanese economy from the volatile world economy and make the country a leading world power. In 1938, Japan therefore proclaimed its New Order in East Asia, and in 1940 and 1941, this widened into the Greater East Asia Co-Prosperity Sphere, a euphemism for Japanese political and economic control over China and Southeast Asia.

Japan's militaristic and colonial enterprises proved to be disastrous at home and abroad. Japanese soldiers inflicted great suffering on civilians throughout the western Pacific. Even now, nations are demanding formal apologies from Japan. In August 1995, on the 50th anniversary of Japan's defeat in World War II, Japan's prime minister did issue a formal general apology for his nation's role in the war. Critics say Japan has not fully atoned for its war crimes and are trying, among other efforts, to block Japan's quest to gain a permanent seat on the United Nations Security Council.

August 1945 found Japan completely defeated. Its overseas territories, acquired during nearly 70 years of successful imperialism, were lost. Its great cities were in ruin. In addition to 1.8 million deaths among its armed services, Japan had suffered some 8 million civilian casualties from the American bombing of the home islands; most of its major cities were over half destroyed. Tokyo's population had fallen from nearly 7 million people to about 3 million, most of them living in shacks. In 1950, 5 years after the beginning of the U.S. military occupation of Japan, national production still stood at only about one-third of its 1931 level, and annual per capita income was $32.

14.3   Japan's Postwar Miracle

Without colonies or empire, Japan has become an economic superpower since World War II. The nation's explosive economic growth after its defeat was one of the most remarkable developments of the late 20th century and is known widely as the Japanese "miracle."

Observers of Japan cite different reasons for the country's economic success. Proponents of dependency theory argue that Japan, never having been colonized, escaped many of the debilitating relationships with Western powers that crippled many potentially wealthy countries. Some analysts believe that the country's postwar economic miracle grew from an intense spirit of achievement and enterprise among the Japanese. Notably, many Japanese attribute this industrious spirit to Japan's geography as a resource-poor island nation. To overcome the constraints nature has placed on them, the Japanese people feel they must work harder.

Still others attribute Japan's postwar achievements to its association with the United States following World War II. Some of the postwar U.S.-imposed reforms worked well economically, and relations between the two countries since the war have also generally stimulated Japan's industrial productivity.

The United States' military occupation of Japan from 1945 to 1952 brought about major changes in Japanese social and economic life. The divine status of the emperor was officially abolished; when Emperor Hirohito announced Japan's surrender on the radio, it was the first time the stunned Japanese public had ever heard the voice of this mythic figure. A new U.S.-written constitution (which has remained unchanged ever since) made Japan a constitutional monarchy with an elective parliamentary government. With the rejection of divine monarchy, the strongly nationalistic Shinto faith lost its status as Japan's official religion, although worship at Shinto shrines was allowed to continue; it still dominates, along with Buddhism, Japan's spiritual life. There was land reform in the countryside to do away with a feudal-style landlord system. Some major Japanese companies were broken up to reduce their monopolistic hold on the economy. Women were enfranchised. A democratic trade union movement was established.

Perhaps most significant for the economy, the United States forbade Japan to rearm, except for small "self-defense" forces. When the occupation ended in 1952, leaving U.S. bases in Japan but returning control of Japanese affairs to the Japanese, Japan was placed under American military protection. Without large military expenditures, much of Japan's capital was freed to invest in economic development.

The U.S. military umbrella over Japan remains controversial today. Many Japanese regard it as an outmoded vestige of colonialism and the war. There have been several incidents of women being raped by U.S. soldiers stationed on the island of Okinawa, where three-quarters of the U.S. bases and more than half of the U.S. troops in Japan are stationed. The resulting protests and diplomatic appeals led the United States to agree to return control of a major airbase in Okinawa to Japan and to relocate some U.S. troops from Okinawa to mainland Japan. In a subsequent referendum, the people of Okinawa voted overwhelmingly in favor of further U.S. military withdrawal from the island.

Japan's special relationship with the United States boosted its economy long after the American occupation ended. The postwar economic recovery accelerated when the United States called on Japanese production of steel, textiles, and clothing to support American forces in the Korean Conflict of 1950–1953. American economic aid continued to flow to Japan. The United States permitted many Japanese products to have free access to the American market and allowed Japan to protect its economy from imports. Japanese firms also had relatively free access to American technologies, which they often improved upon and marketed even before American firms did.

By the 1970s, Japan became an industrial giant with a gross national product far exceeding that of any country except the United States and the Soviet Union. The United States and European nations found themselves at a disadvantage to Japanese competitors in many industries. They busily sought ways to match this competition or to protect themselves from it without doing too much damage to their exporting firms, consumers, and overall trade relations. By the 1980s, the United States and Japan were at odds over Japan's massive trade surplus with the United States.

Observers often point to several unique features of Japanese management and employment to explain the country's meteoric postwar gains. One is recruitment through an extremely challenging (some say brutal) educational system that emphasizes technical training. A rigorous and stressful testing system controls admission to higher education. Japanese management strategies emphasize benevolence toward employees, encouragement of employee loyalty, and participation of workers in decision making. About 20% of Japanese workers enjoy guarantees of lifetime employment in their firms, and many large Japanese companies help provide housing and recreational facilities for their employees.

One essential factor in Japan's postwar economic growth was a high level of investment in new and efficient industrial plants. Investment capital was freed by government policies that cut expenditures on amenities and services such as roads, antipollution measures, parks, housing, and even higher education. Japanese industrial cities grew explosively and became highly polluted areas of dense and inadequate housing, with few public amenities and snarled transportation. The money "saved" by not being spent on amenities and services was made available for investment in industrial growth, but the costs to Japanese society were high.

Some analysts cite elements of Japan's political culture to explain the country's economic successes. One political party, the Liberal Democratic party (LDP), has been repeatedly elected to power since the time Japan regained its sovereignty in 1952. It is a conservative and strongly business-oriented and business-connected organization. The party has promoted Japanese exports with policies to keep the yen (the Japanese unit of currency) — and thus, Japanese goods — inexpensive. The party has also cooperated closely with Japanese business in the development of new products and new industries. However, that very coziness between government and business — the "crony capitalism" seen in several other East and Southeast Asian countries — was partly to blame for the economic malaise that spread over Japan in the 1990s and early 2000s, and so support for the LDP began to erode.

Despite all of these factors in Japan's economic favor, the Japanese miracle did not last. The peak of Japan's postwar success came in the 1980s. A powerful economic boom led to speculative rises in stock prices and land prices. At the end of the 1980s, the total value of all land in Japan was four times greater than that of the United States. The Tokyo Stock Exchange was the world's largest, based on the market value of Japanese shares. Then the "bubble economy" burst, and real estate and stock prices fell by more than 50%. Japan had near-zero economic growth through most of the 1990s and into the 2000s. Japanese industries whose growth had seemed unstoppable experienced an increasing loss of market share to U.S. and European producers. Part of the reason the economy could not regain its footing was official Japanese commitment to prop up faltering industries and farms with huge subsidies and, because of the disproportionate political influence of small towns and rural areas, to modernize remote villages and islands with expensive and underutilized public works projects. Due to such projects, Japan is the world's most indebted country. Overall, the system succeeds in subsidizing its past at the expense of its future.

14.4 Japanese Industry

The gradual evolution of industry over more than three centuries that characterized Europe and the United States has been compressed into little more than a century in Japan. Early iron and steel development was concentrated in north-

Figure 14.11   The port of Tokyo. The island nation of Japan thrives on seaborne trade.

east Kyushu, on and near the country's main coal reserves. The products went mainly into a new railway system and into shipping (Fig. 14.11). Japanese coal, supplemented increasingly by imports, powered the country's factories, railways, and ships. Hydroelectricity was developed on mountain streams and became important in the energy economy.

Japan's petrochemical and other oil-based industries reflect the country's great dependence on imports. Almost devoid of petroleum reserves, the country imports almost all of its oil and gas, mostly from the Persian/Arabian Gulf region. Japan therefore downplays its alignment with the West in most Middle Eastern disputes.

The phenomenal advance of industry since the early 1950s has been marked by a series of booms and declines in various sectors. In the 1960s, electronics, cameras and optical equipment, petrochemicals, synthetic fibers, and automobiles (notably those produced by Mazda, Honda, Toyota, Mitsubishi, and Nissan) became boom industries (Fig. 14.12). In the

Figure 14.12   Japanese-made consumer products are recognized throughout the world for their high quality and competitive prices.

1970s and 1980s, Japan stepped up manufacturing of computers and robots. By the early 1980s, Japan led the world in the robotization of industry, but the United States took and continues to hold the lead in computer manufacturing. For the first time in the postwar era, Japanese industry in the 1990s and early 2000s was not characterized by world supremacy in any single manufacturing sector. Japanese industry must now look for new directions in which to establish leadership. As always, to overcome the severe limitations imposed by small, resource-poor islands, the Japanese must live by their wits.

14.5 The Social Landscape

Japan has one of the world's most homogeneous populations; it is 99.5% ethnic Japanese. The largest non-Japanese group is the 600,000 strong community of ethnic Koreans, the people and their descendents whom Japan imported during World War II to do mainly hard manual labor. These ethnic Koreans have long suffered from discrimination in Japan. Otherwise, few non-Japanese have settled in the country. Only with the country's booming prosperity and growing labor shortages of the 1980s did Japan open its doors to only a few unskilled and low-skilled immigrants from countries such as Pakistan, Bangladesh, Thailand, Peru, and Brazil.

This shortage of racial and ethnic diversity has had mixed results for Japan. Many observers believe that it has helped the country achieve a sense of unity of purpose, allowing the Japanese to persist through periods of adversity, especially the postwar years of reconstruction. However, the Japanese have also earned a reputation for intolerance of ethnic minorities. This attitude may be costly for Japan's future economic development. One way to increase productivity in the face of Japan's declining population would be to import skilled workers from abroad, but the country remains averse to immigration.

Capitalist Japan has a remarkably egalitarian society, with about 80% of the population comprising its middle class. The richest third of the population has a total income just three times as great as that of the poorest third (compared with five times as much in the United States). The Japanese have historically embraced the concept of *wa*, or harmony, based in part on the principle of economic equality. The recessions of the 1990s and early 2000s, however, resulted in growing joblessness and homelessness in Japan, which tarnished the country's self-satisfied image. The country's official unemployment rate reached 5% in 2001. There were loud international appeals for Japan to reform its economic system by allowing inefficient businesses to collapse (rather than prop them up with expensive subsidies) and to resist the temptation to satisfy marginal rural populations with costly education and public works projects. The government balked at these suggestions because of the threat they posed to a socially harmonious Japan.

The legendary Japanese work ethic has had its advantages and drawbacks: It has helped the Japanese create a prosperous country, but it has also created a nation of workaholics beset with the same problems—stress, suicide, depression, and alcoholism—experienced by workers in the world's other MDCs. There is a growing incidence of what Japanese call *karoshi*, or death by overwork. The educational system encourages children to be highly successful but also to conform; critics say this stifles creativity and innovation. College graduates who are not hired during the annual recruiting season face difficult obstacles to entering the job market. And for those it affects, the practice of lifetime employment makes it difficult to change jobs. The country's welfare system, pensions, and social security are inadequate, compelling workers to work harder for savings. In their long hours at work, men are accustomed to spending little time with their spouses and children. Women have not achieved parity with men in the workplace, and they complain increasingly of discrimination and sexism. In struggling to increase their footing in Japanese society, growing numbers of Japanese women are both working longer hours and marrying later, contributing to Japan's remarkably low birthrate (9 per 1,000 annually, one of the lowest in the world). Ironically, Japan's falling birthrates will only increase the burdens of its workers (see Fig. 14.13). The population is both shrinking (it is expected to go from 127 million in 2001 to 100 million in 2050 and 67 million in 2100) and aging (already one in five people was over age 65 in 2001, and 1 million people a year join those ranks). The Japanese of working age will face an increasing load of taxes and family obligations to meet the needs of older citizens.

Crowding and overdevelopment, accompanied by a surprising lack of amenities, are inevitable facts of life in Japan. Officially designated parks comprise 14% of Japan's land area, but many of these areas are developed with roads, houses, golf courses, and resorts. There has been enormous growth in the popularity of winter sports destinations in northern Honshu and Hokkaido and of hot springs and mineral bath resorts throughout the country. Reflecting dissatisfaction with too much growth and development, the conservation ethic is strong and growing in Japan. However, it is difficult for the Japanese to truly "get away from it all" at home.

Daily life for the Japanese is devoid of many of the benefits associated with the MDCs. Apartments and homes are generally very small. Only about half of the country's homes are connected to modern sewage systems. Smog from automobile exhaust and industrial smokestacks is a persistent health hazard. The ever-present traffic jams are known locally as traffic "wars." Japan's much-vaunted rail system, including its high-speed *Shinkansen*, or bullet trains, that carry traffic between northern Honshu and Kyushu at speeds averaging 106 miles (170 km) per hour, is chronically overburdened with passengers (Fig. 14.14). Summarizing these living conditions, Japanese politician Ichiro Ozawa described Japan as having "an ostensibly high-income society with a meager lifestyle." Japan has long emphasized production of manufactured exports to compete with the United States and other large economies rather than raising the material standard of living of its own citizens.

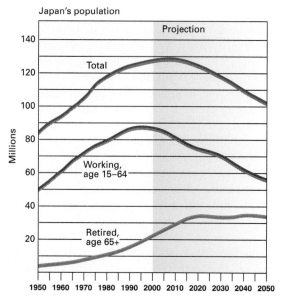

Japan's population

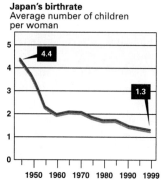

Figure 14.13 Many countries face problems of overpopulation, but Japan is concerned about not having enough people to keep its economic engine running in the future.

Japan has had a troubled period of almost continuous economic recession since the early 1990s. Most analysts expect the country to recover, turning current adversity into opportunity as it has in the past. Japan's highly educated and skilled work force is a major resource. Although natural assets are lacking, the country's favorable geographic location should also aid its recovery. The western Pacific region may resume dynamic economic growth, and Japan is well positioned to benefit from trade with the countries of this region.

Japan will certainly continue to be an important member of the international community. Even during recession, it has maintained its status as the world's second largest donor of foreign aid (after the United States). Its long-term economic well-being is seen as critical to the overall stability of the global economy. The Japanese consume more goods than the 2 billion people of the rest of East Asia combined. Any slowdown in Japanese spending decreases production in the factories of Indonesia, Malaysia, Thailand, and China that make goods for Japanese consumption and reexport; conversely, an economically stronger Japan raises all these economic boats. The United States supplies Japan with a wide range of goods, including semiconductors, software, heavy machinery, movies, apparel, and agricultural goods, and it buys even more goods from Japan, so these two economies — the world's largest — are also extremely interdependent on one another's health.

14.6 Divided Korea

Just 110 miles (177 km) from westernmost Japan is the Korean Peninsula (see Fig. 14.1). Korea's political history during the 20th century, first as a colony of Japan and then as a divided land, has tended to obscure its distinctive culture and contributions to the world. Although the Koreans have been influenced by Chinese culture, and to a smaller extent by Japanese culture, they are ethnically and linguistically a separate people.

Liabilities of Korea's Location

The two Koreas have occupied an unfortunate location in historic geopolitical terms. These small countries are near larger and more powerful neighbors (China and Russia)

Figure 14.14 A bullet train glides through the Ginza district of Tokyo. Japan has an efficient, high-speed rail network.

PETE SEAWARD/STONE/GETTY IMAGES

REGIONAL PERSPECTIVE ::: Nuclear Power and Nuclear Weapons in the Western Pacific

In 1995, on the 50th anniversary of the bombing of Hiroshima and Nagasaki, Americans and Japanese did much soul-searching about the use of nuclear weapons. The surviving U.S. decision makers generally continued to insist that the use of these weapons spared many thousands of lives, both American and Japanese, that would have been lost if the United States had instead undertaken an invasion of the Japanese homeland. For their part, the Japanese continued to condemn the decision, and it remains official policy that Japan will never develop or use atomic weapons.

Despite the unique stigma Japan associates with the power of the atom, the country has come to rely on nuclear power, with 53 plants supplying about 30% of the country's electricity needs. Hydropower supplies 8%, but with fossil fuels almost absent, the critical lack of energy has compelled the nation to develop the world's most energy-efficient economy and to persist with the nuclear technology to which many Japanese are opposed. Small fires, explosions, and other accidents in nuclear power and reprocessing plants in recent years have prompted new domestic concerns about the safety of nuclear energy, forcing a slowdown in Japan's nuclear industry. There are also international fears. Japan imports large quantities of plutonium for use in its nuclear power industry, and the long-distance ocean shipment from Europe of this very hazardous material has contributed to Japan's poor reputation in environmental affairs. International protests against the plutonium shipments have caused Japan to postpone the construction of a series of nuclear breeder reactors, which use and create recyclable plutonium.

Japan continues to worry about the potential nuclear threat from three adversaries, all of which possess or have had programs to develop nuclear weapons: Russia, China, and North Korea. The Japanese feel that the West dismissed such potential threats from Russia too readily when the USSR dissolved. Japan still has territorial disputes with Russia, particularly involving the four Kuril islands of Kunashiri, Etorofu, Shikotan, and Habomai, which the government of Josef Stalin seized at the end of World War II and which are just off the coast of northeast Hokkaido (see Figs. 14.10 and 6.1). Japan and China also have a territorial dispute concerning the East China Sea, and if oil is discovered there, as anticipated, relations between the two countries could deteriorate. China continued to test nuclear weapons through 1996, adding to tensions in Japan. Finally, Japan fears reunification in Korea, which, as a former Japanese colony from 1910–1945, has a particular historic dislike of Japan. There is speculation that Japan may do the unthinkable — develop nuclear weapons — to counter the perceived threat of North Korea's nuclear weapons program.

The world also continues to look nervously at the troubled relations between North and South Korea and between North Korea and the West. A crisis flared in 1994 when North Korea refused to permit full inspection of its nuclear facilities by the International Atomic Energy Agency. The country was suspected of separating plutonium that could be used in making nuclear bombs. Unofficial diplomatic talks between former U.S. President Jimmy Carter and North Korean leader Kim Il Sung helped scale down the tension level. The United States, Japan, and South Korea agreed to assist North Korea in the construction of two nuclear power plants in exchange for a freeze on North Korea's nuclear weapons program and for U.S. deliveries of fuel oil to meet North Korea's winter heating needs.

However, North Korea pressed ahead with a program to build missiles capable of carrying nuclear warheads, and in 1998, launched such a missile (minus the warhead) over Japan. This incident prompted the United States to renew its resolve to establish an antimissile defensive "shield" over the United States — the so called Star Wars initiative dating to the Reagan administration of the 1980s. Some analysts believe the United States is now reluctant to see the two Koreas reunite because it would remove much of the justification on which the missile shield program is based and because it might put pressure on the United States to reduce its military presence in the Western Pacific. They add that Japan, too, would not want to see Korean reunification because it would remove Japan's justification for building up its defenses, which are designed less for confrontation with North Korea than with China.

Late in the Clinton presidency, the United States softened its position on North Korea, even lifting some economic sanctions against the country in 1999 after North Korea suspended its missile building program. President George W. Bush came into office with a harder line against North Korea, effectively suspending dialogue and technical assistance for the still uncompleted nuclear power plants. In 2002, he proclaimed that North Korea was one of three countries making up an "axis of evil," along with Iraq and Iran, thereby signaling that the United States was more interested in confronting than accommodating North Korea. The apparent policy shift aroused great concern in South Korea and Japan.

DEFINITIONS + INSIGHTS

Japan, Korea, and the Law of the Sea

Some 90 miles (145 km) between the shores of Japan and South Korea in the Sea of Japan lay two small, inhospitable islands known to outsiders as the Liancourt Rocks, to Japanese as the Takeshima Islands, and to Koreans as the Tokdo Islands (see Fig. 14.1). South Korea has controlled the islands since 1956, but with only one Korean couple and some Coast Guard personnel living on the islands, Japan has periodically asserted its right to them. In the 1990s, they became the focus of a dispute between Japan and Korea, not because of any riches they contain, but because of a 1970s United Nations treaty known as the **Convention on the Law of the Sea,** which would permit their sovereign power to have greater access to surrounding marine resources.

The Law of the Sea was initiated in an effort to apportion ocean resources as equitably as possible and to avoid precisely the kind of conflict brewing between Japan and South Korea. The treaty gives a coastal nation mineral rights to its own continental shelf, a territorial water limit of 12 miles (19 km) offshore, and the right to establish an exclusive economic zone (EEZ) of up to 200 miles (320 km) offshore (e.g., in which only fishing boats of that country may fish). The power that controls offshore islands such as the Takeshima/Tokdo Islands can extend the area of its exclusive economic zone even further.

By early 1996, 85 nations had ratified the treaty. South Korea ratified the treaty late in 1995. Japan was preparing to ratify it in 1996 when news reached Tokyo that South Korea had plans to build a wharf on the islands. To avoid provoking Japan, North Korea, and China, South Korea avoided declaring an exclusive economic zone off its waters that would include the islands. Japan ratified the treaty. But fears that South Korea may build facilities and station more people on the islands as a step toward establishing an EEZ, and thereby excluding Japanese fishermen from the area, caused Japan to restate its claim to the islands in 1996. South Korean officials answered with military exercises near the islands, and South Korean civilians staged loud demonstrations outside Japan's embassy in Seoul. The issue may have to be settled in the same international legal arena in which it originated.

Even more important than Japan's competition with South Korea over territorial waters and islands is its contest with China. Both lay claims to the islands in the East China Sea that the Japanese know as the Senkakus and the Chinese as the Diaoyu Islands. The islands are uninhabited, but their control is championed by nationalists on both sides and is therefore a potentially explosive issue.

that have frequently been at odds with one another and with the Koreans. North Korea adjoins China along a frontier that follows the Yalu and Tumen Rivers; it faces Japan across the Korea Strait; and in the extreme northeast, it borders Russia for a short distance. For many centuries, the Korean Peninsula has served as a bridge between Japan and the Asian mainland. From an early time, both China and Japan have been interested in controlling this bridge, and Korea was often a subject or vassal state of one or the other. However, from the late 7th century to the mid-20th century, Korea was a unified state, sometimes invaded and forced to pay tribute, but never destroyed as a political entity. The decline of Chinese power in the 19th century was accompanied by the rise of modern Japan, whose influence grew in Korea, and from 1905 until 1945, Korea was firmly under Japanese control. In 1910, it was formally annexed to the Japanese empire. A legacy of hostility and occupation continues to cast a shadow over relations between the Koreas and Japan, and disputes periodically emerge between them over issues such as control of islands and fishing rights (see Definitions & Insights above).

Japan lost Korea at the end of World War II. In accordance with the victorious allies' agreements, Soviet forces occupied Korea north of the 38th parallel, and U.S. forces occupied Korea south of that line. Korea was to have become a unified and independent country, but the occupying powers could not reach agreement on the formation of a government. The occupying powers therefore set up separate governments: the democratic Republic of South Korea in the south, under the auspices of the United Nations, and the People's Republic of Korea in the north, a Chinese Communist satellite. The occupying powers withdrew most of their forces in 1948 and 1949.

Then, in 1950, North Korea attacked South Korea. United Nations units, made up mostly of U.S. forces, entered the peninsula to repel the Communist advance. Late in 1950, when the North Koreans had been driven back almost to the Manchurian border, China entered the war and drove the United Nations forces south. A stalemate then developed just north of the 38th parallel until an armistice was arranged in 1953. The border between the two Koreas, often the tensest boundary on Earth, still follows this armistice

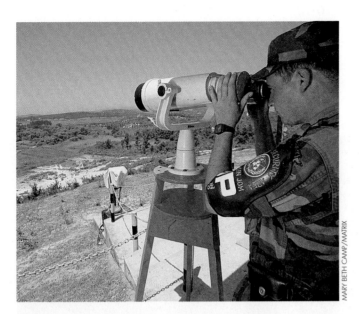

MARY BETH CAMP/MATRIX

Figure 14.15 A United Nations observer peers across the demilitarized zone to North Korea. This is one of the tensest borders on Earth.

line. The 150-mile-long (240-km), 2.5-mile-wide (4-km) demilitarized zone (DMZ) is a virtual no man's land of mines, barbed wire, tank traps, and underground tunnels (Fig. 14.15). The world's largest concentration of hostile troops faces off on either side of it. Remarkably, with the near absence of human activity, rare birds like the Manchurian crane and other endangered animal species have taken refuge in this narrow strip.

Contrasts Between the Two Koreas

The Korean armistice line divides one people, united by their ethnicity and language, into two very different countries (see Fig. 14.16 and Table 10.1). North Korea has an area of about 46,500 square miles (120,400 sq km) inhabited by about 22 million people (2001), with a density approaching 473 per square mile (183 per sq km). South Korea's corresponding figures are 48.8 million people in only 38,000 square miles, averaging 1,287 per square mile (c. 497 per sq km). South Korea is a republic that has fluctuated between attempts at democracy and a repressive military dictatorship. It has a capitalist economy heavily dependent on relationships with the United States and Japan. North Korea is a rigid and very tightly controlled Communist state that promotes what it calls *juche*, or self-reliance, allowing little economic or other exchanges with the outside world. Its principal ties have shifted between China and the Soviet Union; with the collapse of the USSR, China became North Korea's main ally and benefactor.

From World War II until his death in 1994, the dictator Kim Il Sung, whom his citizens called "The Great Leader," governed North Korea. His son and successor, Kim Jong Il, has perpetuated the country's militaristic character. North Korea spends a staggering 30% of its gross domestic product on its military, compared with South Korea's expenditure of only 3.2%, China's at 1.2%, and Japan's at 1%.

There are marked physical contrasts between the two countries. Although both are mountainous or hilly, North Korea is more rugged. North Korea has a humid continental long-summer climate, with hot summers and cold winters. South Korea is mostly a humid subtropical area with shorter and milder winters. Both are monsoonal, with precipitation concentrated in the summer.

Rice is the staple food in both countries. Rice-growing conditions are best in South Korea and diminish northward. Unlike the North, South Korea is able to double crop irrigated rice fields by growing a dry-field winter crop such as barley after the rice has been harvested. The main supplementary crop in the North is corn, which must be grown in the summer and so cannot occupy the same fields as rice. Beginning in 1995, successive waves of flood and drought brought famine to North Korea. The natural calamities only contributed to the already plummeting economic health of the country that came in the wake of the collapse of the Soviet Union, formerly North Korea's main benefactor and trading partner. North Korea's almost impenetrable veil of secrecy made it difficult to calculate the losses, but estimates of the number of people who died in the famine range from 900,000 to 2,400,000, or up to 10% of the country's prefamine population. As one crop after another failed, North Korea gradually and reluctantly sought food aid from abroad. It coaxed emergency supplies of wheat from the United States in part by threatening to withdraw from its 1994 agreement with the United States to halt its nuclear weapons program. Although food aid from the United States, China, and South Korea helped alleviate the immediate crisis, the poor conditions of North Korea's health system, drinking water supplies, and electrical generation have perpetuated malnutrition and disease.

Another physical contrast between the two Koreas is that most of the nonagricultural natural resources are in North Korea. All of the peninsula's major resources — coal, iron ore, some less important metallic ores, hydropower potential, and forests — are more abundant in the North than in the South. North Korea was thus originally the more industrialized state, featuring a typical Communist emphasis on mining and heavy industry, together with hydroelectric production and timber products. However, while the North remained focused on these efforts, the South took over industrial leadership in the 1970s and 1980s by developing a dynamic and diversified capitalist industrial economy. In an effort to begin to close the gap, North Korea opened the Rajin-Sonbong Free Economic and Trade Zone in the 1990s. Centered at the extreme northwest corner of North Korea around the twin cities of Rajin and Sonbong, and ringed by barbed wire, the 288-square-mile (720-sq-km) trade zone is intended to attract foreign investment and industries that will

capitalize on this site's favorable location near the borders of Russia and China.

In the early 2000s, South Korea's main exports were electrical and electronic equipment and appliances, automobiles, textiles, shoes, iron, steel, and ships. Large investments from

a

b

Figure 14.16 The two Koreas are like night and day. A satellite image taken at night shows the profound economic distinctions, with prosperous South Korea awash with urban and industrial lights while North Korea is nearly dark, except for the showcase capital of Pyongyang (a). Affluent shoppers enjoy the fruits of free enterprise in South Korea (b) while military security preoccupies austere Stalinist North Korea (c).

c

The economic and strategic orientations of Australia and New Zealand have been changing, generally away from Britain and toward East Asia and the United States. The recent economic crisis in Asia may dim the prospects for the long-anticipated emergence of the 21st century as the "Pacific Century."

Volcanic high islands and continental islands contain most of the region's economic resources and human populations. The low islands are poorer and less populated, and are vulnerable to anticipated rises in sea level.

Indigenous peoples of the Pacific, including the Aborigines of Australia and the Maori of New Zealand, have unique ethnogeographical perspectives that conflict with the worldviews of the Europeans who have come to dominate their homelands.

A high degree of endemism is characteristic of island flora and fauna, and introduced exotic species have taken a large toll on the native wildlife.

CHAPTER 15

A Geographic Profile of the Pacific World

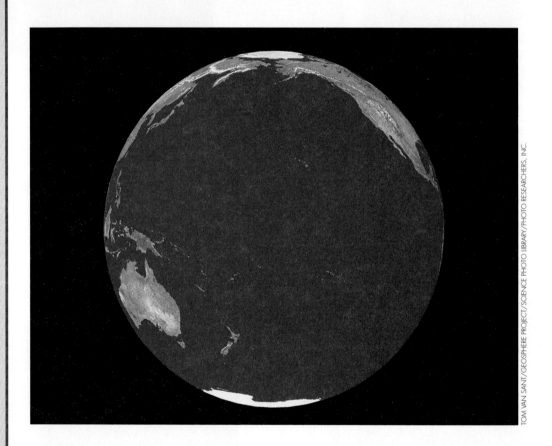

Earth's largest ocean, the vast Pacific. Only about 30 million people — about the population of California — are scattered over this huge portion of the Earth. This is a composite of several thousand satellite images.

Covering fully one-third of Earth's surface, the Pacific World is mostly water. The world's largest ocean, the Pacific, is bigger than all Earth's continents and islands combined. Before World War II, the Western world spawned legends about this ocean and its islands as a kind of utopia. Some islands still do have an idyllic quality for foreign visitors. However, there has long been trouble in this paradise. On many islands, foreign traders, whaling crews, labor recruiters, and other opportunists exploited the indigenous peoples, reducing their numbers and disrupting their cultures. The military battles of World War II further shattered the idyllic qualities of many islands. Yet today, the Pacific mystique, which has been perpetuated in books and films, forms part of the allure for booming tourism development across the region. This chapter presents the cultural and natural histories of the Pacific World, summarizes the impacts of colonial enterprises in the region, and discusses some of the obstacles to development confronting the countries of this watery realm.

15.1   Area and Population

The Pacific World region, often called Oceania, includes Australia, New Zealand, and the islands of the mid-Pacific lying mostly between the Tropics (Table 15.1). The Pacific islands nearer the mainlands of East Asia, Russia, and the Americas are excluded here on the basis of their close ties with the nearby continents. Large areas of the eastern and northern Pacific that contain few islands are also discounted. Australia and New Zealand are included in the Pacific World region because of their strong political and economic interests in the tropical islands, their similar insular character, and the ethnic affiliations of their original inhabitants with the peoples of those islands. But they are also uniquely European in character and dwarf the other countries in economic and political clout. Therefore, they earn separate treatment in Chapter 16.

The Pacific islands are commonly divided into three principal regions: Melanesia, Micronesia, and Polynesia (Fig. 15.1). The islands of **Melanesia** (Greek: "black islands"), bordering Australia on the northeast, are relatively large. New Guinea, the largest, is about 1,500 miles (c. 2,400 km) long and 400 miles (c. 650 km) across at the broadest point. These islands are generally hot, damp, mountainous, and carpeted with dense vegetation. **Micronesia** (Greek: "tiny islands") includes thousands of scattered small islands in the central and western Pacific, mostly north of the equator. **Polynesia** (Greek: "many islands") occupies a greater expanse of ocean than does either Melanesia or Micronesia. It is shaped like a rough triangle with corners at New Zealand, the Hawaiian Islands, and remote Easter Island.

The typical Pacific island country has about 100,000 to 150,000 people in an area of 250 to 1,000 square miles (c. 650 to 2,600 sq km), consists of a number of islands, is poor economically, is an ex-colony of Britain, New Zealand, or Australia, and depends heavily on foreign economic aid. As of 2002, nine fully independent states had emerged, with populations ranging from 5 million in Papua New Guinea, which is exceptionally large in population and area for this region but has a low population density, to 10,000 on tiny Nauru island. The highest population densities are in the smallest island groups, notably the Tuamotus and the Ellices. The region's total population is 31 million. Only Polynesia has an apparent problem of overpopulation, and there is considerable emigration, especially from the Cook Islands, Samoa, Tonga, and Fiji to New Zealand and North America.

15.2   Physical Geography and Human Adaptations

The landscapes, seascapes, and cultural geographies of the islands scattered across the tropical Pacific vary widely. However, there are generally two types of islands: "high" and "low" (Fig. 15.2, p. 400). Most high islands are the result of volcanic eruptions, but high continental islands (islands attached to continents before sea level changes and tectonic activities isolated them) are much larger and more complex in origin. The low islands are made of coral, a material composed of the skeletons and living bodies of small marine organisms that inhabit tropical seas.

These very different settings offer sharply different opportunities for human livelihoods. Overall, however, except in urban, industrialized Australia and New Zealand, the Pacific peoples are rural, living in small farming or fishing villages where they have long depended on root crops (e.g., taro and yams), tree crops (e.g., breadfruit and coconuts), sea fish, and

Political Unit	Area (thousand/sq mi)	Area (thousand/sq km)	Estimated Population (millions)	Annual Rate of Increase (%)	Estimated Population Density (sq mi)	Estimated Population Density (sq km)	Human Development Index	Urban Population (%)	Arable Land (%)	Per Capita GDP PPP (SUS)
Australia and New Zealand										
Australia	2988.9	7741.2	19.4	0.6	6	2	0.936	85	6	23,200
New Zealand	104.5	270.5	3.9	0.8	37	14	0.913	77	9	17,700
Total	**3093.4**	**8011.7**	**23.3**	**0.63**	**8**	**3**	**0.932**	**84**	**6.1**	**22,339**
Other										
Federated States of Micronesia	0.27	0.7	0.1	2.5	444	171	NA	27	NA	2000
Fiji	7.1	18.3	0.8	1.9	119	46	0.757	46	10	7300
French Polynesia	1.5	4.0	0.2	1.6	153	59	NA	54	1	10,800
Guam	0.21	0.55	0.2	2.5	744	287	NA	38	11	21,000
Marshall Islands	0.07	0.18	0.1	2.2	1007	389	NA	65	0	1670
New Caledonia	7.2	18.6	0.2	1.7	30	12	NA	71	0	15,000
Palau	0.18	0.46	0.02	1	107	41	NA	71	NA	7100
Papua New Guinea	178.7	462.8	5	2.3	28	11.0	0.534	15	0.1	2500
Solomon Islands	11.2	28.9	0.5	3.4	41	16	NA	13	1	2000
Vanuatu	4.7	12.2	0.2	3	44	17	NA	21	2	1300
Samoa	1.1	2.8	0.2	2.4	155	60	0.701	33	19	3200
Kiribati	0.28	0.73	0.1	2.4	337	130	NA	37	0	850
Nauru	0.009	0.02	0.01	1.4	1412	545	NA	100	0	5000
Tonga	0.29	0.75	0.1	2.1	349	135	NA	32	24	2200
Tuvalu	0.01	0.026	0.01	2.1	1100	425	NA	18	0	1100
Total	**212.8**	**551.0**	**7.74**	**2.3**	**36**	**14**	**0.569**	**23**	**0.7**	**3937**
Summary Totals	**3306.1**	**8562.8**	**31.0**	**1.1**	**9.4**	**3.6**	**0.857**	**68.6**	**5.8**	**17,729**

Sources: *World Population Data Sheet, Population Reference Bureau, 2001; U.N. Human Development Report, United Nations, 2001; World Factbook, CIA, 2001.*

pigs. Outside Australia and New Guinea, there are only four cities with populations greater than 50,000: Port Moresby in Papua New Guinea, Suva in Fiji, Noumea in New Caledonia, and Papeete in Tahiti.

High Islands

Some **high islands** to the north and northeast of Australia are **continental islands** (once attached to nearby continents), including New Guinea, New Britain, and New Ireland in the Bismarcks; New Caledonia, Bougainville, and smaller islands in the Solomons; the two main islands of Fiji; and a few others. Mountains over 16,000 feet (4,877 m) rise high in New Guinea, and there are lower but still imposing mountains on other islands.

Most of the region's high islands are not continental but volcanic in origin and have a familiar pattern. There is a steep central peak with ridges and valleys radiating outward to the coastline. Permanent streams run through the valleys. A coral reef surrounds the island, and between the shore and the reef is a shallow lagoon. Many of the high volcanic islands are spectacularly scenic (Fig. 15.2a). Classic high islands in the region include the Polynesian groups of Hawaii, Samoa, and the Society Islands.

Some of the volcanic high islands of the Pacific comprise **island chains.** These are formed by the oceanic crust sliding over a stationary **hot spot** in Earth's mantle where molten magma is relatively close to the crust. As the crust slides over the geologic hot spot, magma rises through the crust to form new volcanic islands. The Hawaiian Island chain is an excellent example. The big island of Hawaii is now situated over the hot spot and is still active; it is one of the chain's youngest members. The Pacific Plate of Earth's crust is moving northwestward here; thus, older volcanic islands that were born over the hot spot have moved northwestward, where they have become inactive. Still older islands farther northwest in the chain have submerged to become **seamounts,** or underwater volcanic mountains.

Geographer Tom McKnight explains that the high island's topography historically had a distinct influence on cultural and economic patterns. Each drainage basin formed a relatively distinct unit that was governed by a different chief. Within each unit, there were several subchiefs, each of them controlling a bit of each type of available habitat: coastal land, stream land, forested land, and gardening land. Since available resources were distributed relatively equitably among and between chiefdoms, there were few power struggles or monopolies over resources.

Today, the main cities and seaports of the Pacific World, and largest populations, are in the volcanic high islands and the continental islands. Valuable minerals are scarce, but the rich soils (typically derived from volcanic materials) support a diversity of tropical crops. Generally, these islands are more prosperous than the low islands.

Low Islands

The **low islands** are formed of coral. Most take the shape of an irregular ring surrounding a lagoon; such an island is called an **atoll.** Generally, the coral ring is broken into many pieces, separated by channels leading into the lagoon, but the whole circular group is commonly considered one island. Charles Darwin devised a still widely accepted explanation for the three-stage formation of atolls (Fig. 15.3, p. 400). First, coral builds a fringing reef around a volcanic island. Then, as the island slowly sinks, the coral reef builds upward and forms a barrier reef separated from the shore by a lagoon. Finally, the volcanic island sinks out of sight, and a lagoon occupies the former land area whose outline is reflected in the roughly oval form of the atoll.

The low islands are generally smaller than the volcanic high islands and lack the resources to support dense populations. They are typically fringed by the waving coconut palms that are a mainstay of life and the trademark of the South Sea isles. The islands of the republic of Kiribati (with the unlikely pronunciation KIH-rih-bus) are typical of the picturesque low islands idealized by Hollywood and travel brochures (Fig. 15.2b).

Despite their idyllic appearance, these islands pose many natural hazards to human habitation. The lime-rich soils are often so dry and infertile that trees will not grow, and shortages of drinking water limit permanent settlement. Their low elevation above sea level provides little defense against storm waves and *tsunamis*, the waves generated by earthquakes.

Before the intrusion of outsiders, the island economies were based heavily on subsistence agriculture (with strong reliance on coconuts), gathering, and fishing. Geographer McKnight explains that sparse resources were distributed with relative uniformity among the low islands' populations. The political and social units were smaller and less structured than those of high islands. These resource-poor low islands have a long history of population limiting factors such as war, infanticide, and abortion.

Many low island countries now rely heavily on modern commercialized versions of the coconut and fishing economy. The coconut is so important in the history of their subsistence that these islands have developed a "coconut civilization." The coconut provides food and drink for the islanders, and the dried meat, known as copra, is the only significant export from many islands. The husks and shells of the nuts have many uses, as do the trunks and leaves of the coconut palms. People make baskets and thatching from the leaves, for example, and use timber from the trunk for construction and making furniture.

The Vulnerable Island Ecosystem

Island ecosystems in the Pacific region, like those across the globe, are typically inhabited by **endemic** plant and animal species — those found nowhere else in the world. They result from a process in which ancestral species colonize the islands from distant continents and, over a long period of isolation and successful adaptation to the new environments, evolve to become new species. Gigantic and flightless animals, like the moa bird of New Zealand and the giant tortoises of the Galapagos and Seychelles islands, are typical of endemic island species.

Generally developing in the absence of natural predators and inhabiting relatively small areas, island species have proved to be especially vulnerable to the activities of humankind. People purposely or accidentally introduce to the island new plant and animal species (called **exotic species**) that often prey upon or overtake the endemic species. Habitat destruction and deliberate hunting also lead to extinction. For example, New Zealand's moas and the dodo birds of the Indian Ocean island of Mauritius are now "dead as dodos." Indigenous inhabitants of New Zealand set fires to hunt the giant ostrichlike moa and had already killed off most of the birds by the time the Maori people colonized New Zealand around A.D 1350. By 1800, the moa was extinct.

From one end of the Pacific World to the other, there is heightened concern about such human-induced environmental changes. The U.S. Hawaiian Islands have become known to ecologists as the "extinction capital" of the world because of the irreversible impacts that exotic species, population growth, and development have had on indigenous wildlife there. Environmentalists are also concerned about the impact of commercial logging on the island of New Guinea (Fig. 15.4, p. 400). About 22,000 plant species grow there, fully 90% of them endemic. And ecologists consider nearby New Caledonia to be one of the world's biodiversity hot spots because of the ongoing human impact on its unique flora and fauna.

Human-induced extinctions complement a long list of natural hazards to island species. For example, island animals are particularly prone to natural catastrophes such as typhoons (hurricanes) and volcanic eruptions. Most animals find it difficult or impossible to evacuate in the event of a natural catastrophe, and the island's distance from other lands may hinder or prevent recolonization. So, although island habitats have a higher percentage of endemic species than would be found on mainlands, environmental difficulties often cause the number of different kinds of species (the species diversity) to be lower.

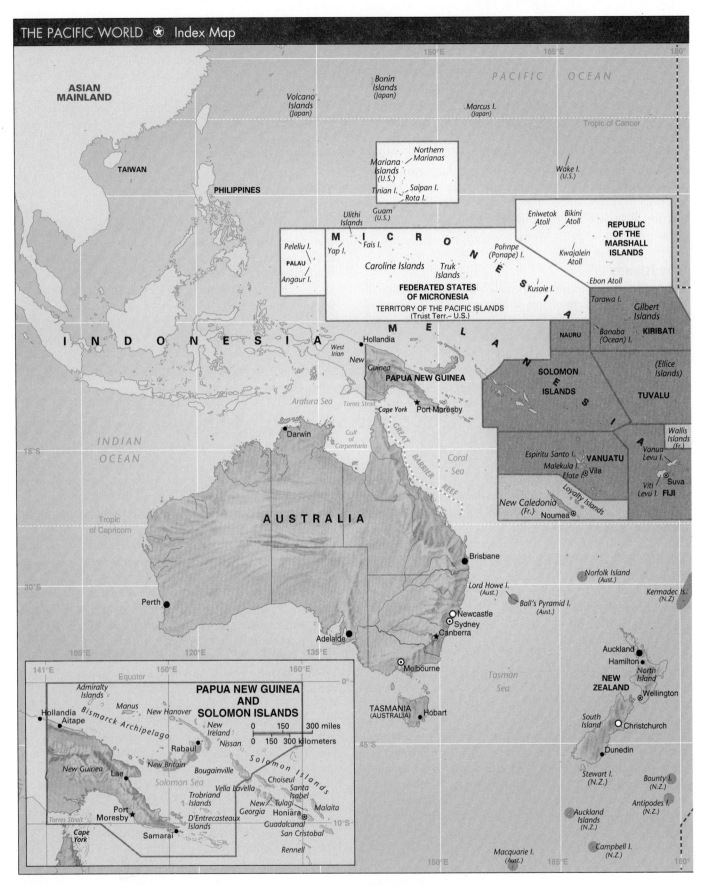

Figure 15.1 General reference map of the Pacific World. In some cases, the tints for political affiliation do not show precise political boundaries. Note the peculiar jog in the International Date Line near Kiribati (along the equator). In a bid to attract more tourists by being the first country in the world to greet the new millennium, Kiribati unilaterally decided in 1997 to shift the date line eastward by more than 2,000 miles.

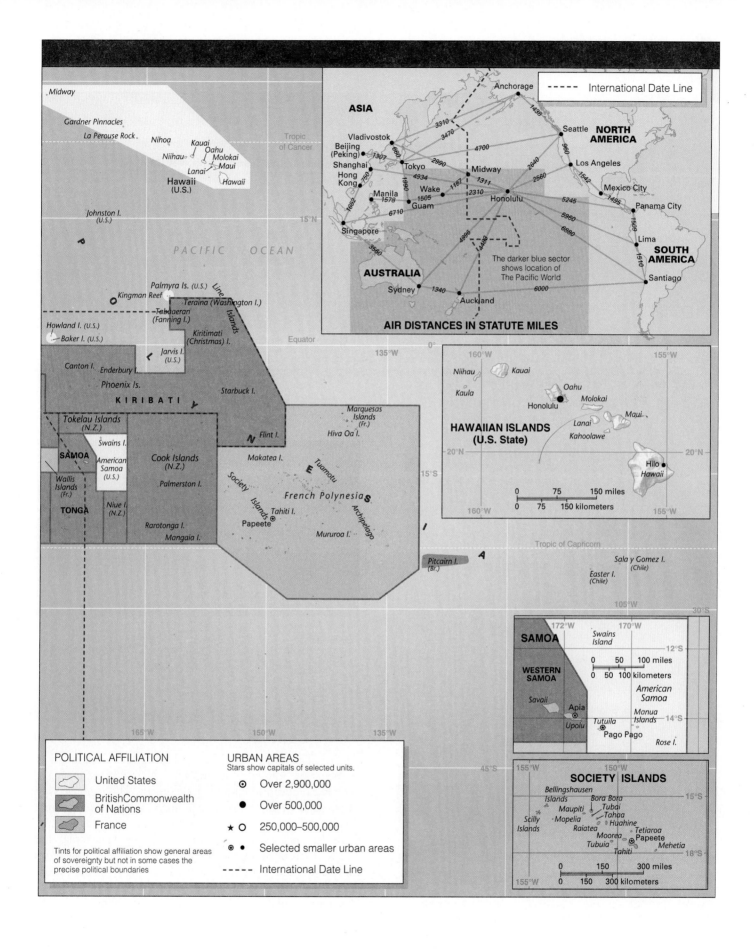

Midway

Gardner Pinnacles
La Perouse Rock
Nihoa
Kauai
Oahu
Niihau
Molokai
Maui
Lanai
Hawaii
Hawaii
(U.S.)

Johnston I.
(U.S.)

P

PACIFIC OCEAN

Palmyra Is. (U.S.)
Kingman Reef
Teraina (Washington I.)
Tabuaeran
(Fanning I.)
O

Howland I. (U.S.)
Baker I. (U.S.)

Kiritimati
(Christmas) I.

Line Islands

Jarvis I.
(U.S.)

Canton I.
Enderbury I.
Equator

Phoenix Is.
K I R I B A T I
Starbuck I.
Y

Tokelau Islands
(N.Z.)
Swains I.
N
Flint I.
SAMOA
American
Samoa
(U.S.)
Cook Islands
(N.Z.)
Marquesas
Islands
(Fr.)
Hiva Oa I.
Makatea I.
Wallis
Islands
(Fr.)
Palmerston I.
E
Tuamotu
Society
Islands
Archipelago
TONGA
Niue I.
(N.Z.)
French Polynesia S
Rarotonga I.
Tahiti I.
Papeete
Mangaia I.
Mururoa I.
I

Pitcairn I.
(Br.)
A

Tropic of Capricorn
Sala y Gomez I.
(Chile)
Easter I.
(Chile)

105°W
30°S

165°W
150°W
135°W
45°S

ASIA

Anchorage
- - - - - International Date Line

Vladivostok
Seattle
NORTH
AMERICA
3310
1438
Beijing
(Peking)
3470
660
1307
4700
960
Los Angeles
Shanghai
Tokyo
2640
Hong
Kong
1990
4934
Midway
1542
Mexico City
750
2560
Manila
Wake
1187
1311
1495
Guam
1505
2310
Honolulu
Panama City
1578
1662
5245
1509
6710
Singapore
5960
Lima
SOUTH
3560
4996
6680
1510
AMERICA
14493
The darker blue sector
Santiago
shows location of
AUSTRALIA
The Pacific World
Sydney
1340
6000
Auckland
AIR DISTANCES IN STATUTE MILES

160°W
155°W
Niihau
Kauai
Kaula
Oahu
Molokai
Honolulu
Maui
HAWAIIAN ISLANDS
Lanai
(U.S. State)
Kahoolawe
20°N
20°N
Hilo
Hawaii

0   75   150 miles
0   75   150 kilometers
160°W
155°W

15°N

135°W
0°
15°S

172°W
170°W
SAMOA
Swains
Island
12°S
WESTERN
SAMOA
0   50   100 miles
0   50   100 kilometers
American
Samoa
Savaii
Apia
Manua
Islands
Upolu
Tutuila
14°S
Pago Pago
Rose I.

155°W
150°W
SOCIETY ISLANDS
Bellingshausen
Islands
Bora Bora
15°S
Maupiti
Tubai
Scilly
Mopelia
Tahaa
Islands
Huahine
Raiatea
Tetiaroa
Moorea
Papeete
Tubuia
Mehetia
Tahiti
18°S

0   150   300 miles
0   150   300 kilometers
155°W
150°W

POLITICAL AFFILIATION

United States	
British Commonwealth of Nations	
France	

Tints for political affiliation show general areas
of sovereignty but not in some cases the
precise political boundaries

URBAN AREAS
Stars show capitals of selected units.

⊙   Over 2,900,000

●   Over 500,000

★ ○   250,000–500,000

⊗ ●   Selected smaller urban areas

- - - - -   International Date Line

ROBERTO ARAKAKI/INTERNATIONAL STOCK

KARL & JILL WALLIN/FPG/GETTY IMAGES

a

b

Figure 15.2 (a) Bora Bora in Tahiti is a classic and much-romanticized high island. (b) Low islands are most associated with the coconut civilizations of the Pacific. Low islands are extremely vulnerable to the dangers that would be posed by global warming.

a Volcanic island with fringing reef

b Slight subsidence barrier reef

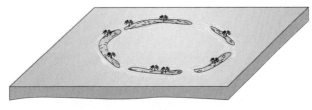

c An atoll

Source: From Gabler et al. 1987. *Essentials of Physical Geography*, p. 562.

Figure 15.3 Charles Darwin's explanation of the development of an atoll. First (a), a fringing coral reef is attached to the volcanic island's shore. Then, as the island subsides (b), a barrier reef forms. With continued subsidence, the coral builds upward, and the volcanic center of the island finally becomes completely submerged, forming an atoll (c).

© MORTON BEEBE, S.F./CORBIS

Figure 15.4 It looks lush, but this verdant landscape of the Markham Valley in New Guinea bears the unmistakable signs of deforestation. The grasslands here were previously cloaked in tropical rain forest.

15.3 Cultural and Historical Context

Their indigenous cultures having been greatly diminished in numbers and influence, Australia and New Zealand today are mainly European in culture and ethnicity. The region's other cultures, however, are overwhelmingly indigenous and quite diverse. Aside from Australia and New Zealand, with their majority European populations, and Fiji, New Caledonia, and Guam, each of which has populations about half indigenous and half foreign, about 80% of the Pacific's people are indigenous. The balance is 13% Asian and 7% European. Of the indigenous populations, Melanesians comprise about 80%, with 14% Polynesian and 6% Micronesian.

The Pacific region began to be settled about 60,000 years ago. Although the archeology is far from definitive on this

DEFINITIONS + INSIGHTS

Deforestation and the Decline of Easter Island

"Easter Island is Earth writ small," wrote naturalist Jared Diamond. Recent archeological and paleobotanical studies suggest that the civilization that built the island's famous monolithic stone statues destroyed itself through overpopulation and abuse of natural resources (Fig. 15.A). **Deforestation,** or the removal of trees by people or their livestock, has serious repercussions wherever it occurs, but perhaps nowhere are these impacts so apparent as on Easter Island.

When the first colonists from eastern Polynesia reached remote Easter Island in about A.D. 400, they found the island cloaked in subtropical forest. Plant foods, especially from the Easter Island palm, and animal foods, notably porpoises and seabirds, were abundant. The human population grew rapidly in

this prolific habitat. The complex, stratified society that emerged on the island grew to an estimated 7,000 to 20,000 people between A.D. 1200 and 1500, when most of the famous statues were built. Apparently in association with their religious beliefs, Easter Islanders erected more than 200 statues, some weighing up to 82 tons and reaching 33 feet (10 m) in height, on gigantic stone platforms. At least 700 more statues were abandoned in their quarry sites and along roads leading to their would-be destinations, "as if the carvers and moving crews had thrown down their tools and walked off the job," Diamond observed.[a]

"Its wasted appearance could give no other impression than of a singular poverty and barrenness," [b] the Dutch explorer Jacob Roggeveen wrote of the island on the day he discovered it— Easter Sunday 1722. Not a single tree stood on the island. The depauperate landscape bears testimony to the fate of the energetic culture that built the great statues. By A.D. 800, people were already exerting considerable pressure on the island's forests for fuel, construction, and ceremonial needs. By 1400, people and the rats they introduced to the island caused the local extinction of the valuable Easter Island palm. Continued deforestation to make room for garden plots and to supply wood to build canoes and to transport and erect the giant statues probably eliminated all of the island's forests by the 15th century. By then, people had hunted to extinction many terrestrial animal species and could no longer hunt porpoises because they lacked the wood needed to build seagoing canoes. Crop yields declined because deforestation led to widespread soil erosion. There is evidence that in the ensuing shortages, people turned on each other as a source of food. By about A.D. 1700, the population began a precipitous decline to only 10–25% of the number who once lived on this isolated Eden.

GEORGE HOLTON/PHOTO RESEARCHERS, INC.

Figure 15.A   The Polynesian people who erected these *moai* statues between A.D. 1200 and 1500 deforested most of Easter Island—in part to supply levers and rollers for the statues—and thus precipitated the demise of their civilization.

[a]Jared Diamond, "Easter's End." *Discover*, 16(8): 1995, p. 64.
[b]Ibid.

point, there is a consensus that this is when the first people settled in Australia. They were the ancestors of today's Aborigines. Some ethnographers refer to the Aborigines as a distinct Australoid race. However, they share racial characteristics with other indigenous groups in Asia, including the Mundas of central India, the Veddahs of Sri Lanka, and even the Ainu of Japan. All of these groups, along with Australia's Aborigines, probably descended from a common ancestral race in Asia, and the Aborigines developed their distinctive features over tens of thousands of years of habitation of Australia. They probably arrived there in several waves either by sea from Timor or by land from what is now

New Guinea. During their migrations, sea levels were much lower, straits were narrower and easier to navigate, and there was a land bridge between New Guinea and Australia until about 10,000 B.C.

New waves of migrants known as Austronesians (previously called Malayo-Polynesians) came after 5000 B.C. Their ancestral stock was southern Mongoloid, probably originating in southern China and Taiwan. Their descendants migrated to the mainland and islands of Southeast Asia. From there, over a period of several thousand years, the Austronesians embarked on voyages to Madagascar, New Zealand, Easter Island, and finally, Hawaii, island-

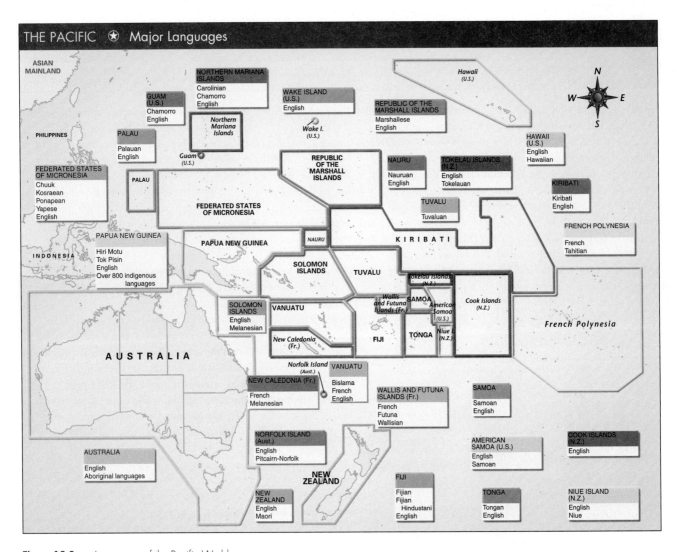

THE PACIFIC ✪ Major Languages

Figure 15.5   Languages of the Pacific World.

hopping all the way and accomplishing extraordinary feats of navigation in their outrigger canoes. Their settlement of Polynesia came rather late in these adventures, within the last 1,500 years. Today's Micronesians and Polynesians are mainly their descendants, but there has been considerable racial mixing over the centuries.

The Pacific region's linguistic diversity is extraordinary (Fig. 15.5). Most of the indigenous peoples of the region speak languages within the Austronesian group. The Melanesian peoples, however, speak Papuan languages, which are not a distinctive linguistic family but an amalgam of often unrelated and mutually unintelligible tongues. There are about 700 Papuan languages. Papua New Guinea alone is home to 20% of the world's spoken languages, making it by far the world's most linguistically diverse country. Such is the diversity that there is an average of one lan-

guage per 5,000 people in Papau New Guinea. There are fewer languages in Micronesia and Polynesia, and more of them are mutually intelligible. Reflecting the colonial past, English and French are official languages in some of the islands and are spoken widely across the region. Another lingua franca is pidgin, comprised of English and other foreign words mixed with indigenous vocabulary and grammar. Pidgin is the official language of Papua New Guinea.

Some ancient indigenous belief systems thrive in parts of the region, often nestled within Christianity and other recently introduced faiths (Fig. 15.6). Overwhelmingly, however, the peoples of the Pacific practice Christianity, with Methodists, Mormons, and Catholics the largest denominations. Where Asians have settled, there are pockets of Hinduism, Buddhism, and Islam.

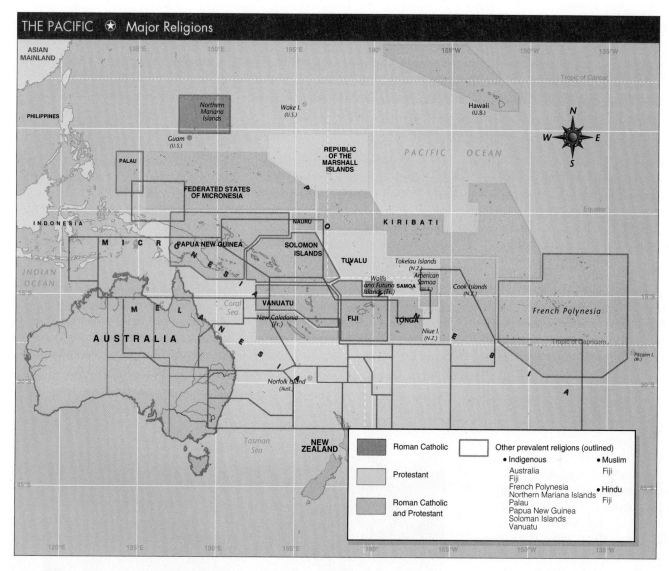

THE PACIFIC ✪ Major Religions

Figure 15.6   Religions of the Pacific World.

Europeans began to visit the Pacific islands in the 16th century and their impact was especially profound after 1800. Spanish and Portuguese voyagers were followed by Dutch, English, French, American, and German explorers. Many famous names are connected with Pacific exploration, including Magellan, Tasman, Bougainville, La Perouse, and the most famous of all, British Captain James Cook, who undertook three great voyages in the 1760s and 1770s, only to be killed by Hawaiian Islanders in an unexpected dispute in 1779. Europeans were interested mainly in the high islands because of their relative resource wealth. They sought sandalwood, pearl shells, whales, and local people to indenture as whaling crews and as manual laborers. They also came as missionaries. On island after island, European penetration decimated the islanders and disrupted their cultures. The intruders introduced venereal and other infectious diseases, alcohol, opium, forced labor, and firearms, which worsened the bloodshed in local wars.

Europeans created new settlement patterns, disrupted old political systems, and rearranged the demographic and natural landscapes. In Polynesia, for example, arriving Europeans typically sought shelter for their vessels on the lee side of an island. That safe place became the island's European port. Whatever chief happened to be in control of that spot typically, because of his association with the Europeans, became more powerful

Perspectives from the Field

With Aborigines and Missionaries in Northwestern Australia

In the austral winter of 1987, I served as guide and lecturer aboard a 30-passenger vessel that took Western tourists, Americans and British mostly, to the remote coastline of the Kimberley region in northwestern Australia. Except for the towns of Wyndham and Broome, respectively on the northeastern and southwestern edges of this region, this is a wilderness, virtually devoid of settlements, roads, and other infrastructure. The major population, quite small in number, is aboriginal, but there are a few European missionaries and cattle ranchers. It is a land of sandstone, spinifex grass, Eucalyptus trees, and abundant wildlife, including the fearsome saltwater crocodile.

My journal from August 10, 1987, reflects on past and present cultural encounters in this tropical wilderness:

Last night we sailed from the Bonaparte Archipelago and made our way to the northern tip of the Kimberley. This part of the mainland is known as the Mitchell Plateau. It has long been an important aboriginal region, and today most of the lands in this coastal area are actually Aboriginal Reserve Lands. Traditionally very resistant to outsiders, these indigenous people succeeded in driving off Malays, Indonesians, and Europeans over a period of several hundred years. Within the past 100 years, various Christian groups successfully established several missions in the area.

Early in the afternoon we entered Napier Broome Bay and did a brief shore excursion. From the beach where our zodiacs [inflatable boats] landed, we climbed up to a rocky sandstone promontory. From there was an excellent view of the beach and the mangrove-lined inlets on either side. In the distance the smoke of bush fires lit by Aborigines could be seen. Returning to the beach, we saw a young green turtle poke its head out of the water for a look around. When he saw us he decided it wasn't safe, and submerged and paddled off rapidly into open water. Some of the crewmen strung out a long net from the beach in the hope of catching some mullet. Almost immediately, a small school blundered into the net. The catch became our gift to the aboriginal people we met later in the day.

Peter [Sartori, the vessel's captain] shifted the ship around Mission Bay, and we went ashore. We were received by a small but very enthusiastic party of aboriginal children, who enjoyed themselves immensely clambering in and out of the zodiacs. Their parents were our hosts for a very interesting couple of hours.

These people are members of the Wunambal tribe. They had come down from the Kalumburu Mission, about 18 miles inland on the King Edward River. One of them, a woman named Mary, was the last child born at the old Pago Mission on the shore of Mission Bay. The Pago Mission was founded by Benedictine (Spanish Catholic) monks in 1908. The missionaries' relationship with the Aborigines was not always peaceful; there were a few hostile encounters in which the missionaries came out the worse. However, the parties came to trust one another, and enjoyed good relations right up until the Pago Mission had to be abandoned in the 1930s, when its water supply ran out. The Kalumburu Mission was established to take its place.

Our hosts were very congenial and went out of their way to make us feel comfortable. We spoke freely with them, and found their English quite good; they had after all been educated at the Mission. The men, who had been wearing Western clothes, disappeared for a while and then returned decked out with loincloths and their finest white ochre dance paint. Just before dusk, a signal was given and a traditional *corroboree* or ceremonial dance began (Fig. 15.B). The women danced, too, but wore Western clothing,

Figure 15.B The Wunambal (with Basil at far left) performing a corroboree by the shore of Mission Bay, Kimberley, Australia.

JOE HOBBS

and had a minor role in the *corroboree*. The dance leader, Basil, was a real ham, boasting about his talents at every opportunity. But mainly he explained how each dance or skit told a story about some aspect of daily or ritual life of the Wunambal. It was quite clear that although they had been in a Christian mission environment for decades, these people held their ancestral beliefs about their origins and the world around them. Of the *Kalinda* dance, Basil said: "It has to do with cyclone Tracy. *Kalinda* is the song spirit. When we see people coming in a line to the *Pruru* or Dancing Place with the totems, we know it is the Song Spirits coming. When we know there are strong winds or cyclones coming, we take the totems to the hills for safety. When we are going to the hill where the Rolling Stone is, and we are halfway up the hill, the big stone rolls down and all the people stop until the stone falls to the ground. One man ventures out to see if it is safe; then all people come out to dance." Then there was the *Palgo*; Basil explained: "The spirits catch fish at night time. They light a bundle of sticks and take them to the waterhole. The fish are attracted by the light, and so we catch them." The last dance that told a story was the *Djarlarimirri* (Witch Doctor): "A man is killed by a spear, and the Witch Doctor is called. He performs his magic, and the man is restored to life." Finally there was the *Djalurru*, a dance without spiritual significance, performed just for fun.

A month later, I visited the ruins of the Pago Mission and the Kalumburu Mission that succeeded it, writing:

From the landing site I followed a dirt track through dry savanna to reach the old mission. Not much is left — some concrete floors, a few posts, and three wells that now have water in them — but its surroundings are attractive. A grove of old mangoes survives untended. The cemetery is overgrown. There is a billabong [pool] here used by water buffalo, and the locals say saltwater crocodiles lurk in it. The first monks who came to this place, in 1907, scrawled their names on a large boab tree still towering over the place. I learned more about their relations with the Aborigines. When the mission was founded, the locals were reluctant to accept it. A few stayed there. Most of the Aborigines in this area though came to get food and supplies, visited with their settled kinsmen, and then disappeared back into the bush.

Missionary records note that the Aborigines of the time practiced formalized warfare, at least once attacking the mission; in 1913, a missionary was speared. Later the place settled down, and the population grew. So did the demand for water, and soon there was not enough, and so the mission had to be abandoned in 1939 in favor of Kalumburu, whose construction had begun 4 years earlier up on the King Edward River.

Later in the day, I visited the Kalumburu Mission at the invitation of one of the nuns working there, the redoubtable Sister Agnes (Fig. 15.C), who sported about on an all-terrain vehicle. Of all things, rather than a cross as her motor vehicle's masthead, she had a bark painting of a *wandjina*, the aboriginal ancestral spirit of typhoons and monsoon rains. The Mission was celebrating the 55th anniversary of its founding. There was a mass going on during much of my brief visit, and I walked around the lifeless grounds. All was tidy. The grass had been mowed close and, in this clearing in the bush, amid tall palms and wide mangoes, stood residences, a clinic, the community's church, and a school. When mass ended, the Mission's entire population poured out. Every one of these aboriginal people, from all ages, wore Western clothes — one young girl even had a Mickey Mouse T-shirt. And each wore distinctive Mission and aboriginal marks on this feast day: a blue headband symbolizing the Mission, and traditional Wunambal white face paint.

Joe Hobbs

JOE HOBBS

Figure 15.C   Sister Agnes, Kalumburu Mission.

and wealthy than other island leaders. Eventually, he became high chief and often expanded his control to other islands; thus, the first Polynesian kings emerged.

Europeans, North Americans, and East Asians changed the natural history of the islands by introducing exotic crops and animals, including arrowroot and cassava; bananas; tropical fruits such as mangoes, pineapples, papayas, and citrus; coffee and cacao; sugarcane; and cattle, goats, and poultry. They established extensive sugar plantations on some islands, notably in the Hawaiian Islands, Fiji Islands (where sugar is still the major export), and Saipan in the Marianas. The valuable Honduran mahogany trees planted by British colonists in Fiji early in the 20th century are only now beginning to mature, and competition for export profits from this resource has become intense.

It is remarkable that as late as 1840 almost all of the Pacific islands had yet to be formally claimed by outside powers; only Spain had taken the Marianas. Then rapidly, Britain, France, Germany, and the United States vied for power. By 1900, the entire Pacific basin was in European and American hands. Germany held Western Samoa (now Samoa), part of New Guinea, the Bismarck Archipelago, Nauru, and Micronesia's most important islands. Britain controlled part of New Guinea, the Solomon Islands, Fiji, and smaller island groups, and held Tonga as a protectorate. France controlled New Caledonia, French Polynesia, and with Britain, the condominium (jointly ruled territory) of the New Hebrides. The United States held Hawaii, Guam, and eastern Samoa. When Germany lost its Pacific territories at the end of World War I, Japan took control of German Micronesia, New Zealand took Western Samoa, and Britain and Australia took the German section of New Guinea and the Bismarcks. Nauru came to be administered jointly by Britain, New Zealand, and Australia. With the outbreak of World War II, Japan quickly overran much of Melanesia. The ensuing ferocious battles between Japan and the Allies for control of New Guinea, the Solomon Islands, and much of Micronesia brought enormous changes to the region. Many of the cultures were in the Stone Age as the war began and by its end had seen the Atomic Age; Micronesia had become a nuclear proving ground. Everywhere, traditional peoples were drawn into cash economies and affected by other global forces originating far beyond their watery horizons.

Most of the colonists are gone, but the seeds of strife they planted are still germinating. Many conflicts in the region today reflect the demographic changes Europeans wrought to meet their economic interests. In many cases, the Europeans introduced laborers from outside the region because local peoples were too few in number or refused to work in the cane fields. Successive infusions of Chinese, Portuguese, Japanese, and Filipinos brought a polyglot character to Hawaii, with little apparent ethnic discord. But ethnic strife is ongoing in Fiji and the Solomon Islands, which became independent from Britain in 1970 and 1978, respectively.

The Solomon Islands' troubles date to 1942, when U.S. Marines fought to push Japanese troops off the island of Guadalcanal. The American forces enlisted the support of thousands of people from the neighboring island of Malaita. After the successful offensive, many of these Malaitans elected to stay on Guadalcanal, where they came to dominate economic and political life. The indigenous people of Guadalcanal resented their dominance and in particular what they saw as the loss of their lands to the Malaitans (who purchased the lands). Conflict between the groups finally erupted in 1999 and persisted for more than 2 years, with the majority Guadalcanalans murdering scores of Malaitans and driving tens of thousands off the island; most returned to Malaita.

The violence in the Solomon Islands was apparently inspired by ethnic unrest in Fiji. In the early 1900s, colonial British plantation owners imported mainly Hindu Indians as indentured laborers to work Fiji's sugarcane plantations. Today, indigenous Fijians now only slightly outnumber ethnic Indians. Here, too, the indigenous inhabitants resent the dominance of the foreign ethnic group; ethnic Indians control most of Fiji's economic life. A coup led by ethnic Fijians in 2000 succeeded in ousting Fiji's ethnic Indian prime minister and throwing out a 1987 constitution that granted wide-ranging rights to ethnic Indians. Accused of lawlessness, the coup leaders were threatened with economic sanctions by the European Union, which pays higher than market prices for most of Fiji's sugar crop, and by the United States, Australia, and New Zealand. The prospect of economic crisis led the army to take control of Fiji's affairs and promise a general election and return to civilian government in 2002. The unrest caused a crash in tourism, Fiji's most valuable industry. Prior to the coup, about 400,000 visitors came each year.

15.4   Economic Geography

A general lack of industrial development is characteristic of Pacific island economies (other than those of Australia and New Zealand), and the poverty typical of less developed countries (LDCs) prevails in the region. There are five major economic enterprises, none of them involving value-added processing of raw materials: exports of plantation crops (especially palm products), exports of fish, exports of minerals, services for Western military interests, and tourism. Significant manufacturing is limited to the Cook Islands and Fiji. Polynesia's Cook Islands, a self-governing territory in free association with New Zealand, has an industrial economy based on exports of labor-intensive clothing manufacturing. Fiji also began exporting manufactured clothes after the mid-1990s, when Australia granted preferential tariff treatment to Fijian-made garments.

A few Pacific countries are beginning to participate, in rather unlikely and sometimes unscrupulous ways, in the information technology revolution that has generated so much global wealth since about 1990. One example is tiny Tuvalu in Polynesia. In 2000, an American entrepreneur agreed to pay the Tuvalu government $50 million to acquire Tuvalu's two-letter Internet suffix: tv. While the new owner of the dot-tv domain has been busily profiting by selling Internet addresses, Tuvalu is delighted with its windfall. It used part of the money to finance the process of joining the United Nations. Vanuatu enjoys significant hard-currency revenue by permitting its country's phone code to be used for 1-900 sex lines. And as described shortly, Nauru and other island nations use telephone and digital technology to practice money laundering.

Mining takes place on only a few islands. The major reserves are on continental islands, notably petroleum, gold, and copper on the main island of Papua New Guinea, copper on Bougainville (an island belonging to Papua New Guinea), and nickel on the French-held island of New Caledonia. The 10,000 people of the tiny oceanic island of Nauru, which gained independence from Australia in 1968, once enjoyed a high average income from the mining of phosphate, the product of thousands of years of accumulation of seabird excrement, or guano. Until recently, Nauru's economy depended totally on annual exports of 2 million tons of phosphate and on revenues earned from overseas investments of profits from phosphate sales (the tallest building in Melbourne, Australia, is a product of Nauruan investments and is known affectionately as Birdshit Tower). This resource is approaching total depletion, however, and because the mining of phosphate has stripped the island of 80% of its soil and vegetation, turning it into a virtual lunar landscape, many people in Nauru are now looking for a new island home (Fig. 15.7).

Others are content with what they see as Nauru's most recent and, they hope, future source of revenue: money laundering. With the phosphate stocks dwindling, Nauruan entrepreneurs have undertaken offshore banking, in which they can operate loosely controlled banks anywhere but in Nauru. In Nauru's case, these are "shell banks," existing only on paper and specializing in turning money earned in illegal ways into money that looks legal. Nauruan law does not require any official oversight of offshore banking transactions, so no official records exist; therefore, any dirty funds moved through a bank registered in Nauru cannot be traced. Russian clients have registered half of Nauru's 400 shell banks, and the assumption is that much of the Russian "Mafia" money is laundered through them. Money laundering and tax haven schemes are also reported in Vanuatu, Palau, the Cook Islands, Marshall Islands, Niue, Tonga, and Samoa.

Issues of control over mineral wealth have brought unrest to New Caledonia and Bougainville. Separatists demanding independence for nickel-rich New Caledonia clashed with French police in the 1980s. An accord signed after the confrontation guaranteed a 1998 referendum for New Caledonia's people (known as Kanaks), who would then choose to become independent or remain under French rule. In 1998, however, apparently to maintain its interests in the colony's nickel, France postponed the vote until 2014. In the meantime, France is allowing New Caledonia to have considerable autonomy; since 1999, a legitimate New Caledonian government has been permitted to look after civil affairs.

In the late 1980s, a crisis shook Bougainville, a former Australian colony that in 1975 became, at Australia's insistence, part of newly independent Papua New Guinea. Most Bougainvilleans were unhappy about their association with Papua New Guinea and wanted to be united again with the Solomon Islands, to whose people they are culturally and ethnically related. Angry landowners calling themselves the Black Rambos destroyed mining equipment and power lines to force the closure of the world's third largest copper mine. These generally younger landholders were receiving none of the royalties and compensation payments made by New Guinean and Australian mining interests to an older generation of the island's inhabitants. Residents of Bougainville concerned about the ruinous environmental effects of copper mining and disturbed that most mine laborers were imported from mainland Papua New Guinea, with little mining revenue remaining on the island, joined the opposition. The growing crisis drove world copper prices to record high levels. Opposition to the Australian-backed mining interests of Papua New Guinea in Bougainville grew after 1988 into a full-fledged independence movement led by the Bougainville Revolutionary Army (BRA). War broke out in 1989 when the BRA leader declared independence from Papua New Guinea. Papua New Guinea responded with an economic blockade around Bougainville to prevent supplies from reaching the rebels, who salvaged World War II-era weapons and ammunition to fight superior troops. In a decade of fighting, Bougainville's 200,000 people suffered huge losses, with nearly 20,000 killed and 40,000 left homeless.

Figure 15.7   The 10 square miles (25 sq km) of land which is Nauru have been devastated by the phosphate mining that once made Nauruans among the wealthiest people per capita on Earth.

REGIONAL PERSPECTIVE   : : :   Foreign Militaries in the Pacific: A Mixed Blessing

The remote locales and nonexistent or small human populations of some Pacific islands have made them irresistible venues for weapons tests by the great powers controlling them (Fig. 15.D). In the 1940s and 1950s, U.S. authorities displaced hundreds of islanders from Kwajalein atoll in the Marshall Islands to make way for intercontinental ballistic missile (ICBM) tests. During those years, the United States also used nearby Bikini atoll as one of its chief nuclear proving grounds (the island's fame at the time led French designers to name a new two-piece bathing suit after it). In 1946, U.S. government authorities relocated the indigenous inhabitants of Bikini to another island, promising they could return when the tests were complete. Over the next 12 years on Bikini, there were 23 nuclear blasts out of a total of 67 conducted in the Marshall Islands. A 15-megaton hydrogen bomb detonation on Bikini in 1954 had the power of 1,000 Hiroshima bombs, blowing a mile wide crater in the island's reef and producing radioactive dust that fell downwind on the Rongelap atoll, where children played in the dust "as though it were snow," one observer wrote.[a] For decades the U.S. Atomic Energy Commission claimed that a sudden shift in winds took the fallout to inhabited Rongelap, but recent records proved authorities knew of the wind shift 3 days prior to the test. The Marshallese today suffer a legacy of cancers and deformities linked to the weapons tests on Bikini and Eniwetok atolls. So far, the United States has provided more than $63 million to compensate Marshall Islanders made ill by the blasts.

The Bikinians have never accepted their exile. But returning to Bikini has been a hazardous prospect given its legacy of radioactive soil. A repatriation of the Bikinians in the 1960s was found to be premature, as radiation persisted, and the nuclear nomads moved again. Recent tests have determined that Bikini has little ambient radioactivity and is essentially habitable, except for the cesium 137 that permeates the soil and becomes concentrated in fruits and coconuts. Bikinians do not want to go home to stay until they have assurances from the United States that the problem of poisoned produce can be resolved and that the island is completely safe. Along with the people of Eniwetok, they are seeking massive funding from the United States for complete environmental decontamination of their homeland and for further compensation for their travails.

Now that Bikini is at least safe to visit, some Bikinians are capitalizing on its formerly off-limits status. The island's lack of human inhabitants for more than 50 years has had the remarkable effect of

Figure 15.D   The nuclear explosion, code-named Seminole, at Eniwetok atoll in the Pacific Ocean on June 6, 1956. This atomic bomb was detonated at ground level and had the same explosive force as 13.7 thousand tons (kilotons) of TNT. It was detonated as part of Operation Redwing, an American program that tested systems for atomic bombs.

A 1988 cease-fire finally brought to an end the longest conflict in the Pacific since World War II.

The most dynamic element of the Pacific economies is the rapid growth of tourism, with most visitors coming from the United States and Japan. The Pacific islands continue to exert an irresistible appeal to people awash in the conveniences and congestion of the industrialized world.

Finally, the military interests of outside powers, particularly the United States and France, generate much revenue for some of the Pacific economies, notably Guam, American Samoa, and French Polynesia. This military presence is a mixed blessing, however, because of its environmental impacts and because it perpetuates dependence on foreign powers. These are among the geopolitical concerns discussed next.

rendering Bikini a true marine wilderness. In the late 1990s, it opened as a tourist destination, with naval ships sunk by the atomic blasts a highlight for many divers. The dive masters and tour guides are Bikinians.

Like the United States, France has had a nuclear stake in the Pacific. In 1995, after a 3-year hiatus, France resumed underground testing of nuclear weapons on the Mururoa ("Place of Deep Secrets") atoll in French Polynesia. There was an immediate backlash from some national governments and from environmental and other organizations. Crew members of the *Rainbow Warrior 2*, the flagship of the environmental organization Greenpeace, confronted French Navy ships near Mururoa in July 1995. French commandos used tear gas to overwhelm the crew and seize the ship. Greenpeace managed to televise the event, inflicting enormous public relations damage to France. Strongly antinuclear New Zealand protested loudly against the nuclear tests. In Australia, customers boycotted imported French goods and local French restaurants, and postal employees refused to deliver mail to the French embassy and consulates. Antinuclear protestors firebombed the French consulate in Perth. Australia joined New Zealand in filing a case at the World Court in The Hague aimed at halting French nuclear tests. Worldwide, informal boycotts diminished sales of French wines. In French Polynesia itself, activists for independence from France used the issue to highlight international attention to their demands. Over several days in September 1995, hundreds of anti-French demonstrators in Tahiti rampaged through French Polynesia's capital city of Papeete, looting and setting parts of the city on fire and causing millions of dollars of damage to the island's international airport, a vital tourist hub. Although French authorities continued to insist that the tests and the atoll itself were safe, France soon bowed to international pressure, ceasing its nuclear tests in the Pacific and signing the Comprehensive Test Ban Treaty in 1996.

Ironically, the greatest fear in the region now is that France will reduce its military — and therefore, economic — presence even more. Despite the nuclear tests, most inhabitants of French Polynesia have long favored continued French rule. Tourism and other local businesses contribute only about 25% of the territory's revenue; the remainder comes from French economic assistance, largely in the military sector. Many locals, however, whose parents and grandparents gave up subsistence fishing and farming for jobs in the military and its service establishments, are worried about their future. Similar fears stalk Palau, the Marshall Islands, and the Federated States of Micronesia (FSM), three independent nations carved from the former U.S. Trust Territory of the Pacific. The United Nations awarded that territory, which includes the Bikini and Eniwetok atolls, to the United States in 1946 as the world's first and only political-military zone. In 1986, the United States granted independence to the Marshall Islands and the FSM in an agreement known as the Compact of Free Association. In this agreement, the countries agreed to rent their military sovereignty to the United States in exchange for tens of millions of dollars in annual payments and access to many federal programs. The United States uses its leases to conduct military tests — for example, on Strategic Defense Initiative (Star Wars) technology. The United States also provides large cash grants and development aid to Palau, which became independent in 1994.

In these cases, the American government handed over the monies with no strings attached. The funds have been embezzled and misspent, and there is little development to show for them. In 2001, the United States began to reassess the payments with an eye to their eventual curtailment. Many islanders fear the bottoms of their economies will subsequently fall out. The people of the Marshall Islands and FSM are especially worried because more than half of the gross domestic product of each country is comprised of economic aid from the United States. The Marshall Islands' leaders, however, are taking a more defiant stand, saying the United States has no right to question what becomes of the aid money and pointing out the United States gets much in return in the form of extensive military rights. The United States is hard-pressed here because the Kwajalein Missile Range in the Marshalls is the only reasonable place in the world where, Pentagon officials say, they can test many of the Star Wars program components.

[a]Nicholas D. Kristof, "An Atomic Age Eden (but Don't Eat the Coconuts)." *New York Times*, March 5, 1997, p. A4.

15.5 Geopolitical Issues

Once entirely colonial, the Pacific World today is a mixture of units still affiliated politically with outside countries and others that have become fully independent. Since the end of World War II, the United States, Britain, Australia, and New Zealand have abandoned most of their colonies in the region. Only France has insisted on holding on to all of its colonies.

In some cases, colonial powers have delayed independence to would-be island nations. They explain that such territories are too small and isolated or are not economically viable enough for independence. Some islands remain dependent because they confer unique military or economic advantages on the governing power. For example, French Polynesia was until recently the venue for French atomic testing (see

Regional Perspective, p. 408), and Guam and American Samoa are still useful to the United States for military purposes. Guam, which has been a U.S. possession since 1898 and a U.S. territory since 1950 (its residents are American citizens), is especially vital in U.S. strategic thinking. It was the jumping-off point for American B-52 jets on bombing missions to Iraq in the 1990s and would be used in any U.S. military action in Korea or elsewhere in East Asia. In the event of such hostilities, the United States would need Japan's permission to operate from bases in Japan, but no such approval would be necessary to act from Guam.

Another set of geopolitical concerns is environmental. These remote islands, so far from the industrial corelands of North America, Europe, and Japan, are actually the places most likely to suffer first and most from the impacts of industrial carbon dioxide emissions. These islands would be the initial victims of any rise in sea level due to global warming. Sea levels have risen in recent years at a rate of 0.08 inches (2 mm) annually, and there are reports of unprecedented tidal surges on Pacific island shores. If the trend continues, the first Pacific islands to be totally submerged would be Kiribati, the Marshall Islands, and Tuvalu, while Tonga, Palau, Nauru, Niue, and the Federated States of Micronesia would lose much of their territories to the sea. These island nations are among the 39 countries comprising the Alliance of Small Island States, which argued forcefully but unsuccessfully at the 1997 Kyoto Climate Change Conference that, by 2005, global greenhouse gas emissions should be reduced to 20% below their 1990 levels. What they did get was a pledge by the industrialized nations to cut these emissions to at least 5% below their 1990 levels by 2012, but political obstacles make even that more modest accomplishment uncertain. As so often in the past, the peoples of the Pacific today are vulnerable to the actions and policies of more powerful nations far from their tranquil shores.

CHAPTER SUMMARY

- The Pacific World region (often called Oceania) encompasses Australia, New Zealand, and the islands of the mid-Pacific lying mostly between the Tropics.
- The Pacific islands are commonly divided into three principal regions: Melanesia, Micronesia, and Polynesia.
- The Pacific World is ethnically complex, having been settled by people with various Asian origins. Polynesia was the last to be populated. Papua New Guinea is the world's most linguistically diverse country. Christianity is the majority faith in this region.
- Although countless islands are scattered across the Pacific Ocean, there are two generally recognizable types: "high islands" and "low islands." Most high islands are volcanic or continental in nature. Low islands are typically made of coral, a material composed of the skeletons and living bodies of small marine organisms that inhabit tropical seas.
- The island ecosystems of the Pacific region are typically inhabited by endemic plant and animal species—species found nowhere else in the world. Island species are especially vulnerable to the activities of humankind such as habitat destruction, deliberate hunting, or the introduction of exotic plant and animal species.
- Europeans began to visit and colonize the Pacific islands early in their Age of Exploration. However, a steady process of decolonization has accompanied a recent surge of Western interest and investment in the region.
- Aside from a few notable exceptions, the poverty typical of LDCs prevails throughout most of the Pacific region.
- In general, the Pacific islands' economic picture is one of nonindustrial economies. Typical economic activities include plantation agriculture, mining, and income derived from activities connected with the military needs of occupying powers. Several countries are profiting from offshore banking and 1-900 telemarketing.
- During the 1940s and 1950s, the United States used the Bikini atoll in the Marshall Islands as one of its chief testing grounds for nuclear weapons. In recent years, strong negative reaction arose throughout the Pacific region as the French resumed underground testing of nuclear weapons on the Mururoa atoll in French Polynesia. The United States relies on the region for testing of its Star Wars missile technology.

REVIEW QUESTIONS

1. Using maps and the text, identify the three principal subregions that comprise the Pacific World.
2. What explains the peculiar jog in the International Date Line?
3. Describe the major differences between high and low islands. According to geographer Tom McKnight, how did the physical characteristics of the islands relate to social and political developments?
4. What is a geologic hot spot and how does it relate to Pacific islands?
5. What unique ecological attributes do islands generally have?
6. What is an atoll? Explain how atolls form.
7. What are the major ethnic groups, languages, and religions of the Pacific World?
8. What are two notable ethnic conflicts on mineral-rich islands in the Pacific World?
9. How has Nauru been able to avoid the poverty typical of most Pacific islands? How long will Nauru be able to enjoy its relative prosperity?
10. Describe the major economic activities that occur in the Pacific islands.
11. What predictions have been issued for the Pacific World if global sea levels rise?

DISCUSSION QUESTIONS

1. Describe some of the natural hazards associated with life on the Pacific islands. How might global warming affect life on the islands?
2. Describe the cultural diversity of the Pacific World.
3. Divide the Pacific World into groups representing high and low islands and compare and contrast them to each other.
4. Why are so many endemic species found on islands such as those of the Pacific World? Why is Hawaii known as the "extinction capital of the world"?
5. How might the story of Easter Island serve as a parable for human activity elsewhere?
6. Describe the positive and negative effects of colonialism on the Pacific islands.

7. How have some Pacific nations taken advantage of economic opportunities in the Information Age?
8. What is the future of the people of Nauru? Is Nauru like Easter Island?
9. Discuss in general terms the pros and cons of foreign military presence in the Pacific World. Why is it described in the text as a mixed blessing? Should the United States pay for the cleanup and repatriation of Bikini?
10. Describe the worldwide reaction to French nuclear weapons testing in the Pacific World. How did various countries show their dissatisfaction with the tests?

chapter 16

Prosperous, Remote Australia and New Zealand

Australia's Aborigines say that spirits "dreamed" the world and all its features into existence. This is a rock painting of one of those creator beings, the *wandjina*, a deity of wind and storms.

Australia and New Zealand are unique. Far removed from Europe, their dominant cultures are nevertheless European. In the less developed Pacific World, they are prosperous. Distinctive plants and animals inhabit the often odd landscapes of these islands, stirring a sense of wonder among indigenous people, European settlers, and foreign tourists alike. Writer Alan Moorehead described the reactions of early European visitors: "Everything was the wrong way about. Midwinter fell in July, and in January, summer was at its height. In the bush there were giant birds that never flew and queer, antediluvian animals that hopped instead of walked or sat munching mutely in the trees. Even the constellations in the sky were upside down."[1]

16.1 Peoples and Populations

These two unusual countries are akin in population, cultural heritage, political problems and orientation, type of economy, and location. Australia and New Zealand are among the world's minority of prosperous countries. Australia's GDP PPP per capita of $23,200 is comparable to that of Germany and Finland, although well below that of the United States and the most prosperous European countries. Australia has been dubbed "The Lucky Country," and a recent World Bank study ranked Australia as the world's wealthiest country based on its natural resource assets divided by the total population. New Zealand is less affluent but prosperous compared with most nations of the world. In both countries, there are relatively few people to spread the wealth among; Australia's 19.4 million people plus New Zealand's 3.9 (as of 2001) amount to only about 70% of the people living in California.

Both countries owe their prosperity to the wholesale transplantation of culture and technology from the industrializing Great Britain to the remote Pacific beginning in the late 18th century. Australia (established originally as a penal colony for British convicts) and New Zealand are products of British colonization and strongly reflect the British heritage in the ethnic composition and culture of their majority populations. They speak English, live under British-style parliamentary forms of government, acknowledge the British sovereign as their own, and attend schools patterned after those of Britain.

Despite their independence, both countries maintain loyalty to Great Britain. They still belong to the British Commonwealth of Nations. In both world wars, the two countries immediately came to the support of Great Britain and lost large numbers of troops on battlefields far from home. In World War I, Australia lost 60,000 of the 330,000 boys it sent to fight in Europe, by far the highest proportion of dead among any of the Allied countries. In World War II, U.S. forces helped to prevent a threatened Japanese assault on Australia. Since that time, Australia and New Zealand have sought closer relations with the United States, and British influence has waned.

Both Australia and New Zealand have minorities of indigenous inhabitants. The native Australian people are known as Aborigines. They have an extraordinarily detailed knowledge of the Australian landscape and its natural history and a complex belief system about how the world came about. According to Aborigines, Earth and all things on it were created by the "dreams" of humankind's ancestors during the period of the Dreamtime. These ancestral beings sang out the names of things, literally "singing the world into existence." The Aborigines call the paths that the beings followed the Footprints of the Ancestors or the Way of the Law; whites know them as Songlines. In a ritual journey called Walkabout, the aboriginal boy on the verge of adulthood follows the pathway of his creator-ancestors. "The Aborigine clings to his native soil with every fiber of his being," wrote the ethnographer Carl Strehlow. "Mountains and creeks and springs and water holes are to the Aborigine not merely interesting or beautiful scenic features. They are the handiwork of ancestors from whom he himself has descended. The whole country is his living, age-old family tree."[2]

The Aborigines numbered between an estimated 300,000 and 1 million when the first Europeans arrived. Colonizing whites slaughtered many of the Aborigines' descendants and drove the majority into marginal areas of the continent (see Problem Landscape, p. 414). The Aborigines today number about 390,000, living mainly in the tropical northern region of the country. Some carry on a traditional way of life in the wilderness, but most live in generally squalid conditions on the fringes of majority white settlements.

When Aboriginal populations began plunging precipitously, many white Australians assumed they would die out

[1] Alan Moorehead, *Cooper's Creek*, 1963, p. 7.

[2] Quoted in Geoffrey Blainey, *Triumph of the Nomads*. Melbourne: Sun Books, 1987, p. 181.

413

PROBLEM LANDSCAPE

Australia's Indigenous People Reclaim Rights to the Land

As in the United States, newcomers to Australia forged a new nation by dispossessing the ancient inhabitants of the land. And as in the United States, in Australia there is a long history of white racism, discrimination, and abuse against people of color, who in Australia's case are the original inhabitants. Aborigines were not mentioned in the 1901 constitution, not allowed to vote until 1962, and not counted in the national census until 1967. The most disadvantaged group in Australian society, Aborigines suffer from high infant mortality (four times that of whites in Australia), low life expectancy (15 to 20 years lower than that of whites), and high unemployment (officially 23%, although aboriginal sources claim 51%, versus 7% for whites). Like Native Americans, a disproportionately large share of Aborigines fall prey to the economic and social costs of alcoholism. Statistics indicate they are 29 times more likely than non-Aborigines to end up in jail. Once in jail, they are more likely than non-Aborigines to commit suicide or be killed by guards. The Australian government has begun a program to improve the justice system for native Australians.

One of the largest issues of contention between Australia's indigenous people and the white majority is land rights. As in the United States, European newcomers pushed the native people off potentially productive ranching, farming, and mining lands into special reservations on inferior lands. The Europeans used a legal doctrine called *terra nullius*,

meaning "the land was unoccupied," to lay claim to the continent. Under the 1976 Land Rights Law, Aborigines were permitted to seek title to vacant state land ("Crown land") to which they could prove a historical relationship, but few succeeded. Following new legislation in 1992, the government began to return titles on a parcel-by-parcel basis — generally, in very marginal lands — to Aborigines who had argued their rights successfully. However, most Aborigines were unhappy with the terms of the returned titles, which allowed the native title to coexist with, but not supersede, the established Crown title. The government retained mineral rights to the land and could therefore lease native land to mining companies.

In 1993, aboriginal leaders expressed their objections to the legislation in a declaration known as the Eva Valley statement, which made four demands:

- Aborigines should have veto rights over mining and pastoral leases on native title land.
- Mining and pastoral leases cannot extinguish native title to the land.
- There should be no access by developers without aboriginal permission.
- The federal government, not the states, should control native title issues.

Prime Minister Paul Keating rejected the Eva Valley statement, arguing

that the government had already made enough concessions. Aboriginal leaders took their case to the United Nations in Geneva, where they argued that existing legislation to validate mineral leases on native land was a breach of international human rights conventions, which Australia had signed.

In 1993, after the longest senate debate in the country's history, Foreign Minister Gareth Evans succeeded in pushing the Native Title Bill through Australia's parliament. The new legislation addressed the Aborigines' major objections, providing the following concessions:

- Aborigines have the right to claim land leased to mining concerns once the lease expires.
- Aborigines have the right to negotiate with mineral leaseholders over development of their land. However, they do not have the right of veto they requested.
- Where their native title has been extinguished, Aborigines are entitled to compensation paid by the government.
- With the help of a Land Acquisitions Fund, impoverished Aborigines can buy land to which they have proven native title.

The Native Title Bill is likely to be of great consequence in the states of South and Western Australia, where there is much vacant Crown land and many

entirely. When they did not, white Australia tried to "assimilate" them by settling them down, particularly in mission stations, and educating them in the service of whites. From 1910 to 1970, Australian authorities abducted about 100,000 aboriginal children (or between a tenth and a third of all aboriginal children) from their parents and, hoping to fully assimilate them, put them in the care of white foster parents. Today, members of the so-called "stolen generation" are seeking to identify their real kin and obtain apologies and

compensation from the government. Thus far, the government has declined to make an official apology because that acceptance of responsibility might mean it would have to pay billions of dollars in reparation to the "stolen children."

In New Zealand, the dominant indigenous culture is the Maori, a Polynesian people living mainly on North Island. Whites (known as *Paheka* to the Maori) broke the Maori hold on the land in a bloody war between 1860 and 1870. The Maori, who in 1900 seemed destined for extinction, have

Aborigines are able to file claims on it. Those states are also heavily dependent on mining. Members of Australia's Mining Industry Council are unhappy with the new legislation and worry that their mine leases will run out before the minerals do, which will require them to negotiate with the aboriginal titleholders.

Potential aboriginal claims to land expanded vastly in 1996 with another Australian High Court ruling on what is known as the Wik case (named after the indigenous people of north Queensland who initiated it). The court concluded that Aborigines could claim title to public lands held by farmers and ranchers under long-term "pastoral leases" granted by state governments. These lands comprise about 42% of Australia's territory. The number of relevant aboriginal land claims awaiting judgment in the courts is 700 and growing. There is bitter debate about what aboriginal claims to these lands would mean. Technically, in light of this court decision, Aborigines have a "right to negotiate," meaning they have a decision-making voice in how non-Aboriginal leaseholders use the land. They also have a right to a share of the profits those leaseholders earn from their land uses. White farmers and ranchers protest that they would be ruined economically if they had to share profits, especially with exports of wheat and beef reduced as a result of recent economic turmoil in Asia. Generally, however, Aborigines have insisted their title rights would not mean running whites out of business but would simply confer nominal recognition of their title and access to the land, especially for visitation to sacred sites.

Under pressure from white farmers, ranchers, and miners, Australia's conservative government under Prime Minister John Howard (elected in 1996) reacted to the Wik case with a 10-point plan that would make it much more difficult for Aborigines to make native title claims and to challenge farming, ranching, and mining activities.

In the Northern Territory, which has its own land-rights legislation, vast tracts of land have already been given back to Aborigines. Aborigines there make up 24% of the population and now control 32% of the land. Aborigines also hold title to Australia's two greatest national parks, Uluru (Ayers Rock) and Kakadu, both in the Northern Territory. In an arrangement that has become a model worldwide for management of national parks where indigenous people reside, the aboriginal owners of the reserves have leased them back to the Australian government park system, which comanages them with the Aborigines. Tourism programs in Uluru and Kakadu now highlight the cultural resources of the parks, in addition to their natural wonders. At both parks, visitors can enjoy interpretive natural history walks that also emphasize aboriginal culture and that are conducted by the land's oldest and most knowledgeable inhabitants (Fig. 16.A). At Uluru, the Anangu Aborigines who own the rock hope that cultural awareness will cut down the number of tourists who climb the rock — estimated at 70% of the 400,000 annual total — because it is sacred to the Anangu.

Although there has been remarkable progress in recent years in this area

and in others involving the Aborigines, the issue of rights to land has not been resolved to the satisfaction of all. "The handover of Ayers Rock is a turning point in Australia's race relations," declared Charles Perkins, then the country's top aboriginal civil servant. "It's a recognition that Aboriginal people were the original owners of this country." But aboriginal activist Galarrwuy Yunupingu declared, "I'd only call it a victory if the governor-general came and gave us title to the whole of Australia, and we leased that back to the Commonwealth."[a]

Figure 16.A   This aboriginal woman of the Anangu clan is a guide at Uluru (Ayers Rock) National Park.

[a]Catherine Foster, "Australia's Ayers Rock: It's Not Just for Climbing Anymore." *Christian Science Monitor,* August 3, 1993, p. 10.

rebounded to make up almost 15% of New Zealand's population. Although their socioeconomic standing is poor, their situation is much better than that of the Australian Aborigines. The Maori are being increasingly integrated, both socially and racially, with New Zealand's white population. At the same time, a Maori cultural and political movement is resisting the loss of traditional culture that this integration may cause. Like the Aborigines, the Maori are increasingly filing claims against the government to recover what they regard as

traditional tribal lands. They assert that New Zealand has violated the spirit and letter of the Treaty of Waitangi, signed in 1840 by Maori tribes and the British government. In the agreement, the Maori gave up political autonomy in exchange for "full and undisturbed" rights to lands, forests, fisheries, and the "national treasures" of New Zealand. In one small but symbolically important concession to such rights, the government has made it legal for Maori to fish without a license.

16.2   The Australian Environment

Australia is truly a world apart. Located in Earth's Southern Hemisphere (Fig. 16.1), where landmasses are few and far between, it is both an island and a continent. Its natural wonders and curiosities have provided endless inspiration, and sometimes consternation, to its inhabitants and to travelers and visiting writers. On arriving in Botany Bay — now Sydney Harbor — in 1788, British major Robert Ross declared, "I do not scruple to pronounce that in the whole world there is not a worse country than what we have yet seen of this. All that is contiguous to us is so very barren and forbidding that it may with truth be said here that nature is reversed; and if not so, she is nearly worn out."[3] Mark Twain marveled at the Land Down Under in his book *Following the Equator:* "To my mind the exterior aspects and character of Australia are fascinating things to look at and think about. They are so strange, so weird, so new, so uncommonplace, such a startling and interesting contrast to the other sections of the planet."[4]

Including the offshore island of Tasmania, Australia has an area of nearly 3 million square miles (7.7 million sq km), about equal to the area of the contiguous United States. On the basis of climate and relief, Australia has four major natural regions: the humid eastern highlands, the tropical savannas of northern Australia, the Mediterranean climate lands of southwestern and southern Australia, and the dry interior (see Fig. 16.2). Mainly because of extreme aridity, most of the continent is sparsely populated, and its people are concentrated where there is adequate fresh water and arable land. Most people live in the humid eastern highlands and coastal plains (especially in the cooler south), in the areas of Mediterranean climate, and in the more humid grasslands adjoining the southern part of the eastern highlands and the Mediterranean areas.

F2.6
29
F2.7
31

The Humid Eastern Highlands

Australia is the world's flattest continent. The only major highlands run down the east coast from just north of the Tropic of Capricorn to southern Tasmania in a belt 100 to 250 miles wide (161 to 410 km). These highlands are generally lower than 3,000 feet (914 m), but their highest summit and the highest point in Australia is Mt. Kosciusko, reaching 7,310 feet (2,228 m). The highlands and the coastal plains at their base are the country's only normally drought-free areas. South of Sydney and at higher elevations to the north, the climate is classified as marine west coast (despite the east coast location). North of Sydney, higher summer temperatures change the classification to humid subtropical, while still farther north, beyond approximately the parallel of 20°S, hotter

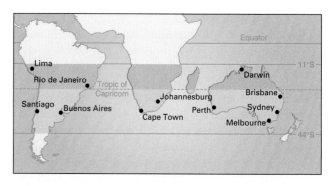

Figure 16.1   Australia and New Zealand compared in latitude and area with southern Africa and South America. The distances separating the continents have been compressed in this figure.

temperatures and greater seasonality of rain create tropical savanna conditions.

Tasmania is a rugged, beautiful island off Australia's southeast coast (Fig. 16.3). About 7% of its land is dedicated to national parks and other protected areas. The state's western region is heavily forested and dotted with lakes. Numerous rivers cut rapid, short courses to the sea, representing potential sources of hydroelectric power. At this latitude — roughly that of the U.S. state of Wyoming — snow sometimes falls on the high peaks even in the height of summer (November–January).

The Tropical Savannas of Northern Australia

Northern Australia, from near Broome on the Indian Ocean to the coast of the Coral Sea, receives heavy rainfall during the season locals call "the wet" — the Southern Hemisphere summer season. The coastal zone is subjected in this season to hurricanes, known as cyclones in Australia. On Christmas Eve in 1974, a powerful cyclone flattened the city of Darwin, which has since been rebuilt. After such deluges, northern Australia experiences almost complete drought during the winter. This highly seasonal distribution of rainfall is the result of monsoonal winds that blow onshore during the summer and offshore during the winter. The seasonality of the rainfall, combined with the tropical heat of the area, has produced a savanna vegetation of coarse grasses with scattered trees and patches of woodland (Fig. 16.4). The long season of drought, the poverty of the soils, and the lack of highlands that would nourish perennial streams all combine to limit agricultural development. The alluvial and volcanic soils that support large populations in some tropical areas are almost absent in northern Australia.

The "Mediterranean" Lands of the South and Southwest

The southwestern corner of Australia and the lands around Spencer Gulf have a Mediterranean or dry-summer subtropical type of climate, with subtropical temperatures, winter

[3] From Howard Ensign Evans and Mary Alice Evans, *Australia: A Natural History.* Washington, D.C.: Smithsonian Institute Press, 1983, p. 12.

[4] Mark Twain, *Following the Equator: A Journey Around the World.* Hartford, Conn.: American Publishing Company, 1897, p. 146.

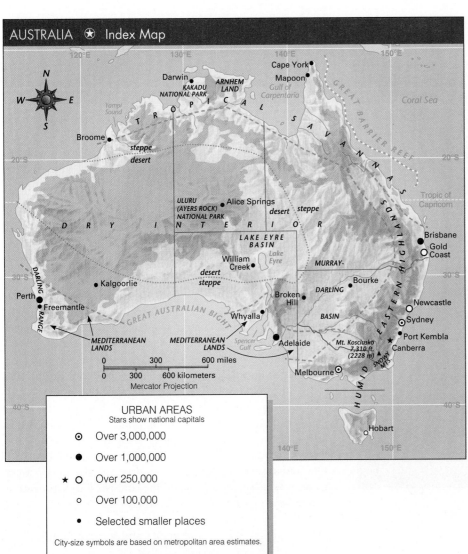

AUSTRALIA ⊛ Index Map

URBAN AREAS
Stars show national capitals

⊙ Over 3,000,000

● Over 1,000,000

★ ○ Over 250,000

○ Over 100,000

• Selected smaller places

City-size symbols are based on metropolitan area estimates.

----- Boundaries of major natural regions

············ Boundary between steppe and desert climates

——— Boundaries between states

⬜ Land over 1000 ft (305 m)

⬛ National parks

Figure 16.2 Reference map of natural features, political units, and cities in Australia.

Figure 16.3 Large parts of Tasmania are rugged and wild. This is Wineglass Bay in Freycinet National Park.

Figure 16.4 Tropical savanna prevails in northern Australia. This is typical vegetation and terrain of the Kimberley region in the continent's extreme northwest.

the "outback" that has lent so much to Australian life and lore. Locals know it as "the bush," the "back of beyond," and the "never-never." Altogether, the interior desert and steppe cover more than half of the continent and extend to the coast in the northwest and along the Great Australian Bight in the south. This huge area of arid and semiarid land is too far south to get much rain from the summer monsoon, too far north to benefit from rainfall brought by the westerlies in winter, and deprived of Pacific winds by the eastern highlands. The region is sparsely populated. The chief city is Alice Springs (population: 25,700; Fig. 16.5), a famous frontier settlement that is the jumping-off point for visitors to Uluru (Ayers Rock).

The western half of Australia is occupied by a vast plateau of ancient igneous, metamorphic, and hardened sedimentary rocks. Its general elevation is only 1,000 to 1,600 feet (c. 300 to 500 m), and its few isolated mountain ranges are too low to influence the climate or to supply perennial streams for irrigation. To the east, between the plateau and the eastern highlands, the land is even lower. The Lake Eyre Basin, the lowest part of this lowland — and of Australia — is the driest part of the continent. However, another part of the lowland, the Murray-Darling Basin, contains Australia's only major river system. It produces half of Australia's agricultural wealth and fully three-quarters of its irrigated crops. Some of the water comes from a massive engineering project, undertaken in the 1950s, to divert 99% of the waters of the Snowy River, which flowed into the Pacific, via underground tunnels to the basin. One unforeseen consequence is a dramatic rise in the water table, which has forced salts to the surface. This **salinization** is proving to be enormously destructive to the basin's cotton and rice. Consideration is now being given to growing more drought-tolerant olives and grapes and to returning more of the Snowy's water to its original drainage.

rain, and summer drought. In winter, the Southern Hemisphere belt of the westerly winds shifts far enough north to bring precipitation to these districts, while in summer this belt lies offshore to the south and the land is dry. Crops introduced from the Mediterranean climate lands of Europe generally do well in these parts of Australia, but the shortage of highlands to catch moisture and supply irrigation water to the lowlands limits agricultural development. Australian wines produced in this region, particularly around Adelaide, are becoming increasingly popular abroad. Approximately 1,300 wineries in Australia now produce about 3% of the world's total wine output.

The Dry Interior

The huge interior of Australia is desert, surrounded by a broad fringe of semiarid grassland (steppe) that is transitional to the more humid areas around the edges of the continent. This is

Figure 16.5 Looking rather like a town in Nevada, this is Australia's frontier town of Alice Springs.

Exotic Species on the Island Continent

Exotic species are nonnative plant and animal species introduced through natural or human-induced means into an ecosystem. Their impact tends to be pronounced and often catastrophic to native species, particularly on island ecosystems where resources and territories are limited. In Ecuador's Galapagos Island National Park, for example, cats, rats, pigs, goats, wasps, and other animals introduced by people have disrupted nests and food supplies and thereby reduced populations of endemic tortoises and other species.

The impacts of exotic species show how Australia truly is an island. Both inadvertently and purposely, people have introduced foreign plants and animals that have multiplied and affected the island, sometimes in biblical plague proportions. English settlers imported the rabbit for sport hunting, confident in the belief that, just as in England, there would be natural checks on the animal's population. They were mistaken (Fig. 16.B). The rabbits bred like rabbits, eating everything green they could find, with ruinous effect on the vegetation. Observers describe the unbelievable concentrations of the lagomorphs as "seething carpets of brown fur." In the mid-20th century, the government mounted a huge eradication effort, employing fences, snares, dogs, guns, fires, and numerous other means, and exhorting combatants with the slogan "The Rabbit War Must Be Won!" An introduced virus called "white blindness" eventually helped reduce the vast numbers, but the animal is still prolific throughout large parts of the country. Other exotic remedies have been applied, with varying success, to correct numerous other problem species, including mice, water buffalo, cane toads, and prickly pear cactuses. Many of the problem populations, notably cats and water buffalo, are **feral animals** — domesticated animals that have abandoned their dependence on people to resume life in the wild. An estimated 12 million feral cats infest the country and are held responsible for causing the extinction or endangerment of 39 native animal species. One Australian politician has been promoting legislation that would kill all cats on the continent by 2020.

Exotic livestock such as sheep and cattle are also a problem in the Australian environment. These hard-hoofed imports cut into the soil, promoting erosion and desertification. Some range-

management scientists and conservationists believe that the future of Australian ranching lies with kangaroos. These soft-hoofed marsupials are adapted to the Australian landscape and do not have such a damaging impact on the soil. Farmers and ranchers have traditionally eradicated them as pests. However, if more consumers at home and abroad could learn to appreciate kangaroo meat, there would be a strong incentive for ranchers to reduce their cattle and sheep herds and allow kangaroos to proliferate and be harvested. South Africa already imports more than 1,000 tons of kangaroo meat yearly from Australia, and nearly 4,000 more tons go each year to countries including Belgium, the Netherlands, Bulgaria, and the Czech Republic. Kangaroo skin is also used to make athletic shoes.

Australia is an unlikely exporter of another quadruped, the dromedary camel. The more than 200,000 camels in Australia are feral descendants of those brought from South Asia and Iran early in the 20th century to provide transport across the arid continent. Australian camels are exported even to Saudi Arabia, where their wild temperaments suit them to racing.

Figure 16.B   Rabbits were one of Australia's many human-caused scourges, and are still a problem, but not nearly as much as in the 1920s when people went on "rabbit drives" like these to chase, corral, and kill the declared vermin.

16.3   Australia's Natural Resource-Based Economy

Due to widespread aridity and the presence of highlands in its only humid area, Australia has little good agricultural land (Fig. 16.6). Although as much as 15% of the total land area is potentially cultivable, only about 6% of Australia is actually cultivated today. Fourteen percent of Australia's area is classified as forest and woodland, mostly of little economic value. More than half is classified as natural grazing land, and about one-fifth as unproductive land. Despite its small proportion of arable area, however, Australia has one of the most favorable ratios of crop-producing land to population in the world.

The sectors of the Australian economy that regularly produce the country's leading exports are mining, ranching, and even with its small proportion of cultivable land, agriculture. Several factors support the country's emphasis on surplus production and export of primary commodities. There are large and dependable markets for Australian primary products in Japan, the United States, Europe, and elsewhere. These products can be taken to market with inexpensive and efficient transportation by sea. The Australian people can supply and import the necessary skills and capital to exploit the continent extensively, utilizing a minimum of labor and a maximum of land and equipment.

Sheep and Cattle Ranching

In founding the first settlement around the excellent natural harbor of Sydney in 1788, the British government intended merely to establish a penal colony. There was to be enough agriculture on this "Fatal Shore" to make the colony self-sufficient in food and perhaps provide a small surplus of some products for export. Although self-sufficiency was elusive in the early days due to the poverty of the leached soils around

Sydney, the colonists soon discovered that sheep did well, particularly the Merino breed that was first imported from Spain in 1796. The market for wool expanded rapidly with the mechanization of Britain's woolen textile industry. Australia thus found its major export staple in the early years of the 19th century, and until recently, sheep ranching was a strong and growing rural industry (Fig. 16.7). Sheep grazers rapidly penetrated the interior of the continent, and by 1850, Australia was already the largest supplier of wool on the world market, a position it has never lost.

People have tried to raise sheep anyplace in Australia that seemed to offer any hope of success. Much of the interior has proven too dry, parts of the eastern mountain belt are too rugged and wet, and most of the north is too hot and wet in summer. The sheep ranching industry has thus become localized, concentrated mainly in a crescent-shaped belt of territory following the gentle western slope of the eastern highlands, from the northern border of New South Wales to the western border of Victoria (see Fig. 16.6). Beyond this concentration, the sheep industry spreads northward on

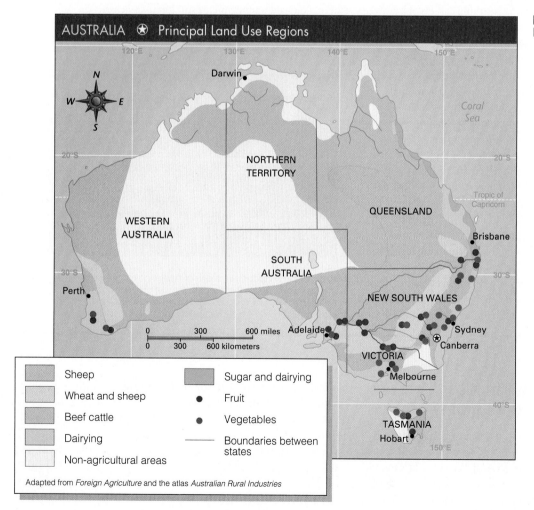

Figure 16.6   Map of agricultural land use in Australia.

Figure 16.7   Sheep are big business in both Australia and New Zealand. These motorized shepherds are mustering Merino sheep at the Narolah Station in southwestern Western Australia, on land that is productive for both sheep and wheat.

poorer pastures into Queensland and westward into South Australia. Still another area of production rims the west coast, especially near Perth. The sheep ranches, or "stations," in Australia vary in size from a few hundred to many thousands of acres.

The main product of the sheep stations is wool, which has been of great importance in Australia's development. Through most of the country's history, wool has been the principal export, amounting to as much as a quarter of all exports as late as the 1960s. Recently, however, it has declined rapidly in importance, partly because of the rise of other commodities in the Australian economy and partly because of wool's slowly losing battle with synthetic fibers. Today, Australian sheep are raised increasingly for meat. Each year, Australia exports large numbers of live animals to Saudi Arabia and other Muslim countries of the Middle East, where the devout slaughter them on the Feast of the Sacrifice commemorating the end of the pilgrimage to Mecca.

Lands in Australia that are unsuitable for sheep or for intensive agriculture are generally devoted to cattle ranching. Cattle stations are spread across the northern lands, with their hot, humid summers and coarse forage. The main belt of cattle ranching extends east-west across the northern part of the country, with the greatest concentration in Queensland (see Fig. 16.6). On many of the remote ranches, especially in the Northern Territory, Aborigines are the principal ranch hands, known as "drovers" and "jackaroos." The beef cattle industry has expanded rapidly in recent decades, spurred by the growing demand for meat that has accompanied increasing affluence in Japan, Europe, and the United States. Construc-

tion of roads for trucking Australian cattle to the coast has also aided this expansion. Australia is now the world's leading beef-exporting nation.

Wheat Farming

Australia is also one of the world's leading wheat exporters (Fig. 16.8). Wheat was the key crop in early attempts to make the settlement at Sydney self-sufficient. In the early years, the colony almost starved as a result of wheat failures, and the problem of grain supply was not solved until the development of the colony (now the state) of South Australia. There, settlers found fertile land and a favorable climate near the port of Adelaide, where wheat could be grown and shipped by sea to other coastal points.

Australian wheat production today is an extensive and highly mechanized form of agriculture, characterized by a small labor force, large acreages, and low yields per acre, but with very high yields per worker. The main belt of wheat production spreads from the eastern coastal districts of South Australia into Victoria and New South Wales, generally following the semiarid and subhumid lands that lie inland from the crests of the eastern highlands (see Fig. 16.6). The main wheat belt has thus come to occupy nearly the same position as the main belt of sheep production. The two types of production are in fact often combined in this area.

Dairy and Sugarcane Farming

Although much less important than grazing and wheat farming for export revenues, dairy farming is Australia's leading type of agriculture in numbers of people employed. It has de-

Figure 16.8   Despite its limited arable land, Australia is a prodigious wheat producer. This is an aerial view of center pivot irrigation for wheat near Wyndham in northwest Australia.

JOE HOBBS

veloped mainly in the humid coastal plains of Queensland, New South Wales, and Victoria (see Fig. 16.6). Dairy farming has grown largely in response to the needs of the country's rapidly growing urban populations.

Australia is a sugar exporter with a unique history. The Queensland sugar-growing area is unusual in that it was a

rare tropical area settled by Europeans doing hard manual work without the benefit of indigenous labor. Sugar production began on the coastal plain of Queensland in the middle of the 19th century to supply the Australian market. Laborers were initially imported from Melanesia to work as virtual slaves in the cane fields. However, when the Com-

Figure 16.9   Mineral map of Australia.

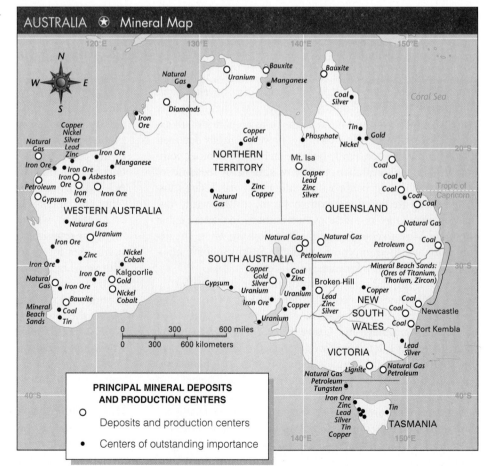

monwealth of Australia was formed in 1901 by the union of the seven states (New South Wales, Victoria, Queensland, Northern Territory, South Australia, Western Australia, and Tasmania), Queensland was required to expel these imported "Kanaka" laborers in the interest of preserving a "white Australia." Anticipating that production costs would soar with the employment of exclusively white labor, the Commonwealth gave Queensland a high protective tariff on sugar. Operating behind this protective wall, the sugar industry has prospered there, though most of the original large estates have been divided into family farms worked by individual farmers.

Australia's Mineral Wealth

Unification of the separate Australian colonies into the independent federal Commonwealth of Australia encouraged the rise of industry, creating a unified internal market and tariff protection for manufacturing as well as for agricultural enterprises. Australia is rich in mineral resources, and that wealth has facilitated industrial development (Fig. 16.9). In addition to supplying most of its own needs for minerals, Australia has long been an important supplier to the rest of the world. In recent years, large new discoveries of iron ore, diamonds, and gold have been expanding this longtime role in international trade.

Mining has long had an impact on Australian life and landscapes. Gold played a large role in Australia's history, as it did in America's; some of the California forty-niners even continued their prospecting careers in Australia. Gold strikes in Victoria in the 1850s set off one of the world's major gold rushes and produced a threefold increase in the continent's population in just 10 years.

Ample supplies of coal also have aided Australian industrialization. Every Australian state has coal, but the major reserves are near the coast in New South Wales and Queensland. Most of Australia's steel-making capacity is located near the coal mines north of Sydney at Newcastle and at Wollongong to the south. Iron ore is brought from distant deposits along and near other sections of the Australian coast (Fig. 16.10). Major new iron ore discoveries and developments in Western Australia have made Australia a leading world supplier.

Australia is by far the world's leading bauxite producer. Major reserves of bauxite are near the north Queensland coast, in the Darling Range inland from Perth, and on the coast of Arnhem Land. Aluminum plants supply the Australian market, and large quantities of the partially processed material called alumina are exported for final processing into aluminum metal at overseas plants.

Among Australia's other important mineral products are copper, lead, titanium, zinc, tin, silver, manganese, nickel, tungsten, uranium, and diamonds. Australia now produces 40% of the world's diamonds, most of them industrial cutting

Figure 16.10 Australia is an important exporter of iron ore. This open-pit mine is in Western Australia.

diamonds rather than gem diamonds. A diamond rush is under way, with prospecting and new finds in the country's northwest, both onshore and at sea. In the 1990s, Australia replaced Russia as the world's third largest gold-producing area, after North America and South Africa. Australia's most famous mining centers are three old settlements in the dry interior — Broken Hill in New South Wales, Kalgoorlie in Western Australia, and Mt. Isa in Queensland — all of which are still active. The Broken Hill Proprietary Company, which grew with the mines of that town, later branched into iron and steel production and other businesses and has become Australia's largest corporation.

A shortage of petroleum has marred Australia's general picture of mineral abundance. There are only small oil fields. Most of these are on the mainland, but the main producing field is offshore Victoria. Although local oil and natural gas supplies have aided the country's economy, Australia continues to import oil and to look for more domestic sources.

Australia and New Zealand have been remote satellites of Great Britain throughout most of their short histories. Even with independence, they pledged loyalty to the British Commonwealth of Nations. The United Kingdom was far more important than any other country in their trade relations, and as recently as the early 1970s, it accounted for about 30% of New Zealand's trade and 15% of Australia's. Their orientation is changing, however, as Australia and New Zealand seek stronger roles in the economic growth projected for the Pacific Basin and its major anchors, Japan and the United States. Great Britain is now a relatively insignificant trading partner of Australia and New Zealand. Already, seven of Australia's ten largest markets are in the Asia-Pacific region, with Japan and the United States its largest and second-largest trading partners, respectively. Australia, Japan, and the United States have become New Zealand's leading trade partners.

Perhaps the best symbol of Australia's new Pacific perspective is the debate over whether or not Australians should convert the country into a republic, ending more than 200 years of formal ties with Britain. Even now, according to Australia's consti-

tution (which was written in 1901 by the English), Australia's head of state cannot be an Australian; it has to be England's king or queen (this continues to be the case in 16 of the 54 countries that make up the Commonwealth). Republic status would change that and allow Australians to have their own elected or appointed president. The monarch's portrait would be removed from Australian currency. The British Union Jack would be lifted from its position in the upper left corner of the Australian flag. In 1998, an Australian Constitutional Convention voted to adopt the republican model, and Australia's avowedly monarchist prime minister, John Howard, reluctantly agreed to put the issue to a referendum in 1999. In essence, the referendum asked Australians where they preferred their economic and political future to rest — either with the other nations of Asia and the Pacific or with Great Britain and the Commonwealth. A narrow majority (55%) voted to keep Australia's lot with Britain and the Commonwealth. The debate has not gone away, however. Polls showed that at the time of the referendum, 60% of Australians favored a republic. Not all of these voted for republic status, however, because the referendum stated that

the republic's president would be chosen by the parliament rather than elected by popular vote. If they were allowed to elect their president, the majority of Australians would probably vote for a republic at the next opportunity.

The monarchy/republic debate has led to something of an identity crisis for Australia and has focused more scrutiny on the sensitive issue of Asian immigration. Between 1945 and 1972, 2 million people, mainly British and continental Europeans, emigrated to Australia. They were allowed in by a "white Australia" policy that excluded Asian and black immigrants because of fears of invasion from Asia and a desire to increase the country's population with white, skilled, English-speaking immigrants. The rate of immigration has slowed since then, but it continues and has been opened considerably to skilled nonwhites. About half of the annual quota of 85,000 legal immigrants are Asian, who now make up about 7% of Australia's population. But Asians will form up to one-fourth of Australia's population by 2025, according to some projections. Recent years have seen a marked surge in racism targeted mainly at Asians. Pauline Hanson, a member of parliament and a candidate for prime minister

Urbanization and Industrialization

Although mining, grazing, and agriculture produce most of Australia's exports, most Australians — about 85% — are city-dwellers. This high degree of urbanization is one of the country's most notable and perhaps surprising characteristics. About 11 million people, or 58% of the population, live in the five largest cities: Sydney (the largest, population: city proper, 3,985,800; Fig. 16.11), Melbourne, Brisbane, Perth, and Adelaide. All of these are seaports, and each is the capital of one of the five mainland states. The country's capital, Canberra (meaning "meeting place" in the aboriginal language), is one of its smaller cities (population: 320,200). Located inland about 190 miles (310 km) southwest of Sydney, Canberra is widely regarded as one of the world's most beautiful modern planned cities.

The Australian government has encouraged industrial development to support the country's growing population, to provide more adequate armaments for defense, and to increase stability through a more diversified and self-contained economy.

A wealthy, urbanized, and resource-rich country such as Australia should, like most more developed countries (MDCs), be a major exporter of finished industrial products. However, approximately three-fourths of the country's exports come directly, or with only early-stage processing, from mining, agriculture, and grazing. Far from being an international industrial power of consequence, Australia does not meet its own demands for manufactured goods and relies heavily on overseas production. Australia is a big importer of manufactured goods from Japan and the United States. The Labor party government of Prime Minister Paul Keating was elected twice in part on its promise to correct this problem and to transform Australia from a protected agriculture- and mining-based economy to a lean, high-technology export economy. Its failure to achieve this goal contributed to the party's defeat in the 1996, 1998, and 2001 elections.

Tourism has become an important industry in Australia, particularly since the worldwide acclaim of the 1986 film *Crocodile Dundee*. (Crocodiles themselves occupy an important

in the 1998 election, said, "I believe we are in danger of being swamped by Asians," and vowed to halt Asian immigration if her One Nation party took power. "A truly multicultural country can never be strong or united," she declared.[a] Ms. Hanson also insisted that Aborigines once practiced cannibalism and that therefore they are unworthy of political sympathy today. Her party was defeated soundly in the election. Since then, there has been a growing tide of illegal immigration from new sources, mainly Afghanistan and Iraq. These boat people are generally trafficked by Indonesian racketeers, who charge a refugee as much as $7,000 for the service.

Ethnic tensions aside, Australia and New Zealand continue to forge important ties with their neighbors around the vast Pacific and with each other. Both countries adopted a free-trade pact in 1990 called the Closer Economic Relations Agreement, which eliminated almost all barriers between them to trade in farm and industrial goods and services. Both countries belong to the 18-member Asia-Pacific Economic Cooperation (APEC) group, which has agreed to establish "free and open" trade and investment between member states by the year 2020. Australia and the United States are the most aggressive member states in arguing for the abolition of trade quotas and the reduction of tariffs imposed on imported goods.

Australians refer to their Asian neighbors not as the "Far East" but as the "Near North." Australia's Northern Territory city of Darwin is well positioned geographically to take advantage of new trading relations with the Near North. Darwin is closer to Jakarta, Indonesia's capital, than it is to Sydney, and Darwin is promoting itself as Australia's "Gateway to Asia." The city has established a Trade Development Zone to provide manufacturing facilities and incentives for overseas companies and for Australian companies wanting to do business overseas. A new deep-water port is under construction, and the regional government is pushing for the completion of a transcontinental Darwin-Adelaide railway to help handle the anticipated boom in freight to be traded between Australia and its neighbors to the northeast.

Australia's Asia-oriented perspective is also changing the nature of one of the country's export staples: beef. Ranchers are shifting to breeds of cattle better suited for live shipment by sea to Indonesia, the Philippines, and Thailand. The country's live-cattle exports tripled in the early 1990s. During that decade, Japan surpassed the United States as Australia's largest recipient of processed beef. But the Asian economic crisis that began in 1997 severely curtailed exports of beef and most other Australian products to Asia. The resulting economic downturn

helped keep the less Asian-oriented Liberal party of John Howard in power for three successive terms.

Australia and New Zealand have Pacific-oriented defense and security agreements. In 1951, they joined the United States to form the security alliance called Australia New Zealand United States (ANZUS). Australian troops supported American forces in the Vietnam War. Since 1987, however, when its liberal government declared that New Zealand would henceforth be a nuclear-free country, New Zealand's relations with the United States have been strained. Legislation banned ships carrying nuclear weapons or powered by nuclear energy from New Zealand's ports. The United States refused to confirm or deny whether its ships violated either restriction, so New Zealand denied port access to them. This action removed New Zealand from ANZUS.

Despite its small army (only 24,000 strong), Australia has begun to provide a selective security blanket over its interests in the region. In 1999, Australian troops were deployed to head up the United Nations peacekeeping force to oversee the referendum that gave independence to East Timor and its subsequent transition to full independence.

[a] From Seth Mydans, "Sea Change Down Under: Drifting to the Orient." *New York Times*, February 7, 1997, p. A4.

Figure 16.11 One great attraction to international tourists is the beautiful port city of Sydney. The white structure is Sydney's famous Opera House.

place in Australia's tourist business. In Kakadu National Park and other sections of northern and western Australia, many tours promise visitors a close-up look at "salties," the great saltwater crocodiles that are ancient and often deadly neighbors of people in the region.) Australia's attractions are sufficiently diverse to draw many different styles of tourists to various destinations on the continent. The 1,200-mile-long (1,920-km) Great Barrier Reef, located off the northeast coast, is one of the world's most prized scuba-diving areas. Hotels and other services on nearby Queensland beaches of the Gold Coast cater to those looking for an idyllic, tropical, restful holiday. The geological features of Uluru (Ayers Rock) and the nearby Olgas in the country's center are on the itinerary of most international tourists. For many visitors, Sydney is as lovely a port city as San Francisco and Venice. Adventurous tourists raft the whitewaters and hike the mountains of Tasmania.

16.4 New Zealand: Pastoral and Urban

Located more than 1,000 miles (c. 1,600 km) southeast of Australia, New Zealand consists of two large islands, North Island and South Island (separated by Cook Strait), and a number of smaller islands (Fig. 16.12). North Island is smaller than South Island but has the majority of the population, which numbered 3.9 million in 2001. With a total area of 104,000 square miles (269,000 sq km), the population density averages only 37 per square mile (14 per sq km). New Zealand is thus a sparsely populated country, although not as sparsely populated as Australia.

Environment and Rural Livelihood

Much of New Zealand is rugged. The Southern Alps, often described as one of the world's most spectacular mountain ranges, dominate South Island (Fig. 16.13). These mountains

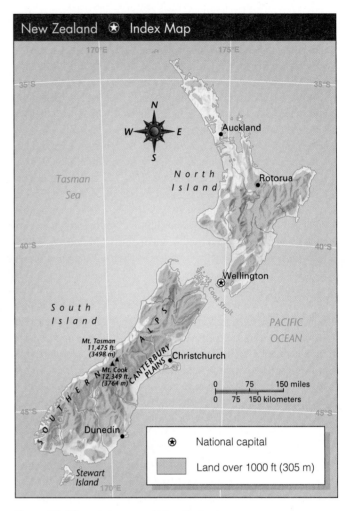

Figure 16.12 Index map of New Zealand.

Figure 16.13 The Southern Alps tower over pastoral landscapes on New Zealand's South Island.

PAUL CHESLEY/STONE/GETTY IMAGES

rise above 5,000 feet (c. 1,500 m), with many glaciated summits over 10,000 feet, and in the southwest, they are cut deeply by coastal fjords. The mountains of North Island are less imposing and extensive, but many peaks exceed 5,000 feet. New Zealand's highlands lie in the Southern Hemisphere belt of westerly winds and receive abundant precipitation. While lowlands generally receive more than 30 inches (76 cm), distributed fairly evenly throughout the year, highlands often receive more than 130 inches (330 cm). Precipitation drops to less than 20 inches (50 cm) in small areas of rain shadow east of South Island's Southern Alps. Temperatures typical of the middle latitudes are moderated by a pervasive maritime influence. The result is a marine west coast climate, with warm-month temperatures generally averaging 60° to 70°F (16° to 21°C) and cool-month temperatures of 40° to 50°F (4° to 10°C). Highland temperatures are more severe, and a few glaciers flow on both islands.

Rugged terrain and heavy precipitation in the Southern Alps and in the mountainous core of North Island account for the almost total absence of people from one-third of New Zealand. About one-fifth of the country is completely unproductive, except for the appeal of the mountains to an expanding tourist industry. New Zealand's attractions for visitors include golfing, fishing, hunting, hiking, and camping in spectacular mountainous national parks, and geyser watching around the Rotorua thermal area of North Island. Despite all this wilderness, New Zealand has suffered the impacts that typically occur as a result of human activities in island ecosystems. Nearly 85% of the country's lowland forests and wetlands have been destroyed since the Maori arrived about 800 years ago. Of New Zealand's presettlement 93 bird species, 43 have become extinct and 37 are endangered (including the kiwi, the national bird and the informal namesake of all New Zealanders). As in Australia, exotic plant and animal species are a major threat to the indigenous varieties.

Well-populated areas are restricted to fringing lowlands around the periphery of North Island and along the drier east and south coasts of South Island. The climate of New Zealand's lowlands is ideal for growing grass and raising livestock. More than half of the country's total area is in pastures and meadows supporting a major sheep and cattle industry. Pastoral industries contribute much to the country's export trade. Meat, wool, dairy products, and hides together account for about half of what New Zealand sells abroad.

The earning power of these pastoral exports has given New Zealand a moderately high standard of living (per capita GDP PPP of $17,700 in 2001, comparable to that of Spain). The United Kingdom was the country's main market until recently. When the U.K. joined the European Common Market in the 1970s, however, a new situation emerged because the Common Market (now the European Union) included countries with surpluses competitive with New Zealand's. In response, New Zealand tried to foster industrial growth and to expand trade with other partners, particularly in Asia. These efforts were not entirely successful. The economy suffered during the 1970s and 1980s, and the country slipped downward in the ranks of the world's affluent countries. A sizable emigration took place, mainly to Australia. These emigrants complained of not only the economic troubles but also boredom in an isolated, placid society in which tax policies made it difficult for almost anyone to make or keep a high income. Today, there is a reverse flow of migrants, particularly of Asians immigrating to New Zealand's cities. Like Australia, New Zealand is seeking to loosen or break its political and economic ties with the British Commonwealth of Nations and to reorient its economic orbit more toward Asia. New Zealand's economy felt the negative impact of the late 1990s Asian economic crisis as its exports to the region declined.

Industrial and Urban Development

In New Zealand, as in Australia, a high degree of urbanization and attempts to develop manufacturing supplement a basic dependence on pastoral industries (Fig. 16.14). The two

Figure 16.14   Queenstown, situated on the South Island, displays both the urbanity and wilderness that are New Zealand.

countries are similar in their conditions and purposes of urban and industrial development. New Zealand's resources for manufacturing do not equal those of Australia, but there are modest reserves of coal, iron, and other minerals. There is much potential for hydropower development. New Zealand has magnificent natural forests, and production and export of forest products have become increasingly important as the country attempts to reduce its dependence on sheep and dairy exports.

New Zealand is not as urbanized as Australia. The country has only six urban areas with estimated metropolitan populations of more than 100,000: Auckland (the largest; population: metropolitan area, 1,119,000; city proper, 381,000), the capital of Wellington (population: metropolitan area, 346,900; city proper, 167,400), at opposite ends of North Island; Napier-Hastings on the southeastern coast of North Island; Hamilton, not far south of Auckland; and Christchurch and Dunedin on the drier east coast of South Island. Christchurch is the main urban center of the Canterbury Plains, which contain the largest concentration of cultivated land in New Zealand. About half of the country's population lives in these six modest-sized cities (in comparison, Australia's Sydney metropolitan area has about the same population as all of New Zealand). Note that all of these cities have English names; as evidence of New Zealand's struggle to reidentify itself, there is a strong movement afoot to replace many of the country's Anglified place names with their original Maori designations.

This concludes our survey of the Pacific World, with its far-flung islands and a single great island-continent. Although islands have an insular quality, this world region is, perhaps more than most, subject to influences from abroad because it depends so much on aid from and trade with greater continental powers. In the years to come, many islands will likely be forced to go it alone as foreign aid is withdrawn, and Australia may forge a new identity as a republic free of Great Britain. Rising sea levels as a result of global warming may also pose dilemmas for the region.

CHAPTER SUMMARY

- Australia and New Zealand are products of British colonization. The British heritage is strongly reflected in the ethnic composition and culture of Australia's and New Zealand's majority populations.
- Both Australia and New Zealand have minorities and indigenous inhabitants. Native Australians are known as Aborigines, and the Maori are the dominant indigenous group in New Zealand.
- In Australia, one of the largest issues of contention between the Aborigines and the white majority is land rights. Aboriginal demands were declared in the Eva Valley statement. In 1993, the Native Title Bill addressed the Aborigines' major concerns and made several land rights concessions.
- On the basis of climate and relief, Australia has four major natural regions: the humid eastern highlands, the tropical savannas of northern Australia, the "Mediterranean" lands of southwestern and southern Australia, and the dry interior.
- Exotic species are nonnative plants and animals introduced through natural or human-induced means into an ecosystem.

- Exotic species often have catastrophic effects on native species and the natural environment.
- The sectors of the Australian economy that regularly produce the country's leading exports are mining, ranching, and agriculture.
- Most Australians — about 85% — live in cities. Nearly 11 million people, or 58% of the population, live in the five largest cities: Sydney, Melbourne, Brisbane, Perth, and Adelaide.
- New Zealand is located more than 1,000 miles southeast of Australia. The country consists of two large islands, North Island and South Island, and a number of smaller islands.
- Pastoral industries, such as meat, wool, dairy products, and hides, contribute greatly to New Zealand's export trade.
- Australia and New Zealand have been reorienting their focus toward the Pacific Rim and away from the United Kingdom. An important expression of this change in focus is the ongoing debate about whether or not Australia will become a republic, thus ending more than 200 years of formal ties with Britain.

REVIEW QUESTIONS

1. What factors foster a basic kinship between Australia and New Zealand?
2. Who were the first nonindigenous settlers of Australia?
3. What is the significance of the Dreamtime to Australia's Aborigines?
4. Using maps and the text, identify the four natural regions of Australia.
5. Using maps and the text, identify the major agricultural regions of Australia.

6. Explain the factors that support Australia's emphasis on surplus production and export of primary commodities.
7. Describe wheat and dairy production in Australia today.
8. List the notable natural resource assets of Australia. What essential natural resources does Australia lack?
9. Using maps and the text, locate Australia's five largest cities. What characteristics do these cities have in common?
10. What are the notable landform features and natural resource assets of New Zealand?

DISCUSSION QUESTIONS

1. What are some popular nicknames for Australia and its regions and what do they tell us about the continent, the landscape, and the people? Describe some of the factors that make Australia unique as a continent.

2. Discuss the influence of Great Britain on the cultures of Australia and New Zealand.

3. Describe some of the similarities between Australia's Aborigines and Native Americans.

4. Describe some of the demands made by aboriginal leaders in the Eva Valley statement. Were these demands met by the Native Title Bill of 1993?

5. What is the history of Australia's wool industry?

6. Provide examples of exotic animals that have been introduced to Australia. What have been the effects of exotic animals on Australia's natural environment?

7. What is notable about Darwin's geographic location and what does the city want to become?

8. What are some indications that Australia and New Zealand's orientation is beginning to focus on the Pacific Rim rather than the United Kingdom?

9. Why is New Zealand no longer a member of ANZUS?

10. What are the pros and cons, and the politics, of Australia's immigration policy? Compare Australia's immigration issues with those of the United States.

11. How do people in New Zealand make a living?

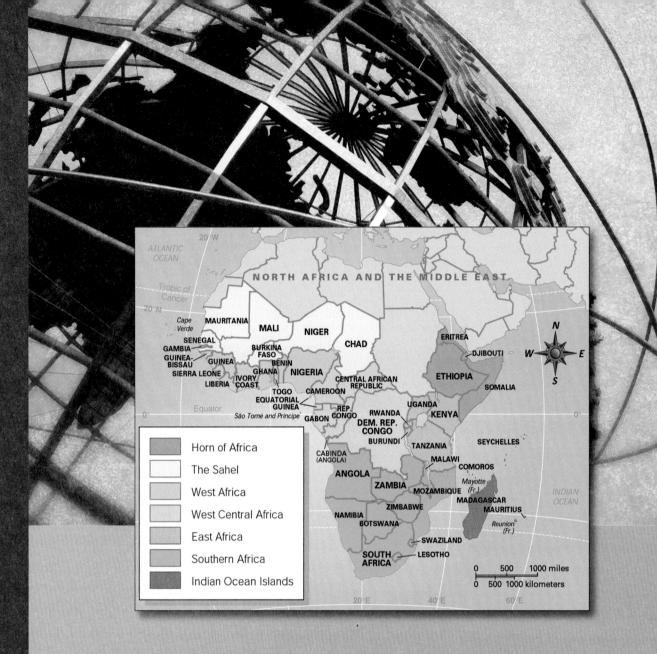

Africa South of the Sahara

So often judged to be marginal in world affairs, Africa south of the Sahara deserves increased international attention because of its humanitarian problems, the global implications of its public health and environmental situations, and problems in the management of its natural resource wealth. It is the poorest world region. Its prospects for economic development trail behind those of other regions due to the legacy of colonialism, indebtedness, corrupt administration, civil and international conflict, and natural hazards.

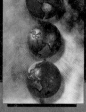

Low overall population density characterizes Africa south of the Sahara, but there are several clusters of dense population. Population growth rates rose rapidly after the mid-20th century, but have recently begun to decline sharply due to falling birth rates and the ravages of HIV/AIDS.

Minerals are the chief economic asset of Africa south of the Sahara, but are distributed unevenly. South Africa enjoys the greatest mineral wealth and by far the highest standard of living on the continent. Abundant mineral resources have in almost all cases failed to yield benefits to most people.

European powers colonized almost all of Africa in the 19th and early 20th centuries. Postcolonial Africa continues, as it did during colonial times, to export mainly raw materials to the industrialized world.

Africa is the cradle of humankind where hominids originated and from where they diffused. The region has seen many indigenous civilizations and empires, and is ethnically and linguistically diverse.

CHAPTER 17

A Geographic Profile of Africa South of the Sahara

This famous Apollo 17 view of Earth from space features almost all of the continent of Africa. Note also that much of the ice-covered continent of Antarctica is visible.

Geographers typically recognize the continent of Africa as having two major divisions: North Africa, which is the predominantly Arab and Berber realm of the continent, and ethnically diverse Africa south of the Sahara, often called sub-Saharan or Black Africa. In this text, North Africa is regarded as essentially Middle Eastern and is discussed in Chapters 8 and 9. This chapter sketches the broad outlines of the geography of Africa south of the Sahara, and Chapter 18 examines the major African subregions. Area and population data are in Table 17.1.

The region of Africa south of the Sahara is culturally complex, physically beautiful, and problem-ridden. In this introductory profile, we focus on the region's diverse environments, peoples and modes of life, major population concentrations, European colonial legacies, and major current problems.

A special section on the continent's economic geography begins on page 454, but it is important to note at the outset that this is by far the world's poorest region. Africa ranks at the very bottom of every statistical indicator of global quality of life. Statistically, 45% of the region's people live in poverty. It is also a continent in strife. Of 27 major armed conflicts around the world in 2001, 11 were in Africa. Africa's plight may seem ironic because by some measures the region has the world's greatest variety of natural resources. But as these chapters will show, some of Africa's recent and current conflicts are not about ideology or ethnicity, but are simply about control over resources such as diamonds, oil, and other minerals.

17.1   Area and Population

Africa south of the Sahara (for convenience, including Madagascar and other nearby Indian Ocean islands) has the largest land area of all the major world regions covered in this book. Its 8.4 million square miles (21.8 million sq km) make it more than twice the size of the United States. "People overpopulation" is apparent in some areas, and yet much of the region is sparsely populated. With a population of 641 million as of 2001, the region's average population density is substantially less than that of the United States. However, barring a catastrophic accelerated loss of population due to AIDS or famine, this density gap should close rapidly; the rate of nat-

ural population increase in Africa south of the Sahara is 2.6% per year, or about four times that of the United States. As in most less developed countries (LDCs), African parents generally want to have large families for several reasons: to have extra hands to perform work; to be looked after when they are old or sick; and in the case of girls, to receive the "bride wealth" a groom pays in a marriage settlement. Large families also convey status.

The majority of this region's people live in a few small, densely populated areas. The main ones are the coastal belt bordering the Gulf of Guinea in West Africa, from the southern part of Africa's most populous country, Nigeria, westward to southern Ghana; the savanna lands of northern Nigeria; the highlands of Ethiopia; the highland region surrounding Lake Victoria in Kenya, Tanzania, Uganda, Rwanda, and Burundi; and the eastern coast and parts of the high interior plateau of South Africa (Fig. 17.1, p. 436, and the world population map, Fig. 1.5b).

This is the world's most rural region (70% in 2001), with rural populations of most countries between 65 and 85%. The most rural are two East African nations with very fertile soils: Rwanda (95%) and Burundi (92%)(see Table 17.1). The most urbanized are Djibouti (83%), where there is no arable land and a small number of people are clustered in a port city, and Gabon (73%), where people are flocking to partake in an oil boom that is benefiting urbanites. Life in villages is the rule, where a typical rural home is a small hut made of sticks and mud, with a dirt floor, a thatched roof, and no electricity or plumbing (Fig. 17.2, p. 437).

This is the world's youngest population. Just under half of the region's people are younger than 15 years old (compared with about a third of the populations of Latin America and Asia). This is another indicator of how rapidly the region's population should grow, barring the vagaries of disease or famine. But the Malthusian scenario does seem to loom over Africa. Analysts fear the consequences of what they call the "1% gap": Since the 1960s, the population of Africa south of the Sahara has grown at a rate of about 3% annually, while food production in the region has grown at only about 2% per year. The wild card in Africa's population deck is the HIV virus, and the disease it causes — AIDS (acquired immunodeficiency syndrome) — has inevitably been identified by some as the Malthusian "check" to the region's population growth.

Table 17.1   Africa: Basic Data

Political Unit	Area (thousand/sq mi)	Area (thousand/sq km)	Estimated Population (millions)	Annual Rate of Increase (%)	Estimated Population Density (sq mi)	Estimated Population Density (sq km)	Human Development Index	Urban Population (%)	Arable Land (%)	Per Capita GDP PPP ($US)
The Horn of Africa										
Ethiopia	426.4	1104.3	65.4	2.9	153	59	0.321	15	12	600
Eritrea	45.4	117.6	4.3	3.0	95	37	0.416	16	12	710
Somalia	246.2	637.7	7.5	3.0	30	12	NA	28	2	600
Djibouti	9.0	23.2	0.6	2.7	71	27	0.447	83	0	1300
Total	727.0	1882.8	77.8	2.9	107	41	0.328	16.8	8.5	611
The Sahel										
Mauritania	396.0	1025.5	2.7	2.8	7	3	0.437	54	0	2000
Senegal	76.0	196.7	9.7	2.8	127	49	0.423	43	12	1600
Mali	478.8	1240.2	11	3.0	23	9	0.378	26	2	850
Niger	489.2	1267.0	10.4	2.9	21	8	0.274	17	3	1000
Burkina Faso	105.8	274.0	12.3	3.0	116	45	0.32	15	13	1000
Gambia	4.4	11.3	1.4	3.0	323	125	0.398	37	18	1100
Cape Verde	1.6	4.0	0.4	3.0	287	111	0.708	53	11	1700
Chad	495.8	1284.0	8.7	3.3	18	7	0.359	21	3	1000
Total	2047.6	5302.7	56.6	3.0	28	11	0.357	25.9	3.1	1129
West Africa										
Guinea	94.9	245.9	7.6	2.3	80	31	0.397	26	2	1300
Côte d'Ivoire	124.5	322.5	16.4	2.0	132	51	0.426	46	8	1600
Togo	21.9	56.8	5.2	2.9	235	91	0.489	31	38	1500
Benin	43.5	112.6	6.6	3.0	152	59	0.42	39	13	1030
Sierra Leone	27.7	71.7	5.4	2.6	196	76	0.258	37	7	510
Ghana	92.1	238.5	19.9	2.2	216	83	0.542	37	12	1900
Nigeria	356.7	923.8	126.6	2.8	355	137	0.455	36	33	950
Guinea-Bissau	13.9	36.1	1.2	2.2	88	34	0.339	22	11	850
Liberia	43.0	111.4	3.2	3.1	75	29	NA	45	1	1100
Total	818.2	2119.3	192.1	2.7	235	91	0.453	36.6	19.5	1125
West Central Africa										
Cameroon	183.6	475.4	15.8	2.7	86	33	0.506	48	13	1700
Gabon	103.3	267.7	1.2	1.6	12	5	0.617	73	1	6300
Congo, Rep. of	132.0	342.0	3.1	3	24	9	0.502	41	0	1100
Central African Republic	240.5	623.0	3.6	2	15	6	0.372	39	3	1700
Congo, Dem. Rep. of	905.4	2344.9	53.6	3.1	59	23	0.429	29	3	600
Equatorial Guinea	10.8	28.0	0.5	3.1	43	17	0.61	37	5	2000
Sao Tome and Principe	0.4	1.0	0.2	3.5	445	172	NA	44	2	1100
Total	1576.0	4082.0	78	2.9	49	19	0.449	34.6	3.8	991
East Africa										
Kenya	224.1	580.4	29.8	2.0	133	51	0.514	20	7	1500
Tanzania	364.9	945.1	36.2	2.8	99	38	0.436	22	3	710
Uganda	93.1	241.0	24	2.9	258	100	0.435	15	25	1100
Rwanda	10.2	26.3	7.3	1.8	719	278	0.395	5	35	900
Burundi	10.7	27.8	6.2	2.5	579	224	0.309	8	44	720
Total	703.0	1820.6	103.5	2.5	147	57	0.448	17.8	8.0	1042
Southern Africa										
Zambia	290.6	752.6	9.8	2.3	34	13	0.427	38	7	880
Malawi	45.7	118.5	10.5	2.3	231	89	0.397	20	34	900

Handwritten annotations: "How powerful" (above Estimated Population); "What does arable" (above Arable Land); "less rural" (beside Burundi)

Table 17.1 Africa: Basic Data (continued)

Political Unit	Area (thousand/sq mi)	Area (thousand/sq km)	Estimated Population (millions)	Annual Rate of Increase (%)	Estimated Population Density (sq mi)	Estimated Population Density (sq km)	Human Development Index	Urban Population (%)	Arable Land (%)	Per Capita GDP PPP ($US)
Zimbabwe	150.9	390.8	11.4	0.9	75	29	0.554	32	7	2500
Angola	481.4	1246.7	12.3	2.4	26	10	0.422	32	2	1000
Mozambique	309.5	801.6	19.4	2.1	63	24	0.323	28	4	1000
South Africa	471.4	1221	43.6	1.2	92	36	0.702	54	10	8500
Namibia	318.3	824.3	1.8	1.9	6	2	0.601	27	1	4300
Botswana	224.6	581.7	1.6	1.0	7	3	0.577	49	1	6600
Lesotho	11.7	30.3	2.2	2.0	186	72	0.541	16	11	2400
Swaziland	6.7	17.4	1.1	2.0	165	64	0.583	25	11	4000
Total	**2310.8**	**5984.9**	**113.7**	**1.7**	**49**	**19**	**0.533**	**38.9**	**5.2**	**4194**
Indian Ocean Islands										
Madagascar	226.7	587.0	16.4	3	71	27	0.462	22	4	800
Mauritius	0.8	2.0	1.2	1	1520	587	0.765	43	49	10400
Mayotte	0.1	0.4	0.2	3.1	1139	440	NA	NA	NA	600
Comoros	0.9	2.2	0.6	3.5	692	267	0.51	29	35	720
Reunion	1.0	2.5	0.7	1.5	744	287	NA	73	17	4800
Seychelles	0.2	0.5	0.1	1.1	449	173	NA	63	2	7700
Total	**229.7**	**594.6**	**19.2**	**2.8**	**84**	**32**	**0.484**	**25.6**	**4.3**	**1577**
Summary Totals	**8412.0**	**21787.0**	**640.9**	**2.6**	**76.2**	**29.4**	**0.444**	**30.1**	**6.3**	**1591**

Sources: *World Population Data Sheet, Population Data Sheet*, 2001; *U.N. Human Development Report*, United Nations, 2001; *World Factbook*, CIA, 2001.

17.2 Physical Geography and Human Adaptations

Africa south of the Sahara is both rich in natural resources and beset with environmental challenges to economic development. It is home to some of the world's greatest concentrations of wildlife and to some of the most degraded habitats.

A Land of Plateaus and Rivers

Most of Africa is a vast plateau, actually a series of plateaus, with a typical elevation of more than 1,000 feet (305 m) (see inset, Fig. 17.1). Near the Great Rift Valley in the Horn of Africa and in southern and eastern Africa, the general elevation rises to 2,000 to 3,000 feet, with many areas at 5,000 feet and higher (Fig. 17.3, p. 438 and Regional Perspective, p. 443). The highest peaks and largest lakes of the continent are located in this belt. The loftiest summits lie within a 250-mile radius (c. 400 km) of Lake Victoria. They include Mt. Kilimanjaro (19,340 ft/5,895 m) and Mt. Kirinyaga (Mt. Kenya; 17,058 ft/5,200 m; Fig. 17.4, p. 439), which are volcanic cones, and the Ruwenzori Range (up to 16,763 ft/5,109 m), a nonvolcanic massif produced by faulting. Lake Victoria, the largest lake in Africa, is surpassed in area among inland waters of the world only by the Caspian Sea and Lake Superior. Other very large lakes in East Africa include Lake Tanganyika and Lake Malawi.

The physical structure of Africa has influenced the character of African rivers. The main rivers, including the Nile, Niger, Congo, Zambezi, and Orange, rise in interior uplands and descend by stages to the sea. At some points, they descend abruptly, particularly at plateau escarpments, so that rapids and waterfalls interrupt their courses. These often block navigation a short distance inland. Helping to offset this problem, Africa's discontinuous inland waterways are interconnected by rail and highways more than on any other continent. The Congo is used more for transportation than is any other river in the region. Unlike the Niger, Zambezi, Limpopo, and Orange, which have built deltas, the Congo has scoured a deep estuary 6 to 10 miles (10 to 16 km) wide that allows ocean vessels to navigate as far as Matadi in the Democratic Republic of Congo, about 85 miles (c. 135 km) inland.

The many waterfalls and rapids do have a positive side: They represent a great potential source of hydroelectric energy. There are major power stations on the Zambezi River at the Cabora Bassa Dam in Mozambique and at the Kariba Dam (Fig. 17.5, p. 439), which Zimbabwe and Zambia share; at the Inga Dam on the Congo River, just upstream from Matadi; at the Kainji Dam on the Niger River in Nigeria; and at the Akosombo Dam on the Volta River in Ghana. But only about 5% of Africa's hydropower potential has been realized (compared to about 60% in North America). Many of the best sites are remote from large markets for power.

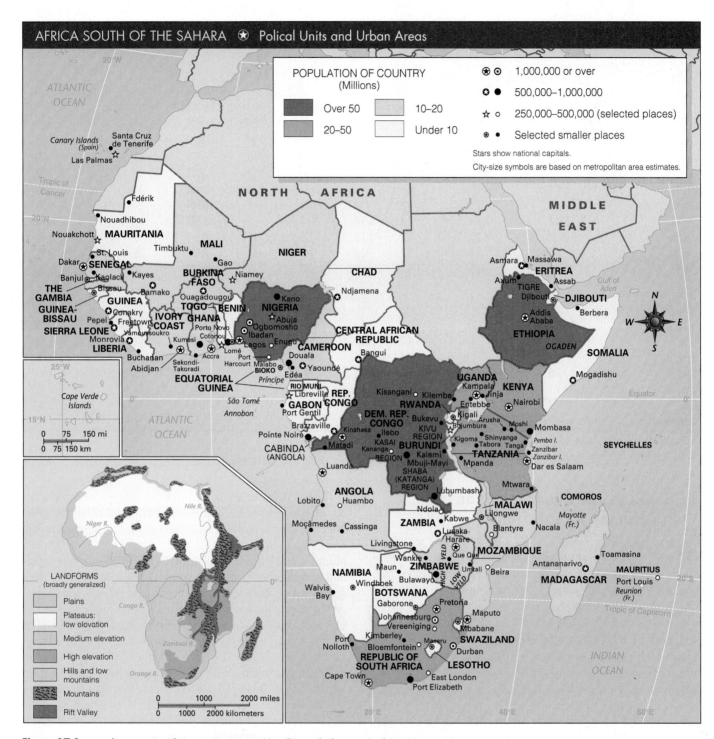

Figure 17.1   Reference map of countries, cities, and landforms of Africa south of the Sahara.

Figure 17.2   This village in northern Madagascar is typical of villages throughout much of Africa south of the Sahara. Homes are elevated above the ground to minimize the risk of flooding during the wet season, and their roofs are thatched. Streets and lanes are unpaved.

Climate, Vegetation, and Water

F2.6
29
F2.7
31

The equator bisects Africa, so about two-thirds of the region lies within the low latitudes and has tropical climates; Africa is the most tropical of the world's continents. One of the most striking characteristics of Africa's climatic pattern is its symmetry or regularity. This is due to the continent's position astride the equator and its generally level surface. Areas of tropical rain forest climate center around the great rain forest of the Congo Basin in central and western Africa. The forest merges gradually into a tropical savanna climate on the north, south, and east. This is the climatic and biotic zone supporting the famous large mammals of Africa. The savanna areas in turn trend into steppe and desert on the north and southwest. A broad belt of drought-prone tropical steppe and savanna bordering the Sahara Desert on the south is known as the Sahel. There is desert on the coasts of Eritrea, Djibouti, and Somalia in the Horn of Africa. In South Africa and Namibia, a coastal desert, the Namib, borders the Atlantic. The Kalahari Desert, which lies inland from the Namib, is better described as steppe or semidesert than as true desert. Along the northwestern and southwestern fringes of the continent are small but productive areas of Mediterranean climate, while eastern coastal sections and adjoining interior areas of South Africa have a humid subtropical climate.

Total precipitation in the region is high but unevenly distributed; some areas are typically saturated, whereas others are perennially bone-dry. Even in many of the rainier parts of the continent, there is a long dry season, and wide fluctuations occur from year to year in the total amount of precipitation. One of the major needs in Africa is better control over water. In the typical village household, women carry water by hand from a stream or lake or a shallow (and often polluted) well. Use of more small dams would help provide water storage throughout the year.

Drought is a persistent problem in most countries. Although all droughts create problems, some last for years with

devastating effects in this heavily agricultural region. Droughts have been particularly severe in recent decades in the Sahel and in the Horn of Africa, respectively in the northwest and northeast parts of the region.

Subsistence Agriculture and Pastoralism

Africa's most productive lands are on river plains, in volcanic regions (especially the East African and Ethiopian highlands), and in some grassland areas of tropical steppes (notably the High Veld in South Africa). Soils of the deserts and regions of Mediterranean climate are often poor. In the tropical rain forests and savannas, there are reddish tropical soils that are infertile once the natural vegetation is removed and can support only shifting agriculture.

Africa's soils favor subsistence agriculture (people farming their own food, but producing little surplus for sale) and pastoralism, and over half of the region's people practice these livelihoods. Women do a large share of the farmwork—they produce 80 to 90% of Africa's food—in addition to household chores and the bearing and nurturing of children (Fig. 17.6, p. 442). Mechanization is rare, fertilizers are expensive, and so crop yields are low. In the steppe of the northern Sahel, both rainfall and cultivation are scarce. The savanna of the southern Sahel, with its more dependable rainfall, is a major area of rainfed cropping. Unirrigated millet, sorghum, corn (maize), and peanuts are major subsistence crops in the savanna. In the tropical savannas south of the equator, corn is a major subsistence crop in most areas, with manioc and millet also widely grown. Corn, manioc, bananas, and yams are the major food crops of the rain forest areas. 457

Many peoples, particularly in the vast tropical grasslands both north and south of the equator, are pastoral. Herding of sheep and hardy breeds of cattle is especially important in the Sahel. Although cattle raising is widespread through the savannas, cattle are largely ruled out over extensive sections both north and south of the equator by the disease called nagana, which is carried by the tsetse flies that also transmit sleeping sickness to humans (see Definitions & Insights, p. 445). In tropical rain forests, tsetse flies are even more prevalent and few cattle are raised, but goats and poultry are common (as they are in tsetse-frequented savanna areas).

Most Africans who live by tilling the soil also keep some animals, even if only goats and poultry. Among African peoples such as the Masai of Kenya and Tanzania and the Tutsi (Watusi) of Rwanda and Burundi, livestock not only contribute to daily diet but are an indispensable part of customary social, cultural, and economic arrangements. Cattle are particularly important, with sheep and goats playing a smaller role. The Masai are probably the most famous African example of close dependence on cattle. They milk and carefully bleed the animals for each day's food and tend them with great care. The Masai give a name to each animal, and herds play a central role in the main Masai social and economic events through the year.

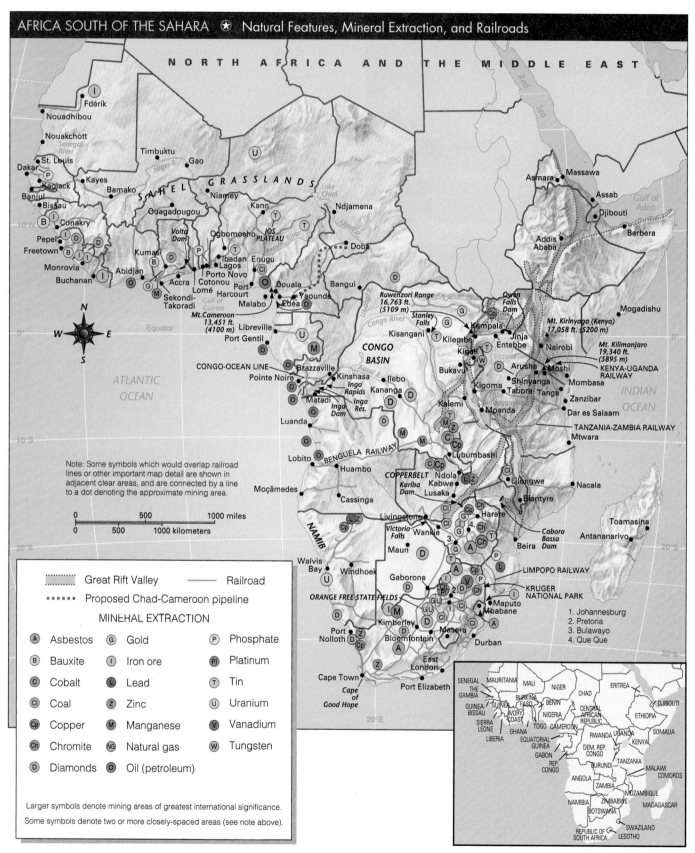

AFRICA SOUTH OF THE SAHARA ✪ Natural Features, Mineral Extraction, and Railroads

Figure 17.3   Map of major natural features, elevations, mineral extraction, and railroads in Africa south of the Sahara.

Figure 17.4 Kenya's Mt. Kirinyaga (Mt. Kenya: 17,058 ft/5,200 m), an extinct volcano, is Africa's second highest mountain.

Figure 17.5 The Kariba Dam straddles the Zambezi River between Zimbabwe and Zambia and provides hydroelectric power to both countries.

REGIONAL PERSPECTIVE   ::: HIV and AIDS in Africa

AIDS is a global problem. It had killed 21 million people worldwide by 2001, and there are an estimated 5 million new infections around the world each year. Africa south of the Sahara is where the virus originated (apparently among chimpanzees, and the first human case was reported in the Congo in 1959) and is today the epicenter of this health crisis. The statistics are startling. In 2001, 70% of the world's estimated 36 million persons infected with HIV lived in Africa south of the Sahara. In this region, 9% of people aged 15 to 49 were HIV-positive in 2001. In 2000, 2.4 million died in the region. Public health authorities predict that 25% of the region's more than 700 million people will have died of AIDS by 2010, leaving behind 35 million orphans and slashing the regional economy by 25%.

The epidemic is most severe in southern Africa. In some countries, including Botswana and Zimbabwe, as much as one-fifth of the general population and more than a third of the adult population is HIV-positive. South Africa

has more HIV cases (4.7 million in 2001) than any other country in the world. One in four adult South Africans, and one in nine of the general population, was HIV-positive in 2001. United Nations officials predict that if the epidemic continued apace, half of South Africa's 15 year olds will die of AIDS in the coming years. Similar impacts are predicted for South Africa's neighbors. The effect will be to change the pyramid-shaped age structure diagram typical of South Africa, Botswana, and Zimbabwe today to "chimney"-shaped structures (Fig. 17.A). In West Africa, more conservative, especially Islamic, mores about sexuality have slowed infection rates. However, AIDS is gaining ground rapidly in that subregion. In Ivory Coast, for example, 11% of adults were HIV-positive in 2001.

Earth has not seen a pandemic like this since the bubonic plague devastated 14th-century Europe and smallpox struck the Aztecs of 16th-century Mexico. This scourge is causing sharp reductions in life expectancy in Africa and, unless contained, will dramatically alter recent pro-

jections of the region's population growth. Life expectancy in Botswana was 61 years in 1993 but, because of the virus, was 41 in 2001 and is projected to be 29 in 2010. Life expectancy in Zimbabwe and Namibia (40 and 46 years, respectively, in 2001) is projected to be 33 in 2010, and in South Africa (53 years in 2001), it is projected to be 35. In Zimbabwe, the population growth rate in 1998 was 1.5% rather than the projected 2.4% that would have been the rate were it not for AIDS-related deaths. By 2001, Zimbabwe's growth rate was down to 0.9%. It is projected that in 2003 the populations of Botswana, Zimbabwe, and South Africa will begin declining.

There is some potentially good news in the frightening picture of AIDS in Africa south of the Sahara. The number of new cases reported has declined slightly. Experts are hopeful this will be a trend and an indication that the epidemic has reached what they call a "plateau." However, these authorities argue that this bit of hope is no reason to temper an all-out war on the problem.

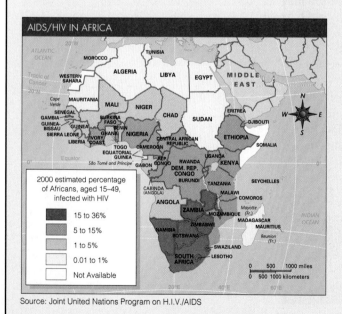

Source: Joint United Nations Program on H.I.V./AIDS

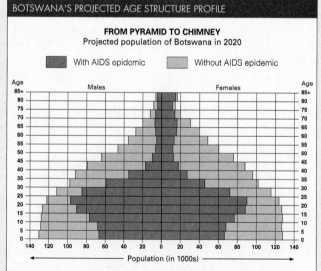

Source: *The Economist*, July 15, 2000.

Figure 17.A   HIV/AIDS has had a devastating impact in Africa south of the Sahara and threatens to redraw the demographic profiles of some countries in unprecedented ways.

This region is already the world's least developed, and HIV/AIDS is only worsening the prospects for development. Although it may seem logical that a lower population would mean more prosperity, the incidence of HIV infection is particularly high among the region's most educated, skilled, and ambitious young urban professionals, including teachers, white-collar workers, and government employees. These are the ones, in addition to truckers and merchants, who travel the most and are therefore the most likely to have many sexual liaisons. Teachers are hit particularly hard. The typical pattern is that a young man trained in a city as a teacher takes his first post in a remote village. Because his salary is low, he often depends on the generosity of villagers for food and shelter. A village headman may assign a girl to cook and clean for him, and the teacher and his caretaker have sex. Girls in the village generally see the teacher as someone with elite status, and they vie for his attention, so he has multiple sexual partners. He will eventually come down with the virus and will die, but not before he has spread the virus further through the village population. In some villages, schools are closed because there are no teachers to replace the ones who have died from AIDS. These kinds of effects are incalculable but probably huge. They affect not only education but corporate profits, research, health care, and other indexes of progress.

Readers may wonder why the HIV infection rate is so high in this region, why people there do not practice safer sex, and why so little is being done to fight the epidemic. A number of factors come into play, and those listed here are generalizations that do not fit every African culture and country. First, attitudes about sexuality are often more relaxed than they are in American culture, for example, with less stigma associated with adultery or multiple partners. Second, both partners are often reluctant to use condoms. Even though she is most at risk of contracting the virus, the woman may insist that the man not use a con-

dom because she does not want to be perceived as lacking virtue and be cast in the same high-risk category as prostitutes. Third, there is a widespread stigma about having AIDS. A person who contracts it is often shunned, even by family members, and often suffers and dies in neglect and isolation. The family or community will insist the person died of tuberculosis or some other ailment, so the real killer problem is never confronted. Fourth, there is a general lack of public awareness and education to fight the disease. Where leadership on this issue is needed most, it is sometimes absent. South Africa's President Thabo Mbeki, for example, was converted to the unusual belief that HIV does not cause AIDS, instead blaming poverty and malnutrition for the virus. Finally, there are inadequate health measures to fight the epidemic. Simple tests to determine the presence of HIV are seldom available, so those who have the virus pass it off unknowingly (the United Nations estimates that 90% of HIV-positive Africans do not know they have the virus). And mainly because of their high costs, anti-AIDS drugs are scarce. The drug nevirapine has a 50% success rate in blocking transmission of the virus from mother to child during childbirth. Its administration as a single pill just prior to delivery would thus have a substantial impact on reducing numbers of new HIV infections, but this drug has been prohibitively expensive.

What can be done about these difficult problems? Two countries that have had remarkable success in stemming the AIDS tide provide some answers. Uganda and Senegal have used relatively inexpensive public health tools, especially education, in their fights against the disease. To reduce the stigma associated with AIDS and to encourage counseling, testing, and condom use, the countries' political and religious leaders have been outspoken about AIDS. The results have been impressive. Uganda's infection rates dropped from 14% in the early 1990s to 8% by 2000. Senegal's president spoke out forcefully about AIDS in the late

1980s, and infection rates have stayed below 2% ever since. Prostitution is legal in Senegal, but a vigorous education campaign about the virus has kept infection rates low among prostitutes and their clients.

The wealthier countries could invest in combating the disease. Harvard University economist Jeffrey Sachs has suggested that if the rich countries combined to contribute just $10 to $20 billion per year to fight AIDS, malaria, and other disease in Africa, there could be huge progress in reducing death rates and thereby a great boost to the region's economic growth rate. This would not be a financial burden on the more developed countries (MDCs), Sachs argues; in fact, they could afford such payments "without breaking a sweat."[a] More modestly, the United Nations says that an annual expenditure of $3 billion would be needed to fight only the African AIDS epidemic. That is about 10 times what is currently spent to combat AIDS in the region.

Drug companies in the West have a unique opportunity to address Africa's AIDS problems. Some of them manufacture life-prolonging "antiretroviral" (ARV) drugs containing protease inhibitors that suppress replication of, but do not kill, the immunodeficiency virus. Importantly, these drugs do reduce the likelihood of the HIV-positive patient transmitting the virus to another person. The companies could help make such drugs more accessible; in 2001, only 10,000 of the 25 million HIV-positive Africans were receiving these medications. That would mean making the drugs inexpensive, which drug companies in the West have until recently been loath to do. An antiretroviral drug "cocktail" can cost $15,000 per patient per year in the United States. However, other countries, notably Brazil and India, have taken advantage of a loophole in international patent rules to manufacture their own cheaper copies of the U.S.-made drugs. Called "compulsory licensing,"

[a]Jeffrey Sachs, "The Best Possible Investment in Africa." *New York Times*, February 10, 2001, p. A27.

(Continued)

the loophole allows the countries to breach patents during national emergencies to manufacture generic versions of patented drugs. Citing AIDS as an emergency, Brazil produces virtually the same drug cocktail for $3,000 per patient per year, and India does it for $1,000 per year.

Early in 2001, the Indian drug manufacturer Cipla offered to sell the drugs to South Africa and other governments for $600 per patient per year and to the organization Doctors Without Borders for $350 per year (Doctors Without Borders would then distribute the drugs free of charge). The pressure was then on U.S. companies to lower their prices on anti-AIDS drugs sold in Africa and elsewhere in the developing world. They hesitated to do so, not only because of the loss of revenue involved, but out of fears that Americans would cry foul and demand lower costs on their own drugs. The tide began to go Africa's way when a German company offered nevirapine free

of charge for 5 years to developing countries. Soon some U.S. drug firms relented. Merck, for example, offered to make one of its key drugs available in Africa at $600 per patient per year, making no profit. The U.S. administration promised to keep the door open for even more, cheaper anti-AIDS drugs by declaring it would not seek sanctions against poor countries stricken with AIDS, even if American patent laws were being broken. Country-by-country decisions on how the multinational drug companies should distribute the drugs are now ongoing. In the meantime, the challenge for MDC governments, aid agencies, and citizens is to help Africans pay for these drugs: Even $600 per year is out of reach. And the drugs will need to be correctly distributed and used; the manufacturers warn that if they are not, new strains of drug-resistant viruses could emerge.

Geographers play an important role in tracking and helping to prevent the

outbreak and spread of HIV and other viruses and diseases. Within geography is the subdiscipline of **medical geography,** which uses a spatial approach to understanding, treating, and preventing disease. Some medical geographers identify locations where a virus outbreak has occurred, track movements of the virus, and warn relevant public health and immigration authorities to take necessary precautions. Others look at how access to health care differs significantly among different socioeconomic groups and advise health-care and political authorities on how to make access more equitable. Medical geographers also serve as environmental watchdogs by monitoring the health impacts of nuclear power plants, chemical factories, and other potentially hazardous facilities. This subfield is likely to grow and diversify even more as issues of globalization, biotechnology, and diseases themselves receive more attention and funding.

Madagascar is a good example of an African country in which cattle represent status and wealth. Malagasy livestock owners tend to want higher numbers of the animals, rather than better quality stock, to enhance their standing (Fig. 17.7). The resulting population of zebu cattle on the island is about 10 million. Their forage needs have grave conse-quences for Madagascar's rain forests and other wild habitats. People clear the forests and repeatedly set fire to the cleared lands to provide a flush of green pasture for their livestock, causing a rapid retreat of the island's natural vegetation.

Wildlife

Africa has the planet's most spectacular and numerous populations of large mammals. The tropical grasslands and open forests of Africa are the habitats of large herbivorous animals, such as the elephant, buffalo, antelope, zebra, and giraffe, as well as of carnivorous and scavenging animals, such as the lion, leopard, and hyena. The tropical rain forests have fewer of these "game" animals (as Africans call them); the most abundant species here are insects, birds, and monkeys, with the hippopotamus, the crocodile, and a great variety of fish present in the streams and rivers draining the forests and wetter savannas.

While film documentaries promote a perception outside Africa that the continent is a vast animal Eden, the reality is less positive. Human population growth, urbanization, and agricultural expansion are taking place in Africa, as elsewhere in the world, at the expense of wildlife. Hunting and competition with domesticated livestock also take their toll. Many species are protected by law. Such laws are difficult to enforce,

Figure 17.6   Mother and child in Zimbabwe. Women plant and harvest most of Africa's food, while also rearing its children.

One of the most spectacular features of Africa's physical geography is the **Great Rift Valley,** a broad, steep-walled trough extending from the Zambezi Valley (on the border between Zimbabwe and Zambia) northward to the Red Sea and the valley of the Jordan River in southwestern Asia (see Fig. 17.3). Its relationship to the tectonic movement of crustal plates is still poorly understood. However, most earth scientists believe it marks the boundary of two crustal plates that are rifting, or tearing apart, causing a central block between two parallel fault lines to be displaced downward, creating a linear valley. This movement will eventually cut much of southern and eastern Africa away from the rest of the continent and allow seawater to fill the valley.

The Great Rift Valley has several branches. Lakes, rivers, seas, and gulfs already occupy much of it. It contains most of the larger lakes of Africa, although Lake Victoria, situated in a depression between two of its principal arms, is an exception. Some, like the 4,823-foot-deep (1,470-m) Lake Tanganyika (the world's second deepest lake after Russia's Lake Baikal), are extremely deep. Most have no surface outlet. Volcanic activity associated with the Great Rift Valley has created Mt. Kilimanjaro, Mt. Kirinyaga, and some of the other great African peaks, along with lava flows, hot springs, and other thermal features. Faulting along the Great Rift Valley in Ethiopia, Kenya, and Tanzania has also exposed remains of the earliest known ancestors of *Homo sapiens.*

Figure 17.7   Zebu cattle are extremely important in the cultures and household economies of rural Madagascar. There are about 10 million cattle in this country of 16 million people.

however, and poaching has devastated some species (see Regional Perspective, p. 446). Still, Africa remains home to some of the world's most extraordinary and successfully managed national parks, such as the Kruger National Park in South Africa and the Ngorogoro Crater National Park in Tanzania. International tourism to these parks is a major source of revenue for some countries, particularly Kenya, Tanzania, South Africa, Namibia, and Botswana.

17.3 Cultural and Historical Context

Many non-Africans are unaware of the achievements and contributions of the cultures of Africa south of the Sahara. The African continent was the original home of humankind. Recent DNA studies suggest that the first modern people (*Homo sapiens*) to inhabit Asia, Europe, and the Americas were descendants of a small group that left Africa via the Isthmus of Suez about 100,000 years ago. After about 5000 B.C., indigenous people were responsible for agricultural innovations in four culture hearths: the Ethiopian Plateau, the West African savanna, the West African forest, and the forest-savanna boundary of West Central Africa. Africans in these areas domesticated important crops such as millet, sorghum, yams, cowpeas, okra, watermelons, coffee, and cotton. From Africa, these diffused to agricultures in other world regions.

Civilizations and empires emerged in Ethiopia, West Africa, West Central Africa, and Southern Africa. In the first century A.D., the Christian empire based in the northern Ethiopian city of Axum controlled the ivory trade from Africa to Arabia. Ethiopian tradition holds that a shrine in Axum still contains the biblical Ark of the Covenant and the tablets of the Ten Commandments, which disappeared from the Temple in

Figure 17.8   Mosque and market in Djenne, Mali. Spirituality is a strong component in each of Africa's many cultures and serves to influence many facets of everyday life.

Jerusalem in 586 B.C. Several Islamic empires, including the Ghana, Mali, and Hausa states, emerged in West Africa between the 9th and 19th centuries. All of these agriculturally based civilizations controlled major trade routes across the Sahara. They profited from the exchange of slaves, gold, and ostrich feathers for weapons, coins, and cloth from North Africa. Three kingdoms arose between the 14th and 18th centuries in what are now the southern Democratic Republic of Congo and northern Angola. These included the Kongo kingdom, which had productive agriculture and was the hub of an interregional trade network for food, metals, and salt. In what is now Zimbabwe, the Karanga kingdom of the 13th to 15th centuries built its capital city at the site known as Great Zimbabwe. Its skilled metalworkers mined and crafted gold, copper, and iron, and merchants traded these metals with faraway India and China.

Some Common Culture Traits

In the 16th century, European colonialism began to overshadow and inhibit the evolution of indigenous African civilization. However, the artistic, technical, and entrepreneurial skills of the region's peoples continued to flourish. These traits are today part of a greater African culture that is poorly understood and often stereotyped in the wider world. In this diverse and often fragmented continent, there are many shared traits that together comprise what geographer Robert Stock describes as "Africanity." He identifies the following eight constituent elements of the African identity:

1. A black skin color.
2. A unique conceptualization of the relationship between people and nature. Indigenous African religions emphasize that spiritual forces are manifested everywhere in the environment, in contrast with the introduced Christian and Muslim faiths, which tend to see nature as separate from God and people as apart from and superior to nature (Fig. 17.8).
3. An identity tied closely to the land, with many people dependent on hunting, herding, and farming. This dependence on the land reinforces the sense of closeness to nature. Africans tend to treat the land as communal rather than individual property.
4. Emphasis on the arts, including sculpture, music, dance, and storytelling, as essential to the expression of African identity.
5. A view of Africans as individuals making up links in a continuing "chain of life," in which reverence of ancestors and nurturing of children are virtues. Parents prefer to have many children to keep the chain growing and strive to educate them in the traditions of the ethnic group.
6. Extended rather than nuclear families, with parents and their children living and interacting with grandparents, cousins, nieces, nephews, and other relatives.
7. Respect for wise and fair authority, with village elders, "big men," and tribal chiefs endowed with powers they are expected to wield to benefit the group.
8. A shared history of colonial occupation that contributes to a unified sense that, in the past, Africans were humiliated and oppressed by outsiders.

DEFINITIONS + INSIGHTS

Africa's Greatest Conservationist

Not much larger than the common housefly, the tsetse fly of sub-Saharan Africa packs a wallop. This insect carries two diseases, both known as trypanosomiasis, which are extremely debilitating to people and their domesticated animals. People contract "sleeping sickness" from the fly's bite, and cattle contract nagana. Since the 1950s, there have been widespread efforts to eradicate tsetse flies so that people can grow crops and herd animals in fly-infested wilderness areas. Where the efforts have been successful, people have cleared, cultivated, and put livestock on the land. The results are mixed. While people have been able to feed growing populations in the process of opening up these lands, they have also eliminated important wild resources and in many cases caused erosion, desertification, and salinization of the land. Where the tsetse fly has been eliminated, so has the wilderness. The diminutive tsetse fly thus may be characterized as a **keystone species** — one that affects many other organisms in an ecosystem. The loss of a keystone species — in this case, a fly that keeps out humans and cattle — can have a series of destructive impacts throughout the ecosystem. For its role in maintaining wilderness in Africa south of the Sahara, some wildlife experts know the tsetse fly as "Africa's greatest conservationist."

Despite the common features of Africanity, Africa south of the Sahara is culturally and ethnically diverse, with complex tribal and ethnic identities.[1]

Languages

There is also great linguistic diversity in Africa south of the Sahara (Fig. 17.9). By one count, the peoples of this region speak more than 1,000 languages. Most of the peoples of this region belong to one of four broad language groupings:

1. The Niger-Congo language family (sometimes listed as a subfamily of the Niger-Kordofanian family) is the largest. It includes the many West African languages and the roughly 400 Bantu subfamily languages. Most of these are spoken south of the equator.
2. The Afro-Asiatic language family, including Semitic branch languages (e.g., Arabic and Amharic) and tongues of the Cushitic (e.g., Oromo and Somali) and Chadic (e.g., Hausa) branches. People living in the area adjoining the Sahara, from West Africa to the Horn of Africa, speak these languages. Even some of the Niger-Congo languages originating south of the Sahara, such as the Swahili (Kiswahili) tongue spoken widely in East Africa, have borrowed much from Arabic and other languages with roots elsewhere. The prominence of Arabic words in Swahili reflects a long history of Arab seafaring along the Indian Ocean coast of Africa; in fact, Swahili means "coastal" in Arabic.

3. The Nilo-Saharan language family of the central Sahel region, the northern region of West Central Africa, and parts of East Africa.
4. The Khoisan languages of the western portion of southern Africa, which the Bushmen (San) and related peoples speak.

The people of Madagascar speak a distinct language without African roots. This is Malagasy, an Austronesian tongue that originated in faraway Southeast Asia.

Religions

The religious landscape of Africa is complex and fluid. Spiritualism is extremely strong, but spiritual affiliations and practices are more interwoven and flexible than in most other world regions. It is not uncommon for parents and children to follow different faiths, for siblings to have different faiths, and even for an individual to change his or her religious beliefs and practices in the course of a lifetime. Broadly, however, there are some dominant patterns of religious geography (Fig. 17.10, p. 449). Islam is the dominant religion in North Africa, although there is a large minority of Christian Copts in Egypt. The Ethiopian Church, closely related to the Coptic faith, makes Ethiopia an exception to the otherwise Islamic Horn of Africa region. Islam is also the prevailing region of the East African coast, where Arab traders introduced the faith. Muslims are a majority or strong minority in rural northern Nigeria and Tanzania, and there are minority Muslim populations in cities and towns across the continent. Christians are a majority in southern Nigeria, Uganda, Lesotho, and parts of South Africa. Both Christianity and Islam are strongest in the cities, whereas traditional religions prevail or overlap with these monotheistic faiths in rural areas. There are Christian and Muslim efforts to dispose of "heretical" notions in African traditional belief systems, but these concepts are very resilient.

These broad strokes only begin to scratch the surface of African spirituality. Even where the large monotheistic faiths

---

[1] Politically and socially, most Africans have traditionally identified themselves by their tribe, recognizing members of the tribe as all those descended in kinship from a single tribal founder, or "eponym." However, because "tribe" and "tribalism" have acquired connotations of primitive feuding between hostile rivals, many Africanists now prefer to use the terms "ethnic group" and "ethnicity" in their place. This blending of meanings can be misleading, however. Sometimes, as in Somalia, there is a single ethnic group made up of numerous tribes. Some countries, like Somalia, Lesotho, Swaziland, and Botswana, are very homogeneous ethnically, whereas Tanzania, Cameroon, and Nigeria have hundreds of ethnic groups.

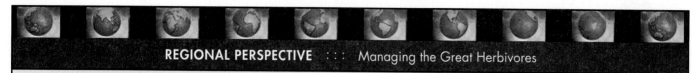

REGIONAL PERSPECTIVE   : : :   Managing the Great Herbivores

Elephants and rhinoceroses are Africa's largest and most endangered herbivores. Their plight has accelerated in recent years and so have local and international efforts to maintain their populations in the wild. The problem has compelled countries with very different wildlife resources to work together toward solutions.

In 1970, there were about 2.5 million African elephants living on the continent. Poaching and habitat destruction reduced their numbers to 1.8 million by 1978. There are now an estimated 350,000. The main reason for the sharp decline is that elephants have something people prize: ivory. Poaching for ivory was reducing African elephants at a rate of 10% annually when delegates of the 112 signatory nations of the Convention on International Trade in Endangered Species (CITES) met in 1989. The organization succeeded in passing a worldwide ban on the ivory trade, and since then, the precipitous decline has halted.

CITES member states won the ivory ban over the strong objections of southern African states led by Zimbabwe. While the East African nations of Uganda, Kenya, and Tanzania were suffering crashing elephant populations, Zimbabwe was experiencing what it regarded as an elephant overpopulation problem (Fig. 17.B). Zimbabwe had 5,000 elephants in 1900. In 2002, there were an estimated 80,000, and they were increasing at a rate of 4% annually. There are also healthy and growing elephant populations in Botswana, Malawi, Namibia, and South Africa. Before the worldwide ivory ban, these countries profited from the sustainable harvest and sale of elephant ivory, hides, and meat. At the 1989 CITES meeting, these countries argued that they should not be punished for their success in protecting the great mammals. They appealed for an exemption from the ivory ban so that they could earn foreign export revenue from a sustainable yield of their elephant populations. The majority of CITES members

rejected this appeal, arguing that any loophole in a complete ban would subject elephants everywhere to illegal poaching, resulting in the loss of the African elephant. The elephant-rich countries reluctantly supported this position.

In 1997, however, CITES reversed its policy and elected to allow Zimbabwe, Namibia, and Botswana to sell ivory as a reward for their positive wildlife policies. These countries were permitted to export a combined total of 60 tons of ivory on a one-time basis only to Japan and only under the most tightly monitored circumstances. As some skeptics had predicted, this loophole in a complete ivory ban promoted increased elephant poaching across the continent. At the following CITES meeting in 2000, these three countries acknowledged the problem and withdrew their previous appeals to sell ivory. While they await better monitoring systems for poaching and trading, the worldwide moratorium on ivory sales is again in effect.

Figure 17.B   Zimbabwe is blessed with elephants. These animals in Hwange National Park are feeding from a box of treats provided by the country's preeminent pachyderm ecologist, Alan Elliot (standing, wearing cap).

Meanwhile, these elephant-rich nations continue to cull (kill) "excess" elephants. Zimbabwean officials argue that their country can support only 45,000 animals. Meat from the cull of about 5,000 elephants yearly goes to needy villagers and crocodile farms. This resource helped the country weather the drought and near-famine of 1992.

If elephant ivory is like gold, rhinoceros horns are like diamonds. Men in the Arabian Peninsula nation of Yemen prize daggers with rhino horn handles (Fig. 17.C). Although Western scientists deny the medicinal efficacy of powdered rhino horn, traditional medicine in East Asia (particularly in Taiwan, China, and Malaysia) makes wide use of it, including as an aphrodisiac. These demands, and the current black-market value of about $25,000 per horn, have led to a precipitous decline in population of black rhinoceroses in Africa. There were an estimated 65,000 black rhinos in Africa in 1982; poaching reduced their numbers to 2,300 by 1992. In 2002, thanks to strenuous antipoaching efforts, there were an estimated 2,700, which is still a critically low number.

Zimbabwe is on the front line in the hard-fought war to protect the black rhino. As recently as 1984, there were as many as 2,000 of the animals in the country. Having reduced the numbers in countries to the north, poachers turned to rhino-rich Zimbabwe. Since 1984, there has been a steady increase in the numbers of poachers crossing international boundaries to kill rhinos in Zimbabwe. Most of them come across the Zambezi River from Zambia. Zambia is a very poor country, and the prospect of making hundreds or thousands of dollars in a night's work is irresistible to many. Even the order to Zimbabwean wildlife rangers to shoot poachers on sight, in effect since 1985, has not slowed the slaughter. Armed with automatic weapons, poachers have killed more than 1,500 rhinos since 1984.

With fewer than 300 rhinos surviving in Zimbabwe in 1991, wildlife officials turned to more desperate measures. They began dehorning rhinos to make them unattractive to poachers. To do this, a sharpshooter tranquilizes the animal and two assistants use a chainsaw to remove the two horns. It is not a permanent solution because the horn grows back at a rate of 3 inches yearly, so each animal must be regularly re-dehorned. Wildlife authorities have admitted that the program has failed. Poachers continue to kill the animals, perhaps out of spite, or because they cannot tell in the dark whether or not the prey has horns, or perhaps because they are after even a few inches of horn stump.

Nevertheless, the combination of efforts to combat the decline in rhino numbers has begun to take hold. By 2002, there were almost 450 rhinos in Zimbabwe. Ever vigilant to discourage poaching, Zimbabwean wildlife officials are now considering the possibility of opening a legal trade in rhino horns. They would raise rhino herds on state farms and regularly harvest their regrowing horns for sale. South Africa supports this idea of sustainable harvest. Like Zimbabwe, South Africa is sitting on a stockpile of tons of confiscated rhino horns and would profit greatly from their legalized trade. In addition, with about 1,100 living animals, South Africa is the last stronghold of the black rhino and does not want to become the next frontline state in the rhinoceros war.

Figure 17.C   Daggers are a nearly universal dress accessory for men in the Arabian Peninsula nation of Yemen. The most prized dagger handles are of rhino horn, a custom that has had a devastating impact on African rhinos thousands of miles distant.

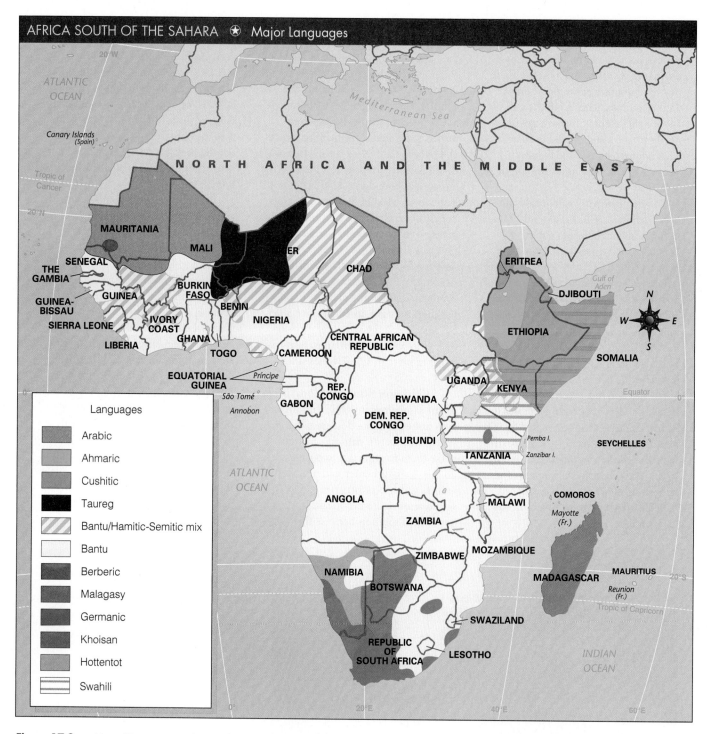

AFRICA SOUTH OF THE SAHARA ⭐ Major Languages

Languages

- Arabic
- Ahmaric
- Cushitic
- Taureg
- Bantu/Hamitic-Semitic mix
- Bantu
- Berberic
- Malagasy
- Germanic
- Khoisan
- Hottentot
- Swahili

Figure 17.9   Map of languages spoken in Africa south of the Sahara.

nominally prevail, and their followers are devout, there is a strong substrate of indigenous beliefs that has persisted and usually comfortably merged with the "official" faith. For example, as Stock mentioned earlier, there is a strong reverence of ancestors, a spiritual belief that runs naturally with the strong social deference to the elderly and respect for preceding generations (see Perspectives from the Field essay on Madagascar,

p. 450). Across the spectrum of African cultures, there is a widespread practice of using mediums to contact spirits, either of deceased ancestors, creator beings, or earth genies that can intercede on one's behalf. Many spirits are tied to particular places on the landscape, affirming Stock's observations that the identities of people themselves are tied closely to the land and that spiritual forces are manifested in the environment. Out-

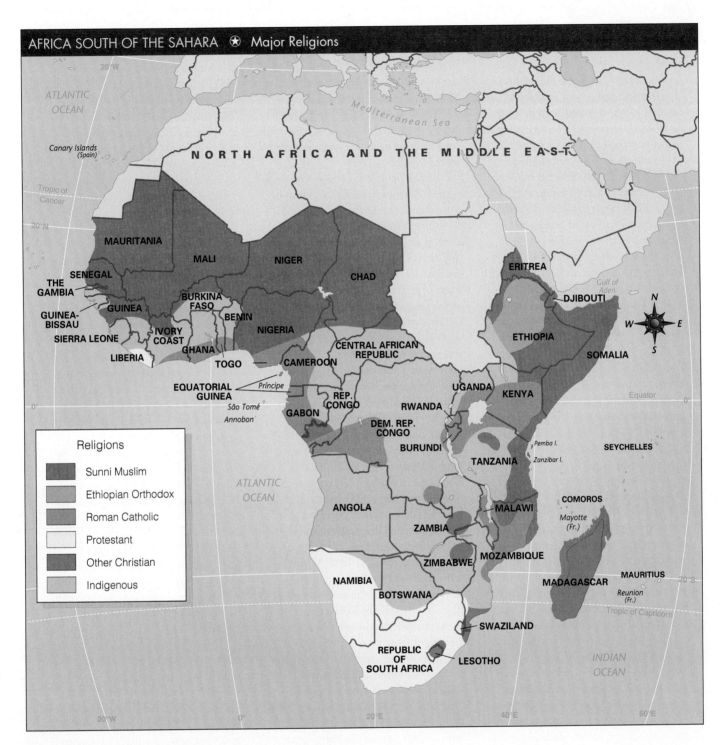

AFRICA SOUTH OF THE SAHARA ✪ Major Religions

Religions

- Sunni Muslim
- Ethiopian Orthodox
- Roman Catholic
- Protestant
- Other Christian
- Indigenous

Figure 17.10   Map of religions practiced in Africa south of the Sahara.

siders have tried, seldom successfully, to give accurate names to indigenous African belief systems, practices, and personnel: "ancestor worship," "animism," *force vitale* (living force), "living dead," and "witch doctor." While it would be gratifying to have accurate names and generalizations for African beliefs, they are unique, and can best be understood by careful reading, observation, and discussion on a case-by-case basis.

The Geographic Impact of Slavery

Until about 1,000 years ago, the cultures of Africa south of the Saharan desert barrier remained largely unknown to the peoples north of the desert. Egyptians, Romans, and Arabs developed contacts with the northern fringes of this region and some trade filtered across the Sahara, but to most outsiders, the "Dark Continent" was a self-contained, tribalized

In the austral autumn of 2000, I went to Madagascar to investigate the relationship between people and caves on the island. This field mission, funded by the American Geographical Society, was part of a larger project I am working on, a book on human-cave interaction around the world since prehistoric times. Of all the countries I hope this project will take me to, Madagascar may have the most complex and fascinating history of human involvement with caves.

The air journey from the United States to Madagascar is one of the longest one may undertake on the planet. It was 42 hours from the time I left home in Columbia, Missouri, until I arrived in Antananarivo (Tana), Madagascar's capital. This included a 12-hour layover in Paris, where an old acquaintance took me out for lunch and showed me some attractions of the legendary city. The fashionable shops and monumental opulence of Paris provided a huge contrast with the poverty and modest architecture of Madagascar's capital.

It is rare in my experience that such a long journey comes off without a hitch, so I was not surprised by the 3-hour delay of the Air Madagascar flight from Paris, during which excess weight, including my baggage, was unloaded from the aircraft. Twelve hours after takeoff, I arrived exhausted and without provisions (except for 30 pounds of camera and video gear in my hand luggage) but very excited about being in this exceptional country.

Fortunately, once in Tana, I was in the capable hands of an old friend and colleague, Steve Goodman. A zoologist based at Chicago's Field Museum, Steve is a pioneer in the natural history study of this extraordinary island, where nearly 90% of all plants and animals are endemic (occurring no where else on Earth). Steve has lived here since 1989 and has virtually become Malagasy. His wife, Asmina, is of the island's coastal Sakalava culture. Steve, Asmina, and their son reside with an extended family network, and as soon as I arrived from the airport, I was im-

mersed in special festivities surrounding the impending wedding of a family member. That afternoon, these included the *fumba fumba*, a ceremony in which the bride's and groom's families seek approvals from their respective ancestors (*razana*) for the impending union.

It is impossible to overstate the significance of the spirits of one's ancestors in this culture. The Malagasy peoples believe them to influence or control virtually every aspect of daily life. Here, in Steve and Asmina's home, the *razana* were appeased with several long oratories and toasts of local rum. I managed an hour's nap before, at 10 P.M., the group headed out for a raucous night of dancing held in, of all places, a prison mess hall (the inmates were tucked in their cells nearby). Here, your author danced to the jailhouse rock into the wee hours, concluding the first day of field work in Madagascar.

But I had come to study caves and people, and when my baggage arrived 2 days later, Steve and I set out across the High Plateau, bound for the caves of northwestern Madagascar's region of Majunga. Not 3 hours into what should have been a 15-hour journey, Steve's vehicle developed a fuel line problem and began to hemorrhage gasoline. There was no choice but to return to Tana. At least this setback allowed me the opportunity to observe in more detail the environmental holocaust wrought by 2,000 years of human activity on the island. Relentless clearing of forest for rice cultivation and cattle pasture has created a virtual moonscape. The rivers run blood red with what is left of the island's precious topsoil.

Steve was drawn away by other obligations, but I had the good fortune to be joined by his sister-in-law, Patty Vavizara, who would be my French and Malagasy interpreter and caving companion for the next 2 weeks. We flew to Majunga, hired a four-wheel-drive vehicle, and made our way to the vast and beautifully decorated caves of Anjohibe and Anjohikely. Rather than describe these, I will tell you of my favorite cave on this journey. We reached

it late at night. We had not set out until dusk, and our two local guides had difficulty getting oriented in the darkness. There was no road to follow, and their views were obscured by grasses nearly twice the height of the vehicle. But they finally located Andoboara Cave, small and uninteresting but for its human and spiritual associations and therefore of huge interest to me.

The people of the nearby village of Ambalakedi consider this cave sacred because on three separate occasions, most recently in 1998, grief-stricken parents whose children had wandered into the forest had recovered them alive here. The children had been abducted by a sometimes malevolent spirit called *kalanoro* who sought to rebuke the parents for not taking proper care of their children. The parents sought guidance from another spirit called *tromba* who possessed a fellow villager and recited a laundry list of demands made by the *kalanoro* that would have to be met for the child to be found. Once the parents fulfilled their obligations by, for example, leaving honey and sacrificing zebu cattle in the place designated by the *tromba*, the *kalanoro* revealed via the *tromba* the lost child's location, and in each case, the joyous reunion took place in the Andoboara Cave. As I was to learn in the coming weeks, there are countless caves in Madagascar bearing rich accounts of this kind.

Near the end of my stay in this exceptional country, Patty and I did a 5-day walk through the Ankarana Special Reserve near the island's northern tip. The park is dominated by a long wall of Jurassic-era limestone eroded by wind and water into razor-sharp formations locally called *tsingy-tsingy* (meaning "ouch ouch," as in how your feet feel when you tread the stuff). The massif is perforated by scores of caves, some so massive that even a powerful Petzl headlamp like mine could not illuminate their ceilings. Others are less grand in scale but graced with lovely calcite formations.

The Ankarana's sinkholes and the perimeter of the limestone massif contain protected remnants of dry tropical forest that are home to abundant wildlife. Each night, members of the primitive primate family of lemurs invaded our camp in search of food, calling so loudly that sleep came only through sheer exhaustion from the day's walking and caving. As a reptile lover, I was in paradise, indulging in the capture, photography, and release of boas and other snakes, riverine turtles, and my favorites, the chameleons.

But the caves and their human connotations were of greatest interest. At one night's campfire, a man from a village just outside the park told me how his ancestors, members of the royal family of the Antakarana tribe, sought refuge in the caves for nearly 3 years during the 1800s. They were escaping would-be slavery at the hands of the Merina tribe of highland Madagascar, whose aspiration was to unite (or pacify and enslave) the island's 17 ethnic groups. The Antakarana had taken advantage of the Merina's hopeless disorientation in the underground labyrinth of the Ankarana massif and survived in its caves while hunting, foraging, and growing some crops in the hidden karst canyons. The first four Antakarana kings were buried in caves here and are still visited on a ceremonial pilgrimage once every 5 years by the current king, Issa, and his subjects. I met with him to seek his permission to enter the royal cave tombs, but he said it was *fady* (taboo) even for him to do so, except on the specified occasion; he would welcome me to take part in the next event.

One particular day in the Ankarana was among the most extraordinary in my life. With an excellent guide named Aurelian Toly, we walked through the massive Andrafiabe Cave for several hundred meters before reaching a giant sunlight chasm containing what botanists call a "sunken forest," essentially a tropical paradise surrounded by karst desert wilderness. Lemurs howled and parrots cried in this lost world. I kept thinking, "Spielberg couldn't top this!" Crossing the forested ravine, we entered Cathedral Cave, whose entrance is littered with the hearths and pots used by the Antakarana in their time of refuge here. Well into the

cave, we crossed a 600-foot-long (200-m) hill of guano built by a massive roost of insectivorous *Hipposideros* bats. Fresh deposits fell on my helmet like raindrops. Our noses were choked by the stench of ammonia and the dust of the guano trodden underfoot. Our headlights shined upon seething masses of cockroaches and centipedes feasting on the treasures dropped by the bats. We couldn't help but crunch this fauna as we walked, and with the shrieking of the annoyed bats, I thought again of a Spielberg movie set.

My last excursion into underground Madagascar was in Crocodile Cave. It is appropriately named, for running through it is a river in which crocodiles seek habitat while the world outside is parched in the long dry season of April–October. On the cave's sand banks we saw abundant crocodile tracks, some very fresh, and we crossed the cave stream with trepidation. Instead of finding the great reptiles (some up to 6 meters long have been seen here), about 1,000 meters into the cave we were astonished to come upon a dugout canoe beached on the bank. Next to the pirogue were three poles, each bearing on either end a strand of 8 to 10 cormorants and other waterfowl tied by their feet. Some had already died and others struggled for life. In the next stretch of cave passage, our headlamps revealed the three poachers responsible, and a confrontation ensued. The two park guides in my party of seven cried out, "We are from the national park service. Come forward. Drop your weapons because the warden has a gun and will use

it against you!" I was surprised to learn that I was the warden, and the GPS pouch on my belt bore the gun. The ruse was enough to convince the poachers, who now pleaded for leniency. They said they had known hunting was forbidden in the park but the need to feed their families had driven them to take this long and strange canoe voyage into the cave.

I still did not understand. Why were there waterfowl in a cave? The park rangers demanded that the hunters return immediately to release the birds where they had taken them, and marching behind the poachers 2 kilometers further into the cave, I had my answer (Fig. 17.D). The cave opened up into another lost world, a great circular cavity in the Ankarana massif, this one filled not with forest but with marshes and lakes teeming with wildfowl. By this time, few of the birds were living, but the hunters did release them. The guides asked me (still feigning as warden) what to do with the criminals and their dead birds. "Would they do this again?" I asked. There was some discussion. Then the three poachers made what seemed a solemn vow that they would never again take wildlife from the park, lest the spirits of the ancestors of the Antakarana seek vengeance upon them. It was a particularly appropriate Malagasy way of dealing with a very understandable problem: the need of poor men to feed their families. For the geographer, the entire incident provided an especially fascinating insight into the relationship between people and caves in an extraordinary country.

Figure 17.D The poachers (with birds on poles) after being apprehended by the Ankarana park staff, here escorting them out of the Crocodile Cave.

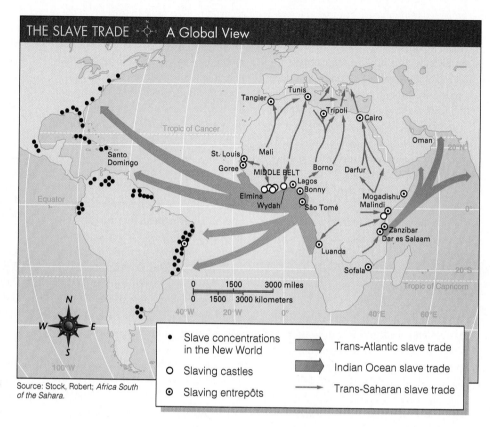

Figure 17.11   Slave export trade routes from Africa south of the Sahara.

Source: Stock, Robert; *Africa South of the Sahara.*

land of mystery. Even at the opening of the 20th century, vast areas of interior tropical Africa were still little known to Westerners.

The tragic impetus for growing contact between Africa and the wider world was slavery (Fig. 17.11). Over a period of 12 centuries, as many as 25 million people from Africa south of the Sahara were forced to become slaves, exported as merchandise from their homelands. The trade began in the seventh century, with Arab merchants using trans-Saharan camel caravan routes to exchange guns, books, textiles, and beads from North Africa for slaves, gold, and ivory from Africa south of the Sahara. As many as two-thirds of the estimated 9.5 million slaves exported between the years 650 and 1900 along this route were young women who became concubines and household servants in North Africa and Turkey. Male slaves usually became soldiers or court attendants (some of whom eventually assumed important political offices). From the 8th to 19th centuries, about 5 million more slaves were exported from East Africa to Arabia, Oman, Persia (modern Iran), India, and China. Again, most were women who became concubines and servants.

The notorious and lucrative traffic in slaves provided the main early motivation for European commerce along the African coasts, and it inaugurated the long era of European exploitation of Africa for profit and political advantage. The European-controlled slave trade was the largest by far. Between the 16th and 19th centuries, the capture, transport, and

sale of slaves was the exclusive preoccupation of trade between the European world and West Africa. Portuguese and Spaniards began the trade in the 15th century, and a century later, English, Danish, Dutch, Swedish, and French slavers were active.

The business boomed with the development of plantations and mines in the New World. Populations of Native Americans in Anglo and Latin America were insufficient for these industries, so the Europeans turned to Africa as a source of labor. The peak of the trans-Atlantic slave trade was between 1700 and 1870, when about 80% of an estimated total of 10 million slaves made the crossing. In escape attempts, in transit, and in the famines and epidemics that followed slave raids in Africa, probably more than 10 million others died.

Slaves were a prized commodity in the triangular trade linking West Africa with Europe and the Americas. European ships carried guns, alcohol, and manufactured goods to West Africa, exchanging them there for slaves. They then transported the slaves to the Americas, exchanging them for gold, silver, tobacco, sugar, and rum to be carried back to Europe. As "raw material" and as the labor working the mines and plantations of Latin America, the West Indies, and Anglo America, slaves generated much of the wealth that made Europe prosperous and helped spark the Industrial Revolution.

While Europeans carried out the trade, their physical presence was limited to coastal shipping points. Africans were the intermediaries who actually raided inland communities to

DEFINITIONS + INSIGHTS

The Modern-Day Slave Trade

Modern slavery is not confined to Africa. The London-based watchdog group Anti-Slavery International defines a slave as someone forced to work under physical or mental threat, who is completely controlled by an owner or employer, and who is bought or sold. By this definition, an estimated 27 million people worldwide were slaves in 2002. Such modern slaves may be found in the United States, where an estimated 50,000 people, mostly women, are trafficked each year to work as sweatshop laborers, strippers, and prostitutes. The victims are struggling to escape poverty in countries such as Thailand, India, and Russia. Promising them legitimate work in a prosperous land, traffickers smuggle them into the United States (particularly into Los Angeles), where they are whisked away to unfamiliar locations and put to their tawdry labors. The slave "master" holds the slave in submission by intimidation, such as reminding her that she is an illegal immigrant and might be raped by police if apprehended, and threatening to have family members in the home country killed if she attempts to escape. Few of the slaves speak English, at least initially, making their escape from servitude more difficult. Their ordeals do not necessarily end when they do find freedom; they may be deported back into the hands of traffickers in their home countries and then be trafficked back to the United States again. New legislation in the U.S. will stiffen penalties for traffickers and make it easier for the victims of trafficking to stay freely in the U.S.

capture the slaves and assemble them at the coast for transit shipment. West African kingdoms initially acquired their own slaves in the course of waging local wars. As the demand for slaves grew, these kingdoms increasingly went to war for the sole purpose of capturing people for the trade. As the exports grew, so did the practice of Africans keeping African slaves. Even after Great Britain (in 1807) and the other European countries abolished slavery — finally bringing an end to the trans-Atlantic trade in 1870 — slavery flourished within Africa. By the end of the century, slaves made up half the populations of many African states.

Slavery has not yet died out in the region. In Mauritania, some light-skinned Moors still enslave blacks, although the national government has outlawed this practice three times. Slavery also exists in the Sudan. Enslavement of children persists in West Africa. The typical pattern is that impoverished parents in one of the region's poorer countries, such as Benin, are approached by an intermediary who promises to take a child from their care and see that the boy or girl is properly educated and employed. This involves a fee (as little as $14) that, unbeknown to the parents, is a sale into slavery. The intermediary sells the child (on average, for $250 to $400) to a trafficker who sees that the child is transported overland or by sea to one of the region's richer countries, such as Gabon. There the child ends up working without wages and under threat of violence as a domestic servant, plantation worker (especially on cocoa and cotton plantations), or prostitute. Benin is the leading slave supplier and trafficking center; Gabon, Ivory Coast, Cameroon, and Nigeria are the main buyers of slaves.

Colonialism

Portugal was the earliest colonial power to build an African empire. The epic voyage of Vasco da Gama to India in 1497–1499 via the Cape route was the culmination of several decades of Portuguese exploration along Africa's western coasts. During the 16th century, Portugal controlled an extensive series of strong points and trading stations along both the Atlantic and Indian Ocean coasts of the continent. European penetration of the African interior began in 1850 with a series of journeys of exploration. Missionaries like David Livingstone, as well as traders, government officials, and now-famous adventurers and scientific explorers such as James Bruce, Richard Burton, and John Speke, undertook these expeditions. By 1881, when Africans still ruled about 90% of the region, their exploits had revealed the main outlines of inner African geography, and the European powers began to scramble for colonial territory in the interior. Much of the carving up of Africa took place at the Conference of Berlin in 1884 and 1885, when the French, British, Germans, Belgians, Portuguese, Italians, and Spanish established their respective spheres of influence in the region. By 1900, only Ethiopia and Liberia had not been colonized. Africa south of the Sahara had became a patchwork of European colonies — a status it retained for more than half a century (Fig. 17.12) — and Europeans in these possessions were a privileged social and economic class.

At the outbreak of World War II in 1939, only three countries — South Africa, Egypt, and Liberia — were independent. The United Kingdom, France, Belgium, Italy, Portugal, and Spain controlled the rest. But after the war, mainly in the 1960s and 1970s, there was a sustained drive for independence. Africa ceased to be a colonial region and came to hold more than one-fourth of the world's independent states. This was a peaceful process in most instances, but bloodshed accompanied or followed independence in several countries, including what are now Angola and the Democratic Republic of Congo.

Dependency theorists often point to Africa as a prime example of how colonialism created lasting disadvantages for

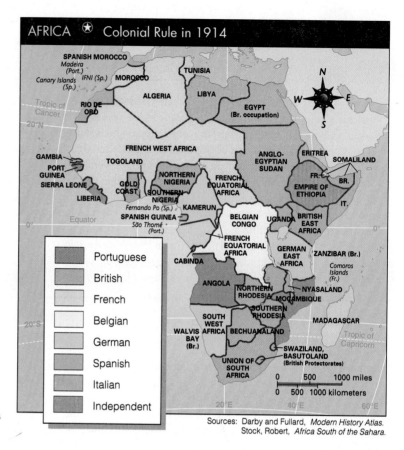

Figure 17.12   Colonial rule in 1914. Germany lost its colonies after World War I.

the colonized. European colonization produced or perpetuated many negative attributes of underdevelopment. These included the marginalization of subsistence farmers, notably those who colonial authorities — intent on cash crop production — displaced from quality soils to inferior land. In addition, European use of indigenous labor to build railways and roads often took a high toll in human lives and disrupted countless families. Furthermore, the colonizers often corrupted traditional systems of political organization to suit their needs, sowing seeds of dissent and interethnic conflict.

The European colonial enterprise did have some positive impacts. The colonies, and the independent nations that succeeded them, were the beneficiaries of new cities and the transport links built with forced or cheap African labor; new medical and educational facilities (often developed through Christian missions); new crops and better agricultural techniques; employment and income provided by new mines and modern industries; new governmental institutions; and government-made maps useful for administration and planning. Such innovations were very helpful, but they were distributed unequally from one colony to another and were inadequate for the needs of modern societies when independence came.

Although formal political colonialism has vanished, most countries still have important links with the colonial powers that formerly controlled them, and many foreign corporations

that operated in colonial days still maintain an important presence. France, but not Britain, has a long history of postindependence intervention in the political and military affairs of its former African colonies. France is the only ex-colonial power to keep troops in Africa (with the highest numbers in Djibouti, Senegal, and the Central African Republic). In postcolonial Africa, France also has taken steps to ensure that most of its former colonies trade almost exclusively with France, and it in turn has supported national currencies with the French treasury. But France has found its paternalistic approach to be extremely expensive and has begun reducing its military presence and other costly assistance to its African clients.

17.4 Economic Geography

Great poverty is characteristic of Africa south of the Sahara; 17 of the world's 20 poorest countries are there (see Table 17.1). All of the economies except South Africa's are underindustrialized. Africa's place in the commercial world is mainly that of a producer of primary products, especially cash crops and raw materials (particularly minerals), for sale outside the region. In most nations, one or two products supply more than two-fifths of all exports — for example, coffee and tea in Kenya (Fig. 17.13). Each country is therefore vulnerable to international oversupply of an export on which it is

JOE HOBBS

Figure 17.13   Kenya's economy is highly dependent on the export of coffee. Cash crops and other raw materials are typical exports of Africa south of the Sahara.

vitally dependent. The value of imports far exceeds that of exports in Africa south of the Sahara, with imports consisting mainly of manufactured goods, oil products, and food. Most African societies lack a substantial middle class and the prospect of upward economic mobility. Instead, most are hierarchical, and what significant income there is flows into the hands of a small elite controlling the lion's share of the nation's wealth.

Despite the overall grim picture of African economies, in recent decades there have been some changes for the better. Improved standards of health and literacy in many areas have resulted from the work of national and international governmental and nongovernmental organizations. The extension of roads, airways, and other transportation facilities has promoted the marketing of farm products, including perishable items, from formerly inaccessible areas. Stores and markets in both rural and urban areas stock a variety of manufactured goods from overseas and African sources. Modern factories have been established in many urban centers, and improved agricultural techniques have been introduced in many areas.

Such changes have affected some peoples and areas more than others, and their total impacts have only begun to lift the region from underdevelopment. There are outstanding problems and potential for development in agriculture, mining, and infrastructure for communication and transportation. There is a pressing need in most countries to deal with a huge burden of debt to international lenders. And perhaps most critically to the resolution of these economic problems, better political leadership is in great demand.

Cash Crops

Per capita food output in most of the countries has declined or has not increased since independence. The average African eats 10% less than he or she did 20 years ago. Rapid population growth and drought are partly responsible for the trend. Many regimes have also invested more in warfare than in getting food to their citizens. Food shortages also relate to government preference for cash crops over subsistence food crops. Coffee, cotton, and cloves provide a means of gaining foreign exchange with which to buy foreign technology, industrial equipment, arms, and consumption items for the elite. The proportion of crops grown for export to overseas destinations and for sale in African urban centers has therefore risen significantly in recent times.

Most export crops are grown on small farms rather than on plantations and estates. Large plantations have never become as established as they are in Latin America and Southeast Asia. Political and economic pressures forced many of them out of business during the period of transition from European colonialism to independence. Governments of the freshly independent states nationalized and subdivided white-owned plantations, and many whites were obliged to sell their land to Africans (this process is still ongoing in Zimbabwe, which became independent in 1980). Now operated by Africans, many of these farms provide vital tax revenue and foreign exchange. One exception is South Africa, where the most important producers and exporters are the ethnic Europeans, who raise livestock, grains, fruit, and sugarcane. But throughout tropical Africa as a whole, there is a growing trend in export production from small, black-owned farms. The most valuable export crops are coffee, cacao, cotton, peanuts, and oil palm prod-

Figure 17.14   A sisal plantation in eastern Kenya.

JOE HOBBS

ucts. Secondary cash crops include sisal (grown for its fibers; Fig. 17.14), pyrethrum (used in insecticides), tea, tobacco, rubber, pineapples, bananas, cloves, vanilla, cane sugar, and cashew nuts.

Although cash crops are important sources of revenue in these poor countries, excessive dependence on them can be harmful to a country's economy. The income they bring often goes almost exclusively to already prosperous farmers and corporations. The prices they fetch are vulnerable to sudden losses amidst changing world market conditions. The plants are susceptible to drought and disease. Finally, they are often grown instead of food crops, and cash crops do not feed hungry people.

Minerals

Mineral exports have had a strong impact on the physical and social geography of Africa south of the Sahara. The three primary mineral source areas are South Africa and Namibia, the Democratic Republic of Congo-Zambia-Zimbabwe region, and West Africa, especially the areas near the Atlantic Ocean (see Fig. 17.3). Notable mineral exports from these regions include precious metals and precious stones (particularly diamonds; see Problem Landscape, p. 459), ferroalloys, copper, phosphate, uranium, petroleum, and high-grade iron ore, all destined principally for Europe and the United States.

Large multinational corporations, financed initially by investors in Europe or America, do most of the mining in Africa. Mining has attracted far more investment capital to Africa than any other economic activity. Money is invested directly in the mines, and many of the transportation lines, port facilities, power stations, housing and commercial areas, manufacturing plants, and other elements in the continent's infrastructure have been developed primarily to serve the needs of the mining industry.

Great numbers of workers in the mines are temporary migrants from rural areas, often hundreds of miles away. The recruitment of migrant workers has had important cultural and public health effects. Millions of Africans who work for mining companies have come into contact with Western ideas as well as those of other ethnic groups and have carried these back to their villages. The growing numbers of radios and televisions appearing in homes across the continent, even in remote villages, also convey an unprecedented wealth of information about the wider world (Fig. 17.15). Unfortunately, the phenomenon of migrant labor has also helped spread HIV, the virus that causes AIDS. Most miners are single or married men who spend extended periods away from their wives or girlfriends, and it is common that they contract the virus from prostitutes and then return home to their villages, where they spread it still further. It was estimated that in 2002, the HIV infection rate among migrant miners working in South Africa was 40%.

Communications and Transportation Infrastructure

Poor transportation hinders development in the region (Fig. 17.16). Few countries can afford to build extensive new road or railroad networks, and much of the colonial infrastructure has deteriorated. Africa south of the Sahara criti-

Many of the important food and export crops of Africa are not native to the continent. For example, corn (maize), manioc, peanuts, cacao, tobacco, and sweet potatoes were introduced from the New World. Cattle, sheep, chickens, and other domesticated animals now characteristic of Africa south of the Sahara also originated outside the region. Such nonnative species are known as **exotic species,** or introduced species.

Some exotic species such as manioc and chickens have adapted well to the environments and human needs in Africa south of the Sahara. Others are more problematic. Known as "mealie-meal," corn has become the major staple food for southern Africans. However, the people of southern Africa came to regard corn as more of a scourge than a blessing during the severe regional drought of 1992. Introduced from relatively well-watered Central America, corn has flourished in African regions as long as rains or irrigation water have been sufficient. But the crop failed altogether during the 1992 drought. The situation was particularly critical in Zimbabwe, where the government had just sold its entire stock of corn reserves in an effort to repay international debts. Zimbabwe suddenly became a large importer of corn. Many in the country proclaimed corn to be "the curse of Africa," lamenting that, with its high water requirements, it should never have been allowed to become the principal crop of this drought-prone region. Agricultural specialists argued that native sorghum and cowpeas are much better adapted to African conditions and should be cultivated as the major foods of the future.

There are also growing questions about whether the cow is an appropriate livestock species for Africa. Cattle frequently overgraze and erode their own rangelands. The traditional approach to improving cattle pastoralism in Africa south of the Sahara, and to boosting livestock exports, has focused on the need to develop new strains of grasses that can feed cattle and other nonindigenous strains of livestock because the native grasses of African savannas suit the wild herbivores but generally not the domesticated livestock.

A new approach turns this problem around by asking whether it might be more appropriate to commercially develop the indigenous animals already suited to the native fodder and soil conditions. Kenya, Zimbabwe, South Africa, and other countries are now maintaining "game ranches" where antelopes — such as eland, impala, sable, waterbuck, and wildebeest — along with zebras, warthogs, ostriches, crocodiles, and even pythons are bred in semicaptive and captive conditions to be slaughtered for their meat, hides, and other useful products (Fig. 17.E). Safari hunting and tourism provide supplemental revenue in some of these operations. So far, these enterprises have demonstrated that indigenous large herbivores produce as much or more protein per area than do domestic livestock under the same conditions. In commercial terms, the systems that have incorporated both wild and domestic stock have proven to be more profitable than systems employing only domestic animals. The World Wildlife Fund and other proponents of this "multispecies" system argue that it is an economically and ecologically sustainable option of land use in Africa. They emphasize that ranchers who keep both wild and domesticated animals will be better able economically to weather periods of drought and will require a much lower capital input than they would expend on cattle or other domesticated livestock alone.

Having convinced many ranchers of the economic value of wildlife, the government of Zimbabwe has begun to enlist peasant farmers in protecting, and thus benefiting from, game animals. In an innovative and so far successful program known as CAMPFIRE (Communal Area Management Program for Indigenous Resources), the government has turned over management of wild animals to local councils, which are permitted to sell quotas of hunting licenses for elephant and other game. Profits from safari hunting return to the local communities to fund a wide range of development projects.

Figure 17.E   Oryx antelopes maintained in a semidomesticated state on the Galana game ranch in Kenya. Husbandry of indigenous wildlife may prove to be a viable alternative to keeping cattle and goats, which have many harmful impacts on soil and vegetation in Africa south of the Sahara.

cally needs a good international transportation network, together with a lowering of trade barriers, to enlarge market opportunities.

Poor communication also slows development. Telephone, fax, email, and other technologies are critical to successful participation in the globalizing economy. Africa cannot take part because it is critically short of most of these assets. For example, while there are more than 500 telephone lines per 1,000 people in the more developed countries, in Africa there are 34 countries with fewer than 10 telephone lines

Figure 17.15 There is no electricity in this village in Niger, but students receive some education and the entire community obtains news and entertainment from a television powered by photovoltaic cells.

per 1,000 people. Put another way, there are more phones in Tokyo than in all of Africa. Getting a phone line installed can be a costly and time-consuming procedure, and then the lines often do not work (a recent survey in 22 African countries showed a 60% faulty rate on phone lines). Calling rates, especially internationally, are very expensive, and few lines are reliable enough to handle Internet traffic. Many of the line problems would be averted with mobile phones. So far, the main obstacle has been that many of the telecommunications companies are government-owned and therefore profitable for governments, which are reluctant to allow competition by the private companies with their cellular technologies. Some governments also like to be able to monitor phone calls and email, and more communication makes that a more challenging task. In some countries, the barriers are falling, however, and mobile phones especially are be-

ginning to flourish in some of the world's most isolated places.

Internet cafés are also springing up, and the Internet may come to be Africa's answer to many of its communications problems. In 2000, the entire continent of Africa had only 3 million Internet users (out of 360 million worldwide), but that number has been growing quickly. Private companies and individuals cooperated to build a 20,000-mile (32,000-km) undersea fiberoptic cable system that, when completed in 2003, will form a ring of connections around Africa. There is much hope that such "digital bridges" will span the "digital divide" between Africa and the more developed world, helping to raise the standard of living and quality of life of many of Africa's peoples.

Debt

Almost all of the countries are heavily in debt to foreign lenders. Particularly after the mid-1970s, nations undertook costly development projects with borrowed money. Western financiers, planners, and contractors gave optimistic assessments of the benefits to be expected from such projects, and African leaders were ready to accept loans as a way to reap quick benefits from newly won independence. Now, however, many countries are having great difficulty in meeting even the interest payments on their debts, and their efforts to do so often create further economic woes and have destructive environmental effects. The International Monetary Fund (IMF), World Bank, and other lenders have generally resisted requests of African debtor nations for rescheduled interest payments and new loans. As a condition of further support, the lenders often demand that African debtors put their finances in better order by such measures as revaluating national currencies downward to make exports more attractive, increasing the prices paid to farmers for their products, and reducing cor-

Figure 17.16 Many African river crossings have no bridges, and it is necessary to use ferries, although some streams can be forded during the dry season. This photo shows vehicles and people boarding a ferry to cross an estuary north of Mombasa on the Kenyan coast. Backup of traffic at these bottlenecks may take hours or even days to clear.

PROBLEM LANDSCAPE

The Trade in Dirty Diamonds

One of the leading sources of revenue for governments and rebels fighting wars in Africa is a thing of great beauty, usually associated with eternity and conjugal bliss: the diamond. An estimated 4 to 15% of the world's total diamond production is made up of the "blood diamonds" that rebels and "rogue" governments sell to buy arms. These gems have financed at least three African wars. In two of these, rebel movements have controlled the mines: In Angola, the UNITA rebels held the diamond-producing area until evicted by government troops in 1999. And in Sierra Leone, for years most of the mines were in the hands of the Revolutionary United Front (RUF) (whose hallmark was the amputation of civilians' limbs). Diamonds have also funded the increasingly isolated government of the Democratic Republic of Congo.

Diplomatic, nongovernmental, and business efforts are underway to stem the tide of Africa's "conflict diamonds." In an effort to promote peace in the region, the United Nations imposes economic sanctions against any country or factions that sell diamonds to finance conflict. In 1998, the United Nations banned trade with UNITA, and 2 years later, the UN banned diamond exports from Sierra Leone. The governments of Belgium, Israel, and India, the leading countries in the diamond cutting and polishing business, are working on laws to prohibit the importation of blood diamonds.

Enforcing a diamond embargo is difficult. Millions of dollars in rough stones can be smuggled in a sock, and expert identification of their place of origin can be a challenge. In addition, neighboring states are often complicit in the trade. Most of Sierra Leone's diamonds are purchased by Liberia, which supports the rebels and profits greatly from the trade (because of its scorned regime, Liberia receives almost no foreign assistance, but earned $290 million from diamond exports in 2000). West African governments have so far stood against an extension of the sanctions to Liberia.

Despite the difficulties, there are indications that sanctions help. Within 6 weeks of the United Nations embargo against Sierra Leone diamonds, for example, the price of diamonds smuggled from Sierra Leone and Angola fell 30% below those of diamonds from reputable mines elsewhere in Africa and abroad. The hope is that even stricter enforcement of the embargo will impoverish the rebels and make them more willing to reach a political settlement of their conflict.

The United Nations General Assembly is working to initiate a global treaty to govern the diamond trade. For it to work, reliable, internationally acceptable tracking standards would have to replace the varying national standards now in place. Special monitoring staff would have to keep their eyes on mines, on government

diamond evaluation offices, on gem polishing and jewelry manufacturing businesses, and on jewelry stores. The challenge is thus to track legitimate diamonds from the moment they are mined until they reach the ring finger.

Diamond merchants are also increasingly anxious to establish an industrywide means of certifying that blood diamonds never make it to the world's jewelry shops. The effort is being spearheaded by the world's leading diamond company (with about two-thirds of the market), a South-African-based multinational named De Beers Consolidated Mines. De Beers was embarrassed by a report that it had bought $14 million worth of diamonds from Angolan rebels in a single year. In 2000, the New York Post ran a photo of a maimed child in Sierra Leone, indicating this was a result of buying diamonds. Perceiving that a public relations disaster could lead to a business disaster, such as happened with an organized boycott against fur products in the 1980s, De Beers seized the initiative. In the name of Africa's welfare, but certainly also a means of increasing demand and profit, De Beers introduced "branded diamonds," or those marked as being from nonconflict areas. The De Beers diamond is sold to the customer with a guarantee that it has not been used by an African ruler or rebel to finance civil war. Other diamond companies have followed suit.

---

ruption, especially among government officials and urban elites. Although many African debtor nations have attempted to meet such demands, internal political factors slow progress. Loss of economic privileges angers urban elites on whom governments depend for support. Relaxation of price controls on food to stimulate production is often met by rioting among city-dwellers who have tight budgets already allocated for other costs of living.

In addition to the problems posed by loans, increased foreign aid in the form of gifts or grants to Africa is controversial. Many argue that foreign aid only increases dependence and deflects national attention away from problems underlying famine. International aid monies often free up funds from national treasuries to be spent on unfortunate uses. In the Horn of Africa, for example, foreign money was available for famine relief, so Ethiopia and Eritrea could more easily afford to purchase weapons used in their war. Aid monies are often simply stolen, and there has been inadequate monitoring by donors of how they are spent. Since about 1990, the donors have proceeded more cautiously, more often funneling money through private and religious groups working in health care, education, and small-business development, for example, through microcredit schemes (see Definitions & Insights on the next page).

A recent promising alternative to the problem of national debt is **microcredit,** or the lending of small sums to poor people to set up or expand small businesses. Typically, poor people have no collateral and therefore cannot borrow from commercial banks. Local loan sharks may charge huge interest rates of 10 to 20% per day. But in the interest of fostering development, microlending organizations encourage poor borrowers to form a small group so as to cross-guarantee one another's loans. One member of the group may take out a loan of $25, for example, to start up a small business such as a restaurant. Only when that individual pays it back may the next person in the group receive a loan. Peer pres-

sure minimizes the default rate. Microlenders generally prefer to lend money to women because women are more likely to use the money in ways that will feed, clothe, and otherwise benefit their children, whereas men may be more likely to spend the money on alcohol or other frivolous pursuits. One fear in Africa is that microloans will not be repaid if the borrower or someone in the borrower's family contracts HIV or some other debilitating illness. Therefore, the loans are often extended only after the borrower receives some amount of public health training, for example, in the use of condoms and in general food preparation hygiene.

Political Leadership

Africa scholars speak of "failed state syndrome," a pernicious process of economic and political decay that has ruined countries like Somalia and Sudan and is eating away at Kenya, Ivory Coast, and Zimbabwe. Some African countries are little more than "shell states." A shell state appears to have all the institutions of a country: a constitution, a parliament, ministries, and more. But most of the positions are occupied by friends and family of the country's ruler, who are happy to line their pockets rather than pursue development or improvement of their fellow citizens. Five of the world's ten most corrupt countries are in this region, according to a leading watchdog group on corruption. One of the worst cases was the government of Zaire (now the Democratic Republic of Congo) under Mobutu. Mobuto's minister of mines was regularly bribed by Zaireans and Europeans to obtain mining permits. He kept as much of the bribe money as he could, and the rest went to Mobuto. Little revenue from the lucrative mining industry ever made it into the state treasury or trickled down through the populace. That is how Zaire functioned, or failed to function, and how a resource-rich country succeeded in achieving one of the worst standards of living on the planet. In Liberia, President Charles Taylor passed a law allowing him to personally sell any of the country's "strategic commodities," including mineral resources, forest products, art, archeological artifacts, fish, and agricultural products. The wealth taken by the countries' leaders generally does not find its way back into the economy but is either spent on grandiose personal accommodations or parked in foreign bank accounts.

One of the assumptions of international lenders and aid agencies is that democracy helps weed out such economic bloodletting. True democracy, they observe, is sorely lacking in the region. According to a 2001 assessment by U.S. intelligence authorities, only 8 of the region's 48 countries

fully embrace pluralism, and all 8 have problems with corruption, weak leadership, and declining education and health services (those 8 are Nigeria, South Africa, Senegal, Mali, Malawi, Tanzania, Botswana, and Namibia). The foreign lenders and donors tell African leaders that if they do not institute democratic reforms, hold fair elections, or improve bad human rights records, they will not receive development aid or loans. Often, the leaders make just enough concessions to win the aid without instituting real reform — a phenomenon known as **donor democracy.** Some Africa observers say that positive development assistance must go where it is needed most, channeled directly into rebuilding destroyed institutions such as the civil service, courts, police, and the military. But it is not clear how foreign powers might do that without becoming guilty of or being accused of neocolonialism.

17.5 Geopolitical Issues

Africa south of the Sahara has lost much of the geopolitical interest it held for the world's great powers after World War II. To boost their competing aims during the Cold War years, the Soviet Union and the United States had played African countries against one another, arming them with weapons with which to wage proxy wars. The superpowers also extended aid generously to many African nations.

The nature of conflict in Africa changed with the end of the Cold War. The great powers withdrew support, but their weapons remained to help smoldering conflicts. The United States helped build the arsenals of eight of the nine countries involved in the recent Democratic Republic of Congo conflict, for example. Cold War–era weapons like AK-47 assault rifles are still being "dumped" — sold at low prices — in Africa, where they are not considered obsolete and where they fan the flames of conflict. Increasingly, fighting has begun to spread across international borders (as in Congo and between Eritrea and Ethiopia). Previously, the United States and the Soviet Union maintained a kind of security balance that kept warfare from becoming internationalized, but that restraint no longer exists. The scenario of regional or even Africa-wide wars began to be feared and was realized with Africa's first "world war" centered on the Democratic Republic of Congo. In the absence of regional or international powers to keep the peace, other countries (e.g., Sierra Leone) are simply imploding, fragmenting into fiefdoms run by factions who claim to be revolutionaries but are essentially profiteers. Meanwhile, some national governments (e.g., Ethiopia) spend vast sums on expensive military aircraft while their people suffer malnutrition.

The end of the Cold War constricted (and in the case of Russia, all but ruptured) aid pipelines. The United States cut its economic assistance to Africa south of the Sahara by 30% between 1985 and 1992 and reduced it another 10% between 1992 and 2001 (to $738 million yearly in 2001 for the entire region, about the same amount given as economic assistance to Egypt alone).

Why the waning concern? American President George W. Bush was quoted as saying, "While Africa may be important, it doesn't fit into the national strategic interests, as far as I can see them."[2] The administration made it clear that it had no interest in sending troops to intervene in any of the continent's many wars. The Bush administration did, however, point out two countries as particularly important: Nigeria and South Africa. Nigeria is the United States' largest trading partner in Africa. Petroleum-rich Nigeria and Angola together supplied 16% of America's oil in 2001 (a figure projected by the U.S. Central Intelligence Agency to rise to 25% by 2015), and South Africa is the region's economic powerhouse and the perceived key to regional stability. South Africa's military strength, which included nuclear weapons until they were dismantled in 1990, contributes to its global significance. It is strategically important for its frontages on both the Atlantic and Indian Oceans, its possession of Africa's finest transport network (vitally important to many African countries that trade with and through South Africa), and its diversified mineral wealth.

Bush's secretary of state, Colin Powell, declared in 2001 that Africa's AIDS epidemic is a national security issue for the United States. He was acknowledging some of the implications of globalization. Despite Africa's apparent remoteness from the United States, the two are linked by hundreds of air traffic routes traveled by thousands of people each day. Hundreds of HIV-positive Africans, perhaps most of them unaware of their virus, pass through U.S. airport gateways daily. There are indications that some African strains (subtypes) of the virus are more easily transmitted by heterosexual contact than the strains prevalent in the United States. This scenario seems, to Powell, to represent a clear and present danger to the United States. There are other dimensions of the potential burden on the United States. As long as the epidemic rages in Africa, there may be large flights of "medical refugees" seeking asylum in the United States. There could be AIDS-related political instability or civil wars that invite U.S. military intervention in the region.

Due to the HIV/AIDS epidemic and the region's other great problems, there will also be humanitarian crises that will tug on American and other Western heartstrings and purse strings. Since the mid-1980s, when television images of famine-wracked Ethiopia prompted American and Western European citizens to give generously, popular sympathy and support during African emergencies have waned. The bitter experience of the United States in Somalia in the early 1990s squelched its appetite to extend humanitarian assistance. This lack of apparent public and official interest in crises abroad is known as **donor fatigue**. How the rest of the world should respond to poverty and recurrent crises is one of the key questions facing the huge, resource-rich, problem-ridden region of Africa south of the Sahara.

CHAPTER SUMMARY

- This chapter examines the culturally complex portion of Africa south of the Sahara, a region often called sub-Saharan or Black Africa.
- The majority of the region's people are rural. There are several clusters of dense population. HIV/AIDS is taking a huge toll, dramatically lowering life expectancy and projections for population growth, especially in southern Africa.
- A number of shared cultural traits may be referred to as comprising "Africanity."
- The relatively low population density of Africa south of the Sahara obscures the fact that a majority of this region's people live in a small number of densely populated areas that together occupy a small share of the region's total area of 8 million square miles.

- Most of Africa consists of a series of plateau surfaces dissected by prominent river systems such as the Nile, Niger, Congo, Zambezi, and Orange.
- One of the most spectacular features of Africa's physical geography is the Great Rift Valley, a broad, steep-walled trough extending from the Zambezi Valley northward to the Red Sea and the valley of the Jordan River in southwestern Asia. In terms of plate tectonics, the feature is believed to mark the boundary of two crustal plates that are rifting, or tearing apart.
- Although about two-thirds of the region lies within the low latitudes and has tropical climates, Africa south of the Sahara contains a great diversity of climate patterns and biomes, some resulting from elevation rather than latitudinal position.

[2] Quoted in Ian Fisher, "Africans Ask If Washington's Sun Will Shine on Them." *New York Times*, February 8, 2001, p. A3.

- Africa's diverse wildlife is often threatened by human population growth, urbanization, and agricultural expansion within the continent. Two of Africa's largest herbivores, elephants and rhinoceroses, are striking examples of Africa's endangered wildlife. Their management and protection require consideration of many issues, including international trade.
- Within the four culture hearths of Africa south of the Sahara — the Ethiopian Plateau, the West African savanna, the West African forest, and the forest-savanna boundary of West Central Africa — early indigenous people were responsible for several agricultural innovations. Within these culture hearths, Africans domesticated important crops such as millet, sorghum, yams, cowpeas, okra, watermelons, coffee, and cotton.
- The tragic impetus for growing contact between Africa and the wider world was slavery. Over a period of 12 centuries, as many as 25 million people from Africa south of the Sahara were forced into slavery.
- Most of Africa south of the Sahara fell under European colonialism after the Conference of Berlin in 1884 and 1885. During this conference, Africa was carved up, as the French, British, Germans, Belgians, Portuguese, Italians, and Spanish established their respective spheres of influence in the region.
- In general, the people of Africa south of the Sahara are poor, live in rural areas, and practice subsistence agriculture as their main occupation. In most countries, per capita food output has declined or has not increased since independence.

- Frequent droughts, lack of education, poor transportation, and serious public health issues have hindered development within Africa south of the Sahara. Most countries are underindustrialized and overly dependent on the export of a few primary products. In addition, while most countries are heavily in debt to foreign lenders, the amount of economic and humanitarian assistance to the region has slowed considerably since the end of the Cold War.
- Although many countries have been under authoritarian governments since independence, there has been some progress toward democracy. Serious political instability is characteristic of many African countries south of the Sahara. A diverse array of political, economic, and social ideas from the West, the Communist bloc, the Muslim world, and Africa south of the Sahara itself has influenced governments since independence. In addition, important links with the colonial powers that formerly controlled them exist in many countries of the region.
- Over half of the people of Africa south of the Sahara depend directly on agriculture or pastoralism for a livelihood. The cultivation of export crops such as coffee, cacao, cotton, peanuts, and oil palm products is increasing throughout the region.
- The export of minerals has had a particularly strong impact on the physical and social geography of Africa south of the Sahara. Notable mineral exports include precious metals and stones, ferroalloys, copper, phosphate, uranium, petroleum, and high-grade iron ore. Income from diamonds in many cases has funded and fueled conflict within and between countries.

REVIEW QUESTIONS

1. Locate the five principal areas of population concentration within Africa south of the Sahara.
2. What factors account for the high HIV infection rates in Africa south of the Sahara? What can be done to fight the epidemic? What is HIV/AIDS doing to life expectancy and the age structure profiles in countries where it is an epidemic?
3. Locate the principal climatic zones of Africa south of the Sahara.
4. Locate and describe the major slave trade routes from Africa south of the Sahara. Where in Africa did they originate? What were some of the major destinations of African slaves?
5. Locate the major colonial possessions of the European powers in Africa south of the Sahara.
6. Explain why African parents typically want to have large families.

7. What is the significance of cattle in many cultures of Africa south of the Sahara?
8. What is a keystone species? Why is the tsetse fly considered a keystone species in Africa south of the Sahara?
9. What is an exotic species? Provide some examples of exotic species that are important to agriculture in Africa south of the Sahara.
10. What is the 1% gap and what are its implications?
11. In African politics, what is a shell state and what is the failed state syndrome?
12. How did the end of the Cold War change the political, military, and economic features of many African countries?
13. What is donor fatigue and why does it exist?

DISCUSSION QUESTIONS

1. Describe the tectonic processes that have led to the development of the Great Rift Valley. What physical features are associated with the Great Rift Valley? What will be the final fate of areas adjacent to the rift?
2. What circumstances make teachers particularly likely to contract the HIV virus? What other segments of the society are hit especially hard and why? What steps could be taken to slow the advance of the epidemic and why haven't they been taken yet? Describe the major impacts AIDS has had on Africa south of the Sahara. Use a PRB data sheet or other sources to update AIDS infection rates, death rates, and population growth rates for countries south of the Sahara. Has the epidemic slowed, accelerated, or stalled since the book went into print? The next time you are in a classroom or a crowd, look around you; count the occupied seats or calculate the chunk of the crowd that would be HIV-positive if you were in Africa.

3. Discuss the main reasons behind the sharp decline in elephants and rhinoceroses in Africa south of the Sahara. What efforts have been taken to preserve these animals? What are the pros and cons of culling elephants? Of the rhino dehorning program?
4. Describe the eight constituent elements of the African identity as identified by Robert Stock.
5. Discuss the pattern of triangular trade. In addition to Africa south of the Sahara, what other regions of the world were involved in this pattern of trade? What goods were traditionally exchanged?
6. Discuss the modern slave trade in Africa and elsewhere, including the United States. How is it that such practices could continue in modern times? What factors explain slavery? Compare and contrast the slavery situation as described here with the discussion of the sex trade in Chapter 7.
7. Describe the positive and negative effects of European colonization of Africa south of the Sahara.

8. Why has corn been called the "curse of Africa" in Zimbabwe? What are some crops that might be better adapted to African conditions?

9. What are the apparent advantages of animals raised in game ranches over cattle and other domesticated livestock?

10. Describe the positive and negative effects of mining on Africa south of the Sahara. What have been the cultural and public health effects of the recruitment of migrant mine workers?

11. Discuss the impacts of foreign debt on African countries. Also discuss what microcredit is and what advantages it offers.

12. What is donor democracy? To what extent should the wealthy countries abroad participate in Africa's economic and political development? How can foreign assistance be put to the best uses possible?

13. Discuss the trade in dirty diamonds. What measures can realistically be taken to ensure that sales of these valuable gems do not finance conflict in the region?

CHAPTER 18

The Assets and Afflictions of Countries South of the Sahara

Sarah Tsialiva, from the village of Mahamasina
in northern Madagascar. Almost half of the people
in Africa south of the Sahara are of her generation,
under the age of 15.

In this chapter, sketches of the major subregions and their countries illustrate some of the characteristic and unique challenges of Africa south of the Sahara. The major subregions are the Horn of Africa, the Sahel, West Africa, West Central Africa, East Africa, southern Africa, and the Indian Ocean islands (see Fig. 18.1). The relevant area and population data are in Table 17.1.

18.1 The Horn of Africa
Environments and Resources

In the extreme northeastern section of the region of Africa south of the Sahara, a great volcanic plateau rises steeply from the desert. This highland and adjacent areas occupy the greater part of the Horn of Africa, named for its projection from the continent into the Indian Ocean. This subregion includes the countries of Ethiopia, Eritrea, Somalia, and Djibouti. Much of the area lies at elevations above 10,000 feet (3,050 m), and one peak in Ethiopia reaches 15,158 feet (4,620 m). The highlands where the Blue Nile, Atbara, and other Nile tributaries rise receive their rainfall during the summer. Temperatures change from tropical to temperate as elevation increases. Bananas, coffee, dates, oranges, figs, temperate fruits, and cereals can be produced without irrigation. Highland pastures support sheep, cattle, and other livestock. The region's people are concentrated in the relatively well-endowed highlands; most of Ethiopia's 65 million people live there.

East of the mountain mass, in Eritrea, Djibouti, and Somalia, lower plateaus and coastal plains descend to the Red Sea, the Gulf of Aden, and the Indian Ocean. Extreme heat and

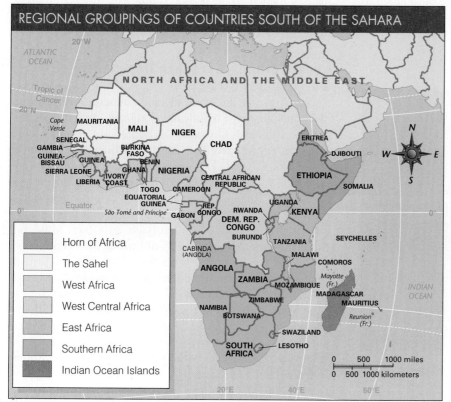

Figure 18.1 Informal regional groupings of countries south of the Sahara.

Source: Robert Stock. *Africa South of the Sahara.*

aridity prevail at these lower levels, and nomadic and seminomadic tribespeople make a living by herding camels, goats, and sheep. Dwellers of scattered oases carry on a precarious agriculture. The arid lowland sections have many characteristics more typical of the Middle East than other areas of Africa south of the Sahara. The populations of these countries are small compared to Ethiopia: Somalia with 7.5 million, Eritrea with 4.3 million, and Djibouti with only 600,000.

Ethiopia has substantial resources, perhaps including significant mineral wealth. However, its resource base remains largely undeveloped, and the country is impoverished. Coffee and animal hides are the main exports. With foreign assistance, a few textile and food-processing factories have been built, and a few dams have been constructed to harness part of the country's large hydroelectric potential. A poor transportation system — whose best feature is the 3,000 miles (c. 4,800 km) of good roads built by the Italians during their occupation in the 1930s — is a major factor keeping the country isolated and underdeveloped. The French-built railroad from Djibouti to Ethiopia's capital, Addis Ababa (population: city proper, 2,562,800), lacks feeder lines and has never been as successful as hoped. Many roads into the Ethiopian heartland are rough tracks, although better connections have gradually been developing. Air services — especially Ethiopia's remarkable fleet of DC-3s, more than 50 years old but kept in top condition — help overcome the deficiencies in ground transportation. Many of Ethiopia's people live in villages on uplands cut off from each other and from the outside world by precipitous chasms (Fig. 18.2). These formidable valleys were cut by streams carrying huge amounts of fertile volcanic silt to the Nile. Before the Aswan High Dam was built, they provided material for the nourishing floods in Egypt.

Eritrea has gold, copper, and marble reserves but lacks oil, and little of its mineral potential has been developed. Its Red Sea coastline has superb coral reefs, and the country is hoping to attract international tourists to dive and snorkel among them. There is some cotton and oilseed agriculture, and fish are a mainstay in the diet. Djibouti supports pastoralists but has almost no arable land, and little to offer economically except its strategic position near the mouth of the Red Sea. It has a fine port also named Djibouti. France maintains a significant military presence there and pumps much-needed economic aid to the country.

Somalia is hyperarid, and only 2% of the country is cultivated. Nomadic and seminomadic pastoralism is the main economic activity. Until an outbreak of livestock disease in the 1990s, Somalia's main exports were live animals (goats, sheep, camels, cattle), shipped mainly to Saudi Arabia, Yemen, and the United Arab Emirates. Fish products and bananas are still shipped, but now charcoal is the main export to the Arabian Peninsula. The resulting environmental impacts on Somalia are severe because trees never were abundant, and there are fights between the tree-cutters and the pastoralists whose camels feed on tree fodder.

Droughts have stricken this region, particularly Ethiopia, almost incessantly since the late 1960s. Ethiopian villagers are especially vulnerable to drought because of their heavy dependence on rainfed agriculture. The volcanic soils are fertile, but the streams are entrenched so far below the upland fields that little irrigation is possible. In 1984 and 1985, drought in Ethiopia reached the point of widespread catastrophe. An estimated 1 million people perished. Only a massive and sustained international relief effort, some of it promoted by British and American rock musicians, helped reduce the suffering.

Local Cultures, European Imperialism, and Regional Struggles

Ethiopia's people are ethnically and culturally diverse. About 45%, including the politically dominant Amhara peoples, practice Ethiopian Orthodox Christianity, an ancient branch of Coptic Christianity that came to Ethiopia in the fourth century from Egypt. The entire area has had important cultural and historical links with Egypt, the Fertile Crescent, and Arabia. The Ethiopian monarchy based its origins and legitimacy on the union of the biblical King Solomon and the Queen of Sheba, who, tradition holds, gave birth to the first Ethiopian emperor, Menelik. Until a Marxist coup brought an end to the emperorship in the 1970s, Ethiopia's rulers were always Christian. Ethiopia has many outstanding Christian artistic and architectural treasures, including the 11 churches of Lalibela, carved from solid rock in the 12th and 13th centuries (Fig. 18.3). Most of the rest of Ethiopia's people are either Muslims (who make up about 40% of the population) or members of Protestant, Evangelical, and Roman Catholic churches. As of 2002, there were still about 17,000 Falashas, or Ethiopian Jews, in Ethiopia, a remnant of a very ancient and isolated Jewish population. The majority, about

Figure 18.2 This highland village in north central Ethiopia, like many in the country, is isolated by deep gorges (not visible in this view). The round structure to which all the village roads lead is the community's Christian church.

JOE HOBBS

Figure 18.3   The churches of Lalibela, carved from volcanic rock in the 12th and 13th centuries, are among Ethiopia's many Christian cultural treasures. Note man at left center for scale.

243 65,000, had fled to Israel in the 1980s and 1990s. Because this mountainous country has long served an isolated refuge for such unique groups, it has been nicknamed the "Galapagos Islands of Religion."

Eritrea's population, like Ethiopia's, is about half Muslim and half Christian. Djibouti is about 95% Muslim and 5% Christian. In Somalia, about 99.8% of the people are Sunni Muslims and 98.3% are ethnic Somalis. They share the same language and the same nomadic culture. This is one of the most homogeneous populations in the world, a fact that would suggest peace and stability, but interclan rivalries have torn the country apart.

Colonialism had a big impact on the region's peoples. European powers seized coastal strips of the Horn of Africa in the latter 19th century (see Fig. 17.12, p. 454). Britain was first with British Somaliland in 1882; then France annexed French Somaliland in 1884; and finally, Italy took control of Italian Somaliland and Eritrea in 1889. These areas along the Suez–Red Sea route have never had much economic importance, but they have had great strategic significance.

Italy tried to extend its domain from Eritrea over more attractive and potentially valuable Ethiopia in 1896, but Ethiopians annihilated the Italian forces at Aduwa. Forty years later, in 1936, the Italians succeeded in taking Ethiopia. In World War II, however, British Commonwealth forces defeated Italian troops throughout the region. Ethiopia regained independence, and Eritrea was federated with it as an autonomous unit in 1952. Ethiopia incorporated Eritrea as a province in 1962.

After World War II, Italian Somaliland was returned to Italian control to be administered as the trust territory of Somalia under the United Nations, pending independence in 1960 as a republic. The present country of Somalia officially includes both the former Italian territory and the former British Somaliland, which chose to unite with Somalia. The people of much smaller French Somaliland remained separate and eventually became independent as the Republic of Djibouti, with its capital at the seaport of the same name.

In 1974, Ethiopia's aging Emperor Haile Selassie (whom Rastafarians regard as their new messiah) was deposed by a civilian and military revolt against the country's feudal order. A Marxist dictatorship, strongly oriented to the former Soviet Union, emerged. American influence, which had been strong under Haile Selassie, was eliminated. Munitions and advisors from the Soviet bloc poured in to equip and train one of Africa's largest armies, and the Soviets gained access to base facilities in Ethiopian ports. The United States, which had previously supported Ethiopia in its struggles against neighboring Somalia, now switched sides to back Somalia. Such foreign interventions and proxy conflicts contributed to the strife that has hindered development in the region.

Ethiopia had many internal troubles following independence. In what was then the province of Eritrea, both Muslims and non-Amhara Christians resented their political subjugation to Ethiopia's Christian Amhara majority. Terrorism and guerrilla actions escalated into civil war in the early 1960s. The Ethiopian army tried but failed to stamp out the revolts of two Eritrean "liberation fronts." At the same time, there was also a rebellion in Tigre Province adjoining Eritrea. In 1991, these combined rebel forces captured Addis Ababa and took over the Ethiopian government. The rebels in Eritrea established an autonomous regime, and Eritrea became independent in 1993. Eritrea's main city and manufacturing center, Asmara (population: metropolitan area, 864,100; city proper, 385,200), is located in highlands at an elevation of more than 7,000 feet (2,130 m) and has a rail link with the Red Sea port of Massawa (see Fig. 17.3, p. 438).

Many geographers subsequently pointed to Eritrea as a model for sustainable development in Africa south of the Sahara. After independence, the country reduced its foreign debt, cut its population growth rate, and took other promising steps toward eliminating the burdens of underdevelopment. However, in 1998, Eritrea dimmed its bright prospects by launching a war against Ethiopia. The conflict emerged from economic rivalries and unresolved territorial issues between the nations. With Eritrea's independence, Ethiopia had agreed to become landlocked, as long as it could freely use Eritrea's Red Sea ports of Massawa and Assab. The concession was in large part Ethiopia's reward to Eritreans for helping with the overthrow of Ethiopia's Marxist regime. The countries' leaders in effect agreed on an economic union in which Eritrea would be the industrial power and Ethiopia the agri-

cultural one. They promised to practice free movement of people and goods between the two states and to share a single currency, the Ethiopian birr.

But Eritrea introduced its own currency and announced steep duties and fees on goods shipped to and from Ethiopia through Eritrean ports. Ethiopia then turned to Djibouti and Kenya to find new outlets to the sea. Deprived of an anticipated major source of revenue, Eritrea attacked Ethiopia. The war finally ended in 2000, when Ethiopia gained the upper hand by securing a border region disputed between the two countries. A demilitarized zone monitored by United Nations troops was established along the border. An estimated 100,000

Figure 18.4   Although ethnically homogenous, Somalia is not unified politically.

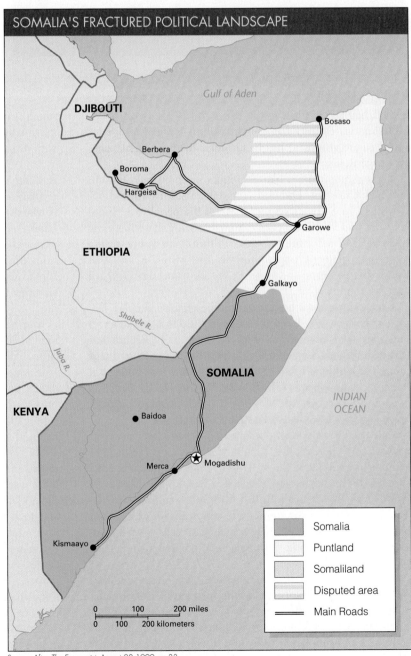

Source: After *The Economist,* August 28, 1999, p. 33.

Eritreans and Ethiopians had perished in the conflict. Eritrea's economy was hit particularly hard. With the Ethiopian market gone, Eritrea's exports declined 80% from their 1997 high. The port of Assab shut down.

Somalia has been in ruin since a rebellion overthrew the government in 1991 and ignited a civil war between regional Somali clans (the country has six major clans and numerous subclans). Contending forces inflicted great damage on the capital, Mogadishu (2001 population: city proper, 1,211,300), from which large numbers of people fled to the countryside as refugees. Massive famine threatened the devastated country in mid-1992. Law and order disintegrated, and humanitarian assistance became difficult. United States military forces then intervened with Operation Restore Hope, a mission to help protect relief workers and aid shipments. However, American policymakers decided to use military force to neutralize the power of one of the country's leading "warlords," Mohamed Farrah Aidid. In a gruesome sequence of events, 18 U.S. servicemen stranded in Aidid's territory in Mogadishu were killed in a single day in 1993. Accused terrorist Osama bin Laden had reportedly equipped Somali forces with the weapons they used to assault the American troops. American public outcry at the tragedy brought an end to the American mission in 1994 (and has made the United States very reluctant to assist in humanitarian crises ever since).

Since then, Somalia has fractured into three de facto states (Fig. 18.4). The southernmost, centered on Mogadishu (which is still much in ruin and without services), is Somalia, the only one with international recognition as a sovereign country. The country is barely functioning as a country, as there is only a nominal government with little authority, there is no legal system, and several warlords (and one warlady) control their virtual territories. To the north is Puntland, which includes the tip of the Horn of Africa and considers itself autonomous but not independent of Somalia. To the northwest, bordering Djibouti and Ethiopia, is Somaliland (the former British Somaliland). It proclaimed itself independent of Somalia in 1991 and has a president and broadly representative government. Its airport at Hargeisa is the only airport in all of greater Somalia that has functioned since 1995. And its port city of Berbera has served as Ethiopia's main trade outlet to the sea since Ethiopia's conflict with Eritrea began. Somaliland's economic mainstay was exports of cattle to the Muslim nations of the Arabian Peninsula, until reports of diseased animals led to a complete cessation of the trade. Now Somaliland is pinning its hopes on exports of gemstones, frankincense, vegetable dyes made from henna, and products made from qasil plants, such as hair conditioners and body cleansers.

18.2 The Sahel

The Sahel region extends eastward from the Cape Verde Islands to the Atlantic shore nations of Mauritania, Senegal, and The Gambia and inland to Mali, Burkina Faso ("Land of the Upright Men"; formerly Upper Volta), Niger, and Chad (see Fig. 18.1). In colonial times, The Gambia was a British dependency; Mauritania, Senegal, Mali, Burkina Faso, Niger, and Chad were French; and Cape Verde was Portuguese. Cape Verde's dry volcanic islands lie well out in the Atlantic off Senegal, but in population, culture, economy, and historical relationships, the islands are so akin to the adjacent mainland that they fit in the Sahelian context.

Environments, Resources, and Settlements

Climatically, the region ranges from desert (in parts of the Sahara) in the north through belts of tropical steppe and dry savanna in the south. The name Sahel in Arabic means "coast" or "shore," referring to the region as a front on the great desert "sea" of the Sahara. Since the late 1960s, the area has been subjected to severe droughts, which, in combination with increased human pressure on resources, have prompted a process of desertification (see Regional Perspective, p. 470).

This area has seen many dramatic changes in climate and vegetation. There is abundant evidence, particularly in the form of prehistoric rock drawings, that much of the now extremely arid northern Sahel and Sahara region was a grassy, well-watered savanna between approximately 6,000 and 10,000 years ago. Many of the large mammals now associated with East Africa frequented the region. Now, even though under much less favorable conditions, the raising of sheep, goats, camels, and cattle, combined where possible with subsistence farming, is the major livelihood for peoples of the Sahel.

One of the most prominent features of the Sahel's landscape is Lake Chad, located north of landlocked Chad's capital, N'Djamena (population: city proper, 593,500). This large, shallow lake (between 7 and 23 ft/2–7 m deep) is situated in a vast open plain. It has no outlets and yet is not very saline because groundwater carries most of the salts away. Its waters are important economically for fish and irrigated agriculture. Characteristic of shallow lakes in arid lands, its surface area historically increased dramatically with rainy periods and shrunk in times of drought. However, the combination of recent decades of persistent drought and the expansion of irrigation along the northward-flowing rivers that feed the lake has caused a huge and enduring shrinkage. Lake Chad's surface area in 2001 was 580 square miles (1,502 sq km), compared with 9,700 square miles (25,123 sq km) in 1963; that is a 95% loss. Four countries share the lake and its drainage area, so decisions to reverse the lake's shrinkage are difficult to reach, and the likelihood of future conflict brought about by water shortages is high. The situation is reminiscent of that of the Aral Sea in Central Asia. *212*

Mining, manufacturing, and urbanization are generally insignificant in this subsistence-oriented region. The most remarkable recent improvement in the region's economy is the discovery of oil in southern Chad (see Fig. 17.3, p. 438). The reserves, with an estimated lifespan of 25 years, are beginning to be tapped, and a controversial World Bank–financed pipeline is being built to carry the crude to Cameroon's

REGIONAL PERSPECTIVE ::: Drought and Desertification in the Sahel

The Sahel region, like much of Africa south of the Sahara, experiences periodic drought. This is a naturally occurring climatic event in which rain fails to fall over an area for an extended period, often years. When rain does return, the arid and semiarid ecosystems of the Sahel come to life with a profusion of flowering plants, insects, and herbivores like gazelles, whose populations climb when foods are abundant. These ecosystems are resilient, meaning they are able to recover from the stress of drought and have mechanisms to cope with a natural cycle that includes periods of dryness and rain.

Desertification is the destruction of that resilience and the biological potential of arid and semiarid ecosystems. It is an unnatural, human-induced condition that has afflicted the Sahel severely since the late 1960s, and thus, it is different from the natural phenomenon of drought (Fig. 18.A). However, the recent history of the Sahel suggests that drought can be the catalyst that initiates the process of desertification. During decades of good rains prior to the late 1960s, the Fulani (Fulbe), Tuareg, and other Sahelian pastoralists allowed their herds of cattle and goats to grow more numerous. The animals represent wealth, and people naturally wanted their numbers to increase beyond the immediate subsistence needs of their families. However, due to declining death rates, the numbers and sizes of families keeping livestock were also very large. Thus, the unprecedented numbers of livestock built up during the rainy years made the Sahelian ecosystem

Figure 18.A   Desertification in progress is shown graphically in this photo from Mali in the Sahel. As growing human populations intensify the impact of naturally occurring drought, the Sahara creeps southward.

STEVE McCURRY/MAGNUM

(Continued)

Atlantic coast. Mauritania exports iron ore and Senegal exports phosphates. Niger exports uranium for France's nuclear power program. Salt is mined on Niger's sterile landscape, but its importance today pales in comparison with centuries ago, when a bar of Niger salt could be traded for a bar of gold. The Sahel region in general was a crossroads for caravan traffic (especially northbound slaves, food, and gold and southbound salt, weapons, and cloth) between West Africa and North Africa. Ancient Mali, with its legendary port of Timbuktu on the Niger River, was an especially important trading intermediary. Today, the traffic focuses on the northward migration of poor people from Nigeria, Niger, Guinea, Ghana, Cameroon and the Democratic Republic of Congo seeking work in oil-rich Libya. Many subsequently use Libya as a jumping-off point for settlement in Europe.

The largest metropolises are Senegal's main port and capital of Dakar (population: 2,294,800) and Bamako (population: metropolitan area, 1,237,300; city proper, 947,100), the capital and main city of Mali. Dakar was once the capital of the immense group of eight colonial territories known as French West Africa. It is the commercial outlet for much of the peanut belt in the Sahelian and West African savanna and is a considerable industrial center by African standards. Dakar is also a tourist and resort destination with fine beaches and an exotic blend of French and Senegalese cultures.

Ethnic Tensions

The Sahel is the border region between the mainly Arab and Berber North Africa and the mainly black Africa south of the Sahara. Relations between the ethnic groups are good in most countries, but there are problems in some. In Mauritania, for example, the government — led by the country's majority Arabic-speaking Moors — began expelling black Mauritanians across the border into Senegal in 1995. Analysts continue to

vulnerable to the impact of drought on an unprecedented scale.

Drought struck the region in 1968 and persisted through 1973. Annual plants failed to grow, so the large herds of cattle and goats turned to acacia trees and other perennial sources of fodder. They ate all of the palatable vegetation.

This destruction was a critical problem because plants play an important role in maintaining soil integrity by helping to intercept moisture and funnel it downward through the root system. Plant litter and decomposer organisms working around the plant contribute to soil fertility and help stabilize the soil. With hungry livestock in the region eating the plants, the landscape changed. No longer anchored and replenished, good soils eroded. "Junk" plants like Sodom apple (*Calotropis*) replaced palatable plants, and an almost impermeable surface formed on the land. This degraded ecosystem lost its resilience so that even when rains returned, vegetation did not recover.

Meteorologists studying the Sahel drought discovered another important link between precipitation and the removal of vegetation. They noticed that when people and livestock reduced the vegetative cover, they increased the **albedo,** or the amount of the sun's energy reflected by the ground. Fewer plants meant more solar energy was de-

flected back into the atmosphere, lowering humidity and reducing local precipitation. This connection is known as the **Charney Effect,** named for the scientist who documented it. It was a tragic sequence: People responding to drought actually perpetuated further drought conditions.

The 1968–1973 drought devastated the great herds of the Sahel. An estimated 3.5 million cattle died. Two million pastoral nomads lost at least half of their herds, and many lost as much as 90%. Farmers dependent on unirrigated cropping were also affected. Harvests were less than half of the usual crop for 15 million farmers, and for many, there was no harvest. The Niger and Senegal Rivers dried up completely, depriving irrigated croplands in Niger, Mali, Burkina Faso, Senegal, and Mauritania. The cost in human lives was estimated at 100,000 to 250,000. Environmental refugees poured into towns like Nouakchatt, the capital of Mauritania, which were unprepared to deal with such a large and sudden influx. Many pastoral nomads gave up herding and settled down to become wage earners and, where possible, farmers, and have never returned to their former ways.

This crisis resulted in an international effort to combat desertification, using recommendations developed by the United Nations in 1977. The U.N. re-

port, however, soon became a case study in how not to deal with an environmental crisis in the developing world. Most of the recommendations were universal, high-technology solutions that proved impossible to implement in the villages and degraded pastures of the Sahel. For example, following the recommended guidelines, many international agencies attempted to relieve the suffering of Sahelian pastoralists using techniques such as digging deep wells to water livestock. But these agencies failed to anticipate that these wells would act like magnets for great numbers of people and animals; although there was plenty to drink, many animals starved to death when they decimated what vegetation remained around the new water supplies.

Since the disappointment of the 1977 U.N. plan, and with the lessons relief organizations learned as a result, there has been greater focus on sustainable development, with its local rather than universal solutions. An example is the technique of "rainwater harvesting," in which villagers use local stones and tools to construct short rock barriers that follow elevation contours. Water striking these barriers backs up to saturate and conserve the soil, allowing more productive farming. The challenge to find such solutions is a continuous one; since 1968, drought has been an unusually frequent visitor to the Sahel.

53

fear that the Mauritanian government aims to "cleanse" the country of all of its 40% black population.

253
475
Chad has north–south tensions that are reminiscent of those of Sudan and Nigeria. The northerners, who now dominate Chad's political life, are mostly Arabic-speaking Muslims whose historic mainstay has been cattle pastoralism. The southerners are non-Arab Christians (mainly of the Sara ethnic group) whose livelihood is farming. The French favored the Christian southerners and empowered them more than the northerners, and now that the tables are turned, the northern elite is discriminating against southerners. Northerners historically have enslaved southerners, and there are reports that they continue to do so. There are frequent land disputes as northern pastoralists overrun southern farms. There are fears that, as in Sudan and Nigeria, the dominant northern power will extract the most benefits from the newly developing oil industry and leave the oil-bearing region in poverty.

18.3 West Africa

West Africa extends from Guinea-Bissau eastward to Nigeria (see Fig. 18.1). Its nine political units make up about 800,000 square miles (c. 2 million sq km), or nearly one-fourth of the area of the United States.

Environments and Resources

Climatic and biotic contrasts within this region are extreme, ranging from tropical steppe, dry savanna, and wetter savanna to areas of tropical rain forest along the southern and southeastern coasts. An estimated 90% of the region's tropical forest has been lost to commercial and subsistence deforestation. Most of the countries' more populous areas are in the zone of tropical savanna climate.

As is typical of less developed countries (LDCs), the West African countries depend heavily on subsistence farming and

F2.6
29
F2.7
31

PROBLEM LANDSCAPE

The Poor, Oil-Rich Delta of Nigeria

Oil is the economic lifeblood of populous Nigeria. Most of the production is concentrated in the Niger River Delta, home to about 7 million mostly Christian people (Fig. 18.B). Some of that production comes from wells drilled in a 350-square-mile (910-sq-km) section of the Niger Delta region inhabited by 500,000 people of the Ogoni tribal group. According to their spokesperson, the Ogoni have derived few benefits and have suffered much from oil development in their homeland. They complain that for more than 30 years, oil spills have tainted their croplands and water, destroying their crops and fisheries, while the flaring off of natural gas has polluted their air and caused acid rain. They note that despite the enormous revenue generated from oil drilled on their land, little money has returned to the area; most goes to the oil companies and to the Muslim-dominated government in the north. Meanwhile, most Ogoni live in palm-roofed mud huts; half of the delta region lacks adequate roads, water supplies, and electricity; schools have few books; and clinics have few medical supplies.

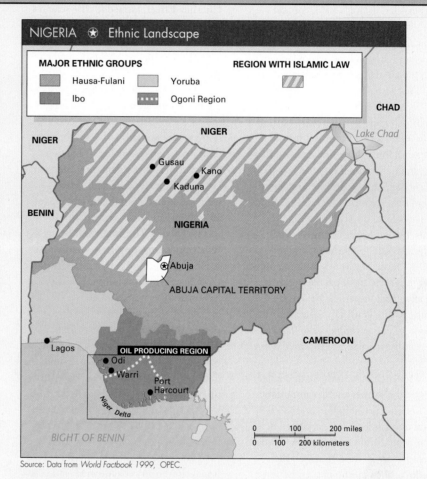

Source: Data from *World Factbook 1999*, OPEC.

Figure 18.B   Nigeria's ethnic landscape.

livestock grazing, along with a few agricultural and/or mineral exports. In the wetter south, subsistence agriculture relies mainly on root crops such as manioc and yams and on maize, the oil palm, and in some areas, irrigated rice. In the drier, seasonally rainy grasslands of the north (e.g., in northern Nigeria and northern Ghana), there is nomadic and seminomadic herding of cattle and goats, along with subsistence farming of millet and sorghum.

The major export specialties of West African agriculture are cacao (from which cocoa and chocolate are made), coffee, and oil-palm products in the wetter, forested south and peanuts and cotton in the drier north. Ivory Coast leads the world in cacao exports. Nigeria was once the world leader in exports of palm oil and peanuts, but these exports almost vanished as Nigerian agriculture was neglected during the oil boom of recent decades.

The other main export industry of West Africa is mining (see Fig. 17.3, p. 438). The most important resource is petroleum, with the OPEC member Nigeria leading the region in production and export. In 2002, when Nigeria stood as the world's sixth-largest oil exporter, petroleum made up over 90% of the country's foreign export earnings. Oil drilling takes place mainly in the mangrove-forested Niger Delta and offshore in the adjacent Gulf of Guinea (see Problem Landscape above).

Other prominent mineral products of West Africa include iron ore, bauxite, diamonds (Fig. 18.5), phosphate, and uranium. Iron ore exports are very important to Liberia. The main bauxite resources lie near the southwestern bulge of the West African coast; Guinea is the main producer and exporter. Togo exports phosphate. Gold, bauxite, and manganese are important to Ghana, as are bauxite, titanium, and diamonds to Sierra Leone.

In an attempt to improve their plight, Ogoni activists founded the Movement for the Survival of the Ogoni People (MOSOP) in 1990. Ogoni chiefs and elders issued an "Ogoni Bill of Rights" in which they declared the right to a safe environment and more federal support of their people. One of the movement's leaders, a popular author, playwright, and television producer named Ken Saro-Wiwa, also called for self-determination for the Ogoni. However, after an antigovernment rally by hundreds of thousands of Ogoni people in 1993, government police razed 27 villages, killing about 2,000 Ogonis and displacing 80,000 more.

In another violent incident, Ogoni activists killed four founding members of MOSOP in 1994. According to family members and supporters of the slain men, Saro-Wiwa incited their murders by stating that they "deserved to die" for not taking a more active position against the government and oil companies. Government authorities arrested Saro-Wiwa and 13 of his associates, charging them with responsibility for the murders. In late 1995, the government court found Saro-Wiwa and eight other defendants guilty and ordered their execution. Despite international protests that Saro-Wiwa and his codefendants were framed, the government had the men hanged. Many analysts regarded the executions as the government's effort to deter actions by other more serious rivals, especially within the army. As Nigeria is a thinly glued collection of ethnic groups, the government may also have seen the Ogoni struggle for minority rights as a Pandora's box that would lead to the breakup of the nation. Redistribution of oil revenue also would have diminished the income of the country's president and senior military officers.

Despite the return to civilian rule in Nigeria that came after Saro-Wiwa's death, dissent in the delta continued to grow. The Ogoni defiance and contempt of government and the oil companies have spread to the region's other major ethnic groups: the Ibo, Yoruba, and Ijaw. Reiterating the Ogoni protest that the region's oil generates $10 billion a year in revenue for the government, the Ijaw (who make up about 5% of Nigeria's population) in particular have pointed to the poverty and squalor of their settlements and demanded that some of the oil money be used to fund roads, electricity, running water, and medical clinics. Ijaw activists have spoken of secession and self-determination if their demands are not met. In 1998, they seized and held about 20 oil stations belonging to Royal Dutch/Shell and Chevron, temporarily cutting Nigeria's oil exports by one-third. Questions of equity in a potentially rich country thus pose a threat to national, regional, and international stability.

Nigerians in the delta and elsewhere hope that their 1999 democratic election of President Olusegun Obasanjo was just the beginning of a long-lasting era of stability and economic reform in their land of promise. Early in his term, Obasanjo vowed to allot the oil-producing states 13% of onshore oil revenues (compared with 5% previously) and considered giving them some portion of offshore receipts, which will soon be the larger.

One of Nigeria's ironies is that in this oil-rich country, there are continuous shortages of gasoline and electricity. Neglect, mismanagement, corruption, and theft are all responsible. In the 1990s, the country's four oil refineries and electricity grid broke down. The Obasanjo government vowed to correct the country's energy problems. One of the cruelest and most repetitive Nigerian news stories is the immolation of scores and even hundreds of poor delta villagers who illegally puncture gasoline or oil pipelines and try to "scoop" and sell the hydrocarbons, only to ignite the fires that consume them.

Population and Settlements

The region has an impressive total population of 192.1 million, or about 92 million fewer people than the United States. The most striking single aspect of West Africa's population distribution is that about two-thirds of the region's population resides within one country, Nigeria (whose 126.6 million also represent 20% of the population of the entire region of Africa south of the Sahara). This unusual concentration of people, which existed even before European contact, is related in part to the region's agricultural productivity and resource wealth.

Only about 37% of West Africa's people are city-dwellers, but the urban population is increasing rapidly. The world's large cities are associated mostly with trade, centralized administration, and large-scale manufacturing, but West Africa's big cities are different. They include national capitals and seaports, but as the region has only minor manufacturing, they have developed mainly in areas of productive commercial agriculture. In nearly every country in the region, the capital is a primate city, far larger than the nation's second largest city.

Nigeria has the most impressive urban development: Of the eleven largest metropolitan cities in West Africa, three are clustered in the ethnic Yoruba-dominated southwestern part of Nigeria, in and near the densely populated belt of commercial cacao production. The largest by far is Lagos (population: metropolitan area, 8,733,100; city proper, 7,720,200; Fig. 18.6). It had long been Nigeria's political capital, but in the 1990s, most governmental functions were moved to a new capital at Abuja in the interior. The development of Lagos surged in the early 20th century when the British colonial administration chose Lagos as the ocean terminus for a trans-Nigerian railway linking the Gulf of Guinea with the north. During the oil boom of recent decades, migrants from all

Figure 18.5 Small-scale diamond mining in Ivory Coast (Côte d'Ivoire), West Africa. The miners will screen for diamonds the small piles of dirt they have extracted by hand from shallow holes. The scrubby vegetation, a secondary growth that is typical of a deforested area, is common in West African tropical rain forest regions.

Many of these port cities have poor connections with their countries' hinterlands. Rapids and lower water during the dry season limit the utility of rivers for transport. Some rivers do carry traffic, notably the Niger, its major tributary the Benue, the Senegal River, and the Gambia River. The Gambia River is the main transport artery for The Gambia — a remarkable country in that it is nearly surrounded by another (Senegal) and consists merely of a 20-mile-wide (32-km) strip of savanna grassland and woodland extending inland for 300 miles (c. 500 km) along either side of the river. Railways are more important than rivers in the region's transport structure, but they are few and widely spaced. Instead of forming a network, the rail pattern is a series of individual fingers that extend inland from different ports without connecting with other rail lines (see Fig. 17.3, p. 438).

Ethnic and Political Complexity and Conflicts

West Africa was an advanced part of Africa south of the Sahara when the European Age of Discovery began. A series of strong pre-European kingdoms and empires developed there, both in the forest belt of the south and in the grasslands of the

parts of Nigeria (and many from outside the country) flooded into the city. With its severe traffic congestion, rampant crime, inflated prices, and inadequate public facilities, Lagos gives the impression of urbanism out of control. It is also indicative of a disorderly national economy.

Close to the border with Niger is Kano, the most important commercial and administrative center of Nigeria's north and a major West African focus of Islamic culture. Immediately west of Nigeria on the Guinea Coast lies Cotonou, the main city and port of Benin. Further west, in southern Ghana — another agricultural area with a high degree of commercialization — are two more of the eleven largest West African metropolises. The larger of the two, Accra (population: metropolitan area, 2,687,800; city proper, 1,551,200), is Ghana's national capital. Until recently, it was the main seaport, despite the lack of deep-water harbor facilities. A modern deep-water port has been developed east of Accra at Tema, and that city is now Ghana's main port and industrial center. Well inland from Accra is the country's second largest city, Kumasi, in the heart of Ghana's cacao belt, which was once the world's greatest cacao-producing area.

Today, the world's leading cacao producer and exporter is Ivory Coast (officially Côte d'Ivoire), with another of West Africa's 11 largest cities, Abidjan (population: metropolitan area, 3,822,400; city proper, 2,900,400 Fig. 18.7). Although the inland city of Yamoussoukro is the official capital, Abidjan is the de facto capital of a country that has vowed to achieve economic prosperity comparable to that of the Asian Tiger countries. Also on the Guinea Coast are three more seaport capitals: Monrovia in Liberia; Freetown in Sierra Leone, on West Africa's finest natural harbor; and Conakry in Guinea.

Figure 18.6 Lagos is Nigeria's vibrant and chaotic capital. The Christian imprint on the country is visible here, but Islamic influences are equally strong.

CHARLES O. CECIL/VISUALS UNLIMITED

Figure 18.7   Only about one-third of the people in Africa south of the Sahara live in cities, but with considerable rural-to-urban migration, that number is growing. This is very modern Abidjan, the leading city of Ivory Coast, with St. Paul's Cathedral in the foreground.

north. In colonial times, it became a French and British realm, except for a small Portuguese dependency and independent Liberia. Of the present countries, The Gambia, Sierra Leone, Ghana, and Nigeria were British colonies; Guinea, Ivory Coast, Togo, and Benin (formerly Dahomey) were French; and Guinea-Bissau was Portuguese. Liberia is unique, becoming Africa's first republic after 1822, when 5,000 freed American slaves sailed there and settled with support from the U.S. Treasury. The country's capital, Monrovia, was named for American President James Monroe.

In West Africa today, the English, French, and Portuguese languages continue in widespread use, are taught in the schools, and are official languages in the respective countries. European languages are a useful means of communication in this region, where hundreds of indigenous languages and dialects, often unrelated, are spoken. Economic, political, and cultural relationships between West African countries and the respective European powers that formerly controlled them continue to be close. France, for instance, supplies economic and military aid to its former colonies in this and other parts of Francophone (French-speaking) Africa. France further cultivates good relations with its former dependencies by inviting their leaders to summit conferences to discuss matters of common interest. Similarly, Britain maintains ties with Anglophone Africa through meetings of the Commonwealth of Nations, to which all its former colonies in Africa south of the Sahara belong.

As is often the case in Africa, the governments of the region must deal with serious problems related to ethnicity. European governments contributed initially to some present troubles by drawing arbitrary boundary lines around colonial units without proper concern for ethnicity. As a result, the typ-

ical West African country today, like most countries elsewhere in Africa south of the Sahara, is a collection of ethnic groups that have little sense of identification with the national unit.

The region's giant, Nigeria, is a good example of an ethnically complex, unnaturally assembled nation. Some Nigerians still refer to the "mistake of 1914," when British colonial cartographers created the country despite its ethnic rifts. Simplistically, the greatest divide is between Muslim north and Christian south. The British colonizers invested more in education and economic development in the south and built army ranks among northerners, whom they thought made better fighters; so the subsequent pattern is that the northern military leaders have tried to blunt the economic clout of southerners.

A total of about 250 ethnic groups are officially recognized, although some sources claim there are as many as 400. The largest are the mostly Muslim Hausa and Fulani (29%) mainly in the north, the mostly Christian Yoruba (21%) in the southwest, the Ibo or Igbo (18%) mainly in the southeast, and the Ijaw of the southern delta region. Recently, there has been escalating tension and violence between some of these groups, especially between the mainly Muslim Hausa and mainly Christian Yoruba, and each has formed militias. As Muslim northerners have traditionally dominated Nigerian politics, the ascendancy of the Christian Yoruba President Obasanjo may be seen as a concession by the northerners to prevent the country from fragmenting. However, mainly as a means of blunting Obasanjo's perceived Christian leanings, most of the Muslim northern states have adopted *shari'a*, or Islamic law, and the Christian minorities of these regions fear persecution and intolerance (see Fig. 18.B). There have been numerous violent riots between Christians and Muslims throughout the country. The prospect of violent devolution hangs over Nigeria, reviving memories of deaths of a million people in 1967–1970 when the country's ethnic Ibo (Igbo) population struggled to establish a separate nation of Biafra.

A diamond purge and civil war have plagued Sierra Leone since 1991. The country's prospects improved when its first democratically elected president, Ahmed Tejan Kabbah, took office in 1996. However, a military junta seized power a year later, forcing Kabbah into exile. Asserting its self-appointed identity as West Africa's regional superpower, and perhaps with an eye on Sierra Leone's mineral wealth, Nigeria responded in 1998 with a military attack on Sierra Leone that in turn forced the junta to flee and restored President Kabbah to office. As of 2002, his hold was firming. Almost two thirds of the country, including all of the diamond-producing area, had been wrested from hands of the rebel Revolutionary United Front (RUF). The RUF had projected its power with a brutal combination of mutilation, rape, and murder. There was an effort to accommodate these rebels when the Lomé Agreement gave the RUF some seats in the national cabinet and established RUF leader Foday Sankoh as the country's vice president. But the RUF broke the peace agreement and took hundreds of United Nations peacekeepers hostage, reigniting the war. As the fighting continued, British troops intervened, capturing Sankoh (who was still imprisoned as of 2002), holding Freetown, and helping Kabbah reassert government authority over this devastated country.

In neighboring Liberia, the regional kingpin Charles Taylor has long provided arms to Sierra Leone's RUF in exchange for diamonds. In 1989, American-educated Taylor ignited a civil war that persisted for 7 years. In 1997, Liberians elected him president because, analysts say, they feared that Taylor would restart the war if not elected. Since then, he has led the country on a path to ruin. Public services such as electricity are all but nonexistent. Schools and hospitals have been downsized or closed. Taylor's family has enriched itself by selling the county's assets, including its tropical hardwoods, and in exchange for Guinean diamonds, Taylor has also supplied guns to rebels in that country. Guinea's struggles created a new humanitarian crisis for West Africa, as by 2002, thousands of Guinean refugees joined the 400,000 war refugees from Sierra Leone and Liberia already huddled in Guinean camps in the border area known as the Parrot's Beak.

There are sectarian tensions in Ivory Coast, too. The political and economic power is concentrated in the hands of mainly Christian southerners. The country grew prosperous from cocoa and coffee exports in the 1990s, developing good roads and electricity grids, working telephones, schools, and a growing middle class. It became attractive to immigrants, rather like the United States is for many Latin Americans, and the government encouraged immigration from neighboring countries to help develop Ivory Coast. The invitation was answered with large numbers of migrants, particularly Muslims from Burkina Faso and Mali, who settled in northern Ivory Coast. They now make up between a third and a half of the country's population. Now that cocoa prices have collapsed and the economy has worsened, the southerners are promoting themselves as the only true Ivoirians (a concept they call *Ivoirité*) and are fomenting antiforeign sentiments directed against the northern immigrants.

18.4   West Central Africa

The subregion of West Central Africa is flanked by Cameroon and the Central African Republic to the north and the Democratic Republic of Congo to the south (see Fig. 17.1). It contains seven countries: Cameroon, Gabon, Central African Republic, Republic of Congo, Democratic Republic of Congo, São Tomé and Principe, and Equatorial Guinea.

Physical Aspects

A narrow band along the immediate coast of West Central Africa is composed of plains below 500 feet (c. 150 m) in elevation. As in West Africa, the coast is fairly straight and has few natural harbors. Most of the region is composed of plateaus with surfaces that are undulating, rolling, or hilly. An exception to the generally uneven terrain is a broad area of flat land in the inner Congo Basin, once the bed of an immense lake. The greater part of West Central Africa lies within the drainage basin of the Congo River. The heart of the region has tropical rain forest climate and vegetation grading into tropical savanna to the north and south. The biological resources are immense. The tropical forests that cover much of the two Congos, Gabon, Cameroon, Equatorial Guinea, and a small corner of the Central African Republic make up one-fifth of the world's remaining tropical rain forests.

Only scattered parts of West Central Africa, primarily along the eastern and western margins, are mountainous. The most prominent mountain ranges lie in the eastern part of the Democratic Republic of Congo near the Great Rift Valley and in the west of Cameroon along and near the border with Nigeria. Vulcanism has been important in mountain building in both instances (see Definitions & Insights, p. 477).

Population, Settlements, and Infrastructure

Within West Central Africa as a whole, population densities are low and populated areas are few and far between. Rough terrain, thick forests, wetlands, and areas with tsetse flies tend to isolate clusters of people from each other. The savanna lands and highlands are generally more densely populated than the rain forests.

Most of the countries have an overall density well below that of Africa south of the Sahara as a whole (see Table 17.1, p. 434). The Democratic Republic of Congo is Africa's second largest country in area (after Sudan); it is about the size of Western Europe, or Alaska, Texas, and Colorado put together, but its estimated population in 2001 was only 54 million, or well under one-half that of the continent's most populous country, Nigeria. Within the subregion, however, it is by far the most populous country. Nearly half of the Democratic Republic of Congo is an area of tropical rain forest in which the population is extremely sparse.

Cities are few in these countries. Most of the larger ones are political capitals and/or seaports, including the Republic of the Congo's capital city of Brazzaville (population: city proper, 1,098,700) and its seaport of Pointe Noire. Brazzaville, located directly across the Congo River from the much larger capital city of Kinshasa in the Democratic Republic of Congo, was once the administrative center of French Equatorial Africa. Kinshasa (population: city proper, 6,069,200) is the Democratic Republic of Congo's capital, largest city, and principal manufacturing center. Another important city is Lubumbashi (formerly Elisabethville) in the copper-mining region of the southeast.

Transportation infrastructure is generally poor due to political mismanagement, warfare, neglect, and poverty. Aside from a handful of rail fingers reaching inland from seaports, West Central Africa is largely devoid of railways, and roads only poorly serve most areas. Isolation and poor transport are

The Cameroonian volcanoes region is also its "deadly lakes" region. Here, in the two volcanic-cone lakes Nyos and Monoun, carbon dioxide gas generated by volcanic activity deep in the earth builds up in the watery depths below a layer of gas-free fresh water. The boundary between these layers is called a **chemocline.** Sometimes an event such as a windstorm or landslide disturbs the top layer deeply enough to puncture the chemocline and release the carbon dioxide from its solution. It erupts explosively, rocketing gas and water upward and outward. On August 21, 1986, such an explosion in Lake Nyos killed 1,700 people who were simply suffocated by a roving cloud of carbon

dioxide. In recent years, U.S. and Cameroonian scientists and engineers have cooperated to install pipes that dissipate the carbon dioxide from these deadly lakes before it builds up to potentially catastrophic levels. The authorities are pleased with the progress but uncertain that enough venting pipes have been installed. Similar steps have yet to be taken in East Africa's Lake Kivu, where the same threat exists.

In 2002, a volcanic eruption near the northern shore of Lake Kivu sent rivers of molten lava into the Congolese city of Goma, incinerating the city and causing almost its entire population of 400,000 to flee as refugees.

major factors limiting economic development. There is not yet a paved road between Brazzaville and Pointe Noire, but the 320-mile (515-km) Congo-Océan Railway does connect the cities. One of the Democratic Republic of Congo's major arteries of transportation is a combination of railroad and river connecting Kinshasa with Lubumbashi (see Fig. 17.3, p. 438). Another pair of cities connected by rail is Cameroon's seaport of Douala (population: metropolitan area, 1,409,200; city proper, 1,204,700) and the country's inland capital, Yaoundé (population: metropolitan area, 1,319,000; city proper, 1,091,300).

Economic Prospects

Subsistence agriculture supports people in most areas. There is small-scale production of export crops, mainly in rain forest areas, and some cotton is exported from savanna areas. Southern Cameroon is an important cash crop area, with coffee and cacao the main export crops, supplemented by bananas, oil-palm products, natural rubber, and tea. The farms occupy an area of volcanic soil at the foot of Mt. Cameroon (13,451 ft/4,100 m), a volcano that is still intermittently active. The mountain has extraordinarily heavy rainfall, averaging around 400 inches (1,000 cm) a year.

The island of Bioko has long been notable for its cacao exports and more recently for its oil boom. Bioko and the adjacent mainland territory of Río Muni, formerly held by the Spanish, received independence in 1968 as the Republic of Equatorial Guinea. Cacao production in Bioko developed on European-owned plantations, but the newly independent country's leaders expelled the plantation owners.

Exports of minerals, principally oil, aid the financing of some countries, notably Gabon, Republic of the Congo, Cameroon, and most recently, Equatorial Guinea, where oil revenues since production began in 1996 have created per capita GDP PPPs that are high for Africa south of the Sahara

(see Table 17.1, p. 434). Cameroon has been engaged in an oil-related dispute with Nigeria over control of the Bakassi Peninsula, which juts into the Gulf of Guinea near the border between the two countries. Control of the peninsula will give the dispute's victor ownership of promising offshore oil reserves of the Gulf of Guinea. Cameroon will also receive royalties and lost-land compensation for the shipment of Chadian petroleum through a pipeline to the Atlantic through Cameroon's territory (see Fig. 17.3, p. 438). Copper from the mineral-rich Katanga (now Shaba) Province in the southeast part of the Democratic Republic of Congo was long the country's main export, but civil war disrupted the industry. There are also rich cobalt deposits there, and diamonds are in the adjoining Kasai region (see Fig. 17.3). Another important resource in Congo is coltan, a rare mineral used in the manufacture of cell phones and fighter jets.

One of the region's most problematic environmental and economic issues is the fate of its extensive tropical rain forests. In 2002, they were being cut, mainly for commercial logging, at a rate of 10 million acres (4 million hectares) yearly. At that pace, they will be gone by 2020. In an effort to slow the destruction, the World Wildlife Fund has reached an agreement with the World Bank whereby the World Bank will not extend loans in this region unless the borrowing governments can demonstrate they are using, or planning to use, their tropical forest resources in a sustainable manner. These forested countries do not like being, as they see it, held hostage by Western environmentalists, but they are making concessions. A good example is Gabon, which has an extensive area of almost-untouched tropical rain forest. As its oil reserves dwindle, Gabon sees its future in selling tropical hardwoods. But in a recent pact, it has promised that part of its pristine rain forest, in the Lopé Reserve of central Gabon, will not be harvested by the Malaysian and French logging companies that have extensive concessions in the country.

Colonialism and Modern Struggles

In colonial days, France held three of the present units — Central African Republic, Republic of the Congo, and Gabon — in the federation of colonies called French Equatorial Africa. Cameroon consisted of two trust territories administered by France and Britain. Belgium held what is now the Democratic Republic of Congo, Spain held Equatorial Guinea, and Portugal held the islands of São Tomé and Principe. During this last quarter of the 19th century, the Congo Basin was virtually a personal possession of Belgium's King Léopold II, whose agents ransacked it ruthlessly for wild rubber, ivory, and other tropical products gathered by Africans. This lawless era inspired Joseph Conrad's famous novel *Heart of Darkness* (1902), in which the trader Kurtz, at the point of death, evokes the ravaged Congo region with the cry "The horror! The horror!" In 1908, the Belgian government formally annexed the greater part of the Congo Basin, creating the colony of Belgian Congo.

Pressure for independence built up rapidly in 1959, and Belgium yielded to the demands of Congolese leaders. In 1960, 3 months before adjacent French Equatorial Africa became the independent People's Republic of the Congo, the Congo colony became the independent Republic of the Congo. In 1971, it took the name Zaire, meaning "river." Following the overthrow of Zaire's government in 1997, the country was renamed Democratic Republic of Congo and is now referred to simply as Congo or in association with its capital as Congo-Kinshasa. The neighboring People's Republic of the Congo is now generally referred to as Congo Republic or Congo-Brazzaville.

Hundreds of ethnic groups speaking many languages and dialects live in West Central Africa. The majority speak Bantu languages of the Niger-Congo family, while Nilo-Saharan language speakers predominate in the drier areas of grassland north of the rain forest. Small bands of Pygmies (Twa) live in the rain forest of the Congo Basin. The population of the Democratic Republic of Congo is especially complex, with some 250 ethnolinguistic groups speaking more than 75 languages.

In recent decades, the region's troubles have focused on its major power, the Democratic Republic of Congo. Then-Zaire's autocratic ruler, Mobutu Sese Seko, had neglected the country's development but acquired a massive fortune during his tenure in office from 1965 to 1997. Political opposition and public discontent with unemployment and inflation led to unrest in the early 1990s. All-out war began after 1997, when Zairean army troops attempted to expel a force of indigenous Tutsis living in the country's far eastern region. The assault backfired when these Tutsis, supported by neighboring Tutsi-dominated governments in Rwanda and Burundi and by other sympathetic countries in the region, took up yet more arms and initiated fierce attacks on government forces. The Tutsi counterattack quickly widened into an antigovernment revolt comprised of many ethnic factions and led by a non-Tutsi guerrilla named Laurent Kabila. Kabila's forces pushed westward virtually unopposed through the huge country and,

in May 1997, occupied Kinshasa with little loss of life. President Mobutu fled Zaire on the eve of Kabila's triumph. Kabila then declared himself president of the country he now called the Democratic Republic of Congo and promised that political reforms would live up to the country's new name. Foreign companies rushed to sign mining contracts with the fledgling government, and throughout Congo, there was hope that decades of neglect and poverty could be reversed in this resource-rich giant of West Central Africa.

But it was not to be. President Laurent Kabila's new regime proved to be a bitter disappointment to most of Congo's citizens and neighbors and to the international community. Like his predecessor Mobutu, he funneled much of the country's wealth and power into the hands of family and friends. He also obstructed international human rights investigations into the deaths and disappearances of thousands of Hutu refugees during the heady days of rebellion against Mobutu. And he failed to bring stability to the far east of his huge country, where an environment of chaos had given birth to his own bid for power.

In 1998, Hutu forces based in the eastern Congo launched successive devastating attacks on Tutsi interests in Rwanda. Although the Tutsis of Rwanda had helped bring Kabila to power in Congo, they now turned against him, forming the backbone of a rebel alliance bent on bringing down Kabila's young regime. The unrest led to something rather rare in the region, a broader fight dubbed "Africa's First World War" that involved numerous countries, some of which are not even Congo's neighbors (Fig. 18.8). Each player became involved to protect its own national, separatist, or economic interests, largely at the expense of anything that would resemble national cohesion and stability in the Democratic Republic of Congo.

On one side was the so-called Rebel Coalition that sent forces to oppose the government in Kinshasa. Rwanda, to avenge Kabila's refusal to crack down on Hutus operating against this Tutsi-led country, and perhaps to establish a security buffer zone on its border with Congo, supported rebels fighting Kabila. Rwanda also wanted protection for the Tutsi minority (known as Banyamulenge) living in extreme eastern Congo. Like Rwanda, Burundi's Tutsi-dominated government fought Congo to protect itself from Hutu attacks launched from inside Congo. Uganda acted against Congo in part to establish a security buffer zone that would help protect it from attacks by Ugandan rebels operating from inside Congo. Angola's rebel UNITA forces also took on Kabila's forces in a bid to retain the bases inside Congo it used to attack the established government in Angola.

On the other side, the so-called Congo Coalition sent forces to support the government in Kinshasa. Zimbabwe supported Kabila, perhaps to divert attention from Zimbabwe's internal problems. By sending forces to aid Kabila, Zimbabwe also may have wanted to establish itself as a regional superpower, akin to the position enjoyed by South Africa. And Zimbabwe acted to protect mining concessions and lucrative contracts that Congo had awarded to Zimbabwean officials.

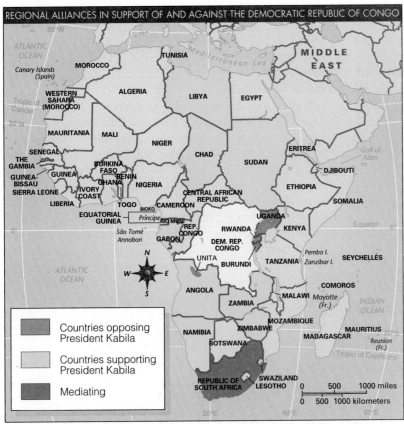

REGIONAL ALLIANCES IN SUPPORT OF AND AGAINST THE DEMOCRATIC REPUBLIC OF CONGO

Countries opposing
President Kabila

Countries supporting
President Kabila

Mediating

Figure 18.8   What began as a localized conflict in Congo spread to neighboring countries, each having a special interest at stake.

Source: Santoro and Dunn. *Africa's War of Nations, with Congo as the Pawn. The Christian Science Monitor,* August 27, 1998: p. 6.

The Angolan government wanted to crack down on UNITA rebel bases inside Congo and so joined the fray on Kabila's side. Zambia's government perceived greater regional stability if Congo remained a single entity governed from Kinshasa, so it joined the coalition. Namibian firms have mining contracts with Congo, so the government in Windhoek supported Kinshasa to protect these interests. The government of Sudan supported Kabila's Congo mainly because of Sudan's bitter enmity with Uganda, which supported southern Sudan's rebellion against the Khartoum government. Finally, Chadian troops were airlifted directly into the diamond-mining region of south central Congo to protect Chad's investments there.

While the fighting was sometime ferocious, it was not warfare that directly claimed most of the estimated 3 million lives lost in this struggle between 1998 and 2001. Most of the casualties were civilians who died from starvation and disease or in widespread massacres directed against ethnic groups, usually Hutus killing Tutsis, but sometimes the reverse.

As the war progressed, Congo effectively split into three entities: the rebel-held east, importing and selling goods through Uganda and Kenya; a south, trading through Zambia and South Africa; and a west, operating from Kinshasa and using the country's outlet to the sea. These geographic divisions and alignments could become formalized should the country dissolve, but as this text went to press, there was hope that a peace treaty signed by the combatants in 1999 would finally stand

firm after many violations. A turning point apparently came when Laurent Kabila was assassinated by a bodyguard in 2001, to be succeeded by his son Joseph, a Congolese leader seemingly, finally, committed to halting the country's decline.

Meanwhile, the expatriate firms that would exploit Congo's mineral wealth and perhaps help bring this country out of dire poverty remained on the sidelines. As the dust of the conflict began to settle, United Nations investigators found evidence of widespread plunder of the country's natural assets during the war years. Uganda and Rwanda in particular were singled out for looting eastern Congo of its gems, minerals, timber, agricultural produce, and wildlife, including elephant ivory from some of the country's national parks. Once allies against Kabila, these powers ended up in a struggle between themselves, focused in part on dividing up Congo's spoils.

18.5   East Africa

Five countries make up East Africa: Kenya and Tanzania, which front the Indian Ocean, and landlocked Uganda, Rwanda, and Burundi (see Fig. 18.1). Their total area is 703,000 square miles (1.8 million sq km), roughly the size of the U.S. state of Texas, with a total population in 2001 of 103.5 million (more than one-third the population of the United States).

Figure 18.9   East Africa is blessed with extensive areas of volcanic soil where commercial and subsistence crops thrive. This is a wheat field in the central highlands of Kenya.

Environments and Resources

East Africa's terrain is mainly plateaus and mountains, with the plateaus generally at elevations of 3,000 to 6,000 feet (c. 900 to 1,800 m). All five countries include sections of the Great Rift Valley. International frontiers follow the floor of the Western Rift Valley for long distances and divide Lakes Tanganyika, Malawi, and Albert and smaller lakes among different countries. The more discontinuous Eastern Rift Valley crosses the heart of Tanzania and Kenya. Lake Victoria, lying in a shallow downwarp between the two major rifts, is divided between Uganda, Kenya, and Tanzania.

The two highest mountains in Africa, Mt. Kilimanjaro (19,340 ft/5,895 m) and Mt. Kirinyaga (formerly Mt. Kenya, 17,058 ft/5,200 m; see Fig. 17.4), are extinct volcanoes. Both are majestic peaks crowned by snow and ice and visible for great distances across the surrounding plains; recent scientific observations, however, suggest that suspected global warming will melt all of Kilimanjaro's legendary snows by 2015. Most of East Africa's productive agricultural districts are in the fertile soils of these volcanic areas (Fig. 18.9). Vulcanism was also im-

portant in building the fertile highlands of Rwanda and Burundi. The Virguna Mountains along the Rwanda-Uganda border include many active volcanoes. One of the region's most active is Mt. Elgon (14,176 ft/4,321 m) on the Uganda-Kenya border. On the Uganda-Democratic Republic of Congo border is Margherita Peak (16,762 ft/5,109 m), part of a nonvolcanic range known as the Ruwenzori, which was commonly identified with the Mountains of the Moon depicted by early Greek geographers on their maps of interior Africa.

The climate of most of East Africa is classified broadly as tropical savanna. In Uganda, Kenya, and Tanzania, park savanna — composed of grasses and scattered flat-topped acacia trees — stretches over broad areas and is the habitat for some of Africa's largest populations of large mammals. Great herds of herbivores and their predators and scavengers migrate seasonally across these countries to follow the changing availability of food and water. The countries manage many parks and wildlife reserves, and in Kenya and Tanzania, these are responsible for a huge share of the nations' hard currency revenues. Among the most famous are the Serengeti Plains and Ngorogoro Crater of Tanzania and the Masai Mara and Amboseli National Parks of Kenya (Fig. 18.10).

Over much of East Africa, moisture is too scarce for nonirrigated agriculture. There are great variations in the amount, effectiveness, and dependability of precipitation and in the length and time of occurrence of the dry season. Long and sometimes catastrophic droughts occur. From 1998 to 2001, Kenya suffered its worst drought in 40 years. More than 3 million people were in need of food aid as crops shriveled and livestock died. As the country's hydropower reserves were depleted, Kenyan cities suffered long periods without electricity.

The main subsistence crops are maize, millets, sorghums, sweet potatoes, plantains, beans, and manioc. Elevation has such a strong effect on temperatures that in some areas midlatitude crops like wheat, apples, and strawberries do well. Kenyan tourist brochures boast that Nairobi (elevation: 5,500 ft/1,676 m) has a springlike climate year-round. Some cultivators depend exclusively on crops for subsistence, and others

Figure 18.10   Savanna grasslands in east and southern Africa host the world's greatest concentrations of large mammals. This is Kenya's Masai Mara National Park in part of the greater Serengeti ecosystem that also extends into northern Tanzania.

also keep livestock. Cattle are the most important livestock animal.

The most valuable export crop is coffee, but tea is also important to Kenya. Kenya and Tanzania also produce sisal for its valuable fiber (see Fig. 17.14). The islands of Zanzibar and Pemba support the millions of clove trees that provide most of world's supply of cloves. The islands' economies have been hurt drastically by a recent fall in world clove prices and by economic and political turmoil in Indonesia, which had long imported most of the clove crop for use in its aromatic *kretek* cigarettes. Seeking a new future, Zanzibar has announced plans to transform itself into a free economic zone modeled after Singapore and Hong Kong.

The East African nations generally lack mineral reserves and production. A project to mine titanium, very controversial because it will require the relocation of thousands of people, is underway near the Kenya coast. Manufacturing is limited mainly to agricultural processing, some textile milling, and the making of simple consumer items. There is almost no petroleum. There is a large cotton-textile mill at Jinja, Uganda, using hydroelectricity from a station at the nearby Owen Stanley Dam on the Nile. The Tana River project in Kenya produces hydropower and supplies irrigation waters.

Populations, Settlements, and Infrastructure

The highest population densities are in a belt along the northern, southern, and eastern shores of Lake Victoria; in south central Kenya around and north of Nairobi; and in Rwanda and Burundi, which have among the world's highest population densities, aside from those of city-states and some islands. As recently as the early 1990s, population growth rates in this region were so high that East Africa was a common academic setting for the Malthusian scenario. But after sharp reductions as a result of falling birth rates and increasing AIDS-related deaths, the population growth rates of the five East African countries were moderate, averaging 2.5% as of 2001. The region's populations are overwhelming rural. Only about 5% of

the populations of Rwanda and Burundi is urban. Kenya and Tanzania are the most urbanized at only about 20%.

The most agriculturally productive parts of East Africa are bound together by railways that form a connected system leading inland from the seaports of Mombasa in Kenya and Dar es Salaam and Tanga in Tanzania (see Fig. 17.3, p. 438). Mombasa is the most important seaport in East Africa (Fig. 18.11). From Mombasa, the main line of the Kenya-Uganda Railway leads inland to Kenya's capital, Nairobi (population: metropolitan area, 2,943,800; city proper, 2,391,600), the largest city and most important industrial center in East Africa. It is also the busiest crossroads of international air traffic and the main outfitting and departure point for safaris into East Africa's world-renowned national parks. Nairobi was founded early in the 20th century as a construction camp on the railway, which continues westward to Kampala (population: metropolitan area, 1,116,900; city proper, 927,900), the capital and largest city of Uganda. Dar es Salaam ("House of Peace," population: city proper, 2,372,000) is the capital, main city, main port, and main industrial center of Tanzania, with rail connections to Lake Tanganyika, Lake Victoria, and Zambia. Rwanda's capital is in the highlands at Kigali (population: city proper, 351,400), and Burundi's capital is on the northernmost point of Lake Tanganyika at Bujumbura (population: city proper, 315,000).

Political and Social Issues

The five East African countries face many social and political dilemmas. None has had much internal peace, order, or prosperity, and none has had easy relations with its neighbors. Ethnic rivalries and conflicts have beset all of them.

Aside from small minorities of Asians, Europeans, and Arabs, the population of East Africa is composed of a large number of African tribes; for example, Tanzania alone has about 120 groups. Among the large groups are the Kikuyu and Luo of Kenya, the Baganda of Uganda, the Sukuma of Tanzania, the pastoral Masai of Kenya and Tanzania, and the Tutsi and Hutu of Rwanda and Burundi. Most East African peoples

Figure 18.11 Naval ships and grain silos in Kenya's Indian Ocean port of Mombasa.

speak Bantu tongues of the Niger-Congo language family, but some in northern Uganda, southern and western Kenya, and northern Tanzania speak Nilo-Saharan languages. Swahili, a Bantu language drawing heavily on Arabic for vocabulary, is a widespread lingua franca (and is the national language of Tanzania), as is English. There are Christian, Muslim, and "animist" faiths in all the countries. The Muslim religion reflects a long history of Arab commercial enterprise in the region, including slave trading.

All of the countries are former European colonies. Rwanda, Burundi, and Tanzania were part of German East Africa until the end of World War I. With the end of the war, Tanzania joined Kenya and Uganda as British dependencies, and Rwanda and Burundi, then known as Ruanda-Urundi, were controlled by Belgium. They became two independent countries in 1962. Now a part of Tanzania, the islands of Zanzibar and Pemba, which lie north of the city of Dar es Salaam and some 25 to 40 miles (40 to 64 km) off the coast, were once a British-protected Arab sultanate. The city of Zanzibar was the major political and slave-trading center and entrepôt of an Arab empire in East Africa. While Arabs formerly controlled political and economic life, the majority of the population was, and is, African. Africans have been in control since 1964, when a revolution overthrew the ruling sultan. That same year, the political union of Tanganyika and Zanzibar was formed with the name United Republic of Tanzania. However, Zanzibar and Pemba have a parliament and president separate from Tanzania's.

Prior to independence, East Africa was home to about 100,000 Europeans, 60,000 Arabs, and 300,000 Asians. Arabs were primarily a mercantile class. Europeans were a professional, managerial, and administrative class. In the "White Highlands" of southwestern Kenya, about 4,000 European families owned and operated estates that were worked by African laborers to produce export crops and cattle. With independence in 1963, Kenya made a point of inviting the British and other Europeans to maintain their presence in the commercial crop sector and to train the black majority in the managerial and technical skills that had served white interests well. Many Europeans left, and Africans acquired their farms, but those who stayed formed the core of the still-influential white Kenyans of today. Collectively known as the "Wahindi" in Swahili, the Asians included Indians, Pakistanis, and Goans (from the former Portuguese colony of Goa on the west coast of India). They dominated retail trade, owned factories, and produced export crops. Some were wealthy, and practically all were more prosperous than most Africans.

Since independence, the Asian minority has lost its commercially predominant position. Ugandan leader Idi Amin abruptly expelled them from Uganda in 1971. Because they were so vital to the country's economy, it nearly collapsed. Some have recently begun to return and have successfully reclaimed lost land and other assets. The recent economic decline of Kenya has led to a slow but steady exodus of Asians, mainly to Australia, Britain, and the United States.

Episodes of bloodshed have punctuated East Africa's history. From 1905 to 1907, when mainland Tanzania was German controlled, German authorities put down a great revolt against colonial control, costing 75,000 African lives. Immediately prior to its independence, Kenya experienced a bloody guerrilla rebellion led by a secret society within the large Kikuyu tribe called the Mau Mau. Antagonism directed against British colonialism and European ways in Kenya cost the lives of about 11,000 Africans and 95 white settlers over the period 1951–1956.

From the 1960s through the 1980s, Kenya was regarded as a relative economic success story for Africa south of the Sahara. However, corruption, ethnic strife, crime, terrorism (including the 1998 bombing of the U.S. Embassy in Nairobi, apparently by associates of Osama bin Laden), drought, and floods then combined to reduce Kenya's prospects for economic development. Recent unrest has threatened the vital business of international tourism in Kenya, which normally contributes about half of the country's hard currency revenue. In addition, the international community has accused President Daniel Arap Moi's government of corruption and human rights abuses against political and tribal opponents. In 1992 and again in 1997, Britain, the United States, and other international lenders reduced and threatened to cut off aid to Kenya unless Moi implemented democratic reforms in advance of national elections. Moi did little, but when he promised a new anticorruption law in 2001, the donors resumed aid, especially to help combat the country's severe drought. The loan terms are extremely stringent and are meant to show other African countries they have to rule the way the international community wants if they wish to receive aid monies.

In postcolonial Uganda, a savage orgy of murder, rape, torture, and looting took place under the notorious regimes of Idi Amin and Milton Obote. The ravaged country had a respite after 1986 when Yoweri Museveni, the leader of a southern guerrilla force that had developed in opposition to Obote, took over the government. Despite some simmering conflicts among Uganda's 40 ethnic groups and despite the country's controversial interventions in regional conflicts, Uganda has come to be regarded widely as a model for democracy, social progress, and economic development in Africa south of the Sahara.

Rwanda and Burundi have had tragic disputes between their majority and minority populations. About 85% of the population in Burundi and 90% in Rwanda are composed of the Hutu (Bahutu), and most of the remainder of the two populations is Tutsi (Watusi). These originally were not separate tribal or ethnic groups; the peoples speak the same language, share a common culture, and often intermarry. The distinction between Hutu and Tutsi rather was based on socioeconomic classification. Historically, the Tutsi were a ruling class who dominated the Hutu majority. Tutsi power and influence were measured especially by the vast numbers of cattle they owned. Tutsi who lost cattle and became poor came to be identified as Hutu, and Hutu who acquired cattle and wealth often became Tutsi.

Colonial rule in East Africa attempted to polarize these groups, thereby increasing antagonism between them. German and Belgian administrators differentiated between them exclusively on the basis of cattle ownership: Anyone with fewer than 10 animals was Hutu and anyone with more was Tutsi. Europeans mythologized the Tutsis as "black Caucasian" conquerors from Ethiopia who were a naturally superior, aristocratic race whose role was to rule the peasant Hutus. The colonists replaced all Hutu chiefs with Tutsi chiefs. These leaders carried out colonial policies that often imposed forced labor and heavy taxation on the Hutus. Education and other privileges were reserved almost exclusively for the Tutsi.

Ferocious violence between these two peoples has marked Rwanda and Burundi intermittently since the states became independent in the early 1960s. After independence, the majority Hutus came to dominate the governments of both Rwanda and Burundi. In 1994, the death of Rwanda's Hutu president in a plane crash (in which Burundi's president also died) sparked civil war in Rwanda between the Hutu-dominated government and Tutsi rebels of the Rwandan Patriotic Front (the RPF, based in Uganda). Hutu government paramilitary troops systematically massacred Tutsi civilians with automatic weapons, grenades, machetes, and nail-studded clubs. Women, children, orphans, and hospital patients were not spared. An estimated 800,000 Tutsis and "moderate" Hutus (those who did not support the genocide) died. International media tracked these horrors, but outside nations did nothing to stop the fighting.

Well-organized and motivated Tutsi rebels seized control of most of the country and took power in the capital, Kigali, in 1994. That precipitated a flood of 2 million Hutu refugees mainly into neighboring Zaire, now the Democratic Republic of Congo (Fig. 18.12). Although the Tutsi-dominated government promised to work with the Hutus to build a new multiethnic democracy, Hutu insurgents, based mainly in eastern Zaire, continued to carry out attacks against Tutsi targets in Rwanda. Rwanda now has a national unity policy aimed at reconciling Hutus and Tutsis; both occupy important government posts and, although the Tutsi still hold the political upper hand, Tutsis have encouraged the resettlement of Hutu refugees who fled the country during its troubles.

In Burundi, the Tutsi minority traditionally dominated politics, the army, and business. In 1993, there was a brief shift of power to the Hutu majority, as Burundians elected the country's first Hutu president. Tutsis murdered him in October 1993, and a large-scale Hutu retaliatory slaughter of Tutsis ensued. An estimated 200,000 died. Since then, the Tutsis have maintained a firm grip on power, but there have been international diplomatic efforts to come to a power-sharing agreement between Hutus and Tutsis leading up to national elections in 2004. Presumably, these elections would restore control to the Hutu. Hutu rebels belonging to the militia known by its French acronym FNL meanwhile continued to attack Tutsi targets around Bujumbura, a mostly Tutsi city. Ostensibly to ensure that Hutu villages near the capital would not provide safe haven for the rebels, Tutsi authorities rounded up perhaps 300,000 Hutu farmers and herded them into squalid, fenced reserves that are essentially concentration camps.

Figure 18.12 Warfare and drought in postcolonial Africa have driven millions of Africans from their homes as refugees. This harrowing scene shows terrorized Rwandans fleeing into Tanzania in 1994 to escape the savage massacres of civilians in the country's civil war.

SCOTT DANIEL PETERSON/GAMMA LIAISON

18.6 Southern Africa

In the southern part of Africa south of the Sahara, five countries — Angola, Mozambique, Zimbabwe, Zambia, and Malawi — share the basin of the Zambezi River and have long had important relations with each other. The first two were former Portuguese colonies, and the latter three were formerly British. Still farther south are South Africa and four countries whose geographical problems are closely linked to South Africa: the three former British dependencies of Botswana (formerly Bechuanaland), Swaziland, and Lesotho (formerly Basutoland), and the former German colony of Namibia. These 10 states are grouped here as the countries of southern Africa (see Fig. 18.1). Other than their geographic location, there is little in the way of economic or political integration that binds these diverse states together. This may change as a result of goals established by the Southern African Development Community (SADC), a regional bloc composed of these 10 countries plus Tanzania and Mauritius. SADC's member nations aim to establish by 2004 an economic free-trade zone in which goods can cross borders without tariffs and taxes. The dominant power in that group, in the region, and on the continent is South Africa, and so it is treated in greatest detail here.

Environment and Resources
Water, Natural Regions, and Agriculture

Water is all-important in this drought-prone region of a dry continent. South Africa is fairly well endowed with water, as two river systems — the Orange (including its tributary, the Vaal) and the Limpopo — drain much of the country's high (3,000 to 6,000 ft/c. 900 to 1,800 m) interior plateaus. Many dams have been constructed on these and other South African rivers to provide irrigation water and hydroelectricity.

The Zambezi is another of the region's great rivers. Rising in Zambia, it flows eastward to empty into the Indian Ocean on the Mozambican coast. On the middle course of the Zambezi River separating Zambia from Zimbabwe is the spectacular Victoria Falls, regarded traditionally as one of the seven Natural Wonders of the World. It is a significant international tourist destination today. Downstream from the falls, hydropower from the Kariba Dam (completed in 1962; see Fig. 17.5, p. 439) flows to both countries.

The driest countries are Namibia and Botswana. They can support pastoralism and, in some areas, limited subsistence agriculture, but commercial agriculture is lacking. The Namib Desert extends the full length of Namibia in the coastal areas (Fig. 18.13). Inland is a broad belt of semiarid plateau country, merging at the east with the Kalahari semi-desert of Botswana. The dry zone extends southward into South Africa's Cape provinces and northward into western Angola. Chronic shortages of fresh water in Namibia have led the government to consider desalinating Atlantic Ocean waters (this option was rejected as too expensive); building the Epupa Dam on the Kunene River, which divides Namibia

Figure 18.13 A surprising abundance and diversity of wild animals and plants are adapted to the harsh conditions of the Namib Desert in Namibia. This is an oryx antelope known as the gemsbok.

from Angola (this project is opposed by environmentalists and by human rights groups, who argue that the homeland of the Himba tribespeople would be inundated); and diverting water by pipeline southward from the Namibian portion of the Okavango River to Windhoek (widely regarded as Africa's cleanest and safest capital; population: city proper, 192,300). Arguing that the World Heritage Site of its Okavango Delta would be damaged or destroyed, Botswana is strongly opposed to this Namibian plan. With a wealth of wildlife attractions in the Okavango Delta and adjacent national parks, Botswana is a growing destination for international nature-loving tourists. Namibia has it own wetlands resource of international importance: Large wildlife populations responding to dramatic fluctuations in wet and dry conditions make Etosha National Park an attraction for the international tourists who represent an important source of hard currency to Namibia.

Not being landlocked, Namibia enjoys an additional resource that Botswana lacks. The arid western coast of southern Africa, swept by the cold, northward-moving Benguela Current, supports Africa's most important fisheries (Fig. 18.14). Upwelling cool water brings up nutrients from the ocean floor and provides a fertile habitat for plankton, the basic food of fish. Namibia has established an economic zone in the Atlantic stretching 200 miles (320 km) offshore, in which fishing and fish processing based at Walvis Bay have become important industries.

Tropical savanna prevails over most of southern Africa, including much of Zimbabwe, Zambia, Malawi, Mozambique, eastern Angola, and northern South Africa. More than 90% of the annual rainfall comes in the cooler 6 months from November to April. The natural vegetation is primarily open woodland, although on the High Veld of Zimbabwe there is tall-grass savanna. The High Veld lies generally at elevations of 4,000 feet (c. 1,200 m) and higher and forms a band about 50 miles (80 km) broad by 400 miles (c. 650 km) long, trending southwest-northeast across the center of Zimbabwe.

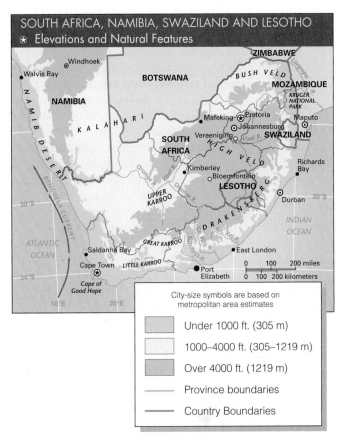

SOUTH AFRICA, NAMIBIA, SWAZILAND AND LESOTHO
⊛ Elevations and Natural Features

City-size symbols are based on metropolitan area estimates

Under 1000 ft. (305 m)

1000–4000 ft. (305–1219 m)

Over 4000 ft. (1219 m)

—— Province boundaries

—— Country Boundaries

Figure 18.14   Major natural features of South Africa and surroundings.

The High Veld was historically the main area of European settlement, and most of Zimbabwe's 70,000 whites still live on it. Among them are about 4,500 farmers whose large holdings, worked by black laborers, produce most of the country's agricultural surplus to feed urban populations and to export. Tobacco from these farms is the most important export, followed by cotton, cane sugar, corn, and roses. As discussed later, the future of these farms and their tenure are uncertain.

South Africa's physical environment has played a major role in its unusual course of development (see Fig. 18.14). Lying almost entirely in the middle latitudes, it is a cooler land than most of Africa south of the Sahara, with annual temperatures generally between 60° and 70°F (16°–21°C). In combination with the country's natural resource wealth, this climate made the area very attractive for European settlement; it was one of the "neo-Europes" of the colonial era and was historically Africa's main center of European settlement and advanced economic development. Europeans favored the extreme east of the country's interior plateau, a region also known as the High Veld. The original vegetation here was a grassy steppe, similar to the midwestern prairies of the United States. This is South Africa's leading area of crop and livestock production. The principal farm animals are beef cattle, dairy cattle, and sheep, and the main crops are corn (maize) and wheat. Corn is the staple diet for South Africa's blacks. Surpluses are exported and used as livestock feed. In KwaZulu/Natal, there is hilly terrain between the Indian Ocean and the Drakensberg, a mountainous escarpment at the edge of the interior plateau. The leading commercial crop of that area is sugarcane.

South Africa's Western Cape Province has a Mediterranean climate. It is the only African area south of the Sahara with this climate type (Fig. 18.15). Average temperatures are

Figure 18.15   A characteristic rural landscape in the Mediterranean climate of South Africa's Western Cape Province. The middle distance is dominated by fields of ripe winter wheat at the base of one of the Cape Ranges.

similar to those of the same climate area in southern California. The early settlers established vineyards in this region, and today, the area's grapes and wines sell both in South Africa and abroad, where they are acquiring a growing share of the world market. The province also produces oranges and other citrus fruits, deciduous fruits, and pineapples. Sheep ranching is a major occupation in the semiarid interior.

Mining and Manufacturing

Minerals, including petroleum, are by far the most valuable material assets of this region. South Africa has the greatest mineral endowment, and that country's relative prosperity owes in large part to that wealth. Immense British and other foreign investments pour into South Africa's mining industry. South Africa's diverse and abundant mineral wealth has fueled its continental leadership in manufacturing. The country has about a third of Africa's manufactures and a third of its electric generation. Its diversified industries include iron and steel, machinery, metal manufacturing, weapons, automobiles, tractors, textiles, chemicals, processed foods, and others. Many thousands of jobs in Western countries are directly linked to mining and manufacturing in South Africa. The mines employ mainly black African workers recruited from South Africa and other African countries, particularly Lesotho, Swaziland, Mozambique, Botswana, and from as far away as Malawi. There is a large illegal immigration to South Africa, too; an estimated 2 to 8 million undocumented migrant workers are in this relatively prosperous country.

South Africa is the world's largest producer of gold (Fig. 18.16), and gold is its most important export. The gold-mining industry is concentrated in the High Veld, especially the Witwatersrand (formerly Rand), an area in the province of Gauteng (Fig. 18.17). Johannesburg, South Africa's largest city (population: metropolitan area, 4,927,200; city proper, 1,617,900), plus the adjacent impoverished and overcrowded black township of Soweto and nearby Pretoria, are located in Gauteng. The Transvaal provinces and adjacent areas make up one of the world's most vital mineral-exporting regions, with exports including copper, asbestos, chromium, platinum, vanadium, antimony, and phosphate.

Recovery of uranium from mine tailings and as a by-product of current gold mining has made South Africa the largest uranium producer on the continent. And since 1867, the country has been an important source of diamonds from alluvial deposits along the Vaal and Orange River valleys and from kimberlite "pipes" (circular rock formations of volcanic origin). Kimberley in Northern Cape is the administrative center for the diamond industry.

South Africa's mining industries require large amounts of electricity. Most of this is supplied by generating plants powered by bituminous coal, which is present in vast quantities near the surface on the interior plateau. These coal deposits, which are the largest in Africa, are immensely important to South Africa, as the country has no petroleum or natural gas output and has only a modest hydroelectric potential. Coal is a major export, and it has facilitated development of Africa's leading iron and steel industry, concentrated at Pretoria and around Johannesburg. Iron ore comes by rail from large deposits in the northeast and central parts of the country.

Some of South Africa's neighbors also have considerable mineral wealth (see Fig. 17.3). Partly for this reason — and because South Africa surrounds Swaziland and Lesotho in particular — South Africa is much engaged in their affairs. Botswana and Lesotho gained independence from Britain in 1966, and Swaziland, the last absolute monarchy in southern

Figure 18.16 Johannesburg, the "City of Gold," is surrounded by huge dumps of waste material from former gold-mining operations. One of these dumps occupies the center of the view. Note the automobile expressway at the bottom of the photo. In the distance lie residential areas occupied by whites.

JESSE H. WHEELER, JR.

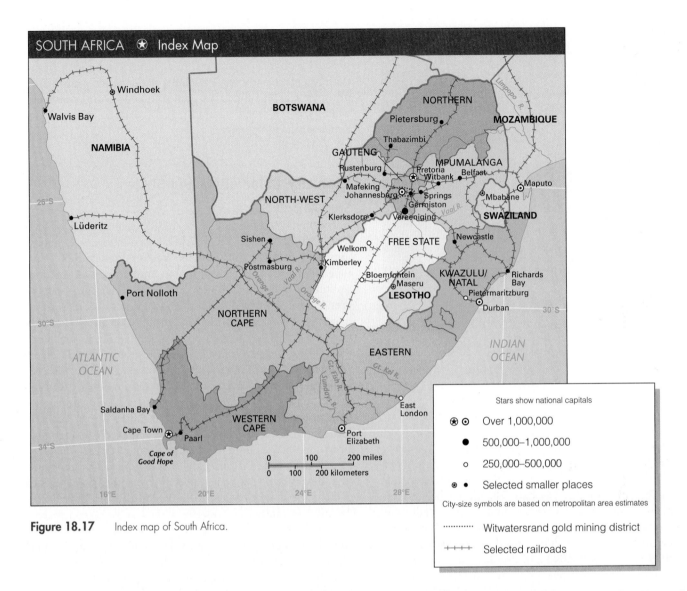

Figure 18.17 Index map of South Africa.

Map legend:

Stars show national capitals

⊛ ⊙ Over 1,000,000

● 500,000–1,000,000

○ 250,000–500,000

⊛ ● Selected smaller places

City-size symbols are based on metropolitan area estimates

·········· Witwatersrand gold mining district

++++ Selected railroads

Africa, became independent in 1968. Since then, however, South Africa has occasionally exerted its power on behalf of its interests in these weaker countries, particularly Lesotho. In 1998, for example, the prime minister of Lesotho perceived that a coup might be imminent in his nation and called on South Africa for assistance. South African troops immediately invaded Lesotho, but rather than quelling the unrest, this intervention led to widespread condemnation of South Africa for meddling in the internal affairs of a sovereign nation. Mining of coal, gold, and tin is an important source of revenue in Swaziland, as is tourism — which features gambling casinos — and sugar production. There is some diamond mining in Lesotho.

Diamonds make up almost three-quarters of the total value of Botswana's exports. South African, British, and U.S. mining companies export Namibian diamonds and various metals. Zimbabwe has begun exporting titanium and is constructing a large aluminum smelting plant. In Zimbabwe, there is an extraordinary geological feature called the Great Dike, an intrusion of younger, mineral-rich material into older rock. Ranging in width from 2 to 7 miles (3 to 11 km), the Great Dike crosses the entire nation and is readily visible to astronauts in space as a giant stripe on the landscape (Fig. 18.18). Along and near the Great Dike are the principal areas of commercial farming and of mines producing gold (Zimbabwe's second most important export, after tobacco), nickel, asbestos, copper, chromium, and other minerals, primarily for export. The Dike thus forms the main axis of economic development in Zimbabwe. Along it lies a railway connecting the largest city and capital, Harare (formerly Salisbury; population: metropolitan area, 2,226,100; city proper, 1,832,008), in the northeast with the second largest city, Bulawayo in the southwest. Iron ore mined at Que Que, about midway between Harare and Bulawayo, is used there to produce steel. The Wankie coal field in western Zimbabwe supplies coal to smelt the ore. Zimbabwe's manufacturing industries are more important, both in value of output and in diversity of production, than those of any other African nation south of the Sahara except Nigeria and South Africa.

NASA/SPL/PHOTO RESEARCHERS, INC.

Figure 18.18 The stripe running top to bottom across this photo is Zimbabwe's Great Dike seen from space. About 300 miles (500 km) long and 6 miles (10 km) wide, the mineral-rich Great Dike was formed about 2.5 billion years ago when molten lava filled and widened a fracture in Earth's crust. Note how faulting has offset the dike in two places.

Zambia's economy depends heavily on copper exported from several large mines developed with British and American capital. An urban area has developed at each mine, and the mining area, known as the Copperbelt, consists of a series of separate population nodes strung close to Zambia's frontier with the adjacent copper mining Shaba region of southeastern Democratic Republic of Congo. The Copperbelt, which contains the majority of Zambia's manufacturing plants and copper mines, is largely an urban region. Copper composed 50% of Zambia's exports in 2001. However, exports are much diminished from their 1960s highs due to years of poor government control of the copper industry. The copper mining industry is now entirely in private hands, and Zambia hopes to rise quickly from its position as the world's 11th largest copper producer (as of 2002). Other mineral production includes cobalt, a by-product of copper mining; zinc and lead mined along the railway between the Copperbelt and Zambia's capital and largest city, Lusaka (population: metropolitan area, 1,669,700; city proper, 1,173,000), and coal production from extreme southern Zambia.

The region's leader in petroleum production is Angola, which is increasingly important on the global energy stage. One of its vital oil producing areas is the coastal exclave of Cabinda, separated from the rest of the country by the Congo River mouth (which is controlled by the Democratic Republic of Congo). Other fields are scattered along the Angolan coast between the Congo mouth and the city of Lobito. Major new reserves have been discovered recently offshore Angola; in fact, more new petroleum reserves were detected there between 1997 and 2001 than in any other country in the world. U.S and other multinational oil companies are at work developing this resource. Crude oil and modest amounts of refined products made up nine-tenths of Angola's exports in 2001, and in that year, Angolan oil comprised 4% of U.S. oil imports. In addition to oil and its products, Angola exports small amounts of diamonds, coffee, and other items. Diamonds are concentrated in the northeast where UNITA rebels mined them for years to finance their antigovernment operations.

Ethnicity, Colonialism, and Strife

South Africa: A Special Case

A country whose people were legally segregated by race, South Africa was one of the world's most controversial nations until its new beginning in 1994. Visitors from Western Europe and North America will find in South Africa most of the institutions and facilities to which they are accustomed. However, the Europeanized cultural landscape does not reflect the majority culture of this unusual country. Whites represent only about 14% of the total population, and blacks make up 75%. There are two other large racial groups: the so-called "coloreds," of mixed origin (8%), and the Asians (3%), most of whom are Indians. Many peoples make up the black African majority. The Zulu and Xhosa, both concentrated in hilly sections of eastern South Africa near the Indian Ocean, are the most populous of the nine officially recognized tribal groups.

A huge economic gulf separates South Africa's impoverished black Africans from about 5.9 million whites of European descent who dominate the economy and enjoy a lifestyle comparable to that of people in Western Europe and North America. Because of the poverty of the non-European majority, South Africa's overall per capita GDP PPP of $8,500 is typical of such moderate income LDCs as Mexico and Oman. For the unskilled and semiskilled labor needed to operate their mines, factories, farms, and services, the whites of South Africa have always relied mainly on low-paid black workers in most of the country and colored workers in the Western Cape Province. The white-dominated economy could not operate without them, and the Africans in turn are extremely dependent on white payrolls. While the races are interlocked economically, however, relations between them have historically been very poor and have been structured to benefit the whites. Following is a summary of how the country's racial structures evolved.

South Africa's white population is split between the British South Africans and the Afrikaners, or Boers (Dutch: "farmers"). The Afrikaners speak Afrikaans, a derivative of Dutch, as their preferred language. They outnumber the British approximately three to two. The Afrikaners are the descendants of Dutch, French Huguenot, and German settlers who began coming to South Africa more than three centuries ago. Their impact on indigenous Africans was catastrophic. White settlers killed, drove out, and through the unintentional introduction of smallpox, decimated most of the original Khoi and San (Bushmen) inhabitants. They put the survivors to work as servants and slaves. When these proved to be insufficient in number, the colonists brought slaves from West Africa, Madagascar, East Africa, Malaya, India, and Ceylon (now Sri Lanka).

Most of the workers in the cane fields of Kwazulu/Natal were indentured laborers brought from India beginning in the 1860s. The majority chose to remain in South Africa when their terms of indenture ended, and some Indians came as free immigrants. About 85% of South Africa's Indian population is still in KwaZulu/Natal, mainly in and near Durban, the province's largest city and main industrial center and port. Most Indians today are employed in commercial, industrial, and service occupations.

Like KwaZulu/Natal, the Cape provinces have a distinctive racial group: South Africa's coloreds (widely known as the "Cape Coloreds"). Nearly nine-tenths live in the provinces, primarily in and near Cape Town, the country's second largest city and one of its four main ports. This group originated in the early days of white settlement as a product of sexual relations between Europeans (Dutch East India Company employees, settlers, and sailors) and non-Europeans, including slaves and Khoi (Hottentot), Malagasy, West African, and various south Asian peoples. The resulting mixed-blood people vary in appearance from those with pronounced African features and others who are physically indistinguishable from Europeans. About nine-tenths speak Afrikaans as a customary language, and most of the others speak English.

Culturally, the coloreds are much closer to Europeans than to Africans, and many with light skins have "passed" into the European community. Coloreds have always had a higher social standing and greater political and economic rights than black South Africans, although they have ranked below the Europeans in these respects. Most coloreds work as domestic servants, factory workers, farm laborers, and fishers and perform other types of unskilled and semiskilled labor. There is a small but growing professional and white-collar class of coloreds. White owners of grape vineyards are donating portions of their fields to an increasing number of the coloreds who once worked as indentured servants on the white lands, leading to a sharp increase in the region's output and exports of wines. Such progress is a sign of how truly different the new South Africa is from the old.

Racial issues in South Africa have a long history of producing strife, even among the ruling white powers. Great Britain occupied South Africa — which it called the Cape Colony — in 1806, during the Napoleonic Wars. Friction developed between many Boers and the British authorities, who imposed tighter administrative and legal controls than the Boers were accustomed to. The British abolished slavery throughout their empire in 1833, contributing to the anti-British sentiments of the slaveholding Boers. Boer discontent resulted in the **Great Trek,** a series of northward migrations by which groups of Boers, primarily from the eastern part of the Cape Colony, sought to find new interior grazing lands and establish new political units beyond British reach. After some earlier ex-

ploratory expeditions, the main trek by horse and ox-wagon began in 1836. It resulted in the founding of the Orange Free State, the Transvaal, and Natal as Boer republics. Britain annexed Natal in 1845, but recognized Boer sovereignty in the Transvaal and the Orange Free State in 1852 and 1854, respectively.

While Boer disaffection with Britain was building, British settlers were coming to South Africa in increasing numbers. The area's natural resource wealth helped shape subsequent events. The British colonies and Boer republics might have developed peaceably side by side if diamonds had not been discovered in the Orange Free State in 1867 and gold in the Transvaal in 1886. These discoveries set off a rush to these republics of prospectors and other fortune hunters and entrepreneurs from outside the region, particularly from Britain. Ill feeling led to the Anglo-Boer War in 1899. British troops defeated the Boer forces decisively, ending the war in 1902 but leaving behind a reservoir of animosity that exists to this day.

The Union of South Africa was organized in 1910 from four British-controlled units as a self-governing constitutional monarchy under the British Crown. Dutch became an official language on par with English (this was later altered to specify Afrikaans rather than Dutch). Pretoria in the Transvaal became the administrative capital as a concession to Boer sentiment (however, the national parliament now meets at Cape Town and the supreme court sits at Bloemfontein in the Free State). In 1961, following a close majority vote by the ruling white population, it became a republic outside the British Commonwealth.

Since 1910, and particularly since the Nationalist (now National) party took power in 1948, Afrikaners have dominated the political life of South Africa by virtue of their greater numbers and cohesion. Racial segregation characterized South African life from 1652 onward, but it was first systematized after 1948 under a comprehensive body of national laws supporting the official governmental policy of "separate development of the races," subsequently called "multinational development" and commonly known as **apartheid.** The new laws imposed racially based restrictions and prohibitions on the entire population, but they weighed most heavily on black Africans. They ruled out significant sharing of power by black people in a unified South Africa. They fragmented and displaced families and ethnic groups, rearranging the country's cultural landscape. Apartheid culminated in the scheme to establish tribal states, or "homelands" (also known as "native reserves," "Bantustans," and "national states"), which were nominally intended to achieve "independence." Whites ejected several million Africans not wanted in "white" South Africa from their homes and transferred them to homelands. They removed great numbers of blacks from urban areas and white farms and dumped them in poverty-stricken and crowded areas unequipped to handle new influxes of people.

The apartheid system drew furious criticism from many other nations. Most of the world community ostracized South

Africa, and many countries, including the United States, imposed economic sanctions against it. Black unrest directed against apartheid and the general underdevelopment of the African majority became so widespread and violent between 1984 and 1986 that the government declared a state of emergency. Blacks killed many other blacks that they perceived to be collaborating with the whites. An intensive government crackdown resulted in the wholesale banning of political activity by antiapartheid groups, intensified restrictions on media, and the jailing of thousands of black Africans perceived as protest leaders. The government lifted the state of emergency in 1990, but widespread violence continued. Much of the fighting was between two large, tribally based rival factions: the Xhosa-dominated African National Congress (ANC), led by Nelson Mandela, and the Zulu-dominated Inkatha Freedom party (IFP), led by Chief Mangosuthu Buthelezi. Tension and violence between these groups have since largely abated.

A complete official turnabout on the issue of apartheid resolved the crisis. It began in 1989 after Frederik W. de Klerk, an insider in the Afrikaner political establishment, became president. Urged on by white South African business leaders, antiapartheid activists, the powerful force of South African and international political opinion, the impact of international economic boycotts against South Africa, and his own stated convictions concerning the injustices and unworkability of apartheid, de Klerk launched a broad-scale program to repeal the apartheid laws and put South Africa on the road to revolutionary governmental changes under a new constitution. Four years of intensive negotiations involving all interested parties resulted in an agreement to hold a countrywide, all-race election to create an interim government that would hold office for not more than 5 years. In 1994, voters turned out in massive numbers to elect a new national parliament, which then chose Nelson Mandela as president. (Only 4 years earlier, the white government had freed Mandela after a 27-year period of political imprisonment.) This historic election ended white European political control in the last bastion of European colonialism on the African continent. The apartheid laws themselves became null and void.

Nelson Mandela, South Africa's "George Washington," retired in 1999 and was succeeded by a second black president, Thabo Mbeki. The country's political landscape is remarkably stable, but challenges remain. The races are continuing on their long path of reconciliation. Some of the healing process has been formalized in a national Truth and Reconciliation Commission, which allows those who were party to racial violence during the apartheid era to confess their misdeeds and, in a sense, be absolved of them. There is still a huge economic gulf between the races to be overcome. South Africa remains, in effect, two countries: a wealthy white one and a poor black one. As in Zimbabwe, there is real need for land reform because the white minority owns about 70% of the productive land. Unlike Zimbabwe, however, South Africa has a plan to peacefully help reduce the imbalance: to redistribute 30% of the country's farmland from white to black hands by 2014, on a willing-seller–willing-buyer basis. A program is also underway to compensate the estimated 3.5 million blacks forcibly displaced by the government to the "homelands" between 1960 and 1982. More broadly, South Africa has an aggressive affirmative action program to help redress the economic imbalance between the races. The Employment Equity Act does not impose quotas but requires employers to move toward "demographic proportionality" based on the national proportions of race and gender.

Troubles for Portugal's Ex-Colonies, and Independent Namibia

Although Portugal was the first colonial power to establish interests in Africa south of the Sahara, stronger European powers came to prevail in the region. However, Portugal managed to retain footholds along both the Atlantic and Indian Ocean coasts. By 1964, Portugal's principal African possessions, Angola (Portuguese West Africa) and Mozambique (Portuguese East Africa), were the most populous European colonies still remaining in the world. Portugal fielded a large military force to combat guerrilla opposition in both areas. In 1974, Portugal's resistance broke, and an army coup in Portugal installed a regime in Lisbon that granted the two colonies independence.

When the Portuguese withdrew from Angola in 1975, a Marxist-Leninist liberation movement supported by the Soviet Union and Cuba took over the central government in the seaport capital and largest city, Luanda (population: metropolitan area, 2,571,600; city proper, 2,094,200; Fig. 18.19). But this movement's authority was not recognized by other liberation movements, one of which — the National Union for the Total Independence of Angola (UNITA) — came to control large parts of the country. South Africa and the United States supplied aid to this anti-Marxist movement headed by Jonas Savimbi, a member of Angola's largest ethnic group. Large quantities of Soviet arms, thousands of Cuban troops, and many Soviet and East German military advisors supported the government in Luanda. The cold war proxy conflict raged for years.

In 1988, representatives of Angola, Cuba, and South Africa signed a U.S.-mediated agreement under which Cuba and South Africa would withdraw militarily from Angola, and South Africa would grant independence to Namibia. But even after the withdrawals, hostilities continued between UNITA and the government until a peace agreement allowing UNITA's participation in all levels of government was reached in 1994. This agreement soon faltered, however, and while some UNITA members took up their legitimate government posts, others, including UNITA leader Savimbi (who was killed in 2002), again took up arms, fighting especially hard to secure the oil fields of Cabinda and using revenue from diamonds to feed the war machine. Like Nigeria, Angola is a resource-rich country that so far has little to show for its

SARAH ERRINGTON/HUTCHISON LIBRARY

Figure 18.19 Luanda, the capital and largest city of Portugal's former colony of Angola, exhibits the common pattern of cities in Africa south of the Sahara: a modest cluster of high-rise buildings in the city center surrounded by a sea of small one-story houses, often on dirt streets and generally roofed with metal sheeting.

oil wealth. The approximate $3 billion in annual revenue (as of 2002) from the oil industry have funded the war against UNITA and apparently made the country's wealthy elite even wealthier but have done little for most Angolans. Desperate for low-interest loans, Angola has promised the international community to triple its expenditures on health, education, and other social services.

Namibia became independent from South Africa quite recently. South African forces overran this former German colony during World War I, and after the war, the League of Nations mandated it to South Africa under the name of South West Africa. Only the war in Angola released South Africa's grip. Backing UNITA, South African military forces had actively intervened in the Angolan struggle in the mid-1970s. Eventually, South African forces withdrew but continued to attack Angolan bases of the South West Africa People's Organization (SWAPO), a guerrilla group fighting for Namibia's independence. In 1988, South Africa agreed to grant independence to Namibia, but only if the would-be Namibian government under SWAPO leadership could convince its main source of support, the government of Angola, to expel Cuban troops from Angola. The South African government granted independence to Namibia in 1990. SWAPO is now the major political party, but its Marxist leanings have not prevented the country from promoting

capitalist investment. White Namibians look anxiously at the confiscation of white farmlands in nearby Zimbabwe; just 4,000 of the country's 80,000 whites own 44% of Namibia's territory.

The anomalous sliver of territory in Namibia's extreme northeast known as the Caprivi Strip, which extends fully 280 miles (450 km) eastward to the waters of the Zambezi River, was important to the German colonists who ruled Namibia early in the 20th century. They insisted on this access to Zambezi in the mistaken belief that it would allow navigation all the way to the Indian Ocean (the Victoria Falls stand in the way). Independent Namibia was permitted to retain the strip, despite protests from neighboring Botswana. The Caprivi is home to the Barotse (sometimes called Lozi) ethnic group, whose tribal lands also extend into Botswana, Angola, and Zambia. Barotse efforts both in the Caprivi Strip and in Zambia to establish an autonomous or independent homeland are strongly opposed by both Namibia and Zambia. With its scattered livestock forage, the Kalahari section of Namibia is peopled mainly by Bantu-speaking herders and by remnants of the Khoi and San (Bushman) populations. Some of the San are among the world's last remaining hunters and gatherers, but the majority have given up the nomadic way of life in recent decades and have settled as farmers and wage laborers.

Portugal's other ex-colony, Mozambique, is one of the world's poorest countries, but it has rather bright prospects. It suffered from years of guerrilla warfare, which first involved African insurgencies against the Portuguese and then warfare by African dissidents against the Marxist government that took over from the Portuguese in Maputo, the country's capital (population: metropolitan area, 1,588,900; city proper, 1,062,700). The highly destructive civil war costing 1 million lives between the Maputo government and the South African- and Rhodesian-backed rebel organization called RENAMO (Mozambique National Resistance) dragged on until 1992 when a peace accord was signed and the United Nations sent a military peacekeeping force to Mozambique. Elections took place peacefully in 1994, and a large share of the 5 million people displaced by the war began returning to their villages. Only periodic disputes between the government and RENAMO have occurred since their reconciliation. Nature has dealt the strongest blows to Mozambique recently; devastating cyclones in 2000 and again in 2001 brought flooding and suffering to much of the country.

Mozambique is being rebuilt. Transportation facilities in the Beira Corridor to Zimbabwe are being upgraded, and a new regional economic link with South Africa has been established. Known as the Maputo Development Corridor, this toll road joins Johannesburg's industrial and mining center with the deep-water port at Maputo. Low-tax zones were created to attract new businesses to the corridor. The country's estimated 30 million land mines are being cleared, and war refugees are returning to farm this fertile land. By 1998, the country was self-sufficient in food production. Major exports are Mozambique's famous huge prawns, cashew nuts, cotton, tea, sugar, and timber. Multinational efforts are underway to process titanium for export and to develop the country's pristine beaches into world-class tourist destinations. Impressed with Mozambique's progress, foreign lenders have agreed to cancel about 80% of the country's debt. Mozambique has acted aggressively on this windfall, with dramatic new investments in health, education, and other social services.

Zimbabwe's Struggles

Prior to the post-World War II movement for African independence, the countries of Zimbabwe, Zambia, and Malawi (then known respectively as Southern Rhodesia, Northern Rhodesia, and Nyasaland) were British colonies. In 1953, the three became linked politically in the Federation of Rhodesia and Nyasaland, or the Central African Federation. It had its own parliament and prime minister but did not achieve full independence. The racial policies and attitudes of Southern Rhodesia's white-controlled government created serious strains within the federation, helping to bring about its dissolution in 1963. Northern Rhodesia then achieved independence as the Republic of Zambia, and Nyasaland became the independent Republic of Malawi.

But the government of Southern Rhodesia, the main area of European settlement in the federation, was unable to come to an agreement with Britain concerning Britain's demand for the political participation of the colony's African majority. In 1965, the government of Southern Rhodesia issued a Unilateral Declaration of Independence (UDI), and in 1970, it took a further step in separation from Britain by declaring itself to be Rhodesia, a republic that would no longer recognize the symbolic sovereignty of the British Crown. Britain and the Commonwealth responded with a trade embargo. Negotiators for Britain and the Rhodesian government unsuccessfully attempted to arrive at a formula to allow political rights and participation for the 95% of Rhodesia's population that was black and thus to achieve reconciliation and world recognition of Rhodesia's independence. Although the economic sanctions were ineffective, pressure on the white government increased in the mid-1970s with black guerrilla warfare against the government and the end of white (Portuguese) control in adjacent Mozambique. Independence for Rhodesia (renamed Zimbabwe) as a parliamentary state within the Commonwealth of Nations finally came in 1980.

This young country faces several ethnic dilemmas. First, the future course of relations between the country's two largest ethnic groups — the majority Shona (75% of the total population) and the Ndebele (20%) — remains uncertain; many Ndebele people want autonomy for the Ndebele stronghold in the south known as Matabeleland. In addition, while most whites left Zimbabwe when it became independent, about 100,000 remained. Now about 70,000 strong, whites participated in Zimbabwe's political life, setting an early example to South Africa of an effective transition to majority black rule. But the apparent harmony has since shattered, particularly over the issue of land ownership. While millions of rural blacks are crowded onto agriculturally inferior "communal lands," whites own about 70% of Zimbabwe's arable land. A poorly handled effort to address this imbalance has recently threatened to tear the country apart.

The problem of disproportionate white land ownership became explosive in 1997, when Zimbabwe's President Robert Mugabe (of the Shona ethnic group) used the issue to boost his party's prospects in impending parliamentary elections. He announced that the government would seize half of Zimbabwe's white-owned farms, "not pay a penny for the soil," and distribute the land to black peasants living in the overcrowded communal areas. Mugabe argued that the former colonial power, Britain, having illegally occupied black lands in the first place, should now pay compensation to whites who lose those lands. Insisting that the seizures violated its agreements with the government, and angry about Zimbabwe's support of the Kinshasa government in Congo's civil war, the International Monetary Fund and most other foreign lenders cut off aid to Zimbabwe in 1999. An economic crisis followed; the value of the Zimbabwean dollar crashed, inflation soared, gold mines shut down, tourism all but disappeared, and unemployment reached an estimated 50% by 2001. Mugabe continued to encourage blacks to challenge whites and their assets in advance of the 2002 presidential election, which Mugabe won, probably fraudulently. Blacks, mostly either Mugabe loyalists or war veterans, moved onto about 1,800 white-owned farms. While there was some violence, an awkward cohabitation settled in on many farms pending a resolution of the crisis. Zimbabwe's Supreme Court ruled that the seizures were illegal, but Mugabe encouraged the squatters to stay. Few had any farming abilities or materials, and the country's agricultural output plummeted. Black militants also descended on white-owned factories. Some blacks have taken to killing game animals on white-owned private nature reserves. They resent that the whites can capitalize by selling hunting licenses for some game animals, while their own hardships only increase.

18.7   The Indian Ocean Islands

The islands and island groups off the Indian Ocean coast of Africa are unique in their cultures and natural histories. Madagascar, the Comoro Islands, Réunion, Mauritius, and the Seychelles have had African, Asian, Arab, European, and even Polynesian ethnic and cultural influences (Fig. 18.20). As island ecosystems, they are home to many endemic plant and animal species.

Madagascar and the smaller Comoro Islands are former colonies of France, and Réunion remains an overseas possession of France. Mauritius became independent of Britain in 1968 as a constitutional monarchy under the British Crown. The Seychelles, which gained independence in 1976, is a republic within the Commonwealth of Nations.

Madagascar (in French, La Grande Ile) is the fourth largest island in the world, nearly 1,000 miles (c. 1,600 km) long and about 350 miles (c. 560 km) wide. It lies off the southeast coast of Africa and has geological formations similar to those of the African mainland. Its distinctive flora and fauna include most of the world's lemur and chameleon species (see Definitions & Insights, p. 494). Some of Madagascar's early human inhabitants migrated from the southwestern Pacific region, bringing with them the cultivation of irrigated rice and other culture traits. There was also a later influx of Africans from the mainland. The official language of the country's 16.4 million people is Malagasy, an Austronesian tongue that has acquired many French, Arabic, and Swahili words.

The east coast of Madagascar rises steeply from the Indian Ocean to heights of over 6,000 feet (1,829 m). Because the island lies in the path of trade winds blowing across the Indian Ocean, the east side receives the heaviest rain and has a natural vegetation of tropical rain forest. The remainder of the island has vegetation consisting primarily of savanna grasses

Figure 18.20   The 16 million people of Madagascar have a unique culture, with roots in both Africa and the Malayo-Polynesian realm.

with scattered woody growth. An eastern coastal strip — low, flat, and sandy — is backed by hills and then by escarpments of the central highlands. Paddy rice is the principal food, and coffee, vanilla, cloves, and sugar represent most of the country's limited exports. There is a growing clothing export industry. The most densely populated part of the island is the central High Plateau, where the economy is a combination of rice growing in valleys and cattle raising on higher lands. Antananarivo (formerly Tananarive; population: metropolitan area, 1,328,900; city proper, 855,500), the capital and largest city, is located in the central highlands. It is connected by rail with Toamasina (formerly Tamatave), the main seaport, located on the east coast.

Madagascar is not as economically developed as other large tropical islands such as Java, Taiwan, and Sri Lanka. Deep economic troubles and political dissension have

DEFINITIONS + INSIGHTS

Madagascar and the Theory of Island Biogeography

Harvard University entomologist Edward Wilson and Stanford University biologist Paul Ehrlich have asserted that if tropical rain forests continue to be cut down at the present rate, a quarter of all of the plant and animal species on Earth will become extinct within the next 50 years.[a] They base their estimate on a model that correlates habitat area with the number of species living in the habitat. This **theory of island biogeography** emerged from observations of island ecosystems in the West Indies and states that the number of species found on an individual island corresponds with the island's area, with a tenfold increase in area normally resulting in a doubling of the number of species. If island A, for example, is 10 square miles in area and has 50 species, 100-square-mile island B may be expected to support 100 species.

What makes the theory useful in projecting species losses is the inverse of this equation: A tenfold diminution in area will result in a halving of the number of species. Therefore, 1-square-mile island C can be expected to hold only 25 species. In applying the model, ecologists treat habitat areas as if they were islands. Thus, if people cut down 90% of the tropical rain forest of the Amazon Basin, for example, the theory of island biogeography suggests they would eliminate half of the species of that ecosystem. Scientists caution that the theory is only a tool meant to help in making rough estimates; the actual number of species lost with habitat removal may be higher or lower.

As a rough guideline, the theory of island biogeography is useful in projecting and attempting to slow the rate of extinction in the world's biodiversity "hot spots," such as Madagascar. Over 90% of Madagascar's plant and animal species are endemic, occurring nowhere else on Earth. Extinction of species was well underway soon after people arrived on the island; the giant, flightless elephant bird (*Aepyornis*) was among the early casualties. But human activities, particularly the clearing of forests to grow rice and provide pasture for zebu cattle, are eliminating habitat areas on the island at a more rapid rate than ever before. Meanwhile, scientists are anxious to learn whether some of Madagascar's remaining plants might be useful in fighting diseases such as AIDS and cancer. Already Madagascar's rosy periwinkle has yielded compounds effective against Hodgkin's disease and lymphocytic leukemia. Other species like the rosy periwinkle could become extinct before their useful properties ever become known.

How urgent is the task to study and attempt to protect plant and animal species in Madagascar? Scientists turn to the theory of island biogeography for an answer. Although people have lived on Madagascar for less than 2,000 years, they have succeeded in removing 90% of the island's forest, setting the stage for some of the most ruinous erosion seen anywhere on Earth (Fig. 18.C). The theory of island biogeography suggests that in the process, they have caused the extinction of roughly half of the island's species. With Madagascar's human population expected to double in 23 years, and with pressure on the island's remaining wild habitats expected to increase accordingly, the task of conservation is extremely urgent.

JOE HOBBS

JOE HOBBS

Figure 18.C   The subsistence needs of a growing human population have had ruinous effects on Madagascar's landscapes and wildlife. Before people came some 2,000 years ago, most of the island was forested, as in the Montagne d'Ambre National Park (top). Today, less than 10% of the island is in forest, and a characteristic landscape feature of its High Plateau is the erosional feature known locally as *lavaka* (bottom).

[a] Edward O. Wilson, "Threats to Biodiversity." *Scientific American*, 261(3): 60–66, (1989).

marked its recent history under a socialist government. Unwise decision making made Madagascar fertile ground in the 1990s for international drug-money launderers, loan sharks, and con artists who purloined much of this poor nation's financial assets. In 2002, the contested results of a presidential election threatened to divide the country into two parts, one ruled from Antananarivo and the other from Toamasina. The political crisis virtually shut down the meager export industries that existed. Given Madagascar's extraordinary wildlife and scenic beauty, the ecotourism sector could grow considerably, but so far, the government has taken little interest in developing an infrastructure to encourage and sustain tourism.

The Comoro Islands, located about midway between northern Madagascar and northern Mozambique, are of volcanic origin. They have a complex mixture of African, Arab, Malayan, and European influences. Aside from the island of Mayotte, which has a Christian majority and which, by its own choice, remains a dependency of France, the Comoro group forms the Federal Islamic Republic of the Comoros (short form: Comoros), in which Islam is the official religion and both Arabic and French are official languages. The capital is Moroni (population: city proper, 40,200) on the island of Grand Comore. France has a defense treaty with the Comoros, which it invoked in 1995 to put down a coup led by a French mercenary. In 1997, two more Comoros islands (Mohéli and Anjouan) declared secession from the national government in Moroni and pleaded to become French dependencies on the hope they could enjoy the somewhat higher standard of living prevailing on nearby Mayotte. The islands' futures remained uncertain as of 2002.

Fishing, subsistence agriculture, and exports of vanilla provide livelihoods to the 600,000 inhabitants of the these islands. The coelacanth, a primitive fish thought to have become extinct soon after the end of the Mesozoic Era (the age of dinosaurs), has been observed alive in Comoro Islands waters since 1938 (and quite recently in Indonesian waters).

Réunion, east of Madagascar, is the tropical island home to 700,000 people, mainly of French and African origin. The island was uninhabited until the French arrived in the 17th century. Elevations reach 10,000 feet (3,048 m). The volcanic soils are fertile, and tropical crops of great variety are grown. Cane sugar from plantations supplies 60% of all exports by value, with rum and molasses (derived from cane) supplying an additional 3%. Réunion also has a significant international tourist industry.

Sugar is the mainstay of the larger island of nearby Mauritius. Grown mainly on plantations, sugarcane occupies most of the cultivated fields on this volcanic landscape. Mauritius has a larger proportion of lowland area than Réunion. The island has developed a clothing industry that supplies more export value than sugar. International tourism also produces valuable revenues (about $485 million annually) for its 1.2 million people. The people of Mauritius are mainly descendants of Indians brought in by the British to work the cane fields, along with descendants of 18th-century French planters as well as some blacks, Chinese, and racial mixtures. Mauritius is also home to about 5,000 Ilois, or "islanders," relocated from their homes in the Chagos Islands, 1,250 miles (2,000 km) to the northwest. The Chagos comprise a British protectorate known as the British Indian Ocean Territory (BIOT). The British administered the islands from Mauritius until 1965 when, in the height of the Cold War, it leased Diego Garcia (the largest island) to the U.S. military and required all the Ilois to move out. Since then, U.S. forces have used Diego Garcia on many military forays, particularly against Iraq during the Gulf War and against Afghanistan in the wake of the terrorist attacks on the United States in 2001. Now many of the Ilois want to return to their home islands, a move opposed by the United States.

The unusual core of the Seychelles island group is composed of granites, which typically are found only on continental mainlands. Many foods have to be imported to the islands, which are generally steep and lack cultivable land. The country's largest exports consist of petroleum products, although the islands have no crude petroleum and no refining industry; the products are simply imported and then reexported. Fish products are the only other important exports. The Aldabra atoll, far to the south of the main group of islands, is home to a population of giant tortoises and other endemic animals (Fig. 18.21).

A tourist boom began when an airport was built in 1971. The exceptional physical beauty of the islands and the renowned friendliness of its people of French, Mauritian, and African descent have made the Seychelles a prime tourist destination among those wealthy enough to afford travel to this remote area. Outsiders are drawn to Seychelles, as they have long been to much of Africa south of the Sahara, as an imagined Eden. The realities of land and life in the region exceed the imagination and are most worthy of the world's attention.

Figure 18.21   A giant Aldabran tortoise on the beach of Aldabra atoll in the Seychelles. As in the Galapagos Islands, unique life forms have evolved through long periods of isolation from other animal populations.

CHAPTER SUMMARY

- Africa south of the Sahara can be divided into seven subregions: West Africa, the Sahel, the Horn of Africa, East Africa, southern Africa, West Central Africa, and the Indian Ocean islands.
- Comprised of a great volcanic plateau and adjacent areas, the Horn of Africa includes the countries of Ethiopia, Eritrea, Somalia, and Djibouti.
- After nearly 30 years of terrorism, guerilla actions, and civil war, the former Ethiopian province of Eritrea became independent in May 1993.
- In addition to the devastation wrought by war with Eritrea, droughts centering in its war-torn north have stricken Ethiopia since the late 1960s. In 1984 and 1985, an estimated 1 million people perished when a drought reached the point of widespread catastrophe.
- The Sahel region extends eastward from the Cape Verde Islands to the Atlantic shore nations of Mauritania, Senegal, and The Gambia and inland to Mali, Burkina Faso, Niger, and Chad. The name Sahel in Arabic means "coast" or "shore," referring to the region as a front on the great desert "sea" of the Sahara.
- Since the late 1960s, the Sahel has been subjected to severe droughts, which, in combination with increased human pressure on resources, have prompted a process of desertification.
- West Africa is defined as the region extending from Guinea-Bissau eastward to Nigeria. Most West African countries maintain strong economic, political, and cultural relationships with the European powers that formerly controlled them.
- The West African countries' heavy dependence on subsistence crop farming, livestock grazing, and a few agricultural and/or export specialties is the customary economic pattern of the world's LDCs.
- Although a number of prominent mineral products are mined in West Africa, the most spectacular and significant mining development has been Nigeria's emergence as a producer and major exporter of oil.
- West Africa's principal cities are also national capitals and seaports, but as the region has only minor manufacturing, its cities have developed primarily in its areas of major commercial agriculture.
- The subregion of West Central Africa is flanked by Cameroon and the Central African Republic to the north and the Democratic Republic of Congo to the south.

- Subsistence agriculture supports people in most areas of West Central Africa. In addition, exports of cash crops and minerals aid the financing of some countries.
- Originally a colony of Belgium, the Democratic Republic of Congo has faced a number of challenges since its independence in 1960. In addition to civil war, the unstable economy of the country has cast the Democratic Republic of Congo into a state of extreme poverty despite its abundant natural resources.
- East Africa includes the countries of Kenya, Tanzania, Uganda, Rwanda, and Burundi.
- A wide array of crops is grown by the subsistence agriculturalists of East Africa. In addition, several export crops, such as coffee and tea, provide a source of income for the region's inhabitants. In comparison with many other African countries, however, the East African nations lack mineral reserves production.
- Ethnic rivalries and conflicts have beset all of the East African countries. Warfare between the Hutus and Tutsis of Rwanda and Burundi serves as a recent example of the tension that can occur between the many ethnic and tribal groups of the region.
- Southern Africa includes the countries of Angola, Mozambique, Zimbabwe, Zambia, Malawi, South Africa, Botswana, Swaziland, Lesotho, and Namibia.
- The dominant power in southern Africa is the country of South Africa. Its economic power, which is based on diversified industries and its mineral wealth, is now comparable to that of the rest of Africa combined.
- Racial segregation characterized South African life from 1652 onward, but it was first systematized after 1948 under a comprehensive body of national laws supporting the official governmental policy and commonly known as apartheid.
- In 1994, Nelson Mandela was elected president of South Africa after an all-race election was held in the country. After years of conflict, this historic election ended white European political control in the last bastion of European colonialism on the African continent. The apartheid laws became null and void. Reconciliation between the races is continuing in postapartheid South Africa.
- The Indian Ocean islands of Madagascar, the Comoro Islands, Réunion, Mauritius, and the Seychelles exhibit African, Asian, Arab, European, and even Polynesian ethnic and cultural influences.

REVIEW QUESTIONS

1. What are the seven subregions of Africa south of the Sahara? Be prepared to identify all of the countries in at least some of these subregions. What is the total number of countries in the region?
2. Why is Ethiopia known as the "Galapagos Islands of Religion"?
3. In what ways is the Sahel both a physical and cultural boundary zone in Africa?
4. What country is the leader in demographic, political, and economic clout in West Africa? What unique difficulties does that country face? Be prepared to discuss its recent turbulent political history.
5. What are the notable natural resource assets of the Democratic Republic of Congo? Why isn't this country richer than it is? What are its relationships with its neighbors?

6. Describe the ethnic diversity of East Africa.
7. What was the original distinction between the Hutus and the Tutsis and what issues brought them into recent conflict?
8. What are the main ethnic groups in South Africa and what have been their traditional social and economic roles?
9. What were the main causes of the Anglo-Boer War of 1899?
10. What are the notable natural resource assets of South Africa?
11. Describe the tensions between blacks and whites over the issue of land tenure in Zimbabwe.
12. Who owns and who uses the island of Diego Garcia? What is its importance in military affairs?

DISCUSSION QUESTIONS

1. Break into groups to study and discuss conflicts in some or all of the subregions. What is a common theme in these conflicts? What factors are unique?

2. What motivated Eritrea to attack Ethiopia in 1998?

3. What was the purpose of Operation Restore Hope? What was the outcome of the United States' intervention in Somalia?

4. What is desertification? What are the major causes of desertification?

5. Discuss the major complaints raised by the Ogoni and Ijaw tribal groups concerning oil development in Nigeria.

6. What is the "mistake of 1914"? Discuss the ethnic geography of Nigeria.

7. Discuss the major reasons for the Democratic Republic of Congo's present economic problems.

8. What African countries became involved in the Democratic Republic of Congo's recent civil war? What reasons motivated each of these countries to enter the conflict?

9. How has the Great Rift Valley affected the physical environment of East Africa?

10. How did European colonialism sow the seeds for violence between Hutus and Tutsis?

11. What are Namibia's difficult choices in trying to increase its access to fresh water?

12. How important is ecotourism in Africa south of the Sahara? Identify some of the major tourist destinations and their attractions.

13. What factors led to South Africa's advanced economic development compared to the rest of Africa?

14. Discuss the evolution and the annulment of the apartheid laws of South Africa. What are the major races in South Africa and how are they getting along today?

15. What is the theory of island biogeography? How exactly is the theory useful in understanding the relationship between habitat loss and species loss?

NORTH
AMERICA
Pittsburgh

to New York (3500 miles)

40°N

EUROPE

to English Channel (3600 miles)

AFRICA

20°N

0°

SOUTH
AMERICA

Natal

to Freetown (1550 miles)

20°S

40°S

RELATIVE POSITION
(Mercator projection)
Distances shown are nautical miles

100°W   80°W   60°W   40°W   20°W   0°   20°E

60°N

Latin America

**From the southern border of the United States all the way
to the South Pole, the region of Latin America extends across
an extraordinary range of latitudes, topographies, ethnicities,
and cultures. From a primary resource for major food and
mineral commodities to a demographic launching for
streams of migrants (both legal and illegal) moving toward
the United States and Canada, to emerging global industrial
centers, Latin America is growing into global significance.**

Before the arrival of Columbus, the peoples of Latin America had been enormously effective in the creation of their cultural landscapes. From Machu Picchu in eastern Peru to Aztec and Mayan ruins in Middle America, Latin America has many scenes that speak elegantly of the high levels of technology and social organization of the pre-Columbian peoples.

Latin America has played the role of major resource supplier for centuries. From the immediacy of coffee in our everyday lives, to exotic as well as common minerals, the world has drawn off the Latin America resource base to great benefit.

Although there are many parts of Latin America that have only the most sparse evidence of human settlement, the cities are densely packed. From the favelas of Brazil to the barrios of Mexico, the urban landscape has striking examples of the tenacity people show in their desire to get to the city and try their luck, even though their accommodations seem so very difficult.

Of the various demographic shifts that have characterized the history of Latin America, one of the most persistent migration streams of the past four or five decades has been the steady northward migration of peoples. These migration streams have been both legal and illegal, but all have been optimistic and represented steady cultural diffusion.

chapter 19

A Geographic Profile of Latin America

TOM VAN SANT/GEOSPHERE PROJECT, SANTA MONICA/SPL/PHOTO RESEARCHERS, INC.

This satellite image of the breadth and length of Latin America shows the rich diversity of environmental settings in this world region. Cultures and countries on this continent express the same diversity.

The land portion of the Western Hemisphere south and southeast of the United States is commonly known as Latin America. This region has a close and long-term relationship with the United States. The name reflects the importance of culture traits inherited primarily from the Latin European colonizing nations of Spain and Portugal (Fig. 19.1). Spanish is the major language, and Portuguese is the language only of the region's largest country, Brazil. Dialects of French are spoken in Haiti and in France's dependencies of French Guiana, Guadeloupe, and Martinique. Some small units in the Caribbean Sea or on the nearby mainland exhibit Latin culture only in minor ways. They speak mainly English or Dutch, and some are still dependencies of the United Kingdom, the United States, or the Netherlands (Table 19.1).

F21.12
567

Figure 19.1   Roman Catholicism is the leading religion in all Latin American countries except some relatively small units colonized by countries other than Spain or Portugal. The old Cathedral of Lima, Peru, is one of the more majestic among great numbers of Catholic church buildings spread through Latin America from Mexico and Cuba to Argentina and Chile.

Their presence in an area of primarily early Spanish control reflects early colonial struggles among European nations initially for territories suitable for sugarcane plantations and other colonial ventures. Important precolonial influences in some Latin American countries are manifest in Native American languages, customs, and costumes.

But Latin influences are predominant today in the region as a whole and can be seen not only in the nearly universal use of Latin-derived (Romance) languages but also in the dominance of the Roman Catholicism that arrived with the first ships from Latin Europe. This religion has been one of the most powerful cultural influences diffused to the so-called New World in the centuries that followed the 1492 arrival of Columbus and the subsequent drive for "Glory, God, and Gold" that led to centuries of exploration, colonization, exploitation, and settlement. However, language and religion are only the most obvious cultural importations, which also included such major elements as land tenure arrangements, governmental practices, legal systems, social structures, and economic systems.

European cultures have been transplanted to Latin America with vital modifications. Underneath a veneer of sameness, promoted by the widespread use of the Spanish language, important distinctions exist between regions, among countries, and within countries. Among the many contributing factors for regional differentiation are diverse pre-European and colonial experiences, different resource bases, distinct percentages of Native American populations, divergent political systems, and differential rates of development. The result is that Latin America presents many faces to the world, including:

1. Extraordinary variety in environmental settings;
2. Population groups of considerable diversity;
3. Local economies that run the scale from technologically advanced to extremely underdeveloped and that range from being very closely linked to North America to being largely self-sufficient at the local level;
4. Diverse governmental philosophies in political control and structures, characterized by relatively frequent shifts;
5. Countries that have distinctive personalities despite enough overall similarity and cohesion to fit together reasonably well in an identifiable major world region; and

501

Table 19.1 Latin America: Basic Data

Political Unit	Area (thousand/sq mi)	Area (thousand/sq km)	Estimated Population (millions)	Estimated Annual Rate of Increase (%)	Estimated Population Density (sq mi)	Estimated Population Density (sq km)	Human Development Index	Urban Population (%)	Arable Land (%)	Per Capita GDP PPP ($US)
Middle America										
Mexico	756.1	1958.2	99.6	1.9	132	51	0.79	74	12	9100
Guatemala	42.0	108.9	13	2.9	309	119	0.626	39	12	3700
Belize	8.9	23.0	0.3	1.9	29	11	0.776	49	10	3200
El Salvador	8.1	21.0	6.4	2.3	788	304	0.701	58	27	4000
Honduras	43.3	112.1	6.7	2.8	155	60	0.634	46	15	2700
Nicaragua	50.2	130.0	5.2	3	104	40	0.635	57	9	2700
Costa Rica	19.7	51.1	3.7	1.8	188	73	0.821	45	6	6700
Panama	29.2	75.5	2.9	2.1	100	39	0.784	56	7	6000
Cuba	42.8	110.9	11.3	0.6	264	102	NA	75	24	1700
Dominican Rep.	18.8	48.7	8.6	2.1	456	176	0.722	61	21	5700
Haiti	10.7	27.7	7	1.7	650	251	0.467	35	20	1800
Jamaica	4.2	11.0	2.6	1.5	624	241	0.738	50	14	3700
Trinidad and Tobago	2.0	5.1	1.3	0.7	656	253	0.798	72	15	9500
Bahamas	5.4	13.9	0.3	1.5	58	22	0.82	84	1	15,000
Barbados	0.2	0.4	0.3	0.5	1620	625	0.864	38	37	14,500
Antigua and Barbuda	0.2	0.4	0.1	1.6	394	152	NA	37	18	8200
Dominica	0.3	0.8	0.1	0.8	262	101	NA	71	9	4000
Grenada	0.1	0.3	0.1	1.3	678	262	NA	34	15	4400
St. Kitts-Nevis	0.1	0.4	0.04	0.9	281	108	NA	43	22	7000
St. Lucia	0.2	0.6	0.2	1.3	656	253	NA	30	8	4500
St. Vincent and the Grenadines	0.2	0.4	0.1	1.2	757	292	NA	44	10	2800
Puerto Rico	3.5	9.0	3.9	0.8	1139	440	NA	71	4	10,000
Guadeloupe	0.7	1.7	0.5	1.2	691	267	NA	48	14	9000
Martinique	0.4	1.1	0.4	0.8	897	346	NA	93	8	11,000
Netherland Antilles	0.3	0.8	0.2	1.1	722	279	NA	70	10	11,400
Total	1047.6	2713.0	174.84	1.9	167	64	0.743	65.5	12.5	6979
South America										
Brazil	3300.2	8547.4	171.8	1.5	52	20	0.75	81	5	6500
Colombia	439.7	1138.9	43.1	1.8	98	38	0.765	71	4	6200
Venezuala	352.1	912.1	24.6	2	70	27	0.765	87	4	6200
Ecuador	109.5	283.6	12.9	2.2	118	46	0.726	62	6	2900
Peru	496.2	1285.2	26.1	1.8	53	20	0.743	72	3	4550
Bolivia	424.2	1098.6	8.5	2.4	20	8	0.648	63	2	2600
Argentina	1073.5	2780.4	37.5	1.1	35	14	0.842	90	9	12,900
Chile	292.1	756.6	15.4	1.3	53	20	0.825	86	5	10,100
Uruguay	68.5	177.4	3.4	0.7	49	19	0.828	92	7	9300
Paraguay	157.0	406.7	5.7	2.7	36	14	0.738	52	6	4750
Guyana	83.0	215	0.7	1.3	8	3	0.704	36	2	4800
French Guiana	34.7	90	0.2	2.3	6	2	NA	79	0	6000
Suriname	63.0	177.4	0.4	1.9	7	3	0.758	69	0	3400
Total	6893.7	17,869.3	350.3	1.6	51	20	0.763	79.1	5.1	6904
Summary Total	7941.4	20,568.2	525.1	1.7	66.1	25.5	0.757	74.6	6.1	6930

Sources: *World Population Data Sheet, Population Data Sheet, 2001; U.N. Human Development Report,* United Nations, 2001; *World Factbook,* CIA, 2001.

6. A mainland region that extends from Mexico south to Argentina and Chile and an island world that is also called Caribbean Middle America.

19.1 Area and Population

With a land area of slightly more than 7.9 million square miles (20.5 million sq km), Latin America — the combination of Middle America and South America — is surpassed in size by Africa, Russia and the Near Abroad, and North America (if you include Greenland) (see Table 1.2). However, its maximum latitudinal extent of more than 85 degrees, or nearly 5,900 miles (c. 9,500 km), is greater than that of any other major world region. Its maximum east–west measurement, amounting to more than 82 degrees of longitude, is also impressive. Yet Latin America is not so large as these statistics might suggest because its two main subregions are offset from each other (Fig. 19.2). The northern part, known as Middle America (see Chapter 20) — which includes Mexico, Central America (Guatemala, Belize, El Salvador, Honduras, Nicaragua, Costa Rica, and Panama), and the islands of the Caribbean — trends sharply northwest — southeast from the north–south orientation of the continent of South America (Fig. 19.3). The latter is thrust much farther into the Atlantic Ocean than is the Caribbean realm or Latin America's northern neighbor, North America. In fact, the meridian of 80°W, which intersects the west coast of South America in Ecuador and Peru, passes through Pittsburgh, Pennsylvania. Brazil also lies less than 2,000 miles (3,200 km) west of Africa (see inset, Fig. 19.2). In area, Latin America totals approximately 7,941,000 square miles (20,568,000 sq km), with South America having 6,893,000 square miles (17,869,000 sq km) and Middle America having 1,047,000 square miles (2,713,000 sq km). This enormous continent possesses a wide range of climates and environments (Fig. 19.4).

Population Geography: Rimland-Mainland Pattern, Diverse Stocks, and Expanding Cities

In mid-2001, world population had grown beyond 6.1 billion people, and Latin Americans totaled approximately 525 million, or almost 8.6% of the world's population. Because Latin America is a large and diverse region with more than twice the land area of the United States, it is not surprising that its population is unevenly distributed and exhibits enormous variations in density. There is, however, some overall order in the fact that most of Latin America's people are packed into two major geographic alignments. The larger of these two areas, called the rim or Rimland, is a discontinuous ring around the margins of South America. The second, a relative highland, extends along a volcanic belt from central Mexico southward into Central America. The two alignments are often defined as the Rimland and Mainland — meaning the coastal regions in contrast to the interior and upland regions — pattern of Latin American population distribution and are outlined by the pattern of population shown in Figure 19.5.

Major Populated Areas

The South American rim of clustered population contains roughly two-thirds of Latin America's people. Two major segments of the rim can be discerned. One is much the larger in population and area. It extends along the eastern margin of the continent from the mouth of the Amazon River in Brazil southward to the humid pampa (subtropical grassland) around Buenos Aires, Argentina. The second segment is located partly on the coast and partly in the high valleys and plateaus of the adjacent Andes Mountains. It stretches around the north end and down the west side of South America. This crescent begins in the vicinity of Caracas, Venezuela, on the Atlantic coast and arches all the way around to the vicinity of Santiago, Chile, on the Pacific side of South America.

This second segment is more fragmented than the first, as it is broken in many places by steep Andean slopes or coastal desert. For example, a strip of hot, rainy coast lies between the Amazon mouth and Caracas in which the thinly populated interior of South America extends to the ocean. In the far south on the Atlantic side of the continent, the population rim is again broken in the rugged, relatively inaccessible, rainswept southern Andes and the dry lands of Argentina's Patagonia that lie in the Andean rain shadow. These latter territories have only a scanty human population, although millions of sheep graze in Patagonia and adjacent Tierra del Fuego (see Fig. 19.5).

It is thought that few Native Americans lived in the Rimland along the Latin American coasts before the European conquest. Major population centers prior to 1492 were located in the Andes in South America, and with the exception of Peru, the coasts were in general only lightly inhabited by migratory native peoples. The high intermontane areas within the Andes already exhibited a pattern of settled communities, ruled by the Incan empire that extended 2,500 miles (4,020 km) from southern Colombia through the western parts of Ecuador, Peru (with Cuzco as the Incan capital), Bolivia, Chile, and into the western edge of Argentina (Fig. 19.6). But when the Europeans began their quest for and settlement of South America in the early 16th century, the coasts largely lay open to whatever settlement pattern the newcomers might devise. Because the Europeans approached from the sea at many separate points, the development of ports as bases for subsequent penetration of the interior was the first major settlement task. Many early ports eventually grew into sizable cities, and a few became major metropolitan centers. In some instances — notably Lima, Caracas, and Santiago — the main city developed a bit inland but retained a close connection with a smaller coastal city that was the actual port.

Around the ports, agricultural districts developed, spread, and shipped an increasing volume of trade products overseas.

Figure 19.2   Index map of Latin America.

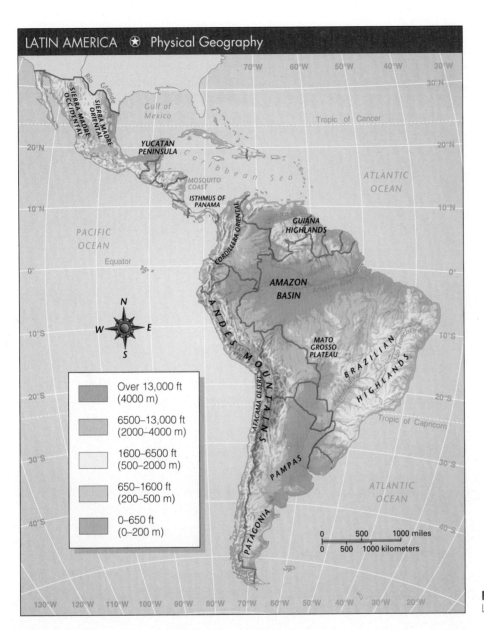

LATIN AMERICA ✪ Physical Geography

Legend:
- Over 13,000 ft (4000 m)
- 6500–13,000 ft (2000–4000 m)
- 1600–6500 ft (500–2000 m)
- 650–1600 ft (200–500 m)
- 0–650 ft (0–200 m)

Figure 19.3   A map of physical features in Latin America.

The ships that took these goods away returned with more from Europe. They also carried passengers, including government officials, in both directions. In the second half of the 1500s, slave ships began to bring the Africans who provided the principal labor on European initiated and owned plantations, primarily in the tropical lowlands with the generally fertile alluvial soils of the coastal plains. Populations of considerable variety along or near the coasts multiplied and continue to do so today.

Meanwhile, the powerful lure of gold and silver stimulated the penetration of the Andes and the Brazilian Highlands in South America. After capturing the stores of precious metals that had been accumulated by the indigenous peoples for ceremonial use, tribute, and trade, the European newcomers opened mines, or in many places took over old mines, and market centers arose to service the mines. Some highland settle-

ments gained a wider importance as centers for new ranching or as plantation areas. Such functions still endure, although many older mines have closed. Mining today is far more diversified and less dependent on gold and silver than was true in the early colonial age. A few highland cities, of which the largest is Bogotá in Colombia, have grown into sizable metropolises. These places are separated from the seaports by difficult terrain that required feats of considerable engineering skill before satisfactory transportation links between the coasts and upland urban centers were developed. Some of the main seaports and highland cities became important centers of colonial government as well as economic nodes, and several are national capitals today, such as La Paz in Bolivia and Quito in Ecuador. Rapid development of manufacturing industries and explosive population growth have characterized such cities since the 1980s (see Fig. 19.2).

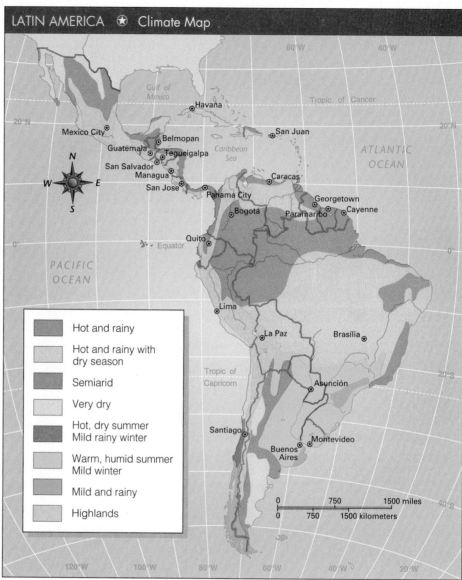

Figure 19.4 A map of climate patterns in Latin America.

Source: Rand McNally, *Atlas of World Geography.*

The second major alignment of populated areas in Latin America lies on the mainland of Middle America to the northwest of the South American Rimland. Composed of the majority of Middle America's people, it extends along an axis of volcanic land that dominates central Mexico and from there reaches southeastward through southern Mexico and along the Pacific side of Central America to Costa Rica (see Fig. 19.5). This belt is characterized by good soils, rainfall adequate but not excessive for crops, and enough elevation in most places to moderate the tropical heat but still permit the raising of significant tropical crops such as coffee. The belt of high population density that extends across central Mexico from the Gulf of Mexico to the Pacific Ocean was already Mexico's center of population when the Spanish conquerors came.

This was the principal domain of the large and technologically advanced Aztec empire whose most important period was approximately A.D. 1300 to 1500. Together with its vassal states, it composed much of Middle America's population in the 16th century (Fig. 19.7). Prior to the Aztecs, the Maya (to the east) had been another dominant and sophisticated Amerindian population. The Maya had their origins as early as 1500 B.C. and experienced their most powerful period from about A.D. 300 to 900 (Fig. 19.8). Then, as today, the soils of this area's high-volcanic plateaus and flat-floored basins provided the foundation for a productive agriculture based primarily on maize. Spanish occupation did not alter the dominance of the area within Mexico, and roughly half of Mexico's 99.6 million people (2001) now live in, around, or between the region's two principal cities of Mexico City and Guadalajara to the north.

LATIN AMERICA ✪ Population Distribution

One dot represents 100,000 people

Figure 19.5 This map shows the power of the Rimland in the distribution of Latin American populations. It highlights the importance of upland locations as well, particularly in the northwest sector of South America and in Mexico. The many open, sparsely settled regions continue to be looked at by government planners as possible frontiers for new agricultural development and human settlement. However, such areas also serve as goal areas for everything from swidden agriculture to forest clearance for beef raising. The area in the breadth of the Amazon Basin in Brazil is particularly susceptible to such transformations.

In Pacific Central America also, the basic pattern of population distribution was established when the Spanish took over. Today, the majority of people live in highland environments, but considerable numbers live in lowlands as well. The ongoing vulcanism of the highlands periodically deposits ash into the highland basins, and this weathers into fertile soils that grow maize, beans, and squash for local food and the coffee that is Central America's largest export crop. Between the highlands and the Pacific Ocean lies a coastal strip of seasonally dry lowlands where cattle and cotton are grown largely for an export market. By contrast, the Atlantic side of Central America is lower, hotter, rainier, and far less populous than the Pacific side. Here, on large corporate-owned plantations, are grown most of the bananas that are Central America's agricultural trademark and second most valuable export after coffee.

Variations in Population Density

Average population densities for entire countries conceal the fact that in nearly every Latin American mainland country

Figure 19.6 One of the most dramatic landscapes of South America is the terraced city of Machu Picchu, also known as the Lost City of the Incas. This highland (8,000 ft/c. 2,400 m) city and ceremonial center of the Inca is high above the Urubama Valley in Peru on the east side of the Andes. This site serves both as a major tourist destination and an archeological treasure of major consequence. It became first known to the West in 1911 when it was "discovered" by a researcher on an exploration funded by the National Geographic Society.

there is a basic spatial configuration consisting of a well-defined population core (or cores) with an outlying, sparsely populated hinterland (see Fig. 19.5). Densities within the cores are often greater than anything to be found within areas of comparable size in the United States. On the other hand, most of the mainland countries have a larger proportion of sparsely populated terrain than is true of the United States. Figures on population density for the greater part of Latin America are extraordinarily low, averaging under two persons per square mile over approximately half of the entire region.

Figure 19.7 The Pyramid of the Sun in the pre-Aztec center of Teotihuacán in the Valley of Mexico is a bold reminder of the architectural skill local Middle American cultures had prior to the arrival of Cortez and the West. This pyramid covered 10 acres (4 hectares) at its base and was the dominant structure in a ceremonial and commercial city complex that was about 7 square miles (18 sq km) in extent and reached its zenith in approximately A.D. 300 to 900. It was virtually abandoned at the time of the Spanish conquest of Mexico in the early 16th century.

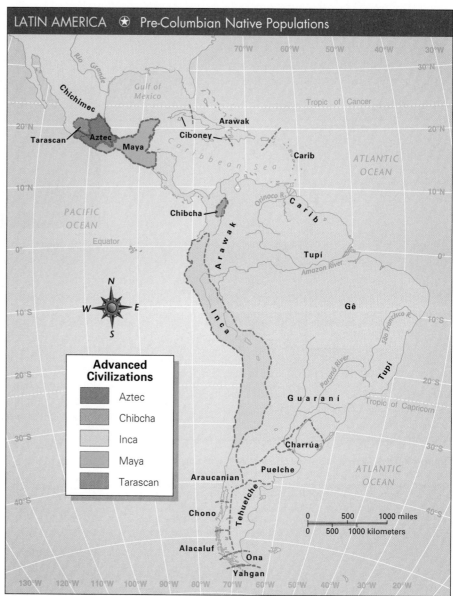

Figure 19.8 This generalized distribution of the pre-Columbian native populations in South America, Meso-America, and the Caribbean shows that a wide range of environments were settled before the very unsettling influences of the Europeans arrived on the scene in 1492.

Source: Modified from Clawson, David L. 1997. *Latin America and the Caribbean*. Wm C. Brown. p. 134.

This pulls down Latin America's average density — and that of most major Latin American countries — to levels somewhat below the average density for the United States (77 people per sq mi/199 people per sq km).

The pattern of core versus the hinterland is particularly evident in Brazil and Argentina. Most Brazilians, for example, live along or very near the eastern seaboard south of Belem, a port city in northeastern Brazil, whereas large areas in the interior are still thinly populated. In Argentina, approximately three-fourths of the population is clustered in Buenos Aires or the adjacent humid pampas — an area containing a little over one-fifth of Argentina's total land. The less densely settled regions in Latin America are generally quite distinctive from the core populations. There is a steady migration from the countryside to the Rimland. However, there is very little reverse migration of the core population going to the hinterland, with its relatively more demanding and less convenient setting for people who have grown accustomed to the environmental and urban amenities of the coastal settings. In Latin America, population patterns in Brazil, Argentina, and Mexico as well as in the Andean countries reflect this demographic imbalance (see Fig. 19.5).

Not every country, however, displays the pattern of core versus hinterland. A very different picture of population density and distribution is evident in El Salvador and Costa Rica (see Definitions & Insights) — both of which are composed of thickly settled volcanic land — and in most Caribbean islands (see Table 19.1). Because the combined population of the islands in the Caribbean is considerably less than that of central Mexico alone, these islands do not contain a concentration of

Imagine a map of the Pacific Ocean and the continents and islands at its eastern and western edge. Now, in your mind, superimpose a large horseshoe atop that map. Have one end of the horseshoe resting on Tierra del Fuego at the very tip of South America. Have the arch of the horseshoe cover the Coastal Ranges of western North America and cover the mountains of Alaska, especially the Aleutian chain that goes northeast-southwest from the mainland down toward East Asia. Then, extend the left side of the horseshoe across the island archipelagos of East Asia, all the way to the islands of Indonesia in Southeast Asia. This horseshoe approximates what is called the Ring of Fire. It is a rough line of approximately 600 active volcanoes and some of the most seismically instable landscape on the surface of Earth.

The Ring of Fire exists because of the particular geography of the tectonic plates — sometimes called lithospheric plates — that define the substructure of Earth's surface. These massive plates are like a jigsaw puzzle that make up the surface on which the continents and ocean surfaces rest. Along the margins of the Pacific Ocean lie major plates that are slipping under continental plates of South and North America, as well as of Eurasia and Indo-Australia.

Earthquakes that include San Francisco (1906), the powerful Alaska (1964) and Mexico City (1965), the Kobe (1995), and

the two in El Salvador (end of 2000 and very beginning of 2001) relate to the instability of this Ring of Fire zone. The enormous explosion of Mt. Saint Helens in 1986, Mt. Pinatubo in the Philippines in 1990, and the 2000 volcano in Hawaii also relate to and derive from this same set of seismic conditions.

The alarming aspect of this natural phenomenon, not only in Latin America but also in the United States and East (and Southeast) Asia, is that humans have made decisions for settlement that fly in the face of this seismic reality. The volcanoes generally provide some warning to humans of forthcoming eruptions and explosion. However, earthquakes are still almost impossible to predict, and as a result, there is always the potential for major human loss of life as quakes move through these heavily settled areas on the various fault systems (e.g., the San Andreas in California) that underlie the dense populations of South, Middle, and North America on the east side of the Pacific Ocean and East Asia and Southeast Asia on the western side of this Ring of Fire.

The political dynamics of warning people about these hazards before houses and other structures are built and then dealing with the crisis of evacuation and reconstruction after earthquakes or volcanic disturbances are all highly complex but also highly important aspects of human settlement all along the Ring of Fire.

people comparable in scale to either of the two major Latin American concentrations — the Rimland and Mainland — discussed earlier. However, they do stand out for exceptionally heavy densities, which have been created over a long period by the accumulation of expanding population on territories of very restricted size. Most often, the crowding is aggravated by high rates of natural population increase and/or the prevalence of steep slopes that restrict possibilities for further farming or settlements.

Ethnic Diversity

Despite the overall dominance in Latin America of culture traits derived from Europe, only three of this region's 38 nations — Argentina, Uruguay, and Costa Rica — have preserved white European racial strains on a large scale with little admixture by native people or blacks. This is due in part to the fact that these nations received relatively large numbers of European immigrants during the late 19th and early 20th centuries. Scattered districts in other countries also are predominantly European or profoundly Amerindian (Native American).

Various sources estimate that 40 to 82% of the population in highland nations of South America are composed of Native Americans. Bolivia at 82% and Ecuador at 60% are estimated to be the highest. Native people are a significant minority

overall in Mexico at 25 to 30%, but they comprise over 50% in some southern states of the country. Guatemala is more than 40% Native American overall.

Native Americans are also a major population element in the basin of the Amazon River and in Panama, where scattered lowland native people maintain their aboriginal cultures. However, the outer world is pushing in on them despite some efforts on the part of governments to protect their ways of life. Over the centuries after Columbus, the encounters of lowland natives with Europeans generally had catastrophic results for the aborigines, who often resisted exploitation fiercely but were no match for the Europeans' diseases, guns, liquor, and relentless hunger for more land, minerals, and religious converts. The more numerous highland native people, who were more advanced in technology, more distant, and better organized than the lowlander Amerindians, were somewhat more successful in dealing with European encroachment and exploitation but still suffered enormous population losses at the hands of their discoverers. Disease was the most active killer because none of the native populations of the so-called New World had any immunity to common European diseases, especially smallpox which was the greatest killer.

Black Latin Americans of relatively unmixed African descent are found in the greatest numbers on the Caribbean islands and along hot, wet Atlantic coastal lowlands in Middle

Box
538

and South America. These are the areas to which African slaves were brought during the colonial period, primarily as a source of labor for sugar plantations. Slavery was gradually abolished during the 19th century, although not until the 1880s in Brazil and Cuba. By that time, slavery had generated large fortunes for many owners of plantations and slave ships. The African peoples it introduced to the region have made cultural contributions that today provide a varied, colorful, and important part of Latin American civilization.

Latin America has escaped many of the racial tensions that grip much of the world. This results in part from the fact that the majority of Latin Americans are themselves of a mixed racial composition. Much of the region exhibits a primary mixture of Spanish and Native Americans, resulting in a heterogeneous group known as **mestizos. Mulatto** is the term meaning a person of African (black) and European (white) ancestry.

Population Growth and Expansive Urbanization

Irrespective of ethnic composition and type of culture, most Latin American countries today find themselves in the second stage of what has been called the demographic transition (see Chapter 2). Startling rates of urban and metropolitan growth have been a major Latin American characteristic during recent times. Recent estimates (2001) give a figure of 74% urban for the population of the region as a whole, compared to a world average of 46%. A major element in the rapid growth of Latin American cities continues to be a huge and continuing migration from the Latin American countryside.

In any consideration of human migration, there is a discussion of the environmental and social factors that cause someone to leave an area (the push factor) and the corresponding factor that attracts (the pull factor) someone toward a specific locale. Generally, economic factors are given the most importance in assessing migration flows, particularly those from the countryside to the city. However, even if there is no certainty about employment, very often a young man or woman will leave a position of chronic underemployment in the rural sector and go to the city hoping for new opportunities.

Very often, this critical migration ends with economic failure, but the pressure to hide such lack of success is so strong that there is very little reverse migration, or return to the countryside. Even where there is active industrial growth and employment opportunity, such as in Mexico City and São Paulo, Brazil (two of the four largest cities in the world), there are many more migrants and applicants than there are factory or service-sector positions. As a result, Latin American slums grow with a steadiness that challenges every urban planner and municipal administrator.

In Latin America, the urban explosion is especially problematic in the major urban centers in the Rimland on the margins of the agricultural lowlands in both South and Middle America. Every large city has big slums, which often take the form of ramshackle shantytowns on the urban periphery. Sometimes they are built on hillsides overlooking the city, as

DAVID R. FRAZIER/PHOTO RESEARCHERS, INC.

Figure 19.9   Many of Latin America's poorest people live in ramshackle housing that clings to nearly vertical slopes. This shantytown overlooks scenic Rio de Janeiro; the locale is the suburb of Santa Marta. Such impromptu settlements serve as another indication of the steady migration stream of rural folk into the city in quest of the alleged good life rumored to exist in Latin American cities.

in Rio de Janeiro, where some of the world's more unsightly housing (Fig. 19.9) — called *favelas* in Brazil and **barrios** in Spanish-speaking countries — commands one of the world's more imposing views. Such shantytowns, which are practically universal in the world's less developed countries (LCDs), are full of underemployed and ill-fed people who still may prefer their present plight to the conditions of the depressed rural landscapes from which they came. Such makeshift settlements do, however, keep new migrants seeking work resident in the hub of economic activity. Thus, these places are sometimes called "slums of hope."

Finally, the overall rate of increase for the population of Latin America averages 1.7% — with the world average being 1.3% — but there is a wide range of rates from country to country, as can be seen in Table 19.1.

19.2 Physical Geography and Human Adaptations

Topographic and Climatic Variations

Latin America is characterized by pronounced differences in elevation and topography from one area to another (see Fig. 19.3). Low-lying plains drained by the Orinoco, Amazon, and Paraná-Paraguay river systems dominate the north and central part of South America and separate older, lower highlands in the east from the rugged Andes of the west. In Mexico, a high interior plateau broken into many basins lies between north–south trending arms of the Sierra Madre Mountains. High mountains within Latin America, largely contained within the Sierra Madre and the Andes, form a nearly continuous landscape feature from northern Mexico and the southern United States south to Tierra del Fuego at the southern tip of South America. Most of the smaller islands of the West Indies are volcanic mountains, although some islands composed of limestone or coral are lower and flatter. The largest islands of the Caribbean have a more diverse topography, including low mountains.

Humid Lowland Climates

Most Latin American lowlands are associated with a dominant form of natural vegetation or ecosystem. The tropical rain forest climate, with its heavy year-round rainfall, monotonous heat and humidity, and associated superabundant vegetation dominated by large broadleaf evergreen trees, lies primarily along or very near the equator, although segments extend to the margin of the tropics in both the Northern and Southern Hemispheres. The largest segment lies in the basin of the Amazon River system. Additional areas are found in southeastern Brazil, eastern Panama, the western coastal plain of Colombia, on the Caribbean side of Central America and southern Mexico, and along the eastern (windward) shores of some Caribbean islands, particularly Hispaniola (Haiti and the Dominican Republic) and Puerto Rico (see Fig. 19.4).

On either side of the principal region of tropical rain forest climate, the tropical savanna climate extends to the vicinity of the Tropic of Capricorn in the Southern Hemisphere and, more discontinuously, to the Tropic of Cancer in the Northern Hemisphere. In this climate zone, the average annual precipitation decreases and becomes more seasonal as one approaches the poles. The mean temperature decreases, and the broadleaf evergreen trees of the rain forest grade into tall savanna grasses, woodlands, or deciduous forests that lose their leaves in the dry season.

Still farther poleward in the eastern portion of South America lies a sizable area of humid subtropical climate, which has cool winters unknown to the tropical zones. Its Northern Hemisphere counterpart is north of the Mexican border in the southeastern United States. In South America, this climate is associated principally with a natural vegetation of prairie grasses in the humid pampas of Argentina, Uruguay, and extreme southern Brazil. On the Pacific side of South America, a small strip of Mediterranean or dry-summer subtropical climate in central Chile is similar to that in southern California.

Dry Climates and the Factors That Produce Them

The humid climates of Latin America have a more or less orderly and repetitive spatial arrangement. One may expect to find similar climates in generally similar positions on all major land masses of the world. But the region's dry climates — desert and steppe — are often the product of, at least in part, local circumstances such as the presence of high landforms that block moisture-bearing air masses or the presence of offshore cool ocean water. In northern Mexico, such climates are partly associated with the global pattern of semipermanent zones of high pressure that create arid conditions in many areas of the world along the Tropics of Cancer and Capricorn. However, the aridity is also due in part to the rain-shadow effect caused by the presence of high mountain ranges on either side of the Mexican plateau.

In Argentina, the extreme height and continuity of the Andean mountain wall are largely responsible for the aridity of large areas, particularly the southern region called Patagonia. Here, the Andes block the path of the prevailing westerly winds, creating heavy orographic precipitation on the Chilean side of the border but leaving Patagonia in rain shadow. But in the west coast tropics and subtropics of South America, the Atacama and Peruvian deserts cannot be explained so simply. Here, shifting winds that parallel the coast, cold offshore currents, and other complexities, as well as the Andes Mountains, are important conditions that combine to create the world's driest area. However, the mountains serve to restrict this area of desert to the coastal strip. Arid and semiarid conditions are also experienced along the northernmost coastal regions of Colombia and Venezuela, and in the Sertão of northeastern Brazil.

Mountains play such a major role in the development of climate patterns and overall demographic patterns of Latin America that a whole pattern of regionally unique zonation has been developed for the region. It is built around **altitudinal zonation** and is characterized by four dominant regional zones: the **tierra caliente, tierra templada, tierra fría,** and **tierra helada.** These are described and shown graphically in the Regional Perspective on page 512.

19.3 Cultural and Historical Context

One of the frustrations that is associated with geographic nomenclature is the traditional use of the term "New World" for the combination of South, Middle, and North America that Columbus found and caused to be explored from 1492 on. It is thought that there were between 12–50 million native "Indians," or aboriginal people, in this world that was

REGIONAL PERSPECTIVE   :::   Altitudinal Zonation in Latin America

One of the most significant features of Latin America's climatic pattern with respect to the distribution of economic activity and population is a series of highland climates arranged into zones by elevation. This zonation results from the fact that air temperature decreases with elevation at a normal rate of approximately 3.6°F (1.7°C) per 1,000 feet (304.8 m). At least four major zones (Fig. 19.A) are commonly recognized in Latin America: the *tierra caliente* (hot country), the *tierra templada* (cool country), the *tierra fría* (cold country), and the *tierra helada* (frost country). At the foot of the highlands, the tierra caliente is a zone embracing the tropical rain forest and tropical savanna climates discussed earlier. The zone reaches upward to approximately 3,000 feet (914.4 m) above sea level at or near the equator and to slightly lower elevations in parts of Mexico and other areas near the margins of the tropics. In this hot, wet environment are grown crops such as rice, sugarcane, bananas, and cacao, often on plantations. Populations of black Africans, brought to the New World as slaves, are concentrated in many of the tierra caliente zones.

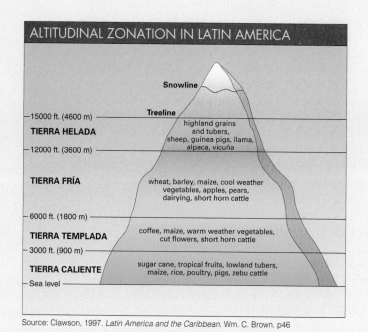

Source: Clawson, 1997. *Latin America and the Caribbean*. Wm. C. Brown. p46

Figure 19.A   A map of altitudinal zonation in Latin America. Because of the profound differences in relief in the landscapes of Latin America, specific patterns of land use have evolved in relation to differing altitudinal zones. This map shows their elevations and general land use patterns.

(Continued)

found by the *Niña*, *Pinta*, and the *Santa Maria*.[1] While it is almost impossible to determine just how many people lived in these two continents before the enormous 16th- and 17th-century population drop began to occur — through disease and terminal response to new mining demands made by Europeans on Native Americans — it is important to have an idea of some of the most important pre-Columbian population groups.

The Maya were an early group who lived mostly on the Yucatán Peninsula — part of Mexico, Belize, and Guatemala today — and who came initially from northern Central America. Their origins go back as far as 3000 B.C., and they practiced agriculture and had a highly organized religious system. They

created monumental religious structures, including pyramids, temples, observatories, and associated highway systems. The contemporary Mayan ruins of Chichen-Itza in the Yucatán Peninsula of Mexico and the Tikal ruins of Belize are today important both as archeological work sites and as valuable tourist destinations. The Maya had highly developed systems of mathematics, astronomy, and engineering. They also had a hieroglyphic writing system that produced many bark books of records found by the Spanish when they first colonized this eastern part of Mexico and Central America. Most of these volumes were put to the torch by Spanish priests who were certain the presence of the Mayan records would retard the pace of religious conversion in the so-called New World. The Mayan civilization peaked several centuries before the arrival of Columbus, and it is still unknown why this advanced civilization seemed to abandon its massive religious and urban

[1] David L. Clawson, *Latin America and the Caribbean: Land and Peoples* (Dubuque, Iowa: William C. Brown, 1997), p. 131.

The tierra caliente merges almost imperceptibly into the tierra templada. Although sugarcane, cacao, bananas, oranges, and other lowland products reach their respective uppermost limits at some point in this higher level, the tierra templada is most notably the zone of the coffee shrub. In the tierra templada, coffee can be grown with relative ease (although by no means are all the soils suitable); at lower elevations, the crop encounters difficulties because of excessive heat and/or moisture. The upper limits of this zone — approximately 6,000 feet (1800 m) above sea level — tend to be the upper limit of European-induced plantation agriculture in Latin America. In its distribution, the tierra templada flanks the rugged western mountain ranges and, in addition, is the uppermost climate in the lower uplands and highlands to the east. Thickly inhabited sections occupy large areas in southeastern Brazil, Colombia, Central America, and Mexico. Although broadleaf evergreen trees characterize the moister, hotter parts of this zone, coniferous evergreens replace them to some degree toward the zone's poleward margins. In places such as the highlands of Brazil or Venezuela where there is less moisture, scrub forest or savanna grasses appear, the latter generally requiring more water.

In brief, the tierra templada is a prominent zone of European-influenced settlement and of commercial agriculture. Six metropolises exceeding 2 million in population — São Paulo, Belo Horizonte, Caracas, Medellín, Cali, and Guadalajara — are in this zone, and two others — Mexico City and Bogotá — lie slightly above it. Four smaller cities that are national capitals and the largest cities in their respective countries are located in the tierra templada: Guatemala City, San Salvador, Tegucigalpa, and San Jose. Others, like Rio de Janeiro, which are situated at lower elevations, have close ties with predominantly residential or resort towns in these cooler temperatures.

The tierra fría (at 6,000 to 12,000 ft) can be distinguished from the other zones by the related facts that it often experiences frost and is often the habitat of a Native American economy with a strong subsistence component. European colonization in Latin America has driven some Native American settlements upslope and into the tierra fría zone, although some major populations (e.g., the Inca of Peru) had already selected upland locations for their settlement before the arrival of Columbus. These upland settlements are most extensive in Ecuador, Peru, and Bolivia, and this type of economy is also very evident in Colombia, Guatemala, and southern Mexico. The tierra fría is comprised of high plateaus, basins, valleys, and mountain slopes within the great mountain chain that extends from northern Mexico to Cape Horn. By far the largest areas are in the Andes, although areas in Mexico are sizable. The upper limit of the zone is generally placed at about 12,000 feet for locations near the equator and at lower elevations toward either pole. This line is usually drawn on the basis of two generalized phenomena: (1) the upper limit of agriculture, as represented by such hardy crops as potatoes and barley, and (2) the upper limit of natural tree growth.

Another zone lies above the other three and consists of the alpine meadows, sometimes called paramos, along with still higher barren rocks and permanent fields of snow and ice. This zone is known as the tierra helada. It occurs above 12,000 feet and goes up to the lower edge of the snow line. It has some grains and animals (llama, alpaca, sheep) but is largely above the mountain flanks that are central to upland native settlement and agriculture. In the highlands of Latin America, the tierra fría tends to be a last retreat and a major home of the indigenous peoples, except in Guatemala and Mexico (see Fig. 19.A), and is characterized by small settlements and what Europeans and Americans might consider rather primitive ways of life. However, certain valuable minerals like tin and copper are located here, attracting modern types of large-scale mining enterprises into the tierra fría and the tierra helada of Bolivia and Peru.

centers and melt back into the forests of the Yucatán Peninsula and points south. By the time the Spanish came to the area, the Maya were only a shadow presence formed by their enormous structures found being slowly reclaimed by jungle growth.

The Aztecs came on the scene in Mexico much later than the Maya, but they — from the time of the founding of their capital, Tenochtitlán, in A.D. 1325 — had a strong impact on the nature of the Spanish role in Mexico and Central America. By the middle of the 15th century, the Aztecs had joined with other strong cultural groups in the Valley of Mexico and formed the Triple Alliance. From this base, the Aztecs continued to develop trade and tribute networks all through central Mexico (see Fig. 19.7).

One of the traits that was characteristic of the Aztecs was their nearly insatiable ceremonial need for human sacrifices. The most imposing structure in their major city of Tenochtitlán was the Pyramid of the Sun on which 20,000 to 50,000 human sacrifices were made annually during the Aztecs' most powerful time prior to collapse in the face of Hernando Cortez in 1519. Even though this habit of bloody sacrifice — the hearts were cut out of slaves and prisoners who were captured in ongoing warfare — the Aztecs left a legacy of much merit. They had, in Tenochtitlán, a city of some 5 to 8 square miles with a population of between 200,000–250,000 when London was one-fifth that size. The city had a sacred center surrounded by active market places, all linked by stone road surfaces to other parts of the urban settlement. There were some outlying districts that provided goods for trade, and craftspeople and merchants came to the city to work. The Aztecs were also skilled metal workers, having learned these skills through diffusion from the Inca of South America. By the time Cortez had completely conquered the Aztecs and had taken over all that they had created, the Aztecs had been dominant barely two centuries. The Inca had an empire of nearly 2,500 miles (c. 4,120 km) in linear extent from Quito, Ecuador, to central Chile, but they were in control for only about a century.

The earliest major Mexican urban center was in full growth long before any European colonization brought either settlement patterns or urban customs to Latin America. The vast city of Teotihuacán, located in the Valley of Mexico, was the first true urban center in the Western Hemisphere. Its construction is thought to have begun approximately 2,000 years ago, and by A.D. 500, it was as large as London in 1500. Located some 30 miles (48 km) from present-day Mexico City, it covered about 8 square miles. The city was laid out on a grid that was aligned to mesh with stars and constellations that were central to the ceremonial timekeeping of the Teotihuacános. As of yet, the origins, language, and culture of these people are not fully understood. The city may have had 200,000 residents at the time of its maximum development as a ceremonial center in A.D. 500. There were some 2,000 apartment compounds, a 20-story Pyramid of the Sun, a Pyramid of the Moon, and a 39-acre civic and religious complex called the Great Compound or the Citadel (Ciudadela). As described by National Geographic Society anthropologist George Stuart, "The whole is a masterpiece of architectural and natural harmony."[2] It still stands as an impressive monument to Native American engineering skills and cultural intensity (see Fig. 19.7). The Aztecs, who came later than the Teotihuacános, honored the city as sacred space and a religious center when they took over the Valley of Mexico in the 1400s.

The Inca, somewhat like the Aztecs, had origins that are not fully known. It is thought that they began to be a presence around Cuzco in the early 1200s and that by 1438 they gained full control of that city. They expanded their empire so that by the end of the 15th century, they had a network of roads and suspension bridges that extended from southern Colombia all the way to Chile and the very western edge of Argentina. That system extended more than 2,500 miles (4,120 km) and was kept active with trade and tribute moving steadily. Machu Picchu became one of the most important centers for the Inca, and its design and engineering serve to demonstrate the skills that the Inca had in stoneworking, terrace construction, irrigation networks, and in the weaving of wool from llama and alpacas and cotton. Although its genesis and full function are not wholely known, it is likely that it was a sacred center, and there are even theories that the varied microgeographies of all of the different terraces might even have served as a kind of "agricultural station" for experimenting with the growth and qualities of the potato.

[2] George E. Stuart, "The Timeless Vision of Teotihuacan," *National Geographic*, Vol. 188, No. 6 (December 1995), pp. 2–35.

There was no written language for the Inca. There was a method of knotting ropes that enabled some recordkeeping. They would post messengers on the road to give out important news to the travelers going north or south on the road network. Warfare was a regular part of the Inca life, and unfortunately for their empire, two brothers had just completed a very bloody and costly series of battles in 1532 when Francisco Pizarro arrived with 170 men and horses. His Spanish army of 102 mounted (nothing even close to the size and power of a horse had ever been seen by any of the 40,000 Incan soldiers facing Pizarro) and another 68 foot soldiers. Although the Spanish had some guns, the real difference in the Spanish military success against the vastly larger numbers of Incan warriors relates to the power of the European swords, daggers, and lances — plus the surprise impact of horses and the occasional shooting of guns. Within a dozen years of the initial stunning Spanish success in 1532, the forces from Spain had completely thwarted the vastly more numerous Inca. Like Mexico, Peru and the Incan kingdom fell under complete control of the European invaders. The map of the Maya, Aztecs, and Inca — and some additional aboriginal populations — is shown in Figure 19.8 and provides some context for the setting and the demography into which the Europeans came in 1492 and the decades following.

19.4 Economic Geography

Latin America is an "emerging" or "developing" region that does not yet provide a good living for most of its people. It is not the worst off of the major world regions; in fact, its overall per capita GNP is almost four times that of Africa (see Table 1.2). But compared with North America or Europe, the Latin American region as a whole is quite poor. This tends to be masked by the glitter of the great metropolises like Mexico City and Rio de Janeiro with their forests of new skyscrapers. In an attempt to overcome their economic deficiencies and constrain popular discontent, many Latin American governments have borrowed heavily from the international banking community. By the late 1990s, unpaid loans had reached staggering proportions, and the Latin American debtor nations were having great difficulty in mustering even the annual interest payments, let alone generating surplus capital to pay toward the principal of these troubling loans. This debtor situation is very often associated with the process of emerging or developing.

Another way countries deal with the demands and dangers of being a developing or emerging nation is through membership in international organizations. In Latin America, there are a number of such linkages that focus on tariff barriers and economic unity (Fig. 19.10). The best known is the North American Free Trade Agreement (NAFTA), of which Canada, the United States, and Mexico are currently members and of which Chile is close to becoming a member as well. In addition to NAFTA is the organization comprised of a

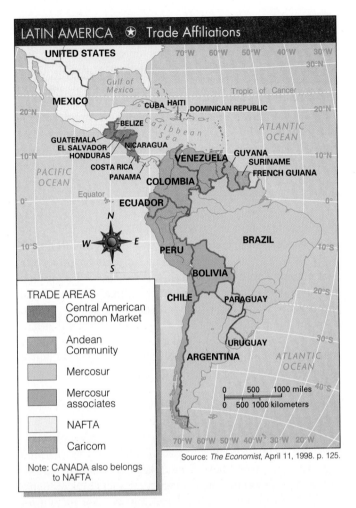

LATIN AMERICA ⊛ Trade Affiliations

TRADE AREAS

Central American Common Market

Andean Community

Mercosur

Mercosur associates

NAFTA

Caricom

Note: CANADA also belongs to NAFTA

Source: *The Economist*, April 11, 1998. p. 125.

Figure 19.10 Latin America is actively experimenting with a variety of alliances in an effort to enhance and expand the economic impact of the region overall and of internal units in different combinations.

cluster of nations in South America — Brazil, Argentina, Uruguay, and Paraguay — called the Southern Cone Common Market, or Mercosur, and is expected to have negotiated a free trade zone with the European Community by the year 2003. There is also the Andean Group, with Venezuela, Colombia, Ecuador, Peru, and Bolivia as current members. Additional groups are shown in Figure 19.10.

There is also considerable interest in a newly planned organization to be called the Free-Trade Area of the Americas (FTAA). This is intended to include every country of the Western Hemisphere. One of the sticking points in this difficult process of such dramatic alliance building is the Latin American disinclination to allow such an alliance to dictate labor and environmental conditions as part of membership in the FTAA. In the United States, the strongest voices in opposition to the current NAFTA are often environmental and labor voices claiming that the only reason U.S. business goes south is that Mexico does not have the same demanding environmental laws held in the United States (and Canada) and that labor is so much cheaper than in the northern two

NAFTA countries. These aspects will take considerable negotiation before FTAA is able to be realized and have all of Latin and North America umbrelled by one free trade alignment.

Minerals and Mining: Spotty Distribution and Unequal Benefits

Latin America is a large-scale producer of a small number of key minerals that are very significant to the outside nations where they are marketed. Only a handful of Latin American nations gain large revenues from such exports; in most countries, mineral production is relatively minor. Even in the countries that do have a large value of mineral output — notably Mexico, Venezuela, Chile, Ecuador, and Brazil — much of the profit appears to be dissipated in the form of showy buildings, corruption, ill-advised development schemes, and enrichment of the upper classes and foreign investors. Even though benefits to the broader population from such mineral production have tended to be rather minimal, significant infrastructural development, including many new highways, power stations, water systems, schools, hospitals, and employment opportunities, has been made possible by mineral revenues. Among Latin America's most important known mineral resources are petroleum, iron ore, bauxite, copper, tin, silver, lead, zinc, and sulfur. However, local use of these minerals for industrialization has been retarded by a lack of good coal, especially the coking coal that is very important in steel production, although Brazil has realized a major improvement in its production and now is the top automobile producer in Latin America.

Most of the extracted ores from Latin America are shipped to overseas consumers in raw or concentrated form, although the region has scattered iron and steel plants, as well as smelters of nonferrous ores, which process metals for use in Latin American industries or for export. Deposits of high-grade iron ore in the eastern highlands of Brazil and Venezuela are the largest known in the Western Hemisphere and are among the largest in the world; the two countries are Latin America's main producers and exporters of ore. Brazil is the main producer of iron and steel by far, with Mexico second. Most of the region's production of bauxite — the major source for aluminum — comes from Jamaica, Brazil, Suriname, and Guyana. The deposits in these countries are located relatively near the sea and are of critical importance to the industrial economies of the United States and Canada. Huge unexploited bauxite deposits in Venezuela are a major resource for the future.

Chile is overwhelmingly the largest copper producer in Latin America and ranks number one in the world. Most of Latin America's known reserves of tin are in Bolivia or Brazil, and these countries produce most of the region's output. The silver of Mexico, Peru, and Bolivia — sought from the very beginning of Spanish colonization — is found principally in mountains or rough plateau country. Mexico and Peru are by

Table 19.2	Primary Exports of Selected Latin American Countries	
Country	Commodity	Percentage of Total Exports
Venezuela	Crude petroleum and petroleum products	77
Cuba	Sugar	63
Mexico	Metallic products, machinery, equipment	58
Chile	Industrial products	45
Ecuador	Food and live animals	48
Panama	Bananas	33
Colombia	Petroleum products	27
Belize	Sugar	30
El Salvador	Coffee	33
Costa Rica	Bananas	24
Dominican Republic	Ferronickel	32
Honduras	Coffee	32
Paraguay	Soybean products	31
Guatemala	Coffee	23
United States	Machinery and transport (for comparison)	47

Source: *Britannica Book of the Year.* Encyclopedia Britannica, Inc, 1999.

far the largest Latin American producers of silver and of lead and zinc as well. Peru's deposits of all three minerals are in the Andes, and Mexico's are mainly in the dry northern and north central sections of the country.

All but a small proportion of Latin America's petroleum is extracted in the Caribbean Sea-Gulf of Mexico area, particularly the central and southern Gulf coast of Mexico and northern Venezuela (see Table 19.2). Other oil fields in Latin America are widely scattered, with the principal ones located along or near the Atlantic coast in Brazil, Argentina (in Patagonia), and Colombia or in sedimentary lowlands along the eastern flanks of the Andes in every country from Trinidad and Tobago to Chile. Natural gas is extracted in many areas that produce oil, but Latin American production is not yet of major world consequence. Currently, Mexico, Venezuela, and Argentina are the largest producers in Latin America. Venezuela is also the only Latin American nation in the global Organization of Petroleum Exporting Countries (OPEC). It was a charter nation in the 1961 founding of OPEC, and Caracas was host to the 2000 winter OPEC meetings.

The Increasing Importance of Manufacturing and Services

Although they are still very spotty in distribution and often lacking in complexity and sophistication, manufacturing industries in Latin America are making an increasingly important contribution to national economies. Most parts of the region are still a long way from full industrialization, but certain districts have an impressive array of factories, including some of the most modern types. Because the Industrial Revolution has come to Latin America very late, the region does not face large

problems of industrial obsolescence. There is a regionwide push to attract industry from abroad with offers of tax exemptions, cheap labor, and other inducements.

The development of the **maquiladora** (assembly operations in Latin America that combine components manufactured from other countries with local labor) movement and NAFTA has done a great deal to keep Latin American labor occupied in locally run factories. In Mexico, there are approximately 3,500 maquiladoras along the Mexico-U.S. border. These shops employ more than 1 million Mexican workers and are considered to be a major factor in a downturn in the flow of illegal immigrants from Mexico and Central America in the year 2000. Mexico estimated the value of the maquiladora movement to be some $40 billion in 1999.

By far the most numerous manufacturing establishments in Latin America are household enterprises or small factories that employ fewer than a dozen workers and sell their products mainly in home markets. Larger operations are almost always located in the larger cities and often are branch plants of overseas companies. The Latin American countries that have the largest output of factory goods are Brazil, Mexico, and Argentina. The industrial scene is dominated by a handful of large metropolises, mostly notably São Paulo, Mexico City, and Buenos Aires. Brazil ranks in the top 10 of the world's nations in production of manufactured goods, although its per capita ranking is much lower than that.

The proportion of Latin Americans in service employment has risen sharply during the past quarter century as people from the countryside have flooded the cities. Expanding government and business bureaucracies, the increasing affluence of urban upper and middle classes, an increase in tourism, and a general increase in the facilities required for even a minimal servicing of hordes of new urban-dwellers have created many millions of new service jobs. Even at general pay scales far below those of Western Europe or Anglo America, such jobs have given new opportunity to desperately poor migrants from overcrowded rural areas, and they have helped provide employment to accommodate the explosive natural increase of population within the Latin American cities themselves.

Two Systems of Agriculture: Latifundia and Minifundia

Although the total number of Latin Americans employed in agriculture has not declined very much in recent decades, there has been a sizable percentage decline in agriculture as a component of both employment and value of product in national economies. The percentages vary a great deal from country to country. Dependence on agricultural exports has also dropped as national economies have become more diversified, but in many countries, more than half of all export revenue is still derived from products of agricultural origin (Table 19.2).

Farms in Latin America are often divided into two major classes by size and system of production. Large estates with a strong commercial orientation are **latifundia** (sing.: latifundio). These estates, whether called haciendas, plantations, or by some other name, are owned by families or corporations. Some have been in the hands of the same family literally for centuries. The desire to own land as a form of wealth and a symbol of prestige and power has always been a strong characteristic of Latin American societies. Huge tracts were granted by Spanish and Portuguese sovereigns to members of the military nobility who led the way in exploration and conquest. Some of this land has been reallocated to small farmers by government action from time to time; the **ejido** — communally farmed land in Mexican Indian villages that gave more land to the farmers and middle class — is a good example, but in many countries, a very large share of the land is still in the hands of a small, wealthy, landowning class.

Minifundia are smaller holdings with a strong subsistence component. The people who farm them generally lack the capital to purchase large and fertile properties; hence, they are relegated to marginal plots, often farmed on a sharecropping basis. Individuals who do own land are frequently burdened by indebtedness, the continued threat of foreclosure of farm loans, and the fragmented nature of farms, which are becoming smaller as they are subdivided through inheritance and the reversal of promised governmental programs of land distribution. Such farms produce food for family use and also for the local market. The crops most commonly raised, especially in Middle America, are maize (Fig. 19.11), beans, and squash, although many other crops are locally important, especially in the different climatic zones of the highland regions.

Although these small farmers make up the bulk of Latin America's agricultural labor force, food has to be imported into many areas, and many of the people are poorly nourished. Productivity is low because of the marginal quality of the land and the generally rudimentary agricultural techniques in peasant farming. There is little capital with which to buy machinery, fertilizers, and improved strains of seeds. Soil erosion and soil depletion are making serious inroads in many areas. As always, there continues to be the image — some would say myth — of a very different lifestyle available to young rural men of life in the city. All through Latin America, the "export" of rural youth, women as well as men increasingly, has been a pattern of demographic shift. In many countries with similar patterns, there has been relatively more capital in the countryside, allowing farming patterns to become more extensive and often more productive. In Latin America, such capital is much scarcer — with the major exception of Brazil — so the rural sector has become increasingly impoverished in relationship to the urban sector.

The Increasing Importance of Tourism

In many Caribbean islands and scattered places on the mainland, tourism has become a major economic asset (Table 19.3). The recent development of major resort complexes in Mexico, the Caribbean islands, Costa Rica, parts of Brazil, and other Latin American countries has expanded tourism as a major player in the generation of foreign

F20.7
535

Figure 19.11 Maize (corn) is a major crop all over Latin America and one of the most important food crops to have been diffused from the New World to the Old World after 1492. In this photo, a government drug inspector in Mexico is about to burn some harvested maize because it is known to contain hidden heroin. Many ingenious methods have been devised for the continuing smuggling of heroin and cocaine into the United States because of the high value of such products on the streets of Latin America's northern neighbor.

Table 19.3	International Tourism to Selected Latin American Destinations			
Country	Total No. Tourists (000s)	Americas	Europe	Receipts ($U.S.) (000,000s)
Argentina	3,030.9	83.0	12.1	3,090
Aruba	541.7	89.6	9.8	443
Bahamas	1,398.9	89.1	8.7	1,244
Barbados	385.4	59.0	40.0	463
Brazil	1,474.9	72.4	23.2	1,307
Chile	1,283.3	89.8	8.0	706
Colombia	1,075.9	92.3	7.1	705
Costa Rica	610.6	83.5	14.8	431
Cuba	460.6	49.1	47.5	382
Dominican Republic	1,523.8	N.A.	N.A.	1,054
Ecuador	403.2	80.2	16.9	192
El Salvador	314.5	92.4	6.6	49
Guatemala	541.0	77.5	19.6	243
Jamaica	909.0	77.1	20.1	858
Martinique	320.7	18.5	80.8	282
Mexico	17,271.0	97.0	2.1	5,997
Panama	311.4	90.2	6.3	207
Puerto Rico	2,639.8	70.2	N.A.	1,511
Uruguay	1,801.7	83.2	2.7	381
U.S. Virgin Islands	390.0	94.9	3.3	792
Venezuela	433.5	54.9	43.2	432

Source: Adapted from David L. Clawson, *Latin America and the Caribbean: Lands and Peoples.* Dubuque, Iowa: Wm. C. Brown, Publishers, 1997: 281.

Perspectives from the Field

The Other Side of Discovery

In 1991, I wrote a short play called *The Other Side of Discovery*. I intended it to be used as a 1992 guide for introducing geography students to the harsh disparities in two polar assessments of the impact of Columbus on the "New World." It has only three characters and, recalling a handful of conversations I have had in fieldwork in Latin America, I am excerpting a segment of the play for this field essay. I do this because there is often an intensity to field encounters, and the lines that follow are in fact a kind of field encounter between Colonio (representing the European presence in Latin America) and Amerinda (representing the native peoples of the same region). In seeing and feeling the distance between their points of view, a geography student can prepare him- or herself for some of the delicate but powerfully interesting field encounters that await you if you step outside the classroom and try to gain a feeling for a sense of place that is distinct from your own home territory.

Time Keeper, who opens the dialogue, is the third character in the play. He represents the forces of politics, culture, and economy that attempt to keep the two opposed perspectives at least adequately supportive of each other to have life carry on. These lines are from the beginning of the play. It takes place sometime around October 12, 1992, the 500th anniversary of the Caribbean island arrival of the three ships of Columbus.

Time Keeper [*He looks beyond the two on stage and gazes out into the audience.*] "We have come here this day to celebrate an event that is, in fact, timeless. We have forced time to stop so that we can reexamine it. If we do not stop and review time, we all rush pell mell to the end of it without taking note of it as it flows by. So, we make it stand still. Today we celebrate a tiny and special segment of such time.

"The event we call to your attention today is the arrival in North America 500 years ago of a small ship with a ragged crew of men ready to kill and abandon their captain. Where they arrived is uncertain and just what their arrival set in motion is also open to continuing argument. But, the captain and his crew completed a voyage that most certainly marked the beginning of a new era of change, of invention, of expansion. We are here to explore and celebrate that voyage of Columbus and the consequent voyage of change that was initiated by their arrival on October 12, 1492. Let me . . . "

Colonio [*Interrupting the Time Keeper*] "Hold on, Old Man, . . . we are not here today to celebrate anything except the voyage of the Niña, the Pinta and the Santa Maria. We have come back to speak of that moment 500 years ago . . . to praise the strength and courage of the men who endured fears and uncertainty like no voyagers ever before in history." [*Colonio squares his shoulders, assuming an even straighter posture in his chair.*]

Time Keeper [*Irritably*] "Wait! Your time is coming, Colonio. Can you not realize that even as your captain was ending darkness over the eastern horizon, he was ushering in the dawn of a new world." [*The Time Keeper looks at the crowd with a little embarrassment, as though to say that the house needs to tolerate this character.*] "You must remember the scope of what your captain launched with his discovery of the New World."

Colonio "Of course I know what my captain did, Viejo, I just don't want to start that New World blather again. I refuse to get into one of those endless discussions of a New World vs. the Old World. I can tell you . . . " [*He is interrupted by Amerinda.*]

Amerinda "Silence! Both of you! It was surely NOT a new world that was found. Here I stand in the shadow of my cities, of my towering pyramids, of my farm fields that now touch all the world, and you two dare to bicker about whether or not Columbus discovered a *New* World. It was not new! It was a world that had been well-settled long before you were able to muster forces and round up *religiosos* to come here. [*Amerinda draws up to full posture. She is strong but not brassy like Colonio. She has an attractive, competent look about her.*]

Colonio "Ah, Amerinda, you miss the point." [*Colonio looks directly at Amerinda with these words. His voice is condescending.*] "I do not mean that there was no world where our captain landed, but I refuse to let this old fellow call it a New World. He knows . . . "

Amerinda "I know what you say, Colonio . . . and I know what you mean. Your European tongue has always found it easy to turn ideas any which way to serve the moment . . . to serve your immediate needs." [*With more boldness*] "To help you conclude one dreadful theft after another."

Colonio [*He is looking directly at Amerinda. As he speaks he begins to turn more to the audience, trying to steal the scene here.*] "Not so, not so. You always stand back, isolated, reserved, remote, and act as though our only goal was to somehow deny you your world. You forget that we risked our very lives in an effort to find land . . . and that by doing that we brought more wonders and improvements to you than you could ever have imagined."

Amerinda "Ah, Señor, what you brought us, we could have done without. With all you brought us to supposedly improve our lives, also came an accompanying horror that brought only pain and destruction." [*She pauses, thinking, then goes on, looking out at the audience.*] "The gift of Columbus was the gift that kept on taking."

Colonio "What are you talking about, woman?"

(Continued)

Amerinda "You gave us religion and took our souls.

"You gave us technology and took our land.

"You gave us whiskey and took our ambition." [*Up to this point, Amerinda has been looking at Colonio. Now she looks beyond him and turns more to the audience.*] "There is nothing that they gave us that we would not be better without. He and his brethren have been only devastation dressed in fancy costume for the past five centuries." [*Clearly distressed, she stops a few seconds. Then she turns to the Time Keeper.*]

"I do not even know why I agreed to come and help you review this story, Viejo."

Time Keeper [*Bothered.*] "Wait, you two. *Wait!* You must let me get the formal introductions and dialogue underway. We have been struggling to get the right forum for centuries and now that this precious time has come and people will — perhaps — listen to you, you both are so tiny-minded that you refuse to let me open the presentation properly. How can these people have any idea of what we are hoping to bring them if you start your endless arguments before I can even tell them who you are?"

Amerinda "Ay, they *know well enough who he is*, Old One. Anyone alive today, anyone who is naive enough to celebrate these 500 years of European theft and destruction knows exactly who he is." [*She looks intently at Colonio.*]

Colonio "And you, Amerinda . . . Do not think yourself unknown! Tales of agony caused by Indians, of Indian attacks, of brutal and bloody Indian rituals abound in the minds and histories of all these people in front of us. Abandon your posture of self-righteousness and let the people see you for all that you were . . . and for all that you are now."

Time Keeper "That's enough! Stop it, both of you." [*He then looks beyond them out to the audience.*]

"What we hope to bring to you this night is neither a burial nor a birth. We gather for this celebration of 500 years of change, of successful effort and dismal failure, of conflict with nature and humankind to settle a world of enormous magnitude and consequence." [*He stops for a second, looks at his notes, and goes on.*]

"We might well have had a cast of literally hundreds of millions who could have played the roles for this small pageant of geography and history. But, partly because of the size of this stage, but chiefly because of the polar positions our two players represent, we have a cast of only two. As you might now understand, Colonio is the sole representative of the Europeans who discovered this world 500 years ago — perhaps not even for the first time — in 1492. Perceiving it not as an unknown world, but rather as the island chain lying off the east of the Spicelands called the Indies, his captain called the people he found 'Indians.' The geography of Columbus was faulty not only in terms of sea charts, but also in terms of peoples." [*The Time Keeper turns from the audience a little and looks at the Indian woman. He goes on.*]

"Amerinda is the sole representative of the myriad native nations that had come themselves to find a new world thousands of years earlier. Like Colonio, her people also stumbled upon this land. Their steady pursuit of herd animals in East Asia kept them moving along the margins of the sea. The rigor of the northern climate hurried them southward toward the equator from their great horseshoe crossing from northeast Asia to northwest North America. They fanned out into hundreds of so-called Indian nations in just as many diverse environments. They . . . "

Amerinda [*Softly*] "You go on too much, Old One. You steal too much of my time. There is no way to put balance in this story that we are wanting to tell these people. Let my story stand for itself and let the muses of the sky and earth determine where the balance is."

Time Keeper [*He looks at the audience, with a faint hand gesture toward Amerinda.*] "Yes. Yes. But I thought . . . "

Colonio [*Interrupting*] "You *do not need* to think, Old Man. You need only to pay attention to your time keeping. Let the people out there think. Let me think. Even let the woman think. *You* do not need to think. Your only task is to move us through time in such a way that we get our story told. My story is clear and to the point." [*At this point, Colonio turns a little more toward the audience.*]

"We came to this land.

"In our eyes, it was virtually unused.

"We used it.

"We made great improvements in its landscape.

"We settled it and we developed it." [*Colonio looks proud of what he has just stated and looks out upon the audience for approval.*]

Amerinda "You are so tiresome, Colonio. I stand at the shore of a world that reaches from pole to pole. It is peopled by countless nations, each of which has found some quiet harmony in their embrace of nature and care of the earth . . . "

Colonio "Oh, spare me the 'quiet harmony in the embrace of nature,' Amerinda." [*Colonio is warming to his role.*] "You who had thousands of prisoners stand in line to march up pyramids in MesoAmerica so that ratty haired medicine men could rip out their hearts with obsidian blades . . . "

Time Keeper "Ay de me! Wait . . . both of you . . . wait." [*With some resignation, the Time Keeper turns once more to the audience.*] "These two represent the world that was found and the world that did the finding. They are the pair who came together 500 years ago when by the light of a faint moon the man on watch spotted the first land seen in 33 days. It was 2:20 A.M. October 12, 1492. From that date until this celebration, these two have been in conflict . . . in relentless, exhausting conflict." [*The Time Keeper stops.*] "Amerinda, are you ready to begin?"

(The dialogue continues for 25 argumentative minutes.)

Kit Salter

Figure 19.12 The landscape, the setting, plays an enormous role in the success of tourist destinations. This major hotel in Nassau in the Bahamas features elaborate water displays, lush vegetation and flowering trees, and comfortable accommodations—all with English as the spoken language. The economic importance of becoming a major tourist destination means a great deal to the economies of many Latin American countries. Since the events of September 11, 2001, tourist activity in the Western Hemisphere has grown much stronger than that of sites in the Middle East and Asia, working to this region's advantage.

exchange (Fig. 19.12). Cancún, Mexico, for example, was recently described in a tourist brochure as "Cancún — Mexico's American Resort . . . Cancún may seem more American than Mexican." The region's proximity to the relatively wealthy and mobile U.S. and Canadian populations, the expansion and improvement of regional air transport, and the increasing inclination of North Americans to spend money for tourism have all brought considerable disposable income to the countries and coastal areas that have made the appropriate landscape investments. Passenger cruise ships now stop at ports of call in the Antilles, on the Pacific coast of Mexico as well as at Cancún, and at Santa Marta and Cartagena in Colombia. The transiting of the Panama Canal continues to be important

as a tourist activity. The Canal continues to be geographically significant for all the ships that can gain passage in it as well because it cuts some 8,000 miles (12,800 km) off the voyage between New York and San Francisco.

Cruise companies are also increasing the number of sailing routes along the east coast of South America with stops in the Amazon Basin and in coastal ports of Brazil, Uruguay, and Argentina. San Juan, Puerto Rico, is now a major port of embarkation for tourist cruises destined for the eastern Antilles, with passengers regularly being flown from U.S. and Canadian cities to San Juan for boarding and embarkation. There is also a growing demand for **ecotourism** destinations that present visitors with a chance to view relatively pristine landscapes with unusual fauna, flora, and native peoples (Fig. 19.13). Countries in which tourist dollars amount to more than 20% of the nation's foreign exchange include Argentina, Dominican Republic, and the Bahamas. Only the export of petroleum generates more foreign exchange for the region than does tourism (see Regional Perspective, p. 521).

19.5 Geopolitical Issues

The more than 40 years of efforts by Fidel Castro in Cuba to introduce a working socialist system of government into Latin America stands as a major example of political innovation. The political strength of Costa Rican democracy for decades — and the minimal percentage of its annual budget spent on a standing military — represents another example of geopolitical idiosyncrasy in Latin America. Everything from persistent dictatorships to major democratic, free-market governments and economies are part of the mosaic of the 38 countries that lie south of the southern border of the United States and reach down to the very edges of Antarctica. The

Figure 19.13 Native American markets of hand woven textiles and local produce are a colorful part of Latin American landscapes.

REGIONAL PERSPECTIVE ::: Ecotourism

The United Nations has selected the year 2002 as the International Year of Ecotourism. This decision has both environmental and economic implications. Economically, it is claimed by Duke University biologist John Terborgh that tourism is a $400 billion industry annually, employing 1 in 15 of workers around the world. He goes on to say that if you are going to preserve a major portion of the fragile global environment — the primary reference is the tropical rain forest — it will be essential to bring a powerful economic perspective to environmental preservation. It is believed by some that ecotourism is the mechanism that can genuinely provide the mix of environmental protection and tourist financial support.

Ecotourism is a variant on the provision of tour facilities that creates a more modest human presence in the environment than does, for instance, classic lodge or resort construction in the midst of a setting that has high touristic significance and draw. For example, some of the most rigorous ecotourism firms create lodges in the midst of the rain forest that have no running water, no electrical outlets, outhouses instead of bathrooms, and kitchen fueled through propane tanks and solar panels. Although younger tour groups might find the rigor of such a setting to be both satisfying and to provide a sense of do-good tourism, they are not the groups that will pay the higher fees necessary if the tourism firms are to generate capital for the broader preservation of their environmental settings. The more costly ecotourism trips are more inclined to require a relatively comfortable lodge settings for after the return from the carefully guided and guarded day trips into the environment. One businessman observed that "unless you can make wildlife [an] asset that

the businessman is going to milk for profit, you're not going to preserve it." Again, ecotourism is the current best hope for such environmental preservation.

In other regions where ecotourism has been given political and economic focus, a sometimes significant portion of the fees for such travel (as in Nepal) are allocated to the sites visited. They are then used for the maintenance and repair of the very environmental elements that make such locales targets for this sort of tourism. The Galapagos Islands off the shore of Ecuador have a long history of very highly controlled ecotourism (an annual tourist flow of some 60,000), but a massive oil spill in January 2001 has forced Ecuador to consider the requirement that all ships that bring petroleum to the islands be double hulled. Enforcing such a law would lead to a considerable increase in the cost of fuel for the islands, and that increase would be passed along to the tourists. Such a development would possibly lead to a decrease in the flow of tourists and tourist dollars, returning the islands once more to a harsh economic pressure for change.

Costa Rica has had a very successful history with the promotion of ecotourism of its well-managed environment. The problem is that it has had too much success and has received too many tourists to be able to control the associated deterioration of the favored environments. Hence, the country is now trying to diminish the number of ecotourist arrivals. The problem with too much success in ecotourism has caused one observer to point out that if the person who takes these tours were "really an ecoperson, [he or she] would let it be and not come!"

Part of the dynamic that has led to the Latin American growth in the interest in ecotourism (as well as

in other parts of the world), is the fact that since 1950 "an area equal to the combined size of India and China has become abandoned wasteland, thanks to deforestation" says John Terborgh of Duke University. Another ecotourist planner says that economic benefits from ecotourism are much more convincing than lectures on the importance of the environment to local businesspeople or farmers.

The issue remains to be one of trade-offs. If there is a wish to utilize an environmental setting through deforestation for ranching, farming, or human settlement, the strongest force retarding such land use is an economic force. Even with governmental policy disallowing forest clearance, there has been a Latin American history of peasant squatters expanding farmland through classic slash-and-burn agriculture or broad scale, heavy machinery forest clearance by people who somehow get permission to work around such governmental policy. It is thereby hoped that ecotourist dollars might be seen as a viable alternative to such deforestation and clearance. Success of these efforts would still mean some environmental change in the creation of trails and lodges with some amenities, but this activity would also provide some employment and perhaps a new source of capital for the preservation of these settings. This struggle will continue into the future for certain. If the pace of deforestation of the past half century continues, it is estimated that all rain forests will be removed before 2050. Such predictions keep people working actively with ecotourist projects.

Sources: Laurent Belsie, "Treading Lightly." *Christian Science Monitor*, February 1, 2001, pp. 11, 13. Larry Rohter, "Isles Rich in Species Are Origin of Much Tension." *New York Times*, January 27, 2001, p. A4.

varied efforts in various systems of government, economy, and global connections all stand in representation of an energetic region of economic ambition and uncertain political format.

One of the most contentious issues of political and social change in Latin America has been land reform. Since the entire region has a long history of major agricultural activity — from the early sugar plantations of the 16th century on through the major banana and coffee plantations of recent decades — it is understandable that land reform would be a geopolitical issue of major concern.

The pattern that has been most problematic is that the most productive lands, and the lands that have benefited most from technologically innovative and modern farming practices, have traditionally been the land owned by colonial powers and now by large, and frequently absentee, landowners. At the same time, the largest sector of the rural population has been made up of *campesinos* — the peasant farmers who sharecrop land owned by others or who have very small holdings — who have a strong image but very little political power in the allocation of national agricultural budgets in Latin America.

In Mexico on New Year's Day in 1994, the campesinos of Chiapas — a southern state in Mexico with a high percentage of *indio* (indigenous people) peasant farmers — staged a very widely reported takeover of San Cristóbal (the state capital). This initial effort has led to arduous negotiations between Mexico's last president and now has become part of the agenda of new President Vicente Fox. Chiapas has become a hallmark issue for much of Latin America, with land reform and improved support of the region's indios as two dominant issues.

Land reform has long been a topic of both political and economic significance in Latin America. The region's history is filled with attempts — sometimes very serious and bloody, sometimes less painful governmental innovations — to achieve a better balance between those who work the land and those who own the land. The main attention has been given to breaking up existing properties or bringing vacant land (whether owned by the public, by private individuals, or by the Catholic church) into cultivation, usually by small farmers. Some new farms have been structured as communal holdings, reflecting indigenous traditions or 20th-century revolutionary plans of agrarian reform. In Cuba, for example, the Communist government that took control in 1959 placed the land from expropriated estates in large farms owned and operated by the state. Workers on these farms are paid wages. The Nicaraguan Sandinistas implemented land reform as part of their revolutionary efforts in the 1980s.

The details and success of land reform policies have varied sharply from one Latin American country to another. Such policies can help relieve poverty in the countryside, but they do little for the majority of Latin America's poor, who live in cities. In Mexico, there was great fanfare at the introduction of the ejido (communally farmed land in Mexican Indian villages) early in the 20th century. It has been so effective that ejidos currently account for 50% of the cultivated land in Mexico and produce nearly 70% of the beans, rice, and corn of this nation. The ejido, by making land available to farming communities, has been the most successful land reform program attempted in Latin America, including Cuba. Puerto Rico has had public battles over "pan, tierra, y libertad" (bread, land, and liberty) because of the difficulty in effecting real distribution of farmland to the farming people. Virtually every nation in Latin America can point to events, martyrs, and programs birthed in protest or blood in an effort to achieve more equitable land ownership. Nevertheless, the great majority of the arable land in Latin America continues to be owned by wealthy farming families, agribusiness operations (both domestically and foreign owned), and the church. There is a world of small farmers in Latin America, but they seldom own as much as 20% of a given nation's arable land.

The Continuing Problem of Natural Hazards and Latin America

Because of the presence of the Ring of Fire all along the western edge of Latin America, the region's history is filled with a major presence of earthquakes and volcanic activity. The city of Antigua, Guatemala, has ruins still captured in lava from a historic eruption. Mt. Popocatepetl in the Valley of Mexico has long been one of the major landscape signatures of that region, and it was again active through much of 2000. At the end of 2000 and the very beginning of 2001, El Salvador was hit by two major earthquakes. In late 1998, Honduras was hit by Hurricane Mitch for a sustained impact, costing thousands of lives, many thousands of homes, and hundreds of millions of dollars. In 2001, local and foreign groups were still making trips to the sites of devastation in an effort to help bring life back to normal. The whole region has a geographic pattern of falling prey to storms, quakes, and even lava flows because of the powerful influences worked on the region by El Niño (see Definitions & Insights) and the seismic instability associated with the Ring of Fire (see Definitions & Insights on p. 509).

These are geopolitical issues because of the settlement patterns that make major cities of Latin America subject to the ravages of these geographic phenomena. But politics also plays a major role in responses to such hazards. For example, in the case of Honduras, there was an international agreement that provided Honduras a forgiveness of two-thirds of its foreign debt of $1.7 billion debt and a 3-year moratorium on debt service payments. To secure these political and international benefits relating to Hurricane Mitch and subsequent flooding, however, the government had to agree to stringent foreign imposed plans for austerity and for dealing with poverty and creating aggressive plans to rebuild infrastructure in the areas hit by Mitch. These environmental hazards have implications that touch all aspects of life in Latin America, even though they may seem to be "tragically and simply" a flood or a quake.

F10.15
296

DEFINITIONS + INSIGHTS

El Niño and Natural Disaster

The years 1997 and 1998 were classic El Niño years in Latin America. Across the Pacific came an unusually warm current, washing up against the western shores of tropical South America, displacing colder normal currents. Not only did this temperature change have a major negative effect on traditional fishing patterns on the west side of Latin America, but the warmer surface water in the Pacific played havoc with normal air movement and weather patterns. Storms lashed a wide band of Middle and northeastern South America, while drought struck broad sections of Brazil, Venezuela, Colombia, and Central America. Fires came after the unseasonably dry periods caused by this cyclic hazard called El Niño, adding considerable forest loss to the fishing loss (Fig. 19.B).

The impact of this weather phenomenon is felt through all bands of society in Latin America and across a great variety of landscapes. In 1997–1998, for example, Bolivia experienced 6 months of highland droughts and lowland floods that were caused in part by the fact that westerly winds that ordinarily would have delivered snow to the Andes came as unusually warm rains; this precipitation washed directly down into the lowlands as floods. Thus, moisture was lost for spring farming, and devastation was created by the floods that brought their own horror to the lowlands. The Panama Canal had to restrict freighter travel through its locks because the unseasonal drought brought water levels to nearly unprecedented lows, leaving larger ships with deeper droughts unable to deal with the artificial waterways of canal passage. The approximate costs of El Niño in these 2 years was put at $20 billion, but in fact the costs will go much higher, for not only must there be enormous repair projects organized and paid for, but the loss of crops and cropland will have a long-term impact on Latin American productivity.

In the same family of natural hazards that plague Latin America was the hurricane called Mitch that wandered leisurely across the Caribbean Sea, the Gulf of Mexico, and the northern sector of Central America in the fall of 1998. It was the rain, not the more common hurricane winds, that made Mitch the worst storm Middle America has known in modern times. In the 4 days that the storm system hovered over the western Gulf of Mexico and the countries of Guatemala, Honduras, Belize, Nicaragua, and El Salvador, it dumped more than 2 feet of rain not only on the coastal regions (the most common focus of hurricane devastation) but also on inland and upland regions. Even the older residents of this Central American region, who had endured the strong hurricanes Fifi in 1974 and Gilbert in 1988, were caught unprepared as Mitch came inland.

More than 15,000 people died as a result of Mitch, and the disease and weakness caused by the loss of potable water, food crops, and shelter were compounded by the disaster worked upon transportation infrastructure, making outside help virtually unable to reach the region except by helicopter. No storm like Mitch has been seen before by anyone now alive in Middle America. In combination with the hardships worked upon Latin America by El Niño, this has been a period of extraordinary demand and difficulty for Latin Americans attempting to attain and maintain the landscapes and lifestyles of aggressively developing countries. The web of influences worked on the people by these natural disasters spills beyond the actual locales visited by these physical systems. Migration is often seen as the only real solution to surviving the disasters associated with these storms and cyclic weather changes, and North America and major cities in Latin America are the general target destinations of these destitute migrants.

Source: *The Economist*, May 9, 1998. p.36.

Figure 19.B   The impact of El Niño in 1997 and 1998 was both far reaching and geographically varied, further supporting the lore that this weather phenomenon is becoming increasingly significant to the region. Consequences of this shift in ocean current patterns and associated air mass movement were felt far beyond Latin America as well.

Staging for the Future

All of the elements for steady development and democratic reform are present in the diversity of Latin America. In many of the 38 countries in this region, there are historical periods of success in both of these realms, but at the same time, these same countries have known great frustration as single-crop or single-mineral economies. They have been depressed by rapid and deep drops in global market prices for these goods, leaving the Latin American export country in major trouble because their base for economic stability is too narrow. The region's populations are varied in stock and not so dense in overall settlement patterns as to be unmanageable, except in cases of overcrowded massive urban centers that have become migration destinations for many rural youth seeking new — often un-realizable — urban opportunities. At the same time, most countries in Latin America have some isolated farming populations, especially in the Tropics, who support themselves by traditional slash-and-burn farming practices. These populations show almost no evidence of the region's aggressive efforts to modernize (see Fig. 19.13). However, the continuing inflow of tourist dollars (see Regional Perspective on Ecotourism) and the steady efforts to slow population growth, even within the context of a Roman Catholic region, are taken as signs of optimism by planners and citizens alike. It is with keen anticipation that the peoples and governments of Latin America will observe the continuing impact of NAFTA during these coming years, trying to assess both the economic and environmental effects of these significant attempts at regional cooperation.

CHAPTER SUMMARY

- Latin America lies south of the United States and includes Mexico, the Central American countries, the island nations of the Caribbean Sea, and all the countries of South America. It is called Latin America because of the nearly universal use of Latin-based languages and the dominance of the Roman Catholic influences that are associated with the colonial period from the early 16th century through the middle of the 19th century. We use the term Middle America for Mexico, Central America, and the Caribbean.

- Major themes in Latin America include extraordinary variety in environmental settings, population groups of great diversity, a wide variety of levels of economic development, countries that have highly diverse cultural and demographic profiles, and diverse governmental philosophies.

- A major tool for the analysis and description of Latin America is altitudinal zonation, which includes tierra caliente, tierra templada, tierra fría, and tierra helada. These zones go from sea level up through various altitudes, each with distinctive climate, soil, agriculture, and settlement patterns.

- Major humid zones are found in the lowlands of Latin America, with tropical rain forests occupying major river basins, especially in Brazil and Venezuela. Grasslands are found poleward of the tropics, and there is a zone of Mediterranean climate on the western coast of Chile.

- Dry climates — desert and steppe — are often located in particular areas because of orographic features that influence the movement of air masses. There are major desert regions on the west coast of both South America and in Northern Mexico.

- Latin America's population amounts to nearly 9% of the world's total population, and its growth rate tends to be above the world average, especially in Central American countries where there continues to be a relatively large Native American population. The general demographic pattern in this region shows a majority of the people living on the rim of the land masses, an area often called the Rimland. There is also a major population band in the mountain valleys and, in some cases, on high plateaus or on the flanks of the Andes Mountains in western South America.

- Before the arrival of the European colonists and settlers in the late 1400s, Latin America had been home to some productive and highly advanced native cultures, with the Aztecs, the Maya, and the Inca the best known and most widely studied. All of these had either declined before European colonization or declined rapidly after 1500. Although their demise is often associated with mining and other land-use demands, the real killer was European disease visited upon a population that had no immunity against Western diseases, especially smallpox.

- Latin America has become one of the most urbanized world regions, with now more than two-thirds of its population living in cities, compared to the world average of 43%. There continues to be steady rural-urban migration as well as a steady stream of both legal and illegal immigrants north into the United States. The North American Free Trade Agreement (NAFTA) has as one of its goals the promotion of industrial and commercial growth in Latin America so that migration to the United States becomes less attractive to the large underemployed Latin American populations.

- Latifundia and minifundia are two major agricultural systems of historical importance that continue to characterize major blocks of regional land use. There is also a strong plantation economy, especially in Central America, the Caribbean, and coastal Brazil, and sugar and fruit all play dominant agricultural roles. Farm landscapes range from peasant households that are largely subsistence farming, to haciendas, to ejidos, to foreign-owned fruit and coffee plantations — all of which have their own distinctive looks and functions.

- Minerals have played a major role in the history, economic development, and trade networks of Latin America and include iron ore, bauxite, copper, tin, silver, lead, zinc, and sulfur. Petroleum has been of dominant importance, especially in Mexico and Venezuela, but has lost some importance recently because of flat or declining world oil prices. Gold has had local importance, and silver is at the center of centuries of European interest in the region.

- In recent years, economic growth has depended increasingly on tourist activity and on manufacturing goods for both domestic and foreign markets. Foreign investment has been influenced by political instability in various parts of Latin America, and the United States military has played a role in promoting and eliminating governments in a number of Middle American nations. The construction of the Panama Canal was enabled, largely, by the quick U.S. recognition of a coup that broke the Panama Isthmus away from Colombia in 1903. Construction of the canal began within a year and was finished a decade later, eliminating some 8,000 miles in the voyage from New York City to San Francisco.

REVIEW QUESTIONS

1. What countries make up the Mainland of Middle America?
2. What are the major island nations of the Caribbean Sea?
3. Besides the Spanish spoken in most of Latin America, what other languages are spoken in this region and where are those languages found?
4. Compare the area and the extent of latitude and longitude of Latin America with the same features in the U.S. and Canada.
5. What are the major climate patterns of Latin America? What are the physical features that play major roles in such patterns?
6. Explain the causes and the significance of hurricanes in Latin America. What other natural hazards are part of the Latin American geography?
7. Explain the four zones discussed in altitudinal zonation in Latin America. What economic and demographic importance do these zones have in the region?
8. What are the major features of the population distribution in Latin America? What features characterize the patterns of population growth in the region? What role does migration play in these patterns?
9. Outline the major agricultural patterns and dominant land use in the region. What role does export play in the overall agricultural pattern of Latin America? What goods and what destinations characterize export activity?
10. Tourism and mineral extraction both relate very closely to geographic conditions and characteristics. What traits are important in each and how are they expressed in Latin America?
11. List the countries included in, and explain the regional breakdown of, Latin America, Middle America, the Central America, Caribbean Sea, and South America.

DISCUSSION QUESTIONS

1. Discuss the possible patterns of development in Latin America had Columbus not come westward in his effort to find East and South Asia. What if no European explorers had found Latin America by 1600? By 1800?
2. Latin America is sometimes described as "a land of contrasts." How would you defend such a broad description of the region?
3. Discuss the central themes from the excerpt "The Other Side of Discovery" in Perspectives from the Field on pp. 518–519. Role play either the position of Amerinda or Colonio and try to determine which attitude seems to make the most sense. What role does the Time Keeper play in this drama?
4. Discuss the positive and the negative impact of NAFTA. Discuss it from the point of view of Latin Americans and from other points of view. Has it been largely beneficial or detrimental for Latin America? For the U.S. and Canada?
5. What major crops were diffused from Latin America to the broader world? What major crops in today's Latin American agriculture came from the Old World when the Europeans began to settle Latin America? What economic and cultural impact has this diffusion had?
6. What are the cultural forces working to reduce population growth rates in Latin America? What forces tend to keep growth rates high? What impact does population growth have on political stability and economic development?
7. Discuss the geography of isolation caused by mountains, vast river systems, islands, and other aspects of Latin America. How do these features influence patterns of political stability and human migration?
8. What were the causes and the impact of 1998's Hurricane Mitch in Middle America? Of El Niño?
9. Consider places as geographically diverse as the Andean Highlands and the Amazon Basin. What are the political and cultural forces that enable such distinct locales to be part of the same world region?
10. Imagine that you are a young Latin American living in a small village or town in Middle America or South America with little business activity. What are the images, history, and knowledge that will cause you to think about migrating? What destinations might you consider and why?

The Many Worlds of Middle America

When the Panama Canal was completed in 1914, it was thought to have been much too large for the shipping that might move through the Isthmus of Panama. The truly massive oil tankers that are now central to global oil transshipment are too wide and too long for these Miraflores locks of the canal system now. Even with such inadequacies, however, the Panama Canal continues to play a critical and major role in shipping patterns between Pacific ports and east U.S. and European ports.

Chapter Outline

Middle America includes (1) Mexico, by far its largest country in area and population, (2) the much smaller Central American states of Guatemala, El Salvador, Belize, Honduras, Nicaragua, Costa Rica, and Panama, and (3) the numerous islands in the Caribbean Sea or near it. Of the many island political units, the largest are Cuba, which is only 90 miles (144 km) away from the United States, Haiti and the Dominican Republic (which share the second largest island, Hispaniola), and Jamaica. The entire region presents a patchwork of physical features, races, cultures, political systems, population densities, and pursuits. Most of the political units are independent, but a few are still dependencies of outside nations (Fig. 20.1). We begin with the most prominent regional unit, Mexico.

20.1   Mexico

The federal republic of Mexico — officially, the Estados Unidos Mexicanos (United Mexican States) — is the largest, most complex, and most influential country in Middle America. Compared to most of the world's countries, Mexico is a spatial giant, but its large size tends to go unrecognized, largely because it lies in the shadow of the United States. Its capital, Mexico City, stands now as the world's second largest city in population (after Tokyo). Triangular in shape and situated next to the compact bulk of the conterminous United States, Mexico looks comparatively small on a world map, but its area of 762,000 square miles (c. 2 million sq km) is nearly eight times that of the United Kingdom, and its elongated territory would stretch from the state of Washington to Florida if superimposed on the United States. The country's huge population (an estimated 99.6 million in 2001, expected to reach 120 million by 2010) makes Mexico by far the largest nation in which Spanish is the main language. Mexico is also a big country economically, ranking quite high among the world's nations in annual output of many different commodities such as metals, oil, gas, sulfur, sugar, coffee, corn, citrus fruit, cacao, and cattle. It has also made substantial gains in its development of manufacturing and component assembly industries, especially since the initiation of the North America Free Trade agreement (NAFTA).

Mexico's remarkable diversity of mineral resources and agricultural products is made possible by great environmental variety. Metal-bearing ores exist in many places, mined particularly in the mountainous terrain of northern and central Mexico. Iron ore from scattered locations supports a fairly sizable iron and steel industry, which also can draw on substantial coal deposits, some suitable for coking. The eastern coastal lowlands along the Gulf of Mexico share the oil, natural gas, and sulfur deposits that are a marked feature of adjacent coastal Texas and Louisiana.

In agriculture, great variety is achieved despite a mountainous environment lacking in tillable land. Only one-eighth of the country is cultivated; good cropland exists mainly (1) on the floors and lower slopes of mountain basins with volcanic soils and (2) on scattered and generally small alluvial plains. Agricultural habitats vary according to elevation, soil types, physiographic history, and atmospheric influences. Mountains too steep to cultivate without excessive erosion dominate the terrain in most areas, but they are flanked by plains, plateaus, basins, and foothills that can be tilled productively if enough water is available. At different times, even the steep slopes are farmed as milpas (slash-and-burn plots) by peasant farmers desperate for access to land.

A little over one-half of Mexico lies north of the Tropic of Cancer and is an area dominated by desert or steppe climates with hot summers and warm to cool winters depending on elevation. South of the Tropic of Cancer, where about four-fifths of all Mexicans live, seasonal temperatures vary less, but elevation differences create conditions ranging from high temperatures in the tierra caliente of the lowlands along the Gulf of Mexico, through moderate heat in the tierra templada of the low highlands, to cool temperatures in the tierra fría of the highlands (See Regional Perspective, Chapter 19, p. 512). Nearly all areas inside the tropics are humid enough to support rain-fed crops, although irrigation is often used to increase the diversity of crops and/or the intensity and productivity of agriculture. Large areas have a dry season in the low-sun period when irrigation becomes a necessity for economically productive cropping.

On the hemispheric and world stage, Mexico speaks with a respected voice, although the country's relative poverty and its great economic dependence on the United States make its influence less potent than might otherwise be true of so large a nation. Mexico has more than one-third as many people as the United States, but in 2000 had a GDP PPP only 9% as large as its neighbor to the north. Economic relations with the

FI9.A
512

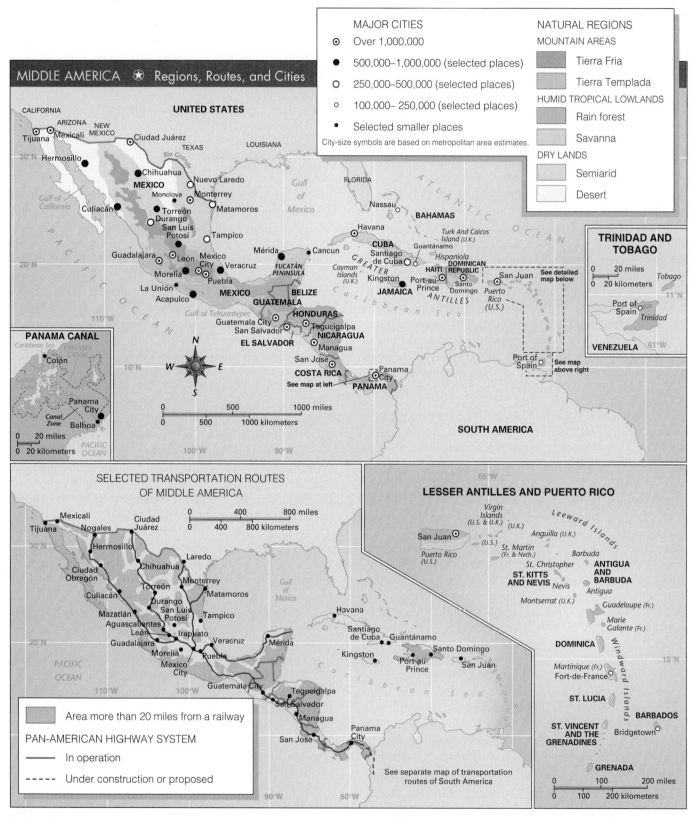

Figure 20.1 Middle America: regions, routes, and cities. General reference maps of Middle America. Curaçao and Bonaire comprise the Netherlands Antilles; Aruba is a separate country. Antigua and Barbuda form one political unit, as do St. Kitts and Nevis, and St. Vincent and the Grenadines. Major highways in the transportation inset are supplemented and interconnected by other surfaced roads not shown.

United States are crucial to Mexico because approximately two-thirds of the country's foreign trade is with the United States, and the latter nation is overwhelmingly the main source of outside capital. The 1994 North American Free Trade Agreement (NAFTA) among the United States, Canada, and Mexico has stimulated considerable economic activity on both sides of the Mexico-U.S. border.

North American Free Trade Agreement: The Significance of NAFTA

One of the most innovative new alliances in Latin America is the coming together of Mexico from Latin America (although Mexico is sometimes assigned linkage with the North American continent), the United States, and Canada in the North American Free Trade Agreement. The origins of NAFTA are varied, but three motivations have been central to this multinational agreement: (1) a wish by all three to promote Mexican economic growth and development; (2) a hope for expanded export markets for the United States and Canada; and (3) a desire to decrease the flow of illegal migrants from Mexico to the United States through the expansion of Mexican manufacturing activity. The earlier success among the European nations in their creation of the European Economic Community (EEC) led NAFTA countries to assume that lower tariff barriers in Latin America and the United States and Canada would enable the affected economies to expand into markets for which they were best suited. Trade between Mexico and the United States has tripled since NAFTA began in 1994. In 2000, trade between the two countries amounted to nearly $50 billion. Mexico produced nearly 1.2 million automobiles in 2001. The migration of the automotive industry southward from the United States has had a strong impact on the major industrial centers in Mexico because of the tariff reduction engineered by NAFTA.

In the United States, there continue to be some populations who are discontent with NAFTA because of the ongoing relocation of textile, shoe, and component assembly industries. Opponents of this agreement say that the flight is simply toward cheaper labor costs, an absence of union activity, and much relaxed environmental constraints on factories. The final impact of this agreement will not be easily determined. The 1994 treaty allows 15 years to pass before all of the targeted tariffs can be fully reduced or eliminated. In the meantime, there continues to be regional economic competition at all levels, as Mexico's economy has devoted considerable capital and energy to climb out of its 1994 near collapse and American industries continue to try to compete with labor costs that approximate $0.80 an hour compared to United States' scales 10 to 15 times that amount. By 2000, Mexico had become the United States' second largest trading partner with some 85% of Mexico's exports coming to the United States. With the election of Vicente Fox as the country's new president in 2000, Mexico has made it clear that it intends to play the role of a cooperative but competitive neighbor in matters both political and economic.

Notwithstanding the linkage spawned by NAFTA, there is inevitably a touchiness in relations between Mexico and the United States due to (1) the disparity between them in wealth, power, and cultural influence; (2) the historic fact that Mexico lost more than one-half its national territory to the United States in the 19th century; and (3) Mexicans' national pride, derived from great cultural longevity. Mexico's aboriginal civilizations were more highly developed than those within the territory of the present United States. It is estimated that in 1519 the city of Tenochtitlán in the Valley of Mexico was larger than any contemporary city in all of Europe. In the early 17th century, when the first tiny English settlements of the future United States were struggling for survival at the edge of a vast wilderness, the court of the Spanish Viceroy in Mexico City was an opulent and far-reaching center of imperial power.

Geographic Signatures from Native American, Spanish, and Mexican Eras

Mexico's human geography exhibits many elements from three major eras: (1) the Native American era prior to the Spanish conquest of the early 16th century; (2) the Spanish colonial era of the 16th to the early 19th centuries; and (3) the Mexican era following Mexican independence in 1821. In the Native American era, the country was the home to aborigines speaking hundreds of languages or distinct dialects and possessing cultures more advanced than any others in the Americas except those within the Inca empire in the Andes of western South America. Great builders in stone, the aboriginal peoples constructed the mighty pyramids near Mexico City (Fig. 20.2) and the huge temples and palaces of the Mayan culture in Yucatán and Guatemala. Today, richly varied native cultures persist, and the native racial component in the population is pronounced; the greater part of the people of Mexico are *mestizos* of mixed Spanish and aboriginal ancestry. The present settlement pattern was powerfully influenced by the distribution of Native Americans at the time the Spanish came. The conquerors sought out concentrations of aborigines as laborers and potential Christian converts; hence, the locations of Spanish-inspired settlement tended to conform to the population pattern already established or led to coastal port cities with access to upland aboriginal settlements and resources.

The Spanish colonial era lasted nearly four centuries. Following the overthrow of the Aztec empire by Hernando Cortez (Hernán Cortés) in 1521, the Spanish Crown established the Viceroyalty of New Spain, which was ruled from Mexico City and eventually encompassed not only the area now included in Mexico but also most of Central America and about one-quarter of the territory now occupied by the 48 conterminous American states. A Hispanic pattern of life, still highly evident today, was established by Spanish administrators, fortune hunters, settlers, and Catholic priests. Racial, ethnic, and cultural residues of their activities are apparent in recent estimates that **mestizos** comprise 60% of Mexico's population, while Caucasians (nearly all of Spanish ancestry) comprise only 9% Spanish is spoken as a first language by 92% of the populace,

Box
597

Figure 20.2   The Pyramid of the Sun (shown in photo) and the Pyramid of the Moon are prominent pre-Aztec structures rising in the midst of maize and maguey fields near Mexico City. They are sited on broad avenues floored by stone and lined with ruined stone temples and priests' quarters. The entire assemblage, known as Teotihuacán, is what remains of a city that may have housed 200,000 people prior to the city's destruction by fires and abandonment 14 centuries ago.

JESSE H. WHEELER, JR.

and religious affiliation in Mexico is 90% Roman Catholic. Other strong Spanish influences are found in architectural styles and urban layouts; towns and cities characteristically have a Spanish-inspired rectangular grid of streets surrounding a plaza at the town center. The plaza functioned both as the major site for periodic markets and as a setting for the dominant Catholic church in the settlement.

Cattle ranching begun by the Spanish also endures as a way of life in large sections of Mexico; the Spanish introduced not only cattle but also wheat, sugarcane, sheep, horses, donkeys, and mules. The Spanish hunger for precious metals greatly expanded mining, particularly of silver. Many of Mexico's larger cities were founded as silver-mining camps or regional service centers for these mining areas. Innumerable towns and cities of today originated essentially as Catholic missions in the midst of clustered native populations. Thus, when the Mexican era began in the early 19th century, Mexico was overwhelmingly rural, illiterate, village centered, church oriented, and poor. In the capital, an educated elite governed and exploited the country with little benefit to the native and mestizo farmers in the countryside.

The bloody and chaotic Mexican Revolution of 1810–1821 ended Spanish control and instituted a short-lived Mexican empire that expired in 1823 and was followed by a federal republic in 1824. Then came a long period of political instability, with personalities of many persuasions contending for power. From the 1830s through the 1840s, huge blocks of the nation's territory were lost, including the Central American part of former New Spain, as well as Texas and areas now comprising California, New Mexico, Arizona, Nevada, Utah, and parts of other states.

During the 1850s and 1860s, a dynamic Native American reformist, Benito Juárez, attained power for a time, but the country then settled into a long period of despotic personal rule by President Porfirio Díaz, which began in 1876 and did not end until 1910. Democratic processes were largely suspended and the rural population sank deeper into poverty even

as heavy foreign investments were made in railroads, the oil industry, metal mining, coal mining, and manufacturing. Political discontent over national disunity and weakness, internal oppression and corruption, foreign exploitation, poverty, and a dynamic need for land reform then led to the Mexican Revolution of 1910–1920, which ousted Díaz and subsequently turned into a destructive struggle among personal armies of regional leaders. In 1917, a new constitution was enacted under which land reform was an urgent priority. Since then, huge acreages have been expropriated from **haciendas** by the government and redistributed or restored to landless farmers and farm workers, most particularly in a program that provided ejidos, or village-level farming units, for village farmers. Large haciendas still exist, especially in the dry north, but they contain only a very small proportion of Mexico's cultivated land.

Agriculture is still a highly important source of livelihood, although recent estimates report Mexico's population to be 74% urban. Maize (corn) is by far the most important crop. Mexico is almost certainly the area where maize was first domesticated and from which it spread to become the premier crop of the New World and many overseas areas. Although not an export of major importance, it is central to the Mexican diet. Beans and squash, also an inheritance from the Native American past, are other universal Mexican crops and foods.

Today, the total value of output from factories and mines is several times that of output from farms. All of Mexico's economic production needs to be seen in a regional context, however, as the output of particular commodities tends to be rather sharply localized within particular areas.

Regional Geography of Mexico

Mexico is strongly regionalized. Although innumerable areas have distinctive personalities, space restricts us to regions of the broadest scale. We focus on (1) Central Mexico — the country's core region; (2) Northern Mexico — dry, mountainous, and increasingly intertwined with the United States through migration patterns and the development of enormously important

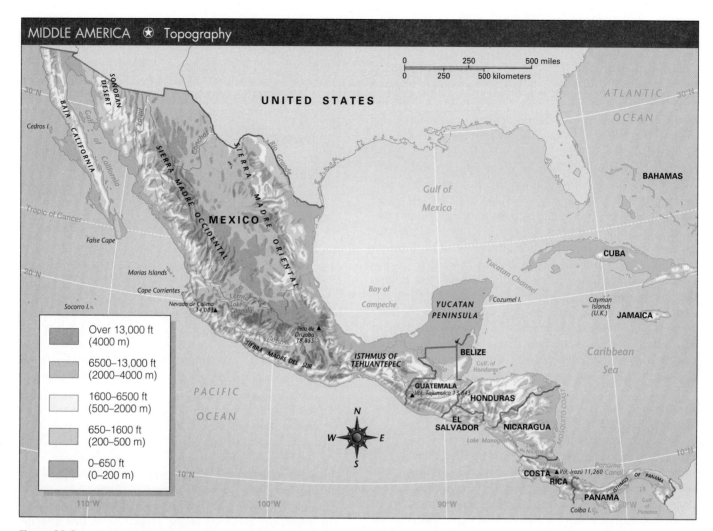

Figure 20.3   The variety of topographic features in Middle America as shown in this map helps illustrate the importance of altitudinal zonation in the region. The significance of such topographic patterns has played a role in everything from mining and economic development to orographic precipitation and associated rain shadows.

maquiladoras; (3) the Gulf Tropics—composed of hot, wet coastal lowlands along the Gulf of Mexico, together with an inland fringe of tierra templada in low highlands; and (4) the Pacific Tropics—a poorly developed southern outlier of humid mountains and narrow coastal plains (Fig. 20.3).

Central Mexico: Volcanic Highland Core Region

It is very common for a Latin American country to have a sharply defined core region where the national life is centered. Mexico is a classic example of this cultural and demographic pattern. About one-half of the population inhabits contiguous highland basins clustered along an axis from Guadalajara (population: metropolitan area, 3,799,000) at the northwest to Puebla (population: metropolitan area, 2,495,000) at the southeast. Toward the eastern end of the axis is Mexico City (population: metropolitan area, 20,750,000; city proper, 8,578,000). The basins comprise Mexico's core region, which is often referred to as Central Mexico or the Central Plateau. The basin floors vary in elevation from about 5,000 feet (1,524 m) to 9,000 feet (2,727 m), with each basin separated from its neighbors by a hilly or

mountainous rim. The centers of the basins contain flat land, which may be marshy. The swampiness has led populations to cluster on higher sloping land, and population pressure has induced cultivation of steep slopes, causing much soil erosion. Most areas are too high and cool for coffee or the other crops of the tierra templada. The prevailing tierra fría environment dictates growth of crops that are less subject to frost damage, such as corn, sorghums, wheat, and potatoes (see Regional Perspective, Chapter 19, p. 512). In addition, cattle (both dairy and beef) and poultry are raised. A productive agriculture that still relies heavily on human labor and animal draft power has been maintained here since the arrival of the Spanish based on fertile and durable soils derived from lava and ash ejected by the many volcanoes.

As mentioned earlier, Mexico City was a major urban and political focus long before its conquest by Cortez in 1521. Near the city are impressive pyramids and other monuments built by aboriginal precursors of the Aztecs. But native political power reached its apex under the Aztec empire in the centuries immediately before the Spanish arrival. The Aztecs, notorious for their practice of ritual human sacrifice, came to the Basin of Mexico from the north and built their capital on an island in a shallow salt lake, with causeways to the shore. This decision to build their most significant ritual and demographic center, Tenochtitlán, in the middle of what was then Lake Texcoco is an example of simple origins confounding complex later histories. Built in the mid-14th century, the city was lined with canals and dike constructions to control flooding and protect water resources for the city. It was here that Cortez finally conquered Tenochtitlán and began the defeat of the Aztecs.

Mexico City was later founded in this same general locale. Most of the lake subsequently was drained, and for the past four and a half centuries, city structures have been built on the drained land. The result has been a gradual subsidence, or settling, of buildings that still continues today as the lake sediments contract. In recent times, this problem has been aggravated by overpumping of ground water to support a metropolitan region of more than 20 million people. But despite the unstable subsurface, including the fact that the city is underlain by a geologic fault, Mexico City now has imposing modern office towers with foundations designed to withstand both subsidence and earthquake shocks. This combination of two such significant environmental hazards underlying what is generally considered the second largest city in the world reminds us that initial settlement decisions seldom anticipate the demographic, political, or cultural magnitude a ritual center can achieve as time passes.

Explosive urbanism has outrun the efforts of Mexico City's planners to cope with the capital city's growth. One mayor of the city commented that administering this monstrous sprawl was like "repairing an airplane in flight." Nonetheless, the city does have some spectacular planning achievements to its credit, not least of which are a modern subway system over 60 miles (100 km) long and a system to pump water to the city through the encircling mountain walls of the Basin of Mexico.

Amid the many memorials to Mexico City's dramatic past, the modern city carries on its functions as the country's leading business and industrial center. It is estimated that approximately one-third of all manufacturing employees in Mexico work in thousands of factories in metropolitan Mexico City. Most factories are small, and manufacturing is devoted largely to miscellaneous consumer items, although some plants carry on heavier manufacturing such as metallurgy, oil refining, and the making of heavy chemicals. The role of high-technology exports has grown from 8% in 1990 to 21% in 2000.

Northern Mexico: Localized Development in a Dry and Rugged Outback

North of the Tropic of Cancer, Mexico is characterized by ruggedness, aridity, and an extensive ranching economy supporting a generally sparse population. There are some widely separated spots of more intense activity based principally on irrigated agriculture, mining, heavy industry, and diversified border industries adjacent to the United States. Most of Northern Mexico lies above 2,000 feet (610 m) in elevation, but there are lowland strips along the Gulf of Mexico, the Gulf of California, and the Pacific. The climate is desert or

Figure 20.4   Trucks from Mexico queue as they wait to enter the United States at a border crossing from Nuevo Larado, Mexico to Laredo, Texas. The Department of Transportation inspectors check trucks for safety violations as they enter the United States from Mexico.

steppe, with scanty xerophytic vegetation except where elevations are high enough to yield more rainfall. In the latter areas, some uplands are carpeted with steppe grasses, and mountains rising still higher have productive coniferous forests. Two major north–south ranges of high mountains — the Sierra Madre Occidental in the west and the Sierra Madre Oriental in the east — tower above the surrounding areas. Surface transport across them is poorly developed. Most forest land is in the Sierra Madre Occidental (see Fig. 20.3). This is Mexico's main source of timber, and lumbering is important in some places. Within Northern Mexico as a whole, ranching continues to be a widespread source of livelihood, although the total number of people supported by it is small. The increasing presence of the maquiladoras that rim the 2,000-mile (3,200-km) border with the United States plays an ever larger social as well as economic role in the region. This economic activity stimulates active trucking networks going into the United States (Fig. 20.4).

In addition to being Mexico's main area of livestock ranching, Northern Mexico also provides the principal output of metals, coal, and products of irrigated agriculture. Iron ore, zinc, lead, copper, manganese, silver, and gold are the main metals produced. The ores, which often contain more than one metal, are smelted with coal mined in Northern Mexico. Some coal is made into coke and used for the manufacture of iron and steel at a number of places, particularly around the city of Monterrey (population: metropolitan area, 3,409,000), which is Mexico's leading center of heavy industry and Northern Mexico's most important business and transportation hub.

Irrigated agriculture in the dry north has expanded greatly since World War II with the building of many multipurpose dams to store the water of rivers originating in the Sierra Madre and emptying into the Gulf of California or the Rio Grande. New or expanded irrigated districts now are widespread, with the main clusters located in the northwest and along or near the Rio Grande. Cotton is a major product of Northern Mexico's irrigated agriculture, and varied fruits and vegetables for shipment to U.S. markets also are grown.

Figure 20.5 Due to the disparity in income and employment opportunities between the United States and Mexico, the 2,000-mile border between the two countries sees attempts by more than 1.5 million Mexicans and Central Americans annually to enter the United States illegally. In this photo, the point of entry is a break in the fence along the Rio Grande River at El Paso, Texas.

In some irrigated areas, the Mexican government has helped develop agricultural colonies of small landowning farmers who have migrated from more crowded parts of the country.

Since World War II, there has been a large upsurge of manufacturing in widely separated Mexican cities immediately adjacent to U.S. cities along the international boundary. The largest are Tijuana (population: city proper, 1,149,000; metropolitan area, probably over 1.35 million) adjacent to San Diego, California, and Ciudad Juárez (population: city proper, 1,107,000) across the Rio Grande from El Paso, Texas.

These Mexican border cities are collecting points for Mexican and Central American migrants who attempt to enter the United States illegally and often succeed despite the efforts of the U.S. Border Patrol to apprehend them (Fig. 20.5). The tension in this zone has also grown because of the smuggling of drugs into the United States through the long border. A U.S. immigration law in 1986 gave amnesty to illegal migrants who could prove residence in the United States for a stated length of time, but the law prescribed tougher measures to try to halt the continuing illegal flow. It remains to be seen how well such measures can succeed along this bicultural border when great poverty lies closely adjacent to affluence and strong demand for low-wage Mexican labor exists on the U.S. side. This tension came to a head in California during the election in 1994, when Proposition 187 — prohibiting the state from providing any welfare or medical support to undocumented aliens (the great majority of those in California who come from Latin America) — brought national attention to the steadily increasing flow of illegal migrants to the United States (Fig. 20.6). One of the most effective means of mitigating the border tensions has been a several-decades-old innovation called the maquiladora.

Maquiladoras: An Effort to Reduce Migration and Increase Industrialization. One of the most significant economic innovations that has been devised to have an impact both on migration patterns and goals of expanded economic development for Mexico has been the *maquiladora*. These are fabrication centers

that stretch along the northern border of Mexico. They are assembly and light-manufacturing businesses that either produce components of products to be exported abroad or assemble products from parts produced abroad and in Mexico. The finished product is then exported, most often to the United States. Maquiladoras began in 1965 and expanded in their economic significance by 70% from 1970 to 1980. The number of these plants has doubled in the 1990s. In 2000, it was estimated that there were 3,500 such plants stretched along the U.S.-Mexican border. They are generally located just south of this border, where there is relatively inexpensive Mexican labor and easy transport of partially assembled goods from north of the border to these free port facilities.

Figure 20.6 In the mid-1990s, a pattern began of makeshift rafts setting sail from Cuba, and later Haiti, for the Florida Peninsula. Often overcrowded and never prepared for bad weather or rough water, these flotillas were more interested in being intercepted by the U.S. Coast Guard than actually crossing the 90 miles of open sea that lies between Cuba and Florida. Once the United States changed its initial policy of bringing the so-called "boat people" to its shores, the flow of rafters decreased. These men have come ashore on the Grand Caymans, having launched themselves from the southern shore of Cuba. Their raft lies just offshore.

In the 2000 election of Vicente Fox as the new president of Mexico came an ambitious outline of dramatic political goals he wanted to accomplish in his first term. One of the most delicate ones pertained to the 2,000-mile border shared by Mexico and the United States. It is estimated that 8 million Mexicans with close ties to family still living south of that border are potential visitors to this largest Middle American nation. The Mexicans who are resident in the United States manifest these continuing links by sending an estimated $6 to $8 billion from the United States to their families still resident in Mexico. President Fox wants to see the U.S.-Mexican border become more permeable so that these millions of Mexican emigrants can more easily return home to reaffirm and maintain family bonds across the border. Fox sees this as a further enhancement of the bonds between these two NAFTA nations. It is probable that his government envisions a considerable cash flow in this travel as well.

In addition, Fox wants to have the United States rethink its closure of the earlier Bracero Program that facilitated the periodic migration of Mexican farm labor into agricultural states. Fox wants to have the new program be a Guest Worker plan that would make it relatively easy for some 350,000 Mexican field workers annually to come to the United States for seasonal, generally agricultural, work. The Mexican government thinks that the explicit expansion of U.S. accommodation of these workers who are generally promised jobs during the fruit and crop harvest seasons would help diminish the flow of the estimated 1.5 million Mexicans who annually try to cross the border illegally. A more welcoming border program might also reduce the average death of one Mexican migrant each day in this continual effort to sneak into the United States.

The 2000 U.S. Census shows that Hispanics are not only the largest minority in the United States, but they are also the fastest growing. Given the dual realities of such a dominant demographic shift and the election of a new Mexican president with energetic concern for changing the dynamics of movement across the Mexican-U.S. border in both directions, there is good precedent for developing another perspective on this continually significant migration stream.

The economic benefits of the maquiladoras to Mexico include the abundance of jobs linked with these factories and their ability to employ Middle American labor that would perhaps otherwise try to migrate to the United States. Since much of the capital supporting the maquiladoras is primarily from the United States and Japan, there is not as much trickle-down economic benefit from these activities as was anticipated, but the importance of the innovation has led it to generate more foreign exchange than tourism in the last decade. There has also been the growth of complementary industries that have grown up to support the maquiladoras, such as small-scale retail and transportation services.

Since 1994–1995, maquiladoras in Mexico have produced more than 14% of Mexico's export trade. In the region around Nuevo Laredo and Matamoros, just across the border from southern Texas, there are more than 2,000 maquiladora operations. The operations have also been effective in providing economic opportunities for the Mexican laborers who settled at the border zone to participate in various legal seasonal labor programs in the American Southwest.

This economic innovation will continue to have a strong impact not only on Mexico's economic development, but on the economies of other nations as they consider this free port arrangement to encourage offshore factory growth. In addition, the manufacturing skills learned by maquiladora workers can be transferred to jobs in the industrial center around Mexico City and Monterrey, where there are solid job openings for skilled labor. Although the maquiladoras seem a highly localized, borderlands economic phenomenon, their impact is in fact much more broadly regional.

The Gulf Tropics: Oil and Gas, Plantations, the Maya, Tourism

On the Gulf of Mexico side, Mexico has tropical lowlands that have a much lower population density than Central Mexico but which are very important to the national economy. Their most valuable product is oil, discovered in huge quantities in the mid-1970s. Just as in coastal Texas and Louisiana, some fields are onshore and others lie underneath the Gulf. The Mexican oil industry is a government monopoly (carried on by a government corporation, Petróleos Mexicanos, or PEMEX), and the United States is its largest foreign customer. The bonanza oil strikes in the 1970s seemed so promising that the government borrowed heavily from foreign banks to finance both oil development and a wave of other economic and welfare projects. Dependence on oil income grew to the point that crude oil represented 67% of all Mexican exports in value in 1983. Subsequently, however, a worldwide oil glut caused prices to skid, and by 1990, crude oil accounted for only 33% of exports. Economic reforms in Mexico during the 1980s reduced inflation and spurred growth, but by 1994, the country's external debt was nearly $128 billion, and a sizable share of export revenue was being spent on debt service. The 1994 collapse of the Mexican peso came in large part because of the combination of massive foreign debt and diminishing oil revenues, a pattern that continued through the late 1990s. Nonetheless, recent restructuring of the economy and a steady increase in the value of oil being exported have created a more hopeful outlook for the future.

The Gulf Tropics are Mexico's main producer of tropical plantation crops, which contribute in a relatively minor way

PROBLEM LANDSCAPE

"Mexico's Green Dream: No More Cancúns"

Mexico City. January 11, 2001. Mexico's Caribbean coast was once a natural paradise. Then a computer program run by the Tourism Ministry spat out the name of a potential gold mine: a spit of sand called Cancun. Today Cancun has nearly 25,000 hotel rooms, roughly three million visitors a year, and nowhere left to build. So for nearly 100 miles down the coast from Cancun, developers are paving paradise as fast as they can. They have built almost 13,000 hotel rooms, most in the last decade, and they intend to double the number in the next five years.[a]

The tension in this Problem Landscape relates to environmental beauty. The eastern coast of Mexico in the Gulf of Mexico is a nearly idyllic tropical environment. Historically, the area has been little settled and developed, yet it is easily accessible by large, relatively affluent, vacation-loving populations in the United States and Canada. From 1970, when Cancún was a hamlet of some 100 Maya, through the decade of the 1980s as the town and its accommodations were built new from the ground up, it has served as a model for a computer search for new, ideal landscapes. The Mexican government (the provinces of Quintana Roo and Yucatán and the federal government of Mexico) had invested 3 years in analysis of computer-determined ideal sites for "an international holiday center." Cancun was selected in Quintana Roo, and it has been a steady moneymaker, although it suffered great damage in Hurricane Gilbert in September 1988.

The tourist landscape elements that have served this site so very well include water (both salt and fresh, although there is little rainfall in Cancún), environmental beauty, tropical climate (Cancún is about 16°N latitude), a service population, and modern accommodations. And increasingly, this setting is being enhanced by the uncovering and opening of Mayan ruins in the Yucatán Peninsula, Belize, and for the more ambitious, in El Peten in northeast Guatemala. This collection of landscape features has been at the heart of Cancún's success.

Mexico's fear now is that not only Cancún but the entire Gulf coast of Quintana Roo and, to the west, Yucatán may endanger their natural beauty through excessive development by well-financed tourism promoters who have come late to the scene but still see an enormous potential for profitable investment in this region. To this scenario comes the cry, "No More Cancúns!" Such a call is a reminder of the yell heard in Oregon four decades ago when California landscape influences were working their way north into Oregon. "Don't Californicate Oregon!" came the cry from environmentalists and long-time settlers alike. Harsh images. Intense feelings!

In any scenario of landscape transformation that becomes a financial success, there is always the danger — even the likelihood — of excessive development. In the Middle American landscapes of the Yucatán Peninsula and the small adjacent islands in the Gulf of Mexico, this excess is particularly problematic because of the very fragile nature of this environment. The dominant landscape question always remains: When is more development too much development? And who is to decide?

[a]Mexico's Green Dream: No More Cancuns." *New York Times*, January 12, 2001, pp. A1, A10.

to the export trade. Cacao, sugarcane, and rubber are leading export crops in the lowlands, and coffee exports come from a strip of tierra templada along the border between the lowlands and Mexico's Central Plateau.

The greater part of Mexico's Gulf Tropics lies in the large Yucatán Peninsula. In pre-Hispanic times, the Mayan people developed a notable civilization there and in adjacent parts of Mexico, Guatemala, and Belize. Its existence was unknown until ruined cities overgrown by dense vegetation were found in the 19th century. Recent translations of Mayan records and ethnographic studies of their descendants reveal an enduring culture with an elaborate past. Why they abandoned their cities is still a mystery. Just as Mayan cities generate an intense curiosity, they also play an increasingly popular role as part of the major tourist flow to the Gulf of Mexico coast (Fig. 20.7). However, even with the magnitude of financial support that tourism gives the Mexican economy, there is always another side to such development. Displeasure with the "Cancúning" of the coast of the Yucatán Peninsula is the focus of the Problem Landscape box above.

Figure 20.7   This boardwalk scene at Mexico's Yucatán Peninsula's Isla Mujeres reflects the ever-growing presence of North American tourists in the Cancún landscape. These vendors on the wharfside depend on the arrival of these traveling populations with their fascination for the luxuries and the simplicity of tourist destinations in Mexico. Even while such travel and traffic bring important dollars to the Mexican economy, they also create cultural awkwardnesses. As noted in the Problem Landscape, the call for "No More Cancúns" is derivative from scenes such as these.

Figure 20.8   The resort hotels that have grown up around the early resort city of Acapulco play a strong role in the expanding importance of tourist dollars in the Mexican economy. The area's mountains, beaches, warm climate, and proximity to reasonably convenient airports have all helped this Mexican city and area grow rapidly. Tourism in Mexico has grown to a nearly $7 billion annual importer of foreign exchange in recent years.

The Pacific Tropics: Mountainous Southwestern Outlier

South of the Tropic of Cancer, rugged mountains with a humid climate lie between the densely populated highland basins of Central Mexico and the Pacific Ocean. This difficult terrain is far more thinly populated than Central Mexico, has a high incidence of relatively unmixed native people living in traditional ways, and has relatively little economic activity. There are two exceptions to the comparative underdevelopment. One is the coastal resort of Acapulco, located on a spectacular bay about 190 miles (c. 300 km) south of Mexico City (Fig. 20.8). This port city of the cliff-diver images continues to play an important role in attracting tourist dollars. There is also a major new iron and steel complex about 150 miles (240 km) up the coast from Acapulco near La Unión. Its operation is based on local deposits of iron ore and is intended to stimulate more general economic growth within this region.

20.2   Central America

Between Mexico and South America, the North American continent tapers southward through an isthmian belt of small and relatively less developed countries (LDCs) known collectively as Central America. Five countries — Guatemala, Honduras, Nicaragua, Costa Rica, and Panama — were the initial members of the Central American Federation. They all have seacoasts on both the Pacific and the Atlantic Oceans (Caribbean Sea; see map, Fig. 20.1). The other two Central American countries have coasts on one ocean only: El Salvador on the Pacific and Belize on the Gulf of Mexico, or the Atlantic. If the seven countries formed one political unit, they would have an area only about one-third that of Mexico (see Table 19.1), with a total population equal to less than one-third that of Mexico.

Geographical fragmentation is a major characteristic of Central America, manifesting itself in the fragmented pattern of its political units. Five of the present countries — Guatemala, El Salvador, Honduras, Nicaragua, and Costa Rica — originally were governed together as a part of New Spain in a unit called the Captaincy-General of Guatemala. Composed of many small settlement nodes isolated from each other by mountains, empty backlands, and poor transportation, the unit never established strong geopolitical cohesion. Separate feelings of nationality became strong enough to fracture overall unity after the end of Spanish imperial rule in 1821. The area gained independence from Mexico in 1825 as the Central American Federation, but in 1838–1839, this loose association segmented into the five republics of today. In recent decades, there have been serious international tensions affecting the five-country area, associated in part with antigovernment guerrilla warfare in various countries. However, by the mid-1990s, tensions had lessened, and the Central American area appeared to be entering a new era of greater political stability.

In Guatemala, El Salvador, Honduras, and Nicaragua, society is composed of groups that differ sharply in wealth, social standing, ethnicity, culture, and opinion. Wealthy owners of large estates — the Latifundia — have formed a social aristocracy since the first land grants were made in early colonial times by the Spanish Crown. Members of this group tend to be of relatively unmixed Spanish descent, although the great majority have some Native American blood. They have, in general, monopolized wealth, defended the status quo, and exercised great political power. Occupying a middle position in society are small landholding farmers, largely mestizos, who own and work their own land. At the bottom of the scale are millions of landless mestizo tenant farmers, hired workers, and the many native peoples (largely in Guatemala) who live essentially outside of white and mestizo society. In the early 1990s, urban-dwellers formed a majority in Nicaragua (62% of its population) but were in the minority in the other three nations. In each country, the national capital is overwhelmingly the largest urban center and serves not only as the demographic hub but also as the economic and political center of the nation, playing the role of primate city quite thoroughly.

Costa Rica has had a rather different historical development from the other four countries of the old Central American Federation. It was distant from the center of government in Guatemala and was left largely to its own devices in developing its economy and polity. Early Spanish settlers eliminated the smaller numbers of indigenous people there, and the ethnic pattern that developed was dominated overwhelmingly by whites of Spanish descent. By one estimate, more than 85% of the present Costa Rican population is white, whereas the same source shows an overwhelming predominance of mestizos and/or native people in El Salvador, Honduras, Nicaragua, and Guatemala. No precious metals of significance were found in Costa Rica, and no large landed aristocracy developed. The basic population of small landowning European immigrant farmers or their descendants managed to create one of Latin America's most democratic societies despite some episodes of dictatorial rule. It has also evolved steadily as a nation undergoing productive economic development.

Panama and Belize (once British Honduras) were not part of the Central American Federation, and they achieved independence much later. Panama was governed from Bogotá as a part of Colombia until 1903, when it broke away as an independent republic. The United States was deeply involved in the political maneuvering that led to Panamanian independence, and in 1904, the United States was granted control over the Panama Canal Zone, site of the famous transoceanic Panama Canal, which was opened in 1914. In 1978, after long agitation by Panama for return of control of the Canal Zone, the United States ratified a treaty under which Panama received full control in stages, ending on December 31, 1999, with the full turnover of the Canal to the government of Panama. Meanwhile, in 1989, a U.S. military invasion of Panama ousted its president, Manuel Noriega, who was captured, brought to the United States, and subsequently convicted of drug-smuggling charges by a U.S. court and jailed in the United States.

Belize, on the Caribbean side of Central America, was held by Great Britain during the colonial era under the name of British Honduras. Remote, poor, underdeveloped, and nearly uninhabited, the colony was used largely as a source of timber, with African slaves brought in to do the woodcutting. It was also claimed by Guatemala as part of its national territory. Not until 1981 did Belize become independent. It is now a parliamentary state governed by a prime minister and bicameral legislature; the British monarch, represented by a governor-general, is the chief of state. English is the official language, and ecotourism (carefully planned tours to environmentally sensitive areas at relatively expensive rates) is steadily more important in the small country as the continuing prohibition of U.S. tourism to Cuba promotes the development of new tourist destinations in and on the margins of Caribbean Latin America.

The commercial production of coffee, bananas, cotton, sugarcane, beef, and a miscellany of other export commodities is very patchy in distribution and not very impressive in overall output, although it does provide the greater part of Central American exports. All the countries except overcrowded El Salvador still have pioneer zones where new agricultural settlement is taking place. There is some manufacturing in the few large cities, but the products are generally simple consumer goods for domestic markets. There are a few factories assembling foreign-made cars, but such bits and pieces of productive enterprise yield scanty returns in relation to the needs of the impoverished population. The situation is worsened by the fact that a large share of the profit from such industrial enterprises makes its way into very few pockets.

An Agricultural Signature: Milpas and Middle and Central America

One of the distinctive landscape signatures of Middle and Central America, and parts of South America as well, is the peasant burning and farming of government land for short-term farming practices. This so-called primitive agricultural system was the system, in all likelihood, that provided support for the Mayan and Aztec societies. Today, it serves as an outlet for peasant frustrations at not being able to gain access to their own land, so they turn to large expanses of land that they can farm in a 1- to 2-year hit and then move on to other lands. Milpa is the term used in Middle America for this pattern that geographers know more commonly as slash-and-burn, or shifting, cultivation. It is not a system well understood, as the following newspaper quote below suggests: "The sky is thickly overcast; the smoke makes eyes water and lungs gasp for breath; the muggy heat is unbearable. This is the Nicaraguan capital, Managua. Life was much the same this week in Mexico City, which was on full smog alert. Honduras reported a continuing sharp rise in respiratory disorders, and until Monday the country's four main airports had all been closed for some days for lack of visibility. At Tikal, in Guatemala, 700 firefighters were struggling to protect historic Mayan temples. For weeks now, fires have raged in scrub and forests from Panama to southern Mexico. Their effects are devastating, and will long outlast the human discomfort."[1]

The records of smoke, haze, and widespread burning associated with Central America in 1998 are thought by some to be just another reflection of the impact of El Niño and associated drought (see Definitions & Insights, Chapter 19, p. 523). However, the real issue is not the cyclic uprising of warm ocean waters linked with El Niño but the more pervasive human trait of attempting to harvest regional forests for economic gain.

There has been a traditional view that the forest lands are the resources of those who come to them, cut down the vegetation, burn the trees to let the light shine in, and plant new crops. This milpa method of land use goes back to the earliest settlements in Middle America (Fig. 20.9). It is more surprising now because all governments have placed legal restrictions on the burning of forest lands. However, as this article shows, the force of law has not generally been as strong as the drive for peasant farmholds or forest products.

In Central America, the article goes on to say, there is a steady incursion of big ranchers clearing land for cattle and of timber companies "getting permission to cut 50 trees and taking 500 . . . " There is, in addition, the ongoing impact of the former Nicaraguan Contras guerrillas staking claim to forest lands in the face of few other economic opportunities. These forces, when totaled, amount to a major dynamic of landscape change. A Nicaraguan ecologist has proclaimed, "Nicaragua will be a desert in 50 years' time." Even making such an assessment has a danger because two

---

[1] "The Burning of Central America." *The Economist*, May 30, 1998, pp. 33–34.

LANDSCAPE IN LITERATURE

And We Sold the Rain

Targeting the financial and political predicament in which many of the Central American countries find themselves, author Carmen Naranjo sets the stage for a satire entitled "And We Sold the Rain" by describing a mythic, very poor Central American country. It is plagued by the classic problems of too little food, too many people, and a crippling gap between poor and rich and city and countryside. In addition, the country has gotten into very deep international debt. Such a situation not only requires enormous debt repayment schedules but also forces the nation to interact with international lenders and fiscal consultants (the "fat cows" in the story). In an effort to take the population's mind off of hunger and these problems, the president decides to sell his nation's major natural resource: the rain that pours down on this impoverished place continually, causing everything to flood and people to grow weary of the wetness. He sells the rain to a Middle Eastern king of a country called the Emirate of the Emirs. The following excerpt begins just after the president has explained to his people how the money gained from the sale of the rain will allow their small Central American country be stronger, more independent, and prosperous.

The people smiled. A little less rain would be agreeable to everyone, and the best part was not having to deal with the six fat cows [the International Monetary Fund, the World Bank, the Agency for International Development, the Embassy, the International Development Bank, and perhaps the European Economic Commission], who were more than a little oppressive. Moreover, one couldn't count on those cows really being fat, since accepting them meant increasing all kinds of taxes, especially those on consumer goods, lifting import restrictions, . . . paying the interest, which was now a little higher, and amortizing the debt that was increasingly at a rate only comparable to the spread of an epidemic. And as if this were not enough, it would be necessary to structure the cabinet in a certain way. . . .

The president added with demented glee, his face garlanded in sappy smiles, that French technicians, those guardians of European meritocracy, would build the rain funnels and the aqueduct, [and provide a] guarantee of honesty, efficiency, and effective transfer of technology.

By then we had already sold, to our great disadvantage, the tuna, the dolphins, and the thermal dome, along with the forests and all Indian artifacts. Also our talent, dignity, sovereignty, and the right to traffic in anything and everything illicit.

The first funnel was located on the Atlantic coast, which in a few months looked worse than the dry Pacific. The first payment from the emir arrived — in dollars! — and the country celebrated with a week's vacation. A little more effort was needed. Another funnel was added in the north and one more in the south. Both zones immediately dried up like raisins. The checks did not arrive. What happened? The IMF garnisheed them for interest payments. Another effort: a funnel was installed in the center of the country, where formerly it had rained and rained. It now stopped raining forever, which paralyzed brains, altered behavior, changed the climate, defoliated the corn, destroyed the coffee, poisoned aromas, devastated canefields, desiccated palm trees, ruined orchards, razed truck gardens, and narrowed faces, making people look and act like rats, ants, and cockroaches, the only animals left alive in large numbers.

To remember what we once had been, people circulated photographs of an enormous oasis with great plantations, parks, and animal sanctuaries full of butterflies and flocks of birds, at the bottom of which was printed, "Come and visit us. The Emirate of Emirs is a paradise."

The first one to attempt it was a good swimmer who took the precaution of carrying food and medicine. Then a whole family left, then whole villages, large and small. The population dropped considerably. One fine day there was nobody left, with the exception of the president and his cabinet. Everyone else, even the deputies, followed the rest by opening the cover of the aqueduct and floating all the way to the cover at the other end, doorway to the Emirate of the Emirs.

In that country we were second-class citizens, something we were already accustomed to. We lived in a ghetto. We got work because we knew about coffee, sugar cane, cotton, fruit trees, and truck gardens. In a short time we were happy and felt as if these things too were ours, or at the very least, that the rain still belonged to us.

A few years passed; the price of oil began to plunge and plunge. The emir asked for a loan, then another, then many; eventually he had to beg and beg for money to service the loans. The story sounds all too familiar. Now the IMF has taken possession of the aqueducts. They have cut off the water because of a default in payments and because the sultan had the bright idea of receiving as a guest of honor a representative of that country that is a neighbor of ours."[a]

From Carmen Naranjo, "And We Sold the Rain," trans. by Jo Anne Engelbert, *Worlds of Fiction*, eds. Roberta Rubenstein and Charles R. Lawson (New York: Macmillan Publishing Co., 1993, 948–952).

Central American ecologists lobbying against the burning of the forests have been murdered in the 1990s, and others have received death threats.

A section of the Nicaraguan coast in the northeast was selected as a United Nations (UNESCO) "biosphere" reserve in February 1998. U.N. scientists in charge of evaluating such landscapes for this U.N. designation have declared that this pristine part of the Caribbean coast lands and forest lands of Nicaragua will have to be given military protection by the government if Nicaragua is serious about preserving this area

PAUL EDMONSON

Figure 20.9 Slash-and-burn agriculture and unauthorized forest burning are very powerful agents of landscape change in all Latin America. Major forest regions are often a long distance from direct governmental protection. A family can turn forest richness into a family resource by cutting, burning, planting, and harvesting an acre or two of forestland. In this process, small populations are given traditional returns on family labor, but primary forest reserves are often lost to unproductive volunteer grasses. This same burning process is also undertaken in a forest-clearing effort to raise cattle for the beef market in the United States and Canada. For whatever reason this burning of forestland is undertaken, it reduces the biotic diversity of the region, and the forest that may ultimately grow back if human intervention is denied will never be as rich a resource as it was as primary forestland.

in its current state. The gloom that these widespread Central America fires evokes is captured by this quote from the newspaper *La Prensa*, printed in Managua: "Desolation and destruction are everywhere; if nothing is done Nicaraguans will be without flora, fauna, or food."

The milpa landscape is evident all through Middle America. In the past, it has been the standard means of support for much of the pre-Columbian civilization, but today, it is a force that makes steady inroads on tropical rain forest lands. A peasant farmer trying to support his family by working an acre or two of unguarded forest land is much less inclined to think about the ecological preservation of a biotic resource base than the potential return on a summer of varied cropping among the fallen timbers of a plot that he and his family have burned and planted for their own benefit. It is a landscape problem that runs the full gamut of forest land expression in Latin America, but especially Middle America.

Physical Belts and Their Activities

The distribution of settlement and the accompanying economic life in Central America strongly relate to three parallel physical belts: the volcanic highlands, the Caribbean lowlands, and the Pacific lowlands. All Central American countries except Belize share the volcanic highlands that reach south from Mexico. Here lie the core regions, including the political capitals, of Guatemala, El Salvador, Honduras, Nicaragua, and Costa Rica. The only cities with over a million people in their metropolitan areas are San Salvador (metropolitan area, 1,729,000), Guatemala City (metropolitan area, 2,476,000), and San José (metropolitan area, 1.4 million)—all national capitals. Tegucigalpa (metropolitan area, 1,309,900) (Honduras) and Managua (metropolitan area, 1,295,300) (Nicaragua) are other significant urban centers. Dotted with majestic volcanic cones and scenic lakes, the highlands are the most densely settled large section of Central America. Climatically, they are classed largely as tierra templada, and most of Central America's coffee is grown there. Only a minor fraction of the land is still forested. The forces that drive such deforestation are in part the human migration into forests that traditionally have been landscapes of opportunity for landless peasants willing to clear forest and farm the ragged lands, generally with milpa or slash-and-burn farming methods.

There is popular sentiment in this region that the government has too much concern for governmental cabinets and high offices and too little for the needs of the common people (see Landscape in Literature, p. 538). Central America historically has had revolutionary movements that embody these fears, and the recent histories of Nicaragua and El Salvador have demonstrated this potential for popular uprising and for U.S. involvement at one level or another. In Nicaragua in 1979, for example, the governing Somoza family—with a long history of U.S. support—was thrown out by the country's Sandinista revolutionaries. After a very contentious 11-year rule characterized by continual battles with the military-backed Contras (also supported by the United States), the Sandinistas were replaced by an elected president, and by the early 1990s, a reasonably stable government was established with some of the earlier Sandinista reforms institutionalized. Nearly one-third of the countries in the broader Latin America have similar histories of U.S. military, economic, or political intervention, most of which stretch into the 20th century, and some which were in the past decades.

By the 17th century, black African slaves were being brought into this region of Central America, providing the initial basis for Central America's relatively small black and mulatto or Creole populations. In the late 19th century, some black workers migrated there from Jamaica to work on the banana plantations being developed by U.S. fruit companies to serve the American market. In the early 1900s, several companies were consolidated into the United Fruit Company (now part of United Brands). An up-and-down history of banana growing saw many plantations eventually shift to the Pacific side of Central America because of devastating

The Panama Canal, providing an interocean passage through the narrow Isthmus of Panama, is of great importance to both Panama and the world. Some 80 shipping routes, or about 5% of the world's cargo volume, use the 50-mile (80-km) shortcut from the Atlantic Ocean to the Pacific Ocean. On an ocean voyage from New York City to San Francisco, nearly 8,000 miles are saved by using the Canal (see Fig. 20.10). The idea of a "shortcut" across the Isthmus of Panama/Central America was long in the minds of people of commerce who saw what an enormous marine cost had to be paid in going around the tip of South America to move goods between the Pacific to the Atlantic Oceans. For the first part of the 1800s, Nicaragua was the primary focus because of natural landscape features that appeared better able to accommodate a crossing without massive construction. The British and the Americans were the two nations most interested in such an accomplishment. In 1846, however, Colombia and the United States signed a treaty supporting a canal project. This was significant because the Isthmus of Panama at that time was in the political control of Colombia.

The pace of interest in a crossing changed profoundly in 1849 with the discovery of gold in California. Prospectors and speculators began to sail to the port of Colón on the east side of Panama, make the 50 miles of land crossing in any way possible, and then reboard a ship in Balboa and sail to San Francisco. By 1855, this pattern was modified by the construction — by a collection of New York business interests — of a railroad from port to port. This method of crossing, which of course required a classic "break-of-bulk" of any goods as well as passengers, was then overshadowed in 1878 by

Colombia's granting of permission to a French firm to build a canal across the isthmus near the railroad line. This was undertaken by Ferdinand De Lesseps, who had been the engineer who had successfully completed the Suez Canal in an engineering feat that took a decade (1859–1869). The French quickly bought control of the railroad line and began digging in 1882, with a plan to make their crossing a sea level transit that would require no locks. This venture failed with bankruptcy in 1889 and was followed by another French attempt, which also failed. As difficult as the construction was in these efforts, failure was also prompted by the high death toll through yellow fever and other tropical diseases. By the beginning of the 20th century, the French had left, and the United States had just been given further evidence of the enormous importance of such a crossing through their naval requirements of the 1898 Spanish-American War.

In 1903, Panama broke away from Colombia, and within weeks, the United States recognized the new government. Within months, an agreement had been struck between the new Panama and the U.S. government for the construction of a transisthmus Panama canal. Construction was begun in 1904 and completed in mid-1914. The first crossing was made under the banner: "The Land Divided; The World United." The Canal has facilitated the transit of an average of 800 to 1,000 vessels a month since its opening.

After decades of tension between Panama and the U.S. government over the control of the Canal, even though the initial treaty gave the Canal land in perpetuity, the Jimmy Carter administration in 1978 agreed to return the Canal Zone to Panama. In 1977, the

control of the Canal had been given over to a Hong Kong firm, Hutchison Whampoa, and in December 1999, the United States removed the last of its troops that had been long stationed in the Canal Zone. It was the end of a very important era in the dominance of the seas in trade and, more specifically, of the centrality of the Panama Canal for trade from East Asia and the west coast of North America to the major market center of the New York area.

The canal had been built by the United States at a cost of approximately $380 million and considerable human costs. Disease was the major killer — yellow fever, bubonic plague, and malaria were the most powerful enemies of the jungle-based project — but it was finally finished. The first ship negotiated the various lakes, including the monumental Gaitun Lake, canals, and completed the passage in August 1914. Work has been done continually on the 50-mile passage, with a 20-year project now underway to widen the Gaillard Cut at the continental divide at the highest point of the Canal to make the passage there a two-way cut. This will not be done until 2012. Currently, approximately one-seventh of all United States trade is carried through it. The Canal generated $486 million in traffic income in 1996. The Panamanian government received $201 million of that total. That balance has been changed completely since the 1999 handover.

There continues to be some tension over the fact that the two access points — Balboa in the Pacific and Cristobal in the Atlantic — are now controlled by a firm with strong Chinese backing. When and if the current locks are widened enough to be able to accommodate the massive oil tankers and container ships that ply the Pacific, a new level of concern is bound to arise.

F4.6 106

plant diseases on the eastern Caribbean side. Today, exports of bananas, grown in part on the Caribbean lowlands and in part on the Pacific lowlands, continue to be very important in the economies of Honduras, Panama, and Costa Rica. The 1999 near trade war between the United States and the European Union related to banana production from this area and was not resolved until early 2001 with a compromise that allowed a return to some steady Central American exporting of bananas to EU countries. The Caribbean lowlands represent the principal area in Central America where agricultural settlement is expanding on a pioneer basis. Settlers generally come from overcrowded highlands and are often aided by government colonization plans featuring land grants and the building of roads to get products to market and bring in supplies. Even with this varied activity, however, there is a continuing problem of sluggish economic development.

More than one-half of Panama's population lives close to the Canal, and until 1997, there were some 10,000 U.S. troops stationed in and around it (Panama City, metropolitan area, 1,017,000). The U.S. citizens and the troops stationed in the Canal Zone were called *Zonians*. The United States handed the Canal over to Panama in December 1999. This plan has generated some anxiety among the shipping nations for whom the Canal is central to their routes. It is also estimated that it will take $7 to $10 billion to modernize the locks and the channels so that the world's largest tankers can use the Panama Canal. Regardless of what the future holds for the Canal, this geographic shortcut has been an innovation of enormous significance in global shipping patterns since its 1914 opening (see Regional Perspective on p. 540 and Fig. 20.10).

The eastern section of Panama adjoining Colombia contains a roadless stretch where the continuity of the Pan-American Highway System is broken in what is called the Darien Gap (see Fig. 20.1). Thick rain forests, mountains, and swamps have presented serious obstacles to the bridging of this last remaining gap in a surfaced highway linkage from northern Canada and the United States through Latin America to southern Argentina and Chile. This region continues to be a home territory of fiercely independent native peoples called the Kuna who occasionally become vocal representatives of the Meso-American aboriginal populations who have made very explicit decisions not to develop and take on the culture and landscape trappings of the Westernizing native peoples.

20.3 The Caribbean Islands

The islands of Caribbean America, commonly known as the West Indies, are very diverse. They encompass a wide range of sizes and physical types, great racial and cultural variations, and a notable variety of political arrangements and economic mainstays. Cuba, the largest island, is primarily lowland, although low mountains exist at the eastern and western ends of the island. The next largest islands — Hispaniola, Jamaica,

and Puerto Rico — are steeply mountainous or hilly. Trinidad and Tobago, just off the northeast corner of South America, has a low mountain range in the north (a continuation of the Andes) and level to hilly areas elsewhere.

Most of the remaining islands, all comparatively small, fall into two broad physical types: (1) low, flat limestone islands rimmed by coral reefs, including the Bahamas (northeast of Cuba) and a few others, and (2) volcanic islands, such as the Virgin Islands, Leeward Islands, and Windward Islands, stretching along the eastern margin of the Caribbean from Puerto Rico toward Trinidad and Tobago (see Fig. 20.1). Each of the volcanic islands consists of one or more volcanic cones (most of which are extinct or inactive), with limited amounts of cultivable land on lower slopes and small plains. Barbados, east of the Windward Islands, is a limestone island with moderately elevated rolling surfaces.

The islands are warm all year, although extremely hot weather is uncommon. Precipitation, however, varies notably, with windward northeastern slopes of mountains often receiving very heavy precipitation, and leeward southwestern slopes and very low islands having rainfall so scanty as to create semiaridity. The natural vegetation varies from luxuriant forests in wet areas to sparse woodland in dry areas. Most areas have two rainy seasons and two dry seasons a year. Hurricanes, approaching from the east and then curving northward, are a scourge of the northern islands and adjacent mainland areas from the Middle Atlantic states of the eastern United States all the way southward through Middle America in late summer and autumn.

Plantation agriculture and tourism are the economic activities for which these islands are best known. Sugarcane, the crop around which the island plantation economies were originally built, is still the main export crop in Cuba, Barbados, and a few other places. Preplantation crops in this region of considerable significance included tobacco, indigo, and cacao. Other commercial crops include coffee, bananas, spices, citrus fruits, and coconuts. Production comes from large plantations or estates worked by tenant farmers or hired laborers, from small farms worked by their owners, or in Cuba, from state-owned farms. Mineral production is absent or unimportant on most islands, the most conspicuous exceptions being Jamaica (bauxite), Trinidad and Tobago (petroleum), and Cuba (metals). Manufacturing has made sizable gains during recent times, and it now provides the largest exports from some islands. Tourism is a major source of revenue in many islands, most notably the Bahamas, Puerto Rico, and the Dominican Republic. Most tourists come from the United States (see Table 19.3).

Emigration has long been an active option for this region's inhabitants, most of whom are economically underprivileged. Millions of West Indians now live in the United States, the United Kingdom, Canada, France and the Netherlands — mainly in cities. The money they send back to relatives is a major source of island income.

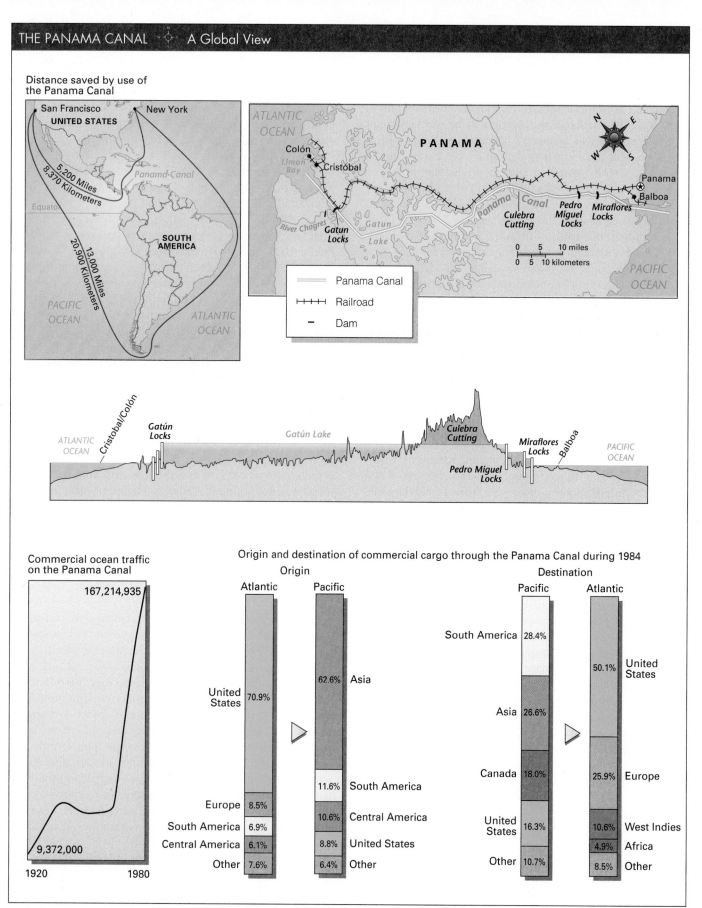

Distance saved by use of
the Panama Canal

Commercial ocean traffic on the Panama Canal

167,214,935

9,372,000

1920   1980

Origin and destination of commercial cargo through the Panama Canal during 1984

Origin

Atlantic   Pacific

United States 70.9%   62.6% Asia

11.6% South America

10.6% Central America

Europe 8.5%
South America 6.9%   8.8% United States
Central America 6.1%
Other 7.6%   6.4% Other

Destination

Pacific   Atlantic

South America 28.4%   50.1% United States

Asia 26.6%

Canada 18.0%   25.9% Europe

United States 16.3%   10.6% West Indies
4.9% Africa
Other 10.7%   8.5% Other

Sources: *Atlas of Central America and the Caribbean.* Macmillan Publishing Co., New York, NY (1985).
The World Book Encyclopedia. World Book, Inc., Chicago, IL (2000).

Figure 20.10   The Panama Canal pulls trade from all over the world not only because of the shipping
time it saves but also because the rounding of the southern end of South America through the Straits of Magellan is a particularly difficult passage. These maps with the trade graphic show the broad range of trade points
that mark the origin and destination of cargo carried through the canal, the vertical profile of the actual canal
route through the Isthmus of Panama, and the distance saved by use of the Canal.

The Greater Antilles: Major Islands of the Caribbean Sea

The four largest Caribbean islands are together called the Greater Antilles; the smaller islands are known as the Lesser Antilles. Of the five present political units in the Greater Antilles, Cuba, the Dominican Republic, and Puerto Rico were Spanish colonies, Haiti was French, and Jamaica was British. This diversity of colonial origins is only one of many ways in which these units have developed and maintained distinct geographic personalities.

By any standard, Cuba is the most important Caribbean island unit. Centrality of location, its large size, a favorable environment for growing sugarcane, some useful minerals, and proximity to the United States are important factors contributing to Cuba's leading role in the Caribbean. Cuba lies only 90 miles (145 km) across the Strait of Florida from the United States. At the western end of the island, Havana developed on a fine harbor as a major colonial Spanish stronghold. The island would stretch from New York to Chicago if superimposed on the United States and is nearly as large in area as the other Caribbean islands combined. Gentle relief, adequate rainfall, warm temperatures, and fertile soils create very favorable agricultural conditions. Agriculture has always been the core of the Cuban economy, and sugarcane has been the dominant crop. Cane sugar still accounted for the largest share of Cuban exports by value in 1999. The 1959 revolution, led by Fidel Castro, shifted the ownership of most Cuban land to the state. About 90% of all farmland is now under state control. However, as in most Communist regimes, state-controlled agriculture has led to significant decreases in crop production in Cuba, and a more than 40-year U.S. trade embargo with Cuba has made it difficult to find effective replacement exports for the island.

Havana (Habana; population: metropolitan area, 2,636,000; city proper, 2,312,000) is Cuba's main industrial center as well as its political capital, largest city, and main seaport. Even with its economic troubles, Cuba's government, aided in a massive way by the former Soviet Union, has had some success in raising the adequacy of medical care, housing, and education over the years, particularly in comparison to the levels of such services during the 25 years of the Batista regime that Castro replaced in 1959 (Fig. 20.11).

In 1962, a great international crisis erupted when the Soviet Union began to build facilities in Cuba that would have accommodated nuclear missiles capable of striking Washington, D.C., New York, and other cities of the eastern United States. The presence of the Soviet missile silos and launchers was discovered through the use of aerial photo interpretation — a geographic research tools of increasing military and commercial utility. Although the missiles were ultimately withdrawn under intense pressure from the U.S. government, led at the time by President John F. Kennedy, Cuba has remained a storm center of political life in the Western Hemisphere. One of the curious geographic outcomes of the tense geopolitics in the Cold War era of the 1960s had Cuba shipping its sugar to mainland

Figure 20.11 Medical graduates listen to a speech given by Cuban President Fidel Castro during a ceremony in Jose Marti Plaza in Havana, Cuba, in August 2000. Some 4,000 graduates took part in this event. The faces of the group reflect well the heterogeneous population base of so much of the Greater Antilles in the Caribbean. Castro has been giving speeches at events like this long before most of the readers of this textbook were even born!

China, while Taiwan (already a close ally of the United States) shipped its sugar to the United States.

Aid to Cuba from the Soviet Union fell sharply when the USSR ended its support for the Soviet bloc in 1989 and the USSR and the Communist party collapsed in 1991. This profound change has placed new strains on the Cuban economy. Castro's efforts in the mid-1990s to have the 1960 U.S. economic sanctions removed were brought to American attention in 1994 when tens of thousands of men, women, and children from Cuba (and thousands more from Haiti at the same time) launched themselves from the northern and western shores of their island. Their hope was either to reach the Florida peninsula or be picked up by the U.S. Coast Guard and ultimately be granted asylum in the United States. Castro argued that if the United States relaxed its sanctions, the Cuban economy would be able to gain enough strength to make flight to the United States relatively less attractive for Cubans (see Fig. 20.6 on p. 533). With the mainland of the United States barely 90 miles distant from Cuba, the political and economic significance of this island continues to be a major thorn in U.S.–Latin American relations.

East of Cuba, the mountainous island of Hispaniola is shared uneasily by the French-speaking Republic of Haiti and the Spanish-speaking Dominican Republic. In Haiti, the black population forms an overwhelming majority, although a small mulatto elite (perhaps 1% of the population) controls a major share of the wealth. Population density is extreme, infrastructure is scanty, and poverty is rampant. The agricultural economy, largely of a subsistence character, reflects little influence from the former French colonial regime, which was driven out at the beginning of the 19th century. The republic, independent since 1804, has a West African flavor in its landscape, crops, and other cultural expressions. Deforestation, erosion, and poor transportation are major problems. Aside from subsistence farming, Haiti's small economy concentrates on light manufacturing (aided by low-wage labor), coffee growing, and tourism. Haiti's annual per capita GDP PPP amounts to only $1,800, the second lowest in all of Latin America (Cuba is lowest at $1,700). These two figures are among the lowest in the entire world.

The Dynamics of Tourism

130   Tourism is one of the classic two-sided coins of economic development. Consider, for the moment, the impact of the international tourist who arrives in a Middle American nation for a short pleasure or business trip. On the one side is the economic impact of the foreign exchange brought in by the traveler. The anticipated arrival of such people leads to the construction of hotel and recreational accommodations and to major investments in the infrastructural investment in ports, highway and rail networks, airports, and restaurants and entertainment facilities. At the same time, these dollars tend to promote the preservation of native crafts and historic landscapes to be showcased to attract visitors.

On the other side of the coin is the argument that much of the capital generated by tourism goes into transnational coffers and that the wages paid to the local people who do everything from hotel work to playing tour guide to entertaining the tourists are at best modest. The sometimes powerful cultural tensions that arise when tour boats or buses arrive are very real (giving rise to the caustic phrase, "If there is a tourist season, why can't we shoot 'em?"). The continued pressure on public space, transportation facilities, and natural resources such as beaches and national parks generated by tourist traffic adds to the tensions. However, looking at the magnitude of the foreign exchange generated annually (Mexico: $7.6 billion; Dominican Republic: $2.1 billion; Puerto Rico: $2.0 billion; Bahamas: $1.4 billion; Jamaica: $1.1 billion), there is a fiscal consequence to being a popular tourist destination. As a geographic theme, the interweaving of the landscapes, the physical settings, the international images of a place, and the network of travel systems that bring people to a popular tourist location present interesting opportunities for study. The dynamics of tourism, although of particular significance to the landscapes of Middle America, are evident all across the globe and this double-edged impact of touring and spending activity is ever part of the fiscal and social scene (see Figs. 20.7 and 20.8).

The Dominican Republic bears chiefly a Spanish cultural heritage. Santo Domingo, its capital, is the oldest European city of continued habitation established in the New World (c. 1496), possessing the first European church, school, university, and hospital. An independent country since 1844, the republic has a history of internal war, foreign intervention (including periods of occupation by U.S. military forces), and misrule, although recent peaceful transfers of power point to-

ward greater governmental stability. The increasingly diversified economy relies on tourism, mining, oil refining, and agricultural products such as sugar, meat, coffee, tobacco, and cacao. Maquiladora export production is playing an increased role in the republic's economy — a pattern developed in the NAFTA world of the Mexico-U.S. borderlands. The Dominican Republic also produces baseballs for the U.S. market and has sent some of the best major-league baseball players to the United States as well. Urbanization is more pronounced than in Haiti, and the capital of Santo Domingo (population: metropolitan area, 2,726,000; city proper, 2,006,000) is considerably larger than Haiti's capital of Port-au-Prince (population: metropolitan area, 1,692,000).

Puerto Rico, smallest and easternmost of the Greater Antilles, is a mountainous island with coastal strips of lowland. The island was ceded to the United States by Spain after the latter's defeat in the Spanish-American War of 1898 and was administered by Congress as a federal territory until 1952. In that year, it became a self-governing commonwealth voluntarily associated with the United States. Puerto Rico's pre–World War II economy was agriculturally based, with a strong emphasis on sugar. There were accompanying wide disparities within the populace in income and living conditions. After the war, a determined effort by Puerto Ricans to raise their living standard brought great changes. Hydroelectric resources were harnessed, a land-classification program provided a basis for sound agricultural planning, and a thriving development of manufacturing industries and tourism began, financed in large measure by capital from the United States. A complex package of tax incentives was organized to attract American businesses and industries, and a steady migration stream sent Puerto Ricans to the United States — primarily to the metropolitan New York area — also providing capital for the island in the form of family remittances.

Today, agriculture has been far outstripped in product value by manufacturing. Petrochemicals, pharmaceuticals, electrical equipment, food products, clothing, and textiles are among the major lines. Sizable copper and nickel deposits have recently been found. Although Puerto Ricans have chosen thus far to remain a commonwealth, there is continuing pressure within the island for statehood, and a small minority has expressed a desire for total independence (Fig. 20.12). San Juan (population: metropolitan area, 1,973,500; city proper, 425,000) is the capital, main city, and port. The relatively successful efforts Puerto Rico has made in its economic development over the past four decades come in good part because of the early efforts to attract American businesses to the island through promises of tax abatements and below market land options. These benefits were not permanent but usually long enough to have the firms decide to continue operations in Puerto Rico even after the most generous dimensions of the enticements expired.

Jamaica, independent from Great Britain since 1962, is located in the Caribbean Sea about 100 miles (c. 160 km) west of Haiti and 90 miles (c. 145 km) south of Cuba. Its population of

The Caribbean as an Engine of Cultural Diffusion

Box
518

One of the fears of geography, or a geographer, is that countries and regions drift into becoming only statistics and bland regional comparisons. For example, in thinking about the relatively small amount of area taken up by the Caribbean countries in contrast to all of South America or Mexico, or even Central America, the danger of the near loss of an area in a world regional geography text is very genuine. If we think, however, about the cultural influence that the Caribbean has worked upon the United States, we are jolted out of the unfair comparison of size with size. It is in the Caribbean that Columbus and his three ships landed on that October night in 1492. It was named for the Carib Indians that Columbus discovered, and it was those same Indians that consumed their prisoners in such a way that the European conquistadors felt justified in all that they did to bring the "cannibals" under control for the church and the flags of Iberia. The Caribbean Sea washes up against the shores of Panama and

brought first French and then American builders to the Isthmus of Panama in quest of a crossing from the Sea to the wide-open Pacific Ocean. In these last 50 years, the Caribbean has been the setting for the profound experimentation with communism and socialism undertaken by Fidel Castro and the Cuban government. From the Cuban Missile Crisis of 1962 to the political tension generated by the safe arrival of a young Cuban boy named Elian in 2000, Cuba has been a major political concern off the flank of the United States for more than half a century. Providing baseballs and baseball players from the islands of the Greater Antilles as well as safe havens for illicit drug money in various islands all through the Caribbean help to make up the engine of cultural diffusion that creates the Caribbean. And that is not even mentioning the power of music from the area. This region is a good reminder of how a small area can have a major influence on the world, both around it and much farther away.

2.6 million (as of 1998) is largely descended from black African slaves transported there by the Spanish and (after 1655) the British. When slavery was abolished in 1838, most blacks left the sugar plantations on the coastal lowlands and migrated to the mountainous interior. Tourism, bauxite, and agriculture have constituted Jamaica's economic base for decades.

The Lesser Antilles: Common Threads but Differing Circumstances

Over 3 million people live on the islands of the Lesser Antilles that are scattered in a 1,400-mile (2,250-km) arc from the United States and British Virgin Islands east of Puerto Rico to

the vicinity of Venezuela. This screen of small islands defining the eastern edge of the Caribbean Sea contains a variety of political entities. The Netherlands Antilles and Aruba are Dutch dependencies evolving toward greater self-government. Guadeloupe and Martinique are French overseas departments, and Anguilla and Montserrat are still British dependencies. The British, however, have withdrawn from these islands in a massive way, starting with independence for Jamaica and Trinidad and Tobago in 1962. Since then, independence from British rule has been gained by Barbados (1966) and six smaller units. The British also granted independence to Bermuda in 1968, to the Bahamas in 1973, and to the mainland unit of Belize in 1981 but still retain their colonies of the British Virgin Islands,

GIULIANO DE PORTU/GAMMA LIAISON

Figure 20.12   Prostatehood demonstrators display an American flag as they participate in a statehood rally. The island of Puerto Rico has had a prostatehood political bloc for more than four decades, but there never has been a popular vote that gave statehood the winning majority in any referendum over the island's political status. The dynamics of the Puerto Rican population that has moved to New York City were given a strong stage presence in *West Side Story* in the 1960s, and the intensity of feelings both for and against statehood continue with the same energy in the present.

Cayman Islands, and Turks and Caicos Islands. All their former colonies have retained membership in the Commonwealth of Nations.

The economies of the islands suffer from many serious problems, including unstable agricultural exports, limited natural resources, a constantly unfavorable trade balance, embryonic industries lacking economies of scale, and chronic unemployment. A shared trait of the islands is their history of European colonialism and plantation slavery. Such generalizations contribute to some understanding of the islands but give little hint of the individuality these small units present when viewed in detail. One example — Trinidad and Tobago — is discussed here.

Trinidad and Tobago comprise the largest, richest in natural resources, and most ethnically varied of the island units in the eastern Caribbean. Trinidad, by far the larger of the two islands, has offshore oil deposits in the south. Yet Trinidad refines more oil than it pumps. Imported crude oil from various sources is refined and then shipped out, mainly to the United States. Some 40% of the population is black, most of whom live mainly in the urban areas and are often employed in the oil industry. Another 40% are the descendants of immigrants from the Indian subcontinent — a population stream that developed for indentured workers after the abolition of slavery. This element lives primarily in the intensively cultivated countryside. About 18% are classified as "mixed," and there are small numbers of Chinese, Madeirans, Syrians, European Jews, Venezuelans, and others. This ethnic complexity of Trinidad is reflected in religion, architecture, language, diet, social class, dress, and politics. It is important to realize that while the mainlands of South America and Central America have maintained major populations of some considerable indigenous homogeneity with the aboriginal populations, the islands of the Caribbean have largely become a blend of many heritages with black Africans playing a very significant role in many of the countries.

The world of Mexico, Central America, and the Caribbean is a prime example of the influence of location and scale. Mexico, in a world distant from the massive presence of the United States as a neighbor, would probably have a more distinctive geographic history and contemporary impact on global affairs. Central America is forever to be in the shadow of the northern neighbors of Mexico, the United States, and Canada. The Caribbean islands, on the other hand, owe a great deal of what financial success they have to their proximity to the United States and Canada, even though their location also throws a shadow over their own efforts to be unique and gain clear geographic identity even among tropical and near-tropical islands. There is seldom any way to escape the influence of geography in the development of national identity.

CHAPTER SUMMARY

- Middle America is made up of Mexico, the Central America countries, and countries in the Greater and the Lesser Antilles in the Caribbean Sea. Cuba, Haiti, the Dominican Republic, Puerto Rico, Jamaica, and Trinidad and Tobago are the largest Caribbean islands.

- Mexico is the largest, most populated, and economically most developed of the Middle America nations. Its area of 762,000 square miles and its population of more than 100 million give Mexico a giant role in Middle America. It is also the largest nation in the world to have Spanish as its dominant language.

- The North American Free Trade Agreement (NAFTA) has played a major role in the pace of economic change in Mexico, particularly in the borderlands between Mexico and the United States. There has been steady demographic and economic growth in the cities lying just south of the U.S. border (maquiladoras), and much of Mexico's light industry occurs in this region. The major industrial manufacturing takes place farther south. NAFTA is seen as a source of both good news and bad news by Mexicans because of the activity that it stimulates and the migrants it collects along the northern edge of the country. Mexico's GNP is 3% that of the United States.

- Mexico's history is composed of three distinct eras: the Native American, the Spanish, and the Mexican. The Aztecs and Mayas were the most important Native American populations, but there were others as well. There is still a strong Spanish presence evident, and the large population of mestizos represents the people who are a racial mix of the Spanish and indigenous groups in Mexico. The Spanish were a major force in establishing land use patterns that continue today. The Mexican era began with the 1810–1821 period of the Mexican Revolution, and the country suffered through many governments and the loss of much land to the United States in the following decades. The second period of revolution in Mexico was 1910 to 1920.

- Central Mexico is a volcanic highland area in which Mexico City, the world's second largest city (15 million or more) is located. There is a steady problem of urban subsidence there, and smog is also a growing problem. It is Mexico's leading industrial center.

- The Gulf Tropics of Mexico have gas, oil, plantations, Mayan ruins, and tourists as major features. Mexican oil has been important in economic development as a nationalized resource, but it has also caused overspending and excessive debt as world oil prices dip. Tourism has been a growing influence on settlement and development in this region as well as on the Pacific Coast and in northern Mexico.

- Central America is made up of Guatemala, Belize, El Salvador, Honduras, Nicaragua, Costa Rica, and Panama. In Guatemala, Honduras, and El Salvador, there are sizable blocks of Native American peoples, and there is a strong European presence in Costa Rica. Panama's history over the past century has been shaped powerfully by the construction of the Panama Canal in 1904–1914 and by the American presence in support of the Canal. In all of the Central American countries — although less in Costa Rica — there is a wide gap in economic prosperity between the very wealthy and the very poor.

- Plantation agriculture continues to be important in the tierra caliente on both coasts of Middle America. The system of altitudinal zonation encompasses the tierra caliente, tierra templada, tierra fría, and tierra helada. Bananas are the major plantation crop, but sugar is also raised in Central and Middle America and the Caribbean. There is new interest in beef raising for the North American market, and this is causing some shift in land use — both legal and illegal. The illegal land use is associated primarily with unauthorized deforestation in both the tierra caliente and tierra templada.

- There are many newly independent countries in Caribbean America that have emerged from colonial connections to Europe in the past four decades. Some islands continue to have colonial relationships with Europe and maintain very strong economic and political affiliation with the countries that colonized them.
- Cuba is the most important Caribbean island because of its size, population, and very unique recent history. Sugar and tourism have long been central to its economy, but since Castro's clash with the United States in the mid-20th century, both tourism and sugar trade have been powerfully constrained. Tourism is slowly gaining economic significance again, but the United States continues to refuse to buy any sugar from the island. The 1962 Cuban Missile Crisis continues to shape the U.S. disinclination to have any open political interaction with Cuba.

- Puerto Rico is the smallest and easternmost of the Greater Antilles and is a self-governing commonwealth of the United States. There continue to be groups on the island seeking either U.S. statehood or complete independence. There is also a steady migration stream of Puerto Ricans from the island to the American Northeast. New York City is a traditional migration destination. Agriculture has been steadily losing ground to industry as Puerto Rico has actively been pursuing stronger economic development.
- Other major islands of settlement and of political and economic significance in the Caribbean include Jamaica and, in the Lesser Antilles, the Netherlands Antilles, Aruba, Guadeloupe, Martinique, Barbados, the Bahamas, and the Virgin Islands. These represent a variety of political affiliations. Banking, tourism, and varied resources are important in the economic profiles of these locales.

REVIEW QUESTIONS

1. Define Middle America and list the countries in Central America and the major islands in Caribbean America.
2. What is the largest country in Middle America and what is the population of its capital city?
3. What does NAFTA stand for? What countries have created this pact and what is its primary significance for Middle America?
4. What are the zones defined in altitudinal zonation in Middle America and what are the most significant settlement and agricultural characteristics of each zone?
5. What are the primary components of the regional geography of Mexico?
6. What is the name of the cooperative manufacturing innovation that brings together Mexican or Middle American labor and in-

dustrial components from all over the world? What is the general market for such activity?
7. Name the countries of Central America and describe the problems with forest burning that takes place in this region.
8. Describe the genesis, development, and significance of the Panama Canal.
9. What is the primary agricultural pattern of the Caribbean islands? What is the role played by tourism in their local economies?
10. Compare and contrast the regional histories of Cuba and Puerto Rico.
11. What role have natural resources played in the development of the Caribbean islands and which have been most important where?

DISCUSSION QUESTIONS

1. Discuss the ways in which proximity to the United States has been a factor in the development of Middle America.
2. What factors have caused Mexico City to grow so rapidly? Discuss the ways in which this growth has been both positive and negative for Mexico.
3. Create a scenario that explains the role that Native American cultures have played in Middle America, bringing their influence into the present as well.
4. Discuss the role of migration in Middle America with particular reference to the U.S.-Mexican border, the maquiladoras, and Mexican economic development.
5. Develop profiles of Guatemala, Costa Rica, and Panama as three distinct Central American nations, assessing how each of them has been influenced by the United States.

6. Discuss the role of the Panama Canal in the 20th century and speculate on its role in this century as well.
7. Lobby for and against the role of tourism in the Caribbean islands and in Middle America overall. What is the impact of this economic activity?
8. Discuss the role of burning forests in Middle America from the perspective of the government, the peasantry, environmentalists, and ranchers looking toward North American markets.
9. Discuss the strongly distinct economic and geographic development of Cuba, Haiti, the Dominican Republic, and Puerto Rico. What has caused such differences?
10. What is the message of the literature excerpt "And We Sold the Rain"?

CHAPTER 21

South America: The Continuing Search for Identity

ROBERT FRIED/TOM STACK & ASSOCIATES

Iguassu Falls, which Argentina shares with Brazil near the Paraguay border, as seen from the Argentine side. These 200-foot falls symbolize the great energy potential available to many South American countries in the form of falling water. In the absence of abundant fossil fuels, resources such as hydroelectric power take on even greater significance.

Chapter Outline

South America comprises the lion's share of Latin America's total land area. It is the fourth largest continent on Earth, amounting to 6,880,706 square miles (17,821,028 sq km). It is approximately 4,500 miles (7,200 km) from north to south and extends to a maximum of 3,200 miles (5,120 km) wide. It includes the world's largest Portuguese-speaking nation, Brazil, which is also home to the largest remaining tropical rain forest in the broad network of the Amazon River. South America has a history of significant Native American development and settlement on the west side of the Andes Mountains, which course south from Colombia in the north all the way to Tierra del Fuego at the very southern tip of the continent (Fig. 21.1). The continent also has a wide range of climates, landscapes, and ecological zones, ranging all the way from the parched dry desert of the Atacama in Chile to the lush tropical rain forests of central Brazil and eastern Peru and Ecuador. The leaders of this region increasingly claim that South America is ready to be fully involved in the economic development of the Western Hemisphere and in the Pacific trade flows that so many feel will play a dominant role in the unfolding of the 21st century. Even though there are enormous areas still distant from transportation hubs and economic centers, and though there are ongoing struggles within its confines, South America sees itself as a giant coming awake (Fig. 21.2).

21.1   The Andean Countries

The South American countries of Venezuela, Colombia, Ecuador, Peru, and Bolivia — all within the tropics and traversed by the Andes — are grouped in this text as the Andean countries (see Fig. 21.1). Common interests among them were recognized in 1969, when five countries signed the Cartagena (Colombia) Agreement and agreed to create a new free-trade zone called the Andean Group. Initially, Chile was part of the group, but it later withdrew. Venezuela came late to the group, which is now made up of Bolivia, Peru, Ecuador, Colombia, and Venezuela. Following is a list of a number of significant traits shared by these countries.

1. *Environmental zonation.* This region's zonation, both vertical and horizontal, is pronounced. The natural setting of each country features local environments and subregions that offer extreme contrasts and that often are very isolated by natural barriers. There are, in addition, demographic patterns that relate to these zones.

2. *Fragmented settlement patterns.* There is widespread fragmentation in patterns of settlement and economic development because of the broken topography that characterizes the Andean countries. This is particularly notable in Peru.

3. *Populous highlands and sparsely settled frontiers.* Every country contains both well-populated highlands and an extensive pioneer fringe of lightly populated interior lowlands or low uplands awaiting development. Some interior areas already yield minerals (notably oil) on an export basis.

4. *Dynamic coastal lowlands.* Every country except Bolivia has an economically productive coastal lowland region. Bolivia originally had a Pacific coastal zone but lost it to Chile during a 19th-century war (1879–1883), thus becoming a landlocked state, as is adjacent Paraguay.

5. *Significant indigenous populations.* All countries had relatively large indigenous populations in the mountains, valleys, and basins at the time of the 16th-century Spanish conquest, and these old settlement areas are today major zones of dense population and strong political influence, especially in the highlands.

6. *Significant Native American landscape signatures.* The Incas of Peru were the most creative and productive pre-Columbian (pre-1492) culture in Latin America. At the time of the Spanish conquest, the Incas had authority over South America from the current border of Ecuador and Colombia south for nearly 2,000 miles (3,200 km) to the Rio Maule in Chile. The Inca clan had taken over from the Quechua civilization, which had been the primary power for centuries in what we now speak of as Inca territory. From their governing center in Cuzco, Peru, the Incas had engineered a system of roads, bridges, and settlements that connected and supported their empire. In building these structures, they demonstrated extraordinary skill, even though they had neither paper nor a writing system. (Records were kept by knotted ropes called *quipus*.) Their power was centered in two cities: Cuzco in the south and Quito, Ecuador, in the north. They used geographic and topographic surveys and extended military control throughout this empire from 1200 until the arrival of Spain's Francisco Pizarro in 1532. When

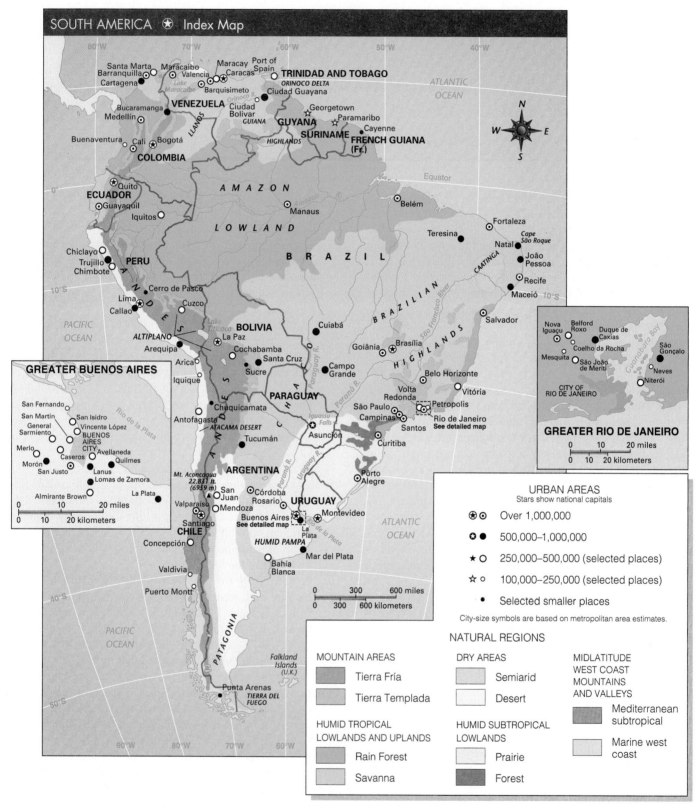

Figure 21.1 Index map of South America. General reference maps of natural areas and major cities in South America. Note the clearly defined arrangement of cities around the rim of the continent and the enormous and relatively empty expanses in the interior. Look particularly at Fig. 21.2 to see the expansiveness of the locations that are more than 20 miles from a railway line.

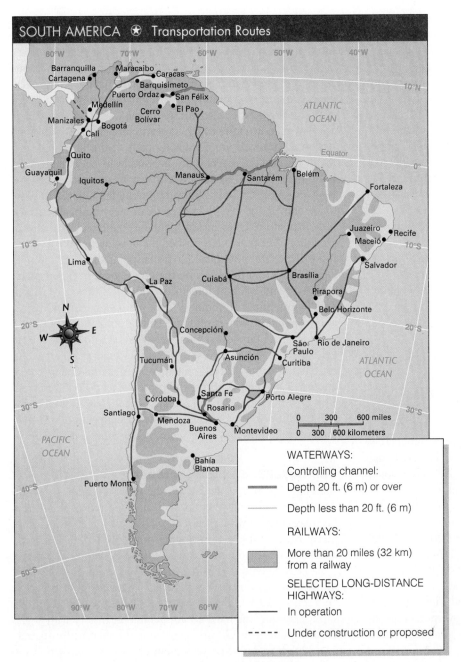

SOUTH AMERICA ✮ Transportation Routes

WATERWAYS:
Controlling channel:
━━━ Depth 20 ft. (6 m) or over

─── Depth less than 20 ft. (6 m)

RAILWAYS:
▨ More than 20 miles (32 km) from a railway

SELECTED LONG-DISTANCE HIGHWAYS:
─── In operation

----- Under construction or proposed

Figure 21.2   Transportation routes in South America. The basic pattern of major highways in the Pan American Highway System is shown, together with new long-distance roads reaching deep into the interior of Brazil, partly because of the move of the national capital from Rio de Janeiro on the coast to the thinly settled interior in the 1960s. Note the enormous size of the regions that are at least 20 miles from a railway line.

Pizarro landed (with a total of only 168 soldiers, including 102 horsemen), he was warmly greeted by an Incan leader, Atahualpa. The Spanish conquistador welcomed the Incan dignitary into his camp and then killed him, launching the destruction of the empire that had ruled a great deal of the western side of the Andes for more than three centuries. Machu Picchu, the grand ceremonial center on the east side of the Andes, was about the farthest east Inca control reached. Of all the riches Pizarro gained from this conquest, perhaps the most important is the potato, a tuber domesticated by predecessors of the Inca and diffused widely to the Old World during the Age of Discovery.

7. *Dominant Hispanic traditions.* Each one of the Andean countries maintains strong Hispanic traditions but at the same time has an important heritage from Andean native populations. The Spanish language is official in all five countries (although Native American languages also are official in Peru and Bolivia), and over nine-tenths of all religious adherents in every country embrace Roman Catholicism. These badges of Iberian culture were imposed by the Spanish on the advanced native cultures within the Inca empire. Today, Bolivia, Peru, and Ecuador have Native American majorities, and the cultural survival of these groups from pre-Spanish times contributes to a keen sense of indigenous identity to each of

these countries. In all the rest of Latin America, only Guatemala has an indigenous majority, although native strains are very pronounced in many countries with important mestizo elements. The latter is also true of Colombia and Venezuela, where Native Americans amount to only 1 or 2% of the total, but mestizos form the largest ethnic element.

8. *Endemic poverty.* All the countries in this region are relatively poor, although oil wealth makes Venezuela considerably better off than the others. However, considerable capital flows into many of these countries through the cultivation, processing, and marketing of illegal drugs, especially cocaine. It is difficult to track this money and accurately discuss its broader significance (see Problem Landscape, p. 554).

9. *Primate cities.* In all countries except Ecuador, the national capital towers above all other cities in size and political and cultural importance, reflecting the characteristics of primate cities.

Venezuela

Venezuela's population is concentrated mainly in Andean valleys and basins not far from the Caribbean coast, as it was when the Spanish arrived in the 1500s. The highlands contain the main city and capital, Caracas (population: metropolitan area, 3,439,000; city proper, 1,784,000). At an elevation of 3,300 feet (c. 1,006 m), Caracas has a tierra templada climate. Around Caracas, the Spanish found a relatively dense native population to draw on for forced labor, as well as some gold to mine and climatic conditions amenable to the commercial production of sugarcane and, later, coffee. Labor needs associated with sugarcane in this region led to the 16th-century introduction of African slaves, but Venezuelan plantation development remained relatively small. Approximately one-tenth of the present population is classified as black. Until the middle of the 20th century, Caracas was a fairly small city, and its valley was still an important and highly productive area of farming. Now, the city fills and overflows the valley, and Venezuelan agriculture is now more generally associated with other highland valleys and basins.

Discovery of oil resources in the coastal area around and under the large Caribbean inlet called Lake Maracaibo have made Venezuela a relatively fortunate Latin American country in recent decades. It was one of the most influential charter nations in the 1961 founding of the Organization of Petroleum Exporting Countries (OPEC). Oil revenues have provided a per capita income that is fairly high for Latin America (see Table 19.1), although Venezuela is far from being a rich nation. The oil-based income has fueled rapid urbanization, attracted foreign immigration, and financed considerable industrialization. However, in recent years, the country's income has suffered because of continuing fluxuation in world oil prices.

While the highlands have supported Venezuela in the past, and the coastal lowland is crucial to the present, the much larger share of the country that lies inland from the Andes is a major hope for the future. This sparsely inhabited section is mostly tropical savanna in the Guiana highlands and in lowlands (often marshy) along the Orinoco River and its tributaries. For centuries, remote interior Venezuela has been Native American country (in the highlands) or poor and very sparsely settled ranching country (in the lowlands), but development is now quickening. Government-aided agricultural settlement projects are occupying parts of the lowlands, and major oil and gas reserves are beginning to be exploited. In the Guiana highlands after World War II, enormous iron-ore reserves south of Ciudad Bolívar and Ciudad Guayana began to be mined for export by American steel companies, and huge bauxite reserves in the highlands continue to await exploitation. Venezuela has pushed the development of a major industrial center at Ciudad Guayana (population: city proper, 748,200). Ocean freighters on the Orinoco carry export products away and bring in foreign coal for industrial fuel. Major steel, aluminum, oil-refining, and petrochemical industries are in operation, with further diversification planned.

In terms of physical setting, Venezuela is also home to Angel Falls (3,212 ft/c. 979 m), the world's highest waterfall. Its existence is part of the reason that Venezuela has dedicated an enormous amount of its area to protected parks and reserves. This governmental decision regarding land use and environmental control also reflects the increasingly important role that tourist dollars play in the country's economic picture.

The Guyana, Suriname, French Guiana Trio

Although this trio of nations on the northeast edge of South America are not classified as Andean, mention of them best fits in association with a discussion of Venezuela. Guyana (formerly British Guiana), Suriname (formerly Dutch Guiana), and French Guiana are a trio of countries (although French Guiana is actually a department of France) that reach from the Atlantic Ocean on the northeast coast of South America between Venezuela on the north all the way to Brazil on the south. The nations have a long plantation history with sugar most important in Guyana and tea and coffee in Suriname. The Dutch traded their settlement of New Amsterdam (later called New York City) for the rights to Suriname and its coastal richness for agricultural exploitation. French Guiana shared the role of plantation agriculture in the production of sugarcane. This country was also home to Devil's Island, an isolated island prison that has a whole world of lore and was still in use until 1954. Today, the three countries have economies that are still based on agriculture, but they also gain a considerable portion of their income from bauxite (especially Guyana and French Guiana) and tourism.

Colombia

Within its wide expanse of the Andes, Colombia includes many populous valleys and basins, some in the *tierra fría* and others in the *tierra templada*. The great majority of Colombia's mestizos and whites are scattered among these upland settlement clusters, which tend to be separated by sparsely settled mountain country with extremely difficult terrain. Lower valleys and basins, along with some coastal districts, earlier developed plantation agriculture employing slave labor. Currently, 14% of the country's population is mulatto and 4% is black.

The Highland Metropolises: Bogotá, Medellín, Cali

Colombia's inland towns and cities have tended to grow as local centers of productive mountain basins. The three largest — Bogotá, Medellín, and Cali — are diverse in geographic character. The largest Andean center in all of South America is Bogotá (population: metropolitan area, 7,594,100; city proper, 6,540,400), Colombia's capital. It lies in the *tierra fría* at an elevation of about 8,700 feet (c. 2,650 m — much higher than Denver, Colorado). The Spanish seized the land in this region, put the native people to work on estates, and improved agricultural productivity by introducing new crops and animals — notably wheat, barley, cattle, sheep, and horses. However, exploitation and disease reduced Native American numbers so drastically that the population took a long time to recover. The Spaniards were not able to develop an export agriculture, as night coolness and frosts precluded such major trading crops as sugarcane and coffee, and early transport to the coast via the Magdalena River and its valley was extremely difficult. Slopes were so formidable that a railroad connection with the Caribbean Sea was not achieved until 1961. In recent decades, modern governmental functions and the provision of air, rail, and highway connections have led to explosive urban growth in upland Colombia. The Bogotá area has even found a large agricultural export market in flowers grown for air shipment to the United States.

Medellín (population: metropolitan area, 2,962,900; city proper, 1,908,600) lies in the *tierra templada* at about 5,000 feet (1,524 m) in a tributary valley above the Cauca River. The area lacked a dense indigenous population to provide estate labor, was difficult to reach, and consequently, remained for centuries a region of shifting subsistence cultivation emphasizing corn, beans, sugarcane, and bananas. The introduction of coffee and the provision of railroad connections changed this situation in the early 20th century. Excellent coffee could be produced under the cover of tall trees on the steep slopes, and the Medellín region became a major producer and exporter. In the United States, a clever advertising campaign designed around Juan Valdez, a fictional representative of Colombian coffee growers, has added landscape and product images of the region to American consciousness. Local enterprise and capital developed the city itself into a considerable textile-manufacturing center, and it grew from about 100,000 in population in the 1920s to its present size approaching 3 million.

Cali (population: metropolitan area, 2,732,300; city proper, 2,161,000) is located in the *tierra templada* some 3,000 feet (915 m) above sea level on a terrace overlooking the floodplain of the Cauca River. Cali had easier connections than did Medellín with the outside world via the Pacific port of Buenaventura. The Spanish developed sugarcane

Figure 21.3   In the area of Putumayo province, south of Bogota and the home of the FARC rebels who contest the regional control of the Colombian government, this coca "factory" exists as one of the nearly countless upslope coca operations. In this photo, a coca leaf worker, or "raspachine," mixes chemicals with coca leaves to help wash out impurities. The U.S. Drug Enforcement Agency estimates that some 60% of the world's cocaine, or about 3 tons a week, is produced in this area. Three tons have a U.S. street value of approximately $400 million.

plantations in the Cali region and also produced tobacco, cacao, and beef cattle. Initial native labor was followed by that of African slaves. From the late 19th century onward, dams on mountain streams provided hydropower for industrial development. This asset, along with cheap labor, attracted foreign manufacturing firms, and Cali grew at a rate comparable to that of Medellín. In the past decades, drug trafficking has further contributed to the growth of this urban center (Fig. 21.3).

Colombian Coasts and Interior Lowlands

The Caribbean and Pacific coasts of Colombia contrast sharply with each other in environment and development. The Pacific coastal strip is the wettest area in all of Latin America, with some places receiving nearly 300 inches (762 cm) of precipitation annually. It has tropical rain forest climate and vegetation. In the north, mountains rise steeply from the sea, but there is a marked coastal plain farther south. The area is thinly populated, with a large proportion of the population of African ancestry. The main settlement is Buenaventura (population: city proper, 228,600), an important port but not a large city. Its principal advantage is that it can be reached from Cali via a low Andean pass.

The Caribbean coastal area is much more populous. Three ports have long competed for the trade of the upland areas lying to the south on either side of the Magdalena and Cauca Rivers. Barranquilla (population: metropolitan area, 1,841,700; city proper, 1,273,000) and Cartagena (population: city proper, 837,600) are the larger and more important cities in terms of legitimate trade. But the smaller city of Santa Marta (population: metropolitan area, 367,900), known as a beach resort as well as a port, may be more important in trade today. It is reputed to be the control center for Colombia's huge illegal export trade in cocaine and marijuana.

PROBLEM LANDSCAPE

The Geography of Coca in Latin America

One of the classic responses of upland farmers to the difficulty and cost of transporting agricultural commodities to the markets in the adjacent lowlands has been to turn grain into liquor. The "white lightning" that is such a part of the mountain lore of the Appalachian region in the United States was a farmer's response to how little money he made carrying bags of grain to the markets compared to the money he could earn selling illegal or untaxed liquor. In Latin America, however, illegal liquor is not the farmer's response

to this geographic isolation from grain markets; it is coca, the common name for the shrub *Erythroxylon* or *E. coca*.

This shrub is found in the upland regions of the Andes of South America, as well as in the mountain regions of Caribbean Middle America. Before the Spanish came to Latin America, coca was limited in its use. It was chewed as a stimulant only by the upper classes of the Inca. The Spanish, however, saw that when it was chewed by the "indios" who were forced into heavy mining labor in the ex-

traction of mountain gold and silver ores, they could endure longer hours and carry heavier loads. The chewing of coca also reduced hunger pangs, diminishing the realization of the inadequate diet so characteristic of the upland native people during the Spanish demands for mine labor.

Today, approximately 90% of the world's cocaine comes from Colombia, Peru, and Bolivia (Figure 21.A). The mountain uplands of these three countries provide good cover, an accommodating environment, and

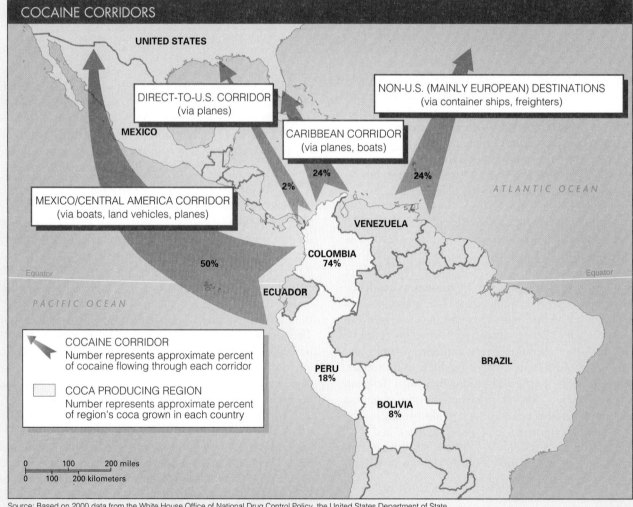

Figure 21.A   The source areas and target destinations of general cocaine flow.

considerable independence from the control of the central government. This native plant can be brought to harvest in under a year and a half, whereas upland fruit trees will take 4 years to produce a harvest. An acre of coca shrubs will generate between $1,000 and $1,300 in cash yield — approximately four to six times the return on any other mountain agricultural crop. In Bolivia from 1960 to 1985, the return on raising coca increased an average of 11% annually. Given the modest return on upland farming and the associated costs of moving bulk commodities to lowland market centers, it is clear why coca has become the upland crop of choice for many subsistence farmers and many export-oriented mountain farmers as well.

The cultural pattern that has developed around the raising of coca includes the involvement of lowland drug traffickers, who contract to receive, process, transport, and sell cocaine from the coca raised by the upland farmers. This trafficking has in turn led to entire networks of transporters, judges, police, customs agents, pilots, and "mules" (the people used to carry the most valuable drugs). The scale of this coca business is such that in 1996 a major drug kingpin, Juan Garcia Abrego, was arrested by Mexican authorities and extradited to Houston for trial. Journalist Howard LaFranchi points out that it is alleged that Abrego was masterminding a trade that brought "100 tons of cocaine annually to United States markets over the past ten years, earning him annual revenues of more than $20 billion."[a]

One researcher on the role of coca in the economy of Latin America pointed out that "the coca industry . . . in less-developed countries . . . is labor intensive, decentralized, growth-pole oriented, cottage-industry promoting, and foreign-exchange earning. If the coca industry were completely licit . . . it could be the final answer to rural development. . . . "[b] Another writer pointed out, "the cocaine industry is expanding swiftly to other Latin American republics — Ecuador, Panama, Venezuela, Brazil, Argentina, and Guatemala — at a speed linked, ironically,

to the intensity of eradication in the Andean countries."[c]

As the United States expands the scale of its antidrug war in Colombia ($300 million and more than 200 military advisors in 1999), it drives the expansion of coca production in the adjacent nations of Peru and Bolivia, especially. In the same way that municipal police cleanups of drug sales at some particular intersections in American cities lead to the same activity popping up in other neighborhoods of the city, the U.S. focus on the growth of coca in Colombia sends ripples of change through the whole Andean region.

Coca represents a truly dramatic and regional problem. Just as the engine that drives the economic development of much of Latin America is the U.S. market, so is it for coca. But coca is also the crop and production activity that the United States most reviles. Even while informal market mechanisms work to promote expansion of *E. coca* in the United States, the government, the military, and the media all make major efforts to force the upland farmers and their associated traffickers to replace this crop with cocoa, cotton, sugarcane, or any other upland crop that could be accommodated by the microgeographies of the tierra templada or the tierra fría where the coca plants are now raised. Environmentalists claim that besides the more than 171,000 hectares of coca cultivated in just Bolivia and Peru, "producers of illicit drugs have deforested as much as 1 million hectares of fragile jungle forest lands in the Amazon region."[d]

Like so many agricultural patterns that hold sway over the peasant farmer, the success of this cropping pattern is dependent on a world beyond the borders and beyond the uplands of the farmer in this problem landscape. The farmer's well-being is not only threatened by a drop in the world price (street price) of this crop, but it is also put under stress by local efforts to eradicate the raising of the crop through everything from military encroachment to governmental efforts to interest the farmers in alternative crops.

Such programs are almost always funded by external (most often United States) sources, so the peasant farmers never know when support for alternative farming and processing will collapse.

This confusion is confounded by the reality that coca leaves play a very regular role in the lives of those in the Andean countries. In Bolivia, for example, the 13,000-foot (4,000-m) upland setting of much of the population has made chewing the coca leaves a normal way to alleviate the influences of elevation. To organize a program to begin to spray coca plants or rigorously enforce prohibitions on growing and processing coca could cause a major disturbance of cultural norms.

The problem of cocaine is in fact a global problem that links the Andean countries and their upland farmers to major blocks of the urban world. To see the global linkage that ties two very disparate parts of the world together, consider what happens when the New York and Los Angeles police departments effect a massive sweep to jail drug sellers in their cities and change the flow of cocaine on the streets of the United States for even a few months. The market forces shift back up the connecting linkages from urban drug dealer to smuggler to Peruvian or other Latin American trafficker to the processor in the mountain highland to, finally, the peasant farmer who raises the actual crop. In a sense, like all farmers, his family's success depends on forces over which he has absolutely no influence or control. While we think of this problem landscape as a major factor in our own urban and suburban instability because of drug abuse, the Andean or Central American or South American upland farmer sees it as a problem that is just as real and just as disturbing to his own patterns of life.

[a]Howard LaFranchi, "Mexico Polishes Images with US," *Christian Science Monitor.* January 17, 1996, p. 7.
[b]Ibid.
[c]Mario de Franco and Richard Godoy, "The Economic Consequences of Cocaine Production in Bolivia: Historical, Local, and Macroeconomic Perspectives." *Journal of Latin American Studies.* Vol. 24, no. 2 (May 1992): 375–406.
[d]U.S. State Department Dispatch, "Fact Sheet: Coca Production and the Environment." Vol. 3, no. 9 (1992), p. 165.

Most of the Caribbean lowland section of Colombia is a large alluvial plain in which agricultural development has been handicapped by floods. Nevertheless, the region has a history of ranching and banana and sugarcane production. The Colombian government is now fostering settlement by small farmers who grow corn and rice.

Colombia's interior lowlands east of the Andes form a resource-rich but sparsely populated region that makes little contribution to the national economy as yet. It is a mixture of plains and hill country, with savanna grasslands in the north and rain forest in the south. The grasslands are used for ranching and the forest for shifting cultivation. Considerable mineral wealth awaits exploitation. As in so many parts of Latin America, and especially in South America, there is intense isolation caused by the difficulty of access to the region. Look at Figure 21.2, especially the symbol for areas that are at least 20 miles (32 km) from a railway.

The Colombian Potential

Colombia has the resources for notable economic development in the future. Energy resources include a large hydropower capacity; coal fields in the Andes, the Caribbean coastal area, and the interior lowlands; oil in the Magdalena Valley as well as in newly discovered and developing fields in the interior; and large supplies of natural gas from the area near Lake Maracaibo. Iron ore from the Andes already supplies a steel mill near Bogotá, and large Andean reserves of nickel are under development. At least one-half of the country is forested, with most of the forest made up of mixed hardwood species in the tropical rain forest. Because of the intermixture of many different kinds of trees, this type of growth has relatively low utility for forest industries other than the making of plywood. Of the foregoing resources, only oil now plays much of a part in the country's export economy. In 2000, petroleum and its products represented approximately 23.5% of all legal Colombian exports by value, with coffee representing 20%. However, illegal cocaine and marijuana may have been the largest category of exports.

Finally, in making any overall assessment of the reality and the potential reality for Colombia, it is essential to realize that there has been a steady, low-grade civil conflict going on for more than 40 years. Kidnappings, assassinations, and hostage taking have become nearly daily matters. The population exists in armed camps — drug lords, conservatives, liberals, other splinter groups — and Colombia speaks of *La Violencia* as a historical period that was allegedly ended in 1957 but in actuality continues to sow uncertainty in the political and economic process of this century as well.

Ecuador

In Ecuador, the Andes form two roughly parallel north-south ranges. Between them lie basins floored with volcanic ash. These basins are mostly in the tierra fría at elevations between 7,000 and 9,500 feet (c. 2,100–2,900 m). Quito (population: metropolitan area, 1,738,600; city proper, 1,610,800), Ecuador's capital and second largest city, is located almost on

the equator, high up on the rim of one of the northern basins. Its elevation of 9,200 feet (2,800 m) yields daytime temperatures averaging 55°F (c. 13°C) every month of the year so that daily fluctuations between warm days and cold nights are the principal temperature variations.

The first Spanish invaders came to Ecuador from Peru in the 1530s. They found dense Native American populations in Ecuador's Andean basins. The conquerors appropriated the land and labor of the native people and built Spanish colonial towns, often on the sites of previous indigenous towns. But they found no great mineral wealth here, and both climate and isolation worked against the development of an export-oriented commercial agriculture. However, from this hearth, the potato (*Solanum tuberosum*) was diffused into Europe and into world trade. Even today, these Andean basins have an agriculture with a strong subsistence component and sales mainly in local markets (Fig. 21.4). Potatoes, grains, dairy cattle, and sheep are the crop and livestock emphases. The isolated Ecuadorian Andes historically attracted relatively few Spaniards, except to Quito, and did not require African slave

JOSEPH J. HOBBS

Figure 21.4   The central market plaza is a landscape feature of all Latin America. The fruits, fresh meat, handicraft goods, and the presence of indio and mestizo populations are all markers of village scenes that are repeated from Chile to Mexico. This particular shot is from Ecuador but has elements that could be found all through the peasant world in both the Eastern and Western Hemisphere.

labor. Consequently, the mountain population has remained predominantly pure Native American, although with some mestizo and white admixture. The isolation has recently decreased somewhat with the growth of a certain amount of industry in Quito and smaller Andean cities.

A drastic shift of Ecuador's economy and population toward the coast took place in the 20th century. Through the 19th century, approximately nine-tenths of the people still lived in the highlands, but that proportion has now been reduced to slightly less than one-half. The growth zone near the coast is a mixture of plains, hills, and low mountains, with a climate varying from rain forest in the north, through savanna along most of the lowland, to steppe in the south. The main focus of development has been the port city of Guayaquil (population: metropolitan area, 2,627,900; city proper, 2,148,600), located at the mouth of the Guayas River, which drains a large alluvial plain where bananas for export (Ecuador is the world's leading banana exporter) and rice for domestic consumption are grown. On adjoining slopes, coffee is cropped for export, and along the Pacific, there is a considerable fishing industry (Figure 21.5). Anchovy harvests have long been a feature of the coastal fishing industry. The Galapagos Islands, located in the Pacific Ocean off the western coast of Ecuador, have been claimed by Ecuador since 1832 and currently are significant in the Latin American promotion of ecotourism — the careful touring of concerned (but well-paying) groups through fragile and often endangered environments. The population of the coastal region is mainly mestizo and white.

In the 1960s, major oil finds in the little-populated trans-Andean interior began to be exploited. The area lies under a tropical rain forest climate in the upper reaches of Amazon drainage. A several-hundred-mile pipeline across the Andes was constructed from the oil fields to a small port (Esmeraldas) on the Pacific, and Ecuador became an oil-exporting country. In 2000, oil and its products provided approximately 27% of the country's foreign-exchange earnings. Other major mineral resources are not known to exist. Possible future mineral discoveries in the interior rain forest add urgency to a border dispute with Peru, which controls considerable territory of this type claimed by Ecuador. In 1981, and again in 1994–1995, a bout of armed conflict between the two countries occurred along their undefined boundary in the mountains of southeastern Ecuador. In late 1998, the two nations signed a treaty resolving this issue for the present.

Peru

This region's native populations have been famous for their extensive road building. The Incas had a network consisting of more than 2,000 miles (3,200 km) of surfaced roadway. Accompanying this road development was the domestication of the potato, one of the most significant crops to diffuse from the Andean uplands to Europe and Asia. It was initially raised by the Incas in Peru and Ecuador and taken to the Old World in around the 1570s. For two centuries, the potato slowly took hold in Britain and Ireland, and by the middle of the 19th century, it was particularly dominant in Ireland. (A potato blight in Ireland in the 1840s led to a decline in productivity and to famine, and millions of Irish fled their country for the east coast of the United States.) The tuber should truly be called the Peruvian or Andean potato rather than the Irish potato, but it got its name to differentiate it from the sweet potato that followed approximately the same Age of Discovery diffusion path. The Irish potato became important because it could be grown in sandy soils, in a wide variety of environmental settings, in high or low moisture, and did reasonably well when kept in the ground until needed. In terms of nutrition, the lowly potato gives a higher caloric yield per unit of

Figure 21.5 Hillside farming has long been a landscape feature of South American agriculture. In this scene from mountainside farming in south central Ecuador, the row crops have been planted in such a way that erosion is diminished. In addition to this contour planting, there have also been eucalyptus trees planted around the margins of the field to serve as a windbreak as they grow.

JOSEPH J. HOBBS

land than wheat, rice, or corn. In fact, it became the most important crop diffused from South America in the centuries following the European arrival and early colonization of Latin America.

Peru is a poor nation with an extraordinarily fragmented distribution of population, induced by an environment that exhibits extremes of ruggedness (in the Andes), dryness (along the coast), or wetness (in the trans-Andean Amazon lowland). More than one-half of the population, predominantly pure Native American, lives in the Andean altiplano. Like the Andean aborigines in other countries, these highlanders are descendants of an ancient civilization that was largely ruled in the three centuries before Spanish times by the Inca empire, which originated around Cuzco (population: city proper, 279,600) in Peru's southern Andes. The highlands provide a notably poor environment for the development of surplus-producing cash crop agriculture. The mountain ranges are considerably higher on the average than those of Ecuador or Colombia. There are cultivable valleys and basins. The agricultural holdings tend to be small, densely populated, and found in the tierra fría, where cool temperatures preclude export crops such as coffee, sugarcane, and cacao. Harvests of potatoes, grains, and livestock (llamas and alpacas) raised for home consumption and minor sale in local markets are the most characteristic household economy. Farmed areas in central and southern sections of the Peruvian Andes tend to be so dry that high productivity is limited to the many fewer irrigated areas. (Long before the Spanish invasion, Incan engineers were adept at constructing irrigation systems.)

Today, considerable cash is earned as tourists travel to varied Inca sites, especially to the monumental city of Machu Picchu in the Andes Mountains (Fig. 21.6 and Definitions & Insights). At the same time, many farmers, as in other Andean countries like Colombia and Bolivia, have turned to the illegal but profitable cultivation of coca leaves from which cocaine is extracted. Peru is reputed to be a major global supplier.

A narrow, discontinuous coastal lowland lies between the Peruvian Andes and the sea. Climatically, it is a desert of remarkable aridity associated with cold ocean waters along the shore. Air moving landward becomes chilled over these waters. Onshore, a relatively cool and heavy layer of surface air underlies warmer air. For this reason, the turbulence needed to generate precipitation is absent, although the surface air does produce heavy fogs and mists during the cooler part of the year, moderating temperatures somewhat and providing some moisture. The great majority of rivers from the Andes that drain the west side of the mountains tend to carry so little water that they disappear before reaching the sea. But irrigated agriculture has been practiced for millennia along these river valleys as they cross the desert coastal plains.

Peru has a wide range of environments but is also powerfully affected by external climatic features that visit the west coast of South America periodically. El Niño is the popular name given to a weather condition that has made major news for decades and that is a cyclic feature of coastal Peru. The name means "The [Christ] Child," referring to a shift in the temperature of ocean currents off the west coast of South America that occurs most often in December with the arrival of warm tropical waters spreading from west to east across the Pacific Ocean. Normal ocean currents give the west coast of South America a resident cold current (historically called the Humboldt Current, but now known as the Peru Current). This cold current is vital not only to the fishing industry but also to temperature patterns on the adjacent lands of Peru,

Figure 21.6   One of the most dramatic landscapes of South America is the remains of Machu Picchu, known also as the Lost City of the Incas. This highland (8,000 ft/2,400 m) city and ceremonial center of the Incas overlooks the Urubamba Valley in Peru and serves as both a major tourist destination and an archeological treasure of major consequence. It became known to the West in 1911. It is sited on the east side of the Andes Mountains and represents the farthest east the Incas reached in their control of an extensive empire in the centuries just prior to the arrival of Columbus and subsequent European conquistadors.

A landscape signature is some aspect of the cultural landscape that plays a role in establishing identity between a place and a specific population. For example, with our students in Missouri, the Gateway Arch in St. Louis is a clear marker that says St. Louis and something about the gateway function of this state in the 19th-century (and earlier) American migration toward the west.

In Peru, there are two very graphic examples of landscape signatures. One is the mountaintop city of Machu Picchu. This city lies at 8,000 feet and represents a crowning achievement of the Incas who used it, in all probability, as a ceremonial and urban center. Its terraces, substantial stone structures, and commanding views over the Urubamba Valley in south central Peru all mark it as an example of the built environment that captured energy, resources, and the imagination of the Incas (see Fig. 21.6). A second stunning example of a landscape signature is the Nazca Lines in the desert of southern Peru. This complex of nearly 200 square miles of massive sculptures in the sandy surface of the desert includes birds, insects, and fish — the smallest of which is an animal figure of some 80 feet in length. What makes them particularly interesting is that the earthen designs can only be appreciated from an aerial perch, and they are believed to have been created some 2,000 years ago by the Nazca people. They are a signature of this place and this people that has led to speculation by some about the possibility of some alien population's presence in the distant past — a population that had the capacity for air flight.

In both cases, these distinctive landscape signatures not only help identify the place as settings for unique landscapes and peoples, but they also bring in tourist dollars and encourage the federal and local governments to give money and attention to the preservation of these landscape creations of earlier peoples.

Ecuador, and Chile in South America. However, once every 4 to 6 years on average, El Niño occurs and changes everything. This thick layer of warmer water comes across the Pacific from west to east in the late summer and fall, often arriving off the coast of South America around Christmas. It caps the usually upwelling cold water associated with the Peru Current. Fishing is changed completely because of the shift in water temperatures. Regional air movement becomes unstable because the warm ocean surface of El Niño sets in motion a whole different set of storm and precipitation patterns than those associated with the normal cold Peru Current. The somewhat mythic power of El Niño is given credit for stimulating weather changes, droughts, hurricanes, tornadoes, and even modifying human behavior. Although it is along the coast of Peru that the impact of El Niño is most frequently experienced, this weather phenomenon has an influence all through Latin America and now is associated with weather abnormalities in East Asia and many others as well.

In 1535, the Spanish founded Lima (population: metropolitan area, 7,451,900; city proper, 6,321,000) in an area of this type a few miles inland from a usable natural harbor. The city proved relatively well located for reaching the Native American communities and mineral deposits of the Andes, especially the Cerro de Pasco silver deposits which, for a time, led the world in silver production. Lima was designated the colonial capital of the Viceroyalty of Peru and became one of the main cities of Spanish America. In the late 19th century, foreign companies became interested in the commercial agricultural possibilities of the Peruvian coastal oases. As a result, commercially oriented oases now spot the coast, producing varying combinations of irrigated cotton, rice, sugarcane, grapes, and olives. Small ports associated with the oases are generally fishing ports and also fish-processing centers, as the cold ocean along the coast is exceptionally rich in marine life. But this resource provides an uncertain livelihood. For a few years in the 1960s and early 1970s, Peru led the world in volume of fish caught (largely anchovies for processing into fishmeal and oil), and the port of Chimbote (population: city proper, 295,300) was the world's leading fishing port. But during the 1970s, a change in water temperature associated with El Niño led to a drastic decline and widespread unemployment in the industry. However, by 1991, the fisheries had recovered to the point that Peru was the world's fourth nation in total fish catch (after China, Russia, and Japan).

Overall, the resources and performance of Peru's agriculture are so poor that the country now imports a large part of its food. Consequently, it remains dependent on mineral resources to supply the exports it needs to gain foreign exchange for such food importation. Oil and metals are the most important. Most of the oil comes from a field developed during the 1970s in the remote Amazonian rain forest of northeastern Peru. Some is transported via pipeline across the Andes and some by river across Brazil from Peru's Amazon port of Iquitos (population: city proper; 340,200). However, the aggregate export value of metallic ores — mainly copper, zinc, lead, and silver — far exceeds that of oil. Metal-mining sites are divided between the Andes and the coast. Peru ranks third globally in production among silver-mining countries.

More than one-half of Peru occupies interior Amazonian plains with a tropical rain forest climate (see the Landscape in

Literature selection in this chapter). Fewer than 2 million people inhabit this large area. The main settlement, Iquitos, originated in the natural-rubber boom of the Amazon Basin during the 19th century and has tended to be more oriented to outside markets via the Amazon than to Andean and Pacific Peru. The Amazon River is open to oceangoing vessels for 1,900 miles (c. 3,060 km) into the interior.

Economic development in Peru has been hampered by governmental instability. The country has gyrated wildly between civilian and military rule. Even the military governments have followed very divergent policies — sometimes leftist and sometimes rightist — with one government undoing the "reforms" of another. The 1980s saw widespread guerrilla activity by such movements as the Shining Path, which purportedly took its inspiration from Communist China's Mao Zedong (Mao Tse-tung). By 1992, this challenge to national stability had resulted in more than 5,000 deaths. Despite the unrest, Peru held a democratic election in 1985, which brought about the first peaceful and democratic change of government in four decades. Then, in April 1992, the elected president, Alberto Fujimori — the son of Japanese immigrants to Peru — and the armed forces dissolved the elected Congress in favor of rule by decree. Fujimori claimed that the new powers would allow him to deal more effectively with economic problems, the illegal drug trade, and guerrilla warfare. By the mid-1990s, Peru had regained reasonable stability and was devoting itself in part to the timely payment of the interest costs on very large foreign debts.

Fujimori was reelected president in 1995, even with his history of suspending certain civil rights to fight the rebels of the Shining Path movement. The 1996–1997 siege of Peru's Japanese Embassy by the Tupac Amaru rebel group is a further example of ongoing tensions between classes in Peru. Five years after the remarkable success of President Fujimori's leadership, there was a commando raid through secretly dug tunnels under the Japanese Embassy in which all of the rebels were killed and all but one of the hostages was brought out alive. The body of the rebel Tupac Amaru leader was dug up and studied by the post-Fujimori government in an attempt to determine whether he died in the rush of battle or had been killed deliberately by the Peruvian forces that rushed up out of the tunnels.

Such a spin on the politics of Peru — for Fujimori in late 2000 returned to Japan from which his parents had immigrated decades earlier and has taken on Japanese citizenship again — reminds us of the uncertainty of politics, the assimilation of widely varied immigrant groups, and the stability of development campaigns that are linked closely with a single political leader. In April 2001, Peru undertook new elections, and Valentin Paniagua became Peru's first Native American president. The new government promised continued economic growth, even as foreign investors shy away from Peru because of the rapidly changing political outlook following the rigid hand of Fujimori at the helm. Such events remind any Latin American observer that the process of economic de-velopment is always highly influenced by political matters that seem unrelated to economic change but are actually profoundly significant.

Landlocked Bolivia

Bolivia is a land of extreme Andean conditions and is by far the poorest Andean country and one of the poorest in all of Latin America. Like the other countries, it has a sparsely populated tropical lowland east of the mountains. But unlike the others, it has no coastal lowland to supplement or surpass the population and production of its mountain basins and valleys. Agriculture still employs two-fifths of Bolivia's people and is limited by extreme tierra fría conditions combined with aridity. La Paz (population: metropolitan area, 1,609,200; city proper, 810,300) is the main city and the de facto capital (Sucre is the legal capital, but the supreme court is the only branch of the government located there). Sited at an elevation of about 12,000 feet (c. 3,700 m), La Paz averages 45°F (8°C) in the daytime in its coolest month and only 53°F (12°C) in its warmest month. Aridity is a further problem; the city averages only 22 inches (56 cm) of precipitation per year, with 5 months essentially without precipitation.

A major rural cluster of people lives at a comparable elevation around Lake Titicaca on the Peruvian border. At 12,507 feet, this lake is the highest navigable lake in the world and boasts regular steamer service among lakeside ports high in the Andes. Smaller settled population clusters are scattered through Andean Bolivia, mainly along the easternmost of the two main Andean ranges and the Altiplano, a very high basin between the two towering ranges. At such high elevations, export-oriented commercial agriculture has not been possible. Not only is the climate unfavorable, but products would have to reach the sea via 13,000-foot (c. 4,000-m) passes to the Pacific coast or perhaps by a long route to the Rio de la Plata to the east. The country has remained predominantly (over 80%) Native American. There continues to be a strong native presence in the upland villages all through Andean South America (Fig. 21.5).

Known as "Upper Peru" by the Spanish in the 16th and 17th centuries, Bolivia was settled by Europeans partly for the land and labor of its native inhabitants but in good part for its minerals. In the late 16th century, the Bolivian mine of Potosí was producing about one-half of the world's silver output. Since the late 19th century, tin has been the mineral most sought in the Bolivian Andes, although exports of tin have now become far less valuable than exports of natural gas from Bolivia's eastern savanna lowlands. Since the 1920s, oil and gas discoveries have been made around the old lowland frontier town of Santa Cruz (population: city proper, 1,089,400). Andean indigenous peoples, Europeans, and Japanese have settled around Santa Cruz in growing agricultural areas. Farther north, where the lowlands have a tropical rain forest climate and where intermediate Andean slopes lie in the tierra templada, other pioneer agricultural settlements have been attracting immigrants from the highlands who come to pan for gold, raise cattle (meat is flown to La Paz), cut wood, and grow coffee, sugar, and illegally, the coca plant. Nevertheless, even with the historic richness of the Potosi mine and the newer lowland settlements, Bolivia today is among the poorest nations in all of Latin America with an annual GNP PPP value of $2,600 compared to $1,800 for Haiti and $6,930 for the 2001 Latin American average (see Table 19.1). However, the coca plant continues to play a major role in the underground Bolivian economy.

In December 2000, Bolivian President Hugo Banzer and U.S. Ambassador Manuel Rocha formally ended a campaign that had "seen 40,000 hectares of coca eradicated in the Chapare [central Bolivia] since 1998." In this effort, the two governments had heavily promoted the raising of bananas and pineapples, but there had been considerable loss of labor and income for the local farmers in the Chapare region. Coffee has also become a replacement crop in regions of the tierra tiemplada and tierra fría, but in nearly all cases, the social and political costs that the local government has had to sustain to undertake these forced land use changes have made the process enormously unpopular. Because this same sort of drug war is fought at a more military level and intensity in Colombia, the market pressure on Bolivia to increase coca production has been strong. This leaves the United States and various Andean governments fighting battles on a number of fronts, all of which are locally unpopular. As noted in the Problem Landscape box in this chapter, the upland farmer finds a crop with relatively high value per unit weight much more attractive than bulk commodities, even if they are relatively valuable bananas and pineapples.[1]

21.2   Brazil

Brazil is a major subdivision of Latin America and South America. It has an area of about 3.3 million square miles (8.5 million sq km) and a population estimated at 172 million as of 2001. Although smaller in area than the entire United States, Brazil is larger than the 48 conterminous states (Fig. 21.7). This giant country had only about 60% as many people as the United States in 2001, but its annual rate of natural increase is so much higher (1.5% as opposed to 0.6% in the United States) that the gap between the two countries is expected to continue to narrow rapidly as we advance in the 21st century.

Rapid population growth is an important facet of the recent dynamism that is giving Brazil an increasingly important role in hemispheric and world affairs. The dynamism is also dramatically expressed in such related phenomena as (1) explosive urbanism that has made São Paulo (population: metropolitan area, 18,167,200; city proper, 9,921,200) and Rio de Janeiro (population: metropolitan area, 11,058,300; city proper,

[1] "Coca's Second Front." *The Economist*, January 6, 2001, pp. 31–32.

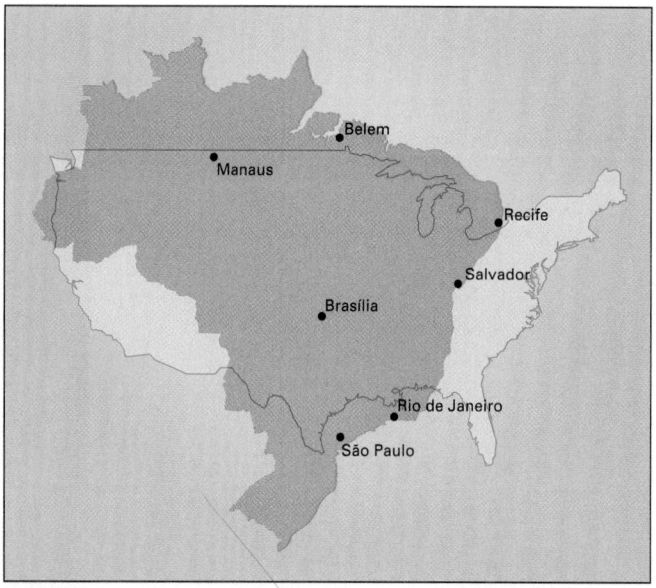

Figure 21.7   Brazil compared in area with the conterminous United States.

5,939,500) two of the world's largest cities (Fig. 21.8); (2) the rise of manufacturing to the point that manufactured exports now exceed agricultural exports in value; and (3) the diversification of export agriculture to such a degree that the traditional overdependence on coffee has disappeared (see Table 19.2).

By providing an expanding domestic market and labor force, rapid population increase has been able to make possible economies of scale and an accelerating economic development in Brazil. But today's burgeoning Brazilian population also poses serious problems for the nation. Brazil is hard-pressed to keep economic output ahead of population increase and to keep education and other social services at adequate levels. The country now has a sizable complement of millionaires, a substantial and growing middle class, and large numbers of technically skilled workers, but the benefits from development have been distributed so unevenly that many millions still live in gross poverty. And Brazil remains deeply mired in a huge foreign debt. However, some portents are favorable: Brazil does have abundant natural resources of many kinds, many friends among the world's advanced economic powers, and a remarkable degree of internal harmony among the diverse racial, ethnic, and cultural elements that make up the Brazilian population.

Environmental Regions

Brazilian development must contend with many problems associated with humid tropical environments. The equator crosses Brazil near the country's northern border, and the Tropic of Capricorn lies just south of Rio de Janeiro and São Paulo. Only a relatively small southern projection extends into midlatitude subtropics. Within the Brazilian tropics, the

Figure 21.8   A dramatic view of one of the most frequently visited cities in Latin America, Rio de Janeiro, Brazil. The decision to create a new capital city in Brasília in the 1950s was an effort, in part, to pull some population away from this setting and into the interior of Brazil.

climate is classed as tropical rain forest or tropical savanna nearly everywhere, although relatively small areas of tierra templada exist in some highlands, and there is one section of tropical steppe climate near the Atlantic in the northeast. The overwhelming predominance of tropical rain forest and tropical savanna ecosystems tends to be characterized by weak soils and, in places, poses severe maintenance and disease problems. An abundance of water sometimes creates general health and sanitation problems, while drought makes the provision of an adequate water supply difficult in the huge areas that have a harsh dry season.

Brazil is not handicapped to any marked degree by excessively high or rugged terrain. Nowhere does it reach the Andes. Mountains exist, but they are limited in extent and elevation. This leaves a country made up largely of relatively low uplands or extensive lowlands, the latter found primarily in the drainage basin of the Amazon River or in zones along the Atlantic coast.

The physiographic diversity of Brazil is very great when viewed in detail, but its major components can be subsumed under three landform divisions: the Brazilian Highlands, the Atlantic Coastal Lowlands, and the Amazon River Lowland. The country also includes a part of the Guiana Highlands that extends across the border from Venezuela, but this region of hills and low mountains is sparsely populated and has comparatively little importance in Brazil's present development.

The Brazilian Highlands

The Brazilian Highlands is a triangular upland that extends from about 200 miles (c. 325 km) south of the lower Amazon River in the north to the border with Uruguay in the south. On the east, it is generally separated from the Atlantic Ocean

by a narrow coastal plain, although it reaches the sea in places. Westward, the Highlands stretch into parts of Paraguay and Bolivia. The northern edge runs raggedly northeastward toward the Atlantic, steadily approaching the Amazon River but never reaching it. The topography is mostly one of hills and river valleys interspersed with tablelands and scattered ranges of low mountains. From Salvador to Pôrto Alegre, the Highlands descend very steeply to the coastal plain or the sea.

One-half of Brazil's metropolitan cities of a million or more population are seaports spotted along this coastal lowland: in the north are Fortaleza (population: metropolitan area, 2,909,100), Salvador (population: metropolitan area, 3,095,100), and Recife (population: metropolitan area, 3,377,600); in the south lie Rio de Janeiro, Santos (population: metropolitan area, 1,515,200), and Pôrto Alegre (population: metropolitan area, 3,576,500). Two of Brazil's three historic capitals are among them: Salvador (formerly Bahia), the colonial capital from the 16th century to 1763, and Rio de Janeiro (Fig. 21.8), which was made the colonial capital in 1763 and then, with independence in 1822, the national capital, until 1960. In that year, the capital was moved to the new inland city of Brasília (population: metropolitan area, 2,020,900) (Fig. 21.9). These locational shifts were motivated by changing economic and political alignments within the country, although the shift to Brasília is most symbolic of the country's hopes for future inland development.

Ironically, Brasília was designed for an estimated 500,000 people when it was conceived and built in the early 1960s. However, by the turn of the century, it was home to nearly 2 million people. It had also been designed to be highly integrated in terms of both socioeconomic classes and ethnic populations. It has, nonetheless, grown to be one of the

Figure 21.9 It is said that Oscar Neimeyer, the primary architect of Brasília, drew his first designs for this new capital city on the back of an envelope. Created in thinly settled savanna in 1956 and assigned the role of national capital in 1960, Brasília has been moderately successful in shifting the seat of national government and the country's demographic center from Rio de Janeiro into the country's interior. A considerable segment of the government population continues to spend the working week in Brasilia, but weekends at second homes in Rio. This photo features the Congress Building, designed by Neimeyer.

country's most segregated cities. With the continued dense occupation of the ring of *favelas* (shantytowns) that were initially thrown together to house the construction crews that were creating the new capital on the open savannas of Brazil's sparsely settled interior, the city now has the classic wide gap in the neighbors of the wealthy and the poor. There are the zones of the professionals and governmental people and the ragged landscapes of the impoverished. Brasília was meant to be a growth pole attracting new settlement, industry, and governmental activity — attempting to break the traditional centrality of Rio de Janeiro and São Paulo on the coast — and it has been successful in that respect. With the region generating approximately 2.4% of the country's GNP in 1959, it had grown to nearly 7.5% by the end of the 1990s, or a threefold increase in economic productivity.

Another exceptional section of the Brazilian Highlands occurs in the northeast, where climatic maps show an interior area of semiarid steppe climate near Brazil's eastern tip. Actually, the steppe area is only a shade drier than a larger surrounding area in which the tropical savanna climate is so dry as to be almost steppe. The whole area, both climatic steppe and the marginal tropical savanna around it, is sometimes called the Caatinga, a word referring to its natural vegetation of sparse and stunted xerophytic forest. Wet enough to attract settlement at an early date, it has been a zone of recurrent disaster because of wide annual fluctuations in rainfall. Years of drought periodically witness agricultural failure, hunger, and extensive mass emigration. This region, also

called the Northeast, is considered the poorest region of Brazil.

The Atlantic Coastal Lowlands

The narrow Atlantic Coastal Lowlands lie between the edge of the Brazilian Highlands and the Atlantic in most places, with the continuity broken occasionally by seaward extensions of the Highlands. Climatically, the northern part of the discontinuous ribbon of plain is tropical savanna, the central part is tropical rain forest, and the southern part is humid subtropical. This was the first part of Brazil to be settled by Europeans, and it has been in agricultural use for a changing variety of crops since the 1500s. It was in fact these alluvial plains that first attracted the Portuguese to introduce and raise sugarcane as the dominant Brazilian export crop for early trans-Atlantic trade. It was for this same plantation economy and trade that African slaves were introduced to Brazil.

The Amazon River Lowland

The other major lowland area of Brazil is radically different from the coastal lowlands in development and size. This Amazon Lowland reaches the Atlantic along a coast that extends some distance on either side of the Amazon's mouth. From this coastal foothold, it extends inland between the Guiana Highlands and the northern edge of the Brazilian Highlands, gradually widening toward the west. Far in the interior, beyond Manaus (population: 1,458,300), the Lowland widens abruptly and extends across the frontiers of Bolivia, Peru, and Colombia to the Andes.

The Amazon Lowland is composed generally of rolling or undulating plains except for the broad floodplains of the Amazon and its many large tributaries. The floodplains are often many miles wide and lie below the general plains level. They are nearly flat, but most of the lowland lies slightly higher, is more uneven, and is not subject to annual flooding. Tropical rain forest climate and vegetation prevail nearly everywhere, although southern fringes are classed as tropical savanna. Manaus and the ocean port of Belém (population: metropolitan area, 1,859,400) are great exceptions to the level of development in the rest of the lowland; most of the Amazon region is still very undeveloped and sparsely populated. However, in recent decades, the Brazilians have opened large areas by building a number of trans-Amazon highways to augment river and air transportation. The result has been a considerable influx of people into Amazonia. These new settlers live, generally, on a semisubsistence basis, employing various forms and combinations of ranching, agriculture, mining, lumbering, and fishing. Many ecologists fear that the push into the Amazon will eventually destroy the trees and the irreplaceable ecology of the rain forest. The number of living species of plants and animals reaches its maximum here (e.g., 600 types of palms, 80,000 plant species, 30 million insect species, and 500 species of fish have been identified). It has been suggested that fewer than 50% of the biota in the Amazon Lowland

ANTONIO MARI/LIAISON/GETTY IMAGES

Figure 21.10   A Yanomani youth shown fishing with a bow and arrow in 1995. In the 1970s, wildcat gold seekers brought diseases that the Yanomani tribe in the Brazilian Amazon jungle had never known, and their numbers were reduced by some 50% in only a decade. This diffusion process that was so costly to this nonliterate people in interior Brazil was parallel to the introduction of disease and enormous change to the "New World" after the arrival of Columbus five centuries ago and the subsequent colonization that took place for centuries. Petes Mathiessen wrote about just such tribes—see Landscape in Literature on p.566.

have been identified at all. The destruction of the rain forest would cause staggering losses to the world's biodiversity as well as to local indigenous peoples (Figs. 21.10 and 21.11). For this reason, international government agencies as well as nongovernmental organizations (NGOs) have taken such a strong and critical interest in attempting to preserve aspects of this, and other, Latin American environments.

Eras of Brazilian Development

In the 1494 Treaty of Tordesillas, Spain and Portugal agreed on a line of demarcation in the so-called New World along a meridian which, as it turned out, intersected the South American coast just south of the mouth of the Amazon (Fig. 21.12). Because of this, the eastward bulge of the South American continent was allocated to Portugal. The Portuguese subsequently expanded beyond this line into the rest of what is now Brazil, with international boundaries drawn in remote and little-populated regions east of the Andes where Portuguese and Spanish penetration met.

The Sugarcane Era

The colonial foundations of Brazil were laid early in the 16th century. A Portuguese fleet, well off-course on a voyage around Africa, discovered the Brazilian coast in 1500, and a sporadic trade with local Indians for Brazilwood (a South American tree that produced a much-desired red dye in early trade) began. Settlement beyond the first small trading posts was promoted by Portugal in the 1530s to combat French and Dutch incursions. Meanwhile, the coastal settlements became increasingly valuable as they rapidly developed exports of cane sugar to Europe and brought shiploads of slaves from Africa to labor in the cane fields. Slave traders in search of Native American slaves penetrated the interior to some degree, and Catholic missions were founded among the native people. Unfortunately, the missions often brought uncontrollable disease epidemics and slaving expeditions in their wake.

Expansion of coastal sugar plantations and increasing penetration of the interior dominated development in 17th-century Brazil. The colony became the world's leading source of sugar. Although some sugarcane was grown as far south as the vicinity of the Tropic of Capricorn, the main plantation areas developed farther north in more thoroughly tropical areas that also lay closer to coastal ports and, consequently, to Europe. Meanwhile, missionaries penetrated well up the Amazon, and a series of wide-ranging expeditions, originating largely from São Paulo, took place in search of either Native Americans to enslave or valuable new mineral deposits. In the late 1600s, sizable gold finds were made in the Brazilian Highlands some hundreds of miles to the north and northwest of São Paulo in what is now Minas Gerais (meaning "mines of many kinds") state.

The Gold and Diamond Era

During the 18th century, Brazil's center of economic gravity shifted southward in response to the expansion of gold and diamond mining in the highlands north and northwest of São Paulo and Rio de Janeiro. By the mid-1700s, Brazil was producing nearly one-half the world's gold. In the latter part of the century, this output declined, but by that time, much of the highland around the present city of Belo Horizonte (founded in 1896 as a state capital) had been settled, along with areas near São Paulo and spots in the far interior. Despite the difficulty of traversing the Great Escarpment between Rio de Janeiro's magnificent harbor and the developing interior, the city became the port and economic focus for the newly settled territories. Gold and diamonds were ideal high-value commodities to move by pack animals over poor roads and trails to the port.

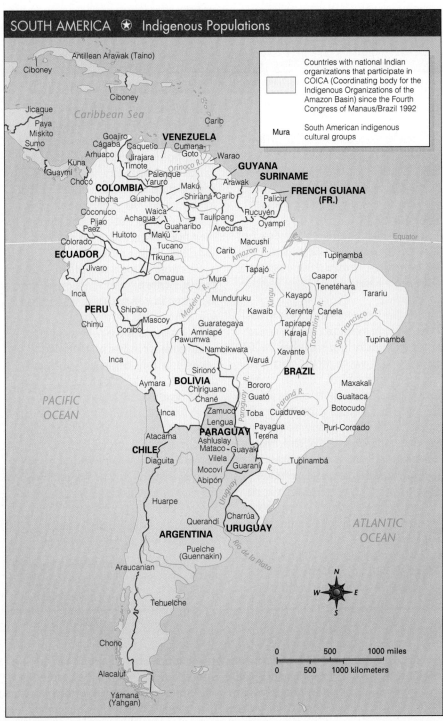

SOUTH AMERICA ✹ Indigenous Populations

Countries with national Indian organizations that participate in COICA (Coordinating body for the Indigenous Organizations of the Amazon Basin) since the Fourth Congress of Manaus/Brazil 1992

Mura — South American indigenous cultural groups

Source: 2000 Britannica *Book of the Year*

Figure 21.11 The great variety of native peoples in South America and part of Central America is shown.

Besides the attraction of mineral wealth, a factor contributing to the rise of the new areas was the economic distress of the sugar coast to the north. This was a result of increasing development of competing sugar production in other New World colonies (notably in the West Indies). One solution was relocation of plantation owners with their slaves to the newly developing areas in Brazil. Another was movement into the Caatinga (Sertão) area inland from the sugar coast. There, a ranching economy began, augmented by cotton growing in the next century.

The following 18th-century events, however, set a pattern of economic hardship for Brazil's Northeast that has never been broken: (1) continued dependence on hard-pressed coastal sugar and inland cotton production, (2) poverty of unusual severity even for Latin America, (3) recurrent disasters brought by drought, and sometimes by

LANDSCAPE IN LITERATURE

At Play in the Fields of the Lord

In 1965, Peter Matthiessen published a novel about the anguishes and seeming pointlessness of missionary work in the depths of the Amazon river system. A North Dakota man, woman, and their grade school son are in the process of landing at the ragtag village of Madre de Dios, the site of their mission work in a refurbished DC-3 as this passage opens. They are all looking out the plane's windows as this excerpt, and their series of life-changing experiences, opens. As you begin to process the images of these paragraphs, use your reaction to think about the extraordinary gaps that exist between the worlds that are common to you and the worlds of challenge, fascination, and fear that lie beyond your horizon. As the three Quarriers make nearly their first trip ever from their home in North Dakota, they are coming to a jungle village that author Matthiessen has dreamed up somewhere in the Spanish-speaking eastern reaches of some Andean country. As they arrive, think of the tensions that had to exist within their own minds as they realized they had come to this post of their own volition (except for Billy). These paragraphs must serve as profound markers of powerful change soon to come to all three of their lives. Read the passage as though you were one of the people in this opening scene of this novel.

Then the window cleared again. They had forsaken the realm of sunlight for a nether world of dark enormous greens, wild strangled greens veined by brown rivers of hot rain, the Andean rain, implacable and mighty as the rain that fell on those sunless days when the earth cooled.

Madre de Dios, already in view, formed a yellow scar in the green waste. With its litter of rust and rotting thatch and mud, the capital of Oriente State resembled a great trash heap, smoking sullenly in the monotony of the rivers.

The three Quarriers at their plane windows were taken aback by the grim prospect; even Billy could not mistake this devastation for an Indian village. The DC-3 slid out into its turn, over the river front, where derelict craft, misshapen and unpainted, nuzzled the bare mud bank; the heat of the place rose to engulf the plane.

Now a few figures came in sight. They stood like sentinels, as if transfixed by a sudden cataclysm. The mud streets, of a yellow-orange color, deeply rutted, were barren of all life but pigs and vultures and a solitary dog. Then these too disappeared, and the mud airstrip, parting a rank depressed savanna, swung into their view. A light plane and an old Mustang fighter were parked at its edge; the fighter plane was belly-deep in weeds.

Quarrier rearranged his legs so that his knees, in the event of a crash, would not bend in the wrong direction. They were certainly a long way from North Dakota, as Hazel had said. Except for their term at Moody Bible Institute in Chicago, they had scarcely set foot out of the Dakotas in all their lives until a month ago, when they had gone to Florida to catch the freighter. And here they were, thousands of miles from the clean kitchens and church suppers of home, ending their second airplane ride, not at the county airport but in a steaming jungle town on the banks of the Rio de las Animas — River of Souls! Such names were common in these parts, and sooner or later Hazel would horrify herself with the translations.

The people and dogs of Madre de Dios awaited them. There was no road into the jungle, and this plane, which crossed the Andes twice a week, was the sole evidence of an outer world. Men and dogs in the foreground, women and children to the rear, they stared at the old machine as it lumbered to a halt. The men looked of a size, and in some hangdog way identical: small brown mustachioed halfbreeds, barefoot in pajama tops and ragged pants and floppy sisal hats. The women, in gingham, were more various and more washed-out; the children were in rags. All stood in ranks behind a large pale man with a pistol on his hip, a small pale man in the white cassock of a priest, and a blond man in rich blond beard, shorts and shortsleeved flowered shirt; this man stood with feet apart, strong fists on hips, and a fine smile on his sunburned face. All three wore shoes. The Quarriers recognized the blond man, from his picture in *Mission Fields,* as Leslie Huben.[a]

[a]Peter Matthiessen, 1965. *At Play in the Fields of the Lord.* New York: Random House-Vintage Books, 1965, pp. 5–6.

---

flood, in the climatically unreliable interior, and (4) flows of population to other parts of Brazil. Such flows turn into strong currents in the worst times.

Rubber and Coffee Booms

Nineteenth-century Brazil was changed by rubber and coffee booms, by much-intensified settlement of the southeastern section, by the beginnings of modern industry, and by extension of the ranching frontier far into the Brazilian Highlands. The country underwent these changes as a colony until 1822, as an independent empire with a strongly federal structure until 1889, as a military dictatorship until 1894, and finally as a civilian-ruled republic. Its political history at this time was turbulent, and not until 1888 did the country emancipate its large slave population.

One of the major economic developments of the 19th century in Brazil was the boom in wild-rubber gathering in the Amazon Basin. This followed the discovery of vulcanization — the process that made rubber more flexible and durable and, consequently, made rubber a much more valuable material — by the American Charles Goodyear. At that time, Amazonia was the only home of the rubber tree (*Hevea brasiliensis*). A "rush" up the rivers followed, and thousands of widely scattered settlers began to tap the rubber trees and send latex downstream. Manaus grew as the interior hub of this traffic and Belém as its port. But some rubber tree seeds were smuggled to England, and seedlings transplanted to Southeast Asia became the basis for competing rubber plantations. By 1920, the Amazon rubber production, which depended on tapping scattered wild trees, was practically dead.

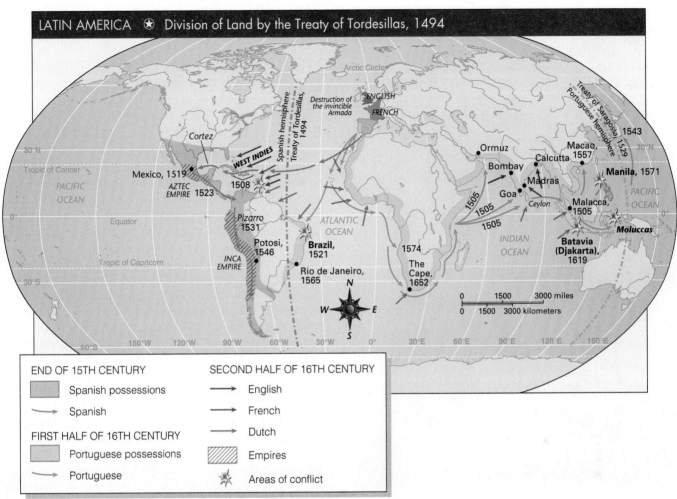

LATIN AMERICA   ✪   Division of Land by the Treaty of Tordesillas, 1494

END OF 15TH CENTURY

Spanish possessions

Spanish

FIRST HALF OF 16TH CENTURY

Portuguese possessions

Portuguese

SECOND HALF OF 16TH CENTURY

English

French

Dutch

Empires

Areas of conflict

Source: *The Harper Atlas of World History*. The Treaty of Tordesillas. Harper Collins Publishers. p. 139. New York. 1992.

Figure 21.12   In 1494, Spain and Portugal met at Tordesillas, Spain, and made decisions about dividing up the non-Christian world. This map shows the Spanish domain lying to the west of the yellow line. The so-called "Spanish Hemisphere" gave Spain virtual control of all the lands of the New World except for the eastern wedge of Brazil. Portugal had its hemisphere lying east of those territories, and that gave the realm of South and East Asia to its sailors and settlers, except for the Philippines, which came under Spanish control in the late 16th century. This 15th-century cartographic division of the surface of Earth played a dominant role in the patterns of colonization and economic control that were operative during the Age of Discovery.

Coffee was the preeminent boom product of Brazil during the 19th century. It had long been produced in Brazil's northeast in small quantities, and by the late 1700s, it was an important crop around and inland from Rio de Janeiro. During the 1800s, production spread in the area around São Paulo and then increased explosively in output during the last decades of the century. This growth was related to the broadening world market for coffee, the emergence of expanded and faster ocean transport, the nearly ideal climatic and soil conditions in this part of the Brazilian Highlands, and the arrival of railway transport to move the product over land. A spectacular and critical transport development was the completion by British interests in 1867 of a railway linking the coffee-collecting city of São Paulo with the ocean port of Santos at the foot of the Great Escarpment. The coffee frontier crossed the Paraná River to

the west and spread widely to the north and south of São Paulo.

Production doubled and redoubled. In the late 1800s, over a million immigrants a year from overseas poured into Brazil, mostly into the region near São Paulo, to work in the coffee industry. São Paulo began its rapid growth as the business center and the collecting and forwarding point for the coffee industry; it also began its productive career as an industrial center. These developments were also facilitated by very early hydroelectric production in the vicinity.

Twentieth-Century Development to the End of World War II

Brazil's development during the 20th century was not spectacular until after World War II. The rubber and coffee booms of the 19th century both collapsed in the early years of the 20th

Figure 21.13   Workers place coffee beans they have just picked into bags. The beans are prepared for roasting at a local coffee mill. Coffee is one of the most important crops of the tierra templada in all of Latin America. However, as is often the case with a heavy dependence on a single agricultural crop, great instability to the region occurs when world coffee prices fall because of high yields and surplus beans on the global market.

century. Brazil became a minor factor in world rubber production. It continued to lead the world in coffee production and export, as it does today, but increasing international competition depressed coffee prices, and the industry struggled economically (Fig. 21.13). The most positive developments of this period were:

1. The continued growth of early-stage industrialization in the São Paulo area;
2. Some expansion of the railway network;
3. The arrival, beginning in 1925, of Japanese immigrants who began to make important contributions to agriculture, especially truck farming;
4. Successful government attacks on two widespread debilitating diseases — malaria and yellow fever — during the 1930s; and
5. A major expansion of irrigated rice production in the Paraíba Valley between Rio de Janeiro and São Paulo. (Rice is a staple of the Brazilian diet, as are corn, beans, and manioc.)

Overall, however, the country struggled through this period without any major export boom. In 1930, at the beginning of the worldwide Great Depression, Brazil's economy practically collapsed, and in that year, democratic government was replaced by a rightist dictatorship that lasted until 1945.

Quickened Development Since World War II

Since 1945, the pace of development in Brazil has generally been very rapid. However, fluctuations have occurred from time to time, including a downturn in the 1980s. Growth took place under elected governments between 1945 and 1964 but occurred most rapidly between 1964 and 1985 under a military dictatorship with technocratic leanings and great ambitions. In 1985, the dictatorship relinquished power to a new elected government, which then had to face the accumulated problems of a stalled "takeoff." Among these were a crushing national debt that the military regime had contracted to finance expansion, and widespread economic misery caused by governmental attempts to keep up with payments on the debt. This pair of constraints (the Latin American plight alluded to directly in the Landscape in Literature selection in Chapter 20) on governmental stability has played a role all through the region. The International Monetary Fund has brokered a major proportion of the loans that have enabled some Latin American governments to stay in authority. Conditions for such loans generally include massive efforts to reduce government spending and control inflation, and monetary policies that support the local currency. Such policies are never popular, although the cash inflows that come from these IMF loans are always welcome.

The provision of a far more adequate internal transportation system has been an essential foundation of Brazil's infrastructural growth in the last few decades. This has primarily involved the building of a network of long-distance paved highways and an emphasis on trucks as movers of goods. Major urban centers have been connected, and additional highways have been pushed into sparsely populated and undeveloped areas. The new roads have attracted ribbons of settlement, have facilitated the flow of pioneer settlers to remote areas, and have furthered the economic development of such areas by connecting them to markets (for rubber, minerals, timber, fish, some agricultural products, and tourism) and sources of supply. This entire move toward more mobility has also stimulated a steady growth in the manufacturing of automobiles and trucks for both domestic and international markets (Fig. 21.14). But the single most spectacular instance of interior development has been the 1960 initial settlement of a new capital in a more interior and central location.

Brasília was a daring effort by the central government to lure people away from the coasts into the largely unsettled Brazilian interior. In 1956, a decision was made to treat the new capital as a growth pole — a term used to describe

Figure 21.14   Brazilians at work on an automobile assembly line in the city of Indiatuba, 70 miles northeast of São Paulo. This plant produces up to 20,000 units a year. Brazil has now grown to an economy of between $800–900 billion annually, and it has moved from one of the lowest to the highest category of the United Nations Human Development Index (see Table 19.1).

something that promotes growth of an industry or of a city through investment of new capital in a specific situation. The whole culture of the government in Rio de Janeiro, the traditional capital, was geared to the environmental amenities of that extraordinary coastal location. In selecting central Brazil for the location of the new federal capital, the government was trying to find the most effective way to prompt a significant demographic shift, impelling migration toward the interior of the country. Built as a totally planned city on largely unsettled savanna in the Brazilian Highlands, Brasília was occupied by the government in 1960. Its growth, accompanied by unwanted adjacent shantytowns, has been such that the new federal district of about 2,300 square miles (6,000 sq km) had more than 2.1 million people by 2001 (see Fig. 21.9). However, even today, Brasília remains only modestly settled, and its metropolitan population pales next to Rio de Janeiro, which continues to grow with over 11 million; São Paulo has about 18 million.

Agricultural Expansion: Sugar, Citrus, and Now, Soybeans

Three major agricultural boosts in older settled territories have contributed to the country's recent economic growth. A program of the government in recent years to modernize and reinvigorate the old sugar industry had considerable success, and Brazil is again a major sugar exporter. In 1962, a severe freeze in Florida gave Brazil an entering wedge into the world market for orange concentrate, and by the 1980s, Brazil was overwhelmingly the world's largest exporter. Also in the 1960s, Brazil became a major soybean producer and exporter, with production centered in the subtropical southeast and in a new

district in the tropical Brazilian Highlands. By the mid-1990s, soy products had become Brazil's biggest agricultural export commodity after sugarcane.

The soybean is the crop of the future for a number of world regions. The plant has a very high percentage of protein — approximately 30–35% — and no starch. It affixes nitrogen to the soil as it grows, making it a crop that is much less demanding on the soil than most. Although a native of China, it has become rapidly adopted by Brazilian farmers, and they have now come to compete with American soybean growers head to head. While nearly all of the U.S. production in soybeans is used for animal feed, the Brazilian crop is widely exported and used domestically for industrial production of chemicals, paints, ink, adhesives, insect sprays, and is now even used as a fuel blend (called soy diesel and akin to ethanol, a blend of corn and gasoline). The world market for soybeans is growing steadily not just because of the industrial uses of the bean but because it also can be blended with other products to create a high protein, modestly priced food substance that is increasingly important in global efforts to diminish world hunger. Soybeans now amount to some 11% of Brazil's annual exports in value.

The Energy in Brazil's New Industrial Age

Even more significant has been the extremely rapid expansion of Brazilian manufacturing. In the 1980s, the value of manufactured goods exported from Brazil began for the first time to exceed the combined value of all other exports. The products involved today are not just the simple ones of early-stage industrialization, although textiles and shoes are still major items. Steel, machinery, automobiles and trucks, ships, chemicals, and plastics are all important exports, as are weapons. Aircraft are also manufactured, although mainly for the extensive domestic market that one would expect to find in a country so large. An industrial core area in Brazil has developed with São Paulo as its core, now producing nearly 60% of Brazil's manufactured goods.

Metals represent the other main contribution of Brazil's resource wealth to Brazilian industry. The Brazilian Highlands are very mineralized with ores yielding numerous metals, outstanding among which are iron, bauxite, manganese, tin, and tungsten. Known reserves give Brazil approximately one-eighth of the world's iron ore, making Brazil a leading producer and exporter. The first modern steel mill began production in 1946 at Volta Redonda, located in the Paraíba Valley between Rio de Janeiro and São Paulo. Others have been built in the industrial core zone since that time, and Brazil has become a considerable exporter of steel.

Brazil's greatest resource deficiency is in fossil fuels. Known supplies of coal are found only in the extreme southeast and are very inadequate for the country's needs, although they do provide some coking coal. Domestic petroleum supplies are known thus far only to exist in the

In 1960, São Paulo, the largest urban concentration in Brazil, had 3.8 million residents. In 1997, it had 16.5 million. In 1970, Rio de Janeiro had 3.4 million people; in 1997, it had more than 10 million. In the quarter century between 1970 and 1995, more than 20 million Brazilians abandoned traditional roots in the countryside and moved to urban centers. On the one hand, these migrants have been pushed out of their rural setting by the steady introduction of new farm machinery that has rendered the traditional roles of the rural underemployed even more marginal. On the other hand, however, the real catalyst in the rural-to-urban migration pattern in Latin America is the image of new opportunities, new landscapes, and wholly new lifestyles associated with the urban scene.[a]

The mayor of Caracas, Venezuela, who saw his city expand from just over a half million in 1975 to nearly 3 million in 1995, commented, "We have to be both firemen and philosophers" to deal with the demands of this unprecedented flow of urban immigrants. Tijuana, Mexico, and Medellín, Colombia, are both

growing at rates of approximately 10% a year. To deal with this growth and its negative as well as positive implications, South American cities are trying various innovations. In São Paulo, the largest city in South America, the Cingapura (Singapore) Project is attempting to replace hundreds of favelas (*barrios* in Spanish) with low-rise blocks of flats. Generally, the favelas in all major Brazilian cities were bootstrapped (built with local scrap material and no governmental help) by self-reliant inhabitants. In the Cingapura Project, residents are moved to temporary barracks while the new flats are put up. Once the flats are finished, these same displaced, often recent rural-to-urban migrants move back in, paying for them with 20-year low-interest loans. The general response, particularly of those who are granted loans, has been positive. This project has been earmarked for 92,000 families, or half a million people. The scale of city-bound migration flows in all of Latin America is such that governments and agencies are looking to see how well the São Paulo effort works.

[a]"Latin America's Drift to the Cities," *Economist,* October 14, 1995.

Northeast. Located offshore for the most part, these fields were located as the result of intense exploration after the drastic rise in world oil prices in 1973. But they supply only a little over half of Brazil's oil consumption (as of 1997), and crude oil has become a major Brazilian import. This is true despite a relatively successful program to develop and use gasohol, whose 20 to 25% alcohol content is currently supplied by sugarcane grown in the Northeast and in the area around São Paulo. Unless the problem of oil supply can be solved, the country's continuing industrial development drive may be seriously hampered.

Another aspect of the industrial future of Brazil relates to the decision, and the need, to capture large loans from the International Monetary Fund in the late 1990s and the first years of this century. Such loans — and recall the conditions outlined in the debtor situation in the Landscape in Literature selection "And We Sold the Rain" in Chapter 20 — require significant internal changes in the management of finance, debt, loan repayment, and inflation. As always, such IMF loans are welcomed with great fanfare, but as the debt burden becomes determined — whether in Brazil, Chile, or Mexico — the debtor nation realizes that the demands of such fiscal situations are very high and may lead to political instability. The scale of the economy in Brazil makes IMF activity in that country of special moment, but the fiscal parallels in most other Latin American nations are quite true to this model of desperation followed by great relief followed by regret and anger at the economic restraints linked with such development loans. The process is all too similar to being overextended in one's own personal credit

line! Protesters against the financial and global web of connections and economic imbalances that have grown so significant since the 1980s follow international meetings from continent to continent in an effort to hold back this tide of global change (Fig. 21.15).

Figure 21.15   Anti-globalization demonstrators throw barricades at police near the site of the Summit of the Americas where leaders of 34 Western Hemisphere nations met in April 2001 to discuss the pending Free Trade Area of the Americas and other regional issues (see Regional Perspective on p. 573). The intensity of protest that globalization has engendered has a great deal to do with the replacement of local crafts and markets with an increasing linkage to international markets and price guidelines. These protests have followed globalization meetings all around the world ever since the mid-1990s.

21.3 Countries of the Southern Midlatitudes

Four countries of southern South America — Argentina, Chile, Uruguay, and Paraguay — differ from all other Latin American countries in that they are essentially midlatitude rather than low-latitude in location and environment. Northern parts of all the countries except Uruguay do extend into tropical latitudes, but the core areas lie south of the Tropic of Capricorn and are climatically subtropical. Argentina barely extends into the tropics in the extreme north, centers in subtropical plains around Buenos Aires, and reaches far southward into latitudes equivalent to those of northern Canada. Chile also extends south to these latitudes. Uruguay's comparatively small territory is entirely midlatitude and subtropical, and most Paraguayans live in the half of their country that lies poleward of the Tropic of Capricorn. The core areas of the three more eastern countries — Argentina, Paraguay, and Uruguay — have the same humid subtropical climatic classification as the southeastern United States. Chile's midsection or core area has a Mediterranean (dry-summer subtropical) climate like that of southern California.

These four Latin American countries share certain other regional characteristics. Agriculturally, for instance, they compete more with other midlatitude countries (in both the Northern and Southern Hemispheres) than with most other Latin American or other tropical countries. Argentina is an important export producer of crops such as wheat, corn, and soybeans and thus competes with North American farmers. Both Uruguay and Argentina are sizable exporters of meat and other animal products. Paraguay exports soybeans and cotton. Chile imports more food than it exports, but it does export fruits, vegetables, and wine.

Because these countries never developed labor-intensive and export-oriented plantation agricultures in colonial times, they have relatively small black populations. Argentina and Uruguay have been major magnets for European immigration and have populations estimated to be 85% "pure" European in descent. Paraguay and Chile, in contrast, attracted fewer Europeans and are about 90% mestizo. Uruguay, Argentina, and Chile are among the most economically developed countries in Latin America. They have relatively low dependence on agricultural employment, with only 10 to 15% of their employed workers still in agriculture. Paraguay, with more than 45% of its labor force still on farms in the 1990s, is less developed and poorer than the other countries, partly because of its history of landlocked isolation from the outside world.

Argentina

Argentina is a major Latin American country whose area of nearly 1.1 million square miles (2.8 million sq km), second in Latin America only to that of Brazil, is about one-third the area of the 48 conterminous American states. Argentina's estimated population of 37.5 million in mid-2001 ranked it far behind Brazil and Mexico within Latin America.

Argentine Agriculture

Agricultural products that Argentina exports through its capital city of Buenos Aires (population: metropolitan area, 12,761,500; city proper, 11,624,000) and smaller ports are sufficient to make the country one of the world's principal sources of surplus food. The leading agricultural exports are wheat, corn, soybeans, and beef. Agriculture is concentrated heavily in the country's core region, known as the humid pampas. (*Pampas* comes from a Native American word for "plains.") This is the Argentine part of the area identified on the map of natural regions as humid subtropical lowland prairie (see Fig. 21.1). The environment is very productive for agriculture. Level to gently rolling plains lie under a climate similar to that of the southeastern United States, featuring hot summers, cool to warm winters, and annual precipitation averaging 20 to 50 inches (c. 50 to 130 cm). But the area is specially advantaged by soils that are more fertile than might be expected from the climatic conditions. The grasses of the original prairie provided an unusually high content of organic material (humus), just as they did in the prairie areas of North America. The soils have a loose structure of fine particles, affording both a high concentration of plant nutrients and the maximum opportunity for plant roots to be nourished by feeding on individual soil particles. The local beef industry is fed from the grasses and grain that flourish in this setting, leading directly to the fact that Argentine per capita beef consumption is the highest in the world.

From the 16th century to the late 1800s, the humid pampas were sparsely settled ranching country, and parts of it remained the domain of native people. Then, in the late 19th and early 20th centuries, Argentina experienced a rapid and dramatic shift to commercial agriculture and urbanization. There was heavy European immigration, and population grew rapidly. Commercial contact with Great Britain contributed greatly to Argentina's new dynamism. At the time, the British market for imported food was increasing rapidly; refrigerated shipping came into use, allowing meat from distant sources such as Argentina to be imported into the United Kingdom; and a British-financed rail network (the densest in Latin America) was built in Argentina to tap the countryside for trade. The network focused on Buenos Aires, which rapidly became a major port and a large city. British cattle breeds, favored by the tastes of British consumers, were introduced to Argentine ranches, or *estancias*, along with alfalfa as a feed crop. European immigrant farmers assumed a major role in agriculture, generally as tenants on the large ranches that continued to dominate the pampas economically and socially. The immigrants came primarily from Italy and Spain but also from a number of other countries, giving present-day Argentina a complex mixture of

Spanish-speaking citizens from various European backgrounds. Soon an export agriculture developed in which wheat, corn, and other crops supplemented and then surpassed the original beef exports.

Argentine areas outside the humid pampas are less environmentally favored, less productive, and less populous. The northwestern lowland just inside the tropics, also called El Gran Chaco, has a tropical savanna climate and supports cattle ranching and some cotton farming. Westward from the lower Paraná River and the southern part of the pampas, the humid plains give way to steppe (see Fig. 21.1) and then desert, with production and population declining accordingly. Population in these western areas is concentrated in oases along the eastern foot of the Andes, where mountain streams emerging onto the eastward-sloping plains have been often impounded for irrigation. Vineyards and, in the hotter north, sugarcane are special emphases in a varied irrigated agriculture. Córdoba (population: metropolitan area, 1,521,700) is the largest oasis service center and the second largest city in Argentina. To the south of the pampas lie thinly populated deserts and steppes in the Patagonian plateau, a bleak region of cold winters, cool summers, and incessant strong winds. Native American country until the early 20th century, Patagonia now has millions of sheep on large ranches but little other agricultural production, although oil production has been gaining importance on the southeast coast.

Urban and Industrial Argentina

Only about one-tenth of Argentina's labor force is directly employed in agriculture and ranching. The great majority of Argentineans are city-dwellers employed in an economy where service occupations are very important and manufacturing employs about twice as many people as does agriculture. Metropolitan Buenos Aires (and the surrounding province) has about one-third of the national population. It was settled in the late 16th century as a port on the broad Paraná estuary called the Rio de la Plata. But until Argentina became reoriented toward British and then world markets, the city remained a rather isolated outpost of a country focused on the north and the western oases, with relatively little trade by sea. In the 19th century, Buenos Aires first became the national capital and then underwent explosive growth as a seaport, rail center, manufacturing city, and recipient of primarily European immigrants.

The growth of industry in Argentina has conformed to a sequence often found in nations emerging from an underindustrialized state:

1. *Local markets.* Industrialization began when the government fixed tariffs to help small-scale industries supply consumer items for sale within the country and to support plants processing agricultural exports (e.g., meat-packing plants). Called a policy of import substitution (encouragement of local firms to produce goods most frequently im-

ported), it restricts consumption of popular foreign goods through the imposition of high tariffs on goods from overseas.

2. *Wartime isolation.* Some consumer industries such as textile manufacturing and shoemaking were expanded with government support when the country was cut off from imported supplies during World War I and again during World War II.

3. *State-led industrialization.* After World War II, a deliberate state-led policy of industrialization was instituted. There have been lapses in this policy and fluctuations in the mechanisms of support, but considerable success has been achieved in the quantitative growth of manufacturing and diversification of industrial output. Imported coal to supply power has been replaced by oil produced within Argentina (mainly from fields on the coast of Patagonia), together with natural gas from the archipelago of Tierra del Fuego and the Andean piedmont. In addition, there is considerable hydroelectric output, and some production of nuclear power generation. Steel plants using imported ore and coal have been built near Rosario on the navigable Paraná River, and diversification has extended into the manufacture of chemicals, machinery, and automobiles. The Iguassu Falls provide considerable additional engineering potential for hydroelectric power, but any government would face enormous resistance if there was an effort to modify the falls for power generation.

However, the economic health of a relatively large urbanized service and industrial economy is open to considerable question in Argentina's case. Manufacturing has never become competitive on the world market, and the country's trade pattern is still dominated by agricultural exports and industrial imports. Manufacturing contributes considerably less to national production than might be expected in a country of Argentina's size and economic status, and recent rapid growth of the small-enterprise service sector of the economy reflects in part a decline in some industries, with resulting unemployment. And despite sizable industrial gains, Argentina has not become a prosperous country. In 2000, its per capita GDP PPP of $12,900 was lower than that of poor European countries such as Portugal ($15,800) or Greece ($17,200) and was a little more than two-thirds that of relatively poor Spain ($18,000).

One of the most alarming shifts in the Argentinean scene is the recent upswing in reverse migration. Children and grandchildren of the hundreds of thousands of European immigrants who came to Argentina after World War II have now begun to return to their European "homes" by the thousands. In the first 6 months of 2000, for example, the Italian embassy in Buenos Aires processed some 7,000 passports for Argentineans wishing to emigrate to Italy. This total is higher than all those who sought the same document in the whole of 1999.

In April 2001, representatives of all 34 Latin American nations except Cuba (uninvited to the summit) convened in Quebec, Canada to try to found the Free-Trade Area of the Americas organization. The discussions were attended in spirit by thousands of protesters who surrounded the Quebec City center and were kept behind a 3-mile (5-km) chain-link fence to prevent their protests from disturbing the state discussions going on within. Tensions in the development of the FTAA relate to U.S concerns focused on environmental and labor conditions in the other member nations. Latin American nations, on the other hand,

claim that the United States complains about those two issues because of the desire to protect U.S. interests. In addition, the United States demands for representative democracy also trouble some of the countries because of the checkered history such political experiments have had in many of the Latin American countries. The drug wars being waged in Colombia, Peru, and Bolivia also are unsettling to these nations as they try to establish a firm political linkage as trade equals in the shadow of the EU, NAFTA, and Mercosur — the economic alliance of Brazil, Argentina, Paraguay, Uruguay, and associate members, Chile and Bolivia.

The protesters in Quebec energetically fought the idea of the continuing globalization of industrial organization because of the heavy impact it has had on small farming operations, smaller businesses, and national autonomy in economic decision making. However, it seems likely that some sort of FTAA-like web of economic interrationship will be in place by or even before the 2005 deadline. All across the economic landscape of today's world are evidences of the financial benefits of lower tariffs, easier migration laws and softer borders, and closer alignment between manufacturing, transporting, and consuming populations.

The steady decline of the economy in Argentina has not only slowed and reversed industrial and commercial growth, but it has rippled through the world of retail commerce as well. That shift in what had been a relatively healthy economy at all levels in the mid-1990s has raised an increasing level of pessimism in the minds of young Argentineans. Those same children and grandchildren of the European immigrants of half a century ago are now seeking the economic setting of the European Union and its more reliable economic growth.

The Disastrous Impact of Argentine Politics

Political turmoil and government mismanagement of the Argentine economy have taken a heavy toll on the country. After initial disturbances following independence, Argentina became a constitutional democracy that was controlled by wealthy landowners. This situation continued for many decades until 1930, when a military coup toppled the constitutional government. Between 1930 and 1983, the country was controlled either by military juntas or by Juan Perón or his party, with brief intervals of semidemocratic government. Although there were variations, the military generals and the Perónists tended to be ultranationalist and semifascist. They distrusted democracy, communism, and free-enterprise capitalism and believed in authoritarianism, violent repression of opposition (which itself was often violent), and state ownership and/or direction of major parts of the economy. They courted the support of a growing urban working class by providing it with jobs, pay, and benefits beyond what was justified by its productivity and output, and

they paid for these things by inflating the currency and borrowing from abroad. Inflation rates eventually reached hundreds of percentage points a year, lowering savings and leading to the flight of investment capital and an economy heavily focused on currency manipulation and speculation rather than on productive investment.

This chapter on Argentina is changing even in the middle of 2002. The country is facing a brutal inflation that comes from having such a great percentage of its national income dedicated to retiring the international debt Argentina has accumulated over the past several decades. All governmental plans for economy and creative infrastructural development are constrained powerfully by a debt burden that overshadows all. In 1990, Argentina expended $6.16 billion dollars in debt service. Less than a decade later, the figure was $25.7 billion.

Chile

Chile's peculiar elongated shape tends to conceal the country's sizable area, which is more than double that of Germany. From its border with Peru to its southern tip at Cape Horn, Chile stretches approximately 2,600 miles (c. 4,200 km), but in most places, its width is only about 100 to 130 miles (c. 160 to 210 km). Given a favorable natural environment, Chile might not find its curious shape a serious obstacle to development, but the country has had to struggle with rugged terrain, zones of extreme aridity, and cool wetness to such a degree that only one relatively small section in the center has any considerable population. The long interior boundary lies mostly near the high Andean crest, with the populous core areas of neighboring countries lying some distance away on the other side of the mountains. Although the Andean barrier has by no means been impassable during Chile's four and a half centuries of colonial and independent existence, traffic across the mountains has been difficult and is still light.

The Middle Chilean Core Region

Chile's populous central region of Mediterranean (dry-summer subtropical) climate occupies lowlands between the Andes and the Pacific from about 31° to 37°S latitude. Mild wet winters and hot dry summers characterize this area, as they do southern California at similar latitudes on the west coast of North America (see Fig. 21.1). This strip of territory has always been the heart of Chile's core area. Topographically, it consists of lower slopes in the Andes, a hilly central valley, and low coastal mountains. The area was the southern outpost of the Inca empire and was occupied in the mid-1500s by a small Spanish army that approached from the north. Members of the army made use of land grants from the king of Spain to institute a rather isolated ranching economy. Grants varied in size by the military rank of the grantees, and this fact, together with the conquered status of the native population, created a society marked by great social and economic inequality. At one end of the scale was the aristocracy, composed of those holding enormous ranches. At the other end was a mass of landless cowboys and subsistence farmers. Men from all ranks of society married native women quite freely, sometimes several at a time. This has created the Chilean nation of today, over 90% of which is classified as mestizo.

The owners of great haciendas dominated the Chilean core area and the country itself well into the 20th century, and ranching remained the primary pursuit in middle Chile. Agriculture, both irrigated and nonirrigated, was carried on but was secondary in importance and generally supplemental to ranching. Wheat, vineyards, and feed crops became agricultural specialties (Fig. 21.16). During the 1960s and early 1970s, population pressure and resulting political demands led to large-scale land reform that somewhat reduced the role of many landowners. Growing population pressure in the core region has contributed to rapid urbanization in recent decades. The national capital of Santiago, founded at the foot of the Andes in the 16th century, now has a metropolitan population of more than 5,500,800. Metropolitan Valparaíso, the core region's main port and coastal resort, has 888,300, and the port of Concepción has about 963,800. Altogether, the area of Mediterranean climate is the home of about 75% of Chile's 15.4 million people. The population of the country was 86% urban by a recent estimate, exceeded among South American republics only by that of Uruguay (92%), Argentina (90%), and Venezuela (87%).

Figure 21.16   The grape harvest in the central valley of Colchagua, Chile, is evocative of images of California. Both locales share a Mediterranean climate, and vineyards and wine production play important economic roles in both regions.

To both north and south, the core area contains transition zones where population density declines until areas of very scanty population are reached. On the north, the population declines irregularly across a narrow band of steppe to the arid and almost empty wastes of the Atacama Desert, which is the driest region on Earth. In the south, the transition zone is much larger and more important economically. South of Concepción, rainfall increases, average temperatures decline, and the Mediterranean type of climate gives way to a notably wet version of the marine west coast climate, whose natural vegetation is impressively dense forests. The southernmost part of the Mediterranean climatic area and the adjacent northern fringe of the marine west coast area as far south as the small seaport of Puerto Montt have become part of Chile's core in the past century, forming a less populous transitional end of the core.

The Araucanian people native to these forests fought off Inca expansion, and they were also able to check most Spanish expansion for three centuries. They were eventually assimilated into Chilean society rather than truly conquered. Their full-blooded native descendants are still numerous in this part of Chile. Also distinctive is a less numerous German element descended from a few thousand immigrants who settled on this wild frontier in the mid-19th century. Today, the German community is economically and culturally very important, even putting its stamp on architectural styles. Most people in the southern transition zone, however, are wholly or partly descendants of mestizo settlers who came from the core of Chile in the past century as population pressure increased. Agriculture in this green landscape of woods and pastures is devoted largely to the raising of beef cattle, dairy cattle, wheat, hay, deciduous fruits, and root crops.

Northern and Southern Chile

North of the core, Chile is desert, except for areas of greater precipitation high in the Andes. This part of Chile, the Atacama Desert, is one of the world's few nearly rainless areas. Mild temperatures, high relative humidity, and numerous fogs are the result of almost constant winds from the cool ocean current offshore. Chile gained most of the Atacama by defeating Bolivia and Peru in the War of the Pacific (1879–1883). Bolivia had previously included a corridor across the Atacama to the sea at the port of Antofagasta, and Peru owned the part of the desert farther north. The war was fought over control of mineral resources, primarily the Atacama's world monopoly on sodium nitrate, a material much in demand at that time for use in fertilizers and the manufacture of smokeless gunpowder. Since then, other mineral resources of the Atacama and the adjacent Andes, principally copper, have eclipsed sodium nitrate in importance.

South of Puerto Montt, the marine west coast section of Chile continues to the country's southern tip in Tierra del Fuego, the cool island region of windswept sheep pastures and wooded mountains divided in ownership between Chile and Argentina. In this little-populated strip, the central valley and coastal mountains of areas to the north continue southward, but here the valley is submerged, and the Pacific reaches the foot of the Andes along a rugged fjorded coastline. The coastal mountains project above sea level only in higher parts that form offshore islands. Extreme wetness, cool temperatures, violent storms, and dense mountain forests are characteristic. Little agriculture beyond some grazing of sheep has been possible, and population is very sparse. Some oil and gas are produced along or near the Strait of Magellan, which separates the South American mainland from Tierra del Fuego.

Chile has been at the edge of the North American Free Trade Agreement (NAFTA) for some years. In partial response to this spatially curious link between Mexico, the United States, Canada, and Chile, in January 1995, the Southern Cone Common Market, known as Mercosur, became operative. It was built around Brazil, Argentina, Uruguay, and Paraguay initially, but later added Bolivia and Chile as associate members. This union was created partly in response to the success that was being realized by the European Common Market (now the European Union) and the 1994 founding of NAFTA. The benefits gained through a common market and its easier trade flows were appreciated by the South American nations, who were all trying aggressively to achieve greater economic development and international stature.

By the end of 1997, Mercosur was seen as an effective economic union of nations with a growing middle class, an expanding base of disposable income, and a particular interest in having Mercosur serve as a southern NAFTA. Relations between Argentina and Chile have improved with the construction of an oil pipeline allowing Argentine gas to flow to Chile. Two additional, and heretofore unprecedented, pipelines are slated for construction in the near future. Electricity is being sold back and forth across the southern Andes, and there is even talk about expansion of the highway network to enable Argentine and Brazilian producers to transport soybeans, mineral resources, and manufactured goods to Chile to get into active Pacific Rim trade.

Although the 1997 slowdown in almost all East Asian economies has slackened the pace of some of these innovations, Mercosur continues to set the stage for a new level of regional economic cooperation in the southern tip of South America. These nations are keenly interested in being more fully integrated into the economic dynamics of the northern end of the hemisphere, especially North America, and of East Asia. For this reason, there is great interest in the potential for the 2005 Free-Trade Area of the Americas (FTAA) that is discussed in this chapter's Regional Perspective box on p. 573.

Uruguay: Costly Dependence on Agriculture in an Urbanized Welfare State

Uruguay, with an area the size of Missouri and bordered only by Argentina and Brazil, is a part of the humid pampas. Its independence from Argentina and Brazil resulted from its historic buffer position between them. In colonial times, there were repeated struggles between the Portuguese in Brazil and the Spanish to their south and west over possession of this territory flanking the mouth of the main Rio de la Plata river system in southern South America. After Argentine independence from Spain, these struggles continued at the same time that a movement for political separation grew in the territory north of the Rio de la Plata. Great Britain eventually intervened, fearing Portuguese expansion south of the river, and the two contenders signed a treaty in 1828 that recognized Uruguay as an independent buffer state between the two larger countries.

Montevideo, Uruguay's national capital, is the only large city in the country, and its metropolitan population of 1,701,600 includes about one-half of all Uruguayans. The city has dominated the country almost from Montevideo's inception in the 17th century, first as a Brazilian (Portuguese) and then as an Argentine (Spanish) fort. Its port facilities handle Uruguay's important waterborne trade. Montevideo directs an economy that developed such a high degree of government ownership that the majority of workers came to be employed directly or indirectly by the government. The city is the main center of Uruguay's manufacturing industries, which employed about one-fifth of the country's labor force in the mid-1980s, but which export little except textiles. Meanwhile, agriculture and ranching, with 4% of the labor force, supply the country's meat-rich diet and the greater part of all Uruguayan exports. This structure began to fail in the 1960s as costs outran production, and the structure was devastated by the high oil prices imposed by oil exporters in the 1970s.

Aside from hydropower, which supplies nearly all of the country's electricity, Uruguay has almost no natural resources to support industrialization and depends heavily on imports of fuels, raw materials, and manufactured goods, which must be purchased with the proceeds from agricultural and textile exports. In the early 1990s, Uruguay was attempting to cope with these problems and to preserve its recently restored democracy. Lower world oil prices were a factor easing the burdens of this conflict-racked country. An extensive program is underway to privatize government-owned industries.

Development Problems and Prospects in Landlocked Paraguay

Landlocked and underdeveloped Paraguay has long been isolated from the main currents of world affairs. When a nation is landlocked, a wide range of geography-based conditions are set in motion. All transit into the landlocked nation depends on the capacity and inclination of the nation's neighbors to build, maintain, and keep open cross-country road networks, mountain passes, bridges, and other elements of a surface transport system. Beyond the transportation difficulties and the political and economic costs that have to be paid to move goods, and even people, from the landlocked nation to an ocean coast—even in this time of expanding air linkages—there is the psychological cost of such a geographic condition. The limits of the map have been the spark of conflict among countries for centuries. To be on the ocean or on major international rivers creates a view of one's place in the world quite distinct from that of a landlocked country.

The Spanish founded Paraguay's capital city of Asunción (population: metropolitan area, 1,427,700) in the 1530s at a site far enough north on the Paraguay River to promise security against the warlike Native Americans of the pampas to the south. In addition, high ground at the riverbank gave security from floods. Asunción became the base from which surrounding areas were occupied. But when independence came, Paraguay had no way to reach the sea except through Argentina or Brazil. The main route that developed reached down the Paraguay and Paraná Rivers to Buenos Aires. But this route is much longer than it appears on maps, owing to meandering of the rivers, and navigation was hindered by difficulties such as shallow and shifting channels, fluctuations in water depth between seasons, snags, and sandbars. Transport costs were so high that Paraguay could not develop an export-oriented commercial economy, and the country remained locked into a predominantly self-contained economy based on the resources of its own territory.

West of the Paraguay River is the region known as the Chaco ("hunting ground"), which extends into Bolivia, Argentina, and Brazil. This is a very dry area, increasingly so toward the west. Precipitation averages 20 to 40 inches (c. 50 to 100 cm) annually, but high temperatures cause rapid evapotranspiration, and porous sandy soils absorb surface moisture. These conditions produce an open xerophytic forest in the east, which thins westward and then gives place to coarse grasses on a vast plain of flat alluvium where wide areas are flooded during the wet season. Paraguayans in the entire western region number only in the neighborhood of 100,000. A striking element in this sparse population is the scattered Mennonite colonies that aggregate approximately 20,000 people.

Paraguay's comparatively low level of development manifests itself in a number of ways, one of which is its total population. Only an estimated 5.7 million people occupied its sizable territory in 2001. This small population results partly from the fact that Paraguay has been a country of emigration for most of its history and has never attracted any large immigration. The latter circumstance has given the country a racial mix containing few whites and a high proportion of native Guaraní blood in the 95% of the population classed officially as mestizo. Urbanization and industrialization have not gone far. Asunción is the only urban place with more than 600,000 people, and about 45% of the country's labor force is still employed directly in agriculture. Per capita income is far below that of Argentina or Uruguay.

Recently, the pace of Paraguay's development has quickened. Modern road and air links with the outside world and between parts of Paraguay itself have been forged. With this access to outside markets, exports of soybeans and cotton have grown rapidly and have come to dominate the country's trade. The future may see larger and more dramatic changes due to three huge new hydroelectric installations along the Paraná River on the southeastern border of the country. Paraguay is now exporting electricity to pay its share of the costs of these projects, which it undertook jointly with Argentina and Brazil. The new hydroelectric output has given Paraguay a far more favorable energy base in the future than this fuel-deficient country has had in the past.

21.4 Antarctica

Antarctica is the world's fifth largest continent, with an area of 5.5 million square miles (14,245,000 sq km) lying south of the tip of South America and virtually filling the Antarctic Circle (Fig. 21.17). It is the setting of enormous human dramas in exploration, bravery, and foolishness as people have crossed the continent with dogsleds, on skis, on foot, in airplanes, and now in tour helicopters. The mystery of the place comes in

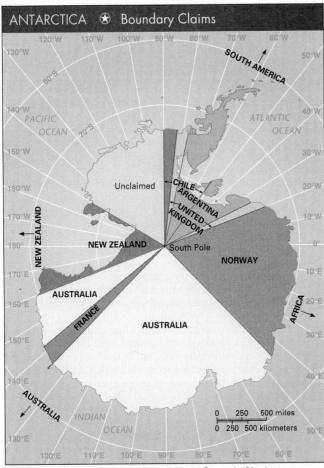

ANTARCTICA ✪ Boundary Claims

Source: Rubenstein, James M. *An Introduction to Human Geography*, 6th ed.,
Upper Saddle River, NJ.: Prentice Hall, 1999. p. 271.

Figure 21.17 International land claims to Antarctica. Antarctica, the continent that surrounds the South Pole and lies at the southern extreme of South America and Tierra del Fuego, has become the focus of considerable global interest in the past four decades. The 1957–1958 International Geophysical Year (IGY) opened up significant international interest in exploration and the settlement of research teams. This map represents the declaration of territorial rights expressed by nations relatively close to Antarctica and by nations wishing to have control of at least some of the resources that are thought to be under the nearly total ice cover.

part from its winter of darkness, with continual problems of "whiteouts" caused by light refraction on an extensive snow and ice surface covering some 95% of the continent. Mirages are common, and average temperatures during the "summer" months barely reach 0°F. Winter averages are the coldest in the world, with winter mean temperatures averaging −70°F (−57°C).

The whole continent is alive with ice formations created by the 2 to 10 inches (5 to 25 cm) of annual precipitation. The formations fed by this seasonal moisture then move toward the open seas of the South Atlantic. Glacial action of all sorts provides a continual breaking away of valley glaciers, ice streams, and tongues all crashing into the sea at the margins of the continent, forming icebergs. There is virtually no human settlement beyond the research teams that have constructed an array of semipermanent structures on the small areas of exposed land that lie near the outer edges of the continent and on the island of Little America in the Ross Sea.

However, there is enormous interest in this distant world. To stay 2 weeks as a "tourist" in varying accommodations costs between $25,000–$43,000. A protocol signed by the 12 countries that signed the 1957 Antarctic Treaty in 1991 states that "the nations participating in the Antarctic Treaty are not to carry out drilling projects or otherwise exploit the continent's natural resources for the next 50 years." And it is the potential for hidden and even unanticipated resources that keep the attention level high over this land of ice.

There has long been speculation about the potential for considerable resource wealth lying beneath the sometimes thousands-of-feet-thick cap of ice. Although many nations now claim some rights to the continent and support the periodic settlements of research scholars, neither the resources nor the fauna are likely to lead to a significant exploitation of this enormous continental mass. It will continue to be, however, another frontier in the geography of the future.

CHAPTER SUMMARY

- South America is made up of a number of geographic subregions, one of which consists of the Andean countries, including Colombia, Venezuela, Ecuador, Peru, and Bolivia. These countries have marked patterns of settlement both along the coasts in the tierra caliente and in the uplands in the tierra templada or tierra fría.

- Major characteristics of the Andean countries include a varied environmental zonation, fragmented topographic patterns, populous highlands and sparsely settled frontiers, significant indigenous populations, significant native landscape signatures, dominant Hispanic traditions, endemic poverty, and primate cities.

- Colombia has three major highland metropolises: Bogotá, Medellín, and Cali. Each has distinct characteristics. Coastal Colombia has less settlement than the uplands, with the Caribbean shores having the larger cities. Bananas and sugarcane

have been the traditional crops, with corn and rice gaining new importance.

- In Venezuela, major settlement is also concentrated in the uplands, with the population in Caracas at more than 4 million. Discovery of oil in the coastal regions a century ago around Lake Maracaibo prompted rapid industrial growth. Iron resources in the Guiana Highlands support an active export flow, utilizing the Orinoco River in part.

- Ecuador also has major settlements in the Andean basins. There has been a steady increase in the proportion of Ecuadoran population locating in the lower slopes and the coastal areas. Oil has begun to be extracted from the eastern side of Ecuador in the tropical lowlands that drain into the Amazon River.

- Peru was the home of the Incas, whose empire extended over 2,000 miles through Andean South America. The Lost City of the Incas, known now as Machu Picchu, represents a bold and well-constructed ceremonial center and city that had reached its zenith well before Columbus began his first voyage. The potato was domesticated in Peru and moved into active Old World trade after the arrival of the Europeans.

- El Niño is a periodic weather phenomenon that brings warm ocean waters from the western Pacific to the west coast of South and Middle America. The normally cold Peru Current is modified by the El Niño waters, influencing not only fishing along the coast but also weather both on the lands lying east of the current and globally.
- Bolivia is South America's poorest country, landlocked high in the Andes with its capital, La Paz, located at 12,000 feet (3,700 m). The Bolivian mine at Potosí produced nearly one-half of the world's silver in the 16th century as a result of the enormous investment of Native American labor and, later, African slaves in mining and refining precious metals for world trade.
- Brazil is a clearly emerging industrial giant in South America. It is larger in area than the 48 conterminous United States, but has more than 100 million fewer people. Settlement tends to cluster more along the coast than in the thinly settled interior. In the late 1950s and early 1960s, Brazil undertook a campaign to promote greater settlement in the interior savanna of the country. A new capital, Brasília, was constructed and, in 1960, the seat of federal government was formally moved from Rio de Janeiro on the eastern coast to the new capital. This move has caused some demographic shift, but it is still along the coast that most of Brazil's population is concentrated.
- Brazil has a large realm of tropical rain forest and tropical savanna climate patterns and does not have the highlands so characteristic of South American Andean countries. There is a broad coastal plain, and the Amazon River drains more than 2 million square miles. Periods of economic change in Brazil can be identified as the sugarcane era, the gold and diamond era, the rubber and coffee booms, and now industrialization, tourism, and export agriculture, with increasing importance on soybeans.
- Countries of the southern midlatitudes include Argentina, Uruguay, Paraguay, and Chile. The first three all possess a humid subtropical climate like that in the southeastern United States. Chile, however, is more similar to southern California since both regions are Mediterranean in climate.

- Argentina is the second largest nation in South America, with an area of more than 1 million square miles (2,776,884 sq km). It has major urban growth along the coast and in the alluvial plain of the Rio de la Plata, but it is the agricultural productivity of the pampas — the dominant grasslands — of the interior that largely defines Argentina. Beef and wheat are major Argentine export commodities.
- Argentina made use of a policy of import substitution in the meat and food processing industries in World War II. The isolation from normal trade flow allowed other industries to expand as well. This led to state promotion of many local industries, but in recent decades, the government has shifted virtually all such manufacturing toward privatization.
- Mercosur is the name of the Southern Cone Common Market. Southern Cone is a term used to represent the cluster of countries that taper like a cone toward Tierra del Fuego. Mercosur includes Brazil, Argentina, Uruguay, and Paraguay, with Bolivia and Chile as associate members.
- Chile has worked to diminish its dependence on the export of copper for its foreign exchange, although it is still very important. Increased trade with East Asia and countries of the Pacific Rim was becoming increasingly important until the mid-1997 collapse of the Thai currency and consequent collapse of export markets to East Asia.
- Uruguay and Paraguay both represent further distinctive national scenes in South America. Uruguay embarked on an unusual experiment as a total welfare state, but in the last two decades, the enormous expense of such a social structure has led the government away from that program. While a number of Latin American nations have been hurt by the low oil prices in the late 1990s, Uruguay has benefited from that decline because it depends on imported oil for its petroleum needs.
- Landlocked Paraguay has been a traditional source of out-migrating youth. The country not only lacks an outlet to the sea that it controls, but it has few mineral resources and lost what proximate land that did have resources in a 19th-century war with Bolivia. Recent efforts to create major hydroelectric installations in cooperation with Argentina and Brazil have changed the outlook for its economic development.
- Antarctica is often perceived as a potential resource outpost of sorts for South America, although current interest in the continent lies primarily in research.

REVIEW QUESTIONS

1. What countries are included among the Andean countries of South America?
2. Outline the most distinctive demographic patterns of South America, making special reference to lowland and highland differentiation.
3. Name some of the most significant pre-Columbian peoples in South America and outline their contributions to culture and economy both in the past and now.
4. What are the most significant Hispanic landscape signatures that exist in South America?
5. Explain how topography has been a major influence in demographic patterns in the Andean countries.
6. What is the resource that has been so important for both Venezuela and Ecuador?
7. Discuss the various dimensions of the Colombian economy and relate these to geography.
8. Discuss the impact of the discovery of petroleum on the eastern flank of the Andes.
9. What are the characteristics of El Niño and where has it had the most impact?
10. List four of the eras that Brazil has experienced in its economic development. What role has geography played in each of these eras?
11. Define growth pole and explain how Brasília can be related to that concept.
12. Locate the major area of tropical rain forest in South America and describe its demography, its biological and environmental significance, and the reasons for its economic instability.
13. Outline the ways in which Antarctica might be a political hot spot in the 21st century.

DISCUSSION QUESTIONS

1. What role have the Andes Mountains had in the development of contemporary South America?

2. What are the geographic factors that make coca and marijuana such popular upland crops in South America?

3. What has caused the Rimland-Upland or Rimland-Mainland demographic patterns in South America?

4. How do you explain the locational and economic patterns of the indigenous people in this region? How and why would you change these patterns it if you were a government planner?

5. Discuss the concept of a growth pole, relate it to a capital city, and also to some new industry, explaining where you would locate a new venture and what impact it might have.

6. Discuss the ways in which Mercosur works and the impact it might have if it were to unfold in ways similar to NAFTA or the European Union.

7. Discuss the ways in which El Niño has changed the face of both economics and short-term weather patterns in South America. Has it had influence elsewhere?

8. Comment on the images of Brazil's tropical rain forest and Antarctica in terms of resources, population densities, and governmental concern.

9. Discuss the significance of being a landlocked nation and suggest ways to modify the locational difficulties associated with such a position. Name examples.

10. What are the reasons Chile or any other South American country might like to be part of NAFTA but not Mercosur?

The United States and Canada

In this last world region, there is the geographic evidence of peoples fortunate in having extraordinary national resource bases in their countries. For the relatively short time that the United States and Canada have been settled, there has been a steady investment in landscape transformation, the construction of transportation networks, and participation in the global economy. As varied as the cultural character of this region is, there is a nearly universal drive to make the human setting ever more productive and satisfying.

The story of landscape design, creation, and utilization has been written at all scales in the United States and Canada. The concept of finding a locale for settlement, reworking its landscape to reflect one's needs and resources, and taking pride (or alarm) in what is created has been repeated continually in the settlement of this world region.

For ages, people living in North America have felt a sense of splendid isolation, at least from much of the warfare that has characterized other regions of the world. The September 11, 2001 terrorist attack let people all across the globe know that they are—whether they wish to be or not—players in a global arena that is political and religious as much as it is economic.

From east to west and, to a lesser degree, from north to south, networks of transportation links have been created from beginnings as simple as animal trails to ones as complex as major river, canal, and lock systems that allow inland wharves to become part of global trade.

The wave of globalization is not seen by all as a welcome change in economic linkage. Even though commodities may be cheaper for many consumers, this change has closed the doors of many plants that traditionally had been a significant cultural as well as economic part of the local scene.

CHAPTER 22

A Geographic Profile of the United States and Canada

North America is not the largest landmass in Earth's collection of continents, but it is fortunate to have one of the most productive agricultural regions of any landmass in the world. In this photograph, the broad, extensive plains regions spread from the central United States northward into Canada, serving as a resource center for the food needs of the entire world.

Chapter Outline

In this final region of the text, we turn to the United States and Canada. They, in combination with Mexico and Central America (known generally as Middle America), make up the continent of North America. However, Mexico is so powerfully linked to Middle and South America through culture, language, and tradition that it belongs to the region called Latin America and is not defined as part of this final region. The United States and Canada as a culture region are sometimes called Anglo America, or sometimes the two nations are simply seen as the region of North America. As part of the so-called New World, the United States and Canada were witness to the presence of active early settlement by numerous Amerindian groups who scattered all across the landscapes of the two countries well before the arrival of Columbus in 1492. Since that date, this region has been a center of continual settlement and of economic and cultural development and diffusion. In fact, this region has served as a major target destination for migrating peoples for the past five centuries. The region is now a complex mosaic of greatly varied geographic and cultural characteristics, with the political, cultural, and economic imprint of the British still at the foundation of the reality of the United States and Canada.

22.1   Area and Population

Possession of a large area assures a country of neither wealth nor power. But it does afford at least the possibility of finding and developing a wider variety of resources and, other things being equal, of supporting a larger population than might be expected in a small country (Table 22.1). The fact that there is no direct relationship between area on the one hand and wealth and power on the other may be seen by comparing these two countries themselves. Canada is slightly the larger in area (3.85 million sq mi/9.97 million sq km compared with 3.72 million sq mi/9.63 million sq km for the United States) but is less wealthy and much less powerful on the global stage than is its neighbor, the United States. In population, Canada had 31 million in 2001, while the United States' total was 284.5 million that same year. Thus, the geographical impact of size and population on either wealth or political power has yet to be fully determined (Fig. 22.1).

Population Characteristics

From the beginning of settlement until the 1960s and 1970s, this region was one of rapidly growing population. By 1800, approximately two centuries after the first European settlements, there were more than 5 million people in the United States and several hundred thousand in Canada. Growth since that time is summarized in Table 22.2. The population of this region long increased at a rate much more rapid than that of the world as a whole. As a result, one element in the mounting importance and power of the United States, where most of this growth occurred, was its possession of an increasing share of the world's population. By the 1980s, however, the population growth rates of both the United States and Canada were well below those of most of the world, except for Europe, Japan, and the former Soviet Union. Together, the United States and Canada now account for about 5% of the estimated world population and about 14% of the world's land area (excluding Antarctica). The majority of this region's

Table 22.1   United States and Canada: Basic Data

Political Unit	Area (thousand/sq mi)	Area (thousand/sq km)	Estimated Population (millions)	Estimated Annual Rate of Increase (%) (actual)	Estimated Population Density (sq mi)	Estimated Population Density (sq km)	Human Development Index	Urban Population (%)	Arable Land (% of total area)	Per Capita GDP PPP ($US)
Canada	3849.7	9970.6	31.0	0.3	8	3	0.936	78	5	24,800
United States	3717.8	9629.1	284.5	0.6	77	30	0.934	75	19	36,200
Total	7567.5	19,559.7	315.5	0.6	42	16	0.934	75	12	35,080

Sources: *World Population Data Sheet*, Population Reference Bureau, 2001; *U.N. Human Development Report*, United Nations, 2001; *World Factbook*, CIA, 2001.

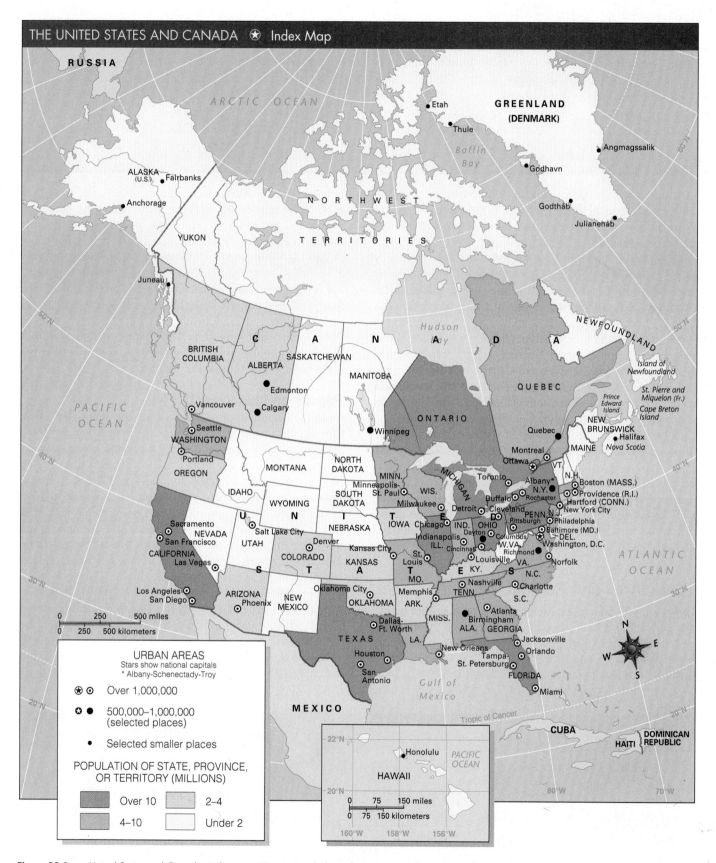

THE UNITED STATES AND CANADA ✪ Index Map

RUSSIA

ARCTIC OCEAN

GREENLAND
(DENMARK)

Etah

Thule

Baffin
Bay

Angmagssalik

ALASKA
(U.S.)   Fairbanks

Godhavn

Anchorage

NORTHWEST

Godthåb

Julianehåb

YUKON

TERRITORIES

Juneau

Hudson
Bay

NEWFOUNDLAND

BRITISH
COLUMBIA

CANADA

Island of
Newfoundland

SASKATCHEWAN

ALBERTA

St. Pierre and
Miquelon (Fr.)

MANITOBA

QUEBEC

Cape Breton
Island

Prince
Edward
Island

Edmonton

PACIFIC
OCEAN

Vancouver

NEW
BRUNSWICK

Calgary

ONTARIO

Quebec

MAINE   Halifax

Seattle

Winnipeg

WASHINGTON

Montreal

Nova Scotia

Portland

Ottawa

VT

N.H.

OREGON

MONTANA

NORTH
DAKOTA

Toronto

Albany*   Boston (MASS.)
N.Y.   Providence (R.I.)

MINN.

MICHIGAN

Buffalo   Rochester   Hartford (CONN.)

IDAHO

Minneapolis-
St. Paul

WIS.

Detroit

Cleveland

New York City

WYOMING

SOUTH
DAKOTA

Milwaukee

PENN.   N.J.

Pittsburgh   Philadelphia

Sacramento

Salt Lake City

NEBRASKA

IOWA   Chicago

IND.   OHIO   Columbus   Baltimore (MD.)

San Francisco

NEVADA

Denver

Indianapolis

Dayton

W.VA.   DEL.

Washington, D.C.

CALIFORNIA

UTAH

ILL.   Cincinnati

Richmond

Las Vegas

COLORADO

Kansas City

St.
Louis

Louisville

KY.

VA.   Norfolk

ATLANTIC
OCEAN

KANSAS

MO.

N.C.

Los Angeles

Nashville   Charlotte

San Diego

ARIZONA   NEW
MEXICO

Oklahoma City

Memphis

TENN.   S.C.

Phoenix

OKLAHOMA

ARK.

MISS.   Birmingham   Atlanta

Dallas-
Ft. Worth

ALA.   GEORGIA

Jacksonville

TEXAS

LA.

New Orleans

Orlando

Houston

Tampa-
St. Petersburg

San
Antonio

Gulf of
Mexico

FLORIDA

Miami

MEXICO

Tropic of Cancer

CUBA

DOMINICAN
REPUBLIC

HAITI

URBAN AREAS
Stars show national capitals
* Albany-Schenectady-Troy

✪ ⊙   Over 1,000,000

✪ ●   500,000–1,000,000
(selected places)

•   Selected smaller places

**POPULATION OF STATE, PROVINCE,
OR TERRITORY (MILLIONS)**

Over 10   2–4

4–10   Under 2

0   250   500 miles
0   250   500 kilometers

22°N

Honolulu

PACIFIC
OCEAN

HAWAII

20°N

0   75   150 miles
0   75   150 kilometers

160°W   158°W   156°W

80°W   70°W

Figure 22.1   United States and Canada: Index map. City-size symbols conform to metropolitan area estimates given in the text.

TABLE 22.2	World's Top 15 Countries in Per Capita GDP PPP
Political Unit	**$U.S.**
1. Luxembourg	36,400
2. United States	36,200
3. Switzerland	28,600
4. Norway	27,700
5. Singapore	26,500
6. Denmark	25,500
7. Belgium	25,300
8. Austria	25,000
9. Japan	24,900
10. Canada	24,800
11. Iceland	24,800
12. France	24,400
13. Netherlands	24,400
14. Germany	23,400
15. Australia	23,200

Source: *World Factbook*, CIA, 2001.

315.5 million people (as of 2001) live in the eastern half of the region, from the St. Lawrence Lowlands and Great Lakes south and east. In Canada, approximately 90% of the population live within 100 miles (160 km) of the border between Canada and the United States (Fig. 22.2). The estimated average population density of the two countries in 2001 was only 80 per square mile (31 per sq km) for the United States and 9 per square mile (3 per sq km) for Canada, with its tremendous expanse of sparsely settled northern lands. And by world standards, even the United States is far from overpopulated (see Fig. 22.1).

The original Native American population was overwhelmed by European settlement and disease and, thus, was reduced to a politically weak and economically underprivileged minority segregated mainly in western and northern areas. The role of the original peoples of the North American continent continues to be one of difficulty and governmental attention.

Both Canada and the United States continue today to each have one exceptionally large ethnic minority that has been

Box *518*

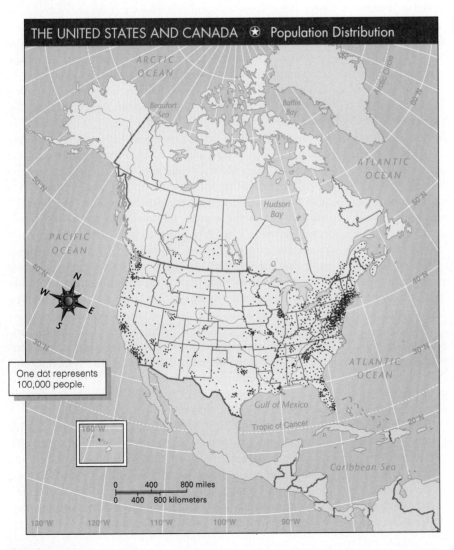

Figure 22.2   Population of the United States and Canada.

the focus of serious problems of national unity. For example, about 26% of Canada's people are French in language and culture. These French Canadians are concentrated mostly in the lowlands along the St. Lawrence River, where they form the majority population in the province of Quebec. Their French ancestors, some 60,000 in number, were left in British hands when France was expelled from Canada in 1763. They did not join the English-speaking colonists of the Atlantic seaboard in the American Revolution, and by their refusal to do so, they laid the basis for the division of the continent of North America into two separate countries. Until the late 20th century, they increased rapidly in numbers, and the French Canadians have clung tenaciously to their distinctive language and culture, attempting twice within the past 20 years to gain independence for the province of Quebec and establish a French-speaking independent country.

At times, major controversies arise between this group and other Canadians, and a provincial government with separatist aims was elected by Quebec's voters in the 1970s. However, the Quebec electorate then refused to approve its government's movement toward secession from Canada, and this possibility temporarily receded, only to flare up with renewed heat in the late 1980s. The Canadian government has attempted to preserve national unity through the recognition of both English and French as official languages and the provision of legal protection for French Canadian institutions. In October 1995, the Québécois, by popular referendum, voted to remain a province within Canada. The vote, however, was 51 to 49% so that the issue of Quebec separatism is still very much alive in Canada and will likely stay on the national agenda for some time to come.

Black people of African descent constitute one-eighth of the U.S. population, and until the 2000 census when Hispanics exceeded the number of African Americans, they had been the principal minority in that country. Unlike the French Canadians, they are not largely confined spatially to only one part of the country, although they do predominate in a number of large cities and particularly in the centers of those cities. Historically, African Americans as a group have not been an active force toward sectionalism in the United States. Nevertheless, a conflict of attitudes toward black slavery was among the important factors that nearly split the United States into two countries in the American Civil War, and race relations are still a major source of friction in American politics. The primary setting for most of the race relations difficulties continues to be urban America. There is a segment of the African American population that has moved into the middle class and above, but a larger segment resides in the largely impoverished and troubled central cities, where problems in housing, jobs, crime, and racial isolation continue to tear at a traditional American sense of unity or at least the hopes for a community of peoples who might thrive within the diversity of the country.

The census of 2000 in the United States has shown a rapid increase in the percentage of Hispanics and Asian Americans in the population, further diminishing the demographic majority of the Anglo population. Even as patterns of immigra-

tion change somewhat toward East Asia and countries south of Mexico from Latin America, the role of the United States and Canada as immigrant destinations has not changed or diminished. It is assumed that this region will continue to play a major role in the global migrant flow. However, it is probable that there will be an increase in the immigration of more people with technical skills and family capital as opposed to the traditional inflow of single male migrants (legal or illegal) making the move in the quest for a new beginning and better economic and political opportunities than they had known in their home countries.

Spanish-speaking people have been a smaller minority in the United States than are African Americans until this recent census. Their growth rate has been more rapid, both domestically and through a major migration stream that flows through all of Middle America toward the long U.S.-Mexico border. A little more than 12.5% of the U.S. population is Hispanic. The majority are of Mexican origin and are strongly concentrated in California, Texas, New Mexico, Arizona, and Colorado. Other important Hispanic groups include people of Puerto Rican origin, most heavily represented in New York City and in the northeast, and people of Cuban origin, who live largely in southern Florida. Estimates vary as to how many million other Hispanic people, mainly Mexicans and Central Americans, are resident in the United States as illegal immigrants. Like African Americans, American Hispanics are not numerically or politically dominant in any state. They are, however, gaining in relative proportion in many of the large urban centers, especially in the Southwest.

Nationally, Asian Americans equal 3.6% of the U.S. population. While initial settlement was concentrated in the distinctive "Chinatown" landscapes that so boldly gave this population (whether Chinese or not) its image in the 19th and early 20th century, Asian Americans have been steadily moving out of the central city and into white and mixed neighborhoods. The closest approach to minority dominance of a state is in multiethnic Hawaii, where Asian Americans have become very powerful politically (although without any hint of separatism). Among the many Asian American groups in Hawaii, Japanese Americans are by far the most numerous. In the United States as a whole in recent decades, there have been large increases in the Filipino, Korean, Vietnamese, and Cambodian populations, thereby further expanding the characteristics of Asian Americans in the North American landscape. Cities throughout the region tend to be the areas where there is the greatest amount of ethnic mixing, particularly in the workplace and in recreation.

The governments of both the United States and Canada have assiduously fostered national unity through the provision of adequate internal transportation. In both countries, transportation networks have had to be built not only over long distances but primarily against the "grain" of the land. Effective transport links between east and west have been the most imperative, but most of the mountains and valleys have a north–south trend, which has presented a series of

REGIONAL PERSPECTIVE ::: Resolution of Native Land Issues

In 1999, Canada ceded nearly one-quarter of its total land area to the Inuit peoples. These 733,600 square miles came from Canada's Northwest Territories and are made up of land lying north of 60°N and surrounding the northwest part of Hudson Bay and extending well into the Arctic Circle. The total population of this new province of Nunavut is approximately 30,000, scattered through fewer than 30 recognized communities or settlements. Approximately nine-tenths of the capital required to keep the new territory viable comes from Ottawa, the federal capital of Canada, making the per capita cost of maintaining Nunavut higher than for any other political unit in Canada.

The bold innovation of ceding such a quantity of land, which may still be found to have a wealth in resources well beyond the current estimation, comes from the ongoing tension in Canada and the United States over honoring 18th-, 19th-, and even 20th-century treaties drawn up between the growing countries of the United States and Canada and the native peoples who had much earlier settlements in landscapes that were increasingly valuable in the eyes of settlers working their way west from more densely settled east coast settlements.

In another dimension of the same issue, museums in both the United States and Canada must deal with native peoples' demands that bone collections be returned to the original and appropriate tribes so that they may be buried with ceremony and solemnity, as opposed to being displayed in museum collections for the public. The institutional response has been slow and is built around the claim that research is being done and still must be done. The bones and ceremonial artifacts associated with these dead are thereby caught in a sort of limbo.

Native peoples have been given certain rights as to fishing, the sale of cigarettes and liquor, and gaming privileges. All across the United States, there are local tensions rising over the perceived economic advantage these unique allowances give native peoples while their nonaboriginal neighbors are disallowed from such fishing or hunting or special economic permits. All of this is evidence of the continuing unsettled nature of the land ownership changes that took place as part of the steady, inexorable wave of Anglo and other settlements that rippled across the United States and Canada from the 17th century on. These issues will be part of the legal and ethical mindset and associated landscapes for decades to come.

---

geographic obstacles to cross-country transport. The coasts of both countries were first tied together effectively by heavily subsidized transcontinental railroads (Fig. 22.3) completed during the latter half of the 19th century. Subsequently, national highway and air networks further enhanced national unity. The populations of both countries have long held mobility to be an important right. The development of affordable automobile transportation in the early decades of the 20th century began what is now more than a century-long fascination with the car that has not abated on either side of the border. The love affair with the car is not only a part of the pattern of regional mobility, but it has served as a significant economic feature of industrial growth near the border as well.

22.2 Physical Geography and Human Adaptations

Even in a broad-scale view that suppresses much detail, the natural environments of the United States and Canada are remarkably diversified. This diversity reflects a number of major circumstances and leads to numerous "microgeographies," or distinctive local landscapes that have their own identities and use patterns.

Major Landform Divisions

The major landform divisions are shown in Figure 22.4. The glacially scoured and thinly populated Canadian Shield is formed of ancient rocks that are rich in metal-bearing ores. The surface is generally rolling or hilly. Agriculture, handicapped by poor soils and a harsh climate, is limited and marginal. Water power, wood, iron, nickel, and uranium are major resources. The United States section bordering Lake Superior is called the Superior Upland.

The Arctic Coastal Plain occurs in two widely separated sections, each of which is largely unpopulated wilderness.

FRED MAROON/PHOTO RESEARCHERS, INC.

Figure 22.3 Although both the U.S. and Canada now use only a portion of their initial railroads, the rail systems continue to be economically and spatially significant.

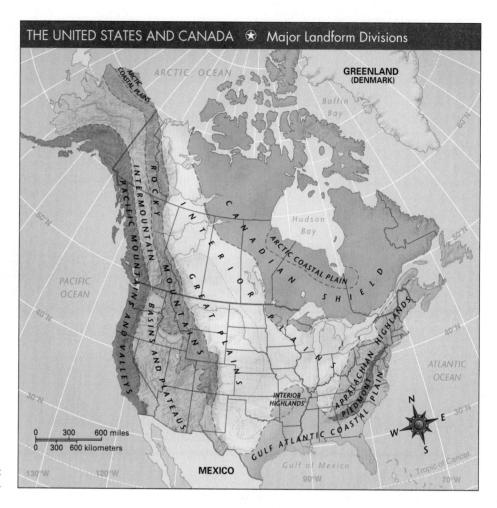

Figure 22.4   United States and Canada: Major landform divisions.

The Alaska section along the Arctic Ocean contains rich oil deposits, exploited on a major scale, but the section along Hudson Bay is economically insignificant. Mountainous Greenland is buried under permanent ice except for thinly inhabited coastal fringes of tundra. Fishing is the main economic activity there.

The Gulf-Atlantic Coastal Plain is low and relatively level, with soils that are generally sandy and relatively infertile. Agriculture involves many different specialties in scattered areas of intensive development. Pine or mixed oak-pine forests cover large areas. Swamps and marshes are abundant, and the coast is indented by river mouths and bays on which many prominent seaports are located. Stands of swamp hardwoods occur in poorly drained areas, and prairie grasslands were the original (but probably fire-modified) vegetation in certain coastal areas in Texas and southwestern Louisiana and in some inland areas underlain by limestone. The Lower Mississippi Valley is a subregion of good and productive soils. Oil, gas, sulfur, and salt are major resources in the Gulf Coast section, where they provide the basis for the location of major oil-refining and chemical industries.

The Piedmont, inland from the Gulf-Atlantic Coastal Plain, is formed of harder and more ancient rocks, is a bit higher and hillier, and has somewhat better soils, although now much depleted and eroded in the center and south because of a past emphasis on growing clean-tilled cotton, tobacco, and corn in

hilly areas with heavy rains. Many cities, such as Richmond, Virginia, Washington, D.C., and Baltimore, Maryland, are spotted along the fall line between the two physiographic regions, where falls and rapids mark the head of navigation on many rivers and have long supplied water power.

The Appalachian Highlands, lying west of the Piedmont, form a complex system in which mountain ranges, ridges, isolated peaks, and rugged dissected plateau areas are interspersed with narrow valleys, lowland pockets, and rolling uplands. Elevations are relatively low, and most land is occupied by coniferous, deciduous, or mixed forests. Coal, water power, and wood are notable resources. The Appalachian region has a surprisingly large population for a highland area, and its importance in the regional economy is significant.

The Interior Highlands share many physical similarities with the Appalachian Highlands, but they are lower and have only minor coal deposits. The Interior Highlands are divided by the Arkansas River valley into the dissected Ozark Plateau on the north and the east-west oriented ridges and valleys of the low Ouachita Mountains on the south.

The Interior Plains lie mostly within the drainage basins of the Mississippi, St. Lawrence, Mackenzie, and Saskatchewan Rivers. The western third of this region is called the Great Plains and is generally delineated from the eastern Interior Plains by the 20-inch (50-cm) isohyet. Huge expanses of gently rolling or flat terrain exhibit some of the world's finest soils and most productive farms. The best soils have developed under prairie or steppe grasslands, but some very good soils in the east have developed under deciduous forest. Large quantities of beef, pork, corn, soybeans, and wheat are produced. Coal, oil, gas, and potash are present in major quantities.

The Rocky Mountains, generally high and rugged, consist of many linear ranges enclosing valleys and basins. Easy passes are almost nonexistent. Ranching, mining (of both metals and fuels), and recreation are all important activities. Many major rivers rise in these mountains, including the Columbia, Colorado, Missouri, and Rio Grande.

The Intermountain Basins and Plateaus consist generally of high plateaus trenched by canyons or, in the extensive Basin and Range Country of Nevada, Utah, and several adjacent states, by hundreds of linear ranges separated by basins of varying size. Shielded from moisture-bearing winds by high mountains on both the west and the east, the Intermountain area is this continent's largest expanse of dry land. Mining of copper, oil, gas, coal, uranium, and other minerals is prominent in scattered spots, and numerous major dams impound irrigation water and generate much hydroelectricity along the Columbia and Colorado Rivers. The importance of water control at all scales is evident in this area.

The Pacific Mountains and Valleys lie parallel to the Pacific shore and contain North America's highest mountains. In California, the Central Valley between the Sierra Nevada mountain system on the east and the Coast Ranges on the west is a huge exhibit of productive irrigated agriculture on flat alluvial land. Farther north, the Willamette-Puget Sound Lowland forms another major valley between the Coast Ranges and the Cascade Range of Oregon and Washington. Extreme western Canada and southern Alaska are dominated by high mountains, with many spectacular glaciers. The economy of the Pacific Mountains and Valleys has grown rapidly on the basis of a major collection of natural resources: abundant forests, fisheries, rivers to produce hydroelectricity and supply irrigation water, oil and gas in California, gold that once drew settlers to California and British Columbia, and both climatic and scenic resources undergirding development in the California subtropics. This wide range of environmental settings has helped stimulate the mobility of particularly the American population.

Major Climatic Regions

The continental climatic pattern in this region exhibits both regional variance and largeness of scale. The United States includes a greater number of major climatic types than does any other country in the world, and even Canada is more varied than is commonly assumed. The variety of economic opportunities and possibilities afforded by this wide range of climates is one of the basic factors underlying the economic strength of this world region.

Each climate region is shown by name and color on Figure 22.5. The tundra climate, characterized by long, cold winters and brief, cool summers, has an associated vegetation of mosses, lichens, sedges, hardy grasses, and low bushes. The subarctic climate has long, cold winters and short, mild summers, with a natural vegetation of coniferous snow forest resembling the Russian taiga. Population is extremely sparse in the tundra and subarctic climates and is practically nonexistent in Greenland's icecap climate. Scattered groups of people engage in trapping, hunting, fishing, mining, logging, and military activities; many live largely on welfare. The humid continental climate with short summers is characterized by long and quite cold winters and short, warm summers. Agriculturally, there is a heavy emphasis on dairy farming except in the extreme west, where spring wheat production is dominant. The humid continental climate has cold winters and long, hot summers; agriculturally, this belt, which includes the agricultural richness of the Midwest, is marked by an emphasis on dairy farming in the east and a corn, soybeans, cattle, and hogs combination in the midwestern (interior) portion. The diverse geographic characteristics that provide this range of environmental settings, however, also give rise to a wide variety of natural hazards. Figure 22.6 shows the hazards that sweep across not only North America but also Middle America as part of a hemispheric characteristic.

In the humid subtropical climate, winters are short and cool, although with cold snaps, and summers are quite long and hot. Agriculture is rather diverse from place to place, with some major specialized emphases on cattle, poultry, soybeans, tobacco, cotton, rice, peanuts, and a wide range of fruits and vegetables. Within each of the three humid climatic regions just described, the pattern of natural vegetation is complex, and each region has areas of coniferous evergreen softwoods, broadleaf deciduous hardwoods, mixed hardwoods and softwoods, and prairie grasses. Extreme southern Florida has a small area of tropical savanna climate, which is a major climate in adjacent Latin America. The state of Hawaii has a tropical rain forest climate. Along the Pacific shore of the United States and Canada, the narrow strip of marine west coast climate is associated with the barrier effect of high mountains near the sea, ocean waters offshore that are warm in winter and cool in summer relative to the land, and winds prevailingly from the west throughout the year. The mild, moist conditions have produced a magnificent growth of giant conifers (most notably the redwood and the Douglas fir) that provides the basis for the large-scale development of lumbering. Lush pastures support dairy farming, as they do in the corresponding climatic region in northwestern Europe.

The Mediterranean or dry-summer subtropical climate of central and southern California, in which nearly all precipitation comes in the winter season, is associated with irrigated production of cattle feeds, vineyards, vegetables, fruits, cotton, and a great range of other crops. These crops, together with associated livestock, dairy, and poultry production, make California the leading U.S. state in total agricultural output. In the semiarid steppe climate, occupying an immense area between the Pacific littoral of the United States and the landward margins of the humid East and extending north into Canada, temperatures range from continental in the north to subtropical in

F5.6
141

Figure 22.5   North America: Major Climatic Regions.

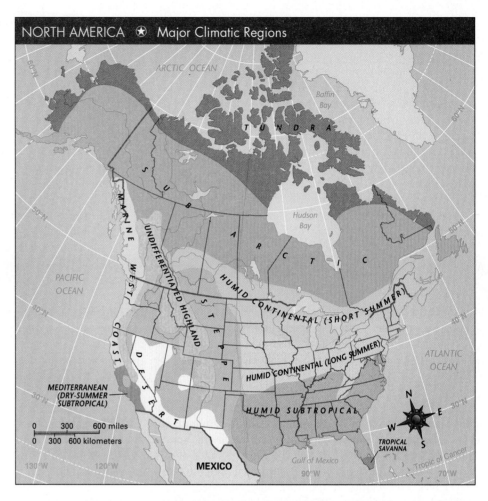

the south. The natural vegetation of short grass, bunch grass, shrubs, and stunted trees supplies forage for cattle ranching, which is the predominant form of agriculture. Areas that are less dry are often used for wheat (both winter wheat and spring wheat are part of the agricultural scene), and other crops are grown in scattered irrigated areas, often associated with major rivers such as the Columbia, Snake, Arkansas, or Rio Grande.

The desert climate of the U.S. Southwest is associated with scattered irrigated districts and settlements emphasizing mining, recreation, and retirement. High, rugged mountains, such as the Rockies and the Sierra Nevada, have undifferentiated highland climates varying with latitude, elevation, and exposure to moisture-bearing winds and to sun. More detailed discussion of climatic regions and major landform divisions and associated landscapes, with many relevant maps and photographs, can be found in Chapters 23 and 24.

22.3   **Cultural and Historical Context**

Both the United States and Canada are wealthy nations, but Canada has much less wealth. Canada's gross domestic product (GDP PPP) is a little less than one-tenth as great as that of the United States. On a per capita 2000 basis, Canada's is

$24,800; in the United States, it is $36,200. The output of goods and services in the United States is approximately equal to that of traditional Europe. In producing these goods and services, the United States uses more than 35% of the world's annual energy consumption. The ranking of the United States among the 15 nations with the highest per capita GDP PPP can be seen in Table 22.2. The relative standing of the North American countries compared with other nations has been declining since World War II because of more rapid economic growth in some other countries.

The development of such a high degree of wealth and power in this region is a topic too complicated to be fully explained in an introductory text and is a phenomenon subject to various interpretations. But it is clear that the United States and Canada possess, or have possessed, a number of specific geographic assets on which they have been able to capitalize. Some of the more important ones include:

1. *Size.* Both countries are large in area and are among the five largest in the world.
2. *Internal unity.* Both possess a comparatively effective internal unity, although Canada has had some difficult problems in recent decades as the province of Quebec continues to have a large percentage of its population that wants to gain independence. There have also been major

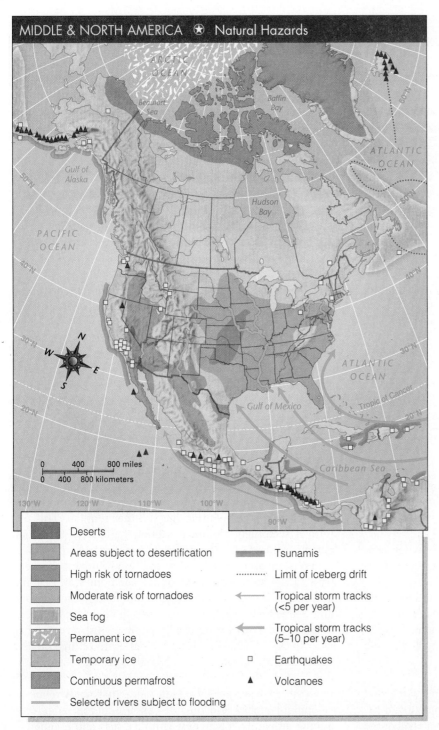

Figure 22.6 This map shows the wide variety of environmental hazards facing Middle and North America. Each of these hazards is caused by natural physical forces, but the impact of these phenomena varies in close relationship to the sorts of human settlement patterns in the location visited by a specific hazard.

MIDDLE & NORTH AMERICA ✪ Natural Hazards

Legend:
- Deserts
- Areas subject to desertification
- High risk of tornadoes
- Moderate risk of tornadoes
- Sea fog
- Permanent ice
- Temporary ice
- Continuous permafrost
- Selected rivers subject to flooding
- Tsunamis
- Limit of iceberg drift
- Tropical storm tracks (<5 per year)
- Tropical storm tracks (5–10 per year)
- Earthquakes
- Volcanoes

negotiations in both countries over Native American demands for expanded land rights and the right to their own self-governing territories.

3. *Resource wealth*. Both are outstandingly rich in natural resources and have well-developed capacities for making use of all levels of resource wealth (see Table 22.3).

4. *Size of populations*. The combined population of both countries is large, yet neither country is overpopulated, although there are images of urban densities and poverty,

especially in the United States, that demonstrate clear problems.

5. *Role of technology*. Both countries developed mechanized economies rather early and under very favorable conditions. Recent innovations in robot technology have led to a continued replacement of human labor with machine labor in, for example, automobile manufacturing in both Canada and the United States. It is likely that these patterns of labor shifting will continue, even though labor

DEFINITIONS + INSIGHTS

Resources

One of the most important definitions you will learn in this text is: "Resources are cultural appraisals." This means that what we call "natural resources" are actually Earth products we define as having value for us. Air, water, and space are, of course, absolutely necessary for life, so they are a special category of resource, and they are "natural" (of nature), although there is a very wide range of qualities of all three elements and differential costs reflect the way in which we value them. Space in downtown Hong Kong has a much higher value than space in the middle of Useful, Missouri, because of factors of location. Bottled water, no matter from which spring it supposedly comes, is more costly than water that flows across a highway in a rain storm and into a storm sewer.

Coal, for example, before it was seen to be an energy source, was perceived as a troublesome rock because it broke up too easily to be useful for building. Brea — the term used in southern California for the tarlike substance that used to seep to the surface in oil-rich areas of the Los Angeles basin — was the reason some land was relatively cheap in the first Los Angeles land boom of the 1880s. Brea lands were the cheapest land a person could buy then because of the difficulty in digging stable footings for new construction in the sticky brea soil. These lands now, of course, are among

the most expensive because of petroleum's resource value to us now. Brea has not changed, but society has now appraised that sticky, black product — petroleum — to be a resource of considerable value. Brea has become a hallmark of very valuable lands because of new uses we have found for this petroleum-rich substance.

Cultural appraisals change because of shifts in technology, in societal desires, and in availability. For example, coal is diminishing in value as a resource not because its characteristics are changing but because virtually all societies currently place a higher value on clean air. Since burning natural gas is less harmful to our atmosphere, the resource value of natural gas is increasing while the relative value of coal is decreasing. The cultural forces that lead to changing appraisals are difficult to predict but are continually at work. Every time you go to a flea market or a garage sale, for example, you are likely to see things that you may consider outdated junk; yet someone else has appraised those old 33-rpm records or that tube radio as having value. Resources are cultural appraisals and therein lies much of the dynamic of the economic and geographic tensions that define our current world. Resources are at the center of a great deal of economic development and differential levels of regional, national, and local prosperity.

unions do what they can to slow the pace of this replacement because of the associated loss of employment.

6. *Neighborly relations.* Despite occasional irritations and quarrels, the relations between the two countries of this region are generally friendly and cooperative.

F24.15
646

7. *Geographic diversity.* A remarkable range of environmental settings has both allowed and stimulated full use of human ingenuity in economic development.

8. *Institutional stability.* In both the United States and Canada, there has been a history of reasonable continuity of political, economic, and cultural institutions. Even in the face of some tensions and institutional evolution, the regional pattern has been one of stability.

The combination of these influences creates a regional base that has a rich history of economic development and relative abundance for its populations. This is not to say that the United States and Canada are not without significant problems of poverty and imbalance in the distribution of the products of this abundance, but taken as a whole, they make up a region of continued strong potential for "the good life."

Mobility and Recreation in the United States and Canada

One of the most dynamic regional images of the United States and Canada culture — although most of the popular imagery seems to focus on U.S. scenes, unless rail travel is

F24.2
628

featured — is that of vast highway networks, motels and roadside commercial strips, and an abundance of motor homes, recreational vehicles, and cars loaded with children and gear, especially during the summer months. This pattern of seemingly continual motion has led to the emergence of abundant car rental facilities, expanding networks of interstate and limited-access highways, hurried weekend trips to not so nearby parks or second homes and cabins, and increasing expansion of government-maintained parks and recreational facilities. Large-scale theme park recreation destinations are especially popular, just as is the network in both Canada and the United States of magnificent national parks created around landscape features of stunning beauty. But there is also a growing world of smaller regional, even municipal, green space that holds retail development at bay and provides environmental and recreational amenities for modest entrance fees and a generally accepted tax burden.

These landscapes of leisure and the pattern of steadily expanding mobility that increases their accessibility are major components of one of the most important culture changes of the past three decades. The construction of new stadiums, aquariums, convention centers, riverboat casinos, auto racing tracks, golf courses, spas, artificial multiuse lakes, and theater complexes of all sizes, combined with a general governmental inclination to support such projects with tax breaks, demonstrates the economic benefits of such mobility and recreation.

TABLE 22.3 United States' and Canada's Share of the World's Known Wealth and Production			
	Approximate Percent of World Total		
Item	U.S.	Canada	Total for Both
Area	7	7	14
Population	4.6	0.5	5.1
Agricultural and Fishery Resources and Production			
Arable land	13	3	16
Meadow and permanent pasture	8	1	9
Production of:			
Wheat	10	5	15
Corn (grain)	51	3	54
Rice	2	0	2
Soybeans	47	2	49
Cotton	18	0	18
Milk	16	2	18
Number of:			
Cattle	7	1	8
Hogs	7	1	8
Fish (commercial catch)	5	1	6
Forest Resources and Production			
Area of forest/woodland	6	8	14
Wood cut	14	5	19
Sawn wood production	28	17	45
Wood pulp production	34	16	50
Minerals, Mining, and Manufacturing			
Coal reserves	23	1	24
Coal production	20	2	22
Coal exports	25	8	33
Published proved oil reserves (tar sands and oil shale)	2	1	3
Crude oil production	10	3	13
Crude oil refinery capacity	21	3	24
Natural gas production	24	7	31
Natural gas reserves	3	1	4
Potential iron ore reserves	5	6	11
Iron ore mined	6	3	9
Pig iron produced	8	2	10
Steel produced	12	2	14
Aluminum produced	15	10	25
Copper reserves	17	5	22
Copper mined	11	5	16
Lead reserves	22	13	35
Lead mined	15	5	20
Zinc reserves	13	15	28
Zinc mined	10	11	21
Nickel mined	0	15	15
Gold mined	14	6	20
Phosphate rock mined	29	0	29
Potash produced	6	23	29
Native sulfur produced	21	0	21
Asbestos mined	1	18	19
Uranium reserves	9	12	21
Uranium produced	16	29	45

(continued)

TABLE 22.3 United States' and Canada's Share of the World's Known Wealth and Production (*continued*)

Item	Approximate Percent of World Total		
	U.S.	Canada	Total for Both
Electricity produced	26	4	30
Potential hydropower	4	3	7
Production of hydropower	4	4	8
Nuclear electricity produced	31	4	35
Sulfuric acid produced	26	3	29

Sources: United Nations Statistics Division, 2000; *Almanac of Politics and Government*, 2000; *World Population Data Sheet*, 2001.

Add to those features the plethora of fast-food restaurants that explode upon the scene as soon as a new highway interchange is opened and you can see the impact of the enormous fascination with staying in motion and spending considerable money and leisure time on recreation, travel, and convenience in such motion.

22.4 Economic Geography

Agricultural and Forest Resources

The United States ranks first among the world's nations in arable land of high quality. A much smaller proportion of Canada is arable, but it still has a total of more arable acreage than all but a handful of countries. Such abundance has helped these two countries become the main food-exporting region in the world. Canada and the United States also have abundant forest resources. Canadian forests are larger in extent than those of the United States but are not as varied. For example, Canada has no forests comparable to the redwood stands of California or the fast-growing pine forests of the southeastern United States. Wood has been an abundant material in the development of both countries. Even now, the United States cuts more wood for lumber each year than any other country in the world. The US and Canada produce one-half of wood pulp and paper production. The United States both exports and imports large quantities, with the exports from private lands going largely to Japan (Fig. 22.7) and the imports coming overwhelmingly from Canada. Canada's timber industry is considerably smaller, although it is still one of the greatest in the world; the country's small population compared to the size of its forest resources allows it to be the world's greatest exporter of wood.

Mineral Resources

The United States and Canada are also outstandingly rich in mineral resources. These resources played a major role in the earlier development of the region's economy, and

Figure 22.7 The economic importance of logging operations at all scales is central to British Columbia. From mountain flanks to forested valleys to rivers punctuated with great floating log rafts, the presence of this economic activity is common in the province. This log carrier is one of the many crafts moving logs into readiness for shipping lanes that go across the Pacific Ocean, destined most often for the Japanese market and other East Asian locales. Canada and the United States both have major resource trade with countries on the Pacific Rim of Asia.

RICHARD KOLAR

mineral output is still huge. Canada is a major exporter of mineral products, largely to the United States. But the United States, although still a major producer, has also become a major importer. This situation reflects the partial depletion of some American resources combined with growing foreign production of all manufactured goods. It also reflects the enormous demands and buying capacity of the U.S. economy.

Abundant energy resources have been basic to the rise of the economy and remain crucial to its operation and expansion. Coal was the largest source of energy in the 19th-century industrialization of the United States. Today, the country ranks with China and Russia as one of the three world leaders in magnitude of estimated coal reserves. In coal production, China and the United States are first and second, with China producing 29% and the United States 20%. The United States is the world's largest coal exporter by a wide margin. Canada is also an important coal exporter, with reserves and production that are small on a world scale but large in terms of Canada's needs.

Petroleum fueled much of the development of the United States in the 20th century, and for decades, the country was the world's greatest producer, with reserves so large that the question of long-term adequacy received little thought. In 1996, the United States was the second largest producer of oil in the world (surpassed by Saudi Arabia and followed by Russia), supplying approximately one-eighth of world output. Even with that production level, the United States still imports approximately half of its own domestic needs. Canada's reserves and production were and are small in comparison but are sufficient to be the basis of a continuing energy boom in its western province of Alberta. The 2001 energy crisis in California reflects the monumental importance electricity has for the industrial economy for California and the rest of the country.

Natural gas became a fuel of rapidly growing significance in the later 20th century. In 1996, the United States was the world's largest consumer of gas (ahead of Russia by a large margin) and was a major importer (from Mexico and Canada). It ranked second in the world in output (behind Russia, but far ahead of any other country), producing one-quarter of the world's gas from 3% of known world reserves. Canada, with smaller but rapidly increasing proven reserves, produced about 7% of world output in 1996 and was a major exporter.

Potential water power in the United States and Canada is small on a world scale, but development has been so intense that the two countries rank with Russia as the world leaders in hydropower capacity and output (Fig. 22.8). Canada generates approximately 62% of its electricity from hydropower; the United States generates 9%. At the same time, nuclear power can draw on the large uranium resources in each country and is an important secondary source of electricity for both.

Nonfuel mineral resources in North America present a similar picture of abundance. Long exploitation on massive scales has had an impact on American reserves, especially of the ores easiest to mine and richest in content, and even large mineral outputs within the United States are insufficient to supply the huge demands of the domestic economy. In many such cases, Canada has been able to assume the role of principal foreign supplier. A major example is iron ore. The United States originally had huge deposits of high-grade ore, principally in Minnesota's Mesabi Range and in less significant ore formations near western Lake Superior, but by the end of World War II, this ore had been seriously depleted. New sources of high-grade ore then were developed in Quebec and Labrador, principally for the American market. Meanwhile, the Lake Superior region somewhat revived its mining industry by extracting lower grade ore, which is

Figure 22.8 Large-scale development of hydropower in the United States and Canada is symbolized by Glen Canyon Dam and Lake Powell, situated on the Colorado River between the states of Arizona (right) and Utah (left). Note the power station at the foot of the dam, the extreme desert environment, the escarpment in the background—separating two levels of the Colorado Plateau—and the mountains rising above the general surface of the plateau in the distance. The rapid pace of settlement in and migration to the western United States has been based in good part on capturing exotic water and bringing it to growing cities, suburbs, and recreation areas for the past century. The system is beginning to show signs of stress because of the steadily increasing demands on water resources.

WILLIAM CAMPBELL/TIME MAGAZINE

present in vast quantities but must be processed in concentrating plants before it can be shipped at a profit to iron and steel plants.

As Table 22.3 shows, there is major production of many other nonfuel minerals by both the United States and Canada. In fact, the mineral wealth of North America makes it easier to summarize gaps in the resource array than to describe the resources that are present. Major minerals in which both countries appear to be truly deficient include chromite, tin, diamonds, high-grade bauxite, and high-grade manganese ore.

Mechanization, Productivity, Services, and Information

The high productivity and national incomes of this region's two countries have come about essentially through the use of machines and mechanical energy on a lavish and ever-increasing scale during the past century and a half. From an early reliance on waterwheels that powered simple machines — especially along the Appalachian Piedmont fall line in the American Northeast — the United States increasingly exploited the power generated from coal by steam engines. Later, the total power used to drive machines increased enormously through the use of oil, internal-combustion engines, and electricity. In achieving this unusually energy-abundant and mechanized economy, the United States was able to take advantage of a unique set of circumstances. For one thing, it had resources so abundant and varied that they attracted foreign capital and also made possible domestic accumulation of capital through large-scale and often wasteful exploitation. There was also a labor shortage, which attracted millions of immigrants as temporary low-wage workers but also promoted higher wages and

Figure 22.9   In San Francisco, as in many of the major urban centers in the United States and Canada, a racial sharing of public space has occurred in recreation areas and in work. This scene from a city park is instructive of such casual interaction among differing population groups.

labor-saving mechanization in the long run. This dynamic continues today, especially at the southwestern border of the United States and through the port of Los Angeles, often spoken of as the current Ellis Island (the New York City port of immigration on the east coast during the 19th and first half of the 20th century) (Fig. 22.9). A relatively free and fluid society encouraged striving for advancement (Fig. 22.10); large segments of this region subscribed with an almost religious fervor to ideals of hard work and economic success. Only occasionally did national energies have to be diverted excessively into defense and war.

Most of the foregoing American conditions were duplicated in Canada, and some were duplicated in a number of other countries, but one American advantage was not: the existence of a large, unified, and growing internal market that

Figure 22.10   One of the most prominent landscape signatures of the general scene in the United States and the southern part of Canada is a busy freeway. This photo of a Los Angeles freeway at dusk illustrates the fascination "Angelenos" have with auto-mobility. While this image perhaps seems understandable at rush hour, it is instructive to know that such traffic density in southern California is common for nearly 15 hours a day. This dependence on the automobile and the independence it allows, even though it means very crowded roadways, are a major characteristic of the urban scene in the United States and Canada.

DEFINITIONS + INSIGHTS

The North American Free Trade Agreement (NAFTA)

The North American Free Trade Agreement (NAFTA) was signed by Mexico, Canada, and the United States in 1993 and made operational in the following year. It was heralded as the most comprehensive trade link between these three nations that has ever been negotiated. It was intended in part to stem the ever-increasing flow of legal and illegal migrants from Mexico and Central America to the United States. It was thought that the economic growth that would take place in Mexico — particularly along the 2,000-mile border between Mexico and the United States — would serve to make Mexico and, ultimately, Central America be seen as a productive place to find jobs. There was considerable reluctance on the part of many labor organizations in the United States to have NAFTA signed because it was felt that jobs in textiles, shoe manufacture, and other industries would be drawn into Mexico. The lower wages there and the less stringent environmental and labor requirements would allow a U.S. or Canadian firm to make a better return on a manufacturing investment.

In its initial design, it was estimated that the United States would end up with a $50 billion trade surplus with Mexico. At the end of the first 6 years, however, the United States realized a trade deficit with Mexico of $93 billion. It is also claimed that the United States has lost nearly 400,000 higher wage, goods-producing jobs and had them replaced by service jobs that pay 38% less. In fact, in the case of the much touted *maquiladora* industries, it is estimated that the United States has lost a total of 500,000 jobs. In terms of wage scales between Mexico and the United States, "Mexican manufacturing wages were 85 percent below those in the US in 1983, 90 percent less in 1998."[a]

The picture is somewhat different in Canada, but the plant closings in the northern two countries and lower cost manufacturing in Mexico are a pattern that seems to have been stimulated to a new pace with the reduction or removal of tariffs that used to regulate the flow of trade between the NAFTA nations. As President Bush now promotes the initiation of the Free-Trade Areas of the Americas (spring 2001), there is considerable concern that the NAFTA pattern will be continued and expanded as a political and economic alliance between these three nations.

[a]David R. Francis, "NAFTA": Off with the Rose-Colored Glasses," *Christian Science Monitor*, October 18, 1999, p.9.

---

allowed producing organizations to specialize in the mechanized mass production of a few items, thus lowering the unit costs of production and further expanding the size both of the producing firms and of the market. In recent decades, however, this enormous American asset has been significantly eroded, along with some others, by the enormous expansion of the global market. Transportation and communication have become so relatively cheap and rapid and so many artificial trade barriers have fallen that the world market, rather than the domestic market, has become the essential one for major industries.

These forces of mechanization and associated productivity were increasingly replaced in the latter decades of the 20th century with a new labor sector: services and information management. The service sector has become a more significant employer than the manufacturing sector in the United States and Canada as the manufacture of shoes, clothes, toys, and cars has migrated to other sites, often outside these two countries. At the same time, a widespread and economically dynamic sector has grown up around the acquisition, manipulation, analysis, and distribution of information. As telecommunication technologies have made all of urban North America become more global in both outlook and operation, the recent economic landscape of this region has seen blast furnaces and conveyor belts steadily replaced by keyboards, monitors, and headsets, as well as other personal service occupations.

22.5 Geopolitical Issues

For many years after the American Revolution, the political division of the United States and Canada between a group of British colonial possessions to the north and the independent United States to the south was accompanied by serious friction between the peoples and governments on either side of the boundary. This heritage of antagonism was a result of (1) the failure of the northern colonies to join the Revolution; (2) the use of those colonies as British bases during that war; (3) a large proportion of the northern population having come from Tory (informally, "the friends of the king," or the name for the colonists who wanted to maintain the political connections with the British government) stock driven from U.S. homes during the Revolution; and (4) uncertainty and rivalry concerning who had ultimate control of the central and western reaches of the continent. The War of 1812 was fought largely as a U.S. effort to conquer Canada, though this intent was not specified at the time. Even after its failure, a series of border disputes occurred, and U.S. ambitions to possess this remaining British territory in North America were openly expressed; suggestions and threats of annexation were made in official quarters throughout the 19th and even into the 20th century.

In fact, Canada's emergence as a unified nation is in good part a result of American pressure. After the American Civil War, the military power of the United States assumed a threatening aspect in Canadian and British eyes.

Suggestions were made in some American quarters that Canadian territory would be a just recompense to the United States for British hostility to the Union during the Civil War. However, the British–North America Act, passed by the British Parliament in 1867, brought an independent Canada into existence by combining Nova Scotia, New Brunswick, and Canada into one dominion under the name of Canada. With that Act, Great Britain sought to establish Canada as an independent nation capable of morally deterring U.S. conquest of the whole of North America. Although hostility between the United States and its northern neighbor did not immediately cease with the establishment of an independent Canada, relations have improved gradually, and the frontier between the two nations has ceased to be a source of insecurity. This frontier, stretching completely unfortified across a continent, has become more a symbol of friendship than of enmity.

The bases of Canadian-American friendship are cultural similarities, the material wealth of both nations, and the mutual need for and advantages of cooperation. A large volume of trade moves across the frontier and strengthens both countries economically and militarily. Each is a vital trading partner of the other, although Canada is much more dependent on the United States in this respect than is the United States on Canada. In 1999, for example, Canada supplied 19% of all U.S. imports by value and took 24% of all U.S. exports, making Canada the leading country in total trade with the United States. On the other hand, in 1999, the United States supplied 68% percent of Canada's imports and took 87% of its exports. The United States also supplies large quantities of capital to Canada, which has been an important factor in Canada's rapid economic development during recent decades but which also has been a frequent source of Canadian disquiet because of the amount of U.S. influence on the Canadian economy that the investment represents. However, Canadians in turn invest heavily in the United States. Except for Canadian exports of automobiles and auto parts to the United States, the

Figure 22.11 The North American Free Trade Agreement (NAFTA) has never had complete support. This demonstration in Austin, Texas, was organized by union leaders fearful of the loss of jobs that would occur as industries left the United States and went south into Mexico and Central America. After more than 7 years of economic changes stimulated by NAFTA, the jury is still out on the ultimate impact of this unprecedented alliance between the United States, Canada, and Mexico. A major complaint in the United States is that although there have been many jobs generated in the United States since the initiation of NAFTA, they have been less well-paying jobs in services versus the decline of better-paying positions in manufacturing.

main pattern of trade between the two countries is the exchange of Canadian raw and intermediate-state materials — primarily ores and metals, timber and newsprint, oil, and natural gas — for American manufactured goods. A free-trade pact to unite the two in one market was signed in 1988. In 1992, Canada joined the United States and Mexico in signing the North American Free Trade Agreement (NAFTA) that was enacted in 1994 (Fig. 22.11). *529*

TABLE 22.4	Population in the United States and Canada, 1850–2000				
				Total Increment	
Year	U.S. Population (millions)	Canadian Population (millions)	Total (millions)	(millions)	(%)
1850	23.3	2.4	25.7	—	—
1900ᵃ	76.1	5.4	81.5	55.8	217
1950ᵃ	151.1	14	165.1	83.6	103
1960ᵃ	179.3	18.2	197.5	32.4	20
1970ᵃ	203.2	21.6	224.8	27.3	14
1980ᵃ	226.5	24.3	250.8	26	12
1990ᵃ	248.7	28.1	276.8	26	10
2000ᵃ	281.4	31.0	312.4	35.6	13

ᵃActual dates for Canada are 1901, 1951, 1961, 1971, 1981, and 1991. The decennial census is taken 1 year later in Canada than in the United States. This table uses census data except for the 2000 midyear estimates.

CHAPTER SUMMARY

- The countries of the United States and Canada comprise what is sometimes called Anglo America or North America. Mexico is defined as part of Middle America, even though the North American Free Trade Agreement (NAFTA) has linked the three countries. Greenland is politically a part of Denmark.
- Both the United States and Canada have sizable areas, although their population statistics are quite dissimilar. The combined population for the two countries amounts to just over 315 million. Greenland, also large, has a population of fewer than 60,000.
- The United States and Canada have a broad range of diverse environmental settings, including the Arctic Coastal Plains, Gulf-Atlantic Coastal Plain, Piedmont, Appalachian Highlands, Interior Highlands, Interior Plains, Rocky Mountains, Intermountain Basins and Plateaus, and Pacific Mountains and Valleys.
- Climatic patterns for the region are also varied. Climate types include tundra, subarctic, humid continental with short summers, humid continental with long summers, tropical savanna, and tropical rain forest. There are also undifferentiated highland and mountain zones. Patterns of vegetation and settlement are associated with these climate patterns.
- A number of factors are central to the geographic richness that characterizes this region. These factors include size, internal unity, resource wealth, population size, technology, intraregional relations, and geographic diversity. Many regions possess some of these features, but this region is particularly fortunate in having all these elements.
- Particularly important resources for this region include agricultural land and forests. Both play a central role in exports of the two countries. Additional major resources include coal, petroleum and natural gas, water power, and minerals such as iron ore. The global importance of these sources in this region is evident in Table 22.3.
- Demographic characteristics of the two countries show similar patterns of early settlement displacing native populations and then subsequent waves of immigration continually changing the demographic, ethnic, and racial characteristics of the resident populations. In Canada, the French are the most significant minority. In the United States, African Americans fill that role. Both countries have an increasingly diverse ethnic blend created by active east, west, and south border immigration flows. Hispanic and Asian American populations are increasingly gaining importance in the United States.
- Both countries have a history of active and significant industrialization in the 19th and 20th centuries. In the past three or four decades, the economic profiles of the countries have changed as manufacturing has become less important and employment in service industries and information management has grown more significant. Small towns have given way to middle-sized and large urban centers in both countries.
- One of the most critical facets of regional success here has been the fundamentally amicable relationship between the two countries. The United States is the major customer for Canadian exports, and Canada plays a similar role for the United States. Ninety percent of the Canadian population lives within 100 miles of the international border. This proximity also causes some frustrations as cultural influences from the United States sometimes seem overwhelming to Canadians.

REVIEW QUESTIONS

1. Explain the reasons for the various traditional names given to this region. How does the name NAFTA fit into this pattern?
2. What are the basic area and population figures for this region? How do they relate to the global picture?
3. List the major landform characteristics of the region. Take three of those characteristics and give some detail on the landscapes and land use patterns associated with them.
4. Do the same for the climatic characteristics in North America. Relate three of the climatic zones to similar zones you have read about in other regions of the world.
5. What are the most important characteristics of the geographic richness of the United States and Canada? List four of the eight traits given in the text and describe your perceptions of them as if you grew up in this region.
6. What aspects of this region's mineral wealth seem most significant to you? Why?
7. Which population characteristics are most significant globally? Which ones seem to be the most important locally? Does this relationship change with time?
8. What role has immigration had in shaping the current demographic scene in the United States and Canada?
9. Give examples from your own knowledge that represent the issues raised in the section on mechanization, productivity, services, and information. How are these likely to change over time?
10. Describe the relations between Canada and the United States, pointing out both positive and negative features of that relationship. How is it likely to change in the next decades?

DISCUSSION QUESTIONS

1. Discuss the geographic factors that might explain the quite distinct history and geographic development of Canada and the United States.
2. Take the topics of area and population size and discuss which patterns of settlement, economics, politics, and culture seem to be dominant in visibility and global consequence. Find exceptions to explain.
3. Utilizing Table 22.3, discuss the regional significance of the United States and Canada during the past 50 years and speculate on what might happen in the next 50 years.
4. Discuss the ways in which the American Revolution and the American Civil War had consequences for Canadian development.
5. Discuss why approximately 90% of the Canadian population lives within 100 miles of the U.S. border.
6. What are the historic and geographic features that have made the United States and Canada exist fundamentally as good neighbors?
7. What factors might change this historical pattern of neighborliness significantly?

Canada

THOMAS KITCHIN/TOM STACK AND ASSOCIATES

The landscapes of farming and particularly wheat cropping are of major importance in the heart of Canada. This scene from Thunder Bay, Ontario, on the northwest shore of Lake Superior shows the grain elevators that are so much a landscape signature along Canadian rail lines and on the shores of the Great Lakes. Thunder Bay is an international port because of the connection to the Atlantic provided by the St. Lawrence Seaway. From this port, a large portion of Canada's massive annual grain export is launched toward global destinations.

Patterns of active migration create regional mosaics that can be quite distinctive. Both the United States and Canada have been largely peopled by migrants moving across the oceans and land in search of a new set of opportunities. The patterns of human development that have emerged from these migration histories are distinctive in that they have provided both urban and rural zones of ethnic continuity. At the same time, the immigrants have endured travails of enormous similarity no matter the place of origin, the ethnic makeup of the migrating population, or the world into which these people came. Migration is truly a process profoundly associated with world regional geography. In this chapter on Canada, it can be seen that one country can be home to quite distinctive mosaics of regional differentiation. Migration is a major element in the creation of such mosaics.

As you learn of Canada and its niche in North America, keep in mind the demands of human migration and the associated demands of being an outsider trying to find a new home and future in a world very different from one's own origins. Not only has this dynamic been important for the cities of Canada, but it has also shaped much of the urban, town, and even rural growth of both countries in this last of our world regions.

Canada is a highly developed country: affluent, industrialized, technologically advanced, and urbanized. But in some ways, it is curiously unlike most developed countries. These differences are closely related to Canada's internal geography and its location adjacent to the United States.

23.1　Canada: Highly Varied and Economically Vital

Canada's prosperity is derived in considerable part from an industrial output sufficiently great to place the country among the world's top dozen manufacturing nations. Products in which it ranks exceptionally high include newsprint, hydroelectricity, commercial motor vehicles, and aluminum. The country is very urbanized (Fig. 23.1), with 78% of its population classed as urban in 2000 (compared to 75% in the United States), with about 35% of its people living in its three largest metropolitan areas: Toronto (population: metropolitan area, 5,411,300; city proper, 2,540,100), Montreal (population: metropolitan area, 3,490,600; city proper, 1,035,100), and Vancouver (population: metropoli-

tan area, 1,922,000; city proper, 530,700). These dynamic metropolises convey a positive image of an advanced and prosperous country. Impressively good national statistics on health, housing, and education provide further evidence of the same. However, the enormous clustering of the Canadian population up against its southern border further demonstrates the country's bittersweet relationship with its neighbor, the United States.

Along with its prosperity, Canada exhibits some striking departures from the expected patterns of developed countries. One of these is an unusual pattern of trade. Despite a fair-sized export market for automobiles and parts and some other fabricated items, Canada is mainly an exporter of raw or semifinished materials, energy, and agricultural products. It is a major importer of manufactured goods and specialty food items. By contrast, most developed nations generally export mainly manufactured goods and import a mixture of manufactured and primary products. This is simply another example of Canada's special geographic characteristics.

A second departure from other developed nations is Canada's overwhelming dependence on trade with a single partner, the United States. In 1999, the United States took 87% of Canada's exports and supplied 68% of Canada's imports. No other developed country comes near these percentages of trade with just one other country.

Still another major divergence from other developed countries lies in the degree to which the Canadian economy is financed and controlled from outside the country, and in this, Canada again has an overwhelming dependence on the United States. More than three-quarters of all foreign investment in Canada is American. The extent of U.S. economic dominance in Canada has long been a sore point with many Canadians, who resent imputations that their country is an American economic "colony" (Fig. 23.2). These irritations tend to be exacerbated by the degree to which American mass-produced culture has permeated Canada and the degree to which Canada is taken for granted in the United States. Few U.S. schools offer courses on Canada, and a high percentage of American College students cannot even name Canada's capital. American news media give relatively scanty coverage to Canada.

For some time, there has been a movement in Canada to achieve a more predominant position for Canadians within their own economy, to strengthen economic relations with industrial countries other than the United States, and to

LANDSCAPE IN LITERATURE

In the Skin of a Lion

Writer Michael Ondaatje's novel about European migrants coming to Canada to find a new life in the early decades of the 20th century provides a window on the difficulties, as well as some of the humor, of this particular migration and transition. The struggles of adjusting to a new home, learning a new language, and finding an adequate job are all illustrated in these images of Nicholas on his move from the Balkans to Toronto.

Nicholas was twenty-five years old when war in the Balkans began. After his village was burned he left with three friends on horseback. They rode one day and a whole night and another day down to Trikala, carrying food and a sack of clothes. Then they jumped on a train that was bound for Athens. Nicholas had a fever, he was delirious, needing air in the thick smoky compartments, waiting to climb up on the roof. In Greece they bribed the captain of a boat a napoleon each to carry them over to Trieste. By now they all had fevers. They slept in the basement of a deserted factory, doing nothing, just trying to keep warm. . . .

Two of Nicholas' friends died on the trip. An Italian showed him how to drink blood in the animal pens to keep strong. It was a French boat called La Siciliana. He still remembered the name, remembered landing in Saint John and everyone thinking how primitive it looked. How primitive Canada was. They had to walk half a mile to the station where they were to be examined. They took whatever they needed from the sacks of the two who had died and walked towards Canada.

Their boat had been so filthy they were covered with lice. The steerage passengers put down their baggage by the outdoor taps near the toilets. They stripped naked and stood in front of their partners as if looking into a mirror. They began to remove the lice from each other and washed the dirt off with cold water and a cloth, working down the body. It was late November. They put on their clothes and went into the Customs sheds.

Nicholas had no passport, he could not speak a word of English. He had ten napoleons, which he showed them to explain he wouldn't be dependent. They let him through. He was in Upper America.

He took a train for Toronto, where there were many from his village; he would not be among strangers. But there was no work. So he took a train north to Copper Cliff, near Sudbury, and worked there in a Macedonian bakery. He was paid seven dollars a month with food and sleeping quarters. After six months he went to Sault Ste. Marie. He still could hardly speak English and decided to go to school, working nights in another Macedonian bakery. If he did not learn the language he would be lost.

The school was free. The children in the class were ten years old and he was twenty-six. He used to get up at two in the morning and make dough and bake until 8:30. At nine he would go to school. The teachers were all young ladies and were very good people. During this time in Sault he had translation dreams — because of his fast and obsessive studying of English. In the dreams trees changed not just their names but their looks and characters. Men started answering in falsettos. Dogs spoke out fast to him as they passed him on the street.

When he returned to Toronto all he needed was a voice for all this language. Most immigrants learned their English from recorded songs or, until the talkies came, through mimicking actors on stage. It was a common habit to select one actor and follow him throughout his career, annoyed when he was given a small part, and seeing each of his plays as often as possible — sometimes as often as ten times during a run. Usually by the end of an east-end production at the Fox or Parrot Theater the actors' speeches would be followed by growing echoes as Macedonians, Finns, and Greeks repeated the phrases after a half-second pause, trying to get the pronunciation right.

This infuriated the actors, especially when a line such as "Who put the stove in the living room, Kristin?" — which had originally brought the house down — was spoken simultaneously by at least seventy people and so tended to lose its spontaneity. When the matinee idol Wayne Burnett dropped dead during a performance, a Sicilian butcher took over, knowing his lines and his blocking meticulously, and money did not have to be refunded.

Certain actors were popular because they spoke slowly. Lethargic ballads, and a kind of blues where the first line of a verse is repeated three times, were in great demand. Sojourners walked out of their accent into regional American voices.[a]

[a]From Michael Ondaatje. *In the Skin of a Lion* (New York: Penguin Books, 1987): 45–47.

F20.5
533

---

heighten cultural self-determination. But there seems to be no way for Canada to pull away from the United States without risking its own prosperity. The country's economy is geared to a close interchange with the gigantic economy next door. In 1988, the two countries moved even closer when they signed a free-trade pact to abolish all tariffs on trade between them and to remove or lower many other restraints on trade. In 1994, this arrangement was expanded when the United States, Canada, and Mexico initiated the North American Free Trade Agreement (NAFTA), but the process was not without dissent.

Borders and Good Neighbors

One of the most frequently cited facts about Canada and the United States is that they are such good neighbors. They share the longest border in the world between two countries — 5,527 miles (8,895 km) — and they have a history that is more generally amicable than confrontational. There is a persistent frustration among Canadians about being in the cultural and material shadow of their southern neighbor; nonetheless, nearly 90% of the Canadian population lives within 100 miles (161 km) of the U.S. border, so the irritation must be bittersweet (90% of the population lives on 12% of the land in these southern reaches). In the patterns of U.S. population distribution, a relatively small percentage of the population lives within 100 miles of the Canadian border. These two facts often serve as a discussion point for the relative power of climate versus culture.

However, there are some specific items that have bred ongoing irritation and that reflect themes comprising the elements of a problem landscape. Here are two.

Figure 23.1   Major regions of Canada. In 1991, the Canadian government undertook to create a new Inuit (Eskimo)-controlled Nunavut Territory taken from portions of the Northwest Territories. In May 1993, Canada's prime minister signed the Nunavut Agreement, setting in motion the final steps to create Nunavut (Inuit: "our land") as a separate territory. The final transfer took place on April 1, 1999.

The Wheat War. Even though less than 5% of Canada's area is suitable for farming and although only about 3% of the Canadian labor force is engaged in agriculture, it is a major world wheat exporter because of the extent of the spring-wheat growing area in the Prairie Region, west of Lake Superior. In this area, farmers have invested heavily in large-scale mechanization. The Great Plains region of the United States is a direct topographic continuation of the immense open plain that courses south-southeast from the

Prairie provinces in Canada. Spring wheat is the major crop in that region, and it is also vital to the Dakotas and adjacent states.

In the 1990s, there was a "wheat war" between the two countries that related to very poor global wheat markets and relatively high yields, particularly in the United States and Canada. With agricultural prices for grains and pork dropping to the lowest levels in more than 50 years, farmers both south and north of the U.S.-Canada border went bankrupt.

Figure 23.2 In Canada, there is a general unease with the presence of American cultural icons all through the Canadian culture. This evidence of cultural diffusion (more often going south to north than north to south) is particularly frustrating to the French Canadians in Montreal and Quebec City. Yet, as is so often the case with the diffusion of popular culture, once one part of a culture complex finds its way to a new destination, everything from blue jeans to fast food finds its way there sooner or later. This is a Wendy's in Montreal.

They left farming, going into local towns and cities or migrating to wholly new lives in new cities. In late 1997, this tension was magnified by American claims that Canada was "dumping" (i.e., selling the product for less than it cost to produce it) its Durham wheat on the United States market. One of the ironies of this tension is the fact that there is not adequate global food production in a world of nearly 6.4 billion people. Yet here is a resource problem that relates much more to marketing and to transportation than to resource availability itself.

The Salmon War. Salmon migrate from the waters at the edge of the Pacific Ocean up and down the coast seeking the rivers of British Columbia (and points north and south) to spawn. With the expiration of a 1985 treaty that had regulated the harvest of such fish, Alaskan fishermen moved in 1997 and 1998 into the shallows and harvested enormous numbers of river-seeking salmon. Canada retaliated by blocking the route of an Alaskan ferry for 3 days, keeping it in Prince Rupert Harbor (on the western edge of British Columbia by the mouth of the Skeena River) as a hostage to negotiations to limit salmon harvest. This act has been followed by other economic boycotts and by Canadian efforts to fully harvest the waters around Vancouver Island so as to deny American fishermen access to these fish. Volleys in this border "skirmish" are still being fired on both sides of the border, reminding us again of the continuing need to carefully chart the use of resources in a border zone.

Being good neighbors is never easy. Virtually every aspect of sharing borders, especially sharing patterns of agricultural or industrial production, means that there are continual areas of friction that crop up, even in a context that is fundamentally agreeable. These two countries, after all, do exchange approximately $1 billion a day in goods and services.

23.2 A Mosaic of Canada's Regional Landscapes

In addition to the problem of maintaining satisfactory relationships with the United States, Canada also has problems posed by its own regional structure. The country is divided between a thinly settled northern wilderness that occupies most of Canada's area and a narrow and discontinuous band of more populous regions stretching from ocean to ocean in the extreme south (see Fig. 23.1). These southern regions are sharply different from each other and are often in varying degrees of political conflict with each other and with the federal government at Ottawa.

About four-fifths of Canada has cold tundra and subarctic climates; more than one-half is in the glacially scoured and agriculturally sterile Canadian Shield (Fig. 23.3), and the part near the Pacific is dominated by rugged mountains. These conditions create the huge expanse of nearly empty land that Canadians call "the North." More than one-half of it lies in territories administered from the country's capital in Ottawa, Ontario, but large parts extend into seven of the ten provinces that stretch across southern Canada and make up the major political subdivisions of Canada's federal system. Even within these provinces, most of the land is lightly inhabited, and nearly all the people are concentrated in limited areas of relatively close and continuous settlement, separated by wedges of thinly populated terrain. These population clusters lie just north of the country's boundary with the 48 contiguous United States. The large percentage of Canadians living close to the border has helped the development of certain areas with strong ties between Canadian regions and adjoining American regions, such as in the American Upper Midwest.

Canada's population clusters can be grouped into four main regions: (1) an Atlantic region of peninsulas and islands at Canada's eastern edge; (2) a culturally divided core region of maximum population and development along the lower Great Lakes and St. Lawrence River in Ontario and Quebec provinces; (3) a Prairie region in the interior plains between the Canadian Shield on the east and the Rocky Mountains on the west; and (4) a Vancouver region on or near the Pacific coast at Canada's southwestern corner (see Fig. 23.1). Each region is marked by a distinctive physical environment and associated cultural landscape; each has a different ethnic mix and distinctive economic emphases. Each plays a different role within Canada and has its own set of relationships with the United States and other countries. All the regions have important links with the Canadian North. (For statistical data on Canadian areas, see Table 22.1.)

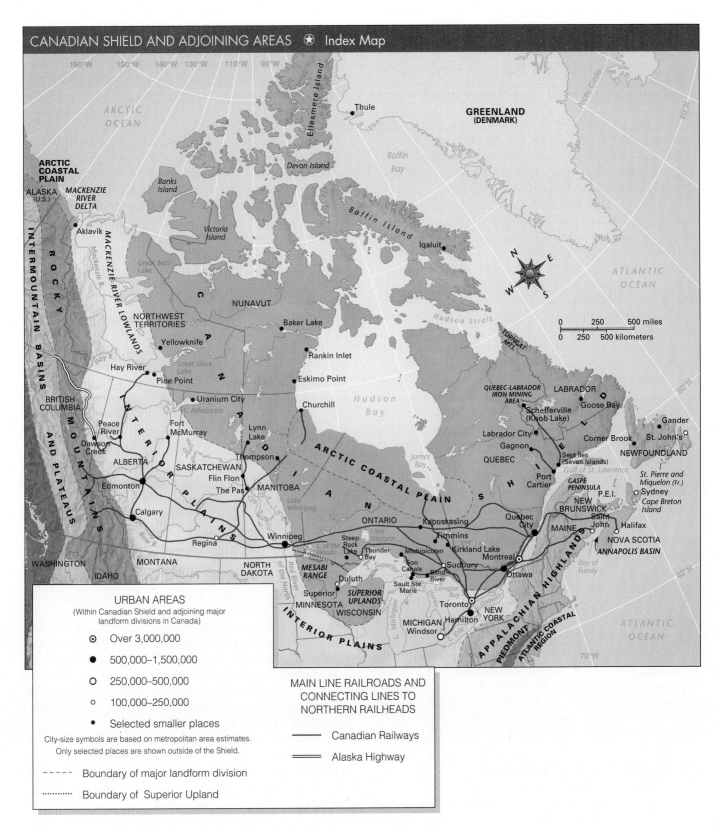

CANADIAN SHIELD AND ADJOINING AREAS ✪ Index Map

URBAN AREAS
(Within Canadian Shield and adjoining major landform divisions in Canada)

◉ Over 3,000,000

● 500,000–1,500,000

○ 250,000–500,000

○ 100,000–250,000

• Selected smaller places

City-size symbols are based on metropolitan area estimates.
Only selected places are shown outside of the Shield.

----- Boundary of major landform division

··········· Boundary of Superior Upland

MAIN LINE RAILROADS AND CONNECTING LINES TO NORTHERN RAILHEADS

——— Canadian Railways

═══ Alaska Highway

Figure 23.3   General reference map of the Canadian Shield and adjoining areas. Only certain portions of Canada's Arctic islands share the rock formation of the Canadian Shield. However, the overall physical conditions of the islands are sufficiently similar to conditions in the Shield that it is convenient for the purposes of this text to regard the entire island group as an extension of the Shield.

23.3 Atlantic Canada: The Frustrations of Fishing Down, Tourism Up

544 The easternmost of Canada's four main populated areas is characterized by a series of strips or nodes of population, largely along or very near the seacoast in the provinces of Newfoundland, New Brunswick, Nova Scotia, and Prince Edward Island. The four provinces are commonly called the Atlantic Provinces, and three, excluding Newfoundland, have long been known as the Maritime Provinces. Their main populated areas are separated from the far more populous and more prosperous Canadian core region in Quebec and Ontario by areas of mountainous wilderness. Prince Edward Island, which is Canada's smallest but most densely populated province (Table 23.1), is a lowland occupied largely by a continuous mat of farms and small settlements, but the main populated areas of New Brunswick, Nova Scotia, and Newfoundland occupy valleys and coastal strips within a frame of lightly settled upland. As a province, Newfoundland includes the dependent territory of Labrador on the mainland, but in common usage, the name Newfoundland is restricted to the large island opposite the mouth of the St. Lawrence River. Formerly a self-governing dominion, Newfoundland was the last province to join Canada, which it did in 1949. Bleak and nearly uninhabited Labrador, situated on the Canadian Shield, is very much a part of Canada's North. It is important to Newfoundland for revenues from iron mining and hydroelectric production.

Isolation from markets afflicts almost all major enterprises in the Atlantic region. The market within the region is small and divided into many small clusters of people separated by sparsely populated land. The largest city in the Atlantic provinces — Halifax, Nova Scotia — has a metropolitan population of about 320,000. Only two other cities — St. John's, Newfoundland, and St. John, New Brunswick — surpass 125,000. The entire area has only 2.4 million people, or fewer than metropolitan Toronto or Montreal. It is also home to a large French population, especially New Brunswick. It is a region of out-migration and the poorest of the four populated regions of Canada. Here the region stands at the mouth of a major navigable river going into the heartland of a very prosperous country, affording the closest link to global markets, and yet this region remains relatively poor and considerably less developed than the rest of the country.

A Long Tradition of Fishing and Its Uncertain Future

The fishing industry has always been of outstanding importance in Canada's Atlantic region. Before the beginning of European settlement in the early 17th century, and probably before the arrival in the Americas of Columbus in 1492, European fishing fleets began operating in the waters, primarily on the "banks," along and near these shores. This area lies relatively close to Europe, and its waters have always been exceptionally rich in fish. However, it has suffered from steady overfishing in recent years, and this has brought about the near disappearance of the cod that have been the mainstay of this region's productivity.

In general, Atlantic Canada provides hard environments for farming. The four provinces lie at the northern end of the Appalachian Highlands and have topographies dominated by hills and low mountains. Soils that are predominantly mediocre to poor exist under a cover of mixed forest. Many soils that once were farmed wore out quickly and were abandoned. Agriculture has now come to be concentrated on relatively small patches of the best land.

Climate also handicaps agriculture. The Maritime Provinces have a humid continental short-summer climate, and Newfoundland has a subarctic climate. The entire region is humid and windy, with much cloud cover and fog. Strong gales are frequent in the winter. Summers are cool, and colder and less hospitable conditions occur as elevation increases; a good part of upland Newfoundland is tundra. Thus, most areas in the Maritimes and Newfoundland have always been agriculturally marginal, although some more favored lowlands with better soils — notably Prince Edward Island and the Annapolis-Cornwallis Valley of Nova Scotia — are exceptions. Potatoes, dairy products, and apples are major agricultural specialties in the Maritimes.

Geographic Profile of a Declining Region

During the first two-thirds of the 19th century, the Maritime Provinces became a relatively prosperous area with a preindustrial economy. Local resources supported this development, including many harbors along indented coastlines, abundant fish for local consumption and export, timber for export and for use in building wooden sailing ships, and land — albeit not very good — for the expansion of an agriculture partly subsistence in character and partly serving local markets. The region carried on extensive commerce with Great Britain and the West Indies. Then followed an era of relative economic decline within an industrializing Canada. The ports were so far from the developing interior of the continent that most shipping bypassed them in favor of the St. Lawrence ports or the Atlantic ports of the United States. Halifax, Nova Scotia, and St. John, New Brunswick, are the main ports of the region today, and although each handles considerable traffic, with an emphasis on containers, neither is in a class with Montreal or the main U.S. ports. St. John's and Gander, in Newfoundland, played a significant role in serving as a refueling stop in the 1960s and 1970s for many trans-Atlantic air flights, but greater fuel efficiency and carrying capacity have now made that stopping point much less popular for general commercial flying.

In the meantime, fishing, forestry, and agriculture suffered various disabilities. Fishing did not prove to be an adequate

Table 23.1 Canadian Provinces and Territories: Basic Data

Political Unit	Land Area (thousand/sq mi)	Land Area (thousand/sq km)	Estimated Population (thousands)	Estimated Population Density (sq mi)	Estimated Population Density (sq km)	Economic Production (% of Value) (farm receipts)	Economic Production (% of Value) (mineral output)	Economic Production (% of Value) (manufacturing)
Atlantic Provinces								
New Brunswick	27.6	71.5	757.1	27	11	1.3	2.5	1.9
Nova Scotia	20.6	53.3	942.7	46	18	1.5	1.5	1.5
Prince Edward Island	2.2	5.7	138.5	63	24	1.1	0.01	0.2
Newfoundland-Labrador	144.4	373.9	533.8	4	1	0.3	2.1	0.4
Total	194.7	504.4	2372.1	12	5	4.2	6.1	4.0
Core Provinces								
Quebec	527.1	1365.1	7410.5	14	5	16.2	7.4	23.7
Ontario	354.3	917.7	11,874.4	34	13	21.5	13.5	52.6
Total	881.4	2282.8	19,284.9	22	8	37.7	20.9	76.3
Prairie Provinces								
Manitoba	213.7	553.6	1150.0	5	2	10.3	3.2	2.3
Saskatchewan	228.5	591.7	1015.8	4	2	20.3	8.6	1.4
Alberta	248.0	642.3	3064.2	12	5	19.8	48.2	8.0
Total	690.2	1787.6	5230.2	8	3	50.4	60	11.7
Pacific Province								
British Columbia	357.2	925.2	4095.9	11	4	7.1	11	7.9
Territories								
Nunavut	747.5	1936.1	28.2	0.04	0.01	—	—	—
Yukon	183.2	474.4	29.9	0.2	0.06	0.01	1	—
Northwest Territories	456.8	1183.1	40.9	0.1	0.03	—	2	—
Total	1387.5	3593.6	99.0	0.1	0.03	0.01	3	—
Summary Total	3511.0	9093.6	31,081.9	9	3	100	100	100

Source: *Statistics Canada* 2001.

Figure 23.4 The fishing villages of Nova Scotia suggest a link with the past that is more a mark of marginal fishing than tourist delight in the contemporary context of Canada's Maritime Provinces. The small craft and family operations of this photo have a very difficult time competing in a world fishing economy that is increasingly high-tech in its design and function. It is often difficult to be the fisherman who needs to begin seeing his setting as part of the tourists' interests rather than a working base to support his family.

THOMAS KITCHIN/TOM STACK & ASSOCIATES

basis for a prosperous modern economy (Fig. 23.4), primarily for the following reasons:

1. *Tariff barriers*. Tariff barriers were imposed by the United States against Canadian fish during the competitive efforts of Canada and the United States to preserve their own domestic markets.

2. *Global competition*. Competition from major world meat exporters and newly developed fishing areas in other parts of the world, particularly after World War II, had grown to such size that it began to disable the Canadian fishing economy.

3. *Foreign competition on the local scene*. The fishing conditions of the Grand Banks and other nearby waters of the Maritime Provinces were so productive that fishing fleets from other nations and regions, many possessing high-tech fishing operations, began to drive Canadian fishermen from their own waters.

4. *Disappearance of the resource*. The combination of expanding numbers of fishermen, more sophisticated and effective fishing technology, and steady overfishing led to a major and continuing decline in the number of fish in

T22.7
593

this traditionally rich area. Such a shift has caused fishing to require much more capital than has been traditional, and this change has also worked against the traditional local fishermen.

The Vanishing Centrality of the Grand Banks

Fishing early became a major aspect of this region's livelihood because for centuries the north Atlantic has been a particularly rich resource base for fishing. Elevated portions of the sea bottom known as "banks" are located off the Atlantic coast from near Cape Cod to the Grand Banks, which lie just off southeastern Newfoundland. The Grand Banks have been at the heart of the fishing economy in Canada's northeastern region for nearly five centuries. The shallowness of the ocean and the mixing of waters from the cold Labrador Current and the warm Gulf Stream foster rich development of the tiny organisms called plankton on which fish feed.

Inshore fisheries supplement the catch from the banks. Although many kinds of fish and shellfish are caught, the early fleets fished mainly for cod, which were cured on land before making the trip to European markets. Both the British and the French established early fishing settlements in Newfoundland, but the French—who had 150 vessels fishing these banks in 1577—were driven out by the British in 1763. The island has remained strongly British in population, as its limited development has attracted very few other immigrants.

In 1977, worried about the effects of overfishing in the Grand Banks, the Canadian government began to enforce a 200-mile (322-km) offshore jurisdiction that prohibited overt foreign competition in the fishing of the Grand Banks. This held off the decline of available fish somewhat, but the greater efficiency of fishing technology propelled this industry toward difficult times. In the 1990s, the overfishing of cod in the banks led to a serious decline in fishing productivity. In Newfoundland, this decrease in yield has led to unemployment that runs higher than 20%, more than three times the overall Canadian level. With the ever more effective catch techniques of the fishing ships that come from all over the world to fish the Grand Banks, this resource is continually threatened. With it slowly departs the fishing tradition and a major component of the historical economic base for Atlantic Canada.

Wood industries also suffered when forests became depleted by overcutting, ships began to be made of iron and steel, and competition set in from new areas of forest exploitation farther west. The position of agriculture in the Maritimes was undermined by the building of railroads, which facilitated settlement of Canada's interior and thus brought cheaper farm products into the Atlantic region. It also provided easier access for local farmers considering a move west. Today, quite a few people in the Maritime Provinces still farm, but they do so largely on a small-scale, part-time basis while earning their living mainly from other employment (Fig. 23.5). The term "hobby farming" has become a descriptor of such an agricultural pattern in the United States. Both in Canada and in the United States, the family farm has been evolving toward such farms if they do not grow through continuing expansion with farm consolidation.

As these economic challenges have arisen, the Atlantic region has attempted to meet them by developing industry. For example, many small cotton-textile factories were built during the 19th century. But these early industrial ventures largely failed when they were undercut by competition from mills in New England, central Canada, and Europe that

Figure 23.5   Broad open plains with checkerboard field patterns, so common in the Great Plains of the United States, are also the farming images that are evoked when one thinks of Canadian agriculture. But in the rolling and even mountainous lands of the Maritime Provinces and the Atlantic Region in Canada, a wholly distinct landscape pattern emerges. This agricultural region on the Gaspe Peninsula in Quebec reminds the reader that settlement and farming began along the Atlantic Coast and only slowly worked west to the more productive and expansive Prairie region of Canada.

FRANCIS LEPINE

were located closer to major markets. In addition, sizable coal reserves existed near the town of Sydney on northern Cape Breton Island in Nova Scotia, and iron ore was present on Newfoundland's Bell Island. On the basis of these resources, a coal mining industry and an iron and steel plant were developed at Sydney, which flourished for some time, largely by supplying steel for Canadian railway construction. But eventually, both the plant and the coal mines declined, with consequent economic depression in the Sydney area. The plant at Sydney was bought by the provincial government after World War II to forestall its closure; it continues to operate at a reduced scale and on a subsidized basis to provide employment. The local coal industry has been handicapped by both the increasing cost of extracting the coal, which is deep lying, and the shrinking market for coal due to its widespread replacement by oil as a fuel. Many coal mines in this area have closed.

Currently, the basic economic activities that support Atlantic Canada's struggling economy are numerous but relatively small:

F24.15
646

1. *Specialized agriculture.* Specialized commercial potato farming is somewhat successful in Prince Edward Island and the St. John River Valley of eastern New Brunswick.
2. *Continued fishing.* Protected from foreign vessels since 1977 by the extension of Canadian control to waters 200 miles (322 km) offshore, commercial fishing continues on a small scale.
3. *Tourism.* Based on the region's scenic beauty, cool summers, and sense of splendid isolation, tourism is also handicapped by some of those same qualities because of the region's distance from population centers with their popular urban amenities.
4. *Forestry.* Pulp and paper manufacturing is making a modest comeback and undergoing steady growth because of the expansion of the international market for paper and wood products.
5. *Energy.* The export of electricity to adjacent provinces and the United States, especially from Labrador's large Churchill Falls hydropower installation and from New Brunswick, has been a steady factor in the modest expansion of economic health in the region.
6. *Minerals.* Atlantic Canada does possess a few large metal mines. Both coal and iron ore have a traditional importance in the region's mineral resources, but they pale compared to current anticipation of petroleum wealth. Offshore oil and gas exploration in the 1970s and 1980s found the Hibernia Field off southeastern Newfoundland. While it is thought that these fields might possess enormous reserves, the extraction is particularly difficult because of local weather conditions and ocean floor characteristics. Production has only moved ahead since the mid-1980s. This potential source of wealth for a region that has been traditionally the poorest in Canada has led to considerable intranational (domestic) bickering over the ownership of the petroleum reserves.

Extremely important to the economy of the Atlantic region is a high level of subsidies from the federal government in the form of (1) pension and welfare payments, (2) the stationing of military forces in the region (Halifax is a major naval base), (3) the presence of many federal administrative offices in the region, (4) direct payments to provincial treasuries, and (5) funding for economic development projects and efforts to attract industry.

23.4 Contrasts in the Two Provinces of Canada's Core Region: Ontario and Quebec

The core region of Canada, with about two-thirds of Canada's total population, has developed along lakes Erie and Ontario and thence seaward along the St. Lawrence River. In this area, the Interior Plains of North America extend northeastward to the Atlantic. They are increasingly constricted seaward, on the north by the edge of the Canadian Shield and on the south by the Appalachian Highlands. The two provinces that divide the core region — Ontario and Quebec — incorporate large and little-populated expanses of the Shield; Quebec has a strip of Appalachian country along its border with the United States. But most of the core region lies in lowlands bordering the Great Lakes or the St. Lawrence River from the vicinity of Windsor, Ontario, to Quebec City, Quebec. The entire lowland area is often loosely termed the St. Lawrence Lowlands, although the Ontario Peninsula between Lakes Huron, Erie, and Ontario is frequently recognized as a separate section.

Contrasts Between the Ontario and Quebec Lowlands

The Ontario and Quebec sections of the core region show marked differences in types of agriculture and cultural features. The Ontario Peninsula, which is strongly British in its roots, echoes the U.S. Midwest's cropbelts on a small scale, with corn and livestock production, dairy farming, and the growth of various specialty crops such as tobacco. Along the St. Lawrence in Quebec, the French heritage is clearly evident in the landscape signatures of the French "long lot" pattern of strip-shaped agricultural landholdings (see Definitions & Insights, p. 610).

In the Quebec landscape along the St. Lawrence, large Catholic churches are prominent landscape features. The Quebec lowlands lie farther north than the Ontario Peninsula and have a harsher winter climate. Dairy farming is the predominant form of agriculture. Together, Ontario and Quebec account for about two-fifths of the value of products marketed from Canadian farms.

The core area contains more than 18 million people. Ontario's population is basically British in origin but has many minority ethnic groups (see Landscape in Literature, p. 602).

DEFINITIONS + INSIGHTS

Long Lots

The French long lot is one of the most distinctive responses to living in a river environment. The geometry of such a land holding is, traditionally, a narrow, long lot that goes upslope from river margin in the floodplain into the higher lands that generally parallel to banks of a river. The concept is said to have been designed to distribute the good and bad aspects of floodplain farming. To have frontage on the river was a benefit for trade, travel, possibly irrigation, and access to the water resource. However, such a location also meant the continual threat of flood. So, the French had the lot extend up to the levee and the higher land adjacent to the flood-prone areas. This meant that if there was a flood — and there has been since our very beginning of human occupancy of the floodplain — then all the farmers living along the close edge of the

river had flooded lands at water's edge, but they also had higher ground to fall back on. It was defined as the best way to give all farmers a variety in the sorts of soils they had and to have farmers suffer in parallel fashion during floods. But at the same time, the long lot meant that no farmers had all of their land inundated. And as noted, each farmer had access to the river "highway." In the United States, this long-lot pattern is part of the landscape along the lower Mississippi River, particularly in Louisiana. In Canada, the same long-lot system is found along the St. Lawrence River, especially in the province of Quebec. It has become a distinctive landscape signature of areas of French settlement not only in Canada but in almost all areas where the French passed a significant period of time as colonizers and settlers.

By contrast, about four-fifths of the people in Quebec are of French origin (Fig. 23.6). This represents the biggest exception to the dominance of English culture in Canada and is the only instance in Canada where a national minority controls a provincial or state government. However, Nunavut Territory on the northwest edge of Hudson Bay and in the eastern reaches of the Northwest Territories became a major political feature of Canada peopled almost entirely by a national minority in 1999.

Within the core, life focuses on Canada's two main cities: Toronto in Ontario (population: metropolitan area, 5,411,300; city proper, 2,540,100) and Montreal in Quebec

Figure 23.6   Quebec City, once the capital of France's North American empire and now the capital of Quebec Province. At the right is the Lower Town along the St. Lawrence River; at the left is the Upper Town. Grain elevators symbolize the city's seaport function. The hotel called the Chateau Frontenac is a reminder of the city's heritage from Imperial France with its distinctive copper sheathing. The city continues to play an important role in Canada's multicultural expressions of settlement and population.

(metropolitan area, 3,490,000; city proper, 1,035,100), the second largest French city in the world (Fig. 23.7). Other large cities of the core include Ottawa (population: metropolitan area, 1,060,400; city proper, 344,300), Quebec City (population: metropolitan area, 705,000; city proper, 170,400), and Hamilton (city proper, 343,200). Ottawa, the federal capital, is located just inside Ontario on the Ottawa River boundary between the two provinces. Quebec City, the original center of French administration in Canada, is the capital of Quebec Province. Hamilton, a Lake Ontario port, is the main center of Canada's steel industry, supplying steel for the automobile and other metal-fabricating industries of Ontario. A good part of the urban development in the core is associated with manufacturing. Three-fourths of Canada's total manufacturing development is here, with somewhat more in Ontario than in Quebec. But almost three times as many people are employed in the service industry, reflecting the region's status as the business and political center of the country.

Historical Evolution

The coreland began as an entry to the interior of North America, and this gateway function continues today. The French founded Quebec City in 1608 at the point where the St. Lawrence estuary leading to the Atlantic narrows sharply. Fortifications on a bold eminence allowed control of the river (Fig. 23.8). From Quebec, fur traders, missionaries, and soldier-explorers soon discovered an extensive network of river and lake routes with connecting portages reaching as far as the Great Plains and the Gulf of Mexico. Montreal, founded later on an island in the St. Lawrence River, became the fur trade's forward post toward the interior native-dominated wilderness. Between Quebec City and Montreal, a thin line of settlement evolved along the St. Lawrence and formed the agricultural base for the colony. Population grew slowly in this

Figure 23.7   In Canadian cities in the eastern part of Canada, there is the refreshing blend of substantial traditional city architecture standing in contrast to new design and modern construction. This scene from Montreal shows the Marie Reine Du Monde Cathedral in the foreground and the Alcan skyscraper as part of the contrasting and contesting urban landscape elements.

northerly outpost where the winter was harsh and only French Catholics were welcome. When finally conquered by Britain in 1759, French Canadians numbered only about 60,000.

A few British settlers came to Quebec after the conquest, and this immigration increased rapidly during and after the American Revolution, thus laying the foundations for the British segment of Canada's core. Many of the early English-speaking immigrants were Loyalist refugees from the newly independent United States, whose rebellion French Canada had refused to join. They were soon joined by more English settlers immigrating directly from Europe. These newcomers formed a sizable minority in French Quebec including a British commercial elite in Montreal that persists to this day (Fig. 23.9), but they also settled west of the French on the Ontario Peninsula. By 1791, the British government found it expedient to separate the two cultural areas into different

political divisions, known at that time as Lower Canada (Quebec) and Upper Canada (Ontario), taking their directional terms from their positions on the St. Lawrence River.

The St. Lawrence Seaway

Very important among the core region's advantages were its lowland and water connections between the Atlantic and the interior of North America. In eastern North America, the only other connection of this sort through the Appalachian barrier is the lowland route connecting New York City with Lake Erie via the Hudson River and Mohawk Valley. Rapids on the St. Lawrence River long barred ship traffic upstream from Montreal, but the lowland along the river facilitated railway construction, and the building of some 19th-century canals bypassed the rapids and allowed small ocean vessels to enter the Great Lakes. Finally, in the 1950s, the river itself was tamed by a series of dams and locks in the St. Lawrence Seaway project (see Fig. 23.8). Since then, good-sized ships have been able to reach the Great Lakes via the river. The Welland Canal, which bypasses Niagara Falls between Lakes Ontario and Erie by means of a series of stairstepped locks, predated the Seaway. The canal admits shipping vessels to the four Great Lakes above the falls. Farther up the lakes, the "Soo" Locks and Canals enable ships to pass between Lake Huron and Lake Superior, and natural channels interconnect Lakes Erie, Huron, and Michigan. The total length of the St. Lawrence Seaway project is 2,342 miles (3,769 km), and cities on the margins of the Great Lakes both in Canada and in the United States now serve as international ports. The Seaway provides an 8-meter-deep waterway in the heart of North America, and it moves wheat, iron ore, petroleum, and durable goods from mid-April until approximately mid-December with the assistance of ice-breakers. Construction was begun in the 1950s, although its full complement of canals, locks, and dams include transportation components that had been built decades earlier. The project has changed commercial geography in the heartland of North America in many ways. Through unwanted events, it has also changed the ecology of the region, particularly with the arrival of zebra mussels (see Problem Landscape, p. 613). The mussels have been diffused into these interior waters from Europe and are now a major blight on shipping and machinery efficiency because of their skyrocketing numbers and persistence.

The Rise of Commerce, Industry, and Urbanism

Until Canada was formed as an independent country in 1867 — an act fomented in good part because of the success of the Union army in the American Civil War — Quebec and Ontario developed largely in a common preindustrial fashion, with economies based primarily on agriculture and forest exploitation. With its larger expanses of good land and a somewhat milder climate, the Ontario Peninsula became more agriculturally productive and prosperous than Quebec. It specialized in the surplus production and export of wheat. On a superior natural harbor, the Lake Ontario port of Toronto developed and became the colonial (now the provincial) capital

353
354

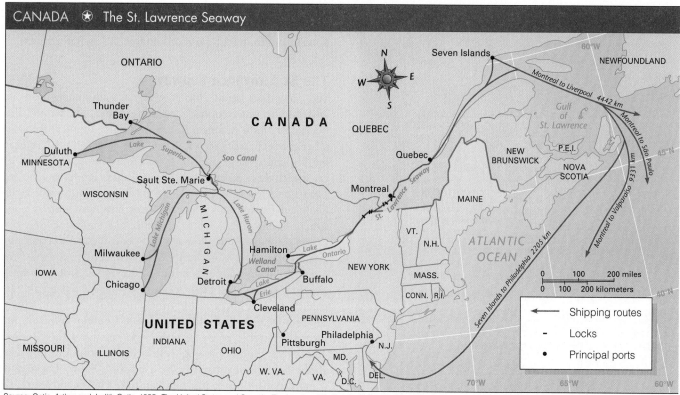

CANADA ✶ The St. Lawrence Seaway

Source: Getis, Arthur and Judith Getis. 1995. *The United States and Canada: The Land and Its People.* Dubuque, Iowa: Wm C. Brown Publishers. p. 231.

Figure 23.8   This waterway, more than 2,300 miles (c. 3,700 km) long, connects the Atlantic Ocean with the Great Lakes in the heart of the continent. This means that cities like Duluth, Minnesota, and Thunder Bay, Ontario, are international ports and serve international vessels. The Seaway was opened in 1959. It is also through this waterway that the zebra mussel came to the Great Lakes in the early 1980s.

and Ontario's largest city. But the bulk of Canada's exports at this time moved eastward to meet ocean shipping, and this movement favored Montreal. Until about 1850, most shipping from the Atlantic went no farther upstream than Quebec City. But subsequent improvements along turbulent sections

Figure 23.9   An October 1995 referendum held in Quebec Province in Canada pitted the French-speaking citizens who wanted to achieve an independent Quebec against people of the same province who wanted to maintain the bilingual lifestyle that had come to characterize this early 18th-century home to French settlers. The move to secede failed by the narrowest of margins, and the issue is likely to be revived in the near future.

of the river enabled ocean ships to navigate safely as far inland as the major rapids (Lachine Rapids) at Montreal. Quebec City was bypassed and its growth began to slow, although it has continued to be an active ocean port. In the later 19th century (see Fig. 23.8) Montreal decisively surpassed its French Canadian rival and became the leading port and largest city in Canada. Only recently has Montreal been exceeded in trade by the Pacific port of Vancouver and surpassed in population by Toronto.

As Montreal's port traffic grew, the city's business interests facilitated it by building a pioneering railway network based in Montreal. This network made use of lowland routes that radiated from the city along the St. Lawrence Valley leading east to the Atlantic and west to the Great Lakes, along the Ottawa River Valley through which Canada's transcontinental railroads found a passage west and north into the Shield; and along the Richelieu Valley-Champlain Lowland leading south toward New York City.

Britain created an independent Canada in 1867 by confederating Quebec, Ontario, New Brunswick, and Nova Scotia. By this time, the core area was already dominant in agriculture and population. In the following century, it was transformed into today's urban-industrial region. In its evolution to industrial dominance, the coreland profited from a number of clear advantages over the rest of Canada: superior position

PROBLEM LANDSCAPE

The Zebra Mussel

One of the most characteristic geographic changes in the contemporary world is the ever-increasing mobility of people, goods, flora, and fauna. The whole process of globalization has meant that landscapes that used to enjoy, or endure, relative isolation are now connected in one way or another with countless other places. One of the most telling outcomes of such expanded interaction is the arrival and frightening success of the zebra mussel (a freshwater shellfish native to the Caspian Sea) into the North American world via the St. Lawrence Seaway.

The process that led to this diffusion is that ships involved in long-haul transport and trade often find that they do not immediately have a replacement cargo after they drop off their cargos. To avoid sailing with an empty hold and having the ship ride too high in the water, they take on water at the time of drop-off and travel with that "ballast" until they get to the next port that has a hold of goods for them. In the early 1980s, zebra mussel–

infested ballast from the region of the Caspian Sea was probably discharged somewhere in the St. Lawrence Seaway. By 1985, the presence of this new shellfish was found in the Great Lakes. This shellfish lives in extremely high densities, and it actively depletes food resources for the sea life that is native to this ecosystem. It also has a special penchant for attaching itself to the intake pipes of irrigation, drinking water, and power generation systems. The presence of highly dense, persistent bivalves in such a location impedes the flow of water and leads to significant machinery damage. It is estimated that $5 billion had been spent by 2000 on efforts to control (there is no longer even a hope of removing them) the zebra mussel and in repairing damage this small shellfish has done.

The Great Flood of 1993, which had a great segment of the Mississippi and Missouri Rivers at extraordinary flood levels all through the Middle West in the United States, further diffused the zebra

mussel. Evidence of this nasty volunteer has now been found all the way from the Hudson River Valley in the east to river systems in Minnesota. The complexity of this specific problem is now confounded by the fact that this mussel has not only found a very accommodating environment all through the St. Lawrence Seaway but has also diffused to hundreds of thousands of square miles of additional territory that has been linked with the Seaway through floods and other diffusion processes.

Figure 23.A The zebra mussel congests pipes with its high density clustering.

DAN SCHLOESSER/GREAT LAKES SCIENCE CTR

---

with respect to transport and trade, the best access to significant resources, accessibility to the largest markets, and the most advantageous labor conditions.

Resource Availability

In addition to the water routes that connect it to the world and to all the shores of the Great Lakes, the Canadian core area has had a number of other major resources to support industrialization:

1. *Agricultural resources.* Farmland good enough for large commercial agriculture, especially in Ontario, has supplied materials for food processing industries and a market for farm equipment industries.

2. *Forest resources.* Forests, mainly in the adjacent southern edge of the Canadian Shield, once supplied lumber for export and now supply wood for a very large pulp and paper industry, mainly in Quebec along the St. Lawrence and its north bank tributaries.

3. *Energy generation.* Especially at the descent from the Shield to the plains, many high-volume, steeply falling rivers drive one of the world's larger concentrations of hy-

droelectric plants. Major hydroelectric plants are also present at Niagara Falls, along the St. Lawrence as part of the Seaway project, and more recently, at far northern sites in the Shield. Electricity is cheap in the coreland and is exported to the United States. Coal requirements are met from nearby Appalachian fields in the United States, and oil comes from Atlantic or Middle Eastern ports by tanker up the St. Lawrence. Both oil and natural gas come by pipeline from western Canada.

4. *Bulk commodity transport.* Metallic minerals are varied and abundant in the Shield to the north and west of the core area. Among them are very large deposits of iron ore located near the western end of Lake Superior on both sides of the international border, and others, more recently developed, are located in northeastern Quebec and adjacent Labrador. A large development of metal processing and fabricating industries, especially in Ontario, uses these Shield minerals, which are also exported.

Through the combination of these resources, their processing, and their distribution to both domestic and foreign markets, the Canadian core region has established a solid base for continuing economic development and relative prosperity.

Market and Labor Advantages

Ever since industrialization began, the population of the core area has represented the main cluster of Canadian consumers and labor. A high-tariff policy to forestall American competition was adopted in Canada shortly after its confederation. Although it fell short of shutting out foreign industrial products, which Canada imports in large quantities, it did favor the growth of a large variety of Canadian industries, mainly in the core region and often including branch plants of U.S. companies "jumping" the tariff wall. However, the policy has long aroused protest from other parts of Canada in which consumers preferred to buy cheaper American goods. It also has fostered many high-cost Canadian industries that lack international competitiveness. The exceptions to this latter phenomenon tend to be industries making products desired in the United States.

Thus, the Canadian pulp and paper and mining industries, with the U.S. market available for their products, tend to be large and efficient. In 1965, the two countries negotiated a free-trade agreement with respect to the production and import and export of automobiles. This sparked rapid expansion of the auto industry into southern Ontario. General Motors, Ford, and Chrysler all have plants there, some of which are in Detroit's Canadian satellite of Windsor (population: 210,500). However, metropolitan Toronto is the main center of the industry. More recently, the inauguration of NAFTA has begun to cause major industrial readjustments — generally characterized by the more labor-intensive firms (clothing especially) moving to Mexican locales — in Canada's core region, just as it has in other parts of North America.

F24.16
647

Industrial Differences Between Ontario and Quebec

Canada's core region produces a wide range of manufactured goods, and for most of these products, it accounts for more than three-fourths of Canada's national output. Within the core, there are marked differences between the industrial emphases of Ontario and those of Quebec. During the formative decades of industrialization, Ontario had cheaper access to U.S. coal. This favored the development of iron and steel capacity and metal products industries there, and these types of industries are still more prominent in Ontario than in Quebec. Montreal also has industries focused on making transport equipment and other metal goods, but these are less typical of Quebec than are more labor-intensive industries such as apparel manufacturing. Such industries were established primarily to deal with a traditional surplus of labor in Quebec created by an unusually rapid natural increase of Quebec's Roman Catholic population.

This labor surplus led to migration into other parts of Canada and the United States and also to the establishment of manufacturing industries that were in need of notably cheap labor. Textiles, shoes, and clothing are typical products. The differences between the two provinces still persist, and they have resulted in lower average wages and incomes in Quebec, which has also experienced a steady decline in its rate of population growth.

Cultural Divergence and Its Consequences

The economic differences between Quebec and Ontario are a serious matter but probably less so than the cultural differences. In Quebec, 82% of the people speak French as a preferred language and 11% speak English. This is the only province in which French speakers predominate and control the provincial government. On the other hand, Ontario, where 77% speak English as a first language and only 5% speak French, is the largest in population of the nine predominantly English-speaking provinces. Major political consequences have arisen from this situation (see Fig. 23.9 and Definitions & Insights, p. 615).

23.5   The Prairie Region: The Tensions of a Strong Resource Base

Manitoba, Saskatchewan, and Alberta are identified as a group under the name the "Prairie Provinces." Isolation is one factor that has led to this grouping. Their populations are separated from other populous areas by hundreds of miles of very thinly inhabited territory. To the east, the Canadian Shield separates them from the populous parts of Ontario, and on the west, the Rocky Mountains and other highlands separate them from the Vancouver region. To the north lies subarctic wilderness, and to the south are lightly populated areas in Montana, North Dakota, and Minnesota.

The prairie environment that gives the provinces their regional name exists only in some of their southern sections, and it is to these sections that we ascribe the name "Prairie region" in this text. The region is a triangle of natural grasslands extending to an apex about 300 miles (c. 500 km) north of the U.S. border. It lies within the Interior Plains between the Shield and the Rockies, with the southern base of the triangle resting on the U.S. boundary. On all sides except the south, the grasslands are bordered by vast reaches of forest. Most of the southern part of the triangle is a northward continuation of steppe grasslands from the Great Plains of the United States. However, toward the forest edges, the soils are moister, and the original settlers found true prairies: taller grasses with scattered clumps and riverine strips of trees. It was to the prairies that settlers were most attracted, and these moister grasslands still form a more densely populated arc-shaped band near the forest edges.

The Quebec Separatist Movement

Increasingly common on the global stage are efforts of specific ethnic, linguistic, or religious population segments to withdraw from a larger body politic and gain partial or full independence. The history of such a movement in Canada is particularly instructive of the types of demographics and political geography that fuel such campaigns. The roots of the tension in Canada can be seen in early Canadian history: The French were early settlers in these lands, but after the 1763 conquest of the British over the French, the broad sweep of language, government, and authority became clearly British in cast. A cultural dichotomy developed that gave the French dominant influence in Quebec but established only modest influence for them beyond the St. Lawrence lowlands region.

After the end of World War II, more widespread French Canadian dissatisfaction developed, possibly because of increasing attention given worldwide to the plight of minorities. In Quebec, this led to the formation of a separatist political party, the Parti Québécois, dedicated to the ultimate full independence of Quebec from Canada. In 1976, the party achieved control of the Quebec provincial government and made French the language of commerce, requiring that all immigrants settling in the province be educated in French. In 1980 and again in 1985, the separatists forced referenda on independence that lost both times. In 1995, the issue was taken to provincial voters again, and the number of voters who favored staying a part of Canada won by a very small margin.

The effects of 40 years of controversy have been considerable. The Ottawa government (the seat of the Canadian federal government) has made French an official language of Canada — along with English — all across the nation. Laws have now stipulated that the children of French Canadians born outside of Quebec may have their children educated in French. And there continues to be political effort in the corridors of power in Quebec as separatist groups try to time another effort for independence. However, a steady influx of non-French-speaking immigrants who seek jobs and settlement in Quebec City and other cities and towns along the St. Lawrence Seaway tends to vote with the pro-English bloc, which continues to thwart the Quebec separatists.

The other implication of this has been that other Canadian provinces have utilized the separatist strategies to gain more control of their own provincial governments. As Ottawa has given French Canadians more authority in Quebec in the hope that they might find the idea of full independence less attractive, other provinces, especially in the Prairie region, have come forward with ever more demanding requests regarding governmental autonomy (See Fig 23.10 to gain a sense of the ethnic variety in that region.). Canadian law supposedly requires that there be no more than one separatist referendum every 5 years, but even that regulation may be overlooked as this process continues. It has the potential for leaving Canada in political turmoil for a long time to come, and the world is watching to see how this pursuit of cultural independence plays itself out.

The Prairie region was settled late, but quite rapidly, in a few decades after 1890. Settlers found reasonably good land, with a climate of long and harsh winters, short and cool summers, and marginal precipitation. Only hardy crops could be grown. Early subsistence farming was succeeded by specialization in spring wheat for export, and the Prairie region became the Canadian part of the North American Spring Wheat Belt (see chapter opening photo). The settlers were predominantly English-speaking people from eastern Canada and the United States, but they included notable minorities of French Canadians, Germans, and Ukrainians. The ethnic composition of the region still reflects these elements (Fig. 23.10 and Table 23.2).

The Downside of Resource Wealth

There is a good deal of luck involved in the discovery of riches in the soil of a specific piece of real estate. This fact applies both at the level of a single miner who strikes it rich on a random piece of land and at the more broad-based and powerful level of settlement across a wide-open expanse, as in the westward movement of pioneers across Canada beyond Lake

Superior in the 19th century. The Prairie Provinces that were settled in this steady movement toward British Columbia's Pacific coast have come to realize that there is considerable wealth in the lands they received in the allocation of provincial territories.

However, frustration has built up, particularly in the Prairie Provinces of western Canada, because of the very wealth these mineral and soil resources represent. It has led to a political posture called *resource separatism*. This occurs when, within the boundaries of a single country, one region has a particularly rich resource base but sees the cash rewards for their good fortune, or regional productivity, go to the federal government and other parts of the country in what the source area considers an unbalanced manner.

In the province of Alberta, the presence of fossil fuels has led to the potential for resource separatism. On a world scale, Canada's oil output is still small; nevertheless, it has led to explosive growth for Calgary and Edmonton, which are the Canadian oil industry's main business centers. Another result of the oil boom has been a running feud between Alberta and the federal government concerning

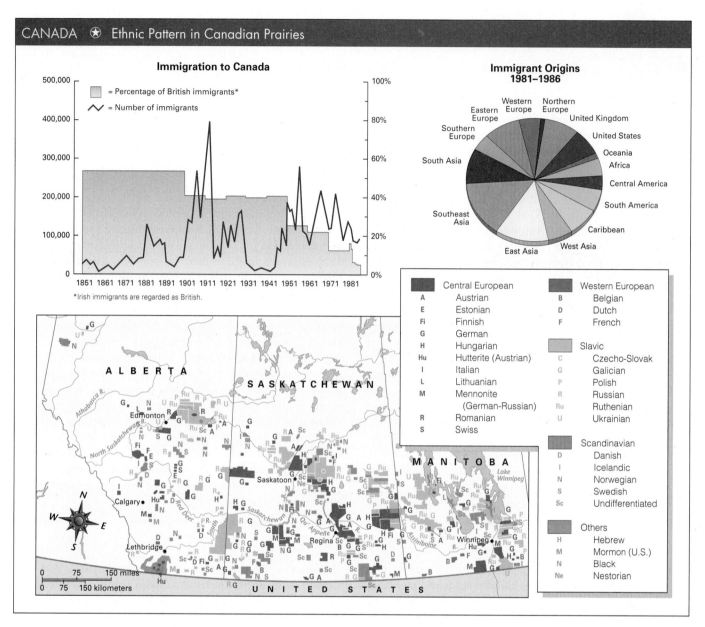

CANADA ★ Ethnic Pattern in Canadian Prairies

Immigration to Canada

= Percentage of British immigrants*

= Number of immigrants

*Irish immigrants are regarded as British.

Immigrant Origins 1981–1986

Eastern Europe · Western Europe · Northern Europe · Southern Europe · United Kingdom · United States · Oceania · Africa · Central America · South Asia · South America · Caribbean · Southeast Asia · East Asia · West Asia

Central European		Western European	
A	Austrian	B	Belgian
E	Estonian	D	Dutch
Fi	Finnish	F	French
G	German		
H	Hungarian		**Slavic**
Hu	Hutterite (Austrian)	C	Czecho-Slovak
I	Italian	G	Galician
L	Lithuanian	P	Polish
M	Mennonite	R	Russian
	(German-Russian)	Ru	Ruthenian
R	Romanian	U	Ukrainian
S	Swiss		

	Scandinavian
D	Danish
I	Icelandic
N	Norwegian
S	Swedish
Sc	Undifferentiated

	Others
H	Hebrew
M	Mormon (U.S.)
N	Black
Ne	Nestorian

ALBERTA · SASKATCHEWAN · MANITOBA · UNITED STATES

Athabasca R. · North Saskatchewan R. · Edmonton · Calgary · Red Deer · Lethbridge · Saskatoon · Saskatchewan R. · South · Qu'Appelle · Regina · Assiniboine · Lake Winnipeg · Winnipeg

0 75 150 miles
0 75 150 kilometers

Figure 23.10   Canada, like the United States, has served as a target destination for human migration for centuries. This figure illustrates the flow patterns, origins, and magnitude of the European migration. The map shows the resultant settlement patterns in Canada's Prairie Provinces.

resource control and, especially, price control of oil piped from Alberta to eastern Canada. Alberta's economic dynamism and aggressive political stance seem likely to continue, as the province is much richer in potential energy than in present energy production. It is estimated that "tar sands" in northern Alberta may contain two to three times the energy equivalent of all the oil now thought to exist in the Middle East. Major extraction must await improved technology and/or a time when greater scarcity of oil yields higher prices.

The oil fields that enrich Alberta and Saskatchewan are extensive and have their origins in Middle America. They course up northward through Texas (where they produce one-sixth of U.S. total oil output) and extend further north through the Canadian Prairie Provinces. Alberta is particularly interested in extracting a greater percentage of the foreign exchange generated by the gas, tar sands, and crude oil of these fossil fuel beds. Resource-based domestic tension such as this reminds us that one can never tell exactly what issues will crop up from the discovery of resource wealth.

Although the economy of the Prairie Provinces has diversified, agriculture is still basic. The region produces about one-half the value of farm products in Canada. Agriculture has been diversified to include other hardy crops such as barley

TABLE 23.2 Ethnic Groups in Canada and Their Origins		
Major Ethnic Groups	**Dates**	**Major Occupations**
Native peoples	pre-1600	Self-contained societies
French	1609–1755	Fishing, farming, fur trading, supporting occupations
Loyalists from the United States	1776–mid-1780s	Farming
English and Scots	mid-1600s on	Farming, skilled crafts
Germans and Scandinavians	1830–50, c. 1900, 1950s	Farming, mining, city jobs
Irish	1840s	Farming, logging, construction
African Americans	1850s–70s	Farming
Mennonites and Hutterites	1870s–80s	Farming
Chinese	1855, 1880s	Gold panning, railway work, mining
Jews	1890–1914	Factory work, skilled trades, small business
Japanese	1890–1914	Logging, service jobs, mining, fishing
East Indians	1890–1914	Logging, service jobs, mining
	1970s	Skilled trades, professions
Ukrainians	1890–1914	Farming
	1940s, 1950s	Variety of jobs
Italians	1890–1914	Railway work, construction, small business
	1950s, 1960s	Construction, skilled trades
Poles	1945–50	Skilled trades, factory jobs, mining
Portuguese	1950s–70s	Factory work, construction, service jobs, farming
Greeks	1955–75	Factory work, small business, skilled trades
Hungarians	1956–57	Professions
West Indians	1950s	Domestic work, nursing
	1967–	Factory work, skilled trades, professional, service jobs
Latin Americans	1970s–	Professions, factory and service jobs
Vietnamese	late 1970s–early 1980s	Variety of occupations

Source: From E. Herberg, *Ethnic Groups in Canada: Adaptations and Transitions.* Copyright © 1989 Nelson Canada, Scarborough, Ontario. Reprinted by Permission.

and rapeseed (for vegetable oil, called canola, and fodder), together with raising livestock.

Urbanization and nonagricultural industries have become increasingly important, especially in Alberta and Manitoba. The three main metropolises are Edmonton (population: metropolitan area, 905,200; city proper, 680,300) and Calgary (population: city proper, 847,900) in Alberta and Winnipeg (population: metropolitan area, 700,100) in Manitoba. These cities have all grown near the corners of the Prairie triangle where connections to the outside world are focused. Winnipeg developed where the rail lines through the Shield crossed the Red River and entered the Prairie region; Calgary settled near a pass over the Rocky Mountains that gave a route toward Vancouver; and Edmonton was established near another Rocky Mountain pass leading to the Pacific. Both Winnipeg and Edmonton have developed additional functions as metropolitan bases for huge sections of Canada's North. None of the three Prairie Provinces has yet become a notable manufacturing center, but the environmental setting has turned this region into an important tourist destination for Canadians and Americans alike, particularly since the horrors of September 11, 2001, and their impact on foreign tourism by American travelers.

Minerals are the foundation of recent advances in the region's economy, with this region now producing more than 60% of Canadian mineral production. Coal deposits underlie both the Rocky Mountain foothills and the plains in Alberta, as well as western Saskatchewan. Coal has recently begun to provide a major export market. Nickel, copper, zinc, and other metals come from the Canadian Shield of Manitoba. The Saskatchewan Shield produces uranium, and southern Saskatchewan produces major quantities of potash. However, petroleum and associated natural gas have had the most economic impact on the Prairie region.

The region's oil industry began near Calgary, Alberta, in the 1940s. As production has multiplied and new fields have been brought in, the industry has remained predominantly in Alberta, with a minor portion in Saskatchewan and a tiny share in southern Manitoba. An important product extracted from natural gas in Alberta is sulfur, of which Canada is a dominant world exporter. The growth of Edmonton has come mostly from petroleum activity.

23.6 The Vancouver Region and British Columbia

Canada's transcontinental belt of population clusters is anchored at the Pacific end by a small region centering on the seaport of Vancouver at the southwestern corner of British Columbia. The region contains no more than 5% of British Columbia's area but has over one-half of the province's population. Metropolitan Vancouver has 1.9 million people, and another 311,000 reside in the metropolitan area of the province's capital city, Victoria. Vancouver is located on a superb natural harbor at the mouth of the Fraser River. This valley provides a natural route for railways and highways across the grain of rugged mountains and plateaus in Alberta and British Columbia. Victoria is located at the southern end of Vancouver Island, the southernmost and largest of a chain of islands along Canada's Pacific coast.

Victoria was the leading commercial and industrial center in British Columbia before the completion of Canada's first transcontinental railroad, the Canadian Pacific, in 1886. Vancouver then became the country's main Pacific port, and it has recently become the largest seaport of the entire country (Fig. 23.11). It has profited from continuing development and a buildup of foreign trade in British Columbia and the Prairie Provinces, particularly Canada's increasing exports of resource commodities to East Asia, primarily Japan. This city served as a major migration destination for Hong Kong Chinese in the years preceding the British handover of Hong Kong to the People's Republic of China. Vancouver now is home to the second largest Chinatown in the United States and Canada, second only to the classic one in San Francisco.

The years of the so-called Asian economic "flu" (recession) in 1997–1999 reduced the Asian tourist flow to British Columbia by about 20%. Japanese tourists have traditionally had the biggest impact on local economies. Korean tourist numbers dropped by 60% in 1998, and those of mainland Chinese dropped by 25%. The only good news in this situation is that the number of American tourists has increased during this period of Asian decline. In early 2002, the numbers of Asian tourists began to rise again.

Vancouver also exports huge quantities of bulk commodities such as grain, wood in various forms, coal, sulfur, metals, potash, and asbestos. Sea trade is fundamental to Vancouver's economy, but the city is far more than just a seaport. Despite its peripheral location within British Columbia, it is the province's main industrial, financial, and corporate administrative center. Thus, it has numerous links with production nodes scattered throughout a sparsely populated and mountainous province larger in area than Texas and California combined. Beyond these industrial functions, however, the real power of Vancouver is the beauty of its setting on the Pacific, the diversity of its resident population, and the growth potential that it radiates.

The Province of British Columbia

British Columbia, along with the Yukon Territory to its north and a narrow fringe of Alberta, is the high part of Canada. The fjorded Coast Ranges along the Pacific and the Rockies of the interior are high and spectacular, and the Intermountain Plateaus between them are largely rugged country. Only the northeastern corner of the province is plains country. This is Peace River Country, the extreme northern outpost of Prairie-region grain and livestock agriculture. The province is also distinctive climatically within Canada by virtue of its coastal strip of marine west coast climate. Here, moderate temperatures and heavy precipitation are characteristic. The

Figure 23.11   The jewel of the Canadian west coast is Vancouver, British Columbia. The growth and expansion of this city in recent years reflects Canada's shifting and expanding trade connections with the nations of the Pacific Rim. The range of design and urban-expression is a hallmark of a city that counts on a future of active trade and tourist activity as the Pacific trade of Canada continues to grow. This view of the Vancouver skyline is from the city's Burrand Inlet.

JEFF GREENBERG/VISUALS UNLIMITED

Figure 23.12   In British Columbia, logging is at the center of the extraction of resources for global trade. This clear cutting on Lyell Island in the Queen Charlotte Islands north of Vancouver shows the intensity with which lumber is pulled off of the local slopes for milling and export. The adjacent uncut forest will be maintained until it makes economic sense to clear cut another portion of the slope.

ART TWOMEY/PHOTO RESEARCHERS, INC.

moisture supports the dense coniferous forests that are one of the province's principal resources, and the mild temperatures, combined with spectacular scenery, attract tourism and retirees. The climate of the interior is subarctic in the north and varies according to elevation and exposure in the south. Like the Coast Ranges, the interior is largely forested, although not so luxuriantly.

Scattered small communities exploit abundant natural resources at increasing scales from an expanding number of locations. Forest products come from both the interior and the coast (Fig. 23.12). Salmon fishing continues along the coast as it has for more than a century. More important than forestry and fishing today is a boom in the mining of metals (copper, molybdenum, silver, lead, and zinc) and coal. Older metal production sites in the south are being supplemented by operations farther and farther north. The driving factor is Japanese and U.S. market demand, with development being supported by foreign capital, largely Japanese. The province is a storehouse of energy resources in the form of both coal and hydropower, with coal shipped to Japan and electricity exported to the United States (Fig. 23.13). Near the coastal town of Kitimat, hydroelectricity is used to produce aluminum from imported alumina (the second-stage material from bauxite ore). Wages tend to be high in the small and isolated production nodes scattered over British Columbia, but so is the cost of living, and cultural amenities are few. Most new residents of British Columbia settle in cosmopolitan Vancouver or in Victoria, known as the most English of Canadian cities.

Figure 23.13   In Sparwood, British Columbia, coal mining is big business. This enormous truck has been designed as the most efficient means of moving vast amounts of coal from the pit to the loading area for both domestic shipping and export. In Canada, major investments in such technology have been made because of the importance of resources in the development of local regions and the overall national importance of such economic activities. This is part of the resource base so important to the Prairie region in Canada.

23.7 The Canadian North: Lots of Land, Few People

The North comprises most of Canada, but where its southern edge lies is arguable. It certainly includes that major section of Canada that is very sparsely populated because of harsh physical environments, but population density fades gradually northward. In this discussion, we use the southern edge of the subarctic climate as, for the most part, the southern edge of the North, but do not include the island of Newfoundland, with its Atlantic associations and somewhat denser population. This definition places large sections of all the provinces except the Maritimes in the North. In addition, we may justify inclusion in the North of parts of the Canadian Shield which project south of the subarctic climate into southeastern Manitoba, Ontario, and Quebec. These areas are so handicapped by isolation (as well as very poor soils) as to be sparsely populated, possessing the demographic patterns that are generally associated with the North (Fig. 23.14).

The southern fringes of the North, often called the "Near North," are spotted with widely scattered islands of development of considerable importance to Canada's economy and, through trade connections, to the economy of the United States. Many of these are mining settlements. The Shield here has a wide variety of metallic ores, although geological processes have left it lacking in fossil fuels. The mining settlements range in size from small hamlets to small metropolitan areas. The largest is Sudbury (population: city proper, 98,000), located in Ontario north of Lake Huron. Particularly noteworthy among the many metals produced in the North are nickel and copper around Sudbury, iron ore from Quebec and Labrador, and uranium mined just north of Lake Huron and in northern Saskatchewan.

Some northern settlements manufacture pulp and paper or smelt metals. A scattering of towns producing wood pulp and paper is spread across the southern fringes of the North. Most are small, but in two places, clusters of mills combine with other functions to produce fairly sizable agglomerations. A string of towns along the Saguenay River tributary of the St. Lawrence in Quebec has a combined population of more than 180,000 people, supported by pulp and paper plants and by the manufacture of aluminum from imported raw materials brought in by ship. The other agglomeration is at Thunder Bay, Ontario (population: city proper, 121,000), on the northwestern shore of Lake Superior (see chapter opening photo). In addition to pulp and paper, Thunder Bay derives support from being at the Canadian head of Great Lakes navigation. Its port links the Prairie Provinces with the Ontario-Quebec core area, the United States, and overseas ports.

Water resources are of major importance in the southern fringe of the North. Hydroelectricity supports local mines and mills and is sent to southern Canada and the United

Figure 23.14   Canada has been negotiating with First Nation peoples and the Inuit peoples for years in an effort to gain resolution of seemingly endless treaty and boundary litigation. In 1999, a new realm of nearly 800,000 square miles (2,072,000 sq km) was deeded to the Inuit peoples. This amounts to nearly one-fifth of Canada's area, and the population resident in these northern lands is approximately 28,000. The Nunavut population shares ancestry with peoples that range all the way from Siberia to Greenland.

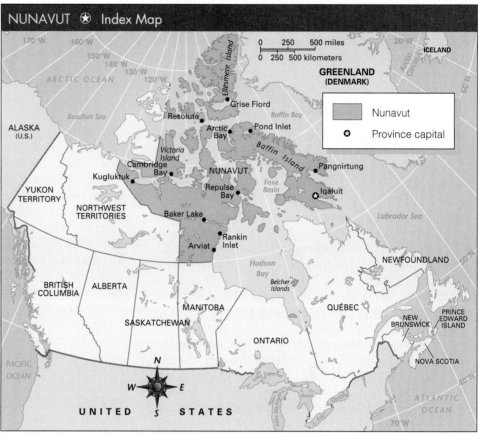

Source: *National Geographic*, 192:3, September 1997. p. 75.

States. Rivers and lakes are so characteristic of the Shield that much of the latter appears from the air to be an amphibious landscape. There is a particular cluster of hydroelectric plants where St. Lawrence tributaries from the north rush down through the Laurentide "Mountains" — the somewhat raised southern edge of the Shield in Quebec. But increasing demand, plus increasing ability to transmit electricity for long distances, has led to major developments much farther north. These are located near James Bay in Labrador and in Manitoba near Hudson Bay. A remarkably small number of people live in this vast area. With agricultural possibilities so limited, a large increase in population seems unlikely in the foreseeable future — with perhaps the exception of the expanding importance of Edmonton and its petroleum base. Agricultural settlements are rare, and most are declining. The port of Churchill on the western edge of Hudson Bay serves as the closest port for movement of Prairie Provinces' wheat to the European market.

Territorial Canada: Bleak and Empty Federal Domain

North of 60° latitude lies the part of Canada not yet considered sufficiently developed and populous for provincial status. This bleak area, known as Territorial Canada, is under the direct administration of the Canadian federal government. Two administrative territories have been established, and a third was formed in 1999 (see Fig. 23.1). In the northwest, the Yukon Territory is cut off from the Pacific by the panhandle of Alaska, once claimed by Canada but acquired by the United States. It is mostly rugged plateau and mountain country. The much larger Northwest Territories lie east of Yukon Territory. They include a fringe of the Rockies, a section of the Interior Plains along which the Mackenzie River flows northward to the Arctic Ocean, a huge area of plains and hills in the Canadian Shield, and the many Arctic islands. Two huge lakes, Great Bear Lake and Great Slave Lake, both drained by the Mackenzie River, mark the western edge of the Shield. The Yukon Territory and the western part of the Northwest Territories have mainly a subarctic climate and taiga vegetation, with tundra near the Arctic Ocean, but in the eastern part of Canada, tundra extends far to the south (see Fig. 23.1). The newly established Nunavut Territory occupies eastern and northern areas withdrawn from the Northwest Territories in 1999.

The population of Territorial Canada was only about 100,800 as of 2001. Approximately one-fifth of the inhabitants in each of the present territories are the native First Nation peoples, and in the Northwest Territories, about one-third — some 15,000 to 20,000 — are Inuit (Eskimos). Most whites are government, corporate, or church employees, often relatively transient. First Nation peoples and Inuit

Box 587

JESSE H. WHEELER, JR.

Figure 23.15 Subsistence hunting and trapping are relatively rare occupations in the present world, although small numbers of people still gain some income from these activities. This photo shows hides of Arctic wolves curing in the sun at the (newly renamed) village of Arviat on Hudson Bay in Canada's Northwest Territories. The month is August, but the Inuit woman is warmly dressed for this tundra climate. Today, Canada's Inuit live in prefabricated houses like the one seen here. Fuel oil for household heating and cooking comes in oil drums brought by ship. The hides came from a hunt by the woman's husband during the preceding winter; pursuing a pack of wolves on his snowmobile, he shot them when they could run no more. All across the landscapes of the Northern Territories and Nunavut, this cultural clash between traditional economies and the modern dependence on canned foods, power snowmobiles, and increasingly sedentary lifestyles continues apace.

receive funding from federal government payrolls in the territorial capitals of Whitehorse (Yukon), Yellowknife (Northwest Territories), and Iqaluit (Nunavut). The nature of the lives of the First Nation peoples has changed immeasurably, however, with the diffusion of things both good and bad from Canada and the United States (see Fig. 23.15).

Nunavut

In April 1999, there was a major shift in Canada's geopolitical map (see Fig. 23.14). An area of approximately 800,000 square miles (2,072,000 sq km; twice the size of the province of Alberta) now divides the Northwest Territories into two realms. The eastern part is known as Nunavut ("our land" in Inuktitut), and the western portion still has not been given an official name. This is the first major change in the map of Canada since the 1949 inclusion of Newfoundland as a province. In addition to this land for the Inuit First Nation peoples, the Canadian government has allocated $1.2 billion to be distributed until 2016 to the Inuit population (an arrangement that averages approximately $38,000 a year per person). This new land, with its population of approximately 28,400, represents another effort to achieve some more equitable alignment between the early treaties signed by the ever-westering settlers and the indigenous peoples resident both in lower and upper Canada and in the North and Northwest Territories. A part of this resolution has been the creation of a number of governmental associations to help guide the use of the land and the capital connected with this agreement in ways that are as positive as possible for the indigenous peoples of this region.

This allocation of land and financial resources is part of the Ottawa government's efforts to cope with the problems of the territorial North. Isolation has been mitigated by the airplane and by telecommunications that now bring telephone service, radio, and even television to remote settlements. A network of schools and medical clinics blankets the area. In Iqaluit, the capital of Nunavut located on remote southern Baffin Island, the school system has undertaken an ambitious Internet linkage with schools all over the world in an effort to diminish the sense of isolation associated with having your capital just south of the Arctic Circle. Air transportation enables the seriously ill to be flown to the few hospitals in the region or to hospitals in cities such as Edmonton or Winnipeg, which serve as metropolitan bases for both the provincial and the territorial North. Meanwhile, the increasingly assertive Native American and Inuit populations are pressing land claims to mineralized areas and pipeline routes and are demanding regulations to prevent further environmental disruptions by mining companies.

still carry on some hunting, gathering, and fishing from homes in small fixed settlements; practically all Inuit inhabit prefabricated government-built housing in widely separated villages along the Arctic Ocean or Hudson Bay (Fig. 23.15). Unemployment among both First Nations folk and Inuit is very high, and a large proportion live essentially on welfare.

Europeans and Anglo Americans have been seeking resources from these far northern reaches since the 17th century, when England's Hudson's Bay Company first established fur-trading posts on the shore of Hudson Bay. The chief 20th-century quest has been for minerals. Mines produce metals or, in one instance, asbestos. Exploration for minerals is active, focusing on oil and gas known to exist in quantity in the Mackenzie Valley and in the Beaufort Sea section of the Arctic Ocean near the Mackenzie delta. In addition to income from the few mining enterprises, the territories also

CHAPTER SUMMARY

- Canada has a vital history of active immigration of people into its vast land. Michael Ondaatje's novel *In the Skin of a Lion* gives a good chronicle of the process of European migrants trying to find work, learn English, and adjust to a very different world. One of the results of this process is the continuing presence of pockets —

both in cities and in the countryside — of ethnic niches that reflect this migration.
- Canada is highly urbanized, with 78% of its population living in cities. It has a major global role in the production of newsprint, hydroelectricity, commercial motor vehicles, and aluminum.

Nearly one-third of its total population lives in Toronto, Montreal, and Vancouver.

- Ten provinces make up the political administration of Canada, but more than one-half of the area of Canada is contained within the North, with its very low population densities. Nearly 90% of the Canadian population is settled within 100 miles (161 km) of the U.S. border, with the largest bloc of population settled in the southeast near the valley of the St. Lawrence River and Seaway. The more than 5,000 miles (8,050 km) of border between the two countries is generally open and allows easy cultural and economic connections between the populations on both sides of the boundary.

- Atlantic Canada consists of the provinces of Newfoundland, New Brunswick, Nova Scotia, and Prince Edward Island. The region is also called the Maritime Provinces, or Atlantic Provinces. Cod fishing has been the mainstay of its economy ever since earliest settlement. The Grand Banks on the east of this region have long been a rich source of fish harvest because of the mixing of warm waters from the south and cold from the north. Recently, the Banks have become so overfished that they have lost their capacity to support the region's fishing population.

- The Atlantic Provinces are now promoting specialized agriculture, tourism, sophisticated fishing gear, forestry, energy, and minerals as the economic activities that will help diminish the importance of federal government support and increase local economic strength.

- Ontario and Quebec are the two provinces that make up the core of Canadian settlement and political influence. Two-thirds of Canada's population lives in this region. The margins of the St. Lawrence River and the Great Lakes make up the areas of primary settlement and industrial activity. The Ontario Peninsula is strongly British, and Quebec is just as powerfully French in its cultural flavor, with 80% of the Quebec population of French origin.

- Industrial development and trade moved upstream on the St. Lawrence steadily until the completion of the St. Lawrence Seaway in the 1950s. Quebec City, the earliest city settlement in Canada, was outgrown by Montreal, and now Montreal has been surpassed by Toronto in size and economic importance, although Vancouver on the west coast has grown to be the biggest port center in Canada.

- The resources of greatest importance in Quebec and Ontario are agriculture and forestry. Energy generation and bulk commodity transport have also been central to the region's economy.

- Hamilton, Ontario, is the center of the steel industry, benefiting from good access to American coal and proximity to automobile manufacturing; apparel manufacturing is centered in Quebec, and Montreal is a node for a variety of metal fabrication industries.

- Quebec separatists have been a political presence for the past 40 years, and in 1995, the push for an independent Quebec was defeated by fewer than 50,000 votes (or less than 1%). In its battle to receive additional accommodations from Ottawa — the head of Canada's federal government — Quebec has gained more provincial autonomy. Other regions, especially the Prairie region, have sought the same benefits by threatening to change their relationship with Ottawa.

- The Prairie region is made up of Manitoba, Saskatchewan, and Alberta. Wheat, petroleum, and coal are the major economic resources. Major urban centers include Edmonton, Calgary, and Winnipeg. These centers provide linkage east and west with other major Canadian regions and also to the North. This region has made continuing efforts to gain more provincial fiscal benefits as a source area for major resources, particularly oil, that flow into export trade. This is sometimes called *resource separatism*.

- The Vancouver region is centered in Vancouver, British Columbia, at the mouth of the Fraser River. More than one-half of the province's population lives in the Vancouver area, which is the region's main industrial, administrative, financial, and cultural center. It is also home to the second largest Chinatown in North America. British Columbia has become a major tourist and retirement destination. Trade is increasingly focused on East Asia, although the economic slowdown in that region has changed trading patterns for this western part of Canada since 1997.

- The Canadian North has an ambiguous southern border. The southern limit of the region of subarctic climate, excluding the island of Newfoundland, can serve to define the southern edge of the North, and population densities tend to decline the farther north one goes. Nickel, copper, and uranium are the major resource metals mined and exported from the North. Forestry, pulp manufacture, and hydroelectricity are additional economic resources. Nunavut is the newest political unit.

REVIEW QUESTIONS

1. Name the provinces in Canada from east to west.
2. Where is the migrant from in the Landscape in Literature selection? Name three different places he tried to find work and learn language in Canada.
3. Name the major import-export patterns for Canada and cite the most important goods and trading partners in the trade.
4. List the political units that make up the Atlantic Provinces, give another name for that region, and outline the reasons for steady economic decline there.
5. What two political units make up the core region of Canada? Outline some of the major similarities and differences in their geographic characteristics.
6. Explain the origins and evolution of the Quebec separatist movement.
7. What role has the St. Lawrence River had in the development of Canada? Describe what has been done in the way of landscape transformation to enhance its utility.
8. What impact has the North American Free Trade Agreement (NAFTA) had on Canadian economic activity?
9. What is the role of automobile manufacturing and wheat production in the economic geography of Canada?
10. What resources have given rise to resource separatism in the Prairie region?
11. What is the major urban center of British Columbia and what trade patterns are of particular importance to that city and that region?
12. What is the name of the new territory turned over to the Inuit in the Northwest Territories in 1999 and what are the reasons for this decision?

DISCUSSION QUESTIONS

1. What geographic characteristics of North America allow Canada and the United States to share a border with basic co-operation? What issues could cause growing friction?

2. What historical events in Michael Ondaatje's literature excerpt have relevance for other parts of the world that you have studied thus far?

3. Discuss the reasons for the Canadian frustration at being so close to the United States and suggest ways in which these feelings might be reversed.

4. Role-play a discussion of Quebec's possible independence, with part of a group representing the French in Quebec and another part representing the English in Ontario. Discuss the reasons for, and the dangers of, independence for Quebec.

5. Discuss the ways in which regions of the United States have experienced similar cultural and economic histories as the Atlantic Provinces, the Prairie region, and the core region.

6. What geographic factors have led to Canada's particular demographic patterns?

7. Speculate on how Canada's development and historical geography would have been different had there been no St. Lawrence River.

8. Discuss the geographic and cultural factors that led to the Nunavut Agreement in 1993 and the subsequent 1999 handing over of approximately 800,000 square miles of land by the Canadian government.

9. Discuss the ways in which the United States has both benefited from and been frustrated by its proximity to Canada. What role has this proximity had in the development of the United States?

10. Speculate on how the map of Canadian settlement will change over the next decade, quarter century, and half century. What will be the role of immigrants? Where are they likely to come from?

The United States: A Landscape Transformed

© BOHEMIAN NOMAD PICTUREMAKERS/CORBIS

In the United States, the pattern of rebuilding earlier landscapes is expressed at all levels. From the massive skyscrapers that stand tall in major cities to the small homemade gardens of city folk who do not want to leave all nature behind, the evidence of a land transformed is writ all across the country. This man, working in a community vegetable garden in Seattle, can be reminded that nature is still close at hand, even in the city.

Chapter Outline

The historical geography of the United States reflects a continuing series of migrations. The country was initially peopled by the earliest migrants coming from northeast Asia across the Bering Strait land bridge and southward into the vastness of North America and beyond. Then, thousands of years later, came immigrants from Europe and slaves from Africa followed by ambitious migrants from all over the world. Those who came of their own volition were seeking some new environment, some new setting that would be safer, more productive, more satisfying. American history and geography have been composed of the outcome of this restlessness since this continent was first found by migrating peoples.

Such steady movement means a continuing change in the country's cultural landscape. To the places peopled by new arrivals there come new languages, new belief systems and customs, new architecture, and new mixes of the cultural personalities created by these demographic blendings. There is also profound change to the landscapes left behind when migrants leave. The nature and magnitude of such change depend in part on the factors that set a particular migration in motion. However, while different migrations have distinct sets of catalysts and migration paths, there are elements of similarity in all the migrations that the United States has known. For example, people move because industrial advances replace human labor, and as a result, people who had worked in that role suddenly have no work. Leaving behind their past and its changing landscape, they head west, which is most often the case in American history, and try to find a place to begin again and anew. These dynamics have traveled with migrants coming to the United States — indeed, to all migration target destinations — since humans began to explore and "hit the road" for new havens and new beginnings. The United States' cultural landscape is rich in the evidence of this human search.

24.1 The Northeast: Original Influences, Contemporary Expression

Of the four major regions of the United States, the Northeast is the most intensively developed, densely populated, ethnically diverse, and culturally intricate. It incorporates the nation's main centers of political and financial activity, and it is the area in which relationships with foreign nations are the most elaborate (e.g., the United Nations in New York City and the country's capital at Washington, D.C., are both located in this region). Strong traditions of intellectual, cultural, scientific, technological, political, and business leadership persist there. The country's political and economic systems had their main beginnings in the Northeast, and traditionally it has been the chief reception center for overseas immigrants.

The Northeast consists of six New England states (Maine, New Hampshire, Vermont, Massachusetts, Connecticut, and Rhode Island), plus five Middle Atlantic states (New York, New Jersey, Pennsylvania, Delaware, and Maryland), and the District of Columbia. Delaware and Maryland have important historical links with the states of the U.S. South, but their present character classifies them more appropriately as Middle Atlantic states; hence, they are included in this region (Fig. 24.1).

As of 2000, about 60 million people, or 21% of the national population, lived within the Northeast, on 5% of the nation's area (Table 24.1). Hence, the regional population density is several times that of the South, Midwest, or West. But the density is much more extreme within a narrow and highly urbanized belt stretching about 500 miles (c. 800 km) along the Atlantic coast from metropolitan Boston (in Massachusetts) through metropolitan Washington, D.C. The belt is often called the Northeastern Seaboard, Northeast Corridor, Boston-to-Washington Axis, "Boswash," or "Megalopolis" (see Fig. 24.1). Here, seven main metropolitan areas contain about 43 million people: Boston (5.8 million), Providence (1.2 million), Hartford (1.2 million), New York (21.2 million), Philadelphia (6.2 million), Washington, D.C. (4.9 million), and Baltimore (2.6 million).[1] Additional population within this belt amounts to roughly 6 million, giving

---

[1] City populations in this chapter are rounded 2000 estimates for Metropolitan Statistical Areas (MSAs), Primary Metropolitan Statistical Areas (PMSAs), or Consolidated Metropolitan Statistical Areas (CMSAs), with the most inclusive figure used (extrapolated from U.S. Bureau of the Census' *Statistical Abstract of the United States*, 2001). In Census Bureau terminology, MSA refers to an urban aggregate composed of a county, or two or more contiguous counties, meeting specified requirements as to metropolitan status. In New England, MSAs are defined by cities and towns rather than by counties. Designation of MSAs represents an attempt to give a realistic picture of functional urban aggregates, each of which often includes more than one incorporated urban place ("political city"), plus unincorporated urbanized areas and stretches of countryside where agriculture may be important. But even the countryside is closely bound to the services, markets, and employment afforded by the urban places. A PMSA is, in general, a larger aggregate, and a CMSA is the most inclusive unit of all. Figures cited in this chapter for the very largest cities, such as New York, Los Angeles, and a considerable list of others, are for CMSAs. Any of the three types of metropolitan units may include counties in more than one state.

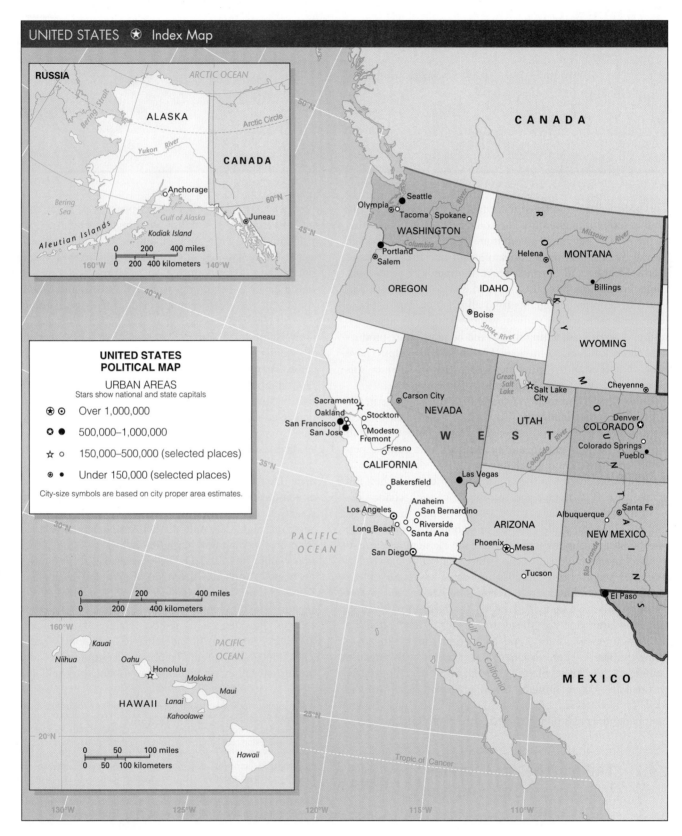

Figure 24.1   General political map of the United States.

Table 24.1	United States Regions: Comparative Data			
	Percent of National Totals			
	Northeast	**South**	**Midwest**	**West**
Total area	5	24	22	49
Total population	21	33	23	22
Land in farms	2	30	37	31
Cropland	3	19	66	12
Value of farm products sold	6	33	36	25
Value of livestock and products sold	8	37	36	20
Value of crops sold	5	26	37	32
Value added by manufacture shipments	16	33	33	18
Value of mineral output	9	29	23	39
Value of retail sales	21	33	23	24
Total personal income	25	30	23	23
Total metropolitan population	24	31	21	24

Source: U.S. Bureau of the Census, *Statistical Abstract of the United States*, 2001.

it a total of about 49 million, or approximately 85% of the people in the Northeast. The leading belt cities are major centers of political decision making, corporations, finance, trade (retailing), and services. An indicator of the centrality of this region is the fact that 37% of all office space in the United States is located within 50 miles (80 km) of New York City. Manufacturing is relatively less important, but the belt still accounts for about one-sixth of the nation's manufacturing. Congested traffic ways and active intercity commercial and cultural linkages bind this mosaic of metropolitan areas together (Fig. 24.2).

Figure 24.2   Rush-hour traffic in downtown Boston, Massachusetts, is representative of the congested Boston-to-Washington "megalopolis".

The Northeastern Environment

Most of the Northeast lies in the Appalachian Highlands, but relatively small sections lie in the Atlantic Coastal Plain, the Piedmont, or the Interior Plains (Fig. 24.3). Atlantic Coastal Plain areas include (1) Cape Cod in Massachusetts, (2) Long Island in New York State, (3) southern New Jersey, (4) the Delmarva Peninsula, which includes nearly all of Delaware plus the eastern shore of Maryland (east of Chesapeake Bay) and parts of Virginia, and (5) the western shore of Maryland inland to the boundary between Baltimore and Washington, D.C. The Plain is low and flat to gently rolling, with many sand dunes and marshes or swamps. Soils are sandy and range from very low to mediocre in fertility. The Plain is indented by broad and deep estuaries ("drowned" lower portions of rivers). The largest northeastern seaports are on the estuaries of the Hudson River (New York), the lower Delaware River (Philadelphia), and the Patapsco River (Baltimore) near the head of Chesapeake Bay.

Between the Coastal Plain and the Appalachians, the northern part of the Piedmont extends across Maryland, Pennsylvania, and New Jersey to New York. The Piedmont is higher than the Coastal Plain but lower than the Appalachians. Where the old, erosion-resistant igneous and metamorphic rocks of the Piedmont meet the younger, softer sedimentary rocks of the Coastal Plain, many streams descend abruptly in rapids or waterfalls, and the boundary between the two physical regions is known as the fall line (see Definitions & Insights). It is marked by a line of cities that have developed along it: Washington, D.C., Baltimore, Wilmington (Delaware), Philadelphia, Trenton (New Jersey), and New Jersey suburbs of New York City (see Fig. 24.1). Most of the Piedmont is rolling, with soils generally better than those of the Coastal Plain or Appalachians.

Aside from a narrow strip of the Interior Plains along the Great Lakes and St. Lawrence River (see Figs. 24.1 and 24.3), the rest of the Northeast lies in the Appalachian Highlands. But in New York, a narrow lowland corridor — the Hudson-Mohawk Trough — breaks the Appalachians into two quite different subdivisions. This Trough is composed of the Hudson Valley from New York City northward to Albany and the Mohawk Valley westward from Albany to the Lake Ontario Plain. Northeast and north of the Trough, the old hard-rock mountains of New England and northern New York form several ranges, characterized by rough terrain, cool summers, snowy winters, and poor soils. The Adirondack Mountains of northern New York rise as a roughly circular mass surrounded by lowlands. Heavily forested and pocked by numerous lakes, these glaciated mountains reach over 5,000 feet (1,524 m) in elevation. They are bounded on the east by the (Lake) Champlain Lowland, which reaches northward into Canada as a continuation of the Hudson Valley. East of it, the relatively low Green Mountains occupy most of Vermont and extend southward to become the Berkshire Hills of western Massachusetts and Connecticut. Farther east across the narrow valley of the upper Connecticut River, the White Mountains occupy north-

DEFINITIONS + INSIGHTS

The Fall Line

The **fall line** is a term that sounds, like so many jargon terms, innocent and of no real consequence. It is defined in most basic terms as the place where rivers flow from relatively resistant rock layers onto more easily eroded rocks. This conjunction means that the rivers cascade down across the hard rocks into the lowland that lies seaward of the falls. Before human settlement, such-fall line locations were of only modest significance. However, with the development of farms, village and town settlement, and increasing patterns of river traffic, these locations became dramatically important. At the actual falls, it became evident that water power could be turned into machine power, and grinding mills and, later, small industrial plants were located so as to use the waterhead (potential power of moving water). River transport patterns saw the fall-line location as requiring the offloading of goods that had been carried up the lowland river to reload the goods on watercraft, or other means

of conveyance, that stayed on the river above the fall line. This **break-of-bulk point** became a logical place to locate settlements because of the human activities associated with such labor, trade, the need for other services, and human interaction. Slowly, cities grew up from these early fall-line settlements. The fall-line cities that are best known along the east coast include Richmond, Virginia (on the James River), Washington, D.C. (on the Potomac River), Baltimore (on the Patapsco River), Philadelphia (at the confluence of the Delaware and Schuylkill Rivers), and Trenton (on the Delaware River). They all play major roles in the economic and demographic character in and at the margins of a megalopolis. Such urban places and their distinctive patterns of growth are a good example of the continuing influence of landscape traits and human response to specific locational settings in decisions for settlement and urban development.

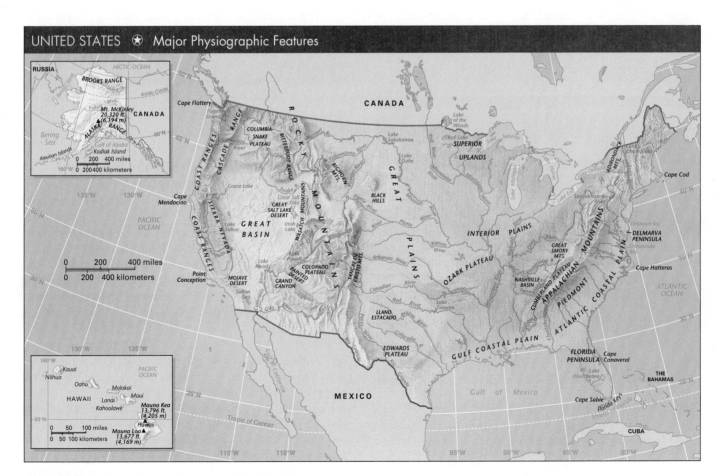

Figure 24.3   Major physiographic features of the United States.

ern New Hampshire and extend into Maine. Here, Mt. Washington in New Hampshire rises to 6,288 feet (1,917 m), the highest elevation in the Northeast. It is in the Presidential Range of the White Mountains, which is characterized by rapid weather changes. Excursions to climb to the top of this regional peak have fooled, sometimes fatally, climbers because of sudden drops in temperature and howling storm winds that come up very rapidly. Wind gusts of greater than 225 miles per hour (360 km/h) have been recorded on Mt. Washington. The Adirondacks and the mountains of New England are major recreation areas for the Northeast's urban populations.

In New England, the hilly areas between the mountains and the sea, sometimes termed the New England Upland, are considered part of the Appalachians, but they are relatively low rather than mountainous. The original soils were persistently stony (caused by the deposition of stones by glaciation), but the early settlers were able to use this area for subsistence agriculture once the stones and trees were laboriously cleared. Boulders from the fields were used to build New England's famous stone fences. Today, many of these have been torn down to facilitate modern farming or urban development, but some derelict fences may still be seen keeping quiet guard in the woodlands where farming has been abandoned (Fig. 24.4). The poor soils of the Upland proved increasingly unable to support stable long-term commercial agriculture, and the late 18th and the 19th centuries saw a large migration of New Englanders from unproductive farms to northeastern

Figure 24.4   Originally, the farmland in New England had to be cleared of boulders left behind by continental glaciers. With the opening of the Erie Canal in 1825, a steady abandonment of New England farmland began. So much farmland has been abandoned that one now often finds stonewalls winding through patches of woodland where crops once grew. Trying to interpret symbols of landscape change like this makes geography a kind of detective work, leading to a clearer understanding of what humans have done in their creation of cultural landscapes.

cities or, especially with the completion in 1825 of the Erie Canal, to more fertile agricultural areas to the west. This migration gave a strong New England cultural flavor to localities scattered all the way to the Pacific coast. Figure 24.5 sets the stage for the settlement patterns created in large part by migrants from the Northeast all across the rest of the country.

South of the Hudson-Mohawk Trough, the Northeastern Appalachians include three distinct sections: the Blue Ridge in the east, the Ridge and Valley Section in the center, and the Appalachian Plateau in the north and west. They lie roughly parallel, with each trending northeast–southwest. The Blue Ridge — long, narrow, and characterized by old igneous and metamorphic rocks — has various local names. Often, it forms the single ridge its name implies, but it broadens into many ridges in the South. West of it, the Ridge and Valley Section, characterized by folded sedimentary rocks, consists of long, narrow, and roughly parallel ridges trending generally north and south and separated by narrow valleys. This valley-dominated eastern strip is essentially a single large valley known as the Great Appalachian Valley. Its limestone floor has decomposed into some of the better soils of the Appalachians. The Appalachian Plateau lies west and north of the Ridge and Valley Section. The northern part is often called the Allegheny Plateau, whereas in eastern Kentucky and farther south (outside the Northeast region), the plateau becomes known as the Cumberland Plateau. Notable east-facing escarpments — the Allegheny Front in the north and the Cumberland Front in the south — mark the eastern edge, from which elevations gradually decline toward the west. Except in New York, the Plateau is underlain by enormous and easily worked deposits of high-quality bituminous coal, which have been extremely important in the economic development of the United States.

Early Development in the Seaboard Cities: The Crucial Role of Commerce

The roots of the Northeast's urban development are deeply set in American history. Boston was founded by English settlers and New Amsterdam (which has become New York City) by the Dutch in the early 1600s, although New York was subsequently annexed and renamed by England in 1664. Philadelphia was founded by the English in the later 1600s and Baltimore in the early 18th century. All four cities were major urban centers by the time of the American Revolution (1775–1783).

Except for Washington, D.C., the largest metropolises of the Northeastern Seaboard gained their initial impetus as seaports, frequently located at river mouths. The federal capital was given its site as part of a political compromise in 1790. It was on the border between the agrarian southern states, with their large slave populations, and the more diversified northern states. The site also placed the capital on the fall line between seaboard and upcountry sectional interests. Many state capitals in the United States have been placed in response to similar locational compromises.

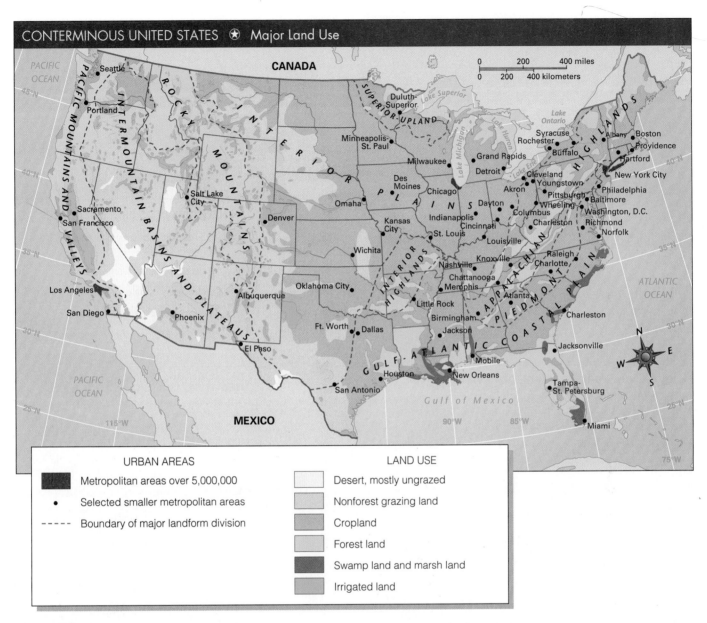

Figure 24.5 This map of generalized land use in the conterminous United States provides a good indication of the variety of land use and also of the importance of moisture and irrigation water.

New York City was founded at the southern tip of Manhattan Island on the great natural harbor of the Upper Bay (Fig. 24.6). It shipped farm produce from lands now occupied by built-up areas of metropolitan New York City and from estates farther up the Hudson River. Philadelphia, also on the fall line, was sited at the point where the Delaware River is joined by a western tributary, the Schuylkill. Philadelphia's local hinterland — the Pennsylvania Piedmont and Great Valley plus parts of southern New Jersey — was productive enough to make the city a major exporter of wheat and helped it become the largest city (population: c. 40,000 in 1776) in the 13 colonies. Baltimore was founded on a Chesapeake Bay harbor at the mouth of the Patapsco River and

shipped tobacco and other farm products from its local hinterland in the Piedmont and the Coastal Plain. Boston had a somewhat different early development because its New England hinterland produced little agricultural surplus for export. Instead, colonial New England placed its main emphasis on activities connected with the sea. Its forests yielded superior timber with which to build ships, and both timber and ships were exported, as were cod and other fish that were initially abundant along this coast. Whaling was important in the 18th and 19th centuries. New Englanders also developed a wide-ranging merchant fleet and trading firms with far-flung interests. Such activities characterized many ports, of which Boston was the largest.

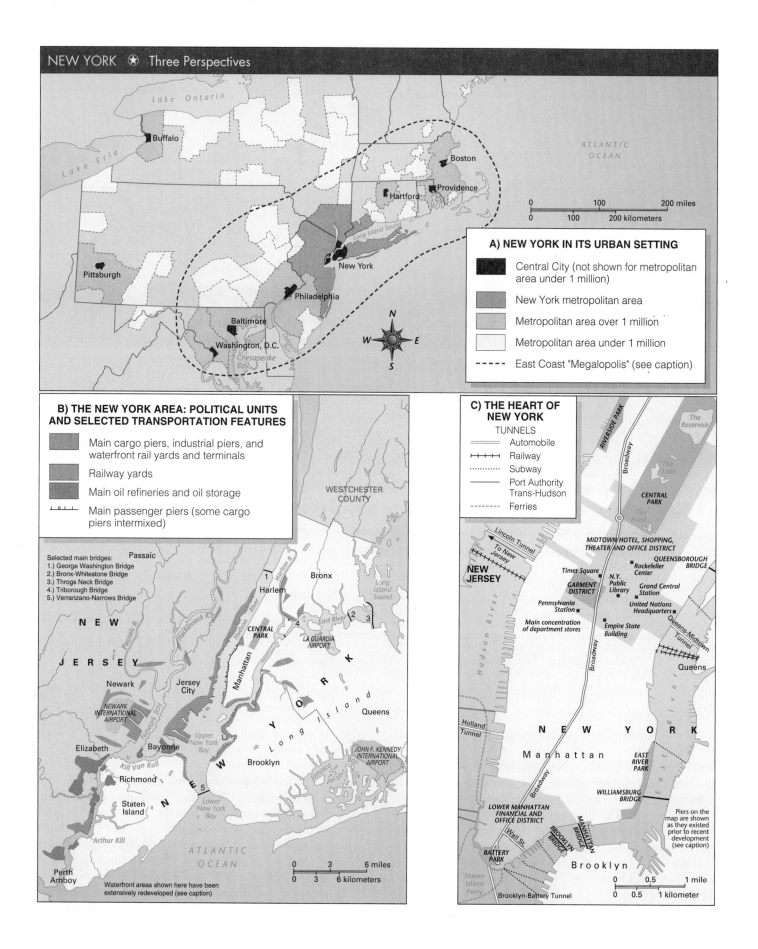

NEW YORK ✦ Three Perspectives

A) NEW YORK IN ITS URBAN SETTING

- ⬛ Central City (not shown for metropolitan area under 1 million)
- New York metropolitan area
- Metropolitan area over 1 million
- Metropolitan area under 1 million
- ----- East Coast "Megalopolis" (see caption)

Buffalo • Boston • Providence • Hartford • New York • Pittsburgh • Philadelphia • Baltimore • Washington, D.C.

Lake Ontario • Lake Erie • ATLANTIC OCEAN • Long Island Sound • Chesapeake Bay

0 100 200 miles
0 100 200 kilometers

B) THE NEW YORK AREA: POLITICAL UNITS AND SELECTED TRANSPORTATION FEATURES

- Main cargo piers, industrial piers, and waterfront rail yards and terminals
- Railway yards
- Main oil refineries and oil storage
- Main passenger piers (some cargo piers intermixed)

Selected main bridges:
1.) George Washington Bridge
2.) Bronx-Whitestone Bridge
3.) Throgs Neck Bridge
4.) Triborough Bridge
5.) Verranzano-Narrows Bridge

WESTCHESTER COUNTY

NEW JERSEY • Passaic • Newark • Jersey City • Newark International Airport • Elizabeth • Bayonne • Richmond • Staten Island • Perth Amboy

Passaic R. • Hackensack R. • Hudson River • Harlem River • Harlem • Bronx • CENTRAL PARK • Manhattan • LA GUARDIA AIRPORT • Long Island Sound • Long Island • Queens • Brooklyn • JOHN F. KENNEDY INTERNATIONAL AIRPORT • Jamaica Bay • Newark Bay • Upper New York Bay • Kill Van Kull • Arthur Kill • Lower New York Bay • ATLANTIC OCEAN

NEW YORK

East River

0 3 6 miles
0 3 6 kilometers

Waterfront areas shown here have been extensively redeveloped (see caption)

C) THE HEART OF NEW YORK

TUNNELS
- Automobile
- ┼┼┼┼ Railway
- ········ Subway
- Port Authority Trans-Hudson
- ----- Ferries

RIVERSIDE PARK • The Reservoir • Broadway • The Lake • CENTRAL PARK • The Pond

Lincoln Tunnel • To New Jersey • NEW JERSEY • MIDTOWN HOTEL, SHOPPING, THEATER AND OFFICE DISTRICT • Times Square • GARMENT DISTRICT • Pennsylvania Station • Main concentration of department stores • QUEENSBOROUGH BRIDGE • Rockefeller Center • N.Y. Public Library • Grand Central Station • United Nations Headquarters • Empire State Building • Queens-Midtown Tunnel • Queens

Holland Tunnel • Hudson River • NEW YORK • Manhattan • EAST RIVER PARK • East River • WILLIAMSBURG BRIDGE • LOWER MANHATTAN FINANCIAL AND OFFICE DISTRICT • Wall St. • BATTERY PARK • BROOKLYN BRIDGE • MANHATTAN BRIDGE • Brooklyn • Staten Island Ferry • Brooklyn-Battery Tunnel

Piers on the map are shown as they existed prior to recent development (see caption)

0 0.5 1 mile
0 0.5 1 kilometer

Figure 24.6   New York: Three Perspectives. New York stands at the center of a strip of highly urbanized land that was christened "megalopolis" by French geographer Jean Gottmann in 1961. The dashed line is not intended to indicate a precise boundary but rather show broadly the area included by Gottmann in his megalopolis concept. Many waterfronts and other areas shown on the two maps of New York are being redeveloped for new commercial, residential, and recreational uses. This is especially true of the piers along both sides of the Hudson River. In the map "The Heart of New York," the shading for the "Midtown hotel, shopping, theater, and office district" covers the areas in which most of Manhattan's well-known hotels, restaurants, department stores, specialty shops, theaters, concert halls, and museums are found. The midtown area also includes numerous office buildings, most notably the cluster of skyscrapers at Rockefeller Center. At the southern end of the area near the main department stores is New York's garment district, with its large work force and numerous workrooms, sweatshops, and showrooms crowded into a remarkably small segment of this densely built-up city. The lower Manhattan financial and office district includes the imposing cluster of office skyscrapers in the Wall Street area and was the home of the World Trade Towers before September 11, 2001. Farther north, City Hall and a large assemblage of other government buildings occupied by city, county, state, and federal offices dominate the cityscape. Note the complex of bridges, tunnels, and ferries that ties Manhattan island to the rest of metropolitan New York.

Transportation Geography and the Race for Midwestern Trade

F23.8
612

As the present Midwest was settled, primarily between 1800 and 1860, its agricultural surpluses and need for manufactured goods greatly expanded the trade of the cities of New York, Philadelphia, and Baltimore. The ports engaged in a race to establish transport connections into the interior. New York City surpassed its rivals, largely because of its access to the Hudson-Mohawk Trough, the only continuous lowland passageway in the United States through the Appalachians (see Figs. 24.1 and 24.3). The Erie Canal was completed along this corridor in 1825. Connecting Lake Erie at Buffalo to the navigable Hudson River near Albany, the canal was a key link in an all-water route from the Great Lakes to New York City. It helped reduce transport costs between New York and the Midwest to a small fraction of their previous level. Such a link stimulated Midwest growth, and a flood of trade was directed through the port of New York. In 1853, using the same route, New York interests became the first to connect a growing Chicago with an east coast port by rail. By the 1850s, New York had become the leading U.S. port and city by a wide margin. The competing ports of Baltimore and Philadelphia also shared in midwestern trade. Major railways were eventually pushed through the Appalachians into the Midwest from both ports. Boston was never a real competitor in this race, although it has remained the largest seaport in New England.

Origins of the Early Industrial Specializations on the Northeastern Seaboard

Industrialization got its first big boost during the same period that the Northeast established connections to the interior. Favorable factors included accumulated capital from commercial profits, access to industrial techniques pioneered in Great Britain, skills from previous commercial operations and handicraft manufacturing, cheap labor supplied by immigrants (see Chapter 23, Landscape in Literature, p. 602), and the presence of energy resources and raw materials that were more important in early industrialization than they are today. Also favorable was a United States high-tariff policy that restricted imports and protected much of the nation's market for its own emerging manufacturers. The first American factory, a water-powered tex-

tile plant using pirated British weaving technology, was established by Samuel Slater on the Blackstone River at Pawtucket (near Providence), Rhode Island, in 1790, and the subsequent growth of factory industry in the Northeast was rapid.

The circumstances and results of industrialization differed in different parts of the Northeast. New England had suitable water power near its ports, with many small and swift streams to turn early waterwheels, but this region lacked coal for metalworking. It specialized in textiles, drawing wool from New England farms, cotton from the South, and abundant labor from recent European immigrants, especially the Irish. However, one New England state, Connecticut, took a different path. It expanded a colonial specialty in metal goods, based on small local ore deposits, by becoming a major manufacturer of machinery, guns, and hardware (which it still is). Mill towns sprang up across southern New England and up the coast into southern Maine.

In the Middle Atlantic states, early industrialization also included textiles, but there was more emphasis on metals, clothing, and chemicals. The emphasis on metals reflected small local deposits of iron ore, especially in southeastern Pennsylvania. Then, in the second quarter of the 19th century, the largest deposits of anthracite coal in the United States became a major source of power. This coal lay deep underneath the Pennsylvania Ridge and Valley Section between Harrisburg (population: metropolitan area, 629,400) and Scranton–Wilkes-Barre (combined population: 624,700). It was made accessible despite the terrain by the determined building of canals and railroads and by large capital investment in the mines themselves. For many decades, anthracite was very significant industrially. Usable in blast furnaces without coking, it fueled large-scale iron and steelmaking in various eastern Pennsylvania cities. Iron and steel from the early plants contributed greatly to the first industrial surge in the Northeast by furnishing material for the manufacture of machinery, steam engines, railway equipment, and other metal goods.

The Urban Dominance of New York City

The United States' largest city is centrally located within the Northeastern Seaboard's urban strip, midway between Boston and Washington, D.C. An enormous harbor, an

PROBLEM LANDSCAPE

The World Trade Towers

On the morning of September 11, 2001, America's sense of geography changed forever. Having been a country bordered by the vast Pacific Ocean on the west, the smaller but stormy Atlantic to the east, the longtime neighbor and vital trade associate, Canada, to the north, and the vibrant and increasingly democratic country of Mexico to the south, the United States was confident in its geographic isolation from proximate enemy forces of any traditional type. We had been called, and had been calling ourselves, "fortress America" because of this geographic distance and isolation from potential enemies. With the stunningly efficient hijacking of four fuel-heavy large commercial passenger aircraft and the effective flying of three of the four into prime commercial and military targets, America was cascaded into a new age, into a new geography. The landscape of all thought now has changed. The relative security that was draped on the structures of distance from Europe, the Middle East, and Southwest Asia now fell away. Suddenly, scores of millions of Americans who had only the most minimal perception of the enmity felt for our country because of our size, wealth, political alliances, and proud independence were thrown face to face with an enemy that was not only clever but also elusive. Even as people lined up daily all around the world, attempting to get legal papers that would enable them to

migrate to the United States and test the rumors of opportunity and reward for diligent efforts, we now saw that there were also perhaps hundreds or thousands lining up to learn how to work destruction on that same country.

This new view of the world around us presented us with a new geography. It is a force that causes us to look more closely at the world directly in front of our eyes. It has led to an excruciating examination and reshaping of the traditional patterns of normalcy relating to air travel, truck transport, shopping mall socializing, and large gatherings. News events that used to feel very far away, which depicted suicide bombers and bloody anger over territorial boundaries, suddenly have taken on a new intimacy in the American landscape. Things that we have never dreamed to be part of our immediate field of concerns have suddenly come as close to us as images of people standing up in a plane aisle to get something out of the overhead bin.

On the spatial side of this equation of change comes the magnitude of the impact that the loss of the office space of the two massive World Trade Towers and some of the adjacent 40- to 50-story skyscrapers will have on the availability of professional office space in Manhattan. New York City has long been the "mother lode" of office space for some foreign as well as American businesses. Up until September 11, 2001, some 40% of the office space in the United States

was located within 50 miles (80 km) of the Statue of Liberty.

However, the impact of the planes on these major towers in lower Manhattan went beyond the physical loss of office space. It has had what appears to be a significant impact on professional decisions about the role of urban centrality for major firms. Classic geographic "laws" about economic and professional interaction have ascribed major importance to easy access to businesspeople and markets. The selection of the World Trade Towers as two of the four targets of the September 11, 2001 terrorism (Fig. 24.7) has led business decision makers to begin to rethink the value and costs of locating in the densely settled lower Manhattan urban nexus of financial activity. This rethinking, coupled with a daily improvement in the technologies of communication and interaction, may also further change the geography of the United States. We may see the thinning out of megaurban space with the associated development of smaller centers of high financial importance with the new offices housed in structures less visually impressive and apparent than the New York City skyscrapers.

The elements in the problem landscape that swirl around Ground Zero and the whole ensemble of things that have become a part of our reality since that September date are more than spatial. But they are profoundly geographic.

active business enterprise, and a central location within the economy of the colonies and the young United States had already made New York City the country's leading seaport before access to the Hudson-Mohawk route gave it an even more decisive advantage. Superior access to the developing Midwest then moved New York rapidly to unquestioned leadership among American cities in size, commerce, and economic impact.

New York City is centered on the less than 24 square miles of Manhattan Island (see Fig. 24.6). The island is narrowly

separated from the southern mainland of New York State on the north by the Harlem River, from Long Island on the east by the East River, and from New Jersey on the west by the Hudson River. Manhattan is one of five boroughs of the city proper. The others are the Bronx on the mainland to the north, Brooklyn and Queens on the western end of Long Island, and Staten Island to the south across Upper New York Bay. In addition, the broader New York metropolitan area includes much of the population and industry of northern and central New Jersey, communities on Long Island east of

Figure 24.7 The ruins seen here are part of the façade of the 110-story twin towers of the World Trade Center. The removal of the thousands of tons of steel, concrete, and all of the material remains of thousands of offices and, tragically, thousands of people, took nearly half a year.

Brooklyn and Queens, southernmost New York State just north and west of the Bronx, and southwestern Connecticut. Expressways and mass-transit lines converge on bridges, tunnels, and ferries that link the boroughs to each other and to northern New Jersey (see Fig. 24.6). Port activity is heavily concentrated in New Jersey along the Upper Bay and Newark Bay, although some traffic continues to be handled in various New York sections of the waterfront. The whole area is held together in part by an amazing 722 miles (1,155 km) of subway system, parts of which are more than a century old. The city itself had a population of 8,008,200 in the 2000 census, and the metropolitan area claims 21,200,000 population. Size, financial and cultural complexity and variety, and national history give New York City a position of primacy that is uncontested in the United States.

Continuing Industrial Emphases of the Northeast

The economic dominance of the Northeast to the whole of the U.S. economy is apparent in a brief look at some of its major specialties:

1. *Clothing design and manufacturing.* Clothing manufacturing is carried on in many locations but is concentrated in metropolitan New York City. The clothing industry in Manhattan was originally an outgrowth of New York's role as an importer of European-made cloth and clothing. It was favored by a large number of skilled immigrant workers in the later 19th century, among them Jewish tailors fleeing from persecution in Tsarist Russia.

2. *Iron and steel manufacturing.* Virtually all iron and steel production has been eliminated from the Northeast. In this region, the powerful iron and steel industries that helped to define American global industrialization from the late 19th century until after World War II were ini-

tially centered. Pittsburgh, Pennsylvania, was the center, but factories and facilities expanded east to Baltimore and west to Chicago at the beginning of the 20th century. In recent years, employment and output have been practically eliminated because of competition from cheaper imported steel, large imports of products made from foreign rather than American steel, and the substitution of aluminum or plastics for steel in industrial processes, especially in automobile manufacturing.

3. *Chemical industries.* Chemical manufacturing is very diversified and widespread and includes production of both industrial chemicals in bulk and a multiplicity of consumer products. The outlying areas of metropolitan New York City form the largest center, although the DuPont Company, which is the world's largest chemical manufacturing concern, has both its headquarters and a major share of its manufacturing in Wilmington, Delaware.

4. *Photographic film and equipment.* Photographic equipment manufacturing is dominated by the world's largest photographic company, Eastman Kodak, which has both its headquarters and main plant in Rochester, New York. The Polaroid Corporation, a major photographic firm for the last half of the 20th century, has now gone bankrupt but was a major employer from the 1960s to the 1990s, based principally in the Boston area.

5. *Electronics manufacturing.* Advanced electronics manufacturing, centering on computers, is widespread in the Northeast. This industry is a major specialty in eastern Massachusetts, being especially concentrated in industrial parks along Route 128 and other beltways that ring the Boston area. Some old textile towns such as Lowell and Lawrence on the Merrimack River have largely converted to electronics or other high-technology industries. They are examples of the widespread adaptive reuse of old mill towns in the Northeast. Eastern Massachusetts was once the nation's leading center of textile milling, but most of this industry has long since moved to the South — and recently, offshore — leaving behind substantial turn-of-the-century building stock that is now being adapted to high-tech industrial use or, increasingly, factory outlet malls.

6. *Electrical equipment manufacturing.* Concentrated in southern New England, New York, and Pennsylvania, electrical equipment manufacturing is dominated by the General Electric Company, headquartered in Stamford, Connecticut, with major production facilities in Schenectady, New York. Westinghouse Electric, headquartered in Pittsburgh, is another major company with plants in the Northeast and other areas.

7. *Aircraft engines.* The manufacture of aircraft engines and helicopters is heavily concentrated near Hartford, Connecticut, where it is carried on by United Technologies Corporation, maker of the famous Pratt and Whitney engines and Sikorsky helicopters.

8. *Nuclear submarines.* The manufacture of Trident nuclear submarines in the New London, Connecticut, area continues New England's long shipbuilding tradition as well as Connecticut's tradition of armaments manufacturing. At a factory near New Haven, Connecticut, in 1798, which supplied muskets under a federal contract, Eli Whitney originated the use of interchangeable product parts in the manufacture of firearms. This practice is central to modern mass production of innumerable types of goods.

9. *Publishing and printing.* Publishing and printing are industries in which the Northeast decisively overshadows the rest of the country. Most of the publishing industry is in the New York metropolis, although the Government Printing Office in Washington, D.C., is the world's largest publishing enterprise and printing plant. Many of the book and magazine printing phases of the industry have been steadily decentralized to other parts of the country, but editorial and management functions remain concentrated in the Northeast.

This assemblage of manufacturing, planning, design, decision making, and management capabilities has also prompted the establishment of tens of thousands of smaller, complementary, and ancillary industries that make their living by supplying materials and information to this concentration of varied dominant businesses. In the geography of commerce, there is great importance placed upon proximity, interaction, timely supply of goods and information, and complex networks of spatial interaction. New York City and the Northeast have historically been very effective in this union.

Demographic and Industrial Shifts and the Migration Gradient Between the "Rust Belt" and the "Sun Belt"

The vernacular term "Rust Belt" has come to describe this region of early urban growth and manufacturing. The region contains the great majority of cities that were already prominent by the early 20th century. Most of the country's old "smokestack industries," a term applied particularly to the steel and other heavy metallurgy industries, are in the Middle Atlantic states and eastern Midwest. Today, these industries are often characterized by outmoded buildings and equipment, depressed sales and profits, high unemployment, abandoned industrial buildings, and landscapes of decline and abandonment.

By contrast, the Sun Belt in the South and West has been enjoying faster growth in population and jobs, including those in manufacturing. "Sun Belt" is a very elastic term, but common usage and geographic logic both suggest an area encompassing most of the South plus the West (in this text's regional system for the United States) at least as far north as Denver, Salt Lake City, and San Francisco. The regional loss of manufacturing employment reflects factors such as (1) closing obsolete plants, (2) shifting production to more profitable locations in other regions or overseas, (3) reduced employment in surviving plants reequipped with new laborsaving machinery, and (4) failure of new plants to locate in the region in sufficient numbers to offset the job losses in existing industries.

Employment in service industries has been expanding quite rapidly in the Northeast, resulting in an expansion of total employment. But despite this, the region has been growing in population relatively slowly, with heavy out-migration from the

Figure 24.8   The Baltimore Inner Harbor Development, called Harborplace, created by the late urban developer James Rouse is one of the most successful reclamations of an old warehouse and wharf area in the country. The current uses of this initial 18th-century harbor facility feature a two-story shopping mall with abundant outdoor restaurants and cafés, a new aquarium, relic brick factory power plant buildings converted into elegant townhouses and condominiums, and easy access to the recently constructed Camden Yards, home of baseball's Baltimore Orioles. Rouse called such (re)developments "festival marketplaces." Berthing for local watercraft and tour ships that cruise Chesapeake Bay are all part of this urban area that, 30 years ago, looked as though it would never be presentable to the public eye or economically useful again. It now serves not only as one of Baltimore's most popular tourist destinations, but it enjoys considerable local popularity as well.

JERRY WATCHER/PHOTO RESEARCHERS, INC.

Middle Atlantic region. The main exceptions to this slow growth are the peripheries of some metropolitan areas into which suburban and exurban (sometimes called satellite development or edge cities) development is expanding. Notable examples include the expansion into New Hampshire from Boston, into Connecticut and New Jersey from New York City, and into Maryland from Washington, D.C. The state of Vermont, whose population growth has recently been high, is not suburban, but much of this quiet and scenic state is exurban, attracting "refugees" from metropolitan areas, owners of second homes and vacation homes, and some diligent commuters, including those traveling by air and, increasingly, professionals who are linked to urban centers through telecommunication networks. This expanding population can thus live in the country and still derive income from offices set right in the heart of the eastern cities. And despite the weight of economic and urban problems, the Northeast recently has been the scene of spectacular new development, such as the Baltimore Inner Harbor area, in the great metropolises, both in old decaying cores and around the peripheries (Fig. 24.8).

For nearly the past half century, these regional differences have stimulated a migration gradient that has meant a continuing movement of people from the Northeast to the Southwest. Although this migration is often described in economic terms, mostly in the phrase "job opportunities," the attraction of the climate and the environmental characteristics of the Southwest has also been a powerful motivating influence in such demographic shifts. With the completion of the 2000 Census in the United States, the national demographic center point has moved from its 1990 location in Sullivan, Missouri, southwest another 35 miles (56 km) to Edgar Springs, Missouri.

Adaptive Reuse

One of the characteristic landscape transformations in the Northeast is the adaptive reuse of older structures. This has become a key factor in any effort to keep or bring new vitality to any urban place. **Adaptive reuse** of earlier building stock means taking abandoned commercial or factory buildings and converting them to some economic function that brings jobs, commercial activity, and people back to a once relatively productive area that has, in the past decades, been abandoned. The process is also associated with **gentrification** — the refurbishment of old building stock by urban professionals for residential or office space. This has been done with particular success in urban places and in the small towns that knew prosperity and economic vitality in the 19th or early 20th centuries. In any downtown that has been losing customers and businesses for some years, the transformation of empty retail stores into specialty shops or trendy restaurants has become a common example of adaptive reuse of older buildings. Although this process is not at all unique to the Northeast, it has been particularly significant in New England. As a process, it is important because (1) it gives new vitality to places and often well-designed buildings that once had economic and social centrality but subsequently lost it; (2) it helps create a sense of the value of the past as old landscapes are visited and

utilized again; and (3) it maintains the structures in their historic context as landscape reference points or pivotal buildings. Adaptive reuse is most often considered in the context of old central business districts; however, the process has become increasingly appropriate in the resettlement of old residential neighborhoods as well. It serves as a bellwether for the revitalization of neighborhoods and helps in the fuller use of at least some sections of a graying and declining (or even abandoned) urban landscape.

24.2 The South: Industrial Efforts to Modify Traditional Images

Various definitions of the South are possible, but none is entirely satisfactory (see Fig. 24.1). The region is discussed here as a block of 14 states: 5 along the Atlantic (Virginia, North Carolina, South Carolina, Georgia, and Florida); 4 along the Gulf of Mexico (Alabama, Mississippi, Louisiana, and Texas); and the 5 interior states of West Virginia, Kentucky, Tennessee, Arkansas, and Oklahoma. Arguments could be made on various grounds for excluding some of these states. For example, West Virginia and Kentucky did not secede from the Union during the Civil War, and the western parts of Oklahoma and Texas lie in dry environments more typical of the West than the South. Although Missouri was a slave state and thus part of the South before the Civil War, it today seems more typically midwestern and thus is not included here.

Patterns for the Utilization of the Physical Environment

Most of the South has a humid subtropical climate, with summers that are long, hot, and wet. January average temperatures range from the 30s (degrees Fahrenheit) (16°–22°C) along the northern fringe to the 50s (27°–33°C) near the Gulf Coast and the 60s (33°–39°C) in southern Florida and the southern tip of Texas. More than 40 inches (c. 100 cm) average annual precipitation is characteristic, rising to over 50 inches in many Gulf Coast and Florida areas and more than 80 inches in parts of the Great Smoky Mountains. High humidity is characteristic, intensifying heat discomfort in summer and chilliness in winter. There are advantages and disadvantages to these climatic conditions. They foster rapid and abundant forest growth, and forest-based industries are of major importance. Agriculturally, they provide long growing seasons, heat, and moisture for a wide variety of crops. But insects and pests also flourish, and the heavy rainfall leaches out soil nutrients and causes severe erosion and flooding.

The Predominance of Plains

Most of the South is classified topographically as plains, but the plains form sections of three major landform divisions: the Gulf-Atlantic Coastal Plain, the Piedmont, and the Interior Plains (see Fig. 24.3). The Coastal Plain occupies the seaward margin from Virginia to the southern tip of Texas, including all of

Florida, Mississippi, and Louisiana and portions of all the other southern states except West Virginia. It is low in elevation, and large areas near the sea and along the Mississippi River are flat, but most inland sections have an irregular surface.

The Interior Plains section of the South occupies central and western Texas, plus all of Oklahoma except the easternmost part. In Texas, the Balcones Escarpment forms an abrupt boundary between the higher Interior Plains and the lower Coastal Plain. Cities such as Dallas, Austin, and San Antonio lie on the Coastal Plain. Westward, the plains rise in elevation and become more level, and the steppe and desert environments of the Interior Plains anticipate major land-use patterns of the West. Irrigation, ranching, and dry farming of wheat are important to the relatively sparse population of this plains region.

The Piedmont extends across central Virginia, the western Carolinas, northern Georgia, and into east central Alabama. Its surface is mainly a rolling plain with some hilly areas. The natural cover is mixed forest in which broadleaved hardwoods, especially oak, tend to predominate over pines. Large areas have reverted to forest from former agricultural use. Soils tend to be poor, and many have suffered serious damage through past cultivation of clean-tilled row crops (cotton, corn, and tobacco) on easily eroded slopes subject to frequent downpours. The region's fall line (often where early industry began) lies between the Piedmont and the Coastal Plain.

Hills, Mountains, and Small Plains of the Upland South

Large parts of the South are hilly or mountainous. In the central and eastern South, two large embayments of rough country project into the South from the Northeast and Midwest. The larger of the two lies mainly in the Appalachian Highlands but includes some rough land west of the Appalachians known as the Unglaciated Southeastern Interior Plain (see Fig. 24.3). Farther west, the second large embayment is the southern part of the Interior Highlands. The three physical areas — Appalachians, Unglaciated Southeastern Interior Plain, and Interior Highlands — are often termed the Upland South.

Southern areas within the Appalachian Highlands extend from western Virginia to eastern Kentucky and southward to northern Alabama. The Highlands include all the major physical subdivisions already distinguished in the Middle Atlantic states, with the total complex narrowing southward (see Figs. 24.1 and 24.3). The Blue Ridge extends from the Potomac River to northern Georgia. Its highest section is called the Great Smoky Mountains. Here, Mt. Mitchell in western North Carolina is the highest point in the eastern United States at 6,684 feet (2,037 m). The Ridge and Valley Section, including the Great Appalachian Valley, runs parallel to the Blue Ridge and inland from it, from western Virginia into Alabama. The southern section of the Appalachian Plateau, known as the Cumberland Plateau in parts of Kentucky and areas to the south, is quite wide in the north, where it occupies eastern Kentucky, most of West Virginia, and a small section of Virginia. It narrows rapidly across eastern Tennessee

and then widens to occupy much of northern Alabama. As in Pennsylvania, it is edged on its east by an abrupt escarpment, is mostly dissected into hills and low mountains, and is underlain by very large deposits of high-quality coal.

Central Kentucky, central Tennessee, and part of northern Alabama are in the Unglaciated Southeastern Interior Plain. This part of the Interior Plain, bounded by the Tennessee River on the west and south, was not subject to the glacial smoothing that occurred on the plains farther north. Hence, a good part of this "plains" area is actually hill country as a result of dissection by streams over a long period of time. True plains are present

a

b

Figure 24.9 (a) Near the junction of Missouri, Iowa, and Nebraska in rich farmland of the Midwest stands this abandoned church structure. Its solitary position in a sea of volunteer grasses coming up to the walls of the sanctuary allows it still to appear central to this scene. It has been unused as a church for at least 30 years, and the late afternoon sun on the old frame building reminds us still of the social and psychic role small churches played in the settlement of the Midwest farm prairies. (b) When you go up to the abandoned church and walk through the doorway that served as the portal to weddings, funerals, church meetings, and probably countless carry-in dinners, your anticipation of interior geography is changed instantly and profoundly. Rather than let the church structure collapse into complete disuse, a local farmer converted the interior space into hog pens. The old wood burning stove, the lath and plaster wallcovering, and the church window trim are the visual cues of early construction and a probable importance to the farms that stretched off from this tristate point in the Midwest.

but not extensive. Two of them — the Kentucky Bluegrass Region around Lexington and the hill-studded Nashville Basin around and south from Nashville, Tennessee — are underlain by limestone that has weathered into good soils. Both areas are famous for their agricultural quality in a region of poor to mediocre soils and difficult slopes.

In the South, the Interior Highlands occupy northwestern Arkansas and most of the eastern margin of Oklahoma. From Arkansas, they extend northward into the southern part of the midwestern state of Missouri. Composed of hill country, low mountains, and some areas of plains, these highlands are split by the broad east–west Arkansas Valley followed by the Arkansas River. North of the valley, the Boston Mountains form the southern edge of the extensive Ozark Plateau South of the valley lie the Ouachita Mountains of west central Arkansas and adjacent Oklahoma. Most of the Ozark Plateau has been dissected into somewhat subdued hill country, but some parts of the Plateau, especially the Boston Mountains, attain mountainous character. The Ouachita Mountains consist of roughly parallel ridges and valleys running east and west.

The Southern Interior Highlands have not been rich in resources. Most soils are quite poor, and many are very stony. Most of the area is still forested, but it has been heavily cut over, and the present timber is generally rather poor. Mineral deposits are varied but often not of sufficient value to justify present exploitation. The main mineral wealth extracted today from the Interior Highlands, primarily lead, comes from the section outside the South in Missouri. A number of large artificial lakes used for recreation are major resources in both Missouri and Arkansas.

Regional patterns of the South in population density, urbanism, ethnicity, and income are distinct from those of the other major regions. Until recently, it was a region of farms, villages, and small cities with low population density compared to that of the Northeast and Midwest (Fig. 24.9a and b). But for the past several decades, the South has been a region of rapid population growth, second only to the West, and its farm population has decreased drastically. It has now surpassed the Midwest in overall population density, although both are far behind the Northeast. Because recent southern growth has been overwhelmingly urban, previous contrasts in this respect also are rapidly fading. The South has more small and medium-sized metropolitan areas than other regions of the country because of the recent rapid growth of many small cities. However, it has fewer centers of truly large size than might be expected from its population. The largest urban complex, Dallas–Ft. Worth, has 5.2 million people, Houston (population: 4.7 million), Atlanta (4.1 million), Miami (3.9 million), and Tampa–St. Petersburg (2.4 million) are among the more than 20 American metropolitan areas with more than 2 million people (Fig. 24.9). But these southern urban complexes continue to grow very rapidly, and some of them have imagery that surpasses their demographic size (Fig. 24.10).

The population of the South has less ethnic and racial variety than the other regions. The overwhelming majority is comprised of white Protestants of British ancestry, African

Figure 24.10   This scene of the Dallas skyline at dusk is effective in illustrating the drama of this Texas financial and commercial center. Dallas, in combination with Ft. Worth, exists as the major commercial and financial center for central Texas. Architects and engineers have given much attention to the clear light of this setting, draping glass over the surfaces of the tall buildings that serve both as office space and as icons or billboards for specific Texas firms.

American Protestants, and Mexican American Catholics. Most striking is the heavy preponderance of white Protestants with Anglo-Saxon origins. The original European settlers along the Atlantic were British. Although small groups of Spanish Catholics settled in Florida and Texas and some French Catholics settled in Louisiana and elsewhere along the Gulf Coast, their numbers were insignificant compared with the tide from Great Britain.

Another way in which the South stands out ethnically from the other major regions is in its high proportion of African American population. This is largely a result of the importation of African slaves for plantation labor in the pre–Civil War South. The five American states that rank highest in percentage of African Americans — from about one-fourth to somewhat over one-third — are Mississippi, Louisiana, South Carolina, Georgia, and Alabama, which together form a Coastal Plain belt. Migration out of the South, mostly in the past 50 years, has given some northern states and California large African American minorities. New York and California both have larger African American populations than any southern state, but there are only 16% in New York and 7% in California. African American populations outside the South are very concentrated in large metropolitan areas. Recently, a sizable reverse migration of African Americans from North to South has occurred as economic and social conditions in the South have improved at the same time economic conditions of many northeastern cities have deteriorated.

Other distinctive racial and ethnic elements in the South's population are less widely distributed. Especially notable are Mexican Americans, Cuban Americans, and Cajuns (of French descent). The 2000 Census reported that the four states leading in percentage of population of Hispanic origin were New Mexico (42%), Texas (32%), California (32%), and Arizona (25%). In Texas, Hispanic people primarily of Mexican descent live mainly in a broad band of territory along the Gulf of Mexico and the Rio Grande, where they form a much higher proportion than the statewide 32%. Natural increase and immigration have rapidly expanded their numbers (in Texas and elsewhere) during recent decades.

South Florida has received intermittent immigration of another Hispanic origin group: Cuban Americans. Almost a million refugees have fled from Communist Cuba since the revolution of 1959, and the majority of them or their descendants now live in metropolitan Miami. Cuban Americans, drawn largely from the middle and upper classes of pre-Communist Cuba, have shown rather rapid upward mobility economically. They have also been assimilating into the general population through marriage and rapid adoption of U.S. culture traits. Meanwhile, they have imprinted aspects of their own culture on south Florida. The country's Mexican American population, on the other hand, largely continues to be a group whose Hispanic identity remains strong. The largest non-Hispanic minority language group in the South are the French-descended Cajuns of southern Louisiana.

Agricultural Geography of the South Today

There is a strong agricultural tradition to landscape and culture in the South. From tobacco to cotton to citrus, there has been a long-term economic dominance gained and held by the products of an active farming culture. But these patterns are changing rapidly. Consider the main features of the South's agricultural geography after a half century of revolutionary departure from regional dependence on cotton, changes in crop and livestock emphases, farm mechanization, farm markets, and the number of farmers. The picture is intricate, but some generalizations are possible:

1. *Declining agricultural importance.* In appearance and land use, the greater part of the South today is not agricultural. Over large areas, less than half the land is in farms, and even where land is farmed, the amount of cropland is apt to be greatly exceeded by woodland and/or pasture. The region's population is only about 2% agricultural, making it just about the same as the U.S. national average.

2. *Farm patterns.* Southern farms tend to be small, low in production, and often part-time. The prevalence of part-time operations and of small full-time farms means that output per farm over most of the South is far below the national average. Within the South as a whole, animal products decisively surpass crops in total farm sales. Beef cattle, broiler chickens, and dairy cattle are the main sources of animal products sold. Rising cattle production has been fostered by (1) the presence of much land that is too poor for crops but amenable to pasture, (2) improvements in pasture grasses and cattle breeds, (3) increased availability of chemicals for pest and disease control, and (4) greater emphasis on soybeans, sorghums, and citrus by-products

for cattle feed. Soybeans have become the most widespread and important southern crop. The leguminous soybean plant adds nitrogen to the soil and can be used for hay. But its greatest value lies in the beans that are pressed for soybean oil (for use in many chemical processes and, now, as a fuel additive as well), with the residue (oilcake) fed to cattle and poultry. Grain sorghums, which have low moisture requirements, have become a mainstay for cattle feeding in drier parts of Texas and Oklahoma. The greater part of Florida citrus fruits are now processed for frozen concentrate, and the residue is fed to cattle; the main market is the large ranch industry in southern Florida. Broiler chicken production has become a specialty of poorer agricultural areas, with major concentrations in the Interior Highlands of Arkansas and the Appalachians of Georgia and Alabama. Dairy farming is growing in importance as southern urban markets grow.

3. *Tobacco and cotton.* The historic staples of tobacco and cotton are still major elements in the agriculture of some parts of the South. Most of the nation's tobacco is grown in the eastern and northern South. In 1996, it was the leading source of income in the agriculture of North and South Carolina, Kentucky, and Tennessee. Nearly all the South's remaining cotton is grown on large mechanized farms in the Mississippi Alluvial Plain or in west Texas.

4. *Specialty crops.* Many islandlike districts of intensive agriculture produce a variety of other specialties. Often, such districts represent the main producing areas of their kind in the United States. Some major specialties and areas include citrus fruits in the central part of the Florida Peninsula; truck crops in the Florida Peninsula; sugarcane in the Louisiana part of the Mississippi Lowland and near Lake Okeechobee in Florida (on a drained section of the northern Everglades); rice in the Mississippi Lowland and the prairies along the coasts of Louisiana and Texas; peanuts in southern Georgia and adjoining parts of Alabama and Florida; and race horses, bred particularly in the Kentucky Bluegrass region around Lexington, where the horse farms are showpieces of productive and conspicuous wealth.

The Major Industrial Belts of the South

The largest cluster of industrial development in the South lies along the Piedmont, with extensions into the adjacent Coastal Plain and Appalachians. Within this concentration, which may be termed the Eastern South Industrial Belt, textiles continue to be the most prominent branch of manufacturing, but clothing, chemical products (including synthetic fibers), furniture, tobacco products, and machinery are also important. In the years since the 1994 signing of the North American Free Trade Agreement (NAFTA), however, textiles and clothing manufacture have experienced significant shifts to Mexico and other offshore manufacturing sites. Much of this region's development, as well as that of adjacent areas in the South, has been fostered by the Tennessee Valley Authority (TVA). The TVA was founded in 1933 by the federal government during

Franklin Roosevelt's New Deal era to promote economic rehabilitation of the deeply depressed Tennessee River Basin. It is a public corporation that built many dams, locks, and hydrostations on both the Tennessee River and its tributaries. These installations controlled floods, allowed barge navigation as far upstream as Knoxville, Tennessee, generated hydroelectricity, and provided recreation at numerous reservoirs. The demand for electricity soon outran the capacity of the hydrostations, and many thermal plants were built, fired by coal or nuclear energy. The success of the TVA in helping industrialize several southern states is a well-publicized facet of the industrial surge that has spread ever more widely across the South in recent decades. Northern industrial development focused on the huge metropolises that arose in the railway age, but in the South, most development came later and took advantage of long-distance power transmission and high-speed truck transportation to spread widely into small cities and towns.

The second main belt of industrial development in the South lies along and near the Gulf Coast from Corpus Christi, Texas, to New Orleans and Baton Rouge, Louisiana. The chemical industry, including oil refining, is the dominant industrial sector within this West Gulf Coast Industrial Belt. It is based on deposits of oil, natural gas, sulfur, and salt. Texas and Louisiana produce nearly one-third of the oil, over three-fifths of the natural gas, and practically all the native sulfur output of the United States, with the Gulf Coast strip the leading area of production for all three. Machinery and equipment for oil and gas extraction, oil refining, and chemical manufacturing are also important products. Major refineries and petrochemical plants are clustered at several locations, most notably metropolitan Houston and Beaumont, Texas, and Baton Rouge, Louisiana.

Many other industries are present in the South, some of which are widespread while others are localized in a few places. Food processing, machinery, furniture, and pulp and paper industries are widely distributed, with food industries especially important in Florida and pulp and paper plants particularly prominent near the Gulf Coast from eastern Texas to Florida and along the Atlantic Coast in Georgia. (For raw material, the latter industry depends largely on the pine trees that grow rapidly in the wet subtropical climate.) Aircraft and aerospace industries are prominent in the Atlanta and Dallas–Ft. Worth areas, and both auto assembly and the manufacture of auto components are expanding, especially in Tennessee and Kentucky, with some in Alabama as well.

Texas-Oklahoma Metropolitan Zone

The Texas-Oklahoma Metropolitan Zone contains the two largest urban clusters in the South—Dallas–Ft. Worth and Houston—plus four other metropolises with populations of roughly 800,000 or more: San Antonio (population: 1.6 million) and Austin (population: 1.2 million), Texas, and Oklahoma City (population: 1.08 million) and Tulsa (population: 803,000), Oklahoma. Houston is a seaport by virtue of the Houston Ship Canal, constructed inland to the city in the late 19th and early 20th centuries (Fig. 24.11 and Definitions & Insights). The city has become a major port serving Texas and the southern Great Plains in addition to being the principal control, supply, and processing center for the oil and gas fields and the chemical industry of the Gulf Coast. Dallas–Ft. Worth, San Antonio, and Austin developed at or near the western edge of the Coastal Plain in a north–south strip of territory known as the Black Land Prairie, where unusually fertile limestone-derived soils supported productive agriculture and towns and cities to service it (Fig. 24.12).

Dallas–Ft. Worth and San Antonio also serve large trading hinterlands stretching far to the west, within which there are no comparable competing centers. In the case of Dallas–Ft. Worth, the same is true in other directions except for the competition of Houston to the south. Dallas–Ft. Worth is the dominant business center for this huge region. San Antonio is

Figure 24.11 The Houston Ship Canal (also called Channel) has linked this rapidly growing manufacturing and food processing city to the Bay of Galveston and the Gulf of Mexico. The Canal was cut through the former Buffalo Bayou in 1914 and has been maintained at a depth of 34 feet and a width of 200 feet and is more than 50 miles long. Through this scale of landscape transformation, Houston has become the country's number one port. In this photo, oil tankers are loading that major Texas product for transport. Just upstream, freighters are being filled with wheat, another major Texas export product.

DEFINITIONS + INSIGHTS

The Houston Ship Canal

"Location, location, location" is one of the classic geographic lessons spoken of in popular culture. It is supposedly the answer to the question: What are the three most important factors in the value of real estate? You have read in prior pages in this text of a number of times that places have had their origins in the location of mountain passes, natural harbors, valuable resources, or bodies of water. In the siting, growth, and development of Houston, Texas, location was a major factor in early history, but it has become even more important due to a major project in landscape transformation.

Houston, founded in 1836 on Buffalo Bayou, was about 25 miles (40 km) distant from Galveston Bay in the Gulf of Mexico and initially had no port function at all. In 1873, engineers began to build the Houston Ship Canal (also called Channel) through the Buffalo Bayou to an inlet on the Bay. With the rapid development of the Houston oil fields near the end of the 19th century, and with the potential growth of Dallas–Ft. Worth and San Antonio, this southeast Texas urban center saw the need to give serious attention to making Houston a port. The Houston Ship Canal was opened in 1914 as an artificial water link some 52 miles long (84 km), approximately 200 feet wide, and maintained at a depth of 34 feet for its whole shipping length. Very unlike the much more complex Panama Canal that opened that same year, it has no locks and has been of dominant importance to the growth and economic change that has characterized Houston's growth in the past century. Before the Houston Ship Canal was opened, the city had a population of approximately 75,000 (1900). By 1930, it had grown to more than 300,000; in 1950, 600,000; and in the 2000 Census, Houston had 1,953,000 with 4.7 million metro. It became the nation's fourth largest city (after New York, Los Angeles, and Chicago) by 1980.

The Ship Canal has facilitated a major shipping role in petroleum, cotton, and other manufactured goods. The sheer force of associated urban growth in the last half century has brought engineering, education, medical and professional centers, and other widely varied economic dimensions to this site as well. The 20 people who died in unprecedented floods in Houston in 2001, with an associated damage tag of many billions of dollars because of those floods, remind us that when landscapes are modified to bring water to locations that have not developed such environmental traits naturally, there is always the risk of the "unintended consequences." On balance, however, Houston has gained enormous economic benefit from the carving of the Houston Ship Canal through the Buffalo Bayou to the active shipping lanes of Galveston Bay, the Gulf of Mexico, and beyond. At the beginning of this century, it had replaced New York as the nation's number one port in terms of shipping volume.

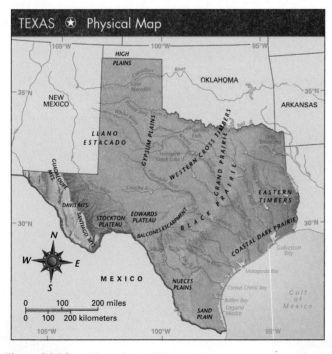

Figure 24.12   Physical map of Texas.

a major military base and retirement center, and Austin is the site of the strongly funded University of Texas, a center for high-technology industries, and the seat of Texas state government. All these cities are involved in the oil and gas industry.

Florida Metropolitan Zone

A second area of exceptional metropolitan development is the Florida Peninsula. Two Florida metropolitan areas have estimated populations of more than 2 million each—Miami (3.9 million) and Tampa (2.4 million)—and three have more than 1 million each: Orlando (1.6 million), Jacksonville (1.1 million), and West Palm Beach (1.1 million). The distribution of Florida's main metropolitan areas reflects the importance of amenities attractive to tourists and retirees. Proximity to the ocean is important, but winter warmth is even more crucial. Miami's marginally tropical winters have undoubtedly helped it to grow faster than Jacksonville in the extreme northeast. But Florida's cities are by no means totally devoted to vacation and retirement functions. Miami has become a major point of contact between the United States and Latin America in regard to air traffic, tourist cruises, and banking, as well as illegal drug traffic. Tampa Bay affords a major harbor well placed for the trade of central

Figure 24.13 In theme parks such as the Disney ones, it is not only the stunning architecture of worlds yet unbuilt but it is also the mystery of the re-creations of landscape features that we all seem to know through stories from our grandparents or from old magazine covers, textbook images, or TV or movie scenes. These features on Tom Sawyer Island remind us of very distinct shelters, transport technologies, and settlement conditions of the past. Experiencing these settings and playing roles as we cross the rope bridge to get there or ride on the old sternwheeler helps us to better appreciate the worlds in which we actually live.

Florida's citrus belt and phosphate mines. High-technology industries and regional corporate offices are increasingly attracted to Florida as the state's population grows and improved transportation and communication reduce the disadvantages of location in places remote from the Old Metropolitan Belt of the Northeast. For example, Orlando's extraordinarily rapid recent growth has been spurred by corporate office development as well as by the presence of Walt Disney World, the world's largest tourist center and tourist attraction (Fig. 24.13).

Southern Piedmont and Coastal Virginia Metropolitan Zone

The third major southern concentration of large metropolises lies along the Piedmont, the fall line, and the nearby Virginia Coastal Plain. Piedmont and fall-line metropolises include Richmond (population: 996,500), Virginia; Charlotte (population: 1.5 million), Greensboro (population: 1.3 million), and Raleigh (population: 1.2 million), North Carolina; and Greenville (population: 962,400), South Carolina. These metropolises have tended to form by the coalescing and integrating of cities located near each other, resulting in multiple-centered city clusters, although each is identified here by the name of its leading central city.

However, the two largest metropolises of this metropolitan zone differ in function from those just named. In the south, Atlanta (population: 4.1 million), Georgia, ranks third in metropolitan population within the entire South. It has diverse industries but stands out as a major inland transport center. Railroads and highways have generally avoided the Great Smoky Mountains, which lie just north of Atlanta. Instead, they skirt the southern end and converge on the city. Its ascendancy as a regional transport center today is seen in its airport, which is the busiest in the United States as measured by vol-

ume of passenger traffic (80 million) and operations size. It is followed by Chicago-O'Hare, Los Angeles International, and Dallas–Ft. Worth. These transport facilities and its superior location have been exploited to make Atlanta the largest southeastern business and governmental center.

In the north on the Coastal Plain is the seaport of Norfolk (population: 1.6 million), Virginia. Located on the spacious Hampton Roads natural harbor at the mouth of the James River, Norfolk ships Appalachian coal by sea and is the main Atlantic base for the U.S. Navy.

Metropolises of the Upland South

The entire Upland South (Appalachians, Unglaciated Southeastern Interior Plain, and Southern Interior Highlands) contains only five metropolitan areas of more than 500,000 people. Louisville (population: 1.02 million), Kentucky, began as a river port at the "Falls of the Ohio," where goods had to be transshipped around this break in transportation. It was also the closest point of contact on the Ohio River for the earliest settled and most productive agricultural area in the state: the Bluegrass Region. By the time a canal had been built to bypass the falls (in 1830), Louisville was a leading presence on the river, and it has continued to grow as a diversified regional center.

Two comparable centers, Nashville (population: 1.2 million) and Knoxville (population: 687,200), Tennessee, also began in locations associated with river transport and limestone soils. Nashville is located where the Cumberland River, a tributary of the Ohio, extends farthest south into the hilly but fertile Nashville Basin of central Tennessee. Knoxville is situated far up the Tennessee River from the Ohio in an area of limestone valleys within the Ridge and Valley Section of the Appalachians.

The same themes of transport advantages, relatively good soils, and growth as the major center for a considerable region are repeated in the case of Little Rock (population: 583,900), Arkansas. In this case, the exceptional land lies mainly in the Mississippi Alluvial Plain to the immediate east and secondarily in the Arkansas Valley to the west. The valley also provides a lowland route channeling traffic through the Interior Highlands, and the Arkansas River flows into the Mississippi.

The major exception, however, to the general pattern of urban location and development in the Upland South just described is Birmingham (population: 921,100), Alabama, which is the only large center of iron and steel industry in the South. Its development in the Ridge and Valley Section of the Appalachians was based on large deposits of both coal (from the nearby Cumberland Plateau) and iron ore.

Other Southern Metropolises

Other southern cities with metropolitan populations of more than 500,000 are New Orleans (population: 1.3 million) and Baton Rouge (population: 602,894), Louisiana; Memphis (population: 1.1 million), Tennessee; Charleston (population: 549,000), South Carolina; and El Paso (population: 679,600), Texas. The first three are Mississippi River ports. New Orleans was founded by the French just upstream from the head of the "bird's foot" delta where the Mississippi divides into separate channels flowing to the Gulf. In the riverboat era before the Civil War, New Orleans was New York's main competitor as a port. It then underwent relative decline as a port in the railway era, but it remains a major seaport, especially for shipping grains and soybeans from the Midwest and the Mississippi Alluvial Plain, as well as a major tourist center. Somewhat upriver, Baton Rouge is accessible to medium-sized ocean tankers and is a major center of the Gulf Coast oil refining and petrochemical industries. Farther upriver, Memphis developed as a river port where the Mississippi flows adjacent to higher ground at the east side of its broad floodplain. The city's function as the main business center for the greater part of the Mississippi Alluvial Plain (and other nearby areas) dates back to antebellum days when a major sector of the "Cotton Kingdom" developed in "The Delta" — a regional name then in use for the large part of the Mississippi Alluvial Plain adjacent to Memphis in northwestern Mississippi.

The remaining metropolises lie at opposite ends of the South. Charleston was the port and commercial capital of the first major plantation area outside Virginia. It was the leading city of the South at the time of the American Revolution, and then languished as the plantation economy moved westward, but has resumed rapid growth with the rise of the modern southern economy. El Paso serves as a major railway center and tourist and business entrance to Mexico. It lies across the Rio Grande from Ciudad Juarez, which also serves as a major transportation center for northern Mexico.

24.3 The Midwest: Agricultural and Industrial Heartland

The existence of a Midwest (Middle West) in the United States is widely recognized, but perceptions differ as to its extent. In this chapter, the term Midwest is applied to 12 states called the North Central states in federal publications. They are commonly subgrouped into the East North Central states of Ohio, Indiana, Illinois, Michigan, and Wisconsin and the West North Central states of Minnesota, Iowa, North Dakota, South Dakota, Nebraska, Kansas, and Missouri. The two subgroups are referred to here as the eastern Midwest and the western Midwest. The Mississippi River separates them except for part of the Wisconsin-Minnesota boundary (see Figs. 24.1 and 24.3). All states in the eastern Midwest border one or more of the Great Lakes, but only Minnesota does so in the western Midwest. In the eastern Midwest, Ohio, Indiana, and Illinois are bounded on the south by the Ohio River; in the

Figure 24.14 Rivers, barges, bridges, ceremonial architecture, urban parkland, persistent centrality, and low-lying sprawl are part of the fabric of the Midwest and urban America. This aerial photograph of St. Louis, Missouri, and the Gateway Arch shows them all, stretching out from the banks of the Mississippi River. The Arch is a 63-story stainless steel structure intended to reveal this city's role as a major departure place in the movement west—beyond the Mississippi, across the state, and eventually across the continent—for the country's ever westward-moving population. The 2000 population center for the United States is some 100 miles (160 km) southwest of this scene.

western Midwest, all states except Minnesota front on the Missouri River (Fig. 24.14).

The use of the term "Midwest" for these 12 states is arbitrary to some degree but can be justified by spatial (mappable) characteristics distinguishing them as a group from the Northeast, South, and West. Among these characteristics are the Midwest's (1) interior "heartland" location; (2) distinctive plains environment; (3) important associations with major regional lakes and rivers; (4) ethnic patterns; (5) outstandingly productive combination of large-scale agriculture, industry, and transportation; and (6) rural landscapes famed for their rectangularity, symmetry, and prosperous appearance.

Midwestern Agriculture and Its Environmental Setting

These vast midwestern plains produce a greater output of foods and feeds than any other area of comparable size in the world. The Midwest commonly accounts for more than two-fifths of all U.S. agricultural production by value. This output is unevenly spread among the midwestern states, with Iowa leading. Within the nation, Iowa is the third-ranking state in value of farm products, exceeded only by much larger California with its irrigation-based agriculture and by Texas. Ordinarily, five midwestern states — Iowa, Nebraska, Illinois, Kansas, and Minnesota — are among the nation's top seven in total agricultural output, and eight midwestern states are among the top fifteen. The lowest ranking midwestern states are Michigan, handicapped by large areas of infertile sandy soil, and the Dakotas, handicapped by dryness, rough land toward the west, and a shorter frost-free season.

Within most of the Midwest, soils are exceptionally good. The best soils developed under grasslands, which supplied abundant humus. Such outstandingly fertile soils normally accompany steppe climates in the middle latitudes, where the fertility is often counterbalanced by precipitation so low as to be marginal for agriculture. However, the remarkable feature of the Midwest is that its soils, which are located in a more humid continental climatic area, still developed under a tall-grass prairie. Such climatic conditions normally produce a natural vegetation of forest, and trees grow well if they are planted. But in the early 19th century, the settlers found large expanses of prairie, or a prairie-forest mixture, as far east as northwestern Indiana. From there, the prairie region extended west in a broadening triangle to the eastern parts of the Dakotas, Nebraska, and Kansas and on into the South in Oklahoma and Texas.

Of the midwestern states, only Michigan and Ohio lacked considerable expanses of prairie. Thus, a very large part of the region had soils of a richness generally found only in semiarid areas but which in the Midwest were combined with humid conditions more favorable to most crops. The existence of the anomalous triangle of prairie bordered by forests both north and south can be accounted for in two main ways. One explanation holds that the prairie was produced by periodic drought conditions, discouraging tree growth in this normally humid area. The other, and more likely, explanation suggests the use

of fire by Native American populations to improve hunting conditions and create grazing range for game. Both factors were assuredly important, and prairie fires set by lightning also may have helped perpetuate the prairie once it was established (grass is less susceptible than tree seedlings to lasting damage by fire). Whatever the causes, the occurrence of grassland soils in a humid area where forests might normally be expected provided an extraordinarily favorable agricultural environment.

Another natural condition contributing to the unusual excellence of many midwestern soils is the widespread presence of loess deposits made of dust particles. These wind-borne soil materials are thought to have accumulated in glacial times when strong winds picked up the particles from dry surfaces where the ice had melted but where no protective mantle of vegetation had yet become established. Because most of the nutrients that nourish plants are contained in these finer soil particles, loess is associated with soils of exceptional fertility in various parts of the world.

In the eastern Midwest, the soils formed under natural broadleaf forests are not exceptionally fertile by nature, but neither are they poor (see Fig. 24.5). The principal areas of poor soil in the Midwest lie in (1) the uplands of the Ozarks and the Appalachian Plateau; (2) sandy areas of needleleaf forest in northern Michigan, Minnesota, and Wisconsin; and (3) rougher western parts of the four westernmost states. After long use, the present quality of midwestern soils depends heavily on how well they have been handled. Where properly fertilized and protected against depletion and erosion, they have proved remarkably durable. Although unwise practices have caused much soil damage, huge expanses remain highly productive. Aided by science and technology, the soils of the Midwest as a whole have higher yields today than ever before.

The Legendary American Corn Belt Becomes the Corn-Soybean Belt

Corn is the single leading product of midwestern farms, and the Corn Belt, where this crop dominates the landscape, has become an agricultural legend for its productivity. Sales of beef cattle in the Midwest are actually greater than receipts from corn, but the cattle are raised primarily on corn. Over four-fifths of American corn production and around two-fifths of world production come from the Midwest. Highly favorable conditions for large-scale corn growing are provided by fertile soils, hot and wet summers, and huge areas of land level enough to permit mechanized operations. The corn plant itself and the methods of production have been greatly improved by extensive research.

Soybeans are another major element in Corn Belt agriculture (Fig. 24.15). Long cultivated in China, they have had a rapidly increasing impact on American and world agriculture during recent decades. In World War II, it was learned that oil of the bean could be used in the manufacture of paints, soap, glycerin, printing ink, and many other chemical-based products. The growth of the bean also adds nitrogen to the soil and provides livestock feed. The oil is now also being experimented

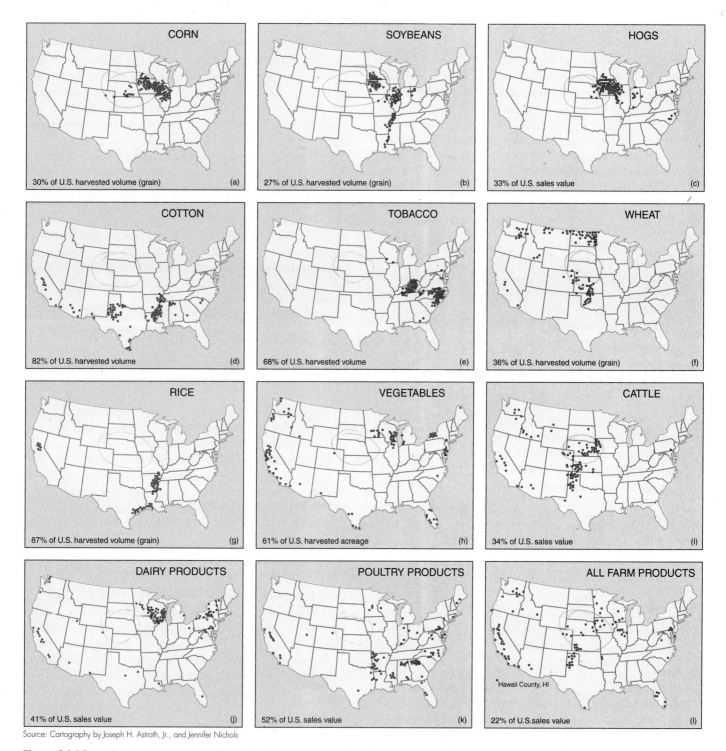

Source: Cartography by Joseph H. Astroth, Jr., and Jennifer Nichols

Figure 24.15 These twelve maps show the 100 leading U.S. counties for selected crops and farm products. Each map also tells the reader what percentage of the U.S. total production of this commodity comes from the aggregate of these 100 counties (except for rice, where only the top 50 counties are shown).

with as a fuel additive (similar to the one derived from corn) for midwestern gas stations. Per capita annual consumption in the United States has grown to 440 pounds, a clear indication of the growing importance of this unusual resource. Soybeans have recently turned the Corn Belt into a region that could more realistically be named the Corn-Soybean Belt.

The Basics of Midwestern Agriculture

Despite its impressive resources and accomplishments, midwestern agriculture is a risky enterprise. Natural hazards such as drought and plant diseases increase the normal business risk, but the main hazards are sharp fluctuations in world supply and demand and therefore in market prices. These forces have driven waves of farmers out of business at various times in the past. Between 1930 and 1997, for example, the number of farms in the Midwest decreased from slightly over 2 million to approximately 749,000. Most of the land of defunct farms was bought or rented by surviving farmers who expanded their acreage. The amount of land in farming decreased only 6%, but the mean size of farms increased from about 180 acres (73 hectares) to about 457 acres (185 hectares). Increased mechanization allowed farmers to work larger acreages but also increased their capital costs and hence increased their borrowing and their household vulnerability to low market prices. Many of the farmers who left farming in the past two decades were victims of excessive indebtedness due to enormous machinery costs. And as U.S. farms become fewer, larger, and more highly capitalized (in the Midwest and elsewhere), corporate forms of farm ownership and enterprise are increasingly evident.

Agricultural Foundations of Midwestern Industry

Midwestern agriculture has been, and remains, one of the main foundations of regional industry. Meat packing, grain milling, and other food processing industries have always been of major importance. Cincinnati first developed as an early hog-packing "Porkopolis" before the Civil War, and Carl Sandburg's poem "Chicago" celebrated that city's role as the "hog butcher for the world." The center of the midwestern packing industry has moved westward with time to a present location in and near east central Nebraska and western Iowa. The same westward migration also has been true of wheat milling, in which Minneapolis–St. Paul is now a leader. In addition to processing agricultural products, midwestern industry supplies many needs of the region's farms. Many technical advances in farm equipment (e.g., Cyrus Hall McCormick's 1840s mechanical reaper for harvesting wheat) originated in the Midwest. The world's largest farm-machinery corporations (John Deere, Caterpillar, Navistar) are headquartered in the region, and many midwestern cities are heavily dependent on the industry. This activity has helped foster the development of a broad range of other types of machine manufacture. By the mid-1990s, St. Louis had become second only to Detroit in the manufacture of automobiles.

The Enormous Automobile Industry

The availability of steel and a central location vis-á-vis markets were factors contributing to a heavy concentration of the U.S. automobile industry in Detroit and nearby midwestern cities. When the industry began to develop during the early 20th century, production on a very large scale was necessary to achieve costs and prices low enough for mass sales. This required that production be centered in a few large plants and the vehicles marketed throughout the country. Steel was needed in huge quantities, as was a location from which distribution could reach the whole national market relatively cheaply (Fig. 24.16). Detroit was nearly optimum for the lat-

Figure 24.16   The process of globalization has been an engine of industrial displacement of the United States. This photo is of a demonstration in Pittsburgh, Pennsylvania, put on by steelworkers and their families from Pennsylvania and Ohio. The goal was to dissuade the owners of the Hazelwood Coke Works of LTV Corporation from closing down this last steel plant in Pittsburgh. At the end of the 19th century, this city was one of the leading cities in the world in the production of steel. A century later, it was found to be in violation of the Clean Air Act, and the Environmental Protection Agency set the conditions for keeping the plant open. The world steel market has so changed and grown to favor new plants and new plant technology that LTV decided not to contest the EPA ruling, and the plant was closed. The woman in this photo was the wife of a man who had been a steelworker for 46 years, and she—and he—knew that there was no other industrial life he could turn to. Seven hundred fifty union employees lost jobs in the closing of the Hazelwood coking plant.

ter, as the national center of population at that time was in southern Indiana (it has now moved west to central Missouri), and buying power was higher toward the north and east. Detroit labor and management had a background in building lake boats and horse carriages, and experiments with internal combustion motors had been carried on by various inventors, including Henry Ford, a Detroit native.

From southeastern Michigan, the auto industry spread over much of the eastern Midwest: north to Flint and Lansing, east to Cleveland, south to Dayton and Indianapolis, and west to the Milwaukee area. Many cities within this circumference became very dependent on supplying auto parts. Assembly itself expanded to St. Louis and Kansas City and eventually to other major regions of the country. The American corporate giants of the industry—General Motors, Ford, and Chrysler (now Daimler-Chrysler)—still have headquarters and major plants in metropolitan Detroit. As in the case of steelmaking, the motor vehicle industry has provided much of the Midwest with a basic industry of great size, vital to employment, and paying high wages. However, like the steel industry, it became a focus of employment problems in the 1980s because of decreasing plant productivity and growing sales of competing operations overseas, especially from Japanese plants. This industry has a history of boom and bust cycles, and it is now having a vigorous resurgence, but its operations have diffused south and west to many additional cities, both in the Midwest and in other regions.

Regional Unity of Midwestern City Complexes

598 A large majority of the Midwest's population is metropolitan. As in other parts of the United States, an increasing amount of commuting has been integrating more and more counties with the cities at or near the centers of metropolitan areas. Hence, most large metropolitan areas are now sprawling multicounty districts. These are essentially urban in employment and function, even though outlying sections may not be highly urban in appearance. Even including the world of small towns and farm communities in the outlying rural landscapes, there is a regional unity of these midwestern urban centers that is both economic and cultural. Farm families go to the regional malls, and city folk go out to the farmers' markets and seasonal fairs.

The larger midwestern cities around which the main metropolitan areas have clustered had their beginnings in the early 19th century or occasionally earlier. Water transport was relatively more important then than it is now, and city locations tend to be related to lake and/or river features that conferred early importance in a developing pattern of water transportation and human settlement. Even today, most major centers still benefit from advantageous locations for freight movement by water. When railroads came, the cities that were already established generally became rail centers as well. Then, in the mid-20th century, highways and airways also focused on the major existing cities.

Great Lakes Metropolises: Chicago, Detroit, Cleveland

The three largest midwestern metropolitan concentrations are around Chicago, Illinois, Detroit, Michigan, and Cleveland, Ohio (see Fig. 24.1). By far the largest is Chicago, the country's third largest city (population: 9.2 million). The city began at the mouth of the small Chicago River, which afforded a harbor near the head of Great Lakes navigation from which the river projected farther into the prime farmlands of the Midwest. A few miles up this river, a portage to tributaries of the Illinois River provided a southward flowing connection via the Illinois to the Mississippi River near St. Louis. A canal built in 1848 along the portage route connected Lake Michigan with the Illinois River and helped Chicago become the major shipping point for farm products moving eastward by lake transport from the rich farming areas in the prairies to the city's south and west. Then, in the 1850s, railways fanned out from Chicago into an ever broader hinterland and also connected the city with Atlantic ports. Chicago became the country's leading center of railway routes and traffic and has remained so. In the 20th century, new harbors were built south of the city center, and new large-scale canal connections facilitated an increasing barge traffic using the Illinois River. Hence, at

this major focus of transport and population growth, Chicago's huge industrial structure rose. Several major industries developed very early: food processing, especially meat packing (which has now almost disappeared); wood products made from white pine and other timber cut in the forests to the north; clothing; agricultural equipment; steel; and a broad range of machinery manufacturing. Today, the most important lines of manufacturing are machinery and steel. Chicago is the leading center for both in the country. The city's banks, corporate offices, commodity exchanges, and other economic institutions make it the overall economic capital of the Midwest. To the east, its metropolitan area reaches into northwestern Indiana, and to the north, its area coalesces with that of Milwaukee (population: 1.7 million), Wisconsin, which also has grown around a natural harbor on Lake Michigan.

The second greatest concentration of people and production in the Midwest is in metropolitan Detroit, together with the smaller adjoining area of Toledo, Ohio. The Detroit area extends along the water passages separating Michigan from the Ontario Peninsula in Canada and connecting Lake Huron with Lake Erie. From north to south, these passages include the St. Clair River, Lake St. Clair, and the Detroit River (see Fig. 24.1). Detroit's metropolitan population is 5.5 million, and the adjacent Toledo metropolis contains 618,200 more. A closely connected Canadian population in metropolitan Windsor, Ontario, just across the Detroit River from Detroit, adds another 278,000 to the metropolitan complex.

Detroit was founded as a French fort and settlement along the Detroit River in 1701. It grew as a service center for a Michigan hinterland that was settled rapidly following completion of the Erie Canal in 1825. It remained modest in size until explosive growth took place during the automobile boom of the 20th century. Toledo was founded on a natural harbor at the western end of Lake Erie. In the 19th century, natural gas and oil from early fields in western Ohio furnished fuel for a glass industry there, and the metropolitan area now produces much of the glass for Detroit automobile production.

The fourth-ranking metropolitan concentration in the Midwest is in northeastern Ohio, where the Lake Erie port of Cleveland has a metropolitan population (including nearby Akron) of 2.9 million. Cleveland developed in the 19th century as a center of heavy industry on a harbor at the mouth of the Cuyahoga River. Ore from the Lake Superior fields was offloaded there for shipment to Pittsburgh, and there was a return flow of Appalachian coal for Great Lakes destinations. Smaller Lake Erie ports also participated in this traffic. Movement of ore one way and coal the other made it logical to develop the iron and steel industry at Cleveland, at other Lake Erie ports, and at Youngstown, Ohio, on the route between Pittsburgh and the lake. Machinery and chemical plants developed, and eventually, the auto industry spread into the district. Part of the latter development was the rise of Akron as the leading center of tire manufacturing in the United States. Today, northeastern Ohio is afflicted by the slow growth, contraction, or closure of many older industrial enterprises.

River Cities: St. Louis, Minneapolis–St. Paul, Cincinnati, Kansas City, Omaha

The remaining major metropolises of the Midwest originated at strategic places on the major rivers that were early arteries of midwestern traffic. These cities still benefit from riverside locations, although other forms of transportation have now relegated river traffic to a subordinate position.

St. Louis, Missouri, is the center of a metropolitan population of about 2.6 million in Missouri and Illinois. It was founded by the French in the 18th century and remained the largest city in the Midwest until the later 19th century. It is located at a natural crossroads of river traffic using the Mississippi to the north and south, the Missouri to the west, the Illinois toward Lake Michigan, and (via a southward connection on the Mississippi) the Ohio toward the east. Early traffic had to adjust to the fact that the Mississippi downstream from the Missouri mouth was notably deeper than the upstream channel or the Missouri. Hence, many riverboats plying the Mississippi below St. Louis drew too much water to use the channels above the city, and St. Louis became a break-of-bulk point where goods were transferred between different kinds of craft. Coal from southern Illinois was available to support industrialization, and the metropolitan area eventually developed an industrial structure noted for its variety (see Fig. 24.14).

Minneapolis–St. Paul (population: 3.0 million), Minnesota (now the third largest metropolitan area in the Midwest), developed at the head of navigation on the Mississippi River. The Falls of St. Anthony, marking this point, supplied water power for early industries, especially sawmilling. Then, transcontinental rail lines were built west from these "Twin Cities" to the Pacific. The spread of business westward along these lines made Minneapolis–St. Paul the economic capital of much of the northern interior United States. No competing center of comparable size emerged between Minneapolis–St. Paul and Seattle. The business hinterland of the Twin Cities thus stretches across the Spring Wheat Belt to the foot of the Montana Rockies. Minneapolis-St. Paul has long been a major focus of the grain trade and flour milling and now has a diversified industrial structure. It is also the home to the nation's largest retail mall — The Mall of America — which has a regular flow of international tourists who come to the Twin Cities largely for shopping there.

A much older river city is Cincinnati (population: 2.0 million) on the Ohio River. Its metropolitan area includes the southwest corner of Ohio and adjacent parts of Indiana and Kentucky. For a time, its situation resembled Minneapolis-St. Paul in that it dominated a large hinterland to the west and north. The Ohio River turns sharply southward from the city, and early 19th-century settlers floating down the river often disembarked at Cincinnati to move overland into Indiana. The city quickly developed a major specialty in slaughtering hogs from Ohio and Indiana farms and shipping salt pork downriver to markets in the South. This early meat-packing function eventually was lost, but coal mining developed in the nearby Appalachian Plateau, and Cincinnati became the economic capital for much of the mining region. It also developed a diversified and relatively stable industrial base, built in part around tool and die machine-working for the region's industrial centers.

Kansas City, Missouri (population: 1.8 million in Missouri and Kansas), strongly resembles Minneapolis–St. Paul in its situation and development. It began at what was essentially the head of navigation westward on the Missouri River. Although river transport was possible upstream, it was subject to navigational hazards and led sharply northwest instead of directly westward. Consequently, two of the major overland trails to the West — the Santa Fe Trail and the Oregon Trail — had their main eastern termini and connection with river transport in the Kansas City area. Later on, cattle drives came to Kansas City from the Great Plains, and railways reached westward from the city. The Pony Express played out its 1 year of history in the early 1860s from St. Joseph, just north of Kansas City. Today, despite a major competitor to the west in Denver, Colorado, Kansas City has a hinterland extending far to the west and is the economic capital of the Winter Wheat Belt. Omaha, Nebraska (population: metropolitan area, 717,000 in Nebraska and Iowa), is a smaller riverside city resembling Kansas City and Minneapolis–St. Paul in its functions and westward reach. Like them, Omaha is in the western edge of the intensively farmed part of the Midwest, with drier, less productive wheat and cattle country to the west.

Two Demographic Shifts of Note: The Middle-sized City and the Small Town

The middle-sized city in the United States (750,000–2,000,000 in population) is taking on a new importance in American migration and development patterns. Cities of this general size — and Cincinnati, Ohio, is a prime example — have emerged in the past decade as target destinations for both industry and finance. The blend of established urban infrastructure and generally well-educated labor pools that are not as expensive as those resident in the really large metropolitan areas has stimulated steady growth. The ever-expanding capacity of telecommunications to link virtually any node with other nodes has reduced the traditional importance of, for example, financial firms locating in the central city. The network afforded by computers, fax machines, the Internet, and expanding local airline links between middle-sized cities and large urban centers has changed the geography of business location.

There are often two critical population blocks resident in or willing to move to middle-sized cities for solid employment. First are the semiskilled high school and college graduates or near graduates who are willing to take beginning work in firms at annual salaries of $20,000–$40,000. As is the case so often in migration to the city, they will take such jobs on the hope that they will either advance in a Proctor & Gamble or CitiCorp or put themselves in a good position to take advantage of

Figure 24.17   With analysis of the results of the 2000 Census, it is clear that a continuing demographic shift of great stability is the ongoing movement of population from the American Northeast to the southwestern corner of the country. Census takers from all across the country gathered in the facts that tell us cities like Las Vegas and Phoenix and even still Los Angeles are growing at rates that are unmatched in the Chicago and Baltimore and New Orleans landscapes.

ANDY NELSON, CHRISTIAN SCIENCE MONITOR. MAY 9, 2000.

another opportunity in that city. Smaller cities often lack the "pull" to bring in growth firms because they do not have the urban amenities in music, museums, shopping, and schools that are found the larger urban places. At the same time, these middle-sized cities do not generate the images of crowding, crime, environmental decline, and gridlock that really large cities often do.

The other population block in middle-sized cities is the well-educated young professionals who find strong enough amenities that they will accept a somewhat lower salary as a tradeoff for the urban characteristics of these cities. Professional sports complexes, energetic renewal of traditional downtown landscapes with trendy cafés and restaurants, and expanding airport facilities are often major aspects of this growth, even in these middle-sized urban centers. These are often complemented by the willingness of city administrators to give incoming firms a sweet deal on land, taxes, and transportation so that business moves to the smaller urban center looks more attractive to the major firms that have grown up generally in much larger cities. These characteristics are increasingly influential in the Midwest, and they are occurring in the Northeast and South as well.

The second demographic shift that is often associated with the Midwest but that in fact is part of population changes all across the United States is the new attention being given to the small town. For example, a statistic of real magnitude is this simple dimension of change: Between 1900 and 2000, the American population more than tripled. That is, from a national population of 76.2 million in 1900, the United States grew to a total population of more than 280 million nine decades later. In that same nine decades, however, 8 of 10 counties all across America had lost population. Or 2,400 counties had fewer people in 1990 than they had in 1900. What can this statistic mean? Where did all of the people who made up the threefold national growth go? They had gone to the major metropolitan centers that we have been talking about in this chapter, and they went to cities lying to the west. What they left was the American small town (Fig. 24.17).

Geography is at the heart of this extraordinary demographic shift. Small-town settlements had emerged as vital clusters of merchants and townsfolk located on some avenue of transportation — the highway, railway, river, or a combination. These clusters were surrounded by the broad, open farmlands being cleared and farmed in the late 18th and 19th centuries. Dirt roads made moving farm produce to local markets, to train stations, and to riverside wharfs difficult, so people quickly learned that there was a geographic wisdom for businesspeople and farmers to locate close together. The small towns that became county seats had a particular advantage because they were the home of local government as well. Particularly fortunate towns became home to food processing firms, shoe or textile factories, or any number of small-scale industries that added a factory payroll to these minor urban centers.

Then, in the steady flow of landscape transformation so central to the expanding settlement and landscape transformation of the United States, roads began to be surfaced so that they could function in all seasons. Some gravel and blacktop roads were replaced by concrete and were made into four lanes, while many were left as two lanes. The larger roads with better surfaces led into hubs of nodality in ever larger settlements. Smaller towns not only lost economic centrality, but they also lost residential utility as more and more people moved to the larger centers where "things seemed to happen." Now, as farm families get into their sedans and drive through the increasingly empty streets of the small towns that lie between their farmsteads and the regional shopping centers or massive discount stores that are their destination for shopping or fast food or entertainment, they become witnesses to a really monumental demographic and cultural change that has left Main Street unused and small towns tired and increasingly empty.

This shift is broadly regional. Distribution of the 80% of American counties that have lost population in the past century is scattered through all of the states. The 20% that have gained—and many have gained enormously—are nested around the major urban centers that have been at the heart of decades of urban growth and sprawl.

However, there is another side to this geographic equation. In the past decade, there has been a significant out-migration of urban professionals from large urban centers. The numbers have not been major yet, but these reverse pioneers are trying to see if they can take up residence in small-town America and still maintain their professional and career links with the major cities in which their offices exist. Can networks of phone lines, fax machines, the Internet, and occasional trips back to the big city substitute for full-time residence in the major metropolitan area? And if so, can residence in small towns with their lower prices for shelter, their homier atmosphere, and their continued link with the American past, a golden age of simpler lifestyles, satisfy the needs of these "refugees" from the big city?

Like so many aspects of the human environment, this demographic shift toward the small town is not quite one thing or another. This new pattern may in fact help return vitality to the small towns that were so much at the heart of American culture a century ago. But the vitality will be of a different nature than that which provides our images of the past. The beauty of this speculation is that you—the very student reading this text—will be able to explore the merits of both the big city and the small town yourselves as you decide where it is you settle after graduation from college and see which built environment works the strongest pull on you. After all, you have been born into a generation that, because of fiberoptic networks, will have a greater range of possible living situations than any prior American generation. Either the middle-sized city or the small town may be the landscape setting you choose for your own career and livelihood!

24.4 The West: Water at the Heart of Everything

Eleven states make up the West (see Figs. 24.1, 24.3, and 24.18). The eight interior states (New Mexico, Arizona, Colorado, Utah, Nevada, Wyoming, Montana, and Idaho) are

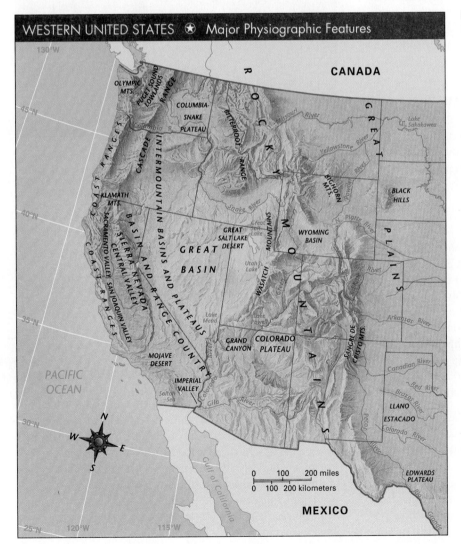

Figure 24.18   Major physiographic features of the western United States.

WESTERN UNITED STATES ✪ Major Physiographic Features

often termed the Mountain states or Mountain West. Farther west are the three Pacific coast states of California, Oregon, and Washington. The latter two states are often termed the Pacific Northwest. Federal publications include Alaska and Hawaii in statistics for the Western states, but these two states are so different that they are discussed in a separate section and are not included in generalizations about the West.

Thus defined, the West comprises 49% of the United States by area, or 38% of the conterminous 48 states. It contains large percentages of land owned by the federal government and accommodates major military facilities all across its predominantly arid landscapes. The region is still lightly settled except for a huge cluster of people and development in central and southern California and a handful of smaller clusters in other widely dispersed locations. In 2000, the West had 61.4 million people (a little more than one-fifth of the nation), 33.9 million of whom lived in California. In 2000, it accounted for 25% of the nation's agricultural output and about 18% of its value added in manufacturing. California is the West's economic colossus, producing almost 57% of the region's manufactured goods and one-half of its farm products. The eight Mountain states contrast sharply with California, with 18.2 million people in 2000 (6.5% of the national population) in eight states with large areas. In 2000, these states produced 13% of the country's agricultural output and 8% of its value added by manufacturing.

Challenges to Development in the West

The generally low intensity of development in the West results principally from dry climates, rough topography, and absence of water transportation except along the Pacific coast. Because settlement is clustered in places where water is adequate, the population pattern is oasislike, with oases separated from each other by little-populated steppe, desert, and/or mountains (Fig. 24.19). The climates that so affect Western settlement are strongly related to topography (see Fig. 24.18). Precipitation in this region is largely orographic, and dryness is largely due to rain shadow. That dryness is caused by being on the relatively dry side (the leeward side) of north–south trending mountains that block the moisture-bearing air masses that come from the west. This is not only a strong regional characteristic but also a source of continuing threats from wildfires, floods, landslides, and droughts. Thus, an understanding of topographic layout is needed to understand the climatic conditions that either handicap or facilitate settlement. If the mountains of the West ran east and west, the environmental settings would be quite different.

Western Topography and Its Influence on Climate

The Pacific shore is bordered by mountains. The relatively low Coast Ranges provide an often spectacular coastline from the Olympic Mountains in Washington to just north of Los Angeles, where the mountains swing inland and then con-

Figure 24.19 This aerial shot of center pivot irrigation in Nebraska provides a stunning visual of the forces of farming and technology. In the arid West and Great Plains, deep wells are sunk, and on the surface of the land, a mobile irrigation rig is attached to a pump. Well water is brought up steadily and run through long pipes of a wheeled rig that pivots around the wellhead, distributing water across parcels of land that are 160 acres (one-quarter section) large. Occasionally, farmers will buy add-on machinery that they put at the end of the wheeler machinery, and these distribute water to the interstitial spaces that do not receive water with the adjacent circular central pivot pattern. This is a capital intensive but labor extensive kind of irrigation, and it makes the arid lands relatively productive. The popularity of this type of irrigation and farming has caused a steady drop in water table levels, causing farmers and nearby urbanites to worry about the future of this extensive irrigation.

tinue south to Mexico. Inland, a line of truly high mountains parallels the Coast Ranges. The northernmost of these inland ranges is the Cascade Mountains, extending southward from British Columbia across Washington and Oregon to northern California. The Cascades are surmounted by many volcanic cones, among which Mt. Hood (near Portland) reaches over 11,000 feet and Mt. Rainier (near Seattle) over 14,000 feet. In northwest Washington is Mt. Saint Helens volcano which, in its major explosion in 1980, killed 57 people and lost 1,300 feet from its height. At its southern end, the Cascade Range joins the rugged Klamath Mountains, a deeply dissected

GEORG GERSTER/PHOTO RESEARCHERS, INC.

LANDSCAPE IN LITERATURE

"There's Nothing Out There"

One of the most vital goals in learning geography is learning to observe, to see what lies before you on the horizon. This selection from William Least Heat-Moon's *Blue Highways* (1982) does a powerful job in making your senses reach out and observe all that is around you. Heat-Moon thinks up this exercise in an arid part of Texas. As you read this passage, think about how you might conduct a similar experiment in your own environment.

Straight as a chief's countenance, the road lay ahead, curves so long and gradual as to be imperceptible except on the map. For nearly a hundred miles due west of Eldorado, not a single town. It was the Texas some people see as barren waste when they cross it, the part they later describe at the motel bar as "nothing." They say, "There's nothing out there."

Driving through the miles of nothing, I decided to test the hypothesis and stopped somewhere in western Crockett County on the top of a broad mesa. . . . I made a list of nothing in particular:

1. mockingbird
2. mourning dove
3. enigma bird (heard not saw)
4. gray flies
5. blue bumblebee
6. two circling buzzards (not yet, boys)
7. orange ants
8. black ants
9. orange-black ants (what's been going on?)
10. three species of spiders
11. opossum skull
12. jackrabbit (chewed on cactus)
13. deer (left scat)
14. coyote (left tracks)
15. small rodent (den full of seed hulls under rock)
16. snake (skin hooked on cactus spine)
17. prickly pear cactus (yellow blossoms)
18. hedgehog cactus (orange blossoms)
19. barrel cactus (red blossoms)
20. devil's pincushions (no blossoms)
21. catclaw (no better name)
22. two species of grass (neither green, both alive)
23. yellow flowers (blossoms smaller than peppercorns)
24. sage (indicates alkali-free soil)
25. mesquite (three-foot plants with eighty-foot roots to reach water that fell as rain two thousand years ago)
26. greasewood (oh, yes)
27. joint fir (steeped stems make Brigham Young tea)
28. earth
29. sky
30. wind (always)

That was all the nothing I could identify then, but had I waited until dark when the desert really comes to life, I could have done better. To say nothing is out here is incorrect; to say the desert is stingy with everything except space and light, stone and earth is closer to the truth.[a]

[a]William Least Heat-Moon, *Blue Highways: A Journey into America.* Boston: Little, Brown and Company, 1982, 148–150.

plateau where the Cascades, Coast Ranges, and the Sierra Nevada Range almost meet. The Sierra Nevada Range increases in elevation southward to the "High Sierra," where Mt. Whitney, at 14,494 feet (4,418 m), is the highest peak in the 48 conterminous states.

Two large lowlands lie between the Coast Ranges and the inner line of Pacific mountains, and a smaller but highly important lowland lies outside the Coast Ranges in the extreme south. In Washington and Oregon, the Willamette-Puget Sound Lowland separates the Cascades from the Coast Ranges. The Oregon sector, extending from the lower Columbia River to the Klamath Mountains, is called the Willamette Valley. The Washington sector, north of the Columbia, is the Puget Sound Lowland, which is deeply penetrated by the deepwater inlet from the Pacific called the Puget Sound. The lowland surface in both states is generally hilly. The second large lowland is the Central Valley, or Great Valley, of California, enclosed by the Klamath, Sierra Nevada, and Coast Ranges or by outliers that seem to weave the systems together. The Central Valley is a practically flat alluvial plain, some 500 miles (c. 800 km) long by 50 miles (80 km) wide. In California's far southwest, a smaller lowland lies between the

coastal mountains and the shore. The wider northern part of this lowland is the Los Angeles Basin, and a narrower coastal strip extends southward to San Diego and Mexico. The Los Angeles Basin is a plain dotted by isolated high hills and overlooked by mountains.

Precipitation is brought to this terrain by air masses moving eastward from the Pacific Ocean. This air movement occurs more frequently in the winter than in the summer, so summer is much the drier season. It also occurs more frequently and generates more precipitation in the north than in the south. Consequently, summer conditions are desertlike in the south, although winter brings some rain and mountain snow. The mountains of the north receive the most precipitation, including the deepest snows in the United States in the higher areas. South of San Francisco, California, the mountains of most areas receive only modest amounts of precipitation at best, and the magnificent Douglas fir and redwood forests of the north give way to the sparse forests and chaparral (drought-tolerant grasses and shrubs) that characterize the southern half of California.

The lowlands of this Pacific strip vary from quite wet to arid. The Coast Ranges of Washington and Oregon are mostly low enough, and the incoming moisture is great enough, that the Willamette–Puget Sound Lowland is not severely affected by its position on the lee side of the mountains. Hence, this lowland is wet and naturally forested. Because its temperatures are moderated in both winter and summer by incoming marine air, this area has a marine west coast climate similar to that of Western Europe. The Central Valley of California is affected more by rain shadow cast by the Coast Ranges. Although generally classed as having a Mediterranean climate — with subtropical temperatures, wet winters, and dry summers — this valley is dry enough that its natural vegetation was steppe grass for the most part, grading into desert shrub in the extreme south. The Los Angeles Basin also has a climate usually classified as Mediterranean subtropical, but the precipitation is so low that a steppe or even a desert classification could be argued.

East of the high mountains rain shadow becomes extreme. A vast area reaching east to the Rockies and around their southern end in New Mexico is predominantly desert or steppe. Little moisture is received from air masses that already have provided precipitation to the mountains near the Pacific. This region is known physiographically as the Intermountain Basins and Plateaus. It includes large parts or all (in the case of Arizona) of nine Western states (see Figs. 24.1 and 24.3).

Three major physical subdivisions of the Intermountain Basins and Plateaus are commonly recognized. The Columbia–Snake River Plateau occupies most of eastern Washington and Oregon and southern Idaho. It is characterized by areas of relatively level terrain formed by past massive lava flows, together with scattered hills and mountains and deep-cut river canyons. The Colorado Plateau occupies northern Arizona, northwestern New Mexico, most of eastern Utah, and western Colorado. Here, rolling uplands lie at varying levels, cut into isolated sections by the deep river canyons of the Colorado and its tributaries. The rest of the Intermountain area is the Basin and Range Country, which continues beyond New Mexico into western Texas. The part between the Wasatch Range of the Rockies and the Sierra Nevada in California has no river reaching the sea and is called the Great Basin. But the whole Basin and Range Country consists of small basins separated from each other by blocklike mountain ranges. This is the driest part of the West.

East of the Intermountain Basins and Plateaus, the high and rugged Rocky Mountains extend southward out of Canada through parts of Idaho, Montana, Wyoming, Utah, and Colorado, and northern Mexico, where a number of peaks rise above 14,000 feet (4,267 m). The Rockies are comprised of many separate ranges with individual names. Valleys and basins separate these ranges. The continuity of the system is almost broken in the Wyoming Basin, which provides the easiest natural passageway through the Rockies and was the route through which the first transcontinental railroad, the Union Pacific, passed. The Rockies also are a formidable barrier to moisture-bearing air masses from either west or east. Hence, they create rain shadows for both the Great Plains at

their eastern foot and the Intermountain area at their western foot. The mountains themselves receive much orographic precipitation and are forested. The highest elevations extend above the tree line to mountain tundra and snowfields.

The higher western section of the Great Plains (part of the Interior Plains) borders the Rockies in Montana, Wyoming, Colorado, and New Mexico. So elevated are these plains that Denver, at their western edge, is famed as the Mile High City. The most common landscape is that of an irregular surface interrupted at intervals by conspicuous valleys. Variations include hilly areas, smooth plains, and isolated mountains, as in the Black Hills of Wyoming and South Dakota. Harney Peak (7,242 ft) is the highest point in the United States east of the Rocky Mountains. The plains are covered by steppe grassland and marked by semiaridity and temperatures that are continental in the north and subtropical in the south.

Water, Agriculture, and Population

"The basin of the Colorado River cuts a huge, wriggling swathe across the western United States. It covers 244,000 square miles [632,000 sq km] and provides water to 25 million people, 16 million in southern California alone. Who gets this water, and why, obsesses the West. So much importance is attached to the distribution of the river's annual surplus of water that the decision is made by the secretary of the interior, no less."[2]

Water for the western United States is a most critical issue. From the crests of the various coastal mountain ranges that course from southern California through the state of Washington to the lands east of the Rocky Mountains, there is general water deficiency. For millennia after the initial arrival of the first human migrants from northeast Asia, water needs were met by keeping settlements small. For centuries after the arrival of Columbus in North America, water was a problem in the West, but it was generally seen as a local concern and, again, settlement size in the most arid areas seldom grew beyond the capacity for landscapes to support local hamlets and towns. But with the arrival of the 20th century, the completion of the transcontinental railway, and the active promotion of real estate development in southern California especially, water needs grew dominant in the whole equation of American demographic shifts to the southwest of the country. What has been a nagging problem for decades may now be growing into a problem of much larger dimensions.

There are two significant "exotic" water systems in the American West. The largest is the flow of the Colorado River coming from the snowmelt of mountain peaks and flanks in north central Colorado and the Green River in Utah. The river flows through Arizona's Grand Canyon, piles up behind Hoover Dam, and from that dam supplies a major portion of the water needs of three lower basin states: California, Arizona, and Nevada (Fig. 24.20). Lake Mead, which is the largest reservoir in the United States in terms of volume of water, also generates a major portion the electricity for the region.

F1.6a
13

---

[2] "Buying a Gulp of the Colorado." *Economist*, January 24, 1998, p. 28.

This water has been a source of litigation and federal and state argument virtually since the dam was completed in the mid-1930s. Because of the fast growth of Arizona in the past decade, that state is pursuing its share of the Colorado River water with more energy than before. The lion's share of the Colorado River water has gone to southern California, where it is critical to both Los Angeles and San Diego and the urban sprawl between the two. The second exotic system is the 223-mile (357-km) gravity flow canal that brings water from the Sierra Nevada snowmelt through the Owens Valley east and north of Los Angeles down to this enormous metropolitan region.

Both of these systems are now in conflict. The Owens Valley has developed such dust storms for the population still resident in the high valley (which sold its water rights piece by piece a century ago) that the California Air Resources Board may require Los Angeles to diminish the amount of water it removes from the area annually. This would enable the valley to expand the area of the water on the Owens lake bed (Mono Lake) so that there would be less potential for dust disturbance. Concerning the Colorado, there is increasing tension for southern California because Arizona and Nevada, which have historically drawn less water from the Colorado flow than they were allowed, are now both seeking their full share because of rapidly increasing urban growth. Las Vegas is the fastest growing large city in the United States, and Arizona continues to grow beyond its water resources as well. There is a complex "water banking" system being devised that will enable Nevada and Arizona to come closer to their goals, but each extra amount they are able to siphon off heightens southern California's anxiety about its own water budget (Fig. 24.21).

Figure 24.21 A view of desert and a major irrigation canal in Arizona's Basin and Range country near Phoenix. The canal brings water from the Colorado River to central and southern Arizona; it is one link in the system of canals, pipelines, tunnels, and pumping stations called the Central Arizona Project. Note the natural spacing of the xerophytic vegetation to take advantage of all available moisture. This photo is an effective reminder of what much of this area's landscape would look like without irrigation playing such a strong role.

There are few geographic features that have stimulated tension, fights, and even warfare the way water rights have. Water problems and associated threatened landscapes spread across a half dozen states in the American West, and there are parallels on a smaller scale that run north, south, east, and west across the country as well. Water is at the heart of settlement in most regions, and its management and availability are often at the very pinnacle of success, or disaster, in human settlement.

Most land east of the coastal states in the West is used for grazing, if at all, and the carrying capacity is so small that only extremely low densities of animals and ranch population are possible. Unirrigated farming of some steppe areas supports only slightly higher population densities. The most important of the latter areas are the wheat-growing areas of the Great Plains and the Columbia Plateau (known also as the Palouse) of Washington. There, yields are relatively low, farms are large and highly mechanized, and population is sparse. Much the greater part of Western agricultural output comes from a few large irrigated oases that grow both food crops and feeds, particularly alfalfa, for dairy and beef cattle kept on feedlots in the oases or for sale as supplementary feed to ranchers outside the oases. Some oases in California and Arizona grow cotton, fruits, vegetables, and sugar beets, and they are highly dependent on irrigation.

Figure 24.20 The Colorado River Basin.

The West is strongly metropolitan in population. Most of these urban westerners are crowded into big clusters located in or near major oases. The Pacific Northwest cluster centering on Seattle and Portland might be considered an exception, but even this rainfed area is a kind of oasis within the larger area of western drylands. These major urban clusters have very diversified functions, among which the servicing of agriculture is generally the oldest and continues to be important.

Southern California: Steps Toward the Creation of a Pacific Megalopolis

The southern California population cluster is by far the West's largest. The name southern California is commonly used for the Los Angeles Basin and the narrow lowlands north to Santa Barbara and south to San Diego, along with their interior frame of mountains. The Native American villagers who originally inhabited this area practiced no agriculture but subsisted on grass seeds, acorns, other wild foods, and some coastal fishing. In the 18th century, small-scale irrigated agriculture and cattle ranching were introduced by Spanish settlers from Mexico. By the 20th century, the Los Angeles area had become a large and important agricultural district using Colorado River water (see Fig. 24.20). Today, agriculture has been largely crowded out of the area by a sprawling urbanization that in 2000 comprised an estimated 19.2 million people in the combined metropolitan areas of Los Angeles (population: 16.4 million) and San Diego (population: 2.8 million). Packed into this small corner of the West are about three-fifths of California's people, one-third of the population of the entire West, and 1 of every 15 people in the nation (Fig. 24.22).

Figure 24.22   Southern California is the hallmark landscape of urban sprawl. The heavy dependence on the automobile and the development of an intricate and far-reaching set of limited access highways have stimulated a 50-year outflowing from the central city. Sometimes the process of replacement of small California bungalows or cottages with new modern or high-rise housing units stimulates civic response. This mural on a home in Venice, California, dramatizes the sense of older folk getting pushed out of their homes by the forces of new residential development. The case is seldom as polar as this, but there is considerable reluctance to see the pattern of 20th-century urban sprawl continue uncontested in many of the urban areas in all parts of western America.

San Diego and Los Angeles were founded as remote footholds on the northern frontier of Spain's empire in the Americas. Catholic priests founded missions in coastal locations as far north as San Francisco, including those in San Diego (founded in 1769) and San Gabriel (founded in 1781) in the Los Angeles area. The settlement at San Diego began on a magnificent natural harbor that is now the main continental base for the U.S. Navy's Pacific Fleet. The San Gabriel Mission and the Pueblo (civil settlement) of Los Angeles were sited near the small Los Angeles River. When Mexico gained independence from Spain in the early 19th century, the California settlements became part of the new country. They subsequently came into the possession of the United States in 1848 at the end of the Mexican-American War. Until the late 1800s, San Diego and Los Angeles were little more than hamlets in one of the world's more isolated corners.

A nearly continuous southern California boom that has lasted to the present began with the arrival of railroad connections to the eastern United States in the 1870s and 1880s. The main lines (the Southern Pacific and the Santa Fe) crossed the area's encircling mountains via the Cajon Pass near San Bernardino. To promote settlement that would create a demand for rail traffic, the railroads offered extremely low fares westward. Oranges shipped in refrigerated cars were the main local product that could stand the transportation cost to far-off eastern markets, and irrigated orchards spread rapidly. Los Angeles was the main transport and business center. In the 1890s, a freeze in Florida put California in the lead in U.S. citrus production, and the state maintained this position until urban sprawl replaced many of the southern California groves in the mid-20th century. (The Disneyland amusement park was constructed in a huge orange grove in 1955.) The climate that favored citrus also attracted a new population seeking health and comfort.

In fact, the southern California climate, water from streams descending the mountains, and other natural resources have continued to support development and attract population. The "brea" of the La Brea Tar Pits was an oil-based resource even for the initial settlers who long antedated the European population and African Americans and Hispanics who formally settled Los Angles in 1781. However, this petroleum resource was not given its full value until early in the 20th century when its role in the expanding automotive industry was becoming important. The region is still a significant producer and refiner of petroleum. Fishing fleets in the Pacific (primarily tuna boats) soon became large and wide-ranging. In the early 1900s, pioneering moviemakers moved from the Northeast to Los Angeles to take advantage of bright, warm weather and varied outdoor locales. The success of the movie industry later led to production of programs for national television when that medium burgeoned after 1950.

In the 1920s and 1930s, aircraft manufacturing was attracted by the favorable climate, the long hours of sunlight year-round, the amount of space still available for factories and runways, and the presence of a considerable labor force.

This industry boomed during World War II and afterward joined with the electronics industry in aerospace development. Prior to the 1990s, the aerospace industry had been southern California's largest manufacturer, heavily supported by federal military expenditures. Thus, beneath the glittery film-industry reputation, Los Angeles had become a major industrial city—an aerospace Detroit with palm trees. In the early 1990s, however, its economy (and that of California generally) slumped, partly as a result of declining federal defense outlays, and for the first time in decades, California saw more people leave the state than enter it. But by 1995, the patterns had reversed, and with the economy growing again—in part from the enormous construction required in recovery from the massive Northridge earthquake in January 1994 and in part from a business shift from defense to industrial operations—population also began to grow through migration. Tourism and a growing regional financial sector supported by the booming Pacific trade have also contributed significantly to the resurging economic dominance of southern California in this region.

In the 2000 U.S. Census, non-Hispanic whites became a minority in California for the first time in over a century. Active in-migration by Hispanics and Asians in the 1990s now has some 16 million whites in the state while there are more than 17 million others in a total population of nearly 34 million. The changes have been particularly dramatic in Los Angeles where the Hispanic population has now reached 45% of the total urban population. There are, in addition, 12% Asian and 9% African Americans. Whites are now 31% of the Los Angeles population. Not only has the ethic mix of the state been changing, but the pace of statewide in-migration has grown again so that now there are 1,700 people a day being born or arriving in California. That amounts to an additional person every 50 seconds. The widely varied ethnic groups that add such a definitive flavor to southern California are not clustered only in homogenous neighborhoods but are also scattered through many different and distinct sections of the metropolitan area.

All this would have been impossible without the supply of vast and increasing amounts of water to this dry area. Streams from the surrounding mountains and wells tapping local groundwater resources became inadequate by late in the 19th century, and water is now brought by aqueduct from the Sierra Nevada and the Colorado River (see Fig. 24.20), both of which are hundreds of miles away. Projects have been proposed to acquire water from still more distant areas as far north as Canada. Economical desalinization of seawater would be one solution if technology could provide a low-cost process.

The San Francisco Metropolis

In 1995, 6.5 million Californians lived in the San Francisco metropolis, which is the fifth largest in the United States. San Francisco is located on a hilly peninsula between the Pacific Ocean and San Francisco Bay, but the metropolitan area includes counties on all sides of the Bay. The climate is unusual. Marine conditions are evidenced in San Francisco's very moderate temperatures, which range from an average of 48°F (9°C) in December to only 64°F (18°C) in September. However, the area is in the border zone between the marine west coast and Mediterranean climates, with the Mediterranean influence evident in the fact that about 80% of the area's rainfall comes in five wet months from November through March. Mean annual precipitation is only 20 inches in San Francisco, but the moderate temperatures and low rates of evaporation have enabled this amount of moisture to produce a naturally forested landscape. There are even groves of redwoods within a short distance both north and south of the Golden Gate strait (Fig. 24.23). East of San Francisco Bay, somewhat warmer temperatures and lower precipitation have produced a scrubby forest (chaparral or dispersed California oaks) or grassland.

San Francisco began as a Spanish fort (presidio) and a mission on the south side of the Golden Gate strait and subsequently was developed, first as a port for the gold mining industry and then for the expanding agricultural economy of central California. The Gold Rush of 1849 and subsequent years centered on the gold bearing alluvial gravels of Sierra Nevada streams entering the Central Valley north and south of the current city of Sacramento. San Francisco, with its spacious natural harbor, was the nearest port to receive immigrants and supplies for the gold camps. Today, San Francisco is a major domestic and international tourist destination for its scenery, wines, and arts and other cultural attractions.

San Francisco has added a sizable industrial sector to its original functions as a port and regional economic center. Manufacturing in the metropolitan area is highly varied, but in recent years, computer-related and other electronics businesses in a strip of land near the southwestern shore of San Francisco Bay have become the outstanding industry. This area is now commonly referred to as Silicon Valley, after the chips that carry microcircuits. The rise of Silicon Valley as a center of high technology is due in part to the attractiveness of the San Francisco area as a place to live and the nearby presence of two outstanding universities: the University of California at Berkeley near the industrial center of Oakland on the east side of the Bay and Stanford University in nearby Palo Alto. Software firms have moved to a number of other regions for development, but Silicon Valley continues to be the dominant U.S. center for this innovation, development, and venture capital.

Seattle and Portland and Their Pacific Northwest Setting

The Puget Sound Lowland of Washington and the Willamette Valley of Oregon to its south are not irrigated oases (Fig. 24.24). They have abundant precipitation and luxuriant natural forests, as do the adjoining Coast Ranges and Cascades. But these

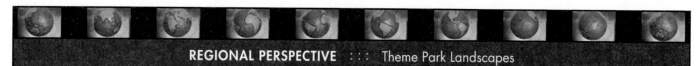

One of the most powerful forces in geography is the human capacity to remake the natural landscape into countless other forms. From turning forest land into cropland, and cropland into industrial space, and industrial space into baseball stadiums, this habit of making and remaking the cultural landscape is central to the human use of the earth. Massive theme parks are an example of this trait that has gained great economic significance in the past half century and in the past two decades particularly. In 1955, Disneyland was opened (see Fig. 24.13). Its 75 acres had been taken from orange orchards that, at that time, more accurately characterized the landscapes of Orange County, adjacent to rapidly growing Los Angeles County. In Disneyland, the first major theme park built (if the 1880s Coney Island on Long Island is not counted), developer Walt Disney introduced themes of Frontierland, Tomorrowland, and created a mythical Main Street that reflected the scenes Disney had grown up with in the small midwestern town of Marceline, Missouri. In this first theme park's Main Street, Disney had architects create a structure that was built at seven-eights scale so that people exploring this space would feel larger and less threatened than he saw the American

city as being in the 1950s. This decision is important in understanding the power of theme parks in not only the American landscape but in the American mind-set about the built environment.

Theme parks have grown to be very important all across the country. The largest complex is the Walt Disney World complex opened on 27,000 acres of swampland near Orlando, Florida, in 1971. In this massive complex, one of the major draws is the EPCOT Center (Experimental Prototype Community of Tomorrow). In this environment, Disney allows the tourist to walk through the future, and the park becomes educational as well as entertaining. Orlando has become the biggest tourist draw in the United States as it now has more than 30 million visitors annually. Las Vegas has just under that number, and the MGM Grand Hotel and Theme Park is the largest hotel and one of the most successful in that city of fantasy. Even The Mall of America in Minneapolis, Minnesota, the nation's largest mall complex, has adopted the theme park trait by placing rides in the center of the mall.

Magic Mountain, Six Flags, Busch Gardens, Sea World, Universal Studios, and many more all reflect this American fascination with the creation of struc-

tured fantasy within the context of safe but frightening rides and entertaining but somewhat educational windows on unusual worlds. The effectiveness of this sort of a built environment is shown in part by the fact that, domestically, the United States has 1.3 billion tourist trips (a trip of more than 100 miles away from home) a year, and this tourism has an economic value of tens of billions of dollars. In counting only the money spent by Mexico to the south and Canada to the north, $11 billion are spent annually by tourists and businesspeople coming to the United States. And major theme parks serve as a draw on these travelers continually. Disney has opened Disneyland Tokyo (1983) and Disneyland Paris (1992) as a reflection of the international fascination with effectively crafted recreation environments.

What is the lesson from this fascination with theme parks? The success of these big "playgrounds" tells us that landscapes, like movies, have the capacity to evoke other worlds, and if such foreign worlds can be explored in safety (even though the explorations are relatively expensive), they are attractive to all ages and to all socioeconomic classes. Build it, and they will come, and many will pay top dollar just to enter these worlds!

humid conditions end abruptly near the crest of the Cascades. Hence, the area is oasislike — a sharply limited area of abundant water and clustered population within the generally dry West. This population is centered in two port cities, each located on a navigable channel extending inland through the Coast Ranges. Seattle (population: 3.6 million), Washington, located on the eastern shore of the Puget Sound, has access to the interior via passes over the Cascades. Even with its pattern of rainfall, the Seattle setting is home to the Boeing Aircraft Company, which is the world's largest aircraft producer. However, to be closer to the decision makers who are critical in the purchases of Boeing products, its administrative center moved to Chicago in 2001. The local early abundance of the Sitka spruce (used for early wooden plane construction) and the availability of cheap electricity led to the 1917 founding of the

firm by William Boeing. Portland (population: 2.3 million), Oregon, is on the navigable lower portion of the Columbia River where the Willamette tributary enters it and has a natural route to the interior via the Columbia River Gorge through the Cascades. Outside these metropolises, the Lowland contains somewhat over a million other people.

Fishing, logging, and trade built the cities of the Lowland in the latter 19th and early 20th centuries. But agricultural land in the Willamette Valley drew the first large contingents of settlers to "Oregon Country" in the 1840s. Agriculture is still of some importance, although it is handicapped by hilly terrain and leached soils. The leading products are milk and truck crops, and a local Oregon wine industry is expanding in importance. Fishing, primarily for salmon, is still carried on, but the Pacific Northwest fish catch is now much smaller than

MICHAEL BISCEGLIE

Figure 24.23   The California redwoods (*Sequoia sempervirens*) are one of the most evocative images of forest in the West. This photo of the Avenue of the Giants just north of the San Francisco Bay Area in California helps provide a visual cue to the size of these trees. Because the wood is so valuable as a building material—valued because of both color and its durability for deck construction—this tree has been a focus of tensions between environmentalists and logging interests for decades.

any one of the three leading states in this industry—Alaska, Massachusetts, and Louisiana (as of 1993). However, Oregon and Washington are still the leading states in production of lumber, and they still have one-quarter of the sawtimber resources of the United States. The trees are mostly coniferous softwoods, and many are enormous in size, especially the famed Douglas fir. The timber industry, now under heavy attack by environmentalists, is still basic to the economies of both states.

Another major industry that has been sited in the Pacific Northwest because of owner environmental preferences is the Microsoft Corporation of Redmond, Washington. The firm was initially founded in Albuquerque, New Mexico, in 1975, but by 1979, the two founders—Bill Gates and Paul Allen—moved headquarters to the Seattle area. It has since grown to be the largest computer software company in the world and is now a source of major payroll and export revenues for the Northwest and the United States.

Population Clusters in the Mountain States

The eight inland states in the Mountain West are drier than those in the Pacific Mountains and Valleys, lack the abundant irrigation water of California, and lack the advantage of ocean transportation. These handicaps have been so constricting

ROBERT HARBISON/CHRISTIAN SCIENCE MONITOR

Figure 24.24   The Dalles Dam on the Columbia River permits salmon to migrate via the fish ladder in the foreground. Note the power-transmission lines in the distance. The dam, located on the border between Oregon and Washington, is one of 56 major dams in the Columbia River watershed.

that the eight states had only 18.2 million people in 2000. About half of these are in a handful of metropolitan areas in irrigated districts. Outside of them, the common pattern in a Mountain West state is that of a few hundred thousand people scattered across a huge area on ranches and farms, on Indian reservations, or in small service centers and mining towns.

Phoenix and Tucson. The largest population cluster in the Mountain West lies in the Basin and Range Country of southern Arizona. Here live about three-quarters of Arizona's people in the adjacent metropolitan areas of Phoenix (population: 3.3 million) and Tucson (population: 843,700). Water from the Salt and Colorado Rivers (see Fig. 24.20), plus groundwater, makes it possible to support this development in a subtropical desert.

Phoenix was originally the business center for an irrigated district along the Salt River. The district expanded after 1912 when the first federal irrigation dam in the West (Roosevelt Dam) was completed on the river. Today, cotton is the area's main crop, although hay to support dairy and beef cattle is important. Growth in both Phoenix and Tucson accelerated after the introduction of air conditioning made them attractive retirement and business centers. Each is the site of a major state university, and Phoenix is the state capital. Tourism to both urban and nonurban destinations, the increasing income level of Sunbelt retirees, and copper mining are also basic to Arizona's economy. The 2000 Census has provided further evidence of the continuing population shift from the American Northeast toward and to the Southwest.

The Colorado Piedmont. Another important population cluster stretches along the Great Plains at the eastern foot of the Rocky Mountains. Within this area, called the Colorado Piedmont, metropolitan Denver contained 2.6 million people in 2000, and more than 516,000 lived in metropolitan Colorado Springs (city proper, 360,900) or smaller cities. These cities developed on streams emerging from the Rockies. The stream valleys provided routes into the mountains, where a series of mining booms began with a gold rush in 1859, a decade after the California gold rush. One mineral rush followed another through the later 19th century, with various minerals being exploited at scattered sites. The Colorado Piedmont towns became supply and service bases for the mining communities, which often were short-lived. Irrigable land along Piedmont rivers afforded a farming base to provide food for the mines, and agricultural servicing became an important function of the towns. In time, agriculture grew into a prosperous business with more stability than mining.

During the 20th century, the Piedmont centers have grown rapidly. Irrigated acreage has been expanded greatly, fostering a boom in irrigated feed to support some of the nation's largest cattle feedlots. The Colorado Rockies have also become an important producer of molybdenum, vanadium, and tungsten, which are comparatively rare alloys much in demand for high-technology industries. High energy prices in

the 1970s led to expanded extraction of fuels from the state's huge coal reserves and scattered oil and gas deposits. And tourism in the Rockies has profited greatly from increasing American affluence, better transportation, and the rise in the popularity of skiing. Finally, Denver has also become a major center for the western regional offices of the federal government, with Colorado Springs the site of the U.S. Air Force Academy and the nearest large city to the headquarters of the North American Air Defense Command, located in a massive cave complex inside a nearby mountain.

The Wasatch Front (Salt Lake) Oasis. Third in size among the population clusters of the interior West is the Wasatch Front, or Salt Lake, Oasis of Utah, composed of metropolitan Salt Lake City–Ogden (population: 1.3 million) and Provo (population: 368,500) to its south. It contains about three-fourths of Utah's people, in a north–south strip between the Wasatch Range of the Rockies and the Great Salt Lake, with a southward extension along the foot of the Wasatch. In the rest of Utah, there are only about 500,000 people. The oasis was developed as a haven for people of the Mormon faith when they were expelled from the eastern United States in the 1840s. After a difficult trek westward, they began irrigation development based on water from the Wasatch Range, and they built at Salt Lake City the temple that is the symbolic center of the Mormon religion and culture. The Mormons established nearly 400 settlements in the West, including some in Canada and Mexico as well. These settlements were successful enough to serve as spatial and cultural models for other parts of the growing settlement patterns in this arid region of the United States. Salt Lake City is a regional religious and cultural capital; in addition, with no large competing metropolis within hundreds of miles, the city is the business capital for a huge although sparsely populated area outside the Mormon core region at the foot of the Wasatch Mountains.

The Less Populous States of the Interior West. The other five states of the interior West are even less populous than Utah. Their aggregate population of about 6.5 million, in states that are very large in area, is less than that of the San Francisco metropolis. The people of New Mexico are largely strung along the north-south valley of the middle and upper Rio Grande, where Albuquerque (population: 712,700) is the main city, although it competes economically with El Paso, Texas, which is located on the Rio Grande just south of New Mexico. New Mexico's economy is a typical Western mix of irrigated agriculture, ranching, mining, tourism, some forestry in the mountains, and federally funded defense contracting. Similarly, most of the population of Idaho is in oases strung along a river — in this case, the Snake River as it crosses the volcanic Snake River Plains in the southern part of the state. Agricultural output is more substantial in Idaho, nationally known for its potatoes. But tourism and federal money yield less income than they do in New Mexico, and the largest city of the state, Boise, has only about 150,000 people in the city proper.

Except for a small section that lies in the Sierra Nevada, the state of Nevada is desert, with very little irrigation. Metropolitan Las Vegas (population: 1.6 million), advertised as the fastest growing major city in the United States with its growth of 83% from 1990–2000, and Reno (population: 339,400) have more than four-fifths of the state's population, and the state's laws allow gambling and related tourism to overshadow the usual Western mix of economic activities. Casinos and associated entertainment and convention business play a dominant role in the state's wealth. The current pace of population growth in the state is also a testimony, again, to the importance of air conditioning.

In Montana and Wyoming, there is little irrigation and no warm-winter climate to encourage year-round tourism or to attract retirees. Economies based on farming, ranching, mining, summer tourism, and even skiing support sparse populations.

24.5 Alaska and Hawaii: Exotic Outliers

In the middle of the 19th century, the United States began to acquire Pacific dependencies. Two of these, Alaska and Hawaii, were admitted to the Union as states in 1959. The United States acquired Alaska by purchase from Russia in 1867, and Hawaii was annexed in 1898. Acquisition of a Pacific empire was motivated partly by defense considerations, and these have continued to play an important role in the settlement and development of both Alaska and Hawaii. Large military installations exist in both states, and defense expenditures are a major element in the economies of these states. In terms of physical geography, these two most recent states are about as different from each other as is possible, and in many ways, they stand apart from the other 48 as well (Fig. 24.25).

Alaska's Difficult Environment

Alaska (from Alyeska, an Aleut word meaning "the great land") makes up about one-sixth of the United States by area, but it is so rugged, cold, and remote that its total population was only 626,900 in 2000. The state's difficult environment can be conveniently assessed in four major physical areas.

1. *Arctic Coastal Plain.* Alaska's Arctic Coastal Plain is an area of tundra along the Arctic Ocean north of the Brooks Range. Known as "the North Slope," it contains very large deposits of oil and natural gas.
2. *The Brooks Range.* The barren Brooks Range forms the northwestern end of the Rocky Mountains. Summit elevations range from about 4,000 to 9,000 feet (c. 1,200 to 2,800 m).
3. *Yukon River Basin.* The Yukon River Basin lies between the Brooks Range on the north and the Alaska Range on the south. This is the Alaskan portion of the Intermountain Basins and Plateaus. The Yukon is a major river, navigable by riverboats for 1,700 miles (c. 2,700 km) from the Bering Sea to Whitehorse in Canada's Yukon Territory. In Alaska, the Basin is generally rolling or hilly, with some sizable areas of flat alluvium that often are swampy. Its subarctic climate is associated with coniferous forest that generally is thin and composed of relatively small trees. Fairbanks (population: city proper, 30,200) is interior Alaska's main town. The Alaska Highway connects it with the "Lower 48" (states) via western Canada, and both paved highways and the Alaska Railroad lead southward to Anchorage and Seward. Fairbanks is a major service center for the Trans-Alaska Pipeline from the North Slope oil area to the port of Valdez in the south. The city is located on the Tanana River tributary of the Yukon and is associated with a small agricultural area in the Tanana Valley.
4. *Pacific Mountains and Valleys.* The Pacific Mountains and Valleys occupy southern Alaska, including the southeastern panhandle. The mainland ranges of the panhandle are an extension of the Cascade Range and the British Columbia Coastal Ranges, and the mountainous offshore islands are an extension of the Coast Ranges of the Pacific Northwest and the islands of British Columbia. The scenic and very popular Inside Passage from Seattle and Vancouver to the Alaskan "panhandle" lies between the islands and the mainland. On the mainland, some mountain peaks exceed 15,000 feet (4,572 m). Fjorded valleys extend long arms of the sea inland. A very cool and wet marine west coast climate prevails, fostering coniferous forests that are a major resource, exploited primarily for the Japanese market. From Juneau (population: city proper, 30,700) northwestward, a majestic display of glaciers defines a long stretch of coast (Fig. 24.24). However, the highest mountains of Alaska do not lie along the coast but are well inland in the Alaska Range, where Mt. McKinley (Denali) rises to 20,320 feet (6,194 m) about 100 miles (160 km) north of Anchorage. This is the highest peak in North America. Anchorage is the state's largest city (population: 260,300) and has more than two-fifths of the state's total population. At the southwest, the Alaska Range merges into the lower mountains occupying the long and narrow Alaska Peninsula and the rough, almost treeless Aleutian Islands.

Immediately south of Anchorage, the Kenai Peninsula lies between Cook Inlet and the Gulf of Alaska, and still farther to the southwest is Alaska's largest island, Kodiak. These features mark the western end of the horseshoe of coastal mountains that curves across the north of the Gulf of Alaska from the southeastern panhandle. The climate of lower areas in this western coastal section is harsher than in the panhandle and is classified as relatively mild subarctic.

Part of the geographic drama of Alaska comes from the fact that it lies within the chain of mountains that comes up from Tierra del Fuego and the Andean Cordillera in South

Box
509

Figure 24.25   Location maps: Alaska, with its agricultural and reindeer herding areas, and Hawaii.

America, through varied mountain chains in Middle and North America, and finally arcs to the west southwest through the Aleutian Islands of Alaska. This chain is part of the **Ring of Fire** that continues to the south and west as it courses through Japan, Taiwan, the Philippines, additional islands in the southwest Pacific, and finally terminates in the North Island of New Zealand. This long arc got its name from the vol-

canoes caused by the intersection of massive plates that sit atop the mantle of Earth's core. Volcanic eruptions at Mt. Saint Helens in Washington in 1986 and Mt. Pinatubo in the Philippines in 1990 and the massive earthquakes in 1964 in Anchorage and in 1995 in Kobe (Japan) all represent recent expressions of the hazards of draping human settlements across the landscape of the Ring of Fire (Fig. 24.26).

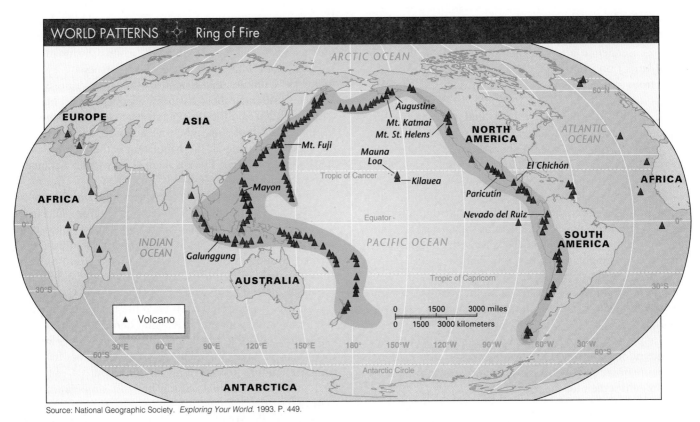

WORLD PATTERNS ✧ Ring of Fire

Source: National Geographic Society. *Exploring Your World.* 1993. P. 449.

Figure 24.26 More than three-quarters of Earth's active volcanoes lie along this arc of mountains from Tierra del Fuego in South America to the islands of New Zealand. The arc is known as the Ring of Fire.

From Fur Trade to Oil Age

After approximately two centuries of occupation by Russia and the United States, the population of Alaska was still only 73,000 in 1940. Eighteenth-century Russian fur traders were spread thinly along the southern and southeastern coasts. Evidence of this era persists in the many Russian place names and in Russian Orthodox churches on the landscape. In 1812, the Russians established an agricultural colony on the California coast at Ft. Ross north of San Francisco Bay. This effort to provision the Alaskan posts more adequately was abandoned in 1841, and then, in 1867, remote and unprofitable Alaska was sold by Russia to the United States. By that time, hunting had nearly wiped out the sea otters, whose expensive pelts had provided the main incentive for Russia's Alaskan fur trade. Only recently, under American and Canadian governmental protection, has a substantial beginning been made at replenishing these mammals. Under the Americans, there was a burst of settlement connected with gold rushes in the 1890s and for a few years thereafter followed by a very slow increase connected largely with fishing and military installations.

Migration patterns to Alaska from the conterminous 48 states have included many more men than women. Such an imbalance has led to the casual female observation (on a popular postcard): "In Alaska, the odds are good . . . but the goods are odd." This imbalance is classic in frontier situations,

and this state has had exactly such population dynamics since its earliest settlement by gold seekers and homesteaders began to come north from the conterminous states.

Beginning with World War II, during which Japanese troops invaded the Aleutian Islands, growth became more rapid. Today, metropolitan Anchorage at the head of Cook Inlet is the main population cluster. The rest of Alaska's people are unevenly distributed over the state but live mainly in or near Fairbanks, in the state capital of Juneau in the panhandle, or in other small and widely spaced towns and villages. Of the total population, about one-sixth is composed of Alaska Natives (Inuit, or Eskimos; Native Americans; and Aleuts). Long the victims of exploitation and neglect by whites, these peoples received some redress in large land grants and cash payments under the Alaska Native Claims Settlement Act passed by Congress in 1971.

The burst of growth since 1940 has been principally due to (1) increased military and governmental employment, (2) the rise of the fishing industry to major importance, with salmon by far the main product, (3) revenues accruing from the development of Arctic oil (discovered in 1965), (4) the rise of air transport as a major carrier of both people and freight, and (5) visits by increasing numbers of tourists, both domestic and foreign. Agriculture consists largely of high-cost dairy farming in a few restricted spots. Both food and other consumer goods come overwhelmingly from the lower 48 states.

New Petroleum Sources?

The petroleum resources became of dominant state and national consequence after the completion of the 800-mile (1,300-km) Trans-Alaska pipeline constructed in the mid-1970s from Prudhoe Bay north of the Arctic Circle southward across the Brooks Range to Valdez on Prince William Sound. The California energy crisis and a strong increase in gasoline prices in mid-2001 led to new consideration of expanded oil exploration in Alaska. Although the environmental tensions associated with such exploration continue to provoke intense arguments, the technology of drilling has changed profoundly since the exploration that led to the 1968 discovery of the petroleum-rich Prudhoe Bay site. There are now strict calendar limitations on when heavy equipment may be moved across the tundra landscape with its fragile peat bogs and shallow ponds. Everything from drilling procedures, to the transport of heavy equipment on temporary ice roads, to the removal of waste products by deep burial has made the impact of oil exploration and productive less intrusive. Although there are many approaches being considered to deal with domestic gasoline prices and regional power shortages, the potential for new petroleum finds in Alaska continues to be on the development screen for the 49th state.

Hawaii: Americanization in a South Seas Setting

The state of Hawaii consists of four major tropical islands (see Fig. 24.25) lying just south of the Tropic of Cancer and somewhat over 2,000 miles (c. 3,200 km) west from California. These islands are all volcanic and largely mountainous, with a combined land area of about 6,400 square miles (c. 17,000 sq km), which is somewhat greater than the combined areas of Connecticut and Rhode Island. The island of Hawaii is by far the largest and the only one to have active volcanic craters. The prevalent trade winds and mountains produce extreme precipitation contrasts within short distances. Windward slopes receive heavy precipitation throughout the year that rise to an annual average of 460 inches (1,168 cm) on Mt. Waialeale on the island of Kauai, the wettest spot on the globe. Nearby leeward and low-lying areas may receive as little as 10 to 20 inches (25 to 50 cm) annually. Some areas are so dry that growing crops requires irrigation. Temperatures are tropical except for colder spots on some mountaintops. Honolulu averages 81°F (27°C) in August and 73°F (23°C) in January; the average annual precipitation in the city is 23 inches (58 cm), with a relatively dry high-sun period ("summer") and a wetter low-sun period ("winter"), bearing some similarity to Mediterranean climate patterns.

Hawaii was more populous than Alaska and was the site of a Polynesian monarchy when it was acquired by the United States. By 2000, its population was 1.2 million, with about 876,200 people concentrated in metropolitan Honolulu. The metropolis occupies the island of Oahu, with the city proper located near the southeastern corner of the island. Immigration from the Pacific Basin has given the state's population a unique mix by ethnic origin: 33% white Caucasian, 25% Japanese, somewhat under 20% native Polynesian, 14% Filipino, 6% Chinese, and a small contingent of Koreans and Samoans. Although there has been ongoing tension between native islanders and the immigrants who come from the U.S. mainland and East Asia, much racial and ethnic mixing has taken place. In general, the population is thoroughly Americanized even through there is a small Polynesian nationalist movement evident as well.

Hawaii's economy is based overwhelmingly on military expenditures, tourism, and commercial agriculture. The main defense installations, including the Pearl Harbor naval base, are on Oahu. The main agricultural products and exports — cane sugar and pineapples — have long been grown on large estates owned and administered by Caucasian families and corporate interests but worked mainly by non-Caucasians. Work on such estates has often been the first employment of new immigrants. However, changing demographic patterns, land values, and the rapidly declining role of agriculture in Hawaii have all led to steadily changing landscapes in the islands. Agriculture now accounts for only about 1% of the annual gross state product. Sugar has lost importance, and tourism has been the great growth industry of recent decades, propelled by the development of air transportation and the growing affluence of the population of mainland America and East Asia. Visitors from Japan and other Pacific Rim countries have become more common with the rising level of economic development since the early 1980s. The main attractions for tourists are the tropical warmth, beaches, spectacular scenery, exotic cultural patterns, and highly developed resorts. However, some of the "South Seas" glamour of the islands has been eroded by the relentless spread of commercial and residential development.

A factor that had strong impact on the importance of the role of tourism in the state economy was the major slowdown in the economic growth of East Asia since 1997. Japan, for example, has been a traditional source of a wealthy tourist population for which Hawaii was a major destination. The still-continuing slowdown in Japanese economic growth has caused a major decline in the Tokyo-Honolulu tourist flow, and this has been of real significance to Hawaii's economy. This broad-scale decline in tourist numbers and subsequent business activity took place in 1998, when Hawaii's unemployment rate also climbed to 33% higher than the national average. Major hotel chains and retailers closed branches or downsized staff and services. It has become apparent that Hawaii's economic vitality is more closely tied to business health in Japan and East Asia than was beneficial. There has been only modest economic return to normal tourist patterns in the first years of the new century. It is assumed that the state will not realize a real turnaround in its own recent declines until East Asian nations, especially Japan, are fully recovered from the "Asian flu."

CHAPTER SUMMARY

- This chapter divides the United States into four regions. The first is the Northeast, comprised of Maine, New Hampshire, Vermont, Massachusetts, Connecticut, Rhode Island, New York, New Jersey, Pennsylvania, Delaware, Maryland, and the District of Columbia. In this region live about 20% of the nation's population, or some 55 million people. The primary center of this population is a region called "Boswash," which is a belt of highly urbanized landscapes running from Boston at the north to Washington, D.C. at the south. There are seven major urban centers in this Boswash region: Boston, Providence, Hartford, New York, Philadelphia, Washington, D.C., and Baltimore.

- The Appalachian Highlands is the overall physiographic term for the Northeast, and the three most significant component parts are the Coastal Plain, the Piedmont, and the Interior Plains. The fall line represents the break in topography between the low-lying Coastal Plain and the inland rise in elevation in the Piedmont. The fall line has been significant as a break-in-bulk point for ships and small freighters coming inland that cannot negotiate the falls that occur. Early manufacturing centers were also built around these drops in elevation. Baltimore, Maryland, Washington, D.C., Philadelphia, Pennsylvania, and Trenton, New Jersey, are examples of fall-line cities.

- Roots of early regional development go back to English settlers in Boston, Dutch settlers in New Amsterdam (New York), and later the English in Baltimore and Philadelphia. The early locations that were favored were on fall lines (Philadelphia) or good harbors (New York).

- The Erie Canal (1825), in conjunction with the Hudson-Mohawk Trough, stimulated a major migration flow to the northwestern territories, now the Midwest. By the 1850s, so much trade had flowed through that corridor that New York harbor became the nation's leading harbor, and it still is of major importance, although Houston has become the country's largest by volume. This passageway also led to significant migration from New England farming as whole farm households traveled to the states at the southern margin of Lake Erie in quest of better soils and farming situations.

- Early industrialization took place along the fall-line cities in the region and then expanded to western Pennsylvania as steel production began along the southern margins of Lake Erie and especially in Pittsburgh. Additional development in flour milling, electrical goods, chemical and textile production, and photographic equipment all took on regional and national importance. However, the real center of major industry, finance, textile manufacturing, and varied manufacturing was in New York City. This city's primacy has been maintained to the present in many of these industries.

- The South includes Virginia, North Carolina, South Carolina, Georgia, Florida, Alabama, Mississippi, Louisiana, Texas, West Virginia, Kentucky, Tennessee, Arkansas, and Oklahoma.

- A humid subtropical climate with long hot summers is common in the South, but rainfall diminishes as you go west in the region. There is considerable regional variety in physiographic units from the Gulf-Atlantic Coastal Plain, the Piedmont, and the Interior Plains, with the fall line again lying between the Piedmont and the Coastal Plain. The Upland South has hills and low mountains, and the subunits of that area include the Appalachians, Unglaciated Southeastern Interior Plain, and the Interior Highlands.

- The South's main agricultural characteristics today include a diminishing role of agriculture overall in the region (only 2% of the population still lives on farms); the reduction of farming to a more part-time occupation, with wages being earned in nearby part-time city jobs; the rise in beef cattle, broiler chickens, and dairy cattle, which have all become more important than crops. Tobacco has been the most recent crop to begin a strong decline. Citrus, peanuts, truck crops, and race horses are some of the special farming activities that now characterize the region.

- New industrial development has included textiles (although this industry is moving on toward Mexico, Middle America, and other offshore sites), and the development of synthetic fibers from local resources has been important. In addition, energy from hydroelectric development, oil and natural gas, and to a lesser degree, coal has been developed. Texas and North Carolina now stand as two major industrial states.

- The industrial belts of the region lie along the Piedmont and in part are connected to the Tennessee Valley Authority of the 1930s. This regional industrial development was unusual because it focused as much on small towns and middle-sized cities as it did on the major metropolitan centers of the Northeast. The second major belt was along and near the Gulf Coast. The Texas-Oklahoma Metropolitan Zone has the two largest urban centers in the South: Dallas–Ft. Worth and Houston. The Houston Ship Canal has made that city a major shipping center. Florida is another major metropolitan zone, but with a focus more on service and tourism than on manufacturing. There are other metropolitan zones in the South, such as cities of the Upland South and the Mississippi Delta.

- The Midwest as a region encompasses 12 states: Ohio, Indiana, Illinois, Michigan, Wisconsin, Minnesota, Iowa, North Dakota, South Dakota, Nebraska, Kansas, and Missouri. As a region, it would rank 13th or 14th in the world in area and population. In the production of pork, farm machinery, and soybeans, no nation in the world other than the United States would surpass the Midwest.

- Plains are the most distinctive topographic feature, with the areas of major relief lying at the margins of the region. Soils are generally very good, and toward the east, rainfall is generally adequate for productive farming. Rainfall diminishes as you go west, making irrigation more an agricultural feature there. Soils also benefit from the presence of loess — windborne dust particles from the west and north, probably glacial in its origin.

- The Corn Belt is a major part of the Midwest, and corn continues to be the single most important crop in Midwest agriculture, although soybeans are gaining in importance each year. Approximately 40% of the world's corn comes from the Midwest.

- Wheat is very important in the Dakotas and south, with spring wheat major in the north and winter wheat most important in Kansas and south. Associated with wheat are sorghums, barley, and livestock production. In the region overall, sales of animal products exceed those of crops. Cattle and hogs are the two primary animals raised in the Midwest. There are also specialty crops that include fruit, sunflowers, sugar beets, and potatoes.

- Detroit has long been the center of automobile manufacturing in the country, related in part to the development of a major steel concentration along the southern shore of Lake Erie, utilizing iron ore from Minnesota, coal from Pennsylvania, and the Great Lakes as the avenue of major movement into the St. Lawrence Seaway and world trade. Slowly, the steel industry has been moving away from its early 20th-century locales, going further south and west. St. Louis has now become second only to Detroit in the manufacture of automobiles because of shifting industrial locations.

- The region of the West includes New Mexico, Arizona, Colorado, Utah, Nevada, Wyoming, Montana, Idaho, California, Oregon, and Washington. Alaska and Hawaii are two very distinct variants on the themes of the West and are the concluding section of this chapter. The region's landscape is predominately arid and amounts to approximately 38% of the conterminous United States. One-fifth of the nation's population lives in the West.

- Aridity is the most pervasive geographic characteristic of the West, and settlement is described as oasislike. Topography plays a strong role in the climate patterns of the region, with windward and leeward locations very important. Major ranges along the west coast (including the Cascades, Klamath Mountains, Coast Ranges, and Sierra Nevadas) have the largest population centers on their windward sides, and population densities generally diminish quickly as you go east from those ranges. The Willamette-Puget Sound Lowland in the north and the Central Valley in California in the south are two significant lowlands in terms of population size.

- Grazing is the most common use of the land in the West, if it is brought into agriculture at all, unless irrigation is possible. Irrigated farmlands raise alfalfa for beef cattle and dairy herds, or they may be used for cotton or specialty crops, particularly in California. In the regions that have access to exotic water, there has been a continual tension between the use of that resource for irrigated farmland or urban settlement and expansion.

- Southern California is the West's largest population cluster. It did not begin to grow significantly until water was brought from the distant Owens Valley to Los Angeles early in the 20th century. As water from the Colorado River Project was also made available to cities in the southwest and in southern California, growth accelerated. Presently, about one-seventh of the United States population lives in southern California, in part because of the networks of water movement and late 19th-century railroad developments. The continual presence of natural hazards, including fire, flood, earthquakes, and intense drought, has done little to retard the growth of this area.

- San Francisco is also a major urban center, sharing many of the same hazards as southern California. It has grown from a different set of origins and has become well-known as the home area of Silicon Valley, the dominant software development center in the West and in the United States. Even with the continuing dispersal of software development (especially to the Seattle area in association with the location of Microsoft), the San Francisco area is dominant in that industrial activity.

- California's Central Valley is a very productive agricultural area, with grapes and cotton as the two leading crops. Irrigation, primarily by waters from the Sierra Nevada range, has long been central to such productivity. Milk and cattle are also major products from this region, and many cities have developed in support of this productivity, Sacramento, Fresno, and Stockton among them.

- Seattle, Washington, its surrounding Puget Sound Lowland, and the Willamette Valley of Oregon have a marine west coast climate that provides abundant natural rainfall. Fishing, logging, and trade built these cities in the late 19th century, and aircraft manufacture (Boeing) and software industries of all sizes (Microsoft is the largest) continue to promote development of this region.

- In the Mountain West — the states that lie east of the Pacific coast states — water is less available than on the coast, and population is clustered around centers that have a reliable water source and transportation lines running at least both east and west. Phoenix and Tucson, Arizona, Denver, Colorado, and Salt Lake City, Utah, are the four largest cities in this region, and they all have distinct development histories and identities.

- Alaska and Hawaii are the two most recent additions to the United States (they both became states in 1959), and they represent very distinctive landscapes and environmental locales. Large military installations exist in both states. Alaska stretches north of the Arctic Circle and has a total population of approximately 600,000. Although there is a variety of landscapes in Alaska, all are characterized by intense winter seasons and a demanding seasonal range of weather. Anchorage is the largest city and has nearly one-half of the state's total population.

- Alaskan history chronicles fur trapping, gold rushes, oil exploration, and military installations. Sports fishing and hunting and general tourism play an increasingly significant role in the state's economy.

- Hawaii lies some 2,000 miles west of California and is comprised of eight major islands. It has a profound range of rainfall patterns, with agriculture playing a very important role in its early development. Today, agriculture amounts to only 1% of its annual gross state product, even though it continues to be a source of sugar and pineapple for mainland markets. Military expenditures have also been important, and during the past two to three decades, tourism has brought major capital to the islands. The Asian economic downturn in 1997 has caused a significant decrease in tourist arrivals.

REVIEW QUESTIONS

1. Name the states that make up the four regions that this text uses to regionalize the United States.
2. Approximately what percentage of the U.S. population lives in the Northeast? In the West?
3. What are the major topographic features that characterize the U.S. Northeast?
4. Explain the U.S. fall-line cities and discuss their role in early industrialization and settlement.
5. When did the Erie Canal open, what route did it follow, and what demographic impact did it have?
6. What are the reasons for the primacy of New York City?
7. What are some of the old images and new images of the U.S. South?
8. Outline the topographic characteristics of the U.S. South and explain how they relate to patterns of urban settlement and agriculture of the region.
9. Explain the significance and the process of the South's efforts to create new regional images of cotton farming in the U.S. South.
10. What are the major aspects of patterns of urbanism in the U.S. South and what are the most important cities?
11. Why are the soils so good in the U.S. Midwest? How do these characteristics relate to topography?
12. Explain the pattern of small-town growth, decline, and potential resurgence in the U.S. Midwest.
13. What are the major characteristics of industrial growth in the U.S. Midwest and what cities have been most influenced by these changes?
14. Why is the term "oasislike" used for urban settlement in the U.S. West?
15. List and give the importance of the major topographic features in the U.S. West. What role do these features have in patterns of agriculture and urban settlement in the region?

16. What is the meaning of "exotic water"? Explain its importance to the U.S. West.
17. Compare Los Angeles, San Francisco, and Seattle in terms of their natural settings, their histories, and their contemporary economic and cultural characteristics.
18. How much of U.S. agricultural production comes from California? Outline what makes this possible and what products it entails.

19. Outline the geographic characteristics of Alaska and Hawaii and relate such features to their individual histories, their contemporary settlement patterns, and their future development.
20. What is the Ring of Fire? Relate it to a global pattern.

DISCUSSION QUESTIONS

1. Discuss criteria that you might use to regionalize a country as diverse and as large as the United States. How successful do you feel this text is in dividing the country into four regions?
2. What has been role of fall-line locations in the urban and economic history of the United States?
3. Discuss the impact changes in transport technology have had on settlement and demographic patterns in the United States.
4. Discuss the distinctive roles of cotton, corn, soybeans, and wheat on the landscapes of the regions of the United States.
5. What roles have rivers played in the patterns of settlement, economic activity, and tourism in the United States?

6. Water has been a key environmental feature of which U.S. regions and how has it been utilized in their economic development and environmental management?
7. Discuss the shifting patterns of American steel production, explaining what geographic, economic, and demographic features have been most influential in today's patterns.
8. What do you think will become of the small town in America? Why?
9. Name the three most highly visible cities in the United States and explain why they have such visibility. What is the significance of their characteristics?

Credits

This page constitutes an extension of the copyright page. We have made every effort to trace the ownership of all copyrighted material and to secure permission from copyright holders. In the event of any question arising as to the use of any material, we will be pleased to make the necessary corrections in future printings. Thanks are due to the following authors, publishers, and agents for permission to use the material indicated.

Chapter 1: Table 1.1, 5: From *Geography for Life: National Geography Standards 1994*. National Geographic Research and Exploration, Washington, D.C., pp. 34–35. Reprinted by permission.

Chapter 2: Figures 2.19, 2.20, 50, 51: Population Reference Bureau. Reprinted by permission.

Chapter 3: Figure 3.A, 64: From *Peoples and Cultures*, by Alisdair Rogers, ed., p.12. Copyright © 1992 Oxford University Press. Reprinted by permission. **Figure 3.16, 84:** From *Planet Management*, by M. Williams, ed. Copyright © 1993 Oxford University Press. Reprinted by permission.

Chapter 4: excerpt, 109: From *The Walking Drum*, by L. L'Amour. Copyright © 1984 Bantam, division of Random House, Inc. Reprinted by permission.

Chapter 5: excerpt, 138: From *The Cloud Sketcher*, by R. Rayner, pp. 162–163. Copyright © 2000 HarperCollins. Reprinted by permission.

Chapter 6: Figure 6.11, 179: From *Goode's World Atlas*, 20th ed., by J. Paul Goode, John C. Hudson (editor), Edward B. Espenshade, Jr. (editor), p. 35. Copyright © 2000 Rand McNally. Reprinted by permission.

Chapter 7: excerpt, 199: From *Doctor Zhivago*, by Boris Pasternak, pp. 232–233, translated by Max Hayward and Manya Harari, English translation copyright © 1958 by William Collins Sons and Co. Ltd. Authorized revisions to the English translation copyright © 1958 by Pantheon Books. Copyright © renewed 1986 by Random House, Inc. Used by permission of Pantheon Books a division of Random House, Inc.

Chapter 8: excerpt, 230: From *Bedouin Life in the Egyptian Wilderness*, by J. Hobbs, pp. 81–82. Copyright © 1989 University of Texas Press. Reprinted by permission.

Chapter 9: Figures 9.1, 9.2, 242, 243: From *The Middle East and North Africa: A Political Geography*, by A. Drysdale and G. Blake. Copyright © 1995 Oxford University Press.

Reprinted by permission. **Figure 9.B, 262:** From *The Middle East and North Africa: A Political Geography*, by A. Drysdale and G. Blake, p. 61. Copyright © 1995 Oxford University Press. Reprinted by permission.

Chapter 10: Figure 10.6, 283: From *Goode's World Atlas*, 20th ed., by J. Paul Goode, John C. Hudson (editor), Edward B. Espenshade, Jr. (editor), p. 15. Copyright © 2000 Rand McNally. Reprinted by permission. **Figure 10.9, 288:** From *Goode's World Atlas*, 20th ed., by J. Paul Goode, John C. Hudson (editor), Edward B. Espenshade, Jr. (editor), p. 35. Copyright © 2000 Rand McNally. Reprinted by permission.

Chapter 12: Figure 12.C, 338: From "Indonesia Cracks Down in Separatist Irian Jaya," by Calvin Sims, *The New York Times*, 12-4-00, A3. Copyright © 2000 The New York Times. Reprinted by permission.

Chapter 17: Figure 17. A (right), 440: From "The Battle With AIDS," *The Economist*, 7-15-00, Copyright © 2000 The Economist. Reprinted by permission.

Chapter 19: Figure 19. 8, 508: Adapted from *Latin America and the Caribbean*, by D. L. Clawson p. 134. Copyright © 1997 Wm. C. Brown Publishers. Reprinted by permission of Times Mirror Higher Education Group, Inc.

Chapter 21: Figure 21.11, 565: From *2000 Britannica Book of the Year*, Encyclopaedia Britannica, Inc. Copyright © 2000 Encyclopaedia Britannica, Inc. Reprinted by permission.

Chapter 23: Figure 23.10, 616: Adapted from *Latin America and the Caribbean*, by D. L. Clawson p. 409. Copyright © 1997 Wm. C. Brown Publishers. Reprinted by permission of Times Mirror Higher Education Group, Inc. (Graph from data in E. N. Herberg, *Ethnic Groups in Canada: Adaptations and Transitions*, Nelson Canada, 1989; Pie chart from data in E. G. Moore, et al, "The Redistribution of Immigrants in Canada," Population Working Paper #12. Dept. of Geography, Queen's University, Kingston, Ontario. Published by Employment and Immig ration Canada, May 1990.

Glossary

Absolute (mathematical) location   Determined by the intersection of lines, such as latitude and longitude, providing an exact point expressed in degrees, minutes, and seconds.

Acid rain   Precipitation that mixes with airborne industrial pollutants, causing the moisture to become highly acidic, and therefore harmful to flora and bodies of water on which it falls. Sulfuric acid is the most common component of this acid precipitation.

Adaptive reuse   Finding new uses for older buildings and stores, often accompanied by a shift from decline to steady renewal in an urban neighborhood.

Age of Discovery   The three to four centuries of European exploration, colonization, and global resource exploitation and trading led largely by European mercantile powers. It began with Columbus at the end of the 15th century and there was still evidence of it late in the 19th century.

Age of Exploration   (also called the **Age of Discovery**) Era of European seafaring lasting from the 15th century through the early 19th century.

Age-structure profile (population pyramid)   Graphic representation of a country's population by gender and 5-year age increments.

Agricultural or Neolithic (New Stone Age) Revolution   The domestication of plants and animals that began about 10,000 years ago.

Albedo   The amount of the sun's energy reflected by the ground. Less vegetation cover correlates with high albedo, and vice-versa.

Alfisol soils   These productive soils found in steppe regions are among the world's more fertile soils. Also known as *chestnut soils*.

Altitudinal zonation   In a highland area, the presence of distinctive climatic and associated biotic and economic zones at successively higher elevations. Ethiopia and Bolivia are examples of these kinds of zones. See *Tierra caliente, Tierra templada, Tierra fría, Tierra helada*.

Anticline   See Folding.

Anticyclone   Atmospheric high-pressure cell. In the cell, the air is descending and becomes warmer. As it warms, its capacity to hold water vapor increases, and the result is minimal precipitation.

Anti-Semitism   Anti-Jewish sentiments and activities.

Apartheid   The Republic of South Africa's former official policy of "separate development of the races," designed to ensure the racial integrity and political supremacy of the white minority.

Arable   Suitable for cultivation.

Archipelago   A chain or group of islands.

Arid China   In a climatic division of China approximately along the 20° isohyet, Arid China lies to the west of the line. It characterizes more than half of the territory of the country but accommodates less than 10% of China's population.

Arms race   Usually associated with the competition between the United States and the Soviet Union, it refers to rival and potential enemy powers increasing their military arsenals — each in an open-ended effort to stay ahead of the other.

Atmospheric pollution   The modification of the blanket of gases that surround the earth largely through the airborne products of industrial production and the human consumption of fossil fuels.

Atoll   Low islands made of coral and usually having an irregular ring shape around a lagoon.

Autonomy   Self-rule, generally with reference to Palestinians' rights to run their own civil (and some security) affairs in portions of the West Bank and Gaza Strip allocated to them in the 1993–2000 peace agreements.

Balkanization   The fragmentation of a political area into many smaller independent units, as in former Yugoslavia in the Balkan peninsula.

Barrios   The name of densely settled neighborhoods in city space that is characteristically inhabited by migrants of latino or hispanic origin.

Barter   The exchange of goods or services in the absence of cash.

Basin irrigation   Ancient Egyptian system of cultivation using fields saturated by seasonal impoundment of Nile floodwaters.

Bazaar (suq)   Central market of the traditional Middle Eastern city, characterized by twisting, close-set lanes, and merchant stalls.

Belief systems   The set of customs that an individual or culture group has relating to religion, social contracts, and other aspects of cultural organization.

Benelux   The name used to collectively refer to the countries of Belgium, Netherlands, and Luxembourg.

Biodiversity hot spots   A ranked list of places scientists believe deserve immediate attention for flora and fauna study and conservation.

Biological diversity (biodiversity)   The number of plant and animal species and the variety of genetic materials these organisms contain.

Biomass   The collective dried weight of organisms in an ecosystem.

Biome   A terrestrial ecosystem type categorized by a dominant type of natural vegetation.

Birth rate   The annual number of live births per 1,000 people in a population.

Black-earth belt   An important area of crop and livestock production spanning parts of Russia, Ukraine, Moldova and Kazakstan. The main soils of this belt are mollisols.

Boat people   In this text, the refugee populations who attempted to flee Cuba and Haiti in the early 1990s by launching themselves from the islands toward Florida in the hope that they would get picked up by the U.S. Coast Guard and given haven. The term had an earlier use by peoples in Southeast Asia making similar efforts to flee Vietnam.

Bourgeoisie   In Marxist doctrine, the capitalist class.

Brain drain   The exodus of educated or skilled persons from a poor to a rich country or from a poor to a rich region within a country.

Break-of-bulk point   A classic geographic term describing a point in transit when bulk goods must be removed from one mode of transport and installed on another. Trainloads of grain carried to a port for transhipment on cargo boats or barges is a common example.

Broadleaf deciduous forest   Forests typical of middle-latitude areas with humid subtropical and humid continental climates. As cool fall temperatures set in, broadleaf trees shed their leaves and cease to grow, thus reducing water loss. They produce new foliage and grow vigorously during the hot, wet summer.

Buffer state   A generally smaller political unit adjacent to a large, or between several large, political units. Such a role often enables the smaller state to maintain its independence because of its mutual use to the larger, proximate nations. Uruguay, between Brazil and Argentina, is an example.

Capital goods   Goods used to produce other goods.

Carrying capacity   The size of a population of any organism that an ecosystem can support.

Carter Doctrine   President Jimmy Carter's declaration, following the Soviet Union's invasion of Afghanistan in 1979, that the United States would use any means necessary to defend its vital interests in the Persian Gulf Region. The "vital interests" were interpreted to mean oil, and "any means necessary" interpreted to mean that the United States would go to war with the Soviet Union if oil supplies were threatened.

Cartogram   A special map in which an area's shape and size are defined by explicit characteristics of population, economy, or distribution of any stated product.

Cartography   The craft of designing and making maps, the basic language of geography. In recent years, this traditional manual art has been changed profoundly through the use of computers and Geographic Information Systems (GIS) and through major improvements in machine capacity to produce detailed, colored, map products.

Cash (commercial) crops   Crops produced generally for export.

Caste   The hierarchy in the Hindu religion that determines a person's social rank. It is determined by birth and cannot be changed.

Charney effect   Observed by an atmospheric scientist named Charney, this states that the less plant cover there is on the ground, the higher the albedo — solar energy deflected back into the atmosphere — and therefore the lower the humidity and precipitation.

Chemocline   The boundary between lower, carbon dioxide-laden waters and higher, gas-free, fresh water in some African lakes. The puncture of this boundary can cause eruptions that are fatal to humans and other life around the lakes.

Chernobyl   The site in the Ukraine where, in April 1986, the worst nuclear power plant accident in history occurred. It is thought that approximately 5,000 people died and a zone with a 20-mile radius is still virtually uninhabitable; 116,000 people were moved from the area and cleanup continues to this day.

Chernozem   A Russian term meaning "black earth." It is a grassland soil that is exceptionally thick, productive, and durable.

Chestnut soils   Productive soils typical of the Russian steppe and North American Great Plains.

China Proper   The relatively well-watered eastern portion of China, which is the home of the great majority of the Chinese population. It is sometimes described as Humid China, as opposed to Arid China that lies to the west, and that has relatively little population.

Chokepoint   A strategic narrow passageway on land or sea that may be closed off by force or threat of force.

Choropleth maps   Maps that are drawn to show the differing distribution of goods or geographic characteristics (including population) across a broad area. Such maps are good for generalizations, but often mask significant local variations in the presence of the item being mapped.

Chunnel (Eurotunnel)   The 23-mile tunnel that links Britain with the European continent. It was completed in 1994 at a cost of more than $15 billion, making it the single most costly project in landscape transformation ever undertaken.

Civilization   The complex culture of urban life.

Climate   The average weather conditions, including temperature, precipitation, and winds, of an area over an extended period of time.

Climatology   The scientific study of patterns and dynamics of climate.

Coal   Residue from organic material compressed for a long period under overlying layers of the earth. Coals vary in hardness and heating value. Anthracite coal burns with a hot flame and almost no smoke. Bituminous coal is used in the largest quantities; some can be used to make coke by baking out the volatile elements. Coke is largely carbon and burns with intense heat when used in blast furnaces to smelt iron ore. Peat, which is coal in the earliest stages of formation, can be burned if dried. Concern over atmospheric pollution by the sulfur in coal smoke has recently increased the demand for low-sulfur coals.

Coke   The residue of coal that has gas burned off under controlled conditions and can then be brought to higher heat in blast furnaces in the process of making iron and steel.

Cold War   The tense but generally peaceful political and military competition between the United States and the Soviet Union, and their respective allies, from the end of World War II until the collapse of the Soviet Union in 1991.

Collective farm   A large-scale farm in the Former Soviet Union that usually incorporated several villages. Workers received shares of the income after the obligations of the collective had been met.

Collectivization   The process of forming collective farms in Communist countries.

Colonization   The European pattern of establishing dependencies abroad to enhance economic development in the home country.

Colored   A South African term referring to persons of mixed racial ancestry.

Command economy   A centrally-planned economy typical of the Soviet Union and its communist allies, in which the government rather than free enterprise determines the production, distribution, and sale of economic goods and services.

Common Market   An earlier name given to the (current) 15 countries that make up the European Union. In 1957, an initial six countries combined to form the European Economic Community (EEC), and this supranational community has grown to have considerable economic and political importance in Europe. *See* European Union.

Compressional fault   *See* Fault.

Computer cartography   Map making using sophisticated software and computer hardware. It is a new, actively growing career field in geography.

Computer-controlled robots   Micro-technology has allowed the creation of highly precise, robotic units that can be used in delicate manufacturing. Although expensive to create, they ultimately reduce per-unit manufacturing costs because of predictable higher quality output and easier maintenance than human work organizations.

Concentric zone model   A generalized model of a city. A city becomes articulated into contrasting zones arranged as concentric rings around its central business district.

Coniferous vegetation   Needleleaf evergreen trees; most bear seed cones.

Consumer goods   Goods that individuals acquire for short-term use.

Consumer organisms   Animals that cannot produce their own food within a food chain.

Consumption overpopulation   The concept that a few persons, each using a large quantity of natural resources from ecosystems across the world, add up to too many people for the environment to support.

Containerization   The prepackaging of items into larger standardized containers for more efficient transport.

Continental islands   Once attached to nearby continents, these are the islands north and northeast of Australia, including New Guinea.

Convectional precipitation   The heavy precipitation that occurs when air is heated by intense surface radiation, then rises and cools rapidly.

Convention on the Law of the Sea   A 1970s United Nations treaty permitting a sovereign power to have greater access to surrounding marine resources.

Coordinate systems   A means of determining exact or absolute location. Latitude and longitude are most often used.

Cottage industry   Handwork that is done in the evenings or in slack seasons in the rural world. Products that are created during such activity can include woven and sewn goods or piecework for local textile or manufacturing industries. Such work brings cash into the rural economy, especially to those not completely involved in full-time farming activity.

Council for Mutual Economic Assistance (COMECON)   A former economic organization consisting of the Soviet Union, Poland, East Germany, Czechoslovakia, Hungary, Romania, Bulgaria, Cuba, Mongolia, and Vietnam, now disbanded.

Crop calendar   The dates by which farmers prepare, plant, and harvest their fields. Farmers who are involved in new agricultural patterns necessitated by greater use of chemical fertilizers and new plant strains are sometimes unable to adjust to the demands of a much more exact crop calendar than has been traditional.

Crop irrigation   Bringing water to the land by artificial methods.

Crusades   A series of European Christian military campaigns between the 11th and 14th centuries aimed at recapturing Jerusalem and the rest of the Holy Land from the Muslims.

Cultural diffusion   One of the most important dynamics in geography, cultural diffusion is the engine of change as crops, languages, culture patterns, and ideas are transferred from one place to other places, often in the course of human migration.

Cultural geography   The study of the ways in which humankind has adopted, adapted to, and modified the face of the earth, with particular attention given to cultural patterns and their associated landscapes. It also includes a culture's influence on environmental perception and assessment.

Cultural landscape   The landscape modified by human transformation, thereby reflecting the cultural patterns of the resident culture at that time.

Cultural mores   The belief systems and customs of a culture group.

Culture   The values, beliefs, aspirations, modes of behavior, social institutions, knowledge, and skills that are transmitted and learned within a group of people.

Culture hearth   An area where innovations develop, with subsequent diffusion to other areas.

Culture System   A system in which Dutch colonizers required farmers in Java to contribute land and labor for the production of export crops under Dutch supervision.

Cyclone   A low-pressure cell that composes an extensive segment of the atmosphere into which different air masses are drawn.

Cyclonic (frontal) precipitation   The precipitation generated in traveling low-pressure cells which bring different air masses into contact.

Death rate   The annual number of deaths per 1,000 people in a population.

Debt-for-nature swap   An arrangement in which a certain portion of international debt is forgiven in return for the borrower's pledge to invest that amount in nature conservation.

Deciduous trees   Broadleaf trees that lose their leaves and cease to grow during the dry or the cold season and resume their foliage and grow vigorously during the hot, wet season.

Deflation   A fall in prices, such as that caused by currency devaluations.

Deforestation   The removal of trees by people or their livestock.

Delta   Landform resulting from the deposition of great quantities of sediment when a stream empties into a larger body of water.

Demographic shift   Major population redistributions as people move from the countryside to the city or from one region to another.

Demographic transition   A model describing population change within a country. The country initially has a high birth rate, a high death rate, and a low rate of natural increase, moves through a middle stage of high birth rate, low death rate, and high rate of natural increase, and ultimately reaches a third stage of low birth rate, low or medium death rate, and low or negative rate of population increase.

Dependency theory   A theory arguing that the world's more developed countries continue to prosper by dominating their former colonies, the now-independent less developed countries.

Desert   An area too dry to support a continuous cover of trees or grass. A desert generally receives less than 10 inches (25 cm) of precipitation per year.

Desert shrub vegetation   Scant, bushy plant life occurring in deserts of the middle and low latitudes where there is not enough rain for trees or grasslands. The plants are generally *xerophytic*.

Desertification   Expansion of a desert brought about by changing environmental conditions or unwise human use.

Detritus   Material shed from rock surfaces in the process of disintegration and erosion.

Development   A process of improvement in the material conditions of people often linked to the diffusion of knowledge and technology.

Devolution   The process by which a sovereign country releases or loses more political and economic control to its constituent elements, such as states and provinces.

Diaspora   The scattering of the Jews outside Palestine beginning in the Roman Era.

Digital divide   The divide between the handful of countries that are the technology innovators and users, and the majority of nations that have little ability to create, purchase, or use new technologies.

Dissection   The carving of a landscape into erosional forms by running water.

Distributary   A stream that results when a river subdivides into branches in a delta.

Domestication   The controlled breeding and cultivation of plants and animals.

Donor democracy   Typical of countries in Africa South of the Sahara, this is a situation in which a government makes just enough concessions on voting rights or human rights to win foreign loans and aid, without instituting any serious democratic reform.

Donor fatigue   Public or official weariness of extending aid to needy people.

Double cropping   The growing of two crops a year on the same field.

Drought avoidance   Adaptations of desert plants and animals to evade dry conditions by migrating (animals) or being active only when wet conditions occur (plants and animals).

Drought endurance Adaptations of desert plants and animals to tolerate dry conditions through water storage and heat loss mechanisms.

Dry farming Planting and harvesting according to the seasonal rainfall cycle.

Ecologically dominant species A species that competes more successfully than others for nutrition and other essentials of life.

Ecology The study of the interrelationships of organisms to one another and to the environment.

Economic development A process that generally includes a major demographic shift toward a more urban population, replacement of major imports with domestically produced goods, expansion of export industries based on manufactured goods, and a general upgrading of domestic patterns of health care, energy use, and literacy.

Economic shock therapy Russia's economic transformation in the early 1990s from a command economy to a free market economy. Overseen by Boris Yeltsin, this transformation was difficult for a country accustomed to government direction in all economic matters, thus the "shock."

Ecosphere (biosphere) The vast ecosystem composed of all of Earth's ecosystems.

Ecosystem A system composed of interactions between living organisms and nonliving components of the environment.

Ecotourism International tourism packages that take small groups willing to pay relatively high fees to see fragile environments and/or unusual or rare fauna and flora. Such groups are often put up in simple, even primitive accommodations to further intensify the experience of making a willing sacrifice to experience such unusual landscapes.

Ejido An agricultural unit in Mexico characterized by communally farmed land, or common grazing land. It is of particular importance to indigenous villages.

Endemic species A species of plant or animal found exclusively in one area.

Energy crisis The petroleum shortages and price surges sparked by the 1973 oil embargo.

Environmental assessment The process of determining the condition and value of a particular environmental setting. Used both in aspects of landscape change and in evaluating environmental perception.

Environmental determinism The belief that the physical environment has played a major role in the cultural development of a people or locale. Also called environmentalism.

Environmental perception The systematic evaluation of environmental characteristics in terms of human patterns of settlement, resources, land use, and attitudes toward the earth and its settings.

Environmental possibilism A philosophy seen in contrast to environmental determinism that declares that although environmental conditions do have an influence on human and cultural development, people have varied possibilities in how they decide to live within a given environment.

Equinox On or about September 23, and again on or about March 20, the Earth reaches the equinox position. Its axis does not point toward or away from the sun, so days and nights are of equal length at all latitudes on the Earth.

Escarpment A steep edge marking an abrupt transition from a plateau to an area of lower elevation.

Estuary A deepened ("drowned") river mouth into which the sea has flooded.

Ethnic cleansing The relocation or killing of members of one ethnic group by another, to achieve some demographic, political or military objective.

Ethnocentrism Regarding one's own group as superior and as setting proper standards for other groups.

Euro Part of the authority of the EU has been the institution of a new currency that has become "coin of the realm" since early 2002. Not all EU nations accept the euro but in more than two-thirds this single currency system is used in electronic and coin and bill exchange.

European Community The name that was replaced in 1993 by the term European Union. The European Community continues to serve as a governing and administrative body for the EU.

European Economic Community An economic organization designed to secure the benefits of large-scale production by pooling resources and markets. The name has been changed to Economic Union. *See* Common Market and Economic Union.

European Free Trade Association (EFTA) An organization that maintains free trade among its members but allows each member to set its own tariffs in trading with the outside world. Members are Iceland, Norway, Sweden, Switzerland, Austria, and Finland.

European Union The current organization begun in the 1950s as the Common Market. It now is made up of 15 nations, including France, Germany, Italy, Belgium, Luxembourg, the Netherlands, the United Kingdom, Denmark, Ireland, Greece, Portugal, Spain, Austria, Finland, and Sweden. *See* Common Market.

Evaporation The loss of moisture from the earth's land surfaces and its water bodies to the air through the ongoing influence of solar radiation and transpiration by plants.

Exotic species A nonnative species introduced into a new area.

Exploration The human fascination with new landscapes that leads to the search for new places, new resources, new cultural patterns, and new routes of travel and trade.

Extensive land use A livelihood, such as hunting and gathering, that requires the use of large land areas.

External costs or **externalities** Consequences of goods and services that are not priced into the initial cost of those goods and services.

Fall line A zone of transition in the eastern United States where rivers flow from the harder rocks of the Piedmont to the softer rocks of the Atlantic and Gulf Coastal Plain. Falls and/or rapids are characteristic features.

Fault A break in a rock mass along which movement has occurred. A break due to rock masses being pulled apart is a tensional or "normal" fault, whereas a break due to rocks being pushed together until one mass rides over the other is a compressional fault. The processes of faulting create these breaks.

Feral animals Domesticated animals that have abandoned their dependence on people to resume life in the wild.

Fertile Crescent The arc-shaped area stretching from southern Iraq through northern Iraq, southern Turkey, Syria, Lebanon, Israel and western Jordan, where plants and animals were domesticated beginning about 10,000 years ago.

Final status issues Issues deferred to the end of the Oslo Peace Process between Israel and the PLO. Finally dealt with at Camp David in 2000, they included the status of Jerusalem, the fate of Palestinian refugees, Palestinian statehood and borders between Israel and a new Palestinian state. The negotiations broke down over these final status issues.

Fjord A long, narrow extension of the sea into the land usually edged by steep valley walls that have been deepened by glaciation.

Flow resource A resource that can be renewed and hence that extracts a lower cost from the environment in its utilization.

Folding The process creating folds (landforms resulting from an intense bending of rock layers). Upfolds are known as anticlines, and downfolds are called synclines.

Food chain The sequence through which energy, in the form of food, passes through an ecosystem.

Formal region *See* Region.

Fossil waters Virtually nonrenewable freshwater supplies, the product of ancient rainfall stored in deep natural underground aquifers, especially in the Middle East and North Africa.

Four Modernizations The effort of the Chinese after the death of Mao Zedong in 1976 to focus Chinese efforts at economic development in agriculture, industry, science and technology, and defense.

Front A contact zone between unlike air masses. A front is named according to the air mass that is advancing (cold front or warm front).

Frontal precipitation *See* Cyclonic precipitation.

Fuelwood crisis Deforestation in the less developed countries caused by subsistence needs.

Functional region *See* Region.

Gaia hypothesis A hypothesis stating that the ecosphere is capable of restoring its equilibrium following any disturbance that is not too drastic.

Gentrification The social and physical process of change in an urban neighborhood by the return of young, often professional, populations to the urban core. These peoples are often attracted by the substantial nature of the original building stock of the place and the proximity to the city center, which generally continues to have major professional opportunities. While this process brings an urban landscape back into a primary role as a tax base, this change does dispossess a considerable number of minority peoples, who tended to remain in these neighborhoods as the white population moved to the suburbs during the past five or six decades.

Geographic analysis By giving attention to the spatial aspects of a distribution, geographic analysis helps to explain distribution, density, and flow of a given phenomenon.

Geographic Information Systems (GIS) The increasingly popular field of computer-assisted geographic analysis and graphic representation of spatial data. It is based on superimposing various data layers that may include everything from soils to hydrology to transportation networks to elevation. Computer software and hardware are steadily improving, enabling GIS to produce ever more detailed and exact output.

Geography The study of the spatial order and associations of things. Also defined as the study of places, the study of relationships between people and environment and the study of spatial organization.

Geomorphology The scientific analysis of the landforms of the earth; sometimes called physiography.

Geopolitics The study of geographical factors in political systems including borders, political unity, and warfare.

Geopolitiks The initial German term for geopolitics that included geographical factors in political systems but that also became associated with concepts of racial superiority.

Glacial deposition In the process of continental and valley glaciation, the deposition of moraines that become lateral or terminal — depending on where they are deposited — in the act of glacial retreat. This same process also leads to glacial scouring as moving ice picks up loose rock and reshapes the landscape as the glacier moves forward or retreats.

Glacial scouring *See* Glacial deposition.

Globalization The growing interaction among worldwide markets, production centers, and labor pools.

Graben (rift valley) A landform created when a segment of Earth's crust is displaced downward between parallel tensional faults or when segments of the crust which border it ride upward along parallel compressional faults.

Gravity flow Water flow in an irrigation or a hydroelectric system that allows for the free flow of water from a source to another area without the addition of motor or animal power. In bringing water to a city, such a characteristic is of particular importance because of its relative cheapness compared to the need to pump water upslope.

Great Rift Valley The result of tectonic processes, this is a broad, steep-walled trough extending from the Zambezi valley in southern Africa northward to the Red Sea and the valley of the Jordan River in southwestern Asia.

Great Trek In what is now South Africa, this was series of northward migrations in the 1830s by which groups of Boers, primarily from the eastern part of the Cape Colony, sought to find new interior grazing lands and establish new political units beyond British reach.

Green Revolution The introduction and transfer of high-yielding seeds, mechanization, irrigation, and massive application of chemical fertilizers to areas where traditional agriculture has been practiced.

Greenhouse effect The observation that increased concentrations of carbon dioxide and other gases in Earth's atmosphere causes a warmer atmosphere.

Gross domestic product (GDP) The value of goods and services produced in a country in a given year. Does not include net income earned outside the country. The value is normally given in current prices for the stated year.

Growth national product (GNP) The value of goods and services produced internally in a given country during a stated year, plus the value resulting from transactions abroad. The value is normally stated in current prices of the stated year. Such data must be used with caution in regard to developing countries because of the broad variance in patterns of data collection and the fact that many people consume a large share of what they produce.

Growth pole A new city, a resource, or some development in a heretofore undersettled area that begins to attract population growth and economic development.

Guano Seabird excrement that is also found in phosphate deposits.

Guest workers Generally young male migrants — often from northern Africa and the eastern Mediterranean — who began to come to western Europe in the 1960s when Germany, France, and England were particularly troubled with labor shortages. The guest worker program did not anticipate permanent residency for these workers but the migrations have, in most cases, led to permanent immigration status for such migrants. Considerable social tension is now associated with the guest worker movement.

Gulf Stream Strong ocean current originating in the tropical Atlantic Ocean that skirts the eastern shore of the United States, curves eastward, and reaches Europe as a part of the broader current called the North Atlantic Drift.

Gulf War The 1990–1991 confrontation between a large military coalition, spearheaded by the United States, and the Iraqi forces that had invaded Kuwait. The allies successfully drove Iraqi forces from Kuwait. Earlier, the "Gulf War" had referred to the 1980–1988 war between Iraq and Iran.

Hacienda A Spanish term for large rural estates owned by the aristocracy in Latin America.

Haj (Hajj) The pilgrimage to Mecca, the principal holy city in the Islamic religion. Every Muslim is required to make this pilgrimage at least once in a lifetime, if possible.

Han Chinese The original Chinese peoples who settled in North China on the margins of the Yellow River (Huang He) and who were central to the development of Chinese culture. Han Chinese make up approximately 94% of China's current population.

Hearth areas Regions of original development in what is now southern Germany and Austria, eventually occupying much of continental Europe and even the British Isles, from which the preliterate Celtic-speaking tribes radiated.

Hierarchichal rule A system by which leadership for a culture group is defined by traditional and sometimes frustrating patterns of age-sex distinction. Evident in most cultures and slowly modified by an increasing role of democratic elections and the establishment of a broader opportunity base for individual advancement.

High islands Generally the result of volcanic eruptions, these are the higher and more agriculturally productive and densely populated islands of the Pacific World.

High Pressure Cell (anticyclone) An air mass descending and warming because of increased pressure and weight of the air above. High pressure typically means low relative humidity and minimal precipitation.

Hinterland The realm of a country that lies away from the capital and largest cities and is most often rural or even unsettled; it is often seen in the minds of economic planners as a realm with a development potential.

Historical geography Concern with the historical patterns of human settlement, migration, town building, and the human use of the earth. Often the subdiscipline that best blends geography and history as a perspective on human activity.

Holocaust Nazi Germany's attempted extermination of Jews, Gypsies, homosexuals, and other minorities during World War II.

Homelands Ten former territorial units in South Africa reserved for native Africans (blacks). They had elected African governments, and some were designated as "independent" republics, although they were not recognized outside of South Africa. Formerly called Bantustans. They were abolished in 1994.

Horizontal migration The movements of pastoral nomads over relatively flat areas to reach areas of pasture and water.

Horst Block mountain formed between two roughly parallel faults; it has a steep face on two sides.

Hot money Short-term and often volatile flows of investment that can cause serious damage to the "emerging market" economies of less developed countries.

Hot spot A small area of Earth's mantle where molten magma is relatively close to the crust. Hot spots are associated with island chains and thermal features.

Household Responsibility System The Chinese system devised in 1978 that enabled their rural population to begin to free itself from the communal structures that had characterized the years of development under Chairman Mao Zedong. This innovation returned much agricultural decision making to the farm household, although there was a required portion of the major crops that had to be sold to the government.

Human Development Index (HDI) A United Nations-devised ranked index of countries' development that evaluates quality of life issues (such as gender equality, literacy, and human rights) in addition to economic performance.

Humboldt (or Peru) Current The cold ocean current that flows northward along the west coast of South and North America. It plays a significant role in regional fish resources.

Humid China The eastern portion of China that is relatively well-watered and where the great majority of the Chinese population has settled. An arc from Kunming in southwest China to Beijing in North China describes the approximate western margin of this zone.

Humid continental Climate with cold winters, warm to hot summers, and sufficient rainfall for agriculture, with the greater part of the precipitation in the summer half-year.

Humid pampa A level to gently rolling area of grassland centered in Argentina with a humid subtropical climate.

Humid subtropical Climate that characteristically occupies the southeastern margins of continents with hot summers, mild to cool winters, and ample precipitation for agriculture.

Humus Decomposed organic soil material. Grasslands characteristically provide more humus than forests do.

Hunting and gathering A mode of livelihood, based on collection of wild plants and hunting of wild animals, generally characterizing preagricultural peoples.

Hydraulic control The ability to control water in irrigation systems, rivers, urban settlements, and for the generation of electricity. It has been central to the development of major civilizations and to culture groups throughout human history.

Ice cap Climate and biome type characterized by permanent ice cover on the ground, no vegetation (except where limited melting occurs), and a severely long, cold winter. Summers are short and cool.

Ideology A system of political and/or economic beliefs such as communism, capitalism, autocracy, or democracy.

Igneous (volcanic) rock Formed by the cooling and solidification of molten materials. Granite and basalt are common types. Such rocks tend to form uplands in areas where sedimentary rocks have weathered into lowlands. Certain types break down into extremely fertile soils. These rocks are often associated with metal-bearing ores.

Industrial Revolution A period beginning in mid-18th-century Britain that saw rapid advances in technology and the use of inanimate power.

Inflation A rise in prices.

Information technology The Internet, wireless telephones, fiber optics, and other technologies characteristic of more developed countries. Generally seen as beneficial for a country's economic prospects, "IT" is also spreading in the less developed countries.

Intensive land use A livelihood requiring use of small land areas, such as farming.

Intertillage The growing of two or more crops simultaneously in alternate rows. Also called interplanting.

Irredentism The demand for an international transfer of territory in order to place a minority population in its alleged homeland.

Irrigation Being able to overcome rainfall deficiencies by the control and transport of water to farmland has historically been pivotal to centers of major agricultural production. This act requires engineering skills, political organization all along a water source, and considerable labor investment in the initial creation and the continuing maintenance of such a system.

Island chain A series of islands formed by ocean crust sliding over a stationary hot spot in the Earth's mantle.

Karst A landscape feature of an area built on limestone or dolomite, which is susceptible to differential erosion, with resultant sinks and mountains that seem to shoot straight up from limestone plains. Although the term comes from the Adriatic Coast in Europe, some of the world's most dramatic karst landscapes occur in southern China.

Keystone species A species which affects many other organisms in an ecosystem.

Kyoto Protocol A treaty on climate change signed by 160 countries in Kyoto, Japan, in 1997. It requires MDCs to reduce their greenhouse gas emissions by more than 5% below their 1990 levels by the year 2012; the U.S. target is 6%.

Lacustrine plain Floor of a former lake where glacial meltwater accumulated and sediments washed in and settled. An extremely flat surface is characteristic.

Landscape Originally, an artist's view of a setting for human activity but now much more broadly used to define the scene that can be observed from any given point and prospective.

Landscape transformation The human process of making over the earth's surface into a setting seen as more productive, more convenient, and more aesthetically pleasing. From initial human shelters to contemporary massive urban centers, humans have been driven to change their environmental settings, causing negative as well as positive environmental outcomes.

Large-scale map A map constructed to show considerable detail in a small area.

Laterite A material found in tropical regions with highly leached soils; composed mostly of iron and aluminum oxides which harden when exposed and make cultivation difficult.

Latifundia (sing. latifundio) Large agricultural Latin American estates with strong commercial orientations.

Latitude Measurement that denotes position with respect to the equator and the poles. Latitude is measured in degrees, minutes, and seconds, which are described as parallels. The low latitudes lie between the Tropics of Cancer and Capricorn; the middle latitudes lie between the Tropics and the Arctic and Antarctic Circles; the high latitudes lie between the Polar Circles and the Poles.

Law of Return An Israeli law permitting all Jews living in Israel to have Israeli citizenship.

Law of the Sea A United Nations treaty or convention permitting coastal nations to have greater access to marine resources.

Less developed countries (LDCs) The world's poorer countries.

Lifeboat ethics Ecologist Garrett Hardin's argument that, for ecological reasons, rich countries should not assist poor countries.

Lithospheric plates The top layer of sediments on the earth and the plates on which human settlement and transport takes place. The *lithic* of Neolithic, for example, comes from the same root meaning stone. Above the lithosphere is the biosphere (layer of life) and above that is the atmosphere (blanket of air).

Location Central to all geographic analysis is the concept of location. Where something "is" relates to all manner of influences, from climate to migration routes. The classic use of this term comes in the response to the question: "What is most important in real estate?" The answer: "Location, location, location." It is a crucial component in trying to understand patterns of historic and economic development.

Loess Fine-grained material that has been picked up, transported, and deposited in its present location by wind; it forms an unusually productive soil.

Long March The 1935–1936 flight of approximately 100,000 Communist troops under Mao Zedong from south central China to the mountains of North China. Mao was fleeing attack by General Chiang Kai-shek, and the success of the Long March led to the establishment of Mao's troops as a force with the capacity to overcome the much larger and better armed troops of Chiang.

Longitude Measurement that denotes a position east or west of the prime meridian (Greenwich, England). Longitude is measured in degrees, minutes, and seconds, and meridians of longitude extend from Pole to Pole and intersect parallels of latitude.

Low islands Made of coral, these are the generally flat, drier, less agriculturally productive, and less densely populated islands of the Pacific World.

Maastricht Treaty of European Union On November 1, 1993 twelve nations (Belgium, Britain, Denmark, France, Germany, Greece, Ireland, Italy, Luxembourg, Netherlands, Portugal, and Spain) met and initiated the European Union allowing unrestricted flow of goods, capital, and services among the member states. In 1996 three additional countries were added (Austria, Finland, and Sweden) and these fifteen have become today's European Union (EU).

Malthusian scenario The model forecasting that human population growth will outpace growth in food and other resources, with a resulting population die-off.

Map projection A way to minimize distortion in one or more properties of a map (direction, distance, shape, or area).

Map scale The actual distance on the earth that is represented by a given linear unit on a map.

Maquiladoras Operations dedicated to the assembly of manufactured goods generally, in Mexico and Central America, from components initially produced in the United States or other places. With the enactment of NAFTA in 1994, there has been a massive expansion of maquiladora operations because of the economic energy generated by this regional economic alliance.

Marginalization A process by which poor subsistence farmers are pushed onto fragile, inferior, or marginal lands which cannot support crops for long and which are degraded by cultivation.

Marine west coast Climate occupying the western sides of continents in the higher middle latitudes; greatly moderated by the effects of ocean currents that are warm in winter and cool in summer relative to the land.

Marshall Plan The plan designed largely by the United States after the conclusion of World War II by which U.S. aid was focused on the rebuilding of the very Germany that had been its enemy in the war just concluded. Secretary of State George Marshall (who had been Chief of Staff of U.S. Army from 1939–1945), was central to the plan's design and implementation.

Material culture The items that can be seen and associated with cultural development, such as house architecture, musical instruments, and tools. In study of the cultural landscape, items of material culture add considerable personality to cultural identity.

Medical geography The study of patterns of disease diffusion, environmental impact on public health, and the interplay of geographic factors, migration, and population. With the increasing ease of international movement, medical geography is becoming more important as the potential for disease diffusion increases.

Medina An urban pattern typical of the Middle Eastern city before the 20th century.

Mediterranean (dry-summer subtropical) Climate that occupies an intermediate location between a marine west coast climate on the poleward side and a steppe or desert climate on the equatorward side. During the high-sun period, it is rainless; in the low-sun period, it receives precipitation of cyclonic or orographic origin.

Mediterranean scrub forest Xerophytic vegetation typical of hot, dry-summer, Mediterranean climate regions. Local names for this vegetation type include *maquis* and *chaparral*.

Meiji Restoration The Japanese revolution of 1868 that restored the legitimate sovereignty of the emperor. *Meiji* means "Enlightened Rule." This event led to a transformation of Japan's society and economy.

Melanesia (Black Islands) A group of relatively large islands in the Pacific Ocean bordering Australia.

Mental maps In every individual's mind is a series of locations, access routes, physical and cultural characteristics of places, and often a general sense of the good or bad of locales. The term mental map is used to define such geographies. Education is often aimed at replacing subjective impressions of places and peoples with more objective information in such personal geographies.

Mercantile colonialism The historical pattern by which Europeans extracted primary products from colonies abroad, particularly in the tropics.

Meridian *See* Longitude.

Mestizo In Latin America, a person of mixed European and Native American ancestry.

Metamorphic rock Formed from igneous or sedimentary rock through changes occurring in the rock structure as a result of heat, pressure, or the chemical action of infiltrating water. Marble, formed of pure limestone, is a common example. Such rocks are generally harder and more closely knit together than sedimentary rocks, are usually more resistant to wearing down by weathering and erosion, and tend to form uplands in areas where the sedimentary rocks have weathered into lowlands. Metal-bearing ores often are found in areas of metamorphic rock.

Metropolitan area A city together with suburbs, satellites, and adjacent territory with which the city is functionally interlocked.

Microcredit The lending of small sums to poor people to set up or expand small businesses.

Micronesia (Tiny islands) Thousands of small and scattered islands in the central and western Pacific Ocean, mainly north of the Equator.

Microstate A political entity that is tiny in area and population and is independent or semi-independent.

Middle Eastern ecological trilogy The model of mostly symbiotic relations among villagers, pastoral nomads, and urbanites in the Middle East.

Migration A temporary, periodic, or permanent move to a new location.

Milpero A Latin American farmer who engages in swidden or shifting cultivation in forest lands. The plot is often called the milpa.

Minifundia (sing. minifundio) Small Latin American agricultural landholdings, usually with a strong subsistence component.

Mixed forest A transitional area where both needleleaf and broadleaf trees are present and compete with each other.

Mobility The pattern of human movement between work and residence, changing because of continual improvement in means of transportation. Increasing mobility has increased the impact of urban space on adjacent farmlands in most developed countries.

Mollisols Thick, productive, and durable soils, such as the *chernozem*, whose fertility comes from abundant humus in the top layer.

Monoculture Single-species cultivation of food or tree crops, usually very economical and productive, but threatening to natural diversity and change.

Monsoon A current of air blowing fairly steadily from a given direction for several weeks or months at a time. Characteristics of a monsoonal climate are a seasonal reversal of wind direction, a strong summer maximum of rainfall, and a long dry season lasting for most or all of the winter months.

Montreal Protocol A 1989 international treaty to ban chlorofluorocarbons.

Moor Rainy, deforested upland, covered with grass or heather and often underlain by water-soaked peat.

Moraine Unsorted material deposited by a glacier during its retreat. Terminal moraines are ridges formed by long-continued deposition at the front of a stationary ice sheet.

More developed countries (MDCs) The world's wealthier countries.

Mulatto A person of mixed European and black ancestry.

Multinational companies Companies that operate, at least in part, outside their home countries.

Multiple use A water system that enables human use of it a number of times, as in stream flow from a river into a turbine and then back into a stream or another, lower, turbine system. Because there is no waste in such a system, it is highly desired and leads to sometimes complex engineering to achieve multiple use of this resource.

Multiple-cropping Farming patterns in which several crops are raised on the same plot of land in the course of a year or even a season. Common examples of this are winter wheat and summer corn or soybeans in North America or several crops of rice in the same plot in Monsoon Asia.

Nation Commonly, a nation is a term describing the citizens of a state — or that state — but it has additional usage as a collective term for peoples of high cultural homogeneity, existing as a separate political entity. *See* Nation-state.

Nation-state A political situation in which high cultural and ethnic homogeneity characterizes the political unit in which such people live. The term ethnic cleansing has been used to define some of the bloody efforts to achieve nation-state status in the collapse of the former Yugoslavia in the 1990s.

Nationalism The drive to expand the identity and strength of a political unit which serves as home to a population interested in greater cohesion and often expanded political power.

Natural levee A strip of land immediately alongside a stream that has been built up by deposition of sediments during flooding.

Natural replacement rate The highest rate at which a renewable resource can be used without decreasing its potential for renewal. Also known as *sustainable yield*.

Natural resource A product of the natural environment that can be used to benefit people. Resources are human appraisals.

Near Abroad Russia's name for the now-independent former republics, other than Russia, of the USSR.

Neocolonialism The perpetuation of a colonial economic pattern in which developing countries export raw materials to, and buy finished goods from, developed countries. This relationship is more profitable for the developed countries.

Neo-Malthusians Supporters of forecasts that resources will not be able to keep pace with the needs of growing human populations.

Newly industrializing countries (NICs) The more prosperous of the world's less developed countries.

Nonblack soil zone Areas of poor soil in cool, humid portions of the Slavic coreland, suitable for cultivation of rye.

North Atlantic Drift A warm current, originating in tropical parts of the Atlantic Ocean, that drifts north and east, moderating temperatures of western Europe. Also called Gulf Stream.

North Atlantic Treaty Organization (NATO) A military alliance formed in 1949 which included the United States, Canada, many European nations, and Turkey.

North European Plain The level to rolling lowlands that extend from the low countries on the west through Germany to Poland. These are areas rich in agricultural development, dense human settlements, and a number of major industrial centers. The plain is broken by a number of rivers flowing from the Alps and other mountain systems in central Europe into the North and Baltic Seas.

Occupied Territories The territories captured by Israel in the 1967 War: the Gaza Strip and West Bank, captured respectively from Egypt and Jordan, and the Golan Heights, captured from Syria.

Oil embargo The 1973 embargo on oil exports imposed by Arab members of the Organization of Petroleum Exporting Countries (OPEC) against the United States and the Netherlands.

Open borders   Political boundaries that have minimal political or structural impediment to easy crossing between two countries.

Organization for Economic Cooperation and Development (OECD)   *See* Organization for European Economic Cooperation.

Organization for European Economic Cooperation (OEEC)   The Organization for European Economic Cooperation was created in 1948 to organize and facilitate the European response to the 1947 Marshall Plan. In the early 1960s, the OEEC became the Organization for Economic Cooperation and Development (OECD), and this served as a multinational base for continued planning in economic and social development. The Common Market came to overshadow this organization, and finally, the European Union, in 1993, became the most powerful and influential multinational organization in Europe.

Orographic precipitation   The precipitation that results when moving air strikes a topographic barrier, such as a mountain, and is forced upward.

Outwash   Glacial material carried by sheets of meltwater underneath a glacier and subsequently deposited.

Ozone hole   Areas of depletion of ozone in Earth's stratosphere, caused by chlorofluorocarbons.

Palestine Liberation Organization   The Palestinian military and civilian organization created in the 1960s to resist Israel, and recognized in the 1990s as the sole legitimate organization representing official Palestinian interests internationally.

Parallel   Latitude line running parallel to the equator.

Patrilineal descent system   Kinship naming system based on descent through the male line.

People overpopulation   The concept that many persons, each using a small quantity of natural resources to sustain life, add up to too many people for the environment to support.

Per capita   For every person or per person.

Per Capita Gross Domestic Product Purchasing Power Parity (GDP PPP)   In this figure, annual per capita gross domestic product (GDP) — the total output of goods and services a country produces for home use in a year — is divided by the country's population. For comparative purposes, that figure is adjusted for purchasing power parity (PPP), which involves the use of standardized international dollar price weights that are applied to the quantities of final goods and services produced in a given economy. The resulting measure, per capita GDP PPP, provides the best available starting point for comparisons of economic strength and well being between countries.

Perennial irrigation   Year-round irrigated cultivation of crops, as in the Nile Valley following construction of barrages and dams.

Periodic markets   Non-daily markets that occur in village centers only on specific days each week or month. Merchants, craftspeople, mobile vendors, and farmers work within a regional schedule and convene in different rural settlements on determined days for selling and buying. If the community gets large, or if the market is unusually successful, the market may become permanent.

Permafrost   Permanently frozen subsoil.

Photosynthesis   The process by which green plants use the energy of the sun to combine carbon dioxide with water to give off oxygen and produce their own food supply.

Physical geography   The subdiscipline in the field of geography most concerned with the climate, landforms, soils, and physiography of the earth's surface.

Piedmont   A belt of country at an intermediate elevation along the base of a mountain range.

Pillars of Islam   The five fundamental tenets of the faith of Islam.

Pivotal countries   Those countries whose collapse would cause international refugee migration, war, pollution, disease epidemics or other international security problems.

Place identity   In geographic analysis of a given locale, the nature of place identity becomes a means of understanding people's response to that particular place. Determination of the environmental and cultural characteristics that are most frequently associated with a certain place helps to establish that "sense of place" for a given location.

Plain   A flat to moderately sloping area, generally of slight elevation.

Planned economy   Government determination of where industrial plants should be located, what products should be produced, and how the labor force should be organized are all aspects of the planned economies found in the Soviet Union, the People's Republic of China, and Eastern Europe from the end of WWII to the early 1990s. From the early 1990s to the present time, this pattern of organization has been replaced with a market economy, or a system that lets markets — both domestic and foreign — determine the organization of resources and industrial activity.

Plantation   Large commercial farming enterprise, generally emphasizing one or two crops and utilizing hired labor. Plantations are commonly found in tropical regions.

Plate tectonics   The dominant force in the creation of the continents, mountain systems, and ocean deeps. The steady, but slow, movement of these massive plates of the earth's mantle and crust has created the positions of the continents that we have and major patterns of volcanic and seismic activity. Areas where plates are being pulled under other plates are called subduction zones.

Plateau   An elevated plain, usually lying above 2,000 feet (610 m) above sea level. Some are known as tablelands. A dissected plateau is a hilly or mountainous area resulting from the erosion of an upraised surface.

Pleistocene overkill   A hypothesis stating that hunters and gatherers of the Pleistocene Era hunted many species to extinction.

Podzol   Soil with a grayish, bleached appearance when plowed, lacking in well-decomposed organic matter, poorly structured, and very low in natural fertility. Podzols are the dominant soils of the taiga.

Polder   An area reclaimed from the sea and enclosed within dikes in the Netherlands and other countries. Polder soils tend to be very fertile.

Polynesia (Many islands)   A Pacific island region that roughly resembles a triangle with its corners at New Zealand, the Hawaiian Islands, and Easter Island.

Population change rate   The birth rate minus the death rate in a population.

Population density   The average number of people living in a square mile or square kilometer. It is a very handy statistic for generalized comparisons but often fails to provide a detailed sense of the real distribution of people.

Population explosion   The surge in Earth's human population which has occurred since the beginning of the Industrial Revolution.

Postindustrial   Just as the Industrial Revolution of the mid-18th century was a monumental time of change in geographic patterns, the current shifts in employment and trade that have taken place in the past two decades are also seen as significant. New importance in postindustrial society is given to information management, financial services, and the service sector.

Prairie   An area of tall grass in the middle latitudes composed of rich soils that have been cleared for agriculture. Original lack of trees may have been due to repeated burnings or periodic drought conditions.

Primary consumers (herbivores)   Consumers of green plants.

Primate city   A city strongly dominant within its country in population, government, and economic activity. Primate cities are generally found in developing countries, although some already developed countries, such as France, have them.

Prime meridian   *See* Latitude.

Producers (Self-feeders)   In a food chain, organisms that produce their own food (mostly, green plants).

Projection   The distortion caused by the transfer of three-dimensional space on Earth's surface to the two dimensions of a flat map.

Push-pull forces of migration   In nonforced human migration, there is generally a series of influences on migrants that tend to push them away from a place. Economics and social conditions are common push factors, but images of other places (often unsubstantiated) also serve to attract the migrants and influence the future use of migration destinations.

Pyramid of biomass   A diagrammatic representation of decreasing biomass in a food chain.

Pyramid of energy flow   A diagrammatic representation of the loss of high quality, concentrated energy as it passes through the food chain.

Qanat   A tunnel used to carry irrigation and drinking water from an underground source by gravity flow; also called foggara or karez.

Rain shadow   A condition creating dryness in an area located on the lee side of a topographic barrier such as a mountain range.

Region   A "human construct" that is often of considerable size and that has substantial internal unity or homogeneity and that differs in significant respects from adjoining areas. Regions can be classed as formal (homogeneous), functional, or vernacular. The formal region, also known as a uniform region, has a unitary quality which derives from a homogeneous characteristic. The United States of America is an example of a formal region. The functional region, also called the nodal region, is a coherent structure of areal units organized into a functioning system by lines of movement or influence that converge on a central node or trunk. A major example would be the trading territory served by a large city and bound together by the flow of people, goods, and information over an organized network of transportation and communication lines. Vernacular or perceptual regions are areas that possess regional identity, such as "The Sun Belt," but share less objective criteria in the use of this regional name. General regions, such as the major world regions in this text, are recognized on the basis of overall distinctiveness.

Remote sensing   Through the use of aerial and satellite imagery, geographers and other scientists have been able to get vast amounts of data describing places all over the face of the earth. Remote sensing is the science of acquiring and analyzing data without being in contact with the subject. It is used in the study of patterns of land use, seasonal change, agricultural activity, and even human movement along transport lines. This process relates closely to GIS. *See also* Geographic Information Systems.

Renewable resource   A resource, such as timber, that is grown or renewed so that a continual supply is available. A finite resource is one that, once consumed, cannot be easily used again. Petroleum products are a good example of such a resource and because it takes too much time to go through the process of creation, they are not seen as renewable.

Resources   Resources become valuable through human appraisal, and their utilization reflects levels of technology, location, and economic ambition. There are few resources that have been universally esteemed (water, land, defensible locales) throughout history, so patterns of resource utilization serve as indexes of other levels of cultural development. *See* Natural resource.

Rift valley.   *See* Graben.

Right of return, Palestinian   Palestinian Arab principle that refugees (and their descendents) displaced from Israel and the Occupied Territories in the 1948 and 1967 wars be allowed to return to the region.

Ring of Fire   The long horseshoe-like chain of volcanoes that goes from the southern Andes Mountains in South America up the west coast of North America and arches over into the northeast Asian island chains of Japan, the Philippines, and Southeast Asia. This zone of seismic instability and erratic vulcanism is caused by the tension built up in plate tectonics. *See* Plate tectonics.

Riparian   A state containing or bordering a river.

Russification   The effort, particularly under the Soviets, to implant Russian culture in non-Russian regions of the former Soviet Union and its Eastern European neighbors.

Sacred space or sacred place   Any locale which people hold in reverence, such as places of worship, cemeteries, and battlefields.

Salinization   The deposition of salts on, and subsequent fertility loss in, soils experiencing a combination of overwatering and high evaporation.

Sand sea   A virtual "ocean" of sand characteristic of parts of the Middle East, where people, plants, and animals are all but nonexistent.

Savanna   Low-latitude grassland in an area with marked wet and dry seasons.

Scale   The size ratio represented by a map; for example, a map with a scale of 1:12,500 is portrayed as 1/12,500 of the actual size.

Scorched earth   The wartime practice of destroying one's own assets to prevent them from falling into enemy hands.

Scrub and thorn forest   Low, sparse vegetation in tropical areas where rainfall is insufficient to support tropical deciduous forest.

Seamount   An underwater volcanic mountain.

Second law of thermodynamics   A natural law stating that high-quality, concentrated energy is increasingly degraded as it passes through the food chain.

Secondary consumers (carnivores)   Consumers of primary consumers.

Sedentarization   Voluntary or coerced settling down, particularly pastoral nomads in the Middle East.

Sedimentary rock   Formed from sediments deposited by running water, wind, or wave action either in bodies of water or on land and which have been consolidated into rock. The main classes are sandstone, shale (formed principally of clay), and limestone, including the pure limestone called chalk. They are extensively associated with lowlands such as the Great Plains of North America but are also common in highlands. In general, they weather into more fertile soils than igneous or metamorphic rocks and are associated with most deposits of coal and petroleum.

Segmentary kinship system   Organization of kinship groups in concentric units of membership, as of lineage, clan, and tribe among Middle Eastern pastoral nomads.

Service sector   The labor sector made up of employees in retail trade and personal services; the sector most likely to increase in employment significance in postindustrial society.

Settler colonization   The historical pattern by which Europeans sought to create new or "neo-Europes" abroad.

Shatter belt   A region having a fragmented and unstable pattern of nationalities and political units.

Shi'a Islam   The branch of Islam regarding male descendants of Ali, the cousin and son-in-law of the Prophet Muhammad, as the only rightful successors to the Prophet Muhammad.

Shifting cultivation   This cultivation involves land rotation between crop and fallow years. In the tropics, secondary vegetation typically takes over fallow fields, while the farmer moves on to clear new mature forest. Where such new lands are not available, fallow periods on old fields are reduced or eliminated, resulting in a soil deterioration. Also known in some contexts as *swidden* or *slash-and-burn* cultivation.

Site and situation   Site is the specific geographic location of a given place, while situation is the accessibility of that site and the nature of the economic and population characteristics of that locale. The combination term deals broadly with the blend of influences of setting and historical development of a place.

Slavic coreland (Fertile Triangle or Agricultural Triangle)   The large area of the western former Soviet region containing most of the region's cities, industries, and cultivated lands.

Small-scale map   A map constructed to give a highly generalized view of a large area.

Soil   The earth mantle made of decomposed rock and decayed organic material.

Sovereign state   A political unit that has achieved political independence and maintains itself as a separate unit.

Soybean   A bean crop that made its way from Asia to the New World during the Age of Discovery and has now become a major crop in the agricultural Midwest of the U.S. It is of particular significance because the plant affixes nitrogen as it grows—making it less demanding on the land—and produces a vegetable crop that is high in protein as well as an oil that can be used as a fuel and as a chemical element in a variety of paint, ink, and other products. It is bound to be a crop of growing significance in the 21st century as both food and petroleum products become ever more costly.

Spatial analysis   *See* Spatial.

Spatial   Geography is concerned at all levels with spatial organization — that is, the way in which space has been compartmentalized, modified, utilized. The patterns that are associated with the distribution of both physical and human features of the landscape are central to all geographic analysis, and such analysis is spatial analysis.

Special Economic Zones (SEZs)   In China, specific urban areas were given special status in the 1980s to enable Chinese planners and entrepreneurs to begin to anticipate economic changes produced by the 1997 return of Hong Kong from Britain. The most important SEZ is Shenzhen, just adjacent to Hong Kong itself, but all of the SEZs have undergone rapid economic and demographic growth since China's current rapid pace of economic growth began in the early 1980s. These locales have served to give Chinese businesspeople a little more latitude in experimentation with free market and joint-venture activities.

Spodosols   Acidic soils that have a grayish, bleached appearance when plowed, lack well-decomposed organic matter, and are low in natural fertility. Also known as *podzols*.

St. Lawrence Seaway   This seaway has a total length of 2,342 miles (3,796 km) between the Atlantic Ocean and its western terminus in Lake Superior. It allows major oceanic vessels to reach the Great Lakes, thus enabling ports as far inland as the west side of Lake Superior to serve as international ports. The seaway was completed in 1959, although Canada and the United States began to anticipate such a waterway project in the last years of the 19th century.

State   A political unit over which an established government maintains sovereign control.

State farm (sovkhoz)   A type of collectivized state-owned agricultural unit in the former Soviet Union; workers receive cash wages in the same manner as industrial workers.

Steppe or **temperate grassland**   A biome composed mainly of short grasses. It occurs in areas of steppe climate, which is a transitional zone between very arid deserts and humid areas.

Subarctic   A high-latitude climate characterized by short, mild summers and long, severe winters.

Subduction zone   *See* Plate tectonics.

Subsidence   The settling of land by the compression of lower layers of rock and soil, usually accelerated by the removal of water from below the surface. Of particular importance in urban centers, with Mexico City having one of the most persistent problems.

Summer solstice   On or about June 22, the first day of summer in the Northern Hemisphere, the northern tip of Earth's axis is inclined toward the sun at an angle of 23 1/2 degrees from a line perpendicular to the plane of the ecliptic. This is the summer solstice in the Northern Hemisphere. On or about December 22, the first day of winter in the Northern Hemisphere, the southern tip of Earth's axis is inclined toward the sun at an angle of 23 1/2 degrees from a line perpendicular to the plane of the ecliptic. This is the summer solstice in the Southern Hemisphere.

Sun Belt   A U.S. area of indefinite extent, encompassing most of the South, plus the West at least as far north as Denver, Salt Lake City, and San Francisco. This region has shown faster growth in population and jobs than the nation as a whole during the past several decades.

Sunni Islam   The branch of Islam regarding successorship to the Prophet Muhammad as a matter of consensus among religious elders.

Sustainable development (ecodevelopment)   Concepts and efforts to improve the quality of human life while living within the carrying capacity of supporting ecosystems.

Sustainable yield (natural replacement rate)   The highest rate at which a renewable resource can be used without decreasing its potential for renewal.

Sweat equity   The actual physical labor expended in remaking some aspect of the cultural landscape. Although it is difficult to give an economic value to this process, the term often reflects the patterns of neighborhood change associated with gentrification.

Swidden   Slash-and-burn agriculture or shifting cultivation. In this process, landless peasant farmers move into remote forest lands and cut down forest growth. After a period of drying out, the forest floor is burned and then varied crops are planted in the ash of the burned forest material. Crops may be raised as little as 1 year, or perhaps several years, and then the farming group moves to adjacent lands to take advantage of the fertilizer afforded by the burned tree growth. While swidden farmers do not make up a large percentage of the world's farm peoples, they have an enormous impact on remaining primary forest growth, especially in tropical rain forests.

Symbolization   The representation of distinct aspects of information shown on a map such as stars for capital cities.

Syncline   *See* Folding.

Taiga   Northern coniferous (needleleaf) forest.

Teaching water   The Chinese phrase used to define the labors that must be accomplished to establish a productive irrigation system.

Technocentrists (Cornucopians)   Supporters of forecasts that resources will keep pace with or exceed the needs of growing human populations.

Technological unemployment   Unemployment caused by the replacement of human labor with highly sophisticated technology, even robots. *See* Postindustrialism.

Tectonic processes   Processes that derive their energy from within the earth's crust and serve to create landforms by elevating, disrupting, and roughening the earth's surface.

Tensional fault   *See* Fault.

Tertiary consumers   Consumers of secondary consumers.

Theory of island biogeography   A theoretical calculation of the relationship between habitat loss and natural species loss, in which a 90% loss of natural forest cover results in a loss of half of the resident species.

Tierra caliente   Latin American climatic zone reaching from sea level upward to approximately 3,000 feet (914 m). In this hot, wet environment are grown such crops as rice, sugarcane, and cacao.

Tierra fría   Latin American climatic zone found between 6000 feet (1800 m) and 12,000 feet (3600 m) above sea level. Frost occurs and the upper limit of agriculture and tree growth is reached.

Tierra helada   The zone above 10,000 feet (3,048 m) that has little vegetation and frequent snow cover.

Tierra templada   A climatic zone extending from approximately 3,000 to 6,000 feet (914–1829 m) above sea level. It is a prominent zone of European-induced settlement and commercial agriculture such as coffee growing.

Togugawa Shogunate   The period 1600 to 1868 in Japan, characterized by a feudal hierarchy and Japan's isolation from the outside world.

Tradable permits   A proposed mechanism for reducing total global greenhouse gases in which each country would be assigned the right to emit a certain quantity of carbon dioxide, according to its population size, setting the total at an acceptable global standard. The assigned quotas could be traded between underproducers and overproducers of carbon dioxide.

Trade winds   Streams of air that originate in semipermanent high-pressure cells on the margins of the tropics and are attracted equatorward by a semipermanent low-pressure cell.

Transhumance   The movement of pastoral nomads and their herd animals between low elevation winter pastures and high-elevation summer pastures. Also known as *vertical migration*.

Transportation infrastructure   In efforts of economic development, it is always seen that the network of transportation facilities — railroads, highways, ports, airports — has a major impact on a government's or business's efforts to achieve higher levels of productivity.

Triangular trade   The 16th- to 19th-century trading links between West Africa, Europe, and the Americas, involving guns, alcohol, and manufactured goods from Europe to West Africa exchanged for slaves. Slaves to the Americans were exchanged for the gold, silver, tobacco, sugar, and rum carried back to Europe.

Trophic (feeding) levels   Stages in the food chain.

Tropical deciduous forest   Vegetation typical of some tropical areas with a dry season. Here, broadleaf trees lose their leaves and are dormant during the dry season and then add foliage and resume their growth during the wet season.

Tropical rain forest   Low-latitude broadleaf evergreen forest found where heat and moisture are continuously, or almost continuously, available.

Tropical savanna climate   A relatively moist low-latitude climate that has a pronounced dry season.

Tundra   A region with a long, cold winter, when moisture is unobtainable because it is frozen, and a very short, cool summer. Vegetation includes mosses, lichens, shrubs, dwarf trees, and some grass.

Underground economy   The "black market" typical of the former Soviet region and many LDCs.

Undifferentiated highland   Climate that varies with latitude, altitude, and exposure to the sun and moisture-bearing winds.

Vernacular region   *See* Region.

Vertical migration   Movements of pastoral nomads between low winter and high summer pastures.

Virgin and idle lands (New lands)   Steppe areas of Kazakstan and Siberia brought into grain production in the 1950s.

Warsaw Pact   A former military alliance consisting of the Soviet Union and the European countries of Poland, East Germany, Czechoslovakia, Hungary, Romania, and Bulgaria (now dissolved).

Waterhead   In attempting to control water for the generation of electricity or a gravity flow irrigation project, it is essential to pond up water in relative quantity so that there is an adequate force — waterhead — to achieve the desired goal. Dam structures are the most common engineering responses to this requirement.

Weather   The atmospheric conditions prevailing at one time and place.

Weathering   The natural process that disintegrates rocks by mechanical or physical means, making soil formation possible.

West   The shorthand term for the countries of the world that have been most influenced by western civilization and that link most closely with the United States and western Europe.

Westerly winds   An airstream located in the middle latitudes that blows from west to east. Also known as westerlies.

Western civilization   The sum of values, practices, and achievements that had roots in ancient Mesopotamia, as well as Palestine, Greece, and Rome, and that subsequently flowered in western and southern Europe.

Westernization   The process whereby non-Western societies acquire Western traits, which are adopted with varying degrees of completeness.

Winter solstice   On or about December 22, the first day of winter in the Northern Hemisphere, the southern tip of Earth's axis is inclined toward the sun at an angle of 23 1/2 degrees from a line perpendicular to the plane of the ecliptic. This is the winter solstice in the Northern Hemisphere. On or about June 22, the first day of summer in the Northern Hemisphere, the northern tip of Earth's axis is inclined toward the sun at an angle of 23 1/2 degrees from a line perpendicular to the plane of the ecliptic. This is the winter solstice in the Southern Hemisphere.

Xerophytic   Literally, "dry plant," referring to desert shrubs having small leaves, thick bark, large root systems, and other adaptations to absorb and retain moisture.

Zero population growth (ZPG)   The condition of equal birth rates and death rates in a population.

Zionist movement   The political effort, beginning in the late 19th century, to establish a Jewish homeland in Palestine.

Zone of transition   An area where the characteristics of one region change gradually to those of another.

Place/Name Pronunciation Guide

The authors acknowledge with thanks the assistance of David Allen, Eastern Michigan University, in preparing an earlier pronunciation guide that the authors expanded and modified for this book. In general, names are those appearing in the text; pronunciations of other names on maps can be found in *Goode's World Atlas* (Rand McNally) or in standard dictionaries and encyclopedias. Vowel sounds are coded as: ă (uh), ā (ay), ĕ (eh), ē (ee), ī (eye), ĭ (ih), o͞o (oo).

Afghanistan (af-*gann'*-ih-stann')
    Helmand R. (*hell'*-mund)
    Hindu Kush Mts. (*hinn'*-doo *koosh'*)
    Kabul (*kah'*-b'l)
    Khyber Pass (*keye'*-ber)
Albania (al-*bay'*-nee-a)
    Tirana (tih-*rahn'*-a)
Algeria (al-*jeer'*-ee-a)
    Ahaggar Mts. (uh-*hahg'*-er)
    Sahara Desert (suh-*hahr'*-a)
    Tanezrouft (*tahn'*-ez-rooft')
American Samoa
    Pago Pago (*pahng'*-oh *pahng'*-oh)
    Tutuila I. (too-too-*ee'*-lah)
Andorra (an-*dohr'*-a)
Angola (ang-*goh'*-luh)
    Benguela R.R. (ben-*gehl'*-a)
    Cabinda (exclave) (kah-*binn'*-duh)
    Luanda (loo-*an'*-duh)
    Lobito (luh-*bee'*-toh)
Anguilla (an-*gwill'*-a)
Antigua and Barbuda (an-*teeg'*-wah,
    bahr-*bood'*-a)
Argentina (ahr'-jen-*teen'*-a)
    Buenos Aires (*bwane'*-uhs *eye'*-ress)
    Córdoba (*kawrd'*-oh-bah)
    Patagonia (region) (patt'-hu-*goh'*-nee-a)
    Rio de la Plata (*ree'*-oh duh lah *plah'*-tah)
    Rosario (roh-*zahr'*-ee-oh)
    Tierra del Fuego (region)
        (tee-ehr'-uh dell *fway'*-goh)
    Tucumán (too'-koo-*mahn'*)
Armenia (ahr'-*meen'*-y-uh)
    Yerevan (yair'-uh-*vahn'*)
Aruba (ah-*roob'*-a)
Australia (aw-*strayl'*-yuh)
    Adelaide (*add'*-el-ade)
    Brisbane (*briz'*-bun)
    Canberra (*can'*-bur-a)
    Melbourne (*mell'*-bern)

    Sydney (*sid'*-nih)
    Tasmania (state) (tazz-*may'*-nih-uh)
Austria (*aw'*-stree-a)
    Graz (*grahts'*)
    Linz (*lints'*)
    Salzburg (*sawlz'*-burg)
    Vienna (vee-*enn'*-a)
Azerbaijan (ah'-zer-*beye'*-jahn)
    Baku (bah-*koo'*)

Bahamas (buh-*hahm'*-uz)
    Nassau (*nass'*-aw)
Bahrain (bah-*rayn'*)
Bangladesh (bahng'-lah-*desh'*)
    Dacca (Dhaka) (*dahk'*-a)
Barbados (bar-*bay'*-dohss)
Belarus (*bell'*-uh-roos)
    Minsk (*mensk'*)
Belgium
    Antwerp (*ant'*-werp)
    Ardennes Upland (ahr-*den'*)
    Brussels (*brus'*-elz)
    Charleroi (*shahr*-luh-*rwah'*)
    Ghent (*gent'*)
    Liège (lee-*ezh'*)
    Meuse R. (*merz'*)
    Sambre R. (*sohm*-bruh)
    Scheldt R. (*skelt'*)
Belize (buh-*leez'*)
Benin (beh-*neen'*)
    Cotonou (koh-toh-*noo'*)
Bhutan (boo-*tahn'*)
    Thimphu (*thim'*-poo')
Bolivia (boh-*liv'*-ee-uh)
    Altiplano (region) (al-tih-*plahn'*-oh)
    La Paz (lah *pahz'*)
    Santa Cruz (sahn-ta *krooz'*)
    Sucre (*soo'*-kray)
    Lake Titicaca (tee'-tee-*kah'*-kah)
Bosnia-Hercegovina
    (*boz'*-nee-uh herts-uh-goh-*veen'*-uh)
    Sarajevo (sahr-uh-*yay'*-voh)
Botswana (baht-*swahn'*-a)
    Gaborone (gahb'-uh-*roh'*-nee)
Brazil (bra-*zill'*)
    Amazonia (region) (am-uh-*zohn'*-ih-a)
    Belo Horizonte (*bay'*-loh haw-ruh-*zonn'*-tee)
    Brasília (bruh-*zeel'*-yuh)
    Caatinga (region) (kay-*teeng'*uh)
    Manaus (mah-*noose'*)
    Minas Gerais (*meen'*-us zhuh-*rice*)

    Natal (nah-*tal'*)
    Paraíba Valley (pah-rah-*ee'*-bah)
    Paraná R. (pah-rah-*nah'*)
    Pôrto Alegre (por-too' al-*egg'*-ruh)
    Recife (ruh-*see'*-fee)
    Rio de Janeiro (*ree'*-oh dih juh-*nair'*-oh)
    Salvador (sal'-vah-*dor'*)
    São Paulo (sau *pau'*-loh)
Brunei (*broo'*-neye)
Bulgaria
    Sofia (*soh'*-fee-a)
Burkina Faso (bur-*kee'*-na *fah'*-soh)
    Ouagadougou (wah'-guh-*doo'*-goo)
Burma (*bur'*-ma)
Burundi (buh-*roon'*-dee) ("oo" as in "wood")
    Bujumbura (boo-jem-*boor'*-a)
        (second "oo" as in "wood")

Cambodia
    Phnom Penh (*nom' pen'*)
Cameroon (kam'-uh-*roon'*)
    Douala (doo-*ahl'*-a)
    Yaoundé (yah-onn-*day'*)
Canada
    Montreal (mon'-tree-*awl'*)
    Nova Scotia (*noh'*-vuh *scoh'*-shuh)
    Ottawa (*aht'*-ah-wah)
    Quebec (kwih-*beck'*; Fr., kay-*beck'*)
    Saguenay R. (*sag'*-uh-nay)
Cape Verde (*verd'*)
Central African Republic
    Bangui (*bahng'*-ee)
Chile (*chill'*-ee)
    Atacama Desert (ah-tah-*kay'*-mah)
    Chuquicamata (choo-kee-*kah'*-mah-tah)
    Concepción (con-sep'-*syohn'*)
    Santiago (sahn'-tih-*ah'*-goh)
    Valparaíso (vahl'-pah-rah-*ee'*-soh)
China
    Altai Mts. (al-*teye*)
    Amur R. (ah-*moor'*) ("oo" as in "wood")
    Anshan (ahn-*shahn'*)
    Baotou (*bau'*-toh)
    Beijing (bay-*zhing'*)
    Canton (kan'-*tahn'*)
    Chang Jiang (chung jee-*ahng'*)
    Chengdu (chung-*doo'*)
    Chongqing (chong-*ching'*)
    Dzungarian Basin (zun-*gair'*-ree-uhn)
    Guangdong (gwahng-*dung'*)
    Guangzhou (gwahng-*joh'*)

Hainan I. (*heye'-nahn'*)
Harbin (*hahr'*-ben)
Huang He (*hwahng' huh'*)
Lanzhou (lahn-*joh'*)
Lhasa (*lah'*-suh)
Li R. (*lee'*)
Liao (lih-ow')
Lüda (*loo'*-dah)
Nanjing (nahn-*zhing'*)
Peking (pea-*king'*)
Shandong (shahn-*dung'*)
Shanghai (shang-*heye'*)
Shantou (shahn-*toh'*)
Shenyang (shun-*yahng'*)
Shenzhen (shun-*zún*)
Sichuan (zehch-*wahn'*)
Taklamakan Desert (*tah'*-kla-mah-*kahn'*)
Tarim Basin (*tah'-reem'*)
Tiananmen (*tyahn'*-an-men)
Tianjin (tyahn-*jeen'*)
Tien Shan (*tih'*-en *shahn'*)
Tientsin (*tinn'-tsinn'*)
Tsinling Shan (*chinn'*-leeng *shahn'*)
Ürümqi (oo-*room'*-chee)
Wuhan (*woo'-hahn'*)
Xian (shee-*ahn'*)
Xinjiang (shin-jee-*ahng'*)
Xizang (shee-*dzahng'*)
Yangtze R. (*yang'*-see)
Yunnan (yoo-*nahn'*) ("oo" as in "wood")
Colombia (koh-*lomm'*-bih-uh)
Barranquilla (bah-rahn-*keel'*-yah)
Bogotá (boh'-ga-*tah'*)
Cali (*kah'*-lee)
Cartagena (kahrt'-a-*hay'*-na)
Cauca R. (*kow'*-kah)
Magdalena R. (mahg'-thah-*lay'*-nah)
Medellín (med-deh-*yeen'*)
Comoros (*kahm'*-o-rohs')
Congo
Brazzaville (*brazz'*-uh-vill'; Fr.: brah-zah-*veel'*)
Pointe Noire (pwahnt *nwahr'*)
Costa Rica (kohs'-tuh *ree'*-kuh)
San José (sahn hoh-*zay'*)
Croatia (kroh-*ay'*-shuh)
Zagreb (*zah'*-grebb)
Cyprus (*seye'*-pruhs)
Nicosia (nik'-o-*see'*-uh)
Czechoslovakia (check'-oh-sloh-*vah'*-kee-a)
Bratislava (bratt'-ih-*slahv'*-a, braht-)
Brno (*ber'*-noh)
Moravia (muh-*rayve'*-ih-uh)
Ostrava (*aw'*-strah-vah)
Prague (*prahg'*)

Denmark
Copenhagen (*koh'*-pen-hahg'-gen)
Djibouti (jih-*boot'*-ee)
Dominica (*dahm'*-i-*nee'*-ka)
Dominican Republic (duh-*min'*-i-kan)
Santo Domingo (sant'-oh d-*min'*-goh)

Ecuador (*ec'*-wa-dawr)
Guayaquil (gweye'-uh-*keel'*)
Quito (*kee'*-toh)

Egypt
Aswan (*ahs'*-wahn)
Cairo (*keye'*-roh)
Port Said (sah-*eed'*)
El Salvador (el *sal'*-va-dor)
Equatorial Guinea (*gin'*-ee)
Estonia (ess-*toh'*-nee-a)
Tallin (*tahl'*-in)
Ethiopia (ee'-thee-*oh'*-pea-a)
Addis Ababa (*ad'*-iss *ab'*-a-ba)
Asmara (az-*mahr'*-a)

Fiji (*fee'*-jee)
Suva (*soo'*-va)
Finland
Helsinki (*hel'*-sing-kee)
Karelian Isthmus (kuh-*ree'*-lih-uhn)
France
Alsace (region) (al-*sass'*)
Ardennes (region) (ahr-*den'*)
Bordeaux (bawr-*doh'*)
Boulogne (boo-*lohn'*-yuh)
Calais (kah-*lay'*)
Carcassonne (kahr-kuh-*suhn'*)
Champagne (sham-*pahn*-yuh)
Garonne R. (guh-*rahn'*)
Jura Mts. (*joo'*-ruh)
Le Havre (luh *ah'*-vruh)
Lille (*leel'*)
Loire R. (*lwahr'*)
Lorraine (loh-*rane'*)
Lyons (Fr., Lyon), both (*lee*-aw)
Marseilles (mahr-*say'*)
Massif Central (upland) (mah-*seef' saw-trahl*)
Meuse R. (*merz'*)
Moselle R. (moh-*zell'*)
Nancy (naw-*see'*)
Nantes (*nawnt'*)
Nice (*neece'*)
Oise R. (*wahz'*)
Pyrenees (peer'-uh-*neez'*)
Riviera (riv'-ih-*ehr'*-a)
Rouen (roo-*aw'*)
Saône R. (*sohn'*)
Seine R. (*senn'*)
Strasbourg (strahz-*boor'*)
Toulon (too-*law'*)
Toulouse (too-*looz'*)
Vosges Mts. (*vohzh'*)
French Guiana (gee-*ahn*-a)
Cayenne (*keye'*-enn)

Gabon (gah-*baw'*)
Gambia (*gamm'*-bih-uh)
Banjul (*bahn*-jool)
Georgia
Abkhazia (ahb-kahz9-zee-uh)
Adjaria (ah-*jahr'*-ree-uh)
Caucasus Mts. (*kaw'*-kuh-suhs)
South Ossetia (oh-*see'*-shee-uh)
Tbilisi (tuh-*blee'*-see)
Germany
Aachen (*ah'*-ken)
Berlin (ber-*lin'*)
Bremen (*bray'*-men)

Chemnitz (*kemm'*-nits)
Cologne (kuh-*lohn'*)
Dresden (*drez'*-den)
Düsseldorf (*doo'*-sel-dorf)
Elbe R. (*ell'*-buh)
Erzgebirge (*ehrts'*-guh-beer-guh)
Frankfurt (*frahngk'*-foort')
Karlsruhe (*kahrls'*-roo-uh)
Leipzig (*leyep'*-sig)
Main R. (*mane'*; Ger., *mine'*)
Mannheim-Ludwigshafen (*man'*-hime *loot'*-vikhs-hah-fen)
Munich (*myoo'*-nik)
Neisse R. (*nice'*-uh)
Oder R. (*oh'*-der)
Ruhr (region) (*roor'*)
Stuttgart (*stuht'*-gahrt)
Weser R. (*vay'*-zer)
Wiesbaden-Mainz (*vees'*-bahd-en *meyents'*)
Wuppertal (*voop'*-er-tahl) ("oo" as in "wood")
Ghana (*gahn'*-a)
Accra (a-*krah'*)
Kumasi (koo-*mahs'*-ee)
Tema (*tay'*-muh)
Volta R. (*vohl'*-tuh)
Greece
Aegean Sea (uh-*jee'*-un)
Piraeus (peye-*ree'*-us)
Thessaloniki (thess'-uh-loh-*nee'*-kih)
Grenada (gruh-*nayd'*-a)
Guadeloupe (gwah'-duh-*loop'*)
Guatemala (gwah'-tuh-*mah'*-luh)
Guinea (*gin'*-ee)
Conakry (*kahn'*-a-kree)
Guinea-Bissau (biw-*ow'*)
Guyana (geye-*an'*-a; geye-*ahn*-a')

Haiti (*hayt'*-ee)
Port-au-Prince (pohrt'-oh-*prants'*)
Honduras (hahn-*dur'*-as)
Tegucigalpa (tuh-goo'-sih-*gahl'*-pah)
Hungary (*hun'*-guh-rih)
Budapest (boo-duh-*pesht'*)

Iceland
Akureyri (ah'-koor-*ray'*-ree)
Reykjavik (*ray'*-kyuh-vik')
India
Agra (*ah'*-gruh)
Ahmedabad (ah'-mud-uh-*bahd'*)
Assam (state) (uh-*sahm'*)
Bengal (region) (benn-*gahl'*)
Bihar (state) (bih-*hahr'*)
Brahmaputra R. (brah'-muh-*poo'*-truh)
Deccan (region) (*deck'*-un)
Delhi (*deh'*-lih)
Ganges R. (*gan'*-jeez)
Eastern, Western Ghats (mountains) (*gahts'*)
Himalaya Mts. (himm-*ah'*-luh-yuh, -ah-*lay*-yuh)
Jaipur (*jeye'*-poor)
Jamshedpur (juhm-*shayd'*-poor)
Kashmir (region) (*cash'*-meer)
Kerala (state) (*kay'*-ruh-luh)
Madras (muh-*drass'*, -*drahs'*)

Punjab (region)
  (*pun*'-jahb, -jab; pun-*jahb*', -*jab*')
Rajasthan (state) (*rah*'-juh-stahn)
Uttar Pradesh (state) (*oo*'-tahr pruh-*desh*')
Indonesia
  Bali (*bah*'-lih)
  Bandung (*bahn*'-doong)
  Celebes (Sulawesi) I.
    (*sell*'-uh-beez; sool-uh-*way*'-see)
  Irian Jaya (ihr'-ee-ahn *jeye*'-uh)
  Jakarta (yah-*kahr*'-tah)
  Java (*jah*'-vuh)
  Kalimantan (kal-uh-*mann*'-tann)
  Medan (muh-*dahn*')
  Palembang (pah-lemm-*bahng*')
  Sumatra (suh-*mah*'-truh)
  Surabaya (soo'-ruh-*bah*'-yah)
  Ujung Pandang (oo'-jung
    pahn-*dahng*')
Iran (ih-*rahn*')
  Abadan (ah-buh-*dahn*')
  Elburz Mts. (ell-*boorz*')
  Isfahan (izz-fah-*hahn*')
  Meshed (muh-*shed*')
  Tabriz (tah-*breez*')
  Tehran (teh-huh-*rahn*')
  Zagros Mts. (*zah*'-gruhs)
Iraq (i-*rahk*')
  Baghdad (*bag*'-dad)
  Basra (*bahs*'-ra)
  Euphrates R. (yoo-frate'-eez)
  Kirkuk (Kihr-*kook*') ("oo" as in "wood")
  Mosul (moh-*sool*')
  Shatt al Arab (R.) (*shaht*' ahl ah-*rahb*')
  Tigris R. (*teye*'-griss)
Israel
  Gaza Strip (*gahz*'-uh)
  Eilat (*ay*'-laht)
  Haifa (*heye*'-fa)
  Tel Aviv (tel' a-*veev*')
Italy
  Adriatic Sea (ay-drih-*at*'-ik)
  Apennines (Mts.) (*app*'-uh-nines)
  Bologna (boh-*lohn*'-ya)
  Genoa (*jen*'-o-a)
  Milan (mih-*lahn*')
  Naples (*nay*'-pels)
  Palermo (pah-*ler*'-moh)
  Turin (*toor*'-in)
Ivory Coast (Côte d'Ivoire) (koht-d'-vwahr)
  Abidjan (abb'-ih-*jahn*')

Jamaica (ja-*may*'-ka)
Japan
  Fukuoka (foo'-koo-*oh*'-kah)
  Hiroshima (heer'-a-*shee*'-mah)
  Hokkaido (island) (hah-*keye*'-doh)
  Honshu (island) (*honn*'-shoo)
  Kitakyushu (kee-tah'-*kyoo*'-shoo)
  Kobe (*koh*'-bay)
  Kyoto (kee-*oht*'-oh)
  Kyushu (island) (kee-*yoo*'-shoo)
  Nagoya (nay-*goy*'-ya)
  Osaka (oh-*sahk*'-a)
  Sapporo (sahp-*pohr*'-oh')
  Shikoku (island) (shih-*koh*'-koo)

Tokyo (*toh*'-kee-oh)
  Yokohama (yoh-ko-*hahm*'-a)
Jordan (*jor*'-dan)
  Amman (a-*mahn*')

Kazakstan (kuh-*zahk*'-stahn)
  Alma Ata (*al*'-mah ah-tah')
  Aral Sea (uh-*rahl*')
  Karaganda (kahr'-rah-*gahn*'-dah)
  Lake Balkhash (bul-*kahsh*')
  Syr Darya (river) (*seer*' dahr-*yah*')
Kenya (*ken*'-ya)
  Mombasa (mahm-*bahs*-a)
  Nairobi (neye-*roh*'-bee)
Kiribati (keer'-ih-*bah*'-tih)
Korea, North
  Pyongyang (pea-ong-*yahng*')
  Yalu River (*yah*'-loo)
Korea, South
  Inchon (*inn*'-chonn)
  Pusan (*poo*'-sahn)
  Seoul (*sohl*')
  Taegu (teye-*goo*')
Kuwait (koo-*wayt*')
Kyrgyzstan (*keer*'-geez-stahn)

Laos (*lah*'-ohs)
  Vientiane (vyen-*tyawn*')
Latvia
  Riga (*reeg*'-a)
Lebanon (*leb*'-a-non)
  Beirut (bay-*root*')
Lesotho (luh-*soh*'-toh)
Libya (*lib*'-yah)
  Benghazi (benn-*gah*'-zee)
  Tripoli (*tripp*'-uh-lee)
Liechtenstein (*lick*'-tun-stine')
Lithuania (lith'-oo-*ane*'-ee-uh)
  Vilnyus (*vill*'-nee-us)
Luxembourg (*luk*'-sem-burg)

Madagascar
  Antananarivo (an-ta-nan'-a-*ree*'-voh)
Malawi (muh-*lah*'-wee)
  Lilongwe (lih-*lawng*'-way)
Malaysia
  Kuala Lumpur (*kwah*'-luh *loom*'-poor)
    ("oo" as in "wood")
  Sabah (*sah*'-bah)
  Sarawak (suh-*rah*'-wahk)
  Strait of Malacca (muh-*lahk*'-uh)
Maldives (*mawl*'-deevz)
Mali (*mah*'-lee)
  Bamako (*bam*'-ah-koh')
Malta (*mawl*'-tuh)
Martinique (mahr'-tih-*neek*')
Mauritania (mahr-ih-*tay*'-nee-a)
Mauritius (muh-*rish*'-us)
Mexico
  Acapulco (a-ca-*pool*-ko)
  Ciudad Juárez (see-yoo-*dahd*'
    wahr'-ez)
  Guadalajara (gwad'-a-la-*har*'-a)
  Monterrey (mahnt-uh-*ray*')
  Puebla (poo-*eb*'-la)
  Tijuana (tee-*hwah*'-nah)

Moldova (mohl-*doh*'-vah)
  Kishinev (kish'-in-*yeff*')
Monaco (*mon*'-a-koh')
Mongolia (mon-*gohl*'-yuh)
  Ulan Bator (oo-*lahn*' *bah*'tor)
Montserrat (mont'-suh-*rat*')
Morocco (moh-*rah*'-koh)
  Casablanca (kas-a-*blang*'-ka)
  Rabat (rah-*baht*')
Mozambique (moh'-zamm-*beek*')
  Beira (*bay*'-rah)
  Cabora Bassa Dam (kah-*bore*'-ah
    bah'-sah)
  Maputo (mah-*poot*'-oh')

Namibia (nah-*mib*'-ee-a)
Nauru (nah-oo-roo)
Nepal (nuh-*pawl*')
  Katmandu (kat'-man-*doo*')
Netherlands
  Ijsselmeer (*eye*'-sul-mahr')
  The Hague (*hayg*')
  Utrecht (*yoo*'-trekt')
  Zuider Zee (*zeye*'-der zay')
Netherlands Antilles
  Curaçao (*koo*'-rahs-ow')
New Zealand
  Auckland (*aw*'-klund)
Nicaragua (nik'-a-*rahg*'-wah)
Niger (*neye*'-jer; Fr., nee'-*zhere*')
  Niamey (nee-*ah*'-may)
Nigeria (neye-*jeer*-ee-a)
  Abuja (a-*boo*'-ja)
  Ibadan (ee-*bahd*'-n)
  Kano (*kahn*'-oh)
  Lagos (*lahg*'-us)
Norway
  Bergen (*berg*'-n)
  Oslo (*ahz*'-loh)
  Stavanger (stah-*vahng*'-er)
  Trondheim (*trahn*'-haym')

Oman (oh-*mahn*')
  Muscat (*mus*'-kat')

Pakistan (pahk-i-*stahn*')
  Hyderabad (*heyed*'-er-a-bahd')
  Islamabad (is-*lahm*-a-bahd')
  Karachi (kah-*rahch*'-ee')
  Karakoram (Mts.) (kahr'-a-*kohr*-um)
  Lahore (lah-*hohr*')
  Peshawar (puh-*shah*'-wahr)
  Punjab (see India)
Papua New Guinea (*pap*'-yoo-uh new *gin*'-ee)
  Port Moresby (*morz*'-bee)
Paraguay (*pair*'-uh-gweye')
  Asuncíon (a-*soon*'-see-*ohn*')
Peru (puh-*roo*')
  Callao (kah-*yow*')
  Cerro de Pasco (*sehr*'-roh day *pahs*'-koh)
  Chimbote (chimm-*boh*'-te)
  Cuzco (*koos*'-koh)
  Iquitos (ee-*kee*'-tohs)
  Lima (*lee*'-ma)
  Machu Picchu (ma-chu *pee*-chu)
Philippines (*fil*-i-peenz')
  Luzon (island) (loo'-*zahn*')

Mindanao (island) (minn'-dah-*nah*'-oh)
Mt. Pinatubo (pea'-nah-*too*'-buh)
Visayan Islands (vih-*sah*'-y'n)
Poland
 Gdansk (ga-*dahntsk*')
 Katowice (kaht-uh-*veet*'-sih)
 Krakow (*krahk*'-ow)
 Lodz (*looj*')
 Posnan (*pohz*'-nan-ya)
 Silesia (region) (sih-*lee*'-shuh)
 Szczecin (*shchet*'-seen')
 Wroclaw (*vrawt*'-slahf)
Portugal (*pohr*'-chi-gal)
 Lisbon (*liz*'-bon)
 Oporto (oh-*pohrt*'-oh)

Qatar (*kaht*'-ar)

Romania (roh-*main*'-yuh)
 Bucharest (boo'-kuh-*rest*')
 Carpathian Mts. (kahr-*pay*'-thih-un)
Russia
 Altai Mts. (*al*'-teye)
 Amur R. (ah-*moor*') ("oo" as in "wood")
 Angara R. (ahn'-guh-*rah*')
 Arkhangelsk (ahr-*kann*'-jelsk)
 Astrakhan (as-trah-*khahn*')
 Baikal, Lake (beye-*kahl*')
 Bashkiria (bahsh-*keer*'-ee-uh)
 Bratsk (*brahtsk*')
 Buryatia (boor-*yaht*'-ee-uh)
 Caucasus (kaw'-kah-suss)
 Chechen-Ingushia
  (*chehch*'-yenn enn-*gooshia*')
 Chelyabinsk (*chell*'-yah-binnsk)
 Cherepovets (*chehr*'-yuh-puh-vyetz')
 Chuvashia (choo-*vahsh*'-i-a)
 Dagestan Rep. (dag'-uh-*stahn*')
 Irkutsk (ihr-*kootsk*')
 Ivanovo (ih-*vahn*'-uh-voh)
 Izhevsk (ih-*zhehvsk*')
 Kama R. (*kah*'-mah)
 Karelia (kah-*reel*'-yuh)
 Kazan (kuh-*zahn*')
 Khabarovsk (kah-*bah*'-rawfsk)
 Kola Pen. (*koh*'-lah)
 Kolyma R. (kuh-*lee*'-muh)
 Krasnoyarsk (kras'-nuh-*yahrsk*')
 Kuril Is. (*koo*'-rill)
 Kuznetsk Basin (kooz'-*netsk*')
 Ladoga, Lake (*lad*'-uh-guh)
 Lena R. (*lee*-nah, *lay*-nah)
 Lipetsk (lyee'-*petsk*')
 Magnitogorsk (mag-*nee*'-toh-gorsk')
 Moscow (mahs'-koh or mahs'-kow)
 Murmansk (moor'-*mahntsk*')
 Nakhodka (nuh-*kawt*-kah)
 Nizhniy Novgorod (*nizh*'-nee nahv-guh-
  rahd')
 Nizhniy Tagil (tah-*geel*)
 Novokuznetsk (naw'-voh-*kooznetsk*')
 Novaya Zemlya (naw'-vuh-yuh zem-*lyah*')
 Novosibirsk (naw-vuh-suh-*beersk*')
 Oka R. (oh-*kah*')
 Okhotsk, Sea of (oh-*kahtsk*')
 Omsk (*ahmsk*')

Pechora R. (peh-*chore*'-a)
Rostov (rahs-*tawf*')
Sakhalin (sahk'-uh-*leen*')
Samara (suh-*mahr*'-a)
Saratov (suh-*raht*'-uff)
Sayan Mts. (sah-*yahn*')
Tatarstan (Tatar Rep.) (tah-*tahr*'-stahn')
Togliatti (tawl-*yah*'-tee)
Tuva Rep. (*too*'-va)
Ufa (oo-*fah*')
Ussuri R. (oo-*soor*'-i)
Vladivostok (vlah'-dih-vahs-*tawk*')
Volga R. (*vahl*-guh, *vohl*'-guh)
Volgograd (vahl-guh-*grahd*', vohl'-)
Voronezh (vah-*raw*'-nyesh)
Yakutsk (yah-*kootsk*')
Yekaterinburg (yeh-*kahta*'-rinn-berg)
Rwanda (roo-*ahn*'-da)

St. Kitts and Nevis (*nee*'-vis)
St. Lucia (*loo*'-sha)
St. Vincent-Grenadines (gren'-a-*deenz*')
São Tomé and Príncipe
 (sau too-*meh*', pren-*see*'-puh)
Saudi Arabia (*saud*'-ee)
 Jiddah (*jid*'-a)
 Riyadh (ree-*adh*')
Senegal (sen'-i-*gahl*')
 Dakar (da'-*kahr*')
Seychelles (say'-*shelz*')
Sierra Leone (see-*ehr*'-a lee-*ohn*')
Slovenia (sloh-*veen*'-i-a)
 Ljubljana (*lyoo*'-b'l-yah-nah)
Somalia (so-mahl'-ee-a)
 Mogadishu (mahg'-uh-*dish*'-oo)
South Africa
 Bloemfontein (*bloom*'-fonn-*tayn*')
 Bophuthatswana (homeland)
  (boh-poot'-aht-*swahna*')
 Drakensberg (escarpment) (*drahk*-n's-burg')
 Durban (*durr*'-bun)
 Johannesburg (joh-*hann*'-iss-burg)
 Kwazulu (homeland) (kwah-zoo'-loo)
 Natal (province) (nuh-*tahl*')
 Pretoria (prih-*tohr*'-ee-uh)
 Transkei (homeland) (trans-*keye*')
 Transvaal (province) (trans-*vahl*')
Spain
 Barcelona (bahr'-suh-*lohn*-uh)
 Bilbao (bill-*bah*'-oh)
 Catalonia (region) (katt'-uh-*lohn*'-ee-a)
 Madrid (muh-*dridd*')
 Meseta (plateau) (muh-*say*'-tuh)
 Seville (suh-*vill*')
 Valencia (vull-*enn*'-shee-uh)
Sri Lanka (sree *lahn*'-kah)
Sudan (*soo*'-dann)
 Gezira (guh-*zeer*'-uh)
 Juba (*joo*'-bah)
 Khartoum (kahr'-*toom*')
 Omdurman (ahm-*durr*'-man)
 Wadi Halfa (wah-di' *hawl*'-fah)
Suriname (*soor*'-i-nahm')
Swaziland (*swah*'-zih-land')
Sweden
 Göteborg (yort'-e-*bor*')

Skåne (*skohn*')
Småland (*smoh*'-lund')
Switzerland
 Basel (*bah*'-z'l)
 Zürich (*zoor*'-ick)
Syria (*seer*'-i-uh)
 Aleppo (a-*lep*-oh')
 Damascus (da-*mas*'-kus)

Tajikistan (tah-*jick*'-ih-stahn')
Tanzania (tan'-za-*nee*'-a)
 Dar es Salaam (dahr' es sa-*lahm*')
Thailand (*teye*'-land')
 Bangkok (*bang*'-kok')
Togo (*toh*'-goh)
 Lomé (loh-*may*')
Tonga (*tawng*'-a)
Trinidad and Tobago (to-*bay*'-goh)
Tunisia (too-*nee*'-zhee-a)
 Tunis (*too*'-nis)
Turkey
 Anatolian Peninsula (ann'-ah-*tohl*'-yun)
 Ankara (*ang*'-kah-rah)
 Bosporus (strait) (*bahs*'-puhr-us)
 Dardanelles (strait) (dahr'-duh-*nelz*')
 Istanbul (iss'-tann-*bool*')
 Izmir (izz'-*meer*')
 Sea of Marmara (*mahr*'-muh-ruh)
Turkmenistan (turk'-menn-ih-*stahn*')
 Ashkhabad (*ash*'-kuh-bahd')

Ukraine (yoo'-*krane*')
 Crimea (cry'-*mee*'-uh)
 Dnepropetrovsk (d'nyepp-pruh-pay-*trawfsk*')
 Dnieper R. (duh-*nyepp*'-er)
 Donetsk (duhn-*yehtsk*')
 Kharkov (*kahr*'-kawf)
 Kiev (*kee*'-yeff)
 Krivoy Rog (*kree*'-voi *rohg*')
 Sevastopol (sih-*vass*'-tuh-*pohl*')
 Yalta (*yahl*'-tuh)
 Zaporozhye (zah-puh-*rawzh*'-yuh)
United Arab Emirates
 Abu Dhabi (*ah*'-boo dah'-bee)
 Dubai (doo-*beye*')
United Kingdom
 Edinburgh (*ed*'-in-buh-ruh)
 Glasgow (*glass*'-goh)
 Thames R. (*timms*')
United States
 Albuquerque (*al*'-buh-ker'-kih)
 Aleutian Is. (uh-*loo*'-sh'n)
 Allegheny R. (*al*'-ih-gay'-nih)
 Appalachian Highlands
  (ap'-uh-*lay*'-chih-un)
 Coeur d'Alene (*kurr*' duh-*layn*')
 Des Moines (duh *moin*')
 Grand Coulee Dam (*koo*'-lee)
 Juneau (*joo*'-noh)
 Mesabi Range (muh-*sahb*'-ih)
 Monongahela R.
  (muh-*nahng*'-guh-hee'-luh)
 Omaha (oh'-mah-*hah*')
 Ouachita Mts. (*wosh*'-ih-taw')
 Palouse (region) (pah-*loose*')
 Phoenix (*fee*'-nix)

Provo (*proh'*-voh)
Rio Grande (*ree'*-oh *grahn'*-day)
San Joaquin R. (sann wah-*keen'*)
Spokane (spoh'-*kann'*)
Tucson (*too'*-sonn)
Valdez (val-*deez'*)
Uruguay (*yoor'*-uh-gweye')
  Montevideo (mahnt'-e-vi-*day'*-oh)
Uzbekistan (ooz-*beck'*-ih-stahn') ("oo" as in
  "wood")
  Amu Darya (river) (ah-moo' *dahr'*-yuh,
    dahr-*yah'*)
  Bukhara (buh-*kah'*-ruh)
  Fergana Valley (fair-*gahn'*-uh)
  Samarkand
    (*samm'*-ur-*kand'*; Rus., suh-mur-*kahnt'*)
  Tashkent (tash-*kent'*, tahsh-)

Vanuatu (van'-oo-*ay'*-too)
  Port Vila (*vee'*-la)
Venezuela (ven'-ez-*way'*-la)
  Caracas (ka-*rak'*-as)

Maracaibo (mar'-a-*keye'*-boh)
Orinoco R. (oh-rih-*noh'*-koh)
Valencia (va-*len'*-see-a)
Vietnam (vee-et'-*nahm'*)
  Annam (region) (uh-*namm'*)
  Cochin China (*koh'*-chin)
  Hanoi (ha-*noi'*)
  Ho Chi Minh City (Saigon)
    (hoe chee *min'*; *seye'*-gahn)
  Mekong R. (may-*kahng'*)

Western Samoa (sa-*moh'*-a)
  Apia (ah-*pea'*-uh)

Yemen
  Aden (*ah*-d'n)
  S'ana (sah-*na'*)
Yugoslavia (yoo'-goh-*slahv'*-ee-a)
  Belgrade (*bell'*-grade, -grahd)
  Kosovo (*kaw'*-suh-voh')
  Montenegro (republic)
    (mahn'-tee-*nay'*-groh)

Zaire (*zeye'*-eer')
  Kananga (kay-*nahng'*-guh)
  Kasai R. (kah-*seye'*)
  Kinshasa (*keen'*-shah-suh)
  Kisangani (kee'-sahn-*gahn'*-ih)
  Lake Kivu (*kee'*-voo)
  Lualaba R. (loo'-ah-*lahb'*-a)
  Lubumbashi (loo-boom-*bah'*-shee)
  Matadi (muh-*tah'*-dee)
  Shaba (region) (*shah'*-bah)
Zambia (*zam'*-bee-a)
  Kariba Dam (kah-*ree'*-buh)
  Lusaka (loo-*sahk'*-a)
  Zambezi R. (zamm-*bee'*-zee)
Zimbabwe (zim-*bahb'*-way)
  Bulawayo (bool'-uh-*way'*-oh)
  Harare (hah-*rahr'*-ee)
  Que Que (*kway' kway'*)
  Wankie (*wahn'*-kee)

Index